Process Dynamics and Control

Process Dynamics and Control

Third Edition

Dale E. Seborg
University of California, Santa Barbara

Thomas F. Edgar
University of Texas at Austin

Duncan A. Mellichamp
University of California, Santa Barbara

Francis J. Doyle III
University of California, Santa Barbara

WILEY

John Wiley & Sons, Inc.

VP & EXECUTIVE PUBLISHER Don Fowley
ACQUISITIONS EDITOR Jennifer Welter
EDITORIAL ASSISTANT Alexandra Spicehandler
DESIGNER RDC Publishing Group Sdn Bhd
MARKETING MANAGER Christopher Ruel
PRODUCTION MANAGER Janis Soo
SENIOR PRODUCTION EDITOR Joyce Poh

This book was set in 10/12 Times Ten by Aptara Corporation and printed and bound by Courier Westford. The cover was printed by Courier Westford.

This book is printed on acid free paper.

Library of Congress Cataloging-in-Publication Data

Seborg, Dale E.
Process dynamics and control / Dale E. Seborg . . . [et al.]. — 3rd ed.
p. cm.
Includes index.
ISBN 978-0-470-12867-1 (cloth)
1. Chemical process control — Data processing. I. Title.
TP155.S35 2011
660′.2815 — dc22

2010000779

Printed in the United States of America

10 9 8 7 6 5 4 3

To our wives, children, and parents

About the Authors

Dale E. Seborg is a Professor and Vice Chair of the Department of Chemical Engineering at the University of California, Santa Barbara. He received his B.S. degree from the University of Wisconsin and his Ph.D. degree from Princeton University. Before joining UCSB, he taught at the University of Alberta for nine years. Dr. Seborg has published over 200 articles and co-edited three books on process control and related topics. He has received the American Statistical Association's Statistics in Chemistry Award, the American Automatic Control Council's Education Award, and the ASEE Meriam-Wiley Award. He was elected to the *Process Automation Hall of Fame* in 2008. Dr. Seborg has served on the Editorial Advisor Boards for control engineering journals and book series, and has been a co-organizer of several major conferences. He is an active industrial consultant who serves as an expert witness in legal proceedings.

Thomas F. Edgar holds the Abell Chair in chemical engineering at the University of Texas at Austin. He earned a B.S. degree in chemical engineering from the University of Kansas and a Ph.D. from Princeton University. Before receiving his doctorate, he was employed by Continental Oil Company. His professional honors include the AIChE Colburn and Lewis Awards, ASEE Meriam-Wiley and Chemical Engineering Division Awards, ISA Education Award, and AIChE Computing in Chemical Engineering Award. He is listed in *Who's Who* in America. He has published over 300 papers in the field of process control, optimization, and mathematical modeling of processes such as separations, combustion, and microelectronics processing. He is co-author of *Optimization of Chemical Processes*, published by McGraw-Hill in 2001. Dr. Edgar was president of AIChE in 1997 and President of the American Automatic Control Council in 1989–91.

Duncan A. Mellichamp is Professor Emeritus and a founding member of the faculty of the chemical engineering department at the University of California, Santa Barbara. He is editor of an early book on data acquisition and control computing and has published more than one hundred papers on process modeling, large scale/plantwide systems analysis, and computer control. He earned a B.S. degree from Georgia Tech and a Ph.D. from Purdue University with intermediate studies at the Technische Universität Stuttgart (Germany). He worked for four years with the Textile Fibers Department of the DuPont Company before joining UCSB. Dr. Mellichamp has headed several organizations, including the CACHE Corporation (1977), the UCSB Academic Senate (1990–92), and the University of California Systemwide Academic Senate (1995–97), where he served on the UC Board of Regents. He presently serves on the governing boards of several nonprofit organizations.

Francis J. Doyle III is the Associate Dean for Research in the College of Engineering at the University of California, Santa Barbara. He holds the Duncan and Suzanne Mellichamp Chair in Process Control in the Department of Chemical Engineering, as well as appointments in the Electrical Engineering Department, and the Biomolecular Science and Engineering Program. He received his B.S.E. from Princeton, C.P.G.S. from Cambridge, and Ph.D. from Caltech, all in Chemical Engineering. Prior to his appointment at UCSB, he has held faculty appointments at Purdue University and the University of Delaware, and held visiting positions at DuPont, Weyerhaeuser, and Stuttgart University. He is a Fellow of IEEE, IFAC, and AIMBE; he is also the recipient of multiple research awards (including the AIChE Computing in Chemical Engineering Award) as well as teaching awards (including the ASEE Ray Fahien Award).

Preface

Process control has become increasingly important in the process industries as a consequence of global competition, rapidly changing economic conditions, faster product development, and more stringent environmental and safety regulations. Process control and its allied fields of process modeling and optimization are critical in the development of more flexible and more complex processes for manufacturing high-value-added products. Furthermore, the rapidly declining cost of digital devices and increased computer speed (doubling every 18 months, according to Moore's Law) have enabled high-performance measurement and control systems to become an essential part of industrial plants.

It is clear that the scope and importance of process control technology will continue to expand during the 21st century. Consequently, chemical engineers need to master this subject in order to be able to design and operate modern plants. The concepts of dynamics, feedback, and stability are also important for understanding many complex systems of interest to chemical engineers, such as in bioengineering and advanced materials. An introductory course should provide an appropriate balance of process control theory and practice. In particular, the course should emphasize dynamic behavior, physical and empirical modeling, computer simulation, measurement and control technology, basic control concepts, and advanced control strategies. We have organized this book so that the instructor can cover the basic material while having the flexibility to include advanced topics. The textbook provides the basis for 10 to 30 weeks of instruction for a single course or a sequence of courses at either the undergraduate or first-year graduate levels. It is also suitable for self-study by engineers in industry. The book is divided into reasonably short chapters to make it more readable and modular. This organization allows some chapters to be omitted without a loss of continuity.

The mathematical level of the book is oriented toward a junior or senior student in chemical engineering who has taken at least one course in differential equations. Additional mathematical tools required for the analysis of control systems are introduced as needed. We emphasize process control techniques that are used in practice and provide detailed mathematical analysis only when it is essential for understanding the material. Key theoretical concepts are illustrated with numerous examples and simulations.

The textbook material has evolved at the University of California, Santa Barbara, and the University of Texas at Austin over the past 40 years. The first edition (SEM1) was published in 1989, adopted by over 80 universities worldwide, and translated into Korean and Japanese. In the second edition (SEM2, 2004), we added new chapters on the important topics of process monitoring (Chapter 21), batch process control (Chapter 22), and plantwide control (Chapters 23 and 24). Even with the new chapters, the length of the second edition was about the same as SEM1. Interactive computer software based on MATLAB® and Simulink® software was extensively used in examples and exercises. The second edition was translated into Chinese in 2004.

For the third edition (SEMD3), we are very pleased to have added a fourth co-author, Professor Frank Doyle (UCSB), and we have made major changes that reflect the evolving field of chemical and biological engineering, as well as the practice of process control, which are described in the following.

The book is divided into five parts. Part I provides an introduction to process control and an in-depth discussion of process modeling. Control system design and analysis increasingly rely on the availability of a process model. Consequently, the third edition includes additional material on process modeling based on first principles, such as conservation equations and thermodynamics. Exercises have been added to several chapters based on MATLAB® simulations of two physical models, a distillation column and a furnace. These simulations are based on the book, *Process Control Modules*, by Frank Doyle, Ed Gatzke, and Bob Parker. Both the book and the MATLAB simulations are available on the book Web site (*www.wiley.com/college/seborg*). National Instruments has provided multimedia modules for a number of examples in the book based on their LabVIEW™ software.

Part II (Chapters 3 through 7) is concerned with the analysis of the dynamic (unsteady-state) behavior of processes. We still rely on the use of Laplace transforms and transfer functions, to characterize the dynamic behavior of linear systems. However, we have kept analytical methods involving transforms at a minimum and prefer the use of computer simulation to determine dynamic responses. In addition, the important topics of empirical models and their development from plant data are presented.

Part III (Chapters 8 through 15) addresses the fundamental concepts of feedback and feedforward control. The topics include an overview of the process instrumentation (Chapter 9) and control hardware and software that are necessary to implement process control (Chapter 8 and Appendix A). Chapter 13 (new) presents the important topic of process control strategies at the unit level, and additional material on process safety has been added to Chapter 10. The design and analysis of feedback control systems still receive considerable attention, with emphasis on industry-proven methods for controller design, tuning, and troubleshooting. The frequency response approach for open and closed-loop processes is now combined into a single chapter (14), because of its declining use in the process industries. Part III concludes with a chapter on feedforward and ratio control.

Part IV (Chapters 16 through 22) is concerned with advanced process control techniques. The topics include digital control, multivariable control, and enhancements of PID control, such as cascade control, selective control, and gain scheduling. Up-to-date chapters on real-time optimization and model predictive control emphasize the significant impact these powerful techniques have had on industrial practice. Other chapters consider process monitoring and batch process control. The two plantwide control chapters that were introduced in SEM2 have been moved to the book Web site, as Appendices G and H. We have replaced this material with two new chapters on biosystems control, principally authored by our recently added fourth author, Frank Doyle. Part V (new Chapters 23 and 24) covers the application of process control in biotechnology and biomedical systems, and introduces basic ideas in systems biology.

The book Web site will contain errata lists for current and previous editions that are available to both students and instructors. In addition, the following resources for instructors (only) are provided: solutions manual, lecture slides, figures from the text, archival material from SEM1 and SEM2, and a link to the authors' Web sites. Instructors need to visit the book Web site to register for a password to access the protected resources. The book Web site is located at *www.wiley.com/college/seborg.*

We gratefully acknowledge the very helpful suggestions and reviews provided by many colleagues in academia and industry: Joe Alford, Anand Asthagiri, Karl Åström, Tom Badgwell, Max Barolo, Larry Biegler, Don Bartusiak, Terry Blevins, Dominique Bonvin, Richard Braatz, Dave Camp, Jarrett Campbell, I-Lung Chien, Will Cluett, Oscar Crisalle, Patrick Daugherty, Bob Deshotels, Rainer Dittmar, Jim Downs, Ricardo Dunia, David Ender, Stacy Firth, Rudiyanto Gunawan, Juergen Hahn, Sandra Harris, Karlene Hoo, Biao Huang, Babu Joseph, Derrick Kozub, Jietae Lee, Bernt Lie, Cheng Ling, Sam Mannan, Tom McAvoy, Greg McMillan, Randy Miller, Samir Mitragotri, Manfred Morari, Duane Morningred, Kenneth Muske, Mark Nixon, Srinivas Palanki, Bob Parker, Michel Perrier, Mike Piovoso, Joe Qin, Larry Ricker, Dan Rivera, Derrick Rollins, Alan Schneider, Sirish Shah, Mikhail Skliar, Sigurd Skogestad, Tyler Soderstrom, Ron Sorensen, Dirk Thiele, John Tsing, Ernie Vogel, Doug White, Willy Wojsznis, Robert Young, and the late Cheng-Ching Yu.

We also gratefully acknowledge the many current and recent students and postdocs at UCSB and UT-Austin who have provided careful reviews and simulation results: Ivan Castillo, Marco Castellani, David Castineira, Dan Chen, Jeremy Cobbs, Jeremy Conner, Eyal Dassau, Doug French, Scott Harrison, John Hedengren, Xiaojiang Jiang, Ben Juricek, Fred Loquastro III, Doron Ronon, Lina Rueda, Ashish Singhal, Jeff Ward, Dan Weber, and Yang Zhang. Eyal Dassau was instrumental in converting the old PCM modules to the version posted to this book's Web site. We revised the solution manual, which was originally prepared for the first and second editions by Mukul Agarwal and David Castineira, with the help of Yang Zhang. We greatly appreciate their careful attention to detail. We commend Chris Bailor for her word processing skill during the numerous revisions for the third edition. We also acknowledge the patience of our editor, Jenny Welter, during the long revision process. Finally, we are deeply grateful for the support and patience of our long-suffering wives (Judy, Donna, Suzanne, and Diana) during the revisions of the book.

In the spirit of continuous improvement, we are interested in receiving feedback from students, faculty, and practitioners who use this book. We hope you find it to be useful.

Dale E. Seborg
Thomas F. Edgar
Duncan A. Mellichamp
Francis J. Doyle III

Contents

PART ONE
INTRODUCTION TO PROCESS CONTROL

PART TWO
DYNAMIC BEHAVIOR OF PROCESSES

PART THREE
FEEDBACK AND FEEDFORWARD CONTROL

Chapter 1

Introduction to Process Control

CHAPTER CONTENTS

In recent years the performance requirements for process plants have become increasingly difficult to satisfy. Stronger competition, tougher environmental and safety regulations, and rapidly changing economic conditions have been key factors in tightening product quality specifications. A further complication is that modern plants have become more difficult to operate because of the trend toward complex and highly integrated processes. For such plants, it is difficult to prevent disturbances from propagating from one unit to other interconnected units.

In view of the increased emphasis placed on safe, efficient plant operation, it is only natural that the subject of *process control* has become increasingly important in recent years. Without computer-based process control systems it would be impossible to operate modern plants safely and profitably while satisfying product quality and environmental requirements. Thus, it is important for chemical engineers to have an understanding of both the theory and practice of process control.

The two main subjects of this book are *process dynamics* and *process control*. The term *process dynamics* refers to unsteady-state (or transient) process behavior. By contrast, most of the chemical engineering curricula emphasize steady-state and equilibrium conditions in such courses as material and energy balances, thermodynamics, and transport phenomena. But process dynamics are also very important. Transient operation occurs during important situations such as start-ups and shutdowns, unusual process disturbances, and planned transitions from one product grade to another. Consequently, the first part of this book is concerned with process dynamics.

The primary objective of process control is to maintain a process at the desired operating conditions, safely and efficiently, while satisfying environmental and product quality requirements. The subject of process control is concerned with how to achieve these goals. In large-scale, integrated processing plants such as oil refineries or ethylene plants, thousands of process variables such as compositions, temperatures, and pressures are measured and must be controlled. Fortunately, large numbers of process variables (mainly flow rates) can usually be manipulated for this purpose. Feedback control systems compare measurements with their desired values and then adjust the manipulated variables accordingly.

As an introduction to the subject, we consider representative process control problems in several industries.

1.1 REPRESENTATIVE PROCESS CONTROL PROBLEMS

The foundation of process control is *process understanding*. Thus, we begin this section with a basic question: what is a process? For our purposes, a brief definition is appropriate:

> ***Process:*** *The conversion of feed materials to products using chemical and physical operations. In practice, the term* process *tends to be used for both the processing operation and the processing equipment.*

Note that this definition applies to three types of common processes: continuous, batch, and semi-batch. Next, we consider representative processes and briefly summarize key control issues.

1.1.1 Continuous Processes

Four continuous processes are shown schematically in Figure 1.1:

(a) ***Tubular heat exchanger.*** A process fluid on the tube side is cooled by cooling water on the shell side. Typically, the exit temperature of the process fluid is controlled by manipulating the cooling water flow rate. Variations in the inlet temperatures and the process fluid flow rate affect the heat exchanger operation. Consequently, these variables are considered to be disturbance variables.

(b) ***Continuous stirred-tank reactor (CSTR).*** If the reaction is highly exothermic, it is necessary to control the reactor temperature by manipulating the flow rate of coolant in a jacket or cooling coil. The feed conditions (composition, flow rate, and temperature) can be manipulated variables or disturbance variables.

(c) ***Thermal cracking furnace.*** Crude oil is broken down ("cracked") into a number of lighter petroleum fractions by the heat transferred from a burning fuel/air mixture. The furnace temperature and amount of excess air in the flue gas can be controlled by manipulating the fuel flow rate and the fuel/air ratio. The crude oil composition and the heating quality of the fuel are common disturbance variables.

(d) ***Multicomponent distillation column.*** Many different control objectives can be formulated for distillation columns. For example, the distillate composition can be controlled by adjusting the reflux flow rate or the distillate flow rate. If the composition cannot be measured on-line, a tray temperature near the top of the column can be controlled instead. If the feed stream is supplied by an upstream process, the feed conditions will be disturbance variables.

For each of these four examples, the process control problem has been characterized by identifying three important types of process variables.

- *Controlled variables (CVs):* The process variables that are controlled. The desired value of a controlled variable is referred to as its *set point*.
- *Manipulated variables (MVs):* The process variables that can be adjusted in order to keep the controlled variables at or near their set points. Typically, the manipulated variables are flow rates.
- *Disturbance variables (DVs):* Process variables that affect the controlled variables but cannot be manipulated. Disturbances generally are related to changes in the operating environment of the process: for example, its feed conditions or ambient temperature. Some disturbance variables can be measured on-line, but many cannot such as the crude oil composition for Process (c), a thermal cracking furnace.

The specification of CVs, MVs, and DVs is a critical step in developing a control system. The selections should

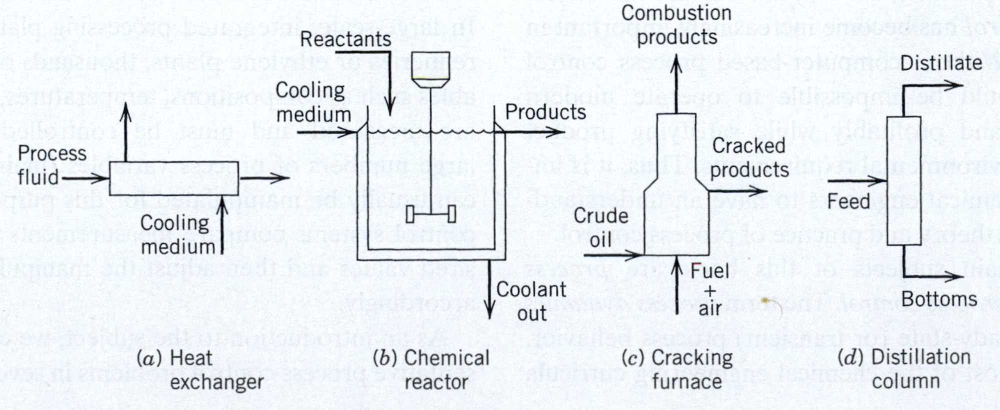

Figure 1.1 Some typical continuous processes.

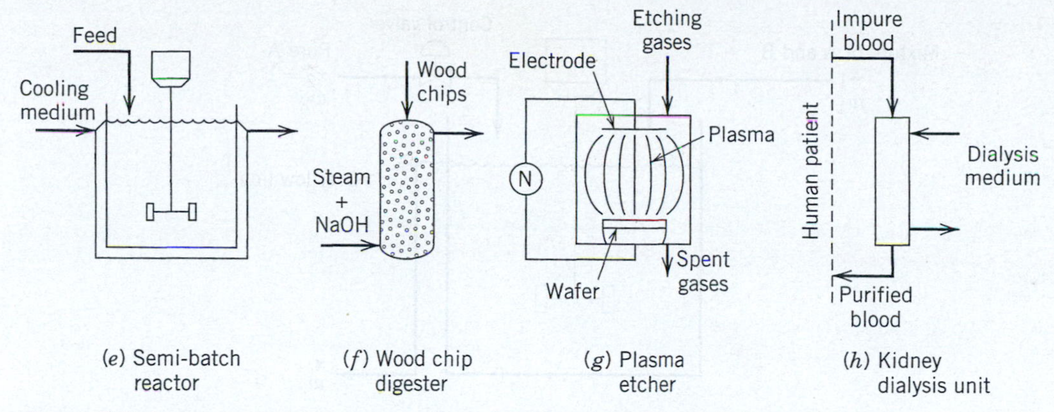

Figure 1.2 Some typical processes whose operation is noncontinuous.

be based on process knowledge, experience, and control objectives.

1.1.2 Batch and Semi-Batch Processes

Batch and semi-batch processes are used in many process industries, including microelectronics, pharmaceuticals, specialty chemicals, and fermentation. Batch and semi-batch processes provide needed flexibility for multiproduct plants, especially when products change frequently and production quantities are small. Figure 1.2 shows four representative batch and semi-batch processes:

(e) ***Batch or semi-batch reactor.*** An initial charge of reactants is brought up to reaction conditions, and the reactions are allowed to proceed for a specified period of time or until a specified conversion is obtained. Batch and semi-batch reactors are used routinely in specialty chemical plants, polymerization plants (where a reaction byproduct typically is removed during the reaction), and in pharmaceutical and other bioprocessing facilities (where a feed stream, e.g., glucose, is fed into the reactor during a portion of the cycle to feed a living organism, such as a yeast or protein). Typically, the reactor temperature is controlled by manipulating a coolant flow rate. The end-point (final) concentration of the batch can be controlled by adjusting the desired temperature, the flow of reactants (for semi-batch operation), or the cycle time.

(f) ***Batch digester in a pulp mill.*** Both continuous and semi-batch digesters are used in paper manufacturing to break down wood chips in order to extract the cellulosic fibers. The end point of the chemical reaction is indicated by the kappa number, a measure of lignin content. It is controlled to a desired value by adjusting the digester temperature, pressure, and/or cycle time.

(g) ***Plasma etcher in semiconductor processing.*** A single wafer containing hundreds of printed circuits is subjected to a mixture of etching gases under conditions suitable to establish and maintain a plasma (a high voltage applied at high temperature and extremely low pressure). The unwanted material on a layer of a microelectronics circuit is selectively removed by chemical reactions. The temperature, pressure, and flow rates of etching gases to the reactor are controlled by adjusting electrical heaters and control valves.

(h) ***Kidney dialysis unit.*** This medical equipment is used to remove waste products from the blood of human patients whose own kidneys are failing or have failed. The blood flow rate is maintained by a pump, and "ambient conditions," such as temperature in the unit, are controlled by adjusting a flow rate. The dialysis is continued long enough to reduce waste concentrations to acceptable levels.

Next, we consider an illustrative example in more detail.

1.2 ILLUSTRATIVE EXAMPLE—A BLENDING PROCESS

A simple blending process is used to introduce some important issues in control system design. Blending operations are commonly used in many industries to ensure that final products meet customer specifications.

A continuous, stirred-tank blending system is shown in Fig. 1.3. The control objective is to blend the two inlet streams to produce an outlet stream that has the desired composition. Stream 1 is a mixture of two chemical species, A and B. We assume that its mass flow rate w_1 is constant, but the mass fraction of A, x_1, varies with time. Stream 2 consists of pure A and thus $x_2 = 1$. The mass flow rate of Stream 2, w_2, can be manipulated using a control valve. The mass

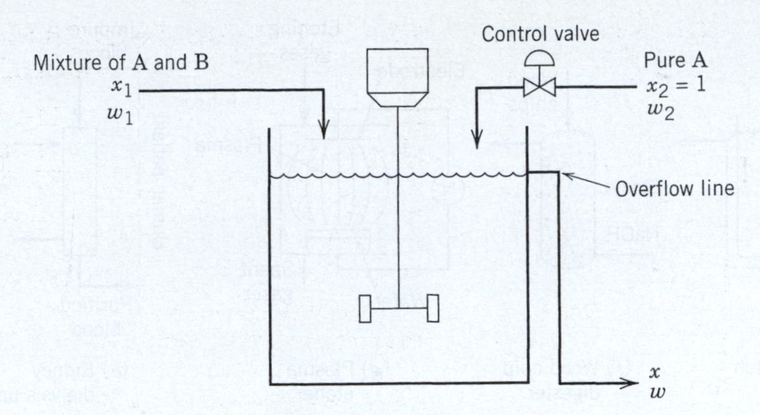

Figure 1.3 Stirred-tank blending system.

fraction of A in the exit stream is denoted by x and the desired value (set point) by x_{sp}. Thus for this control problem, the controlled variable is x, the manipulated variable is w_2, and the disturbance variable is x_1.

Next we consider two questions.

Design Question. *If the nominal value of x_1 is $\overline{x}_1$, what nominal flow rate $\overline{w}_2$ is required to produce the desired outlet concentration, x_{sp}?*

To answer this question, we consider the steady-state material balances:

Overall balance:

$$0 = \overline{w}_1 + \overline{w}_2 - \overline{w} \tag{1-1}$$

Component A balance:

$$0 = \overline{w}_1\overline{x}_1 + \overline{w}_2\overline{x}_2 - \overline{w}\,\overline{x} \tag{1-2}$$

The overbar over a symbol denotes its nominal steady-state value, for example, the value used in the process design. According to the process description, $\overline{x}_2 = 1$ and $\overline{x} = x_{sp}$. Solving Eq. 1-1 for $\overline{w}$, substituting these values into Eq. 1-2, and rearranging gives:

$$\overline{w}_2 = \overline{w}_1 \frac{x_{sp} - \overline{x}_1}{1 - x_{sp}} \tag{1-3}$$

Equation 1-3 is the design equation for the blending system. If our assumptions are correct and if $x_1 = \overline{x}_1$, then this value of w_2 will produce the desired result, $x = x_{sp}$. But what happens if conditions change?

Control Question. *Suppose that inlet concentration x_1 varies with time. How can we ensure that the outlet composition x remains at or near its desired value, x_{sp}?*

As a specific example, assume that x_1 increases to a constant value that is larger than its nominal value, $\overline{x}_1$. It is clear that the outlet composition will also increase due to the increase in inlet composition. Consequently, at this new steady state, $x > x_{sp}$.

Next we consider several strategies for reducing the effects of x_1 disturbances on x.

Method 1. *Measure x and adjust w_2.* It is reasonable to measure controlled variable x and then adjust w_2 accordingly. For example, if x is too high, w_2 should be reduced; if x is too low, w_2 should be increased. This control strategy could be implemented by a person (*manual control*). However, it would normally be more convenient and economical to automate this simple task (*automatic control*).

Method 1 can be implemented as a simple control algorithm (or control law),

$$w_2(t) = \overline{w}_2 + K_c[x_{sp} - x(t)] \tag{1-4}$$

where K_c is a constant called the *controller gain*. The symbols, $w_2(t)$ and $x(t)$, indicate that w_2 and x change with time. Equation 1-4 is an example of *proportional control,* because the change in the flow rate, $w_2(t) - \overline{w}_2$, is proportional to the deviation from the set point, $x_{sp} - x(t)$. Consequently, a large deviation from set point produces a large corrective action, while a small deviation results in a small corrective action. Note that we require K_c to be positive because w_2 must increase when x decreases, and vice versa. However, in other control applications, negative values of K_c are appropriate, as discussed in Chapter 8.

A schematic diagram of Method 1 is shown in Fig. 1.4. The outlet concentration is measured and transmitted to the controller as an electrical signal. (Electrical signals are shown as dashed lines in Fig. 1.4.) The controller executes the control law and sends the calculated value of w_2 to the control valve as an electrical signal. The control valve opens or closes accordingly. In Chapters 8 and 9 we consider process instrumentation and control hardware in more detail.

Method 2. *Measure x_1, adjust w_2.* As an alternative to Method 1, we could measure disturbance variable x_1 and adjust w_2 accordingly. Thus, if $x_1 > \overline{x}_1$, we would decrease w_2 so that $w_2 < \overline{w}_2$. If $x_1 < \overline{x}_1$, we would

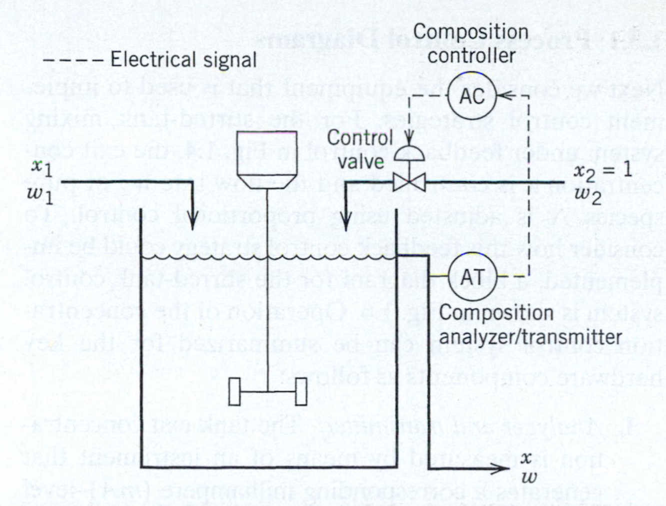

Figure 1.4 Blending system and Control Method 1.

increase w_2. A control law based on Method 2 can be derived from Eq. 1-3 by replacing $\overline{x}_1$ with $x_1(t)$ and $\overline{w}_2$ with $w_2(t)$:

$$w_2(t) = \overline{w}_1 \frac{x_{sp} - x_1(t)}{1 - x_{sp}} \tag{1-5}$$

The schematic diagram for Method 2 is shown in Fig. 1.5. Because Eq. 1-3 is valid only for steady-state conditions, it is not clear just how effective Method 2 will be during the transient conditions that occur after an x_1 disturbance.

Method 3. *Measure x_1 and x, adjust w_2.* This approach is a combination of Methods 1 and 2.

Method 4. *Use a larger tank.* If a larger tank is used, fluctuations in x_1 will tend to be damped out as a result of the larger volume of liquid. However, increasing tank size is an expensive solution due to the increased capital cost.

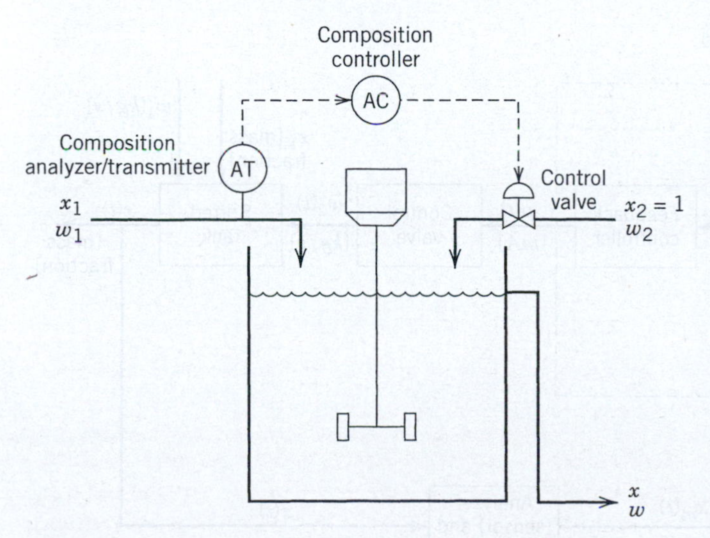

Figure 1.5 Blending system and Control Method 2.

1.3 CLASSIFICATION OF PROCESS CONTROL STRATEGIES

Next, we will classify the four blending control strategies of the previous section and discuss their relative advantages and disadvantages. Method 1 is an example of a *feedback control* strategy. The distinguishing feature of feedback control is that the controlled variable is measured, and that the measurement is used to adjust the manipulated variable. For feedback control, the disturbance variable is *not* measured.

It is important to make a distinction between *negative feedback* and *positive feedback*. In the engineering literature, negative feedback refers to the desirable situation in which the corrective action taken by the controller forces the controlled variable toward the set point. On the other hand, when positive feedback occurs, the controller makes things worse by forcing the controlled variable farther away from the set point. For example, in the blending control problem, positive feedback takes place if $K_c < 0$, because w_2 will increase when x increases.[1] Clearly, it is of paramount importance to ensure that a feedback control system incorporates negative feedback rather than positive feedback.

An important advantage of feedback control is that corrective action occurs regardless of the source of the disturbance. For example, in the blending process, the feedback control law in (1-4) can accommodate disturbances in w_1, as well as x_1. Its ability to handle disturbances of unknown origin is a major reason why feedback control is the dominant process control strategy. Another important advantage is that feedback control reduces the sensitivity of the controlled variable to unmeasured disturbances and process changes. However, feedback control does have a fundamental limitation: no corrective action is taken until after the disturbance has upset the process, that is, until after the controlled variable deviates from the set point. This shortcoming is evident from the control law of (1-4).

Method 2 is an example of a *feedforward control strategy*. The distinguishing feature of feedforward control is that the disturbance variable is measured, but the controlled variable is not. The important advantage of feedforward control is that corrective action is taken *before* the controlled variable deviates from the set point. Ideally, the corrective action will cancel the effects of the disturbance so that the controlled variable is not affected by the disturbance. Although ideal cancelation is generally not possible, feedforward control can significantly

[1] Note that social scientists use the terms negative feedback and positive feedback in a very different way. For example, they would say that teachers provide "positive feedback" when they compliment students who correctly do assignments. Criticism of a poor performance would be an example of "negative feedback."

Table 1.1 Concentration Control Strategies for the Blending System

Method	Measured Variable	Manipulated Variable	Category
1	x	w_2	FB
2	x_1	w_2	FF
3	x_1 and x	w_2	FF/FB
4	—	—	Design change

FB = feedback control; FF = feedforward control; FF/FB = feedforward control and feedback control.

reduce the effects of measured disturbances, as discussed in Chapter 15.

Feedforward control has three significant disadvantages: (i) the disturbance variable must be measured (or accurately estimated), (ii) no corrective action is taken for unmeasured disturbances, and (iii) a process model is required. For example, the feedforward control strategy for the blending system (Method 2) does not take any corrective action for unmeasured w_1 disturbances. In principle, we could deal with this situation by measuring both x_1 and w_1 and then adjusting w_2 accordingly. However, in industrial applications it is generally uneconomical to attempt to measure all potential disturbances. A more practical approach is to use a combined feedforward–feedback control system, in which feedback control provides corrective action for unmeasured disturbances, while feedforward control reacts to eliminate measured disturbances before the controlled variable is upset. Consequently, in industrial applications feedforward control is normally used in combination with feedback control. This approach is illustrated by Method 3, a combined feedforward-feedback control strategy because both x and x_1 are measured.

Finally, Method 4 consists of a process design change and thus is not really a control strategy. The four strategies for the stirred-tank blending system are summarized in Table 1.1.

1.3.1 Process Control Diagrams

Next we consider the equipment that is used to implement control strategies. For the stirred-tank mixing system under feedback control in Fig. 1.4, the exit concentration x is controlled and the flow rate w_2 of pure species A is adjusted using proportional control. To consider how this feedback control strategy could be implemented, a block diagram for the stirred-tank control system is shown in Fig. 1.6. Operation of the concentration control system can be summarized for the key hardware components as follows:

1. *Analyzer and transmitter:* The tank exit concentration is measured by means of an instrument that generates a corresponding milliampere (mA)–level signal. This time-varying signal is then sent to the controller.
2. *Feedback controller:* The controller performs three distinct calculations. First, it converts the actual set point x_{sp} into an equivalent internal signal $\tilde{x}_{sp}$. Second, it calculates an error signal $e(t)$ by subtracting the measured value $x_m(t)$ from the set point $\tilde{x}_{sp}$, that is, $e(t) = \tilde{x}_{sp} - x_m(t)$. Third, controller output $p(t)$ is calculated from the proportional control law similar to Eq. 1-4.
3. *Control valve:* The controller output $p(t)$ in this case is a DC current signal that is sent to the control valve to adjust the valve stem position, which in turn affects flow rate $w_2(t)$. Because many control valves are pneumatic, i.e., are operated by air pressure, the controller output signal may have to be converted to an equivalent air pressure signal capable of adjusting the valve position. For simplicity, we do not show such a transducer in this diagram.

The block diagram in Fig. 1.6 provides a convenient starting point for analyzing process control problems. The physical units for each input and output signal are also

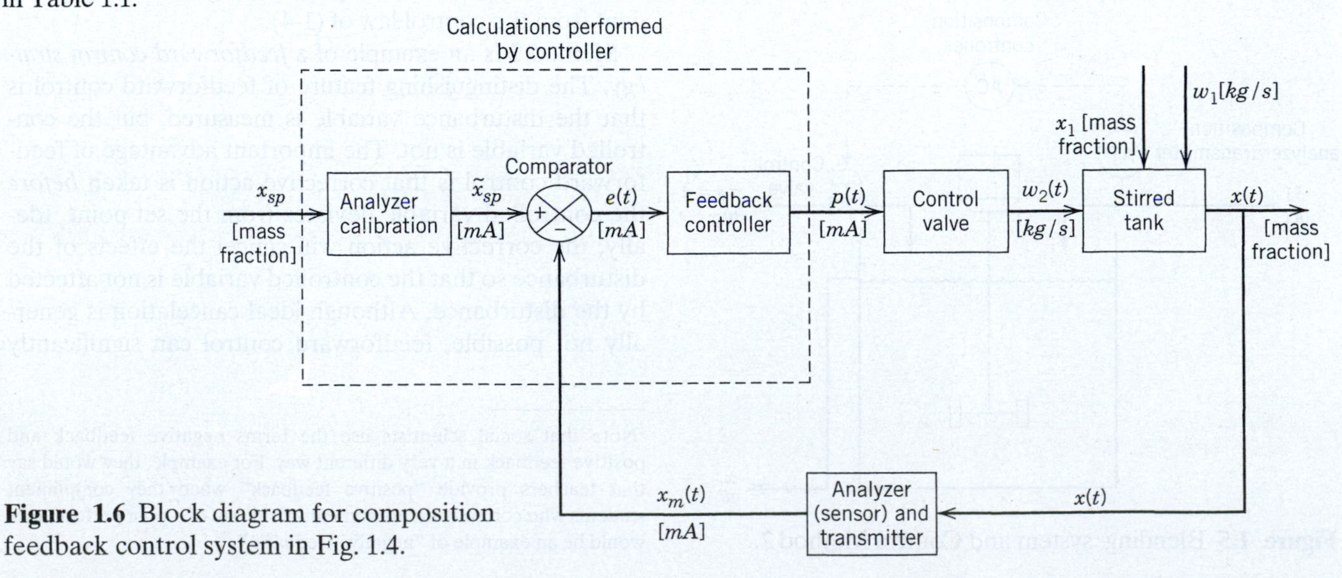

Figure 1.6 Block diagram for composition feedback control system in Fig. 1.4.

shown. Note that the schematic diagram in Fig. 1.4. shows the *physical connections* between the components of the control system, while the block diagram shows the *flow of information* within the control system. The block labeled "control valve" has $p(t)$ as its input signal and $w_2(t)$ as its output signal, which illustrates that the signals on a block diagram can represent either a physical variable such as $w_2(t)$ or an instrument signal such as $p(t)$.

Each component in Fig. 1.6 exhibits behavior that can be described by a differential or algebraic equation. One of the tasks facing a control engineer is to develop suitable mathematical descriptions for each block; the development and analysis of such dynamic mathematical models are considered in Chapters 2–7.

Other elements in the block diagram (Fig. 1.6) are discussed in detail in future chapters. Sensors and control valves are presented in Chapter 9, and the feedback controller is covered in Chapter 8.

1.4 A MORE COMPLICATED EXAMPLE—A DISTILLATION COLUMN

The blending control system in the previous section is quite simple, because there is only one controlled variable and one manipulated variable. For most practical applications, there are multiple controlled variables and multiple manipulated variables. As a representative example, we consider the distillation column in Fig. 1.7, which has five controlled variables and five manipulated variables. The controlled variables are product compositions, x_D and x_B, column pressure, P, and the liquid levels in the reflux drum and column base, h_D and h_B. The five manipulated variables are product flow rates, D and B, reflux flow, R, and the heat duties for the condenser and reboiler, Q_D and Q_B. The heat duties are adjusted via the control valves on the coolant and heating medium lines. The feed stream is assumed to come from an upstream unit. Thus, the feed flow rate cannot be manipulated, but it can be measured and used for feedforward control.

A conventional *multiloop control* strategy for this distillation column would consist of five feedback control loops. Each control loop uses a single manipulated variable to control a single controlled variable. But how should the controlled and manipulated variables be paired? The total number of different multiloop control configurations that could be considered is 5!, or 120. Many of these control configurations are impractical or unworkable, such as any configuration that attempts to control the base level h_B by manipulating distillate flow D or condenser heat duty Q_D. However, even after the infeasible control configurations are eliminated, there are still many reasonable configurations left. Thus, there is a need for systematic techniques that can identify the most promising configurations. Fortunately, such tools are available; these are discussed in Chapter 18.

In control applications, for which conventional multiloop control systems are not satisfactory, an alternative approach, *multivariable control*, can be advantageous. In multivariable control, each manipulated variable is adjusted based on the measurements of all the controlled variables rather than only a single controlled variable, as in multiloop control. The adjustments are based on a dynamic model of the process that indicates how the manipulated variables affect the controlled variables. Consequently, the performance of multivariable control, or any model-based control technique, will depend heavily on the accuracy of the process model. A specific type of multivariable control, *model predictive control*, has had a major impact on industrial practice, as discussed in Chapter 20.

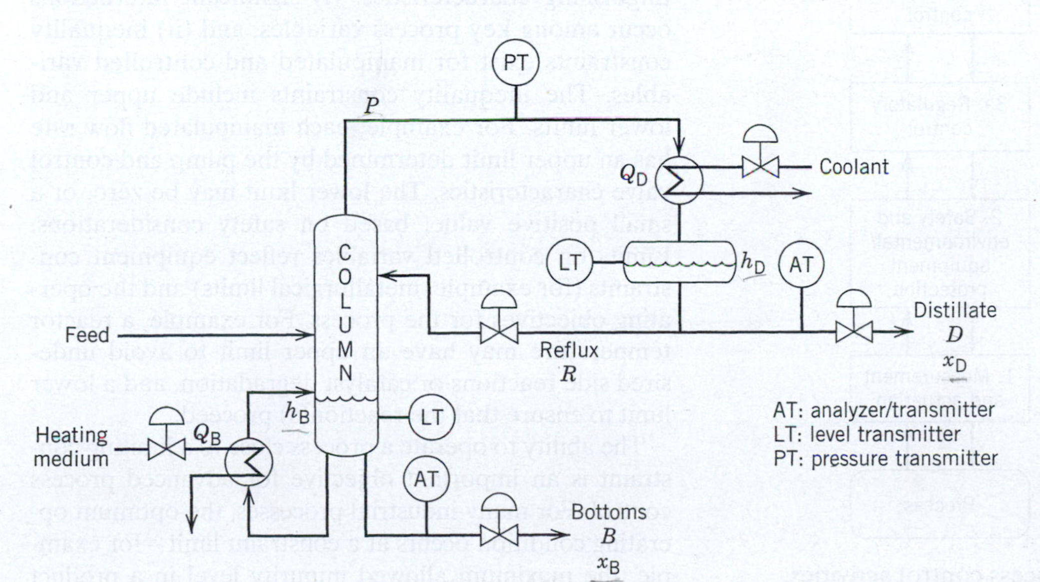

Figure 1.7 Controlled and manipulated variables for a typical distillation column.

1.5 THE HIERARCHY OF PROCESS CONTROL ACTIVITIES

As mentioned earlier, the chief objective of process control is to maintain a process at the desired operating conditions, safely and efficiently, while satisfying environmental and product quality requirements. So far, we have emphasized one process control activity, keeping controlled variables at specified set points. But there are other important activities that we will now briefly describe.

In Fig. 1.8 the process control activities are organized in the form of a hierarchy with required functions at the lower levels and desirable but optional functions at the higher levels. The time scale for each activity is shown on the left side. Note that the frequency of execution is much lower for the higher-level functions.

Measurement and Actuation (Level 1)

Measurement devices (sensors and transmitters) and actuation equipment (for example, control valves) are used to measure process variables and implement the calculated control actions. These devices are interfaced to the control system, usually digital control equipment such as a digital computer. Clearly, the measurement and actuation functions are an indispensable part of any control system.

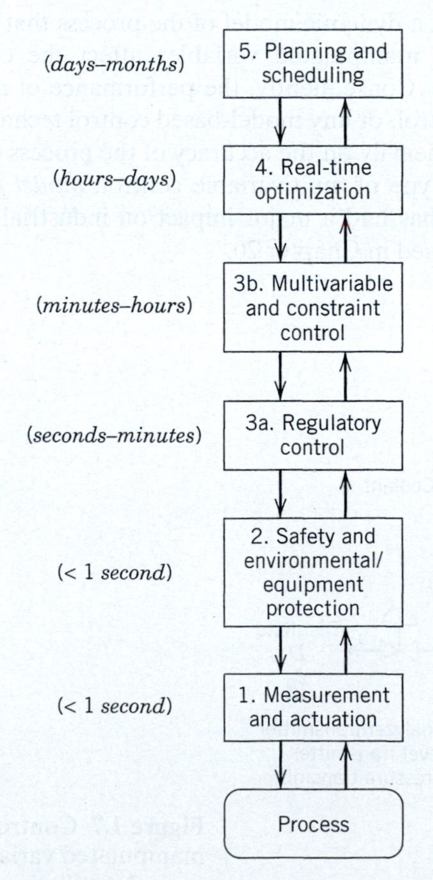

Figure 1.8 Hierarchy of process control activities.

Safety and Environmental/Equipment Protection (Level 2)

The Level 2 functions play a critical role by ensuring that the process is operating safely and satisfies environmental regulations. As discussed in Chapter 10, process safety relies on the principle of *multiple protection layers* that involve groupings of equipment and human actions. One layer includes process control functions, such as alarm management during abnormal situations, and *safety instrumented systems* for emergency shutdowns. The safety equipment (including sensors and control valves) operates independently of the regular instrumentation used for regulatory control in Level 3a. Sensor validation techniques can be employed to confirm that the sensors are functioning properly.

Regulatory Control (Level 3a)

As mentioned earlier, successful operation of a process requires that key process variables such as flow rates, temperatures, pressures, and compositions be operated at or close to their set points. This Level 3a activity, *regulatory control*, is achieved by applying standard feedback and feedforward control techniques (Chapters 11–15). If the standard control techniques are not satisfactory, a variety of advanced control techniques are available (Chapters 16–18). In recent years, there has been increased interest in monitoring control system performance (Chapter 21).

Multivariable and Constraint Control (Level 3b)

Many difficult process control problems have two distinguishing characteristics: (i) significant interactions occur among key process variables, and (ii) inequality constraints exist for manipulated and controlled variables. The inequality constraints include upper and lower limits. For example, each manipulated flow rate has an upper limit determined by the pump and control valve characteristics. The lower limit may be zero, or a small positive value, based on safety considerations. Limits on controlled variables reflect equipment constraints (for example, metallurgical limits) and the operating objectives for the process. For example, a reactor temperature may have an upper limit to avoid undesired side reactions or catalyst degradation, and a lower limit to ensure that the reaction(s) proceed.

The ability to operate a process close to a limiting constraint is an important objective for advanced process control. For many industrial processes, the optimum operating condition occurs at a constraint limit—for example, the maximum allowed impurity level in a product

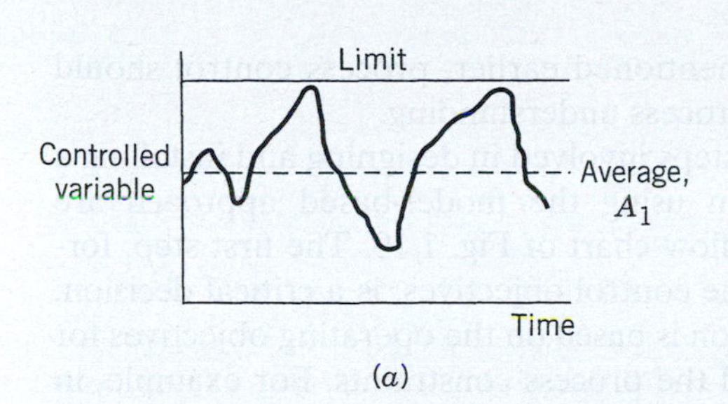

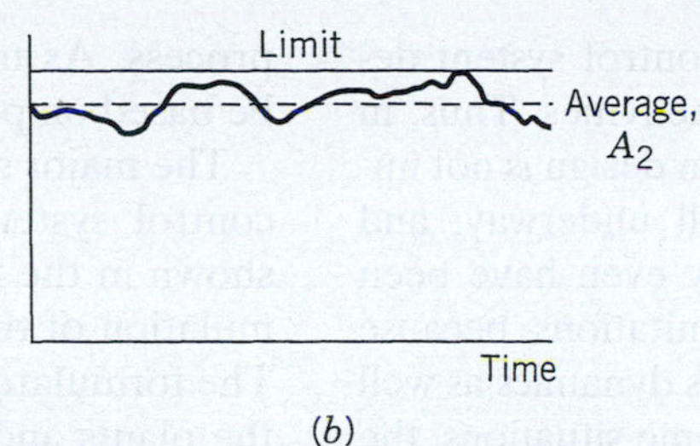

Figure 1.9 Process variability over time: (*a*) before improved process control; (*b*) after.

stream. For these situations, the set point should not be the constraint value, because a process disturbance could force the controlled variable beyond the limit. Thus, the set point should be set conservatively, based on the ability of the control system to reduce the effects of disturbances. This situation is illustrated in Fig. 1.9. For (*a*), the variability of the controlled variable is quite high, and consequently, the set point must be specified well below the limit. For (*b*), the improved control strategy has reduced the variability; consequently, the set point can be moved closer to the limit, and the process can be operated closer to the optimum operating condition.

The standard process control techniques of Level 3a may not be adequate for difficult control problems that have serious process interactions and inequality constraints. For these situations, the advanced control techniques of Level 3b, *multivariable control* and *constraint control*, should be considered. In particular, the *model predictive control* (*MPC*) strategy was developed to deal with both process interactions and inequality constraints. MPC is the subject of Chapter 20.

Real-time Optimization (Level 4)

The optimum operating conditions for a plant are determined as part of the process design. But during plant operations, the optimum conditions can change frequently owing to changes in equipment availability, process disturbances, and economic conditions (for example, raw material costs and product prices). Consequently, it can be very profitable to recalculate the optimum operating conditions on a regular basis. This Level 4 activity, *real-time optimization* (*RTO*), is the subject of Chapter 19. The new optimum conditions are then implemented as set points for controlled variables.

The RTO calculations are based on a steady-state model of the plant and economic data such as costs and product values. A typical objective for the optimization is to minimize operating cost or maximize the operating profit. The RTO calculations can be performed for a single process unit and/or on a plantwide basis.

The Level 4 activities also include data analysis to ensure that the process model used in the RTO calculations is accurate for the current conditions. Thus, *data reconciliation* techniques can be used to ensure that steady-state mass and energy balances are satisfied. Also, the process model can be updated using parameter estimation techniques and recent plant data (Chapter 7).

Planning and Scheduling (Level 5)

The highest level of the process control hierarchy is concerned with planning and scheduling operations for the entire plant. For continuous processes, the production rates of all products and intermediates must be planned and coordinated, based on equipment constraints, storage capacity, sales projections, and the operation of other plants, sometimes on a global basis. For the intermittent operation of batch and semi-batch processes, the production control problem becomes a batch scheduling problem based on similar considerations. Thus, planning and scheduling activities pose difficult optimization problems that are based on both engineering considerations and business projections.

Summary of the Process Control Hierarchy

The activities of Levels 1, 2, and 3a in Fig. 1.8, are required for all manufacturing plants, while the activities in Levels 3b–5 are optional but can be very profitable. The decision to implement one or more of these higher-level activities depends very much on the application and the company. The decision hinges strongly on economic considerations (for example, a cost/benefit analysis), and company priorities for their limited resources, both human and financial. The immediacy of the activity decreases from Level 1 to Level 5 in the hierarchy. However, the amount of analysis and the computational requirements increase from the lowest level to the highest level. The process control activities at different levels should be carefully coordinated and require information transfer from one level to the next. The successful implementation of these process control activities is a critical factor in making plant operation as profitable as possible.

1.6 AN OVERVIEW OF CONTROL SYSTEM DESIGN

In this section, we introduce some important aspects of control system design. However, it is appropriate first to describe the relationship between process design and process control.

Traditionally, process design and control system design have been separate engineering activities. Thus, in the traditional approach, control system design is not initiated until after plant design is well underway, and when major pieces of equipment may even have been ordered. This approach has serious limitations, because the plant design determines the process dynamics as well as the operability of the plant. In extreme situations, the process may be uncontrollable, even though the design appears satisfactory from a steady-state point of view. A more desirable approach is to consider process dynamics and control issues early in the process design. The interaction between process design and control is analyzed in more detail in Chapters 13, 25, and 26.

Next, we consider two general approaches to control system design:

1. ***Traditional Approach.*** The control strategy and control system hardware are selected based on knowledge of the process, experience, and insight. After the control system is installed in the plant, the controller settings (such as controller gain K_c in Eq. 1-4) are adjusted. This activity is referred to as *controller tuning*.
2. ***Model-Based Approach.*** A dynamic model of the process is first developed that can be helpful in at least three ways: (i) it can be used as the basis for model-based controller design methods (Chapters 12 and 14), (ii) the dynamic model can be incorporated directly in the control law (for example, model predictive control), and (iii) the model can be used in a computer simulation to evaluate alternative control strategies and to determine preliminary values of the controller settings.

In this book, we advocate the philosophy that, for complex processes, a dynamic model of the process should be developed so that the control system can be properly designed. Of course, for many simple process control problems, controller specification is relatively straightforward and a detailed analysis or an explicit model is not required. For complex processes, however, a process model is invaluable both for control system design and for an improved understanding of the process. As mentioned earlier, process control should be based on process understanding.

The major steps involved in designing and installing a control system using the model-based approach are shown in the flow chart of Fig. 1.10. The first step, formulation of the control objectives, is a critical decision. The formulation is based on the operating objectives for the plants and the process constraints. For example, in the distillation column control problem, the objective might be to regulate a key component in the distillate stream, the bottoms stream, or key components in both streams. An alternative would be to minimize energy consumption (e.g., heat input to the reboiler) while meeting product quality specifications on one or both product streams. The inequality constraints should include upper and lower limits on manipulated variables, conditions that lead to flooding or weeping in the column, and product impurity levels.

After the control objectives have been formulated, a dynamic model of the process is developed. The dynamic model can have a theoretical basis, for example, physical and chemical principles such as conservation laws and rates of reactions (Chapter 2), or the model can be developed empirically from experimental data (Chapter 7). If experimental data are available, the dynamic model should be validated, with the data and the model accuracy characterized. This latter information is useful for control system design and tuning.

The next step in the control system design is to devise an appropriate control strategy that will meet the control objectives while satisfying process constraints. As indicated in Fig. 1.10, this design activity is both an art and a science. Process understanding and the experience and preferences of the design team are key factors. Computer simulation of the controlled process is used to screen alternative control strategies and to provide preliminary estimates of appropriate controller settings.

Finally, the control system hardware and instrumentation are selected, ordered, and installed in the plant. Then the control system is tuned in the plant using the preliminary estimates from the design step as a starting point. Controller tuning usually involves trial-and-error procedures as described in Chapter 12.

SUMMARY

In this chapter we have introduced the basic concepts of process dynamics and process control. The process dynamics determine how a process responds during transient conditions, such as plant start-ups and shutdowns, grade changes, and unusual disturbances. Process control enables the process to be maintained at the desired operating conditions, safely and efficiently, while satisfying environmental and product quality requirements. Without effective process control, it would be impossible to operate large-scale industrial plants.

Two physical examples, a continuous blending system and a distillation column, have been used to introduce basic control concepts, notably, feedback and feedforward control. We also motivated the need for a systematic approach for the design of control systems for complex processes. Control system development consists of a number of separate activities that are shown in Fig. 1.10. In this book we advocate the design philosophy that for complex processes, a dynamic model of the process should be developed so that the control system can be properly designed.

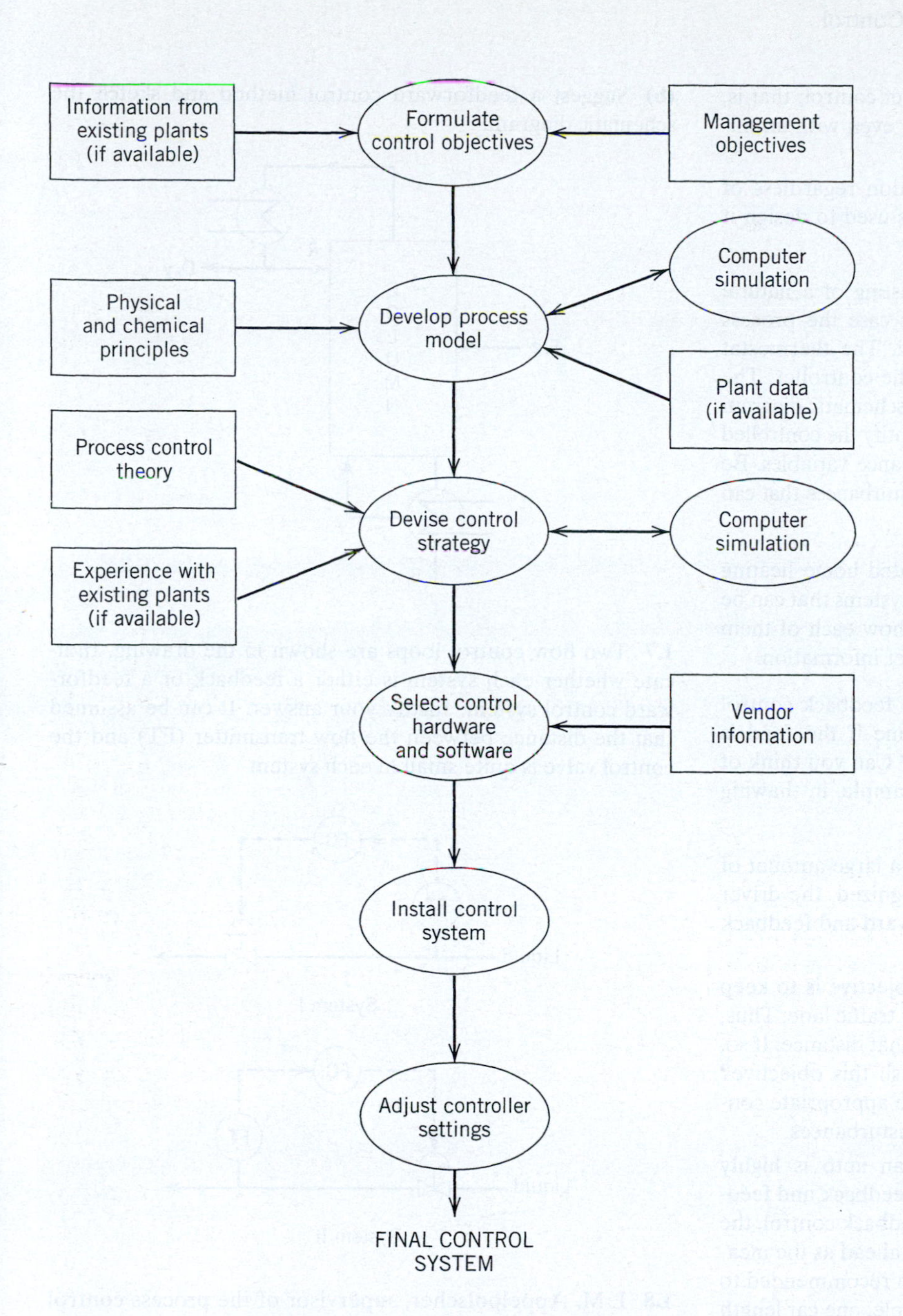

Figure 1.10 Major steps in control system development.

A hierarchy of process control activities was presented in Fig. 1.8. Process control plays a key role in ensuring process safety and protecting personnel, equipment, and the environment. Controlled variables are maintained near their set points by the application of regulatory control techniques and advanced control techniques such as multivariable and constraint control. Real-time optimization can be employed to determine the optimum controller set points for current operating conditions and constraints. The highest level of the process control hierarchy is concerned with planning and scheduling operations for the entire plant. The different levels of process control activity in the hierarchy are related and should be carefully coordinated.

EXERCISES

1.1 Which of the following statements are true?

(a) *Feedback* and *feedforward control* both require a measured variable.

(b) The process variable to be controlled is measured in *feedback control.*

(c) *Feedforward control* can be perfect in the theoretical sense that the controller can take action via the manipulated variable even while the controlled variable remains equal to its desired value.

(d) *Feedforward control* can provide perfect control; that is, the output can be kept at its desired value, even with an imperfect process model.

(e) *Feedback control* will always take action regardless of the accuracy of any process model that was used to design it and the source of a disturbance.

1.2 Consider a home heating system consisting of a natural gas-fired furnace and a thermostat. In this case the process consists of the interior space to be heated. The thermostat contains both the measuring element and the controller. The furnace is either on (heating) or off. Draw a schematic diagram for this control system. On your diagram, identify the controlled variables, manipulated variables, and disturbance variables. Be sure to include several possible sources of disturbances that can affect room temperature.

1.3 In addition to a thermostatically operated home heating system, identify two other feedback control systems that can be found in most residences. Describe briefly how each of them works; include sensor, actuator, and controller information.

1.4 Does a typical microwave oven utilize feedback control to set cooking temperature or to determine if the food is "cooked"? If not, what mechanism is used? Can you think of any disadvantages to this approach, for example, in thawing and cooking foods?

1.5 Driving an automobile safely requires a large amount of individual skill. Even if not generally recognized, the driver needs an intuitive ability to utilize feedforward and feedback control methods.

(a) In the process of steering a car, the objective is to keep the vehicle generally centered in the proper traffic lane. Thus, the controlled variable is some measure of that distance. If so, how is feedback control used to accomplish this objective? Identify the sensor(s), the actuator, how the appropriate control action is determined, and some likely disturbances.

(b) The process of braking/accelerating an auto is highly complex, requiring the skillful use of both feedback and feedforward mechanisms to drive safely. For feedback control, the driver normally uses distance to the vehicle ahead as the measured variable. The "set point" then is often recommended to be some distance related to speed, for example, one car length separation for each 10 mph. If this assertion is correct, how does feedforward control come into the accelerating/braking process when one is attempting to drive in traffic at a constant speed? In other words, what other information—in addition to distance separating the two vehicles, which obviously should never equal zero—does the driver utilize to avoid colliding with the car ahead?

1.6 The distillation column shown in the drawing is used to distill a binary mixture. Symbols *x*, *y*, and *z* denote mole fractions of the more volatile component, while *B*, *D*, *R*, and *F* represent molar flow rates. It is desired to control distillate composition *y* despite disturbances in feed flow rate *F*. All flow rates can be measured and manipulated with the exception of *F*, which can only be measured. A composition analyzer provides measurements of *y*.

(a) Propose a feedback control method and sketch the schematic diagram.

(b) Suggest a feedforward control method and sketch the schematic diagram.

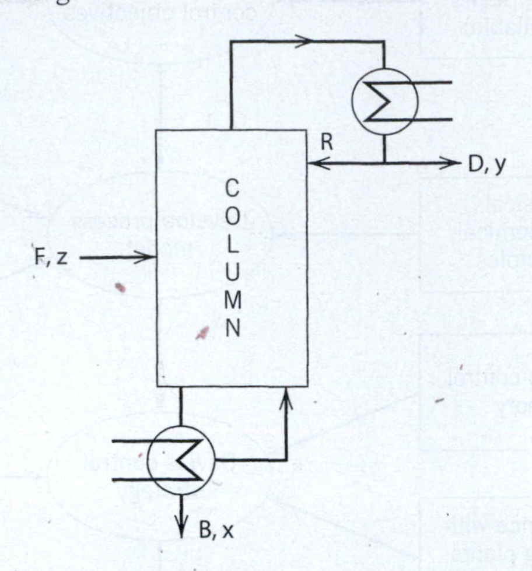

1.7 Two flow control loops are shown in the drawing. Indicate whether each system is either a feedback or a feedforward control system. Justify your answer. It can be assumed that the distance between the flow transmitter (FT) and the control valve is quite small in each system.

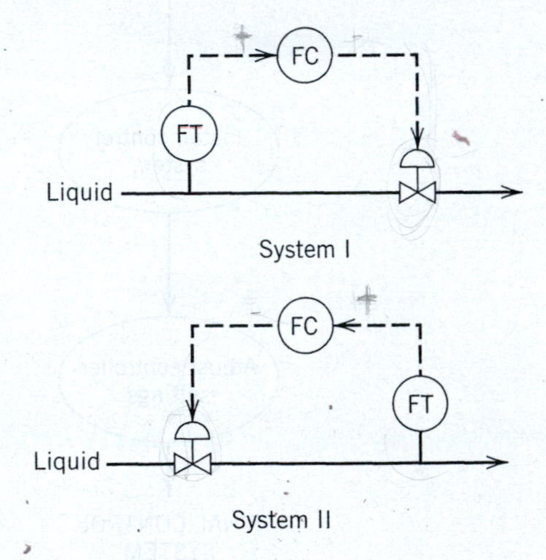

1.8 I. M. Appelpolscher, supervisor of the process control group of the Ideal Gas Company, has installed a 25 × 40 × 5-ft swimming pool in his backyard. The pool contains level and temperature sensors used with feedback controllers to maintain the pool level and temperature at desired values. Appelpolscher is satisfied with the level control system, but he feels that the addition of one or more feedforward controllers would help maintain the pool temperature more nearly constant. As a new member of the process control group, you have been selected to check Appelpolscher's mathematical analysis and to give your advice. The following information may or may not be pertinent to your analysis:

(i) Appelpolscher is particular about cleanliness and thus has a high-capacity pump that continually recirculates the water through an activated charcoal filter.

(ii) The pool is equipped with a natural gas-fired heater that adds heat to the pool at a rate $Q(t)$ that is directly proportional to the output signal from the controller $p(t)$.

(iii) There is a leak in the pool, which Appelpolscher has determined is constant equal to F (volumetric flow rate). The liquid-level control system adds water from the city supply system to maintain the level in the pool exactly at the specified level. The temperature of the water in the city system is T_w, a variable.
(iv) A significant amount of heat is lost by conduction to the surrounding ground, which has a constant, year-round temperature T_G. Experimental tests by Appelpolscher showed that essentially all of the temperature drop between the pool and the ground occurred across the homogeneous layer of gravel that surrounded his pool. The gravel thickness is Δx, and the overall thermal conductivity is k_G.
(v) The main challenge to Appelpolscher's modeling ability was the heat-loss term accounting for convection, conduction, radiation, and evaporation to the atmosphere. He determined that the heat losses per unit area of open water could be represented by

$$\text{losses} = U(T_p - T_a)$$

where

T_p = temperature of pool
T_a = temperature of the air, a variable
U = overall heat transfer coefficient

Appelpolscher's detailed model included radiation losses and heat generation due to added chemicals, but he determined that these terms were negligible.

(a) Draw a schematic diagram for the pool and all control equipment. Show all inputs and outputs, including all disturbance variables.

(b) What additional variable(s) would have to be measured to add feedforward control to the existing pool temperature feedback controller?

(c) Write a steady-state energy balance. How can you determine which of the disturbance variables you listed in part (a) are most/least likely to be important?

(d) What recommendations concerning the prospects of adding feedforward control would you make to Appelpolscher?

1.9 In a thermostat control system for a home heating system
(a) Identify the manipulated variable
(b) Identify the controlled variable
(c) How is a valve involved in the control system? What does it manipulate?
(d) Name one important disturbance (it must change with respect to time).

1.10 Identify and describe three automatic control systems in a modern automobile (besides cruise control).

1.11 In Figure 1.2 (*h*), identify the controlled, manipulated, and disturbance variables (there may be more than one of each type). How does the length of time for the dialysis treatment affect the waste concentration?

1.12 For the steam-heated tank shown below, identify manipulated, controlled, and disturbance variables. What disturbances are measured for feedforward control? How would the control system react to an increase in feed temperature in order to keep the tank temperature at its setpoint?

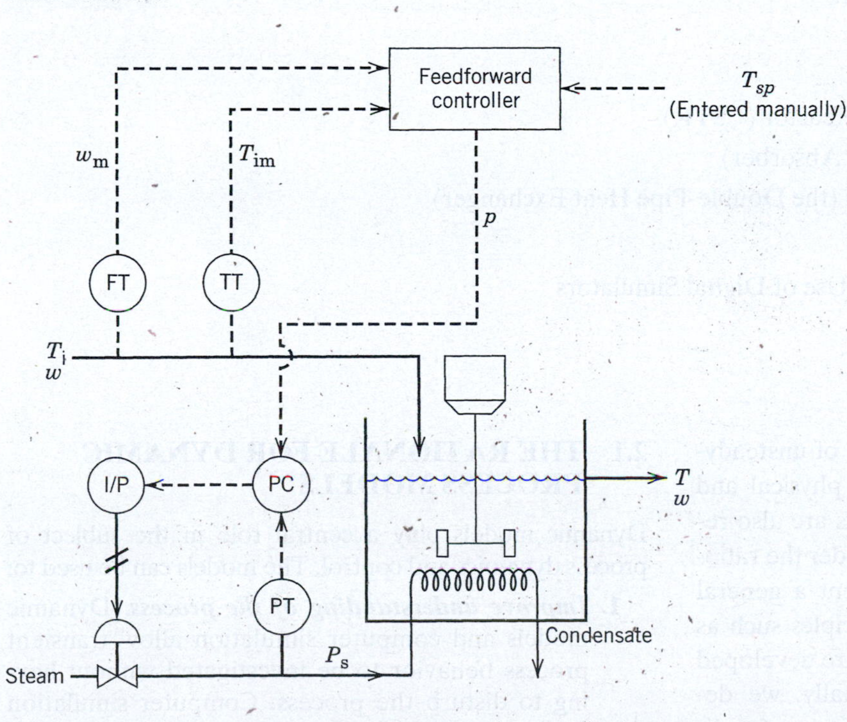

Figure E1.12. Feedforward control with a feedback control loop for outlet temperature.

Chapter 2

Theoretical Models of Chemical Processes

CHAPTER CONTENTS

In this chapter we consider the derivation of unsteady-state models of chemical processes from physical and chemical principles. Unsteady-state models are also referred to as *dynamic models*. We first consider the rationale for dynamic models and then present a general strategy for deriving them from first principles such as conservation laws. Then dynamic models are developed for several representative processes. Finally, we describe how dynamic models that consist of sets of ordinary differential equations and algebraic relations can be solved numerically using computer simulation.

2.1 THE RATIONALE FOR DYNAMIC PROCESS MODELS

Dynamic models play a central role in the subject of process dynamics and control. The models can be used to:

1. ***Improve understanding of the process.*** Dynamic models and computer simulation allow transient process behavior to be investigated without having to disturb the process. Computer simulation allows valuable information about dynamic and steady-state process behavior to be acquired, even before the plant is constructed.

2. ***Train plant operating personnel.*** Process simulators play a critical role in training plant operators to run complex units and to deal with emergency situations. By interfacing a process simulator to standard process control equipment, a realistic training environment is created.
3. ***Develop a control strategy for a new process.*** A dynamic model of the process allows alternative control strategies to be evaluated. For example, a dynamic model can help identify the process variables that should be controlled and those that should be manipulated. For model-based control strategies (Chapters 16 and 20), the process model is part of the control law.
4. ***Optimize process operating conditions.*** It can be advantageous to recalculate the optimum operating conditions periodically in order to maximize profit or minimize cost. A steady-state process model and economic information can be used to determine the most profitable operating conditions (see Chapter 19).

For many of the examples cited above—particularly where new, hazardous, or difficult-to-operate processes are involved—development of a suitable process model can be crucial to success. Models can be classified based on how they are obtained:

(a) *Theoretical models* are developed using the principles of chemistry, physics, and biology.

(b) *Empirical models* are obtained by fitting experimental data.

(c) *Semi-empirical models* are a combination of the models in categories (a) and (b); the numerical values of one or more of the parameters in a theoretical model are calculated from experimental data.

Theoretical models offer two very important advantages: they provide physical insight into process behavior, and they are applicable over wide ranges of conditions. However, there are disadvantages associated with theoretical models. They tend to be expensive and time-consuming to develop. In addition, theoretical models of complex processes typically include some model parameters that are not readily available, such as reaction rate coefficients, physical properties, or heat transfer coefficients.

Although empirical models are easier to develop than theoretical models, they have a serious disadvantage: *empirical models typically do not extrapolate well.* More specifically, empirical models should be used with caution for operating conditions that were not included in the experimental data used to fit the model. The range of the data is typically quite small compared to the full range of process operating conditions.

Semi-empirical models have three inherent advantages: (i) they incorporate theoretical knowledge, (ii) they can be extrapolated over a wider range of operating conditions than empirical models, and (iii) they require less development effort than theoretical models. Consequently, semi-empirical models are widely used in industry. Interesting industrial case studies that involve semi-empirical models have been reported by Foss et al. (1998).

This chapter is concerned with the development of theoretical models from first principles such as conservation laws. Empirical dynamic models are considered in Chapter 7.

2.1.1 An Illustrative Example: A Blending Process

In Chapter 1 we developed a steady-state model for a stirred-tank blending system based on mass and component balances. Now we develop an unsteady-state model that will allow us to analyze the more general situation where process variables vary with time. Dynamic models differ from steady-state models because they contain additional accumulation terms.

As an illustrative example, we consider the isothermal stirred-tank blending system in Fig. 2.1. It is a more general version of the blending system in Fig. 1.3 because the overflow line has been omitted and inlet stream 2 is not necessarily pure A (that is, $x_2 \neq 1$). Now the volume of liquid in the tank V can vary with time, and the exit flow rate is not necessarily equal to the sum of the inlet flow rates. An unsteady-state mass balance for the blending system in Fig. 2.1 has the form

$$\left\{\begin{matrix}\text{rate of accumulation} \\ \text{of mass in the tank}\end{matrix}\right\} = \left\{\begin{matrix}\text{rate of} \\ \text{mass in}\end{matrix}\right\} - \left\{\begin{matrix}\text{rate of} \\ \text{mass out}\end{matrix}\right\} \tag{2-1}$$

The mass of liquid in the tank can be expressed as the product of the liquid volume V and the density ρ.

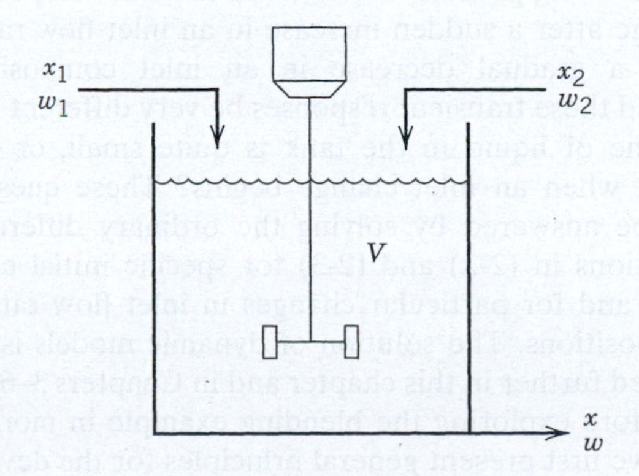

Figure 2.1 Stirred-tank blending process.

Consequently, the rate of mass accumulation is simply $d(V\rho)/dt$, and (2-1) can be written as

$$\frac{d(V\rho)}{dt} = w_1 + w_2 - w \tag{2-2}$$

where w_1, w_2, and w are mass flow rates.

The unsteady-state material balance for component A can be derived in an analogous manner. We assume that the blending tank is perfectly mixed. This assumption has two important implications: (i) there are no concentration gradients in the tank contents and (ii) the composition of the exit stream is equal to the tank composition. The perfect mixing assumption is valid for low-viscosity liquids that receive an adequate degree of agitation. In contrast, the assumption is less likely to be valid for high-viscosity liquids such as polymers or molten metals. Nonideal mixing is modeled in books on reactor analysis (e.g., Fogler, 1999).

For the perfect mixing assumption, the rate of accumulation of component A is $d(V\rho x)/dt$, where x is the mass fraction of A. The unsteady-state component balance is

$$\frac{d(V\rho x)}{dt} = w_1x_1 + w_2x_2 - wx \tag{2-3}$$

Equations 2-2 and 2-3 provide an unsteady-state model for the blending system. The corresponding steady-state model was derived in Chapter 1 (cf. Eqs. 1-1 and 1-2). It also can be obtained by setting the accumulation terms in Eqs. 2-2 and 2-3 equal to zero,

$$0 = \overline{w}_1 + \overline{w}_2 - \overline{w} \tag{2-4}$$

$$0 = \overline{w}_1\overline{x}_1 + \overline{w}_2\overline{x}_2 - \overline{w}\,\overline{x} \tag{2-5}$$

where the nominal steady-state conditions are denoted by $\overline{x}$ and $\overline{w}$, and so on. In general, a steady-state model is a special case of an unsteady-state model that can be derived by setting accumulation terms equal to zero.

A dynamic model can be used to characterize the transient behavior of a process for a wide variety of conditions. For example, some relevant concerns for the blending process: How would the exit composition change after a sudden increase in an inlet flow rate or after a gradual decrease in an inlet composition? Would these transient responses be very different if the volume of liquid in the tank is quite small, or quite large, when an inlet change begins? These questions can be answered by solving the ordinary differential equations in (2-2) and (2-3) for specific initial conditions and for particular changes in inlet flow rates or compositions. The solution of dynamic models is considered further in this chapter and in Chapters 3–6.

Before exploring the blending example in more detail, we first present general principles for the development of dynamic models.

2.2 GENERAL MODELING PRINCIPLES

It is important to remember that a process model is nothing more than a mathematical abstraction of a real process. The model equations are at best an approximation to the real process as expressed by the adage that "all models are wrong, but some are useful." Consequently, the model cannot incorporate all of the features, whether macroscopic or microscopic, of the real process. Modeling inherently involves a compromise between model accuracy and complexity on one hand, and the cost and effort required to develop the model and verify it on the other hand. The required compromise should consider a number of factors, including the modeling objectives, the expected benefits from use of the model, and the background of the intended users of the model (for example, research specialists versus plant engineers).

Process modeling is both an art and a science. Creativity is required to make simplifying assumptions that result in an appropriate model. The model should incorporate all of the important dynamic behavior while being no more complex than is necessary. Thus, less important phenomena are omitted in order to keep the number of model equations, variables, and parameters at reasonable levels. The failure to choose an appropriate set of simplifying assumptions invariably leads to either (1) rigorous but excessively complicated models or (2) overly simplistic models. Both extremes should be avoided. Fortunately, modeling is also a science, and predictions of process behavior from alternative models can be compared, both qualitatively and quantitatively. This chapter provides an introduction to the subject of theoretical dynamic models and shows how they can be developed from first principles such as conservation laws. Additional information is available in the books by Bequette (1998), Aris (1999), and Cameron and Hangos (2001).

A systematic procedure for developing dynamic models from first principles is summarized in Table 2.1. Most of the steps in Table 2.1 are self-explanatory, with the possible exception of Step 7. The *degrees of freedom analysis* in Step 7 is required in model development for complex processes. Because these models typically contain large numbers of variables and equations, it is not obvious whether the model can be solved, or whether it has a unique solution. Consequently, we consider the degrees of freedom analysis in Sections 2.3 and 10.3.

Dynamic models of chemical processes consist of ordinary differential equations (ODE) and/or partial differential equations (PDE), plus related algebraic equations. In this book we will restrict our discussion to ODE models, with the exception of one PDE model considered in Section 2.4. For process control problems, dynamic models are derived using unsteady-state conservation laws.

Table 2.1 A Systematic Approach for Developing Dynamic Models

1. State the modeling objectives and the end use of the model. Then determine the required levels of model detail and model accuracy.
2. Draw a schematic diagram of the process and label all process variables.
3. List all of the assumptions involved in developing the model. Try to be parsimonious: the model should be no more complicated than necessary to meet the modeling objectives.
4. Determine whether spatial variations of process variables are important. If so, a partial differential equation model will be required.
5. Write appropriate conservation equations (mass, component, energy, and so forth).
6. Introduce equilibrium relations and other algebraic equations (from thermodynamics, transport phenomena, chemical kinetics, equipment geometry, etc.).
7. Perform a degrees of freedom analysis (Section 2.3) to ensure that the model equations can be solved.
8. Simplify the model. It is often possible to arrange the equations so that the output variables appear on the left side and the input variables appear on the right side. This model form is convenient for computer simulation and subsequent analysis.
9. Classify inputs as disturbance variables or as manipulated variables.

In this section we first review general modeling principles, emphasizing the importance of the mass and energy conservation laws. Force-momentum balances are employed less often. For processes with momentum effects that cannot be neglected (e.g., some fluid and solid transport systems), such balances should be considered. The process model often also includes algebraic relations that arise from thermodynamics, transport phenomena, physical properties, and chemical kinetics. Vapor-liquid equilibria, heat transfer correlations, and reaction rate expressions are typical examples of such algebraic equations.

2.2.1 Conservation Laws

Theoretical models of chemical processes are based on conservation laws such as the conservation of mass and energy. Consequently, we now consider important conservation laws and use them to develop dynamic models for representative processes.

Conservation of Mass

$$\begin{Bmatrix}\text{rate of mass}\\ \text{accumulation}\end{Bmatrix} = \begin{Bmatrix}\text{rate of}\\ \text{mass in}\end{Bmatrix} - \begin{Bmatrix}\text{rate of}\\ \text{mass out}\end{Bmatrix} \tag{2-6}$$

Conservation of Component i

$$\begin{Bmatrix}\text{rate of component } i\\ \text{accumulation}\end{Bmatrix} = \begin{Bmatrix}\text{rate of component } i\\ \text{in}\end{Bmatrix} - \begin{Bmatrix}\text{rate of component } i\\ \text{out}\end{Bmatrix} + \begin{Bmatrix}\text{rate of component } i\\ \text{produced}\end{Bmatrix} \tag{2-7}$$

The last term on the right-hand side of (2-7) represents the rate of generation (or consumption) of component i as a result of chemical reactions. Conservation equations can also be written in terms of molar quantities, atomic species, and molecular species (Felder and Rousseau, 2000).

Conservation of Energy

The general law of energy conservation is also called the First Law of Thermodynamics (Sandler, 2006). It can be expressed as

$$\begin{Bmatrix}\text{rate of energy}\\ \text{accumulation}\end{Bmatrix} = \begin{Bmatrix}\text{rate of energy in}\\ \text{by convection}\end{Bmatrix} - \begin{Bmatrix}\text{rate of energy out}\\ \text{by convection}\end{Bmatrix} + \begin{Bmatrix}\text{net rate of heat addition}\\ \text{to the system from}\\ \text{the surroundings}\end{Bmatrix} + \begin{Bmatrix}\text{net rate of work}\\ \text{performed on the system}\\ \text{by the surroundings}\end{Bmatrix} \tag{2-8}$$

The total energy of a thermodynamic system, U_{tot}, is the sum of its internal energy, kinetic energy, and potential energy:

$$U_{tot} = U_{int} + U_{KE} + U_{PE} \tag{2-9}$$

For the processes and examples considered in this book, it is appropriate to make two assumptions:

1. Changes in potential energy and kinetic energy can be neglected, because they are small in comparison with changes in internal energy.
2. The net rate of work can be neglected, because it is small compared to the rates of heat transfer and convection.

For these reasonable assumptions, the energy balance in Eq. 2-8 can be written as (Bird et al., 2002)

$$\frac{dU_{int}}{dt} = -\Delta(w\hat{H}) + Q \tag{2-10}$$

where U_{int} is the internal energy of the system, $\hat{H}$ is the enthalpy per unit mass, w is the mass flow rate, and Q is the rate of heat transfer to the system. The Δ operator denotes the difference between outlet conditions and inlet conditions of the flowing streams. Consequently, the $-\Delta(w\hat{H})$ term represents the enthalpy of the inlet stream(s) minus the enthalpy of the outlet stream(s). The analogous equation for molar quantities is

$$\frac{dU_{int}}{dt} = -\Delta(\tilde{w}\tilde{H}) + Q \tag{2-11}$$

where $\tilde{H}$ is the enthalpy per mole and $\tilde{w}$ is the molar flow rate.

Note that the conservation laws of this section are valid for batch and semi-batch processes, as well as for continuous processes. For example, in batch processes, there are no inlet and outlet flow rates. Thus, $w = 0$ and $\tilde{w} = 0$ in (2-10) and (2-11).

In order to derive dynamic models of processes from the general energy balances in Eqs. 2-10 and 2-11, expressions for U_{int} and $\hat{H}$ or $\tilde{H}$ are required, which can be derived from thermodynamics. These derivations and a review of related thermodynamics concepts are included in Appendix B.

2.2.2 The Blending Process Revisited

Next, we show that the dynamic model of the blending process in Eqs. 2-2 and 2-3 can be simplified and expressed in a more appropriate form for computer simulation. For this analysis, we introduce the additional assumption that the density of the liquid, ρ, is a constant. This assumption is reasonable because often the density has only a weak dependence on composition. For constant ρ, Eqs. 2-2 and 2-3 become

$$\rho\frac{dV}{dt} = w_1 + w_2 - w \tag{2-12}$$

$$\rho\frac{d(Vx)}{dt} = w_1x_1 + w_2x_2 - wx \tag{2-13}$$

Equation 2-13 can be simplified by expanding the accumulation term using the "chain rule" for differentiation of a product:

$$\rho\frac{d(Vx)}{dt} = \rho V\frac{dx}{dt} + \rho x\frac{dV}{dt} \tag{2-14}$$

Substitution of (2-14) into (2-13) gives

$$\rho V\frac{dx}{dt} + \rho x\frac{dV}{dt} = w_1x_1 + w_2x_2 - wx \tag{2-15}$$

Substitution of the mass balance in (2-12) for $\rho dV/dt$ in (2-15) gives

$$\rho V\frac{dx}{dt} + x(w_1 + w_2 - w) = w_1x_1 + w_2x_2 - wx \tag{2-16}$$

After canceling common terms and rearranging (2-12) and (2-16), a more convenient model form is obtained:

$$\frac{dV}{dt} = \frac{1}{\rho}(w_1 + w_2 - w) \tag{2-17}$$

$$\frac{dx}{dt} = \frac{w_1}{V\rho}(x_1 - x) + \frac{w_2}{V\rho}(x_2 - x) \tag{2-18}$$

The dynamic model in Eqs. 2-17 and 2-18 is quite general and is based on only two assumptions: perfect mixing and constant density. For special situations, the liquid volume V is constant (that is, $dV/dt = 0$), and the exit flow rate equals the sum of the inlet flow rates, $w = w_1 + w_2$. For example, these conditions occur when

1. An overflow line is used in the tank as shown in Fig. 1.3.
2. The tank is closed and filled to capacity.
3. A liquid-level controller keeps V essentially constant by adjusting a flow rate.

In all three cases, Eq. 2-17 reduces to the same form as Eq. 2-4, not because each flow rate is constant, but because $w = w_1 + w_2$ at all times.

The dynamic model in Eqs. 2-17 and 2-18 is in a convenient form for subsequent investigation based on analytical or numerical techniques. In order to obtain a solution to the ODE model, we must specify the inlet compositions (x_1 and x_2) and the flow rates (w_1, w_2 and w) as functions of time. After specifying initial conditions for the dependent variables, $V(0)$ and $x(0)$, we can determine the transient responses, $V(t)$ and $x(t)$. The derivation of an analytical expression for $x(t)$ when V is constant is illustrated in Example 2.1.

EXAMPLE 2.1

A stirred-tank blending process with a constant liquid holdup of 2 m^3 is used to blend two streams whose densities are both approximately 900 kg/m^3. The density does not change during mixing.

(a) Assume that the process has been operating for a long period of time with flow rates of $w_1 = 500$ kg/min and $w_2 = 200$ kg/min, and feed compositions (mass fractions) of $x_1 = 0.4$ and $x_2 = 0.75$. What is the steady-state value of x?

(b) Suppose that w_1 changes suddenly from 500 to 400 kg/min and remains at the new value. Determine an expression for $x(t)$ and plot it.

(c) Repeat part **(b)** for the case where w_2 (instead of w_1) changes suddenly from 200 to 100 kg/min and remains there.

(d) Repeat part **(c)** for the case where x_1 suddenly changes from 0.4 to 0.6.

(e) For parts **(b)** through **(d)**, plot the normalized response $x_N(t)$,

$$x_N(t) = \frac{x(t) - x(0)}{x(\infty) - x(0)}$$

where $x(0)$ is the initial steady-state value of $x(t)$ and $x(\infty)$represents the final steady-state value, which is different for each part.

SOLUTION

(a) Denote the initial steady-state conditions by $\overline{x}$, $\overline{w}$, and so on. For the initial steady state, Eqs. 2-4 and 2-5 are applicable. Solve (2-5) for $\overline{x}$:

$$\overline{x} = \frac{\overline{w}_1\overline{x}_1 + \overline{w}_2\overline{x}_2}{\overline{w}} = \frac{(500)(0.4) + (200)(0.75)}{700} = 0.5$$

(b) The component balance in Eq. 2-3 can be rearranged (for constant V and ρ) as

$$\tau\frac{dx}{dt} + x = \frac{w_1x_1 + w_2x_2}{w} \qquad x(0) = \overline{x} = 0.5 \tag{2-19}$$

where $\tau \triangleq \overline{V}\rho/\overline{w}$. In each of the three parts, **(b)**–**(d)**, $\tau = 3$ min and the right side of (2-19) is constant for this example. Thus, (2-19) can be written as

$$3\frac{dx}{dt} + x = C^* \qquad x(0) = 0.5 \tag{2-20}$$

where

$$C^* \triangleq \frac{\overline{w}_1\overline{x}_1 + \overline{w}_2\overline{x}_2}{\overline{w}} \tag{2-21}$$

The solution to (2-20) can be obtained by applying standard solution methods (Kreyszig, 1999):

$$x(t) = 0.5e^{-t/3} + C^*(1 - e^{-t/3}) \tag{2-22}$$

For case **(b)**:

$$C^* = \frac{(400 \text{ kg/min})(0.4) + (200 \text{ kg/min})(0.75)}{600 \text{ kg/min}} = 0.517$$

Substituting C^* into (2-22) gives the desired solution for the step change in w_1:

$$x(t) = 0.5e^{-t/3} + 0.517(1 - e^{-t/3}) \tag{2-23}$$

(c) For the step change in w_2,

$$C^* = \frac{(500 \text{ kg/min})(0.4) + (100 \text{ kg/min})(0.75)}{600 \text{ kg/min}} = 0.458$$

and the solution is

$$x(t) = 0.5e^{-t/3} + 0.458(1 - e^{-t/3}) \tag{2-24}$$

(d) Similarly, for the simultaneous changes in x_1 and w_2, Eq. 2-21 gives $C^* = 0.625$. Thus, the solution is

$$x(t) = 0.5e^{-t/3} + 0.625(1 - e^{-t/3}) \tag{2-25}$$

(e) The individual responses in (2-22)–(2-24) have the same normalized response:

$$\frac{x(t) - x(0)}{x(\infty) - x(0)} = 1 - e^{-t/3} \tag{2-26}$$

The responses of **(b)**–**(e)** are shown in Fig. 2.2.

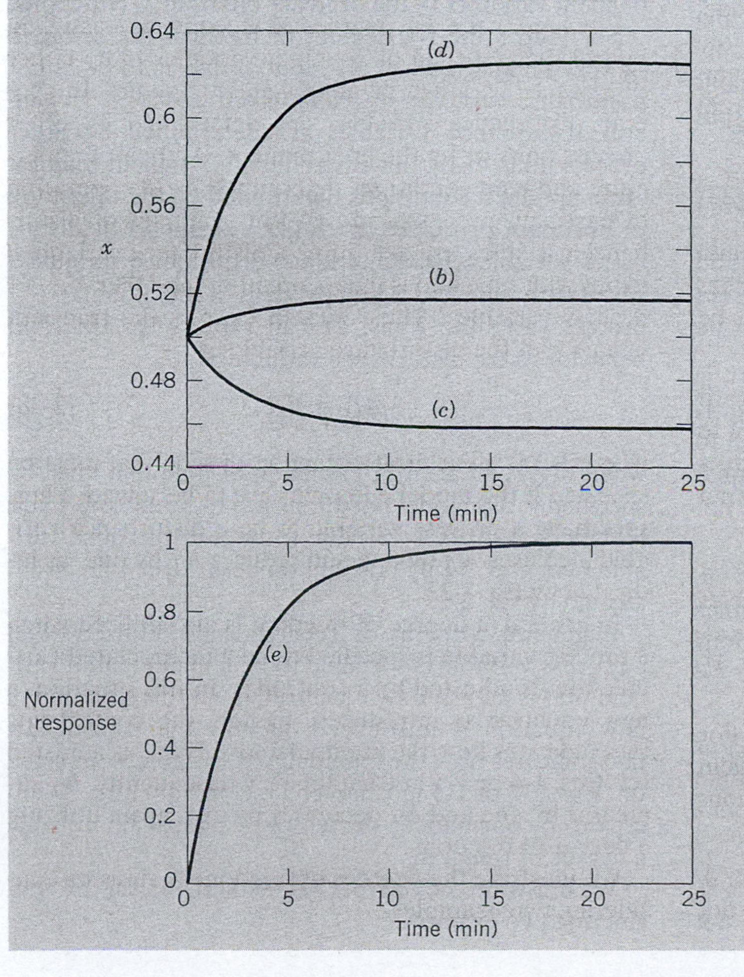

Figure 2.2 Exit composition responses of a stirred-tank blending process to step changes in
(b) Flow rate w_1
(c) Flow rate w_2
(d) Flow rate w_2 and inlet composition x_1
(e) Normalized response for parts (b)–(d)

The individual responses and normalized response have the same time dependence for cases **(b)**—**(d)** because $\tau = \overline{V}\rho/\overline{w} = 3$ min for each part. Note that τ is the mean residence time of the liquid in the blending tank. If $\overline{w}$ changes, then τ and the time dependence of the solution also change. This situation would occur, for example, if $\overline{w}_1$ changed from 500 kg/min to 600 kg/min. These more general situations will be addressed in Chapter 4.

2.3 DEGREES OF FREEDOM ANALYSIS

To simulate a process, we must first ensure that its model equations (differential and algebraic) constitute a solvable set of relations. In other words, the output variables, typically the variables on the left side of the equations, can be solved in terms of the input variables on the right side of the equations. For example, consider a set of linear algebraic equations, $\mathbf{y} = \mathbf{A}\mathbf{x}$. In order for these equations to have a unique solution for $\mathbf{x}$, vectors $\mathbf{x}$ and $\mathbf{y}$ must contain the same number of elements and matrix $\mathbf{A}$ must be nonsingular (that is, have a nonzero determinant).

It is not easy to make a similar evaluation for a large, complicated steady-state or dynamic model. However, there is one general requirement. In order for the model to have a unique solution, the number of unknown variables must equal the number of independent model equations. An equivalent statement is that all of the available *degrees of freedom* must be utilized. The number of degrees of freedom, N_F, can be calculated from the expression

$$N_F = N_V - N_E \tag{2-27}$$

where N_V is the total number of process variables and N_E is the number of independent equations. A degrees of freedom analysis allows modeling problems to be classified according to the following categories:

1. $N_F = 0$: The process model is *exactly specified*. If $N_F = 0$, then the number of equations is equal to the number of process variables and the set of equations has a solution. (However, the solution may not be unique for a set of nonlinear equations.)
2. $N_F > 0$: The process is *underspecified*. If $N_F > 0$, then $N_V > N_E$, so there are more process variables than equations. Consequently, the N_E equations have an infinite number of solutions, because N_F process variables can be specified arbitrarily.
3. $N_F < 0$: The process model is *overspecified*. For $N_F < 0$, there are fewer process variables than equations, and consequently the set of equations has no solution.

Note that $N_F = 0$ is the only satisfactory case. If $N_F > 0$, then a sufficient number of input variables have not been assigned numerical values. Then additional independent model equations must be developed in order for the model to have an exact solution.

Table 2.2 Degrees of Freedom Analysis

1. List all quantities in the model that are *known* constants (or parameters that can be specified) on the basis of equipment dimensions, known physical properties, and so on.
2. Determine the number of equations N_E and the number of process variables, N_V. Note that time t is *not* considered to be a process variable, because it is neither a process input nor a process output.
3. Calculate the number of degrees of freedom, $N_F = N_V - N_E$.
4. Identify the N_E output variables that will be obtained by solving the process model.
5. Identify the N_F input variables that must be specified as either disturbance variables or manipulated variables, in order to utilize the N_F degrees of freedom.

A structured approach to modeling involves a systematic analysis to determine the number of degrees of freedom and a procedure for assigning them. The steps in the degrees of freedom analysis are summarized in Table 2.2. In Step 4 the output variables include the dependent variables in the ordinary differential equations.

For Step 5 the N_F degrees of freedom are assigned by specifying a total of N_F input variables to be either *disturbance variables* or *manipulated variables*. In general, disturbance variables are determined by other process units or by the environment. Ambient temperature and feed conditions determined by the operation of upstream processes are typical examples of disturbance variables. By definition, a disturbance variable d varies with time and is independent of the other $N_V - 1$ process variables. Thus, we can express the transient behavior of the disturbance variable as

$$d(t) = f(t) \tag{2-28}$$

where $f(t)$ is an arbitrary function of time that must be specified if the model equations are to be solved. Thus, specifying a process variable to be a disturbance variable increases N_E by one and reduces N_F by one, as indicated by Eq. 2-27.

In general, a degree of freedom is also utilized when a process variable is specified to be a manipulated variable that is adjusted by a controller. In this situation, a new equation is introduced, namely the control law that indicates how the manipulated variable is adjusted (cf. Eqs. 1-4 or 1-5 in Chapter 1). Consequently, N_E increases by one and N_F decreases by one, again utilizing a degree of freedom.

We illustrate the degrees of freedom analysis by considering two examples.

EXAMPLE 2.2

Analyze the degrees of freedom for the blending model of Eq. (2-3) for the special condition where volume V is constant.

SOLUTION

For this example, there are

2 parameters:	V, ρ
4 variables ($N_V = 4$):	x, x_1, w_1, w_2
1 equation ($N_E = 1$):	Eq. 2-3

The degrees of freedom are calculated as $N_F = 4 - 1 = 3$. Thus, we must identify three input variables that can be specified as known functions of time in order for the equation to have a unique solution. The dependent variable x is an obvious choice for the output variable in this simple example. Consequently, we have

1 output:	x
3 inputs:	x_1, w_1, w_2

The three degrees of freedom can be utilized by specifying the inputs as

2 disturbance variables:	x_1, w_1
1 manipulated variable:	w_2

Because all of the degrees of freedom have been utilized, the single equation is exactly specified and can be solved.

EXAMPLE 2.3

Analyze the degrees of freedom of the blending system model in Eqs. 2-17 and 2-18. Is this set of equations linear, or nonlinear, according to the usual working definition?[1]

SOLUTION

In this case, volume is now considered to be a variable rather than a constant parameter. Consequently, for the degrees of freedom analysis we have

1 parameter:	ρ
7 variables ($N_V = 7$):	$V, x, x_1, x_2, w, w_1, w_2$
2 equations ($N_E = 2$):	Eqs. 2-17 and 2-18

Thus, $N_F = 7 - 2 = 5$. The dependent variables on the left side of the differential equations, V and x, are the model outputs. The remaining five variables must be chosen as inputs. Note that a physical output, effluent flow rate w, is classified as a mathematical input, because it can be specified arbitrarily. Any process variable that can be specified arbitrarily should be identified as an input. Thus, we have

2 outputs:	V, x
5 inputs:	w, w_1, w_2, x_1, x_2

[1]A linear model cannot contain any nonlinear combinations of variables (for example, a product of two variables) or any variable raised to a power other than one.

Because the two outputs are the only variables to be determined in solving the system of two equations, no degrees of freedom are left. The system of equations is exactly specified and hence solvable.

To utilize the degrees of freedom, the five inputs are classified as either disturbance variables or manipulated variables. A reasonable classification is

3 disturbance variables:	w_1, x_1, x_2
2 manipulated variables:	w, w_2

For example, w could be used to control V and w_2 to control x.

Note that Eq. 2-17 is a linear ODE, while Eq. 2-18 is a nonlinear ODE as a result of the products and quotients.

2.4 DYNAMIC MODELS OF REPRESENTATIVE PROCESSES

For the simple process discussed so far, the stirred-tank blending system, energy effects were not considered due to the assumed isothermal operation. Next, we illustrate how dynamic models can be developed for processes where energy balances are important.

2.4.1 Stirred-Tank Heating Process: Constant Holdup

Consider the stirred-tank heating system shown in Fig. 2.3. The liquid inlet stream consists of a single component with a mass flow rate w_i and an inlet temperature T_i. The tank contents are agitated and heated using an electrical heater that provides a heating rate, Q. A dynamic model will be developed based on the following assumptions:

1. Perfect mixing; thus, the exit temperature T is also the temperature of the tank contents.

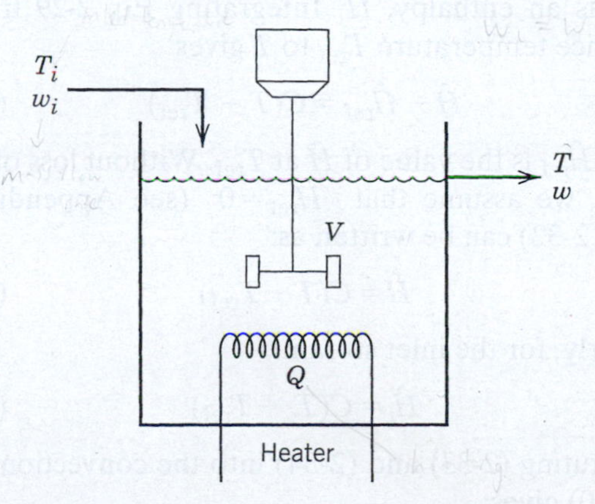

Figure 2.3 Stirred-tank heating process with constant holdup, V.

2. The inlet and outlet flow rates are equal; thus, $w_i = w$ and the liquid holdup V is constant.
3. The density ρ and heat capacity C of the liquid are assumed to be constant. Thus, their temperature dependence is neglected.
4. Heat losses are negligible.

In general, dynamic models are based on conservation laws. For this example, it is clear that we should consider an energy balance, because thermal effects predominate. A mass balance is not required in view of Assumptions 2 and 3.

Next, we show how the general energy balance in Eq. 2-10 can be simplified for this particular example. For a pure liquid at low or moderate pressures, the internal energy is approximately equal to the enthalpy, $U_{int} \approx H$, and H depends only on temperature (Sandler, 2006). Consequently, in the subsequent development, we assume that $U_{int} = H$ and $\hat{U}_{int} = \hat{H}$ where the caret (^) means per unit mass. As shown in Appendix B, a differential change in temperature, dT, produces a corresponding change in the internal energy per unit mass, $d\hat{U}_{int}$,

$$d\hat{U}_{int} = d\hat{H} = C\,dT \tag{2-29}$$

where C is the constant pressure heat capacity (assumed to be constant). The total internal energy of the liquid in the tank can be expressed as the product of $\hat{U}_{int}$ and the mass in the tank, ρV:

$$U_{int} = \rho V \hat{U}_{int} \tag{2-30}$$

An expression for the rate of internal energy accumulation can be derived from Eqs. 2-29 and 2-30:

$$\frac{dU_{int}}{dt} = \rho V C \frac{dT}{dt} \tag{2-31}$$

Note that this term appears in the general energy balance of Eq. 2-10.

Next, we derive an expression for the enthalpy term that appears on the right-hand side of Eq. 2-10. Suppose that the liquid in the tank is at a temperature T and has an enthalpy, $\hat{H}$. Integrating Eq. 2-29 from a reference temperature T_{ref} to T gives

$$\hat{H} - \hat{H}_{ref} = C(T - T_{ref}) \tag{2-32}$$

where $\hat{H}_{ref}$ is the value of $\hat{H}$ at T_{ref}. Without loss of generality, we assume that $\hat{H}_{ref} = 0$ (see Appendix B). Thus, (2-32) can be written as:

$$\hat{H} = C(T - T_{ref}) \tag{2-33}$$

Similarly, for the inlet stream:

$$\hat{H}_i = C(T_i - T_{ref}) \tag{2-34}$$

Substituting (2-33) and (2-34) into the convection term of (2-10) gives:

$$-\Delta(w\hat{H}) = w[C(T_i - T_{ref})] - w[C(T - T_{ref})] \tag{2-35}$$

Finally, substitution of (2-31) and (2-35) into (2-10) gives the desired dynamic model of the stirred-tank heating system:

$$V\rho C \frac{dT}{dt} = wC(T_i - T) + Q \tag{2-36}$$

Note that the T_{ref} terms have canceled, because C was assumed to be constant, and thus independent of temperature.

A degrees of freedom analysis for this model gives

3 parameters:	V, ρ, C
4 variables:	T, T_i, w, Q
1 equation:	Eq. 2-36

Thus, the degrees of freedom are $N_F = 4 - 1 = 3$. The process variables are classified as

1 output variable:	T
3 input variables:	T_i, w, Q

For control purposes, it is reasonable to classify the three inputs as

2 disturbance variables:	T_i, w
1 manipulated variable:	Q

2.4.2 Stirred-Tank Heating Process: Variable Holdup

Now we consider the more general situation in which the tank holdup can vary with time. This analysis also is based on assumptions 1, 3 and 4 of the previous section. Now an overall mass balance is needed, because the holdup is not constant. The overall mass balance is

$$\frac{d(V\rho)}{dt} = w_i - w \tag{2-37}$$

The energy balance for the current stirred-tank heating system can be derived from Eq. 2-10 in analogy with the derivation of Eq. 2-36. We again assume that $U_{int} = H$ for the liquid in the tank. Thus, for constant ρ:

$$\frac{dU_{int}}{dt} = \frac{dH}{dt} = \frac{d(\rho V\hat{H})}{dt} = \rho\frac{d(V\hat{H})}{dt} \tag{2-38}$$

From the definition of $-\Delta(w\hat{H})$ and Eqs. 2-33 and 2-34, it follows that

$$-\Delta(w\hat{H}) = w_i\hat{H}_i - w\hat{H} = w_iC(T_i - T_{ref}) - wC(T - T_{ref}) \tag{2-39}$$

where w_i and w are the mass flow rates of the inlet and outlet streams, respectively. Substituting (2-38) and (2-39) into (2-10) gives

$$\rho\frac{d(V\hat{H})}{dt} = w_iC(T_i - T_{ref}) - wC(T - T_{ref}) + Q \tag{2-40}$$

Next we simplify the dynamic model. Because ρ is constant, (2-37) can be written as

$$\rho \frac{dV}{dt} = w_i - w \quad (2\text{-}41)$$

The chain rule can be applied to expand the left side of (2-40) for constant C and ρ:

$$\rho \frac{d(V\hat{H})}{dt} = \rho V \frac{d\hat{H}}{dt} + \rho \hat{H} \frac{dV}{dt} \quad (2\text{-}42)$$

From Eq. 2-29 or 2-33, it follows that $d\hat{H}/dt = CdT/dt$. Substituting this expression and Eqs. 2-33 and 2-41 into Eq. 2-42 gives

$$\rho \frac{d(V\hat{H})}{dt} = C(T - T_{\text{ref}})(w_i - w) + \rho CV \frac{dT}{dt} \quad (2\text{-}43)$$

Substituting (2-43) into (2-40) and rearranging gives

$$C(T - T_{\text{ref}})(w_i - w) + \rho CV \frac{dT}{dt} = w_i C(T_i - T_{\text{ref}}) - wC(T - T_{\text{ref}}) + Q \quad (2\text{-}44)$$

Rearranging (2-41) and (2-44) provides a simpler form for the dynamic model:

$$\frac{dV}{dt} = \frac{1}{\rho}(w_i - w) \quad (2\text{-}45)$$

$$\frac{dT}{dt} = \frac{w_i}{V\rho}(T_i - T) + \frac{Q}{\rho CV} \quad (2\text{-}46)$$

This example and the blending example in Section 2.2.2 have demonstrated that process models with variable holdups can be simplified by substituting the overall mass balance into the other conservation equations.

Equations 2-45 and 2-46 provide a model that can be solved for the two outputs (V and T) if the two parameters (ρ and C) are known and the four inputs (w_i, w, T_i, and Q) are known functions of time.

2.4.3 Electrically Heated Stirred Tank

Now we again consider the stirred-tank heating system with constant holdup (Section 2.4.1), but we relax the assumption that energy is transferred instantaneously from the heating element to the contents of the tank. Suppose that the metal heating element has a significant thermal capacitance and that the electrical heating rate Q directly affects the temperature of the element rather than the liquid contents. For simplicity, we neglect the temperature gradients in the heating element that result from heat conduction and assume that the element has a uniform temperature, T_e. This temperature can be interpreted as the average temperature for the heating element.

Based on this new assumption, and the previous assumptions of Section 2.4.1, the unsteady-state energy balances for the tank and the heating element can be written as

$$mC \frac{dT}{dt} = wC(T_i - T) + h_e A_e (T_e - T) \quad (2\text{-}47)$$

$$m_e C_e \frac{dT_e}{dt} = Q - h_e A_e (T_e - T) \quad (2\text{-}48)$$

where $m = V\rho$ and $m_e C_e$ is the product of the mass of metal in the heating element and its specific heat. The term $h_e A_e$ is the product of the heat transfer coefficient and area available for heat transfer. Note that mC and $m_e C_e$ are the thermal capacitances of the tank contents and the heating element, respectively. Q is an input variable, the thermal equivalent of the instantaneous electrical power dissipation in the heating element.

Is the model given by Eqs. 2-47 and 2-48 in suitable form for calculation of the unknown output variables T_e and T? There are two output variables and two differential equations. All of the other quantities must be either model parameters (constants) or inputs (known functions of time). For a specific process, m, C, m_e, C_e, h_e, and A_e are known parameters determined by the design of the process, its materials of construction, and its operating conditions. Input variables w, T_i, and Q must be specified as functions of time for the model to be completely determined—that is, to utilize the available degrees of freedom. The dynamic model can then be solved for T and T_e as functions of time by integration after initial conditions are specified for T and T_e.

If flow rate w is constant, Eqs. 2-47 and 2-48 can be converted into a single second-order differential equation. First, solve Eq. 2-47 for T_e and then differentiate to find dT_e/dt. Substituting the expressions for T_e and dT_e/dt into Eq. 2-48 yields

$$\frac{mm_e C_e}{wh_e A_e} \frac{d^2T}{dt^2} + \left(\frac{m_e C_e}{h_e A_e} + \frac{m_e C_e}{wC} + \frac{m}{w}\right)\frac{dT}{dt} + T = \frac{m_e C_e}{h_e A_e}\frac{dT_i}{dt} + T_i + \frac{1}{wC}Q \quad (2\text{-}49)$$

The reader should verify that the dimensions of each term in the equation are consistent and have units of temperature. In addition, the reader should consider the steady-state versions of (2-36) and (2-49). They are identical, which is to be expected. Analyzing limiting cases is one way to check the consistency of a more complicated model.

The model in (2-49) can be simplified when $m_e C_e$, the thermal capacitance of the heating element, is very small compared to mC. When $m_e C_e = 0$, Eq. 2-49 reverts to the first-order model, Eq. 2-36, which was derived for the case where the heating element has a negligible thermal capacitance.

It is important to note that the model of Eq. 2-49 consists of only a single equation and a single output variable, T. The intermediate variable, T_e, is less important than T and has been eliminated from the earlier model (Eqs. 2-47 and 2-48). Both models are exactly specified; that is, they have no unassigned degrees of freedom. To integrate Eq. 2-49, we require initial conditions for both T and dT/dt at $t = 0$, because it is a second-order differential equation. The initial condition for dT/dt can be found by evaluating the right side of Eq. 2-47 when $t = 0$, using the values of $T_e(0)$ and $T(0)$. For both models, the inputs (w, T_i, Q) must be specified as functions of time.

EXAMPLE 2.4

An electrically heated stirred-tank process can be modeled by Eqs. (2-47) and (2-48) or, equivalently, by Eq. (2-49) alone. Process design and operating conditions are characterized by the following four parameter groups:

$$\frac{m}{w} = 10 \text{ min} \qquad \frac{m_e C_e}{h_e A_e} = 1.0 \text{ min}$$

$$\frac{m_e C_e}{wC} = 1.0 \text{ min} \qquad \frac{1}{wC} = 0.05 \text{ °C min/kcal}$$

The nominal values of Q and T_i are

$$\overline{Q} = 5000 \text{ kcal/min} \qquad \overline{T}_i = 100^\circ\text{C}$$

(a) Calculate the nominal steady-state temperature, $\overline{T}$.

(b) Assume that the process is initially at the steady state determined in part (**a**). Calculate the response, $T(t)$, to a sudden change in Q from 5000 to 5400 kcal/min using Eq. (2-49). Plot the temperature response.

(c) Suppose that it can be assumed that the term $m_e C_e/h_e A_e$ is small relative to other terms in (2-49). Calculate the response $T(t)$ for the conditions of part (**b**), using a first-order differential equation approximation to Eq. (2-49). Plot $T(t)$ on the graph for part (**b**).

(d) What can we conclude about the accuracy of the approximation for part (**c**)?

SOLUTION

(a) The steady-state form of Eq. 2-49 is

$$\overline{T} = \overline{T}_i + \frac{1}{wC}\overline{Q}$$

Substituting parameter values gives $\overline{T} = 350$ °C.

(b) Substitution of the parameter values in (2-49) gives

$$10\frac{d^2T}{dt^2} + 12\frac{dT}{dt} + T = 370$$

The following solution can be derived using standard solution methods (Kreyszig, 1999):

$$T(t) = 350 + 20\left[1 - 1.089\, e^{-t/11.099} + 0.0884\, e^{-t/0.901}\right]$$

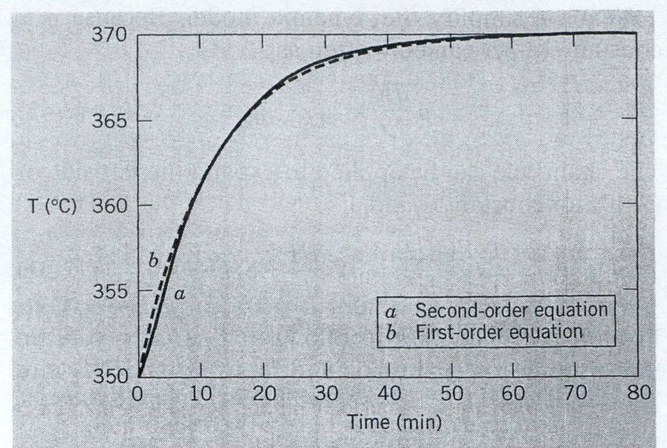

Figure 2.4 Responses of an electrically-heated stirred-tank process to a sudden change in the heater input.

This response is plotted in Fig. 2.4 as the slightly "s-shaped" curve (a).

(c) If we assume that $m_e C_e$ is small relative to other terms, then Eq. 2-49 can be approximated by the first-order differential equation:

$$12\frac{dT}{dt} + T = 370, \quad T(0) = 350°C$$

The solution is

$$T(t) = 350 + 20\,(1 - e^{-t/12})$$

(d) Figure 2.4 shows that the approximate solution (b) is quite good, matching the exact solution very well over the entire response. For purposes of process control, this approximate model is likely to be as useful as the more complicated, exact model.

2.4.4 Steam-Heated Stirred Tank

Steam (or some other heating medium) can be condensed within a coil or jacket to heat liquid in a stirred tank, and the inlet steam pressure can be varied by adjusting a control valve. The condensation pressure P_s then fixes the steam temperature T_s through an appropriate thermodynamic relation or from tabular information such as the steam tables (Sandler, 2006):

$$T_s = f(P_s) \tag{2-50}$$

Consider the stirred-tank heating system of Section 2.4.1 with constant holdup and a steam heating coil. We assume that the thermal capacitance of the liquid condensate is negligible compared to the thermal capacitances of the tank liquid and the wall of the heating coil. This assumption is reasonable when a steam trap is used to remove the condensate from the coil as it is produced. As a result of this assumption, the dynamic model consists of energy balances on the liquid and the heating coil wall:

$$mC\frac{dT}{dt} = wC(T_i - T) + h_pA_p(T_w - T) \quad (2\text{-}51)$$

$$m_wC_w\frac{dT_w}{dt} = h_sA_s(T_s - T_w) - h_pA_p(T_w - T) \quad (2\text{-}52)$$

where the subscripts w, s, and p refer, respectively, to the wall of the heating coil and to its steam and process sides. Note that these energy balances are similar to Eqs. 2-47 and 2-48 for the electrically heated example.

The dynamic model contains three output variables (T_s, T, and T_w) and three equations: an algebraic equation with T_s related to P_s (a specified function of time or a constant) and two differential equations. Thus, Eqs. 2-50 through 2-52 constitute an exactly specified model with three input variables: P_s, T_i, and w. Several important features are noted.

1. Usually $h_sA_s \gg h_pA_p$, because the resistance to heat transfer on the steam side of the coil is much lower than on the process side.
2. The change from electrical heating to steam heating increases the complexity of the model (three equations instead of two) but does not increase the model order (number of first-order differential equations).
3. As models become more complicated, the input and output variables may be coupled through certain parameters. For example, h_p may be a function of w, or h_s may vary with the steam condensation rate; sometimes algebraic equations cannot be solved explicitly for a key variable. In this situation, numerical solution techniques have to be used. Usually, implicit algebraic equations must be solved by iterative methods at each time step in the numerical integration.

We now consider some simple models for liquid storage systems utilizing a single tank. In the event that two or more tanks are connected in series (cascaded), the single-tank models developed here can be easily extended, as shown in Chapter 5.

2.4.5 Liquid Storage Systems

A typical liquid storage process is shown in Fig. 2.5 where q_i and q are volumetric flow rates. A mass balance yields

$$\frac{d(\rho V)}{dt} = \rho q_i - \rho q \quad (2\text{-}53)$$

Assume that liquid density ρ is constant and the tank is cylindrical with cross-sectional area, A. Then the volume of liquid in the tank can be expressed as $V = Ah$, where h is the liquid level (or *head*). Thus, (2-53) becomes

$$A\frac{dh}{dt} = q_i - q \quad (2\text{-}54)$$

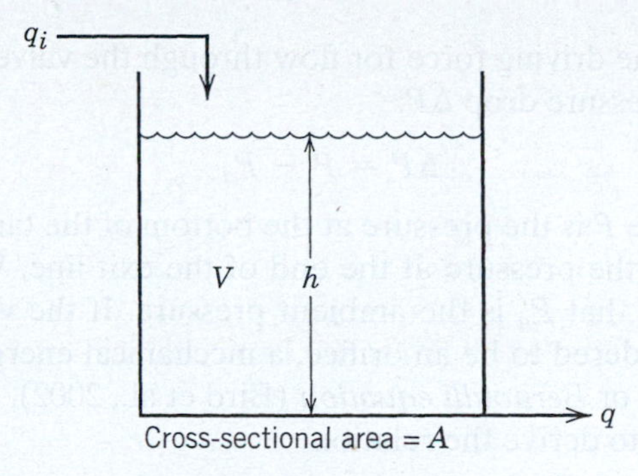

Figure 2.5 A liquid-level storage process.

Note that Eq. 2-54 appears to be a *volume balance*. However, in general, volume is *not* conserved for fluids. This result occurs in this example due to the constant density assumption.

There are three important variations of the liquid storage process:

1. The inlet or outlet flow rates might be constant; for example, exit flow rate q might be kept constant by a constant-speed, fixed-volume (metering) pump. An important consequence of this configuration is that the exit flow rate is then completely independent of liquid level over a wide range of conditions. Consequently, $q = \overline{q}$ where $\overline{q}$ is the steady-state value. For this situation, the tank operates essentially as a flow *integrator*. We will return to this case in Section 5.3.

2. The tank exit line may function simply as a resistance to flow from the tank (distributed along the entire line), or it may contain a valve that provides significant resistance to flow at a single point. In the simplest case, the flow may be assumed to be linearly related to the driving force, the liquid level, in analogy to Ohm's law for electrical circuits ($E = IR$)

$$h = qR_v \quad (2\text{-}55)$$

where R_v is the resistance of the line or valve. Rearranging (2-55) gives the following *flow-head equation*:

$$q = \frac{1}{R_v}h \quad (2\text{-}56)$$

Substituting (2-56) into (2-54) gives a first-order differential equation:

$$A\frac{dh}{dt} = q_i - \frac{1}{R_v}h \quad (2\text{-}57)$$

This model of the liquid storage system exhibits dynamic behavior similar to that of the stirred-tank heating system of Eq. 2-36.

3. A more realistic expression for flow rate q can be obtained when a fixed valve has been placed in the exit line and turbulent flow can be assumed.

The driving force for flow through the valve is the pressure drop ΔP:

$$\Delta P = P - P_a \tag{2-58}$$

where P is the pressure at the bottom of the tank and P_a is the pressure at the end of the exit line. We assume that P_a is the ambient pressure. If the valve is considered to be an orifice, a mechanical energy balance, or *Bernoulli equation* (Bird et al., 2002), can be used to derive the relation

$$q = C_v^* \sqrt{\frac{P - P_a}{\rho}} \tag{2-59}$$

where C_v^* is a constant. The value of C_v^* depends on the particular valve and the valve setting (how much it is open). See Chapter 9 for more information about control valves.

The pressure P at the bottom of the tank is related to liquid level h by a force balance

$$P = P_a + \frac{\rho g}{g_c} h \tag{2-60}$$

where the acceleration of gravity g is constant. Substituting (2-59) and (2-60) into (2-54) yields the dynamic model

$$A\frac{dh}{dt} = q_i - C_v\sqrt{h} \tag{2-61}$$

where $C_v \triangleq C_v^*\sqrt{g/g_c}$. This model is nonlinear due to the square root term.

The liquid storage processes discussed above could be operated by controlling the liquid level in the tank or by allowing the level to fluctuate without attempting to control it. For the latter case (operation as a surge tank), it may be of interest to predict whether the tank would overflow or run dry for particular variations in the inlet and outlet flow rates. Thus, the dynamics of the process may be important even when automatic control is not utilized.

2.4.6 The Continuous Stirred-Tank Reactor (CSTR)

Continuous stirred-tank reactors have widespread application in industry and embody many features of other types of reactors. CSTR models tend to be simpler than models for other types of continuous reactors such as tubular reactors and packed-bed reactors. Consequently, a CSTR model provides a convenient way of illustrating modeling principles for chemical reactors.

Consider a simple liquid-phase, irreversible chemical reaction where chemical species A reacts to form species B. The reaction can be written as A → B. We assume that the rate of reaction is first-order with respect to component A,

$$r = kc_A \tag{2-62}$$

where r is the rate of reaction of A per unit volume, k is the reaction rate constant (with units of reciprocal time), and c_A is the molar concentration of species A. For single-phase reactions, the rate constant is typically a strong function of reaction temperature given by the Arrhenius relation,

$$k = k_0 \exp(-E/RT) \tag{2-63}$$

where k_0 is the frequency factor, E is the activation energy, and R is the gas constant. The expressions in (2-62) and (2-63) are based on theoretical considerations, but model parameters k_0 and E are usually determined by fitting experimental data. Thus, these two equations can be considered to be *semi-empirical* relations, according to the definition in Section 2.2.

The schematic diagram of the CSTR is shown in Fig. 2.6. The inlet stream consists of pure component A with molar concentration, c_{Ai}. A cooling coil is used to maintain the reaction mixture at the desired operating temperature by removing heat that is released in the exothermic reaction. Our initial CSTR model development is based on three assumptions:

1. The CSTR is perfectly mixed.
2. The mass densities of the feed and product streams are equal and constant. They are denoted by ρ.
3. The liquid volume V in the reactor is kept constant by an overflow line.

For these assumptions, the unsteady-state mass balance for the CSTR is:

$$\frac{d(\rho V)}{dt} = \rho q_i - \rho q \tag{2-64}$$

Because V and ρ are constant, (2-64) reduces to

$$q = q_i \tag{2-65}$$

Thus, even though the inlet and outlet flow rates may change due to upstream or downstream conditions,

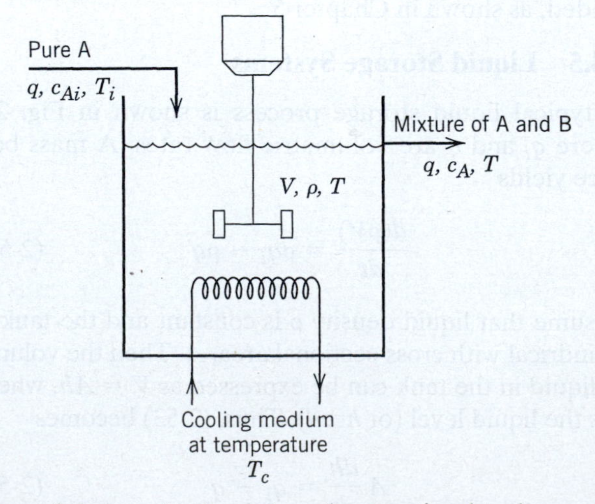

Figure 2.6 A nonisothermal continuous stirred-tank reactor.

Eq. 2-65 must be satisfied at all times. In Fig. 2.6, both flow rates are denoted by the symbol q.

For the stated assumptions, the unsteady-state component balances for species A (in molar units) is

$$V\frac{dc_A}{dt} = q(c_{Ai} - c_A) - Vkc_A \tag{2-66}$$

This balance is a special case of the general component balance in Eq. 2-7.

Next, we consider an unsteady-state energy balance for the CSTR. But first we make five additional assumptions:

4. The thermal capacitances of the coolant and the cooling coil wall are negligible compared to the thermal capacitance of the liquid in the tank.
5. All of the coolant is at a uniform temperature, T_c. (That is, the increase in coolant temperature as the coolant passes through the coil is neglected.)
6. The rate of heat transfer from the reactor contents to the coolant is given by

$$Q = UA(T_c - T) \tag{2-67}$$

   where U is the overall heat transfer coefficient and A is the heat transfer area. Both of these model parameters are assumed to be constant.
7. The enthalpy change associated with the mixing of the feed and the liquid in the tank is negligible compared with the enthalpy change for the chemical reaction. In other words, the heat of mixing is negligible compared to the heat of reaction.
8. Shaft work and heat losses to the ambient can be neglected.

The following form of the CSTR energy balance is convenient for analysis and can be derived from Eqs. 2-62 and 2-63 and Assumptions 1–8 (Fogler, 2006; Russell and Denn, 1972),

$$V\rho C\frac{dT}{dt} = wC(T_i - T) + (-\Delta H_R)Vkc_A + UA(T_c - T) \tag{2-68}$$

where ΔH_R is the heat of reaction per mole of A that is reacted.

In summary, the dynamic model of the CSTR consists of Eqs. 2-62 to 2-64, 2-66, 2-67, and 2-68. This model is nonlinear as a result of the many product terms and the exponential temperature dependence of k in Eq. 2-63. Consequently, it must be solved by numerical integration techniques (Fogler, 2006). The CSTR model will become considerably more complex if

1. More complicated rate expressions are considered. For example, a mass action kinetics model for a second-order, irreversible reaction, $2A \rightarrow B$, is given by

$$r = k_2c_A^2 \tag{2-69}$$

2. Additional species or chemical reactions are involved. If the reaction mechanism involved production of an intermediate species, $2A \rightarrow B^* \rightarrow B$, then unsteady-state component balances for both A and B* would be necessary (to calculate c_A and c_B^*), or balances for both A and B could be written (to calculate c_A and c_B). Information concerning the reaction mechanisms would also be required.

Reactions involving multiple species are described by high-order, highly coupled, nonlinear reaction models, because several component balances must be written.

EXAMPLE 2.5

To illustrate how the CSTR can exhibit nonlinear dynamic behavior, we simulate the effect of a step change in the coolant temperature T_c in positive and negative directions. Table 2.3 shows the parameters and nominal operating condition for the CSTR based on Eqs. 2–66 and 2–68 for the exothermic, irreversible first-order reaction $A \rightarrow B$. The two-state variables of the ODEs are the concentration of A (c_A) and the reactor temperature T. The manipulated variable is the jacket water temperature, T_c.

Two cases are simulated, one based on increased cooling by changing T_c from 300 K to 290 K and one reducing the cooling rate by increasing T_c from 300 K to 305 K.

These model equations are solved in MATLAB with a numerical integrator (ode15s) over a 10 min horizon. The decrease in T_c results in an increase in c_A. The results are displayed in two plots of the temperature and reactor concentration as a function of time (Figs. 2.7 and 2.8).

At a jacket temperature of 305 K, the reactor model has an oscillatory response. The oscillations are characterized by apparent reaction run-away with a temperature spike. However, when the concentration drops to a low value, the reactor then cools until the concentration builds, then there is another temperature rise. It is not unusual for chemical reactors to exhibit such widely different behaviors for different directional changes in the operating conditions.

Table 2.3 Nominal Operating Conditions for the CSTR

Parameter	Value	Parameter	Value
q	100 L/min	E/R	8750 K
c_{Ai}	1 mol/L	k_0	7.2×10^{10} min^{-1}
T_i	350 K	UA	5×10^4 J/min K
V	100 L	$T_c(0)$	300 K
ρ	1000 g/L	$c_A(0)$	0.5 mol/L
C	0.239 J/g K	$T(0)$	350 K
$-\Delta H_R$	5×10^4 J/mol		

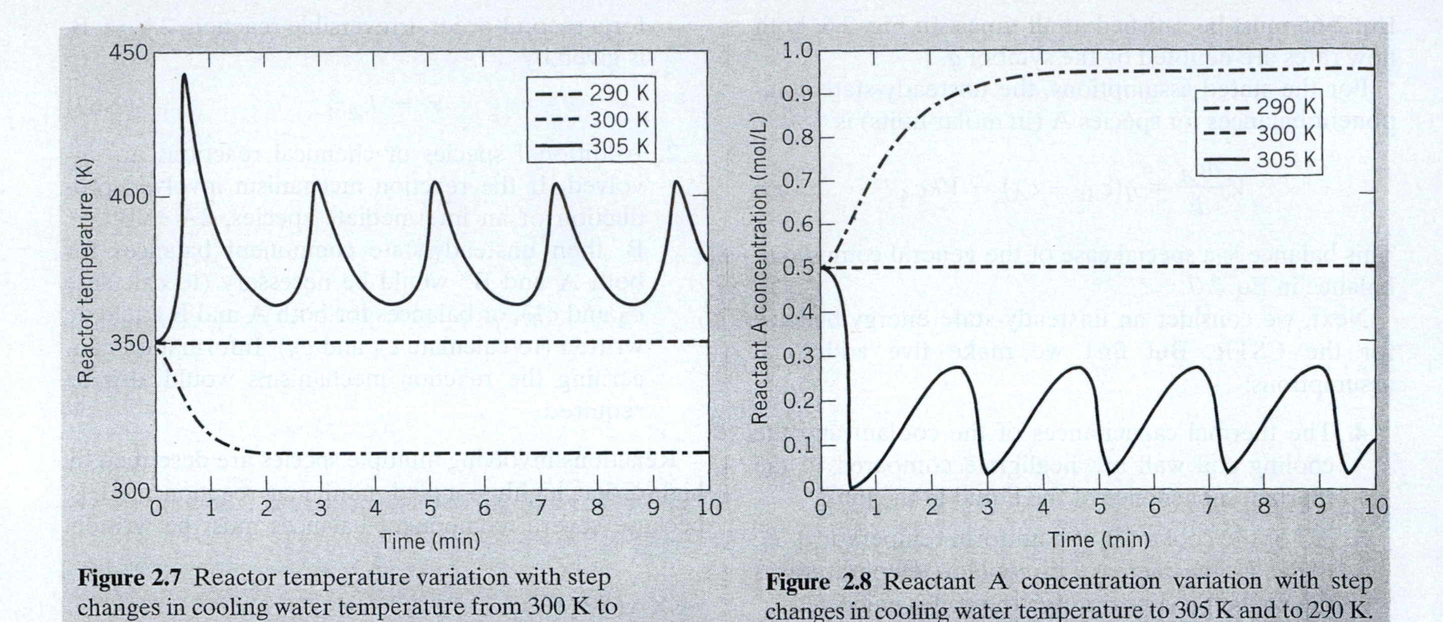

Figure 2.7 Reactor temperature variation with step changes in cooling water temperature from 300 K to 305 K and from 300 K to 290 K.

Figure 2.8 Reactant A concentration variation with step changes in cooling water temperature to 305 K and to 290 K.

Although the modeling task becomes much more complex, the same principles illustrated above can be extended and applied. We will return to the simple CSTR model again in Chapter 4.

2.4.7 Staged Systems (a Three-Stage Absorber)

Chemical processes, particularly separation processes, often consist of a sequence of stages. In each stage, materials are brought into intimate contact to obtain (or approach) equilibrium between the individual phases. The most important examples of staged processes include distillation, absorption, and extraction. The stages are usually arranged as a cascade with immiscible or partially miscible materials (the separate phases) flowing either cocurrently or countercurrently. Countercurrent contacting, shown in Fig. 2.9, usually permits the highest degree of separation to be attained in a fixed number of stages and is considered here.

The feeds to staged systems may be introduced at each end of the process, as in absorption units, or a single feed may be introduced at a middle stage, as is usually the case with distillation. The stages may be physically connected in either a vertical or horizontal configuration, depending on how the materials are transported, that is, whether pumps are used between stages, and so forth. Below we consider a gas-liquid absorption process, because its dynamics are somewhat simpler to develop than those of distillation and extraction processes. At the same time, it illustrates the characteristics of more complicated countercurrent staged processes (Seader and Henley, 2005).

For the three-stage absorption unit shown in Fig. 2.10, a gas phase is introduced at the bottom (molar flow rate G) and a single component is to be absorbed into a liquid phase introduced at the top (molar flow rate L, flowing countercurrently). A practical example of such a process is the removal of sulfur dioxide (SO_2) from combustion gas by use of a liquid absorbent. The gas passes up through the perforated (sieve) trays and contacts the liquid cascading down through them. A series of weirs and downcomers typically are used to retain a significant holdup of liquid on each stage while forcing the gas to flow upward through the perforations. Because of intimate mixing, we can assume that the component to be absorbed is in equilibrium between the gas and liquid streams leaving each stage i. For example, a simple linear relation is often assumed. For stage i

$$y_i = ax_i + b \tag{2-70}$$

where y_i and x_i denote gas and liquid concentrations of the absorbed component. Assuming constant liquid

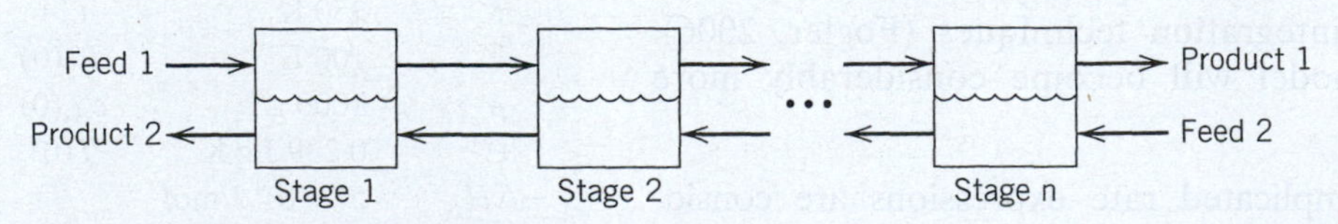

Figure 2.9 A countercurrent-flow staged process.

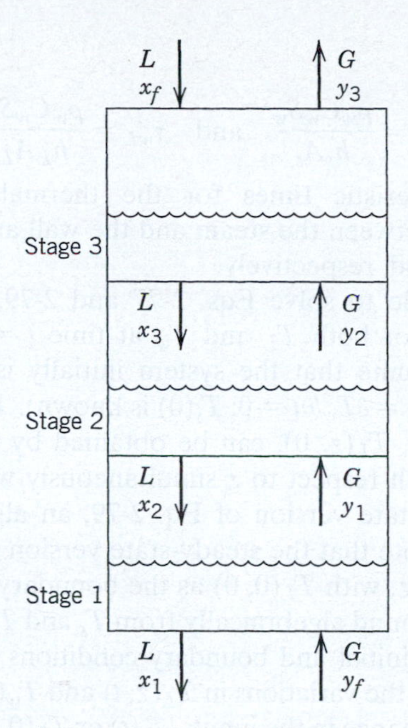

Figure 2.10 A three-stage absorption unit.

holdup H and perfect mixing on each stage, and neglecting the holdup of gas, the component material balance for any stage i is

$$H\frac{dx_i}{dt} = G(y_{i-1} - y_i) + L(x_{i+1} - x_i) \quad (2\text{-}71)$$

In Eq. 2-71 we also assume that molar liquid and gas flow rates L and G are unaffected by the absorption, because changes in concentration of the absorbed component are small, and L and G are approximately constant. Substituting Eq. 2-70 into Eq. 2-71 yields

$$H\frac{dx_i}{dt} = aGx_{i-1} - (L + aG)x_i + Lx_{i+1} \quad (2\text{-}72)$$

Dividing by L and substituting $\tau = H/L$ (the stage liquid residence time), $\mathcal{S} = aG/L$ (the *stripping factor*), and $K = G/L$ (the gas-to-liquid ratio), the following model is obtained for the three-stage absorber:

$$\tau\frac{dx_1}{dt} = K(y_f - b) - (1 + \mathcal{S})\,x_1 + x_2 \quad (2\text{-}73)$$

$$\tau\frac{dx_2}{dt} = \mathcal{S}x_1 - (1 + \mathcal{S})x_2 + x_3 \quad (2\text{-}74)$$

$$\tau\frac{dx_3}{dt} = \mathcal{S}x_2 - (1 + \mathcal{S})x_3 + x_f \quad (2\text{-}75)$$

In the model of (2-73) to (2-75) note that the individual equations are linear but also coupled, meaning that each output variable—x_1, x_2, x_3—appears in more than one equation. This feature can make it difficult to convert these three equations into a single higher-order equation in one of the outputs, as was done in Eq. 2-49.

2.4.8 Distributed Parameter Systems (the Double-Pipe Heat Exchanger)

All of the process models discussed up to this point have been of the *lumped parameter* type, meaning that any dependent variable can be assumed to be a function only of time and not of spatial position. For the stirred-tank systems discussed earlier, we assumed that any spatial variations of the temperature or concentration within the liquid could be neglected. Perfect mixing in each stage was also assumed for the absorber. Even when perfect mixing cannot be assumed, a lumped or average temperature may be taken as representative of the tank contents to simplify the process model.

While lumped parameter models are normally used to describe processes, many important process units are inherently *distributed parameter*; that is, the output variables are functions of both time and position. Hence, their process models contain one or more partial differential equations. Pertinent examples include shell-and-tube heat exchangers, packed-bed reactors, packed columns, and long pipelines carrying compressible gases. In each of these cases, the output variables are a function of distance down the tube (pipe), height in the bed (column), or some other measure of location. In some cases, two or even three spatial variables may be considered; for example, concentration and temperature in a tubular reactor may depend on both axial and radial positions, as well as time.

Figure 2.11 illustrates a double-pipe heat exchanger where a fluid flowing through the inside tube with velocity v is heated by steam condensing in the outer tube. If the fluid is assumed to be in plug flow, the temperature of the liquid is expressed as $T_L(z, t)$ where z denotes distance from the fluid inlet. The fluid heating process is truly distributed parameter; at any instant in time there is a temperature profile along the inside

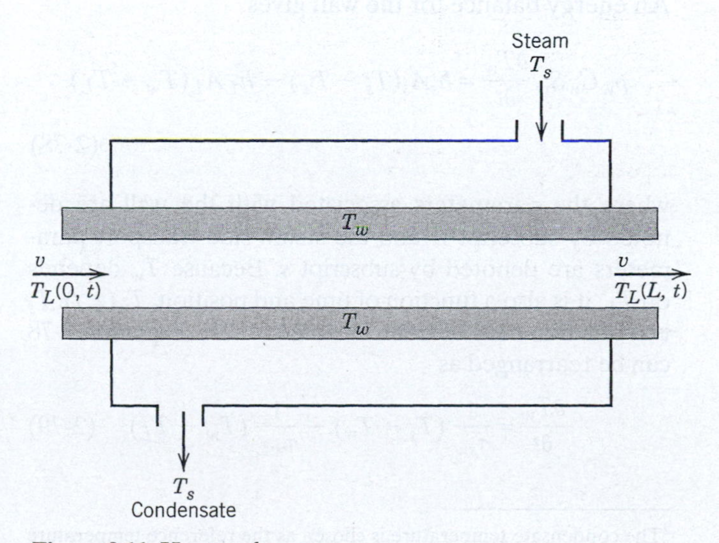

Figure 2.11 Heat exchanger.

tube. The steam condensation, on the other hand, might justifiably be treated as a lumped process, because the steam temperature $T_s(t)$ can be assumed to be a function of the condensation pressure, itself presumably a function only of time and not a function of position. We also assume that the wall temperature $T_w(z, t)$ is different from T_L and T_s due to the thermal capacitance and resistances.

In developing a model for this process, assume that the liquid enters at temperature $T_L(0, t)$—that is, at $z = 0$. Heat transfer coefficients (steam-to-wall h_s and wall-to-liquid h_L) can be used to approximate the energy transfer processes. We neglect the effects of axial energy conduction, the resistance to heat transfer within the metal wall, and the thermal capacitance of the steam condensate.[2] A distributed parameter model for the heat exchanger can be derived by applying Eq. 2-8 over a differential tube length Δz of the exchanger. In such a *shell* energy balance, the partial differential equation is obtained by taking the limit as $\Delta z \to 0$ (Bird et al., 2002). Using the conservation law, Eq. 2-8, the following PDE results (Coughanowr, 1991).

$$\rho_L C_L S_L \frac{\partial T_L}{\partial t} = -\rho_L C_L S_L v \frac{\partial T_L}{\partial z} + h_L A_L (T_w - T_L) \tag{2-76}$$

where the following parameters are constant: ρ_L = liquid density, C_L = liquid heat capacity, S_L = cross-sectional area for liquid flow, h_L = liquid heat transfer coefficient, and A_L = wall heat transfer area of the liquid. This equation can be rearranged to yield

$$\frac{\partial T_L}{\partial t} = -v \frac{\partial T_L}{\partial z} + \frac{1}{\tau_{HL}} (T_w - T_L) \tag{2-77}$$

where $\tau_{HL} = \rho_L C_L S_L / h_L A_L$ has units of time and is called the characteristic time for heating of the liquid. An energy balance for the wall gives

$$\rho_w C_w S_w \frac{\partial T_w}{\partial t} = h_s A_s (T_s - T_w) - h_L A_L (T_w - T_L) \tag{2-78}$$

where the parameters associated with the wall are denoted by subscript w and the steam-side transport parameters are denoted by subscript s. Because T_w depends on T_L, it is also a function of time and position, $T_w(z, t)$. T_s is a function only of time, as noted above. Equation 2-78 can be rearranged as

$$\frac{\partial T_w}{\partial t} = \frac{1}{\tau_{sw}} (T_s - T_w) - \frac{1}{\tau_{wL}} (T_w - T_L) \tag{2-79}$$

where

$$\tau_{sw} = \frac{\rho_w C_w S_w}{h_s A_s} \quad \text{and} \quad \tau_{wL} = \frac{\rho_w C_w S_w}{h_L A_L} \tag{2-80}$$

are characteristic times for the thermal transport processes between the steam and the wall and the wall and the liquid, respectively.

To be able to solve Eqs. 2-77 and 2-79, boundary conditions for both T_L and T_w at time $t = 0$ are required. Assume that the system initially is at steady state ($\partial T_L/\partial t = \partial T_w/\partial t = 0$; $T_s(0)$ is known). The steady-state profile, $T_L(z, 0)$, can be obtained by integrating Eq. 2-77 with respect to z simultaneously with solving the steady-state version of Eq. 2-79, an algebraic expression. Note that the steady-state version of (2-77) is an ODE in z, with $T_L(0, 0)$ as the boundary condition. $T_w(z, 0)$ is found algebraically from T_s and $T_L(z, 0)$.

With the initial and boundary conditions completely determined, the variations in $T_L(z, t)$ and $T_w(z, t)$ resulting from a change in the inputs, $T_s(t)$ or $T_L(0, t)$, can now be obtained by solving Eqs. 2-77 and 2-79 simultaneously using an analytical approach or a numerical procedure (Hanna and Sandall, 1995). Because analytical methods can be used only in special cases, we illustrate a numerical procedure here. A numerical approach invariably requires that either z, t, or both z and t be *discretized*. Here we use a finite difference approximation to convert the PDEs to ODEs. Although numerically less efficient than other techniques such as those based on weighted residuals (Chapra and Canale, 2010), finite difference methods yield more physical insight into both the method and the result of physical lumping.

To obtain ODE models with time as the independent variable, the z dependence is eliminated by discretization. In Fig. 2.12 the double-pipe heat exchanger has been redrawn with a set of grid lines to indicate points at which the liquid and wall temperatures will be evaluated. We now rewrite Eqs. 2-77 and 2-79 in terms of the liquid and wall temperatures $T_L(0)$, $T_L(1)$, . . . , $T_L(N)$ and $T_w(0)$, $T_w(1)$, . . . , $T_w(N)$. Utilizing the backward difference approximation for the derivative $\partial T_L/\partial z$ yields

$$\frac{\partial T_L}{\partial z} \approx \frac{T_L(j) - T_L(j-1)}{\Delta z} \tag{2-81}$$

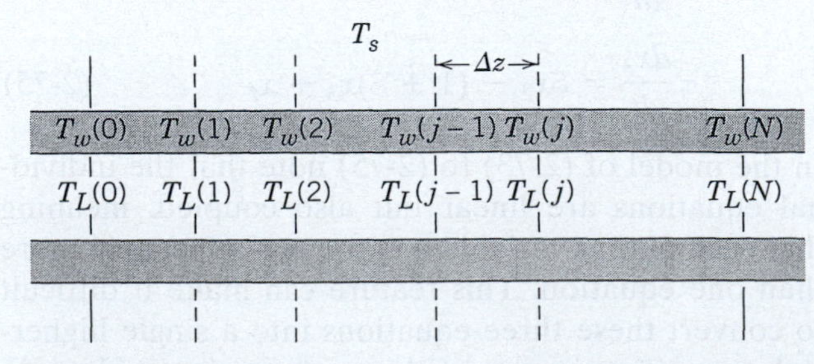

Figure 2.12 Finite-difference approximations for double-pipe heat exchanger.

[2]The condensate temperature is chosen as the reference temperature for energy balances.

where $T_L(j)$ is the liquid temperature at the jth node (discretization point). Substituting Eq. 2-81 into Eq. 2-77, the equation for the jth node is

$$\frac{dT_L(j)}{dt} = -v\frac{T_L(j) - T_L(j-1)}{\Delta z} + \frac{1}{\tau_{HL}}[T_w(j) - T_L(j)]\ (j = 1, \ldots, N) \quad (2\text{-}82)$$

The boundary condition at $z = 0$ becomes

$$T_L(0, t) = T_F(t) \quad (2\text{-}83)$$

where $T_F(t)$ is a specified forcing (input) function. Rearranging Eq. 2-82 yields

$$\frac{dT_L(j)}{dt} = \frac{v}{\Delta z}T_L(j-1) - \left(\frac{v}{\Delta z} + \frac{1}{\tau_{HL}}\right)T_L(j) + \frac{1}{\tau_{HL}}T_w(j) \quad (j = 1, \ldots, N) \quad (2\text{-}84)$$

Similarly, for the wall equation,

$$\frac{dT_w(j)}{dt} = -\left(\frac{1}{\tau_{sw}} + \frac{1}{\tau_{wL}}\right)T_w(j) + \frac{1}{\tau_{wL}}T_L(j) + \frac{1}{\tau_{sw}}T_s(j) \quad (j = 1, \ldots, N) \quad (2\text{-}85)$$

Note that Eqs. 2-84 and 2-85 represent $2N$ linear ordinary differential equations for N liquid and N wall temperatures. There are a number of anomalies associated with this simplified approach compared to the original PDEs. For example, it is clear that heat transfer from steam to wall to liquid is not accounted for at the zeroth node (the entrance), but is accounted for at all other nodes. Also, a detailed analysis of the discrete model will show that the steady-state relations between $T_L(j)$ and either input, T_s or T_F, are a function of the number of grid points and thus the grid spacing, Δz. The discrepancy can be minimized by making N large, that is, Δz small, The lowest-order model for this system that retains some distributed nature would be for $N = 2$. In this case, four equations result:

$$\frac{dT_{L1}}{dt} = \frac{v}{\Delta z}T_F(t) - \left(\frac{v}{\Delta z} + \frac{1}{\tau_{HL}}\right)T_{L1} + \frac{1}{\tau_{HL}}T_{w1} \quad (2\text{-}86)$$

$$\frac{dT_{L2}}{dt} = \frac{v}{\Delta z}T_{L1} - \left(\frac{v}{\Delta z} + \frac{1}{\tau_{HL}}\right)T_{L2} + \frac{1}{\tau_{HL}}T_{w2} \quad (2\text{-}87)$$

$$\frac{dT_{w1}}{dt} = -\left(\frac{1}{\tau_{sw}} + \frac{1}{\tau_{wL}}\right)T_{w1} + \frac{1}{\tau_{wL}}T_{L1} + \frac{1}{\tau_{sw}}T_s(t) \quad (2\text{-}88)$$

$$\frac{dT_{w2}}{dt} = -\left(\frac{1}{\tau_{sw}} + \frac{1}{\tau_{wL}}\right)T_{w2} + \frac{1}{\tau_{wL}}T_{L2} + \frac{1}{\tau_{sw}}T_s(t) \quad (2\text{-}89)$$

where the node number has been denoted by the second subscript on the output variables to simplify the notation. Equations 2-86 to 2-89 are coupled, linear, ordinary differential equations.

2.4.9 Fed-Batch Bioreactor

Biological reactions that involve microorganisms and enzyme catalysts are pervasive and play a crucial role in the natural world. Without such bioreactions, plant and animal life as we know it simply could not exist. Bioreactions also provide the basis for production of a wide variety of pharmaceuticals and healthcare and food products. Other important industrial processes that involve bioreactions include fermentation and wastewater treatment. Chemical engineers are heavily involved with biochemical and biomedical processes. In this section we present a dynamic model for a representative process, a bioreactor operated in a semi-batch mode. Additional biochemical and biomedical applications appear in other chapters.

In general, bioreactions are characterized by the conversion of feed material (or *substrate*) into products and cell mass (or *biomass*). The reactions are typically catalyzed by enzymes (Bailey and Ollis, 1986; Fogler, 1999). When the objective is to produce cells, a small amount of cells (*inoculum*) is added to initiate subsequent cell growth. A broad class of bioreactions can be represented in simplified form as

$$\text{substrate} \xrightarrow{\text{cells}} \text{more cells} + \text{products} \quad (2\text{-}90)$$

The stoichiometry of bioreactions can be very complex and depends on many factors that include the environmental conditions in the vicinity of the cells. For simplicity we consider the class of bioreactions where the substrate contains a single limiting nutrient and only one product results. The following *yield coefficients* are based on the reaction stoichiometry:

$$Y_{X/S} = \frac{\textit{mass of new cells formed}}{\textit{mass of substrate consumed to form new cells}} \quad (2\text{-}91)$$

$$Y_{P/S} = \frac{\textit{mass of product formed}}{\textit{mass of substrate consumed to form product}} \quad (2\text{-}92)$$

Many important bioreactors are operated in a semicontinuous manner that is referred to as *fed-batch* operation, which is illustrated in Figure 2.13. A feed stream containing substrate is introduced to the fed-batch reactor continually. The mass flow rate is denoted by F and the substrate mass concentration by S_f. Because there is no exit stream, the volume V of the bioreactor contents increases during the batch. The advantage of fed-batch operation is that it allows the substrate

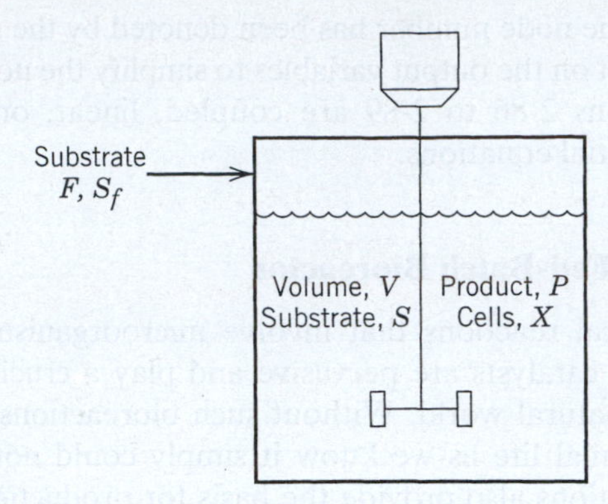

Figure 2.13 Fed-batch reactor for a bioreaction.

concentration to be maintained at a desired level, in contrast to batch reactors where the substrate concentration varies continually throughout the batch (Shuler and Kargi, 2002).

Fed-batch operation is used to manufacture many important industrial products, including antibiotics and protein pharmaceuticals. In batch and fed-batch reactors, cell growth occurs in different stages after the inoculum is introduced. We will consider only the exponential growth stage where the cell growth rate is autocatalytic and is assumed to be proportional to the cell concentration. A standard reaction rate expression to describe the rate of cell growth with a single limiting substrate is given by (Bailey and Ollis, 1986; Fogler, 2006)

$$r_g = \mu X \tag{2-93}$$

where r_g is the rate of cell growth per unit volume, X is the cell mass, and μ is the *specific growth rate,* which is well described by the *Monod equation*:

$$\mu = \mu_{max} \frac{S}{K_S + S} \tag{2-94}$$

Note that μ has units of reciprocal time—for example, h^{-1}. Model parameter μ_{max} is referred to as the *maximum growth rate,* because μ has a maximum value of μ_{max} when $S \gg K_S$. The second model parameter, K_S, is called the *Monod constant.* The Monod equation has the same form as the Michaelis-Menten equation, a standard rate expression for enzyme reactions (Bailey and Ollis, 1986; Fogler, 2006).

A dynamic model for the fed-batch bioreactor in Fig. 2.13 will be derived based on the following assumptions:

1. The cells are growing exponentially.
2. The fed-batch reactor is perfectly mixed.
3. Heat effects are small so that isothermal reactor operation can be assumed.
4. The liquid density is constant.
5. The *broth* in the bioreactor consists of liquid plus solid material (i.e., cell mass). This heterogeneous mixture can be approximated as a homogenous liquid.
6. The rate of cell growth r_g is given by (2-93) and (2-94).
7. The rate of product formation per unit volume r_p can be expressed as

$$r_P = Y_{P/X} r_g \tag{2-95}$$

where the *product yield coefficient* $Y_{P/X}$ is defined as:

$$Y_{P/X} = \frac{\textit{mass of product formed}}{\textit{mass of new cells formed}} \tag{2-96}$$

8. The feed stream is sterile and thus contains no cells.

The dynamic model of the fed-batch reactor consists of individual balances for substrate, cell mass, and product, plus an overall mass balance. The general form of each balance is

$$\{\textit{Rate of accumulation}\} = \{\textit{rate in}\} + \{\textit{rate of formation}\} \tag{2-97}$$

The individual component balances are

$$\textit{Cells:} \quad \frac{d(XV)}{dt} = V r_g \tag{2-98}$$

$$\textit{Product:} \quad \frac{d(PV)}{dt} = V r_p \tag{2-99}$$

$$\textit{Substrate:} \quad \frac{d(SV)}{dt} = F S_f - \frac{1}{Y_{X/S}} V r_g \tag{2-100}$$

where P is the mass concentration of the product and V is reactor volume. Reaction rates r_g and r_p and yield coefficients were defined in Eqs. 2-91 through 2-96. The overall mass balance (assuming constant density) is

$$\textit{Mass:} \quad \frac{dV}{dt} = F \tag{2-101}$$

The dynamic model is simulated for two different feed rates (0.02 L/hr and 0.05 L/hr). Figure 2.14 shows the profile of cell, product, and substrate concentration, together

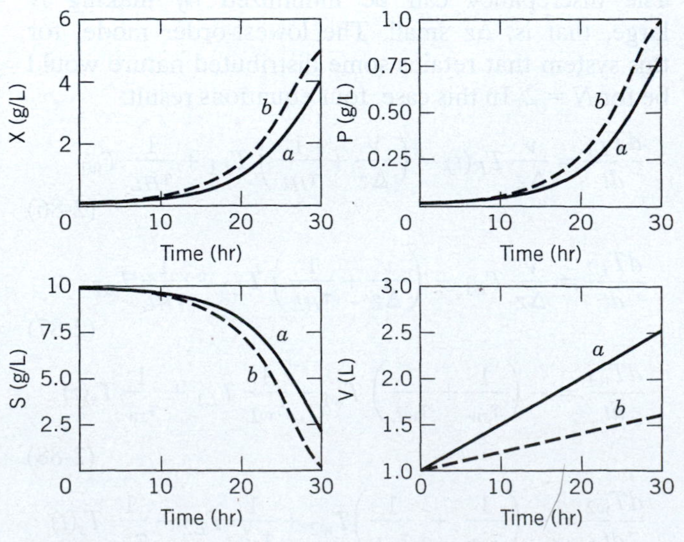

Figure 2.14 Fed-batch reaction profile (*a*: F = 0.05 L/hr; b: F = 0.02 L/hr).

Table 2.3 Model Parameters and Simulation Conditions for Bioreactor

Model Parameters			Simulation Conditions		
μ_{max}	0.20	hr^{-1}	S_f	10.0	g/L
K_S	1.0	g/L	$X(0)$	0.05	g/L
$Y_{X/S}$	0.5	g/g	$S(0)$	10.0	g/L
$Y_{P/X}$	0.2	g/g	$P(0)$	0.0	g/L
			$V(0)$	1.0	L

with liquid volume in the reactor. The model parameters and simulation conditions are given in Table 2.3. For different feed rates, the bioreactor gives different responses; thus, the product can be maximized by varying F.

2.5 PROCESS DYNAMICS AND MATHEMATICAL MODELS

Once a dynamic model has been developed, it can be solved for a variety of conditions that include changes in the input variables or variations in the model parameters. The transient responses of the output variables as functions of time are calculated by numerical integration after specifying the initial conditions, the inputs and the time interval at which the system is to be integrated.

A large number of numerical integration techniques are available, ranging from simple techniques (e.g., the Euler and Runge-Kutta methods) to more complicated ones (e.g., the implicit Euler and Gear methods). All of these techniques represent some compromise between computational effort (computing time) and accuracy. Although a dynamic model can always be solved in principle, for some situations it may be difficult to generate useful numerical solutions. Dynamic models that exhibit a wide range of time scales (*stiff equations*) are quite difficult to solve accurately in a reasonable amount of computation time. Software for integrating ordinary and partial differential equations is readily available. Websites for the following popular software packages are given at the end of the chapter: *MATLAB, Mathematica, POLYMATH, ACSL, IMSL, Mathcad* and *GNU Octave.*

For solving dynamic models that contain large numbers of algebraic and ordinary differential equations, standard programs have been developed to assist in this task. A *graphical-user interface (GUI)* allows the user to enter the algebraic and ordinary differential equations and related information, such as the total integration period, error tolerances, the variables to be plotted, and so on. The simulation program then assumes responsibility for:

1. Checking to ensure that the set of equations is exactly specified.
2. Sorting the equations into an appropriate sequence for iterative solution.
3. Integrating the equations.
4. Providing numerical and graphical output.

Examples of equation-oriented simulators include *ACSL, gPROMS,* and *Aspen Custom Modeler* (Luyben, 2002).

One disadvantage of equation-oriented packages is the amount of time and effort required to develop all of the equations for a complex process. An alternative approach is to use modular simulation, in which prewritten subroutines provide models of individual process units, such as distillation columns or chemical reactors. Consequently, this type of simulator has a direct correspondence to the process flowsheet. The modular approach has the significant advantage that plant-scale simulations only require the user to identify the appropriate modules and to supply the numerical values of model parameters and initial conditions, which is easily accomplished via a graphical user inteface. This activity requires much less effort than writing all of the equations, and it is also easier to program and debug than sets of equations. Furthermore, the software is responsible for all aspects of the solution. Because each module is rather general in form, the user can simulate alternative flowsheets for a complex process—for example, different configurations of distillation towers and heat exchangers, or different types of chemical reactors. Similarly, alternative process control strategies can be quickly evaluated. Some software packages allow the user to add custom modules for novel applications.

In many modeling applications, it may be desirable to develop a simulation using vendor-provided software packages involving different modules or functionalities (for example, software packages for thermodynamic properties, simulation, optimization, and control system design). Historically, it has been difficult to establish communication between software packages developed by different sources, such as software and equipment vendors, universities, and user companies. Fortunately, through worldwide efforts such as Global CAPE-OPEN, standard software protocols have been developed (*open standards*) to accommodate *plug-and-play* software. A list of websites for simulation software packages is given at the end of the chapter.

Modular dynamic simulators have been available since the early 1970s. Several commercial products are available from Aspen Technology (ASPEN PLUS and HYSYS), Honeywell (UniSim), Chemstations (ChemCAD), and Invensys (PRO/II). *Modelica* is an example of a collaborative effort that provides modeling capability for a number of application areas. These packages also offer equation-oriented capabilities. *Modular* dynamic simulators have achieved a high degree of acceptance in process engineering and control studies because they allow plant dynamics, real-time optimization, and alternative control configurations to be evaluated for an existing or new plant, sometimes in the context of operator training. Current open systems utilize *OLE* (*Object Linking and Embedding*), which allows dynamic simulators to be integrated with software for other applications, such as control system design and optimization. A more recent and widely used standard is *OPC* (*OLE for Process Control*), which is a worldwide standard of application interface in industrial automation software and enterprise systems. The OPC Foundation provides the standard specifications for exchange of process control data between data sources and hardware, databases, calculation engines (such as process simulators), spreadsheets, and process historians.

While a dynamic simulator can incorporate some features of control loops, sequences, and the operator interface (e.g., displays and historian), a more practical approach embeds the simulation in the *Distributed Control System (DCS)* and has an adjustable real-time factor. The process simulator reads the DCS outputs for the modulation of final control elements (e.g., control valves and variable speed drives), on/off control of motors (e.g., agitators, fans, and pumps), and the open-close control of automated valves (e.g., isolation and interlock valves). Such simulations for DCS configuration checkout and operator training can significantly reduce the time to commission new equipment and automation systems and to achieve the desired process performance.

Alternatively, process models and the DCS can reside in an off-line personal computer, to provide a more portable, accessible, and maintainable dynamic representation of the plant. Such a virtual plant can be used to enhance process understanding and testing, investigate startups and transitions, diagnose and prevent abnormal operation, improve process automation, and prototype advanced control systems.

SUMMARY

In this chapter we have considered the derivation of dynamic models from first principles, especially conservation equations. Model development is an art as well as a science. It requires making assumptions and simplifications that are consistent with the modeling objectives and the end use of the model. A systematic approach for developing dynamic models is summarized in Table 2.1. This approach has been illustrated by deriving models for representative processes. Although these illustrative examples are rather simple, they demonstrate fundamental concepts that are also valid for more complicated processes. Finally, we have discussed how the development and solution of dynamic models continues to be made easier by commercial simulation software and open standards.

In Chapters 3–6, we will consider analytical solutions of linear dynamic models using Laplace transforms and transfer functions. These useful techniques allow dynamic response characteristics to be analyzed on a more systematic basis. They also provide considerable insight into common characteristics shared by complex processes.

REFERENCES

Aris, R., *Mathematical Modeling: A Chemical Engineer's Perspective*, Academic Press, New York, 1999.

Bailey, J. E. and D. F. Ollis, *Biochemical Engineering Fundamentals*, 2nd ed., McGraw-Hill, New York, 1986.

Bequette, B. W., *Process Dynamics: Modeling, Analysis, and Simulation*, Prentice Hall, Upper Saddle River, NJ, 1998.

Bird, R. B., W. E. Stewart, and E. N. Lightfoot, *Transport Phenomena*, 2nd ed., John Wiley, New York, 2002.

Braunschweig, B. L., C. C. Pantelides, H. I. Britt, and S. Sama, Process Modeling: The Promise of Open Software Architectures, *Chem. Eng. Progr.*, **96**(9), 65 (2000).

Cameron, I. T., and K. Hangos, *Process Modeling and Model Analysis*, Academic Press, New York, 2001.

Chapra, S. C., and R. P. Canale, Numerical Methods for Engineers, 6th ed., McGraw-Hill, New York, 2010.

Coughanowr, D. R., *Process Systems Analysis and Control*, 2nd ed. McGraw-Hill, New York, 1991.

Felder, R. M., and R. W. Rousseau, *Elementary Principles of Chemical Processes*, 3d ed., John Wiley, New York, 2005.

Fogler, H. S., *Elements of Chemical Reactor Engineering*, 3rd ed., Prentice-Hall, Upper Saddle River, NJ, 2006.

Foss, B. A., A. B. Lohmann, and W. Marquardt, A Field Study of the Industrial Modeling Process, *J. Process Control*, **8**, 325 (1998).

Kreyszig, E., *Advanced Engineering Mathematics*, 8th ed., John Wiley, New York, 1998.

Luyben, W. L., *Plantwide Dynamic Simulators in Chemical Processing and Control*, Marcel Dekker, New York, 2002.

Marquardt, W., Trends in Computer-Aided Modeling, *Comput. Chem. Engng.*, **20**, 591 (1996).

Russell, T. F. W. and M. M. Denn, *Introduction to Chemical Engineering Analysis*, John Wiley, New York, 1972.

Sandler, S. I., *Chemical and Engineering Thermodynamics*, 3rd ed., John Wiley, New York, 2006.

Seader, J. D., and E. J. Henley, *Separation Process Principles*, 2d ed., John Wiley, New York, 2005.

Shuler, M. L. and F. Kargi, *Bioprocess Engineering*, 2nd ed., Prentice-Hall, Upper Saddle River, NJ, 2002.

Simulation Software: Web Sites

Advanced Continuous Simulation Language (ACSL), Mitchell and Gauthier Associates Inc., www.mga.com.

Aspen Custom Modeler, www.aspentech.com.

CHEMCAD, www.chemstations.com.

Global Cape Open, www.global-cape-open.org.

GNU-Octave, www.octave.org.

gPROMS, www.psenterprise.com.

HYSYS, www.aspentech.com.

IMSL, www.vni.com.

Mathcad, www.mathcad.com.

Mathematica, www.wolfram.com.

MATLAB, www.mathworks.com.

Modelica, www.modelica.org.

POLYMATH, www.cache.org/polymath.html.

PRO II, www.simsci.com

Unisim, www.honeywell.com/ps.

EXERCISES

2.1 A perfectly stirred, constant-volume tank has two input streams, both consisting of the same liquid. The temperature and flow rate of each of the streams can vary with time.

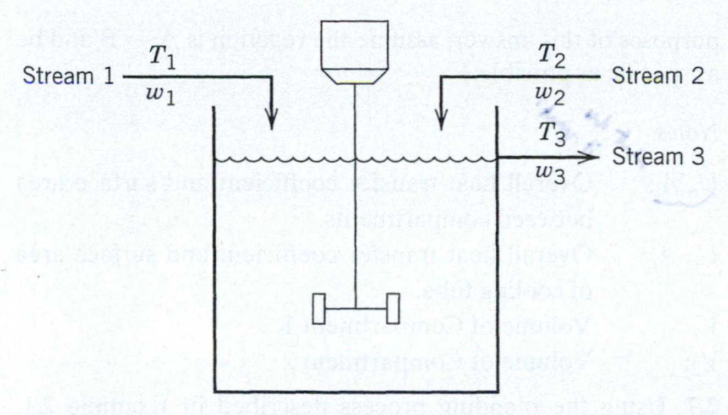

Figure E2.1

(a) Derive a dynamic model that will describe transient operation. Make a degrees of freedom analysis assuming that both Streams 1 and 2 come from upstream units (i.e., their flow rates and temperatures are known functions of time).

(b) Simplify your model, if possible, to one or more differential equations by eliminating any algebraic equations. Also, simplify any derivatives of products of variables.

Notes:

w_i denotes mass flow rate for stream *i*.
Liquid properties are constant (not functions of temperature).

2.2 A completely enclosed stirred-tank heating process is used to heat an incoming stream whose flow rate varies.

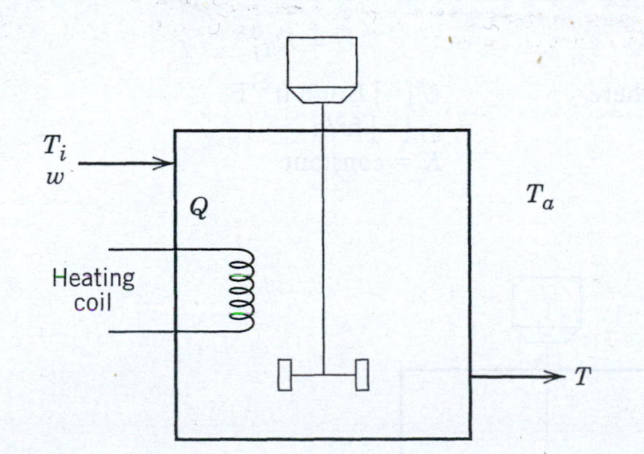

Figure E2.2

The heating rate from this coil and the volume are both constant.

(a) Develop a mathematical model (differential and algebraic equations) that describes the exit temperature if heat losses to the ambient occur and if the ambient temperature (T_a) and the incoming stream's temperature (T_i) both can vary.

(b) Discuss qualitatively what you expect to happen as T_i and w increase (or decrease). Justify by reference to your model.

Notes:

ρ and C_ρ are constants.
U, the overall heat transfer coefficient, is constant.
A_s is the surface area for heat losses to ambient.
$T_i > T_a$ (inlet temperature is higher than ambient temperature).

2.3 Two tanks are connected together in the following unusual way in Fig. E2.3.

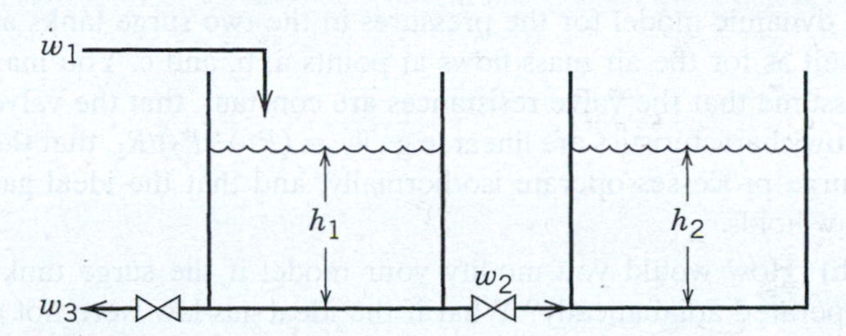

Figure E2.3

(a) Develop a model for this system that can be used to find h_1, h_2, w_2, and w_3 as functions of time for any given variations in inputs.

(b) Perform a degrees of freedom analysis. Identify all input and output variables.

Notes:

The density of the incoming liquid, ρ, is constant.
The cross-sectional areas of the two tanks are A_1 and A_2.
w_2 is positive for flow from Tank 1 to Tank 2.
The two valves are linear with resistances R_2 and R_3.

2.4 Consider a liquid flow system consisting of a sealed tank with noncondensible gas above the liquid as shown in Fig. E2.4. Derive an unsteady-state model relating the liquid level h to the input flow rate q_i. Is operation of this system independent of the ambient pressure P_a? What about for a system open to the atmosphere?

You may make the following assumptions:

(i) The gas obeys the ideal gas law. A constant amount of m_g/M moles of gas are present in the tank.

(ii) The operation is isothermal.

(iii) A square root relation holds for flow through the valve.

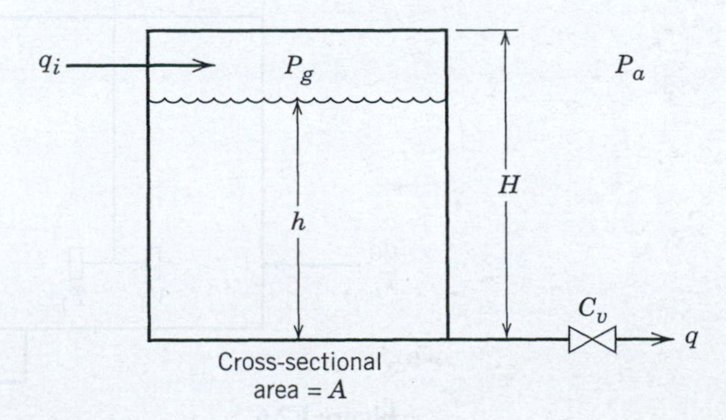

Figure E2.4

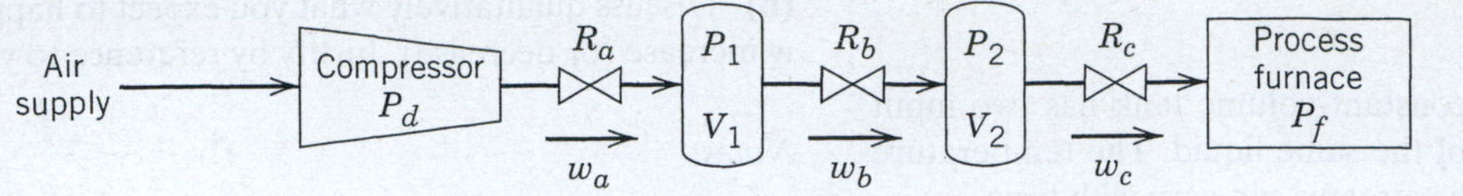

Figure E2.5

2.5 Two surge tanks are used to dampen pressure fluctuations caused by erratic operations of a large air compressor. (See Fig. E2.5.)

(a) If the discharge pressure of the compressor is $P_d(t)$ and the operating pressure of the furnace is P_f (constant), develop a dynamic model for the pressures in the two surge tanks as well as for the air mass flows at points a, b, and c. You may assume that the valve resistances are constant, that the valve flow characteristics are linear, e.g., $w_b = (P_1 - P_2)/R_b$, that the surge processes operate isothermally, and that the ideal gas law holds.

(b) How would you modify your model if the surge tanks operated adiabatically? What if the ideal gas law were not a good approximation?

2.6 A closed stirred-tank reactor with two compartments is shown in Fig. E2.6. The basic idea is to feed the reactants continuously into the first compartment, where they will be preheated by energy liberated in the exothermic reaction, which is anticipated to occur primarily in the second compartment. The wall separating the two compartments is quite thin, thus allowing heat transfer; the outside of the reactor is well insulated; and a cooling coil is built into the second compartment to remove excess energy liberated in the reaction.

Tests are to be conducted initially with a single-component feed (i.e., no reaction) to evaluate the reactor's thermal characteristics.

(a) Develop a dynamic model for this process under the conditions of no reaction. Assume that q_0, T_i, and T_c all may vary.

(b) Make a degrees of freedom analysis for your model—identifying all parameters, outputs, and inputs that must be known functions of time in order to obtain a solution.

(c) In order to estimate the heat transfer coefficients, the reactor will be tested with T_i much hotter than the exit temperature. Explain how your model would have to be modified to account for the presence of the exothermic reaction. (For purposes of this answer, assume the reaction is A → B and be as specific as possible.)

Notes:

U_t, A_t: Overall heat transfer coefficient and surface area between compartments.

U_c, A_c: Overall heat transfer coefficient and surface area of cooling tube.

V_1: Volume of Compartment 1.

V_2: Volume of Compartment 2.

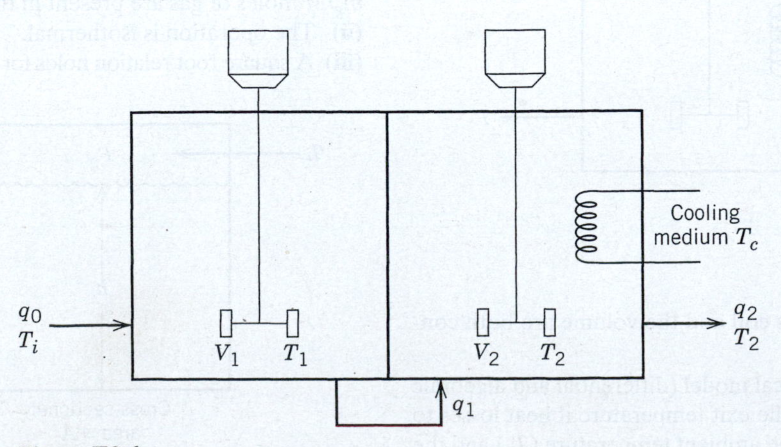

Figure E2.6

2.7 Using the blending process described in Example 2.1, calculate the response of x to a change in x_1 (the disturbance from 0.4 to 0.5 and a change in w_2 from 200 to 100 kg/min. Plot the response using appropriate software for $0 \leq t \leq$ 25 minutes. Explain physically why the composition increases or decreases, compared to case (d) in Fig. 2.2.

2.8 A jacketed vessel is used to cool a process stream as shown in Fig. E2.8. The following information is available:

(i) The volume of liquid in the tank V and the volume of coolant in the jacket V_J remain constant. Volumetric flow rate q_F is constant, but q_J varies with time.

(ii) Heat losses from the jacketed vessel are negligible.

(iii) Both the tank contents and the jacket contents are well mixed and have significant thermal capacitances.

(iv) The thermal capacitances of the tank wall and the jacket wall are negligible.

(v) The overall heat transfer coefficient for transfer between the tank liquid and the coolant varies with coolant flow rate:

$$U = Kq_J^{0.8}$$

where U [=] Btu/h ft^2 °F
q_J [=] ft^3/h
K = constant

Derive a dynamic model for this system. (State any additional assumptions that you make.)

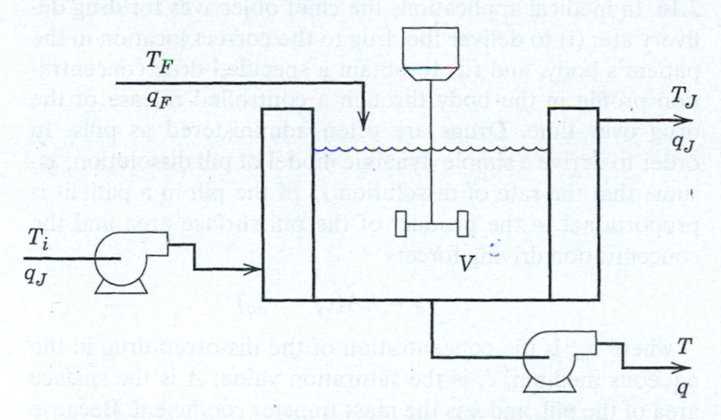

Figure E2.8

2.9 Solve the nonlinear differential equation (2–61) for $q_i = 0$, either analytically or numerically, to obtain $h(t)$. Assume $A = 2$, $C_v{}^* = 0.5$, $\rho = 60$, $g/gc = 1$, and $h(0) = 10$, and that units of these parameters are consistent.

2.10 Irreversible consecutive reactions $A \xrightarrow{k_1} B \xrightarrow{k_2} C$ occur in a jacketed, stirred-tank reactor as shown in Fig. E2.10. Derive a dynamic model based on the following assumptions:

(i) The contents of the tank and cooling jacket are well mixed. The volumes of material in the jacket and in the tank do not vary with time.

(ii) The reaction rates are given by

$$r_1 = k_1 e^{-E_1/RT} c_A \,[=]\, \text{mol A/h L}$$

$$r_2 = k_2 e^{-E_2/RT} c_B \,[=]\, \text{mol B/h L}$$

(iii) The thermal capacitances of the tank contents and the jacket contents are significant relative to the thermal capacitances of the jacket and tank walls, which can be neglected.

(iv) Constant physical properties and heat transfer coefficients can be assumed.

Note:

All flow rates are volumetric flow rates in L/h. The concentrations have units of mol/L. The heats of reaction are ΔH_1 and ΔH_2.

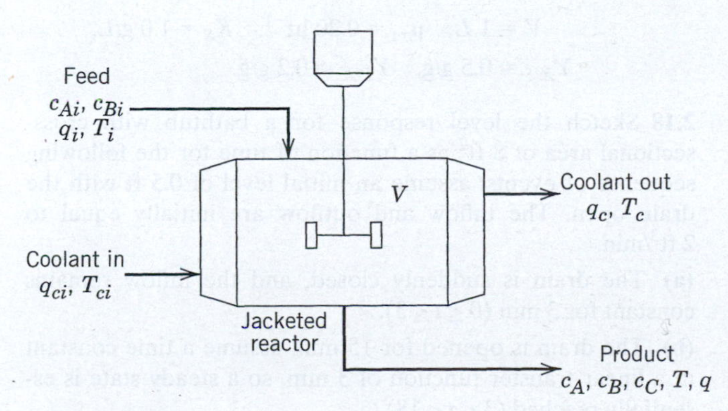

Figure E2.10

2.11 Example 2.1 plots responses for changes in input flows for the stirred tank blending system. Repeat part (b) and plot it. Next, relax the assumption that V is constant, and plot the response of $x(t)$ and $V(t)$ for the change in w_1 for $t = 0$ to 15 minutes. Assume that w_2 and w remain constant.

2.12 A process tank has two input streams—Stream 1 at mass flow rate w_1 and Stream 2 at mass flow rate w_2. The tank's effluent stream, at flow rate w, discharges through a fixed valve to atmospheric pressure. Pressure drop across the valve is proportional to the flow rate squared. The cross-sectional area of the tank, A, is 5 m^2, and the mass density of all streams is 940 kg/m^3.

(a) Draw a schematic diagram of the process and write an appropriate dynamic model for the tank level. What is the corresponding steady-state model?

(b) At initial steady-state conditions, with $w_1 = 2.0$ kg/s and $w_2 = 1.2$ kg/s, the tank level is 2.25 m. What is the value of the valve constant (give units)?

(c) A process control engineer decides to use a feedforward controller to hold the level approximately constant at the set-point value ($h_{sp} = 2.25$ m) by measuring w_1 and manipulating w_2. What is the mathematical relation that will be used in the controller? If the w_1 measurement is not very accurate and always supplies a value that is 1.1 times the actual flow rate, what can you conclude about the resulting level control? (*Hint:* Consider the process initially at the desired steady-state level and with the feedforward controller turned on. Because the controller output is slightly in error, $w_2 \neq 1.2$, so the process will come to a new steady state. What is it?) What conclusions can you draw concerning the need for accuracy in a steady-state model? for the accuracy of the measurement device? for the accuracy of the control valve? Consider all of these with respect to their use in a feedforward control system.

2.13 The liquid storage tank shown in Fig. E2.13 has two inlet streams with mass flow rates w_1 and w_2 and an exit stream with flow rate w_3. The cylindrical tank is 2.5 m tall and 2 m in diameter. The liquid has a density of 800 kg/m^3. Normal operating procedure is to fill the tank until the liquid level reaches a nominal value of 1.75 m using constant flow rates: $w_1 = 120$ kg/min, $w_2 = 100$ kg/min, and $w_3 = 200$ kg/min. At that point, inlet flow rate w_1 is adjusted so that the level remains constant. However, on this particular day, corrosion of the tank has opened up a hole in the wall at a height of 1 m, producing

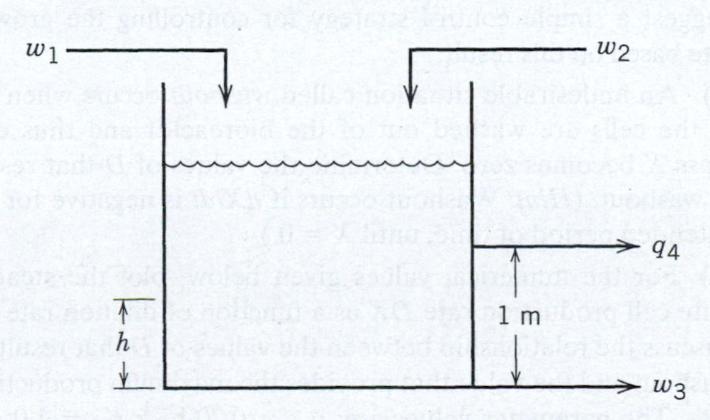

Figure E2.13

a leak whose volumetric flow rate q_4 (m^3/min) can be approximated by

$$q_4 = 0.025\sqrt{h - 1}$$

where h is height in meters.

(a) If the tank was initially empty, how long did it take for the liquid level to reach the corrosion point?

(b) If mass flow rates w_1, w_2, and w_3 are kept constant indefinitely, will the tank eventually overflow? Justify your answer.

2.14 Consider a blending tank that has the same dimensions and nominal flow rates as the storage tank in Exercise 2.13 but that incorporates a valve on the outflow line that is used to establish flow rate w_3. (For this exercise, there is no leak in the tank as in Exercise 2.13.) In addition, the nominal inlet stream mass fractions of component A are $x_1 = x_2 = 0.5$.

The process has been operating for a long time with constant flow rates and inlet concentrations. Under these conditions, it has come to steady state with exit mass fraction $x = 0.5$ and level $h = 1.75$ m. Using the information below, answer the following questions:

(a) What is the value of w_3? the constant, C_v?

(b) If x_1 is suddenly changed from 0.5 to 0.6 without changing the inlet flow rates (of course, x_2 must change as well), what is the final value of x_3? How long does it take to come within 1% of this final value?

(c) If w_1 is changed from 120 kg/min to 100 kg/min without changing the inlet concentrations, what will be the final value of the tank level? How long will it take to come within 1% of this final value?

(d) Would it have made any difference in part (c) if the concentrations had changed at the same time the flow rate was changed?

Useful information: The tank is perfectly stirred.

$$w_3 = C_v\sqrt{h}$$

2.15 Suppose that the fed-batch bioreactor in Fig. 2.11 is converted to a continuous, stirred-tank bioreactor (also called a *chemostat*) by adding an exit stream. Assume that the inlet and exit streams have the same mass flow rate F and thus the volume of liquid V in the chemostat is constant.

(a) Derive a dynamic model for this chemostat by modifying the fed-batch reactor model in Section 2.4.9.

(b) Derive the steady-state relationship between growth rate μ in Eq. 2-93 and *dilution rate* D where by definition, $D = F/V$. Suggest a simple control strategy for controlling the growth rate based on this result.

(c) An undesirable situation called *washout* occurs when all of the cells are washed out of the bioreactor and thus cell mass X becomes zero. Determine the values of D that result in washout. (*Hint:* Washout occurs if dX/dt is negative for an extended period of time, until $X = 0$.)

(d) For the numerical values given below, plot the steady-state cell production rate DX as a function of dilution rate D. Discuss the relationship between the values of D that result in washout and the value that provides the maximum production rate. The parameter values are: $\mu_m = 0.20\ h^{-1}$; $K_S = 1.0$ g/l, and $Y_{X/S} = 0.5$ g/g. The steady-state condition is $\overline{D} = 0.1\ h^{-1}$, $\overline{X} = 2.25$ g/L, $\overline{S} = 1.0$ g/L, and $\overline{S}_f = 10$ g/L.

2.16 In medical applications the chief objectives for drug delivery are: (i) to deliver the drug to the correct location in the patient's body, and (ii) to obtain a specified drug concentration profile in the body through a controlled release of the drug over time. Drugs are often administered as pills. In order to derive a simple dynamic model of pill dissolution, assume that the rate of dissolution r_d of the pill in a patient is proportional to the product of the pill surface area and the concentration driving force:

$$r_d = kA(c_s - c_{aq})$$

where c_{aq} is the concentration of the dissolved drug in the aqueous medium, c_s is the saturation value, A is the surface area of the pill, and k is the mass transfer coefficient. Because $c_s .. c_{aq}$, even if the pill dissolves completely, the rate of dissolution reduces to $r_d = kAc_s$.

(a) Derive a dynamic model that can be used to calculate pill mass M as a function of time. You can make the following simplifying assumptions:

(i) The rate of dissolution of the pill is given by $r_d = kAc_s$.

(ii) The pill can be approximated as a cylinder with radius r and height h. It can be assumed that $h/r .. 1$. Thus the pill surface area can be approximated as $A = 2\pi rh$.

(b) For the conditions given below, how much time is required for the pill radius r to be reduced by 90% from its initial value of r_0?

$\rho = 1.2$ g/ml $\quad r_0 = 0.4$ cm $\quad h = 1.8$ cm
$c_s = 500$ g/L $\quad k = 0.016$ cm/min

2.17 Bioreactions are often carried out in batch reactors. The fed-batch bioreactor model in Section 2.4.9 is also applicable to batch reactors if the feed flow rate F is set equal to zero. Using the available information shown below, determine how much time is required to achieve a 90% conversion of the substrate. Assume that the volume V of the reactor contents is constant.

Available information:

(i) *Initial conditions:*

$$X(0) = 0.05 \text{ g/L}, \quad S(0) = 10 \text{ g/L}, \quad P(0) = 0 \text{ g/L}.$$

(ii) *Parameter values:*

$$V = 1\,L, \quad \mu_m = 0.20\ \text{hr}^{-1}, \quad K_S = 1.0 \text{ g/L},$$
$$Y_{X/S} = 0.5 \text{ g/g}, \quad Y_{P/X} = 0.2 \text{ g/g}.$$

2.18 Sketch the level response for a bathtub with cross-sectional area of 8 ft^2 as a function of time for the following sequence of events; assume an initial level of 0.5 ft with the drain open. The inflow and outflow are initially equal to 2 ft^3/min.

(a) The drain is suddenly closed, and the inflow remains constant for 3 min ($0 \le t \le 3$).

(b) The drain is opened for 15 min; assume a time constant in a linear transfer function of 3 min, so a steady state is essentially reached ($3 \le t \le 18$).

(c) The inflow rate is doubled for 6 min ($18 \le t \le 24$).

(d) The inflow rate is returned to its original value for 16 min ($24 \le t \le 40$).

2.19 Perform a degrees of freedom analysis for the model in Eqs. 2-64 through 2-68. Identify parameters, output variables, and inputs (manipulated and disturbance variables).

2.20 Surge and storage tanks are important dynamic processes in a chemical plant. We can investigate their behavior by using simple experiments at home. Obtain a translucent paper cup (available at fine fast-food restaurants) approximately 6 to 8 in high (or more). Puncture the cup on the side near the bottom with a small hole (~ 1/8 in).

(a) Fill the cup to the top and record how long it takes for the cup to empty. Try other heights (h) and record the time to empty (t_e) (and repeat some of the trials due to experimental error). Plot the results (h vs. t_e). Is the relationship between h and t_e linear or nonlinear? Note this data is related to the case if you measure the height vs. time in a single experiment.

(b) Is the outflow rate constant with respect to time? Explain why, or why not.

(c) Develop a nonlinear dynamic model (ODE) for the process that describes the height vs. time:

$$\frac{dh}{dt} = f(h)$$

(d) A linear model would be $\frac{dh}{dt} = a_1 h$. What is its solution for an initial condition $h(0)$? Can you estimate a_1 from the data in part (a)?

2.21 Plot the level response for a tank with constant cross-sectional area of 4 ft^2 as a function of time for the following sequence of events; assume an initial level of 1.0 ft with the drain open, and that level and outflow rate are linearly related. The steady-state inflow and outflow are initially equal to 2 ft^3/min. The graph should show numerical values of level vs. time.

(a) The drain is suddenly closed, and the inflow remains constant for 3 min ($0 \le t \le 3$).

(b) The drain is opened for 15 min, keeping the inflow at 2 ft^3/min, where a steady state is essentially reached ($3 \le t \le 18$).

(c) The inflow rate is doubled to 4 ft^3/min for 15 min ($18 \le t \le 33$).

(d) The inflow rate is returned to its original value of 2 ft^3/min for 17 min ($33 \le t \le 50$).

Chapter 3

Laplace Transforms

CHAPTER CONTENTS

In Chapter 2 we developed a number of mathematical models that describe the dynamic operation of selected processes. Solving such models—that is, finding the output variables as functions of time for some change in the input variable(s)—requires either analytical or numerical integration of the differential equations. Sometimes considerable effort is involved in obtaining the solutions. One important class of models includes systems described by linear ordinary differential equations (ODEs). Such *linear systems* represent the starting point for many analysis techniques in process control.

In this chapter we introduce a mathematical tool, the *Laplace transform*, which can significantly reduce the effort required to solve and analyze linear differential equation models. A major benefit is that this transformation converts ordinary differential equations to algebraic equations, which can simplify the mathematical manipulations required to obtain a solution or perform an analysis.

First, we define the Laplace transform and show how it can be used to derive the Laplace transforms of simple functions. Then we show that linear ODEs can be solved using Laplace transforms, along with a technique called *partial fraction expansion*. Some important general properties of Laplace transforms are presented, and we illustrate the use of these techniques with a series of examples.

3.1 THE LAPLACE TRANSFORM OF REPRESENTATIVE FUNCTIONS

The Laplace transform of a function $f(t)$ is defined as

$$F(s) = \mathcal{L}[f(t)] = \int_0^\infty f(t)e^{-st}\,dt \tag{3-1}$$

where $F(s)$ is the symbol for the Laplace transform, s is a complex independent variable, $f(t)$ is some function of time to be transformed, and $\mathcal{L}$ is an operator, defined by the integral. The function $f(t)$ must satisfy mild conditions that include being piecewise continuous for $0 < t < \infty$ (Churchill, 1971); this requirement almost always holds for functions that are useful in process modeling and control. When the integration is performed, the transform becomes a function of the Laplace transform variable s. The *inverse Laplace transform* ($\mathcal{L}^{-1}$) operates on the function $F(s)$ and converts it to $f(t)$. Notice that $F(s)$ contains no information about $f(t)$ for $t < 0$. Hence, $f(t) = \mathcal{L}^{-1}\{F(s)\}$ is not defined for $t < 0$ (Schiff, 1999).

One of the important properties of the Laplace transform and the inverse Laplace transform is that they are linear operators; a linear operator satisfies the *superposition principle*:

$$\mathcal{F}(ax(t) + by(t)) = a\mathcal{F}(x(t)) + b\mathcal{F}(y(t)) \quad (3\text{-}2)$$

where $\mathcal{F}$ denotes a particular operation to be performed, such as differentiation or integration with respect to time. If $\mathcal{F} \equiv \mathcal{L}$, then Eq. 3-2 becomes

$$\mathcal{L}(ax(t) + by(t)) = aX(s) + bY(s) \quad (3\text{-}3)$$

Therefore, the Laplace transform of a sum of functions $x(t)$ and $y(t)$ is the sum of the individual Laplace transforms $X(s)$ and $Y(s)$; in addition, multiplicative constants can be factored out of the operator, as shown in (3-3).

In this book we are more concerned with operational aspects of Laplace transforms—that is, using them to obtain solutions or the properties of solutions of *linear differential equations*. For more details on mathematical aspects of the Laplace transform, the texts by Churchill (1971) and Dyke (1999) are recommended.

Before we consider solution techniques, the application of Eq. 3-1 should be discussed. The Laplace transform can be derived easily for most simple functions, as shown below.

Constant Function. For $f(t) = a$ (a constant),

$$\mathcal{L}(a) = \int_0^\infty ae^{-st}\,dt = -\frac{a}{s}e^{-st}\Big|_0^\infty \quad (3\text{-}4)$$

$$= 0 - \left(-\frac{a}{s}\right) = \frac{a}{s}$$

Step Function. The unit step function, defined as

$$S(t) = \begin{cases} 0 & t < 0 \\ 1 & t \geq 0 \end{cases} \quad (3\text{-}5)$$

is an important input that is used frequently in process dynamics and control. The Laplace transform of the unit step function is the same as that obtained for the constant above when $a = 1$:

$$\mathcal{L}[S(t)] = \frac{1}{s} \quad (3\text{-}6)$$

If the step magnitude is a, the Laplace transform is a/s. The step function incorporates the idea of *initial time*, *zero time*, or *time zero* for the function, which refers to the time at which $S(t)$ changes from 0 to 1. To avoid any ambiguity concerning the value of the step function at $t = 0$ (it is discontinuous), we will consider $S(t = 0)$ to be the value of the function approached from the positive side, $t = 0^+$.

Derivatives. The transform of a first derivative of f is important because such derivatives appear in dynamic models:

$$\mathcal{L}(df/dt) = \int_0^\infty (df/dt)e^{-st}\,dt \quad (3\text{-}7)$$

Integrating by parts,

$$\mathcal{L}(df/dt) = \int_0^\infty f(t)e^{-st}s\,dt + f(t)e^{-st}\Big|_0^\infty \quad (3\text{-}8)$$

$$= s\mathcal{L}(f(t)) - f(0) = sF(s) - f(0) \quad (3\text{-}9)$$

where $F(s)$ is the Laplace transform of $f(t)$. Generally, the point at which we start keeping time for a solution is arbitrary. Model solutions are most easily obtained assuming that time *starts* (i.e., $t = 0$) at the moment the process model is first perturbed. For example, if the process initially is assumed to be at steady state and an input undergoes a unit step change, *zero time* is taken to be the moment at which the input changes in magnitude. In many process modeling applications, functions are defined so that they are zero at initial time—that is, $f(0) = 0$. In these cases, (3-9) simplifies to $\mathcal{L}(df/dt) = sF(s)$.

The Laplace transform for higher-order derivatives can be found using Eq. 3-9. To derive $\mathcal{L}[f''(t)]$, we define a new variable ($\phi = df/dt$) such that

$$\mathcal{L}\left(\frac{d^2f}{dt^2}\right) = \mathcal{L}\left(\frac{d\phi}{dt}\right) = s\phi(s) - \phi(0) \quad (3\text{-}10)$$

$$\phi(s) = sF(s) - f(0) \quad (3\text{-}11)$$

Substituting into Eq. 3-10

$$\mathcal{L}\left(\frac{d^2f}{dt^2}\right) = s[sF(s) - f(0)] - \phi(0) \quad (3\text{-}12)$$

$$= s^2F(s) - sf(0) - f'(0) \quad (3\text{-}13)$$

where $f'(0)$ denotes the value of df/dt at $t = 0$. The Laplace transform for derivatives higher than second order can be found by the same procedure. An nth-order derivative, when transformed, yields a series of $(n + 1)$ terms:

$$\mathcal{L}\left(\frac{d^nf}{dt^n}\right) = s^nF(s) - s^{n-1}f(0) - s^{n-2}f^{(1)}(0) - \cdots$$
$$- sf^{(n-2)}(0) - f^{(n-1)}(0) \quad (3\text{-}14)$$

where $f^i(0)$ is the ith derivative evaluated at $t = 0$. If $n = 2$, Eq. 3-13 is obtained.

Exponential Functions. The Laplace transform of an exponential function is important because exponential functions appear in the solution to all linear differential

Table 3.1 Laplace Transforms for Various Time-Domain Functions[a]

$f(t)$	$F(s)$
1. $\delta(t)$ (unit impulse)	1
2. $S(t)$ (unit step)	$\dfrac{1}{s}$
3. t (ramp)	$\dfrac{1}{s^2}$
4. t^{n-1}	$\dfrac{(n-1)!}{s^n}$
5. e^{-bt}	$\dfrac{1}{s+b}$
6. $\dfrac{1}{\tau}e^{-t/\tau}$	$\dfrac{1}{\tau s+1}$
7. $\dfrac{t^{n-1}e^{-bt}}{(n-1)!}$ $(n>0)$	$\dfrac{1}{(s+b)^n}$
8. $\dfrac{1}{\tau^n(n-1)!}t^{n-1}e^{-t/\tau}$	$\dfrac{1}{(\tau s+1)^n}$
9. $\dfrac{1}{b_1-b_2}(e^{-b_2t}-e^{-b_1t})$	$\dfrac{1}{(s+b_1)(s+b_2)}$
10. $\dfrac{1}{\tau_1-\tau_2}(e^{-t/\tau_1}-e^{-t/\tau_2})$	$\dfrac{1}{(\tau_1 s+1)(\tau_2 s+1)}$
11. $\dfrac{b_3-b_1}{b_2-b_1}e^{-b_1t}+\dfrac{b_3-b_2}{b_1-b_2}e^{-b_2t}$	$\dfrac{s+b_3}{(s+b_1)(s+b_2)}$
12. $\dfrac{1}{\tau_1}\dfrac{\tau_1-\tau_3}{\tau_1-\tau_2}e^{-t/\tau_1}+\dfrac{1}{\tau_2}\dfrac{\tau_2-\tau_3}{\tau_2-\tau_1}e^{-t/\tau_2}$	$\dfrac{\tau_3 s+1}{(\tau_1 s+1)(\tau_2 s+1)}$
13. $1-e^{-t/\tau}$	$\dfrac{1}{s(\tau s+1)}$
14. $\sin \omega t$	$\dfrac{\omega}{s^2+\omega^2}$
15. $\cos \omega t$	$\dfrac{s}{s^2+\omega^2}$
16. $\sin(\omega t+\phi)$	$\dfrac{\omega\cos\phi+s\sin\phi}{s^2+\omega^2}$
17. $e^{-bt}\sin\omega t$ } b, ω real	$\dfrac{\omega}{(s+b)^2+\omega^2}$
18. $e^{-bt}\cos\omega t$	$\dfrac{s+b}{(s+b)^2+\omega^2}$
19. $\dfrac{1}{\tau\sqrt{1-\zeta^2}}e^{-\zeta t/\tau}\sin(\sqrt{1-\zeta^2}\,t/\tau)$ $(0\le\lvert\zeta\rvert<1)$	$\dfrac{1}{\tau^2s^2+2\zeta\tau s+1}$
20. $1+\dfrac{1}{\tau_2-\tau_1}(\tau_1e^{-t/\tau_1}-\tau_2e^{-t/\tau_2})$ $(\tau_1\neq\tau_2)$	$\dfrac{1}{s(\tau_1 s+1)(\tau_2 s+1)}$
21. $1-\dfrac{1}{\sqrt{1-\zeta^2}}e^{-\zeta t/\tau}\sin[\sqrt{1-\zeta^2}\,t/\tau+\psi]$ $\psi=\tan^{-1}\dfrac{\sqrt{1-\zeta^2}}{\zeta}$, $(0\le\lvert\zeta\rvert<1)$	$\dfrac{1}{s(\tau^2s^2+2\zeta\tau s+1)}$
22. $1\ e^{-\zeta t/\tau}[\cos(\sqrt{1-\zeta^2}\,t/\tau)+\dfrac{\zeta}{\sqrt{1-\zeta^2}}\sin(\sqrt{1-\zeta^2}\,t/\tau)]$ $(0\le\lvert\zeta\rvert<1)$	$\dfrac{1}{s(\tau^2s^2+2\zeta\tau s+1)}$

Table 3.1 (*Continued*)

$f(t)$	$F(s)$
23. $1 + \frac{\tau_3 - \tau_1}{\tau_1 - \tau_2} e^{-t/\tau_1} + \frac{\tau_3 - \tau_2}{\tau_2 - \tau_1} e^{-t/\tau_2}$ $(\tau_1 \neq \tau_2)$	$\frac{\tau_3 s + 1}{s(\tau_1 s + 1)(\tau_2 s + 1)}$
24. $\frac{df}{dt}$	$sF(s) - f(0)$
25. $\frac{d^n f}{dt^n}$	$s^n F(s) - s^{n-1} f(0) - s^{n-2} f^{(1)}(0) - \cdots - sf^{(n-2)}(0) - f^{(n-1)}(0)$
26. $f(t - t_0)S(t - t_0)$	$e^{-t_0 s} F(s)$

[a]Note that $f(t)$ and $F(s)$ are defined for $t \geq 0$ only.

equations. For a negative exponential, e^{-bt}, with $b > 0$,

$$\mathcal{L}(e^{-bt}) = \int_0^\infty e^{-bt} e^{-st} dt = \int_0^\infty e^{-(b+s)t} dt \quad (3\text{-}15)$$

$$= \frac{1}{b+s}\left[-e^{-(b+s)t}\right]\Big|_0^\infty = \frac{1}{s+b} \quad (3\text{-}16)$$

The Laplace transform for $b < 0$ is unbounded if $s < b$; therefore, the real part of s must be restricted to be larger than $-b$ for the integral to be finite. This condition is satisfied for all problems we consider in this book.

Trigonometric Functions. In modeling processes and in studying control systems, there are many other important time functions, such as the trigonometric functions, $\cos \omega t$ and $\sin \omega t$, where ω is the frequency in radians per unit time. The Laplace transform of $\cos \omega t$ or $\sin \omega t$ can be calculated using integration by parts. An alternative method is to use the Euler identity[1]

$$\cos \omega t = \frac{e^{j\omega t} + e^{-j\omega t}}{2}, \quad j \triangleq \sqrt{-1} \quad (3\text{-}17)$$

and to apply (3-1). Because the Laplace transform of a sum of two functions is the sum of the Laplace transforms,

$$\mathcal{L}(\cos \omega t) = \tfrac{1}{2}\mathcal{L}(e^{j\omega t}) + \tfrac{1}{2}\mathcal{L}(e^{-j\omega t}) \quad (3\text{-}18)$$

Using Eq. 3-16 gives

$$\mathcal{L}(\cos \omega t) = \frac{1}{2}\left(\frac{1}{s - j\omega} + \frac{1}{s + j\omega}\right)$$

$$= \frac{1}{2}\left(\frac{s + j\omega + s - j\omega}{s^2 + \omega^2}\right) = \frac{s}{s^2 + \omega^2} \quad (3\text{-}19)$$

Note that the use of imaginary variables above was merely a device to avoid integration by parts; imaginary numbers do not appear in the final result. To find $\mathcal{L}(\sin \omega t)$, we can use a similar approach.

[1]The symbol j, rather than i, is traditionally used for $\sqrt{-1}$ in the control engineering literature.

Table 3.1 lists some important Laplace transform pairs that occur in the solution of linear differential equations. For a more extensive list of transforms, see Dyke (1999).

Note that in all the transform cases derived above, $F(s)$ is a ratio of polynomials in s, that is, a *rational form*. There are some important cases when nonpolynomial (nonrational) forms occur. One such case is discussed next.

The Rectangular Pulse Function. An illustration of the rectangular pulse is shown in Fig. 3.1. The pulse has height h and width t_w. This type of signal might be used to depict the opening and closing of a valve regulating flow into a tank. The flow rate would be held at h for a duration of t_w units of time. The area under the curve in Fig. 3.1 could be interpreted as the amount of material delivered to the tank ($= ht_w$). Mathematically, the function $f(t)$ is defined as

$$f(t) = \begin{cases} 0 & t < 0 \\ h & 0 \leq t < t_w \\ 0 & t \geq t_w \end{cases} \quad (3\text{-}20)$$

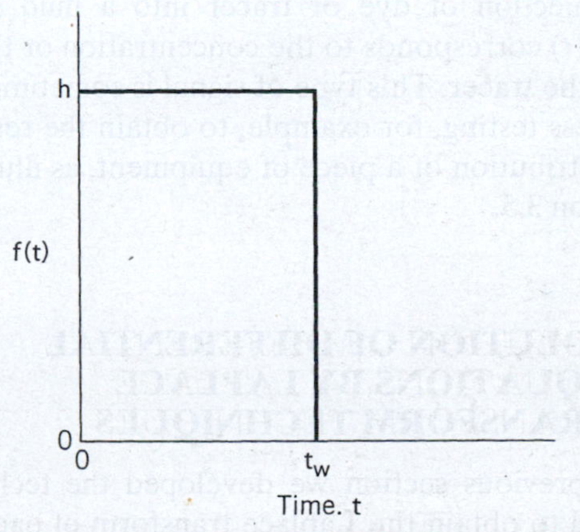

Figure 3.1 The rectangular pulse function.

The Laplace transform of the rectangular pulse can be derived by evaluating the integral (3-1) between $t = 0$ and $t = t_w$ because $f(t)$ is zero everywhere else:

$$F(s) = \int_0^{\infty} f(t)e^{-st}\,dt = \int_0^{t_w} he^{-st}\,dt \qquad (3\text{-}21)$$

$$F(s) = -\frac{h}{s}e^{-st}\bigg|_0^{t_w} = \frac{h}{s}(1 - e^{-t_w s}) \qquad (3\text{-}22)$$

Note that an exponential term in $F(s)$ results. For a *unit rectangular pulse*, $h = 1/t_w$ and the area under the pulse is unity.

Impulse Function. A limiting case of the unit rectangular pulse is the *impulse* or *Dirac delta function*, which has the symbol $\delta(t)$. This function is obtained when $t_w \rightarrow 0$ while keeping the area under the pulse equal to unity. A pulse of infinite height and infinitesimal width results. Mathematically, this can be accomplished by substituting $h = 1/t_w$ into (3-22); the Laplace transform of $\delta(t)$ is

$$\mathcal{L}(\delta(t)) = \lim_{t_w \rightarrow 0} \frac{1}{t_w s}(1 - e^{-t_w s}) \qquad (3\text{-}23)$$

Equation 3-23 is an indeterminate form that can be evaluated by application of L'Hospital's rule (also spelled L'Hôpital), which involves taking derivatives of both numerator and denominator with respect to t_w:

$$\mathcal{L}(\delta(t)) = \lim_{t_w \rightarrow 0} \frac{se^{-t_w s}}{s} = 1 \qquad (3\text{-}24)$$

If the impulse magnitude (i.e., area $t_w h$) is a constant a rather than unity, then

$$\mathcal{L}(a\delta(t)) = a \qquad (3\text{-}25)$$

as given in Table 3.1. The unit impulse function may also be interpreted as the time derivative of the unit step function $S(t)$. The response of a process to a unit impulse is called its *impulse response*, which is illustrated in Example 3.7.

A physical example of an impulse function is the rapid injection of dye or tracer into a fluid stream, where $f(t)$ corresponds to the concentration or the flow rate of the tracer. This type of signal is sometimes used in process testing, for example, to obtain the residence time distribution of a piece of equipment, as illustrated in Section 3.5.

3.2 SOLUTION OF DIFFERENTIAL EQUATIONS BY LAPLACE TRANSFORM TECHNIQUES

In the previous section we developed the techniques required to obtain the Laplace transform of each term in a linear ordinary differential equation. Table 3.1 lists important functions of time, including derivatives, and their Laplace transform equivalents. Because the Laplace transform converts any function $f(t)$ to $F(s)$ and the inverse Laplace transform converts $F(s)$ back to $f(t)$, the table provides an organized way to carry out these transformations.

The procedure used to solve a differential equation is quite simple. First Laplace transform both sides of the differential equation, substituting values for the initial conditions in the derivative transforms. Rearrange the resulting algebraic equation, and solve for the transform of the dependent (output) variable. Finally, find the inverse of the transformed output variable. The solution method is illustrated by means of several examples.

EXAMPLE 3.1

Solve the differential equation,

$$5\frac{dy}{dt} + 4y = 2 \quad y(0) = 1 \qquad (3\text{-}26)$$

using Laplace transforms.

SOLUTION

First take the Laplace transform of both sides of Eq. 3-26:

$$\mathcal{L}\left(5\frac{dy}{dt} + 4y\right) = \mathcal{L}(2) \qquad (3\text{-}27)$$

Using the principle of superposition, each term can be transformed individually:

$$\mathcal{L}\left(5\frac{dy}{dt}\right) + \mathcal{L}(4y) = \mathcal{L}(2) \qquad (3\text{-}28)$$

$$\mathcal{L}\left(5\frac{dy}{dt}\right) = 5\mathcal{L}\left(\frac{dy}{dt}\right) = 5(sY(s) - 1) = 5sY(s) - 5 \qquad (3\text{-}29)$$

$$\mathcal{L}(4y) = 4\mathcal{L}(y) = 4Y(s) \qquad (3\text{-}30)$$

$$\mathcal{L}(2) = \frac{2}{s} \qquad (3\text{-}31)$$

Substitute the individual terms:

$$5sY(s) - 5 + 4Y(s) = \frac{2}{s} \qquad (3\text{-}32)$$

Rearrange (3-32) and factor out $Y(s)$:

$$Y(s)(5s + 4) = 5 + \frac{2}{s} \qquad (3\text{-}33)$$

or

$$Y(s) = \frac{5s + 2}{s(5s + 4)} \qquad (3\text{-}34)$$

Take the inverse Laplace transform of both sides of Eq. 3-34:

$$\mathcal{L}^{-1}[Y(s)] = \mathcal{L}^{-1}\left[\frac{5s + 2}{s(5s + 4)}\right] \tag{3-35}$$

The inverse Laplace transform of the right side of (3-35) can be found by using Table 3.1. First, divide the numerator and denominator by 5 to put all factors in the $s + b$ form corresponding to the table entries:

$$y(t) = \mathcal{L}^{-1}\left(\frac{s + 0.4}{s(s + 0.8)}\right) \tag{3-36}$$

Because entry 11 in the table, $(s + b_3)/[(s + b_1)(s + b_2)]$, matches (3-36) with $b_1 = 0.8$, $b_2 = 0$, and $b_3 = 0.4$, the solution can be written immediately:

$$y(t) = 0.5 + 0.5e^{-0.8t} \tag{3-37}$$

Note that in solving (3-26) both the forcing function (the constant 2 on the right side) and the initial condition have been incorporated easily and directly. As for any differential equation solution, (3-37) should be checked to make sure it satisfies the initial condition and the original differential equation for $t \geq 0$.

Next we apply the Laplace transform solution to a higher-order differential equation.

EXAMPLE 3.2

Solve the ordinary differential equation

$$\frac{d^3y}{dt^3} + 6\frac{d^2y}{dt^2} + 11\frac{dy}{dt} + 6y = 1 \tag{3-38}$$

with initial conditions $y(0) = y'(0) = y''(0) = 0$.

SOLUTION

Take Laplace transforms, term by term, using Table 3.1:

$$\mathcal{L}\left(\frac{d^3y}{dt^3}\right) = s^3Y(s)$$

$$\mathcal{L}\left(6\frac{d^2y}{dt^2}\right) = 6s^2Y(s)$$

$$\mathcal{L}\left(11\frac{dy}{dt}\right) = 11sY(s)$$

$$\mathcal{L}(6y) = 6Y(s)$$

$$\mathcal{L}(1) = \frac{1}{s}$$

Rearranging and factoring $Y(s)$, we obtain

$$Y(s)(s^3 + 6s^2 + 11s + 6) = \frac{1}{s} \tag{3-39}$$

$$Y(s) = \frac{1}{s(s^3 + 6s^2 + 11s + 6)} \tag{3-40}$$

To invert (3-40) to find $y(t)$, we must find a similar expression in Table 3.1. Unfortunately, no formula in the table has a fourth-order polynomial in the denominator. This example will be continued later, after we develop the techniques necessary to generalize the solution method in Section 3.3.

In general, a transform expression may not exactly match any of the entries in Table 3.1. This problem always arises for higher-order differential equations, because the order of the denominator polynomial (characteristic polynomial) of the transform is equal to the order of the original differential equation, and no table entries are higher than third order in the denominator. It is simply not practical to expand the number of entries in the table ad infinitum. Instead, we use a procedure based on elementary transform building blocks. This procedure, called *partial fraction expansion*, is presented in the next section.

3.3 PARTIAL FRACTION EXPANSION

The high-order denominator polynomial in a Laplace transform solution arises from the differential equation terms (its *characteristic polynomial*) plus terms contributed by the inputs. The factors of the characteristic polynomial correspond to the roots of the characteristic polynomial set equal to zero. The input factors may be quite simple. Once the factors are obtained, the Laplace transform is then expanded into *partial fractions*. As an example, consider

$$Y(s) = \frac{s + 5}{s^2 + 5s + 4} \tag{3-41}$$

The denominator can be factored into a product of first-order terms, $(s + 1)(s + 4)$. This transform can be expanded into the sum of two partial fractions:

$$\frac{s + 5}{(s + 1)(s + 4)} = \frac{\alpha_1}{s + 1} + \frac{\alpha_2}{s + 4} \tag{3-42}$$

where α_1 and α_2 are unspecified coefficients that must satisfy Eq. 3-42. The expansion in (3-42) indicates that the original denominator polynomial has been factored into a product of first-order terms. In general, for every partial fraction expansion, there will be a unique set of α_i that satisfy the equation.

There are several methods for calculating the appropriate values of α_1 and α_2 in (3-42):

Method 1. Multiply both sides of (3-42) by $(s + 1)(s + 4)$:

$$s + 5 = \alpha_1(s + 4) + \alpha_2(s + 1) \tag{3-43}$$

Equating coefficients of each power of s gives

$$s^1: \quad \alpha_1 + \alpha_2 = 1 \tag{3-44a}$$

$$s^0: \quad 4\alpha_1 + \alpha_2 = 5 \tag{3-44b}$$

Solving for α_1 and α_2 simultaneously yields $\alpha_1 = \frac{4}{3}$, $\alpha_2 = -\frac{1}{3}$.

Method 2. Because Eq. 3-42 must be valid for all values of s, we can specify two values of s and solve for the two constants:

$$s = -5\text{:}\ 0 = -\tfrac{1}{4}\alpha_1 - \alpha_2 \tag{3-45a}$$

$$s = -3\text{:}\ -\tfrac{2}{2} = -\tfrac{1}{2}\alpha_1 + \alpha_2 \tag{3-45b}$$

Solving, $\alpha_1 = \frac{4}{3}$, $\alpha_2 = -\frac{1}{3}$.

Method 3. The fastest and most popular method is called the *Heaviside expansion*. In this method multiply both sides of the equation by one of the denominator terms $(s + b_i)$ and then set $s = -b_i$, which causes all terms except one to be multiplied by zero. Multiplying Eq. 3-42 by $s + 1$ and then letting $s = -1$ gives

$$\alpha_1 = \left.\frac{s+5}{s+4}\right|_{s=-1} = \frac{4}{3}$$

Similarly, after multiplying by $(s + 4)$ and letting $s = -4$, the expansion gives

$$\alpha_2 = \left.\frac{s+5}{s+1}\right|_{s=-4} = -\frac{1}{3}$$

As seen above, the coefficients can be found by simple calculations.

For a more general transform, where the factors are real and distinct (no complex or repeated factors appear), the following expansion formula can be used:

$$Y(s) = \frac{N(s)}{D(s)} = \frac{N(s)}{\prod_{i=1}^{n}(s+b_i)} = \sum_{i=1}^{n}\frac{\alpha_i}{s+b_i} \tag{3-46}$$

where $D(s)$, an nth-order polynomial, is the denominator of the transform. $D(s)$ is the characteristic polynomial. The numerator $N(s)$ has a maximum order of $n - 1$. The ith coefficient can be calculated using the Heaviside expansion

$$\alpha_i = (s + b_i)\left.\frac{N(s)}{D(s)}\right|_{s=-b_i} \tag{3-47}$$

Alternatively, an expansion for real, distinct factors may be written as

$$Y(s) = \frac{N'(s)}{D'(s)} = \frac{N'(s)}{\prod_{i=1}^{n}(\tau_i s+1)} = \sum_{i=1}^{n}\frac{\alpha_i'}{\tau_i s+1} \tag{3-48}$$

Using Method 3, calculate the coefficients by

$$\alpha_i' = (\tau_i s + 1)\left.\frac{N'(s)}{D'(s)}\right|_{s=-\frac{1}{\tau_i}} \tag{3-49}$$

Note that several entries in Table 3.1 have the $\tau s + 1$ format.

We now can use the Heaviside expansion to complete the solution of Example 3.2.

EXAMPLE 3.2 *(Continued)*

First factor the denominator of Eq. 3-40 into a product of first-order terms ($n = 4$ in Eq. 3-46). Simple factors, as in this case, rarely occur in actual applications.

$$s(s^3 + 6s^2 + 11s + 6) = s(s + 1)(s + 2)(s + 3) \tag{3-50}$$

This result determines the four terms that will appear in the partial fraction expansion—namely,

$$\begin{aligned} Y(s) &= \frac{1}{s(s+1)(s+2)(s+3)} \\ &= \frac{\alpha_1}{s} + \frac{\alpha_2}{s+1} + \frac{\alpha_3}{s+2} + \frac{\alpha_4}{s+3} \end{aligned} \tag{3-51}$$

The Heaviside expansion method gives $\alpha_1 = 1/6$, $\alpha_2 = -1/2$, $\alpha_3 = 1/2$, $\alpha_4 = -1/6$.

After the transform has been expanded into a sum of first-order terms, invert each term individually using Table 3.1:

$$\begin{aligned} y(t) &= \mathcal{L}^{-1}[Y(s)] \\ &= \mathcal{L}^{-1}\left(\frac{1/6}{s} - \frac{1/2}{s+1} + \frac{1/2}{s+2} - \frac{1/6}{s+3}\right) \\ &= \frac{1}{6}\mathcal{L}^{-1}\left(\frac{1}{s}\right) - \frac{1}{2}\mathcal{L}^{-1}\left(\frac{1}{s+1}\right) \\ &\quad + \frac{1}{2}\mathcal{L}^{-1}\left(\frac{1}{s+2}\right) - \frac{1}{6}\mathcal{L}^{-1}\left(\frac{1}{s+3}\right) \\ &= \frac{1}{6} - \frac{1}{2}e^{-t} + \frac{1}{2}e^{-2t} - \frac{1}{6}e^{-3t} \end{aligned} \tag{3-52}$$

Equation 3-52 is thus the solution $y(t)$ to the differential equation (3-38). The α_i's are simply the coefficients of the solution. Equation 3-52 also satisfies the three initial conditions of the differential equation. The reader should verify the result.

3.3.1 General Procedure for Solving Differential Equations

We now state a general procedure to solve ordinary differential equations using Laplace transforms. The procedure consists of four steps, as shown in Fig. 3.2.

Note that solution for the differential equation involves use of Laplace transforms as an intermediate step. Step 3 can be bypassed if the transform found in Step 2 matches an entry in Table 3.1. In order to factor $D(s)$ in Step 3, software such as MATLAB, Mathematica, or Mathcad can be utilized (Chapra and Canale, 2010).

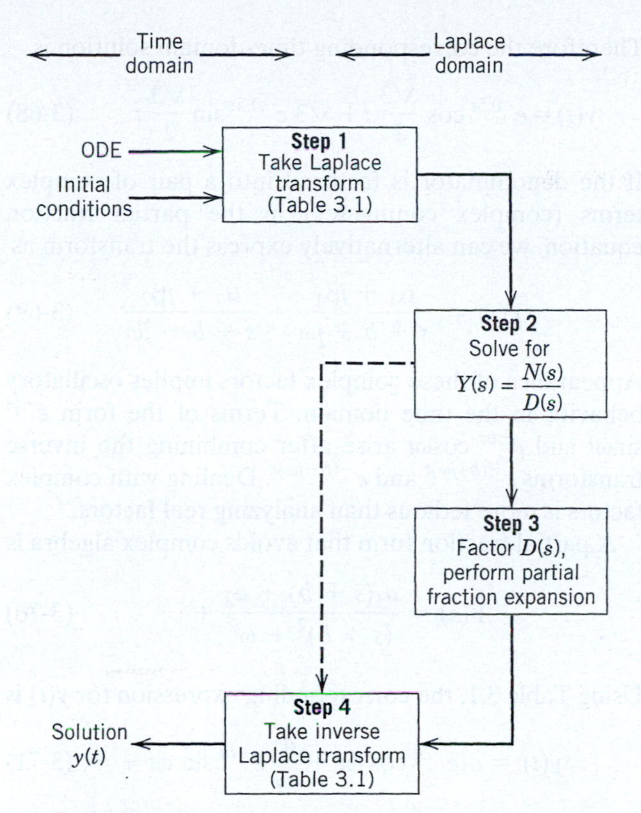

Figure 3.2 The general procedure for solving an ordinary differential equation using Laplace transforms.

In Step 3, other types of situations can occur. Both repeated factors and complex factors require modifications of the partial fraction expansion procedure.

Repeated Factors

If a term $s + b$ occurs r times in the denominator, r terms must be included in the expansion that incorporate the $s + b$ factor

$$Y(s) = \frac{\alpha_1}{s+b} + \frac{\alpha_2}{(s+b)^2} + \cdots + \frac{\alpha_r}{(s+b)^r} + \cdots \tag{3-53}$$

in addition to the other factors. Repeated factors arise infrequently in process models of real systems, mainly for a process that consists of a series of identical units or stages.

EXAMPLE 3.3

For

$$Y(s) = \frac{s+1}{s(s^2+4s+4)} = \frac{\alpha_1}{s+2} + \frac{\alpha_2}{(s+2)^2} + \frac{\alpha_3}{s} \tag{3-54}$$

evaluate the unknown coefficients α_i.

SOLUTION

To find α_1 in (3-54), the Heaviside rule cannot be used for multiplication by $(s + 2)$, because $s = -2$ causes the second term on the right side to be unbounded, rather than 0 as desired. We therefore employ the Heaviside expansion method for the other two coefficients (α_2 and α_3) that can be evaluated normally and then solve for α_1 by arbitrarily selecting some other value of s. Multiplying (3-54) by $(s + 2)^2$ and letting $s = -2$ yields

$$\alpha_2 = \left.\frac{s+1}{s}\right|_{s=-2} = \frac{1}{2} \tag{3-55}$$

Multiplying (3-54) by s and letting $s = 0$ yields

$$\alpha_3 = \left.\frac{s+1}{s^2+4s+4}\right|_{s=0} = \frac{1}{4} \tag{3-56}$$

Substituting the value $s = -1$ in (3-54) gives

$$0 = \alpha_1 + \alpha_2 - \alpha_3 \tag{3-57}$$

$$\alpha_1 = -\frac{1}{4} \tag{3-58}$$

An alternative approach to find α_1 is to use differentiation of the transform. Equation 3-54 is multiplied by $s(s + 2)^2$,

$$s + 1 = \alpha_1(s+2)s + \alpha_2 s + \alpha_3(s+2)^2 \tag{3-59}$$

Then (3-59) is differentiated twice with respect to s,

$$0 = 2\alpha_1 + 2\alpha_3; \quad \text{so that } \alpha_1 = -\alpha_3 = -\frac{1}{4} \tag{3-60}$$

Note that differentiation in this case is tantamount to equating powers of s, as demonstrated earlier.

The differentiation approach illustrated above can be used as the basis of a more general method to evaluate the coefficients of repeated factors. If the denominator polynomial $D(s)$ contains the repeated factor $(s + b)^r$, first form the quantity

$$Q(s) = \frac{N(s)}{D(s)}(s+b)^r = (s+b)^{r-1}\alpha_1 + (s+b)^{r-2}\alpha_2 + \ldots$$
$$+ \alpha_r + (s+b)^r[\text{other partial fractions}] \tag{3-61}$$

Setting $s = -b$ will generate α_r directly. Differentiating $Q(s)$ with respect to s and letting $s = -b$ generates α_{r-1}. Successive differentiation a total of $r - 1$ times will generate all α_i, $i = 1, 2, \ldots, r$ from which we obtain the general expression

$$\alpha_{r-i} = \left.\frac{1}{i}\frac{d^{(i)}Q(s)}{ds^{(i)}}\right|_{s=-b} \quad i = 0, \ldots, r-1 \tag{3-62}$$

For $i = 0$ in (3-62), 0! is defined to be 1 and the zeroth derivative of $Q(s)$ is defined to be simply $Q(s)$ itself.

Returning to the problem in Example 3.3,

$$Q(s) = \frac{s+1}{s} \tag{3-63}$$

from which

$$i = 0:\ \alpha_2 = \left.\frac{s+1}{s}\right|_{s=-2} = \frac{1}{2} \tag{3-64a}$$

$$i = 1:\ \alpha_1 = \left.\frac{d\left(\frac{s+1}{s}\right)}{ds}\right|_{s=-2} = \left.\frac{-1}{s^2}\right|_{s=-2} = -\frac{1}{4} \tag{3-64b}$$

Complex Factors

An important case occurs when the factored characteristic polynomial yields terms of the form

$$\frac{c_1 s + c_0}{s^2 + d_1 s + d_0}$$

where

$$\frac{d_1^2}{4} < d_0$$

Here the denominator cannot be written as the product of two real factors, which can be determined by using the quadratic formula.

For example, consider the transform

$$Y(s) = \frac{s+2}{s^2+s+1} \tag{3-65}$$

To invert (3-65) to the time domain, we complete the square of the first two terms in the denominator:

$$Y(s) = \frac{s+2}{(s+0.5)^2 - 0.25 + 1} = \frac{(s+0.5)+1.5}{(s+0.5)^2 + \left(\frac{\sqrt{3}}{2}\right)^2} \tag{3-66}$$

Dividing the numerator of $Y(s)$ into two terms,

$$Y(s) = \frac{s+0.5}{(s+0.5)^2 + \left(\frac{\sqrt{3}}{2}\right)^2} + \frac{1.5}{(s+0.5)^2 + \left(\frac{\sqrt{3}}{2}\right)^2} \tag{3-67}$$

To determine $y(t)$, we invert each term separately. Note that in Table 3.1, $\frac{s+b}{(s+b)^2+\omega^2}$ transforms to $e^{-bt}\cos\omega t$, while $\frac{\omega}{(s+b)^2+\omega^2}$ transforms to $e^{-bt}\sin\omega t$.

Therefore the corresponding time-domain solution is

$$y(t) = e^{-0.5t}\cos\frac{\sqrt{3}}{2}t + \sqrt{3}\,e^{-0.5t}\sin\frac{\sqrt{3}}{2}t \tag{3-68}$$

If the denominator is factored into a pair of complex terms (complex conjugates) in the partial fraction equation, we can alternatively express the transform as

$$Y(s) = \frac{\alpha_1 + j\beta_1}{s+b+j\omega} + \frac{\alpha_2 + j\beta_2}{s+b-j\omega} \tag{3-69}$$

Appearance of these complex factors implies oscillatory behavior in the time domain. Terms of the form $e^{-bt}\sin\omega t$ and $e^{-bt}\cos\omega t$ arise after combining the inverse transforms $e^{-(b+j\omega)t}$ and $e^{-(b-j\omega)t}$. Dealing with complex factors is more tedious than analyzing real factors.

A partial fraction form that avoids complex algebra is

$$Y(s) = \frac{a_1(s+b) + a_2}{(s+b)^2 + \omega^2} + \cdots \tag{3-70}$$

Using Table 3.1, the corresponding expression for $y(t)$ is

$$y(t) = a_1 e^{-bt}\cos\omega t + \frac{a_2}{\omega}e^{-bt}\sin\omega t + \cdots \tag{3-71}$$

However, the coefficients a_1 and a_2 must be found by solving simultaneous equations, rather than by the Heaviside expansion, as shown as follows in Example 3.4.

EXAMPLE 3.4

Find the inverse Laplace transform of

$$Y(s) = \frac{s+1}{s^2(s^2+4s+5)} \tag{3-72}$$

SOLUTION

The roots of the denominator term $(s^2 + 4s + 5)$ are imaginary $(s + 2 + j, s + 2 - j)$, so we know the solution will involve oscillatory terms (sin, cos). The partial fraction form for (3-72) that avoids using complex factors or roots is

$$Y(s) = \frac{s+1}{s^2(s^2+4s+5)} = \frac{\alpha_1}{s} + \frac{\alpha_2}{s^2} + \frac{\alpha_5 s + \alpha_6}{s^2+4s+5} \tag{3-73}$$

Multiply both sides of Eq. 3-73 by $s^2(s^2 + 4s + 5)$ and collect terms:

$$s + 1 = (\alpha_1 + \alpha_5)s^3 + (4\alpha_1 + \alpha_2 + \alpha_6)s^2 + (5\alpha_1 + 4\alpha_2)s + 5\alpha_2 \tag{3-74}$$

Equate coefficients of like powers of s:

$$s^3:\ \alpha_1 + \alpha_5 = 0 \tag{3-75a}$$

$$s^2:\ 4\alpha_1 + \alpha_2 + \alpha_6 = 0 \tag{3-75b}$$

$$s^1: 5\alpha_1 + 4\alpha_2 = 1 \quad (3\text{-}75c)$$

$$s^0: 5\alpha_2 = 1 \quad (3\text{-}75d)$$

Solving simultaneously gives $\alpha_1 = 0.04$, $\alpha_2 = 0.2$, $\alpha_5 = 0.04$, $\alpha_6 = -0.36$. The inverse Laplace transform of $Y(s)$ is

$$y(t) = \mathcal{L}^{-1}\left(\frac{0.04}{s}\right) + \mathcal{L}^{-1}\left(\frac{0.2}{s^2}\right) + \mathcal{L}^{-1}\left(\frac{-0.04s - 0.36}{s^2 + 4s + 5}\right) \quad (3\text{-}76)$$

Before using Table 3.1, the denominator term $(s^2 + 4s + 5)$ must be converted to the standard form by completing the square to $(s + 2)^2 + 1^2$; the numerator is $-0.04(s + 9)$. In order to match the expressions in Table 3.1, the argument of the last term in (3-76) must be written as

$$\frac{-0.04s - 0.36}{(s + 2)^2 + 1} = \frac{-0.04(s + 2)}{(s + 2)^2 + 1} + \frac{-0.28}{(s + 2)^2 + 1} \quad (3\text{-}77)$$

This procedure yields the following time-domain expression:

$$y(t) = 0.04 + 0.2t - 0.04e^{-2t}\cos t - 0.28e^{-2t}\sin t$$

It is clear from this example that the Laplace transform solution for complex or repeated roots can be quite cumbersome for transforms of ODEs higher than second order. In this case, using numerical simulation techniques may be more efficient to obtain a solution, as discussed in Chapters 5 and 6.

3.4 OTHER LAPLACE TRANSFORM PROPERTIES

In this section, we consider several Laplace transform properties that are useful in process dynamics and control.

3.4.1 Final Value Theorem

The asymptotic value of $y(t)$ for large values of time $y(\infty)$ can be found from (3-78), providing that $\lim_{s\to 0}[sY(s)]$ exists for all $Re(s) \geq 0$:

$$\lim_{t\to\infty} y(t) = \lim_{s\to 0}[sY(s)] \quad (3\text{-}78)$$

Equation 3-78 can be proved using the relation for the Laplace transform of a derivative (Eq. 3-9):

$$\int_0^\infty \frac{dy}{dt} e^{-st}\, dt = sY(s) - y(0) \quad (3\text{-}79)$$

Taking the limit as $s \to 0$ and assuming that dy/dt is continuous and that $sY(s)$ has a limit for all $Re(s) \geq 0$,

$$\int_0^\infty \frac{dy}{dt}\, dt = \lim_{s\to 0}[sY(s)] - y(0) \quad (3\text{-}80)$$

Integrating the left side and simplifying yields

$$\lim_{t\to\infty} y(t) = \lim_{s\to\infty}[sY(s)] \quad (3\text{-}81)$$

If $y(t)$ is unbounded for $t \to \infty$, Eq. 3-81 gives erroneous results. For example, if $Y(s) = 1/(s - 5)$, Eq. 3-81 predicts $y(\infty) = 0$. Note that Eq. 3-79, which is the basis of (3-79), requires that $\lim y(t \to \infty)$ exists. In this case, $y(t) = e^{5t}$, which is unbounded for $t \to \infty$. However, Eq. 3-79 does not apply here, because $sY(s) = s/(s - 5)$ does not have a limit for some real value of $s \geq 0$, in particular, for $s = 5$.

3.4.2 Initial Value Theorem

Analogous to the final value theorem, the initial value theorem can be stated as

$$\lim_{t\to 0} y(t) = \lim_{s\to\infty}[sY(s)] \quad (3\text{-}82)$$

The proof of this theorem is similar to the development in (3-78) through (3-81). It also requires that $y(t)$ is continuous. The proof is left to the reader as an exercise.

EXAMPLE 3.5

Apply the initial and final value theorems to the transform derived in Example 3.1:

$$Y(s) = \frac{5s + 2}{s(5s + 4)}$$

SOLUTION

Initial Value

$$y(0) = \lim_{s\to\infty}[sY(s)] = \lim_{s\to\infty} \frac{5s+2}{5s+4} = 1 \quad (3\text{-}83a)$$

Final Value

$$y(\infty) = \lim_{s\to 0}[sY(s)] = \lim_{s\to 0} \frac{5s+2}{5s+4} = 0.5 \quad (3\text{-}83b)$$

The initial value of 1 corresponds to the initial condition given in Eq. 3-26. The final value of 0.5 agrees with the time-domain solution in Eq. 3-37. Both theorems are useful for checking mathematical errors that may occur in obtaining Laplace transform solutions.

EXAMPLE 3.6

A process is described by a third-order ODE:

$$\frac{d^3y}{dt^3} + 6\frac{d^2y}{dt^2} + 11\frac{dy}{dt} + 6y = 4u \quad (3\text{-}84)$$

with all initial conditions on y, dy/dt, and dy^2/dt^2 equal to zero. Show that for a step change in u of 2 units, the steady-state result in the time domain is the same as applying the final value theorem.

SOLUTION

If $u = 2$ the steady-state result for y can be found by setting all derivatives to zero and substituting for u. Therefore

$$6y = 8 \quad \text{or} \quad y = 4/3 \tag{3-85}$$

The transform of (3-84) is

$$(s^3 + 6s^2 + 11s + 6)Y(s) = 8/s \tag{3-86}$$

$$Y(s) = \frac{8}{s^4 + 6s^3 + 11s^2 + 6s} \tag{3-87}$$

One of the benefits of the final value theorem is that we do not have to solve for the analytical solution of $y(t)$. Instead, simply apply Eq. 3-81 to the transform $Y(s)$ as follows:

$$\lim_{s\to 0} sY(s) = \lim_{s\to 0} \frac{8}{s^3+6s^2+11s+6} = \frac{8}{6} = \frac{4}{3} \tag{3-88}$$

This is the same answer as obtained in Eq. 3-85. The time-domain solution obtained from a partial fraction expansion is

$$y = 4/3 - 2e^{-t} + 2e^{-2t} - 2/3e^{-3t} \tag{3-89}$$

As $t \to \infty$, only the first term remains, which is the same result as in Eq. 3-90 (using the final value theorem).

3.4.3 Transform of an Integral

Occasionally, it is necessary to find the Laplace transform of a function that is integrated with respect to time. By applying the definition (Eq. 3-1) and integrating by parts,

$$\mathcal{L}\left\{\int_0^t f(t^*)\,dt^*\right\} = \int_0^\infty \left\{\int_0^t f(t^*)\,dt^*\right\} e^{-st}\,dt \tag{3-90}$$

$$= -\frac{1}{s}\left[e^{-st}\int_0^t f(t^*)\,dt^*\right]_0^\infty + \frac{1}{s}\int_0^\infty e^{-st} f(t)\,dt \tag{3-91}$$

The first term in (3-91) yields 0 when evaluated at both the upper and lower limits, as long as $f(t^*)$ possesses a transform (is bounded). The integral in the second term is simply the definition of the Laplace transform of $f(t)$. Hence,

$$\mathcal{L}\left\{\int_0^t f(t^*)\,dt^*\right\} = \frac{1}{s}F(s) \tag{3-92}$$

Note that Laplace transformation of an integral function of time leads to division of the transformed function by s.

We have already seen in (3-9) that transformation of time derivatives leads to an inverse relation—that is, multiplication of the transform by s.

3.4.4 Time Delay (Translation in Time)

Functions that exhibit time delay play an important role in process modeling and control. Time delays commonly occur as a result of the transport time required for a fluid to flow through piping. Consider the stirred-tank heating system example presented in Chapter 2. Suppose one thermocouple is located at the outflow point of the stirred tank, and a second thermocouple is immersed in the fluid a short distance (L = 10 m) downstream. The heating system is off initially, and at time zero it is turned on. If there is no fluid mixing in the pipe (the fluid is in plug flow) and if no heat losses occur from the pipe, the shapes of the two temperature responses should be identical. However, the second sensor response will be translated in time; that is, it will exhibit a *time delay*. If the fluid velocity is 1 m/s, the time delay ($t_0 = L/v$) is 10 s. If we denote $f(t)$ as the transient temperature response at the first sensor and $f_d(t)$ as the temperature response at the second sensor, Fig. 3.3 shows how they are related. The function $f_d = 0$ for $t < t_0$. Therefore, f_d and f are related by

$$f_d(t) = f(t - t_0)S(t - t_0) \tag{3-93}$$

Note that f_d is the function $f(t)$ delayed by t_0 time units. The unit step function $S(t - t_0)$ is included to denote explicitly that $f_d(t) = 0$ for all values of $t < t_0$. If $\mathcal{L}(f(t)) = F(s)$, then

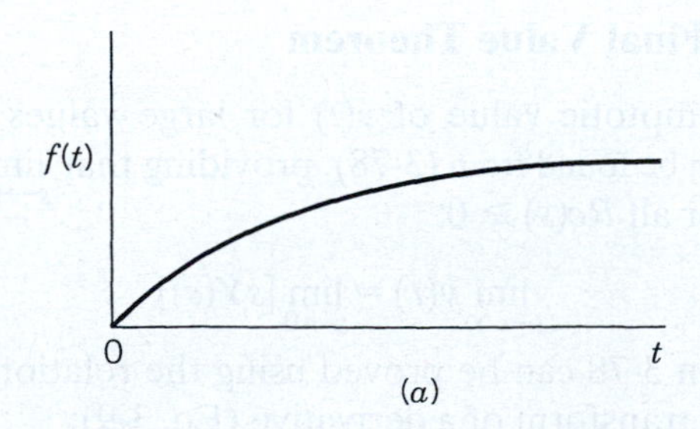

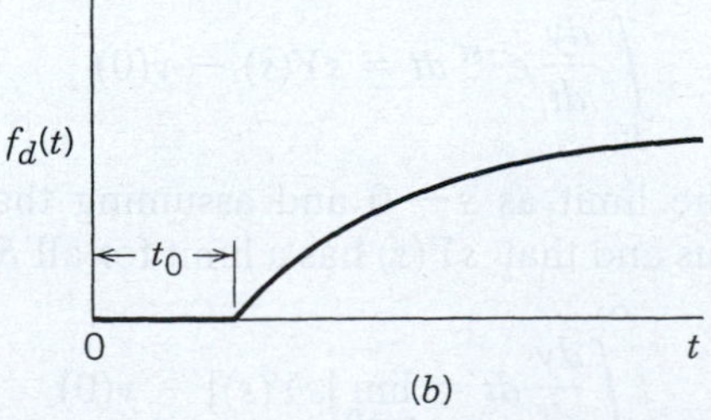

Figure 3.3 A time function with and without time delay. (*a*) Original function (no delay); (*b*) function with delay t_0.

$$\begin{aligned}\mathcal{L}(f_d(t)) &= \mathcal{L}(f(t-t_0)S(t-t_0)) \\ &= \int_0^\infty f(t-t_0)S(t-t_0)e^{-st}\,dt \\ &= \int_0^{t_0} f(t-t_0)(0)e^{-st}\,dt + \int_{t_0}^\infty f(t-t_0)e^{-st}\,dt \\ &= \int_{t_0}^\infty f(t-t_0)e^{-s(t-t_0)}e^{-st_0}\,d(t-t_0) \quad (3\text{-}94)\end{aligned}$$

Because $(t - t_0)$ is now the artificial variable of integration, it can be replaced by t^*.

$$\mathcal{L}(f(t)) = e^{-st_0}\int_0^\infty f(t^*)e^{-st^*}\,dt^* \qquad (3\text{-}95)$$

yielding the *Real Translation Theorem:*

$$F_d(s) = \mathcal{L}(f(t-t_0)S(t-t_0)) = e^{-st_0}F(s) \quad (3\text{-}96)$$

In inverting a transform that contains an e^{-st_0} element (time-delay term), the following procedure will easily yield results and also avoid the pitfalls of dealing with translated (shifted) time arguments. Starting with the Laplace transform

$$Y(s) = e^{-st_0}F(s) \qquad (3\text{-}97)$$

1. Invert $F(s)$ in the usual manner; that is, perform partial fraction expansion, and so forth, to find $f(t)$.
2. Find $y(t) = f(t - t_0)S(t - t_0)$ by replacing the argument t, wherever it appears in $f(t)$, by $(t - t_0)$; then multiply the entire function by the shifted unit step function, $S(t - t_0)$.

EXAMPLE 3.6

Find the inverse transform of

$$Y(s) = \frac{1 + e^{-2s}}{(4s+1)(3s+1)} \qquad (3\text{-}98)$$

SOLUTION

Equation 3-98 can be split into two terms:

$$Y(s) = Y_1(s) + Y_2(s) \qquad (3\text{-}99)$$

$$= \frac{1}{(4s+1)(3s+1)} + \frac{e^{-2s}}{(4s+1)(3s+1)} \qquad (3\text{-}100)$$

The inverse transform of $Y_1(s)$ can be obtained directly from Table 3.1:

$$y_1(t) = e^{-t/4} - e^{-t/3} \qquad (3\text{-}101)$$

Because $Y_2(s) = e^{-2s}Y_1(s)$, its inverse transform can be written immediately by replacing t by $(t - 2)$ in (3-101), and then multiplying by the shifted step function:

$$y_2(t) = [e^{-(t-2)/4} - e^{-(t-2)/3}]S(t-2) \qquad (3\text{-}102)$$

Thus, the complete inverse transform is

$$y(t) = e^{-t/4} - e^{-t/3} + [e^{-(t-2)/4} - e^{-(t-2)/3}]S(t-2) \qquad (3\text{-}103)$$

Equation 3-103 can be numerically evaluated without difficulty for particular values of t, noting that the term in brackets is multiplied by 0 (the value of the unit step function) for $t < 2$, and by 1 for $t \geq 2$. An equivalent and simpler method is to evaluate the contributions from the bracketed time functions only when the time arguments are nonnegative. An alternative way of writing Eq. 3-105 is as two equations, each one applicable over a particular interval of time:

$$0 \leq t < 2: \quad y(t) = e^{-t/4} - e^{-t/3} \qquad (3\text{-}104)$$

and

$$\begin{aligned} t \geq 2: \quad y(t) &= e^{-t/4} - e^{-t/3} + [e^{-(t-2)/4} - e^{-(t-2)/3}] \\ &= e^{-t/4}(1 + e^{2/4}) - e^{-t/3}(1 + e^{2/3}) \\ &= 2.6487e^{-t/4} - 2.9477e^{-t/3} \qquad (3\text{-}105)\end{aligned}$$

Note that (3-104) and (3-105) give equivalent results for $t = 2$, because in this case, $y(t)$ is continuous at $t = 2$.

3.5 A TRANSIENT RESPONSE EXAMPLE

In Chapter 4 we will develop a standardized approach for using Laplace transforms to calculate transient responses. That approach will unify the way process models are manipulated after transforming them, and it will further simplify the way initial conditions and inputs (forcing functions) are handled. However, we already have the tools to analyze an example of a transient response situation in some detail. Example 3.7 illustrates many features of Laplace transform methods in investigating the dynamic characteristics of a physical process.

EXAMPLE 3.7

The Ideal Gas Company has two fixed-volume, stirred-tank reactors connected in series as shown in Fig. 3.4. The three IGC engineers who are responsible for reactor operations—Kim Ng, Casey Gain, and Tim Delay—are concerned about the adequacy of mixing in the two tanks and want to run a tracer test on the system to determine whether dead zones and/or channeling exist in the reactors.

Their idea is to operate the reactors at a temperature low enough that reaction will not occur, and to apply a rectangular pulse in the reactant concentration to the first stage for test purposes. In this way, available instrumentation on the second-stage outflow line can be used without modification to measure reactant (tracer) concentration.

Before performing the test, the engineers would like to have a good idea of the results that should be expected if perfect mixing actually is accomplished in the reactors. A rectangular pulse input for the change in reactant concentration will be used with the restriction that the resulting output concentration changes must be large enough to be measured precisely.

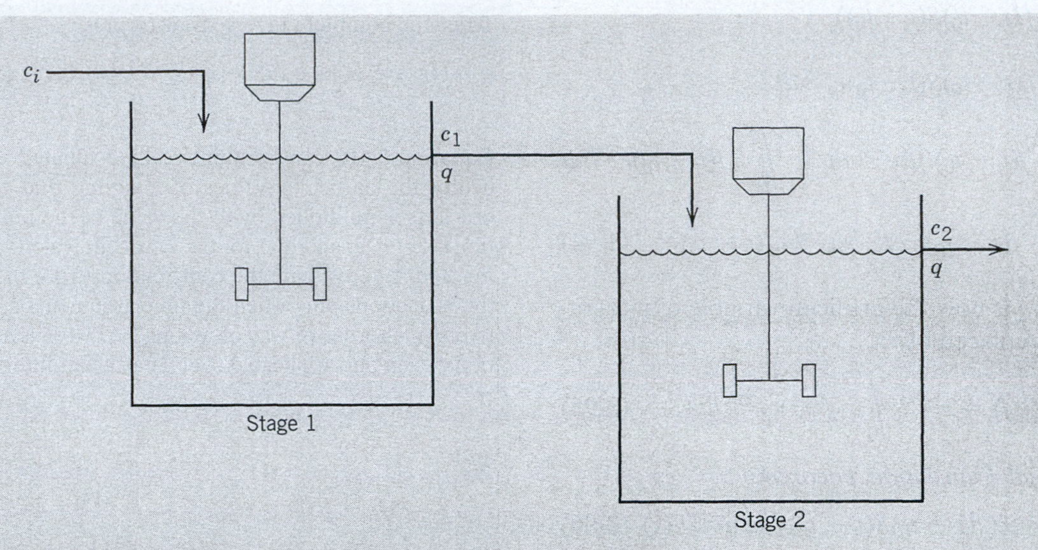

Figure 3.4 Two-stage stirred-tank reactor system.

Table 3.2 Two-Stage Stirred-Tank Reactor Process and Operating Data

Volume of Stage 1	$= 4\ \text{m}^3$
Volume of Stage 2	$= 3\ \text{m}^3$
Total flow rate q	$= 2\ \text{m}^3/\text{min}$
Nominal feed reactant concentration (c_i)	$= 1$ kg mol/m^3

The process data and operating conditions required to model the reactor tracer test are given in Table 3.2. Figure 3.5 shows the proposed pulse change of 0.25 min duration that can be made while maintaining the total reactor input flow rate constant. As part of the theoretical solution, Kim, Casey, and Tim would like to know how closely the rectangular pulse response can be approximated by the system response to an impulse of equivalent magnitude. Based on all of these considerations, they need to obtain the following information:

(a) The magnitude of an impulse input equivalent to the rectangular pulse of Fig. 3.5.

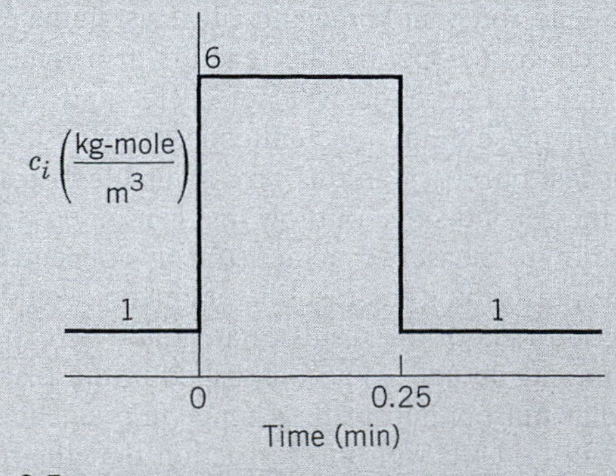

Figure 3.5 Proposed input pulse in reactant concentration.

(b) The impulse and pulse responses of the reactant concentration leaving the first stage.

(c) The impulse and pulse responses of the reactant concentration leaving the second stage.

SOLUTION

The reactor model for a single-stage CSTR was given in Eq. 2-66 as

$$V\frac{dc}{dt} = q(c_i - c) - Vkc$$

where c is the reactant concentration of component A. Because the reaction term can be neglected in this example ($k = 0$), the stages are merely continuous-flow mixers. Two material balance equations are required to model the two stages:

$$4\frac{dc_1}{dt} + 2c_1 = 2c_i \qquad (3\text{-}106)$$

$$3\frac{dc_2}{dt} + 2c_2 = 2c_1 \qquad (3\text{-}107)$$

If the system initially is at steady state, all concentrations are equal to the feed concentration:

$$c_2(0) = c_1(0) = c_i(0) = 1\ \text{kg mol/m}^3 \qquad (3\text{-}108)$$

(a) The pulse input is described by

$$c_i^P = \begin{cases} 1 & t < 0 \\ 6 & 0 \le t < 0.25\ \text{min} \\ 1 & t \ge 0.25\ \text{min} \end{cases} \qquad (3\text{-}109)$$

A convenient way to interpret (3-109) is as a constant input of 1 added to a rectangular pulse of height = 5 kg mol/m^3:

$$c_i^P = 6 \quad \text{for} \quad 0 \le t < 0.25 \text{ min} \tag{3-110}$$

The magnitude of an impulse input that is equivalent to the time-varying portion of (3-110) is simply the integral of the rectangular pulse:

$$M = 5\frac{\text{kg mol}}{\text{m}^3} \times 0.25 \text{ min} = 1.25\frac{\text{kg mol} \cdot \text{min}}{\text{m}^3}$$

Therefore, the equivalent impulse input is

$$c_i^\delta(t) = 1 + 1.25\delta(t) \tag{3-111}$$

Although the units of M have little physical meaning, the product

$$qM = 2\frac{\text{m}^3}{\text{min}} \times 1.25\frac{\text{kg mol} \cdot \text{min}}{\text{m}^3} = 2.5 \text{ kg mol}$$

can be interpreted as the amount of additional reactant fed into the reactor as either the rectangular pulse or the impulse.

(b) The impulse response of Stage 1 is obtained by Laplace transforming (3-106), using $c_1(0) = 1$:

$$4sC_1(s) - 4(1) + 2C_1(s) = 2C_i(s) \tag{3-112}$$

By rearranging (3-112), we obtain $C_1(s)$:

$$C_1(s) = \frac{4}{4s+2} + \frac{2}{4s+2}C_i(s) \tag{3-113}$$

The transform of the impulse input in feed concentration in (3-111) is

$$C_i^\delta(s) = \frac{1}{s} + 1.25 \tag{3-114}$$

Substituting (3-114) into (3-113), we have

$$C_1^\delta(s) = \frac{2}{s(4s+2)} + \frac{6.5}{4s+2} \tag{3-115}$$

Equation 3-115 does not correspond exactly to any entries in Table 3.1. However, putting the denominator in $\tau s + 1$ form yields

$$C_1^\delta(s) = \frac{1}{s(2s+1)} + \frac{3.25}{2s+1} \tag{3-116}$$

which can be directly inverted using the table, yielding

$$c_1^\delta(t) = 1 - e^{-t/2} + 1.625e^{-t/2} = 1 + 0.625e^{-t/2} \tag{3-117}$$

The rectangular pulse response is obtained in the same way. The transform of the input pulse (3-109) is given by (3-22), so that

$$C_i^P(s) = \frac{1}{s} + \frac{5(1 - e^{-0.25s})}{s} \tag{3-118}$$

Substituting (3-118) into (3-113) and solving for $C_1^P(s)$ yields

$$C_1^P(s) = \frac{4}{4s+2} + \frac{12}{s(4s+2)} - \frac{10e^{-0.25s}}{s(4s+2)} \tag{3-119}$$

Again, we have to put (3-119) into a form suitable for inversion:

$$C_1^P(s) = \frac{2}{2s+1} + \frac{6}{s(2s+1)} - \frac{5e^{-0.25s}}{s(2s+1)} \tag{3-120}$$

Before inverting (3-120), note that the term containing $e^{-0.25s}$ will involve a translation in time. Utilizing the procedure discussed above, we obtain the following inverse transform:

$$c_1^P(t) = e^{-t/2} + 6(1 - e^{-t/2}) - 5[1 - e^{-(t-0.25)/2}]S(t - 0.25) \tag{3-121}$$

Note that there are two solutions; for $t < 0.25$ min (or t_w) the rightmost term, including the time delay, is zero in the time solution. Thus, for

$$t < 0.25 \text{ min:} \quad c_1^P(t) = e^{-t/2} + 6(1 - e^{-t/2}) = 6 - 5e^{-t/2} \tag{3-122}$$

$$\begin{aligned} t \ge 0.25 \text{ min:} \quad c_1^P(t) &= e^{-t/2} + 6(1 - e^{-t/2}) \\ &\quad - 5(1 - e^{-(t-0.25)/2}) \\ &= 1 - 5e^{-t/2} + 5e^{-t/2}e^{+0.25/2} \\ &= 1 + 0.6657e^{-t/2} \end{aligned} \tag{3-123}$$

Plots of (3-117), (3-122), and (3-123) are shown in Fig. 3.6. Note that the rectangular pulse response approximates the impulse response fairly well for $t > 0.25$ min. Obviously, the approximation cannot be very good before $t = 0.25$ min, because the full effect of the rectangular pulse is not felt until that time, while the full effect of the hypothetical impulse begins immediately at $t = 0$.

(c) For the impulse response of Stage 2, Laplace transform (3-107), using $c_2(0) = 1$:

$$3sC_2(s) - 3(1) + 2C_2(s) = 2C_1(s) \tag{3-124}$$

Rearrange to obtain $C_2(s)$:

$$C_2(s) = \frac{3}{3s+2} + \frac{2}{3s+2}C_1(s) \tag{3-125}$$

For the input to (3-125), substitute the Laplace transform of the output from Stage 1—namely, (3-116):

$$C_2^\delta(s) = \frac{3}{3s+2} + \frac{2}{3s+2}\left[\frac{1}{s(2s+1)} + \frac{3.25}{2s+1}\right] \tag{3-126}$$

which can be rearranged to

$$C_2^\delta(s) = \frac{1.5}{1.5s+1} + \frac{1}{s(1.5s+1)(2s+1)} + \frac{3.25}{(1.5s+1)(2s+1)} \tag{3-127}$$

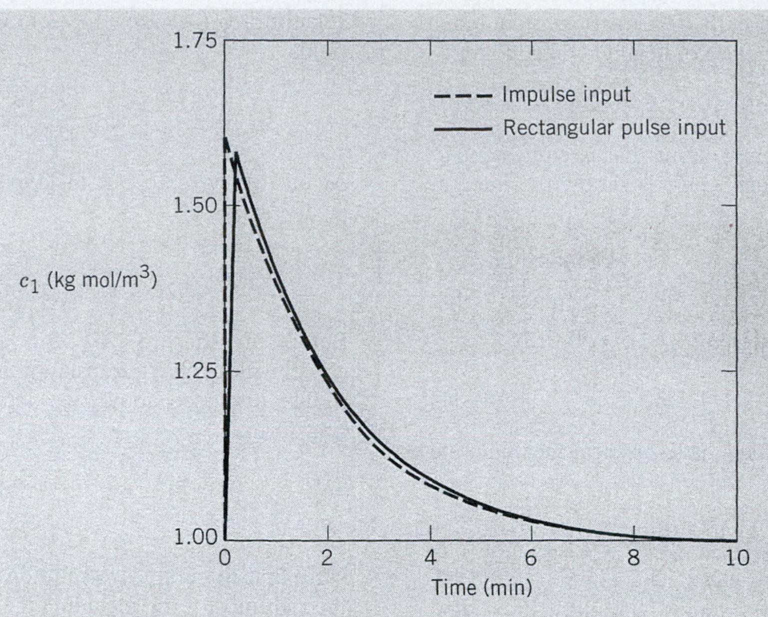

Figure 3.6 Reactor Stage 1 response.

Because each term in (3-127) appears as an entry in Table 3.1, partial fraction expansion is not required:

$$c_2^{\delta}(t) = e^{-t/1.5} + \left[1 + \frac{1}{0.5}(1.5e^{-t/1.5} - 2e^{-t/2})\right]$$
$$+ \frac{3.25}{0.5}[e^{-t/2} - e^{-t/1.5}]$$
$$= 1 - 2.5e^{-t/1.5} + 2.5e^{-t/2} \tag{3-128}$$

For the rectangular pulse response of Stage 2, substitute the Laplace transform of the appropriate stage output, Eq. 3-120, into Eq. 3-125 to obtain

$$C_2^p(s) = \frac{1.5}{1.5s+1} + \frac{2}{(1.5s+1)(2s+1)}$$
$$+ \frac{6}{s(1.5s+1)(2s+1)} - \frac{5e^{-0.25s}}{s(1.5s+1)(2s+1)} \tag{3-129}$$

Again, the rightmost term in (3-129) must be excluded from the inverted result or included, depending on whether or not $t < 0.25$ min. The calculation of the inverse transform of (3-129) gives

$$t < 0.25: \quad c_2^p(t) = 6 + 15e^{-t/1.5} - 20e^{-t/2} \tag{3-130}$$
$$t \geq 0.25: \quad c_2^p(t) = 1 - 2.7204e^{-t/1.5} + 2.663e^{-t/2} \tag{3-131}$$

Plots of Eqs. 3-128, 3-130, and 3-131 are shown in Fig. 3.7. The rectangular pulse response is virtually indistinguishable from the impulse response. Hence, Kim, Casey, and Tim can use the simpler impulse response solution to compare with real data obtained when the reactor is forced by a rectangular pulse. The maximum expected value of $c_2(t)$ is approximately 1.25 kg mol/m^3. This value should be compared with the nominal concentration before and after the test ($\bar{c}_2$ = 1.0 kg mol/m^3) to determine if the instrumentation is precise enough to record the change in concentration. If the change is too small, then the pulse amplitude, pulse width, or both must be increased.

Because this system is linear, multiplying the pulse magnitude (h) by a factor of four would yield a maximum concentration of reactant in the second stage of about 2.0 (the difference between initial and maximum concentration will be four times as large). On the other hand, the solutions obtained above strictly apply only for t_w = 0.25 min. Hence, the effect of a fourfold increase in t_w can be predicted only by resolving the model response for t_w = 1 min. Qualitatively, we know that the maximum value of c_2 will increase as t_w increases. Because the impulse response

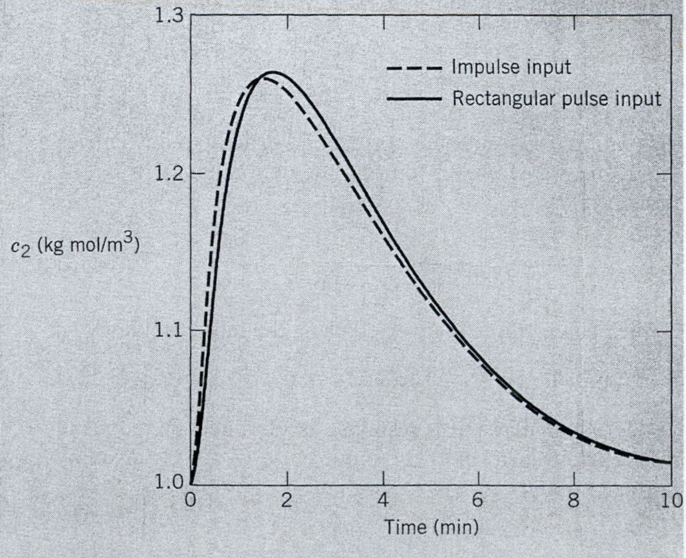

Figure 3.7 Reactor Stage 2 response.

model is a reasonably good approximation with $t_w = 0.25$ min, we expect that *small changes* in the pulse width will yield an approximately proportional effect on the maximum concentration change. This argument is based on a proportional increase in the approximately equivalent impulse input. A quantitative verification using numerical simulation is left as an exercise.

SUMMARY

In this chapter we have considered the application of Laplace transform techniques to solve linear differential equations. Although this material may be a review for some readers, an attempt has been made to concentrate on the important properties of the Laplace transform and its inverse, and to point out the techniques that make manipulation of transforms easier and less prone to error.

The use of Laplace transforms can be extended to obtain solutions for models consisting of simultaneous differential equations. However, before addressing such extensions, we introduce the concept of input-output models described by transfer functions. The conversion of differential equation models into transfer function models, covered in the next chapter, represents an important simplification in the methodology, one that can be exploited extensively in process modeling and control system design.

REFERENCES

Churchill, R. V., *Operational Mathematics*, 3d ed., McGraw-Hill, New York, 1971.

Chapra, S. C., and R. P. Canale, *Numerical Methods for Engineers*, 6th ed., McGraw-Hill, New York, 2010.

Dyke, P. R. G., *An Introduction to Laplace Transforms and Fourier Series*, Springer-Verlag, New York, 1999.

Schiff, J. L., *The Laplace Transform: Theory and Application*, Springer, New York, 1999.

EXERCISES

3.1 The differential equation (dynamic) model for a chemical process is as follows:

$$\frac{d^2y}{dt^2} + 5\frac{dy}{dt} + 3y = 2u(t)$$

where $u(t)$ is the single input function of time. $y(0)$ and $dy/dt\,(0)$ are both zero.

What are the functions of the time (e.g., $e^{-t/\tau}$) in the solution to the ODE for output $y(t)$ for each of the following cases?

(a) $u(t) = be^{-2t}$

(b) $u(t) = ct$

b and c are constants.

Note: You do not have to find $y(t)$ in these cases. Just determine the functions of time that will appear in $y(t)$.

3.2 Solve the ODE

$$\frac{d^4y}{dt^4} + 16\frac{d^3y}{dt^3} + 86\frac{d^2y}{dt^2} + 176\frac{dy}{dt} + 105y = 1$$

using partial fraction expansion. Note you need to calculate the roots of a fourth-order polynomial in s. All initial conditions on y and its derivatives are zero.

3.3 Figure E3.3 shows a pulse function.

(a) From details in the drawing, calculate the pulse width, t_w.

(b) Construct this function as the sum of simpler time elements, some perhaps translated in time, whose transforms can be found directly from Table 3.1.

(c) Find $U(s)$.

(d) What is the area under the pulse?

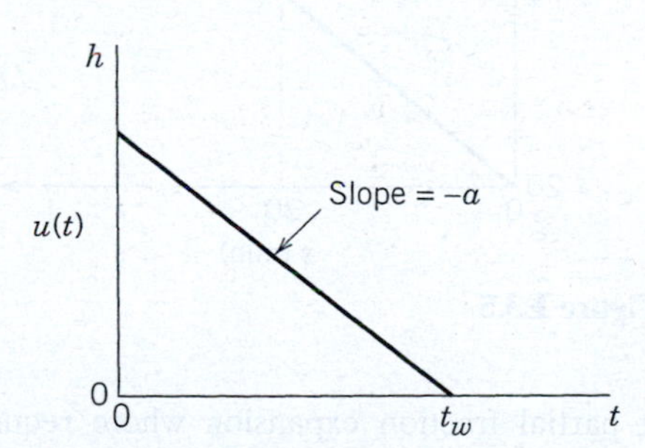

Figure E3.3 Triangular pulse function.

3.4 Derive Laplace transforms of the input signals shown in Figs. E3.4*a* and E3.4*b* by summing component functions found in Table 3.1.

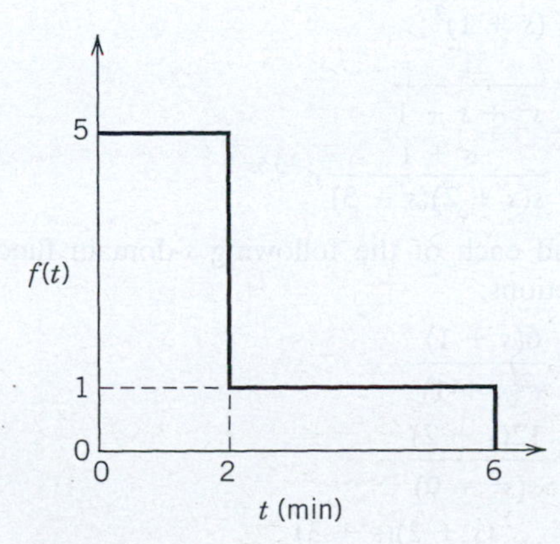

Figure E3.4*a*

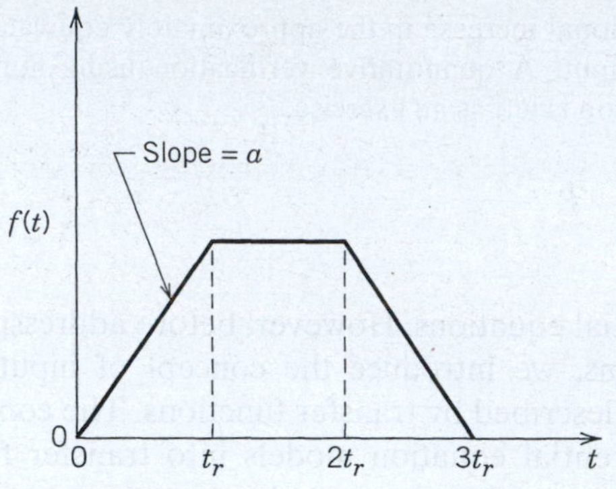

Figure E3.4b

3.5 The start-up procedure for a batch reactor includes a heating step where the reactor temperature is gradually heated to the nominal operating temperature of 75°C. The desired temperature profile $T(t)$ is shown in Fig. E3.5. What is $T(s)$?

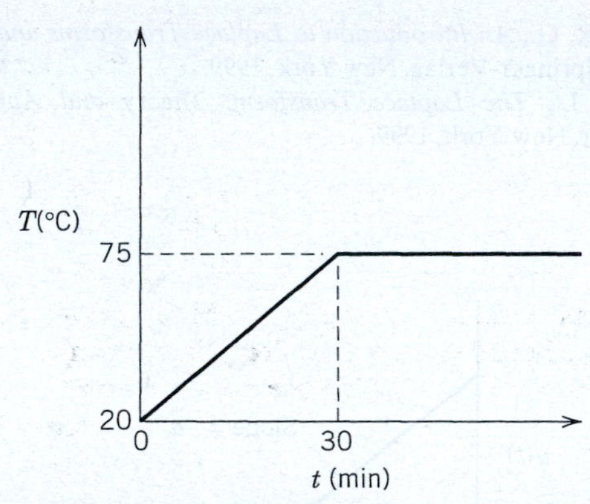

Figure E3.5

3.6 Using partial fraction expansion where required, find $x(t)$ for

(a) $X(s) = \dfrac{s(s+1)}{(s+2)(s+3)(s+4)}$

(b) $X(s) = \dfrac{s+1}{(s+2)(s+3)(s^2+4)}$

(c) $X(s) = \dfrac{s+4}{(s+1)^2}$

(d) $X(s) = \dfrac{1}{s^2+s+1}$

(e) $X(s) = \dfrac{s+1}{s(s+2)(s+3)}e^{-0.5s}$

3.7 Expand each of the following s-domain functions into partial fractions:

(a) $Y(s) = \dfrac{6(s+1)}{s^2(s+1)}$

(b) $Y(s) = \dfrac{12(s+2)}{s(s^2+9)}$

(c) $Y(s) = \dfrac{(s+2)(s+3)}{(s+4)(s+5)(s+6)}$

(d) $Y(s) = \dfrac{1}{[(s+1)^2+1]^2(s+2)}$

3.8 **(a)** For the integro-differential equation

$$\ddot{x} + 3\dot{x} + 2x = 2\int_0^t e^{-\tau}\,d\tau$$

find $x(t)$. Note that $\dot{x} = dx/dt$, etc.

(b) What is the value of $x(t)$ as $t \to \infty$?

3.9 For each of the following functions $X(s)$, what can you say about $x(t)$ $(0 \le t \le \infty)$ without solving for $x(t)$? In other words, what are $x(0)$ and $x(\infty)$? Is $x(t)$ converging, or diverging? Is $x(t)$ smooth, or oscillatory?

(a) $X(s) = \dfrac{6(s+2)}{(s^2+9s+20)(s+4)}$

(b) $X(s) = \dfrac{10s^2-3}{(s^2-6s+10)(s+2)}$

(c) $X(s) = \dfrac{16s+5}{s^2+9}$

3.10 For each of the following cases, determine what functions of time, e.g., $\sin 3t$, e^{-8t}, will appear in $y(t)$. (Note that you do not have to find $y(t)$!) Which $y(t)$ are oscillatory? Which exhibit a constant value of $y(t)$ for large values of t?

(i) $Y(s) = \dfrac{2}{s(s^2+4s)}$

(ii) $Y(s) = \dfrac{2}{s(s^2+4s+3)}$

(iii) $Y(s) = \dfrac{2}{s(s^2+4s+4)}$

(iv) $Y(s) = \dfrac{2}{s(s^2+4s+8)}$

(v) $Y(s) = \dfrac{2(s+1)}{s(s^2+4)}$

3.11 Which solutions of the following equations will exhibit convergent behavior? Which are oscillatory?

(a) $\dfrac{d^3x}{dt^3} + 2\dfrac{d^2x}{dt^2} + 2\dfrac{dx}{dt} + x = 3$

(b) $\dfrac{d^2x}{dt^2} - x = 2e^t$

(c) $\dfrac{d^3x}{dt^3} + x = \sin t$

(d) $\dfrac{d^2x}{dt^2} + \dfrac{dx}{dt} = 4$

Note: All of the above differential equations have one common factor in their characteristic equations.

3.12 The differential equation model for a particular chemical process has been found by testing to be as follows:

$$\tau_1\tau_2\frac{d^2y}{dt^2} + (\tau_1+\tau_2)\frac{dy}{dt} + y = Ku(t)$$

where τ_1 and τ_2 are constant parameters and $u(t)$ is the input function of time.

What are the functions of time (e.g., e^{-t}) in the solution for each output $y(t)$ for the following cases? (Optional: find the solutions for $y(t)$.)

(a) $u(t) = aS(t)$ unit step function

(b) $u(t) = be^{-t/\tau}$ $\tau \neq \tau_1 \neq \tau_2$

(c) $u(t) = ce^{-t/\tau}$ $\tau = \tau_1 \neq \tau_2$

(d) $u(t) = d \sin \omega t$ $\tau_1 \neq \tau_2$

3.13 Find the complete time-domain solutions for the following differential equations using Laplace transforms:

(a) $\frac{d^3x}{dt^3} + 4x = e^t$ with $x(0) = 0$, $\frac{dx(0)}{dt} = 0$, $\frac{d^2x(0)}{dt^2} = 0$

(b) $\frac{dx}{dt} - 12x = \sin 3t$ $x(0) = 0$

(c) $\frac{d^2x}{dt^2} + 6\frac{dx}{dt} + 25x = e^{-t}$ $x(0) = 0$, $\frac{dx(0)}{dt} = 0$

(d) A process is described by two differential equations:

$$\frac{dy_1}{dt} + y_2 = x_1$$

$$\frac{dy_2}{dt} - 2y_1 + 3y_2 = 2x_2$$

If $x_1 = e^{-t}$ and $x_2 = 0$, what can you say about the form of the solution for y_1? for y_2?

3.14 The dynamic model between an output variable y and an input variable u can be expressed by

$$\frac{d^2y(t)}{dt^2} + 3\frac{dy(t)}{dt} + y(t) = 4\frac{du(t-2)}{dt} - u(t-2)$$

(a) Will this system exhibit an oscillatory response after an arbitrary change in u?

(b) What is the steady-state gain?

(c) For a step change in u of magnitude 1.5, what is $y(t)$?

3.15 Find the solution of

$$\frac{dx}{dt} + 4x = f(t)$$

$$\text{where } f(t) = \begin{cases} 0 & t < 0 \\ h & 0 \leq t < 1/h \\ 0 & t \geq 1/h \end{cases}$$

$$x(0) = 0$$

Plot the solution for values of $h = 1$, 10, 100, and the limiting solution ($h \to \infty$) from $t = 0$ to $t = 2$. Put all plots on the same graph.

3.16 **(a)** The differential equation

$$\frac{d^2y}{dt^2} + 6\frac{dy}{dt} + 9y = \cos t$$

has initial conditions $y(0) = 1$, $y'(0) = 2$. Find $Y(s)$ and, without finding $y(t)$, determine what functions of time will appear in the solution.

(b) If $Y(s) = \frac{s+1}{s(s^2 + 4s + 8)}$, find $y(t)$.

3.17 A stirred-tank blending system initially is full of water and is being fed pure water at a constant flow rate, q. At a particular time, an operator shuts off the pure water flow and adds caustic solution at the same volumetric flow rate q but with concentration $\bar{c}_i$. If the liquid volume V is constant, the dynamic model for this process is

$$V\frac{dc}{dt} + qc = q\bar{c}_i$$

with $c(0) = 0$.

What is the concentration response of the reactor effluent stream, $c(t)$? Sketch it as a function of time.

Data: $V = 2$ m^3; $q = 0.4$ m^3/min; $\bar{c}_i = 50$ kg/m^3

3.18 For the dynamic system

$$2\frac{dy}{dt} = -y + 5u$$

y and u are deviation variables—y in degrees, u in flowrate units.

(a) u is changed from 0.0 to 2.0 at $t = 0$. Sketch the response and show the value of y_{ss}. How long does it take for y to reach within 0.1 degree of the final steady state?

(b) If u is changed from 0.0 to 4.0 at $t = 0$, how long does it take to cross the same steady state that was determined in part (a)? What is the new steady state?

(c) Suppose that after step (a) that the new temperature is maintained at 10 degrees for a long time. Then, at $t = t_1$, u is returned to zero. What is the new steady-state value of y? Use Laplace transformation to show how to obtain the analytical solution to the above ODE for this case. (Hint: select a new time, $t = 0$, where $y(0) = 10$).

3.19 Will the solution to the ODE that follows reach a steady state? Will it oscillate?

$$\frac{d^2x}{dt^2} + \frac{dx}{dt} = 4$$

Show appropriate calculations using partial fraction expansion and Laplace transforms.

3.20 Three stirred-tanks in series are used in a reactor train (see Fig. E3.20). The flow rate into the system of some inert species is maintained constant while tracer test are conducted. Assuming that mixing in each tank is perfect and volumes are constant:

(a) Derive model expressions for the concentration of tracer leaving each tank, c_i is the concentration of tracer entering the first tank.

(b) If c_i has been constant and equal to zero for a long time and an operator suddenly injects a large amount of tracer material in the inlet to tank 1, what will be the form of $c_3(t)$ (i.e., what kind of time functions will be involved) if

1. $V_1 = V_2 = V_3$
2. $V_1 \neq V_2 \neq V_3$.

(c) If the amount of tracer injected is unknown, is it possible to back-calculate the amount from experimental data? How?

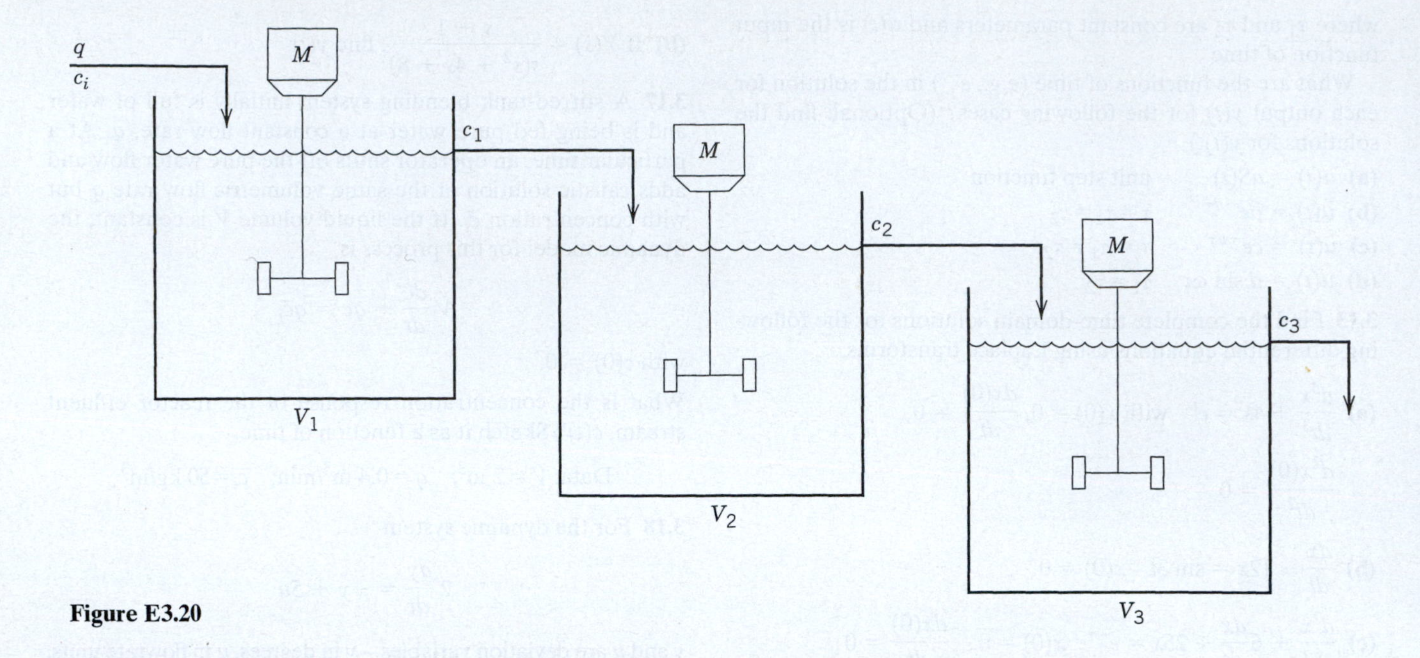

Figure E3.20

3.21 A stirred-tank reactor is operated with a feed mixture containing reactant A at a mass concentration C_{Ai}. The feed flow rate is w_i, as shown in Fig. E3.21. Under certain conditions the system operates according to the model

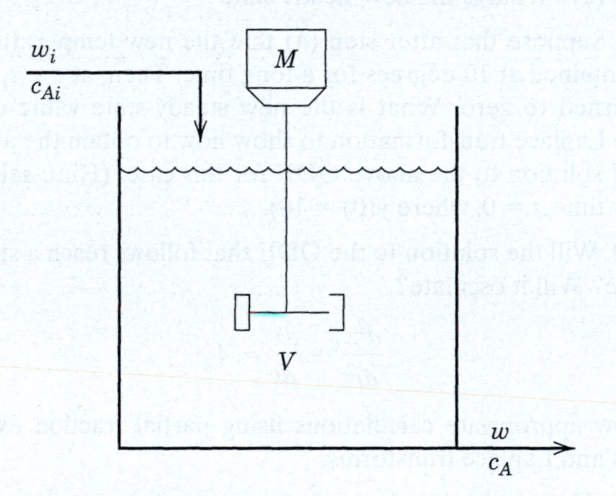

Figure E3.21

$$\frac{d(\rho V)}{dt} = w_i - w$$

$$\frac{d(\rho V c_A)}{dt} = w_i c_{At} - w c_A - \rho V k c_A$$

(a) For cases where the feed flow rate and feed concentration may vary and the volume is not fixed, simplify the model to one or more equations that do not contain product derivatives. The density may be assumed to be constant. Is the model in a satisfactory form for Laplace transform operations? Why or why not?

(b) For the case where the feed flow rate has been steady at $\overline{w}_i$ for some time, determine how c_A changes with time if a step change in c_{Ai} is made from c_{A1} to c_{A2}. List all assumptions necessary to solve the problem using Laplace transform techniques.

Chapter 4

Transfer Function Models

CHAPTER CONTENTS

Chapters 2 and 3 have considered dynamic models in the form of ordinary differential equations (ODE). In this chapter, we introduce an alternative model form based on Laplace transforms: the *transfer function model*. Both types of models can be used to determine the dynamic behavior of a process after changes in input variables. The transfer function also plays a key role in the design and analysis of control systems, as will be considered in later chapters.

A transfer function model characterizes the dynamic relationship of two process variables, a dependent variable (or *output variable*) and an independent variable (or *input variable*). For example, in a continuous chemical reactor, the output variable could be the exit concentration and the input variable a feed flow rate. Thus the input can be considered to be a "cause" and the output an "effect." Transfer function models are only directly applicable to processes that exhibit linear dynamic behavior, such as a process that can be modeled by a linear ODE. If the process is nonlinear, a transfer function can provide an approximate linear model, as described in Section 4.4.

4.1 AN ILLUSTRATIVE EXAMPLE: A CONTINUOUS BLENDING SYSTEM

Consider the continuous blending process of Section 2.2.2. For simplicity, we make the following assumptions:

1. Liquid density ρ and volume V are constant.
2. Flow rates w_1, w_2, and w are constant.

Then the component balance in Eq. 2-3 becomes

$$\rho V\frac{dx}{dt} = w_1x_1 + w_2x_2 - wx \tag{4-1}$$

Case (i): Inlet concentration x_1 varies while x_2 is constant
We will derive a transfer function model between exit composition x and inlet composition x_1, starting with Eq. 4-1. The steady-state version of (4-1) is

$$0 = w_1\bar{x}_1 + w_2x_2 - w\bar{x} \tag{4-2}$$

where the bar over a symbol denotes a nominal steady-state value. Subtracting (4-2) from (4-1) gives

$$\rho V\frac{dx}{dt} = w_1x_1' - wx' \tag{4-3}$$

where the two *deviation variables* (sometimes called *perturbation variables*) are defined as

$$x_1' \triangleq x_1 - \bar{x}_1 \tag{4-4}$$

$$x' \triangleq x - \bar{x}$$

Because $\bar{x}$ is a constant, it follows that

$$\frac{dx}{dt} = \frac{d(x-\bar{x})}{dt} = \frac{dx'}{dt} \tag{4-5}$$

Substituting (4-5) into (4-3) gives the solute component balance in deviation variable form:

$$\rho V\frac{dx'}{dt} = w_1x_1' - wx' \tag{4-6}$$

Assume that the blending system is initially at the nominal steady state. Thus, $x(0) = \bar{x}$ and $x'(0) = 0$. Taking the Laplace transform of Eq. 4-6 gives

$$\rho V s[X'(s) - x'(0)] = w_1 X_1'(s) - wX'(s) \tag{4-7}$$

where $X'(s) \triangleq \mathcal{L}[x'(t)]$ and $X_1'(s) \triangleq \mathcal{L}[x_1'(t)]$. Rearranging gives the *transfer function* $G(s)$ between the exit and inlet and compositions:

$$\frac{X'(s)}{X_1'(s)} = \frac{w_1}{\rho V s + w} \tag{4-8}$$

It is useful to place the transfer function in a standard form by dividing both the numerator and the denominator by w:

$$\frac{X'(s)}{X_1'(s)} = G(s) \triangleq \frac{K_1}{\tau s + 1} \tag{4-9}$$

where constants K_1 and τ are defined as

$$K_1 \triangleq \frac{w_1}{w} \tag{4-10}$$

$$\tau \triangleq \frac{\rho V}{w} \tag{4-11}$$

Later, useful physical interpretations of K and τ are provided in Section 5.2.

Case (ii): Both inlet concentrations, x_1 and x_2, vary
For the case of two input variables, x_1 and x_2, two transfer functions are needed to describe their effects on output variable x. Their derivation is analogous to the derivation for Case (i).

For this case, the steady-state version of (4-1) can be written as

$$0 = w_1\bar{x}_1 + w_2\bar{x}_2 - w\bar{x} \tag{4-12}$$

Subtracting (4-12) from (4-1) and introducing deviation variables gives

$$\rho V\frac{dx'}{dt} = w_1 x_1' + w_2 x_2' - wx' \tag{4-13}$$

where $x_2' \triangleq x_2 - \bar{x}_2$. Again assuming that the blending system is initially at the nominal steady state, taking the Laplace transform of Eq. 4-13 gives

$$\rho V s[X'(s) - x'(0)] = w_1 X_1'(s) + w_2 X_2'(s) - wX'(s) \tag{4-14}$$

which can be rearranged as

$$X'(s) = \frac{K_1}{\tau s + 1}X_1'(s) + \frac{K_2}{\tau s + 1}X_2'(s) \tag{4-15}$$

where K_2 is defined as

$$K_2 \triangleq \frac{w_2}{w} \tag{4-16}$$

and K_1 and τ are defined in (4-10) and (4-11). In order to derive the transfer function between x and x_1, assume that x_2 is constant at its nominal steady-state value, $x_2 = \bar{x}_2$. Therefore, $x_2'(t) = 0$, $X_2'(s) = 0$, and (4-15) reduces to the previous transfer function relating x and x_1 (see Eq. 4-9).

$$\frac{X'(s)}{X_1'(s)} \triangleq G_1(s) \triangleq \frac{K_1}{\tau s + 1} \tag{4-17}$$

Similarly, the transfer function between x and x_2 can be derived from (4-14) and the assumption that x_1 is constant at its nominal steady-state value, $x_1 = \bar{x}_1$.

$$\frac{X'(s)}{X_2'(s)} \triangleq G_2(s) \triangleq \frac{K_2}{\tau s + 1} \tag{4-18}$$

The models in (4-17) and (4-18) are referred to as *first-order transfer functions*, because the denominators are first-order in the Laplace variable s.

Three important aspects of these derivations are

1. A comparison of (4-15) to (4-18) shows that the effects of the individual input variables on the output variable are additive. This result is a consequence of the Principle of Superposition for linear models (see Section 3.1).
2. The assumption of an input being constant in the derivations of Eqs. 4-17 and 4-18 seems restrictive but actually is not, for the following reason. Because a transfer function concerns the effect of a *single input* on an output, it is not restrictive to assume that the other independent inputs are constant for purposes of the derivation. Simultaneous changes in both inputs can be analyzed, as indicated by Eq. 4-15.
3. A transfer function model allows the output response to be calculated for a specified input change. For example, Eq. 4-17 can be rearranged as

$$X'(s) = G_1(s)\, X_1'(s) \tag{4-19}$$

After specifying $x_1'(t)$, its Laplace transform $X_1'(s)$ can be determined using Table 3.1. Then the output response $x'(t)$ can be derived from (4-19), as illustrated by Example 4.1.

EXAMPLE 4.1

Consider the stirred-tank blending process for Case (i) and Eqs. 4-1 and 4-2. The nominal steady-state conditions are $w_1 = 600$ kg/min, $w_2 = 2$ kg/min, $x_1 = 0.050$, and $x_2 = 1$ (for pure solute). The liquid volume and density are constant: $V = 2\ \text{m}^3$ and $\rho = 900\ \text{kg/m}^3$, respectively.

(a) Calculate the nominal exit concentration, $\bar{x}$.

(b) Derive an expression for the response, $x(t)$, to a sudden change in x_1 from 0.050 to 0.075 that occurs at time, $t = 0$. Assume that the process is initially at the nominal steady state.

SOLUTION

(a) Exit flow rate w can be calculated from an overall mass balance

$$w = w_1 + w_2 = 600 + 2 = 602 \text{ kg/min}$$

and $\bar{x}$ can be determined from (4-2):

$$\bar{x} = \frac{w_1\bar{x}_1 + w_2 x_2}{w} = \frac{(600)(0.05) + (2)(1)}{602} = 0.053$$

(b) To determine $x(t)$ for a sudden change in x_1, we first derive an expression for $x'(t)$ and then obtain $X'(s)$. Thus, the appropriate starting point for the derivation is the transfer function in (4-9) where $K_1 = 600/602 = 0.997$ and $\tau = \rho V/w = (900)(2)/(602) = 2.99$ min. The sudden change in x_1 can be expressed in deviation variable form as

$$x_1'(t) = x_1 - \bar{x}_1 = 0.075 - 0.050$$

$$= 0.025 \quad \Rightarrow \quad X_1'(s) = \frac{0.025}{s}$$

Rearranging (4-9) and substituting numerical values gives

$$X'(s) = \left(\frac{K_1}{\tau s + 1}\right)X_1'(s) = \left(\frac{0.997}{2.99s + 1}\right)\left(\frac{0.025}{s}\right)$$

$$= \frac{0.0249}{s(2.99s + 1)} \tag{4-20}$$

Using Item 13 in Table 3.1, the inverse Laplace transform is

$$x'(t) = 0.0249\,(1 - e^{-t/2.99}) \tag{4-21}$$

From (4-4),

$$x(t) = \bar{x} + x'(t) = 0.053 + 0.0249\,(1 - e^{-t/2.99}) \tag{4-22}$$

Example 4.1 has shown how an expression can be derived for the response $x(t)$ to a step change in x_1. Analogous derivations could be made for other types of x_1 or x_2 changes, such as a sinusoidal change, or for simultaneous changes in x_1 and x_2. The starting point for the latter derivation would be Eq. 4-15. In order to derive $x(t)$ for a flow rate change, the process model must first be *linearized*, a technique considered in Section 4.4.

4.2 TRANSFER FUNCTIONS OF COMPLICATED MODELS

In the next example, we extend the concept of a transfer function model based on a single differential equation model to a model consisting of two differential equations. A more complicated transfer function results, but the approach remains the same.

EXAMPLE 4.2

Consider the model of the electrically heated stirred-tank system in Section 2.4.3. Subscript e refers to the heating element:

$$mC\frac{dT}{dt} = wC(T_i - T) + h_eA_e(T_e - T) \tag{2-47}$$

$$m_eC_e\frac{dT_e}{dt} = Q - h_eA_e(T_e - T) \tag{2-48}$$

(a) Derive transfer functions relating changes in outlet temperature T to changes in the two input variables: heater input Q (assuming no change in inlet temperature), and inlet temperature T_i (for no change in heater input).

(b) Show how these transfer functions are simplified when negligible thermal capacitance of the heating element ($m_eC_e \to 0$) is assumed.

SOLUTION

(a) First write the steady-state equations:

$$0 = wC(\bar{T}_i - \bar{T}) + h_eA_e(\bar{T}_e - \bar{T}) \tag{4-23}$$

$$0 = \bar{Q} - h_eA_e(\bar{T}_e - \bar{T}) \tag{4-24}$$

Next subtract (4-23) from (2-47), and (4-24) from (2-48):

$$mC\frac{dT}{dt} = wC[(T_i - \bar{T}_i) - (T - \bar{T})] + h_eA_e[(T_e - \bar{T}_e) - (T - \bar{T})] \tag{4-25}$$

$$m_eC_e\frac{dT_e}{dt} = (Q - \bar{Q}) - h_eA_e[(T_e - \bar{T}_e) - (T - \bar{T})] \tag{4-26}$$

Note that $dT/dt = dT'/dt$ and $dT_e/dt = dT_e'/dt$. Substitute deviation variables; then multiply (4-25) by $1/wC$ and (4-26) by $1/h_eA_e$:

$$\frac{m}{w}\frac{dT'}{dt} = -(T' - T_i') + \frac{h_eA_e}{wC}(T_e' - T') \tag{4-27}$$

$$\frac{m_eC_e}{h_eA_e}\frac{dT_e'}{dt} = \frac{Q'}{h_eA_e} - (T_e' - T') \tag{4-28}$$

The Laplace transform of each equation, after rearrangement, and assuming $T'(0) = T_e'(0) = 0$, is:

$$\left(\frac{m}{w}s + 1 + \frac{h_eA_e}{wC}\right)T'(s) = T_i'(s) + \frac{h_eA_e}{wC}T_e'(s) \tag{4-29}$$

$$\left(\frac{m_eC_e}{h_eA_e}s + 1\right)T_e'(s) = \frac{Q'(s)}{h_eA_e} + T'(s) \tag{4-30}$$

We can eliminate one of the output variables, $T'(s)$ or $T_e'(s)$, by solving (4-30) for it, and substituting into (4-29). Because $T_e'(s)$ is the intermediate variable, remove it. Then rearranging gives

$$\left[\frac{mm_eC_e}{wh_eA_e}s^2 + \left(\frac{m_eC_e}{h_eA_e} + \frac{m_eC_e}{wC} + \frac{m}{w}\right)s + 1\right]T'(s)$$

$$= \left(\frac{m_eC_e}{h_eA_e}s + 1\right)T_i'(s) + \frac{1}{wC}Q'(s) \tag{4-31}$$

By inspection, it is clear that Eq. 2-49, obtained by time-domain analysis is equivalent to (4-31).

Because both inputs influence the dynamic behavior of T', it is necessary to develop two transfer functions for the model. The effect of Q' on T' can be derived by assuming that T_i is constant at its nominal steady-state value, $\overline{T}_i$. Thus, $T_i' = 0$ and (4-31) can be rearranged as

$$\frac{T'(s)}{Q'(s)} = \frac{1/wC}{b_2s^2 + b_1s + 1} = G_1(s) \qquad (T_i'(s) = 0) \quad (4\text{-}32)$$

Similarly, the effect of T_i' on T' is obtained by assuming that $Q = \overline{Q}$ (that is, $Q' = 0$):

$$\frac{T'(s)}{T_i'(s)} = \frac{\dfrac{m_eC_e}{h_eA_e}s + 1}{b_2s^2 + b_1s + 1} = G_2(s) \qquad (Q'(s) = 0) \quad (4\text{-}33)$$

where

$$b_1 \triangleq \frac{m_eC_e}{h_eA_e} + \frac{m_eC_e}{wC} + \frac{m}{w} \quad (4\text{-}34)$$

$$b_2 \triangleq \frac{mm_eC_e}{wh_eA_e} \quad (4\text{-}35)$$

By the Superposition Principle, the effect of simultaneous changes in both inputs is given by

$$T'(s) = G_1(s)Q'(s) + G_2(s)T_i'(s) \quad (4\text{-}36)$$

This expression can also be derived by rearranging (4-31).

(b) The limiting behavior of $m_eC_e \rightarrow 0$ has $b_2 = 0$ and $b_1 = m/w$ and simplifies (4-36) to

$$T'(s) = \frac{1/wC}{\dfrac{m}{w}s + 1}Q'(s) + \frac{1}{\dfrac{m}{w}s + 1}T_i'(s) \quad (4\text{-}37)$$

4.3 PROPERTIES OF TRANSFER FUNCTIONS

One important property of the transfer function is that the steady-state output change for a sustained input change can be calculated directly. Very simply, setting $s = 0$ in $G(s)$ gives the steady-state gain of a process if the gain exists.[1] This feature is a consequence of the final value theorem presented in Chapter 3. If a unit step change in input is assumed, the corresponding output change for $t \rightarrow \infty$ is lim $G(s)$ as $s \rightarrow 0$.

The steady-state gain is the ratio of the output variable change to an input variable change when the input is adjusted to a new value and held there, thus allowing the process to reach a new steady state. Stated another way, the steady-state gain K of a process corresponds to the following expression:

$$K = \frac{\overline{y}_2 - \overline{y}_1}{\overline{u}_2 - \overline{u}_1} \quad (4\text{-}38)$$

where 1 and 2 indicate different steady states and $(\overline{y}, \overline{u})$ denote the corresponding steady-state values of the output and input variables. The steady-state gain is constant for linear processes regardless of the operating conditions. This is not true for a nonlinear process, as discussed in Section 4.4.

Another important property of the transfer function is that the order of the denominator polynomial (in s) is the same as the order of the equivalent differential equation. A general linear nth-order differential equation has the form

$$a_n\frac{d^ny}{dt^n} + a_{n-1}\frac{d^{n-1}y}{dt^{n-1}} + \cdots + a_1\frac{dy}{dt} + a_0y$$
$$= b_m\frac{d^mu}{dt^m} + b_{m-1}\frac{d^{m-1}u}{dt^{m-1}} + \cdots + b_1\frac{du}{dt} + b_0u \quad (4\text{-}39)$$

where u and y are input and output deviation variables, respectively. The transfer function obtained by Laplace transformation of (4-39) with $y(0) = 0$ and all initial conditions for the derivatives of u and y set equal to zero is

$$G(s) = \frac{Y(s)}{U(s)} = \frac{\sum_{i=0}^{m} b_is^i}{\sum_{i=0}^{n} a_is^i}$$
$$= \frac{b_ms^m + b_{m-1}s^{m-1} + \cdots + b_o}{a_ns^n + a_{n-1}s^{n-1} + \cdots + a_o} \quad (4\text{-}40)$$

Note that the numerator and denominator polynomials of the transfer function have the same orders (m and n, respectively) as the differential equation. In order for the model in (4-40) to be physically realizable, $n \geq m$.

The steady-state gain of $G(s)$ in (4-40) is b_o/a_o, obtained by setting $s = 0$. If both the numerator and denominator of (4-40) are divided by a_o, the characteristic (denominator) polynomial can be factored into a product $\Pi(\tau_is + 1)$ where τ_i denotes a time constant.

$$G(s) = \frac{Y(s)}{U(s)} = \frac{KB(s)}{(\tau_1s + 1)(\tau_2s + 1)\ (\tau_ns + 1)} \quad (4\text{-}41)$$

where gain K and *m-th* order polynomial $B(s)$ are obtained from the numerator of (4-40).

In this *time constant form*, inspection of the individual time constants provides information about the speed and qualitative features of the system response. This important point is discussed in detail in Chapters 5 and 6, after some additional mathematical tools have been developed.

---

[1]Some processes do not exhibit a steady-state gain, for example, the integrating elements discussed in Chapter 5.

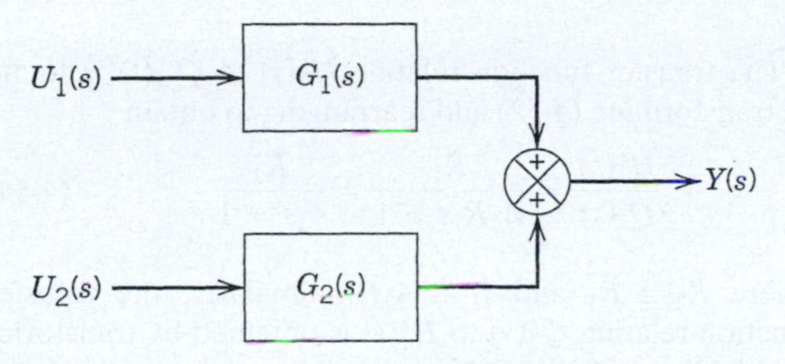

Figure 4.1 Block diagram of additive transfer function model.

The orders of the numerator and denominator polynomials in Eq. 4-40 are restricted by physical reasons so that $n \geq m$. Suppose that a real process could be modeled by

$$a_0 y = b_1 \frac{du}{dt} + b_0 u \tag{4-42}$$

That is, $n = 0$ and $m = 1$ in (4-39). This system will respond to a step change in $u(t)$ with an impulse at time zero, because dx/dt is infinite at the time the step change occurs. The ability to respond infinitely quickly to a sudden change in input is impossible to achieve with any real (physical) process, although it is approximated in some instances—for example, in an explosion. Therefore, we refer to the restriction $n \geq m$ as a *physical realizability* condition. It provides a diagnostic check on transfer functions derived from a high-order differential equation or from a set of first-order differential equations. Those transfer functions where $m > 0$, such as (4-33), are said to exhibit *numerator dynamics*. There are, however, many important cases where m is zero.

We have already illustrated the important *additive* property of transfer functions in deriving Eqs. 4-15 and 4-36, which is depicted in Fig. 4.1. Observe that a single process output variable (Y) can be influenced by more than one input (U_1 and U_2) acting individually or together.

EXAMPLE 4.3

The stirred-tank heating process described in Eq. 4-37 operates at steady state with an inlet temperature of 70 °F and a heater input of 1920 Btu/min. The liquid flow rate is 200 lb/min, the liquid has constant density ($\rho = 62.4$ lb/ft^3) and specific heat (0.32 Btu/lb °F), and the liquid volume is constant at 1.60 ft^3. Then the inlet temperature is changed to 90 °F, and the heater input is changed to 1,600 Btu/min. Calculate the output temperature response.

SOLUTION

The steady-state energy balance for the nominal conditions can be written as

$$wC(\overline{T} - \overline{T}_i) = \overline{Q} \tag{4-43}$$

Substituting numerical values gives $\overline{T} = 100$ °F.

$$T_i'(s) = \frac{90 - 70}{s} = \frac{20}{s}$$

$$Q'(s) = \frac{1600 - 1920}{s} = -\frac{320}{s}$$

The time constant τ and process gain K are

$$\tau = \frac{(62.4)(1.6)}{200} = 0.5 \text{ min}$$

$$K = \frac{1}{(200)(0.32)} = 1.56 \times 10^{-2} \frac{°\text{F}}{\text{Btu/min}}$$

Substituting in Eq. 4-37 yields

$$T'(s) = \frac{0.0156}{0.5s + 1}\left(-\frac{320}{s}\right) + \frac{1}{0.5s + 1}\left(\frac{20}{s}\right) \tag{4-44}$$

After simplification,

$$T'(s) = \frac{-5}{s(0.5s + 1)} + \frac{20}{s(0.5s + 1)} = \frac{15}{s(0.5s + 1)} \tag{4-45}$$

The corresponding time-domain solution is

$$T(t) = 100 + 15(1 - e^{-2t}) \tag{4-46}$$

Equation (4-43) shows the individual effects of the two input changes. At steady state, the reduction in the heater input lowers the temperature 5 °F, while the inlet temperature change increases it by 20 °F, for a net increase of 15 °F.

Transfer functions also exhibit a *multiplicative* property for sequential processes or process elements. Suppose two processes with transfer functions G_1 and G_2 are in a series configuration (see Fig. 4.2). The input $U(s)$ to G_1 produces an output $Y_1(s)$, which is the input to G_2. The output from G_2 is $Y_2(s)$. In equation form,

$$Y_1(s) = G_1(s)U(s) \tag{4-47}$$

$$Y_2(s) = G_2(s)Y_1(s) = G_2(s)G_1(s)U(s) \tag{4-48}$$

In other words, the transfer function between the original input U and the output Y_2 can be obtained by multiplying G_2 by G_1, as shown by the block diagram in Fig. 4.2.

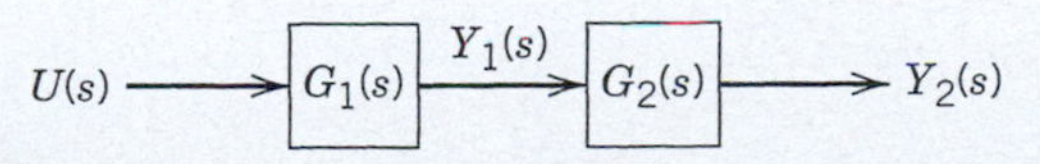

Figure 4.2 Block diagram of multiplicative (series) transfer function model.

EXAMPLE 4.4

Suppose that two liquid surge tanks are placed in series so that the outflow from the first tank is the inflow to the second tank, as shown in Fig. 4.3. If the outlet flow rate from each tank is proportional to the height of the liquid (head) in that tank, find the transfer function relating changes in flow rate from the second tank, $Q_2'(s)$, to changes in flow rate into the first tank, $Q_i'(s)$. Show how this transfer function is related to the individual transfer functions, $H_1'(s)/Q_i'(s)$, $Q_1'(s)/H_1'(s)$, $H_2'(s)/Q_1'(s)$, and $Q_2'(s)/H_2'(s)$. $H_1'(s)$ and $H_2'(s)$ denote the deviations in Tank 1 and Tank 2 levels, respectively. Assume that the two tanks have different cross-sectional areas A_1 and A_2, and that the valve resistances are R_1 and R_2.

SOLUTION

Equations 2-56 and 2-57 are valid for each tank; for Tank 1,

$$A_1\frac{dh_1}{dt} = q_i - q_1 \tag{4-49}$$

$$q_1 = \frac{1}{R_1}h_1 \tag{4-50}$$

Substituting (4-50) into (4-49) eliminates q_1:

$$A_1\frac{dh_1}{dt} = q_i - \frac{1}{R_1}h_1 \tag{4-51}$$

Putting (4-50) and (4-51) into deviation variable form gives

$$A_1\frac{dh_1'}{dt} = q_i' - \frac{1}{R_1}h_1' \tag{4-52}$$

$$q_1' = \frac{1}{R_1}h_1' \tag{4-53}$$

The transfer function relating $H_1'(s)$ to $Q_i'(s)$ is found by transforming (4-52) and rearranging to obtain

$$\frac{H_1'(s)}{Q_i'(s)} = \frac{R_1}{A_1R_1s+1} = \frac{K_1}{\tau_1 s+1} \tag{4-54}$$

where $K_1 \triangleq R_1$ and $\tau_1 \triangleq A_1R_1$. Similarly, the transfer function relating $Q_1'(s)$ to $H_1'(s)$ is obtained by transforming (4-53).

$$\frac{Q_1'(s)}{H_1'(s)} = \frac{1}{R_1} = \frac{1}{K_1} \tag{4-55}$$

The same procedure leads to the corresponding transfer functions for Tank 2,

$$\frac{H_2'(s)}{Q_1'(s)} = \frac{R_2}{A_2R_2s+1} = \frac{K_2}{\tau_2 s+1} \tag{4-56}$$

$$\frac{Q_2'(s)}{H_2'(s)} = \frac{1}{R_2} = \frac{1}{K_2} \tag{4-57}$$

where $K_2 \triangleq R_2$ and $\tau_2 \triangleq A_2R_2$. Note that the desired transfer function relating the outflow from Tank 2 to the inflow to Tank 1 can be derived by forming the product of (4-54) through (4-57).

$$\frac{Q_2'(s)}{Q_i'(s)} = \frac{Q_2'(s)}{H_2'(s)}\frac{H_2'(s)}{Q_1'(s)}\frac{Q_1'(s)}{H_1'(s)}\frac{H_1'(s)}{Q_i'(s)} \tag{4-58}$$

or

$$\frac{Q_2'(s)}{Q_i'(s)} = \frac{1}{K_2}\frac{K_2}{\tau_2 s+1}\frac{1}{K_1}\frac{K_1}{\tau_1 s+1} \tag{4-59}$$

which can be simplified to yield

$$\frac{Q_2'(s)}{Q_i'(s)} = \frac{1}{(\tau_1 s+1)(\tau_2 s+1)} \tag{4-60}$$

which is a second-order transfer function (does the unity gain make sense on physical grounds?). Figure 4.4 is a block diagram showing the *information flow* for this system.

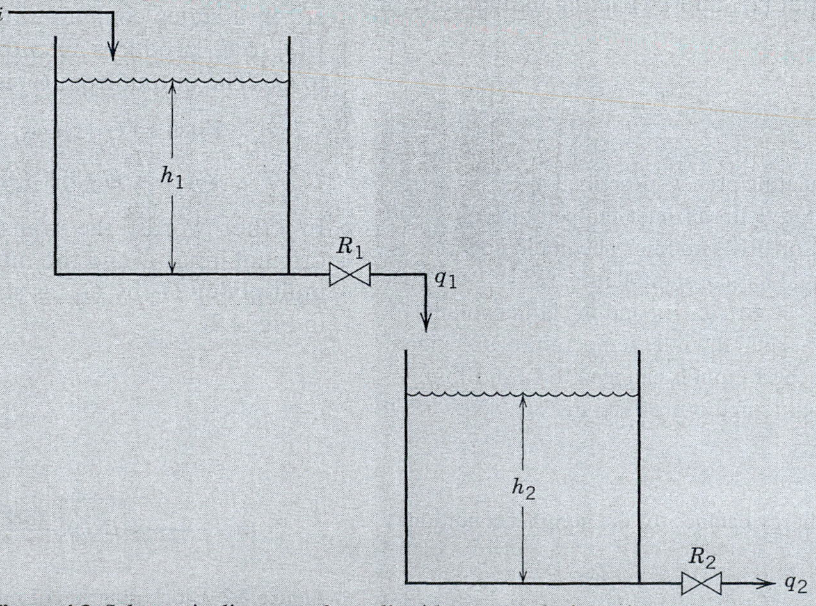

Figure 4.3 Schematic diagram of two liquid surge tanks in series.

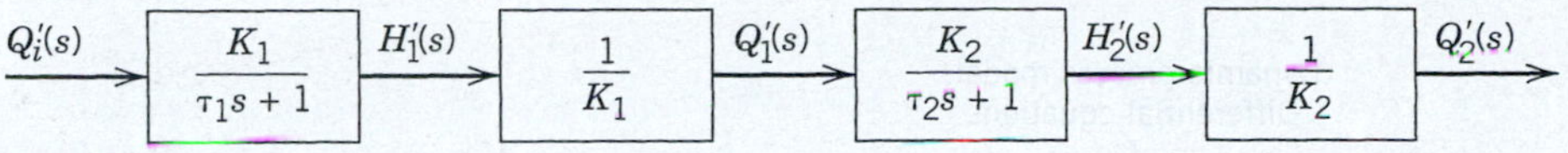

$$Q_i'(s) \rightarrow \boxed{\frac{K_1}{\tau_1 s + 1}} \xrightarrow{H_1'(s)} \boxed{\frac{1}{K_1}} \xrightarrow{Q_1'(s)} \boxed{\frac{K_2}{\tau_2 s + 1}} \xrightarrow{H_2'(s)} \boxed{\frac{1}{K_2}} \xrightarrow{} Q_2'(s)$$

Figure 4.4 Input–output model for two liquid surge tanks in series.

The multiplicative property of transfer functions proves to be quite valuable in designing process control systems because of the series manner in which process units are connected.

4.4 LINEARIZATION OF NONLINEAR MODELS

In the previous sections, we have limited the discussion to those processes that can be modeled by linear ordinary differential equations. However, there is a wide variety of processes for which the dynamic behavior depends on the process variables in a nonlinear manner.

Prominent examples include the exponential dependence of reaction rate on temperature (considered in Chapter 2), the nonlinear behavior of pH with flow rate of acid or base, and the asymmetric responses of distillate and bottoms compositions in a distillation column to changes in feed flow. Classical process control theory has been developed for linear processes, and its use, therefore, is restricted to linear approximations of the actual nonlinear processes. A linear approximation of a nonlinear steady-state model is most accurate near the point of linearization. The same is true for dynamic process models. Large changes in operating conditions for a nonlinear process cannot be approximated satisfactorily by linear expressions.

In many instances, however, nonlinear processes remain in the vicinity of a specified operating state. For such conditions, a linearized model of the process may be sufficiently accurate. Suppose a nonlinear dynamic model has been derived from first principles (material, energy, or momentum balances):

$$\frac{dy}{dt} = f(y, u) \tag{4-61}$$

where y is the output and u is the input. A linear approximation of this equation can be obtained by using a Taylor series expansion and truncating after the first-order terms. The reference point for linearization is the nominal steady-state operating point $(\bar{y}, \bar{u})$.

$$f(y, u) \cong f(\bar{y}, \bar{u}) + \left.\frac{\partial f}{\partial y}\right|_{\bar{y}, \bar{u}}(y - \bar{y}) + \left.\frac{\partial f}{\partial u}\right|_{\bar{y}, \bar{u}}(u - \bar{u}) \tag{4-62}$$

By definition, the steady-state condition corresponds to $f(\bar{y}, \bar{u}) = 0$. In addition, note that deviation variables arise naturally out of the Taylor series expansion—namely, $y' = y - \bar{y}$ and $u' = u - \bar{u}$. Hence, the linearized differential equation in terms of y' and u' is (after substituting, $dy'/dt = dy/dt$)

$$\frac{dy'}{dt} = \left.\frac{\partial f}{\partial y}\right|_s y' + \left.\frac{\partial f}{\partial u}\right|_s u' \tag{4-63}$$

where $(\partial f/\partial y)|_s$ is used to denote $(\partial f/\partial y)|_{\bar{y}, \bar{u}}$. If another input variable, z, is in the physical model, then Eq. 4-62 must be generalized further:

$$\frac{dy'}{dt} = \left.\frac{\partial f}{\partial y}\right|_s y' + \left.\frac{\partial f}{\partial u}\right|_s u' + \left.\frac{\partial f}{\partial z}\right|_s z' \tag{4-64}$$

where $z' = z - \bar{z}$.

In order to develop a transfer function of a nonlinear model, it is useful to summarize the general procedure, as is shown in Fig. 4.5. We use this procedure in the next example.

EXAMPLE 4.5

Again consider the stirred-tank blending system in Eqs. 2-17 and 2-18, written as

$$\rho \frac{dV}{dt} = w_1 + w_2 - w \tag{4-65}$$

$$\rho V \frac{dx}{dt} = w_1(x_1 - x) + w_2(x_2 - x) \tag{4-66}$$

Assume that volume V remains constant (due to an overflow line that is not shown) and consequently, $w = w_1 + w_2$. Inlet composition x_1 and inlet flow rates w_1 and w_2 can vary, but stream 2 is pure solute so that $x_2 = 1$.

Derive transfer functions that relate the exit composition to the three input variables (w_1, w_2, and x_1) using the steps shown in the flow chart of Fig. 4.5.

SOLUTION

The nonlinearities in Eq. 4-66 are due to the product terms, $w_1 x_1$, and so forth. The right side of (4-66) has the functional form $f(x, x_1, w_1, w_2)$. For Step 1 of Fig. 4.5, find the steady-state values of x and w by settings the derivatives of (4-65) and (4-66) equal to zero and substituting the steady-state values. For Step 2, linearize (4-66) about the nominal steady-state values to obtain

$$\rho V \frac{dx}{dt} = \rho V \frac{dx'}{dt} = \left(\frac{\partial f}{\partial x}\right)_s (x - \bar{x}) + \left(\frac{\partial f}{\partial x_1}\right)_s (x_1 - \bar{x}_1) + \left(\frac{\partial f}{\partial w_1}\right)_s (w_1 - \bar{w}_1) + \left(\frac{\partial f}{\partial w_2}\right)_s (w_2 - \bar{w}_2) \tag{4-67}$$

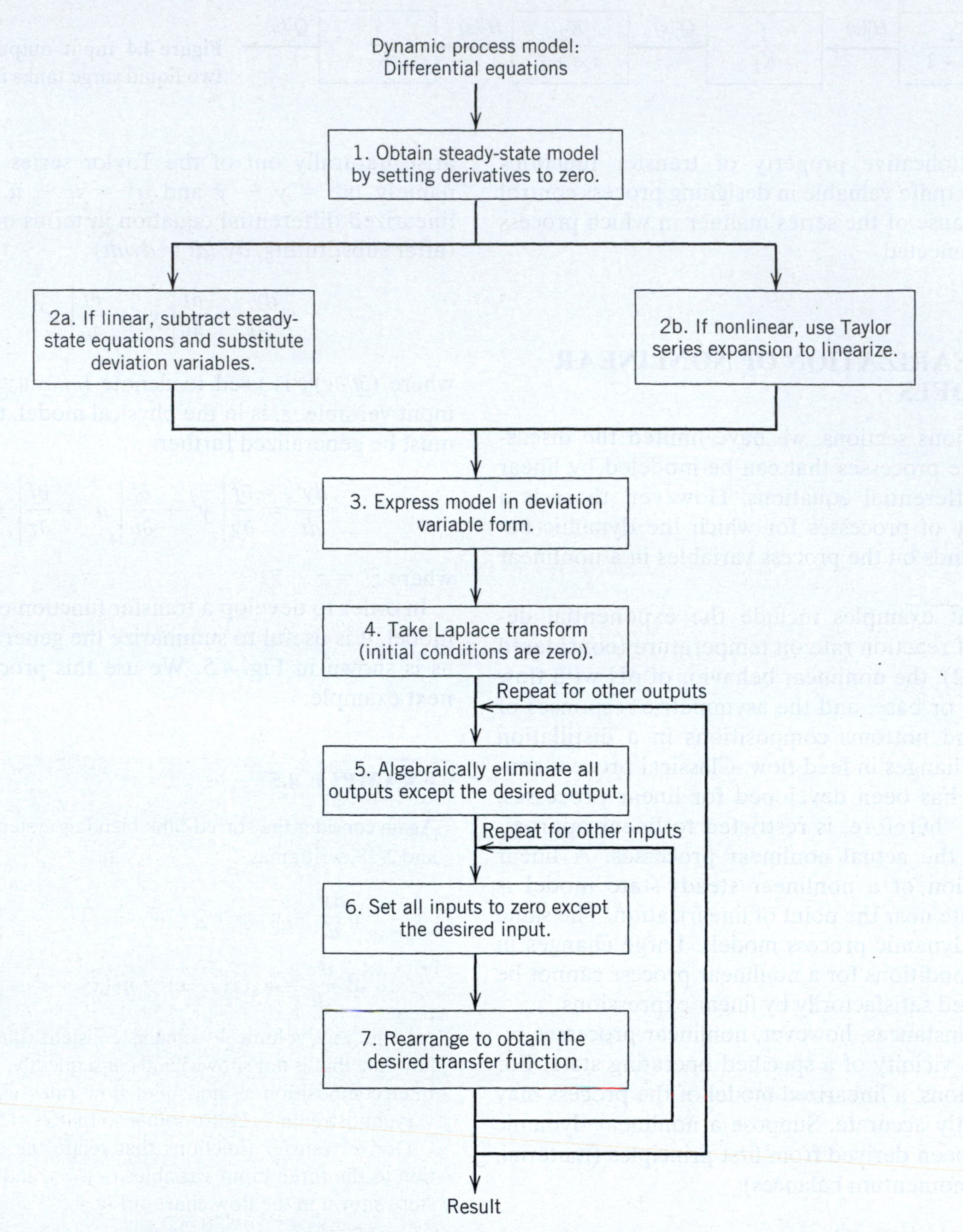

Figure 4.5 General procedure for developing transfer function models.

The partial derivatives are as follow:

$$\left(\frac{\partial f}{\partial x}\right)_s = -\overline{w}_1 - \overline{w}_2$$

$$\left(\frac{\partial f}{x_1}\right)_s = \overline{w}_1 \qquad (4\text{-}68)$$

$$\left(\frac{\partial f}{w_1}\right)_s = \overline{x}_1 - \overline{x}$$

$$\left(\frac{\partial f}{w_2}\right)_s = 1 - \overline{x}$$

Substitute (4-68) and introduce deviation variables (Step 3):

$$\rho V \frac{dx'}{dt} = -\overline{w}x' + \overline{w}_1 x_1' + (\overline{x}_1 - \overline{x})w_1' + (1 - \overline{x})w_2' \qquad (4\text{-}69)$$

The above equation is general in that it applies to any specified operating point.

For Step 4 take the Laplace transform of both sides of Eq. 4-69 with the initial condition, $x'(0) = 0$:

$$\rho V s X'(s) = -\overline{w}X'(s) + \overline{w}_1 X_1'(s) + (\overline{x}_1 - \overline{x})W_1'(s) + (1 - \overline{x})W_2'(s)$$

Rearranging and dividing by $\overline{w}$ yields

$$\left(\frac{V\rho}{\overline{w}}s+1\right)X'(s)=\frac{\overline{w}_1}{\overline{w}}X_1'(s)+\frac{\overline{x}_1-\overline{x}}{\overline{w}}W_1'(s)+\frac{1-\overline{x}}{\overline{w}}W_2'(s)$$

Define

$$\tau=\frac{V\rho}{\overline{w}}$$

$$K_1=\frac{\overline{w}_1}{\overline{w}},\quad K_2=\frac{1-\overline{x}}{\overline{w}},\quad \text{and}\quad K_3=\frac{\overline{x}_1-\overline{x}}{\overline{w}}$$

Applying Step 5 gives the relationship for the single output and three inputs:

$$X'(s)=\frac{K_1}{\tau s+1}X_1'(s)+\frac{K_2}{\tau s+1}W_2'(s)+\frac{K_3}{\tau s+1}W_1'(s) \tag{4-70}$$

Three input-output transfer functions can be derived from Steps 6 and 7:

$$G_1(s)=\frac{X'(s)}{X_1'(s)}=\frac{K_1}{\tau s+1} \tag{4-71}$$

$$G_2(s)=\frac{X'(s)}{W_2'(s)}=\frac{K_2}{\tau s+1} \tag{4-72}$$

$$G_3(s)=\frac{X'(s)}{W_1'(s)}=\frac{K_3}{\tau s+1} \tag{4-73}$$

This example shows that individual transfer functions for a model with several inputs can be obtained by linearization of the nonlinear differential equation model. Note that all three transfer functions have the same time constant τ but different gains.

$$K_1=\frac{\overline{w}_1}{\overline{w}}>0$$

$$K_2=\frac{1-\overline{x}}{\overline{w}}>0$$

$$K_3=\frac{\overline{x}_1-\overline{x}}{\overline{w}}<0$$

If a gain is positive, a steady-state increase in its input produces a steady-state increase in the output. A negative gain (e.g., K_3) has just the opposite effect.

Note that the gains of this nonlinear process depend on the nominal steady-state conditions. Thus, if these conditions were changed to improve process performance, the numerical values of the gains and time constant would also change.

EXAMPLE 4.6

Consider a single tank liquid-level system where the outflow passes through a valve. Recalling Eq. 2-56, assume now that the valve discharge rate is related to the square root of liquid level:

$$q=C_v\sqrt{h} \tag{4-74}$$

where C_v depends on the fixed opening of the valve (see Chapter 9). Derive an approximate dynamic model for this process by linearization and compare with the results in Example 4.4.

SOLUTION

The material balance for the process (Eq. 2-54) after substituting (4-74) is

$$A\frac{dh}{dt}=q_i-C_v\sqrt{h} \tag{4-75}$$

To obtain the system transfer function, linearize (4-75) about the steady-state conditions $(\overline{h},\overline{q}_i)$. The deviation variables are

$$h'=h-\overline{h}$$
$$q_i'=q_i-\overline{q}_i$$

Applying (4-63) where $y=h$ and $x=q_i$, and $f(h,q_i)$ is the right side of (4-75), the linearized differential equation is

$$A\frac{dh'}{dt}=q_i'-\frac{C_v}{2\sqrt{\overline{h}}}h' \tag{4-76}$$

If we define the valve resistance R using the relation

$$\frac{1}{R}=\frac{C_v}{2\sqrt{\overline{h}}} \tag{4-77}$$

the resulting dynamic equation is analogous to the linear model presented earlier in (4-52):

$$A\frac{dh'}{dt}=q_i'-\frac{1}{R}h' \tag{4-78}$$

The transfer function corresponding to (4-77) was derived earlier as (4-54).

EXAMPLE 4.7

A horizontal cylindrical tank shown in Fig. 4.6*a* is used to slow the propagation of liquid flow surges in a processing line. Figure 4.6*b* illustrates an end view of the tank and w_t is the width of the liquid surface, which is a function of its height, both of which can vary with time. Develop a model for the height of liquid h in the tank at any time with the inlet and outlet volumetric flow rates as model inputs. Linearize the model assuming that the process initially is at steady state and that the liquid density ρ is constant.

SOLUTION

Note that the primary complication in modeling this process is that the liquid surface area A varies as the level

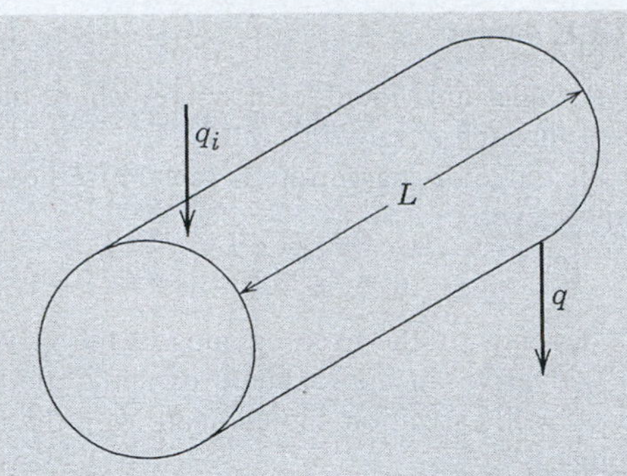

Figure 4.6a A horizontal cylindrical liquid surge tank.

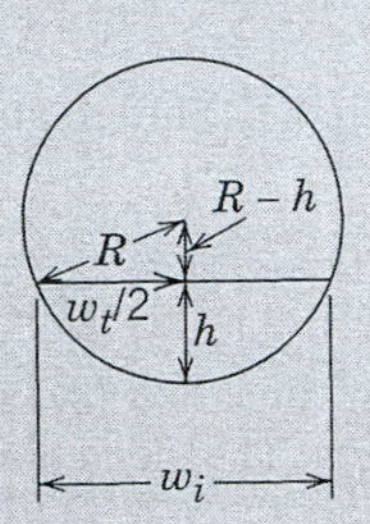

Figure 4.6b The end view of a cylindrical surge tank.

varies. The accumulation term must represent this feature. For constant density, a mass balance yields

$$\frac{dm}{dt} = \rho q_i - \rho q \tag{4-79}$$

The mass accumulation term in (4-79) can be written, noting that $dV = A dh = w_t L\, dh$, as

$$\frac{dm}{dt} = \rho\frac{dV}{dt} = \rho w_t L\frac{dh}{dt} \tag{4-80}$$

where $w_t L$ represents the changing surface area of the liquid. Substituting (4-80) in (4-79) and simplifying gives

$$w_t L\frac{dh}{dt} = q_i - q \tag{4-81}$$

The geometric construction in Fig. 4.6*b* indicates that $w_t/2$ is the length of one side of a right triangle whose hypotenuse is R. Thus, $w_t/2$ is related to the level h by

$$\frac{w_t}{2} = \sqrt{R^2 - (R - h)^2} \tag{4-82a}$$

After rearrangement,

$$w_t = 2\sqrt{(D - h)h} \tag{4-82b}$$

with $D = 2R$ the diameter of the tank. Substituting (4-81b) into (4-81) yields a nonlinear dynamic model for the tank with q_i and q as inputs:

$$\frac{dh}{dt} = \frac{1}{2L\sqrt{(D - h)h}}(q_i - q) \tag{4-83}$$

To linearize (4-83) about the operating point ($h = \bar{h}$), let

$$f = \frac{q_i - q}{2L\sqrt{(D - h)h}}$$

Then

$$\left(\frac{\partial f}{\partial q_i}\right)_s = \frac{1}{2L\sqrt{(D - \bar{h})\bar{h}}}$$

$$\left(\frac{\partial f}{\partial q}\right)_s = \frac{-1}{2L\sqrt{(D - \bar{h})\bar{h}}}$$

$$\left(\frac{\partial f}{\partial h}\right)_s = (\bar{q}_i - \bar{q})\left[\frac{\partial}{\partial h}\left(\frac{1}{2L\sqrt{(D - \bar{h})\bar{h}}}\right)\right]_s = 0$$

The last partial derivative is zero, because $\bar{q}_i - \bar{q} = 0$ from the steady-state relation, and the derivative term in brackets is finite for all $0 < h < D$. Consequently, the linearized model of the process, after substitution of deviation variables, is

$$\frac{dh'}{dt} = \frac{1}{2L\sqrt{(D - \bar{h})\bar{h}}}(q_i' - q') \tag{4-84}$$

Recall that the term $2L\sqrt{(D - h)h}$ in (4-84) represents the variable surface area of the tank. The linearized model (4-84) treats this quantity as a constant $(2L\sqrt{(D - \bar{h})\bar{h}}$ that depends on the nominal (steady-state) operating level. Consequently, operation of the horizontal cylindrical tank for small variations in level around the steady-state value would be much like that of any tank with equivalent but constant liquid surface. For example, a vertical cylindrical tank with diameter D' has a surface area of liquid in the tank $= \pi(D')^2/4 = 2L\sqrt{(D - \bar{h})\bar{h}}$. Note that the coefficient $1/2L\sqrt{(D - \bar{h})\bar{h}}$ is infinite for $\bar{h} = 0$ or $\bar{h} = D$ and is a minimum at $\bar{h} = D/2$. Thus, for large variations in level, Eq. 4-84 would not be a good approximation, because dh/dt is independent of h in the linearized model. In these cases, the horizontal and vertical tanks would operate very differently.

Finally, we examine the application of linearization methods when the model involves more than one nonlinear equation.

EXAMPLE 4.8

As shown in Chapter 2, a continuous stirred-tank reactor with a single first-order chemical reaction has the following material and energy balances:

$$V\frac{dc_A}{dt} = q(c_{Ai} - c_A) - Vkc_A \tag{2-66}$$

$$V\rho C\frac{dT}{dt} = wC(T_i - T) + (-\Delta H_R)Vkc_A + UA(T_c - T) \tag{2-68}$$

If the reaction rate coefficient k is given by the Arrhenius equation,

$$k = k_0 e^{-E/RT} \tag{2-63}$$

this model is nonlinear. However, it is possible to find approximate transfer functions relating the inputs and outputs. For the case where the flow rate (q or w) and inlet conditions (c_{Ai} and T_i) are assumed to be constant, calculate the transfer function relating changes in the reactor concentration c_A to changes in the coolant temperature T_c.

SOLUTION

For this situation, there is a single input variable T_c and two output variables c_A and T. First, the steady-state operating point must be determined (Step 1 in Fig. 4.5). Note that such a determination will require iterative solution of two nonlinear algebraic equations; this can be done using a Newton-Raphson method or similar algorithm (Chapra and Canale, 2010). Normally, we would specify $\overline{T}_i$, $\overline{c}_{Ai}$, and $\overline{c}_A$ and then determine $\overline{T}$ and $\overline{T}_c$ that satisfy (2-66) and (2-68) at steady state. Then we can proceed with the linearization of (2-66) and (2-68). Defining deviation variables c'_A, T', and T'_c, we obtain the following equations:

$$\frac{dc'_A}{dt} = a_{11}c'_A + a_{12}T' \tag{4-85}$$

$$\frac{dT'}{dt} = a_{21}c'_A + a_{22}T' + b_2T'_c \tag{4-86}$$

where

$$a_{11} = -\frac{q}{V} - k_0e^{-E/R\overline{T}}$$

$$a_{12} = -k_0e^{-E/R\overline{T}}\overline{c}_A\left(\frac{E}{R\overline{T}^2}\right)$$

$$a_{21} = \frac{(-\Delta H_R)k_0e^{-E/R\overline{T}}}{\rho C}$$

$$a_{22} = \frac{1}{V\rho C}\left[-(wC + UA) + (-\Delta H_R)V\overline{c}_Ak_0e^{-E/R\overline{T}}\left(\frac{E}{R\overline{T}^2}\right)\right]$$

$$b_2 = \frac{UA}{V\rho C}$$

Note that Eq. 2-66 does not contain input variable T_c, so no T'_c term appears in (4-85). We can convert (4-85) and (4-86) into a transfer function between the coolant temperature $T'_c(s)$ and the tank outlet concentration $C'_A(s)$ via Laplace transformation:

$$(s - a_{11})C'_A(s) = a_{12}T'(s) \tag{4-87}$$

$$(s - a_{22})T'(s) = a_{21}C'_A(s) + b_2T'_c(s) \tag{4-88}$$

Substituting for $T'(s)$, (4-87) becomes

$$(s - a_{11})(s - a_{22})C'_A(s) = a_{12}a_{21}C'_A(s) + a_{12}b_2T'_c(s) \tag{4-89}$$

yielding

$$\frac{C'_A(s)}{T'_c(s)} = \frac{a_{12}b_2}{s^2 - (a_{11} + a_{22})s + a_{11}a_{22} - a_{12}a_{21}} \tag{4-90}$$

which is a second-order transfer function. The a and b coefficients can be evaluated for a particular operating condition.

SUMMARY

In this chapter, we have introduced an important concept, the transfer function. It relates changes in a process output to changes in a process input and can be derived from a linear differential equation model using Laplace transformation methods. The transfer function contains key information about the steady-state and dynamic relations between input and output variables, namely, the process gain and time constants, respectively. Transfer functions are usually expressed in terms of deviation variables, that is, deviations from nominal steady-state conditions.

REFERENCES

Chapra, S. C., and R. P. Canale, *Numerical Methods for Engineers*, 6th ed., McGraw-Hill, New York, 2010.

Henson, M. A. and D. E. Seborg (eds.), *Nonlinear Process Control*, Prentice Hall, Upper Saddle River, NJ, 1997.

EXERCISES

4.1 Consider a transfer function:

$$\frac{Y(s)}{U(s)} = \frac{a}{bs + c}$$

(a) What is the steady-state gain?

(b) For a step change of magnitude M in the input, will the output response be bounded for all values of constants a, b, and c? Briefly justify your answer.

4.2 Consider the following transfer function:

$$G(s) = \frac{Y(s)}{U(s)} = \frac{5}{10s + 1}$$

(a) What is the steady-state gain?

(b) What is the time constant?

(c) If $U(s) = 2/s$, what is the value of the output $y(t)$ when $t \rightarrow \infty$?

(d) For the same $U(s)$, what is the value of the output when $t = 10$? What is the output when expressed as a fraction of the new steady-state value?

(e) If $U(s) = (1 - e^{-s})/s$, that is, the unit rectangular pulse, what is the output when $t \rightarrow \infty$?

(f) If $u(t) = \delta(t)$, that is, the unit impulse at $t = 0$, what is the output when $t \rightarrow \infty$?

(g) If $u(t) = 2 \sin 3t$, what is the value of the output when $t \rightarrow \infty$?

4.3 The dynamic behavior of a pressure sensor/transmitter can be expressed as a first-order transfer function (in deviation variables) that relates the measured value P_m to the actual pressure, P:

$$\frac{P'_m(s)}{P'(s)} = \frac{1}{30s + 1}$$

Both P'_m and P' have units of psi and the time constant has units of seconds. Suppose that an alarm will sound if P_m exceeds 45 psi. If the process is initially at steady state, and then P suddenly changes from 35 to 50 psi at 1:10 PM, at what time will the alarm sound?

4.4 Consider the first-order transfer function model in Exercise 4.2 where y and u are deviation variables. For an initial condition of $y(0) = 1$ and a step change in u of magnitude 2 (at $t = 0$), calculate the response, $y(t)$.
Hint: First determine the corresponding differential equation model by using the inverse Laplace tranform.

4.5 For the process modeled by

$$2\frac{dy_1}{dt} = -2y_1 - 3y_2 + 2u_1$$

$$\frac{dy_2}{dt} = 4y_1 - 6y_2 + 2u_1 + 4u_2$$

Find the four transfer functions relating the outputs (y_1, y_2) to the inputs (u_1, u_2). The u_i and y_i are deviation variables.

4.6 A stirred-tank blending system can be described by a first-order transfer function between the exit composition x and the inlet composition x_1 (both are mass fractions of solute):

$$\frac{X'(s)}{X'_i(s)} = \frac{K}{\tau s + 1}$$

where $K = 0.6$ (dimensionless) and $\tau = 10$ min. When the blending system is at steady state ($\bar{x} = 0.3$), the dynamic behavior is tested by quickly adding a large amount of a radioactive tracer, thus approximating an impulse function with magnitude 1.5.

(a) Calculate the exit composition response $x(t)$ using Laplace transforms and sketch $x(t)$. Based on this analytical expression, what is the value of $x(0)$?

(b) Using the initial value theorem of Section 3.4, determine the value of $x(0)$.

(c) If the process is initially at a steady state with $\bar{x} = 0.3$, what is the value of $x(0)$?

(d) Compare your answer for parts (a)–(c) and briefly discuss any differences.

4.7 A single equilibrium stage in a distillation column is shown in Fig. E4.7. The model that describes this stage is

$$\frac{dH}{dt} = L_0 + V_2 - (L_1 + V_1)$$

$$\frac{dHx_1}{dt} = L_0x_0 + V_2y_2 - (L_1x_1 + V_1y_1)$$

$$y_1 = a_0 + a_1x_1 + a_2x_1^2 + a_3x_1^3$$

(a) Assuming that the molar holdup H in the stage is constant and that equimolal overflow holds, for a mole of vapor that condenses, one mole of liquid is vaporized, simplify the model as much as possible.

(b) Linearize the resulting model and introduce deviation variables.

(c) For constant liquid and vapor flow rates, derive the four transfer functions relating outputs x_1 and y_1 to inputs x_0 and y_2. Put in standard form.

4.8 A surge tank in Fig. E4.8 is designed with a slotted weir so that the outflow rate, w, is proportional to the liquid level to the 1.5 power; that is,

$$w = Rh^{1.5}$$

where R is a constant. If a single stream enters the tank with flow rate w_i, find the transfer function $H'(s)/W'(s)$. Identify the gain and all time constants. Verify units.
The cross-sectional area of the tank is A. Density ρ is constant.

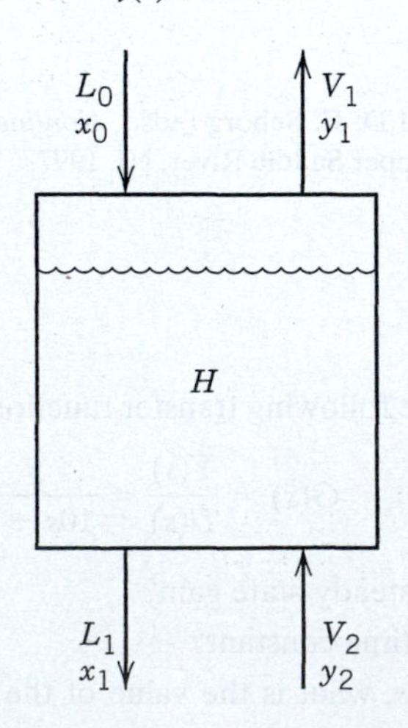

Figure E4.7

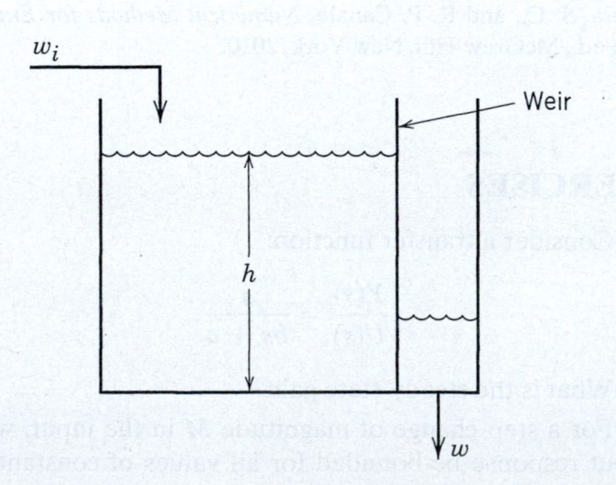

Figure E4.8

4.9 For the steam-heated stirred-tank system modeled by Eqs. 2-51 and 2-52, assume that the steam temperature T_s is constant.

(a) Find a transfer function relating tank temperature T to inlet liquid temperature T_i.

(b) What is the steady-state gain for this choice of input and output?

(c) Based on physical arguments only, should the gain be unity? Justify your answer.

4.10 The contents of the stirred-tank heating system shown in Figure E4.10 are heated at a constant rate of Q(Btu/h) using a gas-fired heater. The flow rate w(lb/h) and volume V(ft^3) are constant, but the heat loss to the surroundings Q_L(Btu/h) varies with the wind velocity v (ft/s) according to the expressions

$$Q_L = UA(T - T_a)$$
$$U(t) = \overline{U} + bv(t)$$

where $\overline{U}$, A, b, and T_a are constants. Derive the transfer function between exit temperature T and wind velocity v. List any additional assumptions that you make.

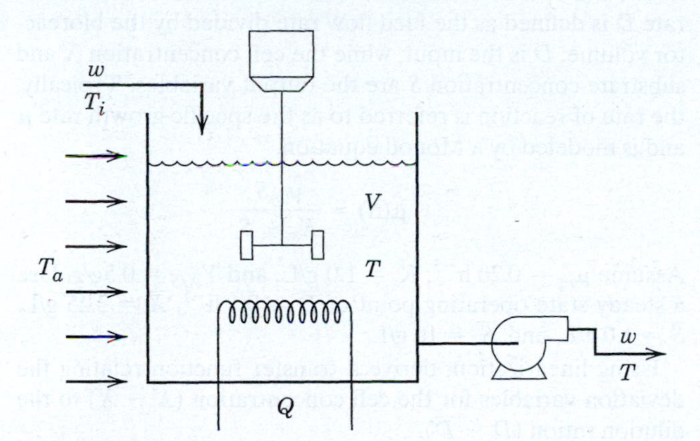

Figure E4.10

4.11 Consider a pressure surge system to reduce the effect of pressure variations at a compressor outlet on the pressure in a compressed gas header. We want to develop a two-tank model and evaluate the form of the resulting transfer function for the two-tank process shown in Fig. E4.11.

(a) Develop a dynamic model that can be used to solve for the gas flow rate, $w_3(t)$, to the header given known pressures at the compressor, $P_c(t)$, and in the header, $P_h(t)$. Determine the degrees of freedom.

Available Information:

(i) The three valves operate linearly with resistances R_1, R_2, R_3. e.g., $w_1 = (P_c - P_1)R_1$

(ii) The tank volumes (V_1 and V_2) are constant.

(iii) The Ideal Gas Law holds.

(iv) The molecular weight of the gas is M.

(v) Operation is isothermal.

(b) Develop the model (linearize, Laplace transform, etc.) just to the point where you can identify the following characteristics of the transfer function

$$\frac{W_3'(s)}{P_c'(s)}$$

(i) What is the order of the denominator?

(ii) What is the order of the numerator?

(iii) Are any integrating elements present?

(iv) Does the gain equal one?

Note: There is no need to derive the actual transfer function. On the other hand, you should justify your answer to each question.

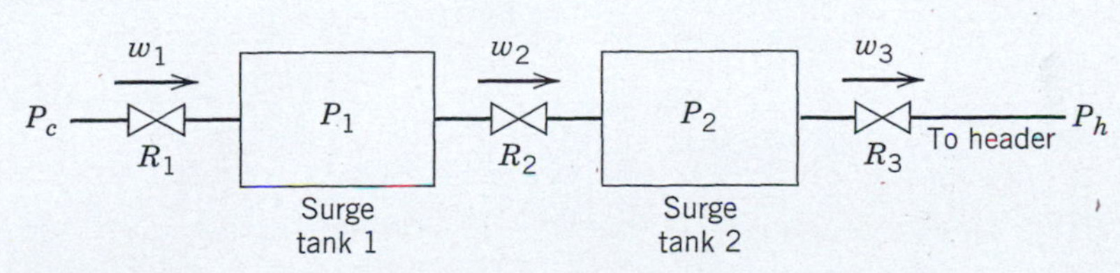

Figure E4.11

4.12 A simple surge tank with a valve on the exit line is illustrated in Figure E4.12. If the exit flow rate is proportional to the square root of the liquid level, an unsteady-state model for the level in the tank is given by

$$A\frac{dh}{dt} = q_i - C_v h^{1\backslash 2}$$

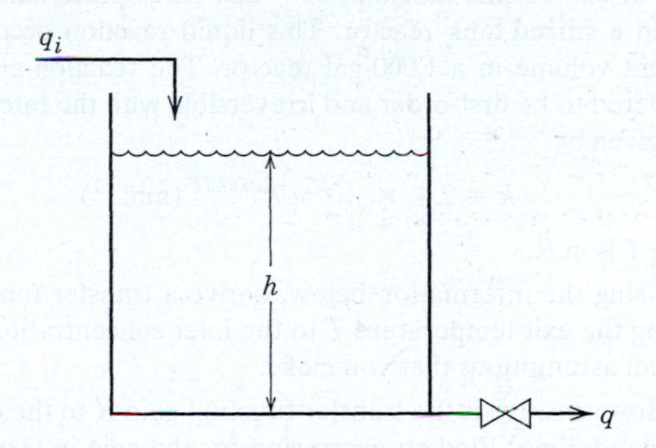

Figure E4.12

As usual, you can assume that the process initially is at steady state:

$$\overline{q_i} = \overline{q} = C_v \overline{h}^{0.5}$$

(a) Find the transfer function $H'(s)/Q_i'(s)$. Put the transfer function in standard gain/time constant form.

(b) Find the transfer function $Q'(s)/Q_i'(s)$ and put it in standard form.

(c) If the algebraic relation for the exit flow rate is linear instead of square root, the level transfer function can be put into a first-order form,

$$\frac{H'(s)}{Q_i'(s)} = \frac{K^*}{\tau^* s + 1}$$

with $K^* = \bar{h}/\bar{q}_i$, $\tau^* = \bar{V}/\bar{q}_i$, and $\bar{V} = A\bar{h}$ is the initial steady-state volume.

When written this way, τ^* is easily interpreted as the liquid residence time at the nominal operating conditions. What are equivalent expressions for K and τ in the part (a) level transfer function, that is, for the square root outflow relation?

4.13 Liquid flow out of a spherical tank discharging through a valve can be described approximately by the following nonlinear differential equation:

$$\frac{dh}{dt} = \frac{1}{\pi(D-h)h}\left(q_i - C_v\sqrt{h}\right)$$

where the variables used are consistent with other liquid level models we have developed.

(a) Derive a linearized model (in deviation variables) of the form

$$\frac{dh'}{dt} = ah' + bq_i'$$

(b) Develop a transfer function relating the liquid level to the volumetric flow of liquid into the tank. Give the final expression in terms of model coefficients, *a* and *b*.

4.14 An exothermic reaction, A → 2B, takes place adiabatically in a stirred-tank reactor. This liquid reaction occurs at constant volume in a 1,000-gal reactor. The reaction can be considered to be first-order and irreversible with the rate constant given by

$$k = 2.4 \times 10^{15} e^{-20{,}000/T}\ (\text{min}^{-1})$$

where T is in R.

(a) Using the information below, derive a transfer function relating the exit temperature T to the inlet concentration c_{Ai}. State all assumptions that you make.

(b) How sensitive is the transfer function gain K to the operating conditions? Find an expression for the gain in terms of $\bar{q}$, $\bar{T}$, and $\bar{c}_{Ai}$ and evaluate the sensitivities (that is, $\partial K/\partial \bar{q}$, etc.)

Available Information

(i) Nominal steady-state conditions are

$$\bar{T} = 150°\text{F},\ \bar{c}_{Ai} = 0.8\ \text{mol/ft}^3$$

$$\bar{q} = 20\ \text{gal/min} = \text{flow in and out of the reactor}$$

(ii) Physical property data for the mixture at the nominal steady state:

$$C = 0.8\frac{\text{Btu}}{\text{lb°F}},\quad \rho = 52\ \text{lb/ft}^3,\quad -\Delta H_R = 500\text{kJ/mol}$$

4.15 A chemostat is a continuous stirred tank bioreactor that can carry out fermentation of a plant cell culture. Its dynamic behavior can be described by the following equations:

$$\dot{X} = \mu(S)X - DX$$

$$\dot{S} = -\mu(S)X/Y_{X/S} - D(S_f - S)$$

X and S are the cell and substrate concentrations, respectively, and S_f is the substrate feed concentration. The dilution rate D is defined as the feed flow rate divided by the bioreactor volume. D is the input, while the cell concentration X and substrate concentration S are the output variables. Typically, the rate of reaction is referred to as the specific growth rate μ and is modeled by a Monod equation,

$$\mu(S) = \frac{\mu_m S}{K_s + S}$$

Assume $\mu_m = 0.20\ \text{h}^{-1}$, $K_s = 1.0$ g/L, and $Y_{X/S} = 0.5$g/g. Use a steady-state operating point of $\bar{D} = 0.1\ \text{h}^{-1}$, $\bar{X} = 2.25$ g/L, $\bar{S} = 1.0$ g/L, and $\bar{S}_f = 10$ g/L.

Using linearization, derive a transfer function relating the deviation variables for the cell concentration $(X - \bar{X})$ to the dilution ration $(D - \bar{D})$.

Chapter 5

Dynamic Behavior of First-Order and Second-Order Processes

CHAPTER CONTENTS

In Chapter 2 we derived dynamic models for several typical processes, and in Chapter 4 we showed how these models can be put into standard transfer function form. Now we investigate how processes respond to typical changes in their environment, that is, to changes in their inputs. We have already seen in Chapter 1 that process inputs fall into two categories:

1. Inputs that can be manipulated to control the process.
2. Inputs that are not manipulated, classified as disturbance variables.

The transfer function representation makes it easy to compare the effects of different inputs. For example, the dynamic model for the constant-flow stirred-tank blending system was derived in Section 4.1. Rewriting Eq. 4-15 in terms of process parameters yields

$$X'(s) = \frac{w_1/w}{\frac{V\rho}{w}s+1}X_1'(s) + \frac{w_2/w}{\frac{V\rho}{w}s+1}X_2'(s) \tag{5-1}$$

The resulting first-order transfer functions,

$$\frac{X'(s)}{X_1'(s)} = \frac{w_1/w}{\frac{V\rho}{w}s+1} \quad \text{and} \quad \frac{X'(s)}{X_2'(s)} = \frac{w_2/w}{\frac{V\rho}{w}s+1} \tag{5-2}$$

relate changes in outlet mass fraction $X'(s)$ to changes in inlet mass fractions $X_1'(s)$ and $X_2'(s)$.

A second advantage of the transfer function representation is that the dynamic behavior of a given process can be generalized easily. Once we analyze the response of the process to an input change, the response of any other process described by the same generic transfer function is then known.

For a general first-order transfer function with output $Y(s)$ and input $U(s)$,

$$Y(s) = \frac{K}{\tau s + 1} U(s) \tag{5-3}$$

a general time-domain solution can be found once the nature of the input change is specified (e.g., step or impulse change). This solution applies to any other process with a first-order transfer function, for example, the liquid surge tanks of Eqs. 4-53 and 4-55. Another benefit of transfer function form (e.g., (5-3)) is that it is not necessary to re-solve the ODE when K, τ, or $U(s)$ changes.

We will exploit this ability to develop general process dynamic formulas as much as possible, concentrating on transfer functions that commonly arise in describing the dynamic behavior of industrial processes. This chapter covers the simplest transfer functions: first-order processes, integrating units, and second-order processes. In Chapter 6 the responses of more complicated transfer functions will be discussed. To keep the results as general as possible, we now consider several standard process inputs that are used to characterize the behavior of many actual processes.

5.1 STANDARD PROCESS INPUTS

We have previously discussed outputs and inputs for process models; we now introduce more precise working definitions. The word *output* generally refers to a controlled variable in a process, a process variable to be maintained at a desired value (set point). For example, the output from the stirred blending tank just discussed is the mass fraction x of the effluent stream. The word *input* refers to any variable that influences the process output, such as the flow rate of the stream flowing into the stirred blending tank. The characteristic feature of all inputs, whether they are disturbance variables or manipulated variables, is that they influence the output variables that we wish to control.

In analyzing process dynamics and in designing control systems, it is important to know how the process outputs will respond to changes in the process inputs. There are six important types of input changes used in industrial practice for the purposes of modeling and control.

1. *Step Input.* One characteristic of industrial processes is that they can be subjected to sudden and sustained input changes; for example, a reactor feedstock may be changed quickly from one supply to another, causing a corresponding change in important input variables such as feed concentration and feed temperature. Such a change can be approximated by the step change

$$u_S(t) = \begin{cases} 0 & t < 0 \\ M & t \geq 0 \end{cases} \tag{5-4}$$

where *zero time*, as noted earlier, is taken to be the time at which the sudden change of magnitude M occurs. Note that $u_S(t)$ is defined as a deviation variable—that is, the change from the current steady state. Suppose the heat input to a stirred-tank heating unit suddenly is changed from 8,000 to 10,000 kcal/h, by changing the electrical heater input. Then

$$Q(t) = 8000 + 2000\, S(t) \tag{5-5a}$$

$$Q'(t) = 2000\, S(t) \tag{5-5b}$$

where $S(t)$ is the unit step function. The Laplace transform of a step of magnitude M (cf. Eq. 3-4) is

$$u_S(s) = \frac{M}{s} \tag{5-6}$$

2. *Ramp Input.* Industrial processes often are subjected to inputs that *drift*—that is, they gradually change upward or downward for some period of time with a roughly constant slope. For example, ambient conditions (air temperature and relative humidity) can change slowly during the day so that the plant cooling tower temperature also changes slowly. Set points are sometimes ramped from one value to another rather than making a step change. We can approximate such a change in an input variable by means of the ramp function:

$$u_R(t) = \begin{cases} 0 & t < 0 \\ at & t \geq 0 \end{cases} \tag{5-7}$$

where $u_R(t)$ is a deviation variable. The Laplace transform of a ramp input with a slope of 1 is given in Table 3.1 as $1/s^2$. Hence, transforming Eq. 5-7 yields

$$u_R(s) = \frac{a}{s^2} \tag{5-8}$$

3. *Rectangular Pulse.* Processes sometimes are subjected to a sudden step change that then returns to its original value. Suppose that a feed to a reactor is shut off for a certain period of time or a natural-gas-fired furnace experiences a brief interruption in fuel gas. We might approximate this type of input change as a rectangular pulse:

$$u_{RP}(t) = \begin{cases} 0 & t < 0 \\ h & 0 \leq t < t_w \\ 0 & t \geq t_w \end{cases} \tag{5-9}$$

where the pulse width t_w can range from very short (approximation to an impulse) to very long. An alternative way of expressing (5-9) utilizes the shifted unit step input $S(t - t_w)$, which is equal to unity for $t \geq t_w$ and equal to zero for $t < t_w$ (cf. Eq. 3-23). Equation 5-9

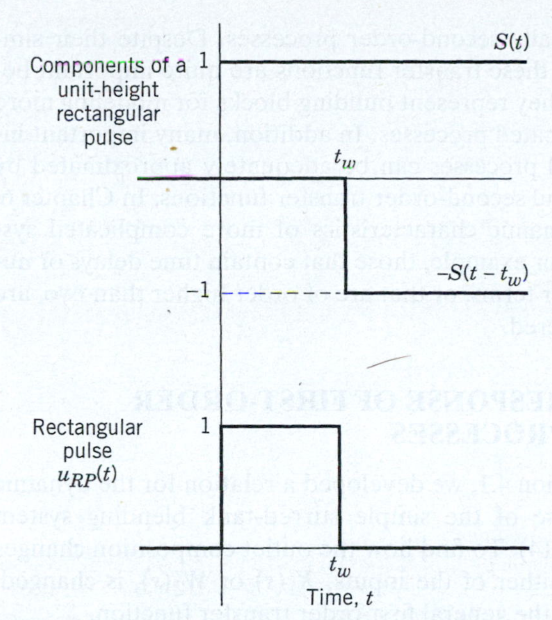

Figure 5.1 How two step inputs can be combined to form a rectangular pulse.

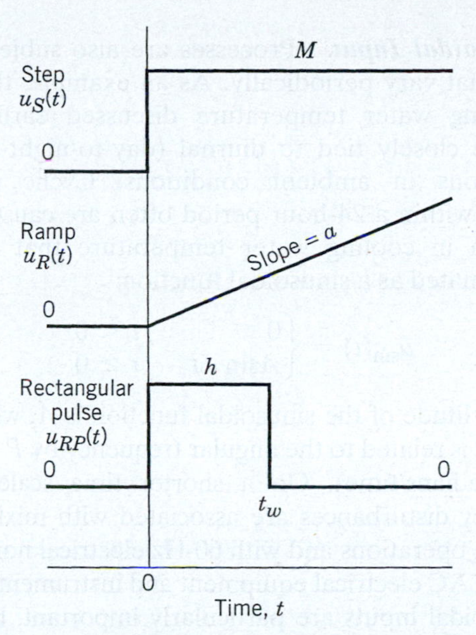

Figure 5.2 Three important examples of deterministic inputs.

is depicted in Fig. 5.1 as the sum of two steps, one step of magnitude equal to 1 occurring at $t = 0$ combined with a second step of magnitude equal to -1 occurring at $t = t_w$. Mathematically, this combination can be expressed as

$$u_{RP}(t) = h[S(t) - S(t - t_w)]$$

Because the Laplace transform is only defined for $t \geq 0$, this expression can be simplified to

$$u_{RP}(t) = h[1 - S(t - t_w)] \qquad t \geq 0 \qquad (5\text{-}10)$$

which can be Laplace transformed to yield

$$u_{RP}(s) = \frac{h}{s}(1 - e^{-t_w s}) \qquad (5\text{-}11)$$

which is the same result given in (3-22).

The three important inputs discussed above—step, ramp, rectangular pulse—are depicted in Fig. 5.2. Note that many types of inputs can be represented as combinations of step and ramp inputs. For example, a unit height (isosceles) triangular pulse of width t_w can be constructed from three ramp inputs, as shown in Fig. 5.3. In this case, we write a single expression for the triangular pulse function

$$u_{TP}(t) = \frac{2}{t_w}[tS(t) - 2(t - t_w/2)S(t - t_w/2) + (t - t_w)S(t - t_w)]$$

$$= \frac{2}{t_w}[t - 2(t - t_w/2)S(t - t_w/2) + (t - t_w)S(t - t_w)] \qquad t \geq 0 \qquad (5\text{-}12)$$

where the second relation is valid only for $t \geq 0$. Equation 5-12 can be Laplace transformed term-by-term to obtain

$$u_{TP}(s) = \frac{2}{t_w}\left(\frac{1 - 2e^{-t_w s/2} + e^{-t_w s}}{s^2}\right) \qquad (5\text{-}13)$$

Note that Eq. 5-12 written without the unit step function multipliers is incorrect.

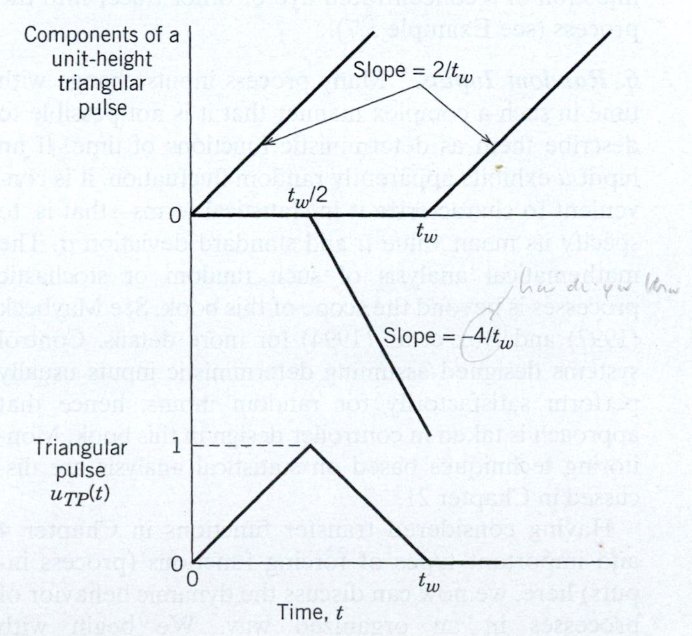

Figure 5.3 How three ramp inputs can be combined to form a triangular pulse.

4. *Sinusoidal Input.* Processes are also subjected to inputs that vary periodically. As an example, the drift in cooling water temperature discussed earlier can often be closely tied to diurnal (day-to-night-to-day) fluctuations in ambient conditions. Cyclic process changes within a 24-hour period often are caused by a variation in cooling water temperature that can be approximated as a sinusoidal function:

$$u_{\sin}(t) = \begin{cases} 0 & t < 0 \\ A\sin \omega t & t \geq 0 \end{cases} \tag{5-14}$$

The amplitude of the sinusoidal function is A, while the period P is related to the angular frequency by $P = 2\pi/\omega$ (ω in radians/time). On a shorter time scale, high-frequency disturbances are associated with mixing and pumping operations and with 60-Hz electrical noise arising from AC electrical equipment and instrumentation.

Sinusoidal inputs are particularly important, because they play a central role in frequency response analysis, which is discussed in Chapter 14. The Laplace transform of the sine function in Eq. 5-14 can be obtained by multiplying entry 14 in Table 3.1 by the amplitude A to obtain

$$U_{\sin}(s) = \frac{A\omega}{s^2 + \omega^2} \tag{5-15}$$

5. *Impulse Input.* The unit impulse function discussed in Chapter 3 has the simplest Laplace transform, $U(s) = 1$ (Eq. 3-24). However, true impulse functions are not encountered in normal plant operations. To obtain an impulse input, it is necessary to inject a finite amount of energy or material into a process in an infinitesimal length of time, which is not possible. However, this type of input can be approximated through the injection of a concentrated dye or other tracer into the process (see Example 3.7).

6. *Random Inputs.* Many process inputs change with time in such a complex manner that it is not possible to describe them as deterministic functions of time. If an input u exhibits apparently random fluctuation, it is convenient to characterize it in statistical terms—that is, to specify its mean value $\overline{u}$ and standard deviation σ. The mathematical analysis of such random or stochastic processes is beyond the scope of this book. See Maybeck (1997) and Box et al. (1994) for more details. Control systems designed assuming deterministic inputs usually perform satisfactorily for random inputs; hence that approach is taken in controller design in this book. Monitoring techniques based on statistical analysis are discussed in Chapter 21.

Having considered transfer functions in Chapter 4 and important types of forcing functions (process inputs) here, we now can discuss the dynamic behavior of processes in an organized way. We begin with processes that can be modeled as first-order transfer functions. Then integrating elements are considered and finally second-order processes. Despite their simplicity, these transfer functions are quite important because they represent building blocks for modeling more complicated processes. In addition, many important industrial processes can be adequately approximated by first- and second-order transfer functions. In Chapter 6, the dynamic characteristics of more complicated systems, for example, those that contain time delays or numerator terms, or that are of order higher than two, are considered.

5.2 RESPONSE OF FIRST-ORDER PROCESSES

In Section 4.1, we developed a relation for the dynamic response of the simple stirred-tank blending system (Eq. 4-14). To find how the outlet composition changes when either of the inputs, $X_1'(s)$ or $W_2'(s)$, is changed, we use the general first-order transfer function,

$$\frac{Y(s)}{U(s)} = \frac{K}{\tau s + 1} \tag{5-16}$$

where K is the process gain and τ is the time constant. Now we investigate some particular forms of input $U(s)$, deriving expressions for $Y(s)$ and the resulting response, $y(t)$.

5.2.1 Step Response

For a step input of magnitude M, $U(s) = M/s$, and (5-16) becomes

$$Y(s) = \frac{KM}{s(\tau s + 1)} \tag{5-17}$$

Using Table 3.1, the time-domain response is

$$y(t) = KM(1 - e^{-t/\tau}) \tag{5-18}$$

The plot of this equation in Fig. 5.4 shows that a first-order process does not respond instantaneously to a

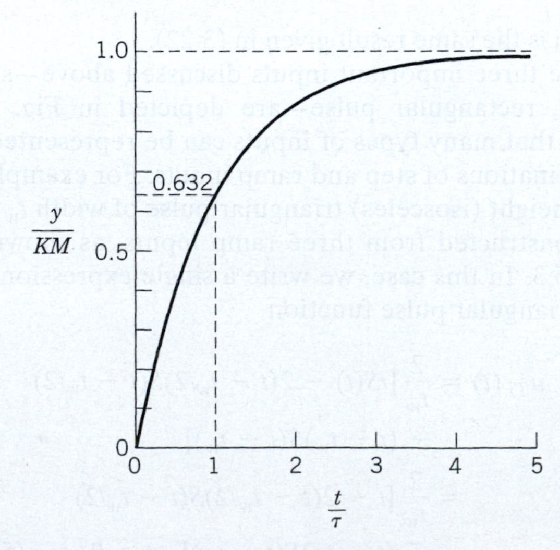

Figure 5.4 Step response of a first-order process.

Table 5.1 Response of a First-Order Process to a Step Input

t	$y(t)/KM = 1 - e^{-t/\tau}$
0	0
τ	0.6321
2τ	0.8647
3τ	0.9502
4τ	0.9817
5τ	0.9933

sudden change in its input. In fact, after a time interval equal to the process time constant ($t = \tau$), the process response is still only 63.2% complete. Theoretically, the process output never reaches the new steady-state value except as $t \rightarrow \infty$; it does approximate the final steady-state value when $t \approx 5\tau$, as shown in Table 5.1. Notice that Fig. 5.4 has been drawn in dimensionless or normalized form, with time divided by the process time constant and the output change divided by the product of the process gain and magnitude of the input change. Now we consider a more specific example.

EXAMPLE 5.1

A stirred-tank heating system described by Eq. 4-37 is used to preheat a reactant containing a suspended solid catalyst at a constant flow rate of 1000 kg/h. The volume in the tank is 2 m^3, and the density and specific heat of the suspended mixture are, respectively, 900 kg/m^3 and 1 cal/g °C. The process initially is operating with inlet and outlet temperatures of 100 and 130 °C. The following questions concerning process operations are posed:

(a) What is the heater input at the initial steady state and the values of K and τ?

(b) If the heater input is suddenly increased by +30%, how long will it take for the tank temperature to achieve 99% of the final temperature change?

(c) Assume the tank is at its initial steady state. If the inlet temperature is increased suddenly from 100 to 120 °C, how long will it take before the outlet temperature changes from 130 to 135 °C?

SOLUTION

(a) First calculate the process steady-state operating conditions and then the gain and time constant in Eq. 4-37. Assuming no heat losses, the energy input from the heater at the initial steady state is equal to the enthalpy increase between the inlet and outlet streams. Thus, the steady-state energy balance provides the answer:

$$\overline{Q} = \overline{w}C(\overline{T} - \overline{T}_i)$$
$$= \left(10^6 \frac{g}{h}\right)\left(\frac{1 \text{ cal}}{g\ °C}\right)(130\ °C - 100\ °C)$$
$$= 3 \times 10^7 \text{ cal/h}$$

Using Eq. 4-37, the gain and time constants can be determined (the disturbance gain is unity):

$$K = \frac{1}{wC} = \frac{1}{\left(10^6 \frac{g}{h}\right)\left(\frac{1 \text{ cal}}{g\ °C}\right)}$$
$$= 10^{-6} \frac{°C}{\text{cal/h}}$$
$$\tau = \frac{V\rho}{w} = \frac{(2 \text{ m}^3)\left(9 \times 10^5 \frac{g}{m^3}\right)}{10^6 \frac{g}{h}} = 1.8 \text{ h}$$

(b) According to Table 5.1, the time required to attain the 99% response following a step change of any magnitude in heater input will be 5 process time constants—that is, 9 h. The steady-state change in temperature due to a change of +30% in Q (9×10^6 cal/h) can be found from the Final Value Theorem, Eq. 3-94:

$$T'(t \rightarrow \infty) = \lim_{s \rightarrow 0} s\left(\frac{10^{-6}}{1.8s + 1}\frac{9 \times 10^6}{s}\right) = 9\ °C$$

Note that we have calculated the outlet temperature *change* as a result of the input change; hence, the outlet temperature at the final steady state will be 130 °C + 9 °C = 139 °C. However, use of the Final Value Theorem is an unnecessary formality when a process transfer function is written in the standard form with gain and time constants. The input change need only be multiplied by the process gain to obtain the ultimate change in the process output, assuming that the final value does in fact exist and is finite. In this case $T'(t \rightarrow \infty) = K\,\Delta Q = (10^{-6}\ °C/\text{cal} \cdot \text{h})(9 \times 10^6 \text{ cal/h}) = 9\ °C$.

(c) Because the gain of the appropriate transfer function (that relates T' to T_i') is one, an input temperature change of 20 °C causes an outlet temperature of 20 °C. The time required for the output to change by 5 °C, or 25% of the ultimate steady-state change, can be estimated from Fig. 5.4 as $t/\tau = 0.3$ or $t = 0.54$ h. Equation 5-18 furnishes a more accurate way to calculate this value:

$$\frac{y(t)}{KM} = 1 - e^{-t/\tau}$$
$$\frac{5\ °C}{(1)(20\ °C)} = 1 - e^{-t/\tau}$$
$$e^{-t/\tau} = 0.75$$
$$-\frac{t}{\tau} = \ln 0.75 = -0.288$$
$$t = 0.52 \text{ h}$$

5.2.2 Ramp Response

We now evaluate how a first-order system responds to the ramp input, $U(s) = a/s^2$ of Eq. 5-8. Performing a partial fraction expansion yields

$$Y(s) = \frac{Ka}{(\tau s + 1)s^2} = \frac{\alpha_1}{\tau s + 1} + \frac{\alpha_2}{s} + \frac{\alpha_3}{s^2} \quad (5\text{-}19)$$

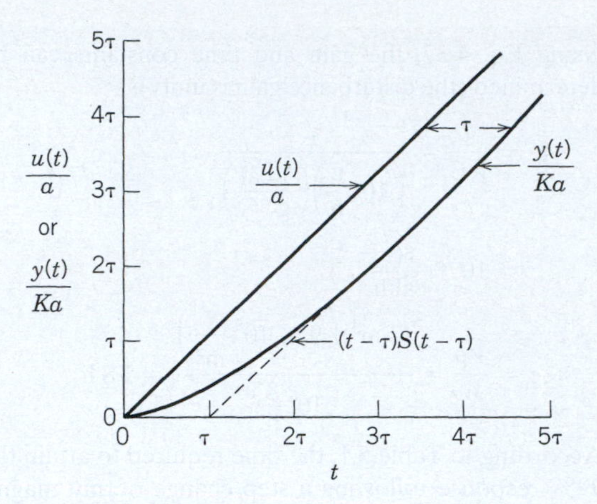

Figure 5.5 Ramp response of a first-order process (comparison of input and output).

The Heaviside expansion (Chapter 3) gives

$$Y(s) = \frac{Ka\tau^2}{\tau s + 1} - \frac{Ka\tau}{s} + \frac{Ka}{s^2} \tag{5-20}$$

Using Table 3.1

$$y(t) = Ka\tau(e^{-t/\tau} - 1) + Kat \tag{5-21}$$

The above expression has the interesting property that for large values of time ($t >> \tau$)

$$y(t) \approx Ka(t - \tau) \tag{5-22}$$

Equation 5-22 implies that after an initial transient period, the ramp input yields a ramp output with slope equal to Ka, but shifted in time by the process time constant τ (see Fig. 5.5). An unbounded ramp input will ultimately cause some process component to saturate, so the duration of the ramp input ordinarily is limited. A process input frequently will be *ramped* from one value to another in a fixed amount of time so as to avoid the sudden change associated with a step change. Ramp inputs of this type are particularly useful during the start-up of a continuous process or in operating a batch process.

5.2.3 Sinusoidal Response

As a final example of the response of first-order processes, consider a sinusoidal input $u_{sin}(t) = A \sin \omega t$, with transform given by Eq. (5-15):

$$y(s) = \frac{KA\omega}{(\tau_s + 1)(s^2 + \omega^2)} \tag{5-23}$$

$$= \frac{KA}{\omega^2\tau^2 + 1}\left(\frac{\omega\tau^2}{\tau s + 1} - \frac{s\omega\tau}{s^2 + \omega^2} + \frac{\omega}{s^2 + \omega^2}\right) \tag{5-24}$$

Inversion gives

$$y(t) = \frac{KA}{\omega^2\tau^2 + 1}(\omega\tau e^{-t/\tau} - \omega\tau \cos \omega t + \sin \omega t) \tag{5-25}$$

or, by using trigonometric identities,

$$y(t) = \frac{KA\omega\tau}{\omega^2\tau^2 + 1}e^{-t/\tau} + \frac{KA}{\sqrt{\omega^2\tau^2 + 1}}\sin(\omega t + \phi) \tag{5-26}$$

where

$$\phi = -\tan^{-1}(\omega\tau) \tag{5-27}$$

Notice that in both (5-25) and (5-26) the exponential term goes to zero as $t \to \infty$, leaving a pure sinusoidal response. This property is exploited in Chapter 14 for frequency response analysis.

Students often have difficulty imagining how a real process variable might change sinusoidally. How can the flow rate into a reactor be negative as well as positive? Remember that we have defined the input u and output y in these relations to be deviation variables. An actual input might be

$$q(t) = 0.4\frac{\text{m}^3}{\text{s}} + \left(0.1\frac{m^3}{s}\right)\sin \omega t \tag{5-28}$$

where the amplitude of the deviation input signal A is 0.1 m^3/s. After a long period of time, the output response (5-26) also will be a sinusoidal deviation, similar to that given in Eq. 5-28.

EXAMPLE 5.2

A liquid surge tank similar to the one described by Eq. 4-50 has the transfer function form of Eq. 4-53:

$$\frac{H'(s)}{Q_i'(s)} = \frac{10}{50s + 1}$$

where h is the tank level (m), q_i is the flow rate (m^3/s), the gain has units m/m^3/s, or s/m^2, and the time constant has units of seconds. The system is operating at steady state with $\bar{q} = 0.4$ m^3/s and $\bar{h} = 4$ m when a sinusoidal perturbation in inlet flow rate begins with amplitude = 0.1 m^3/s and a cyclic frequency of 0.002 cycles/s. What are the maximum and minimum values of the tank level after the flow rate disturbance has occurred for 6 min or more? What are the largest level perturbations expected as a result of sinusoidal variations in flow rate with this amplitude? What is the effect of high-frequency variations, say, 0.2 cycles/s?

SOLUTION

Note that the actual input signal $q(t)$ is given by Eq. 5-28, but only the amplitude of the input deviation (0.1 m^3/s) is required. From Eq. 5-26 the value of the exponential term 6 min after the start of sinusoidal forcing is $e^{-360/50} = e^{-7.2} < 10^{-3}$. Thus, the effect of the exponential transient term is less than 0.1% of the disturbance amplitude and can be safely neglected. Consequently, from Eq. 5-26 the amplitude of the output (level) perturbation is

$$\frac{KA}{\sqrt{\omega^2\tau^2 + 1}}$$

where A is the input amplitude and ω is the frequency (in radians) = (2π) (cyclic frequency) = (6.28)(0.002) radians/s. The amplitude of the perturbation in the liquid level is

$$\frac{10(\text{s/m}^2)(0.1\ \text{m}^3/\text{s})}{\sqrt{[(6.28\ \text{rad/cycles})(0.002\ \text{cycles/s})(50\ \text{s})]^2 + 1}}$$

or 0.85 m. Hence, the actual tank level varies from a minimum of 3.15 m to a maximum of 4.85 m.

The largest deviations that can result from sinusoidal variations of amplitude 0.1 m^3/s occur for $\omega \to 0$—that is, for very low frequencies. In this case, the deviations would be $\pm KA = \pm(10\ \text{s/m}^2)\ (0.1\ \text{m}^3/\text{s}) = \pm 1$ m. Hence, the minimum and maximum values of level would be 3 and 5 m, respectively.

For high-frequency variations (0.2 cycles/s), the amplitude will approach zero. This occurs because the rapid variations of flow rate are averaged in the tank when the residence time is sufficiently large, giving a relatively constant level.

5.3 RESPONSE OF INTEGRATING PROCESSES

In Section 2.4 we briefly considered a liquid-level system with a pump attached to the outflow line. Assuming that the outflow rate q can be set at any time by the speed of the pump, Eq. 2-54 becomes

$$A\frac{dh(t)}{dt} = q_i(t) - q(t) \quad (5\text{-}29)$$

Suppose at $t = 0$, the process is at the nominal steady state where $q_i = \bar{q}$ and $h = \bar{h}$. After subtracting the steady-state equation ($0 = \bar{q}_i - \bar{q}$) from (5-29) and noting that $dh(t)/dt = dh'(t)/dt$,

$$A\frac{dh'(t)}{dt} = q_i'(t) - q'(t) \quad (5\text{-}30)$$

where the primed deviation variables are all zero at $t = 0$. Taking Laplace transforms

$$sAH'(s) = Q_i'(s) - Q'(s) \quad (5\text{-}31)$$

and rearranging gives

$$H'(s) = \frac{1}{As}[Q_i'(s) - Q'(s)] \quad (5\text{-}32)$$

Both transfer functions, $H'(s)/Q_i'(s) = 1/As$ and $H'(s)/Q'(s) = -1/As$, represent *integrating models*, characterized by the term $1/s$. The integral of (5-29) is

$$\int_{\bar{h}}^{h} dh^* = \frac{1}{A}\int_0^t [q_i(t^*) - q(t^*)]\,dt^*$$

or

$$h(t) - \bar{h} = \frac{1}{A}\int_0^t [q_i(t^*) - q(t^*)]dt^* \quad (5\text{-}33)$$

hence the term *integrating process*. Integrating processes do not have a steady-state gain in the usual sense. For such a process operating at steady state, any positive step change in q_i (increase in q_i above q) will cause the tank level to increase linearly with time in proportion to the difference, $q_i(t) - q(t)$, while a positive step change in q will cause the tank level to decrease linearly. Thus, no new steady state will be attained, unless the tank overflows or empties. In contrast, a tank with an exit line valve rather than a pump will reach a steady state when the outflow rate becomes equal to the inflow rate. This process is described by a first-order transfer function rather than an integrator (cf. Example 4.6).

EXAMPLE 5.3

A vented cylindrical tank is used for storage between a tank car unloading facility and a continuous reactor that uses the tank car contents as feedstock (Fig. 5.6). The reactor feed exits the storage tank at a constant flow rate of 0.02 m^3/s. During some periods of operation, feedstock is simultaneously transferred from the tank car to the feed tank and from the tank to the reactor. The operators have to be particularly careful not to let the feed tank overflow or empty. The feed tank is 5 m high (distance to the vent) and has an internal cross-sectional area of 4 m^2.

(a) Suppose after a long period of operation, the tank level is 2 m at the time the tank car empties. How long can the reactor be operated before the feed tank is depleted?

(b) Another tank car is moved into place and connected to the tank, while flow continues into the reactor at 0.02 m^3/s. If flow is introduced into the feed tank just as the tank level reaches 1 m, how long can the transfer pump from the tank car be operated? Assume that it pumps at a constant rate of 0.1 m^3/s when switched on.

SOLUTION

(a) For such a system, there is no unique steady-state level corresponding to a particular value of input and output flow rate. Suppose the initial level is $h = 2$ m and the constant flow rate from the feed pump to the reactor, $q = 0.02\ \text{m}^3/\text{s}$, is the basis for defining deviation variables for h, q, and q_i. Then

$$\bar{h} = 2\ \text{m}$$
$$\bar{q}_i = \bar{q} = 0.02\ \text{m}^3/\text{s}$$

and, from Eq. 5-32, the process model for the tank is

$$H'(s) = \frac{1}{4s}[Q_i'(s) - Q'(s)]$$

At the time the tank car empties

$$q_i = 0 \Rightarrow q_i' = -0.02 \Rightarrow Q_i'(s) = -\frac{0.02}{s}$$
$$q = 0.02 \Rightarrow q' = 0 \Rightarrow Q'(s) = 0$$

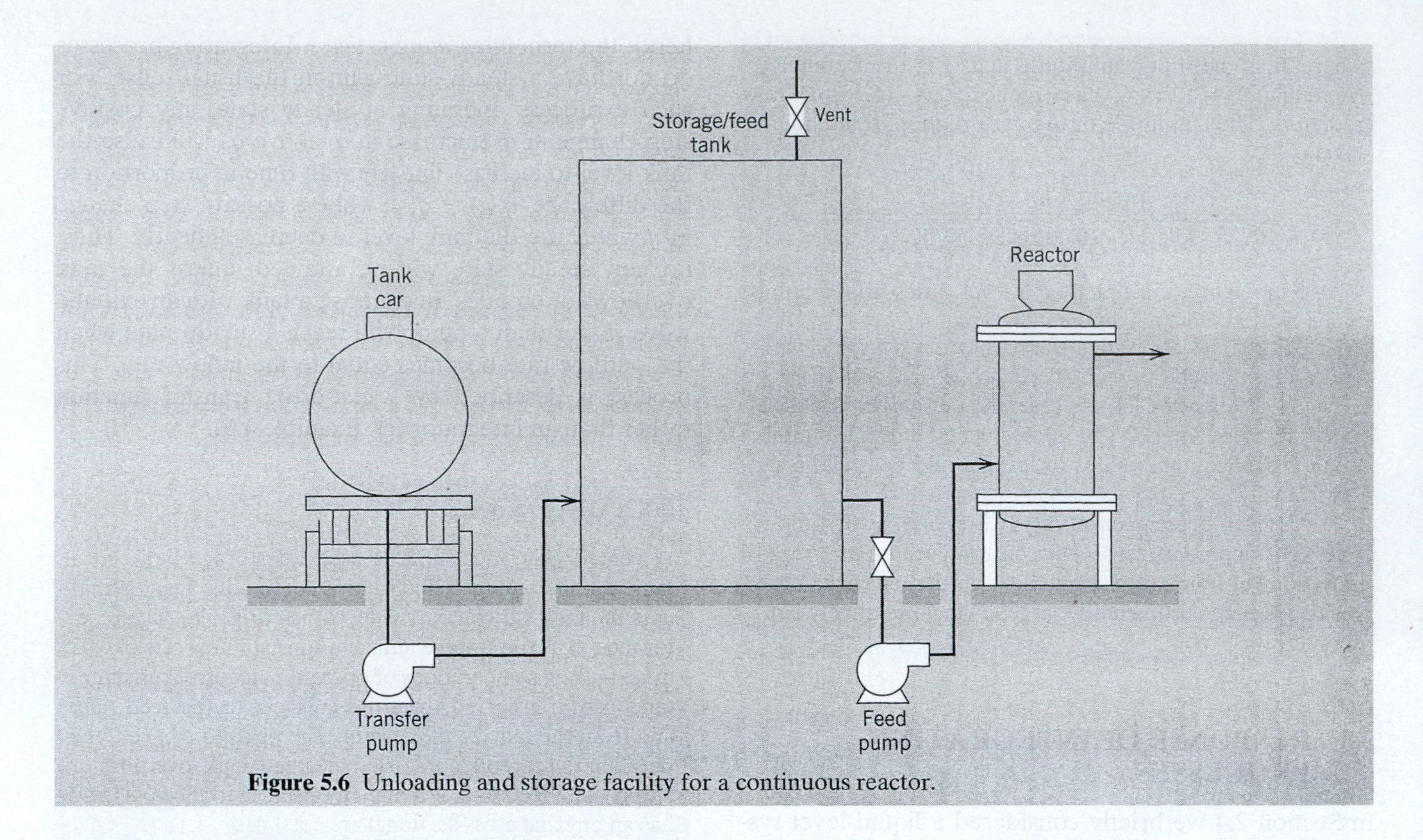

Figure 5.6 Unloading and storage facility for a continuous reactor.

Thus

$$H'(s) = \frac{1}{4s}\left(-\frac{0.02}{s} - 0\right) = -\frac{0.005}{s^2}$$

Inversion to the time domain gives $h'(t) = -0.005t$ and $h(t) = 2 - 0.005t$. The length of time for $h(t)$ to go to zero is $t = 2/0.005 = 400$ s.

(b) For the tank-filling period and using the same basis for deviation variables,

$$q_i = 0.1 \Rightarrow q_i' = +0.08 \Rightarrow Q_i'(s) = \frac{0.08}{s}$$
$$q = 0.02 \Rightarrow q' = 0 \Rightarrow Q'(s) = 0$$

Consequently, from (5-32), the tank model is

$$H'(s) = \frac{1}{4s}\left(\frac{0.08}{s} - 0\right) = \frac{0.02}{s^2}$$

Inversion to the time domain yields $h(t) = 1 + 0.02t$. Thus, the transfer pump can be operated for 200 s until $h(t) = 5$ m, when the tank would overflow. Note that this time (as well as the time to empty the tank in (a)) can be calculated without using Laplace transforms, simply by using the constant rate of inflow (or outflow) and the tank volume.

This example illustrates that integrating process units do not reach a new steady state when subjected to step changes in inputs, in contrast to first-order processes (cf. Eq. 5-18). Integrating systems represent an example of *non-self-regulating* processes. Closed pulse inputs, where the initial and final values of the input are equal, do lead to a new steady state. For example, the rectangular pulse with height h given in Eq. 5-9 has the Laplace transform given in Eq. 5-10. The response of an integrating process with transfer function

$$\frac{Y(s)}{U(s)} = \frac{K}{s} \tag{5-34}$$

to a rectangular pulse input is

$$Y(s) = \frac{Kh(1 - e^{-t_w s})}{s^2} = Kh\left(\frac{1}{s^2} - \frac{e^{-t_w s}}{s^2}\right) \tag{5-35}$$

There are two regions for the solution of (5-35), depending on the value of t compared to the pulse width t_w. For $0 \le t < t_w$, the second term in the parentheses of (5-35) is 0, hence

$$y(t) = Kht \tag{5-36}$$

corresponding to a linear increase with respect to time. For $t \ge t_w$, taking the inverse Laplace transform of (5-35) gives

$$y(t) = Kh\,[t - (t - t_w)] = Kht_w \tag{5-37}$$

which is a constant value. Combining the solutions yields

$$y(t) = \begin{cases} Kht & t < t_w \\ Kht_w & t \ge t_w \end{cases} \tag{5-38}$$

Equation 5-38 shows that the change in y at any time is proportional to the area under the input pulse curve (the integral), an intuitive result.

5.4 RESPONSE OF SECOND-ORDER PROCESSES

As noted in Chapter 4, a second-order transfer function can arise physically whenever two first-order processes are connected in series. For example, two stirred-tank blending processes, each with a first-order transfer function relating inlet to outlet mass fraction, might be physically connected so that the outflow stream of the first tank is used as the inflow stream of the second tank. Figure 5.7 illustrates the signal flow relation for such a process. Here

$$G(s) = \frac{Y(s)}{U(s)} = \frac{K_1K_2}{(\tau_1 s + 1)(\tau_2 s + 1)} = \frac{K}{(\tau_1 s + 1)(\tau_2 s + 1)} \tag{5-39}$$

where $K = K_1K_2$. Alternatively, a second-order process transfer function will arise upon transforming either a second-order differential equation process model such as the one given in Eq. 4-29 for the electrically heated stirred-tank unit, or two coupled first-order differential equations, such as for the CSTR (cf. Eqs. 4-84 and 4-85). In this chapter we consider the case where the second-order transfer function has the standard form

$$G(s) = \frac{K}{\tau^2 s^2 + 2\zeta\tau s + 1} \tag{5-40}$$

We defer discussion of the more general cases with a time-delay term in the numerator or other numerator dynamics present until Chapter 6.

In Eq. 5-40, K and τ have the same importance as for a first-order transfer function. K is the process gain, and τ determines the speed of response (or, equivalently, the response time) of the system. The damping coefficient ζ (zeta) is dimensionless. It provides a measure of the amount of *damping* in the system—that is, the degree of oscillation in a process response after a perturbation. Small values of ζ imply little damping and a large amount of oscillation, as, for example, in an automobile suspension system with ineffective shock absorbers. Hitting a bump causes such a vehicle to bounce up and down dangerously. In some textbooks, Eq. 5-40 is written in terms of $\omega_n = 1/\tau$, the undamped natural frequency of the system. This name arises because it represents the frequency of oscillation of the system when there is no damping ($\zeta = 0$).

There are three important classes of second-order systems as shown in Table 5.2 The case where $\zeta < 0$ is omitted here because it corresponds to an unstable second-order system that would have an unbounded response to any input (effects of instability are covered in Chapter 11). The overdamped and critically damped forms of the second-order transfer function most often appear when two first-order systems occur in series (see Fig. 5.7). The transfer functions given by Eqs. 5-39 and 5-40 differ only in the form of the denominators. Equating the denominators yields the relation between the two alternative forms for the *overdamped* second-order system:

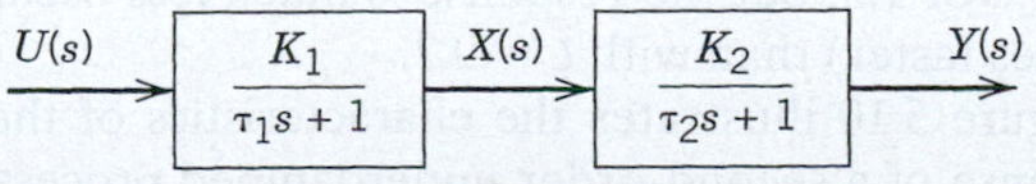

$$\tau^2 s^2 + 2\zeta\tau s + 1 = (\tau_1 s + 1)(\tau_2 s + 1) \tag{5-41}$$

Expanding the right side of (5-41) and equating coefficients of the s terms,

$$\tau^2 = \tau_1\tau_2$$
$$2\zeta\tau = \tau_1 + \tau_2$$

from which we obtain

$$\tau = \sqrt{\tau_1\tau_2} \tag{5-42}$$

$$\zeta = \frac{\tau_1 + \tau_2}{2\sqrt{\tau_1\tau_2}} \tag{5-43}$$

Alternatively, the left side of (5-41) can be factored:

$$\tau^2 s^2 + 2\zeta\tau s + 1 = \left(\frac{\tau s}{\zeta - \sqrt{\zeta^2 - 1}} + 1\right) \times \left(\frac{\tau s}{\zeta + \sqrt{\zeta^2 - 1}} + 1\right) \tag{5-44}$$

from which expressions for τ_1 and τ_2 are obtained:

$$\tau_1 = \frac{\tau}{\zeta - \sqrt{\zeta^2 - 1}} \quad (\zeta \geq 1) \tag{5-45}$$

$$\tau_2 = \frac{\tau}{\zeta + \sqrt{\zeta^2 - 1}} \quad (\zeta \geq 1) \tag{5-46}$$

Table 5.2 The Three Classes of Second-Order Transfer Functions

Damping Coefficient	Characterization of Response	Roots of Characteristic Equation[1]
$\zeta > 1$	Overdamped	Real and unequal
$\zeta = 1$	Critically damped	Real and equal
$0 < \zeta < 1$	Underdamped	Complex conjugates (of the form $a + jb$ and $a - jb$)

[1]This equation is $\tau^2 s^2 + 2\zeta\tau s + 1 = 0$.

$U(s)$ → $\frac{K_1}{\tau_1 s + 1}$ → $X(s)$ → $\frac{K_2}{\tau_2 s + 1}$ → $Y(s)$

Figure 5.7 Two first-order systems in series yield an overall second-order system.

EXAMPLE 5.4

An overdamped system consists of two first-order processes operating in series ($\tau_1 = 4$, $\tau_2 = 1$). Find the equivalent values of τ and ζ for this system.

SOLUTION

From Eqs. 5-42 and 5-43,

$$\tau = \sqrt{(4)(1)} = 2$$

$$\zeta = \frac{4+1}{(2)(2)} = 1.25$$

Equations 5-45 and 5-46 provide a check on these results:

$$\tau_1 = \frac{2}{1.25 - \sqrt{(1.25)^2 - 1}} = \frac{2}{1.25 - 0.75} = 4$$

$$\tau_2 = \frac{2}{1.25 + \sqrt{(1.25)^2 - 1}} = \frac{2}{1.25 + 0.75} = 1$$

The underdamped form of (5-40) can arise from some mechanical systems, from flow or other processes such as a pneumatic (air) instrument line with too little line capacity, or from a mercury manometer, where inertial effects are important.

For process control problems the underdamped form is frequently encountered in investigating the properties of processes under feedback control. Control systems are sometimes designed so that the controlled process responds in a manner similar to that of an underdamped second-order system (see Chapter 12). Next we develop the relation for the step response of all three classes of second-order processes.

5.4.1 Step Response

For the step input ($U(s) = M/s$) to a process described by (5-40),

$$Y(s) = \frac{KM}{s(\tau^2 s^2 + 2\zeta\tau s + 1)} \tag{5-47}$$

After inverting to the time domain, the responses can be categorized into three classes:

Overdamped ($\zeta > 1$)
If the denominator of Eq. 5-47 can be factored using Eqs. 5-45 and 5-46, then the response can be written

$$y(t) = KM\left(1 - \frac{\tau_1 e^{-t/\tau_1} - \tau_2 e^{-t/\tau_2}}{\tau_1 - \tau_2}\right) \tag{5-48}$$

The response can also be written in the equivalent form

$$y(t) = KM\left\{1 - e^{-\zeta t/\tau}\left[\cosh\left(\frac{\sqrt{\zeta^2 - 1}}{\tau}t\right) + \frac{\zeta}{\sqrt{\zeta^2 - 1}}\sinh\left(\frac{\sqrt{\zeta^2-1}}{\tau}t\right)\right]\right\} \tag{5-49}$$

Critically Damped ($\zeta = 1$)

$$y(t) = KM\left[1 - \left(1 + \frac{t}{\tau}\right)e^{-t/\tau}\right] \tag{5-50}$$

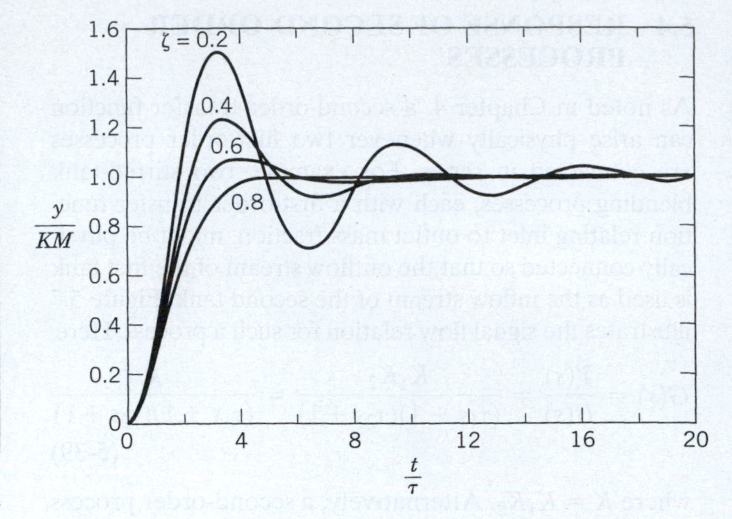

Figure 5.8 Step response of underdamped second-order processes.

Underdamped ($0 \leq \zeta < 1$)

$$y(t) = KM\left\{1 - e^{-\zeta t/\tau}\left[\cos\left(\frac{\sqrt{1-\zeta^2}}{\tau}t\right) + \frac{\zeta}{\sqrt{1-\zeta^2}}\sin\left(\frac{\sqrt{1-\zeta^2}}{\tau}t\right)\right]\right\} \tag{5-51}$$

Plots of the step responses for different values of ζ are shown in Figs. 5.8 and 5.9, where the time axis is normalized with respect to τ. Thus, when τ is small, a rapid response is signified, implying a large value for the undamped natural frequency, $\omega_n = 1/\tau$.

Several general remarks can be made concerning the responses shown in Figs. 5.8 and 5.9:

1. Responses exhibit a higher degree of oscillation and overshoot ($y/KM > 1$) as ζ approaches zero.
2. Large values of ζ yield a sluggish (slow) response.
3. The fastest response without overshoot is obtained for the critically damped case ($\zeta = 1$).

Control system designers sometimes attempt to make the response of the controlled variable to a set-point change approximate the ideal step response of an underdamped second-order system, that is, make it exhibit a prescribed amount of overshoot and oscillation as it settles at the new operating point. When damped oscillation is desirable, values of ζ in the range 0.4 to 0.8 may be chosen. In this range, the controlled variable y reaches the new operating point faster than with $\zeta = 1.0$ or 1.5, but the response is much less oscillatory (settles faster) than with $\zeta = 0.2$.

Figure 5.10 illustrates the characteristics of the step response of a second-order underdamped process. The following terms are used to describe the dynamics of underdamped processes:

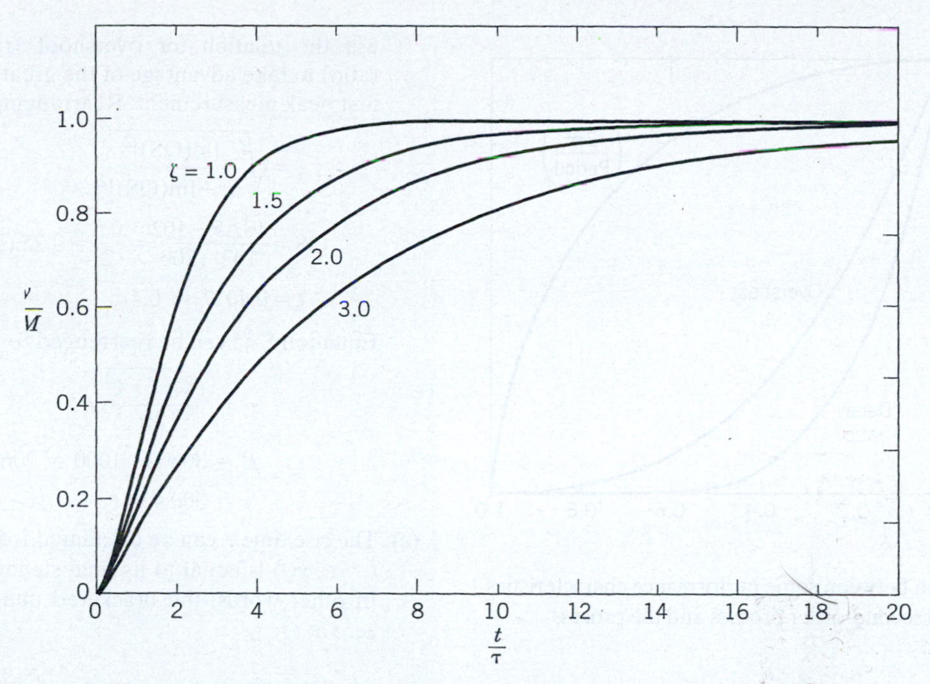

Figure 5.9 Step response of critically damped and overdamped second-order processes.

1. ***Rise Time.*** t_r is the time the process output takes to first reach the new steady-state value.
2. ***Time to First Peak.*** t_p is the time required for the output to reach its first maximum value.
3. ***Settling Time.*** t_s is the time required for the process output to reach and remain inside a band whose width is equal to ±5% of the total change in y for 95% response time (99% response time is also used for some applications).
4. ***Overshoot.*** OS = a/b (% overshoot is 100 a/b).
5. ***Decay Ratio.*** DR = c/a (where c is the height of the second peak).
6. ***Period of Oscillation.*** P is the time between two successive peaks or two successive valleys of the response.

Note that the above definitions generally apply to the step response of any underdamped process. If the process does not exhibit overshoot, the rise time definition is modified to be the time to go from 10% to 90% of the steady-state response (Åström and Hägglund, 2006). For the particular case of an underdamped second-order process, we can develop analytical expressions for some of these characteristics. Using Eq. 5-51

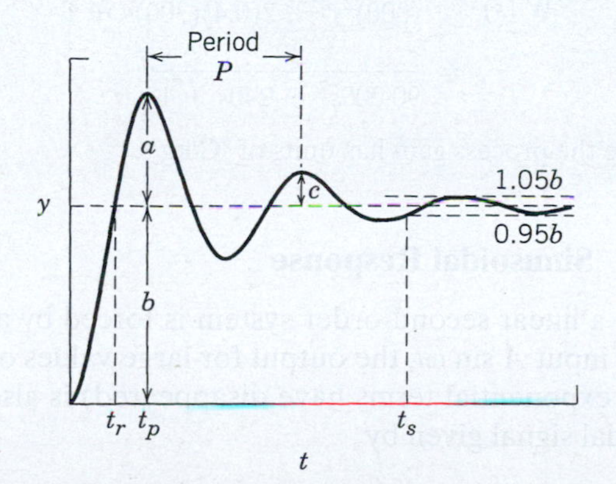

Figure 5.10 Performance characteristics for the step response of an underdamped process.

Time to first peak: $$t_p = \pi\tau/\sqrt{1-\zeta^2} \tag{5-52}$$

Overshoot: $$\text{OS} = \exp\left(-\pi\zeta/\sqrt{1-\zeta^2}\right) \tag{5-53}$$

Decay ratio: $$\text{DR} = (\text{OS})^2 = \exp(-2\pi\zeta/\sqrt{1-\zeta^2}) \tag{5-54}$$

Period: $$P = \frac{2\pi\tau}{\sqrt{1-\zeta^2}} \tag{5-55}$$

Note that OS and DR are functions of ζ only. For a second-order system, the decay ratio is constant for each successive pair of peaks. Figure 5.11 illustrates the dependence of overshoot and decay ratio on damping coefficient.

For an underdamped second-order transfer function, Figs. 5.8 and 5.11 and Eq. 5-55 can be used to obtain estimates of ζ and τ based on step response characteristics.

EXAMPLE 5.5

A stirred-tank reactor has an internal cooling coil to remove heat liberated in the reaction. A proportional controller is used to regulate coolant flow rate so as to keep the reactor temperature reasonably constant. The controller has been designed so that the controlled reactor exhibits typical underdamped second-order temperature

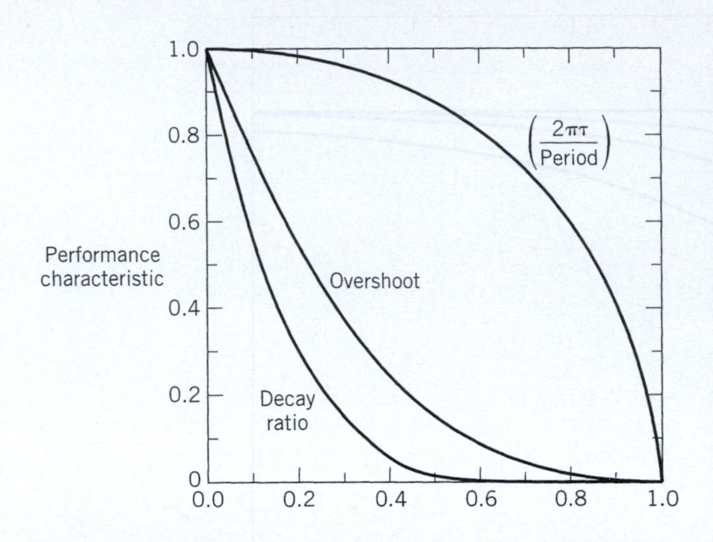

Figure 5.11 Relation between some performance characteristics of an underdamped second-order process and the process damping coefficient.

response characteristics when it is disturbed, either by feed flow rate or by coolant temperature changes.

(a) The feed flow rate to the reactor changes suddenly from 0.4 to 0.5 kg/s, and the temperature of the reactor contents, initially at 100 °C, changes eventually to 102 °C. What is the gain of the transfer function (under feedback control) that relates changes in reactor temperature to changes in feed flow rate? (Be sure to specify the units.)

(b) The operator notes that the resulting response is slightly oscillatory with maxima estimated to be 102.5 and 102.0 °C occurring at times 1000 and 3060 s after the change is initiated. What is the complete process transfer function?

(c) The operator failed to note the rise time. Predict t_r based on the results in (a) and (b).

SOLUTION

(a) The gain is obtained by dividing the steady-state change in temperature by the feed flow rate (disturbance) change:

$$K = \frac{102 - 100}{0.5 - 0.4} = 20\,\frac{°\text{C}}{\text{kg/s}}$$

(b) The oscillatory characteristics of the response can be used to find the dynamic elements in the transfer function relating temperature to feed flow rate. Assuming the step response is due to an underdamped second-order process, Figs. 5.8 and 5.11 can be used to obtain estimates of ζ and τ. Alternatively, analytical expressions can be used, which is the approach taken here. Either Eq. 5-53 or 5-54 can be employed to find ζ independently of τ. Because the second peak value of temperature (102.0 °C) is essentially the final value (102 °C), the calculated value of peak height c will be subject to appreciable measurement error. Instead, we use the relation for overshoot (rather than decay ratio) to take advantage of the greater precision of the first peak measurement. Rearranging (5-53) gives

$$\zeta = \sqrt{\frac{[\ln(\text{OS})]^2}{\pi^2 + [\ln(\text{OS})]^2}}$$

$$\text{OS} = \frac{102.5 - 102}{102 - 100} = \frac{0.5}{2} = 0.25 \text{ (i.e., 25\%)} \qquad (5\text{-}56)$$

$$\zeta = 0.4037 \;\approx 0.4$$

Equation 5-55 can be rearranged to find τ:

$$\tau = \frac{\sqrt{1 - \zeta^2}}{2\pi} P$$

$$P = 3060 - 1000 = 2060 \text{ s} \qquad (5\text{-}57)$$

$$\tau = 300 \text{ s}$$

(c) The rise time t_r can be calculated from Eq. 5-51. When $t = t_r$, $y(t)$ is equal to its final steady-state value, KM. In other words, the bracketed quantity is identically zero at $t = t_r$:

$$\cos\left(\frac{\sqrt{1 - \zeta^2}}{\tau} t_r\right) + \frac{\zeta}{\sqrt{1 - \zeta^2}} \sin\left(\frac{\sqrt{1 - \zeta^2}}{\tau} t_r\right) = 0 \qquad (5\text{-}58)$$

The general solution has multiple values of t that satisfy $y(t) = KM$:

$$t = \frac{\tau}{\sqrt{1 - \zeta^2}}(n\pi - \cos^{-1}\zeta) \qquad n = 1, 2, \ldots \qquad (5\text{-}59)$$

The rise time corresponds to the first time ($n = 1$) that $y(t) = KM = y(\infty)$. Solving for the rise time gives

$$t_r = \frac{\tau}{\sqrt{1 - \zeta^2}}(\pi - \cos^{-1}\zeta) \qquad (5\text{-}60)$$

where the result of the inverse cosine computation must be in radians. Because $\tau = 300$ s and $\zeta = 0.40$

$$t_r = 649 \text{ s}$$

In summary, the disturbance transfer function between feed flow rate and outlet temperature while under feedback control is

$$\frac{T'(s)}{W'(s)} = \frac{20}{(300)^2 s^2 + 2(0.4)(300)s + 1} = \frac{20}{90{,}000s^2 + 240s + 1}$$

where the process gain has units of °C/kg/s.

5.4.2 Sinusoidal Response

When a linear second-order system is forced by a sinusoidal input $A \sin \omega t$, the output for large values of time (after exponential terms have disappeared) is also a sinusoidal signal given by

$$y(t) = \frac{KA}{\sqrt{[1 - (\omega\tau)^2]^2 + (2\zeta\omega\tau)^2}} \sin(\omega t + \phi) \qquad (5\text{-}61)$$

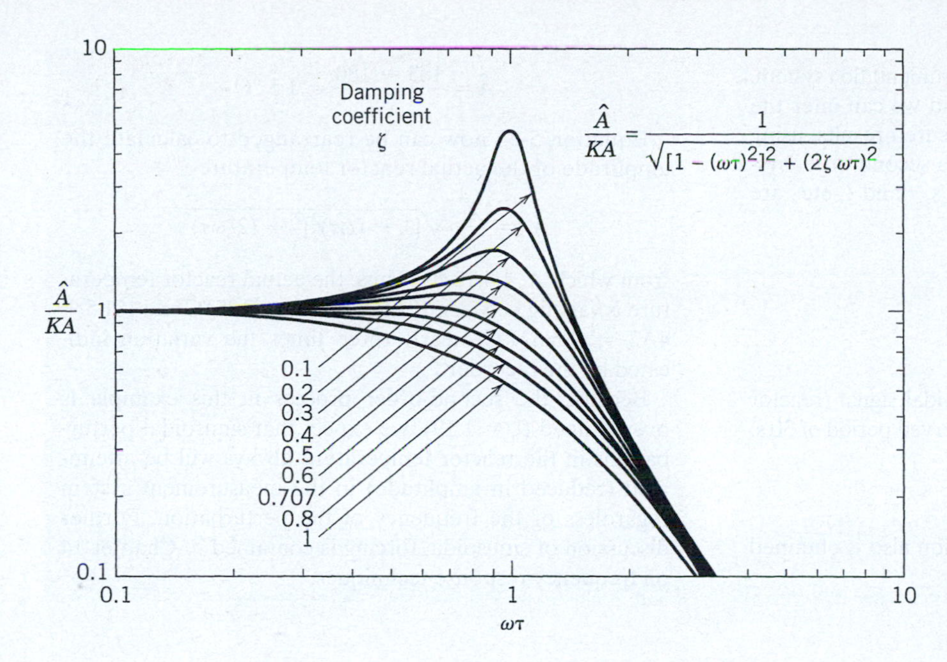

Figure 5.12 Sinusoidal response amplitude of a second-order system after exponential terms have become negligible.

where

$$\phi = -\tan^{-1}\left[\frac{2\zeta\omega\tau}{1-(\omega\tau)^2}\right] \quad (5\text{-}62)$$

The output amplitude $\hat{A}$ is obtained directly from Eq. 5-61:

$$\hat{A} = \frac{KA}{\sqrt{[1-(\omega\tau)^2]^2 + (2\zeta\omega\tau)^2}} \quad (5\text{-}63)$$

The ratio of output to input amplitude is the *amplitude ratio* AR $(= \hat{A}/A)$. When normalized by the process gain, it is called the *normalized amplitude ratio* AR_N

$$AR_N = \frac{\hat{A}}{KA} = \frac{1}{\sqrt{[1-(\omega\tau)^2]^2 + (2\zeta\omega\tau)^2}} \quad (5\text{-}64)$$

AR_N represents the effect of the dynamic model parameters (ζ, τ) on the sinusoidal response; that is, AR_N is independent of steady-state gain K and the amplitude of the forcing function, A. The maximum value of AR_N can be found (if it exists) by differentiating (5-64) with respect to ω and setting the derivative to zero. Solving for ω_{max} gives

$$\omega_{max} = \frac{\sqrt{1-2\zeta^2}}{\tau} \quad \text{for } 0 < \zeta < 0.707 \quad (5\text{-}65)$$

For $\zeta \geq 0.707$, there is no maximum, as Fig. 5.12 illustrates. Substituting (5-65) into (5-64) yields an expression for the maximum value of AR_N:

$$AR_N\bigg|_{max} = \frac{\hat{A}}{KA}\bigg|_{max} = \frac{1}{2\zeta\sqrt{1-\zeta^2}}$$
$$\text{for } 0 < \zeta < 0.707 \quad (5\text{-}66)$$

We see from (5-66) that the maximum output amplitude for a second-order process that has no damping $(\zeta = 0)$ is undefined. Small values of ζ are invariably avoided in the design of processes, as well as in designing control systems. Equation 5-66 indicates that a process with little damping can exhibit very large output oscillations if it is perturbed by periodic signals with frequency near ω_{max}.

EXAMPLE 5.6

An engineer uses a temperature sensor mounted in a thermowell to measure the temperature in a CSTR. The temperature sensor/transmitter combination operates approximately as a first-order system with time constant equal to 3 s. The thermowell behaves like a first-order system with time constant of 10 s. The engineer notes that the measured reactor temperature has been cycling approximately sinusoidally between 180 and 183 °C with a period of 30 s for at least several minutes. What can be concluded concerning the actual temperature in the reactor?

SOLUTION

First, note that the sensor/transmitter and the transmission line act as two first-order processes in series (Eq. 5-39) with overall gain K equal to 1, with the approximate transfer function

$$\frac{T'_{meas}(s)}{T'_{reactor}(s)} = \frac{1}{(3s+1)(10s+1)} \quad (5\text{-}67)$$

From the reported results, we conclude that some disturbance has caused the actual reactor temperature (and its deviation) to vary sinusoidally, which, in turn, has caused the recorded output to oscillate. The cycling has continued for a period of time that is much longer than the time constants of the process—that is, the instrumentation system. Hence, the transients have died out and we can infer the conditions in the reactor from the measured results, using Eq. 5-63 for the sinusoidal response of a second-order system. From (5-67), $\tau_1 = 3$ s and $\tau_2 = 10$ s; τ and ζ, etc., are calculated from Eqs. 5-42 and 5-43:

$$\tau = \sqrt{(3)(10)} = 5.48 \text{ s}$$

$$\zeta = \frac{13}{2(5.477)} = 1.19$$

The frequency of the perturbing sinusoidal signal (reactor temperature) is calculated from the observed period of 30 s:

$$\omega = \frac{2\pi}{P} = \frac{6.28}{30} = 0.2093 \text{ s}^{-1}$$

The amplitude of the output perturbation also is obtained from observed results as

$$\hat{A} = \frac{183 - 180}{2} = 1.5\ ^\circ\text{C}$$

Equation 5-63 now can be rearranged to calculate the amplitude of the actual reactor temperature

$$A = \frac{\hat{A}}{K}\sqrt{[1 - (\omega\tau)^2]^2 + (2\zeta\omega\tau)^2}$$

from which $A = 4.12$ °C. Thus, the actual reactor temperature is varying between 181.5 − 4.12 = 177.38 °C and 181.5 + 4.12 = 185.62 °C, nearly three times the variation indicated by the recorder.

Because the second-order process in this example is overdamped ($\zeta = 1.19$), we expect that sinusoidal perturbations in the reactor temperature always will be attenuated (reduced in amplitude) in the measurement system regardless of the frequency of the perturbation. Further discussion of sinusoidal forcing is contained in Chapter 14 on frequency response techniques.

SUMMARY

Transfer functions can be used conveniently to obtain output responses to any type of input change. In this chapter we have focused on first- or second-order transfer functions and integrating processes. Because a relatively small number of input changes have industrial or analytical significance, we have considered in detail the responses of these basic process transfer functions to the important types of inputs, such as step, ramp, impulse, and sine inputs.

If a process can be modeled as a first-order or second-order transfer function, the process response to any standard input change can be found analytically or numerically. When a theoretical model is not available, as occurs in many plant situations, data can be used to obtain an approximate process transfer function if the input is known, as discussed in Chapter 7. A model permits predictions of how a process will react to other types of disturbances or input changes.

Unfortunately, not all processes can be modeled by such simple transfer functions. Hence, in Chapter 6 several additional transfer function elements are introduced in order to construct more complicated transfer functions. However, the emphasis there is to show how complex process behavior can be explained with combinations of simple transfer function elements.

REFERENCES

Åström, K. J., and T. Hägglund, Advanced PID Control, 3d ed., Instrument Society of America, Research Triangle Park, NC, 2006.

Box, G.E.P., G. M. Jenkins, and G. C. Reinsel, *Time Series Analysis, Forecasting, and Control*, 3d ed., Prentice-Hall, Englewood Cliffs, NJ, 1994.

Maybeck, P. S., *Stochastic Models, Estimation, and Control*, 2d ed., Academic Press, New York, 1997.

EXERCISES

5.1 In addition to the standard inputs discussed in Section 5.1, other input functions occasionally are useful for special purposes. One, the so-called doublet pulse, is shown in Fig. E5.1.

(a) Find the Laplace transform of this function by first expressing it as a composite of functions whose transforms you already know.

(b) What is the response of a process having a first-order transfer function $K/(\tau s + 1)$ to this input? of the integrating process K/s?

(c) From these results, can you determine what special property this input offers?

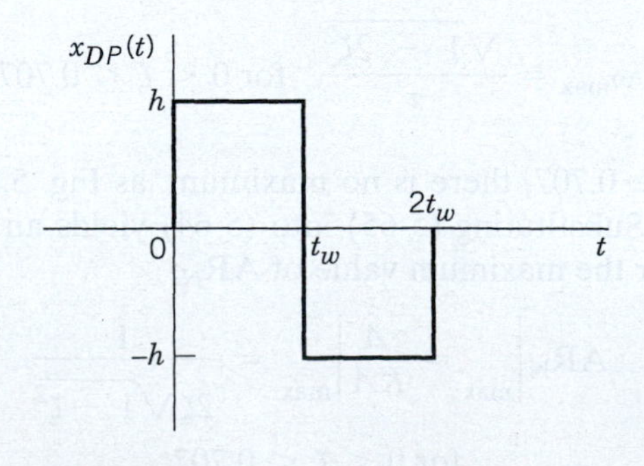

Figure E5.1

5.2 A heater for a semiconductor wafer has first-order dynamics, that is, the transfer function relating changes in temperature T to changes in the heater input power level P is

$$\frac{T'(s)}{P'(s)} = \frac{K}{\tau s + 1}$$

where K has units [°C/Kw] and τ has units [min].

The process is at steady state when an engineer changes the power input stepwise from 1 to 1.5 Kw. She notes the following:

(i) The process temperature initially is 80 °C.

(ii) Four minutes after changing the power input, the temperature is 230 °C.

(iii) Thirty minutes later the temperature is 280 °C.

(a) What are K and τ in the process transfer function?

(b) If at another time the engineer changes the power input linearly at a rate of 0.5 kW/min, what can you say about the maximum rate of change of process temperature: When will it occur? How large will it be?

5.3 A composition sensor is used to continually monitor the contaminant level in a liquid stream. The dynamic behavior of the sensor can be described by a first-order transfer function with a time constant of 10 s,

$$\frac{C'_m(s)}{C'(s)} = \frac{1}{10s + 1}$$

where C' is the actual contaminant concentration and C'_m is the measured value. Both are expressed as deviation variables (e.g., $C' = C - \overline{C}$). The nominal concentration is $\overline{C} =$ 5 ppm. Both C and C_m have values of 5 ppm initially (i.e., the values at $t = 0$).

An alarm sounds if the measured value exceeds the environmental limit of 7 ppm. Suppose that the contaminant concentration C gradually increases according to the expression $C(t) = 5 + 0.2t$, where t is expressed in seconds. After the actual contaminant concentration exceeds the environmental limit, what is the time interval, Δt, until the alarm sounds?

5.4 The dynamic response of a stirred-tank bioreactor can be represented by the transfer function

$$\frac{C'(s)}{C'_F(s)} = \frac{4}{2s + 1}$$

where C' is the exit substrate concentration, mol/L, and C'_F is the feed substrate concentration, mol/L.

(a) Derive an expression for $c'(t)$ if $c_F(t)$ is a rectangular pulse (Fig. 5.2) with the following characteristics:

$$c_F(t) = \begin{cases} 2 & t < 0 \\ 4 & 0 \le t < 2 \\ 2 & 2 \le t < \infty \end{cases}$$

(b) What is the maximum value of $c'(t)$? When does it occur? What is the final value of $c'(t)$?

(c) If the initial value is $c(0) = 1$, how long does it take for $c(t)$ to return to a value of 1.05 after it has reached its maximum value?

5.5 A thermocouple has the following characteristics when it is immersed in a stirred bath:

Mass of thermocouple = 1 g
Heat capacity of thermocouple = 0.25 cal/g °C
Heat transfer coefficient = 20 cal/cm^2h °C (for thermocouple and bath)
Surface area of thermocouple = 3 cm^2

(a) Derive a transfer function model for the thermocouple relating the change in its indicated output T to the change in the temperature of its surroundings T_s assuming uniform temperature (no gradients in the thermocouple bead), no conduction in the leads, constant physical properties, and conversion of the millivolt-level output directly to a °C reading by a very fast meter.

(b) If the thermocouple is initially out of the bath and at room temperature (23 °C), what is the maximum temperature that it will register if it is suddenly plunged into the bath (80 °C) and held there for 20 s?

5.6 Consider the transfer function

$$G(s) = \frac{Y(s)}{U(s)} = \frac{10}{(5s + 1)(3s + 1)}$$

What is $y(t \to \infty)$ for the following inputs:

(a) step input of height M

(b) unit impulse input ($\delta(t)$)

(c) $\sin t$

(d) unit rectangular pulse (Eq. 3-20, $h = 1$)

5.7 Appelpolscher has just left a meeting with Stella J. Smarly, IGC's vice-president for process operations and development. Smarly is concerned about an upcoming extended plant test of a method intended to improve the yields of a large packed-bed reactor. The basic idea, which came from IGC's university consultant and was recently tested for feasibility in a brief run, involves operating the reactor cyclically so that nonlinearities in the system cause the time-average yield at the exit to exceed the steady-state value. Smarly is worried about the possibility of sintering the catalyst during an extended run, particularly in the region of the "hotspot" (axially about one-third of the way down the bed and at the centerline) where temperatures invariably peak. Appelpolscher, who plans to leave the next day on a two-week big game photo safari, doesn't want to cancel his vacation. On the other hand, Smarly has told him he faces early, unexpected retirement in Botswana if the measurement device (located near the hot spot) fails to alert operating people and the reactor catalyst sinters. Appelpolscher likes Botswana but doesn't want to retire there. He manages to pull together the following data and assumptions before heading for the airport and leaves them with you for analysis with the offer of the use of his swimming pool while he is gone. What do you report to Smarly?

Data:

Frequency of cyclic operation = 0.1 cycles/min
Amplitude of thermal wave (temperature) at the measurement point obtained experimentally in the recent brief run = 15 °C
Average operating temperature at the measurement point, T_{meas} = 350 °C

Time constant of temperature sensor and thermowell = 1.5 min

Temperature at the reactor wall = 200 °C

Temperature at which the catalyst sinters if operated for several hours = 700 °C

Temperature at which the catalyst sinters instantaneously = 715 °C

Assumptions:

The reactor operational cycle is approximately sinusoidal at the measurement point.

The thermowell is located near the reactor wall so as to measure a "radial average" temperature rather than the centerline temperature.

The approximate relation is

$$T = \frac{T_{\text{center}} + 2T_{\text{wall}}}{3}$$

which also holds during transient operation.

5.8 A liquid storage system is shown below. The normal operating conditions are $\bar{q}_1 = 10$ ft^3/min, $\bar{q}_2 = 5$ ft^3/min, $\bar{h} = 4$ ft. The tank is 6 ft in diameter, and the density of each stream is 60 lb/ft^3. Suppose that a pulse change in q_1 occurs as shown in Fig. E5.8.

(a) What is the transfer function relating H' to Q_1'?

(b) Derive an expression for $h(t)$ for this input change.

(c) What is the new steady-state value of liquid level $\bar{h}$?

(d) Repeat (b) and (c) for the doublet pulse input of Exercise 5.1 where the changes in q_1 are from 10 to 15 to 5 to 10 ft^3/min.

5.9 Two liquid storage systems are shown in Fig. E5.9. Each tank is 4 feet in diameter. For System I, the valve acts as a linear resistance with the flow–head relation $q = 8.33\,h$, where q is in gal/min and h is in feet. For System II, variations in liquid level h do not affect exit flow rate q. Suppose that each system is initially at steady state with $\bar{h} = 6$ ft and $\bar{q}_i = 50$ gal/min and that at time $t = 0$ the inlet flow rate suddenly changes from 50 to 70 gal/min. For each system, determine the following information:

(a) The transfer function $H'(s)/Q_i'(s)$ where the primes denote deviation variables.

(b) The transient response $h(t)$.

(c) The new steady-state levels.

(d) If each tank is 8 ft tall, which tank overflows first? when?

5.10 The dynamic behavior of the liquid level in each leg of a manometer tube, responding to a change in pressure, is given by

$$\frac{d^2h'}{dt^2} + \frac{6\mu}{R^2\rho}\frac{dh'}{dt} + \frac{3}{2}\frac{g}{L}h' = \frac{3}{4\rho L}p'(t)$$

where $h'(t)$ is the level of fluid measured with respect to the initial steady-state value, $p'(t)$ is the pressure change, and R, L, g, ρ, and μ are constants.

(a) Rearrange this equation into standard gain-time constant form and find expressions for K, τ, ζ in terms of the physical constants.

(b) For what values of the physical constants does the manometer response oscillate?

(c) Would changing the manometer fluid so that ρ (density) is larger make its response more oscillatory, or less? Repeat the analysis for an increase in μ (viscosity).

5.11 A process is described by the following transfer function:

$$\frac{Y(s)}{U(s)} = \frac{K}{s(\tau s + 1)}$$

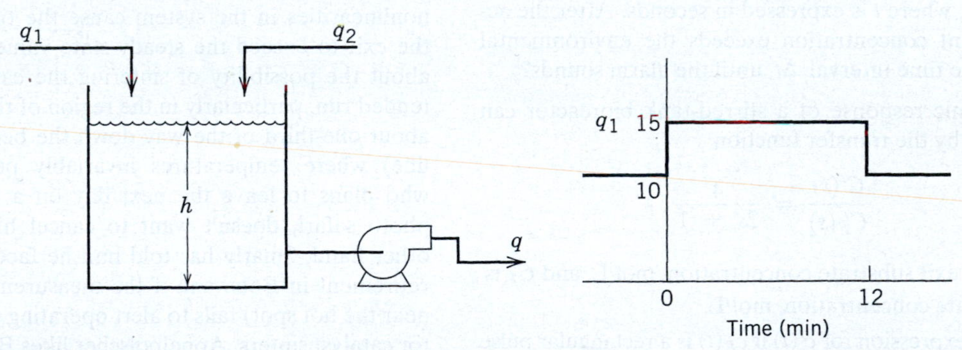

Figure E5.8

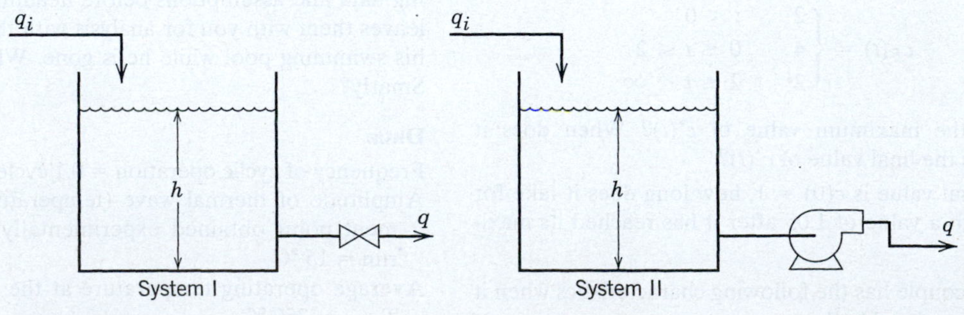

Figure E5.9

Thus, it exhibits characteristics of both first-order and integrating processes.

How could you utilize a step change in the input of magnitude M to find quickly the two parameters K and τ? (Be sure to show all work and sketch the anticipated process response.)

5.12 For the equation

$$\frac{d^2y}{dt^2} + K\frac{dy}{dt} + 4y = u$$

(a) Find the transfer function and put it in standard gain/time constant form.

(b) Discuss the qualitative form of the response of this system (independent of the input forcing) over the range $-10 \leq K \leq 10$.

Specify values of K where the response will converge and where it will not. Write the form of the response without evaluating any coefficients.

5.13 A second-order critically damped process has the transfer function

$$\frac{Y(s)}{U(s)} = \frac{K}{(\tau s + 1)^2}$$

(a) For a step change in input of magnitude M, what is the time (t_S) required for such a process to settle to within 5% of the total change in the output?

(b) For $K = 1$ and a ramp change in input, $u(t) = at$, by what time period does $y(t)$ lag behind $u(t)$ once the output is changing linearly with time?

5.14 A step change from 15 to 31 psi in actual pressure results in the measured response from a pressure-indicating element shown in Fig. E5.14.

(a) Assuming second-order dynamics, calculate all important parameters and write an approximate transfer function in the form

$$\frac{R'(s)}{P'(s)} = \frac{K}{\tau^2 s^2 + 2\zeta\tau s + 1}$$

where R' is the instrument output deviation (mm), P' is the actual pressure deviation (psi).

(b) Write an equivalent differential equation model in terms of actual (not deviation) variables.

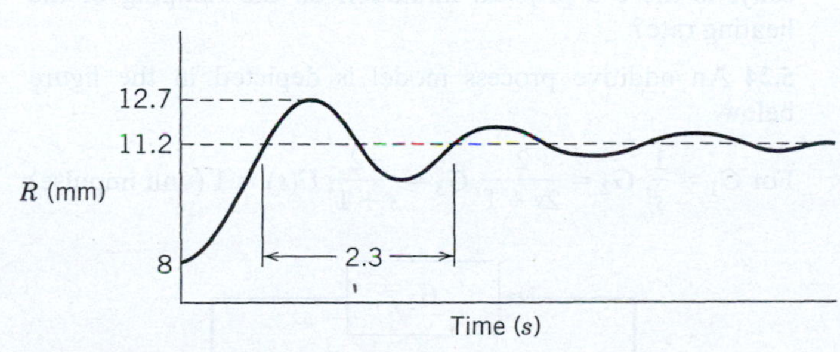

Figure E5.14

5.15 An electrically heated process is known to exhibit second-order dynamics with the following parameter values: $K = 3$ °C/kW, $\tau = 3$ min, $\zeta = 0.7$. If the process initially is at steady state at 70 °C with heater input of 20 kW and the heater input is suddenly changed to 26 kW and held there,

(a) What will be the expression for the process temperature as a function of time?

(b) What will be the maximum temperature observed? When will it occur?

5.16 Starting with Eq. 5-51, derive expressions for the following response characteristics of the underdamped second-order system.

(a) The time to first peak t_p (Eq. 5-52).

(b) The fraction overshoot (Eq. 5-53).

(c) The decay ratio (Eq. 5-54).

(d) The settling time (t_s, defined in Fig. 5.10). Can a *single* expression be used for t_s over the full range of ζ, $0 < \zeta < 1$?

5.17 A tank used to dampen liquid flow rate surges is known to exhibit second-order dynamics. The input flow rate changes suddenly from 120 to 140 gal/min. An operator notes that the tank level changes as follows:

Before input change: level = 6 ft and steady

Four minutes later: level = 11 ft

Forty minutes later: level = 10 ft and steady

(a) Find a transfer function model that describes this process, at least approximately. Evaluate all parameters in your model, including units.

(b) Is your model unique? Why or why not?

5.18 A process has the transfer function

$$G(s) = \frac{2}{s^2 + s + 1} = \frac{Y(s)}{U(s)}$$

(a) For a step change in the input $U(s) = 2/s$, sketch the response $y(t)$ (you do not need to solve the differential equation). Show as much detail as possible, including the steady-state value of $y(t)$, and whether there is oscillation.

(b) What is the decay ratio?

5.19 A surge tank system is to be installed as part of a pilot plant facility. The initial proposal calls for the configuration shown in Fig. 4.3. Each tank is 5 ft high and 3 ft in diameter. The design flow rate is $q_i = 100$ gal/min. It has been suggested that an improved design will result if the two-tank system is replaced by a single tank that is 4 ft in diameter and has the same total volume (i.e., $V = V_1 + V_2$).

(a) Which surge system (original or modified) can handle larger step disturbances in q_i? Justify your answer.

(b) Which system provides the best damping of step disturbances in q_i? (Justify your answer).

In your analysis you may assume that:

(i) The valves on the exit lines act as linear resistances.

(ii) The valves are adjusted so that each tank is half full at the nominal design condition of $q_i = 100$ gal/min.

5.20 The caustic concentration of the mixing tank shown in Fig. E5.20 is measured using a conductivity cell. The total volume of solution in the tank is constant at 7 ft^3 and the density ($\rho = 70$ lb/ft^3) can be considered to be independent of concentration. Let c_m denote the caustic concentration measured

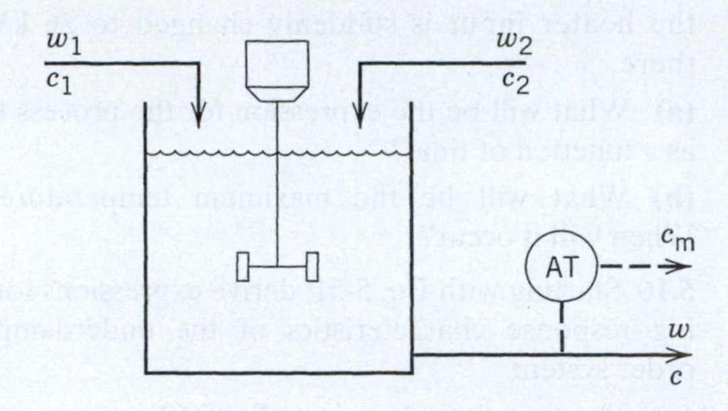

Figure E5.20

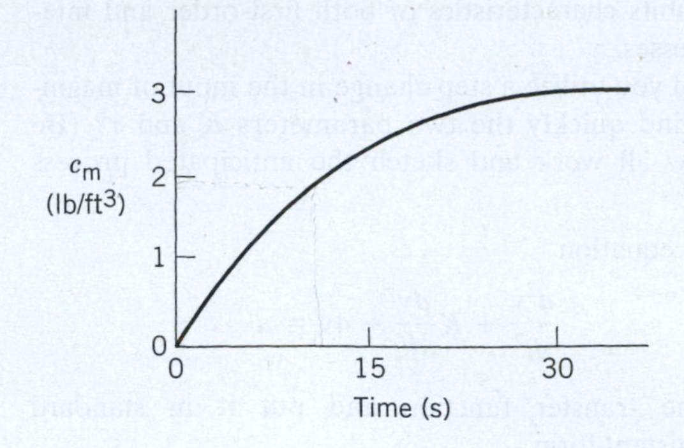

by the conductivity cell. The dynamic response of the conductivity cell to a step change (at $t = 0$) of 3 lb/ft^3 in the actual concentration (passing through the cell) is also shown in Fig. E5.20.

(a) Determine the transfer function $C'_m(s)/C'_1(s)$ assuming the flow rates are equal and constant: ($w_1 = w_2 = 5$ lb/min):

(b) Find the response for a step change in c_1 from 14 to 17 lb/ft^3.

(c) If the transfer function $C'_m(s)/C'(s)$ were approximated by 1 (unity), what would be the step response of the system for the same input change?

(d) By comparison of (b) and (c), what can you say about the dynamics of the conductivity cell? Plot both responses, if necessary.

5.21 An exothermic reaction, A → 2B, takes place adiabatically in a stirred-tank system. This liquid phase reaction occurs at constant volume in a 100-gal reactor. The reaction can be considered to be first order and irreversible with the rate constant given by

$$k = 2.4 \times 10^{15} e^{-20{,}000/T} \ (\text{min}^{-1})$$

where T is in °R. Using the information below, derive a transfer function relating the exit temperature T to the inlet concentration c_{Ai}. State any assumptions that you make. Simplify the transfer function by making a first-order approximation and show that the approximation is valid by comparing the step responses of both the original and the approximate models.

Available Information

(i) Nominal steady-state conditions are:

$$\overline{T} = 150\,°\text{F}, \quad \overline{c}_{Ai} = 0.8 \text{ lb mole/ft}^3$$

$$q = 20 \text{ gal/min} = \text{flow rate in and out of the reactor}$$

(ii) Physical property data for the mixture at the nominal steady state: $C_p = 0.8$ Btu/lb °F,

$$\rho = 52 \text{ lb/ft}^3, \quad -\Delta H_R = 500 \text{ kJ/lb mole}$$

5.22 Using the step responses of (1) an integrating element and (2) a first-order process to an input change of magnitude M.

(a) Show that the step response for an input change M of a first-order process

$$G_1(s) = \frac{K_1}{\tau s + 1}$$

can be approximately modeled by the step response of an integrator.

$$G_0(s) = \frac{K_0}{s}$$

for low values of t—i.e., $t << \tau$. (Hint: you can use a first-order Taylor series approximation of $e^{-t/\tau}$.)

(b) What is the relation between K_0 and K_1 if the two responses match for $t << \tau$?

(c) This relationship motivates the use of an integrator model to approximate a first-order process by means of a single-parameter model. Explain how you would analyze a single step test to find K_0 and a time delay (if one exists).

5.23 For a stirred-tank heater, assume the transfer function between the heater input change $u(t)$ (cal/sec) and the tank temperature change $y(t)$ (°C) can be modeled as

$$G(s) = \frac{K}{\tau s + 1}$$

(a) Using the Final Value Theorem, find the steady-state response for a unit rectangular pulse change in the heating rate $\left(U(s) = \dfrac{1 - e^{-s}}{s} \right)$.

(b) Repeat the calculation in (a) for a unit ramp $\left(U(s) = \dfrac{1}{s^2} \right)$.

(c) For both cases (a) and (b), explain your answer physically. Is there a physical limitation on the ramping of the heating rate?

5.24 An additive process model is depicted in the figure below.

For $G_1 = \dfrac{1}{s}$, $G_2 = \dfrac{2}{2s+1}$, $G_3 = \dfrac{-2}{s+1}$, $U(s) = 1$ (unit impulse)

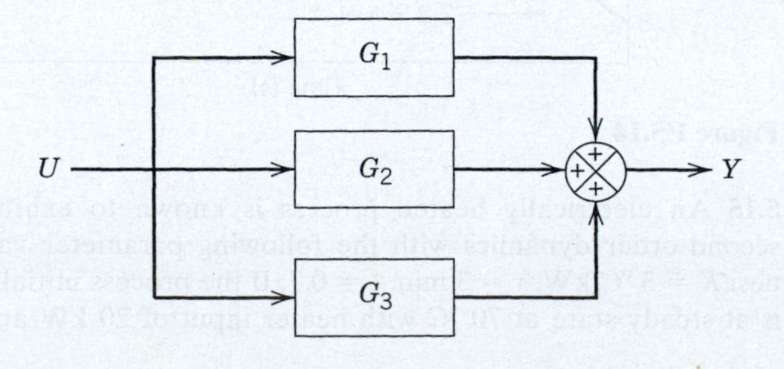

(a) Derive the response $Y(s)$ and describe $y(t)$ quantitatively.

(b) Sketch the response and show its major characteristics.

5.25 Can a tank with the outflow rate fixed by a constant speed pump reach a steady state if the inlet flow rate undergoes a step change? Why, or why not? If the transfer function is $G(s) = K/s$, is it possible to calculate a steady-state gain?

5.26 A thermometer with first-order time constant = 0.1 min and gain = 1.0 is placed in a temperature bath (25 °C). After the thermometer comes to equilibrium with the bath, the temperature of the bath is increased linearly at a rate of 1°/min.

(a) What is the difference between the measured temperature T_m and the bath temperature T at $t = 0.1$ min and $t = 1.0$ min after the change in temperature?

(b) What is the maximum deviation between T_m (t) and $T(t)$? When does it occur?

(c) Plot both $T(t)$ and $T_m(t)$ to 3 mins. For large values of t, determine the time lag between T_m and T.

5.27 A thermometer has first-order dynamics with a time constant of 1 sec and is placed in a temperature bath at 120°F. After the thermometer reaches steady state, it is suddenly placed in a bath at 140 °F for $0 \le t \le 10$ sec. Then it is returned to the bath at 100 °F.

(a) Sketch the variation of the measured temperature $T_m(t)$ with time.

(b) Calculate $T_m(t)$ at $t = 0.5$ sec and at $t = 15.0$ sec.

Chapter 6

Dynamic Response Characteristics of More Complicated Processes

CHAPTER OF CONTENTS

In Chapter 5 we discussed the dynamics of relatively simple processes, those that can be modeled as either first- or second-order transfer functions or as an integrator. Now we consider more complex transfer function models that include additional time constants in the denominator and/or functions of s in the numerator. We show that the forms of the numerator and denominator of the transfer function model influence the dynamic behavior of the process. We also introduce a very important concept, the *time delay,* and consider the approximation of complicated transfer function models by simpler, low-order models. Additional topics in this chapter include interacting processes, state-space models, and processes with multiple inputs and outputs.

6.1 POLES AND ZEROS AND THEIR EFFECT ON PROCESS RESPONSE

An important feature of the simple process elements discussed in Chapter 5 is that their response characteristics are determined by the factors of the transfer function denominator. For example, consider a transfer function,

$$G(s) = \frac{K}{s(\tau_1 s + 1)(\tau_2^2 s^2 + 2\zeta\tau_2 s + 1)} \tag{6-1}$$

where $0 \leq \zeta < 1$. Using partial fraction expansion followed by the inverse transformation operation, we know that the response of system (6-1) to *any* input will contain the following functions of time:

- A constant term resulting from the s factor
- An e^{-t/τ_1} term resulting from the $(\tau_1 s + 1)$ factor
- $e^{-\zeta t/\tau_2} \sin \dfrac{\sqrt{1-\zeta^2}}{\tau_2} t$ and $e^{-\zeta t/\tau_2} \cos \dfrac{\sqrt{1-\zeta^2}}{\tau_2} t$ — terms resulting from the $(\tau_2^2 s^2 + 2\zeta\tau_2 s + 1)$ factor

Additional terms determined by the specific input forcing will also appear in the response, but the intrinsic dynamic features of the process, the so-called *response modes* or *natural modes*, are determined by the process itself. Each of the above response modes is determined from the factors of the denominator polynomial, which is also called the characteristic polynomial (cf. Section 3.3). The roots of these factors are

$$s_1 = 0$$
$$s_2 = -\frac{1}{\tau_1}$$
$$s_3 = -\frac{\zeta}{\tau_2} + j\frac{\sqrt{1-\zeta^2}}{\tau_2} \tag{6-2}$$
$$s_4 = -\frac{\zeta}{\tau_2} - j\frac{\sqrt{1-\zeta^2}}{\tau_2}$$

Roots s_3 and s_4 are obtained by applying the quadratic formula.

Control engineers refer to the values of s that are roots of the denominator polynomial as the *poles* of transfer function $G(s)$. Sometimes it is useful to plot the roots (poles) and to discuss process response characteristics in terms of root locations in the complex s plane. In Fig. 6.1 the ordinate expresses the imaginary part of each root; the abscissa expresses the real part. Figure 6.1 is based on Eq. 6-2 and indicates the presence of four poles: an integrating element (pole at the origin), one real pole (at $-1/\tau_1$), and a pair of complex poles, s_3 and s_4. The real pole is closer to the imaginary axis than the complex pair, indicating a slower response mode (e^{-t/τ_1} decays slower than $e^{-\zeta t/\tau_2}$). In general, the speed of response for a given mode increases as the pole location moves farther away from the imaginary axis.

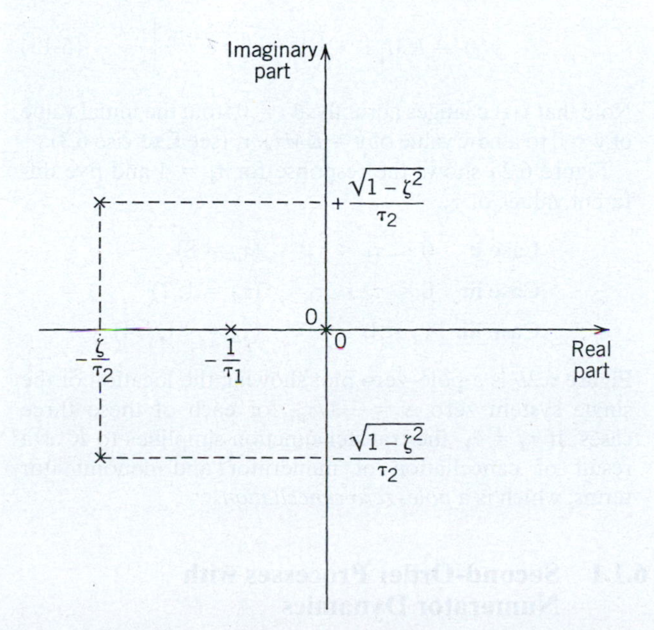

Figure 6.1 Poles of $G(s)$ (Eq. 6-1) plotted in the complex s plane (× denotes a pole location).

Historically, plots such as Fig. 6.1 have played an important role in the design of mechanical and electrical control systems, but they are rarely used in designing process control systems. However, it is helpful to develop some intuitive feeling for the influence of pole locations. A pole to the right of the imaginary axis (called a *right-half plane pole*), for example, $s = +1/\tau$, indicates that one of the system response modes is $e^{t/\tau}$. This mode grows without bound as t becomes large, a characteristic of unstable systems. As a second example, a complex pole always appears as part of a conjugate pair, such as s_3 and s_4 in Eq. 6-2. The complex conjugate poles indicate that the response will contain sine and cosine terms; that is, it will exhibit oscillatory modes.

All of the transfer functions discussed so far can be extended to represent more complex process dynamics simply by adding numerator terms. For example, some control systems contain a *lead–lag element*. The differential equation for this element is

$$\tau_1\frac{dy}{dt} + y = K\left(\tau_a\frac{du}{dt} + u\right) \tag{6-3}$$

In Eq. 6-3 the standard first-order dynamics have been modified by the addition of the du/dt term multiplied by a time constant τ_a. The corresponding transfer function is

Lead-lag element

$$G(s) = \frac{K(\tau_a s + 1)}{\tau_1 s + 1} \tag{6-4}$$

Transfer functions with numerator terms such as $\tau_a s + 1$ above are said to exhibit *numerator dynamics*. Suppose that the integral of u is included in the input terms:

$$\tau_1\frac{dy}{dt} + y = K\left(u + \frac{1}{\tau_a}\int_0^t u(t^*)\,dt^*\right) \tag{6-5}$$

The transfer function for Eq. 6-5, assuming zero initial conditions, is

$$G(s) = \frac{K(\tau_a s + 1)}{\tau_a s(\tau_1 s + 1)} \tag{6-6}$$

In this example, integration of the input introduces a pole at the origin (the $\tau_a s$ term in the denominator), an important point that will be discussed later.

The dynamics of a process are affected not only by the poles of $G(s)$, but also by the values of s that cause the numerator of $G(s)$ to become zero. These values are called the *zeros* of $G(s)$.

Before discussing zeros, it is useful to show several equivalent ways in which transfer functions can be written. In Chapter 4, a standard transfer function form was discussed:

$$G(s) = \frac{\sum_{i=0}^{m} b_i s^i}{\sum_{i=0}^{n} a_i s^i} = \frac{b_m s^m + b_{m-1} s^{m-1} + \cdots + b_0}{a_n s^n + a_{n-1} s^{n-1} + \cdots + a_0} \tag{4-40}$$

which can also be written as

$$G(s) = \frac{b_m}{a_n} \frac{(s - z_1)(s - z_2) \ldots (s - z_m)}{(s - p_1)(s - p_2) \ldots (s - p_n)} \tag{6-7}$$

where the z_i and p_i are zeros and poles, respectively. Note that the poles of $G(s)$ are also the roots of the characteristic equation. This equation is obtained by setting the denominator of $G(s)$, the characteristic polynomial, equal to zero.

It is convenient to express transfer functions in *gain/time constant form*; that is, b_0 is factored out of the numerator of Eq. 4-40 and a_0 out of the denominator to show the steady-state gain explicitly ($K = b_0/a_0 = G(0)$). Then the resulting expressions are factored to give

$$G(s) = K \frac{(\tau_a s + 1)(\tau_b s + 1) \cdots}{(\tau_1 s + 1)(\tau_2 s + 1) \cdots} \tag{6-8}$$

for the case where all factors represent real roots. Thus, the relationships between poles and zeros and the time constants are

$$z_1 = -1/\tau_a, \quad z_2 = -1/\tau_b, \quad \cdots \tag{6-9}$$

$$p_1 = -1/\tau_1, \quad p_2 = -1/\tau_2, \quad \cdots \tag{6-10}$$

The presence or absence of system zeros in Eq. 6-7 has no effect on the number and location of the poles and their associated response modes unless there is an exact cancellation of a pole by a zero with the same numerical value. However, the zeros exert a profound effect on the coefficients of the response modes (i.e., how they are weighted) in the system response. Such coefficients are found by partial fraction expansion. For practical control systems the number of zeros in Eq. 6-7 is less than or equal to the number of poles ($m \leq n$). When $m = n$, the output response is discontinuous after a step input change, as illustrated by Example 6.1.

EXAMPLE 6.1

Calculate the response of the lead–lag element (Eq. 6-4) to a step change of magnitude M in its input.

SOLUTION

For this case,

$$Y(s) = \frac{KM(\tau_a s + 1)}{s(\tau_1 s + 1)} \tag{6-11}$$

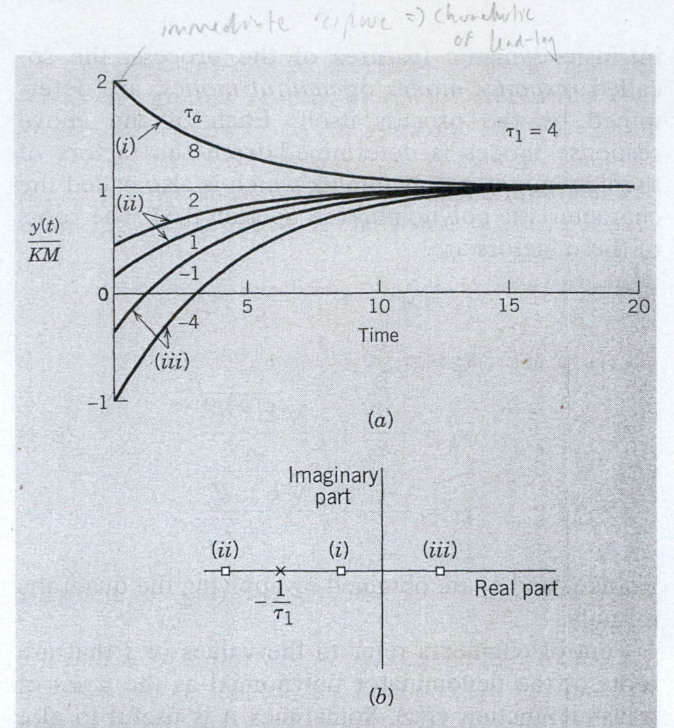

Figure 6.2. (*a*) Step response of a lead–lag process (Eq. 6-13) for five values of a single zero [$y(t = 0) = KM\,\tau_a/\tau_1$]. (*b*) Pole–zero plot for a lead–lag process showing alternative locations of the single zero. × is a pole location; □ is a location of single zero.

which can be expanded into partial fractions

$$Y(s) = KM\left(\frac{1}{s} + \frac{\tau_a - \tau_1}{\tau_1 s + 1}\right) \tag{6-12}$$

yielding the response

$$y(t) = KM\left[1 - \left(1 - \frac{\tau_a}{\tau_1}\right) e^{-t/\tau_1}\right] \tag{6-13}$$

Note that $y(t)$ changes abruptly at $t = 0$ from the initial value of $y = 0$ to a new value of $y = KM\tau_a/\tau_1$ (see Exercise 6.3).

Figure 6.2*a* shows the response for $\tau_1 = 4$ and five different values of τ_a.

Case i: $0 < \tau_1 < \tau_a$ $(\tau_a = 8)$

Case ii: $0 < \tau_a < \tau_1$ $(\tau_a = 1, 2)$

Case iii: $\tau_a < 0$ $(\tau_a = -1, -4)$

Figure 6.2*b* is a pole–zero plot showing the location of the single system zero, $s = -1/\tau_a$, for each of these three cases. If $\tau_a = \tau_1$, the transfer function simplifies to K as a result of cancellation of numerator and denominator terms, which is a *pole–zero cancellation*.

6.1.1 Second-Order Processes with Numerator Dynamics

From inspection of Eq. 6-13 and Fig. 6.2*a*, the presence of a zero in the first-order system causes a jump discontinuity in $y(t)$ at $t = 0$ when the step input is applied. Such an instantaneous step response is possible only when the numerator and denominator polynomials have the same order, which includes the case $G(s) = K$. Industrial processes have higher-order dynamics in the denominator, causing them to exhibit some degree of *inertia*. This feature prevents them from responding instantaneously to any input, including an impulse input. Thus, we say that $m \leq n$ for a system to be *physically realizable*.

EXAMPLE 6.2

For the case of a single zero in an overdamped second-order transfer function,

$$G(s) = \frac{K(\tau_a s + 1)}{(\tau_1 s + 1)(\tau_2 s + 1)} \tag{6-14}$$

calculate the response to a step input of magnitude M and plot the results for $\tau_1 = 4$, $\tau_2 = 1$ and several values of τ_a.

SOLUTION

The response of this system to a step change in input is (see Table 3.1)

$$y(t) = KM\left(1 + \frac{\tau_a - \tau_1}{\tau_1 - \tau_2} e^{-t/\tau_1} + \frac{\tau_a - \tau_2}{\tau_2 - \tau_1} e^{-t/\tau_2}\right) \tag{6-15}$$

Note that $y(t \to \infty) = KM$ as expected; thus, the effect of including the single zero does not change the final value, nor does it change the number or locations of the poles. But the zero does affect how the response modes (exponential terms) are weighted in the solution, Eq. 6-15.

Mathematical analysis (see Exercise 6.3) shows that three types of responses are involved here, as illustrated for eight values of τ_a in Fig. 6.3:

Case i:	$\tau_a > \tau_1$	$(\tau_a = 8, 16)$
Case ii:	$0 < \tau_a \leq \tau_1$	$(\tau_a = 0.5, 1, 2, 4)$
Case iii:	$\tau_a < 0$	$(\tau_a = -1, -4)$

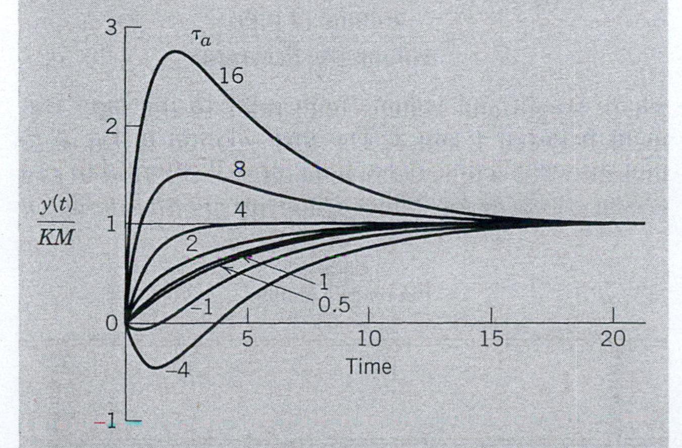

Figure 6.3 Step response of an overdamped second-order system (Eq. 6-14) for different values of τ_a ($\tau_1 = 4$, $\tau_2 = 1$).

where $\tau_1 > \tau_2$ is arbitrarily chosen. Case (i) shows that *overshoot* can occur if τ_a is sufficiently large. Case (ii) is similar to a first-order process response. Case (iii), which has a *positive zero*, also called a *right-half plane zero*, exhibits an *inverse response*, an infrequently encountered yet important dynamic characteristic. An inverse response occurs when the initial response to a step input is in one direction but the final steady state is in the opposite direction. For example, for case (iii), the initial response is in the negative direction while the new steady state $y(\infty)$ is in the positive direction in the sense that $y(\infty) > y(0)$. Inverse responses are associated with right-half plane zeros.

The phenomenon of overshoot or inverse response results from the zero in the above example and will not occur for an overdamped second-order transfer function containing two poles but no zero. These features arise from competing dynamic effects that operate on two different time scales (τ_1 and τ_2 in Example 6.2). For example, an inverse response can occur in a distillation column when the steam pressure to the reboiler is suddenly changed. An increase in steam pressure ultimately will decrease the reboiler level (in the absence of level control) by boiling off more of the liquid. However, the initial effect usually is to increase the amount of frothing on the trays immediately above the reboiler, causing a rapid spillover of liquid from these trays into the reboiler below. This initial increase in reboiler liquid level, is later overwhelmed by a decrease due to the increased vapor boil-up. See Buckley et al. (1985) for a detailed analysis of this phenomenon.

As a second physical example, tubular catalytic reactors with exothermic chemical reactions exhibit an inverse response in exit temperature when the feed temperature is increased. Initially, increased conversion in the entrance region of the bed momentarily depletes reactants at the exit end of the bed, causing less heat generation there and decreasing the exit temperature. Subsequently, higher reaction rates occur, leading to a higher exit temperature, as would be expected. Conversely, if the feed temperature is decreased, the inverse response initially yields a higher exit temperature.

Inverse response or overshoot can be expected whenever two physical effects act on the process output variable in different ways and with different time scales. For the case of reboiler level mentioned above, the *fast* effect of a steam pressure increase is to spill liquid off the trays above the reboiler immediately as the vapor flow increases. The *slow* effect is to remove significant amounts of the liquid mixture from the reboiler through increased boiling. Hence, the relationship between reboiler level and reboiler steam pressure can be represented approximately as an overdamped second-order transfer function with a right-half plane zero.

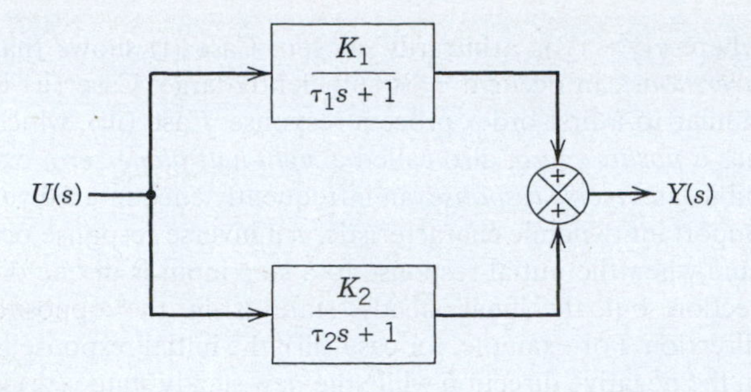

Figure 6.4 Two first-order process elements acting in parallel.

Next, we show that inverse responses can occur for two first-order transfer functions in a parallel arrangement, as shown in Fig. 6.4. The relationship between $Y(s)$ and $U(s)$ can be expressed as

$$\frac{Y(s)}{U(s)} = \frac{K_1}{\tau_1 s + 1} + \frac{K_2}{\tau_2 s + 1} = \frac{K_1(\tau_2 s + 1) + K_2(\tau_1 s + 1)}{(\tau_1 s + 1)(\tau_2 s + 1)} \tag{6-16}$$

or, after rearranging the numerator into standard gain/time constant form, we have

$$\frac{Y(s)}{U(s)} = \frac{(K_1 + K_2)\left(\frac{K_1\tau_2 + K_2\tau_1}{K_1 + K_2} s + 1\right)}{(\tau_1 s + 1)(\tau_2 s + 1)} \tag{6-17}$$

$$K = K_1 + K_2 \tag{6-18}$$

and

$$\tau_a = \frac{K_1\tau_2 + K_2\tau_1}{K_1 + K_2} \tag{6-19}$$

$$= \frac{K_1\tau_2 + K_2\tau_1}{K} \tag{6-20}$$

The condition for an inverse response to exist is $\tau_a < 0$, or

$$\frac{K_1\tau_2 + K_2\tau_1}{K} < 0 \tag{6-21}$$

For $K > 0$, Eq. 6-21 can be rearranged to the more convenient form

$$-\frac{K_2}{K_1} > \frac{\tau_2}{\tau_1} \tag{6-22}$$

For $K > 0$, the inequality in Eq. 6-22 is reversed. Note that Eq. 6-22 indicates that K_1 and K_2 have opposite signs, because $\tau_1 > 0$ and $\tau_2 > 0$. It is left to the reader to show that $K > 0$ when $K_1 > 0$ and that $K < 0$ when $K_1 < 0$. In other words, the sign of the overall transfer function gain is the same as that of the slower process. Exercise 6.5 considers the analysis of a right-half, plane zero in the transfer function.

The step response of the process described by Eq. 6-14 will have a negative slope initially (at $t = 0$) if the product of the gain and step change magnitude is positive ($KM > 0$), τ_a is negative, and τ_1 and τ_2 are both positive. To show this, let $U(s) = M/s$:

$$Y(s) = G(s)U(s) = \frac{KM(\tau_a s + 1)}{s(\tau_1 s + 1)(\tau_2 s + 1)} \tag{6-23}$$

Because differentiation in the time domain corresponds to multiplication by s in the Laplace domain (cf. Chapter 3), we let $z(t)$ denote dy/dt. Then

$$Z(s) = sY(s) = G(s)M = \frac{KM(\tau_a s + 1)}{(\tau_1 s + 1)(\tau_2 s + 1)} \tag{6-24}$$

Applying the Initial Value Theorem,

$$z(0) = \left.\frac{dy}{dt}\right|_{t=0} = \lim_{s\to\infty}\left[s\frac{KM(\tau_a s + 1)}{(\tau_1 s + 1)(\tau_2 s + 1)}\right]$$

$$= \lim_{s\to\infty}\left[\frac{KM(\tau_a + 1/s)}{(\tau_1 + 1/s)(\tau_2 + 1/s)}\right] = \frac{KM\tau_a}{\tau_1\tau_2} \tag{6-25}$$

which has the sign of τ_a if the other constants (KM, τ_1, and τ_2) are positive. Note that if τ_a is zero, the initial slope is zero. Evaluation of Eq. 5-48 for $t = 0$ yields the same result.

6.2 PROCESSES WITH TIME DELAYS

Whenever material or energy is physically moved in a process or plant, there is a *time delay* associated with the movement. For example, if a fluid is transported through a pipe in plug flow, as shown in Fig. 6.5, then the transportation time between points 1 and 2 is given by

$$\theta = \frac{\text{length of pipe}}{\text{fluid velocity}} \tag{6-26}$$

or equivalently, by

$$= \frac{\text{volume of pipe}}{\text{volumetric flowrate}}$$

where length and volume both refer to the pipe segment between 1 and 2. The first relation in Eq. 6-26 indicates why a time delay sometimes is referred to as a *distance–velocity lag*. Other synonyms are *transportation*

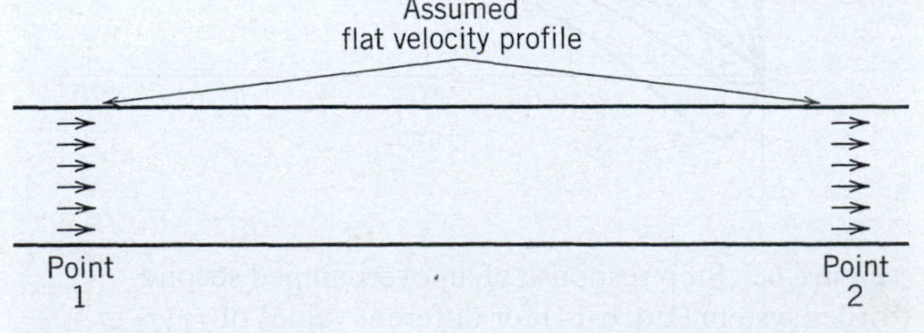

Figure 6.5 Transportation of fluid in a pipe for turbulent flow.

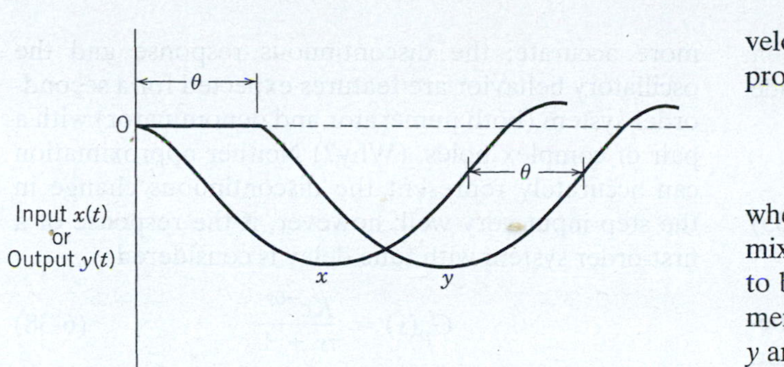

Figure 6.6 The effect of a time delay is a translation of the function in time.

lag, transport delay, and *dead time*. If the plug flow assumption does not hold, for example, with laminar flow or for non-Newtonian liquids, approximation of the bulk transport dynamics using a time delay still may be useful, as discussed below.

Suppose that x is some fluid property at point 1, such as concentration, and y is the same property at point 2 and that both x and y are deviation variables. Then they are related by a time delay θ

$$y(t) = \begin{cases} 0 & t < \theta \\ x(t - \theta) & t \geq \theta \end{cases} \tag{6-27}$$

Thus, the output $y(t)$ is simply the same input function shifted backward in time by θ. Figure 6.6 shows this *translation in time* for an arbitrary $x(t)$.

Equation 3-97 shows that the Laplace transform of a function shifted in time by t_0 units is simply $e^{-t_0 s}$. Thus, the transfer function of a time delay of magnitude θ is given by

$$\frac{Y(s)}{X(s)} = G(s) = e^{-\theta s} \tag{6-28}$$

Besides the physical movement of liquid and solid materials, there are other sources of time delays in process control problems. For example, the use of a chromatograph to measure concentration in liquid or gas stream samples taken from a process introduces a time delay, the analysis time. One distinctive characteristic of chemical processes is the common occurrence of time delays.

Even when the plug flow assumption is not valid, transportation processes usually can be modeled approximately by the transfer function for a time delay given in Eq. 6-28. For liquid flow in a pipe, the plug flow assumption is most nearly satisfied when the radial velocity profile is nearly flat, a condition that occurs for Newtonian fluids in turbulent flow. For non-Newtonian fluids and/or laminar flow, the fluid transport process still might be modeled by a time delay based on the average fluid velocity. A more general approach is to model the flow process as a *first-order plus time-delay* transfer function

$$G(s) = \frac{e^{-\theta_m s}}{\tau_m s + 1} \tag{6-29}$$

where τ_m is a time constant associated with the degree of mixing in the pipe or channel. Both τ_m and θ_m may have to be determined from empirical relations or by experiment. Note that the process gain in (6-29) is unity when y and x are material properties such as composition.

Next we demonstrate that analytical expressions for time delays can be derived from the application of conservation equations. In Fig. 6.5 suppose that a very small cell of liquid passes point 1 at time t. It contains $Vc_1(t)$ units of the chemical species of interest where V is the volume of material in the cell and c_1 is the concentration of the species. At time $t + \theta$, the cell passes point 2 and contains $Vc_2(t + \theta)$ units of the species. If the material moves in plug flow, not mixing at all with adjacent material, then the amount of species in the cell is constant:

$$Vc_2(t + \theta) = Vc_1(t) \tag{6-30}$$

or

$$c_2(t + \theta) = c_1(t) \tag{6-31}$$

An equivalent way of writing (6-31) is

$$c_2(t) = c_1(t - \theta) \tag{6-32}$$

if the flow rate is constant. Putting (6-32) in deviation form (by subtracting the steady-state version of (6-32)) and taking Laplace transforms gives

$$\frac{C_2'(s)}{C_1'(s)} = e^{-\theta s} \tag{6-33}$$

When the fluid is incompressible, flow rate changes at point 1 propagate instantaneously to any other point in the pipe. For compressible fluids such as gases, the simple expression of (6-33) may not be accurate. Note that use of a constant time delay implies constant flow rate.

6.2.1 Polynomial Approximations to $e^{-\theta s}$

The exponential form of Eq. 6-28 is a nonrational transfer function that cannot be expressed as a rational function, a ratio of two polynominals in s. Consequently, (6-28) cannot be factored into poles and zeros, a convenient form for analysis, as discussed in Section 6.1. However, it is possible to approximate $e^{-\theta s}$ by polynomials using either a Taylor series expansion or a Padé approximation.

The Taylor series expansion for $e^{-\theta s}$ is:

$$e^{-\theta s} = 1 - \theta s + \frac{\theta^2 s^2}{2!} - \frac{\theta^3 s^3}{3!} + \frac{\theta^4 s^4}{4!} - \frac{\theta^5 s^5}{5!} + \cdots \tag{6-34}$$

The Padé approximation for a time delay is a ratio of two polynomials in s with coefficients selected to match the terms of a truncated Taylor series expansion of $e^{-\theta s}$. The simplest pole–zero approximation is the 1/1 Padé approximation:

$$e^{-\theta s} \approx G_1(s) = \frac{1 - \frac{\theta}{2}s}{1 + \frac{\theta}{2}s} \tag{6-35}$$

Equation 6-35 is called the 1/1 Padé approximation because it is first-order in both numerator and denominator.

Performing the indicated long division in (6-35) gives

$$G_1(s) = 1 - \theta s + \frac{\theta^2 s^2}{2} - \frac{\theta^3 s^3}{4} + \cdots \tag{6-36}$$

A comparison of Eqs. 6-34 and 6-36 indicates that $G_1(s)$ is correct through the first three terms. There are higher-order Padé approximations, for example, the 2/2 Padé approximation:

$$e^{-\theta s} \approx G_2(s) = \frac{1 - \frac{\theta s}{2} + \frac{\theta^2 s^2}{12}}{1 + \frac{\theta s}{2} + \frac{\theta^2 s^2}{12}} \tag{6-37}$$

Figure 6.7*a* illustrates the response of the 1/1 and 2/2 Padé approximations to a unit step input. The first-order approximation exhibits the same type of discontinuous response discussed in Section 6.1 in connection with a first-order system with a right-half plane zero. (Why?) The second-order approximation is somewhat more accurate; the discontinuous response and the oscillatory behavior are features expected for a second-order system (both numerator and denominator) with a pair of complex poles. (Why?) Neither approximation can accurately represent the discontinuous change in the step input very well; however, if the response of a first-order system with time delay is considered,

$$G_p(s) = \frac{Ke^{-\theta s}}{\tau s + 1} \tag{6-38}$$

Figure 6.7*b* shows that the approximations are satisfactory for a step response, especially if $\theta \ll \tau$, which is often the case.

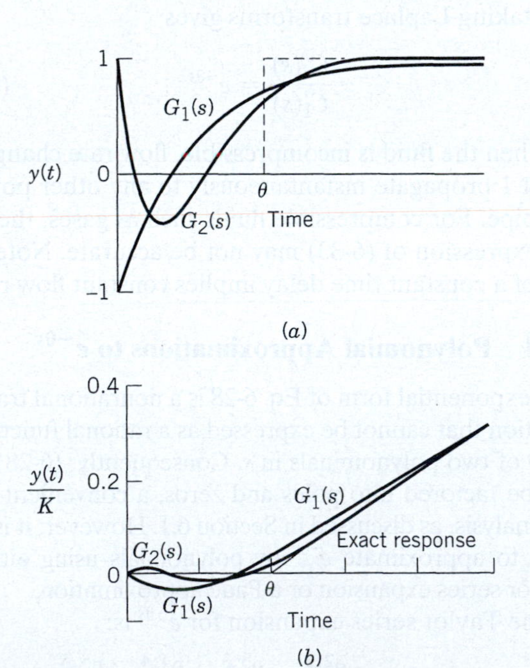

Figure 6.7 (*a*) Step response of 1/1 and 2/2 Padé approximations of a time delay ($G_1(s)$ and $G_2(s)$, respectively). (*b*) Step response of a first-order plus time-delay process ($\theta = 0.25\tau$) using 1/1 and 2/2 Padé approximations of $e^{-\theta s}$.

EXAMPLE 6.3

The trickle-bed catalytic reactor shown in Fig. 6.8 utilizes product recycle to obtain satisfactory operating conditions for temperature and conversion. Use of a high recycle rate eliminates the need for mechanical agitation. Concentrations of the single reactant and the product are measured at a point in the recycle line where the product stream is removed. A liquid phase first-order reaction is involved.

Under normal operating conditions, the following assumptions may be made:

(i) The reactor operates isothermally with a reaction rate given by $r = kc$, where $-r$ denotes the rate of disappearance of reactant per unit volume, c is the concentration of reactant, and k is the rate constant.

(ii) All flow rates and the liquid volume V are constant.

(iii) No reaction occurs in the piping. The dynamics of the exit and recycle lines can be approximated as constant time delays θ_1 and θ_2, as indicated in the figure. Let c_1 denote the reactant concentration at the measurement point.

(iv) Because of the high recycle flow rate, mixing in the reactor is complete.

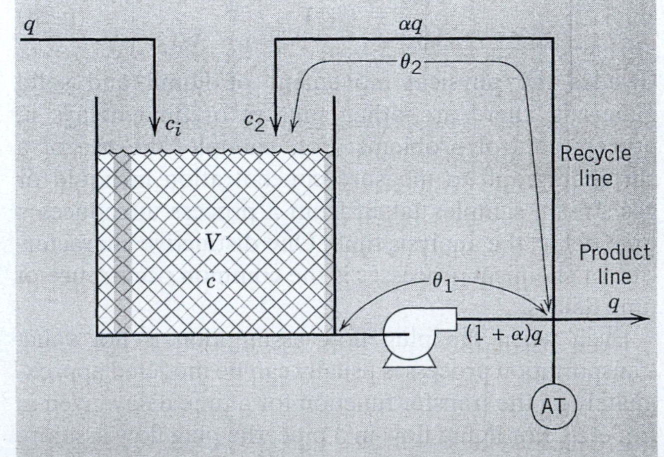

Figure 6.8 Schematic diagram of a trickle-bed reactor with recycle line. (AT: analyzer transmitter; θ_1: time delay associated with material flow from reactor outlet to the composition analyzer; θ_2: time delay associated with material flow from analyzer to reactor inlet.)

(a) Derive an expression for the transfer function $C_1'(s)/C_i'(s)$.

(b) Using the following information, calculate $c_1'(t)$ for a step change in $c_i'(t) = 2000\ \text{kg/m}^3$

Parameter Values

$V = 5\ \text{m}^3$	$\alpha = 12$
$q = 0.05\ \text{m}^3/\text{min}$	$\theta_1 = 0.9$ min
$k = 0.04\ \text{min}^{-1}$	$\theta_2 = 1.1$ min

SOLUTION

(a) In analogy with Eq. 2-66; the component balance around the reactor is,

$$V\frac{dc}{dt} = qc_i + \alpha q c_2 - (1+\alpha)qc - Vkc \quad (6\text{-}39)$$

where the concentration of the reactant is denoted by c. Equation 6-39 is linear with constant coefficients. Subtracting the steady-state equation and substituting deviation variables yields

$$V\frac{dc'}{dt} = qc_i' + \alpha q c_2' - (1+\alpha)qc' - Vkc' \quad (6\text{-}40)$$

Additional relations are needed for $c_2'(t)$ and $c_1'(t)$. Because the exit and recycle lines can be modeled as time delays,

$$c_1'(t) = c'(t-\theta_1) \quad (6\text{-}41)$$

$$c_2'(t) = c_1'(t-\theta_2) \quad (6\text{-}42)$$

Equations 6-40 through 6-42 provide the process model for the isothermal reactor with recycle. Taking the Laplace transform of each equation yields

$$sVC'(s) = qC_i'(s) + \alpha q C_2'(s) - (1+\alpha)qC'(s) - VkC'(s) \quad (6\text{-}43)$$

$$C_1'(s) = e^{-\theta_1 s}C'(s) \quad (6\text{-}44)$$

$$C_2'(s) = e^{-\theta_2 s}C_1'(s) = e^{-(\theta_1+\theta_2)s}C'(s) = e^{-\theta_3 s}C'(s) \quad (6\text{-}45)$$

where $\theta_3 \triangleq \theta_1 + \theta_2$. Substitute (6-45) into (6-43) and solve for $C'(s)$:

$$C'(s) = \frac{q}{sV - \alpha q e^{-\theta_3 s} + (1+\alpha)q + Vk}C_i'(s) \quad (6\text{-}46)$$

Equation 6-46 can be rearranged to the following form:

$$C'(s) = \frac{K}{\tau s + 1 + \alpha K(1 - e^{-\theta_3 s})}C_i'(s) \quad (6\text{-}47)$$

where

$$K = \frac{q}{q+Vk} \quad (6\text{-}48)$$

$$\tau = \frac{V}{q+Vk} \quad (6\text{-}49)$$

Note that, in the limit as $\theta_3 \to 0$, $e^{-\theta_3 s} \to 1$ and

$$C'(s) = \frac{K}{\tau s + 1}C_i'(s) \quad (6\text{-}50)$$

So K and τ can be interpreted as the process gain and time constant, respectively, of a recycle reactor with no time delay in the recycle line, which is equivalent to a stirred isothermal reactor with no recycle.

The desired transfer function $C_1'(s)/C_i'(s)$ is obtained by combining Eqs. 6-47 and 6-44 to obtain

$$\frac{C_1'(s)}{C_i'(s)} = \frac{Ke^{-\theta_1 s}}{\tau s + 1 + \alpha K(1 - e^{-\theta_3 s})} \quad (6\text{-}51)$$

(b) To find $c_1'(t)$ when $c_i'(t) = 2000\ \text{kg/m}^3$, we multiply (6-51) by $2000/s$

$$C_1'(s) = \frac{2000Ke^{-\theta_1 s}}{s[\tau s + 1 + \alpha K(1 - e^{-\theta_3 s})]} \quad (6\text{-}52)$$

and take the inverse Laplace transform. From inspection of (6-52) it is clear that the numerator time delay can be inverted directly; however, there is no transform in Table 3.1 that contains a time-delay term in the denominator. To obtain an analytical solution, the denominator time-delay term must be eliminated by introducing a rational approximation, for example, the 1/1 Padé approximation in (6-35). Substituting (6-35) and rearranging yields

$$C_1'(s) \approx \frac{2000K\left(\frac{\theta_3}{2}s + 1\right)e^{-\theta_1 s}}{s\left[\tau\frac{\theta_3}{2}s^2 + \left(\tau + \frac{\theta_3}{2} + \alpha K\theta_3\right)s + 1\right]} \quad (6\text{-}53)$$

This expression can be written in the form

$$C_1'(s) = \frac{2000K(\tau_a s + 1)e^{-\theta_1 s}}{s(\tau_1 s + 1)(\tau_2 s + 1)} \quad (6\text{-}54)$$

where $\tau_a = \theta_3/2$ and τ_1 and τ_2 are obtained by factoring the expression in brackets. For $\alpha K\theta_3 > 0$, τ_1 and τ_2 are real and distinct.

The numerical parameters in (6-53) are

$$K = \frac{q}{q+Vk} = \frac{0.05}{0.05 + (5)(0.04)} = 0.2$$

$$\tau = \frac{V}{q+Vk} = 20\ \text{min}$$

Substituting these values in (6-53) gives

$$C_1'(s) = \frac{400(s+1)e^{-0.9s}}{s[20s^2 + (20 + 1 + (24)(0.2)(1))s + 1]} = \frac{400(s+1)e^{-0.9s}}{s(25s+1)(0.8s+1)} \quad (6\text{-}55)$$

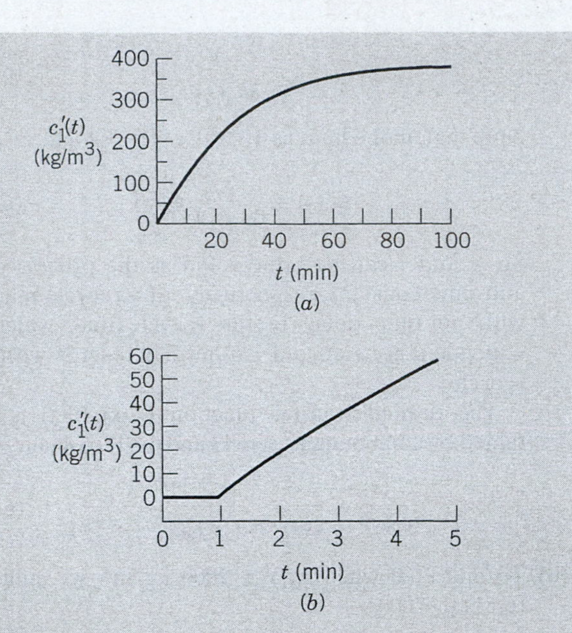

Figure 6.9 Recycle reactor composition measured at analyzer: (*a*) complete response; (*b*) detailed view of short-term response.

Taking the inverse Laplace and introducing the delayed unit step function $S(t - 0.9)$ gives

$$c_1'(t) = 400(1 - 0.99174e^{-(t-0.9)/25} - 0.00826e^{-(t-0.9)/0.8})S(t - 0.9) \quad (6\text{-}56)$$

which is plotted in Fig. 6.9. A numerical solution of Eqs. 6-40 through 6-42 that uses no approximation for the total recycle delay is indistinguishable from the approximate solution. Note that in obtaining (6-56), we did not approximate the numerator delay. It is dealt with exactly and appears as a time delay of 0.9 min in several terms.

6.3 APPROXIMATION OF HIGHER-ORDER TRANSFER FUNCTIONS

In this section, we present a general approach for approximating higher-order transfer function models with lower-order models that have similar dynamic and steady-state characteristics. The low-order models are more convenient for control system design and analysis, as discussed in Chapter 12.

In Eq. 6-34 we showed that the transfer function for a time delay can be expressed as a Taylor series expansion. For small values of s, truncating the expansion after the first-order term provides a suitable approximation:

$$e^{-\theta s} \approx 1 - \theta s \quad (6\text{-}57)$$

Note that this time-delay approximation is a right-half plane (RHP) zero at $s = +\theta$. An alternative first-order approximation consists of the transfer function,

$$e^{-\theta s} = \frac{1}{e^{\theta s}} \approx \frac{1}{1 + \theta s} \quad (6\text{-}58)$$

which is based on the approximation, $e^{\theta s} \approx 1 + \theta s$. Note that the time constant has a value of θ.

Equations 6-57 and 6-58 were derived to approximate time-delay terms. However, these expressions can be reversed to approximate the pole or zero on the right-hand side of the equation by the time-delay term on the left side. These pole and zero approximations will be demonstrated in Example 6.4.

6.3.1 Skogestad's "Half Rule"

Skogestad (2003) has proposed a related approximation method for higher-order models that contain multiple time constants. He approximates the largest neglected time constant in the denominator in the following manner. One-half of its value is added to the existing time delay (if any), and the other half is added to the smallest retained time constant. Time constants that are smaller than the largest neglected time constant are approximated as time delays using (6-58). A right-half plane zero is approximated by (6-57). The motivation for this "half rule" is to derive approximate low-order models that are more appropriate for control system design. Examples 6.4 and 6.5 illustrate Skogestad's half rule.

EXAMPLE 6.4

Consider a transfer function:

$$G(s) = \frac{K(-0.1s + 1)}{(5s + 1)(3s + 1)(0.5s + 1)} \quad (6\text{-}59)$$

Derive an approximate first-order-plus-time-delay model,

$$\tilde{G}(s) = \frac{Ke^{-\theta s}}{\tau s + 1} \quad (6\text{-}60)$$

using two methods:

(a) The Taylor series expansions of Eqs. 6-57 and 6-58.

(b) Skogestad's half rule.

Compare the normalized responses of $G(s)$ and the approximate models for a unit step input.

SOLUTION

(a) The dominant time constant (5) is retained. Applying the approximations in (6-57) and (6-58) gives

$$-0.1s + 1 \approx e^{-0.1s} \quad (6\text{-}61)$$

and

$$\frac{1}{3s + 1} \approx e^{-3s} \qquad \frac{1}{0.5s + 1} \approx e^{-0.5s} \quad (6\text{-}62)$$

Substitution into (6-59) gives the Taylor series approximation, $\tilde{G}_{TS}(s)$:

$$\tilde{G}_{TS}(s) = \frac{Ke^{-0.1s}e^{-3s}e^{-0.5s}}{5s + 1} = \frac{Ke^{-3.6s}}{5s + 1} \quad (6\text{-}63)$$

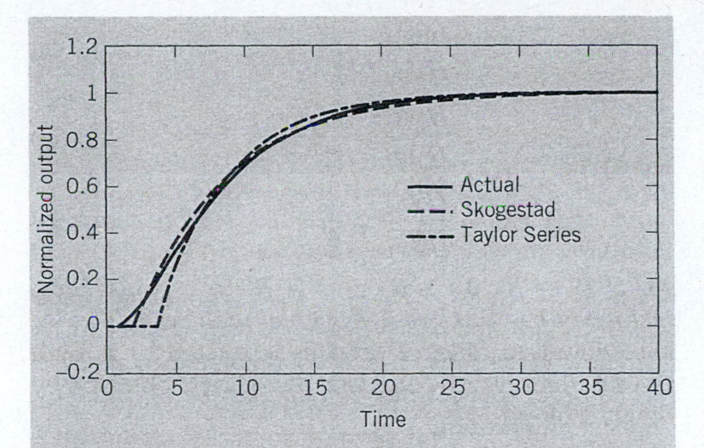

Figure 6.10 Comparison of the actual and approximate models for Example 6.4.

(b) To apply Skogestad's method, we note that the largest neglected time constant in (6-59) has a value of three. According to his "half rule," half of this value is added to the next largest time constant to generate a new time constant, $\tau = 5 + 0.5(3) = 6.5$. The other half provides a new time delay of $0.5(3) = 1.5$. The approximation of the RHP zero in (6-61) provides an additional time delay of 0.1. Approximating the smallest time constant of 0.5 in (6-59) by (6-58) produces an additional time delay of 0.5. Thus, the total time delay in (6-60) is

$$\theta = 1.5 + 0.1 + 0.5 = 2.1$$

and $G(s)$ can be approximated as

$$\tilde{G}_{Sk}(s) = \frac{Ke^{-2.1s}}{6.5s + 1} \tag{6-64}$$

The normalized step responses for $G(s)$ and the two approximate models are shown in Fig. 6.10. Skogestad's method provides better agreement with the actual response.

EXAMPLE 6.5

Consider the following transfer function:

$$G(s) = \frac{K(1 - s)e^{-s}}{(12s + 1)(3s + 1)(0.2s + 1)(0.05s + 1)} \tag{6-65}$$

Use Skogestad's method to derive two approximate models:

(a) A first-order-plus-time-delay model in the form of (6-60).

(b) A second-order-plus-time-delay model in the form

$$\tilde{G}(s) = \frac{Ke^{-\theta s}}{(\tau_1 s + 1)(\tau_2 s + 1)} \tag{6-66}$$

Compare the normalized output responses for $G(s)$ and the approximate models to a unit step input.

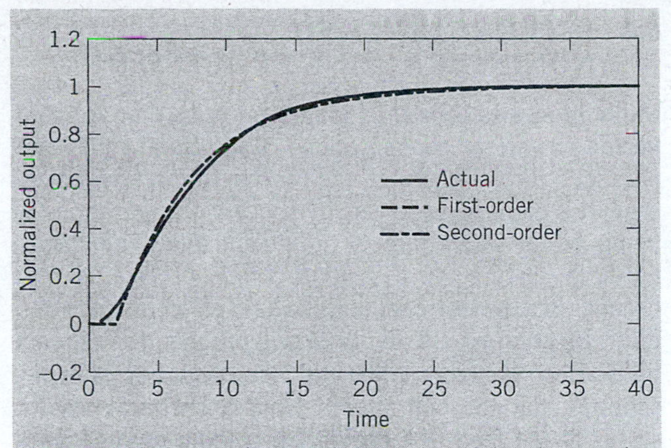

Figure 6.11 Comparison of the actual model and approximate models for Example 6.5. The actual and second-order model responses are almost indistinguishable.

SOLUTION

(a) For the first-order-plus-time-delay model, the dominant time constant (12) is retained. One-half of the largest neglected time constant (3) is allocated to the retained time constant and one-half to the approximate time delay. Also, the small time constants (0.2 and 0.05) and the zero (1) are added to the original time delay. Thus, the model parameters in (6-60) are

$$\theta = 1 + \frac{3}{2} + 0.2 + 0.05 + 1 = 3.75$$

$$\tau = 12 + \frac{3}{2} = 13.5$$

(b) An analogous derivation for the second-order-plus-time-delay model gives

$$\theta = 1 + \frac{0.2}{2} + 0.05 + 1 = 2.15$$

$$\tau_1 = 12, \qquad \tau_2 = 3 + 0.1 = 3.1$$

In this case, the half rule is applied to the third largest time constant, 0.2.

The normalized step responses of the original and approximate transfer functions are shown in Fig. 6.11. The second-order model provides an excellent approximation, because the neglected time constants are much smaller than the retained time constants. The first-order-plus-time-delay model is not as accurate, but it does provide a suitable approximation of the actual response.

Skogestad (2003) has also proposed approximations for left-half plane zeros of the form, $\tau_a s + 1$, where $\tau_a > 0$. However, these approximations are more complicated and beyond the scope of this book. In these situations, a simpler model can be obtained by empirical fitting of the step response using the techniques in Chapter 7.

6.4 INTERACTING AND NONINTERACTING PROCESSES

Many processes consist of individual units that are connected in various configurations that include series and parallel structures, as well the recycle of material or energy. It is convenient to classify process configurations as being either *interacting* or *noninteracting*. The distinguishing feature of a *noninteracting process* is that changes in a downstream unit have no effect on upstream units. By contrast, for an *interacting process*, downstream units affect upstream units, and *vice versa*. For example, suppose that the exit stream from a chemical reactor serves as the feed to a distillation column used to separate product from unreacted feed. Changes in the reactor affect column operation but not *vice versa*—a noninteracting process. But suppose that the distillate stream from the column contains largely unreacted feed; then, it could be beneficial to increase the reactor yield by recycling the distillate to the reactor where it would be added to the fresh feed. Now, changes in the column affect the reactor, and *vice versa*—an interacting process.

An example of a system that does *not* exhibit interaction was discussed in Example 4.4. The two storage tanks were connected in series in such a way that liquid level in the second tank did not influence the level in the first tank (Fig. 4.3). The following transfer functions relating tank levels and flows were derived:

$$\frac{H'_1(s)}{Q'_i(s)} = \frac{K_1}{\tau_1 s + 1} \tag{4-53}$$

$$\frac{Q'_1(s)}{H'_1(s)} = \frac{1}{K_1} \tag{4-54}$$

$$\frac{H'_2(s)}{Q'_1(s)} = \frac{K_2}{\tau_2 s + 1} \tag{4-55}$$

$$\frac{Q'_2(s)}{H'_2(s)} = \frac{1}{K_2} \tag{4-56}$$

where $K_1 = R_1$, $K_2 = R_2$, $\tau_1 = A_1R_1$, $\tau_2 = A_2R_2$. Each tank level has first-order dynamics with respect to its inlet flow rate. Tank 2 level h_2 is related to q_i by a second-order transfer function that can be obtained by simple multiplication:

$$\frac{H'_2(s)}{Q'_i(s)} = \frac{H'_2(s)}{Q'_1(s)}\frac{Q'_1(s)}{H'_1(s)}\frac{H'_1(s)}{Q'_i(s)}$$

$$= \frac{K_2}{(\tau_1 s + 1)(\tau_2 s + 1)} \tag{6-67}$$

A simple generalization of the dynamic expression in Eq. 6-67 is applicable to n tanks in series shown in Fig. 6.12:

$$\frac{H'_n(s)}{Q'_i(s)} = \frac{K_n}{\prod_{i=1}^{n}(\tau_i s + 1)} \tag{6-68}$$

and

$$\frac{Q'_n(s)}{Q'_i(s)} = \frac{1}{\prod_{i=1}^{n}(\tau_i s + 1)} \tag{6-69}$$

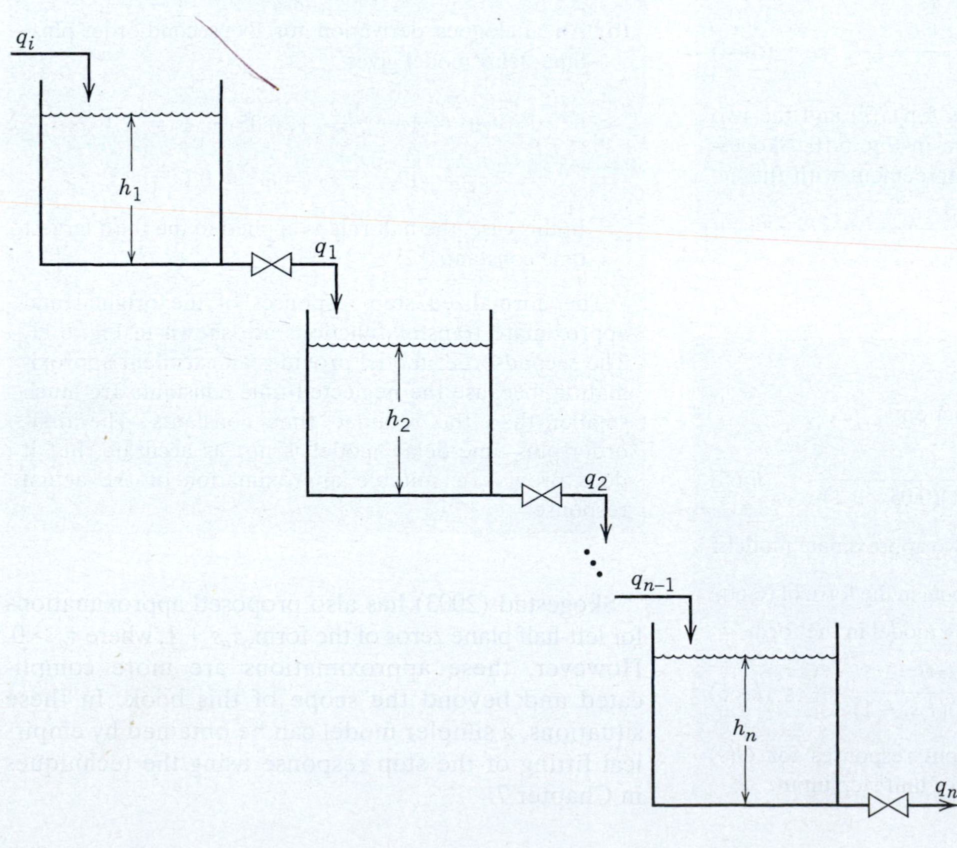

Figure 6.12 A series configuration of n noninteracting tanks.

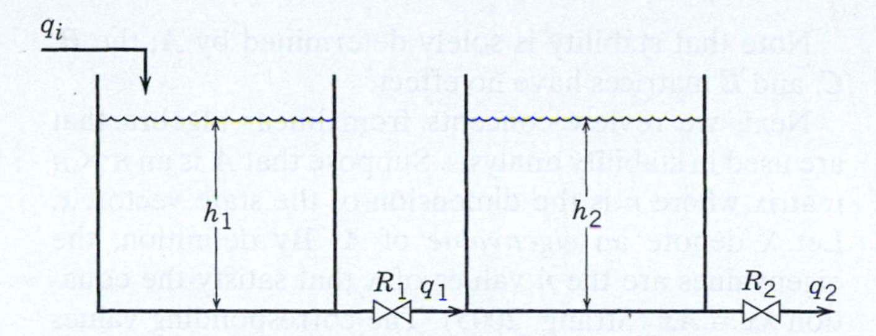

Figure 6.13 Two tanks in series whose liquid levels interact.

Next consider an example of an interacting process that is similar to the two-tank process in Chapter 4. The process shown in Fig. 6.13 is called an *interacting system* because h_1 depends on h_2 (and vice versa) as a result of the interconnecting stream with flow rate q_1. Therefore, the equation for flow from Tank 1 to Tank 2 must be written to reflect that physical feature:

$$q_1 = \frac{1}{R_1}(h_1 - h_2) \tag{6-70}$$

For the Tank 1 level transfer function, a much more complicated expression than (4-53) results:

$$\frac{H_1'(s)}{Q_i'(s)} = \frac{(R_1 + R_2)\left(\dfrac{R_1R_2A_2}{R_1 + R_2}s + 1\right)}{R_1R_2A_1A_2s^2 + (R_2A_2 + R_1A_1 + R_2A_1)s + 1} \tag{6-71}$$

It is of the form

$$\frac{H_1'(s)}{Q_i'(s)} = \frac{K_1'(\tau_a s + 1)}{\tau^2 s^2 + 2\zeta\tau s + 1} \tag{6-72}$$

In Exercise 6.15, the reader can show that $\zeta > 1$ by analyzing the denominator of (6-71); hence, the transfer function is overdamped and second-order, and has a negative zero at $-1/\tau_a$, where $\tau_a = R_1R_2A_2/(R_1 + R_2)$. The transfer function relating h_1 and h_2,

$$\frac{H_2'(s)}{H_1'(s)} = \frac{\dfrac{R_2}{R_1 + R_2}}{\dfrac{R_1R_2A_2}{R_1 + R_2}s + 1} \tag{6-73}$$

is of the form $K_2'/(\tau_a s + 1)$. Consequently, the overall transfer function between H_2' and Q_i' is

$$\frac{H_2'(s)}{Q_i'(s)} = \frac{R_2}{\tau^2 s^2 + 2\zeta\tau s + 1} \tag{6-74}$$

The above analysis of the interacting two-tank system is more complicated than that for the noninteracting system of Example 4.4. The denominator polynomial can no longer be factored into two first-order terms, each associated with a single tank. Also, the numerator of the first tank transfer function in (6-72) contains a zero that modifies the dynamic behavior along the lines suggested in Section 6.1.

6.5 STATE-SPACE AND TRANSFER FUNCTION MATRIX MODELS

Dynamic models derived from physical principles typically consist of one or more ordinary differential equations (ODEs). In this section, we consider a general class of ODE models referred to as *state-space models*, that provide a compact and useful representation of dynamic systems. Although we limit our discussion to linear state-space models, nonlinear state-space models are also very useful and provide the theoretical basis for the analysis of nonlinear processes (Henson and Seborg, 1997; Khalil, 2002).

Consider a linear state-space model,

$$\dot{\boldsymbol{x}} = \boldsymbol{Ax} + \boldsymbol{Bu} + \boldsymbol{Ed} \tag{6-75}$$

$$\boldsymbol{y} = \boldsymbol{Cx} \tag{6-76}$$

where $\boldsymbol{x}$ is the *state vector*; $\boldsymbol{u}$ is the input vector of manipulated variables (also called *control variables*); $\boldsymbol{d}$ is the disturbance vector; and $\boldsymbol{y}$ is the output vector of measured variables. (**Boldface symbols** are used to denote vectors and matrices, and plain text to represent scalars.) The elements of $\boldsymbol{x}$ are referred to as *state variables*. The elements of $\boldsymbol{y}$ are typically a subset of $\boldsymbol{x}$, namely, the state variables that are measured. In general, $\boldsymbol{x}$, $\boldsymbol{u}$, $\boldsymbol{d}$ and $\boldsymbol{y}$ are functions of time. The time derivative of $\boldsymbol{x}$ is denoted by $\dot{\boldsymbol{x}}(=d\boldsymbol{x}/dt)$; it is also a vector. Matrices $\boldsymbol{A}$, $\boldsymbol{B}$, $\boldsymbol{C}$, and $\boldsymbol{E}$ are constant matrices. The vectors in (6-75) can have different dimensions (or "lengths") and are usually written as deviation variables.

Because the state-space model in Eqs. (6-75) and (6-76) may seem rather abstract, it is helpful to consider a physical example.

EXAMPLE 6.6

Show that the linearized CSTR model of Example 4.8 can be written in the state-space form of Eqs. 6-75 and 6-76. Derive state-space models for two cases:

(a) Both c_A and T are measured

(b) Only T is measured

SOLUTION

The linearized CSTR model in Eqs. 4-84 and 4-85 can be written in vector-matrix form using deviation variables:

$$\begin{bmatrix} \dfrac{dc_A'}{dt} \\ \dfrac{dT'}{dt} \end{bmatrix} = \begin{bmatrix} a_{11} & a_{12} \\ a_{21} & a_{22} \end{bmatrix} \begin{bmatrix} c_A' \\ T' \end{bmatrix} + \begin{bmatrix} 0 \\ b_2 \end{bmatrix} T_c' \tag{6-77}$$

Let $x_1 \triangleq c'_A$ and $x_2 \triangleq T'$, and denote their time derivatives by $\dot{x}_1$ and $\dot{x}_2$. In (6-77) the coolant temperature T_c is considered to be a manipulated variable. For this example, there is a single control variable, $u \triangleq T'_c$, and no disturbance variable. Substituting these definitions into (6-77) gives

$$\begin{bmatrix} \dot{x}_1 \\ \dot{x}_2 \end{bmatrix} = \underbrace{\begin{bmatrix} a_{11} & a_{12} \\ a_{21} & a_{22} \end{bmatrix}}_{A} \begin{bmatrix} x_1 \\ x_2 \end{bmatrix} + \underbrace{\begin{bmatrix} 0 \\ b_2 \end{bmatrix}}_{B} u \tag{6-78}$$

which is in the form of Eq. 6-75 with $x = \text{col}[x_1, x_2]$. ("col" denotes a column vector.)

(a) If both T and c_A are measured, then $y = x$ and $C = I$ in Eq. 6-76, where I denotes the 2×2 identity matrix. A and B are defined in (6-78).

(b) When only T is measured, output vector y is a scalar, $y = T'$, and C is a row vector, $C = [0, 1]$.

Note that the state-space model for Example 6.6 has $d = 0$, because disturbance variables were not included in (6-77). By contrast, suppose that the feed composition and feed temperature are considered to be disturbance variables in the original nonlinear CSTR model in Eqs. 2-66 and 2-68. Then the linearized model would include two additional deviation variables c'_{Ai} and T'_i, which would also be included in (6-77). As a result, (6-78) would be modified to include two disturbance variables, $d_1 \triangleq c'_{Ai}$ and $d_2 \triangleq T'_i$.

The state-space model in Eq. 6-75 contains both dependent variables, the elements of x, and independent variables, the elements of u and d. But why is x referred to as the "state vector"? This term is used because $x(t)$ uniquely determines the state of the system at any time, t. Suppose that at time t, the initial value $x(0)$ is specified and $u(t)$ and $d(t)$ are known over the time period $[0, t]$. Then $x(t)$ is unique and can be determined from the analytical solution or by numerical integration. Analytical techniques are described in control engineering textbooks (e.g., Franklin et al., 2005; Ogata, 2008), while numerical solutions can be readily obtained using software packages such as *MATLAB* or *Mathematica*.

6.5.1 Stability of State-Space Models

A detailed analysis of state-space models is beyond the scope of this book but is available elsewhere (e.g., Franklin et al., 2005; Ogata, 2008). One important property of state-space models is *stability*. A state-space model is said to be stable if the response $x(t)$ is bounded for all $u(t)$ and $d(t)$ that are bounded. The stability characteristics of a state-space model can be determined from a necessary and sufficient condition:

Stability Criterion for State-Space Models

The state-space model in Eq. (6-75) will exhibit a bounded response $x(t)$ for all bounded $u(t)$ and $d(t)$ if and only if all of the eigenvalues of A have negative real parts.

Note that stability is solely determined by A; the B, C, and E matrices have no effect.

Next, we review concepts from linear algebra that are used in stability analysis. Suppose that A is an $n \times n$ matrix where n is the dimension of the state vector, x. Let λ denote an *eigenvalue* of A. By definition, the eigenvalues are the n values of λ that satisfy the equation $\lambda x = Ax$ (Strang, 2005). The corresponding values of x are the *eigenvectors* of A. The eigenvalues are the roots of the characteristic equation.

$$|\lambda I - A| = 0 \tag{6-79}$$

where I is the $n \times n$ identity matrix and $|\lambda I - A|$ denotes the determinant of the matrix $\lambda I - A$.

EXAMPLE 6.7

Determine the stability of the state-space model with the following A matrix:

$$A = \begin{bmatrix} -4.0 & 0.3 & 1.5 \\ 1.2 & -2.0 & 1.0 \\ -0.5 & 2.0 & -3.5 \end{bmatrix}$$

SOLUTION

The stability criterion for state-space models indicates that stability is determined by the eigenvalues of A. They can be calculated using the MATLAB command, *eig*, after defining A:

```
A = [-4.0 0.3 1.5; 1.2 -2  1.0; -0.5 2.0 -3.5]
eig(A)
```

The eigenvalues of A are -0.83, $-4.33 + 1.18j$, and $-4.33 - 1.18j$ where $j \triangleq \sqrt{-1}$. Because all three eigenvalues have negative real parts, the state-space model is stable, although the dynamic behavior will exhibit oscillation due to the presence of imaginary components in the eigenvalues.

6.5.2 The Relationship between State-Space and Transfer Function Models

State-space models can be converted to equivalent transfer function models. Consider again the CSTR model in (6-78), which can be expanded as

$$\dot{x}_1 = a_{11}x_1 + a_{12}x_2 \tag{6-80}$$

$$\dot{x}_2 = a_{21}x_1 + a_{22}x_2 + b_2u \tag{6-81}$$

Apply the Laplace transform to each equation (assuming zero initial conditions for each deviation variable, x_1 and x_2):

$$sX_1(s) = a_{11}X_1(s) + a_{12}X_2(s) \tag{6-82}$$

$$sX_2(s) = a_{21}X_1(s) + a_{22}X_2(s) + b_2U(s) \tag{6-83}$$

Solving (6-82) for $X_2(s)$ and substituting into (6-83) gives the equivalent transfer function model relating X_1 and U:

$$\frac{X_1(s)}{U(s)} = \frac{a_{12}b_2}{s^2 - (a_{11} + a_{22})s + a_{11}a_{22} - a_{12}a_{21}} \tag{6-84}$$

Equation 6-82 can be used to derive the transfer function relating X_2 and U:

$$\frac{X_2(s)}{U(s)} = \frac{b_2(s - a_{11})}{s^2 - (a_{11} + a_{22})s + a_{11}a_{22} - a_{12}a_{21}} \tag{6-85}$$

Note that these two transfer functions are also the transfer functions for $C'_A(s)/T'_c(s)$ and $T'(s)/T'_c(s)$, respectively, as a result of the definitions for x_1, x_2, and u. Furthermore, the roots of the denominator of (6-84) and (6-85) are also the eigenvalues of $\mathbf{A}$ in (6-78).

EXAMPLE 6.8

To illustrate the relationships between state-space models and transfer functions, again consider the electrically heated, stirred tank model in Section 2.4.3. First, equations (2-47) and (2-48) are converted into state-space form by dividing (2-47) by mC and (2-48) by m_eC_e, respectively:

$$\frac{dT}{dt} = \frac{w}{m}(T_i - T) + \frac{h_eA_e}{mC}(T_e - T) \tag{6-86}$$

$$\frac{dT_e}{dt} = \frac{Q}{m_eC_e} - \frac{h_eA_e}{m_eC_e}(T_e - T) \tag{6-87}$$

The nominal parameter values are the same as in Example 2.4:

$$\frac{m}{w} = 10 \text{ min} \qquad \frac{m_eC_e}{h_eA_e} = 1.0 \text{ min}$$

$$\frac{m_eC_e}{wC} = 1.0 \text{ min} \qquad \frac{1}{wC} = 0.05°\text{C min/kcal}$$

Consequently,

$$m_eC_e = 20 \text{ kcal/°C}$$
$$mC = 200 \text{ kcal/°C}$$
$$h_eA_e = 20 \text{ kcal/°C min}$$

(a) Using deviation variables (T', T'_e, Q') determine the transfer function between temperature T' and heat input Q'. Consider the conditions used in Example 2.4: $\overline{Q} = 5000$ kcal/min and $\overline{T}_i = 100$°C; at $t = 0$, Q is changed to 5,400 kcal/min. Compare the expression for $T'(s)$ with the time domain solution obtained in Example 2.4.

(b) Calculate the eigenvalues of the state-space model.

SOLUTION

(a) Substituting numerical values for the parameters gives

$$\frac{dT}{dt} = 0.1\,(T_i - T) + 0.1\,(T_e - T) \tag{6-88}$$

$$\frac{dT_e}{dt} = 0.05\,Q - (T_e - T) \tag{6-89}$$

The model can be written in deviation variable form (note that the steady-state values can be calculated to be $\overline{T} = 350$°C and $\overline{T}_e = 640$°C):

$$\frac{dT'}{dt} = 0.1\,(T'_i - T') + 0.1\,(T'_e - T') \tag{6-90}$$

$$\frac{dT'_e}{dt} = 0.05\,Q' - (T'_e - T') \tag{6-91}$$

$T'_i = 0$ because the inlet temperature is assumed to be constant. Taking the Laplace transform gives

$$sT'(s) = -0.2\,T'(s) + 0.1\,T'_e(s) \tag{6-92}$$

$$sT'_e(s) = 0.05\,Q'(s) - T'_e(s) + T'(s) \tag{6-93}$$

Using the result derived earlier in (6-84) (see also (4-32)), the transfer function is

$$\frac{T'(s)}{Q'(s)} = \frac{0.05}{10s^2 + 12s + 1} = \frac{0.005}{s^2 + 1.2s + 0.1} \tag{6-94}$$

For the step change of 400° kcal/min, $Q'(s) = \dfrac{400}{s}$ kcal/min, then

$$T'(s) = \frac{2}{s(s^2 + 12s + 0.1)}$$

The reader can verify that the inverse Laplace transform is

$$T'(t) = 20\,[1 - 1.089e^{-0.09t} + 0.0884e^{-1.11t}] \tag{6-95}$$

which is the same solution as obtained in Example 2.4.

(b) The state-space model in 6-90 and 6-91 can be written as

$$\begin{bmatrix} \dot{T}' \\ \dot{T}'_e \end{bmatrix} = \begin{bmatrix} -0.2 & 0.1 \\ 1 & -1 \end{bmatrix} \begin{bmatrix} T' \\ T'_e \end{bmatrix} + \begin{bmatrix} 0 \\ 0.05 \end{bmatrix} Q' \tag{6-96}$$

The 2 × 2 state matrix for this linear model is the same when either deviation variables (T', T'_e, Q') or the original physical variables (T, T_e, Q) are employed. The eigenvalues λ_i of the state matrix can be calculated from setting the determinant of $\mathbf{A} - \boldsymbol{\lambda}\mathbf{I}$ equals zero.

$$\det \begin{bmatrix} -0.2 - \lambda & 0.1 \\ 1 & -1 - \lambda \end{bmatrix} = 0$$

$$(-0.2 - \lambda)(-1 - \lambda) - 0.1 = 0$$

$$\lambda^2 + 1.2\lambda + 0.1 = 0$$

Solving for λ using the quadratic formula gives

$$\lambda_{1,2} = \frac{-1.2 + \sqrt{1.44 - 0.4}}{2} = -1.11, -0.09$$

which is the same result that was obtained using the transfer function approach. Because both eigenvalues are real, the response is non-oscillating, as shown in Figure 2.4.

The state space form of the dynamic system is not unique. If we are principally interested in modeling the dynamics of the temperature T, the state variables of the process model can be defined as

$$x = \begin{bmatrix} T' \\ \dfrac{dT'}{dt} \end{bmatrix} \text{ (note } T_e' \text{ is not an explicit variable).}$$

The resulting state-space description analogous to (6-88) and (6-89) would be

$$\frac{dx_1}{dt} = x_2 \tag{6-97}$$

$$\frac{dx_2}{dt} = -1.2x_2 - 0.1x_1 + 0.05u \tag{6-98}$$

Note that if (6-97) is differentiated once and we substitute the right hand side of (6-98) for dx_2/dt, then the same second-order model for T' is obtained. This is left as an exercise for the reader to verify. In addition, it is possible to derive other state space descriptions of the same second-order ODE, because the state-space form is not unique.

A general expression for the conversion of a state-space model to the corresponding transfer function model will now be derived. The starting point for the derivation is the standard state-space model in Eqs. 6-75 and 6-76. Taking the Laplace transform of these two equations gives

$$s\boldsymbol{X}(s) = \boldsymbol{A}\boldsymbol{X}(s) + \boldsymbol{B}\boldsymbol{U}(s) + \boldsymbol{E}\boldsymbol{D}(s) \tag{6-99}$$

$$\boldsymbol{Y}(s) = \boldsymbol{C}\boldsymbol{X}(s) \tag{6-100}$$

where $\boldsymbol{Y}(s)$ is a column vector that is the Laplace transform of $\boldsymbol{y}(t)$. The other vectors are defined in an analogous manner. After applying linear algebra and rearranging, a transfer function representation can be derived (Franklin et al., 2005):

$$\boldsymbol{Y}(s) = \boldsymbol{G}_p(s)\boldsymbol{U}(s) + \boldsymbol{G}_d(s)\boldsymbol{D}(s) \tag{6-101}$$

where the *process transfer function matrix*, $\boldsymbol{G}_p(s)$ is defined as

$$\boldsymbol{G}_p(s) \triangleq \boldsymbol{C}[s\boldsymbol{I} - \boldsymbol{A}]^{-1}\boldsymbol{B} \tag{6-102}$$

and the *disturbance transfer function matrix* $\boldsymbol{G}_d(s)$ is defined as

$$\boldsymbol{G}_d(s) \triangleq \boldsymbol{C}[s\boldsymbol{I} - \boldsymbol{A}]^{-1}\boldsymbol{E} \tag{6-103}$$

Note that the dimensions of the transfer function matrices depend on the dimensions of $\boldsymbol{Y}$, $\boldsymbol{U}$, and $\boldsymbol{D}$.

Fortunately, we do not have to perform tedious evaluations of expressions such as (6-102) and (6-103) by hand. State-space models can be converted to transfer function form using the MATLAB command *ss2tf*.

EXAMPLE 6.9

Determine $G_p(s)$ for temperature T' and input Q' for Example 6.8 using Equations 6-102 and 6-103.

SOLUTION

For part (a) of Example 6.8, $Y(s) = X_1(s)$, and there is one manipulated variable and no disturbance variable. Consequently, (6-101) reduces to

$$Y(s) = \boldsymbol{C}(s\boldsymbol{I} - \boldsymbol{A})^{-1}\boldsymbol{B}U(s) \tag{6-104}$$

where $\boldsymbol{G}_p(s)$ is now a scalar transfer function.

The calculation of the inverse matrix can be numerically challenging, although for this 2×2 case it can be done analytically by recognizing that

$$(s\boldsymbol{I} - \boldsymbol{A})^{-1} = \frac{\text{adjoint } (s\boldsymbol{I} - \boldsymbol{A})}{\det (s\boldsymbol{I} - \boldsymbol{A})} \tag{6-105}$$

The adjoint matrix is formed by the transpose of the cofactors of $\boldsymbol{A}$, so that

$$(s\boldsymbol{I} - \boldsymbol{A})^{-1} = \frac{\begin{bmatrix} s+1 & 0.1 \\ 1.0 & s+0.2 \end{bmatrix}}{s^2 + 1.2s + 0.1} \tag{6-106}$$

Note that the denominator polynomial formed by the determinant is the same one derived earlier in Example 6.8 using transfer functions and algebraic manipulation. You should verify that the inverse matrix when multiplied by $(s\boldsymbol{I} - \boldsymbol{A})$ yields the identity matrix.

To find the multivariable transfer function for $T'(s)/Q'(s)$, we use the following matrices from the state-space model:

$$\boldsymbol{B} = \begin{bmatrix} 0 \\ 0.05 \end{bmatrix} \quad \boldsymbol{C} = [1 \quad 0]$$

Then the product

$$\boldsymbol{C}(s\boldsymbol{I} - \boldsymbol{A})^{-1}\boldsymbol{B} = [1 \quad 0]\begin{bmatrix} \dfrac{s+1}{s^2+1.2s+1} & \dfrac{0.1}{s^2+1.2s+1} \\ \dfrac{1.0}{s^2+1.2s+1} & \dfrac{s+0.2}{s^2+1.2s+1} \end{bmatrix}\begin{bmatrix} 0 \\ 0.05 \end{bmatrix} \tag{6-107}$$

$$G_p(s) = \frac{0.005}{s^2 + 1.2s + 0.1} \quad (6\text{-}108)$$

which is the same result as in Eq. 6-95. The reader can also derive $G_d(s)$ relating $T'(s)$ and $T_i'(s)$ using the matrix-based approach in Eq. 6-103; see Eq. (4-33) for the solution.

It is also possible to convert a transfer function matrix in the form of Eq. 6-102 to a state-space model, and vice versa, using a single command in MATLAB. Using such software is recommended when the state matrix is larger than 2 × 2.

6.6 MULTIPLE-INPUT, MULTIPLE-OUTPUT (MIMO) PROCESSES

Most industrial process control applications involve a number of input (manipulated) variables and output (controlled) variables. These applications are referred to as *multiple-input/multiple-output (MIMO)* systems to distinguish them from the *single-input/single-output (SISO)* systems that have been emphasized so far. Modeling MIMO processes is no different conceptually than modeling SISO processes. For example, consider the thermal mixing process shown in Figure 6.14. The level h in the stirred tank and the temperature T are to be controlled by adjusting the flow rates of the hot and cold streams, w_h and w_c, respectively. The temperatures of the inlet streams T_h and T_c are considered to be disturbance variables. The outlet flow rate w is maintained constant by the pump, and the liquid properties are assumed to be constant (not affected by temperature) in the following derivation.

Noting that the liquid volume can vary with time, the energy and mass balances for this process are

$$\rho C \frac{d[V(T - T_{\text{ref}})]}{dt} = w_h C(T_h - T_{\text{ref}}) + w_c C(T_c - T_{\text{ref}}) - wC(T - T_{\text{ref}}) \quad (6\text{-}109)$$

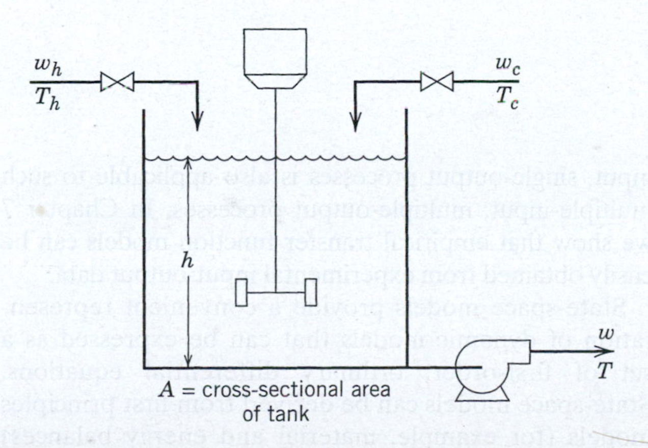

Figure 6.14 A multi-input, multi-output thermal mixing process.

$$\rho \frac{dV}{dt} = w_h + w_c - w \quad (6\text{-}110)$$

The energy balance includes a thermodynamic reference temperature T_{ref} (see Section 2.4.1). Expanding the derivative gives

$$\frac{d[V(T - T_{\text{ref}})]}{dt} = (T - T_{\text{ref}})\frac{dV}{dt} + V\frac{dT}{dt} \quad (6\text{-}111)$$

Equation 6-111 can be substituted for the left side of Eq. 6-109. Following substitution of the mass balance (6-110), a simpler set of equations results with $V = Ah$

$$\frac{dT}{dt} = \frac{1}{\rho A h}[w_h T_h + w_c T_c - (w_h + w_c)T] \quad (6\text{-}112)$$

$$\frac{dh}{dt} = \frac{1}{\rho A}(w_h + w_c - w) \quad (6\text{-}113)$$

After linearizing (6-112) and (6-113), putting them in deviation form, and taking Laplace transforms, we obtain a set of eight transfer functions that describe the effect of each input variable (w_h', w_c', T_h', T_c') on each output variable (T' and h'):

$$\frac{T'(s)}{W_h'(s)} = \frac{(\overline{T}_h - \overline{T})/\overline{w}}{\tau s + 1} \quad (6\text{-}114)$$

$$\frac{T'(s)}{W_c'(s)} = \frac{(\overline{T}_c - \overline{T})/\overline{w}}{\tau s + 1} \quad (6\text{-}115)$$

$$\frac{T'(s)}{T_h'(s)} = \frac{\overline{w}_h/\overline{w}}{\tau s + 1} \quad (6\text{-}116)$$

$$\frac{T'(s)}{T_c'(s)} = \frac{\overline{w}_c/\overline{w}}{\tau s + 1} \quad (6\text{-}117)$$

$$\frac{H'(s)}{W_h'(s)} = \frac{1/A\rho}{s} \quad (6\text{-}118)$$

$$\frac{H'(s)}{W_c'(s)} = \frac{1/A\rho}{s} \quad (6\text{-}119)$$

$$\frac{H'(s)}{T_h'(s)} = 0 \quad (6\text{-}120)$$

$$\frac{H'(s)}{T_c'(s)} = 0 \quad (6\text{-}121)$$

where $\tau = \rho A\overline{h}/\overline{w}$ is the average residence time in the tank and an overbar denotes a nominal steady-state value.

Equations 6-114 through 6-117 indicate that all four inputs affect the tank temperature through first-order transfer functions and a single time constant that is the nominal residence time of the tank τ. Equations 6-118 and 6-119 show that the inlet flow rates affect level through integrating transfer functions that result from the pump on the exit line. Finally, it is clear from Eqs. 6-120 and 6-121, as well as from physical intuition, that inlet temperature changes have no effect on liquid level.

A very compact way of expressing Eqs. 6-114 through 6-121 is by means of a *transfer function matrix*:

$$\begin{bmatrix} T'(s) \\ H'(s) \end{bmatrix} = \begin{bmatrix} \dfrac{(\overline{T}_h - \overline{T})/\overline{w}}{\tau s + 1} & \dfrac{(\overline{T}_c - \overline{T})/\overline{w}}{\tau s + 1} & \dfrac{\overline{w}_h/\overline{w}}{\tau s + 1} & \dfrac{\overline{w}_c/\overline{w}}{\tau s + 1} \\ \dfrac{1/A\rho}{s} & \dfrac{1/A\rho}{s} & 0 & 0 \end{bmatrix} \cdot \begin{bmatrix} W_h'(s) \\ W_c'(s) \\ T_h'(s) \\ T_c'(s) \end{bmatrix} \tag{6-122}$$

Equivalently, two transfer function matrices can be used to separate the manipulated variables, w_h and w_c, from the disturbance variables, T_h and T_c:

$$\begin{bmatrix} T'(s) \\ H'(s) \end{bmatrix} = \begin{bmatrix} \dfrac{(\overline{T}_h - \overline{T})/\overline{w}}{\tau s + 1} & \dfrac{(\overline{T}_c - \overline{T})/\overline{w}}{\tau s + 1} \\ \dfrac{1/A\rho}{s} & \dfrac{1/A\rho}{s} \end{bmatrix} \begin{bmatrix} W_h'(s) \\ W_c'(s) \end{bmatrix} + \begin{bmatrix} \dfrac{\overline{w}_h/\overline{w}}{\tau s + 1} & \dfrac{\overline{w}_c/\overline{w}}{\tau s + 1} \\ 0 & 0 \end{bmatrix} \begin{bmatrix} T_h'(s) \\ T_c'(s) \end{bmatrix} \tag{6-123}$$

The block diagram in Figure 6.15 illustrates how the four input variables affect the two output variables.

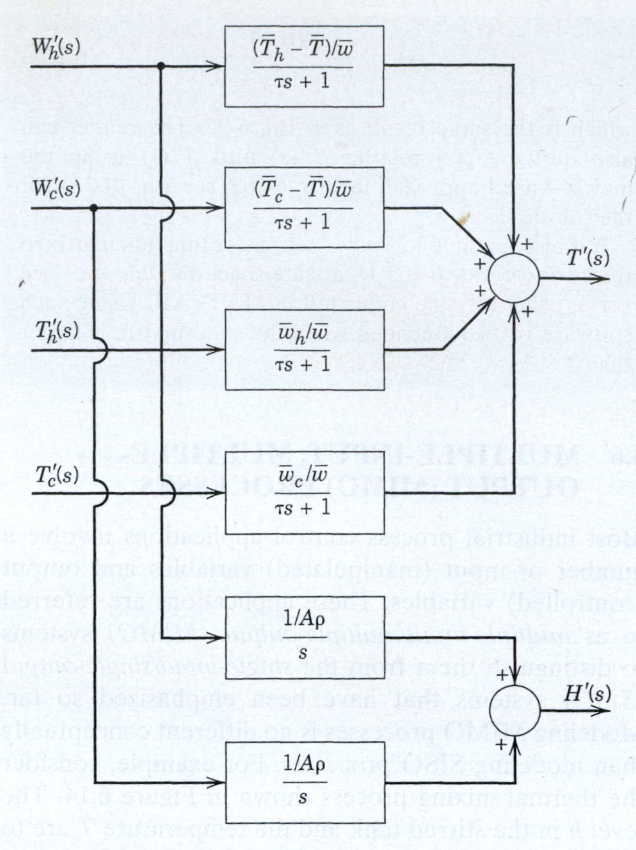

Figure 6.15 Block diagram of the MIMO thermal mixing system with variable liquid level.

Two points are worth mentioning in conclusion:

1. A transfer function matrix, or, equivalently, the set of individual transfer functions, facilitates the design of control systems that deal with the interactions between inputs and outputs. For example, for the thermal mixing process in this section, control strategies can be devised that minimize or eliminate the effect of flow changes on temperature and level. This type of multivariable control system is considered in Chapters 18 and 20.
2. The development of physically-based MIMO models can require a significant effort. Thus, empirical models rather than theoretical models often must be used for complicated processes. Empirical modeling is the subject of the next chapter.

SUMMARY

In this chapter we have considered the dynamics of processes that cannot be described by simple transfer functions. Models for these processes include numerator dynamics such as time delays or process zeros. An observed time delay is often a manifestation of higher-order dynamics; consequently, a time-delay term in a transfer function model provides a way of approximating high-order dynamics (for example, one or more small time constants). Important industrial processes typically have several input variables and several output variables. Fortunately, the transfer function methodology for single-input, single-output processes is also applicable to such multiple-input, multiple-output processes. In Chapter 7 we show that empirical transfer function models can be easily obtained from experimental input-output data.

State-space models provide a convenient representation of dynamic models that can be expressed as a set of first-order, ordinary differential equations. State-space models can be derived from first principles models (for example, material and energy balances) and used to describe both linear and nonlinear dynamic systems.

REFERENCES

Buckley, P. S., W.L. Luyben, and J. P. Shunta, *Design of Distillation Column Control Systems*, Instrument Society of America, Research Triangle Park, NC, 1985.

Franklin, G. F., J. D. Powell, and A. Emani-Naeini, *Feedback Control of Dynamic Systems*, 5th ed., Prentice-Hall, Upper Saddle River, NJ, 2005.

Henson, M. A. and D. E. Seborg, (eds.), *Nonlinear Process Control*, Prentice-Hall, Upper Saddle River, NJ, 1997.

Khalil, H. K., *Nonlinear Systems*, 3rd ed., Prentice-Hall, Upper Saddle River, NJ, 2002.

Ogata, K., *Modern Control Engineering*, 5th ed., Prentice-Hall, Upper Saddle River, NJ, 2008.

Skogestad, S., Simple Analytic Rules for Model Reduction and PID Controller Tuning, *J. Process Control*, 13, 291 (2003).

Strang, G., *Linear Algebra and Its Applications*, 4th ed., Brooks Cole, Florence, KY, 2005.

EXERCISES

6.1 Consider the transfer function:

$$G(s) = \frac{0.7(s^2 + 2s + 2)}{s^5 + 5s^4 + 2s^3 - 4s^2 + 6}$$

(a) Plot its poles and zeros in the complex plane. A computer program that calculates the roots of the polynomial (such as the command *roots* in MATLAB) can help you factor the denominator polynomial.

(b) From the pole locations in the complex plane, what can be concluded about the output modes for any input change?

(c) Plot the response of the output to a unit step input. Does the form of your response agree with your analysis for part (b)? Explain.

6.2 The following transfer function is not written in a standard form:

$$G(s) = \frac{2(s + 0.5)}{(s + 2)(2s + 1)} e^{-5s}$$

(a) Put it in standard gain/time constant form.

(b) Determine the gain, poles and zeros.

(c) If the time-delay term is replaced by a 1/1 Padé approximation, repeat part (b).

6.3 For a lead–lag unit,

$$\frac{Y(s)}{X(s)} = \frac{K(\tau_a s + 1)}{\tau_1 s + 1}$$

show that for a step input of magnitude M:

(a) The value of y at $t = 0^+$ is given by $y(0^+) = KM\tau_a/\tau_1$.

(b) Overshoot occurs only for $\tau_a > \tau_1$, in which case $dy/dt < 0$.

(c) Inverse response occurs only for $\tau_a < 0$.

6.4 A second-order system has a single zero:

$$\frac{Y(s)}{X(s)} = \frac{K(\tau_a s + 1)}{(\tau_1 s + 1)(\tau_2 s + 1)} \quad (\tau_1 > \tau_2)$$

For a step input, show that:

(a) $y(t)$ can exhibit an extremum (maximum or minimum value) in the step response only if

$$\frac{1 - \tau_a/\tau_2}{1 - \tau_a/\tau_1} > 1$$

(b) Overshoot occurs only for $\tau_a/\tau_1 > 1$.

(c) Inverse response occurs only for $\tau_a < 0$.

(d) If an extremum in y exists, the time at which it occurs can be found analytically. What is it?

6.5 A process has the transfer function of Eq. 6-14 with $K = 2$, $\tau_1 = 10$, $\tau_2 = 2$. If τ_a has the following values,

Case i: $\tau_a = 20$
Case ii: $\tau_a = 4$
Case iii: $\tau_a = 1$
Case iv: $\tau_a = -2$

Simulate the responses for a step input of magnitude 0.5 and plot them in a single figure. What conclusions can you make, about the effect of the zero location? Is the location of the pole corresponding to τ_2 important so long as $\tau_1 > \tau_2$?

6.6 A process consists of an integrating element operating in parallel with a first-order element (Fig. E6.6).

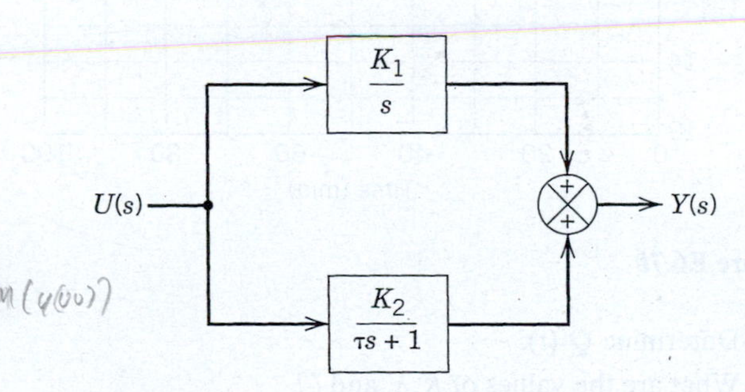

Figure E6.6

(a) What is the order of the overall transfer function, $G(s) = Y(s)/U(s)$?

(b) What is the gain of $G(s)$?

(c) What are the poles of $G(s)$? Where are they located in the complex s-plane?

(d) What are the zeros of $G(s)$? Where are they located? Under what condition(s) will one or more of the zeros be located in the right-half s-plane?

(e) For what conditions, will this process exhibit both a negative gain and a right-half plane zero?

(f) For any input change, what functions of time (response modes) will be included in the response, $y(t)$?

(g) Is the output bounded for any bounded input change, for example, $u(t) = M$?

6.7 A pressure-measuring device has been analyzed and can be described by a model with the structure shown in Fig. E6.7*a*. In other words, the device responds to pressure changes as if it were a first-order process in parallel with a second-order process. Preliminary tests have shown that the gain of the first-order process is -3 and the time constant equals 20 min, as shown in Fig. E6.7*a*. An additional test is made on the overall system. The actual output P_m (not P'_m) resulting from a step change in P from 4 to 6 psi is plotted in Fig. E6.7*b*.

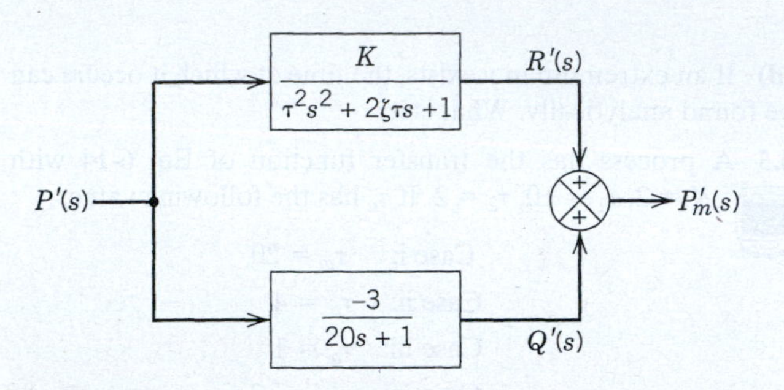

Figure E6.7*a*

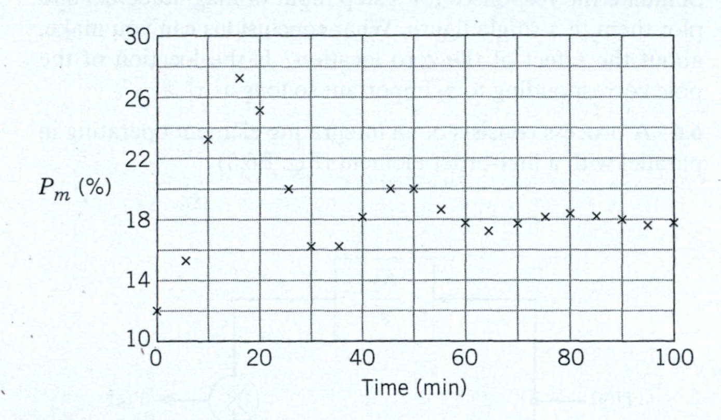

Figure E6.7*b*

(a) Determine $Q'(t)$.

(b) What are the values of K, τ, and ζ?

(c) What is the overall transfer function for the measurement device $P'_m(s)/P'(s)$?

(d) Find an expression for the overall process gain.

6.8 A blending tank that provides nearly perfect mixing is connected to a downstream unit by a long transfer pipe. The blending tank operates dynamically like a first-order process.

The mixing characteristics of the transfer pipe, on the other hand, are somewhere between plug flow (no mixing) and perfectly mixed. A test is made on the transfer pipe that shows that it operates as if the total volume of the pipe were split into five equal-sized perfectly stirred tanks connected in series.

The process (tank plus transfer pipe) has the following characteristics:

$$V_{tank} = 2 \text{ m}^3$$
$$V_{pipe} = 0.1 \text{ m}^3$$
$$q_{total} = 1 \text{ m}^3/\text{min}$$

where q_{total} represents the sum of all flow rates into the process.

(a) Using the information provided above, what would be the most accurate transfer function $C'_{out}(s)/C'_{in}(s)$ for the process (tank plus transfer pipe) that you can develop? *Note:* c_{in} and c_{out} are inlet and exit concentrations.

(b) For these particular values of volumes and flow rate, what approximate (low-order) transfer function could be used to represent the dynamics of this process?

(c) What can you conclude concerning the need to model the transfer pipe's mixing characteristics very accurately for this particular process?

(d) Simulate approximate and full-order model responses to a step change in c_{in}.

6.9 *By inspection* determine which of the following process models can be approximated reasonably accurately by a first-order-plus-time-delay model. For each acceptable case, give your best estimate of θ and τ.

(a) $\dfrac{K}{(10s + 1)(10s + 1)}$

(b) $\dfrac{K}{(10s + 1)(8s + 1)(s + 1)}$

(c) $\dfrac{K}{(10s + 1)(s + 1)^2}$

(d) $\dfrac{K(20s + 1)}{10s + 1}$

(e) $\dfrac{K(0.5s + 1)}{(10s + 1)(s + 1)}$

(f) $\dfrac{K}{10s^2 + 11s + 1}$

(g) $\dfrac{K}{100s^2 + 10s + 1}$

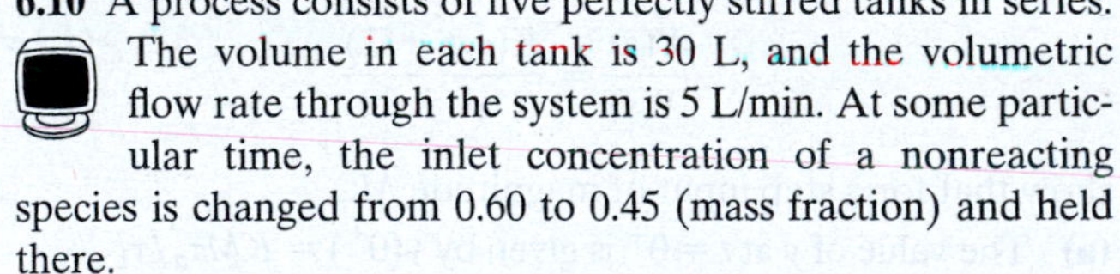

6.10 A process consists of five perfectly stirred tanks in series. The volume in each tank is 30 L, and the volumetric flow rate through the system is 5 L/min. At some particular time, the inlet concentration of a nonreacting species is changed from 0.60 to 0.45 (mass fraction) and held there.

(a) Write an expression for c_5 (the concentration leaving the fifth tank) as a function of time.

(b) Simulate and plot $c_1, c_2, \ldots, c_5$. Compare c_5 at $t = 30$ min with the corresponding value for the expression in part (a).

6.11 A composition analyzer is used to measure the concentration of a pollutant in a wastewater stream. The relationship between the measured composition C_m and the actual composition C is given by the following transfer function (in deviation variable form):

$$\frac{C'_m(s)}{C'(s)} = \frac{e^{-\theta s}}{\tau s + 1}$$

where $\theta = 2$ min and $\tau = 4$ min. The nominal value of the pollutant is $\overline{C} = 5$ ppm. A warning light on the analyzer turns on whenever the measured concentration exceeds 25 ppm.

Suppose that at time $t = 0$, the actual concentration begins to drift higher—$C(t) = 5 + 2t$, where C has units of ppm and t has units of minutes. At what time will the warning light turn on?

6.12 For the process described by the exact transfer function

$$G(s) = \frac{5}{(10s + 1)(4s + 1)(s + 1)(0.2s + 1)}$$

(a) Find an approximate transfer function of second-order-plus-time-delay form that describes this process.

(b) Simulate and plot the response $y(t)$ of both the approximate model and the exact model on the same graph for a unit step change in input $x(t)$.

(c) What is the maximum error between the two responses? Where does it occur?

6.13 Find the transfer functions $P_1'(s)/P_d'(s)$ and $P_2'(s)/P_d'(s)$ for the compressor-surge tank system of Exercise 2.5 when it is operated isothermally. Put the results in standard (gain/time constant) form. For the second-order model, determine whether the system is overdamped or underdamped.

6.14 A process has the block diagram

$$U(s) \rightarrow \boxed{\frac{0.8}{(0.4s + 1)^2}} \rightarrow \boxed{\frac{5e^{-\theta s}}{2s^2 + 3s + 1}} \rightarrow Y(s)$$

Derive an approximate first-order-plus-time-delay transfer function model.

6.15 Show that the liquid-level system consisting of two interacting tanks (Fig. 6.13) exhibits overdamped dynamics; that is, show that the damping coefficient in Eq. 6-72 is larger than one.

6.16 An open liquid surge system (ρ = constant) is designed with a side tank that normally is isolated from the flowing material as shown in Fig. E6.16.

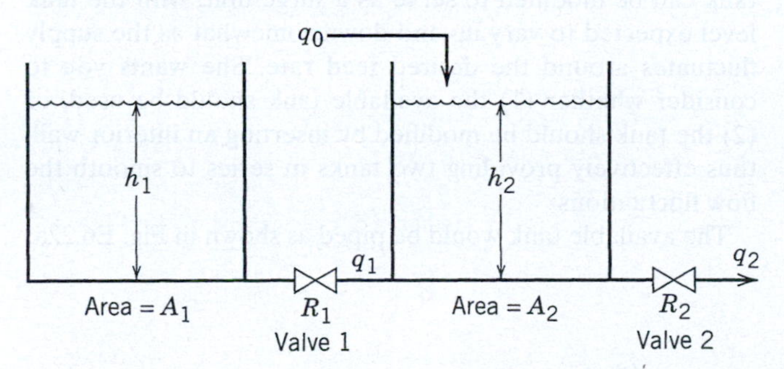

Figure E6.16

(a) In normal operation, Valve 1 is closed ($R_1 \rightarrow \infty$) and $q_1 = 0$. What is the transfer function relating changes in q_0 to changes in outflow rate q_2 under these conditions?

(b) Someone inadvertently leaves Valve 1 partially open ($0 < R_1 < \infty$). What is the differential equation model for this system?

(c) What do you know about the form of the transfer function $Q_2'(s)/Q_0'(s)$ for Valve 1 partially open? Discuss but do not derive.

(d) Is the response to changes in q_0 faster or slower for Case (b) compared to Case (a)? Explain why but do not derive the response.

6.17 The dynamic behavior of a packed-bed reactor can be approximated by a transfer function model

$$\frac{T'(s)}{T_i'(s)} = \frac{3(2 - s)}{(10s + 1)(5s + 1)}$$

where T_i is the inlet temperature, T is the outlet temperature (°C), and the time constants are in hours. The inlet temperature varies in a cyclic fashion due to the changes in ambient temperature from day to night.

As an approximation, assume that T_i' varies sinusoidally with a period of 24 hours and amplitude of 12°C. What is the maximum variation in the outlet temperature, T?

6.18 Example 5.1 derives the gain and time constant for a first-order model of a stirred-tank heating process.

(a) Simulate the response of the tank temperature to a step change in heat input from 3×10^7 cal/h to 5×10^7 cal/h.

(b) Suppose there are dynamics associated with changing the heat input to the system. If the dynamics of the heater itself can be expressed by a first-order transfer function with a gain of one and a time constant of 10 min, what is the overall transfer function for the heating system (tank plus heater)?

(c) For the process in **(b)**, simulate the step increase in heat input.

6.19 Distributed parameter systems such as tubular reactors and heat exchangers often can be modeled as a set of lumped parameter equations. In this case an alternative (approximate) physical interpretation of the process is used to obtain an ODE model directly rather than by converting a PDE model to ODE form by means of a lumping method

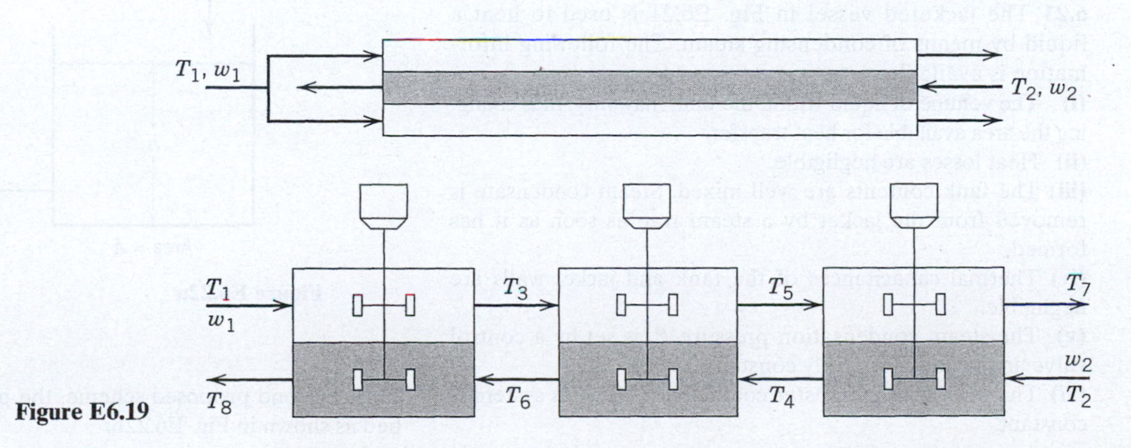

Figure E6.19

such as finite differences. As an example, consider a single concentric-tube heat exchanger with energy exchange between two liquid streams flowing in opposite directions, as shown in Fig. E6.19. We might model this process as if it were three small, perfectly stirred tanks with heat exchange. If the mass flow rates w_1 and w_2 and the inlet temperatures T_1 and T_2 are known functions of time, derive transfer function expressions for the exit temperatures T_7 and T_8 in terms of the inlet temperature T_1. Assume that all liquid properties (ρ_1, ρ_2, Cp_1, Cp_2) are constant, that the area for heat exchange in each stage is A, that the overall heat transfer coefficient U is the same in each stage, and that the wall between the two liquids has negligible thermal capacitance.

6.20 A two-input/two-output process involving simultaneous heating and liquid-level changes is illustrated in Fig. E6.20. Find the transfer function models and expressions for the gains and the time constant τ for this process. What is the output response for a unit step change in Q? for a unit step change in w? *Note:* Transfer function models for a somewhat similar process depicted in Fig. 6.15 are given in Eqs. 6-80 through 6-87. They can be compared with your results. For this exercise, T and h are the outputs and Q and w are the inputs.

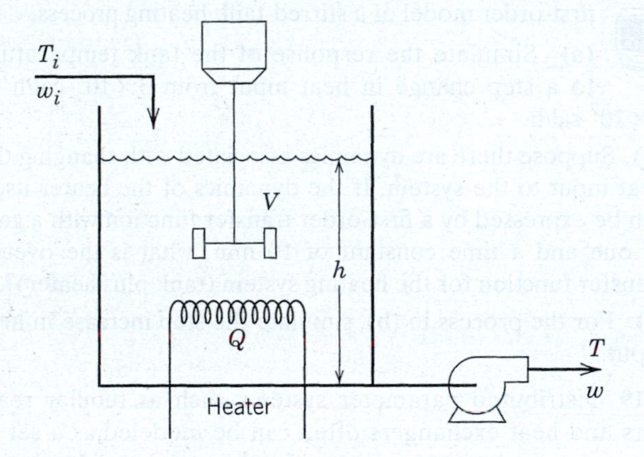

Figure E6.20

6.21 The jacketed vessel in Fig. E6.21 is used to heat a liquid by means of condensing steam. The following information is available:

(i) The volume of liquid within the tank may vary, thus changing the area available for heat transfer.

(ii) Heat losses are negligible.

(iii) The tank contents are well mixed. Steam condensate is removed from the jacket by a steam trap as soon as it has formed.

(iv) Thermal capacitances of the tank and jacket walls are negligible.

(v) The steam condensation pressure P_s is set by a control valve and is not necessarily constant.

(vi) The overall heat transfer coefficient U for this system is constant.

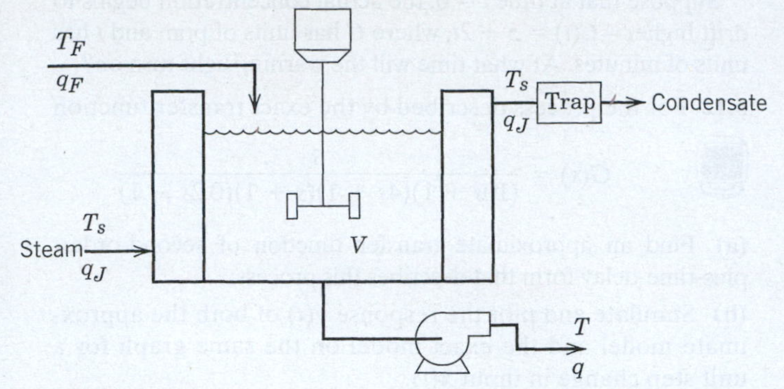

Figure E6.21

(vii) Flow rates q_F and q are independently set by external valves and may vary.

Derive a dynamic model for this process. The model should be simplified as much as possible. State any additional assumptions that you make.

(a) Find transfer functions relating the two primary output variables h (level) and T (liquid temperature) to inputs q_F, q, and T_S. You should obtain six separate transfer functions.

(b) Briefly interpret the form of each transfer function using physical arguments as much as possible.

(c) Discuss qualitatively the form of the response of each output to a step change in each input.

6.22 Your company is having problems with the feed stream to a reactor. The feed must be kept at a constant mass flow rate ($\overline{w}$) even though the supply from the upstream process unit varies with time, $w_i(t)$. Your boss feels that an available tank can be modified to serve as a surge unit, with the tank level expected to vary up and down somewhat as the supply fluctuates around the desired feed rate. She wants you to consider whether (1) the available tank should be used, or (2) the tank should be modified by inserting an interior wall, thus effectively providing two tanks in series to smooth the flow fluctuations

The available tank would be piped as shown in Fig. E6.22a:

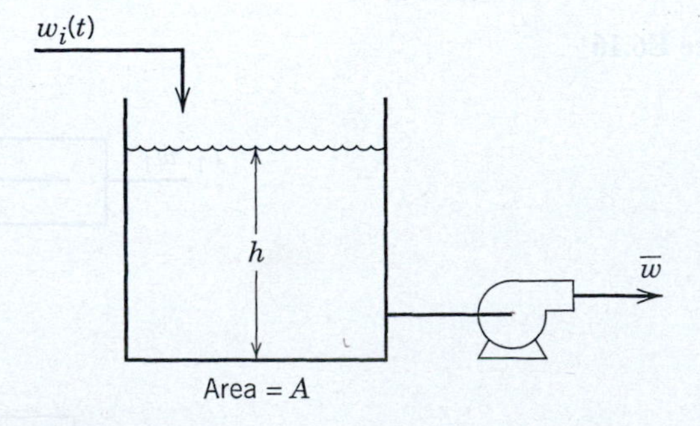

Figure E6.22a

In the second proposed scheme, the process would be modified as shown in Fig. E6.22b:

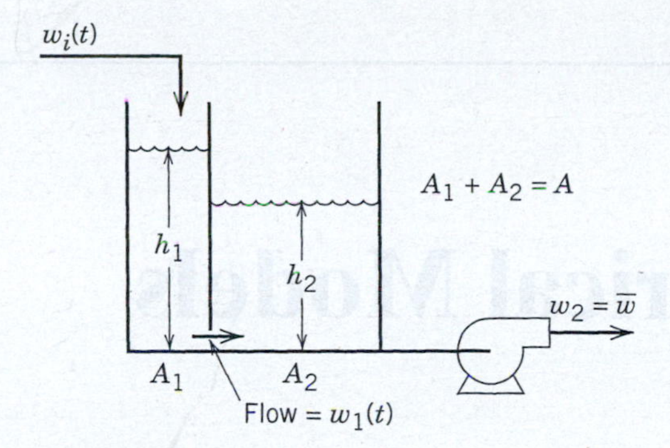

Figure E6.22b

In this case, an opening placed at the bottom of the interior wall permits flow between the two partitions. You may assume that the resistance to flow $w_1(t)$ is linear and constant (R).

(a) Derive a transfer function model for the two-tank process $[H_2'(s)/W_i'(s)]$ and compare it to the one-tank process $[H'(s)/W_i'(s)]$. In particular, for each transfer function indicate its order, presence of any zeros, gain, time constants, presence or absence of an integrating element, whether it is interacting or noninteracting, and so on.

(b) Evaluate how well the two-tank surge process would work relative to the one-tank process for the case $A_1 = A_2 = A/2$ where A is the cross-sectional area of the single tank. Your analysis should consider whether h_2 will vary more or less rapidly than h for the same input flow rate change, for example, a step input change.

(c) Determine the best way to partition the volume in the two-tank system to smooth inlet flow changes. In other words, should the first tank contain a larger portion of the volume than the second, and so on.

(d) Plot typical responses to corroborate your analysis. For this purpose, you should size the opening in the two-tank interior wall (i.e., choose R) such that the tank levels will be the same at whatever nominal flow rate you choose.

6.23 A process has the following block diagram representation:

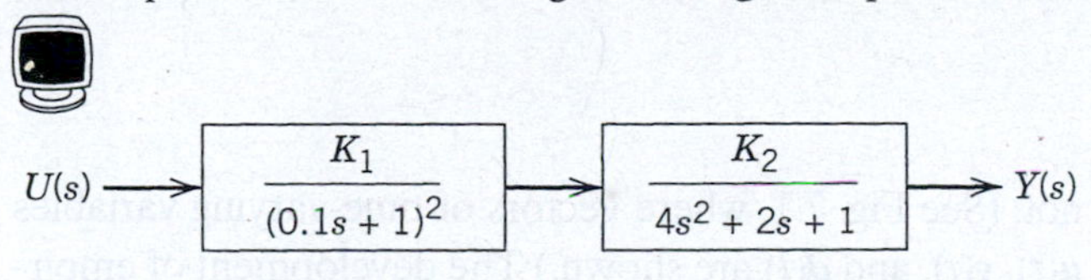

(a) Will the process exhibit overshoot for a step change in u? Explain/demonstrate why or why not.

(b) What will be the approximate maximum value of y for $K = K_1K_2 = 1$ and a step change, $U(s) = 3/s$?

(c) Approximately when will the maximum value occur?

(d) Simulate and plot both the actual fourth-order response and a second-order-plus-time-delay response that approximates the critically damped element for values of $\tau_1 = 0.1$, 1, and 5. What can you conclude about the quality of the approximation when τ_1 is much smaller than the underdamped element's time scale? about the order of the underdamped system's time scale?

6.24 The transfer function that relates the change in blood pressure y to a change in u the infusion rate of a drug (sodium nitroprusside) is given by[1]

$$G_p(s) = \frac{Ke^{-\theta_1 s}(1 + \alpha e^{-\theta_2 s})}{\tau s + 1}$$

The two time delays result from the blood recirculation that occurs in the body, and α is the recirculation coefficient. The following parameter values are available:

$$K = -1.0 \frac{\text{mm Hg}}{\text{ml/h}},$$

$$\alpha = 0.4,\ \theta_1 = 30\text{ s},\ \theta_2 = 45\text{ s, and } \tau = 40\text{ s}$$

Simulate the blood pressure response to a unit step change ($u = 1$) in sodium nitroprusside infusion rate. Is it similar to other responses discussed in Chapters 5 or 6?

6.25 In Example 4.4, a two-tank system is presented. Using state-space notation, determine the matrices A, B, C, and E, assuming that the level deviations is h_1' and h_2' are the state variables, the input variable is q_1', and the output variable is the flow rate deviation, q_2'.

6.26 The staged system model for a three-stage absorber is presented in Eqs. (2-73)–(2-75), which are in state-space form. A numerical example of the absorber model suggested by Wong and Luus[2] has the following parameters: $H = 75.72$ lb, $L = 40.8$ lb/min, $G = 66.7$ lb/min, $a = 0.72$, and $b = 0.0$. Using the MATLAB function *ss2tf*, calculate the three transfer functions (Y_1'/Y_f', Y_2'/Y_f', Y_3'/Y_f') for the three state variables and the feed composition deviation Y_f' as the input.

[1]Hahn, J., T. Edison, and T. F. Edgar, Adaptive IMC Control for Drug Infusion for Biological Systems, *Control Engr. Practice*, **10**, 45 (2002).
[2]Wong, K. T., and R. Luus, Model Reduction of High-order Multistage Systems by the Method of Orthogonal Collocation, *Can. J. Chem. Eng.* **58**, 382 (1980).

Chapter 7

Development of Empirical Models from Process Data

CHAPTER CONTENTS

Several modeling approaches are used in process control applications. Theoretical models based on the chemistry and physics of the process represent one alternative. However, the development of rigorous theoretical models may not be practical for complex processes if the model requires a large number of equations with a significant number of process variables and unknown parameters (e.g., chemical and physical properties). An alternative approach is to develop an empirical model directly from experimental data. Empirical models are sometimes referred to as *black box* models, because the process being modeled can be likened to an opaque box. Here the input and output variables (u and y, respectively) are known, but the inner workings of the box are not. (See Fig. 7.1, where vectors of time-varying variables $\boldsymbol{u}(t)$, $\boldsymbol{y}(t)$, and $\boldsymbol{d}(t)$ are shown.) The development of empirical steady-state and dynamic models is the subject of this chapter. This activity is referred to as *process* or *system identification* (Ljung and Glad, 1994; Ljung, 1999). In general, empirical dynamic models are simpler than theoretical models and offer the advantage that they can be solved in "real time." In other words, the computational time required for the model solution (e.g., transient response) is much shorter than the actual process response time. However, this may not be true for complex models with many variables and equations.

The key differences between process simulation and process identification can be summarized with the aid

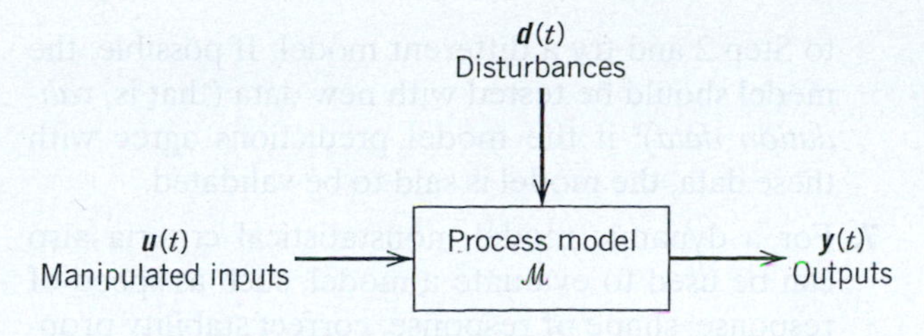

Figure 7.1 Input-output process model.

of Fig. 7.1. In simulation, the process model $\mathcal{M}$ is known, and we wish to generate the response $\boldsymbol{y}(t)$ for a specified input $\boldsymbol{u}(t)$ and a specified disturbance $\boldsymbol{d}(t)$. If $\mathcal{M}$ is a linear dynamic model and $\boldsymbol{u}(t)$ and $\boldsymbol{d}(t)$ are expressed analytically, $\boldsymbol{y}(t)$ can be derived using Laplace transforms (see Chapter 4). Alternatively, $\boldsymbol{y}(t)$ can be calculated numerically using software packages such as MATLAB (Ljung, 2007). If $\mathcal{M}$ is a nonlinear dynamic model, $\boldsymbol{y}(t)$ can be obtained by numerical integration (cf. Chapter 2) after $\boldsymbol{u}(t)$ and $\boldsymbol{d}(t)$ are specified. By contrast, in process identification the model $\mathcal{M}$ is determined from data for $\boldsymbol{u}(t)$, $\boldsymbol{y}(t)$, and $\boldsymbol{d}(t)$, if $\boldsymbol{d}$ can be measured. If the model structure is postulated but contains unknown model parameters, then the model parameters can be obtained using regression techniques. This parameter estimation can be done with commercially available software regardless of whether the process model is linear or nonlinear, or whether it is theoretically-based or empirical in nature.

Steady-state empirical models can be used for instrument calibration, process optimization, and specific instances of process control. Single-input, single-output (SISO) models typically consist of simple polynomials relating an output to an input. Dynamic empirical models can be employed to understand process behavior during upset conditions. They are also used to design control systems and to analyze their performance. Empirical dynamic models typically are low-order differential equations or transfer function models (e.g., first- or second-order model, perhaps with a time delay), with unspecified model parameters to be determined from experimental data. However, in some situations more complicated models are valuable in control system design, as discussed later in this chapter.

The concept of a *discrete-time model* will now be introduced. These models are generally represented by difference equations rather than differential equations. Most process control tasks are implemented via digital computers, which are intrinsically discrete-time systems. In digital control, the continuous-time process variables are sampled at regular intervals (e.g., every 0.1 s); hence, the computer calculations are based on sampled data rather than continuous measurements. If process variables are observed only at the sampling instants, the dynamic behavior can be modeled using a discrete-time model in the form of a difference equation. The selection of discrete-time models over continuous-time models is becoming commonplace, especially for advanced control strategies.

Several methods for determining steady-state and dynamic empirical models for both continuous-time and discrete-time model types will now be presented. We first consider general model-fitting techniques based on linear and nonlinear regression that can be used to calculate model parameters for any type of model. Then simple but very useful methods are presented for obtaining first-order and second-order dynamic models from step response data using analytical solutions. These methods yield models suitable for the design of control systems; however, the resulting models are usually accurate only for a narrow range of operating conditions close to the nominal steady state, where the process exhibits linear behavior. We also show the relationship between continuous-time and discrete-time models. Finally, we present several methods for developing linear discrete-time models for dynamic processes.

7.1 MODEL DEVELOPMENT USING LINEAR OR NONLINEAR REGRESSION

Before developing an empirical model for two variables, a single input u and a single output y, it is instructive first to plot the available data (e.g., y vs. u for steady-state data and y and u vs. time for transient response data). From these plots it may be possible to visualize overall trends in the data and to select a reasonable form for the model. After the model form is selected, the unknown model parameters can be calculated and the model accuracy evaluated. This parameter calculation procedure is referred to as *parameter estimation* or regression (Ljung, 1999; Montgomery and Runger, 2007). These calculations are usually based on model fitting, that is, minimizing a measure of the differences between model predictions and data. However, the problem of fitting a model to a set of input-output data becomes complicated when the model relation is not simple or involves multiple inputs and outputs.

First, we consider steady-state models. Suppose that a set of steady-state input-output data is available and shown as circles in Fig. 7.2. Variable y represents a process output (e.g., a reactor yield), whereas u represents an input variable (e.g., an operating condition such as temperature). Although a straight-line model (Model 1) provides a reasonable fit, higher-order polynomial relations (Models 2 and 3) result in smaller

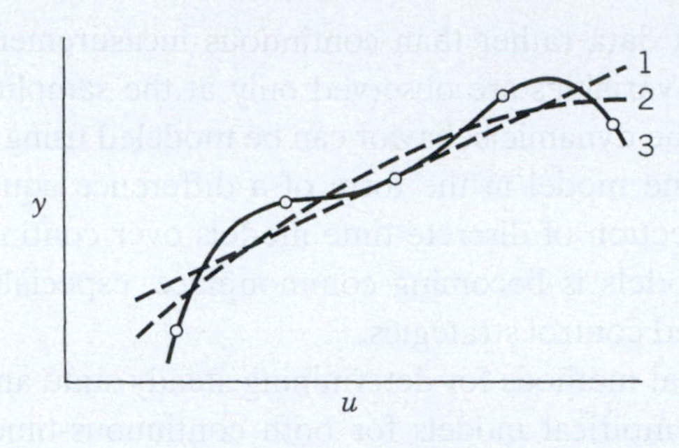

Figure 7.2 Three models for scattered data.

errors between the data and the curve representing the empirical model. Models 2 and 3 provide better agreement with the data at the expense of greater complexity because more model parameters must be determined. Sometimes the model form may be known from theoretical considerations or past experience with the process.

In Fig. 7.2, if the actual process behavior is linear, the differences (or residuals) between Model 1 and the data could be due to process disturbances or measurement errors. In empirical modeling, it is preferable to choose the simplest model structure that yields a good fit of the data, providing that the model is physically reasonable. Note that in Fig. 7.2, if Model 3 is extrapolated beyond the data range, it would apparently yield significantly different model predictions than Model 1 or 2. The selection of the best model might require collecting more data, perhaps outside the range shown in Fig. 7.2, which then could be used to *validate* each model.

7.1.1 Model Building Procedure

In this section we present a systematic procedure for developing empirical dynamic models (Ljung, 1999). The procedure consists of the following steps:

1. Formulate the model objectives; that is, how will the model be used, and who will be the user?
2. Select the input and output variables for the model.
3. Evaluate available data and develop a plan to acquire additional data. A testing plan would specify the values of u or the form of $u(t)$, for example, a step change or some other input sequence (see Section 7.2).
4. Select the model structure and level of model complexity (e.g., steady-state vs. dynamic model, linear vs. nonlinear model).
5. Estimate the unknown model parameters using linear or nonlinear regression.
6. Using input and output data, evaluate model accuracy based on statistical considerations. It is desirable to use new data (if available) as well as the "old" data that were used to develop the model. If the model does not provide a satisfactory fit, return to Step 2 and try a different model. If possible, the model should be tested with new data (that is, *validation data*); if the model predictions agree with these data, the model is said to be validated.
7. For a dynamic model, nonstatistical criteria also can be used to evaluate a model, such as speed of response, shape of response, correct stability properties, and correct steady-state gain. The utility of a model for designing controllers is also important in process control, where an overly complex model can be a disadvantage. Thus *control-relevant models* are desirable (Rivera and Jun, 2000).

7.1.2 Linear Regression

Statistical analysis can be used to estimate unknown model parameters and to specify the uncertainty associated with the empirical model. It can also be used to compare several candidate models (Draper and Smith, 1998; Montgomery and Runger, 2007). For linear models, the *least-squares* approach is widely used to estimate model parameters. Consider the linear (or straight-line) model in Fig. 7.2 (Model 1) and let Y_i represent the data point where $\hat{y}_i$ is the model prediction for $u = u_i$. Then for the model, $y = \beta_1 + \beta_2 u + \epsilon$, the individual data points can be expressed as

$$Y_i = \beta_1 + \beta_2 u_i + \epsilon_i \tag{7-1}$$

where β_1 and β_2 are the model parameters to be estimated. ϵ_i is the random error for the particular data point.

The least-squares method is the standard approach for calculating the values of β_1 and β_2 that minimize the sum of the squares of the errors S for an arbitrary number of data points, N:

$$S = \sum_{i=1}^{N} \epsilon_i^2 = \sum_{i=1}^{N} (Y_i - \beta_1 - \beta_2 u_i)^2 \tag{7-2}$$

In (7-2), note that the values of Y_i and u_i are known, while β_1 and β_2 are to be calculated so as to minimize S, the objective function. The optimal estimates of β_1 and β_2 calculated for a specific data set are designated as $\hat{\beta}_1$ and $\hat{\beta}_2$. The model predictions are given by the regression model:

$$y = \hat{\beta}_1 + \hat{\beta}_2 u \tag{7-3}$$

and the *residuals* e_i are defined as

$$e_i \triangleq Y_i - \hat{y}_i \tag{7-4}$$

These least-squares estimates will provide a good fit if the errors ϵ_i are statistically independent and normally distributed.

For a linear model and N data points, values of $\hat{\beta}_1$ and $\hat{\beta}_2$ that minimize (7-2) are obtained by first setting the derivatives of S with respect to β_1 and β_2 equal to zero. Because S is a quadratic function, this approach leads to two linear equations in two unknowns $\hat{\beta}_1$ and $\hat{\beta}_2$. The analytical solution (Edgar et al., 2001) is

$$\hat{\beta}_1 = \frac{S_{uu}S_y - S_{uy}S_u}{NS_{uu} - (S_u)^2} \tag{7-5}$$

$$\hat{\beta}_2 = \frac{NS_{uy} - S_uS_y}{NS_{uu} - (S_u)^2} \tag{7-6}$$

where

$$S_u \triangleq \sum_{i=1}^{N} u_i \quad S_{uu} \triangleq \sum_{i=1}^{N} u_i^2$$

$$S_y \triangleq \sum_{i=1}^{N} Y_i \quad S_{uy} \triangleq \sum_{i=1}^{N} u_i Y_i$$

These calculations can be made using statistical packages or spreadsheets such as Excel.

This least-squares estimation approach (also called linear regression) can be extended to more general models with

1. More than one input or output variable
2. Functions of the input variables u, such as polynomials and exponentials, providing that the unknown parameters appear linearly.

A general nonlinear steady-state model which is linear in the parameters has the form

$$y = \sum_{j=1}^{p} \beta_j X_j + \epsilon \tag{7-7}$$

The p unknown parameters (β_j) are to be estimated, and the X_j are the p specified functions of u. Note that the unknown parameters β_j appear linearly in the model, and a constant term can be included by setting $X_1 = 1$.

The sum of the squares analogous to (7-2) is

$$S = \sum_{i=1}^{N}\left(Y_i - \sum_{j=1}^{p} \beta_j X_{ij}\right)^2 \tag{7-8}$$

For X_{ij} the first subscript corresponds to the ith data point, and the second index refers to the jth function of u. This expression can be written in matrix notation as

$$S = (\boldsymbol{Y} - \boldsymbol{X}\beta)^T(\boldsymbol{Y} - \boldsymbol{X}\beta) \tag{7-9}$$

where the superscript T denotes the transpose of a vector or matrix and

$$\boldsymbol{Y} = \begin{bmatrix} Y_1 \\ \vdots \\ Y_n \end{bmatrix} \quad \beta = \begin{bmatrix} \beta_1 \\ \vdots \\ \beta_p \end{bmatrix} \quad \boldsymbol{X} = \begin{bmatrix} X_{11} & X_{12} & \cdots & X_{1p} \\ X_{21} & X_{22} & \cdots & X_{2p} \\ \vdots & \vdots & & \vdots \\ X_{n1} & X_{n2} & \cdots & X_{np} \end{bmatrix}$$

The least-squares estimate $\hat{\beta}$ is given by Draper and Smith (1998), and Montgomery and Runger (2007),

$$\hat{\beta} = (\boldsymbol{X}^T\boldsymbol{X})^{-1}\boldsymbol{X}^T\boldsymbol{Y} \tag{7-10}$$

providing that matrix $\boldsymbol{X}^T\boldsymbol{X}$ is nonsingular so that its inverse exists. Note that the matrix $\boldsymbol{X}$ is comprised of functions of u; for example, if $y = \beta_1 + \beta_2 u + \beta_3 u^2 + \epsilon$, then $X_1 = 1$, $X_2 = u$, and $X_3 = u^2$.

If the number of data points is equal to the number of model parameters (i.e., $N = p$), Eq. 7-10 provides a unique solution to the parameter estimation problem, one that provides perfect agreement with the data points, as long as $\boldsymbol{X}^T\boldsymbol{X}$ is nonsingular. For $N > p$, a *least-squares solution* results that minimizes the sum of the squared deviations between each of the data points and the model predictions.

The least-squares solution in Eq. 7-10 provides a *point estimate* for the unknown model parameters β_i but does indicate how accurate the estimates are. The degree of accuracy is expressed by confidence intervals that have the form, $\hat{\beta}_i \pm \Delta\beta_i$. The $\Delta\beta_1$ are calculated from the (u, y) data for a specified *confidence level* (Draper and Smith, 1998).

Next we consider the development of a steady-state performance model, such as might be used in optimizing the operating conditions of an electrical power generator (see Chapter 19).

EXAMPLE 7.1

An experiment has been performed to determine the steady-state power delivered by a gas turbine-driven generator as a function of fuel flow rate. The following normalized data were obtained:

Fuel Flow Rate u_i	Power Generated Y_i
1.0	2.0
2.3	4.4
2.9	5.4
4.0	7.5
4.9	9.1
5.8	10.8
6.5	12.3
7.7	14.3
8.4	15.8
9.0	16.8

The linear model in (7-1) should be satisfactory because the data reveal a monotonic trend. Compare the best linear and quadratic models.

SOLUTION

To solve for the linear model, Eqs. 7-5 and 7-6 could be applied directly. However, to illustrate the use of Eq. 7-7, first define the terms in the linear model: $X_1 = 1$ and $X_2 = u$. The following matrices are then obtained:

$$\boldsymbol{X}^T = \begin{bmatrix} 1 & 1 & 1 & 1 & 1 & 1 & 1 & 1 & 1 & 1 \\ 1.0 & 2.3 & 2.9 & 4.0 & 4.9 & 5.8 & 6.5 & 7.7 & 8.4 & 9.0 \end{bmatrix}$$

$$\boldsymbol{Y}^T = [2.0 \;\; 4.4 \;\; 5.4 \;\; 7.5 \;\; 9.1 \;\; 10.8 \;\; 12.3 \;\; 14.3 \;\; 15.8 \;\; 16.8]$$

$$\mathbf{X}^T\mathbf{X} = \begin{bmatrix} \sum_{i=1}^{10} (1)^2 & \sum_{i=1}^{10} u_i \\ \sum_{i=1}^{10} u_i & \sum_{i=1}^{10} u_i^2 \end{bmatrix}$$

$$\mathbf{X}^T\mathbf{Y} = \begin{bmatrix} \sum_{i=1}^{10} Y_i \\ \sum_{i=1}^{10} u_i Y_i \end{bmatrix}$$

Solving for $\hat{\beta}_1$ and $\hat{\beta}_2$ using Eq. 7-10 yields the same results given in Eqs. 7-5 and 7-6, with $\hat{\beta}_1 = 0.0785 \pm 0.0039$ and $\hat{\beta}_2 = 1.859 \pm 0.093$ (95% confidence limits are shown).

To determine how much the model accuracy can be improved by using a quadratic model, Eq. 7-10 is again applied, this time with $X_1 = 1$, $X_2 = u$, and $X_3 = u^2$. The estimated parameters and 95% confidence limits for this quadratic model are

$\hat{\beta}_1 = 0.1707 \pm 0.0085$, $\hat{\beta}_2 = 1.811 \pm 0.096$, and $\hat{\beta}_3 = 0.0047 \pm 0.0002$

The predicted values of $y(\hat{y})$ are compared with the measured values (actual data) in Table 7.1 for both the linear and quadratic models. It is evident from this comparison that the linear model is adequate and that little improvement results from the more complicated quadratic model.

Table 7.1 A Comparison of Model Predictions from Example 7.1

u_i	y_i	Linear Model Prediction $\hat{y}_{1i} = \hat{\beta}_1 + \hat{\beta}_2 u_i$	Quadratic Model Prediction $\hat{y}_{2i} = \hat{\beta}_1 + \hat{\beta}_2 u_i + \hat{\beta}_3 u_i^2$
1.0	2.0	1.94	1.99
2.3	4.4	4.36	4.36
2.9	5.4	5.47	5.46
4.0	7.5	7.52	7.49
4.9	9.1	9.19	9.16
5.8	10.8	10.86	10.83
6.5	12.3	12.16	12.14
7.7	14.3	14.40	14.40
8.4	15.8	15.70	15.72
9.0	16.8	16.81	16.85
		($S = 0.0613$)	($S = 0.0540$)

7.1.3 Nonlinear Regression

If the empirical model is nonlinear with respect to the model parameters, then nonlinear regression rather than linear regression must be used. For example, suppose that a reaction rate expression of the form $r_A = kc_a^n$ is to be fit to experimental data, where r_A is the reaction rate of component A, c_A is the reactant concentration, and k and n are model parameters.

This model is *linear* with respect to rate constant k but is *nonlinear* with respect to reaction order n. A general nonlinear model can be written as

$$y = f(u_1, u_2, u_3, \ldots \beta_1, \beta_2, \beta_3 \ldots) \quad (7\text{-}11)$$

where y is the model output, u_j are inputs, and β_j are the parameters to be estimated. In this case, the β_j do not appear linearly in the model. However, we can still define a sum of squares error criterion to be minimized by selecting parameters β_j:

$$S = \sum_{i=1}^{N} (Y_i - \hat{y}_i)^2 \quad (7\text{-}12)$$

where Y_i and $\hat{y}_i$ denote the ith output measurement and model prediction corresponding to the ith data point, respectively. The least-squares estimates are again denoted by $\hat{\beta}_j$.

Consider the problem of estimating the time constants for first-order and overdamped second-order dynamic models based on the measured output response to a step input change of magnitude M. Analytical expressions for these step response were developed in Chapter 5.

Transfer Function	***Step Response***	
$\dfrac{Y(s)}{U(s)} = \dfrac{K}{\tau s + 1}$	$y(t) = KM(1 - e^{-t/\tau})$	(5-18)
$\dfrac{Y(s)}{U(s)} = \dfrac{K}{(\tau_1 s + 1)(\tau_2 s + 1)}$	$y(t) = KM\left(1 - \dfrac{\tau_1 e^{-t/\tau_1} - \tau_2 e^{-t/\tau_2}}{\tau_1 - \tau_2}\right)$	(5-48)

In the step response equations, t is the independent variable instead of the input u used earlier, and y is the dependent variable expressed in deviation form. Although steady-state gain K appears linearly in both response equations, the time constants are contained in a nonlinear manner, which means that linear regression cannot be used to estimate them.

Sometimes a variable transformation can be employed to transform a nonlinear model so that linear regression can be used (Montgomery and Runger, 2007). For example, if K is assumed to be known, the first-order step response can be rearranged:

$$\ln\left(1 - \frac{y(t)}{KM}\right) = -\frac{t}{\tau} \quad (7\text{-}13)$$

Because $\ln(1 - y/KM)$ can be evaluated at each time t_i, this model is linear in the parameter $1/\tau$. Thus, this model has the standard linear form as Eq. 7-1, where the left-hand side of (7-13) is Y_i, $\beta_1 = 0$, and $u_i = t_i$.

The transformation in Eq. 7-13 leads to the *fraction incomplete response* method of determining first-order models discussed in the next section. However, for step responses of higher-order models, such as Eq. 5-48, the transformation approach is not feasible. For these calculations, we must use an iterative optimization method to find the least-squares estimates of the time constants (Edgar et al., 2001).

As an alternative to nonlinear regression, a number of graphical correlations can be used quickly to find approximate values of τ_1 and τ_2 in second-order models. The accuracy of models obtained in this way is often sufficient for controller design. In the next section, we present several shortcut methods for estimating transfer function parameters based on graphical analysis.

7.2 FITTING FIRST- AND SECOND-ORDER MODELS USING STEP TESTS

A plot of the output response of a process to a step change in input is sometimes referred to as the *process reaction curve*. If the process of interest can be approximated by a first-order or second-order linear model, the model parameters can be obtained by inspection of the process reaction curve. For example, recall the first-order model expressed in deviation variables,

$$\tau \frac{dy}{dt} + y = Ku \tag{7-14}$$

where the system is initially at a steady state, with $u(0) = 0$ and $y(0) = 0$. If the input u is abruptly changed from 0 to M at time $t = 0$, the step response in Eq. 5-18 results. The normalized step response is shown in Fig. 7.3. The response $y(t)$ reaches 63.2% of its final value at $t = \tau$. The steady-state change in y, Δy, is given by $\Delta y = KM$. From Eq. 5-18 or 7-13, after rearranging and evaluating the limit at $t = 0$, the initial slope of the normalized step response is

$$\frac{d}{dt}\left(\frac{y}{KM}\right)_{t=0} = \frac{1}{\tau} \tag{7-15}$$

Thus, as shown in Fig. 7.3, the intercept of the tangent at $t = 0$ with the horizontal line, $y/KM = 1$, occurs at $t = \tau$. As an alternative, τ can be estimated from a step response plot using the value of t at which the response is 63.2% complete, as shown in the following example.

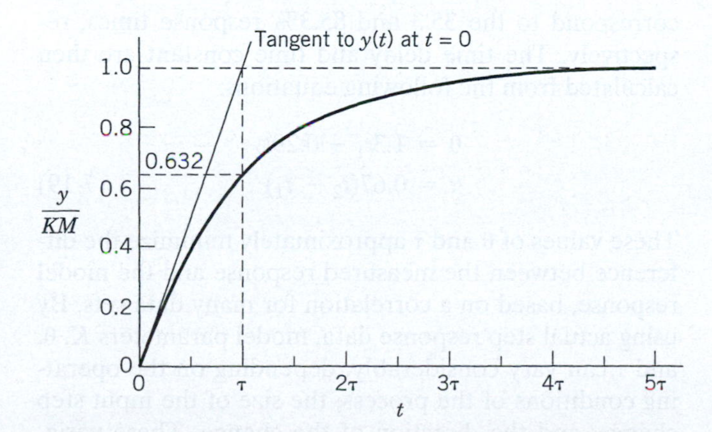

Figure 7.3 Step response of a first-order system and graphical constructions used to estimate the time constant, τ.

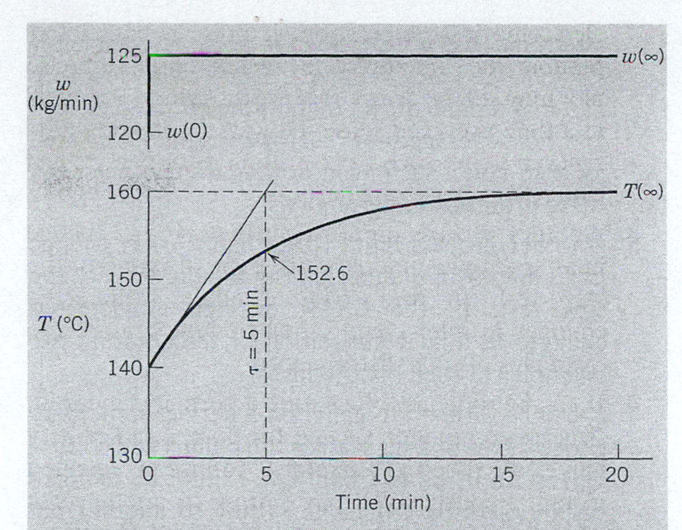

Figure 7.4 Temperature response of a stirred-tank reactor for a step change in feed flow rate.

EXAMPLE 7.2

Figure 7.4 shows the response of the temperature T in a continuous stirred-tank reactor to a step change in feed flow rate w from 120 to 125 kg/min. Find an approximate first-order model for the process and these operating conditions.

SOLUTION

First note that $\Delta w = M = 125 - 120 = 5$ kg/min. Because $\Delta T = T(\infty) - T(0) = 160 - 140 = 20$ °C, the steady-state gain is

$$K = \frac{\Delta T}{\Delta w} = \frac{20°\text{C}}{5 \text{ kg/min}} = 4 \frac{°\text{C}}{\text{kg/min}}$$

The time constant obtained from the tangent construction shown in Fig. 7.4 is $\tau = 5$ min. Note that this result is consistent with the "63.2% method," because

$$T = 140 + 0.632(20) = 152.6 \text{ °C}$$

Consequently, the resulting process model is

$$\frac{T'(s)}{W'(s)} = \frac{4}{5s + 1}$$

where the steady-state gain is 4 °C/kg/min.

Very few experimental step responses exhibit exactly first-order behavior, because

1. The true process model is usually *neither first-order nor linear*. Only the simplest processes exhibit such ideal dynamics.
2. The output data are usually corrupted with noise; that is, the measurements contain a random component. Noise can arise from normal operation of the process, for example, inadequate mixing that produces eddies of higher and lower concentration (or temperature), or from

electronic instrumentation. If noise is completely random (i.e., uncorrelated), a first-order response plot may still be drawn that fits the output data well in a time-averaged sense. However, autocorrelated random noise, such as in drifting disturbances, can cause problems in the analysis.

3. Another process input (disturbance) may change in an unknown manner during the duration of the step test. In the CSTR example, undetected changes in inlet composition or temperature are examples of such disturbances.
4. It can be difficult to generate a perfect step input. Process equipment, such as the pumps and control valves discussed in Chapter 9, cannot be changed instantaneously from one setting to another but must be ramped over a period of time. However, if the ramp time is small compared to the process time constant, a reasonably good approximation to a step input may be obtained.

In summary, departures from the ideal response curve in Fig. 7.3 are common.

In order to account for higher-order dynamics that are neglected in a first-order model, a time-delay term can be included. This modification can improve the agreement between model and experimental responses. The fitting of a first-order plus time-delay model (FOPTD),

$$G(s) = \frac{Ke^{-\theta s}}{\tau s + 1} \tag{7-16}$$

to the actual step response requires the following steps, as shown in Fig. 7.5:

1. The process gain K is found by calculating the ratio of the steady-state change in y to the size of the input step change, M.
2. A tangent is drawn at the point of inflection of the step response; the intersection of the tangent line and the time axis (where $y = 0$) is the time delay.
3. If the tangent is extended to intersect the steady-state response line (where $y = KM$), the point of intersection corresponds to time $t = \theta + \tau$. Therefore, τ can be found by subtracting θ from the point of intersection.

The tangent method presented here for obtaining the time constant suffers from using only a single point to estimate the time constant. Use of several points from the response may provide a better estimate. Again consider Eq. 7-13, but now introduce the time shift $t - \theta$ and rearrange to give the expression

$$\ln\left[\frac{y(\infty) - y_i}{y(\infty)}\right] = -\frac{t_i - \theta}{\tau} \tag{7-17}$$

The final steady-state value, $y(\infty)$, equals KM. In 7-17, $y(\infty) - y_i$ can be interpreted as the *incomplete response*

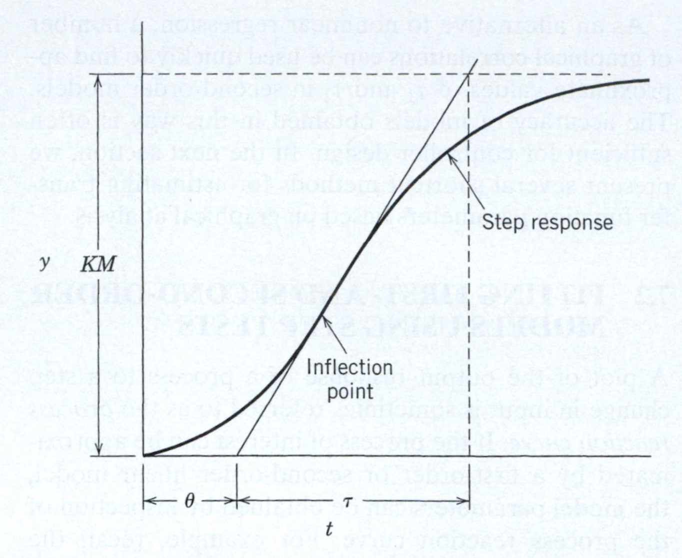

Figure 7.5 Graphical analysis of the process reaction curve to obtain parameters of a first-order-plus-time-delay model.

for data point i; dividing by $y(\infty)$ yields the *fraction incomplete response*: a semilog plot of $[y(\infty) - y_i]/y(\infty)$ vs. $(t_i - \theta)$ will then yield a straight line with a slope of $-1/\tau$, from which an average value of τ is obtained. An equation equivalent to 7-17 for the variables of Example 7.2 is

$$\ln\left[\frac{T(\infty) - T(t)}{T(\infty) - T(0)}\right] = -\frac{t - \theta}{\lambda\tau} \tag{7-18}$$

The major disadvantage of the time-delay estimation method in Fig. 7.5 is that it is difficult to find the point of inflection, as a result of measurement noise and small-scale recorder charts or computer displays. The method of Sundaresan and Krishnaswamy (1978) avoids use of the point of inflection construction entirely to estimate the time delay. They proposed that two times t_1 and t_2 be estimated from a step response curve. These times correspond to the 35.3 and 85.3% response times, respectively. The time delay and time constant are then calculated from the following equations:

$$\begin{aligned} \theta &= 1.3t_1 - 0.29t_2 \\ \tau &= 0.67(t_2 - t_1) \end{aligned} \tag{7-19}$$

These values of θ and τ approximately minimize the difference between the measured response and the model response, based on a correlation for many data sets. By using actual step response data, model parameters K, θ, and τ can vary considerably, depending on the operating conditions of the process, the size of the input step change, and the direction of the change. These variations usually can be attributed to process nonlinearities and unmeasured disturbances.

7.2.1 Graphical Techniques for Second-Order Models

In general, a better approximation to an experimental step response can be obtained by fitting a second-order model to the data. Figure 7.6 shows the range of step response shapes that can occur for the second-order model,

$$G(s) = \frac{K}{(\tau_1 s + 1)(\tau_2 s + 1)} \tag{5-39}$$

Figure 7.6 includes two limiting cases: $\tau_2/\tau_1 = 0$, where the system becomes first-order, and $\tau_2/\tau_1 = 1$, the critically damped case. The larger of the two time constants, τ_1, is called the *dominant time constant*. The S-shaped response becomes more pronounced as the ratio of τ_2/τ_1 becomes closer to one.

Model parameters for second-order systems which include time delays can be estimated using graphical or numerical methods. A method due to Smith (1972) utilizes a model of the form

$$G(s) = \frac{Ke^{-\theta s}}{\tau^2 s^2 + 2\zeta\tau s + 1} \tag{7-20}$$

which includes both overdamped and underdamped cases. Smith's method requires the times (with apparent time delay removed) at which the normalized response reaches 20% and 60%, respectively. Using Fig. 7.7, the ratio of t_{20}/t_{60} gives the value of ζ. An estimate of τ can be obtained from the plot of t_{60}/τ vs. t_{20}/t_{60}.

When graphically fitting second-order models, some caution must be exercised in estimating θ. A second-order model with no time delay exhibits a point-of-inflection (see Fig. 7.6 when $\tau_1 \approx \tau_2$). If the tangent to the point-of-inflection shown in Fig. 7.5 is applied to this case, however, a nonzero time delay is indicated. To avoid this conflict, visual determination of θ is recommended for graphical estimation, but estimation of θ by trial and error may be required to obtain a good fit. In

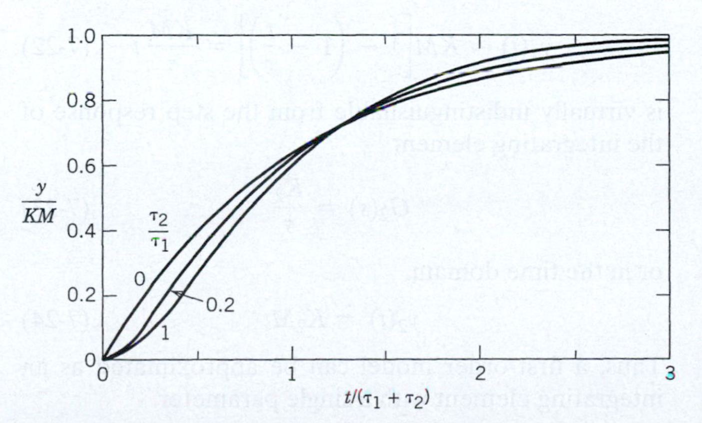

Figure 7.6 Step response for several overdamped second-order systems.

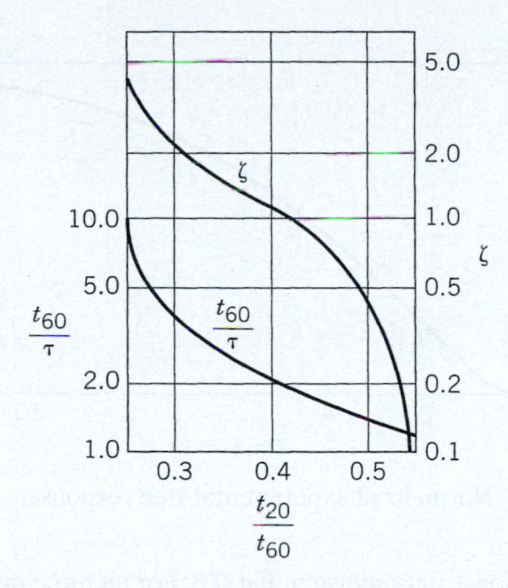

Figure 7.7 Smith's method: relationship of ζ and τ to τ_{20} and τ_{60}.

the following examples, the time delay is subtracted from the actual time value; then the adjusted time, $t' = t - \theta$, is employed for the actual graphical analysis. An alternative approach to fitting the parameters of the second-order model utilizes three points in the step response. Rangaiah and Krishnaswamy (1994, 1996).

7.2.2 Regression of Step Response Data

Model parameters of transfer function models can be estimated using nonlinear regression and standard software such as Excel and MATLAB. To use Excel, the measured data must be placed in one column. The model predictions to be compared with the measured data are placed in a second column. The sum of squares of the errors is calculated and put into a cell, called the target cell. The target cell value can be minimized using the built-in *Solver* in the *Tools* menu. The window of the Solver allows the user to select the cell to minimize/maximize, the range of cells to be adjusted (the model parameters), and the restrictions, if any, that apply. Clicking on ⟨solve⟩ will calculate the parameter values that minimize the sum of squares. The optimization method used by Excel is based on the generalized reduced gradient technique (Edgar et al., 2001).

In order to use MATLAB, it is necessary to write an M-file that defines the sum of squares of errors. Then the command *fminu* is used to calculate the minimum. The default algorithm in MATLAB is the BFGS quasi-Newton method (Ljung, 2007).

EXAMPLE 7.3

Step test data have been obtained for the off-gas CO_2 concentration response obtained from changing the feed rate to a bioreactor. Use Smith's method as well as nonlinear regression based on Excel and MATLAB to estimate parameters in a second-order model from experimental

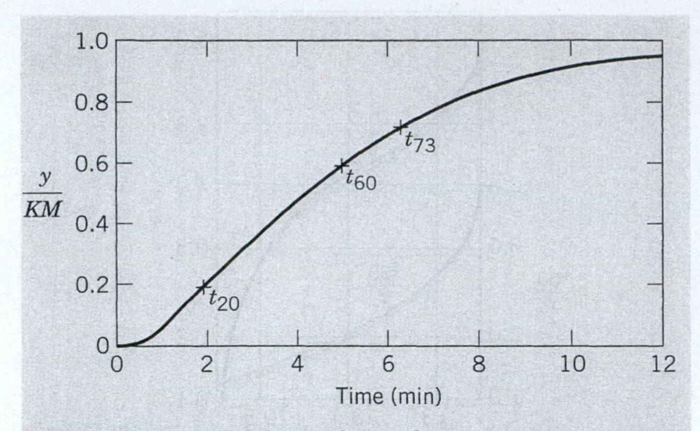

Figure 7.8 Normalized experimental step response.

step response data shown in Fig. 7.8. For all three methods, assume $\theta = 0$ because the response curve becomes nonzero immediately after $t = 0$. Compare the results with a first-order-plus-time-delay (FOPTD) model that is fit using the 63.2% response method to estimate the time constant.

SOLUTION

Smith's Method

The two points of interest are the 20% response time, $t_{20} = 1.85$ min, and the 60% response time, $t_{60} = 5.0$ min. Hence, $t_{20}/t_{60} = 0.37$. From Fig. 7.7, $\zeta = 1.3$ and $t_{60}/\tau = 2.8$; thus, $\tau = 5.0/2.8 = 1.79$ min. Because the model is overdamped, the two time constants can be calculated from the following expressions:

$$\tau_1 = \tau\zeta + \tau\sqrt{\zeta^2 - 1}, \quad \tau_2 = \tau\zeta - \tau\sqrt{\zeta^2 - 1}$$

Solving gives $\tau_1 = 3.81$ min and $\tau_2 = 0.84$ min.

For Fig. 7.8 the 63.2% response is estimated to occur at $t = 5.3$ min. Using the slope at the point of inflection, we can estimate the time delay to be $\theta = 0.7$ min. Note that $\tau = 4.6$ min, which is approximately equal to the sum of τ_1 and τ_2 for the second-order model.

Nonlinear Regression

Using Excel and MATLAB, we calculate the time constants in Eq. 5-48 that minimize the sum of the squares of the errors between experimental data and model predictions (see Eq. 7-12). The data consisted of 25 points between $t = 0$ and $t = 12$ min with a sampling period of 0.5 min. A comparison of the model parameters and the sum of squared errors for each method is shown below; the time delay is set to zero for the three second-order methods.

	τ_1 (min)	τ_2 (min)	Sum of Squares
Smith	3.81	0.84	0.0769
First order ($\theta = 0.7$ min)	4.60	—	0.0323
Excel and MATLAB	3.34	1.86	0.0057

Clearly, the nonlinear regression method is superior in terms of the goodness of fit, as measured by the sum of squares of the prediction error, but the required calculations are more complicated. Note that the nonlinear regression methods employed by Excel and MATLAB produce identical results.

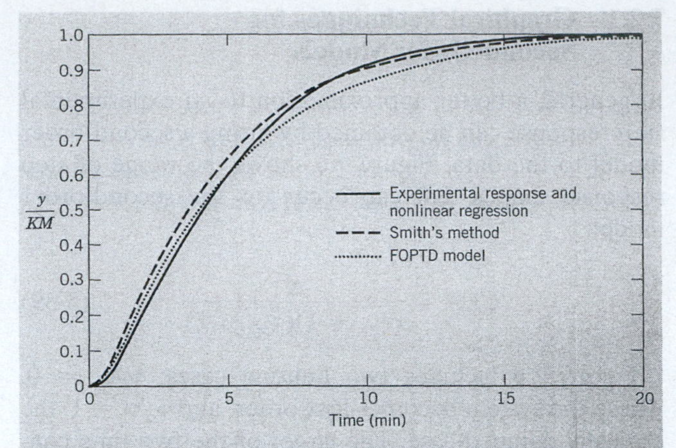

Figure 7.9 Comparison of step responses of fitted models with the original response data.

The step responses are plotted in Fig. 7.9; all three calculated models give an acceptable fit to the original step response curve. In fact, the nonlinear regression model is indistinguishable from the experimental response. Nonlinear regression does not depend on graphical correlations and always provides a better fit to the data. It also permits the experimental step test to be terminated before the final steady state is reached; however, sufficient response data must be obtained for the regression method to be effective.

7.2.3 Fitting an Integrator Model to Step Response Data

In Chapter 5, we considered the response of a first-order process to a step change in input of magnitude M:

$$y_1(t) = KM(1 - e^{-t/\tau}) \tag{5-18}$$

For short times, $t < \tau$, the exponential term can be approximated by a truncated Taylor Series expansion

$$e^{-t/\tau} \approx 1 - \frac{t}{\tau} \tag{7-21}$$

so that the response

$$y_1(t) \approx KM\left[1 - \left(1 - \frac{t}{\tau}\right)\right] = \frac{KM}{\tau}t \tag{7-22}$$

is virtually indistinguishable from the step response of the integrating element

$$G_2(s) = \frac{K_2}{s} \tag{7-23}$$

or in the time domain,

$$y_2(t) = K_2Mt \tag{7-24}$$

Thus, a first-order model can be approximated as an integrating element with a single parameter

$$K_2 = \frac{K}{\tau} \tag{7-25}$$

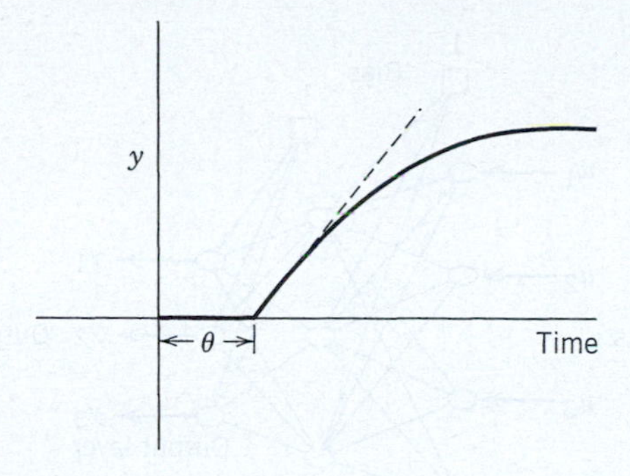

Figure 7.10 Comparison of step responses for a FOPTD model (solid line) and the approximate integrator plus time delay model (dashed line).

that matches the early ramp-like response to a step change in input.

Similarly, the approximate short-term response of the FOPTD model in (7-16) to a step input of magnitude M is

$$y(t) = \frac{KM}{\tau}(t - \theta)S(t - \theta)$$

that is, a ramp shifted by the time delay, θ. Thus, an approximate integrator plus time delay model consisting of a constant $K_2 = K/\tau$ and time delay θ is obtained. As shown in Chapter 12, whenever τ is large compared to θ, a control system can be designed using a two-parameter (K_2 and θ) model that is equally as effective as when using a three-parameter (K, τ, and θ) FOPTD model. Figure 7.10 shows that the two responses match well for relatively short times.

7.2.4 Other Types of Input Excitation

Sometimes a step change in a process input is not permissible owing to safety considerations or the possibility of producing off-specification (*off-spec*) material as a result of the process output deviating significantly from the desired value. In these situations, other types of input changes that do not move the process to a new steady state can be selected. They include rectangular pulses (see Fig. 7.11), pulses of arbitrary shape, or even white (Gaussian) noise. Such "plant-friendly" inputs should be as short as possible, stay within actuator limits, and cause minimum disruption to the controlled variables (Rivera and Jun, 2000). For pulse forcing, the input is suddenly changed, left at its new value for a period of time, and then returned to its original value. Consequently, the process output also returns to its initial steady state, unless the process has an integrating mode (e.g., Eq. 7-23).

Random Binary Sequence (RBS) forcing involves a series of pulses of fixed height and random duration. At each sampling instant, a random number generator determines whether the input signal is set at its maximum or minimum value. However, it is more convenient to

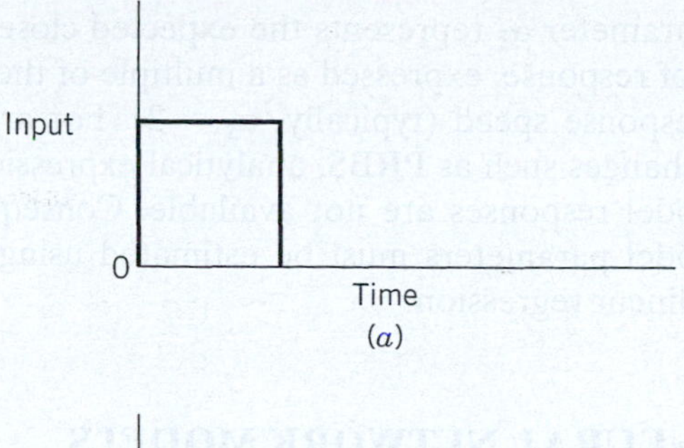

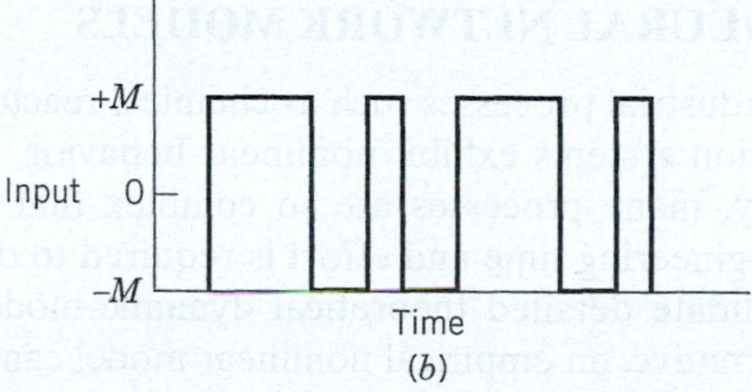

Figure 7.11 (*a*) Pulse and (*b*) PRBS inputs (one cycle).

implement a *pseudo random binary sequence (PRBS)*, which is a two-level, periodic, deterministic signal of a specified length, shown in Fig. 7.11. The actual sequence of inputs can be repeated multiple times. The term *pseudo random* indicates the input is a repeating sequence that has the spectral characteristics of a random signal (Godfrey, 1993). The advantage of a PRBS is that the input excitation can be concentrated in particular frequency ranges that correspond to the process dynamics and that are important for control system design (see Chapter 14 for more information on frequency response analysis).

A PRBS sequence is characterized by two parameters: the duration of the switching sequence (N_s) and the switching time or clock period T_{sw}, which is the minimum time between changes in the level of the signal. N_s is a positive integer value, while T_{sw} is an integer multiple of the sampling period Δt. The signal repeats itself after $N_s T_{sw}$ units of time. The actual input sequence is generated by a set of n shift registers such that $N_s = 2^n - 1$. This means that N_s assumes specific values such as 3, 7, 15, 31, etc. For example, for $n = 4$ and $N_s = 15$, the input binary sequence is [0, 1, 1, 1, 1, 0, 1, 0, 1, 1, 0, 0, 1, 0, 0, 0], where "0" represents the lower input value and "1" represents the higher input value. N_s and T_{sw} can be determined from *a priori* information about the process. Rivera and Jun (2000) have recommended guidelines for specifying T_{sw} and N_s,

$$T_{sw} \leq \frac{2.78\tau^L_{dom}}{\alpha_s} \qquad N_s \geq \frac{2\pi\beta_s\tau^H_{dom}}{T_{sw}}$$

where τ^H_{dom} and τ^L_{dom} are high and low estimates of the dominant time constant. β_s is an integer corresponding to the settling time of the process (e.g., for $t_{95\%}$, $\beta_s = 3$; for $t_{99\%}$, $\beta_s = 5$), which determines the length of the test. Parameter α_s represents the expected closed-loop speed of response, expressed as a multiple of the open-loop response speed (typically, $\alpha_s \approx 2$). For arbitrary input changes such as PRBS, analytical expressions for the model responses are not available. Consequently, the model parameters must be estimated using linear or nonlinear regression.

7.3 NEURAL NETWORK MODELS

Most industrial processes such as chemical reactors and separation systems exhibit nonlinear behavior. Unfortunately, many processes are so complex that significant engineering time and effort is required to develop and validate detailed theoretical dynamic models. As an alternative, an empirical nonlinear model can be obtained from experimental data. *Neural networks (NN)* or artificial neural networks are an important class of empirical nonlinear models. Neural networks have been used extensively in recent years to model a wide range of physical and chemical phenomena and to model other nonengineering situations such as stock market analysis, chess strategies, speech recognition, and medical diagnoses. Neural networks are attractive whenever it is necessary to model complex or little understood processes with large input-output data sets, as well as to replace models that are too complicated to solve in real time (Su and McAvoy, 1997; Himmelblau, 2008).

The exceptional computational abilities of the human brain have motivated the concept of an NN. The brain can perform certain types of computation, such as perception, pattern recognition, and motor control, much faster than existing digital computers (Haykin, 2009). The operation of the human brain is complex and nonlinear and involves massive parallel computation. Its computations are performed using structural constituents called *neurons* and the synaptic interconnections between them (that is, a *neural network*). The development of artificial neural networks is an admittedly approximate attempt to mimic this biological neural network, in order to achieve some of its computational advantages.

A multilayer feedforward network, one of the most common NN structures, is shown in Fig. 7.12. The *neurons* (or *nodes*) are organized into layers (input, output, hidden); each neuron in the hidden layer is connected to the neurons in adjacent layers via connection weights. These weights are unknown parameters that are estimated based on the input/output data from the process to be modeled. The number of unknown parameters can be quite large (e.g., 50 to 100), and powerful nonlinear programming algorithms are required to fit the parameters to the data using the least-squares objective function (Edgar et al., 2001). If enough neurons are utilized, an input-output process can be accurately modeled by a neural net model.

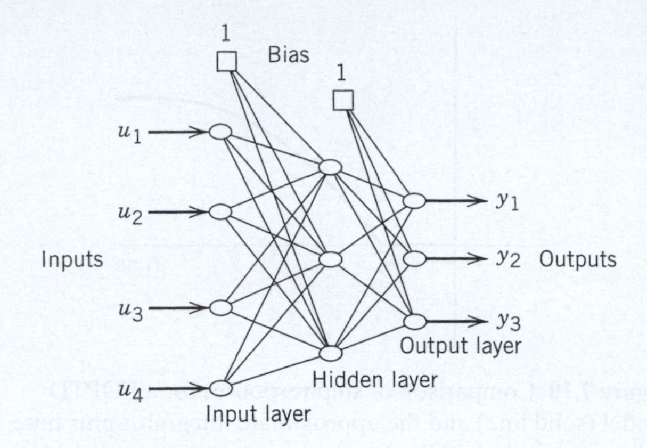

Figure 7.12 Multilayer neural network with three layers.

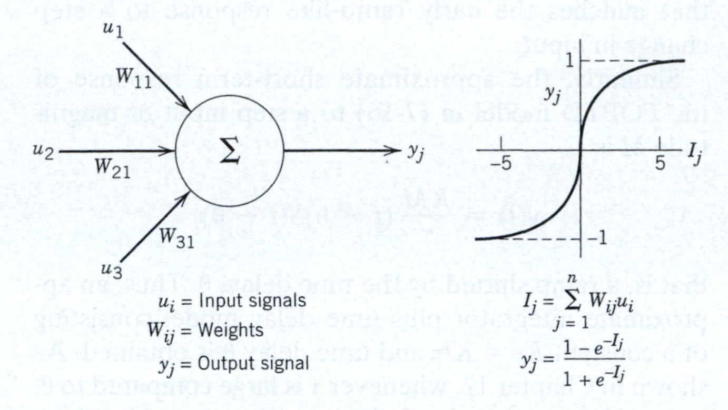

Figure 7.13 Signal diagram for a neuron.

As shown in Fig. 7.13, at each neuron inputs are collected from other neurons or from bias terms, and their strength or magnitude is evaluated. These inputs are then summed and compared with a threshold level, and the appropriate output is determined. The connection weight (W_{ij}) determines the relative importance of that input. The sum of the weighted inputs is then passed to a nonlinear transformation, as shown in Fig. 7.13. One type of transformation has a sigmoidal shape as is shown in the figure, although many options are available.

The *training* of a neural network involves estimating the unknown parameters; this procedure generally utilizes normal operating data (often large data sets) taken in the operating region where the model is intended to be used. After the parameters are estimated (the network is *trained*), another large set of data can be used to validate that the model is adequate. Sometimes the resulting NN model is not satisfactory, and changes in the model structure must be made, often by trial and error. Commercial software packages are available that make automatic empirical modeling of complex processes feasible.

Advanced applications of neural nets have been commercially implemented in the areas of fault detection and diagnosis, sensor errors, and dynamic modeling and control (Su and McAvoy, 1997). In some cases, neural nets have been used to determine controller settings in advanced control systems.

7.3.1 Soft Sensors

A common problem shared by many industrial processes is the inability to measure key process variables noninvasively and in real time, especially the compositions of process streams and product properties. The development of improved sensors, based on new techniques of analytical chemistry and modern electronic devices using fiber optics and semiconductors, has been an active area (cf. Appendix A). As an alternative, the use of easily measured secondary variables to infer values of unmeasured process variables is now receiving great interest; the term *soft sensors* is often used to denote this approach. *Chemometrics* is a term related to soft sensors that describes how data from process analyzers (e.g., spectra) can be analyzed and modeled for use in process monitoring and control (Brown, 1998).

Soft sensors have become an attractive alternative to the high cost of accurate on-line measurements for applications where empirical models can accurately infer (that is, predict) unmeasured variables. For example, the environmental regulatory agency in Texas permits NN models to be used for monitoring emissions from various process units such as power boilers. The NN models use measurements of selected input and output variables to predict pollutants at the parts per billion level (Martin, 1997). In materials manufacturing, the real-time detection of cracks, inclusions, porosity, dislocations, or defects in metallurgical or electronic materials would be highly desirable during processing, rather than after processing is completed and defective products are shipped. Use of virtual sensor models to predict quality control measures, such as the formation and location of defects, can greatly reduce the stringent requirements imposed on hardware-based sensors.

7.4 DEVELOPMENT OF DISCRETE-TIME DYNAMIC MODELS

A digital computer by its very nature deals internally with discrete-time data or numerical values of functions at equally spaced intervals determined by the sampling period. Thus, discrete-time models such as *difference equations* are widely used in computer control applications. One way a continuous-time dynamic model can be converted to discrete-time form is by employing a finite difference approximation (Chapra and Canale, 2010). Consider a nonlinear differential equation,

$$\frac{dy(t)}{dt} = f(y, u) \tag{7-26}$$

where y is the output variable and u is the input variable. This equation can be numerically integrated (although with some error) by introducing a finite difference approximation for the derivative. For example, the first-order, backward difference approximation to the derivative at $t = k\Delta t$ is

$$\frac{dy(t)}{dt} \cong \frac{y(k) - y(k-1)}{\Delta t} \tag{7-27}$$

where Δt is the integration interval specified by the user and $y(k)$ denotes the value of $y(t)$ at $t = k\Delta t$. Substituting Eq. 7-26 into (7-27) and evaluating $f(y, u)$ at the previous values of y and u (i.e., $y(k-1)$ and $u(k-1)$) gives

$$\frac{y(k) - y(k-1)}{\Delta t} \cong f(y(k-1)), u(k-1)) \tag{7-28}$$

or

$$y(k) = y(k-1) + \Delta t\, f(y(k-1), u(k-1)) \tag{7-29}$$

Equation 7-29 is a first-order difference equation that can be used to predict $y(k)$ based on information at the previous time step $(k-1)$. This type of expression is called a *recurrence relation*. It can be used to numerically integrate Eq. 7-26 by successively calculating $y(k)$ for $k = 1, 2, 3, \ldots$ starting from a known initial condition $y(0)$ and a specified input sequence, $\{u(k)\}$. In general, the resulting numerical solution becomes more accurate and approaches the correct solution $y(t)$ as Δt decreases. However, for extremely small values of Δt, computer roundoff can be a significant source of error (Chapra and Canale, 2010).

EXAMPLE 7.4

For the first-order differential equation,

$$\tau \frac{dy(t)}{dt} + y(t) = Ku(t) \tag{7-30}$$

derive a recursive relation for $y(k)$ using a first-order backwards difference for $dy(t)/dt$.

SOLUTION

The corresponding difference equation after approximating the first derivative is

$$\frac{\tau(y(k) - y(k-1))}{\Delta t} + y(k-1) = Ku(k-1) \tag{7-31}$$

Rearranging gives

$$y(k) = \left(1 - \frac{\Delta t}{\tau}\right) y(k-1) + \frac{K\,\Delta t}{\tau} u(k-1) \tag{7-32}$$

The new value $y(k)$ is a weighted sum of the previous value $y(k-1)$ and the previous input $u(k-1)$. Equation 7-32 can also be derived directly from (7-29).

As shown in numerical analysis textbooks, the accuracy of Eq. 7-32 is influenced by the integration interval. However, discrete-time models involving no approximation errors can be derived for any linear differential equation under the assumption of a piecewise constant input signal, that is, the input variable u is held constant over Δt. Next, we develop discrete-time modeling methods that introduce no integration error for piecewise constant inputs, regardless of the size of Δt.

Such models are important in analyzing computer-controlled processes where the process inputs are piecewise constant.

7.4.1 Exact Discrete-Time Models

For a process described by a linear differential equation, the corresponding discrete-time model can be derived from the analytical solution for a piecewise constant input. This analytical approach eliminates the discretization error inherent in finite-difference approximations. Consider a first-order model in Eq. 7-30 with previous output $y[(k-1)\Delta t]$ and a constant input $u(t) = u[(k-1)\Delta t]$ over the time interval, $(k-1)\Delta t \le t < k\Delta t$. The analytical solution to Eq. 7-30 at $t = k\Delta t$ is

$$y(k\Delta t) = (1 - e^{-\Delta t/\tau})Ku[(k-1)\Delta t] + e^{-\Delta t/\tau}y[(k-1)\Delta t] \qquad (7\text{-}33)$$

Equation 7-33 can be written more compactly as

$$y(k) = e^{-\Delta t/\tau}y(k-1) + K(1 - e^{-\Delta t/\tau})u(k-1) \qquad (7\text{-}34)$$

Equation 7-34 is the exact solution to Eq. 7-30 at the sampling instants provided that $u(t)$ is constant over each sampling interval of length Δt. Note that the continuous output $y(t)$ is not necessarily constant between sampling instants, but (7-33) and (7-34) provide an exact solution for $y(t)$ at the sampling instants, $k = 1, 2, 3, \ldots$.

In general, when a linear differential equation of order p is converted to discrete time, a linear difference equation of order p results. For example, consider the second-order model:

$$G(s) = \frac{Y(s)}{U(s)} = \frac{K(\tau_a s + 1)}{(\tau_1 s + 1)(\tau_2 s + 1)} \qquad (7\text{-}35)$$

The analytical solution for a constant input provides the corresponding difference equation, which is also referred to as an autogressive model with external (or exogenous) input, or *ARX model* (Ljung, 1999):

$$y(k) = a_1 y(k-1) + a_2 y(k-2) + b_1 u(k-1) + b_2 u(k-2) \qquad (7\text{-}36)$$

where

$$a_1 = e^{-\Delta t/\tau_1} + e^{-\Delta t/\tau_2} \qquad (7\text{-}37)$$

$$a_2 = -e^{-\Delta t/\tau_1}e^{-\Delta t/\tau_2} \qquad (7\text{-}38)$$

$$b_1 = K\left(1 + \frac{\tau_a - \tau_1}{\tau_1 - \tau_2}e^{-\Delta t/\tau_1} + \frac{\tau_2 - \tau_a}{\tau_1 - \tau_2}e^{-\Delta t/\tau_2}\right) \qquad (7\text{-}39)$$

$$b_2 = K\left(e^{-\Delta t(1/\tau_1 + 1/\tau_2)} + \frac{\tau_a - \tau_1}{\tau_1 - \tau_2}e^{-\Delta t/\tau_2} + \frac{\tau_2 - \tau_a}{\tau_1 - \tau_2}e^{-\Delta t/\tau_1}\right) \qquad (7\text{-}40)$$

In Eq. 7-36 the new value of y depends on the values of y and u at the two previous sampling instants; hence, it is a second-order difference equation. If $\tau_2 = \tau_a = 0$ in Eqs. 7-36 through 7-40, the first-order difference equation in (7-33) results.

The steady-state gain of the second-order difference equation model can be found by considering steady-state conditions. Let $\overline{u}$ and $\overline{y}$ denote the new steady-state values after a step change in u. Substituting these values into Eq. 7-36 gives

$$\overline{y} = a_1\overline{y} + a_2\overline{y} + b_1\overline{u} + b_2\overline{u} \qquad (7\text{-}41)$$

Because y and u are deviation variables, the steady-state gain is simply $\overline{y}/\overline{u}$, the steady-state change in y divided by the steady-state change in u. Rearranging Eq. 7-41 gives

$$\text{Gain} = \frac{\overline{y}}{\overline{u}} = \frac{b_1 + b_2}{1 - a_1 - a_2} \qquad (7\text{-}42)$$

Substitution of Eqs. 7-37 through 7-40 into (7-42) gives K, the steady-state gain for the transfer function model in Eq. 7-35.

Higher-order linear differential equations can be converted to a discrete-time, difference equation model using a state-space analysis (Åström and Wittenmark, 1997).

7.5 IDENTIFYING DISCRETE-TIME MODELS FROM EXPERIMENTAL DATA

If a linear discrete-time model is desired, one approach is to fit a continuous-time model to experimental data (cf. Section 7.2) and then to convert it to discrete-time form using the above approach. A more attractive approach is to estimate parameters in a discrete-time model directly from input-output data based on linear regression. This approach is an example of *system identification* (Ljung, 1999). As a specific example, consider the second-order difference equation in (7-36). It can be used to predict $y(k)$ from data available at times, $(k-1)\Delta t$ and $(k-2)\Delta t$. In developing a discrete-time model, model parameters a_1, a_2, b_1, and b_2 are considered to be unknown. They are estimated by applying linear regression to minimize the error criterion in Eq. 7-8 after defining

$$\beta^T = [a_1\ a_2\ b_1\ b_2], \quad X_1 = y(k-1), \quad X_2 = y(k-2),$$

$$X_3 = u(k-1), \quad \text{and} \quad X_4 = u(k-2)$$

EXAMPLE 7.5

Consider the step response data $y(k)$ in Table 7.2, which were obtained from Example 7.3 and Fig. 7.8 for $\Delta t = 1$. At $t = 0$ a unit step change in u occurs, but the first output change is not observed until the next sampling instant.

Estimate the model parameters in the second-order difference equation from the input-output data. Compare this model with the models obtained in Example 7.3 using nonlinear regression.

Table 7.2 Step Response Data

k	$y(k)$
1	0.058
2	0.217
3	0.360
4	0.488
5	0.600
6	0.692
7	0.772
8	0.833
9	0.888
10	0.925

*$\Delta t = 1$; for $k < 0$, $y(k) = 0$ and $u(k) = 0$.

Table 7.3 Data Regression for Example 7.5

$$\boldsymbol{Y} = \begin{bmatrix} 0.058 \\ 0.217 \\ 0.360 \\ 0.488 \\ 0.600 \\ 0.692 \\ 0.722 \\ 0.833 \\ 0.888 \\ 0.925 \end{bmatrix} \qquad \boldsymbol{X} = \begin{bmatrix} 0 & 0 & 1 & 0 \\ 0.058 & 0 & 1 & 1 \\ 0.217 & 0.058 & 1 & 1 \\ 0.360 & 0.217 & 1 & 1 \\ 0.488 & 0.360 & 1 & 1 \\ 0.600 & 0.488 & 1 & 1 \\ 0.692 & 0.600 & 1 & 1 \\ 0.772 & 0.692 & 1 & 1 \\ 0.833 & 0.772 & 1 & 1 \\ 0.888 & 0.833 & 1 & 1 \end{bmatrix}$$

SOLUTION

For linear regression, there are four independent variables, $y(k-1)$, $y(k-2)$, $u(k-1)$, $u(k-2)$, one dependent variable $y(k)$, and four unknown parameters (a_1, a_2, b_1, b_2). We structure the data for regression as shown in Table 7.3 and solve Eq. 7-10.

Table 7.4 compares the estimated parameters obtained by the two approaches. The linear regression results were obtained from Eq. 7-10. The results labeled nonlinear regression were obtained by fitting a continuous-time model (overdamped second-order with time constants τ_1 and τ_2 and gain K) to the data using nonlinear regression. The continuous-time model was then converted to the corresponding discrete-time model using Eqs. 7-36 to 7-40.

Table 7.4 Comparison of Estimated Model Parameters for Example 7.5

	Linear Regression	Nonlinear Regression
a_1	0.975	0.984
a_2	−0.112	−0.122
b_1	0.058	0.058
b_2	0.102	0.101
K	1.168	1.159

Table 7.5 Comparison of Simulated Responses for Various Difference Equation Models[a]

n	y	$\hat{y}_L$	$\hat{y}_N$
1	0.058	0.058	0.058
2	0.217	0.217	0.216
3	0.360	0.365	0.366
4	0.488	0.487	0.487
5	0.600	0.595	0.596
6	0.692	0.690	0.690
7	0.772	0.768	0.767
8	0.833	0.835	0.835
9	0.888	0.886	0.885
10	0.925	0.933	0.932

[a] y, experimental data; $\hat{y}_L$, linear regression; $\hat{y}_N$, nonlinear regression

The parameters obtained from linear regression in Table 7.4 are slightly different from those for nonlinear regression. This result occurs because for linear regression, four parameters were estimated; with nonlinear regression, three parameters were estimated. The estimated gain for linear regression, $K = 1.168$, is about 1% higher than the value obtained from nonlinear regression.

Table 7.5 compares the simulated responses for the two empirical models. Linear regression gives slightly better predictions, because it fits more parameters. However, in this particular example, it is difficult to distinguish graphically among the three model step responses.

Example 7.5 has shown how we can fit a second-order difference equation model to data directly. The linear regression approach can also be used for higher-order models, provided that the parameters still appear linearly in the model. It is important to note that the estimated parameter values depend on the sampling period Δt for the data collection.

An advantage of the regression approach is that it is not necessary to make a step change in u in order to estimate model parameters. An arbitrary input variation such as a PRBS signal (see Fig. 7.11) over a limited period of time would suffice. In fact, a PRBS has certain advantages in forming the $\boldsymbol{X}^T\boldsymbol{X}$ matrix in Eq. 7-10 (Rivera and Jun, 2000). In particular, it is not necessary to force the system to a new steady state, a beneficial feature for industrial applications. Other advantages of PRBS are that the input is not correlated with other process trends, and that the test can be run for a longer time period than for a step change in the input.

7.5.1 Impulse and Step Response Models

Another type of discrete-time model, the *finite impulse response (FIR)* or *convolution* model, has become important in computer control. This model can be written as

$$y(k+1) = y(0) + \sum_{i=1}^{N} h_i u(k-i+1) \quad (7\text{-}43)$$

Integer N is selected so that $N\Delta t \geq t_s$, the settling time of the process (see Chapter 5). Note that an equivalent version of Eq. 7-43 can be written with $y(k)$ instead of $y(k+1)$ on the left-hand side (as in Section 7.4) by shifting the index backward one sampling period. A related discrete-time model can be derived from Eq. 7-43 and is called the *finite step response model*, or just the *step response model*. To illustrate this relationship, we consider a simple (finite) impulse response model where $N = 3$. Expanding the summation in Eq. 7-43 gives

$$y(k+1) = y(0) + h_1 u(k) + h_2 u(k-1) + h_3 u(k-2) \quad (7\text{-}44)$$

The *step response coefficients* S_i are related to the impulse response coefficients h_i as shown in Fig. 7.14. By definition, the step response coefficients are simply the values of the response y at the sampling instants. Note that the impulse response coefficients are equal to the differences between successive step response coefficients, $h_i = S_i - S_{i-1}$. If we substitute for h_i in terms of S_i in (7-44), then

$$y(k+1) = y(0) + (S_1 - S_0)u(k) + (S_2 - S_1)u(k-1) + (S_3 - S_2)u(k-2) \quad (7\text{-}45)$$

Recognizing that $S_0 = 0$ (see Fig. 7.14) and rearranging gives

$$y(k+1) = y(0) + S_1[u(k) - u(k-1)) + S_2(u(k-1) - u(k-2)] + S_3 u(k-2) \quad (7\text{-}46)$$

After defining $\Delta u(k) \triangleq u(k) - u(k-1)$, Eq. 7-46 becomes

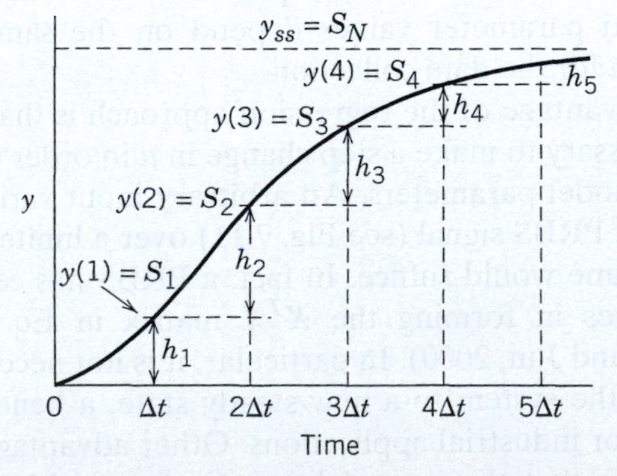

Figure 7.14 The relationship between the step response (S_i) and impulse response (h_i) coefficients for the situation where $y(0) = 0$.

$$y(k+1) = y(0) + S_1 \Delta u(k) + S_2 \Delta u(k-1) + S_3 u(k-2) \quad (7\text{-}47)$$

Similarly, the step response model that corresponds to the full impulse response model in Eq. 7-43 is given by

$$y(k+1) = y(0) + \sum_{i=1}^{N-1} S_i \Delta u(k-i+1) + S_N u(k-N+1) \quad (7\text{-}48)$$

This derivation is left to the reader.

A generalized framework for using step response models in model predictive control is presented in Chapter 20. Note that Fig. 7.14 illustrates the case where there is no time delay in the process model. When a time delay is present, the initial step (or impulse) coefficients are zero. For example, if there is a time delay of d sampling periods, then $S_0, S_1, \ldots, S_d$ are zero in Eq. 7-48. Similarly, $h_i = 0$ for $0 \leq i \leq d$ in Eq. 7-43.

A discrete-time impulse or step response model can be developed from a transfer function model or a linear differential (or difference) equation model. For example, consider a first-order transfer function with $\tau = 1$ min and $K = 1$, and a unit step input change. The first-order difference equation corresponding to Eq. 7-34 with $y(0) = 0$ and $\Delta t = 0.2$ is

$$y(k) = 0.8187y(k-1) + 0.1813u(k-1),$$

or, equivalently,

$$y(k+1) = 0.8187y(k) + 0.1813u(k).$$

For $u(0) = u(1) = \cdots = u(k-1) = 1.0$ and $\Delta t = 0.2$, the step response for $t = 0$ to $t = 10$, consists of the data points in Table 7.6. The values of the step response are

Table 7.6 Selected Step Response Coefficients for First-Order Model

Time Step, k	S_k
0	0.0000
1	0.1813
2	0.3297
3	0.4513
4	0.5507
5	0.6322
6	0.6989
7	0.7535
8	0.7982
9	0.8348
10	0.8647
15	0.9502
20	0.9817
25	0.9933
30	0.9975
35	0.9991
40	0.9997
45	0.9999
49	1.0000

shown for $0 \le k \le 10$, and selected values are included for $10 \le k \le 49$.

This modeling approach leads to a step response model that is equivalent to the first-order difference equation in Eq. 7-32 but that has many more parameters. Given the two alternatives, it is clear that the number of model parameters (*model parsimony*) is an issue in selecting the appropriate model.

When should a step response or impulse response model be selected? First, this type of model is useful when the actual model order or time delay is unknown, because this information is not required for step response models. The model parameters can be calculated directly using linear regression. Second, step or impulse response models are appropriate for processes that exhibit unusual dynamic behavior that cannot be described by standard low-order models. We consider such an example next.

EXAMPLE 7.6

The industrial data shown in Fig. 7.15 were obtained for a step test of a distillation column in a gas recovery unit. The input is the column pressure, the output is in analyzer composition, and the sampling period is $\Delta t = 1$ min (120 data points). Obtain the following models for the unit step change in the input:

(1) Step response model with 50 coefficients

(2) Discrete-time ARX model ($N = 50$):

$$y(k) = a_1 y(k-1) + a_2 y(k-2) + a_3 y(k-3) + a_4 y(k-4) + b_1 u(k-1) + b_2 u(k-2) + b_3 u(k-3)$$

(3) First-order-plus-time-delay model (cf. Eq. 7-16)

(4) Second-order-plus-time-delay model with inverse response:

$$G(s) = \frac{K(1 - \tau_a s)e^{-\theta s}}{(\tau_1 s + 1)(\tau_2 s + 1)}$$

SOLUTION

Figure 7.15 compares the four model responses with the experimental data. Excel was used to fit Models 3 and 4, while linear regression was used for Models 1 and 2. A step response model with 50 coefficients (and $\Delta t = 2$ min) provides a predicted response that is indistinguishable from the experimental data (solid line); it is shown as Model 1. A step response model with 120 coefficients and $\Delta t = 1$ min would provide an exact fit of the 120 data points.

Models 2, 3, and 4 are as follows:

(2) $$y(k) = 3.317y(k-1) - 4.033y(k-2) + 2.108y(k-3) + 0.392y(k-4) - 0.00922u(k-1) + 0.0322u(k-2) - 0.0370u(k-3) + 0.0141u(k-4)$$

(3) $$G(s) = \frac{0.082e^{-44.8s}}{7.95s + 1}$$

(4) $$G(s) = \frac{0.088(1 - 12.2s)e^{-25.7s}}{109.2s^2 + 23.1s + 1}$$

Model 2 gives an adequate fit except for the initial inverse response (an artifact near $t = 0$). Models 3 and 4 provide poor approximations of the response for $t \le 25$. However, Models 3 and 4 may be adequate for designing simple controllers.

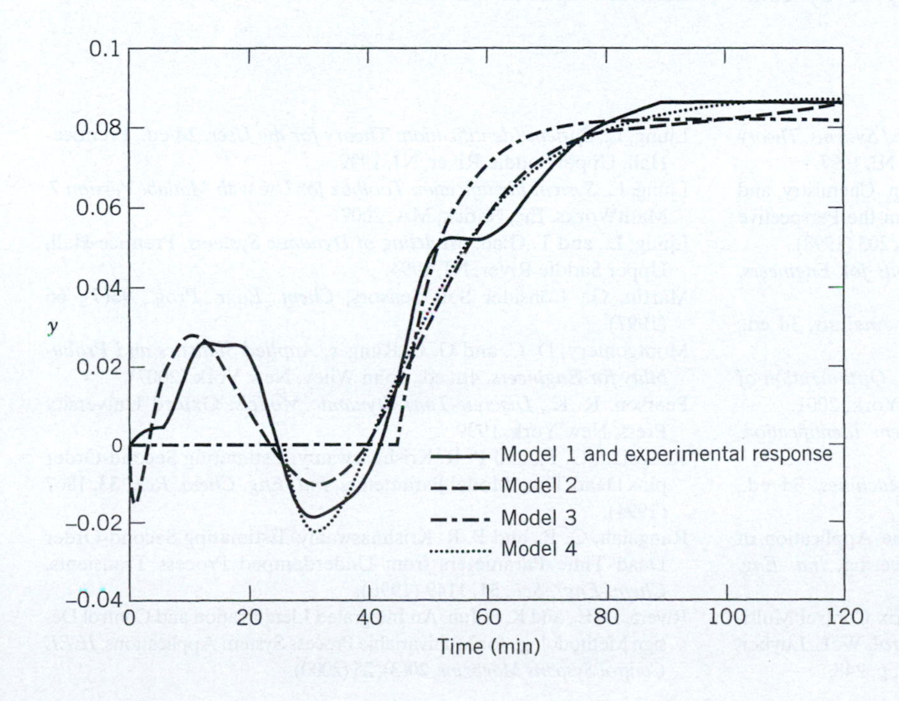

Figure 7.15 Comparison of model predictions for industrial column step responses, Example 7.7.

7.5.2 Process Identification of More Complicated Models

In this section, we briefly consider three classes of more complicated process models: MIMO models, stochastic, models, and nonlinear discrete-time models.

MIMO (multiple input, multiple output) process modeling is inherently more complicated than SISO modeling. For linear systems, the Principle of Superposition holds, which allows MIMO models to be developed through a series of single step tests for each input, while holding the other inputs constant. For a process with three inputs ($\boldsymbol{u}$) and three outputs ($\boldsymbol{y}$), we can introduce a step change in u_1, and record the responses for y_1, y_2, and y_3. The three transfer functions involving u_1, namely

$$\frac{Y_1}{U_1} = G_{11}, \quad \frac{Y_2}{U_1} = G_{21}, \quad \frac{Y_3}{U_1} = G_{31}$$

can be obtained using the techniques described in Section 7.2. In a similar fashion, step changes in U_2 and U_3 can be introduced in order to determine the other six G_{ij}. Alternatively, discrete-time models can be developed for each output, as discussed earlier in this section, using linear regression techniques. See Chapter 20 for a discussion of how such models are developed and used in model predictive controller calculations.

It is also possible to use PRBS forcing to obtain MIMO models. To generate a multi-input PRBS signal, it is desirable that the input changes be independent. One way to accomplish this is to implement "shifted" or delayed versions of a single PRBS signal in each input. This means that if a single-input PRBS test requires 20 hours, a three-input test is designed to take three times as long. Hokanson and Gerstle (1992) suggest that about 20 total moves in each independent variable should be made.

Linear discrete-time models can also be developed that include the effects of unmeasured stochastic disturbances. For example, separate process models and disturbance models, also called *noise models*, can be obtained from data (Ljung, 1999). In this case the error term ϵ in Eq. 7-1 is not white noise but *colored noise* that is autocorrelated. In other words, there are underlying disturbance dynamics that causes ϵ to depend on previous values.

A variety of nonlinear discrete-time models have also been used in process control (Pearson, 1999). They include the neural net models discussed in Section 7.3 as well as nonlinear models obtained by adding nonlinear terms to the linear models of the previous section.

SUMMARY

When theoretical models are not available or are very complicated, empirical process models provide a viable alternative. In these situations, a model that is sufficiently accurate for control system design can often be obtained from experimental input/output data. Step response data can be analyzed graphically or by computer (nonlinear regression) to obtain a first- or second-order transfer function model. Discrete-time models in the form of linear difference equations are frequently used in process control. These models can be readily obtained by least-squares fitting of experimental response data.

REFERENCES

Åström, K. J., and B. Wittenmark, *Computer-Controlled Systems: Theory and Design*, 3d ed., Prentice-Hall, Englewood Cliffs, NJ, 1997.

Brown, S. D., Information and Data Handling in Chemistry and Chemical Engineering: The State of the Field from the Perspective of Chemometrics, *Computers and Chem. Engr.*, **23,** 203 (1998).

Chapra, S. C. and R. P. Canale, *Numerical Methods for Engineers,* 6th ed, McGraw-Hill, New York, 2010.

Draper, N. R., and H. Smith, *Applied Regression Analysis*, 3d ed., Wiley, New York, 1998.

Edgar, T. F., D. M. Himmeblau, and L. S. Lasdon, *Optimization of Chemical Processes,* 2d ed., McGraw-Hill, New York, 2001.

Godfrey, K. (Ed.), *Perturbation Signals for System Identification*, Prentice-Hall, Englewood Cliffs, NJ (1993).

Haykin, S. S., *Neural Networks and Learning Machines,* 3d ed., Prentice-Hall, Upper Saddle River, NJ, 2009.

Himmelblau, D. M., Accounts of Experience in the Application of Artificial Neural Networks in Chemical Engineering, *Ind. Eng. Chem. Res.,* **47,** 5782 (2008).

Hokanson, D. A., and J. G. Gerstle, Dynamic Matrix Control Multivariable Controllers, in *Practical Distillation Control*, W. L. Luyben (Ed.), Van Nostrand Reinhold, New York (1992), p. 248.

Ljung, L., *System Identification: Theory for the User,* 2d ed., Prentice-Hall, Upper Saddle River, NJ, 1999.

Ljung, L., *System Identification Toolbox for Use with Matlab. Version 7.* MathWorks, Inc, Natick, MA, 2007.

Ljung, L., and T. Glad, *Modeling of Dynamic Systems*, Prentice-Hall, Upper Saddle River, NJ, 1994.

Martin, G., Consider Soft Sensors, *Chem. Engr. Prog.*, **93**(7), 66 (1997).

Montgomery, D. C. and G. C. Runger, *Applied Statistics and Probability for Engineers*, 4th ed., John Wiley, New York (2007).

Pearson, R. K., *Discrete-Time Dynamic Models*, Oxford University Press, New York, 1999.

Rangaiah, G. P., and P. R. Krishnaswamy, Estimating Second-Order plus Dead Time Model Parameters, *Ind. Eng. Chem. Res.*, **33,** 1867 (1994).

Rangaiah, G. P., and P. R. Krishnaswamy, Estimating Second-Order Dead Time Parameters from Underdamped Process Transients, *Chem. Engr. Sci.*, **51,** 1149 (1996).

Rivera, D. E., and K. S. Jun, An Integrated Identification and Control Design Methodology for Multivariable Process System Applications, *IEEE Control Systems Magazine*, **20**(3), 25 (2000).

Smith, C. L., *Digital Computer Process Control*, Intext, Scranton, PA, 1972.

Su, H. T., and T. J. McAvoy, Artificial Neural Networks for Nonlinear Process Identification and Control, Chapter 7 in *Nonlinear Process Control*, M. A. Henson and D. E. Seborg (Eds.), Prentice Hall, Upper Saddle River, NJ, 1997.

Sundaresan, K. R., and R. R. Krishnaswamy, Estimation of Time Delay, Time Constant Parameters in Time, Frequency, and Laplace Domains, *Can. J. Chem. Eng.*, **56**, 257 (1978).

EXERCISES

7.1 An operator introduces a step change in the flow rate q_i to a particular process at 3:05 A.M., changing the flow from 500 to 540 gal/min. The first change in the process temperature T (initially at 120 °F) occurs at 3:09 A.M. After that, the response in T is quite rapid, slowing down gradually until it appears to reach a steady-state value of 124.7 °F. The operator notes in the logbook that there is no change after 3:34 A.M. What approximate transfer function might be used to relate temperature to flow rate for this process in the absence of more accurate information? What should the operator do next time to obtain a better estimate?

7.2 A single-tank process has been operating for a long period of time with the inlet flow rate q_i equal to 30.4 ft^3/min. After the operator increases the flow rate suddenly at $t = 0$ by 10%, the liquid level in the tank changes as shown in Table E7.2.

Table E7.2

t (min)	h (ft)	t (min)	h (ft)
0	5.50	1.4	6.37
0.2	5.75	1.6	6.40
0.4	5.93	1.8	6.43
0.6	6.07	2.0	6.45
0.8	6.18	3.0	6.50
1.0	6.26	4.0	6.51
1.2	6.32	5.0	6.52

Assuming that the process dynamics can be described by a first-order model, calculate the steady-state gain and the time constant using three methods:

(a) From the time required for the output to reach 63.2% of the total change

(b) From the initial slope of the response curve

(c) From the slope of the fraction incomplete response curve

(d) Compare the data and the three models by simulating their step responses.

7.3 A process consists of two stirred tanks with input q and outputs T_1 and T_2 (see Fig. E7.3). To test the hypothesis that the dynamics in each tank are basically first-order, a step change in q is made from 82 to 85 L/min, with output responses given in Table E7.3.

(a) Find the transfer functions $T'_1(s)/Q'(s)$ and $T'_2(s)/T'_1(s)$. Assume that they are of the form $K_i/(\tau_i s + 1)$.

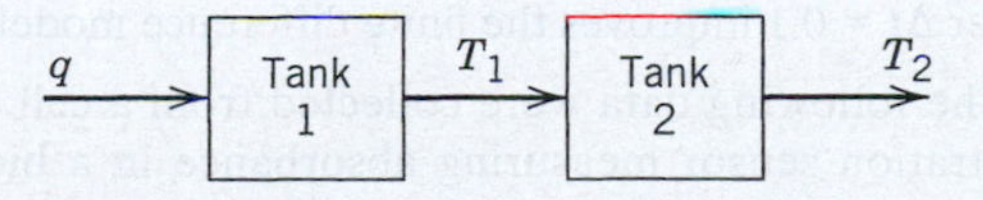

Figure E7.3

Table E7.3

t(min)	T_1 (°C)	T_2 (°C)	t(min)	T_1 (°C)	T_2 (°C)
0	10.00	20.00	11	17.80	25.77
1	12.27	20.65	12	17.85	25.84
2	13.89	21.79	13	17.89	25.88
3	15.06	22.83	14	17.92	25.92
4	15.89	23.68	15	17.95	25.94
5	16.49	24.32	16	17.96	25.96
6	16.91	24.79	17	17.97	25.97
7	17.22	25.13	18	17.98	25.98
8	17.44	25.38	19	17.99	25.98
9	17.60	25.55	20	17.99	25.99
10	17.71	25.68	50	18.00	26.00

(b) Calculate the model responses to the same step change in q and plot with the experimental data.

7.4 For a multistage bioseparation process described by the transfer function,

$$G(s) = \frac{2}{(5s + 1)(3s + 1)(s + 1)}$$

calculate the response to a step input change of magnitude, 1.5.

(a) Obtain an approximate first-order-plus-delay model using the fraction incomplete response method.

(b) Find an approximate second-order model using a method of Section 7.2.

(c) Calculate the responses of both approximate models using the same step input as for the third-order model. Plot all three responses on the same graph. What can you conclude concerning the approximations?

7.5 Fit an integrator plus time-delay model to the unit step response in Figure E7.5 for $t \leq 15$. The step response has been normalized by the steady-state gain. Compare the experimental response with the response predicted from the model.

7.6 For the unit step response shown in Fig. E7.5, estimate the following models using graphical methods:

(a) First-order plus time-delay.

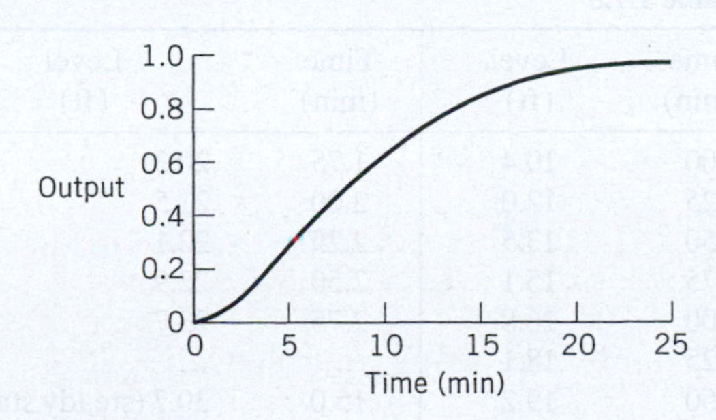

Figure E7.5

(b) Second-order using Smith's method and nonlinear regression.

Plot all three predicted model responses on the same graph.

7.7 A heat exchanger used to heat a glycol solution with a hot oil is known to exhibit FOPTD behavior, $G_1(s) = T'(s)/Q'(s)$, where T' is the outlet temperature deviation and Q' is the hot oil flow rate deviation. A thermocouple is placed 3 m downstream from the outlet of the heat exchanger. The average velocity of the glycol in the outlet pipe is 0.5 m/s. The thermocouple also is known to exhibit first-order behavior; however, its time constant is expected to be considerably smaller than the heat exchanger time constant.

(a) Data from a unit step test in Q' on the complete system are shown in Fig. E7.7. Using a method of your choice, calculate the time constants of this process from the step response.

(b) From your empirical model, find transfer functions for the heat exchanger, pipe, and thermocouple. Think of the model as the product of three transfer functions: process, pipe flow, and sensor. What assumptions do you have to make to obtain these individual transfer functions from the overall transfer function?

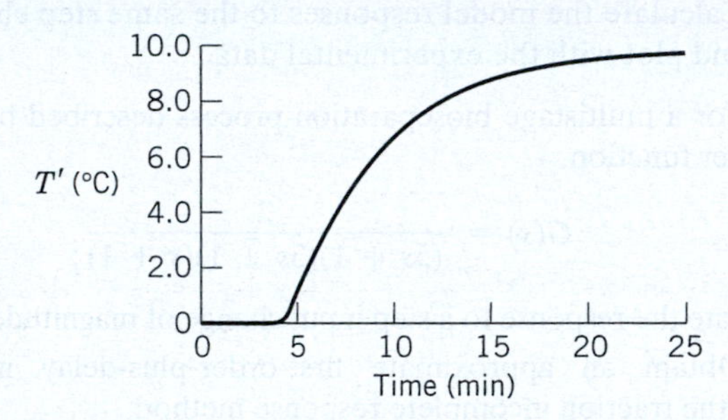

Figure E7.7

7.8 The level in a tank responds as a first-order system to changes in its inlet flow. The data shown below were gathered after the inlet flow was increased quickly from 1.5 to 4.8 gal/min.

(a) Determine the transfer function by estimating the time constant using one of the methods of Section 7.2. Be sure to use deviation variables and include units for the model parameters.

(b) Repeat part (a) using nonlinear regression (e.g., Excel) and the liquid level data.

(c) Graphically compare the two model responses with the data. Which model is superior? (Justify your answer)

Table E7.8

Time (min)	Level (ft)	Time (min)	Level (ft)
0.00	10.4	1.75	20.3
0.25	12.0	2.00	21.5
0.50	13.5	2.25	22.1
0.75	15.1	2.50	22.9
1.00	16.8	2.75	23.7
1.25	18.1	…	…
1.50	19.2	15.0	30.7 (steady state)

Table E7.9

t	y	t	y
0	0	7	1.8
1	0	8	2.4
2	0	9	2.7
3	0.3	10	2.8
4	0.6	11	2.9
5	0.9	12	3.0
6	1.3	13	3.0

7.9 The output response data y shown above were generated from a step change in input u from 2 to 4 at time $t = 0$. Develop a transfer function model of the form

$$\frac{Y(s)}{U(s)} = \frac{Ke^{-\theta s}}{(\tau_1 s + 1)(\tau_2 s + 1)}$$

7.10 Noisy data for the step response of a boiler temperature T to a decrease in air flow rate q from 1000 to 950 cfm are shown below. Develop a FOPTD model using a method from Chapter 7. Be sure to use deviation variables and report units for the model parameters.

Table E7.11

t (min)	q (cfm)	T (°C)
0	1000	849
1	1000	851
2	1000	850
3	950	851
4	950	849
5	950	860
6	950	867
7	950	873
8	950	878
9	950	882
10	950	886
11	950	888
12	950	890
13	950	890

7.11 Consider the first-order differential equation

$$5\frac{dy}{dt} + y(t) = 6u(t) \qquad y(0) = 3$$

where $u(t)$ is piecewise constant and has the following values:

$u(0) = 0 \quad u(3) = 3$

$u(1) = 1 \quad u(4) = 0$

$u(2) = 2 \quad u(t) = 0 \quad \text{for } t > 4$

Derive a difference equation for this ordinary equation using $\Delta t = 1$ and

(a) Exact discretization

(b) Finite difference approximation

Compare the integrated results for $0 \leq t \leq 10$. Examine whether $\Delta t = 0.1$ improves the finite difference model.

7.12 The following data were collected from a cell concentration sensor measuring absorbance in a biochemical stream. The input x is the flow rate deviation (in

Table E7.12

Time (*s*)	*x*	*y*
0	0	3.000
1	3	2.456
2	2	5.274
3	1	6.493
4	0	6.404
5	0	5.243
6	0	4.293
7	0	3.514
8	0	2.877
9	0	2.356
10	0	1.929

dimensionless units) and the sensor output y is given in volts. The flow rate (input) is piecewise constant between sampling instants. The process is not at steady state initially, so y can change even though $x = 0$.

Fit a first-order model, $y(k) = a_1 y(k-1) + b_1 x(k-1)$, to the data using the least-squares approach. Plot the model response and the actual data. Can you also find a first-order continuous transfer function $G(s)$ to fit the data?

7.13 Obtain a first-order discrete-time model from the response data in Table E7.12. Compare your results with the first-order graphical method for step response data, fitting the gain and time constant. Plot the two simulated step responses for comparison with the observed data.

7.14 Data for a person with type 1 diabetes are available as both MATLAB and Excel data files on the book web site.[1] Glucose measurements (y) were recorded every five minutes using a wearable sensor that measures subcutaneous glucose concentration. The insulin infusion rate (u) from a wearable subcutaneous insulin pump was also recorded every five minutes. The data files consist of experimental data for two step changes in the insulin infusion rate. The data are reported as deviations from the initial values that are considered to be the nominal steady-state values.

It is proposed that the relationship between the glucose concentration y and the insulin infusion rate u can be described by a discrete-time, dynamic model of the form:

$$y(k) = a_1 y(k-1) + a_2 y(k-2) + b_1 u(k-1) + b_2 u(k-2)$$

Do the following:

(a) Use the least squares approach to estimate the model parameters from the *basal1* dataset. This data will be referred to as the *calibration data*. Graphically compare the model response and this data.

(b) In order to assess the accuracy of the model from part (a), calculate the model response $\hat{y}$ to the u step changes in the *validation data* (*basal2*). Then graphically compare the model response $\hat{y}$ *with* the validation data y.

(c) Repeat Steps (a) and (b) using an alternative transfer function model:

$$\frac{Y(s)}{U(s)} = \frac{K}{\tau s + 1}$$

Estimate the model parameters using graphical techniques and the *basal1* dataset. Then compare the model and experimental response data for both datasets.

(d) Which model is superior? Justify your answer by considering the least squares index for the one-step-ahead prediction errors,

$$S = \sum_{k=1}^{N} [y(k) - \hat{y}(k)]^2$$

where N is the number of data points.

7.15 Consider the PCM furnace module of Appendix E. Assume that hydrocarbon temperature T_{HC} is the output variable and that air flow rate F_A is the input variable.

Do the following:

(a) Develop a FOPTD model from response data for a step change in F_A at $t = 10$ min from 17.9 to 20.0 m^3/min. Summarize your calculated model parameters in a table and briefly describe the method used to calculate them.

(b) Repeat (a) for a second-order plus time-delay (SOPTD) model.

(c) Plot the actual T_{HC} response and the two model responses for the F_A step change of part (a).

(d) Are the two models reasonably accurate? Which model is superior? Justify your answer by considering the least squares index for the prediction errors,

$$S = \sum_{k=1}^{N} [y(k) - \hat{y}(k)]^2$$

where N is the number of data points.

7.16 Consider the PCM distillation column module of Appendix E. Assume that distillate MeOH composition x_D is the output variable and that reflux ratio R is the input variable.

Do the following:

(a) Develop a first-order plus time-delay (FOPTD) transfer function model from response data for a step change in R at $t = 10$ min from 1.75 to 2.0. Summarize your calculated model parameters in a table and briefly describe the method used to calculate them.

(b) Repeat (a) for a second-order plus time-delay (SOPTD) model.

(c) Plot the actual x_D response and the two model responses for the R step change of part (a).

(d) Are the two models reasonably accurate? Which model is better? Justify your answer by considering the least squares index for the prediction errors,

$$S = \sum_{k=1}^{N} [y(k) - \hat{y}(k)]^2$$

where N is the number of data points.

[1] Book web site: www.wiley.com/college/seborg

Chapter 8

Feedback Controllers

CHAPTER CONTENTS

In previous chapters, we considered the dynamic behavior of representative processes and developed mathematical tools required to analyze process dynamics. We are now prepared to consider the important topic of feedback control.

The standard feedback *control algorithms* (also called *control laws*) are presented in this chapter, with emphasis on control algorithms that are widely used in the process industries. *Proportional-integral-derivative* (PID) control and *on-off control* are the predominant types of feedback control. Consequently, features and options for PID controllers are discussed in detail. Finally, we introduce digital PID control algorithms to emphasize the strong parallels between digital and analog (continuous) versions of feedback control. The remaining elements in the feedback control loop—sensors, transmitters, and control valves—will be considered in Chapter 9.

8.1 INTRODUCTION

We introduce feedback control systems by again considering the stirred-tank blending process of Chapters 2 and 4.

8.1.1 Illustrative Example: The Continuous Blending Process

A schematic diagram of a stirred-tank blending process is shown in Fig. 8.1. The control objective is to keep the tank exit composition x at the desired value set point by adjusting w_2, the flow rate of species A, via the control valve. The composition analyzer-transmitter (AT)

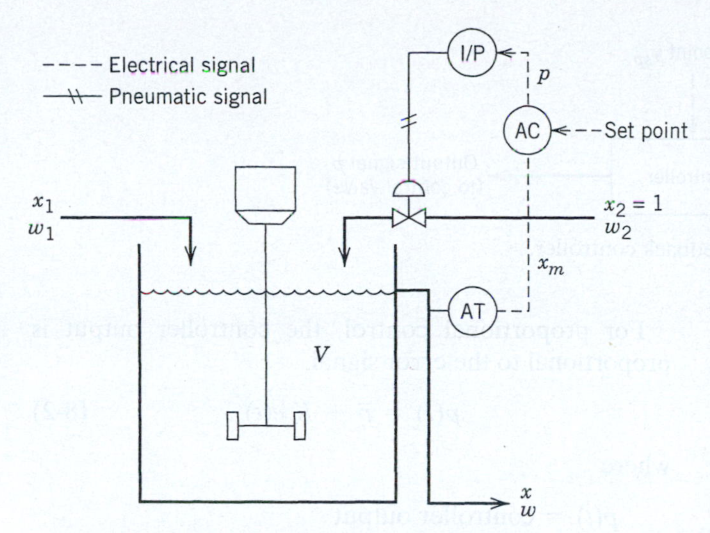

Figure 8.1 Schematic diagram for a stirred-tank blending system.

measures the exit composition and transmits it as an electronic signal to the feedback controller (AC). The controller compares the measured value x_m to the desired value (set point) and calculates an appropriate output signal p, an electronic signal that is sent to a *current-to-pressure transducer* (I/P) where it is converted to an equivalent pneumatic (air) signal that is compatible with the control valve. The symbols of Fig. 8.1 are examples of the standard instrumentation symbols published by the Instrumentation, Systems and Automation (ISA) Society. In particular, an electronic signal is denoted by a dashed line and a pneumatic signal by a solid line with crosshatches. A compilation of common instrumentation symbols appears in Appendix D.

This example illustrates that the basic components in a feedback control loop are:

- Process being controlled (blending system)
- Sensor-transmitter combination (AT)
- Feedback controller (AC)
- Current-to-pressure transducer (I/P)
- Final control element (control valve)
- Transmission lines between the various instruments (electrical cables and pneumatic tubing)

A current-to-pressure (or voltage-to-pressure) transducer is required if the control loop contains both electronic instruments and a pneumatic control valve. The term *final control element* refers to the device that is used to adjust the manipulated variable. It is usually a control valve but could be some other type of device, such as a variable speed pump or an electrical heater. The operation of this blending control system has been described in Section 1.2.

The blending system in Fig. 8.1 involves *analog* instrumentation. For an analog device, the input and output signals are continuous (analog) rather than discontinuous (digital or discrete time). Analog devices can be either electronic or pneumatic. For electronic devices such as sensors and controllers, the standard ranges for input and output signals are 4–20 mA and 1–5V (DC). Pneumatic instruments continue to be used, particularly in older plants or hazardous areas where electronic instruments are not intrinsically safe. For a pneumatic instrument, the input and output signals are air pressures in the range of 3 to 15 psig. Metal or plastic tubing (usually 1/4 or 3/8 OD) is used to interconnect the various pneumatic instruments. As indicated in Fig. 8.1, both electronic and pneumatic devices can be used in the same feedback control loop.

Most new control systems utilize digital technology with the control algorithms implemented via digital computers and with digital signal pathways (networks) used (see Appendix A) for data transmission. Consequently, we consider digital control algorithms. Instrumentation for process control, including computer hardware and software, are considered in greater detail in Chapter 9 and Appendix A.

Now we consider the heart of a feedback control system, the controller itself.

8.1.2 Historical Perspective

We tend to regard automatic control devices as a modern development. However, ingenious feedback control systems for water-level control were used by the Greeks as early as 250 B.C. (Mayr, 1970), with their mode of operation being very similar to that of the level regulator in the modern flush toilet. The fly-ball governor, which was first applied by James Watt to the steam engine in 1788, played a key role in the development of steam power.

During the 1930s, *three-mode controllers* with proportional, integral, and derivative (PID) feedback control action became commercially available (Ziegler, 1975). The first theoretical papers on process control were published during this same period. Pneumatic PID controllers gained widespread industrial acceptance during the 1940s, and their electronic counterparts entered the market in the 1950s. The first computer control applications in the process industries were reported in the late 1950s and early 1960s. Since the 1980s, digital hardware has been used on a routine basis and has had a tremendous impact on process control.

As a simple example of feedback control, consider the flow control loop in Fig. 8.2 where the flow rate of a

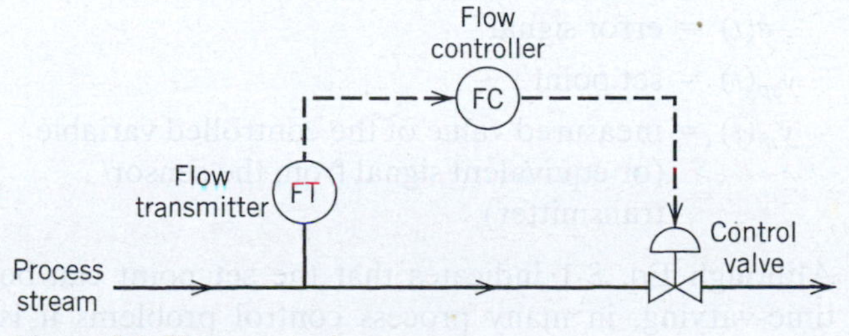

Figure 8.2 Flow control system.

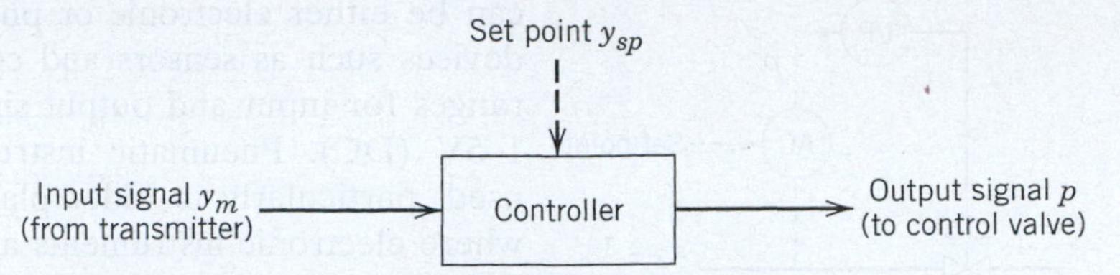

Figure 8.3 Simple diagram of a feedback controller.

process stream is measured and transmitted electronically to a flow controller. The controller compares the measured value to the set point and takes the appropriate corrective action by calculating the controller output and transmitting it as an electronic signal to the control valve.

The block diagram for the feedback controller of Fig. 8.2 is shown in Fig. 8.3. The set point is shown as a dashed line. For digital control systems, the set point would be entered by an operator using a computer terminal. For an analog controller, the set point would be specified via a dial setting on the equipment. In addition to this local set point, some controllers have a remote set-point option that permits them to receive an external set-point from another controller or a computer. The input and output signals for analog controllers are continuous signals that are either electrical or pneumatic. For digital control systems, the input signals are first converted from analog to digital form prior to the control calculations. Then, the calculated value of the controller output is converted from a digital signal to an analog signal for transmission to the control valve (or some other type of final control element). These types of signal conversions are described in Appendix A.

8.2 BASIC CONTROL MODES

Next we consider the three basic feedback control modes starting with the simplest mode, *proportional control.*

8.2.1 Proportional Control

In feedback control, the objective is to reduce the error signal to zero where

$$e(t) = y_{sp}(t) - y_m(t) \tag{8-1}$$

and

$e(t)$ = error signal

$y_{sp}(t)$ = set point

$y_m(t)$ = measured value of the controlled variable (or equivalent signal from the sensor/transmitter)

Although Eq. 8-1 indicates that the set point can be time-varying, in many process control problems it is kept constant for long periods of time.

For proportional control, the controller output is proportional to the error signal,

$$p(t) = \bar{p} + K_c e(t) \tag{8-2}$$

where

$p(t)$ = controller output

$\bar{p}$ = bias (steady-state) value

K_c = controller gain (usually dimensionless)

The key concepts behind proportional control are that (1) the controller gain can be adjusted to make the controller output changes as sensitive as desired to deviations between set point and controlled variable, and that (2) the sign of K_c can be chosen to make the controller output increase (or decrease) as the error signal increases. For example, for the blending process in Fig. 8.1, we want w_2 to decrease as x increases; hence, K_c should be a positive number.

For proportional controllers, bias $\bar{p}$ can be adjusted, a procedure referred to as *manual reset*. Because the controller output equals $\bar{p}$ when the error is zero, $\bar{p}$ is adjusted so that the controller output, and consequently the manipulated variable, are at their nominal steady-state values when the error is zero. For example, if the final control element is a control valve, $\bar{p}$ is adjusted so that the flow rate through the control valve is equal to the nominal, steady-state value when $e = 0$. The controller gain K_c is adjustable and is usually *tuned* (i.e., adjusted) after the controller has been installed.

For general-purpose controllers, K_c is dimensionless. This situation occurs when p and e in Eq. 8-2 have the same units. For example, the units could be associated with electronic or pneumatic instruments (mA, volts, psi, etc.). For digital implementation, p and e are often expressed as numbers between 0 and 100%. The latter representation is especially convenient for graphical displays using computer control software.

On the other hand, in analyzing control systems it can be more convenient to express the error signal in engineering units such as °C or mol/L. For these situations, K_c will not be dimensionless. As an example, consider the stirred-tank blending system. Suppose that e [=] mass fraction and p [=] mA; then Eq. 8.2 implies that K_c [=] mA because mass fraction is a dimensionless quantity. If a controller gain is not dimensionless, it

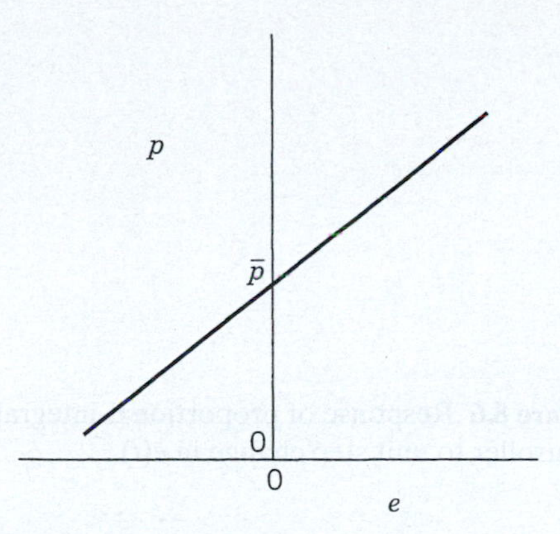

Figure 8.4 Proportional control: ideal behavior (slope of line = K_c).

includes the steady-state gain for another component of the control loop such as a transmitter or control valve. This situation is discussed in Chapter 11.

Some controllers have a proportional band setting instead of a controller gain. The *proportional band PB* (in %) is defined as

$$PB \triangleq \frac{100\%}{K_c} \tag{8-3}$$

This definition applies only if K_c *is* dimensionless. Note that a small (narrow) proportional band corresponds to a large controller gain, whereas a large (wide) *PB* value implies a small value of K_c.

The ideal proportional controller in Eq. 8-2 and Fig. 8.4 does not include physical limits on the controller output, *p*. A more realistic representation is shown in Fig. 8.5, where the controller *saturates* when its output reaches a physical limit, either p_{max} or p_{min}. In order to derive the transfer function for an ideal proportional controller (without saturation limits), define a deviation variable $p'(t)$ as

$$p'(t) \triangleq p(t) - \bar{p} \tag{8-4}$$

Then Eq. 8-2 can be written as

$$p'(t) = K_c e(t) \tag{8-5}$$

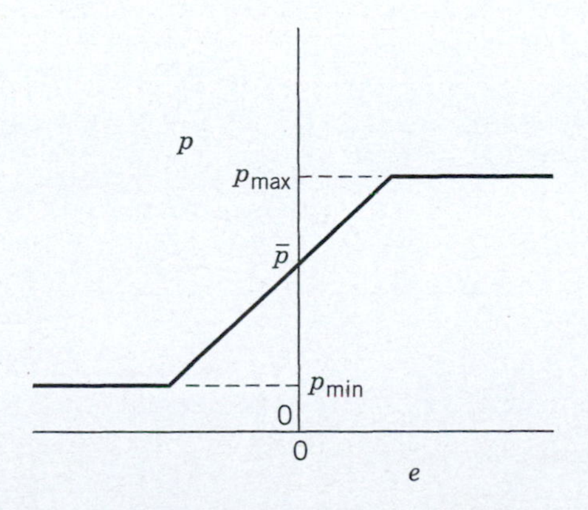

Figure 8.5 Proportional control: actual behavior.

It is unnecessary to define a deviation variable for the error signal, because *e* is already in deviation form, and its nominal steady-state value is $\bar{e} = 0$. Taking Laplace transforms and rearranging (8-5) gives the transfer function for proportional-only control:

$$\frac{P'(s)}{E(s)} = K_c \tag{8-6}$$

An inherent disadvantage of proportional-only control is that a steady-state error (or *offset*) occurs after a set-point change or a sustained disturbance. In Chapter 11 we demonstrate that offset will occur for proportional-only control regardless of the value of K_c that is employed. Fortunately, the addition of the integral control mode facilitates offset elimination, as discussed in the next section.

For control applications where offsets can be tolerated, proportional-only control is attractive because of its simplicity. For example, in some level control problems, maintaining the liquid level close to the set point is not as important as merely ensuring that the storage tank does not overflow or run dry.

8.2.2 Integral Control

For integral control action, the controller output depends on the integral of the error signal over time,

$$p(t) = \bar{p} + \frac{1}{\tau_I}\int_0^t e(t^*)\, dt^* \tag{8-7}$$

where τ_I, an adjustable parameter referred to as the *integral time* or *reset time*, has units of time. In the past, integral control action has been referred to as *reset* or *floating control*, but these terms are seldom used anymore.

Integral control action is widely used because it provides an important practical advantage, the elimination of offset. To understand why offset is eliminated, consider Eq. 8-7. In order for the controlled process to be at steady state, the controller output *p* must be constant so that the manipulated variable is also constant. Equation 8-7 implies that *p* changes with time unless $e(t^*) = 0$. Thus, when integral action is used, *p* automatically changes until it attains the value required to make the steady-state error zero. This desirable situation always occurs unless the controller output or final control element saturates and thus is unable to bring the controlled variable back to the set point. Controller saturation occurs whenever the disturbance or set-point change is so large that it is beyond the range of the manipulated variable.

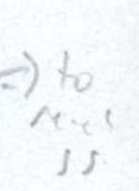

Although elimination of offset is usually an important control objective, the integral controller in Eq. 8-7 is seldom used by itself, because little control action takes place until the error signal has persisted for some

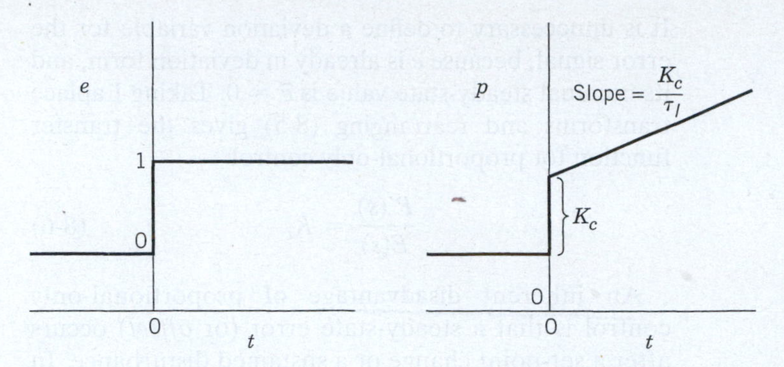

Figure 8.6 Response of proportional-integral controller to unit step change in $e(t)$.

time. In contrast, proportional control action takes immediate corrective action as soon as an error is detected. Consequently, integral control action is normally used in conjunction with proportional control as the *proportional-integral (PI)* controller:

$$p(t) = \bar{p} + K_c\left(e(t) + \frac{1}{\tau_I}\int_0^t e(t^*)\,dt^*\right) \quad (8\text{-}8)$$

The corresponding transfer function for the PI controller in Eq. 8-8 is given by

$$\frac{P'(s)}{E(s)} = K_c\left(1 + \frac{1}{\tau_I s}\right) = K_c\left(\frac{\tau_I s + 1}{\tau_I s}\right) \quad (8\text{-}9)$$

The response of the PI controller to a unit step change in $e(t)$ is shown in Fig. 8.6. At time zero, the controller output changes instantaneously due to the proportional action. Integral action causes the ramp increase in $p(t)$ for $t > 0$. When $t = \tau_I$, the integral term has contributed the same amount to the controller output as the proportional term. Thus, the integral action has *repeated* the proportional action once. Some commercial controllers are calibrated in terms of $1/\tau_I$ (repeats per minute) rather than τ_I (minutes, or minutes per repeat). For example, if $\tau_I = 0.2$ min, this corresponds to $1/\tau_I$ having a value of 5 repeats/minute.

One disadvantage of using integral action is that it tends to produce oscillatory responses of the controlled variable and, as we will see in Chapter 11, it reduces the stability of the feedback control system. A limited amount of oscillation can usually be tolerated, because it often is associated with a faster response. The undesirable effects of too much integral action can be avoided by proper tuning of the controller or by including derivative control action (Section 8.2.3), which tends to counteract the destabilizing effects.

Reset Windup

An inherent disadvantage of integral control action is a phenomenon known as *reset windup*. Recall that the integral mode causes the controller output to change as long as $e(t^*) \neq 0$ in Eq. 8-8. When a sustained error occurs, the integral term becomes quite large and the controller output eventually saturates. Further buildup of the integral term while the controller is saturated is referred to as reset windup or *integral windup*. Figure 8.7 shows a typical response to a step change in set point when a PI controller is used. Note that the indicated areas under the curve provide either positive or negative contributions to the integral term depending on whether the measurement of the controlled variable y_m is below or above the set point y_{sp}. The large overshoot in Fig. 8.7 occurs because the integral term continues to increase until the error signal changes sign at $t = t_1$. Only then does the integral term begin to decrease. After the integral term becomes sufficiently small, the controller output moves away from

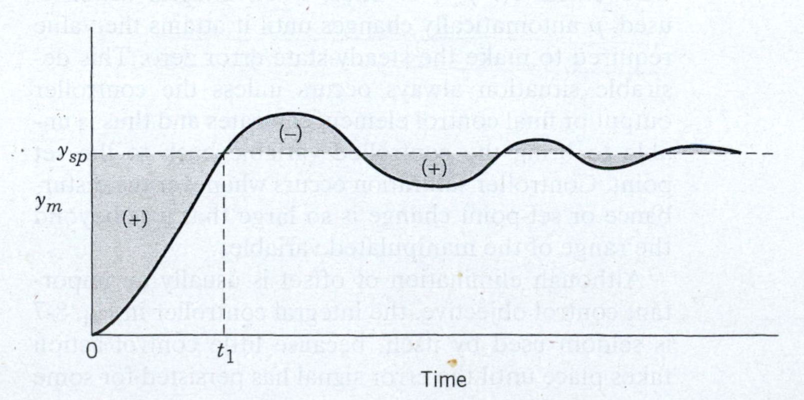

Figure 8.7 Reset windup during a set-point change.

the saturation limit and has the value determined by Eq. 8-8.

Reset windup occurs when a PI or PID controller encounters a sustained error, for example, during the start-up of a batch process or after a large set-point change. It can also occur as a consequence of a large sustained disturbance that is beyond the range of the manipulated variable. In this situation, a physical limitation (control valve fully open or completely shut) prevents the controller from reducing the error signal to zero. Clearly, it is undesirable to have the integral term continue to build up after the controller output saturates, because the controller is already doing all it can to reduce the error. Fortunately, commercial controllers provide *anti-reset windup*. In one approach, reset windup is reduced by temporarily halting the integral control action whenever the controller output saturates. The integral action resumes when the output is no longer saturated. The anti-reset windup feature is sometimes referred to as a *batch unit,* because it is required when batch processes are started up automatically (see Chapter 22).

8.2.3 Derivative Control

The function of derivative control action is to anticipate the future behavior of the error signal by considering its rate of change. In the past, derivative action was also referred to as *rate action*, *pre-act*, or *anticipatory control*. For example, suppose that a reactor temperature increases by 10 °C in a short period of time, say, 3 min. This clearly is a more rapid increase in temperature than a 10 °C rise in 30 min, and it could indicate a potential *runaway* situation for an exothermic reaction. If the reactor were under manual control, an experienced plant operator would anticipate the consequences and quickly take appropriate corrective action to reduce the temperature. Such a response would not be obtainable from the proportional and integral control modes discussed so far. Note that a proportional controller reacts to a deviation in temperature only, making no distinction as to the time period over which the deviation develops. Integral control action is also ineffective for a sudden deviation in temperature, because the corrective action depends on the duration of the deviation.

The anticipatory strategy used by the experienced operator can be incorporated in automatic controllers by making the controller output proportional to the rate of change of the error signal or the controlled variable. Thus, for *ideal* derivative action,

$$p(t) = \bar{p} + \tau_D \frac{de(t)}{dt} \tag{8-10}$$

where τ_D, the *derivative time*, has units of time. Note that the controller output is equal to the nominal value $\bar{p}$ as long as the error is constant (that is, as long as $de/dt = 0$). Consequently, derivative action is never used alone; it is always used in conjunction with proportional or proportional-integral control. For example, an ideal PD controller has the transfer function:

$$\frac{P'(s)}{E(s)} = K_c(1 + \tau_D s) \tag{8-11}$$

By providing anticipatory control action, the derivative mode tends to stabilize the controlled process. Thus, it is often used to counteract the destabilizing tendency of the integral mode (Chapters 11 and 14).

Derivative control action also tends to improve the dynamic response of the controlled variable by the settling time, the time it takes reducing to reach steady state. But if the process measurement is *noisy*, that is, if it contains high-frequency, random fluctuations, then the derivative of the measured variable will change wildly, and derivative action will amplify the noise unless the measurement is *filtered*, as discussed in Chapter 17. Consequently, derivative action is seldom used for flow control, because flow control loops respond quickly and flow measurements tend to be noisy.

Unfortunately, the ideal proportional-derivative control algorithm in Eq. 8-11 is *physically unrealizable* because it cannot be implemented exactly using either analog or digital controllers. For analog controllers, the transfer function in (8-11) can be approximated by

$$\frac{P'(s)}{E(s)} = K_c\left(1 + \frac{\tau_D s}{\alpha\tau_D s + 1}\right) \tag{8-12}$$

where the constant α typically has a value between 0.05 and 0.2, with 0.1 being a common choice. In Eq. 8-12 the denominator term serves as a *derivative mode filter* (or a *derivative filter*) that reduces the sensitivity of the control calculations to noisy measurements. Derivative filters are used in virtually all commercial PD and PID controllers.

8.2.4 Proportional-Integral-Derivative Control

Now we consider the combination of the proportional, integral, and derivative control modes as a PID controller. PI and PID control have been the dominant control techniques for process control for many decades. For example, a survey has indicated that large-scale continuous processes typically have between 500 and 5,000 feedback controllers for individual process variables such as flow rate and liquid level (Desborough and Miller, 2001). Of these controllers, 97% utilize some form of PID control.

Many variations of PID control are used in practice; next, we consider the three most common forms.

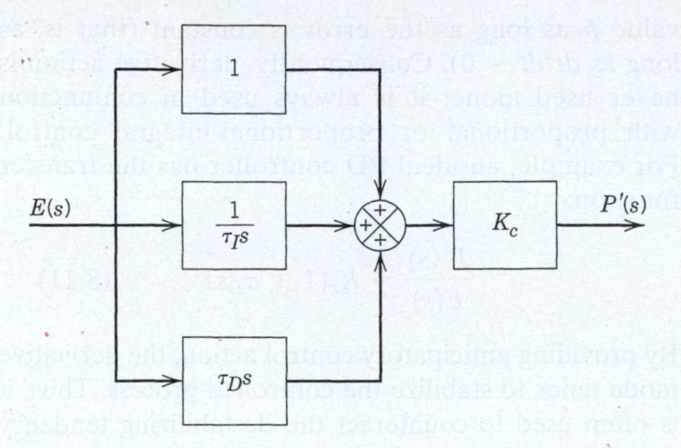

Figure 8.8 Block diagram of the parallel form of PID control (without a derivative filter).

Parallel Form of PID Control

The *parallel form* of the PID control algorithm (without a derivative filter) is given by

$$p(t) = \bar{p} + K_c\left[e(t) + \frac{1}{\tau_I}\int_0^t e(t^*)\,dt^* + \tau_D\frac{de(t)}{dt}\right] \quad (8\text{-}13)$$

The corresponding transfer function is

$$\frac{P'(s)}{E(s)} = K_c\left[1 + \frac{1}{\tau_I s} + \tau_D s\right] \quad (8\text{-}14)$$

Figure 8.8 illustrates that this controller can be viewed as three separate elements operating in parallel on $E(s)$.

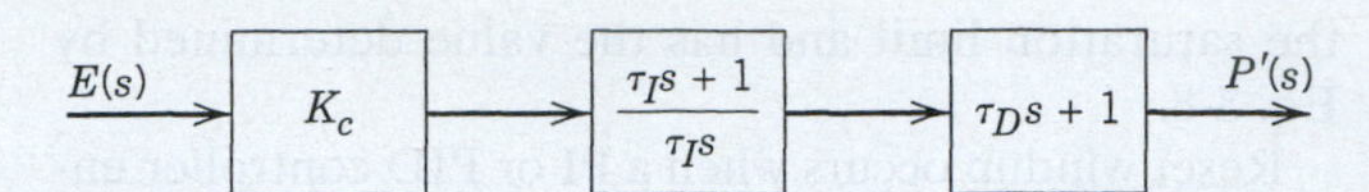

Figure 8.9 Block diagram of the series form of PID control (without a derivative filter).

The parallel-form PID controller with and without a derivative filter are shown in Table 8.1.

Series Form of PID Control

Historically, it was convenient to construct early analog controllers (both electronic and pneumatic) so that a PI element and a PD element operated in series. The *series form* of PID control without a derivative filter is shown in Fig. 8.9. In principle, it makes no difference whether the PD element or the PI element comes first. Commercial versions of the series-form controller have a derivative filter that is applied to either the derivative term, as in Eq. 8-12, or to the PD term, as in Eq. 8-15:

$$\frac{P'(s)}{E(s)} = K_c\left(\frac{\tau_I s + 1}{\tau_I s}\right)\left(\frac{\tau_D s + 1}{\alpha\tau_D s + 1}\right) \quad (8\text{-}15)$$

The consequences of adding a derivative filter are analyzed in Exercise 14.16.

Table 8.1 Common PID Controllers

Controller Type	Other Names Used	Controller Equation	Transfer Function
Parallel	Ideal, additive, ISA form	$p(t) = \bar{p} + K_c\left(e(t) + \frac{1}{\tau_I}\int_0^t e(t^*)\,dt^* + \tau_D\frac{de(t)}{dt}\right)$	$\frac{P'(s)}{E(s)} = K_c\left(1 + \frac{1}{\tau_I s} + \tau_D s\right)$
Parallel with derivative filter	Ideal, realizable, ISA standard	*See Exercise 8.10(a)*	$\frac{P'(s)}{E(s)} = K_c\left(1 + \frac{1}{\tau_I s} + \frac{\tau_D s}{\alpha\tau_D s + 1}\right)$
Series	Multiplicative, interacting	*See Exercise 8.11*	$\frac{P'(s)}{E(s)} = K_c\left(\frac{\tau_I s + 1}{\tau_I s}\right)(\tau_D s + 1)$
Series with derivative filter	Physically realizable	*See Exercise 8.10(b)*	$\frac{P'(s)}{E(s)} = K_c\left(\frac{\tau_I s + 1}{\tau_I s}\right)\left(\frac{\tau_D s + 1}{\alpha\tau_D s + 1}\right)$
Expanded	Noninteracting	$p(t) = \bar{p} + K_c e(t) + K_I\int_0^t e(t^*)\,dt^* + K_D\frac{de(t)}{dt}$	$\frac{P'(s)}{E(s)} = K_c + \frac{K_I}{s} + K_D s$
Parallel, with proportional and derivative weighting	Ideal β, γ controller	$p(t) = \bar{p} + K_c\left(e_P(t) + \frac{1}{\tau_I}\int_0^t e(t^*)\,dt^* + \tau_D\frac{de_D(t)}{dt}\right)$ where $e_P(t) = \beta y_{sp}(t) - y_m(t)$, $e(t) = y_{sp}(t) - y_m(t)$, $e_D(t) = \gamma y_{sp}(t) - y_m(t)$	$P'(s) = K_c\left(E_P(s) + \frac{1}{\tau_I s}E(s) + \tau_D s E_D(s)\right)$ where $E_P(s) = \beta Y_{sp}(s) - Y_m(s)$, $E(s) = Y_{sp}(s) - Y_m(s)$, $E_D(s) = \gamma Y_{sp}(s) - Y_m(s)$

Expanded Form of PID Control

The *expanded form* of PID control is:

$$p(t) = \bar{p} + K_c e(t) + K_I \int_0^t e(t^*)\, dt^* + K_D \frac{de(t)}{dt} \quad (8\text{-}16)$$

Note that the controller parameters for the expanded form are three "gains," K_c, K_I, and K_D, rather than the standard parameters, K_c, τ_I, and τ_D. The expanded form of PID control is used in MATLAB. This form might appear to be well suited for controller tuning, because each gain independently influences only one control mode. But the well-established controller tuning relations presented in Chapters 12 and 14 were developed for the series and parallel forms. Thus, there is little advantage in using the expanded form in Eq. 8-16.

8.3 FEATURES OF PID CONTROLLERS

Next, we consider common extensions of the basic PID controllers that greatly enhance their performance.

8.3.1 Elimination of Derivative and Proportional Kick

One disadvantage of the previous PID controllers is that a sudden change in set point (and hence the error, e) will cause the derivative term momentarily to become very large and thus provide a *derivative kick* to the final control element. This sudden "spike" is undesirable and can be avoided by basing the derivative action on the measurement, y_m, rather than on the error signal, e. To illustrate the elimination of derivative kick, consider the parallel form of PID control in Eq. 8-13. Replacing de/dt by $-dy_m/dt$ gives

$$p(t) = \bar{p} + K_c\left[e(t) + \frac{1}{\tau_I}\int_0^t e(t^*)\, dt^* - \tau_D \frac{dy_m(t)}{dt}\right] \quad (8\text{-}17)$$

This method of eliminating derivative kick is a standard feature in most commercial controllers. For a series-form PID controller, it can be implemented quite easily by placing the PD element in the feedback path, as shown in Fig. 8.10. Note that the elimination of derivative kick for set-point changes does not affect the controller performance when the y_{sp} is constant. Thus, Eqs. (8-13) and (8-17) provide identical responses to process disturbances when the set point is constant.

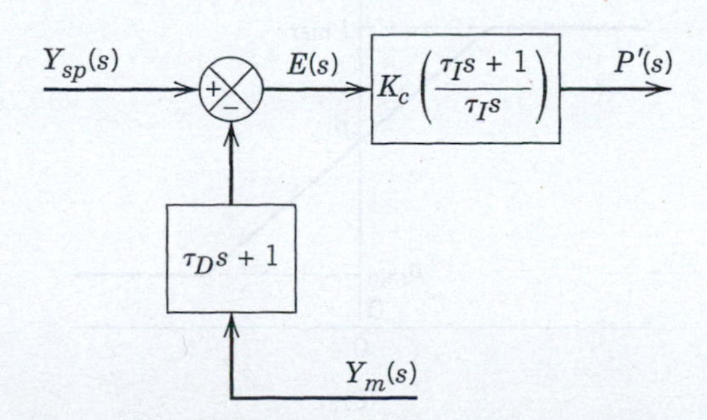

Figure 8.10 Block diagram of the series form of PID control that eliminates derivative kick.

A more flexible PID control algorithm can be obtained by weighting the set point in both the proportional and the derivative terms. This modification eliminates the *proportional kick* that also occurs after a step change in set point. For this modified PID algorithm, a different error term is defined for each control mode:

$$p(t) = \bar{p} + K_c\left(e_P(t) + \frac{1}{\tau_I}\int_0^t e(t^*)\, dt^* + \tau_D \frac{de_D(t)}{dt}\right) \quad (8\text{-}18)$$

with:

$$e_P(t) \triangleq \beta y_{sp}(t) - y_m(t) \quad (8\text{-}19)$$

$$e(t) \triangleq y_{sp}(t) - y_m(t) \quad (8\text{-}20)$$

$$e_D(t) \triangleq \gamma y_{sp}(t) - y_m(t) \quad (8\text{-}21)$$

where β and γ are nonnegative constants. This control algorithm is known as the *parallel PID controller with proportional and derivative mode weighting*, or the *beta-gamma controller*. The modified PID control algorithm in Eq. 8-18 allows for independent set-point weighting in the proportional and derivative terms. Thus, to eliminate derivative kick, γ is set to zero; to eliminate proportional kick, β is set to zero. The β weighting parameter can be used to tune this PID controller performance for set-point changes, as discussed in Chapter 12. Note that the definition of the integral mode error in (8-20) is the same as for the standard control law in (8-13); this error term is essential in order to eliminate offset after a set-point change or sustained disturbance.

Finally, it should be noted that, although digital controller settings can be specified exactly, analog controller settings represent only nominal values. Although it would be desirable to be able to specify K_c, τ_I, and τ_D accurately and independently for analog controllers, in practice there are interactions among the control modes owing to hardware limitations. Consequently, the actual controller settings may differ from the dial settings by as much as 30%.

Table 8.1 shows the most important forms of PID controllers, controller equations, and transfer functions. The derivation of several controller equation forms is left as an exercise for the reader. The table is organized by the descriptive names used in this book, but common synonyms are also included. However, all these terms should be used with caution as a result of the inconsistent terminology that occurs in the literature. For example, referring to the parallel form (the first line of Table 8.1) as an "ideal controller" is misleading, because its derivative

Table 8.2 Key Characteristics of Commercial PID Controllers

Controller Feature	Controller Parameter	Symbol	Units	Typical Range*
Proportional mode	*Controller gain*	K_c	Dimensionless [%/%, mA/mA]	0.1–100
	Proportional band	$PB = 100\%/K_c$	%	1–1000%
Integral mode	*Integral time* (or *reset time*)	τ_I	Time [min, s]	0.02–20 min 1–1000 s
	Reset rate	$1/\tau_I$	Repeats/time [min^{-1}, s^{-1}]	0.001–1 repeats/s 0.06–60 repeats/min
	Integral mode "gain"	K_I	Time^{-1} [min^{-1}, s^{-1}]	0.1–100
Derivative mode	*Derivative time*	τ_D	Time [min, s]	0.1–10 min. 5–500 s
	Derivative mode "gain"	K_D	Time [min, s]	0.1–100
	Derivative filter parameter	α	Dimensionless	0.05–0.2
Control interval (Digital controllers)		Δt	Time [s, min]	0.1 s–10 min

*Based on McMillan (2006).

mode amplifies noise, an undesirable characteristic. In addition, the terms *interacting* and *noninteracting* can be quite confusing, because a controller's modes can be noninteracting in the time domain (controller equation) but interacting in the Laplace domain (transfer function) and vice versa. Some of these idiosyncrasies are evident from the exercises and from the frequency response analysis of Chapter 14.

Table 8.2 summarizes important characteristics of representative commercial PID controllers. The operating interval (sampling period/sampling frequency) information applies to the digital controllers of Section 8.6.

8.3.2 Reverse or Direct Action

The controller gain can be either negative or positive.[1] For proportional control, when $K_c > 0$, the controller output $p(t)$ increases as its input signal $y_m(t)$ decreases, as is apparent after combining Eqs. 8-2 and 8-1:

$$p(t) - \bar{p} = K_c[y_{sp}(t) - y_m(t)] \quad (8\text{-}22)$$

Thus if $K_c > 0$, the controller is called a *reverse-acting* controller. When $K_c < 0$, the controller is said to be *direct acting*, because p increases as y_m increases. Note that these definitions are based on the measurement, $y_m(t)$, rather than the error, $e(t)$. Direct-acting and reverse-acting proportional controllers are compared in Fig. 8.11.

[1]For some computer control software, K_c must be positive. The user enters the designation of reverse or direct action as a separate binary parameter.

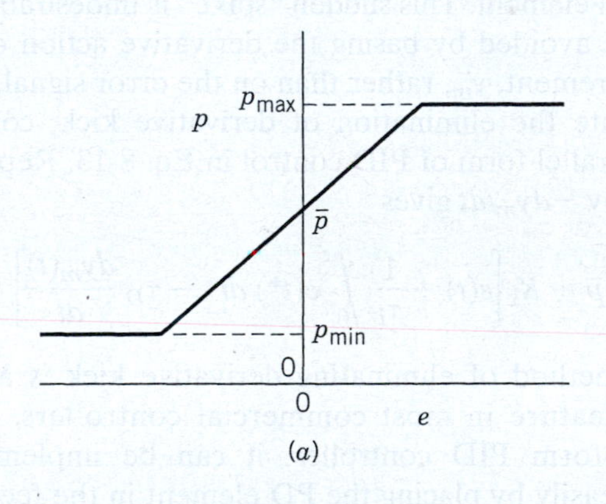

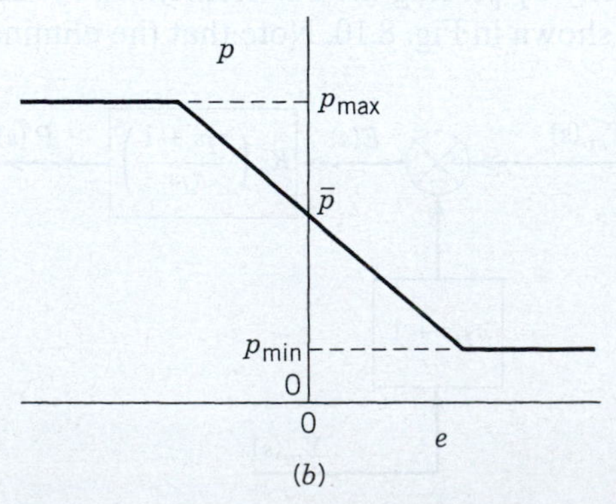

Figure 8.11 Reverse and direct-acting proportional controllers: (*a*) reverse acting ($K_c > 0$), (*b*) direct acting ($K_c < 0$).

To illustrate why both direct-acting and reverse-acting controllers are needed, again consider the flow control loop in Fig. 8.2. Suppose that the flow transmitter is designed to be direct-acting so that its output signal increases as the flow rate increases. Most transmitters are designed to be direct-acting. Also assume that the control valve is designed so that the flow rate through the valve increases as the signal to the valve, $p(t)$, increases. In this case the valve is designated as *air-to-open* (or *fail close*). The question is: should the flow controller have direct or reverse action? Clearly, when the measured flow rate is higher than the set point, we want to reduce the flow by closing the control valve. For an air-to-open valve, the controller output signal should be decreased. Thus, the controller should be *reverse-acting*.

But what if the control valve is air-to-close (or *fail open*) rather than air-to-open? Now when the flow rate is too high, the controller output should increase to further close the valve. Here, a *direct-acting* controller is required.

It is extremely important that the controller action be specified correctly, because an incorrect choice usually results in loss of control. For the flow control example, having the wrong controller action would force the control valve to stay fully open or fully closed (why?). Thus, the controller action must be carefully specified when a controller is installed or when a troublesome control loop is being analyzed. The following guideline is very useful and can be justified by the stability analysis techniques of Chapter 11.

General Guideline for Specifying the Controller Action (Direct or Reverse): *The overall product of the gains for all of the components in the feedback control loop must be positive.*

For example, the blending control system in Fig. 8.1 has five components in the feedback control loop: the process, the sensor, the controller, the I/P transducer, and the control valve.

8.3.3 Automatic/Manual Control Modes

Equations 8-2 to 8-16 describe how controllers perform during the *automatic mode* of operation. However, in certain situations, the plant operator may decide to override the automatic mode and adjust the controller output manually.

This *manual mode* of controller operation is very useful during a plant start-up, shutdown, or emergency situation. A manual/automatic switch, or the software equivalent, is used to transfer the controller from the automatic mode to the manual mode, and vice versa. During these transfers, it is important that the controller output not change abruptly and "bump" the process. Consequently, most controllers facilitate *bumpless transfers*.

A controller may be left in manual for long periods of time (or indefinitely) if the operator is not satisfied with its performance in the automatic mode. Consequently, if a significant percentage of the controllers in a plant is in manual, it is an indication that the control systems are not performing well or that the plant operators do not have much confidence in them. The topic of troubleshooting poorly performing control loops is considered in Chapter 12.

8.4 ON-OFF CONTROLLERS

On-off controllers are simple, inexpensive feedback controllers that are commonly used as thermostats in home heating systems and domestic refrigerators. They are also used in noncritical industrial applications such as some level control loops and heating systems. However, on-off controllers are less widely used than PID controllers, because they are not as versatile or as effective.

For ideal on-off control, the controller output has only two possible values:

$$p(t) = \begin{cases} p_{max} & \text{if } e \geq 0 \\ p_{min} & \text{if } e < 0 \end{cases} \qquad (8\text{-}23)$$

where $p_{\max}$ and $p_{\min}$ denote the on and off values, respectively (for example, for a typical digital computer implementation, $p_{\max} = 100\%$ and $p_{\min} = 0\%$; for a current-based electronic controller, $p_{\max} = 20$ mA and $p_{\min} = 4$ mA). On-off controllers can be modified to include a dead band for the error signal to reduce sensitivity to measurement noise (Shinskey, 1996). Equation 8-23 also indicates why on-off control is sometimes referred to as *two-position* or *bang-bang* control. Note that on-off control can be considered a special case of proportional control with a very high controller gain (see Fig. 8.5).

The disadvantages of on-off control are that it results in continual cycling of the controlled variable and produces excessive wear on the control valve (or other final control element). The latter disadvantage is significant if a control valve is used, but less of a factor for solenoid valves or solenoid switches that are normally employed with on-off controllers.

8.5 TYPICAL RESPONSES OF FEEDBACK CONTROL SYSTEMS

The responses shown in Fig. 8.12 illustrate the typical behavior of a controlled process after a step change in a disturbance variable occurs. The controlled variable y represents the deviation from the initial steady-state value. If feedback control is not used, the process slowly reaches a new steady state. Proportional control speeds up the process response and reduces the offset.

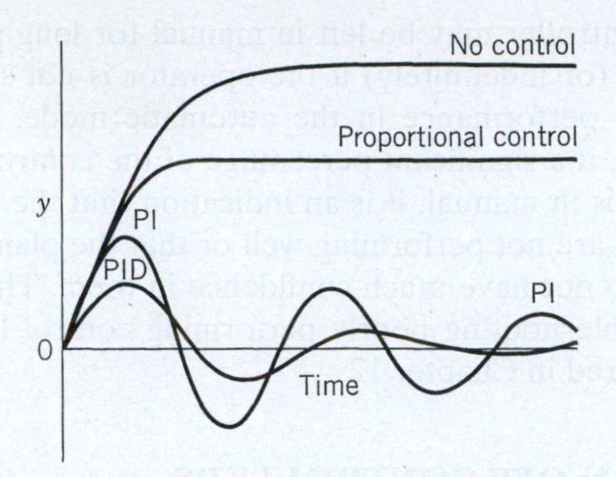

Figure 8.12 Typical process responses with feedback control.

The addition of integral control action eliminates offset but tends to make the response more oscillatory. Adding derivative action reduces both the degree of oscillation and the response time. The use of P, PI, and PID controllers does not always result in oscillatory process responses; the nature of the response depends on the choice of the controller settings (K_c, τ_I, and τ_D) and the process dynamics. However, the responses in Fig. 8.12 are representative of what occurs in practice.

The qualitative effects of changing individual controller settings are shown in Figs. 8.13 to 8.15. In general, increasing the controller gain tends to make the process response less sluggish; however, if too large a value of K_c is used, the response may exhibit an undesirable degree of oscillation or even become unstable. Thus, an intermediate value of K_c usually results in the best control. These guidelines are also applicable to PI and PID control, as well as to the proportional controller shown in Fig. 8.13.

Increasing the integral time, τ_I, usually makes PI and PID control more conservative (sluggish) as shown in Fig. 8.14. Theoretically, offset will be eliminated for all positive values of τ_I. But for very large values of τ_I, the controlled variable will return to the set point very slowly after a disturbance or set-point change occurs.

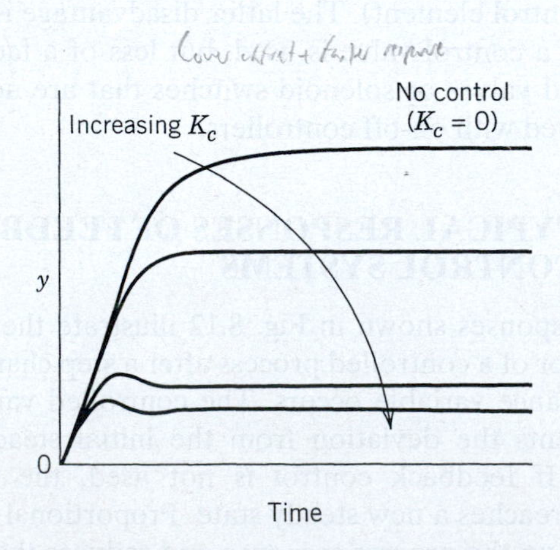

Figure 8.13 Proportional control: effect of controller gain.

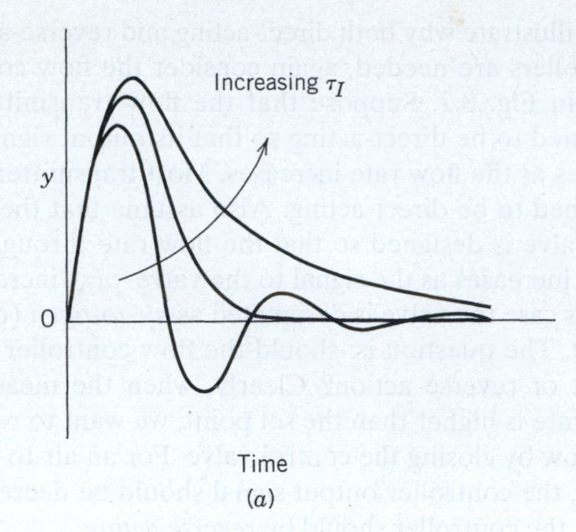

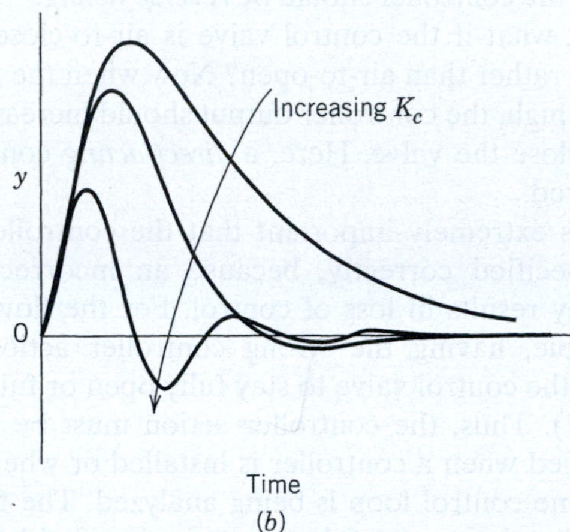

Figure 8.14 Proportional-integral control: (*a*) effect of integral time, (*b*) effect of controller gain.

It is more difficult to generalize about the effect of the derivative time τ_D. For small values, increasing τ_D tends to improve the response by reducing the maximum deviation, response time, and degree of oscillation, as shown in Fig. 8.15. However, if τ_D is too large, measurement noise is amplified and the response may become oscillatory. Thus, an intermediate value of τ_D is desirable. More detailed discussions of how PID controller settings should be specified are presented in Chapters 11, 12, and 14.

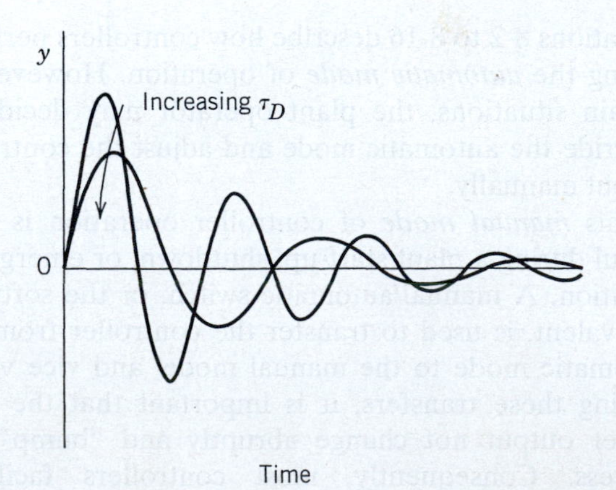

Figure 8.15 PID control: effect of derivative time.

8.6 DIGITAL VERSIONS OF PID CONTROLLERS

So far we have assumed that the input and output signals of the controller are continuous functions of time. However, there has also been widespread application of digital control systems due to their flexibility, computational power, and cost effectiveness. In this section we briefly introduce digital control techniques by considering digital versions of PID control. A more complete discussion of digital computer control is presented in Chapter 17 and Appendix A.

When a feedback control strategy is implemented digitally, the controller input and output are digital (or discrete-time) signals rather than continuous (or analog) signals. Thus, the continuous signal from the measurement device (sensor/transmitter) is sampled and converted to a digital signal by an *analog-to-digital converter* (ADC). A digital control algorithm is then used to calculate the controller output, a digital signal. Because most final control elements are analog devices, the digital output signal is usually converted to a corresponding analog signal by a *digital-to-analog converter* (DAC). However, some electronic final control elements can receive digital signals directly, as discussed in Chapter 9.

8.6.1 Position and Velocity Algorithms for Digital PID Control

There are two alternative forms of the digital PID control equation, the *position form* and the *velocity form*. A straightforward way of deriving a digital version of the parallel form of the PID controller (Eq. 8-13) is to replace the integral and derivative terms by finite difference approximations,

$$\int_0^t e(t^*)\,dt^* \approx \sum_{j=1}^{k} e_j \Delta t \tag{8-24}$$

$$\frac{de}{dt} \approx \frac{e_k - e_{k-1}}{\Delta t} \tag{8-25}$$

where

Δt = the sampling period (the time between successive measurements of the controlled variable)

e_k = error at the kth sampling instant for $k = 1, 2, \ldots$

Substituting Eqs. 8-24 and 8-25 into (8-13) gives the *position form*,

$$p_k = \bar{p} + K_c\left[e_k + \frac{\Delta t}{\tau_I}\sum_{j=1}^{k} e_j + \frac{\tau_D}{\Delta t}(e_k - e_{k-1})\right] \tag{8-26}$$

where p_k is the controller output at the kth sampling instant. The other symbols in Eq. 8-26 have the same meaning as in Eq. 8-13. Equation 8-26 is referred to as the position form, because the actual value of the controller output is calculated.

In the *velocity form*, the change in controller output is calculated. The velocity form can be derived by writing Eq. 8-26 for the $(k-1)$ sampling instant:

$$p_{k-1} = \bar{p} + K_c\left[e_{k-1} + \frac{\Delta t}{\tau_I}\sum_{j=1}^{k-1} e_j + \frac{\tau_D}{\Delta t}(e_{k-1} - e_{k-2})\right] \tag{8-27}$$

Note that the summation still begins at $j = 1$, because it is assumed that the process is at the desired steady state for $j \leq 0$, and thus $e_j = 0$ for $j \leq 0$. Subtracting Eq. 8-27 from (8-26) gives the velocity form of the digital PID algorithm:

$$\Delta p_k = p_k - p_{k-1} = K_c\left[(e_k - e_{k-1}) + \frac{\Delta t}{\tau_I}e_k + \frac{\tau_D}{\Delta t}(e_k - 2e_{k-1} + e_{k-2})\right] \tag{8-28}$$

The velocity form has three advantages over the position form:

1. It inherently contains antireset windup, because the summation of errors is not explicitly calculated.
2. This output is expressed in a form, Δp_k, that can be utilized directly by some final control elements, such as a control valve driven by a pulsed stepping motor.
3. For the velocity algorithm, transferring the controller from manual to automatic model does not require any initialization of the output ($\bar{p}$ in Eq. 8-26). However, the control valve (or other final control element) should be placed in the appropriate position prior to the transfer.

Certain types of advanced control strategies, such as cascade control and feedforward control, require that the actual controller output p_k be calculated explicitly. These strategies are discussed in Chapters 15 and 16. However, p_k can easily be calculated by rearranging Eq. 8-28:

$$p_k = p_{k-1} + K_c\left[(e_k - e_{k-1}) + \frac{\Delta t}{\tau_I}e_k + \frac{\tau_D}{\Delta t}(e_k - 2e_{k-1} + e_{k-2})\right] \tag{8-29}$$

A minor disadvantage of the velocity form is that the integral mode *must* be included. When the set point is constant, it cancels out in both the proportional and derivative error terms. Consequently, if the integral mode were omitted, the process response to a disturbance would tend to drift away from the set point.

The position form of the PID algorithm (Eq. 8-26) requires a value of $\bar{p}$, while the velocity form in Eq. 8-28 does not. Initialization of either algorithm is straightforward, because manual operation of the control system usually precedes the transfer to automatic control.

Hence, $\bar{p}$ (or p_{k-1} for the velocity algorithm) is simply set equal to the signal to the final control element at the time of transfer. As noted previously, the velocity form is less prone to reset windup problems.

8.6.2 Modifications of the Basic PID Control Algorithms

We now consider several modifications of the basic PID control algorithms that are widely used in industry.

1. ***Elimination of Reset Windup.*** For controllers that contain integral control action, *reset windup* can occur when the error summation grows to a very large value. Suppose the controller output saturates at an upper or lower limit, as the result of a large sustained error signal. Even though the measured variable eventually reaches its set point (where $e_k = 0$), the controller may be wound up because of the summation term. Until the error changes sign for a period of time, thereby reducing the value of the summation, the controller will remain at its saturation limit.

   For the position algorithm, several modifications can be made to reduce reset windup:

   a. Place an upper limit on the value of the summation. When the controller saturates, suspend the summation until the controller output moves away from the limit.

   b. Back-calculate the value of e_k that just causes the controller to saturate. When saturation occurs, use this value as the error term, e_{k-1}, in the next controller calculation.

   Experience has indicated that approach (b) is superior to (a), although it is somewhat more complicated.

   For the velocity form in Eqs. 8-28 or 8-29, no summation appears, and thus the reset windup problem is avoided. However, the control algorithm must be implemented so that Δp_k is disregarded if p_k is at a saturation limit, implying that p_k should be monitored at all times. In general, the velocity form is preferred over the position form.

2. ***Elimination of Derivative Kick.*** When a sudden set-point change is made, the PID control algorithms in Eq. 8-26 or Eq. 8-28 will produce a large immediate change in the output due to the derivative control action. For digital control algorithms, several methods are available for eliminating derivative kick:

   a. In analogy with Eq. 8-17, derivative action can be applied to the measurement, $y_{m,k}$, rather than the error signal. Thus, for the position form in Eq. 8-26, e_k is replaced by $-y_{m,k}$ in the derivative term:

$$p_k = \bar{p} + K_c\left[e_k + \frac{\Delta t}{\tau_I}\sum_{j=1}^{k} e_j - \frac{\tau_D}{\Delta t}(y_{m,k} - y_{m,k-1})\right] \quad (8\text{-}30)$$

   The velocity form in Eq. 8-28 can be modified in an analogous fashion.

   b. Change the set point gradually by ramping it to the new value. This strategy limits the rate of change of the set point and thus reduces the derivative kick.

   If measurement noise combined with a large ratio of derivative time to sampling period ($\tau_D/\Delta t$) causes an overactive derivative mode, then the error signal must be filtered before calculating the derivative action (see Chapter 17).

3. ***Effect of Saturation on Controller Performance.*** Another difficulty that can occur for a digital controller equation such as Eq. 8-30 is that a small change in the error can cause the controller output to saturate for certain values of the controller settings. Suppose that $K_c\tau_D/\Delta t = 100$ due to a small sampling period, and that e_k and p_k are both scaled from 0 to 100%. A 1% change in $\Delta e_k = e_k - e_{k-1}$ will cause a 100% change in p_k, thus exceeding its upper limit. Therefore, the values of the controller settings and Δt should be checked to ensure that they do not cause such *overrange* problems. For the velocity algorithm, the change in the controller output can be constrained by using *rate limits* or *clamps*, that is, lower and upper bounds on the change, Δp_k.

4. ***Other Optional features.*** For some control applications, it is desirable that the controller output signal not be changed when the error is small, within a specified tolerance. This optional feature is referred to as *gap action*. Finally, in *gain scheduling*, the numerical value of K_c depends on the value of the error signal. These controller options are discussed in more detail in Chapter 16.

For a more detailed discussion of digital control algorithms, see Chapter 17.

SUMMARY

In this chapter we have considered the most common types of feedback controllers. Although there are potentially many forms of feedback control, the process industries rely largely on variations of PID control and on-off control. The remaining important elements within the control loop—sensors, transmitters, and final control elements—are discussed in detail in the next chapter. Once the steady-state and dynamic characteristics of these elements are understood, we can investigate the dynamic characteristics of the controlled process (Chapter 11).

REFERENCES

Åström, K. J., and T. Hägglund, *Advanced PID Control, 3rd ed.*, ISA, Research Triangle Park, NC, 2006.

Desborough, L., and R. Miller, Increasing Customer Value of Industrial Control Performance Monitoring—Honeywell, Experience, *Proc. 6th Internat. Conf. on Chemical Process Control (CPC VI)*, p. 169, AIChE, NY (2002).

Edgar, T. F., C. L. Smith, F. G. Shinskey, G. W. Gassman, P. J. Schafbuch, T. J. McAvoy, and D. E. Seborg, Process Control, Section 8 in *Perry's Chemical Engineers' Handbook, 8th edition*, D. W. Green and R. H. Perry (ed.), McGraw-Hill, New York, 2008.

Instrumentation Symbols and Identification, Standard ISA-5.1-1984 (R1992), International Society of Automation (ISA), Research Triangle Park, NC (1992).

Mayr, O., *The Origins of Feedback Control*, MIT Press, Cambridge, MA, 1970.

McMillan, G. M., *Good Tuning: A Pocket Guide, 2nd ed.*, ISA, Research Triangle Park, NC, 2006.

Shinskey, F. G., *Process Control Systems*, 4th ed., McGraw-Hill, New York, 1996.

Ziegler, J. G., Those Magnificent Men and Their Controlling Machines, *J. Dynamic Systems, Measurement and Control, Trans. ASME*, **97**, 279 (1975).

EXERCISES

8.1 An electronic PI temperature controller has an output p of 12 mA when the set point equals the nominal process temperature. The controller response to step change in the temperature set point of 3 mA (equivalent to a change of 5°F) is shown below:

t, s	p, mA
0−	12
0+	10
20	9
60	7
80	6

Determine the controller gain K_c (mA/mA) and the integral time, τ_I. Is the controller reverse-acting or direct-acting?

8.2 A physically realizable form of the ideal PD controller transfer function in Eq. 8-11 is given by

$$\frac{P'(s)}{E(s)} = \frac{K_c(\tau_D s + 1)}{\alpha\tau_D s + 1}$$

where $0.05 < \alpha < 0.2$.

(a) Show how to obtain this transfer function with a parallel arrangement of two much simpler functions in Fig. E8.2:

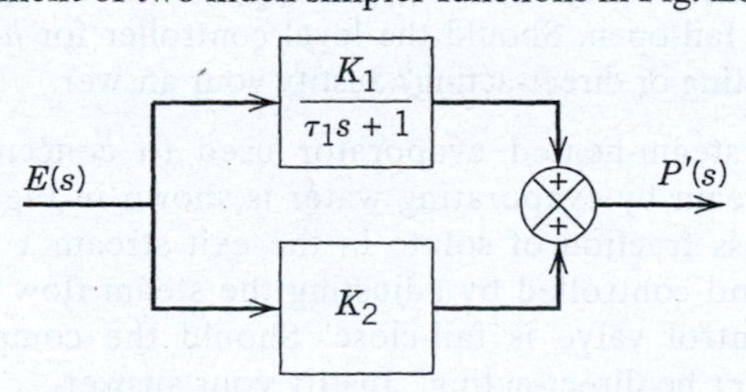

Figure E8.2

(b) Find expressions for K_1, K_2, and τ_1 that can be used to obtain desired values of K_c, τ_D, and α.

(c) Verify the relations for $K_c = 3$, $\tau_D = 2$, $\alpha = 0.1$.

8.3 The parallel form of the PID controller has the transfer function given by Eq. 8-14. Many commercial analog controllers can be described by the series form given by Eq. 8-15.

(a) For the simplest case, $\alpha \to 0$, find the relations between the settings for the parallel form ($K_c^\dagger$, $\tau_I^\dagger$, $\tau_D^\dagger$) and the settings for the series form (K_c, τ_I, τ_D).

(b) Does the series form make each controller setting (K_c, τ_I, or τ_D) larger or smaller than would be expected for the parallel form?

(c) What are the magnitudes of these interaction effects for $K_c = 4$, $\tau_I = 10$ min, $\tau_D = 2$ min?

(d) What can you say about the effect of nonzero α on these relations? (Discuss only first-order effects.)

8.4 Exercise 1.7 shows two possible ways to design a feedback control loop to obtain a desired rate of liquid flow. Assume that in both Systems I and II, the flow transmitter is direct-acting (i.e., the output increases as the actual flow rate increases). However, the control valve in System I is "air-to-open," meaning that an increasing pressure signal from the controller will open the valve more, thus increasing the flow rate (See Chapter 9). On the other hand, the control valve in System II is "air-to-close." The dynamics for both of the valves are negligible.

(a) For each of these valves, what is the sign of its gain, K_v?

(b) Which controller must be direct-acting? reverse-acting? Use physical arguments to support your answers.

(c) What sign should the controller gain have for each case?

8.5 A liquid-level control system can be configured in either of two ways: with a control valve manipulating flow of liquid into the holding tank (Fig. E8.5*a*), or with a control valve

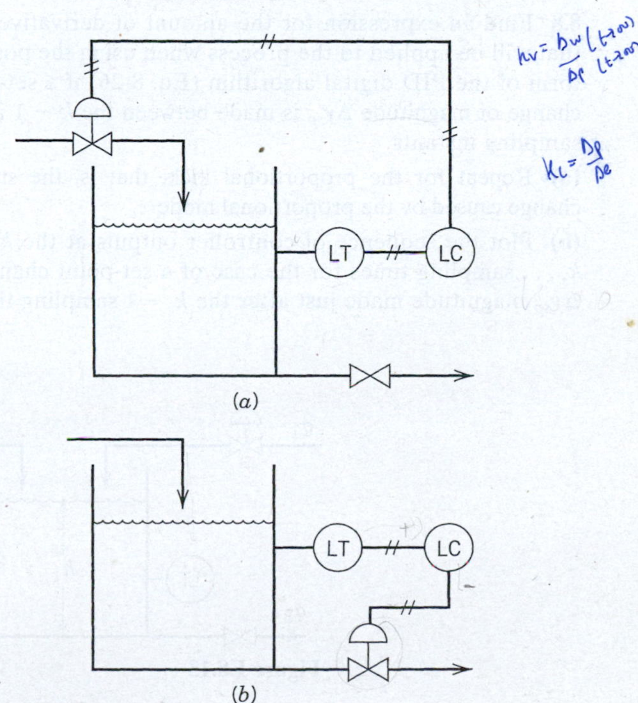

Figure E8.5

manipulating the flow of liquid from the tank (Fig. E8.5*b*). Assuming that the liquid-level transmitter always is direct-acting,

(a) For each configuration, what control action should a proportional pneumatic controller have if the control valve is air-to-open?

(b) If the control valve is air-to-close?

8.6 If the measured input to a PI controller changes stepwise ($Y_m(s) = 2/s$) and the controller output changes initially as in Fig. E8.6, what are the values of the controller gain and integral time?

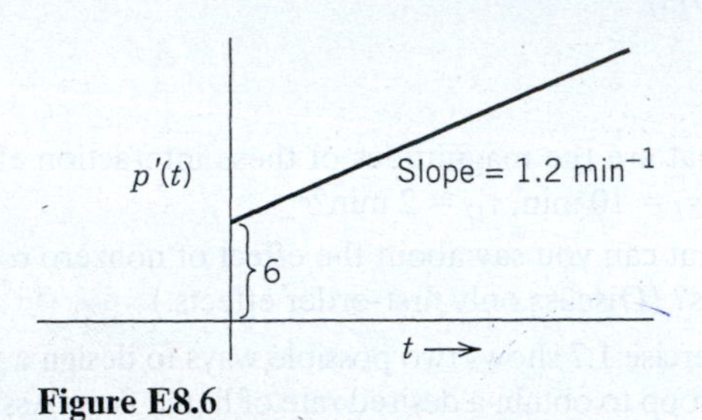

Figure E8.6

8.7 An electronic PID temperature controller is at steady state with an output of 12 mA. The set point equals the nominal process temperature initially. At $t = 0$, the set point is increased at the rate of 0.5 mA/min (equivalent to a rate of 2°F/min). If the current settings are

$$K_c = 2 \text{ (dimensionless)}$$
$$\tau_I = 1.5 \text{ min}$$
$$\tau_D = 0.5 \text{ min}$$

(a) Derive an expression for the controller output $p(t)$.

(b) Repeat (a) for a PI controller.

(c) Plot the two controller outputs and qualitatively discuss their differences.

8.8 Find an expression for the amount of derivative kick that will be applied to the process when using the position form of the PID digital algorithm (Eq. 8-26) if a set-point change of magnitude Δy_{sp} is made between the $k - 1$ and k sampling instants.

(a) Repeat for the proportional kick, that is, the sudden change caused by the proportional mode.

(b) Plot the sequence of controller outputs at the $k - 1$, $k, \ldots$ sampling times for the case of a set-point change of Δy_{sp} magnitude made just after the $k - 1$ sampling time if the controller receives a constant measurement $\bar{y}_m$ and the initial set point is $\bar{y}_{sp} = \bar{y}_m$. Assume that the controller output initially is $\bar{p}$.

(c) How can Eq. 8-26 be modified to eliminate derivative kick?

8.9 (a) For the case of the digital velocity P and PD algorithms, show how the set point enters into calculation of Δp_k on the assumption that it is not changing, that is, y_{sp} is a constant.

(b) What do the results indicate about use of the velocity form of P and PD digital control algorithms?

(c) Are similar problems encountered if the integral mode is present, that is, with PI and PID forms of the velocity algorithm? Explain.

8.10 What differential equation model represents the parallel PID controller with a derivative filter? (**Hint:** Find a common denominator for the transfer function first.)

(a) Repeat for the series PID controller with a derivative filter.

(b) Simulate the time response of each controller for a step change in $e(t)$.

8.11 What is the corresponding control law for the series PID controller? Qualitatively describe its response to a step change in $e(t)$.

8.12 Consider a standard feedback control system where each component is functioning properly. Briefly indicate whether you agree or disagree with the following statements:

(a) For proportional-only control, the controller output is always proportional to the error signal.

(b) A PI controller always eliminates offset after a sustained, unmeasured disturbance.

8.13 Consider the liquid storage system in Fig. E8.13. Suppose that q_2 must be kept constant, and, consequently, h_2 is to be controlled by adjusting q_1. Suppose that the q_1 control valve is fail-open. Should the level controller for h_2 be reverse acting or direct-acting? Justify your answer.

8.14 A steam-heated evaporator used to concentrate a feed stream by evaporating water is shown in Fig. E8.14. The mass fraction of solute in the exit stream x is measured and controlled by adjusting the steam flow rate, S. The control valve is fail-close. Should the composition controller be direct-acting? Justify your answer.

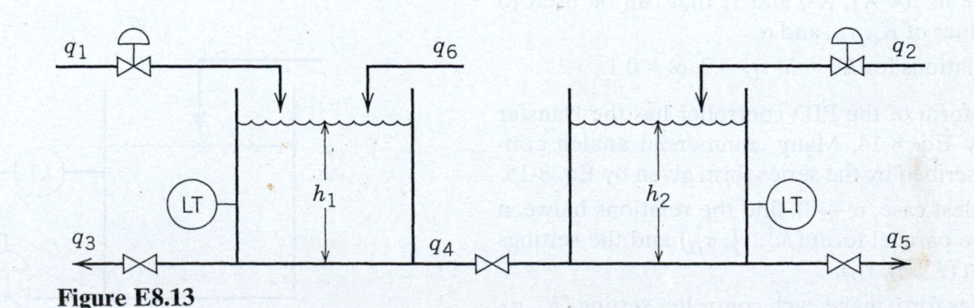

Figure E8.13

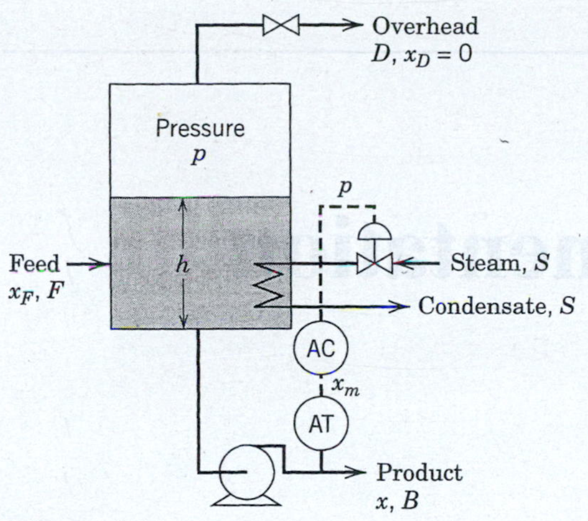

Figure E8.14

8.15 A very hot stream is cooled by cold water in a counter-current heat exchanger: shown in Fig. E8.15:

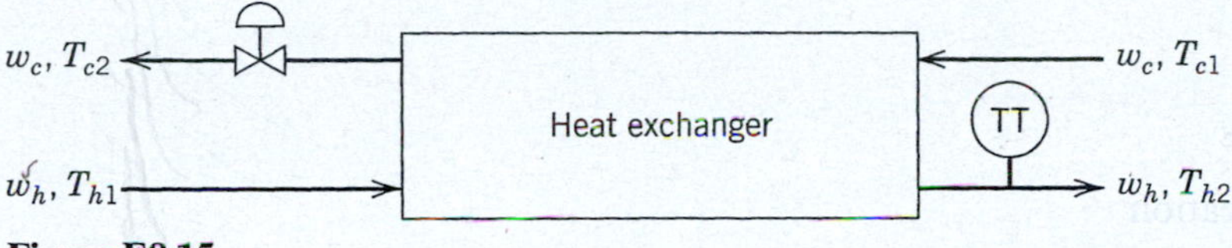

Figure E8.15

Temperature T_{h2} is to be controlled by adjusting flow rate, w_c. The temperature sensor/transmitter (TT) is direct-acting. Should the feedback controller be direct-acting or reverse-acting?

8.16 Consider the schematic diagram of a controlled blending process shown in Fig. 8.1. The control objective is to control the mass fraction of the exit stream, x, by adjusting inlet flow rate, w_2, using a feedback controller. The mass fractions of a key chemical component in the inlet streams, x_1 and x_2, are constant, and mass flow rate w_1 is a disturbance variable. The liquid volume V is constant. The composition sensor/transmitter (AT) and the current-to-pressure transducer (I/P) are both direct-acting devices.

What is the *minimum* amount of information you would need in order to decide whether the feedback controller, AC, should be reverse-acting or direct-acting?

Chapter 9

Control System Instrumentation

CHAPTER CONTENTS

Having considered PID controllers in Chapter 8, we now consider the other components of the feedback control loop. As an illustrative example, consider the stirred-tank heating system in Fig. 9.1. A thermocouple measures the liquid temperature and converts it to a millivolt-level electrical signal. This signal is then amplified to a voltage level and transmitted to the electronic controller. The feedback controller performs the control calculations and sends the calculated value as an output signal to the final control element, an electrical heater that adjusts the rate of heat transfer to the liquid. This example illustrates the three important functions of a feedback control loop: (1) measurement of the controlled variable (CV), (2) adjustment of the manipulated variable (MV), and (3) signal transmission between components.

The interconnection between the process and the controller in Fig. 9.1 can be considered to be an *interface* (analog or digital). The interconnection is required for a single controller Fig. 9.2 or for a number of controllers in a computer control system Fig. 9.3. In each case, the interface consists of all measurement, manipulation, and transmission instruments. The interface elements in Fig. 9.3 all contain a common feature. Each involves the conversion of a variable, for example, temperature to a voltage-level signal. Final control elements, or *actuators*, are used to manipulate process variables (usually flow rates).

This chapter introduces key instrumentation concepts and emphasizs how the choice of measurement and manipulation hardware affects the characteristics of the control system. Many of the assumptions that are commonly used to simplify the design of control systems—linear behavior of instruments and actuators, negligible instrumentation and signal transmission dynamics—depend on the proper design and specification of control loop instrumentation. A number of general references and handbooks can be utilized for specification of instrumentation (e.g., Connell, 1996; Lipták, 2003; Johnson, 2008; Edgar et al., 2008; Scott, 2008).

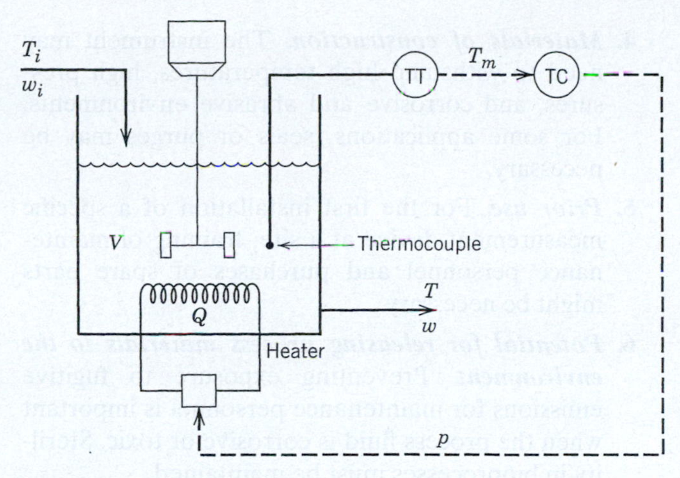

Figure 9.1 Schematic diagram for a stirred-tank heating control system.

Appendix A describes digital computer control and digital instrumentation systems. A significant amount of instrumentation used today is based on digital technology, although traditional analog instrumentation is still-viable used. Consequently, we consider both in this chapter.

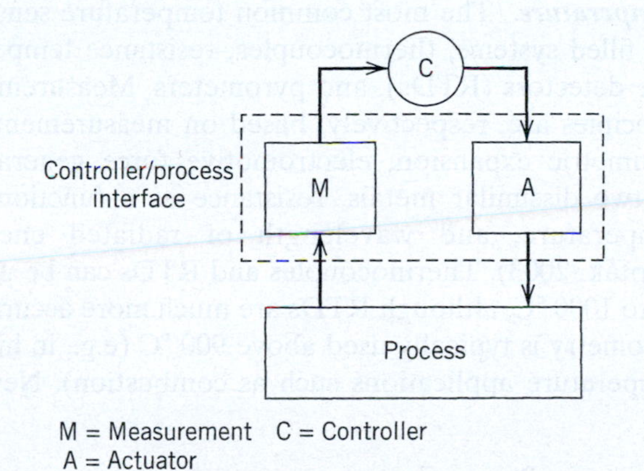

Figure 9.2 A controller/process interface.

9.1 SENSORS, TRANSMITTERS, AND TRANSDUCERS

The operation of complex industrial plants would be difficult, if not impossible, without the measurement and control of critical process variables. Large plants typically have hundreds or thousands of process variables that are repetitively measured on-line every few seconds or minutes. In addition, important product properties are measured in quality control labs less frequently—e.g., once per hour, once an eight-hour shift, or daily. Consequently, the design and maintenance of accurate, reliable measurement systems is a critical aspect of process control. The lack of a reliable, cost—effective on-line sensor can be a key limitation on the effectiveness of a process control system.

A physical variable is measured by a *sensor* which produces a physical response (e.g., electrical or mechanical) that is related to the value of the process variable. For example, in the stirred-tank heating system in Fig. 9.1, the thermocouple generates a millivolt electrical signal that increases as the temperature of the liquid increases. However, for this temperature measurement to be used in the control calculations, the millivolt-level signal must be converted to an appropriate voltage or current signal in a standard input range for the controller (see Section 9.1.1). This conversion is done by a *transmitter*.

In the process control literature, the terms *sensor*, *transmitter*, and *sensor-transmitter*, are used more or less interchangeably; we follow suit in this book.

It is often necessary to convert an instrumentation signal from one form to another. A device that performs this conversion is referred to as a *transducer*. One common application is when the controller output signal is a current signal and the final control element is a pneumatic control valve (see Section 9.2.1). The required conversion is performed by a current-to-pressure (I/P) transducer. Voltage-to-pressure (E/P) transducers are also quite common.

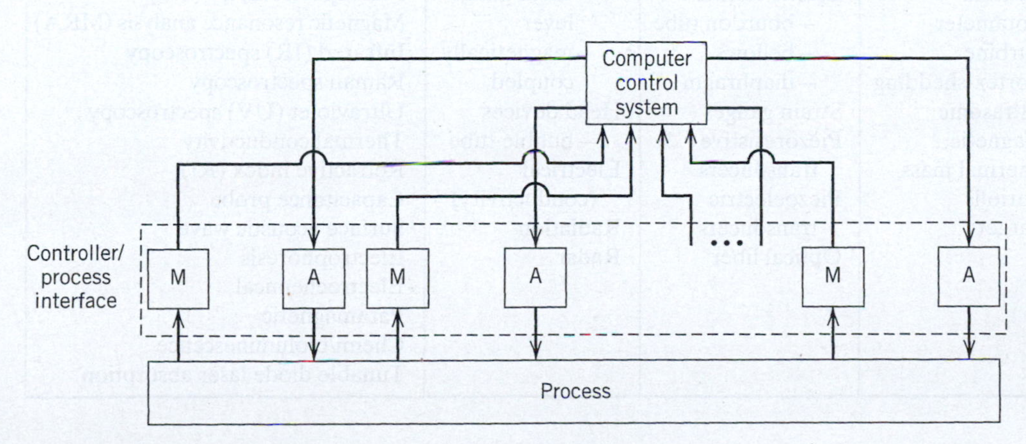

Figure 9.3 Computer control system with multiple measurements and multiple actuators.

9.1.1 Standard Instrumentation Signal Levels

Before 1960, instrumentation in the process industries utilized pneumatic (air pressure) signals to transmit measurement and control information almost exclusively. These devices make use of mechanical force-balance elements to generate signals in the range of 3 to 15 psig, an industry standard. Since about 1960, electronic instrumentation has become predominant. The standard signal ranges for analog instruments are 4 to 20 mA and 1 to 5 V, direct current (VDC).

9.1.2 Sensors

We now briefly discuss commonly used sensors for the most important process variables. Additional information is available in Soloman (1999), Lipták (2003), Shuler and Kargi (2002), Scott (2008), and Connell (1996). The main categories of measurements used in process control are temperature, pressure, flow rate, liquid level, and composition. Table 9.1 lists sensor options for each of these five categories.

Selection Criteria. The selection of a measurement device should consider the following factors:

1. ***Measurement range (span).*** The required measurement range for the process variable must lie entirely within the instrument's range of performance.
2. ***Performance.*** Depending on the application, accuracy, repeatability, or some other measure of performance is appropriate. For closed-loop control, speed of response is also important.
3. ***Reliability.*** Manufacturers provide baseline conditions. Previous experience with the measurement device is very important.
4. ***Materials of construction.*** The instrument may need to withstand high temperatures, high pressures, and corrosive and abrasive environments. For some applications, seals or purges may be necessary.
5. ***Prior use.*** For the first installation of a specific measurement device at a site, training of maintenance personnel and purchases of spare parts might be necessary.
6. ***Potential for releasing process materials to the environment.*** Preventing exposure to fugitive emissions for maintenance personnel is important when the process fluid is corrosive or toxic. Sterility in bioprocesses must be maintained.
7. ***Electrical classification.*** If the sensor is not inherently compatible with possible exposure to hazards, suitable enclosures must be purchased and included in the installation costs.
8. ***Invasive or non-invasive.*** The insertion of a probe (invasive) can cause fouling, which leads to inaccurate measurements. Probe location must be selected carefully to ensure measurement accuracy and minimize fouling.

Temperature. The most common temperature sensors are filled systems, thermocouples, resistance temperature detectors (RTDs), and pyrometers. Measurement principles are, respectively, based on measurement of volumetric expansion, electromotive force generated by two dissimilar metals, resistance as a function of temperature, and wavelength of radiated energy (Lipták, 2003). Thermocouples and RTDs can be used up to 1000 °C, although RTDs are much more accurate. Pyrometry is typically used above 900 °C (e.g., in high-temperature applications such as combustion). Newer

Table 9.1 On-Line Measurement Options for Process Control

Temperature	Flow	Pressure	Level	Composition
Thermocouple Resistance temperature detector (RTD) Filled-system thermometer Bimetal thermometer Pyrometer —total radiation —photoelectric —ratio Laser Surface acoustic wave Semiconductor	Orifice Venturi Rotameter Turbine Vortex-shedding Ultrasonic Magnetic Thermal mass Coriolis Target	Liquid column Elastic element —bourdon tube —bellows —diaphragm Strain gauges Piezoresistive transducers Piezoelectric transducers Optical fiber	Float-activated —chain gauge, lever —magnetically coupled Head devices —bubble tube Electrical (conductivity) Radiation Radar	Gas-liquid chromatography (GLC) Mass spectrometry (MS) Magnetic resonance analysis (MRA) Infrared (IR) spectroscopy Raman spectroscopy Ultraviolet (UV) spectroscopy Thermal conductivity Refractive index (RI) Capacitance probe Surface acoustic wave Electrophoresis Electrochemical Paramagnetic Chemi/bioluminescence Tunable diode laser absorption

options include surface acoustic wave (SAW), which measures attenuation and frequency shift as a function of temperature on a solid surface, and semiconductors, whose resistance varies with temperature.

Differential Pressure. For pneumatic instrumentation, pressure sensing is quite straightforward. A bellows, bourdon tube, or diaphragm isolates process liquid or gas from the instrument, at the same time furnishing a deflection to a force-balance element that generates a proportional signal in the 3 to 15 psig range. For electronic instrumentation, a strain gauge often is used to convert pressure into an elongation of resistance wires, which changes a millivolt-level emf. This signal can be amplified to an appropriate voltage or current range.

A pressure difference can be measured similarly by placing the two process pressure connections on either side of a diaphragm. Electronic measurements typically use a strain gauge to convert the diaphragm deflection in differential pressure instruments in the same way as in pressure measurement instruments. For many proceses, the liquid or gas streams cannot be brought into direct contact with the sensing element (diaphragm) because of high temperature or corrosion considerations. In these cases an inert fluid, usually an inert gas, is used to isolate the sensing element.

More recently, fiber-optic sensors have been developed to measure pressure in high-temperature environments (Krohn, 2000). Multivariable transmitters are available that measure several process variables (see Fig. 9.4).

Liquid or Gas Flow Rate. Selection of a flow transmitter should consider the following factors: nature of the flowing material (liquid/gas/solid), corrosiveness, mass vs. volume measurement, nature of the signal, cost, accuracy, current plant practice, space available, and necessary maintenance (Spitzer, 2001).

Flow rate can be measured indirectly, using the pressure drop across an orifice or venturi as the input signal to conventional differential pressure instrumentation. In this case, the volumetric flow rate is proportional to the square root of the pressure drop. The orifice plate is normally sized to provide a pressure drop in the range of 20 to 200 in of water, but this approach is not very accurate (±5%) compared to venturi meters (±2%). Volumetric flow rates can also be measured using turbine flowmeters. The pulse output signal can be modulated to give an electronic signal, which can be *totalized* in a counter and sent periodically to a controller. Deflection of a vane inserted in the pipe or channel also can be used as a flow sensor, such as in a target meter or a vortex shedding meter. Magnetic flow meters can be used to measure volumetric flow rates of conducting fluids.

Mass flow meters that are independent of changes in pressure, temperature, viscosity, and density include the thermal mass meter and the coriolis meter. Thermal mass meters are widely used in semiconductor manufacturing and in bioprocessing for control of low flow rates (called mass flow controllers, or MFCs). MFCs measure the heat loss from a heated element, which varies with flow rate, with an accuracy of ±1%. Coriolis meters use a vibrating flow loop that undergoes a twisting action due to the coriolis effect. The amplitude of the deflection angle is converted to a voltage that is nearly proportional to the liquid mass flow rate, with an accuracy of ±0.5%. Sufficient space must be available to accommodate the flow loop, and pressure losses of 10 psi should be allowable (Henry et al., 2000). Capacitance probes measure the dielectric constant of the fluid and are useful for flow measurements of slurries and other two-phase flows. The accuracy of ultrasonic meters has been improved during recent years owing to better sensors and software for analysis of the wave patterns; they are attractive because of their noninvasive nature and the absence of moving parts that can wear out (Baker, 2000).

Liquid Level. The position of a free float or the buoyancy effects on a fixed float can be detected and converted to level if the liquid density is known. The difference in pressure between the vapor above the liquid and the bottom of the liquid can be similarly used. Pressure taps (tubes connected from the transmitter to

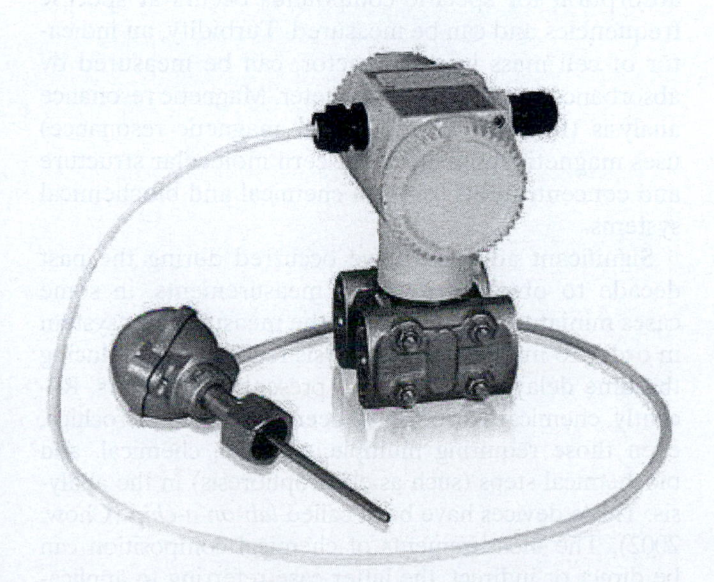

Figure 9.4 A multivariable pressure transmitter that measures absolute pressure, differential pressure, and temperature (Courtesy of ABB).

the appropriate process locations) can be kept from plugging by maintaining very low flows of inert gas through the taps to the process. The attenuation of high-energy radiation (e.g., from nuclear sources) by the liquid also can be used when solid material or gas streams cannot be in contact with process liquids.

Chemical Composition. Chemical composition is generally the most challenging on-line measurement. Before the era of on-line analyzers, messengers were required to deliver samples to the laboratory for analysis and to return the results to the control room. The long time delay involved prevented process adjustment from being made, affecting product quality. The development of on-line analyzers has automated this approach and reduced the analysis time. However, manual sampling is still frequently employed, especially in the specialty chemical industry, where few instruments are commercially available. Because a chemical composition analysis system can be very expensive, it is important to assess the payback of such an investment vs. the cost of manual sampling. Potential quality improvements can be an important consideration.

In order to obtain quantitative composition measurements, specific instruments must be chosen depending on the nature of the species to be analyzed. Measuring a specific concentration requires a unique chemical or physical attribute. In infrared (IR) spectroscopy, the vibrational frequency of specific molecules like CO and CO_2 can be probed by absorbing electromagnetic radiation. Ultraviolet radiation analyzers operate similarly to infrared analyzers in that the degree of absorption for specific compounds occurs at specific frequencies and can be measured. Turbidity, an indicator of cell mass in a bioreactor, can be measured by absorbance in a spectrophotometer. Magnetic resonance analysis (formerly called nuclear magnetic resonance) uses magnetic moments to discern molecular structure and concentrations for both chemical and biochemical systems.

Significant advances have occurred during the past decade to obtain lower cost measurements, in some cases miniaturizing the size of the measurement system in order to make on-line analysis feasible and reducing the time delays that often are present in analyzers. Recently, chemical sensors have been placed on microchips, even those requiring multiple, physical, chemical, and biochemical steps (such as electrophoresis) in the analysis. These devices have been called *lab-on-a-chip* (Chow, 2002). The measurements of chemical composition can be direct or indirect, the latter case referring to applications where some property of the process stream is measured (such as refractive index) and then related to composition of a particular component.

In gas chromatography (GC), usually the thermal conductivity is used to measure concentration. The GC can measure many components in a mixture at the same time, whereas most other analyzers can only detect one component; hence, GC is widely employed. A gas sample (or a vaporized liquid sample) is carried through the GC by an inert gas, and components of the sample are separated by a packed bed. Because each component has a different affinity for the column packing, it passes through the column at a different time during the sample analysis, allowing individual concentrations to be measured. Typically, all components can be analyzed in a five- to ten-minute time period (although miniaturized GCs are faster). The GC can measure concentrations ranging from parts per billion (ppb) to tens of percent, depending on the compound (Nichols, 1988). High-performance liquid chromatography (HPLC) can be used to measure dissolved solute levels, including proteins (Shuler and Kargi, 2002).

Mass spectroscopy (MS) determines the partial pressures of gases in a mixture by directing ionized gases into a detector under a vacuum (10^{-6} torr), and the gas phase composition is then monitored more or less continuously based on the molecular weight of the species (Nichols, 1988). Sometimes GC is combined with MS in order to obtain a higher level of discrimination of the components present. As an example, complete analysis of a combustion gas requires multiple on-line analyzers as follows:

O_3	UV photometer
SO_2	UV fluorescence
NO_x	Chemiluminescence
CO, CO_2, SO_2	Infrared
O_2	Paramagnetic
Trace hydrocarbons	GC/MS

Fiber-optic sensors are attractive options (but more expensive) for acquiring measurements in harsh environments such as high temperature or pressure. The transducing technique used by these sensors is optical and does not involve electrical signals, so they are immune to electromagnetic interference. Raman spectroscopy uses fiber-optics and involves pulsed light scattering by molecules. It has a wide variety of applications in process control (Dakin and Culshaw, 1997).

Many composition measurements are both difficult and expensive to obtain. Indirect means of measuring concentrations are often less expensive and faster for example, relating the mole or mass fraction of a liquid component to pH or conductivity, or the concentration of one component in a vapor stream to its IR or UV absorption. Often an indirect measure is used to *infer* composition; for example, the liquid temperature on a plate near the top of a distillation column might be used to indicate composition of the distillate stream.

A related approach is to use a process model as a soft sensor (see Section 7.3) to estimate process variables that cannot be measured in real time. For example, *predictive emissions monitoring systems (PEMS)* relate trace pollutant concentrations to operating conditions such as temperature, pressure, and excess air; see Section 7.3.

Physical Properties. This category includes such measurements as density, moisture, turbidity, viscosity, refractive index, pH, dielectric constant, and thermal conductivity. See Lipták (2003) and Shuler and Kargi (2002) for more details on various on-line instrument alternatives.

9.1.3 Static and Dynamic Characteristics

As noted above, the output signal from a sensor-transmitter (or transmitter) must be compatible with the input range of the controller that receives the signal. Transmitters are generally designed to be *direct-acting*; that is, the output signal increases as the measured variable increases. In addition, most commercial transmitters have an adjustable input range. For example, a temperature transmitter might be adjusted so that the input range of a platinum resistance element (the sensor) is 50–150 °C. In this case, the following correspondence is obtained:

Input	Output
50 °C	4 mA
150 °C	20 mA

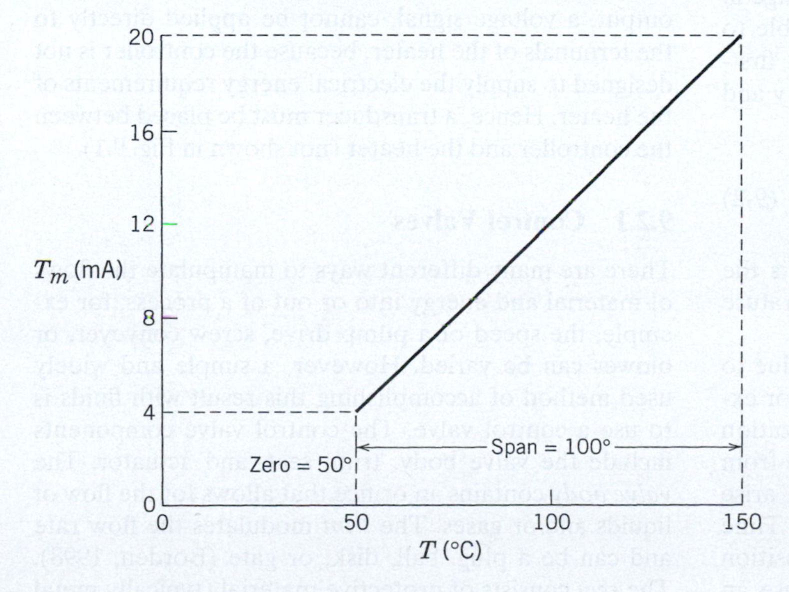

Figure 9.5 A linear instrument calibration showing its zero and span.

This instrument has a lower limit, or *zero*, of 50 °C and a range, or *span*, of 150 ° − 50 ° = 100 °C. Note that the transmitter is designed for a specific type of sensor; hence, the zero and span of the overall sensor/transmitter are adjustable. Figure 9.5 illustrates the concepts of zero and span. In this example, the relation between temperature and the transmitted (measured) signal is linear. If the sensor power supply fails, the transmitter output signal has a value of 0 mA, which would move the controller output and final control element to their minimum or maximum values. If this action could lead to an unsafe condition, the transmitter output signal could be inverted to give the highest value in the operating range.

For this temperature transmitter, the relation between the output and input is

$$T_m(\text{mA}) = \left(\frac{20\text{ mA} - 4\text{ mA}}{150\text{ °C} - 50\text{ °C}}\right)(T - 50\text{°C}) + 4\text{ mA}$$

$$= \left(0.16\,\frac{\text{mA}}{\text{°C}}\right) T(\text{°C}) - 4\text{ mA}$$

The gain of the measurement element K_m is 0.16 mA/°C. For any linear instrument

$$K_m = \frac{\text{output range}}{\text{span}} \tag{9-1}$$

For a nonlinear instrument, the gain at an operating point is the tangent to the characteristic input-output relation at the operating point. Figure 9.6 illustrates a typical case. Note that the gain changes whenever the operating point changes; hence, it is preferable to utilize instruments that exhibit nearly linear behavior. Gain K_m changes when the span is changed but is invariant to changes in the zero.

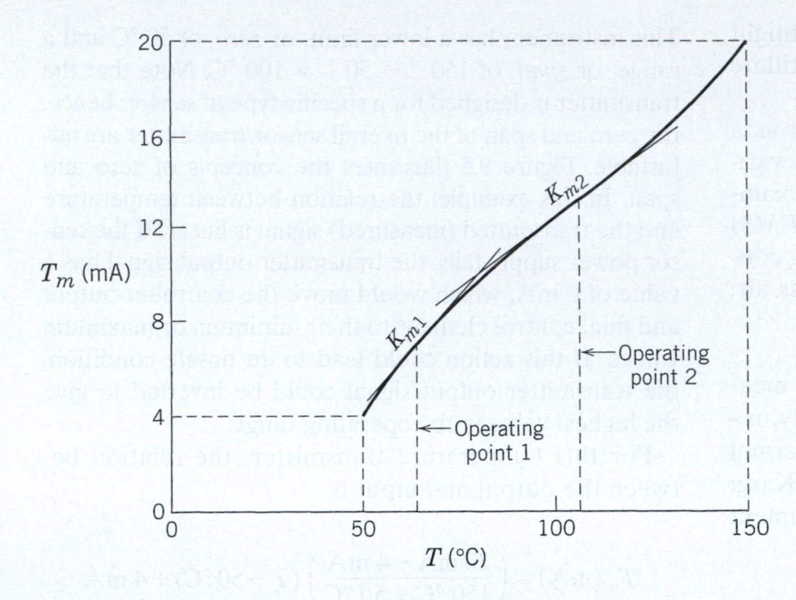

Figure 9.6 Gain of a nonlinear transmitter as a function of operating point.

Dynamic Characteristics of Sensor-Transmitters. Many sensor-transmitters respond quickly and have measurement dynamics that are negligible compared to slower process dynamics. For other applications where measurement dynamics are not negligible, significant dynamic errors can occur, that is, large differences between the true values and the measured values for transient conditions. For example, a bare thermocouple will have a rapid response to a changing fluid temperature. But a thermocouple placed in a protective thermowell with a large mass and large specific heat, can have a significant measurement time constant (see Section 9.4). Representative time constants for a variety of sensors have been reported (Riggs and Karim, 2006).

Many sensor-transmitters have overdamped dynamics and exhibit monotonic responses to a step change in the variable being measured. Thus, it is reasonable to model this type of measurement dynamics as a first-order transfer function between the actual value y and the measured value y_m:

$$\frac{Y_m(s)}{Y(s)} = \frac{K_m}{\tau_m s + 1} \tag{9-2}$$

where K_m is the gain given by Eq. 9-1 and τ_m is the measurement time constant. For the temperature transmitter example, the units of K_m are mA/°C.

Significant measurement dynamics can occur due to a poor sensor location or a long sampling line. For example, if a pH sensor for a continuous neutralization process is located in the exit line, a long distance from the process vessel, a significant time delay can arise due to the distance-velocity lag (see Section 6.2). Time delays can also result when an on-line composition measurement requires a long sample line because an expensive analyzer in a protected environment is located a long distance from the sample location near the process unit. This common situation can produce a significant distance-velocity lag.

9.2 FINAL CONTROL ELEMENTS

Every process control loop contains a final control element (or actuator), the device that enables a process variable to be manipulated. For most chemical and petroleum processes, the final control elements (usually control valves) adjust the flow rates of materials—solid, liquid, and gas feeds and products—and, indirectly, the rates of energy transfer to and from the process. Figure 9.1 illustrates the use of an electrical resistance heater as the final control element. In this case, the controller output, a voltage signal, cannot be applied directly to the terminals of the heater, because the controller is not designed to supply the electrical energy requirements of the heater. Hence, a transducer must be placed between the controller and the heater (not shown in Fig. 9.1).

9.2.1 Control Valves

There are many different ways to manipulate the flows of material and energy into or out of a process; for example, the speed of a pump drive, screw conveyer, or blower can be varied. However, a simple and widely used method of accomplishing this result with fluids is to use a control valve. The control valve components include the valve body, trim, seat, and actuator. The *valve body* contains an orifice that allows for the flow of liquids and/or gases. The *trim* modulates the flow rate and can be a plug, ball, disk, or gate (Borden, 1998). The *seat* consists of protective material (typically metal

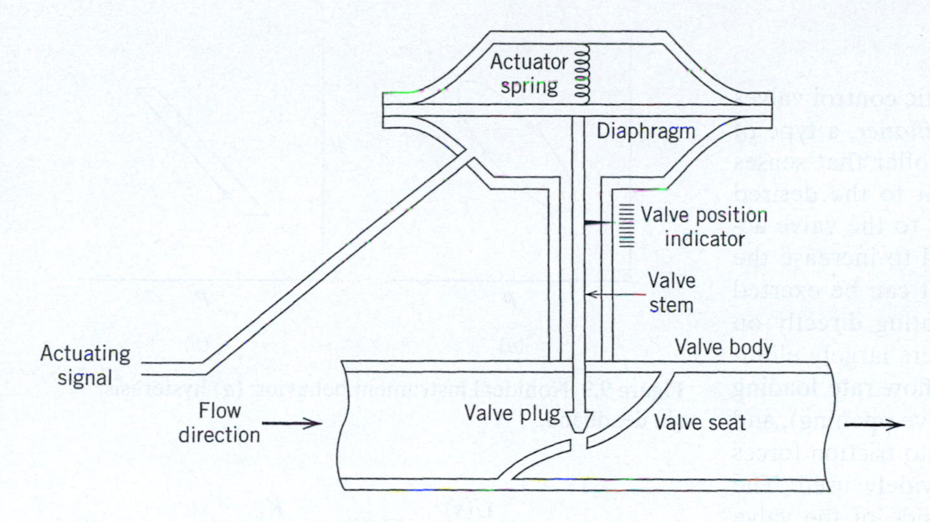

Figure 9.7 A pneumatic control valve (air-to-open).

or soft polymer) inserted around the orifice to provide a tight shutoff and to increase the life of the valve when corrosive or solid materials pass through it. Control valves are either linear (rising stem) or rotary in design. Linear valves are usually *globe valves* that open and close the valve by moving a plug vertically away from the orifice and seat. This movement changes the cross-sectional area available for fluid flow. *Rotary valves* are closed by a 90° turn of the closing element (also called *quarter-turn valves*); these valves are used for both on-off and flow modulating control valves. The *actuator* provides the force for opening and closing the valve. Rotary valves are more compact, less expensive, and easier to maintain. The primary types of quarter-turn valves are the plug valve, the butterfly valve, the ball valve, and the rotary globe valve. For more information concerning control valves, there is extensive literature (Borden, 1998; Fitzgerald, 1995; Lipták, 2003, 2006; Emerson Process Management, 2005; Edgar et al., 2008).

Control valves typically utilize some type of mechanical driver to move the valve plug into and out of its seat, thus opening or closing the area for fluid flow. The mechanical driver can be either (1) a DC motor or a stepping motor that screws the valve stem in and out in much the same way as a hand valve would be operated or (2) a pneumatically operated diaphragm device that moves the stem vertically against the opposing force of a fixed spring, called a *rising stem valve*. Motor drivers are used for very large valves and with some electronic controllers. The stepping motor is particularly useful for control valves using digital control, because the valve rotates a small fraction of a turn (2 or 3°) for each pulse it receives from the controller.

Despite the growing use of motor-driven valves, most control applications utilize pneumatically driven control valves of the rising stem type shown schematically in Fig. 9.7. As the pneumatic controller output signal increases, increased pressure on the diaphragm compresses the spring, thus pulling the stem out and opening the valve further. This type of control valve is referred to as *air-to-open* (A–O). By reversing either the plug/seat or the spring/air inlet orientation, the valve becomes *air-to-close* (A–C). For example, if the spring is located below the diaphragm and the air inlet is placed above the diaphragm, an air-to-close valve results. Normally, the choice of A–O or A–C valve is based on safety considerations. We choose the way the valve should operate (full flow or no flow) based on the desired response in an emergency situation. Hence, A–C and A–O valves often are referred to as *fail-open* (FO) and *fail-close* (FC), respectively.

EXAMPLE 9.1

Pneumatic control valves are to be specified for the applications listed below. State whether an A–O or A–C valve should be specified for the following manipulated variables, and give reason(s):

(a) Steam pressure in a reactor heating coil.

(b) Flow rate of reactants into a polymerization reactor.

(c) Flow of effluent from a wastewater treatment holding tank into a river.

(d) Flow of cooling water to a distillation condenser.

SOLUTION

(a) A–O (fail-close) to make sure that a transmitter failure will not cause the reactor to overheat, which is usually more serious than having it operate at too low a temperature.

(b) A–O (fail-close) to prevent the reactor from being flooded with excessive reactants.

(c) A–O (fail-close) to prevent excessive and perhaps untreated waste from entering the stream.

(d) A–C (fail-open) to ensure that overhead vapor is completely condensed before it reaches the receiver.

9.2.2 Valve Positioners

For important control loops, pneumatic control valves should be equipped with a *valve positioner*, a type of mechanical or digital feedback controller that senses the actual stem position, compares it to the desired position, and adjusts the air pressure to the valve accordingly. Valve positioners are used to increase the relatively small mechanical force that can be exerted by a 3–15 psig pressure signal operating directly on the valve diaphragm. Valve positioners largely eliminate valve deadband and hysteresis, flow rate loading (the effect of back pressure on the valve opening), and other undesirable characteristics due to friction forces in the valve unit; hence, they are widely used. The valve positioner is mounted on the side of the valve actuator and can reduce the valve deadband from about ±5% to ±0.5%, a significant enhancement. Details on mechanical valve positioners are given in *Perry's Handbook* (Edgar et al., 2008). A photo of a control valve with a valve positioner is shown in Fig. 9.8.

Control Valve Dynamics. Control valve dynamics tend to be relatively fast compared to the dynamics of the process itself. However, the overall behavior of pneumatic control valves can include nonlinear phenomena such as dead band, stick-slip phenomena, backlash and hysteresis (Blevins et al., 2003; Edgar et al., 2008). Dead band and hysteresis are illustrated in Fig. 9.9. Fortunately, their effects can be reduced significantly by employing valve positioners.

For purposes of control system analysis using transfer functions, the dynamic behavior of the control valve (and valve positioner) can be approximated by a first-order transfer function $G_v(s)$ between the manipulated variable $u(t)$ and the signal to the control valve $p(t)$,

Figure 9.8 Modern control valves using digital valve controllers (Courtesy Emerson Process Management).

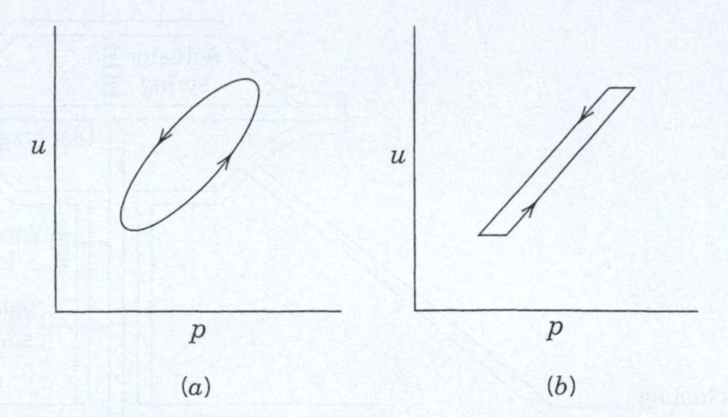

Figure 9.9 Nonideal instrument behavior: (*a*) hysteresis; (*b*) deadband.

$$\frac{U(s)}{P(s)} = G_v(s) = \frac{K_v}{\tau_v s + 1} \tag{9-3}$$

where $\tau_v << \tau_p$ and τ_p is the largest process time constant.

9.2.3 Specifying and Sizing Control Valves

Control valves are specified by first considering both properties of the process fluid and the desired flow characteristics in order to choose the valve body material and type. Then the desired characteristics for the actuator are considered. The choice of construction material depends on the corrosive properties of the process fluid at operating conditions. Commercial valves made of brass, carbon steel, and stainless steel can be ordered off the shelf, at least in smaller sizes. For large valves and more exotic materials of construction, special orders usually are required.

A design equation used for sizing control valves relates valve lift ℓ to the actual flow rate q by means of the *valve coefficient* C_v, the proportionality factor that depends predominantly on valve size or capacity:

$$q = C_v f(\ell) \sqrt{\frac{\Delta P_v}{g_s}} \tag{9-4}$$

Here q is the flow rate, $f(\ell)$ is the *valve characteristic*, ΔP_v is the pressure drop across the valve, and g_s is the specific gravity of the fluid. This relation is valid for nonflashing liquids. See Edgar et al. (2008) for other cases such as flashing liquids.

Specification of the valve size is dependent on the valve characteristic f. Three control valve characteristics are mainly used. For a fixed pressure drop across the valve, the valve characteristic $f(0 \le f \le 1)$ is related to the lift ℓ $(0 \le \ell \le 1)$ that is, the extent of valve opening, by one of the following relations:

$$\begin{aligned} &\text{Linear:} && f = \ell \\ &\text{Quick opening:} && f = \sqrt{\ell} \\ &\text{Equal percentage:} && f = R^{\ell-1} \end{aligned} \tag{9-5}$$

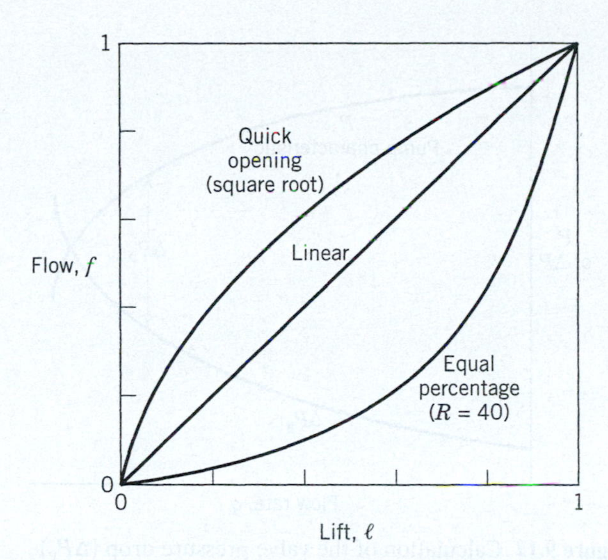

Figure 9.10 Inherent control valve characteristics.

where R is a valve design parameter that is usually in the range 20 to 50. Figure 9.10 illustrates these three flow/lift characteristics. The quick-opening valve above is referred to as a *square root* valve (valves with quicker-opening characteristics are available). The equal percentage valve is given that name because the slope of the f vs. ℓ curve, $df/d\ell$, is proportional to f, leading to an equal percentage change in flow for a particular change in ℓ, anywhere in the range.

Unfortunately, sizing of control valves depends on the fluid processing units, such as pumps, heat exchangers, or filters, that are placed in series with the valve. Considering only *control* objectives, the valve would be sized to take most of the pressure drop in the line. This choice would give the valve maximum influence over process changes that disturb the flow rate, such as upstream (supply) pressure changes. It also would yield the smallest (least expensive) valve. However, the most *economical operating conditions* require the valve to introduce as little pressure drop as possible, thus minimizing pumping costs (electrical power).

The following compromise serves as a guideline:

Guideline. *In general, a control valve should be sized so that it takes approximately one-quarter to one-third of the total pressure drop in the line at the design flow rate.*

To illustrate the tradeoffs, consider the following example adapted from Luyben and Luyben (1997). A control valve with linear characteristics is placed in series with a heat exchanger, both supplied by a pump with a constant discharge pressure at 40 psi (although its discharge flow rate varies). If the heat exchanger has already been sized to give a 30-psi pressure drop (ΔP_{he}) for a 200-gal/min flow of liquid (specific gravity equal to one), then the valve will take a 10-psi drop (ΔP_v). Figure 9.11 shows the equipment configuration. The linear control valve is sized so that it is half open ($f = \ell = 0.5$) at these conditions. Hence

$$C_v = \frac{200}{0.5\sqrt{10}} = 127$$

which, using manufacturers' data books, would require a 4-in control valve.

The *rangeability* of a control valve is defined as the ratio of maximum to minimum input signal level. For control valves, rangeability translates to the need to operate the valve within the range $0.05 \le f \le 0.95$ or a rangeability of $0.95/0.05 = 19$. For the case where the flow is reduced to 25% of design, (50 gal/min) the heat exchanger pressure drop will be reduced approximately to 1.9 psi $[30 \times (0.25)^2]$, leaving the control valve to supply the remaining 38.1 psi. The valve operating value of f, obtained by rearranging Eq. 9-4, is $(50/127)/\sqrt{38.1}$ or 0.06; hence, the valve is barely open. If the valve exhibits any hysteresis or other undesirable behavior due to internal *stiction* (combination of mechanical sticking and friction), it likely will *cycle* (close completely, then open too wide) as the controller output signal attempts to maintain this reduced flow setting. A valve positioner will reduce the cycling somewhat. If a centrifugal pump is used, the pump discharge pressure will actually increase at the lower flow rate, leading to a lower value of f and even worse problems.

Choice of Valve Characteristics. The choice of valve characteristic and its effect on valve sizing deserve some discussion at this point. A valve with linear behavior would appear to be the most desirable. However, the designer's objective is to obtain an *installed* valve characteristic that is as linear as possible that is, to have the flow through the valve and all connected process units vary linearly with ℓ. Because ΔP_v usually

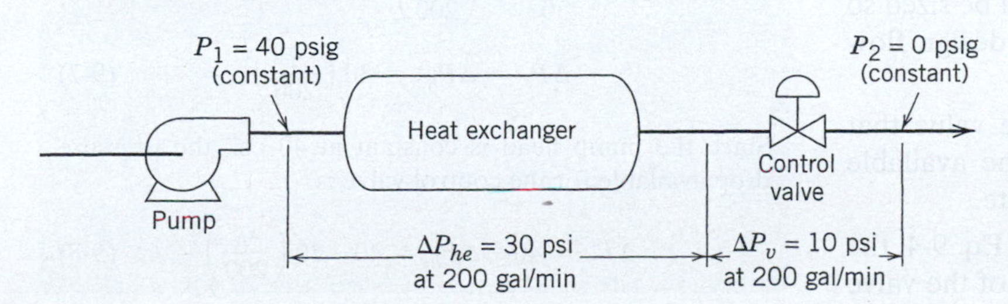

Figure 9.11 A control valve placed in series with a pump and a heat exchanger. Pump discharge pressure is constant.

varies with flow rate, a nonlinear valve often will yield a more linear flow relation *after installation* than will a linear valve characteristic. In particular, the equal percentage valve is designed to compensate, at least approximately, for changes in ΔP_v with flow rate. In the heat exchanger case, the valve coefficient C_v should be selected to match the choice of $\Delta P_{he}/\Delta P_v$ at design operating conditions. The objective is to obtain a nearly linear relation for q with respect to ℓ over the normal operating range of the valve.

Control valves must be sized very carefully. This can be particularly difficult, because many of the published recommendations are ambiguous or conflicting, or do not match control system objectives. For example, one widely used guideline is that the valve be half open at nominal operating conditions, while some vendors recommend that the required C_v not exceed 90% of the valve's rated C_v. The latter approach may result in poor control of the process if the controller output often exceeds the "required" (design) conditions as a result of disturbances. However, this recommendation does reduce the valve size considerably compared to the use of Luyben and Luyben's criterion. A more conservative criterion is that the required valve C_v be sized at 70% the valve's rated C_v and that the C_v required to accommodate the maximum expected flow rate (not the design flow rate) equal to 90% of the valve's rated C_v. If energy costs are high, it may be more economical to have the valve take 33% of the pressure drop at nominal operating condition. The lower figures will yield larger valves, and therefore higher equipment costs, but lower pumping costs (energy costs) due to lower pressure loss.

Some general guidelines for valve characteristic selection are as follows:

1. If the pump characteristic (discharge pressure vs. flow rate) is fairly flat and system frictional losses are quite small over the entire operating region, choose a linear valve. However, this situation occurs infrequently, because it results from an overdesigned process (pump and piping capacity too large).
2. To select an equal percentage valve;
   a. Plot the pump characteristic curve and ΔP_s, the system pressure drop curve without the valve, as shown in Fig. 9.12. The difference between these two curves is ΔP_v. The pump should be sized so that $\Delta P_v/\Delta P_s$ is 25 to 33% at the design flow rate q_d.
   b. Calculate the valve's rated C_v, the value that yields at least 100% of q_d with the available pressure drop at that higher flow rate.
   c. Compute q as a function of ℓ using Eq. 9-4, the rated C_v, and ΔP_v from (a). A plot of the valve characteristic (q vs. ℓ) should be reasonably linear in the operating region of interest (at least around the design flow rate). If it is not suitably linear, adjust the rated C_v and repeat.

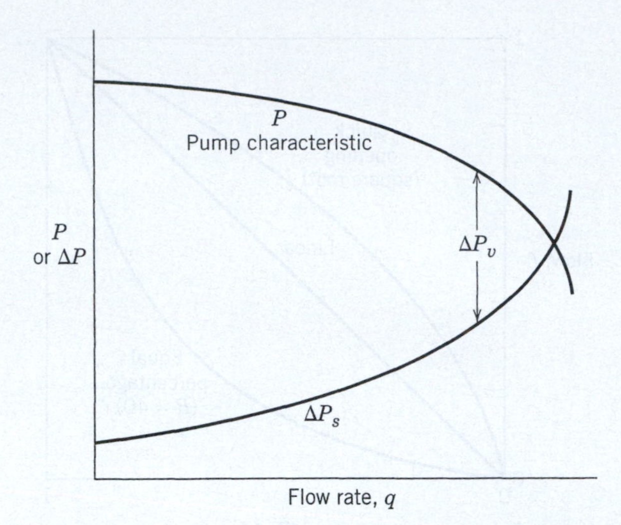

Figure 9.12 Calculation of the valve pressure drop (ΔP_v) from the pump characteristic curve and the system pressure drop without the valve (ΔP_s).

EXAMPLE 9.2

A pump furnishes a constant head of 40 psi over the entire flow rate range of interest. The heat exchanger pressure drop is 30 psig at 200 gal/min (q_d) and can be assumed to be proportional to q^2. Select the rated C_v of the valve and plot the installed characteristic for the following cases:

(a) A linear valve that is half open at the design flow rate.

(b) An equal percentage valve ($R = 50$ in Eq. 9-5) that is sized to be completely open at 110% of the design flow rate.

(c) Same as in **(b)**, except with a C_v that is 20% higher than calculated.

(d) Same as in **(b)**, except with a C_v that is 20% lower than calculated.

SOLUTION

First we write an expression for the pressure drop across the heat exchanger

$$\frac{\Delta P_{he}}{30} = \left(\frac{q}{200}\right)^2 \tag{9-6}$$

$$\Delta P_s = \Delta P_{he} = 30\left(\frac{q}{200}\right)^2 \tag{9-7}$$

Since the pump head is constant at 40 psi, the pressure drop available for the control valve is

$$\Delta P_v = 40 - \Delta P_{he} = 40 - 30\left(\frac{q}{200}\right)^2 \tag{9-8}$$

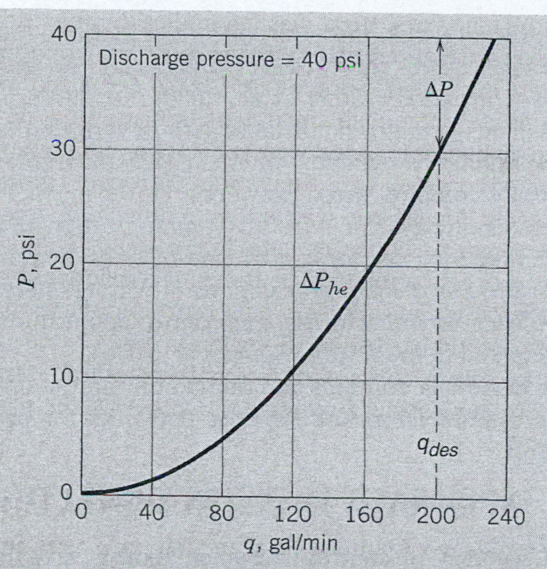

Figure 9.13 Pump characteristic and system pressure drop for Example 9.2.

Figure 9.13 illustrates these relations. Note that in all four design cases $\Delta P_v/\Delta P_s = 10/30 = 33\%$ at q_d.

(a) First calculate the rated C_v using (9-4).

$$C_v = \frac{200}{0.5\sqrt{10}} = 126.5$$

We will use $C_v = 125$. For a linear characteristic valve, use the relation between ℓ and q from Eq. 9-4:

$$\ell = \frac{q}{C_v\sqrt{\Delta P_v}} \tag{9-9}$$

Using Eq. 9-9 and values of ΔP_v from Eq. 9-8, the installed valve characteristic curve can be plotted (Fig. 9.14).

(b) From (9-7) and (9-8), $\Delta P_v = 3.7$ psi. The rated Cv at 110% of q_d can be calculated.

$$C_v = \frac{220}{\sqrt{3.7}} = 114.4$$

Use a value of $C_v = 115$. For the equal percentage valve, rearrange Eq. 9-2 as follows:

$$\frac{q}{C_v\sqrt{\Delta P_v}} = R^{\ell-1} \tag{9-10}$$

or

$$\ell = 1 + \log\left(\frac{q}{C_v\sqrt{\Delta P_v}}\right)\Big/\log R \tag{9-11}$$

Substituting $C_v = 115$, $R = 50$, and values of q and ΔP_v yields the installed characteristic curve in Fig. 9.14.

(c) $C_v = 1.2(115) = 138$

(d) $C_v = 0.8(115) = 92$

Using the installed characteristics in Fig. 9.14, note that the maximum flow rate that could be achieved in this system (negligible pressure drop across the valve) would have a pressure drop of 40 psi across the heat exchanger:

$$\left(\frac{q_{\max}}{200}\right)^2 = \frac{40}{30} \tag{9-12}$$

From these results we conclude that an equal percentage valve with $C_v \approx 115$ would give a reasonably linear installed characteristic over a large range of flows and have sufficient capacity to accommodate flows as high as 110% of the design flow rate.

Sometimes it is advantageous to use two control valves, even though there is only a single controlled variable. As an example, consider a jacketed batch reactor. For an exothermic reaction, it may be necessary to initially heat the batch to a temperature at which the reaction rate is significant by passing a hot liquid through the jacket. But as the exothermic reaction proceeds, it is necessary to remove heat by passing coolant through the jacket in order to control the batch temperature. One solution is to use a *split range control strategy* (Chapter 16) with the heating and cooling valves operating in parallel.

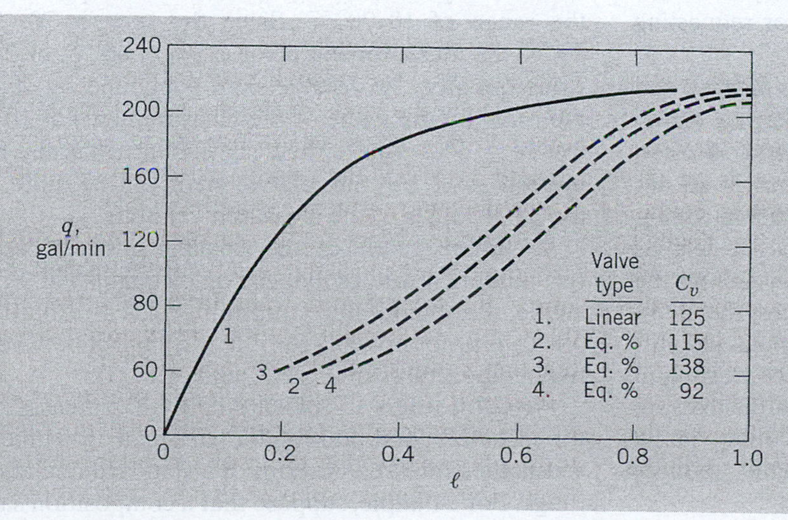

Figure 9.14 Installed valve characteristics for Example 9.2.

For flow control problems where the nominal flow rate is very large, two control valves, one large and one small, can be used to provide tight flow control (see Exercise 9.8).

9.3 SIGNAL TRANSMISSION AND DIGITAL COMMUNICATION

Electronic controllers (digital or analog) can be located relatively far from their instruments with little concern for the impedance of the intervening transmission lines or for the time of transmission, which for all practical purposes is instantaneous. Multipair shielded cable has traditionally been used for this purpose. An advantage of such two-wire systems is that the power supply can be located *in the loop*; thus, separate wiring is not required. Most transmitter analog signals are in the form of current rather than voltage, because voltage is affected by wire and connector resistances, which change with wire length, temperature and aging. Voltage-level control and instrumentation signals (e.g., 1 to 5 VDC) are better restricted to situations where short distances are involved. Very careful practice must be followed in wiring and terminating analog signals to prevent biasing, attenuation, or inducing noise (e.g., 60-Hz noise) in the transmission lines. Usually, shielded coaxial cable is required to minimize these effects.

Pneumatic pressure signals between instruments are transmitted by means of tubing, usually 1/4- or 3/8 in. diameter. The propagation of a signal changing in time through such a medium is limited by dynamic accuracy considerations (Section 9.4) to 100 or 200 meters at most.

Signals from digital instruments and controllers are usually transmitted in digital format as a sequence of on-off pulses. Digital transmission is carried out over a single *data highway* that is linked in serial or daisy-chain fashion to all instruments and controllers. A microcomputer built into each instrument or controller is responsible for communicating periodically over the highway, either directing information to or requesting information from some other device.

Various field network protocols such as *fieldbus* and *profibus* provide the capability of transferring digital information and instructions among field devices, instruments, and control systems. Fieldbus is an all-digital, serial, two-wire communications system configured like a local area network (Berge, 2002). The fieldbus software mediates the flow of information among the components. Fieldbus technology replaces the dedicated set of wires required for each instrument and mitigates the problem of electrical interference existing in 4–20 mA signal transport. Multiple digital devices can be connected and communicate with each other via the digital communication line, which greatly reduces wiring cost for a typical plant.

In recent years, there has been considerable interest in using wireless networks for process control applications (Caro, 2008; Song et al., 2006). For large industrial plants, a major financial benefit of wireless devices is the savings associated with the design and implementation of complex wired networks. However, there are major challenges for wireless communication that include security concerns and line-of-sight limitations. Most process control applications of wireless networks have been for monitoring, but control applications are anticipated in the future (Song et al., 2006).

Appendix A contains additional information on digital communication and wireless networks for process control.

9.4 ACCURACY IN INSTRUMENTATION

The accuracy of control instrumentation is very important with accuracy requirements inherently related to control system objectives. For example, cooling water flow errors on the order of 10% (of the measured flow rate) might be acceptable in a control loop regulating the temperature of a liquid leaving a condenser, as long as the measurements are simply biased from the true value by this constant amount. On the other hand, errors in the feed flow rate to a process on the order of 1 or 2% might be unacceptable if throughput/inventory calculations must be made with these data.

9.4.1 Terms Used to Describe Instrument Accuracy

We now introduce several terms that are commonly associated with the accuracy of process instruments. The *measurement error* (or *error*) is the difference between the true value and the measured value. Instrument vendors often express accuracy as a *percentage of full scale* (% FS) where the term *full scale* refers to the span of the instrument. Suppose that the % FS error of a temperature transmitter is reported as 1% and the zero and span are adjusted so that the instrument operates over the range of 10–70 °C. Since the span is 70−10 = 60 °C, the measurement error is 1% of 60 °C, or 0.6 °C. Consequently, the *relative error* (obtained by dividing the error by the value of the measurement) at 10 °C is 0.6/10 = 6%. Thus, when instrument accuracy is expressed as % FS, the relative error can be quite large for small values of the measured variable.

Resolution refers to the smallest interval between two numerical values that can be distinguished. For example, if a temperature transmitter has a resolution of 0.1 °C, it is not possible for it to distinguish between actual temperatures of 21.62 °C and 21.67 °C.

Precision refers to the variability of a measurement for specified conditions and a particular instrument. It is usually expressed in terms of a standard deviation or range. For example, suppose that a composition sample

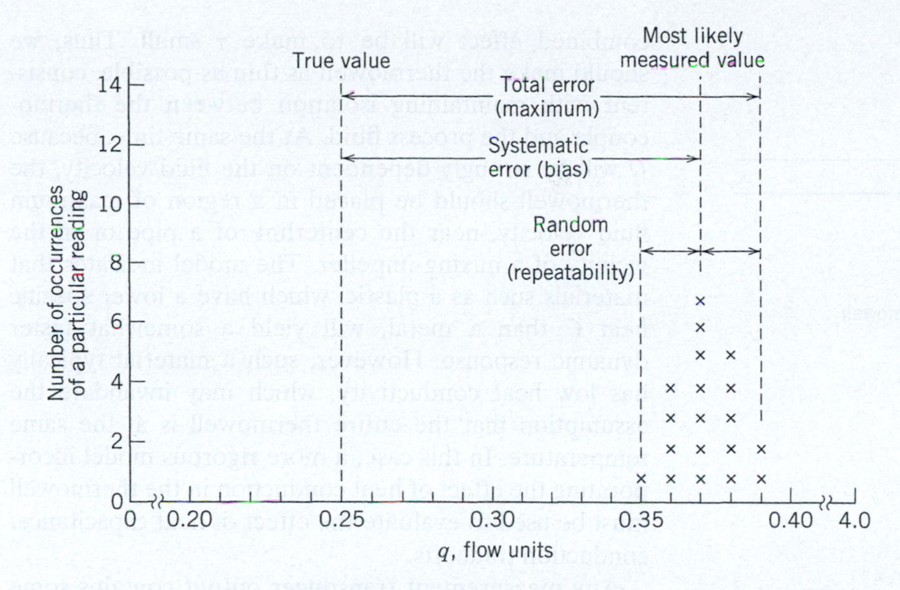

Figure 9.15 Analysis of types of error for a flow instrument whose range is 0 to 4 flow units.

was carefully prepared and divided into four parts, and that the composition of each part was measured using an analyzer. If the four measurements of a key component are 21.3, 22.7, 20.6, and 21.5%, the analyzer precision could be expressed as the range, 22.7−20.6=2.1%, or as the standard deviation, 0.87%.

Repeatability is similar to precision but refers to the variability of replicate measurements in a set of data. The variability is due to random errors from both known and unknown sources. The variability can be expressed as a range or standard deviation.

Bias refers to a constant error in the data due to a deterministic cause rather than random variations. A thermocouple measurement in a vessel could be consistently lower than the actual fluid temperature because of conduction heat losses, in which the thermocouple is in contact with the vessel wall.

Figure 9.15 displays an error analysis for a dataset consisting of replicate measurements at known flow rates. The dataset contains both a systematic error (bias) and random errors.

Manufacturers of measurement devices always state the accuracy of the instrument. However, these statements specify reference conditions at which the measurement device will perform with the stated accuracy, with temperature and pressure most often appearing in the reference conditions. When the measurement device is used at other conditions, the accuracy is affected. Manufacturers usually provide statements indicating how accuracy is affected when the conditions of use deviate from the reference conditions. Whenever a measurement device provides data for real-time optimization, accuracy is very important (see Chapter 19).

9.4.2 Calibration of Instruments

Any important instrument should be calibrated both initially (before commissioning) and periodically (as it remains in service). In recent years, the use of *smart sensors* has become more widespread. These devices incorporate a microcomputer as part of the sensor/transmitter, which can greatly reduce the need for in-service calibration and checkout. Their key features are:

1. Checks on the internal electronics, such as verifying that the voltage levels of internal power supplies are within specifications.
2. Checks on environmental conditions within the instruments.
3. Compensation of the measured value for conditions such as temperature and pressure within the instrument.
4. Linearizing the output of the transmitter if necessary (e.g., square root extraction of the differential pressure for a head-type flow transducer) can be done within the instrument instead of within the control system.
5. Configuring the transmitter from a remote location, such as changing its zero or span.
6. Automatic recalibration of the transmitter.

9.4.3 Dynamic Measurement Errors

In this section, the dynamic measurement error associated with a temperature sensor is analyzed. Figure 9.16 shows a thermocouple placed in a metal thermowell with mass m and specific heat C. The dynamic lag introduced by the thermowell/thermocouple combination can be easily estimated if several simplifying assumptions are made. In particular, assume that the well and thermocouple are always at the same temperature T_m, which can be different from the surrounding fluid temperature T. Further assume that heat is transferred only between the fluid and the well (there are no ambient

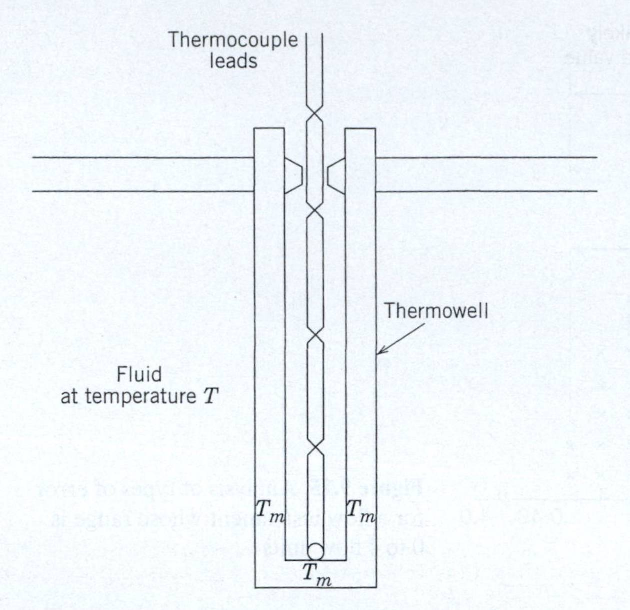

Figure 9.16 Schematic diagram of a thermowell/thermocouple.

losses from the thermowell due to conduction along its length to the environment). An energy balance on the thermowell gives

$$mC\,\frac{dT_m}{dt} = UA(T - T_m) \tag{9-13}$$

where U is the heat transfer coefficient and A is the heat transfer area. Rearranging gives

$$\frac{mC}{UA}\frac{dT_m}{dt} + T_m = T \tag{9-14}$$

Converting to deviation variables and taking the Laplace transform gives

$$\frac{T'_m(s)}{T'(s)} = \frac{1}{\tau s + 1} \tag{9-15}$$

with $\tau \triangleq mC/UA$.

The transfer function in Eq. 9-15 indicates that the dynamic measurement lag of the sensor will be minimized if the thermal capacitance of the well (mC) is made as small as possible and if UA is made large. The combined effect will be to make τ small. Thus, we should make the thermowell as thin as possible, consistent with maintaining isolation between the thermocouple and the process fluid. At the same time, because U will be strongly dependent on the fluid velocity, the thermowell should be placed in a region of maximum fluid velocity, near the centerline of a pipe or in the vicinity of a mixing impeller. The model indicates that materials such as a plastic, which have a lower specific heat C than a metal, will yield a somewhat faster dynamic response. However, such a material typically has low heat conductivity, which may invalidate the assumption that the entire thermowell is at the same temperature. In this case, a more rigorous model incorporating the effect of heat conduction in the thermowell must be used to evaluate the effect of heat capacitance/conduction tradeoffs.

Any measurement transducer output contains some dynamic error; an estimate of the error can be calculated if transducer time constant τ and the maximum expected rate of change of the measured variable are known. For a ramp input, $x(t) = at$, and a first-order dynamic model (see Eq. 9-15), the transducer output y is related to x by:

$$Y(s) = \frac{1}{\tau s + 1}X(s) = \frac{1}{\tau s + 1}\frac{a}{s^2} \tag{9-16}$$

The ramp response $y(t)$ of a first-order system was obtained in Eqs. 5-19 through 5-21. The maximum deviation between input and output is $a\tau$ (obtained when $t >> \tau$), as shown in Fig. 5.5. Hence, as a general result, we can say that the maximum dynamic error that can occur for any instrument with *first-order* dynamics is

$$\epsilon_{max} = |y(t) - x(t)|_{max} = a\tau$$

Clearly, by reducing the time constant, the dynamic error can be made negligibly small.

In general, measurement and transmission time constants should be less than one-tenth the largest process time constant, preferably much less, to reduce dynamic measurement errors. The dynamics of measurement, transmission, and final control elements also significantly limit the speed of response of the controlled process. Thus, it is important that the dynamics of these components be made as fast as is practical or economical.

SUMMARY

In this chapter we have considered the instrumentation required for process control applications. Sensors provide information about process output variables (in a form that can be transmitted to the controllers), and the final control elements are used to manipulate process input variables based on signals from the controllers. The technology trend is for more microcomputer-based instrumentation and digital transmission of information, which are considered in more detail in Appendix A.

Another major trend is the increasing integration of sensing elements into silicon chip microcircuitry.

Using this approach, we can now measure pressure, temperature, ion and gas concentration, radiation level, and other important process variables with sensors that directly incorporate all circuitry needed to self-compensate for environmental changes and to yield a linear output that is suitably amplified for transmission to standard electronic or digital controllers. These new sensors offer the advantage of small size, greatly reduced prices, and virtually no mechanical parts to wear out.

REFERENCES

Baker, R. C., *Flow Measurement Handbook*, Cambridge University Press, New York, 2000.

Berge, J., *Fieldbuses for Process Control Engineering, Operation and Maintenance*, ISA, Research Triangle Park, NC, 2002.

Blevins, T. L., G. K. McMillan, W. K. Wojsznis, and M. W. Brown, *Advanced Control Unleashed*, ISA, Research Triangle Park, NC, 2003.

Borden, G. (ed.), *Control Valves*, ISA, Research Triangle Park, NC, 1998.

Caro, D., *Wireless Networks for Industrial Automation* 3rd ed., ISA, Research Triangle Park, NC, 2008.

Chow, A. W., Lab-on-a-Chip: Opportunities for Chemical Engineering, *AIChE J.*, **48**, 1590 (2002).

Connell, R., *Process Instrumentation Applications Manual*, McGraw-Hill, New York, 1996.

Dakin, J., and B. Culshaw (eds.), *Optical Fiber Sensors: Applications, Analysis, and Future Trends*, Vol. IV, Artech House, Norwood, MA, 1997.

Edgar, T. F., C. L. Smith, F. G. Shinskey, G. W. Gassman, P. J. Schafbuch, T. J. McAvoy, and D. E. Seborg, Process Control, Section 8 in *Perry's Chemical Engineers Handbook*, 8th ed., D. W. Green and R. H. Perry (ed.), McGraw-Hill, New York, 2008.

Emerson Process Management, *Control Valve Handbook*, 4th ed., http://www.documentation.emersonprocess.com/groups/public/documents/book/cvh99.pdf, 2005.

Fitzgerald, B., *Control Valves for the Chemical Process Industries*, McGraw-Hill, New York, 1995.

Henry, M. P., D. W. Clarke, N. Archer, J. Bowles, M. J. Leahy, R. P. Liu, J. Vignos, and F. B. Zhou, A Self-Validating Digital Control Mass-Flow Meter: An Overview, *Control Engr. Practice*, **8**, 487 (2000).

Instrumentation Symbols and Identification, Standard *ISA-5.1-1984 (R1992)*, International Society of Automation (ISA), Research Triangle Park, NC, 1992.

Johnson, C. D., *Process Control Instrumentation Technology*, 8th ed., Prentice-Hall, Upper Saddle River, NJ, 2008.

Krohn, D. A., *Fiber Optic Sensors: Fundamentals and Applications*, 3d ed., ISA, Research Triangle Park, NC, 2000.

Lipták, B. (ed.), *Instrument Engineers' Handbook,* 4th ed., *Vol. I: Process Measurement and Analysis*; *Vol. 2: Process Control*, Radnor, PA, 2003.

Lipták, B., How to Select Control Valves, Parts 1-3, *Control Magazine*, available at www.controlglobal.com (July 14, 2006).

Luyben, W. L., and M. L. Luyben, *Essentials of Process Control*, McGraw-Hill, New York, 1997.

Nichols, G. D., *On-Line Process Analyzers*, Wiley, New York, 1988.

Riggs, J. B., and M. N. Karim, *Chemical and Biochemical Process Control*, 3rd ed., Ferret Press, Lubbock, TX, 2006.

Scott, D. M., *Industrial Process Sensors*, CRC Press, Boca Raton, FL, 2008.

Shuler, M. P., and F. Kargi, *Bioprocess Engineering*, Prentice-Hall, Upper Saddle River, NJ, 2002.

Soloman, S., *Sensors Handbook*, McGraw-Hill, New York, 1999.

Song, J. A. K. Mok, D. Chen, and M. Nixon, Challenges of Wireless Control in Process Industry, *Workshop on Research Directions for Security and Networking in Critical Real-Time and Embedded Systems*, San Jose, CA, 2006.

Spitzer, D. W., *Flow Measurement*, 3rd ed., ISA, Research Triangle Park, NC, 2004.

EXERCISES

9.1 Several linear transmitters have been installed and calibrated as follows:

Flow rate:	400 gal/min	$\rightarrow$ 15 psig	pneumatic
	0 gal/min	$\rightarrow$ 3 psig	transmitter
Pressure:	30 in Hg	$\rightarrow$ 20 mA	current
	10 in Hg	$\rightarrow$ 4 mA	transmitter
Level:	20 m	$\rightarrow$ 5 VDC	voltage
	0.5 m	$\rightarrow$ 1 VDC	transmitter
Concentration:	20 g/L	$\rightarrow$ 10 VDC	voltage
	2 g/L	$\rightarrow$ 1 VDC	transmitter

(a) Develop an expression for the output of each transmitter as a function of its input. Be sure to include appropriate units.

(b) What is the gain of each transmitter? zero? span?

9.2 A process instrumentation diagram for a flash drum is shown in Fig. E9.2. Steam is condensed in a steam coil to vaporize a portion of the liquid feed, and the liquid product is removed by a pump. There are control valves for the steam flow, vapor product, liquid product, feed flow, and steam chest (which allows the steam chest to be rapidly evacuated in emergency situations). Determine whether the five valves should be fail-close (F/C) or fail-open (F/O) for safe operation, for each of three cases:

(a) The safest conditions are achieved by the lowest temperature and pressure in the flash vessel.

(b) Vapor flow to downstream equipment can cause a hazardous situation.

(c) Liquid flow to downstream equipment can cause a hazardous situation.

Discuss various scenarios of air failure (or power failure).

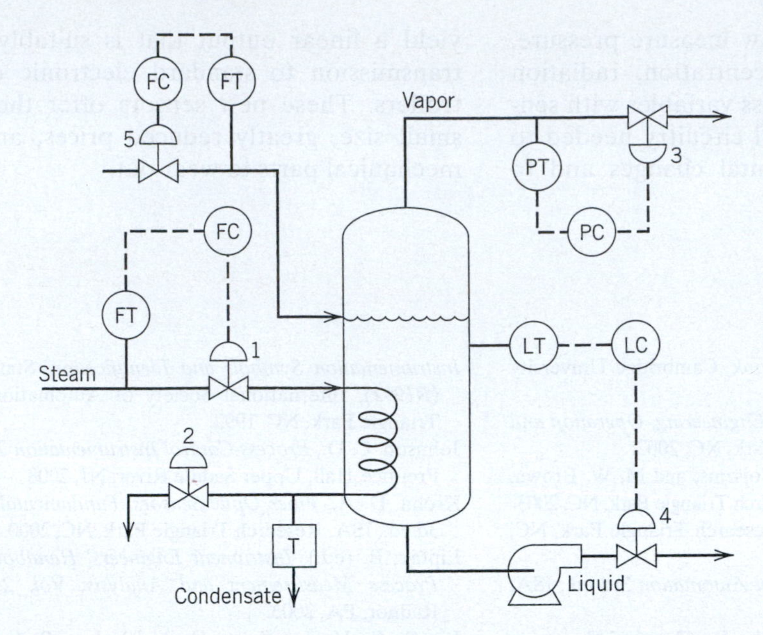

Figure E9.2

9.3 Suppose that the temperature in an exothermic continuous stirred-tank reactor is controlled by manipulating the coolant flow rate using a control valve. A PID controller is used and is well-tuned. Which of these changes could adversely affect the stability of the closed-loop system? Briefly justify your answers.

(a) The span of the temperature transmitter is increased from 20 °C to 40 °C.

(b) The zero of the temperature transmitter is increased from 15 °C to 20 °C.

(c) The control valve trim is changed from linear to equal percentage.

(d) The feed flow rate is doubled.

9.4 Chilled ethylene glycol (sp gr = 1.11) is pumped through the shell side of a condenser and a control valve at a nominal flow rate of 200 gal/min. The total pressure drop over the entire system is constant. The pressure drop over the condenser is proportional to the square of the flow rate and is 30 psi at the nominal flow rate. Make plots of flow rate versus vs. stem position ℓ for linear and equal percentage control valves, assuming that the valves are set so that $f(\ell) = 0.5$ at the nominal flow rate. Prepare these plots for the situations where the pressure drop over the control valve at the design flow is

(a) 5 psi

(b) 30 psi

(c) 90 psi

What can you conclude concerning the results from these three sets of design conditions? In particular, for each case, comment on linearity of the installed valve, ability to handle flow rates greater than nominal, and pumping costs.

9.5 A pneumatic control valve is used to adjust the flow rate of a petroleum fraction (specific gravity = 0.9) that is used as fuel in a cracking furnace. A centrifugal pump is used to supply the fuel, and an orifice meter/differential pressure transmitter is used to monitor flow rate. The nominal fuel rate to the furnace is 320 gal/min. Select an equal percentage valve that will be satisfactory to operate this system. Use the following data (all pressures in psi; all flow rates in gal/min):

(a) Pump characteristic (discharge pressure):

$$P = (1 - 2.44 \times 10^{-6} q^2) P_{de}$$

where P_{de} is the pump discharge pressure when the pump is dead ended (no flow).

(b) Pressure drop across the orifice:

$$\Delta P_O = 1.953 \times 10^{-4} q^2$$

(c) Pressure drop across the furnace burners:

$$\Delta P_b = 40$$

(d) R for the valve: 50

(e) Operating region of interest:

$$250 \leq q \leq 350$$

This design attempt should attempt to minimize pumping costs by keeping the pump capacity (related to P_{de}) as low as possible. In no case should $\Delta P_v / \Delta P_s$ be greater than 0.33 at the nominal flow rate. Show, by means of a plot of the installed valve characteristic (q vs. ℓ), just how linear the final design is.

9.6 Consider the evaporator and control system in Figure 13.6. Briefly answer the following questions:

(a) Should each control valve be fail-open (FO) or fail-close (FC)?

(b) Should each PI controller be direct-acting or reverse-acting?

9.7 A theoretical force balance for the control valve shown in Fig. 9.7 can be expressed as

$$PA_D + M\frac{g}{g_c} - Kx - P_f A_p - R\frac{dx}{dt} = \frac{M}{g_c}\frac{d^2x}{dt^2}$$

where

M = mass of movable stem = 10 lb_m
P = valve air pressure input
A_D = diaphragm area
g, g_c = gravity, conversion constants
K = spring constant = 3,600 lb_f/ft
P_f = fluid pressure
A_p = valve plug area
R = coefficient of friction (stem to packing) = 15,000 lb_f/ft/s
x = valve position

Assuming the second-order differential equation is linear, find values of the coefficients of the equation (in deviation variable form) and determine whether the valve dynamic behavior is overdamped or underdamped.

9.8 It has been suggested that the liquid flow rate in a large diameter pipeline could be better regulated by using two control valves instead of one. Suppose that one control valve has a large C_v value, that the other has a small C_v value, and that the flow controller will primarily adjust the smaller valve while also making occasional adjustments to the large valve, as needed. Which of the two alternative configurations seems to be the more promising: placing the control valves in series (Configuration I), or in parallel (Configuration II)? Briefly justify your answer.

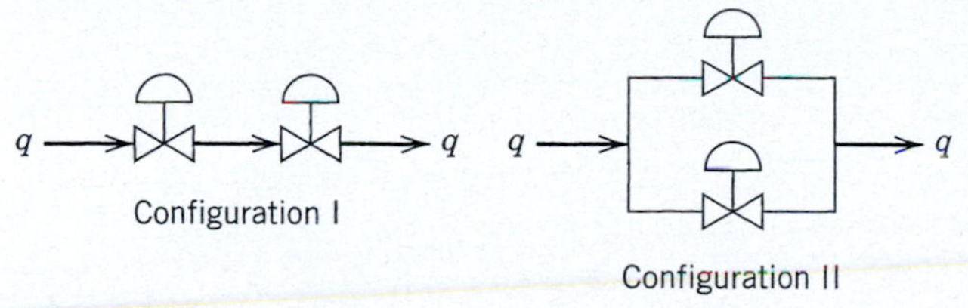

Figure E9.8

9.9 A temperature transmitter is used to measure the liquid temperature in a bioreactor. A steady-state calibration of this instrument yields the following data:

Temperature, °C	Measurement, mA
0	4.0
100	8.1
200	11.9
300	16.1
400	20.0

A process engineer runs a test on the reactor under controlled conditions in which its temperature is changed by +3 °C/min. The transmitter output was recorded during this test, converted to °C, and compared with a standard thermometer which is known to be accurate and to have a time constant of 20 s. The test data are

Time from	Temperature (°C)	
Start of Test, min	Std. Thermometer	T/C Transmitter
2.0	111.8	107.8
3.0	115.1	111.0
4.0	117.9	114.1
5.0	121.1	117.0

For steady-state conditions, the standard thermometer and thermocouple-transmitter outputs are identical. Assuming that the transmitter/thermocouple can be modeled by a first-order transfer function, find K and τ.

9.10 An engineer sets the pressure in a supply tank using a very accurate manometer as a guide and then reads the output of a 20-psig pressure gauge attached to the tank as 10.2 psig. Sometime later she repeats the procedure and obtains values of 10.4 and 10.3 psig. Discuss the following gauge properties:

(a) Precision
(b) Accuracy
(c) Resolution
(d) Repeatability

Express these answers on a percentage of full-scale basis.

9.11 A process temperature sensor/transmitter in a fermentation reactor exhibits second-order dynamics with time constants of 1 s and 0.1 s. If the quantity being measured changes at a constant rate of 0.1 °C/s, what is the maximum error that this instrument combination will exhibit? What is the effect of neglecting the smaller time constant? Plot the response.

Chapter 10

Process Safety and Process Control

CHAPTER CONTENTS

Process safety has been a primary concern of the process industries for decades. But in recent years, safety issues have received increased attention as a result of increased public awareness of potential risks, stricter legal requirements, and the increased complexity of modern industrial plants. Chemical engineers have a special role to perform in assuring process safety. As Turton et al. (2008) have noted, "As the professional with the best knowledge of the risks of a chemical processing operation, the chemical engineer has a responsibility to communicate those risks to employers, employees, clients, and the general public." Furthermore, in the American Institute of Chemical Engineers (AIChE) Code of Ethics, the first responsibility of chemical engineers is to "hold paramount the safety, health, and welfare of the public in performance of their professional duties." Professional societies that include the AIChE, the Institution of Chemical Engineers (London), and the International Society of Automation (ISA) have played a leading role in developing safety standards and reference materials. For example, the AIChE Center for Chemical Process Safety (CCPS) has published a number of books (AIChE, 1993, 2001, 2007) and a journal devoted to safety, *Process Safety Progress.*

The overall safety record of the process industries has been quite good, despite several highly publicized plant incidents. In fact, the accident and loss statistics for the chemicals and allied products industries are among the best of the manufacturing sectors (Crowl and Louvar, 2002). But it is not possible to eliminate risk entirely, and unfortunate accidents occasionally occur (Kletz, 1995; Mannan and Lees, 2005; Crowl and Louvar, 2002; Banerjee, 2003).

Process safety is considered at various stages during the development and operation of a process:

1. An initial safety analysis is performed during the preliminary process design.
2. A very thorough safety review is conducted during the final stage of the process design, using techniques such as hazard and operability (HAZOP) studies, failure mode and effect analysis, and fault tree analysis (AIChE, 1993; Kletz, 1999; Crowl and Louvar, 2002).
3. After plant operation begins, HAZOP studies are conducted on a periodic basis in order to identify and eliminate potential hazards.
4. Many companies require that any proposed plant change or change in operating conditions requires formal approval via a Management-of-Change (MOC) process that considers the potential impact of the change on the safety, environment, and health of the workers and the nearby communities. Furthermore, facilities that process significant quantities of hazardous materials must comply with government regulations from agencies such as the Environmental Protection Agency (EPA) and the Occupational Safety and Health Administration (OSHA).
5. After a serious industrial plant accident or incident in the United States, a thorough investigation is conducted by an independent government agency, the Chemical Safety Board (CSB), to determine the root cause of the incident, assess responsibility, and suggest safety improvements. The subsequent accident report is made available on an Internet site: www.chemsafety.gov.

The process control system, instrumentation, and alarms play critical roles in ensuring plant safety. But if they do not function properly, they can be a contributing factor or even a root cause of a serious incident (Kletz et al., 1995). In this chapter, we provide an overview of the influence of process dynamics and control on process safety.

10.1 LAYERS OF PROTECTION

In modern industrial plants, process safety relies on the principle of multiple layers of protection (AIChE, 1993, 2001; ISA, 1996). A typical configuration is shown in Figure 10.1. Each layer of protection consists of a grouping of equipment and/or human actions. The layers of protection are shown in the order of activation that occurs as a plant incident develops, with the most effective layers used first. The basic concept is that an incident should be handled at the lowest possible layer. In the interior of the diagram, the process design itself provides the first level of protection. The next two layers consist of the *basic process control system* (BPCS), augmented with two levels of alarms and operator supervision or intervention. An alarm indicates that a measurement has exceeded its specified limits and may require either operator intervention or an automated response.

The fourth layer consists of a *safety instrumented system* (SIS) and/or an *emergency shutdown* (ESD) *system*. The SIS, formerly referred to as a safety interlock system, automatically takes corrective action when the process and BPCS layers are unable to handle an emergency. For example, the SIS could automatically turn off the reactant and catalyst pumps for a chemical reactor after a high temperature alarm occurs. The SIS is described in Section 10.1.4.

In the fifth layer of protection, passive relief devices, such as rupture disks and relief valves, provide physical protection by preventing excessive pressures from being generated within process vessels and pipelines. If overpressurization occurs, the relief valve or rupture disk opens and the fluid is vented to an appropriate location, such as a combustion flare stack, incinerator, scrubber, waste treatment facility, or the atmosphere. Such passive devices operate independently of the SIS system.

Finally, dikes are located around process units and storage tanks to contain liquid spills. Emergency response plans are used to address extreme situations, inform the nearby community, and implement evacuation plans, if necessary.

The functioning of the multiple-layer protection system can be summarized as follows (AIChE, 1993): "Most failures in well-designed and operated chemical processes are contained by the first one or two protection layers. The middle levels guard against major releases and the outermost layers provide mitigation response to very unlikely major events. For major hazard potential, even more layers may be necessary."

It is evident from Figure 10.1 that process control and automation play important roles in ensuring process safety. In particular, many of the protective layers in Figure 10.1 involve instrumentation and control equipment. Furthermore, the process and instrument dynamics are key considerations in safety analysis. For example, after a major incident develops, how much time elapses before the sensors detect the new conditions and corrective action is taken? If the incident remains undetected, how long will it take for an emergency situation to result?

Next, we consider the role of process control and instrumentation in the protection layers of Fig. 10.1.

10.1.1 The Role of the Basic Process Control System

The basic process control system (BPCS) consists of feedback and feedforward control loops that regulate process variables such as temperatures, flow rates, liquid levels, and pressures. Although the BPCS typically provides

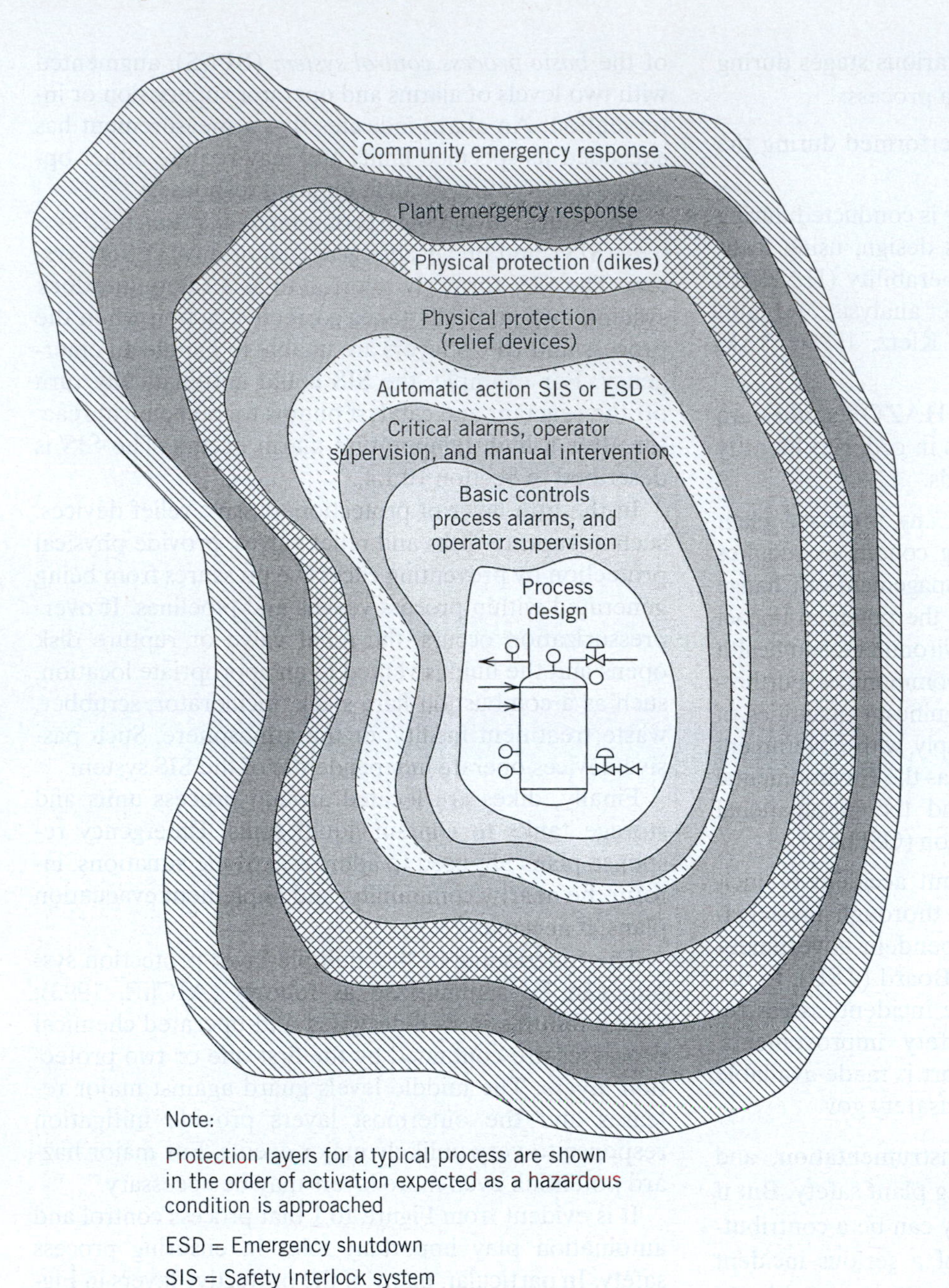

Figure 10.1 Typical layers of protection in a modern chemical plant (AIChE, 1993).

satisfactory control during routine process operation, it may not do so during abnormal conditions. For example, if a controller output or manipulated variable saturates, the controlled variable may exceed allowable limits. Similarly, the failure or malfunction of a component in the feedback loop, such as a sensor, control valve, or transmission line, could cause the process operation to enter an unacceptable region. In this case, the operation of the process is transferred to the SIS system. Typical component failure rates are shown in Table 10.1, expressed as the average number of faults per year.

10.1.2 Process Alarms

The second and third layers of protection rely on process alarms to inform operators of abnormal situations. A block diagram for an alarm system is shown in Fig. 10.2. An alarm is generated automatically when a measured variable exceeds a specified high or low limit. The logic block is programmed to take appropriate corrective action when one or more alarm switches are triggered. After an alarm occurs, the logic block activates an annunciator, either a visual display or an audible sound such as a horn or bell. For example, if a reactor temperature exceeds a high alarm limit, a light might flash on a computer screen, with the color indicating the alarm priority (e.g., yellow for a less serious situation, red for a critical situation). An alarm continues until it is acknowledged by an operator action, such as pressing a button or a key on a computer keyboard. If the alarm indicates a potentially hazardous situation, an automated corrective action is initiated by the SIS. Two

Table 10.1 Failure Rates for Selected Components (Mannan and Lees, 2005)

Instrument	Failure Frequency (faults per year)
Control valve	0.60
Valve positioner	0.44
Current/pressure transducer	0.49
Pressure measurement	1.41
Flow measurement (fluids)	
Orifice plate & D/P transmitter	1.73
Magnetic flowmeter	2.18
Temperature measurement	
Thermocouple	0.52
Mercury-in-steel thermometer	0.027
Controller (electronic)	0.29
Flow switch	1.12
Pressure switch	0.34
Alarm indicator lamp	0.044
Gas-liquid chromatograph	30.6

types of high and low alarm limits are widely employed. Warning limits are used to denote minor excursions from nominal values, whereas alarm limits indicate larger, more serious excursions.

Connell (1996) has proposed the following classification system for process alarms:

Type 1 Alarm: Equipment status alarm. Equipment status indicates, for example, whether a pump is on or off, or whether a motor is running or stopped.

Type 2 Alarm: Abnormal measurement alarm. If a measurement is outside specified limits, a high alarm or low alarm signal is triggered.

Type 3 Alarm: An alarm switch without its own sensor. This type of an alarm is directly activated by the process, rather than by a sensor signal; thus it is utilized in lieu of a sensor. Type 3 alarms are used for situations where it is not necessary to know the actual value of the process variable, only whether it is above or below a specified limit.

Figure 10.3 shows typical configurations for Type 2 and 3 alarms. In the Type 2 alarm system, the flow sensor/transmitter (FT) signal is sent to both a flow controller (FC) and a flow switch (FSL refers to "flow-switch-low"). When the measurement is below the specified low limit, the flow switch sends a signal to an alarm that activates an annunciator in the control room (FAL refers to "flow-alarm-low"). By contrast, for the Type 3 alarm system in Fig. 10.3b, the flow switch is self-actuated and thus does not require a signal from a flow sensor/transmitter. Type 3 alarms are preferred because they still function in case the sensor is out of service. Type 3 alarms are also used to indicate that an automatic shutdown system has tripped.

Type 4 Alarm: An alarm switch with its own sensor. A Type 4 alarm system has its own sensor, which serves as a backup in case the regular sensor fails. The alarm sensor also allows sensor drifts and failures to be detected more readily than a switch does.

Type 5 Alarm: Automatic shutdown or startup system. These important and widely used systems are described in the next section on Safety Instrumented Systems.

It is tempting to specify tight alarm limits for a large number of process variables. But this temptation should be resisted, because an excessive number of unnecessary alarms can result. Furthermore, too many alarms can be as detrimental as too few, for several reasons. First, frequent nuisance alarms tend to make the plant operators less responsive to important alarms. For example, if a tank is being filled during a plant startup, low-level alarms could occur repeatedly but be of no value to the plant operator. Second, in an actual emergency, a large number of unimportant alarms can obscure the root cause of the problem. Third, the relationships between alarms need to be considered. Thus, the design of an appropriate alarm management system is a challenging task.

10.1.3 Safety Instrumented System (SIS)

The SIS in Figure 10.1 serves as an emergency backup system for the BPCS. The SIS starts automatically when a critical process variable exceeds specified alarm limits that define its allowable operating range. Its initiation results in a drastic action, such as starting or stopping a pump or shutting down a process unit. Consequently, it is used only as a last resort to prevent injury to people or equipment. The term *Safety Interlock System* was previously used, but the newer term *Safety*

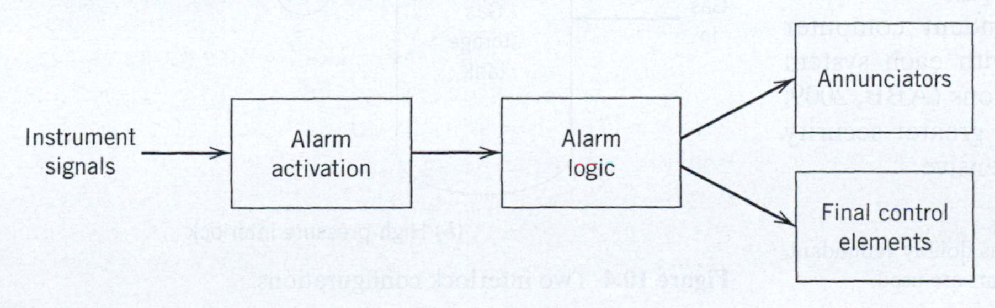

Figure 10.2 A general block diagram for an alarm system

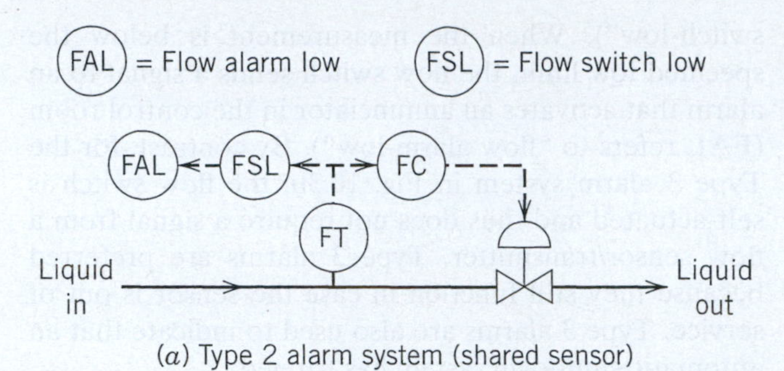

(a) Type 2 alarm system (shared sensor)

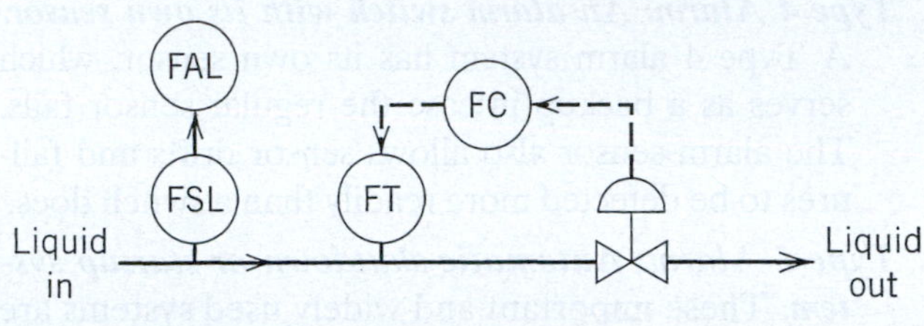

(b) Type 3 alarm system (based on a switch)

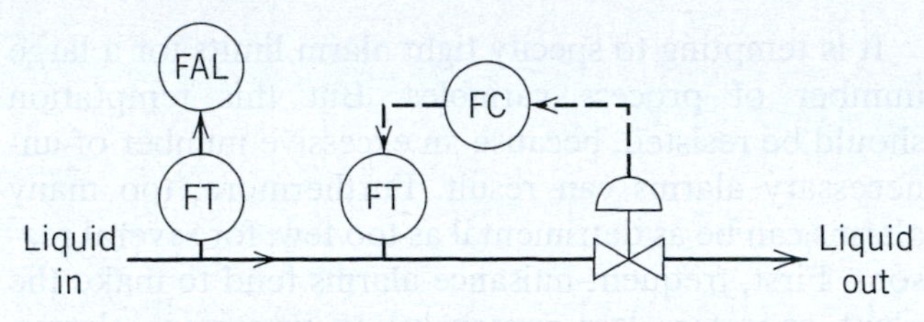

(c) Type 4 alarm system (separate sensor)

Figure 10.3 Alternative flow alarm configurations

Instrumented System is now preferred because it is more general.

It is very important that the SIS function independently of the BPCS; otherwise, emergency protection will be unavailable during periods when the BPCS is not operating (e.g., as a result of a malfunction or power failure). Thus, it is recommended that the SIS be physically separated from the BPCS and have its own sensors and actuators (AIChE, 1993).

Sometimes redundant sensors and actuators are utilized. For example, triply redundant[1] sensors are used for critical measurements, with SIS actions based on the median of the three measurements. This strategy prevents a single sensor failure from crippling SIS operation. The SIS also has a separate set of alarms so that the operator can be notified when the SIS initiates an action (e.g., turning on an emergency cooling pump), even if the BPCS is not operational.

As an alternative approach, redundant computer control systems can be employed, with each system having the BPCS, SIS, and ESD functions (ABB, 2009; Camp, 2009). This approach provides greater security but tends to be more complex and expensive.

[1] Arguably, this strategy should be referred to as doubly redundant, rather than triply redundant, because three sensors are used.

10.1.4 Interlocks and Emergency Shutdown Systems

The SIS operation is designed to provide automatic responses after alarms indicate potentially hazardous situations. The objective is to have the process reach a safe condition. The automatic responses are implemented via interlocks and via automatic shutdown and startup systems. Distinctions are sometimes made between safety interlocks and process interlocks; the latter are used for less critical situations to provide protection against minor equipment damage and undesirable process conditions, such as the production of off-specification product.

Two simple interlock systems are shown in Fig. 10.4. For the liquid storage system, the liquid level must stay above a minimum value in order to avoid pump damage such as cavitation. If the level drops below the specified limit, the low-level switch (LSL) triggers both an alarm (LAL) and a solenoid switch (S) (or solenoid) that turns the pump off. For the gas storage system in Fig. 10.4b, the solenoid valve is normally closed. But if the pressure of the hydrocarbon gas in the storage tank exceeds a specified limit, the high pressure switch (PSH) activates an alarm (PAH) and causes the solenoid valve to open fully,

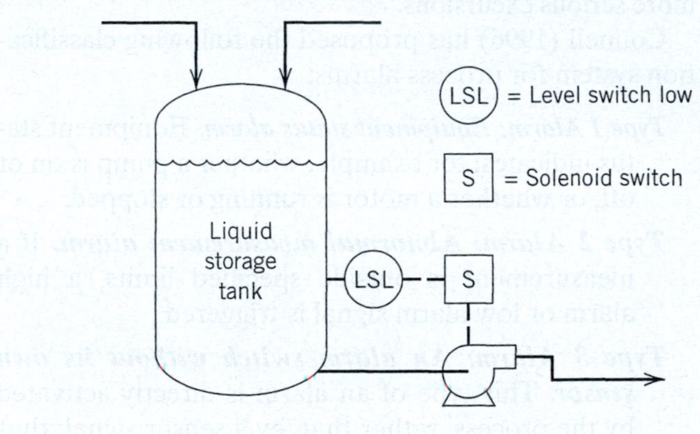

(a) Low-level interlock

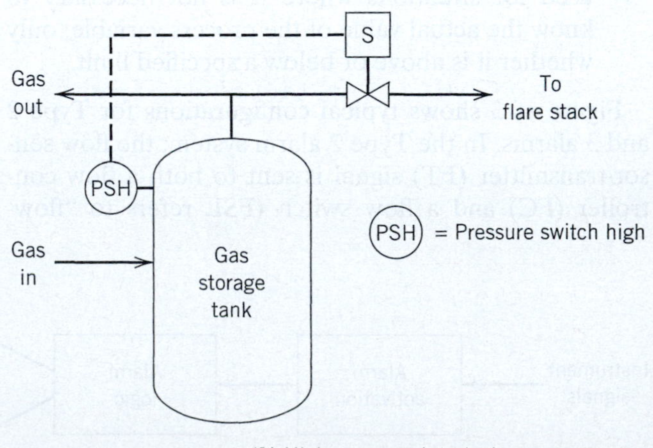

(b) High-pressure interlock

Figure 10.4 Two interlock configurations

thus reducing the pressure in the tank. For interlock and other safety systems, a solenoid switch can be replaced by a sensor transmitter if the measurement is required.

Another common interlock configuration is to locate a solenoid switch between a controller and a control valve. When an alarm is actuated, the solenoid trips and causes the air pressure in the pneumatic control valve to be vented; consequently, the control valve reverts to either its fail-open or fail-close position. Interlocks have traditionally been implemented as hard-wired systems that are independent of the control hardware. But, for most applications, software implementation of the interlock logic via a digital computer or a programmable logic controller is a viable alternative. Programmable logic controllers (PLCs) used for batch processes are considered in Chapter 22 and Appendix A.

If a potential emergency situation is very serious, the ESD system automatically shuts down or starts up equipment. For example, a pump would be turned off (or tripped) if it overheats or loses lubricant pressure. Similarly, if an exothermic chemical reaction starts to run away, it may be possible to add a quench material that stops the reaction quickly. For some emergency situations, the appropriate response is an automatic startup of equipment, rather than an automatic shutdown. For example, a backup generator or a cooling water pump could be started if the regular unit shuts down unexpectedly.

Although the ESD function is essential for safe process operation, unnecessary plant shutdowns and startups should be avoided, because they result in loss of production and generate off-specification product during the subsequent plant startup. Also, the emergency shutdowns and startups for a process unit involve risks and may activate additional safety systems that also shutdown other process units. Such nuisance shutdowns can create additional hazards. The use of redundant sensors can reduce unnecessary shutdowns.

10.2 ALARM MANAGEMENT

As industrial processes have become more complex and integrated, the topics of alarm management and abnormal situation management have become increasingly important, both from safety and economic considerations. For example, it has been estimated that preventable abnormal situations have an annual impact of over $10 billion on the operations of the U.S.-based petrochemical industry because of production loss, equipment damage, etc. (ASM, 2009). Alarm management and the occurrence of an excessive number of alarms during an abnormal situation (an alarm flood) have often been cited as contributing factors in investigations of major industrial plant accidents. In this section, we consider key elements of effective alarm management and response to abnormal situations.

Before computer control systems were available, a process alarm was connected to a light box and an audible alarm on a panel board. When an alarm occurred, the light box flashed and/or a horn sounded, thus attracting the operator's attention. The operator then acknowledged the alarm by pressing a button. Because panel board space was quite limited, only key process variables could be alarmed. But the introduction of modern computer control systems, beginning in the 1970s, drastically changed this situation. It became feasible to alarm virtually any measured variable and to display the alarms on a computer monitor. As a result, engineers are able to alarm very large numbers of process variables, which can inadvertently generate many spurious or nuisance alarms, an alarm overload. For example, during the Three Mile Island accident in 1979, the nuclear power plant operators were overwhelmed with information, much of which was irrelevant or incorrect. Fortunately, no one was injured in the accident (Nuclear Regulatory Commission, 2008).

Representative alarm experience for the process industries is compared with recommended guidelines in Table 10.2. A standing (or stale) alarm is one that remains in an alarm state for an extended period of time, such as 24 h. Unfortunately, the actual alarm rates in Table 10.2 exceed the guidelines by large margins. This alarm overload can distract or confuse operators, making it difficult for them to determine which alarms are most important for the current situation. Dire consequences can result, such as incorrect decisions, loss production, and unsafe process operation. Consequently, alarm management and abnormal situation management

Table 10.2 Alarm Rates Per Plant Operator in the Process Industries[†]

	EEMUA Guideline	Oil & Gas	Petrochemical	Power	Other
Average alarms per day	144	1200	1500	2000	900
Average standing alarms	9	50	100	65	35
Peak alarms (per 10-min interval)	10	220	180	350	180
Average alarms (per 10-min interval)	1	6	9	8	5
Alarm distribution % (low/medium/high)	80/15/5	25/40/35	25/40/35	25/40/35	25/40/35

[†]Engineering Equipment & Materials Users' Association (2007).

have become increasingly important issues in recent years. Alarm requirements are also the subject of government regulations (OSHA, 1996).

10.2.1 Alarm Guidelines

The design and maintenance of alarm management systems has been the subject of numerous articles, standards, and books (AIChE, 2007; Hollifield and Habibi, 2007; Katzel, 2007; ASM, 2009). Some important guidelines are shown in Table 10.3.

These alarm guidelines are illustrated by two common situations.

Tank-Filling Operation

A tank is filled using a pump on an inlet line. In order to avoid nuisance alarms, the low-level alarm should be suppressed (or the alarm limit changed) during this operation. Similarly, the low-flow alarm should be suppressed (or its limit changed) when the pump is turned off. Automatic adjustment of the alarm limits for routine tank filling is an example of state-based alarming (Guideline 11). It also satisfies Guideline 1 by not generating an alarm for a routine and known situation.

Table 10.3 Guidelines for Alarm Design and Management

1. Each alarm should have two important characteristics; it should result from an abnormal situation and require a specific operator response.
2. Alarm systems and displays should be designed with the user (the plant operator) in mind.
3. Each alarm should have a priority level to indicate its level of importance. Typically, two to four priority levels are employed. For example, the four levels could be designated as critical, emergency, high, and low (Hollifield and Habibi, 2007).
4. Protective alarms related to process safety or alarms that require an immediate response should be assigned the highest priority.
5. An alarm should continue until it is acknowledged.
6. Operators should respond to every alarm, regardless of priority, but alarm overload should be avoided.
7. Alarm suppression should only be allowed when there is a legitimate reason (e.g., an instrument is out of service). Also, automated reminders should be employed so that it is not possible for an operator to suppress an alarm and then forget about it (i.e., alarm storing is preferable to alarm suppression).
8. Each alarm should be logged with a time/date stamp and a record of the corresponding operator actions. Suppressed alarms should be included in the log.
9. Alarm histories should be reviewed on a regular basis in order to reduce standing alarms and nuisance alarms.
10. Alarm limit changes should be carefully considered and well-documented. Changes for critical protective alarms should be approved via an MOC order.
11. State-based alarming can greatly improve alarm management by eliminating nuisance alarms. It involves automatic adjustment of alarm limits to accommodate different process conditions, such as startups, shutdowns, changes in feedstocks or production rates (Hollifield and Habibi, 2007).

Regular and Backup Pumps

Consider a situation where there is a regular pump and a spare pump (Hollifield and Habibi, 2007). During typical operations, there could be zero, one, or two pumps running, with the latter situation occurring during a routine switch from one pump to the other. Alarms can be used to indicate the status of each pump. If the alarm strategy is to have an alarm activate when a pump is not running, one of the two pumps is always in an alarm state.

An alternative approach is to allow the operator to specify the number of pumps that are supposed to be running. Then digital logic could be used to activate an alarm only when the actual number of operating pumps differs from the desired number. This approach conforms with Guideline 1, because nuisance alarms are avoided during routine changes in process operations.

EXAMPLE 10.1

Consider the liquid surge system shown in Figure 10.5. A high-level alarm is used to prevent tank overflow. After a high alarm sounds, the operator has ten minutes to respond before the ESD system turns off the pumps on the inlet streams. The tank is 6 ft in diameter, 8 ft tall, and half-full at the nominal conditions. The density of the liquid is constant. The nominal values and ranges of the flow rates are shown as follows:

Flow rate	Nominal Value (ft^3/min)	Range (ft^3/min)
q_1	10	8–12
q_2	5	4–6
q	15	13–17

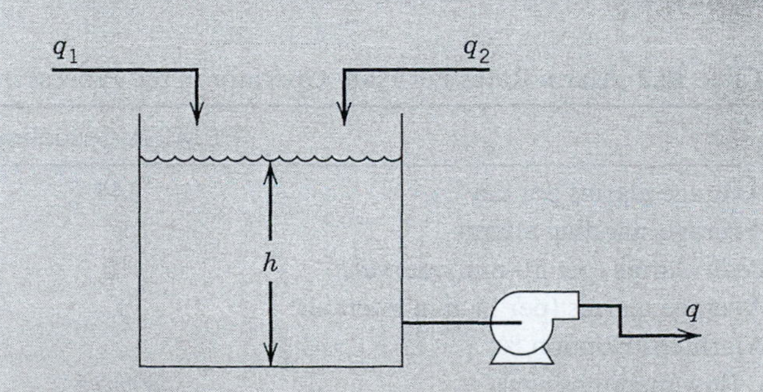

Figure 10.5 Liquid surge system.

(a) Assuming that the pumps are operating properly, determine an appropriate set-point value for the high-level alarm, that is, the numerical value of liquid level h that triggers the alarm.

(b) What additional safety features are required to handle unanticipated pump shutdowns?

SOLUTION

(a) For constant liquid density, the mass balance for the liquid in the tank is

$$\rho A \frac{dh}{dt} = \rho(q_1 + q_2 - q) \tag{10-1}$$

In order to specify the high alarm limit, consider the worst-case conditions that result in the largest rate of change, dh/dt. This situation occurs when the inlet flow rates are at their maximum values and the exit flow rate is a minimum. Substituting $q_1 = 12$ ft^3/min, $q_2 = 6$ ft^3/min, and $q = 13$ ft^3/min gives:

$$\frac{dh}{dt} = \frac{5 \text{ ft}^3/\text{min}}{\pi(3\text{ ft})^2} = 0.176 \text{ ft/ min} \tag{10-2}$$

Consequently, the maximum increase in liquid level that can occur during any 10-min period is,

$$\Delta h_{max} = (10 \text{ min})\ (0.176 \text{ ft/min}) = 1.76 \text{ ft} \tag{10-3}$$

Thus, the alarm limit should be at least 1.76 ft below the tank height. As a safety margin, choose 2 ft and set the high alarm limit at 6 ft.

(b) A pump failure could result in tank overflow or a tank level so low that the exit pump could be damaged. Thus, high and low level alarms should be connected to interlocks on the inlet and exit pumps, respectively.

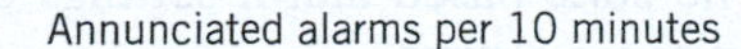

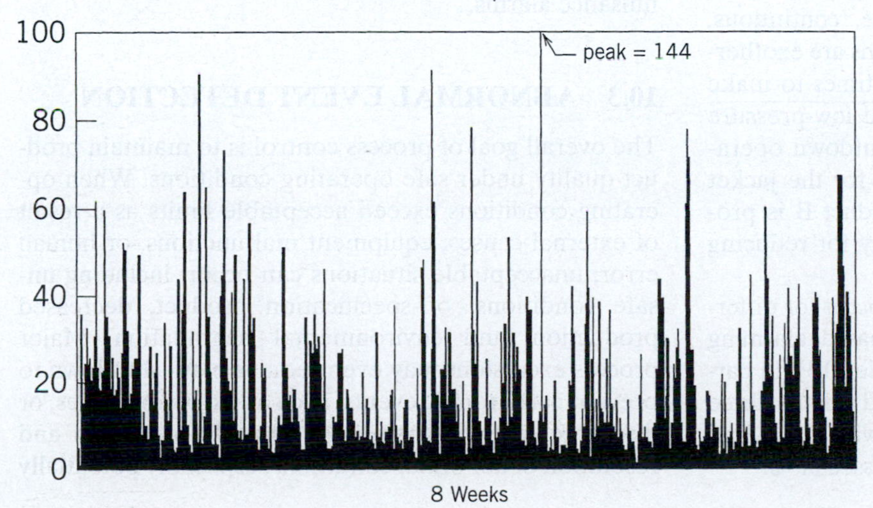

Figure 10.6 Example of annunciated graphs for one operator station over a period of 8 weeks (Hollifield and Habibi, 2007).

Table 10.4 Recommended Alarm Event Priority Distribution†

Alarm Priority	% of Total Alarms
Emergency	3–7% (5%)
High	15–25% (15%)
Low	70–80% (80%)

†Hollifield and Habibi (2007).

10.2.2 Alarm Rationalization

Because processes and process conditions change over time, it is necessary to review alarm system performance on a periodic basis. The review should be based on the alarm history and a number of key performance metrics. Some of the most important alarm metrics are the following (Jofriet, 2005):

(a) Frequency of alarm activations (e.g., alarms per day)

(b) Priority distribution (e.g., numbers of high, medium, and low alarms)

(c) Alarm performance during upset conditions (e.g., the numbers of alarms activated during the first 10 min and the subsequent 10-min periods)

(d) Identification of bad actors—individual alarm points that generate a large fraction of the total alarms

These metrics are illustrated in Table 10.2 for a set of industrial plant data (Hollifield and Habibi, 2007). A recommended alarm priority distribution is shown in Table 10.4 with the Table 10.2 guidelines in parentheses.

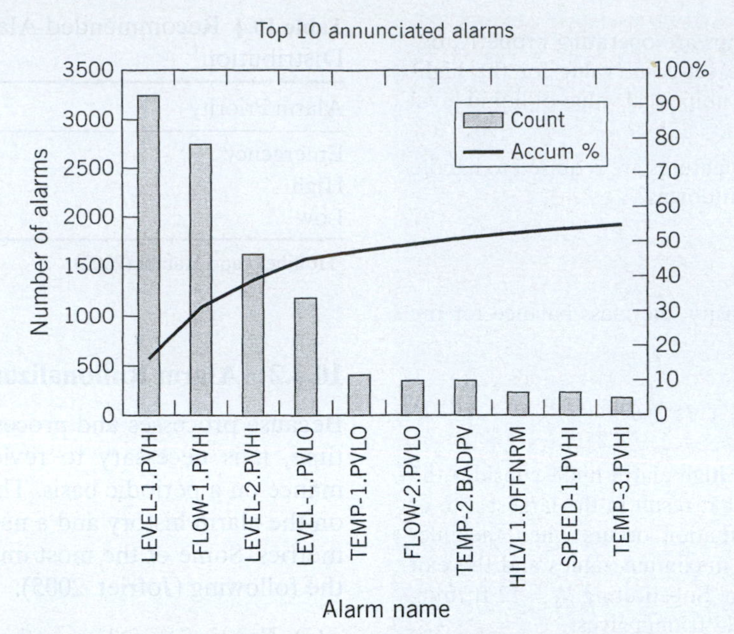

Figure 10.7 An example of the 10 most frequent alarms (bad actors) over an 8-week period (Hollifield and Habibi, 2007).

A critical analysis of the alarm metrics can result in significant improvements in alarm management. In particular, it can identify potentially dangerous alarm floods where operators are overwhelmed by a large number of alarms in a small time interval. It can also delineate bad actors. An industrial example of an alarm flood is shown in Fig. 10.6; identification of the associated bad actors is shown in Fig. 10.7. Analysis of bad actor alarm data can identify process problems or alarm limits that are inappropriate (e.g., too aggressive). Figure 10.7 indicates that only four alarms generate over 40% of the total number of alarms.

Case Study

A review of an alarm history has identified that the bad actors include level, pressure, and temperature alarms that are associated with a liquid-phase, continuous, stirred-tank reactor. The chemical reactions are exothermic, and the CSTR is used at different times to make two products, A or B. The low-level and low-pressure alarm violations occur mainly during shutdown operations, whereas high temperature alarms for the jacket cooling water occur primarily when product B is produced. It is desirable to devise a strategy for reducing these bad actor alarms.

Because these nuisance alarms tend to occur for different known-process conditions, a state-based alarming strategy is warranted (Guideline 11 of Table 10.3). In particular, different alarm levels are required for the three reactor conditions. For example, the following alarm limits could be employed with variables expressed in %.

Condition	Level (high)	Level (low)	Pressure (high)	Cooling water temperature (high)
Product A	90%	10%	85%	75%
Product B	90%	10%	85%	85%
Shutdown/ empty	5%	0%	10%	30%

Production of B involves a more exothermic reaction and thus tends to result in a higher cooling water temperature. When the reactor is shut down and evacuated, the level alarm settings are reduced to very low values. In particular, if reactor is not completely evacuated, the alarm sounds if the liquid level exceeds 5%. The state-based alarm settings eliminate many of the nuisance alarms.

10.3 ABNORMAL EVENT DETECTION

The overall goal of process control is to maintain product quality under safe operating conditions. When operating conditions exceed acceptable limits as a result of external causes, equipment malfunctions, or human error, unacceptable situations can occur, including unsafe conditions, off-specification product, decreased production, and environmental degradation. Major process excursions may even require plant shutdown to prevent catastrophic events, such as explosions, fires, or discharge of toxic chemicals. Thus the detection and resolution of abnormal conditions, especially potentially disastrous situations, are very important plant activities that have received considerable attention in recent years, especially from government organizations (e.g., OSHA, 1996), engineering societies (AIChE 2007), and industrial organizations (EEMUA, 1997; ASM, 2009).

In this section, we consider the early detection of abnormal conditions, an activity referred to as *abnormal event detection* (AED). It is important to note the crucial role of plant operating personnel, especially plant operators, in AED. Their process knowledge and experience can provide both early detection of abnormal situations and remedial action to either return the plant to normal levels of operation or to shut down the process. However, an operator's response depends on many factors: the number of alarms and the frequency of occurrence of abnormal conditions; how information is presented to the operator; the complexity of the plant; and the operator's intelligence, training, experience, and reaction to stress. Consequently, computational tools that assist plant personnel are crucial to the success of operating complex manufacturing plants. These computational tools can be embedded in the process control system. In this section, we consider three general approaches for AED.

10.3.1 Fault Detection Based on Sensor and Signal Analysis

Analysis of past and current values of measured variables can provide valuable diagnostic information for the detection of abnormal events. As described in Section 10.1, on-line measurements are routinely checked to ensure that they are between specified high and low limits and to ensure that the rate of change is consistent with the physical process. Also, a simple calculation can identify a common type of sensor malfunction or fault.

Suppose that the sample variance of a measured variable, x, is calculated from n consecutive measurements with a constant sample period (e.g., $\Delta t = 1$ min). The sample variance s^2 is defined as (see Appendix F):

$$s^2 = \frac{1}{n-1}\sum_{i=1}^{n} x_i^2 \tag{10-4}$$

Based on the expected process variability, measurement noise, and past experience, a lower limit for s^2, s^2_{min}, can be specified. For subsequent on-line monitoring, if $s^2 < s^2_{\text{min}}$, there is reason to be believe that the measurement may be essentially constant, because of a "dead" or "frozen" sensor.

Other simple signal analyses can be used to detect common problems, such as sticking control valves (Shoukat Choudhury et al., 2008) and oscillating control loops (Thornhill and Horch, 2007). For AED applications that involve many process variables, multivariable statistical techniques, such as Principal Component Analysis (PCA) and Partial Least Squares (PLS), can be very effective (see Chapter 21).

10.3.2 Model-Based Methods

Both steady-state and dynamic process models can provide useful information for AED. The model can be either a physical model (Chapter 2) or an empirical model (Chapter 7). For example, equipment performance calculations based on mass and energy balances can be used to calculate thermal efficiencies for energy-intensive processes such as furnaces. An unusually low value could indicate a potentially hazardous situation, such as a burner malfunction.

If a chemical composition or a product quality variable cannot be measured on-line, it may be possible to predict it from measured process variables, such as flow rates, temperatures, and pressures. When the predictions are used for feedback control, this strategy is referred to as inferential control (see Chapter 16). On-line prediction of future values of controlled variables based on a dynamic model is a key element of a widely used, advanced control strategy: model predictive control (see Chapter 20).

Periodic evaluations of steady-state conservation equations provide a powerful tool for abnormal event detection. The error of closure, E_c, of a steady-state mass or component balance for a continuous process can be defined as

$$E_c = Rate\ In - Rate\ Out \tag{10-5}$$

An unusually large error of closure (in absolute value) suggests an abnormal event, for example, a malfunctioning sensor or an equipment problem, such as heat exchanger fouling, as illustrated by Example 10.2.

EXAMPLE 10.2

A feed mixture of two components is separated in a distillation column, as shown in Fig. 10.8 where F, D, and B are molar flow rates and z, y, and x are mole fractions of the more volatile component. Assume that the three flow rates are constant: $F = 4$ mol/min, $D = B = 2$ mol/min. During normal operation, each composition measurement contains random errors. Based on previous experience, it

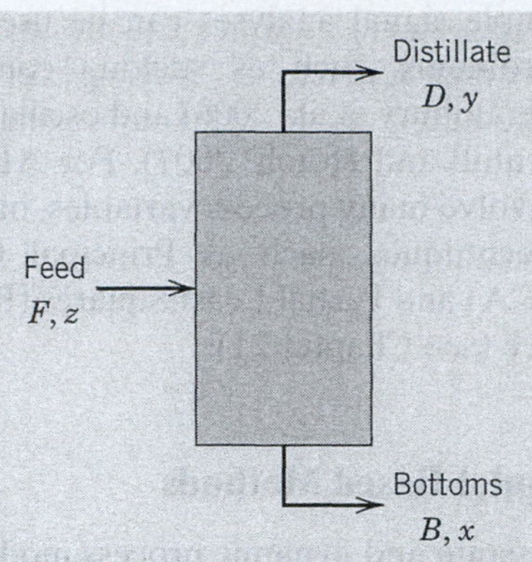

Figure 10.8 Schematic diagram of a distillation column.

can be assumed that the population means μ and population standard deviations σ (see Appendix F for definitions) have the following values:

	z	y	x
μ	0.500	0.800	0.200
σ	0.010	0.020	0.005

At steady state, the measured compositions were:

$$\bar{z} = 0.485, \quad \bar{y} = 0.825, \quad \bar{x} = 0.205$$

Is there statistically significant evidence to support the assertion that one or more of the composition sensors are not working properly?

SOLUTION

The error of closure for the steady-state component balance is:

$$E_c = Fz - Dy - Bx$$

Substituting the measured values gives,

$$E_c = 4(0.485) - 2(0.825) - 2(0.205) = -0.\,120 \text{ mol/min}$$

As shown in Appendix F, the standard deviation of E_c is:

$$\sigma_{Ec} = \sqrt{F^2\sigma_z^2 + D^2\sigma_y^2 + B^2\sigma_x^2}$$

$$\sigma_{Ec} = \sqrt{(4)^2(0.01)^2 + (2)^2(0.02)^2 + (2)^2(0.005)^2}$$
$$= 0.0574 \text{ mol/min}$$

Because the absolute value of the observed value of E_c is more than two standard deviations away from the expected value of zero, there is strong statistical evidence that an abnormal event has occurred. For example, if E_c is normally distributed, with a mean of $\mu_{Ec} = 0$ and a standard deviation of $\sigma_{Ec} = 0.0574$, the corresponding Z value (see Appendix F) is

$$Z = \frac{E_c - \mu_{Ec}}{\sigma_{Ec}} = \frac{-0.120 - 0}{0.0574} = -2.09$$

The probability that an abnormal event has occurred is the probability that $|Z| > 2.09$. From the table for the standard normal probability distribution (Montgomery and Runger, 2007), this probability is 0.963. Thus, there is significant statistical evidence to include that an abnormal event has occurred.

This error analysis can be extended to the situation where the flow measurements also contain random errors.

10.3.3 Knowledge-Based Methods

In the previous section, quantitative models were used to detect abnormal events. As an alternative, the AED assessment can utilize *qualitative* information based on process knowledge and past experience. This general strategy is referred to as a *knowledge-based approach.* It relies on specific methods, such as causal analysis, fault-tree diagrams, fuzzy logic, logical reasoning, and expert systems (Chiang et al., 2001; Dash and Venkatasubramanian, 2003).

To illustrate the use of qualitative knowledge in AED, consider the liquid storage system in Fig. 10.5. Using qualitative concepts such as low, normal and high, an AED would be indicated if the following logical IF-THEN statement is true based on measurements of q_1, q_2, q_3, and dh/dt:

IF ("q_1 is normal or low") AND ("q_2 is normal or low") AND ("q is normal or high") AND ("dh/dt is high"), THEN ("an AED has occurred"). This type of analysis requires thresholds for each qualitative concept and process variable.

For this simple example, a similar analysis could be based on a quantitative approach, namely, a dynamic model based on an unsteady-state mass balance. The level could be changing as a result of process changes or a sensor failure. However, for more complicated processes, a reasonably accurate dynamic model may not be available, and thus a qualitative approach can be used to good advantage. The diagnosis of the abnormal event could lead to a subsequent root cause analysis, where the source of the abnormality is identified and appropriate corrective actions are taken.

10.4 RISK ASSESSMENT

Risk assessment considers the consequences of potential hazards, faults, failures, and accident scenarios. In particular, it provides a quantitative assessment of risk in contrast to other approaches (e.g., a HAZOP study) that provide qualitative assessments. Because industrial processes are complex and interconnected, risk assessment determines overall failure probabilities from those of individual components. For example, the *failure rate* of a typical temperature control loop depends on the failure rates of the individual components: sensor/transmitter, controller, I/P transducer (if required), and the control valve. Consequently, in this section, we consider relevant reliability concepts based on probability theory and statistics. Appendix F reviews concepts from probability and statistics that are needed for this analysis.

10.4.1 Reliability Concepts

A typical failure rate curve for process equipment and other hardware has the "bathtub shape" shown in Fig. 10.9. For much of its lifetime, the failure rate is approximately constant, with a value denoted by μ. In the subsequent analysis, a constant failure rate is assumed for each component; however, extensions can be made to include time-varying failure rates.

The probability that the component does not fail during the interval, $(0, t)$, is defined to be the *reliability*, R. For a component with a constant failure rate μ, R is given by the exponential probability distribution (Crowl and Louvar, 2002),

$$R = e^{-\mu t} \tag{10-6}$$

and the corresponding failure probability P is:

$$P = 1 - R \tag{10-7}$$

Both P and R are bounded by $[0, 1]$. The expected *mean time between failures* (MTBF) is given by

$$MTBF = \frac{1}{\mu} \tag{10-8}$$

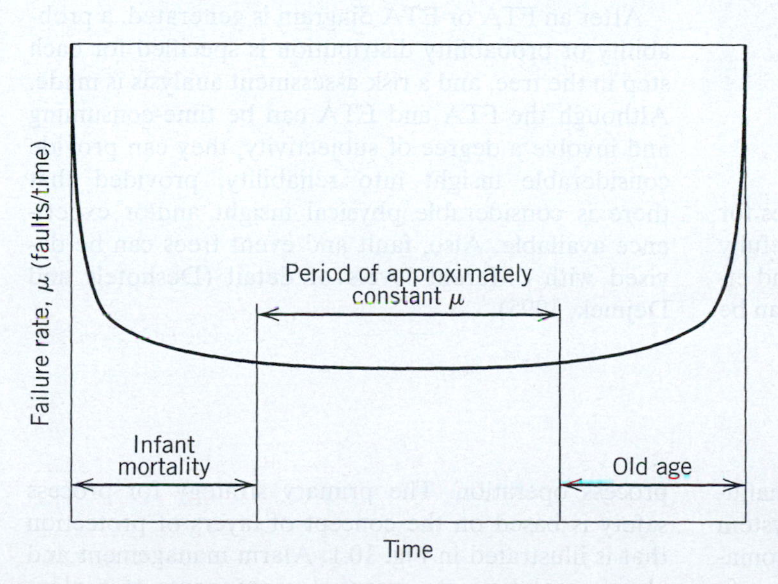

Figure 10.9 The "bathtub curve," showing a typical failure rate plot for hardware and instruments.

10.4.2 Overall Failure Rates

We consider overall failure rates for two common situations.

Components in series: Consider a system of n independent components, each with an individual failure rate, μ_i. Suppose that all of the components must operate properly in order for the overall system to function. The reliability of the overall system R is given by,

$$R = \prod_{i=1}^{n} R_i \tag{10-9}$$

where R_i is the reliability of the i-th component. Then the overall failure probability P is,

$$P = 1 - R = 1 - \prod_{i=1}^{n} (1 - P_i) \tag{10-10}$$

where $P_i = 1 - R_i$

Components in parallel (redundant components): Improved system reliability can be achieved by using redundant components. For example, in Section 10.1, redundant sensors and control loops were used in SIS and ESD systems. Suppose that m independent sensors are available and only one needs to be operational in order for the feedback control loop to operate properly. The probability that all m sensors fail is:

$$P = \prod_{i=1}^{m} P_i \tag{10-11}$$

and the reliability of the overall system is:

$$R = 1 - P = 1 - \prod_{i=1}^{m} (1 - R_i) \tag{10-12}$$

The calculation of overall failure rates from individual failure rates is illustrated in the following example.

EXAMPLE 10.3

A flow control loop consists of a differential pressure flow sensor/transmitter, a digital PI controller, an I/P transducer, a control valve, and a valve positioner. Determine the reliability, failure rate, and MTBF for this control loop.

SOLUTION

Using the failure rate data in Table 10.1, the calculated values are

Component	Failure rate, μ (failures/yr)	Reliability (per yr) $R = e^{-\mu t}$	Failure probability $P = 1 - R$
D/P flowmeter	1.73	0.18	0.82
Digital controller†	0.05	0.95	0.05
I/P transducer	0.49	0.61	0.39
Control valve	0.60	0.55	0.45
Valve positioner	0.44	0.64	0.36

† Assumed value (not in Table 10.1).

All of the components must function properly in order for the control loop to operate. Thus, the components can be considered to be in a series configuration. The overall reliability of components in series is the probability that no failures occur. It can be calculated using Eq. (10-9):

$$R = \prod_{i=1}^{5} R_i = (0.18)(0.95)(0.61)(0.55)(0.64) = 0.037$$

The overall failure probability per year is,

$$P = 1 - 0.037 = 0.96$$

The overall failure rate is calculated from Eq. (10-6):

$$0.037 = e^{-\mu}$$

$$\mu = -\ln(0.037) = 3.3 \text{ failures/yr}$$

The mean time between failures is:

$$\text{MTBF} = \frac{1}{\mu} = 0.3 \text{ yr}$$

10.4.3 Fault and Event Tree Analysis

In order to evaluate the reliability and failure rates for complex processes, it is necessary to consider carefully the available physical information about causes and effects of possible failure modes. This information can be incorporated into the risk assessment using two types of analyses based on logic diagrams: fault tree analysis (FTA) and event tree analysis (ETA). Fault trees display all of the component failures that can lead to a very serious situation, such as an accident or an explosion, and the subsequent chain of events. Event trees are similar to fault trees but focus on a single initiating event, such as a component failure, and then evaluate the consequences, classified according to how serious they are. Both FTA and ETA are used to analyze proposed designs for new processes and to diagnose the reliability of existing or retrofit processes (Crowl and Louvar, 2002; Banerjee, 2003).

For fault tree analysis, the starting point is to specify an undesirable serious situation, called the top effect, and then to consider all possible causes that could produce it. For example, the specified top effect could be the over-pressurization of a chemical reactor. Possible causes could include a reduction or loss of coolant, excess catalyst, an ineffective pressure control loop, etc. Each possible cause is analyzed further to determine why it occurred. For example, the pressure control loop problem could be to the result of a sensor or control valve malfunction. Thus, FTA is a top-down approach that generates a tree of causal relations, starting with the specified top event and working backward. Standard logic concepts, such as AND and OR, are used in the logic diagrams.

Event tree analysis is a similar technique, but its starting point is a single cause, rather than a single outcome. For the previous chemical reactor example, the specified starting point might be a blockage in the coolant piping or a sensor failure in the pressure control loop. Logic diagrams are used to illustrate how the initial root cause propagates through different levels in the event tree and produces one or more undesirable outcomes. Thus ETA is called a *bottom-up approach*.

After an FTA or ETA diagram is generated, a probability or probability distribution is specified for each step in the tree, and a risk assessment analysis is made. Although the FTA and ETA can be time-consuming and involve a degree of subjectivity, they can provide considerable insight into reliability, provided that there is considerable physical insight and/or experience available. Also, fault and event trees can be devised with different levels of detail (Deshotels and Dejmek, 1995).

SUMMARY

Process safety is both a paramount concern in manufacturing plants and a primary issue in control system design. In this chapter, it has been shown that automation and process control play key roles in ensuring safe process operation. The primary strategy for process safety is based on the concept of layers of protection that is illustrated in Fig. 10.1. Alarm management and alarm resolution are essential components of a plant safety system. The selection of the variables to be alarmed and their alarm limits are based on hazard identification and risk assessment. Abnormal event detection and risk assessment can play key roles in enhancing plant safety.

Finally, as Rinard (1990) has poignantly noted, "The regulatory control system affects the size of your paycheck; the safety control system affects whether or not you will be around to collect it."

REFERENCES

ABB, Industrial IT System 800xA, http://www.abb.com (2009).

AIChE Center for Chemical Process Safety, *Guidelines for Safe and Reliable Instrumented Protective Systems*, AIChE, NY, 2007.

AIChE Center for Chemical Process Safety, *Guidelines for Safe Automation of Chemical Processes*, AIChE, NY, 1993.

AIChE Center for Chemical Process Safety, *Layer of Protection Analysis: Simplified Process Risk Assessment*, AIChE, NY, 2001.

Abnormal Situation Management (ASM) Consortium, http://www.asmconsortium.com 2009.

Banerjee, S., *Industrial Hazards and Plant Safety*, Taylor and Francis, NY, 2003.

Camp, D., Private Communication 2009.

Chiang, L. H., E. L. Russell, and R. D. Bratz, *Fault Detection and Diagnosis in Industrial Systems*, Springer-Verlag, London, U.K., 2001.

Connell, B., *Process Instrumentation Applications Manual*, McGraw-Hill, NY, 1996.

Connelly, C. S., Lack of Planning in Alarm System Configuration Is, in Essence, Planning to Fail, *ISA Trans.*, **36**, 219 (1997).

Crowl, D. A. and J. F. Louvar, *Chemical Process Safety: Fundamentals with Applications, 2nd ed.*, Prentice Hall, Englewood Cliffs, NJ, 2002.

Dash, S. and V. Venkatasubramanian, Integrated Framework for Abnormal Event Management and Process Hazard Analysis, *AIChE J.*, **49**, 124 (2003).

Deshotels, R. and M. Dejmek, Choosing the Level of Detail for Hazard Identification, *Process Safety Progress*, **14**, 218 (1995).

Engineering Equipment & Materials Users' Association (EEMUA), *Alarm Systems - A Guide to Design, Management and Procurement*, 2nd ed., Publication 191, London, 2007.

Hollifield, B. R. and E. Habibi, *Alarm Management: Seven Effective Methods for Optimum Performance*, ISA, Research Triangle Park, NC, 2007.

International Society of Automation, *ANSI/ISA-84.01-1996: Application of Safety Instrumented Systems for the Process Industries*, Research Triangle Park, NC (1996).

Jofriet, P., Alarm Management, *Chem. Eng.*, **112** (2), 36 (2005).

Katzel, J., Managing Alarms, *Control Eng.*, **54** (2), 50 (2007).

Kletz, T. A., P. Chung, E. Broomfield, and C. Shen-Orr, *Computer Control and Human Error*, Gulf Publishing Co., Houston, TX, 1995.

Kletz, T. A., *HAZOP and HAZAN: Identifying and Assessing Process Industry Hazards*, Taylor & Francis, London, 1999.

Mannan, S., and F. P. Lees, *Lees' Loss Prevention in the Process Industries: Hazard Identification, Assessment, and Control, Vol. 1*, 3rd ed., Butterworth-Heinemann Elsevier, NY, 2005.

Montgomery, D.C. and G.C. Runger, *Applied Statistics and Probability for Engineers, 4th ed.*, John Wiley, Hoboken, N.J., 2007.

Nuclear Regulatory Commission, Fact Sheet on the Three Mile Island Accident, http://www.nrc.gov/ 2008.

Occupational Safety and Health Administration (OSHA), *Process Safety Management of Highly Hazardous Chemicals*, CFR 29, Regulation 1910.119 (1996).

Perry, R. H. and D. W. Green (ed.), *Chemical Engineers' Handbook*, 8th ed., Section 23, *Process Safety*, McGraw-Hill, NY, 2008.

Rinard, I., Discussion, *Chem. Eng. Educ.*, **24**, Spring Issue, pg. 76 (1990).

Shoukat Choudhury, M. A. A., M. Jain, and S. L. Shah, Stiction—Definition, Modelling, Detection and Quantification, *J. Process Control*, **18**, 232–243 (2008).

Thornhill, N. F. and A. Horch, Advances and New Directions in Plant-Wide Controller Performance Assessment, *Control Eng. Practice*, **15**, 1196–1206 (2007).

Turton, R., R. C. Bailie, W. B. Whiting, and J. A. Shaeiwitz, *Analysis, Synthesis and Design of Chemical Processes*, 3rd ed., Prentice-Hall PTR, Upper Saddle River, NJ, 2008.

EXERCISES

10.1 Air samples from a process area are continuously drawn through a ¼-in diameter tube to an analytical instrument that is located 40 m away. The tubing has an outside diameter of 6.35 mm and a wall thickness of 0.762 mm. The flow rate through the transfer line is 10 cm^3/s for ambient conditions of 20°C and 1 atm. The pressure drop in the transfer line is negligible. Because chlorine gas is used in the process, a leak can poison workers in the area. It takes the analyzer 5 s to respond after chlorine first reaches it. Determine the amount of time that is required to detect a chlorine leak in the processing area. State any assumptions that you make. Would this amount of time be acceptable if the hazardous gas were carbon monoxide, instead of chlorine?

(Adapted from: *Student Problems for Safety, Health, and Loss Prevention in Chemical Processes,* AIChE Center for Chemical Process Safety, NY, 1990).

10.2 The two-phase feed stream for the gas-liquid separator (or flash drum) shown in Fig. E10.2 consists of a hydrocarbon mixture. Because the pressure in the vessel is significantly lower than the feed pressure, part of the liquid feed flashes to form a gas phase. The hydrocarbons are flammable and somewhat hazardous. Discuss the process safety issues and propose an alarm/SIS strategy.

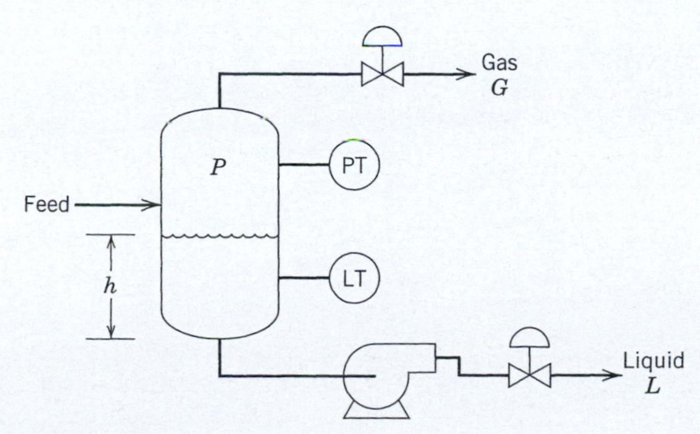

Figure E10.2

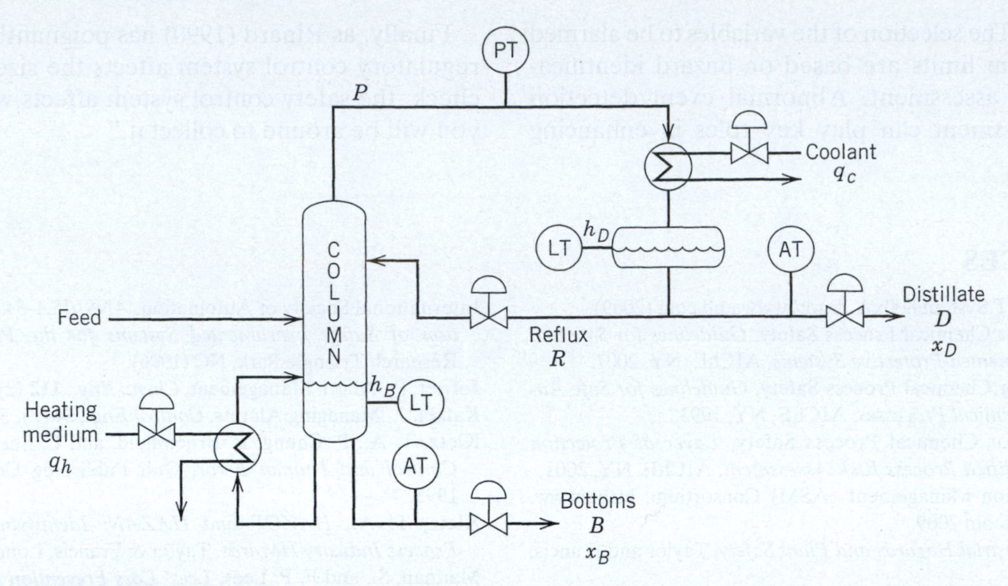

Figure E10.3

10.3 The loss of the coolant to a process vessel can produce an unacceptably high pressure in the vessel. As a result, a pressure relief valve is used to reduce the pressure by releasing the vapor mixture to the atmosphere. But if the mixture is toxic or flammable, the release can be hazardous. For the distillation column in Fig. E10.3, which operates at above ambient temperature, propose an alarm/SIS system that will reduce the number of releases to the environment, even though the occasional loss of coolant flow to the condenser is unavoidable. (*Note:* The pressure relief valve at the top of the column is not shown in Fig. E10.3.)

10.4 The probability of a particular type of sensor functioning properly is 0.99. Consequently, a triply redundant sensor system has been proposed for a critical measurement. Thus, three independent sensors will be installed, and the median of the three measurements will be used for the alarms and control calculations. What is the probability that at least two of the sensors will be working at any time?

10.5 Consider the liquid storage tank with a low-level interlock, as shown in Fig. 10.4. Suppose that an independent low-level alarm is added, with its set-point value above the value for the low-level switch. If both the low-level alarm and the low-level interlock system fail simultaneously, the pump could be seriously damaged. What is the probability that this occurs? What is the mean time between failures?

Failure rates (faults per year):

Solenoid switch: $\mu_S = 0.01$

Level switch: $\mu_{LS} = 0.4$

Level alarm: $\mu_A = 0.2$

10.6 For the reliability analysis of the flow control loop in Example 10.3, the D/P flowmeter is the least reliable component. Suppose that a second, identical flowmeter is used in a backup mode so that it could be automatically and immediately employed if the first flowmeter failed. How much would the overall system reliability improve by adding the second sensor?

10.7 Using the failure rate data in Table 10.1, evaluate the reliability and mean time between failures for the high-pressure interlock in Fig. 10.4. Assume that the failure rate for the solenoid switch and valve is $\mu = 0.42$ faults per year.

Chapter 11

Dynamic Behavior and Stability of Closed-Loop Control Systems

CHAPTER CONTENTS

In this chapter we consider the dynamic behavior of processes that are operated using feedback control. This combination of the process, the feedback controller, and the instrumentation is referred to as a *feedback control loop* or a *closed-loop system*. Thus, the term *closed-loop system* is used to denote the controlled process. We begin by demonstrating that block diagrams and transfer functions provide a useful description of closed-loop systems. We then use block diagrams to analyze the dynamic behavior of several simple closed-loop systems.

Although feedback control yields many desirable characteristics, it also has one undesirable characteristic.

If the controller is poorly designed or the process dynamic characteristics change after the controller is implemented, the resulting closed-loop system can be unstable. This means that the controller can produce a growing oscillation in the controlled variable rather than keeping it at the set point. Understanding the source of this unstable behavior, and how to prevent it, are important issues. In this chapter, several mathematical stability criteria are introduced, and practical methods for analyzing closed-loop stability are considered.

11.1 BLOCK DIAGRAM REPRESENTATION

In Chapters 1 and 8 we have shown that a block diagram provides a convenient representation of the flow of information around a feedback control loop. The previous discussion of block diagrams was qualitative rather than quantitative, because the blocks were labeled but did not indicate the relationships between process variables. However, quantitative information can also be included by showing the transfer function for each block.

To illustrate the development of a block diagram, we return to a previous example, the stirred-tank blending process considered in earlier chapters. The schematic diagram in Fig. 11.1 shows the blending tank with the flow rate of pure component A, w_2, as the manipulated variable. The control objective is to regulate the tank composition, x, by adjusting the mass flow rate w_2. The primary disturbance variable is assumed to be inlet composition x_1. The tank composition is measured by a sensor/transmitter whose output signal x_m is sent to an electronic controller. Because a pneumatic control valve is used, the controller output (an electrical signal in the range of 4 to 20 mA) must be converted to an equivalent pneumatic signal p_t (3 to 15 psig) by a current-to-pressure transducer. The transducer output signal is then used to adjust the valve.

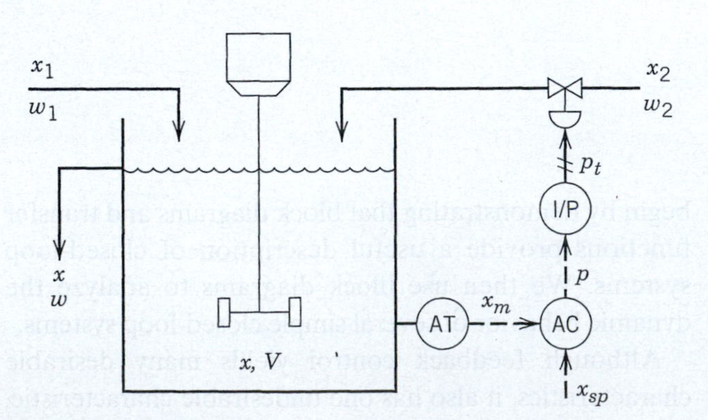

Figure 11.1 Composition control system for a stirred-tank blending process.

Next, we develop a transfer function for each of the five elements in the feedback control loop. For the sake of simplicity, flow rate w_1 is assumed to be constant, and the system is initially operating at the nominal steady rate. Later, we extend this analysis to more general situations.

11.1.1 Process

In Example 4.5 the approximate dynamic model of a stirred-tank blending system was developed:

$$X'(s) = \left(\frac{K_1}{\tau s + 1}\right)X_1'(s) + \left(\frac{K_2}{\tau s + 1}\right)W_2'(s) \tag{11-1}$$

where

$$\tau = \frac{V\rho}{\overline{w}}, \quad K_1 = \frac{\overline{w}_1}{\overline{w}}, \quad \text{and} \quad K_2 = \frac{1-\overline{x}}{\overline{w}} \tag{11-2}$$

Figure 11.2 provides a block diagram representation of the information in Eqs. 11-1 and 11-2 and indicates the units for each variable. In the diagram, the deviation variable, $X_d'(s)$, denotes the change in exit composition due to a change in inlet composition $X_1'(s)$ (the disturbance). Similarly, $X_u'(s)$ is a deviation variable that denotes the change in $X'(s)$ due to a change in the manipulated variable (the flow rate of pure A, $W_2'(s)$). The effects of these changes are additive because $X'(s) = X_d'(s) + X_u'(s)$ as a direct consequence of the Superposition Principle for linear systems discussed in Chapter 3. Recall that this transfer function representation is valid only for linear systems and for nonlinear systems that have been linearized, as is the case for the blending process model.

11.1.2 Composition Sensor-Transmitter (Analyzer)

We assume that the dynamic behavior of the composition sensor-transmitter can be approximated by a first-order transfer function:

$$\frac{X_m'(s)}{X'(s)} = \frac{K_m}{\tau_m s + 1} \tag{11-3}$$

This instrument has negligible dynamics when $\tau >> \tau_m$. For a change in one of the inputs, the measured composition $x_m'(t)$ rapidly follows the true composition $x'(t)$, even

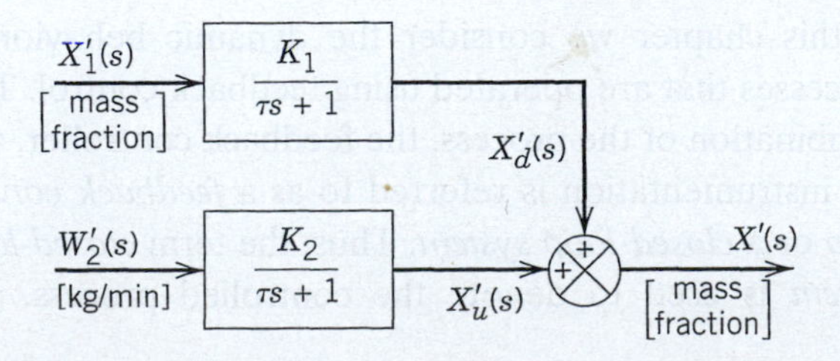

Figure 11.2 Block diagram of the blending process.

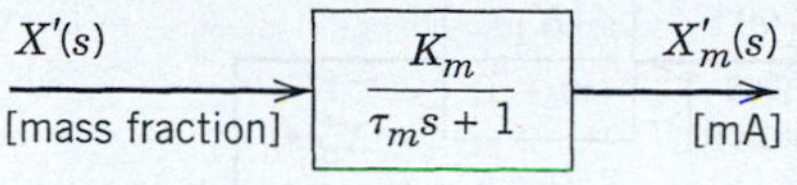

Figure 11.3 Block diagram for the composition sensor–transmitter (analyzer).

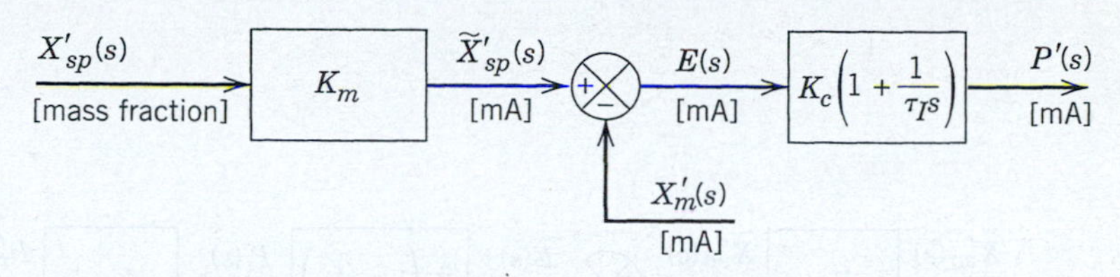

Figure 11.4 Block diagram for the controller.

while $x'(t)$ is slowly changing with time constant τ. Hence, the dynamic error associated with the measurement can be neglected (cf. Section 9.4). A useful approximation is to set $\tau_m = 0$ in Eq. 11-3. The steady-state gain K_m depends on the input and output ranges of the composition sensor-transmitter combination, as indicated in Eq. 9-1. The block diagram for the sensor-transmitter is shown in Fig. 11.3.

11.1.3 Controller

Suppose that an electronic proportional plus integral controller is used. From Chapter 8, the controller transfer function is

$$\frac{P'(s)}{E(s)} = K_c\left(1 + \frac{1}{\tau_I s}\right) \tag{11-4}$$

where $P'(s)$ and $E(s)$ are the Laplace transforms of the controller output $p'(t)$ and the error signal $e(t)$. Note that p' and e are electrical signals that have units of mA, while K_c is dimensionless. The error signal is expressed as

$$e(t) = \tilde{x}'_{sp}(t) - x'_m(t) \tag{11-5}$$

or after taking Laplace transforms,

$$E(s) = \tilde{X}'_{sp}(s) - X'_m(s) \tag{11-6}$$

The symbol $\tilde{x}'_{sp}(t)$ denotes the *internal set-point* composition expressed as an equivalent electrical current signal. This signal is used internally by the controller. $\tilde{x}'_{sp}(t)$ is related to the actual composition set point $x'_{sp}(t)$ by the sensor-transmitter gain K_m (which is the slope of the calibration curve):

$$\tilde{x}'_{sp}(t) = K_m x'_{sp}(t) \tag{11-7}$$

Thus

$$\frac{\tilde{X}'_{sp}(s)}{X'_{sp}(s)} = K_m \tag{11-8}$$

The block diagram representing the controller in Eqs. 11-4 through 11-8 is shown in Fig. 11.4. The symbol that represents the subtraction operation is called a *comparator*.

In general, if a reported controller gain is not dimensionless, it includes the gain of at least one other device (such as an actuator) in addition to the dimensionless controller gain.

11.1.4 Current-to-Pressure (I/P) Transducer

Because transducers are usually designed to have linear characteristics and negligible (fast) dynamics, we assume that the transducer transfer function merely consists of a steady-state gain K_{IP}:

$$\frac{P'_t(s)}{P'(s)} = K_{IP} \tag{11-9}$$

In Eq. 11-9, $P'_t(s)$ denotes the output signal from the I/P transducer in deviation form. The corresponding block diagram is shown in Fig. 11.5.

11.1.5 Control Valve

As discussed in Section 9.2, control valves are usually designed so that the flow rate through the valve is a nearly linear function of the signal to the valve actuator. Therefore, a first-order transfer function usually provides an adequate model for operation of an installed valve in the vicinity of a nominal steady state. Thus, we assume that the control valve can be modeled as:

$$\frac{W'_2(s)}{p'_t(s)} = \frac{K_v}{\tau_v s + 1} \tag{11-10}$$

The block diagram for an I/P transducer plus pneumatic control valve is shown in Fig. 11.6.

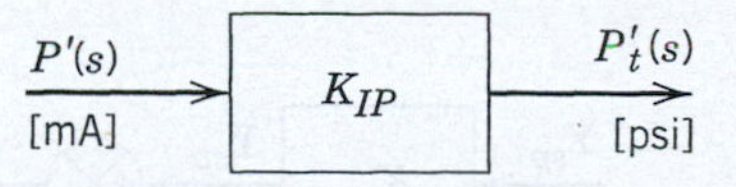

Figure 11.5 Block diagram for the I/P transducer.

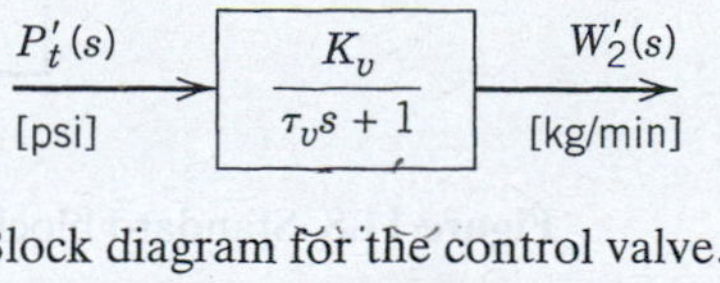

Figure 11.6 Block diagram for the control valve.

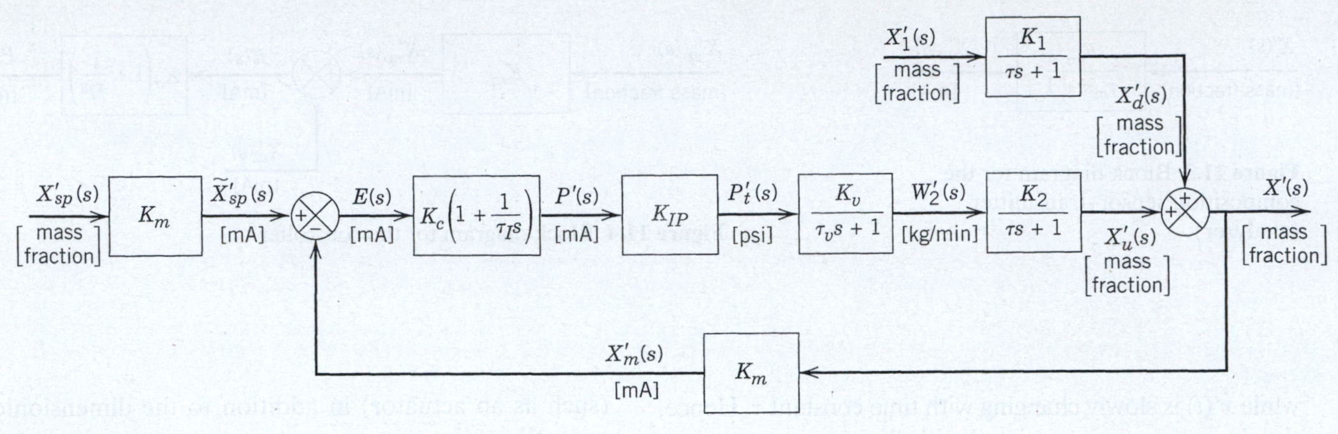

Figure 11.7 Block diagram for the entire blending process composition control system.

Now that transfer functions and block diagrams in Figs. 11.2 to 11.6 have been developed for the individual components of the feedback control system, we can combine this information to obtain the composite block diagram of the controlled system shown in Fig. 11.7.

11.2 CLOSED-LOOP TRANSFER FUNCTIONS

The block diagrams considered so far have been specifically developed for the stirred-tank blending system. The more general block diagram in Fig. 11.8 contains the standard notation:

Y = controlled variable
U = manipulated variable
D = disturbance variable (also referred to as the *load variable*)
P = controller output
E = error signal
Y_m = measured value of Y
Y_{sp} = set point
$\widetilde{Y}_{sp}$ = internal set point (used by the controller)
Y_u = change in Y due to U
Y_d = change in Y due to D
G_c = controller transfer function
G_v = transfer function for the final control element
G_p = process transfer function
G_d = disturbance transfer function
G_m = transfer function for sensor-transmitter
K_m = steady-state gain for G_m

In Fig. 11.8 each variable is the Laplace transform of a deviation variable. To simplify the notation, the primes and s dependence have been omitted; thus, Y is used rather than $Y'(s)$. Because the final control element is often a control valve, its transfer function is denoted by G_v. Note that the process transfer function G_p indicates the effect of the manipulated variable on the controlled variable. The disturbance transfer function G_d represents the effect of the disturbance variable on the controlled variable. For the

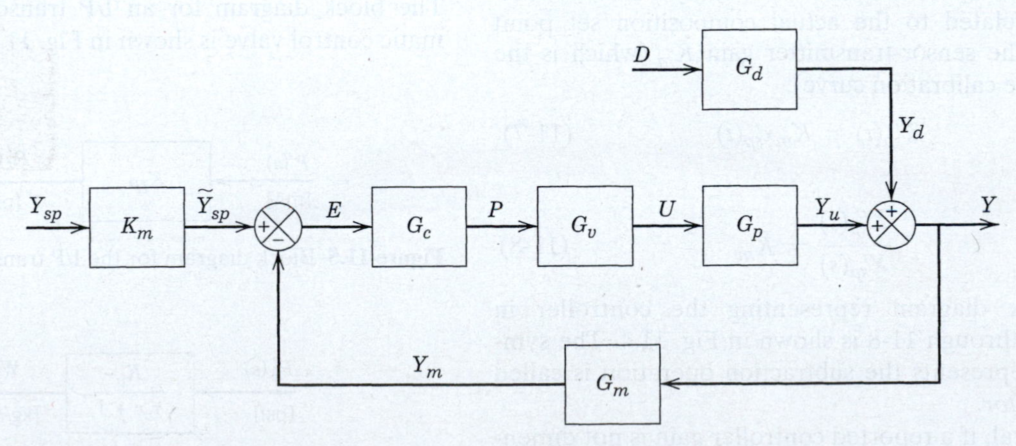

Figure 11.8 Standard block diagram of a feedback control system.

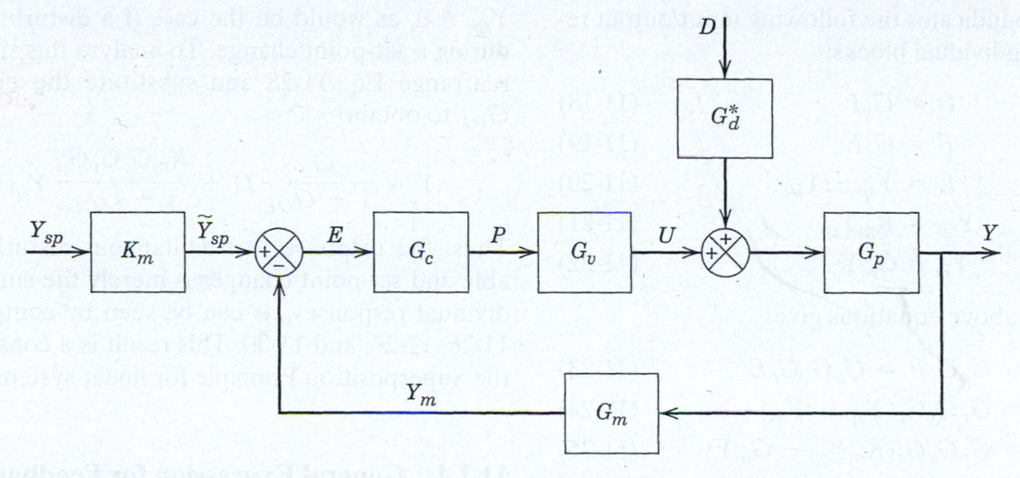

Figure 11.9 Alternative form of the standard block diagram of a feedback control system.

stirred-tank blending system, G_d and G_p are given in Eq. 11-1.

The standard block diagram in Fig. 11.8 can be used to represent a wide variety of practical control problems. Other blocks can be added to the standard diagram to represent additional elements in the feedback control loop such as the current-to-pressure transducer in Fig. 11.7. In Fig. 11.8, the signal path from E to Y through blocks G_c, G_v, and G_p is referred to as the *forward path*. The path from Y to the comparator through G_m is called the *feedback path*.

Figure 11.9 shows an alternative representation of the standard block diagram that is also used in the control literature. Because the disturbance transfer functions appear in different locations in Figs. 11.8 and 11.9, different symbols are used. For these two block diagrams to be equivalent, the relation between Y and D must be preserved. Thus, G_d and G_d^* must be related by the expression $G_d = G_p G_d^*$.

Note that Y_{sp} and D are the independent input signals for the controlled process because they are not affected by operation of the control loop. By contrast, U and D are the independent inputs for the uncontrolled process. To evaluate the performance of the control system, we need to know how the controlled process responds to changes in D and Y_{sp}. In the next section, we derive expressions for the *closed-loop transfer functions*, $Y(s)/Y_{sp}(s)$ and $Y(s)/D(s)$. But first, we review some block diagram algebra.

11.2.1 Block Diagram Reduction

In deriving closed-loop transfer functions, it is often convenient to combine several blocks into a single block. For example, consider the three blocks in series in Fig. 11.10. The block diagram indicates the following relations:

$$\begin{aligned} X_1 &= G_1 U \\ X_2 &= G_2 X_1 \\ X_3 &= G_3 X_2 \end{aligned} \tag{11-11}$$

By successive substitution,

$$X_3 = G_3 G_2 G_1 U \tag{11-12}$$

or

$$X_3 = GU \tag{11-13}$$

where $G \triangleq G_3 G_2 G_1$. Equation 11-13 indicates that the block diagram in Fig. 11.10 can be reduced to the equivalent block diagram in Fig. 11.11.

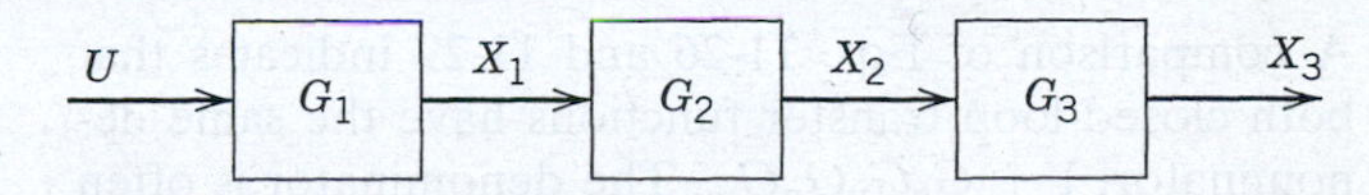

Figure 11.10 Three blocks in series.

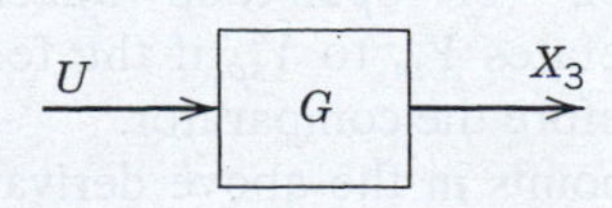

Figure 11.11 Equivalent block diagram.

11.2.2 Set-Point Changes

Next we derive the closed-loop transfer function for set-point changes. The closed-loop system behavior for set-point changes is also referred to as the servomechanism (*servo*) *problem* in the control literature, because early applications were concerned with positioning devices called servomechanisms. We assume for this case that no disturbance change occurs and thus $D = 0$. From Fig. 11.8, it follows that

$$Y = Y_d + Y_u \tag{11-14}$$

$$Y_d = G_d D = 0 \text{ (because } D = 0) \tag{11-15}$$

$$Y_u = G_p U \tag{11-16}$$

Combining gives

$$Y = G_p U \tag{11-17}$$

Figure 11.8 also indicates the following input/output relations for the individual blocks:

$$U = G_v P \tag{11-18}$$
$$P = G_c E \tag{11-19}$$
$$E = \widetilde{Y}_{sp} - Y_m \tag{11-20}$$
$$\widetilde{Y}_{sp} = K_m Y_{sp} \tag{11-21}$$
$$Y_m = G_m Y \tag{11-22}$$

Combining the above equations gives

$$Y = G_p G_v P = G_p G_v G_c E \tag{11-23}$$
$$= G_p G_v G_c(\widetilde{Y}_{sp} - Y_m) \tag{11-24}$$
$$= G_p G_v G_c(K_m Y_{sp} - G_m Y) \tag{11-25}$$

Rearranging gives the desired closed-loop transfer function,

$$\frac{Y}{Y_{sp}} = \frac{K_m G_c G_v G_p}{1 + G_c G_v G_p G_m} \tag{11-26}$$

In both the numerator and denominator of Eq. (11-26) the transfer functions have been rearranged to follow the order in which they are encountered in the feedback control loop. This convention makes it easy to determine which transfer functions are present or missing in analyzing subsequent problems.

11.2.3 Disturbance Changes

Now consider the case of disturbance changes, which is also referred to as the *regulator problem* since the process is to be *regulated* at a constant set point. From Fig. 11.8,

$$Y = Y_d + Y_u = G_d D + G_p U \tag{11-27}$$

Substituting (11-18) through (11-22) gives

$$Y = G_d D + G_p U$$
$$= G_d D + G_p G_v G_c(K_m Y_{sp} - G_m Y) \tag{11-28}$$

Because $Y_{sp} = 0$ we can rearrange (11-28) to give the closed-loop transfer function for disturbance changes:

$$\frac{Y}{D} = \frac{G_d}{1 + G_c G_v G_p G_m} \tag{11-29}$$

A comparison of Eqs. 11-26 and 11-29 indicates that both closed-loop transfer functions have the same denominator, $1 + G_c G_v G_p G_m$. The denominator is often written as $1 + G_{OL}$ where G_{OL} is the *open-loop transfer function*, $G_{OL} \triangleq G_c G_v G_p G_m$. The term open-loop transfer function (or open-loop system) is used because G_{OL} relates Y_m to $\widetilde{Y}_{sp}$ if the feedback loop is opened just before the comparator.

At different points in the above derivations, we assumed that $D = 0$ or $Y_{sp} = 0$, that is, that one of the two inputs was constant. But suppose that $D \neq 0$ and $Y_{sp} \neq 0$, as would be the case if a disturbance occurs during a set-point change. To analyze this situation, we rearrange Eq. 11-28 and substitute the definition of G_{OL} to obtain

$$Y = \frac{G_d}{1 + G_{OL}} D + \frac{K_m G_c G_v G_p}{1 + G_{OL}} Y_{sp} \tag{11-30}$$

Thus, the response to simultaneous disturbance variable and set-point changes is merely the sum of the individual responses, as can be seen by comparing Eqs. 11-26, 11-29, and 11-30. This result is a consequence of the Superposition Principle for linear systems.

11.2.4 General Expression for Feedback Control Systems

Closed-loop transfer functions for more complicated block diagrams can be written in the general form

$$\frac{Z}{Z_i} = \frac{\Pi_f}{1 + \Pi_e} \tag{11-31}$$

where

Z is the output variable or any internal variable within the control loop

Z_i is an input variable (e.g., Y_{sp} or D)

Π_f = product of the transfer functions in the *forward* path from Z_i to Z

Π_e = product of *every* transfer function in the feedback loop

Thus, for the previous servo problem, we have $Z_i = Y_{sp}$, $Z = Y$, $\Pi_f = K_m G_c G_v G_p$, and $\Pi_e = G_{OL}$. For the regulator problem, $Z_i = D$, $Z = Y$, $\Pi_f = G_d$, and $\Pi_e = G_{OL}$. It is important to note that Eq. 11-31 is applicable only to portions of a block diagram that include a *feedback loop* with a negative sign in the comparator.

EXAMPLE 11.1

Find the closed-loop transfer function Y/Y_{sp} for the complex control system in Fig. 11.12. Notice that this block diagram has two feedback loops and two disturbance variables. This configuration arises when the cascade control scheme of Chapter 16 is employed.

SOLUTION

Using the general rule in (11-31), we first reduce the inner loop to a single block as shown in Fig. 11.13. To solve the servo problem, set $D_1 = D_2 = 0$. Because Fig. 11.13 contains a single feedback loop, use (11-31) to obtain Fig. 11.14*a*. The final block diagram is shown in Fig. 11.14*b* with $Y/Y_{sp} = K_{m1} G_5$. Substitution for G_4 and G_5 gives the desired closed-loop transfer function:

$$\frac{Y}{Y_{sp}} = \frac{K_{m1} G_{c1} G_{c2} G_1 G_2 G_3}{1 + G_{c2} G_1 G_{m2} + G_{c1} G_2 G_3 G_{m1} G_{c2} G_1}$$

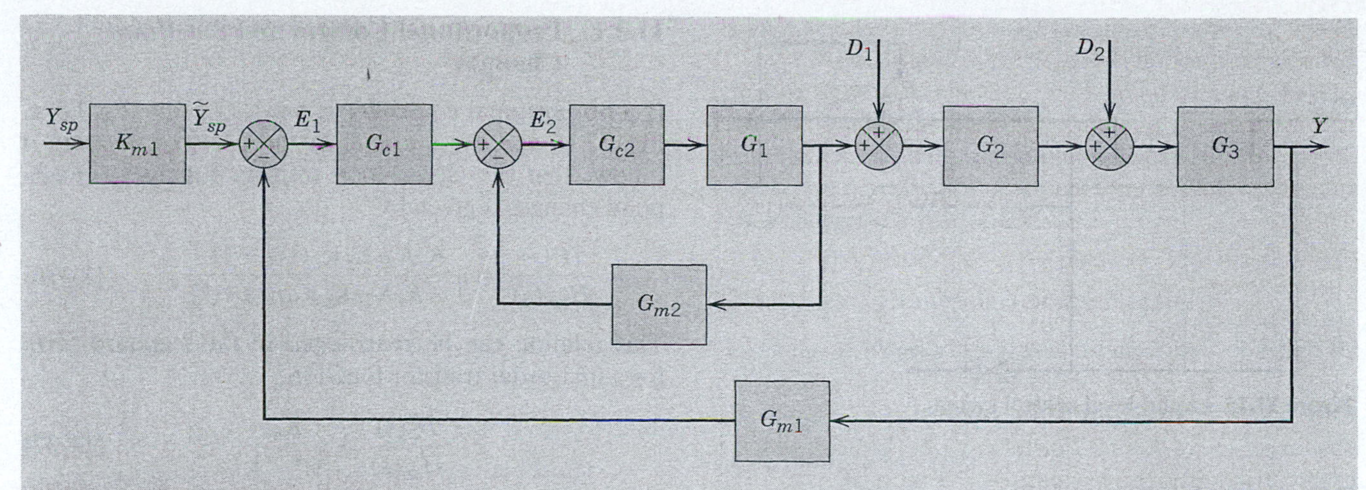

Figure 11.12 Complex control system.

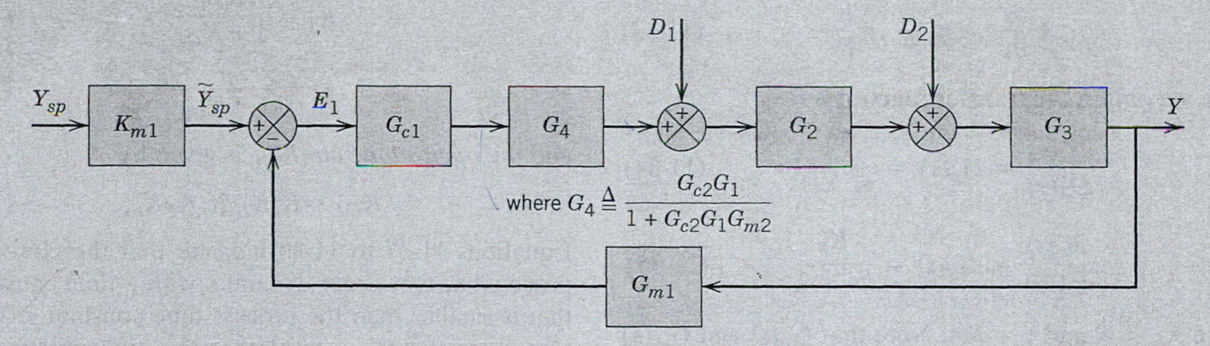

Figure 11.13 Block diagram for reduced system.

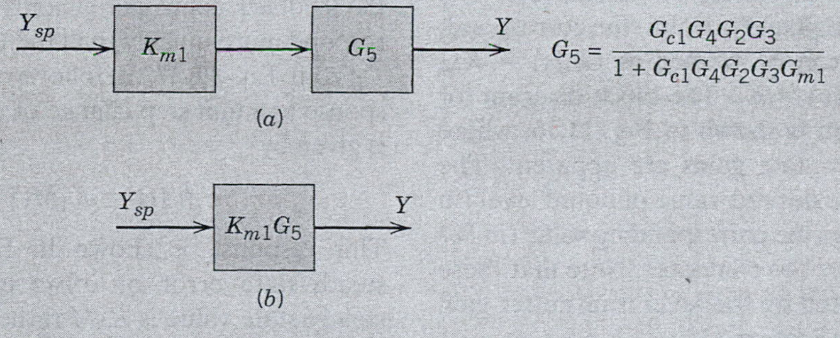

Figure 11.14 Final block diagrams for Example 11.1.

11.3 CLOSED-LOOP RESPONSES OF SIMPLE CONTROL SYSTEMS

In this section, we consider the dynamic behavior of several elementary control problems for disturbance variable and set-point changes. The transient responses can be determined in a straightforward manner if the closed-loop transfer functions are available.

Consider the liquid-level control system shown in Fig. 11.15. The liquid level is measured and the level transmitter (LT) output is sent to a feedback controller (LC) that controls liquid level h by adjusting volumetric flow rate q_2. A second inlet flow rate, q_1, is the disturbance variable. Assume that

1. The liquid density ρ and the cross-sectional area of the tank A are constant.
2. The flow-head relation is linear, $q_3 = h/R$.
3. The level transmitter, I/P transducer, and pneumatic control valve have negligible dynamics.
4. An electronic controller with input and output in % is used (full scale = 100%).

Derivation of the process and disturbance transfer functions directly follows Example 4.4. Consider the unsteady-state mass balance for the tank contents:

$$\rho A \frac{dh}{dt} = \rho q_1 + \rho q_2 - \rho q_3 \tag{11-32}$$

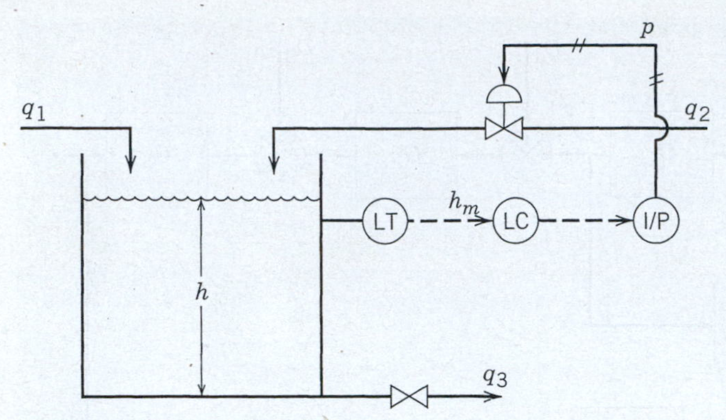

Figure 11.15 Liquid-level control system.

Substituting the flow-head relation, $q_3 = h/R$, and introducing deviation variables gives

$$A\frac{dh'}{dt} = q'_1 + q'_2\frac{h'}{R} \tag{11-33}$$

Thus, we obtain the transfer functions

$$\frac{H'(s)}{Q'_2(s)} = G_p(s) = \frac{K_p}{\tau s + 1} \tag{11-34}$$

$$\frac{H'(s)}{Q'_1(s)} = G_d(s) = \frac{K_p}{\tau s + 1} \tag{11-35}$$

where $K_p = R$ and $\tau = RA$. Note that $G_p(s)$ and $G_d(s)$ are identical, because q_1 and q_2 are both inlet flow rates and thus have the same effect on h.

Because the level transmitter, I/P transducer, and control valve have negligible dynamics, the corresponding transfer functions can be written as $G_m(s) = K_m$, $G_{IP}(s) = K_{IP}$, and $G_v(s) = K_v$. The block diagram for the level control system is shown in Fig. 11.16, where the units of the steady-state gains are apparent. The symbol H'_{sp} denotes the desired value of liquid level (in meters), and $\widetilde{H}'_{sp}$ denotes the corresponding value (in %) that is used internally by the computer. Note that these two set-points are related by the level transmitter gain K_m, as was discussed in Section 11.1.

The block diagram in Fig. 11.16 is in the alternative form of Fig. 11.9 with $G^*_d(s) = 1$.

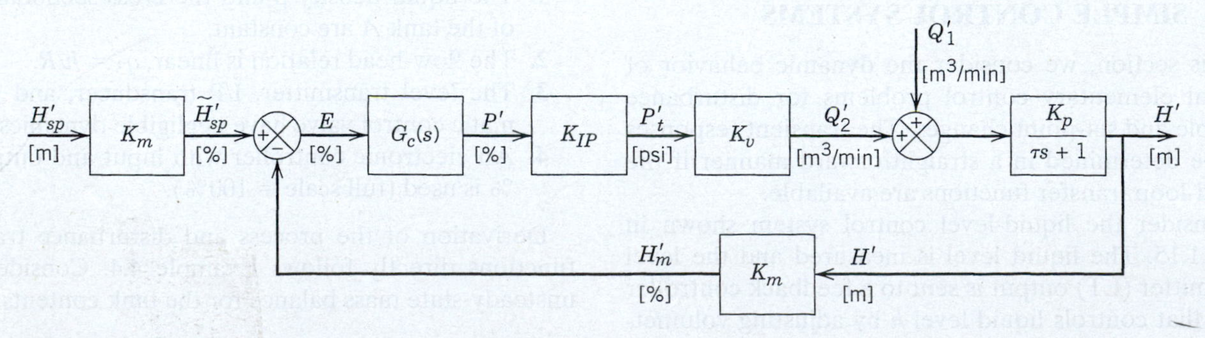

Figure 11.16 Block diagram for level control system.

11.3.1 Proportional Control and Set-Point Changes

If a proportional controller is used, $G_c(s) = K_c$. From Fig. 11.16 and the material in the previous section, it follows that the closed-loop transfer function for set-point changes is given by

$$\frac{H'(s)}{H'_{sp}(s)} = \frac{K_c K_{IP} K_v K_m/(\tau s + 1)}{1 + K_c K_{IP} K_v K_m/(\tau s + 1)} \tag{11-36}$$

This relation can be rearranged in the standard form for a first-order transfer function,

$$\frac{H'(s)}{H'_{sp}(s)} = \frac{K_1}{\tau_1 s + 1} \tag{11-37}$$

where

$$K_1 = \frac{K_{OL}}{1 + K_{OL}} \tag{11-38}$$

$$\tau_1 = \frac{\tau}{1 + K_{OL}} \tag{11-39}$$

and the *open-loop gain* K_{OL} is given by

$$K_{OL} = K_c K_{IP} K_v K_P K_m \tag{11-40}$$

Equations 11-37 to 11-40 indicate that the closed-loop process has first-order dynamics with a time constant τ_1 that is smaller than the process time constant τ. We assume here that $K_{OL} > 0$; otherwise, the control system would not function properly, as will be apparent from the stability analysis later in this chapter. Because $\tau_1 < \tau$, the feedback controller enables the controlled process to respond more quickly than the uncontrolled process.

From Eq. 11-37, it follows that the closed-loop response to a unit step change of magnitude M in set point is given by

$$h'(t) = K_1 M(1 - e^{-t/\tau_1}) \tag{11-41}$$

This response is shown in Fig. 11.17. Note that a steady-state error or *offset* exists, because the new steady-state value is $K_1 M$ rather than the desired value $M (K_1 < 1)$. The offset is defined as

$$\text{offset} \triangleq h'_{sp}(\infty) - h'(\infty) \tag{11-42}$$

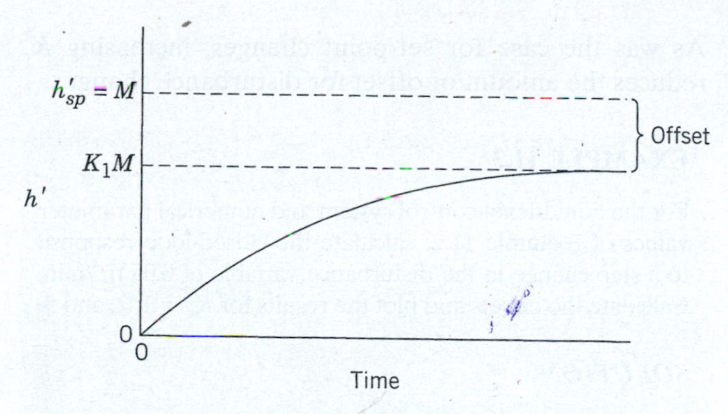

Figure 11.17 Step response for proportional control (set-point change).

For a step change of magnitude M in set point, $h'_{sp}(\infty) = M$. From (11-41), it is clear that $h'(\infty) = K_1M$. Substituting these values and (11-38) into (11-42) gives

$$\text{offset} = M - K_1M = \frac{M}{1 + K_{OL}} \tag{11-43}$$

EXAMPLE 11.2

Consider the level control system shown in Fig. 11.15, implemented with a computer whose inputs and outputs are calibrated in terms of full range (100%). The tank is 1 m in diameter, while the valve on the exit line acts as a linear resistance with $R = 6.37$ min/m^2. The level transmitter has a span of 2.0 m and an output range of 0 to 100%. The valve characteristic f of the equal percentage control valve is related to the fraction of lift ℓ by the relation $f = (30)^{\ell-1}$. The air-to-open control valve receives a 3 to 15 psi signal from an I/P transducer, which, in turn, receives a 0 to 100% signal from the computer-implemented proportional-only controller. When the control valve is fully open ($\ell = 1$), the flow rate through the valve is 0.2 m^3/min. At the nominal operating condition, the control valve is half-open ($\ell = 0.5$). Using the dynamic model in the block diagram of Fig. 11.16, calculate the closed-loop responses to a step change in the set point of 0.3 m for three values of the controller gain: $K_c = 4$, 8, and 20.

SOLUTION

From the given information, we can calculate the cross-sectional area of the tank A, the process gain K_p, and the time constant:

$$A = \pi\,(0.5\text{ m})^2 = 0.785\text{ m}^2$$

$$K_p = R = 6.37\text{ min/m}^2 \tag{11-44}$$

$$\tau = RA = 5\text{ min}$$

The sensor-transmitter gain K_m can be calculated from Eq. 9-1:

$$K_m = \frac{\text{output range}}{\text{input range}} = \frac{100 - 0\%}{2\text{ m}} = 50\%/\text{m} \tag{11-45}$$

The gain for the IP transducer is given by

$$K_{IP} = \frac{15 - 3\text{ psi}}{100 - 0\%} = 0.12\text{ psi}/\% \tag{11-46}$$

Next, we calculate the gain for the control valve K_v. The valve relation between flow rate q and fraction of lift ℓ can be expressed as (cf. Eqs. 9-4 and 9-5)

$$q = 0.2(30)^{\ell-1} \tag{11-47}$$

Thus

$$\frac{dq}{d\ell} = 0.2\ln 30\,(30)^{\ell-1} \tag{11-48}$$

At the nominal condition, $\ell = 0.5$ and

$$\frac{dq}{d\ell} = 0.124\text{ m}^3/\text{min} \tag{11-49}$$

The control valve gain K_v can be expressed as

$$K_v = \frac{dq}{dp_t} = \left(\frac{dq}{d\ell}\right)\left(\frac{d\ell}{dp_t}\right) \tag{11-50}$$

If the valve actuator is designed so that the fraction of lift ℓ varies linearly with the IP transducer output p_t, then

$$\frac{d\ell}{dp_t} = \frac{\Delta\ell}{\Delta p_t} = \frac{1 - 0}{15 - 3\text{ psi}} = 0.0833\text{ psi}^{-1} \tag{11-51}$$

Then, from Eqs. 11-48, 11-50, and 11-51,

$$K_v = 1.03 \times 10^{-2}\text{ m}^3/\text{min psi} \tag{11-52}$$

An alternative method for estimating K_v is to use the tangent to the valve characteristic curve (see Chapter 9). Now that all of the gains and the time constant in Fig. 11.16 have been calculated, we can calculate the closed-loop gain K_1 and time constant τ_1 in Eq. 11-41. Substituting these numerical values into Eqs. 11-38 and 11-39 for the three values of K_c gives

K_c	τ_1 (min)	K_1
4	1.94	0.612
8	1.20	0.759
20	0.56	0.887

The closed-loop responses are shown in Fig. 11.18. Increasing K_c reduces both the offset and the time required to reach the new steady state.

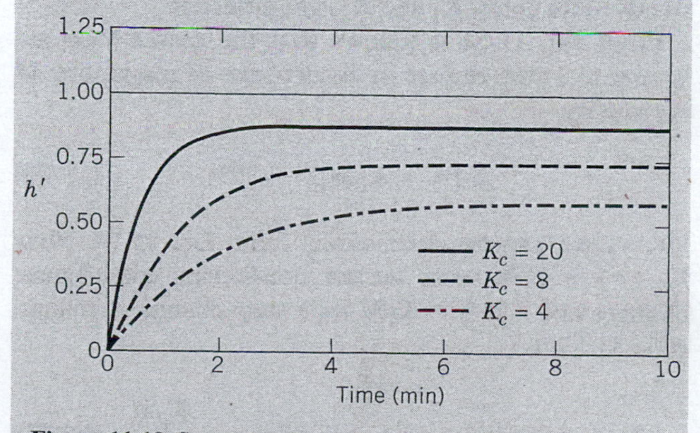

Figure 11.18 Set-point responses for Example 11.2.

Equation 11-43 suggests that offset can be reduced by increasing K_c. However, for most control problems, making K_c too large can result in oscillatory or unstable responses due to the effect of additional lags and time delays that have been neglected in the present analysis. For example, we have neglected the dynamics associated with the control valve, level transmitter, and pneumatic transmission line between the I/P transducer and control valve. A more rigorous analysis, including the dynamics of these components, would reveal the possibility of oscillations or instability. Stability problems associated with feedback control systems are analyzed in Section 11.4.

For many liquid-level control problems, a small offset can be tolerated, because the vessel serves as a surge capacity (or intermediate storage volume) between processing units.

11.3.2 Proportional Control and Disturbance Changes

From Fig. 11.16 and Eq. 11-29, the closed-loop transfer function for disturbance changes with proportional control is

$$\frac{H'(s)}{Q_1'(s)} = \frac{K_p/(\tau s+1)}{1+K_{OL}/(\tau s+1)} \tag{11-53}$$

Rearranging gives

$$\frac{H'(s)}{Q_1'(s)} = \frac{K_2}{\tau_1 s+1} \tag{11-54}$$

where τ_1 is defined in (11-39) and K_2 is given by

$$K_2 = \frac{K_p}{1+K_{OL}} \tag{11-55}$$

A comparison of (11-54) and (11-37) indicates that both closed-loop transfer functions are first-order, and they have the same time constant. However, the steady-state gains, K_1 and K_2, are different.

From Eq. 11-54 it follows that the closed-loop response to a step change in disturbance of magnitude M is given by

$$h'(t) = K_2 M(1-e^{-t/\tau_1}) \tag{11-56}$$

The offset can be determined from Eq. 11-56. Now $h'_{sp}(\infty) = 0$, because we are considering disturbance changes and $h'(\infty) = K_2 M$ for a step change of magnitude M. Thus,

$$\text{offset} = 0 - h'(\infty) = -K_2 M = -\frac{K_p M}{1+K_{OL}} \tag{11-57}$$

As was the case for set-point changes, increasing K_c reduces the amount of offset for disturbance changes.

EXAMPLE 11.3

For the liquid-level control system and numerical parameter values of Example 11.2, calculate the closed-loop response to a step change in the disturbance variable of 0.05 m^3/min. Calculate the offsets and plot the results for K_c = 1, 2, and 5.

SOLUTION

The closed-loop responses in Fig. 11.19 indicate that increasing K_c reduces the offset and speeds up the closed-loop response. The offsets are

K_c	Offset
4	−0.124
8	−0.077
20	−0.036

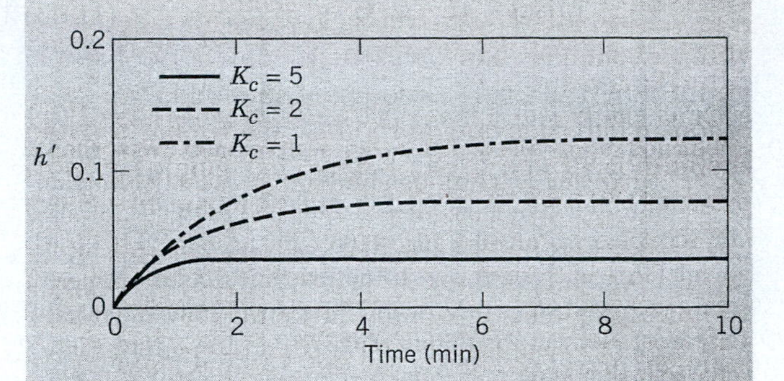

Figure 11.19 Disturbance responses for Example 11.3.

The negative values of offset indicate that the controlled variable is greater than the set point. For K_c = 0 (no control), the offset is –0.318.

11.3.3 PI Control and Disturbance Changes

For PI control, $G_c(s) = K_c(1 + 1/\tau_I s)$. The closed-loop transfer function for disturbance changes can then be derived from Fig. 11.16:

$$\frac{H'(s)}{Q_1'(s)} = \frac{K_p/(\tau s+1)}{1+K_{OL}(1+1/\tau_I s)/(\tau s+1)} \tag{11-58}$$

Clearing terms in the denominator gives

$$\frac{H'(s)}{Q_1'(s)} = \frac{K_p \tau_I s}{\tau_I s(\tau s+1)+K_{OL}(\tau_I s+1)} \tag{11-59}$$

Further rearrangement allows the denominator to be placed in the standard form for a second-order transfer function:

$$\frac{H'(s)}{Q_1'(s)} = \frac{K_3 s}{\tau_3^2 s^2 + 2\zeta_3 \tau_3 s + 1} \tag{11-60}$$

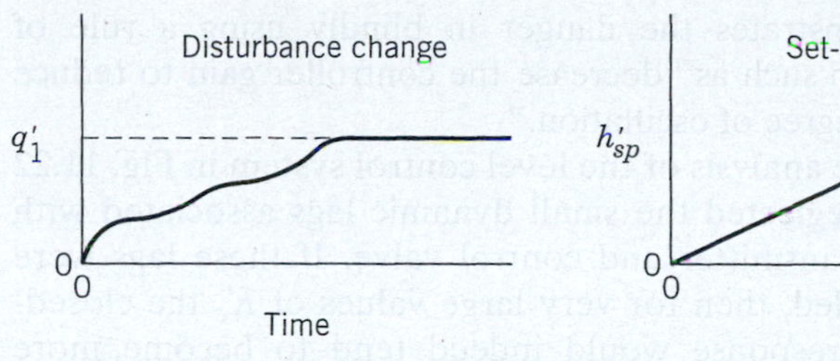

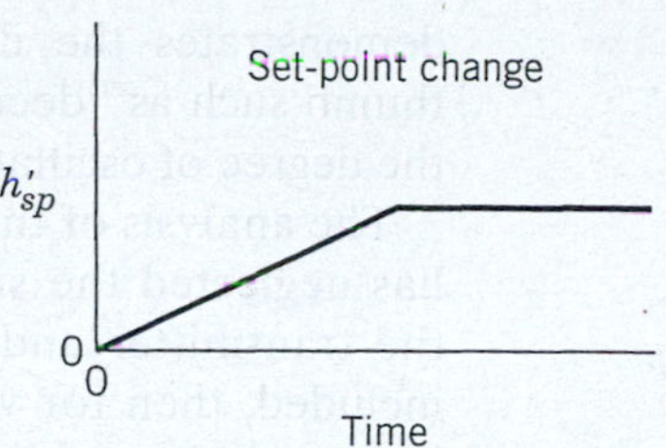

Figure 11.20 Sustained changes in disturbance and set point.

$$K_3 = \tau_I / K_c K_{IP} K_v K_m \tag{11-61}$$

$$\zeta_3 = \frac{1}{2}\left(\frac{1+K_{OL}}{\sqrt{K_{OL}}}\right)\sqrt{\frac{\tau_I}{\tau}} \tag{11-62}$$

$$\tau_3 = \sqrt{\tau\tau_I/K_{OL}} \tag{11-63}$$

For a unit step change in disturbance, $Q_1'(s) = 1/s$, and (11-59) becomes

$$H'(s) = \frac{K_3}{\tau_3^2 s^2 + 2\zeta_3\tau_3 s + 1} \tag{11-64}$$

For $0 < \zeta_3 < 1$, the response is a damped oscillation that can be described by

$$h'(t) = \frac{K_3}{\tau_3\sqrt{1-\zeta_3^2}}\, e^{-\zeta_3 t/\tau_3} \sin[\sqrt{1-\zeta_3^2}\; t/\tau_3] \tag{11-65}$$

It is clear from (11-65) that $h'(\infty) = 0$ because of the negative exponential term. Thus, the addition of integral action eliminates offset for a step change in disturbance. It also eliminates offset for step changes in set point. In fact, integral action eliminates offset not only for step changes, but also for any type of sustained change in disturbance or set point. By a sustained change, we mean one that eventually settles out at a new steady-state value, as shown in Fig. 11.20. However, integral action does not eliminate offset for a ramp disturbance.

Equation (11-63) and Fig. 11.21 indicate that increasing K_c or decreasing τ_I tends to speed up the response. In addition, the response becomes more oscillatory as either K_c or τ_I decreases. But in general, closed-loop responses become more oscillatory as K_c is *increased* (see Example 11.4). These anomalous results occur because the small dynamic lags associated with the control valve and transmitter were neglected. If these lags are included, the transfer function in (11-60) is no longer second-order, and then increasing K_c makes the response more oscillatory.

11.3.4 PI Control of an Integrating Process

Consider the electronic liquid-level control system shown in Fig. 11.22. This system differs from the previous example in two ways: (1) the exit line contains a pump, and (2) the manipulated variable is the exit flow rate rather than an inlet flow rate. In Section 5.3, we saw that a tank with a pump in the exit stream can act as an integrator with respect to flow rate changes, because

$$\frac{H'(s)}{Q_3'(s)} = G_p(s) = -\frac{1}{As} \tag{11-66}$$

$$\frac{H'(s)}{Q_1'(s)} = G_d(s) = \frac{1}{As} \tag{11-67}$$

If the level transmitter and control valve in Fig. 11.22 have negligible dynamics, then $G_m(s) = K_m$ and $G_v(s) = K_v$. For PI control, $G_c(s) = K_c(1 + 1/\tau_I s)$.

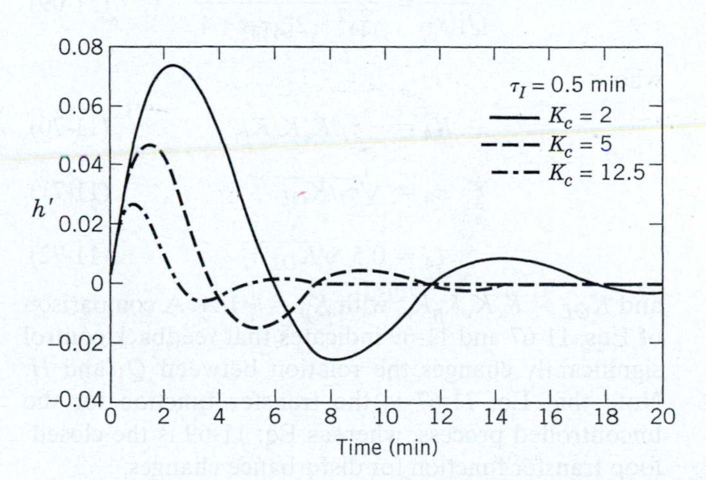

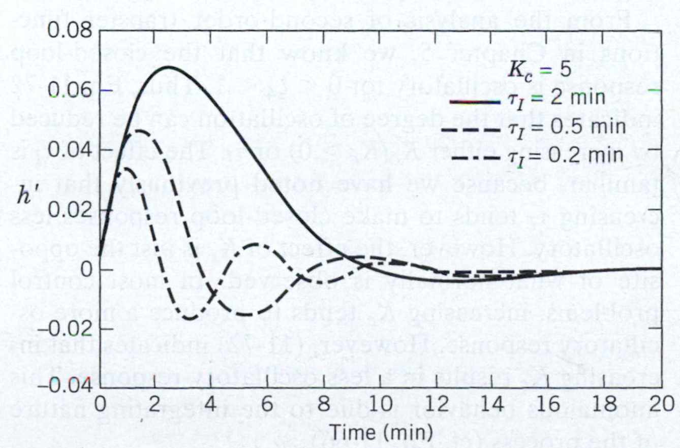

Figure 11.21 Effect of controller settings on disturbance responses.

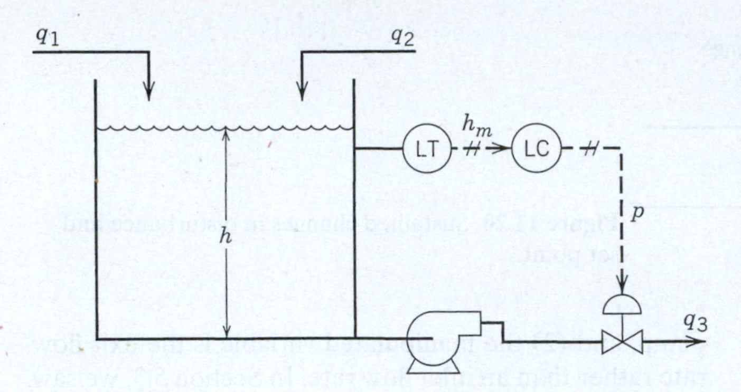

Figure 11.22 Liquid-level control system with pump in exit line.

Substituting these expressions into the closed-loop transfer function for disturbance changes

$$\frac{H'(s)}{Q_1'(s)} = \frac{G_d}{1 + G_c G_v G_p G_m} \tag{11-68}$$

and rearranging gives

$$\frac{H'(s)}{Q_1'(s)} = \frac{K_4 s}{\tau_4^2 s^2 + 2\zeta_4 \tau_4 s + 1} \tag{11-69}$$

where

$$K_4 = -\tau_I / K_c K_v K_m \tag{11-70}$$

$$\tau_4 = \sqrt{\tau_I / K_{OL}} \tag{11-71}$$

$$\zeta_4 = 0.5\sqrt{K_{OL}\tau_I} \tag{11-72}$$

and $K_{OL} = K_c K_v K_p K_m$ with $K_p = -1/A$. A comparison of Eqs. 11-67 and 11-69 indicates that feedback control significantly changes the relation between Q_1 and H. Note that Eq. 11-67 is the transfer function for the uncontrolled process, whereas Eq. 11-69 is the closed-loop transfer function for disturbance changes.

From the analysis of second-order transfer functions in Chapter 5, we know that the closed-loop response is oscillatory for $0 < \zeta_4 < 1$. Thus, Eq. 11-72 indicates that the degree of oscillation can be reduced by increasing either $K_c (K_c > 0)$ or τ_I. The effect of τ_I is familiar, because we have noted previously that increasing τ_I tends to make closed-loop responses less oscillatory. However, the effect of K_c is just the opposite of what normally is observed. In most control problems, increasing K_c tends to produce a more oscillatory response. However, (11-72) indicates that increasing K_c results in a less oscillatory response. This anomalous behavior is due to the integrating nature of the process (cf. Eq. 11-66).

This liquid-level system illustrates the insight that can be obtained from block diagram analysis. It also demonstrates the danger in blindly using a rule of thumb such as "decrease the controller gain to reduce the degree of oscillation."

The analysis of the level control system in Fig. 11.22 has neglected the small dynamic lags associated with the transmitter and control valve. If these lags were included, then for very large values of K_c the closed-loop response would indeed tend to become more oscillatory. Thus, if τ_I is held constant, the effect of K_c on the higher-order system can be summarized as follows:

Value of K_c	Closed-Loop Response
Small	Oscillatory
Moderate or large	Overdamped (nonoscillatory)
Very large	Oscillatory or unstable

Because the liquid-level system in Fig. 11.22 acts as an integrator, the question arises whether the controller must also contain integral action to eliminate offset. This question is considered further in Exercise 11.6.

In the previous examples, the denominator of the closed-loop transfer function was either a first- or second-order polynomial in s. Consequently, the transient responses to specified inputs were easily determined. In many control problems, the order of the denominator polynomial is three or higher, and the roots of the polynomial have to be determined numerically. Furthermore, for higher-order ($n > 2$) systems, feedback control can result in unstable responses if inappropriate values of the controller settings are employed.

11.4 STABILITY OF CLOSED-LOOP CONTROL SYSTEMS

An important consequence of feedback control is that it can cause oscillatory responses. If the oscillation has a small amplitude and damps out quickly, then the control system performance is generally considered to be satisfactory. However, under certain circumstances, the oscillations may be undamped or even have an amplitude that increases with time until a physical limit is reached, such as a control valve being fully open or completely shut. In these situations, the closed-loop system is said to be *unstable*.

In the remainder of this chapter, we analyze the stability characteristics of closed-loop systems and present several useful criteria for determining whether a system will be stable. Additional stability criteria based on frequency response analysis are discussed in Chapter 14. But first we consider an illustrative example of a closed-loop system that can become unstable.

EXAMPLE 11.4

Consider the feedback control system shown in Fig. 11.8 with the following transfer functions:

$$G_c = K_c \qquad G_v = \frac{1}{2s+1} \tag{11-73}$$

$$G_p = G_d = \frac{1}{5s+1} \qquad G_m = \frac{1}{s+1} \tag{11-74}$$

Show that the closed-loop system produces unstable responses if controller gain K_c is too large.

SOLUTION

To determine the effect of K_c on the closed-loop response $y(t)$, consider a unit step change in set point, $Y_{sp}(s) = 1/s$. In Section 11.2 we derived the closed-loop transfer function for set-point changes (cf. Eq. 11-26):

$$\frac{Y}{Y_{sp}} = \frac{K_m G_c G_v G_p}{1 + G_c G_v G_p G_m} \tag{11-75}$$

Substituting (11-73) and (11-74) into (11-75) and rearranging gives

$$Y(s) = \frac{K_c(s+1)}{10s^3 + 17s^2 + 8s + 1 + K_c}\,\frac{1}{s} \tag{11-76}$$

After K_c is specified, $y(t)$ can be determined from the inverse Laplace transform of Eq. 11-76. But first the roots of the cubic polynomial in s must be determined before performing the partial fraction expansion. These roots can be calculated using standard root-finding techniques (Chapra and Canale, 2010). Figure 11.23 demonstrates that as K_c increases, the response becomes more oscillatory and is unstable for $K_c = 15$. More details on the actual stability limit of this control system are given in Example 11.10.

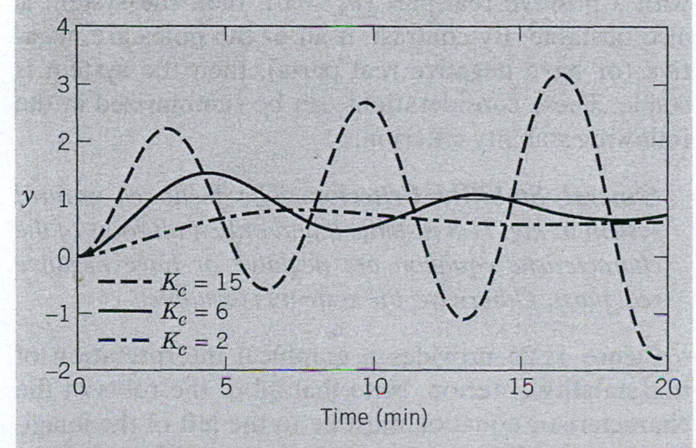

Figure 11.23 Effect of controller gains on closed-loop response to a unit step change in set point (Example 11.4).

The unstable response for Example 11.4 is oscillatory, with the amplitude growing in each successive cycle. In contrast, for an actual physical system, the amplitudes will increase until a physical limit is reached or an equipment failure occurs. Because the final control element usually has saturation limits (see Chapter 9), the unstable response will manifest itself as a sustained oscillation with a constant amplitude instead of a continually increasing amplitude. Sustained oscillations can also occur without having the final control element *saturate*, as was mentioned in Section 11.3.

Clearly, a feedback control system must be stable as a prerequisite for satisfactory control. Consequently, it is of considerable practical importance to be able to determine under what conditions a control system becomes unstable. For example, for what values of the PID controller parameters K_c, τ_I, and τ_D is the controlled process stable?

11.4.1 General Stability Criterion

Most industrial processes are stable without feedback control. Thus, they are said to be *open-loop stable,* or *self-regulating*. An open-loop stable process will return to the original steady state after a transient disturbance (one that is not sustained) occurs. By contrast, there are a few processes, such as exothermic chemical reactors, that can be *open-loop unstable*. These processes are extremely difficult to operate without feedback control.

Before presenting various stability criteria, we introduce the following definition for unconstrained linear systems. We use the term *unconstrained* to refer to the ideal situation where there are no physical limits on the input and output variables.

Definition of Stability. *An unconstrained linear system is said to be stable if the output response is bounded for all bounded inputs. Otherwise it is said to be unstable.*

By a bounded input, we mean an input variable that stays within upper and lower limits for all values of time. For example, consider a variable $u(t)$ that varies with time. If $u(t)$ is a step or sinusoidal function, then it is bounded. However, the functions $u(t) = t$ and $u(t) = e^{3t}$ are not bounded.

EXAMPLE 11.5

A liquid storage system is shown in Fig. 11.24. Show that this process is not self-regulating by considering its response to a step change in inlet flow rate.

SOLUTION

The transfer function relating liquid level h to inlet flow rate q_i was derived in Section 5.3:

$$\frac{H'(s)}{Q_i'(s)} = \frac{1}{As} \tag{11-77}$$

where A is the cross-sectional area of the tank. For a step change of magnitude M_0, $Q_i'(s) = M_0/s$, and thus

$$H'(s) = \frac{M_0}{As^2} \tag{11-78}$$

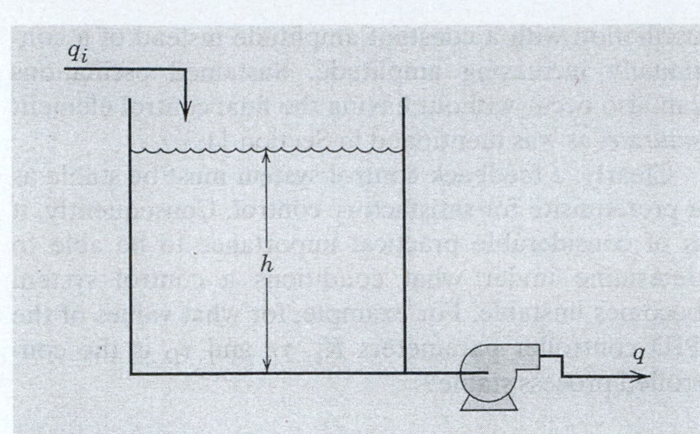

Figure 11.24 A liquid storage system that is not self-regulating.

Taking the inverse Laplace transform gives the transient response,

$$h'(t) = \frac{M_0}{A}t \tag{11-79}$$

We conclude that the liquid storage system is open-loop unstable (or *non-self-regulating*) because a bounded input has produced an unbounded response. However, if the pump in Fig. 11.24 were replaced by a valve, then the storage system would be self-regulating (cf. Example 4.4).

Characteristic Equation

As a starting point for the stability analysis, consider the block diagram in Fig. 11.8. Using block diagram algebra that was developed earlier in this chapter, we obtain

$$Y = \frac{K_m G_c G_v G_p}{1 + G_{OL}} Y_{sp} + \frac{G_d}{1 + G_{OL}} D \tag{11-80}$$

where G_{OL} is the open-loop transfer function, $G_{OL} = G_c G_v G_p G_m$.

For the moment consider set-point changes only, in which case Eq. 11-80 reduces to the closed-loop transfer function

$$\frac{Y}{Y_{sp}} = \frac{K_m G_c G_v G_p}{1 + G_{OL}} \tag{11-81}$$

If G_{OL} is a ratio of polynomials in s (i.e., a rational function), then the closed-loop transfer function in Eq. 11-81 is also a rational function. After rearrangement, it can be factored into poles (p_i) and zeroes (z_i) as

$$\frac{Y}{Y_{sp}} = K' \frac{(s - z_1)(s - z_2) \ldots (s - z_m)}{(s - p_1)(s - p_2) \ldots (s - p_n)} \tag{11-82}$$

where K' is a multiplicative constant that gives the correct steady-state gain. To have a physically realizable system, the number of poles must be greater than or equal to the number of zeroes; that is, $n \geq m$ (Kuo, 2002). Note that a *pole-zero cancellation* occurs if a zero and a pole have the same numerical value.

Comparing Eqs. 11-81 and 11-82 indicates that the poles are also the roots of the following equation, which is referred to as the *characteristic equation* of the closed-loop system:

$$1 + G_{OL} = 0 \tag{11-83}$$

The characteristic equation plays a decisive role in determining system stability, as discussed below.

For a unit change in set point, $Y_{sp}(s) = 1/s$, and Eq. 11-82 becomes

$$Y = \frac{K'}{s} \frac{(s - z_1)(s - z_2) \ldots (s - z_m)}{(s - p_1)(s - p_2) \ldots (s - p_n)} \tag{11-84}$$

If there are no repeated poles (i.e., if they are all *distinct poles*), then the partial fraction expansion of Eq. 11-84 has the form considered in Section 6.1,

$$Y(s) = \frac{A_0}{s} + \frac{A_1}{s - p_1} + \frac{A_2}{s - p_2} + \cdots + \frac{A_n}{s - p_n} \tag{11-85}$$

where the $\{A_i\}$ can be determined using the methods of Chapter 3. Taking the inverse Laplace transform of Eq. 11-85 gives

$$y(t) = A_0 + A_1 e^{p_1 t} + A_2 e^{p_2 t} + \cdots + A_n e^{p_n t} \tag{11-86}$$

Suppose that one of the poles is a positive real number; that is, $p_k > 0$. Then it is clear from Eq. 11-86 that $y(t)$ is unbounded, and thus the closed-loop system in Fig. 11.8 is unstable. If p_k is a complex number, $p_k = a_k + jb_k$, with a positive real part ($a_k > 0$), then the system is also unstable. By contrast, if all of the poles are negative (or have negative real parts), then the system is stable. These considerations can be summarized in the following stability criterion:

General Stability Criterion. *The feedback control system in Fig. 11.8 is stable if and only if all roots of the characteristic equation are negative or have negative real parts. Otherwise, the system is unstable.*

Figure 11.25 provides a graphical interpretation of this stability criterion. Note that all of the roots of the characteristic equation must lie to the left of the imaginary axis in the complex plane for a stable system to exist. The qualitative effects of these roots on the transient response of the closed-loop system are shown in Fig. 11.26. The left portion of each part of Fig. 11.26 shows representative root locations in the complex plane. The corresponding figure on the right shows the contributions these poles make to the closed-loop response for a step change in set point. Similar responses would occur for a step change in a disturbance. A system that has all negative real roots will have a stable,

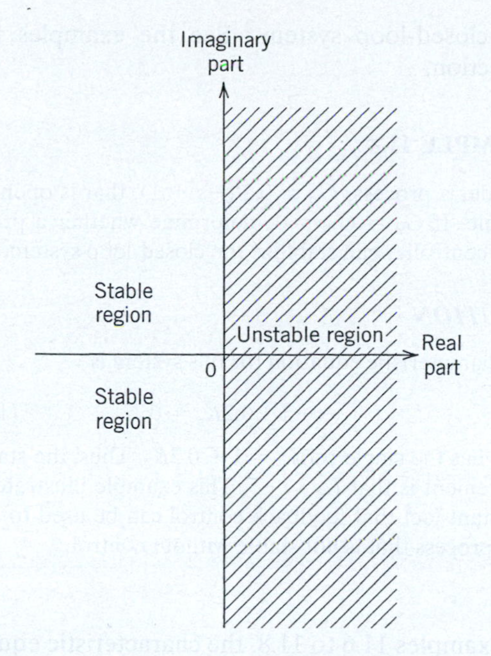

Figure 11.25 Stability regions in the complex plane for roots of the characteristic equation.

nonoscillatory response, as shown in Fig. 11.26*a*. On the other hand, if one of the real roots is positive, then the response is unbounded, as shown in Fig. 11.26*b*. A pair of complex conjugate roots results in oscillatory responses as shown in Figs. 11.26*c* and 11.26*d*. If the complex roots have negative real parts, the system is stable; otherwise it is unstable. Recall that complex roots always occur as complex conjugate pairs.

The root locations also provide an indication of how rapid the transient response will be. A real root at $s = -a$ corresponds to a closed-loop time constant of $\tau = 1/a$, as is evident from Eqs. 11-85 and 11-86. Thus, real roots close to the imaginary (vertical) axis result in slow responses. Similarly, complex roots near the imaginary axis correspond to slow response modes. The farther the complex roots are away from the real axis, the more oscillatory the transient response will be (see Example 11.14). However, the process zeros also influence the response, as discussed in Chapter 6.

Note that the same characteristic equation occurs for both disturbance and set-point changes because the term, $1 + G_{OL}$, appears in the denominator of both terms in Eq. 11-80. Thus, if the closed-loop system is stable for disturbances, it will also be stable for set-point changes.

The analysis in Eqs. 11-80 to 11-86 that led to the general stability criterion was based on a number of assumptions:

1. Set-point changes (rather than disturbance changes) were considered.
2. The closed-loop transfer function was a ratio of polynomials.
3. The poles in Eq. 11-82 were all distinct.

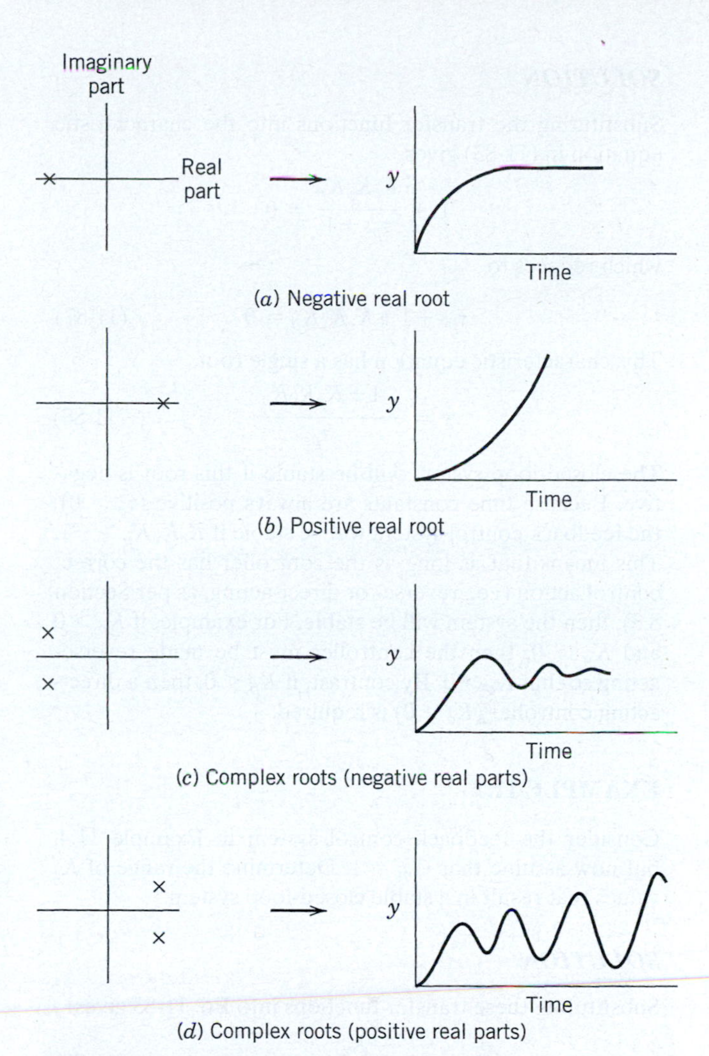

Figure 11.26 Contributions of characteristic equation roots to closed-loop response.

However, the general stability criterion is valid even if these assumptions are removed. In fact, this stability criterion is valid for *any* linear control system (comprised of linear elements described by transfer functions). By contrast, for nonlinear systems rigorous stability analyses tend to be considerably more complex and involve special techniques such as Liapunov and Popov stability criteria (Khalil, 2001). Fortunately, a stability analysis of a linearized system using the techniques presented in this chapter normally provides useful information for nonlinear systems operating near the point of linearization.

From a mathematical point of view, the general stability criterion presented above is a *necessary and sufficient condition*. Thus, linear system stability is completely determined by the roots of the characteristic equation.

EXAMPLE 11.6

Consider the feedback control system in Fig. 11.8 with $G_v = K_v$, $G_m = 1$, and $G_p = K_p/(\tau_p s + 1)$. Determine the stability characteristics if a proportional controller is used, $G_c = K_c$.

SOLUTION

Substituting the transfer functions into the characteristic equation in (11-83) gives

$$1 + \frac{K_c K_v K_p}{\tau_p s + 1} = 0$$

which reduces to

$$\tau_p s + 1 + K_c K_v K_p = 0 \tag{11-87}$$

This characteristic equation has a single root,

$$s = -\frac{1 + K_c K_v K_p}{\tau_p} \tag{11-88}$$

The closed-loop system will be stable if this root is negative. Because time constants are always positive ($\tau_p > 0$), the feedback control system will be stable if $K_c K_v K_p > -1$. This means that as long as the controller has the correct control action (i.e., reverse- or direct-acting, as per Section 8.3), then the system will be stable. For example, if $K_p > 0$ and $K_v > 0$, then the controller must be made reverse-acting so that $K_c > 0$. By contrast, if $K_p < 0$, then a direct-acting controller ($K_c < 0$) is required.

EXAMPLE 11.7

Consider the feedback control system in Example 11.4, but now assume that $G_m = 1$. Determine the range of K_c values that result in a stable closed-loop system.

SOLUTION

Substituting these transfer functions into Eq. 11-83 gives

$$1 + \frac{K_c}{(2s+1)(5s+1)} = 0 \tag{11-89}$$

which can be rearranged as

$$10s^2 + 7s + K_c + 1 = 0 \tag{11-90}$$

Applying the quadratic formula yields the roots,

$$s = \frac{-7 \pm \sqrt{49 - 40(K_c + 1)}}{20} \tag{11-91}$$

To have a stable system, both roots of this characteristic equation must have negative real parts. Equation 11-91 indicates that the roots will be negative if $40(K_c + 1) > 0$, because this means that the square root will have a value less than 7. If $40(K_c + 1) > 0$, then $K_c + 1 > 0$ and $K_c > -1$. Thus, we conclude that the closed-loop system will be stable if $K_c > -1$.

The stability analyses for Examples 11.6 and 11.7 have indicated that these closed-loop systems will be stable for all positive values of K_c, no matter how large. However, this result is *not* typical, because it occurs only for the special case where the open-loop system is stable and the open-loop transfer function G_{OL} is first- or second-order with no time delay. In more typical problems, K_c must be below an upper limit to have a stable closed-loop system.[2] See the examples in the next section.

EXAMPLE 11.8

Consider a process, $G_p = 0.2/(-s + 1)$, that is open-loop unstable. If $G_v = G_m = 1$, determine whether a proportional controller can stabilize the closed-loop system.

SOLUTION

The characteristic equation for this system is

$$-s + 1 + 0.2K_c = 0 \tag{11-92}$$

which has the single root $s = 1 + 0.2K_c$. Thus, the stability requirement is that $K_c < -5$. This example illustrates the important fact that feedback control can be used to stabilize a process that is not stable without control.

In Examples 11.6 to 11.8, the characteristic equations were either first- or second-order, and thus we could find the roots analytically. For higher-order polynomials, this is not possible, and numerical root-finding techniques (Chapra and Canale, 2010), also available in MATLAB and Mathematica, must be employed. An attractive alternative, the Routh stability criterion, is available to evaluate stability without requiring calculation of the roots of the characteristic equation.

11.4.2 Routh Stability Criterion

Routh (1905) published an analytical technique for determining whether any roots of a polynomial have positive real parts. According their general stability criterion, a closed-loop system will be stable only if all of the roots of the characteristic equation have negative real parts. Thus, by applying Routh's technique to analyze the coefficients of the characteristic equation, we can determine whether the closed-loop system is stable. This approach is referred to as the *Routh stability criterion*. It can be applied only to systems whose characteristic equations are polynomials in *s*. Thus, the Routh stability criterion is not directly applicable to systems containing time delays, because an $e^{-\theta s}$ term appears in the characteristic equation where θ is the time delay. However, if $e^{-\theta s}$ is replaced by a Padé approximation (see Section 6.2), then an *approximate* stability analysis can be performed (cf. Example 11.11). An exact stability analysis of systems containing time delays can be performed by direct root-finding or by using a frequency response analysis and the Bode or Nyquist stability criterion presented in Chapter 14.

[2]If a direct-acting controller is used (i.e., $K_c < 0$), then stability considerations place an upper limit on $-K_c$ rather than on K_c.

The Routh stability criterion is based on a characteristic equation that has the form

$$a_n s^n + a_{n-1}s^{n-1} + \cdots + a_1 s + a_0 = 0 \qquad (11\text{-}93)$$

We arbitrarily assume that $a_n > 0$. If $a_n < 0$, simply multiply Eq. 11-93 by -1 to generate a new equation that satisfies this condition. A *necessary* (but not sufficient) condition for stability is that all of the coefficients ($a_0, a_1, \ldots, a_n$) in the characteristic equation be positive. If any coefficient is negative or zero, then at least one root of the characteristic equation lies to the right of, or on, the imaginary axis, and the system is unstable. If all of the coefficients are positive, we next construct the following Routh array:

Row				
1	a_n	a_{n-2}	a_{n-4}	$\cdots$
2	a_{n-1}	a_{n-3}	a_{n-5}	$\cdots$
3	b_1	b_2	b_3	$\cdots$
4	c_1	c_2		
$\vdots$	$\vdots$			
$n+1$	z_1			

The Routh array has $n + 1$ rows, where n is the order of the characteristic equation, Eq. 11-93. The Routh array has a roughly triangular structure with only a single element in the last row. The first two rows are merely the coefficients in the characteristic equation, arranged according to odd and even powers of s. The elements in the remaining rows are calculated from the formulas

$$b_1 = \frac{a_{n-1}a_{n-2} - a_n a_{n-3}}{a_{n-1}} \qquad (11\text{-}94)$$

$$b_2 = \frac{a_{n-1}a_{n-4} - a_n a_{n-5}}{a_{n-1}} \qquad (11\text{-}95)$$

$$\vdots$$

$$c_1 = \frac{b_1 a_{n-3} - a_{n-1}b_2}{b_1} \qquad (11\text{-}96)$$

$$c_2 = \frac{b_1 a_{n-5} - a_{n-1}b_3}{b_1} \qquad (11\text{-}97)$$

$$\vdots$$

Note that the expressions in the numerators of Eqs. 11-94 to 11-97 are similar to the calculation of a determinant for a 2×2 matrix except that the order of subtraction is reversed. Having constructed the Routh array, we can now state the Routh stability criterion:

Routh Stability Criterion. *A necessary and sufficient condition for all roots of the characteristic equation in Eq. 11-93 to have negative real parts is that all of the elements in the left column of the Routh array are positive.*

Next we present three examples that show how the Routh stability criterion can be applied.

EXAMPLE 11.9

Determine the stability of a system that has the characteristic equation

$$s^4 + 5s^3 + 3s^2 + 1 = 0 \qquad (11\text{-}98)$$

SOLUTION

Because the s term is missing, its coefficient is zero. Thus, the system is unstable. Recall that a necessary condition for stability is that all of the coefficients in the characteristic equation must be positive.

EXAMPLE 11.10

Find the values of controller gain K_c that make the feedback control system of Example 11.4 stable.

SOLUTION

From Eq. 11-76, the characteristic equation is

$$10s^3 + 17s^2 + 8s + 1 + K_c = 0 \qquad (11\text{-}99)$$

All coefficients are positive provided that $1 + K_c > 0$ or $K_c > -1$. The Routh array is

10	8
17	$1 + K_c$
b_1	b_2
c_1	

To have a stable system, each element in the left column of the Routh array must be positive. Element b_1 will be positive if $K_c < 12.6$. Similarly, c_1 will be positive if $K_c > -1$. Thus, we conclude that the system will be stable if

$$-1 < K_c < 12.6 \qquad (11\text{-}100)$$

This example illustrates that stability limits for controller parameters can be derived analytically using the Routh array; in other words, it is not necessary to specify a numerical value for K_c before performing the stability analysis.

EXAMPLE 11.11

Consider a feedback control system with $G_c = K_c$, $G_v = 2$, $G_m = 0.25$, and $G_p = 4e^{-s}/(5s + 1)$. The characteristic equation is

$$1 + 5s + 2K_c e^{-s} = 0 \qquad (11\text{-}101)$$

Because this characteristic equation is not a polynomial in s, the Routh criterion is not directly applicable. However, if a polynomial approximation to e^{-s} is introduced, such as a Padé approximation (see Chapter 6), then the Routh criterion can be used to determine approximate stability limits. For simplicity, use the 1/1 Padé approximation,

$$e^{-s} \approx \left(\frac{1-0.5s}{1+0.5s}\right) \tag{11-102}$$

and determine the stability limits for the controller gain.

SOLUTION

Substituting Eq. 11-102 into 11-101 gives

$$1 + 5s + 2K_c\left(\frac{1 - 0.5s}{1 + 0.5s}\right) = 0 \tag{11-103}$$

Multiplying both sides by $1 + 0.5s$ and rearranging gives

$$2.5s^2 + (5.5 - K_c)s + (1 + 2K_c) = 0 \tag{11-104}$$

The necessary condition for stability is that each coefficient in this characteristic equation must be positive. This situation occurs if $-0.5 < K_c < 5.5$. The Routh array is

2.5	$1 + 2K_c$
$5.5 - K_c$	0
$1 + 2K_c$	

In this example, the Routh array provides no additional information but merely confirms that the system with the Padé approximation is stable if $-0.5 < K_c < 5.5$.

An exact time-delay analysis, without the Padé approximation and based on the Bode stability criterion (see Example 14.6), indicates that the actual upper limit on K_c is 4.25, which is 23% lower than the approximate value of 5.5 from the Routh stability criterion and the 1/1 Padé approximation. If the 2/2 Padé approximation in Eq. 6-37 is used with the Routh stability criterion, an approximate maximum controller gain $K_{cm} = 4.29$ will be obtained, much closer to the correct value of 4.25. This derivation is left as an exercise for the reader.

11.4.3 Direct Substitution Method

The imaginary axis divides the complex plane into stable and unstable regions for the roots of the characteristic equation, as indicated in Fig. 11.26. On the imaginary axis, the real part of s is zero, and thus we can write $s = j\omega$. Substituting $s = j\omega$ into the characteristic equation allows us to find a stability limit such as the maximum value of K_c (Luyben and Luyben, 1997). As the gain K_c is increased, the roots of the characteristic equation cross the imaginary axis when $K_c = K_{cm}$.

EXAMPLE 11.12

Use the direct substitution method to determine K_{cm} for the system with the characteristic equation given by Eq. 11-99.

SOLUTION

Substitute $s = j\omega$, $\omega = \omega_m$, and $K_c = K_{cm}$ into Eq. 11-99:

$$-10j\omega_m^3 - 17\omega_m^2 + 8j\omega_m + 1 + K_{cm} = 0$$

or

$$(1 + K_{cm} - 17\omega_m^2) + j(8\omega_m - 10\omega_m^3) = 0 \tag{11-105}$$

Equation 11-105 is satisfied if both the real and imaginary parts are identically zero:

$$1 + K_{cm} - 17\omega_m^2 = 0 \tag{11-106a}$$

$$8\omega_m - 10\omega_m^3 = \omega_m(8 - 10\omega_m^2) = 0 \tag{11-106b}$$

Therefore,

$$\omega_m^2 = 0.8 \Rightarrow \omega_m = \pm 0.894 \tag{11-107}$$

and from (11-106a),

$$K_{cm} = 12.6$$

Thus, we conclude that $K_c < 12.6$ for stability. Equation 11-107 indicates that at the stability limit (where $K_c = K_{cm} = 12.6$), a sustained oscillation occurs that has a frequency of $\omega_m = 0.894$ radian/min if the time constants have units of minutes. (Recall that a pair of complex roots on the imaginary axis, $s = \pm j\omega$, results in an undamped oscillation of frequency ω.) The corresponding period P is $2\pi/0.894 = 7.03$ min.

The direct-substitution method is related to the Routh stability criterion in Section 11.4.2. If the characteristic equation has a pair of roots on the imaginary axis, equidistant from the origin, and all other roots are in the left-hand plane, the single element in the next-to-last row of the Routh array will be zero. Then the location of the two imaginary roots can be obtained from the solution of the equation.

$$Cs^2 + D = 0$$

where C and D are the two elements in the $(n - 1)$ row of the Routh array, as read from left to right.

The direct-substitution method is also related to the frequency response approach of Chapter 14, because both techniques are based on the substitution $s = j\omega$.

11.5 ROOT LOCUS DIAGRAMS

In the previous section we have seen that the roots of the characteristic equation play a crucial role in determining system stability and the nature of the closed-loop responses. In the design and analysis of control systems, it is instructive to know how the roots of the characteristic equation change when a particular system parameter such as a controller gain changes. A root locus diagram provides a convenient graphical display of this information, as indicated in the following example.

EXAMPLE 11.13

Consider a feedback control system that has the open-loop transfer function,

$$G_{OL}(s) = \frac{4K_c}{(s+1)(s+2)(s+3)} \tag{11-108}$$

Plot the root locus diagram for $0 \le K_c \le 20$.

SOLUTION

The characteristic equation is $1 + G_{OL} = 0$ or

$$(s+1)(s+2)(s+3) + 4K_c = 0 \tag{11-109}$$

The root locus diagram in Fig. 11.27 shows how the three roots of this characteristic equation vary with K_c. When $K_c = 0$, the roots are merely the poles of the open-loop transfer function, -1, -2, and -3. These are designated by an $\times$ symbol in Fig. 11.27. As K_c increases, the root at -3 decreases monotonically. The other two roots converge and then form a complex conjugate pair when $K_c = 0.1$. When $K_c = K_{cm} = 15$, the complex roots cross the imaginary axis and enter the unstable region. This illustrates why the substitution of $s = j\omega$ (Section 11.3) determines the unstable controller gain. Thus, the root locus diagram indicates that the closed-loop system is unstable for $K_c > 15$. It also indicates that the closed-loop response will be nonoscillatory for $K_c < 0.1$.

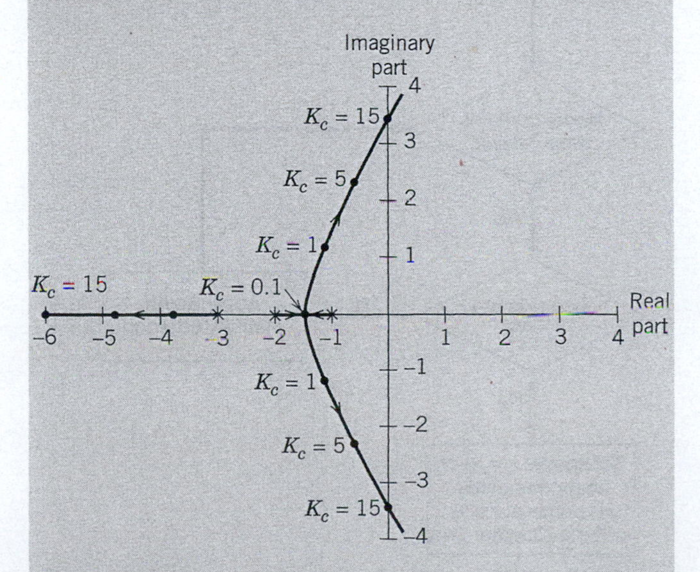

Figure 11.27 Root locus diagram for third-order system.

The root locus diagram can be used to provide a quick estimate of the transient response of the closed-loop system. The roots closest to the imaginary axis correspond to the slowest response modes. If the two closest roots are a complex conjugate pair (as in Fig. 11.28), then the closed-loop system can be approximated by an underdamped second-order system as follows.

Consider the standard second-order transfer function of Chapter 6,

$$G(s) = \frac{K}{\tau^2 s^2 + 2\zeta\tau s + 1} \tag{11-110}$$

which has the following roots when $0 < \zeta < 1$:

$$s = -\frac{\zeta}{\tau} \pm j\frac{\sqrt{1-\zeta^2}}{\tau} \tag{11-111}$$

These roots are shown graphically in Fig. 11.28. Note that the length d in Fig. 11.28 is given by

$$d = \sqrt{\left(\frac{\zeta}{\tau}\right)^2 + \frac{1-\zeta^2}{\tau^2}} = \sqrt{\frac{1}{\tau^2}} = \frac{1}{\tau} \tag{11-112}$$

Consequently,

$$\cos\psi = \frac{\zeta/\tau}{1/\tau} = \zeta \tag{11-113}$$

and

$$\psi = \cos^{-1}(\zeta) \tag{11-114}$$

This information provides the basis for a second-order approximation to a higher-order system, as illustrated in Example 11.14.

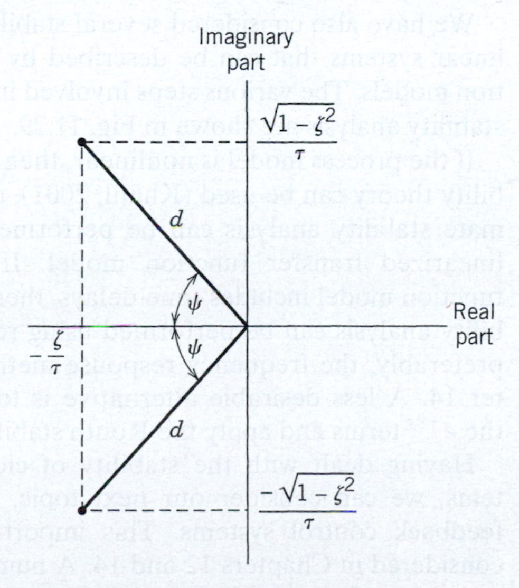

Figure 11.28 Root locations for underdamped second-order system.

EXAMPLE 11.14

Consider the root locus diagram in Fig. 11.27 for the third-order system of Example 11.13. For $K_c = 10$, determine values of ζ and τ that can be used to characterize the transient response approximately.

SOLUTION

For $K_c = 10$, there is one real root and two complex roots. By measuring the angle ψ and the distance d to the complex root, we obtain

$$\psi = \cos^{-1}\zeta = 75^\circ$$

$$d = 2.3$$

Then it follows from Eqs. 11-113 and 11-114 that

$$\zeta = 0.25 \quad \text{and} \quad \tau = 0.43.$$

Thus, the third-order system can be approximated by an underdamped second-order system with ζ and τ values given above. This information (and the material in Chapter 6) provide a useful characterization of the transient response.

The utility of root locus diagrams has been illustrated by the third-order system of Examples 11.13 and 11.14. The major disadvantage of root locus analysis is that time delays cannot be handled conveniently, and they require iterative solution of the nonlinear and nonrational characteristic equation. Nor is it easy to display simultaneous changes in more than one parameter (e.g., controller parameters K_c and τ_I). For this reason, the root locus technique has not found much use as a design tool in process control.

Root locus diagrams can be quickly generated by using a hand calculator or a computer with root-finding techniques such as are provided in MATLAB.

SUMMARY

This chapter has considered the dynamic behavior of processes that are operated under feedback control. A block diagram provides a convenient representation for analyzing control system performance. By using block diagram algebra, expressions for closed-loop transfer functions can be derived and used to calculate the closed-loop responses to disturbance and set-point changes. Several liquid-level control problems have been considered to illustrate key features of proportional and proportional-integral control. Proportional control results in offset for sustained disturbance or set-point changes; however, these offsets can be eliminated by including integral control action.

We have also considered several stability criteria for linear systems that can be described by transfer function models. The various steps involved in performing a stability analysis are shown in Fig. 11.29.

If the process model is nonlinear, then advanced stability theory can be used (Khalil, 2001), or an approximate stability analysis can be performed based on a linearized transfer function model. If the transfer function model includes time delays, then an exact stability analysis can be performed using root-finding or, preferably, the frequency response methods of Chapter 14. A less desirable alternative is to approximate the $e^{-\theta s}$ terms and apply the Routh stability criterion.

Having dealt with the stability of closed-loop systems, we can consider our next topic, the design of feedback control systems. This important subject is considered in Chapters 12 and 14. A number of prominent control system design and tuning techniques are based on stability criteria.

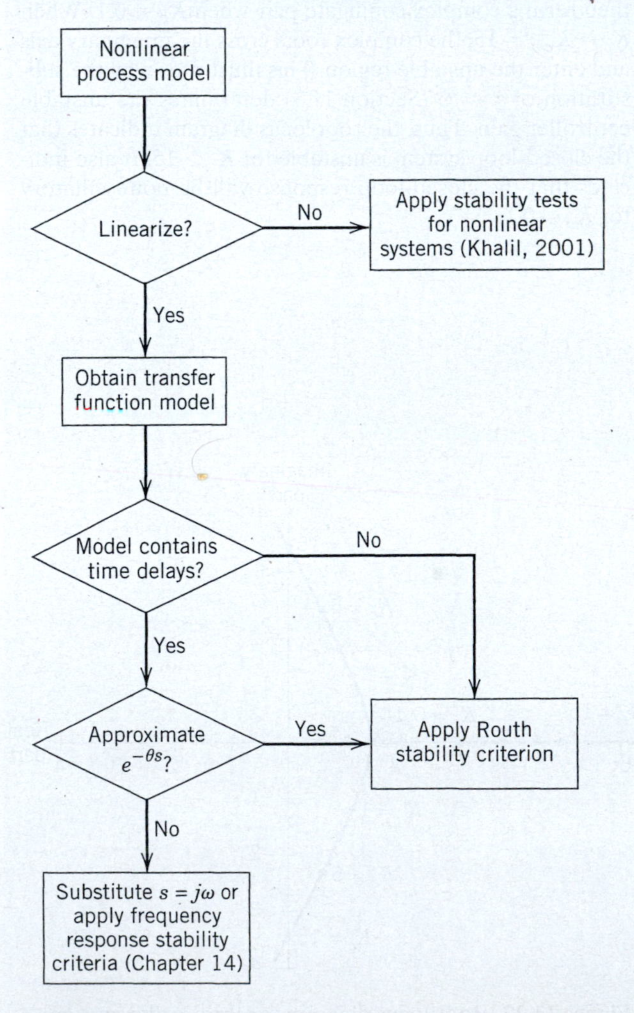

Figure 11.29 Flow chart for performing a stability analysis.

REFERENCES

Chapra, S. C. and R. P. Canale, *Numerical Methods for Engineers*, 6th ed., McGraw-Hill, New York, 2010.

Khalil, H. K., *Nonlinear Systems*, 3d ed., Prentice Hall, Englewood Cliffs, NJ, 2001.

Kuo, B. C., *Automatic Control Systems*, 8th ed., Prentice Hall, Englewood Cliffs, NJ, 2002.

Luyben, W. L. and M. L. Luyben, *Essentials of Process Control*, McGraw-Hill, New York, 1997.

Routh, E. J., *Dynamics of a System of Rigid Bodies, Part 11*, Macmillan, London, 1905.

EXERCISES

11.1 A temperature control system for a distillation column is shown in Fig. E11.1. The temperature T of a tray near the top of the column is controlled by adjusting the reflux flow rate R. Draw a block diagram for this feedback control system. You may assume that both feed flow rate F and feed composition x_F are disturbance variables and that all of the instrumentation, including the controller, is pneumatic.

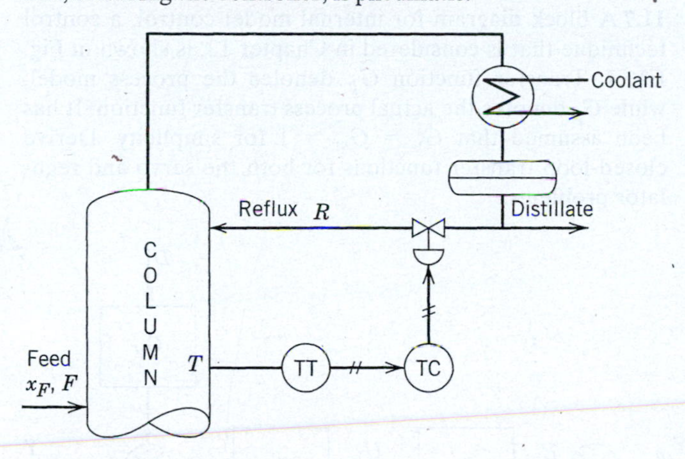

Figure E11.1

11.2 Consider the liquid-level, PI control system similar to Fig. 11.16 with the following parameter values: $A = 3$ ft^2, $R = 1.0$ min/ft^2, $K_v = 0.2$ ft^3/min psi, $K_m = 4$ mA/ft, $K_c = 5.33$, $K_{IP} = 0.75$ psi/mA, and $\tau_I = 3$ min. Suppose that the system is initially at the nominal steady state with a liquid level of 2 ft. If the set point is suddenly changed from 2 to 3 ft, how long will it take the system to reach (a) 2.5 ft and (b) 3 ft?

11.3 Consider proportional-only control of the stirred-tank heater control system in Fig. E11.3. The temperature transmitter has a span of 50 °F and a zero of 55 °F. The nominal design conditions are $\overline{T} = 80$ °F and $\overline{T}_i = 65$ °F. The controller has a gain of 5, while the gains for the control valve and current-to-pressure transducer are $K_v = 1.2$ (dimensionless) and $K_{IP} = 0.75$ psi/mA, respectively. The time constant for the tank is $\tau = 5$ min. The control valve and transmitter dynamics are negligible. After the set point is changed from 80 to 85 °F, the tank temperature eventually reaches a new steady-state value of 84.14 °F, which was measured with a highly accurate thermometer.

(a) What is the offset?

(b) What is the process gain K_2?

(c) What is the pressure signal p_t to the control valve at the final steady state?

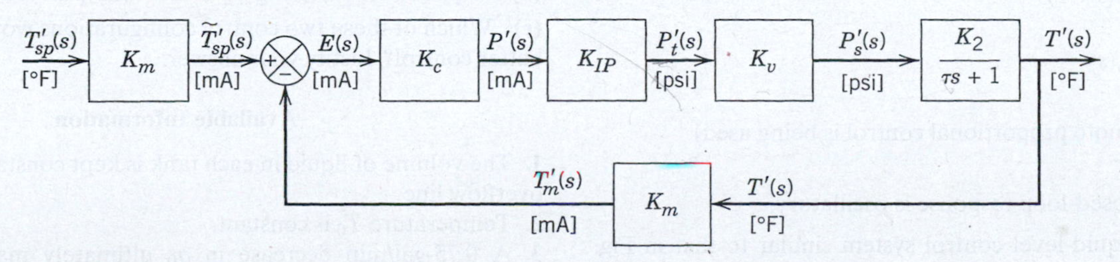

Figure E11.3

11.4 It is desired to control the exit concentration of c_3 of the liquid blending system shown in Fig. E11.4. Using the information given below, do the following:

(a) Draw a block diagram for the composition control scheme, using the symbols in Fig. E11.4.

(b) Derive an expression for each transfer function and substitute numerical values.

(c) Suppose that the PI controller has been tuned for the nominal set of operating conditions below. Indicate whether the controller should be retuned for each of the following situations. (Briefly justify your answers).

(i) The nominal value of c_2 changes to $\overline{c}_2 = 8.5$ lb solute/ft^3.

(ii) The span of the composition transmitter is adjusted so that the transmitter output varies from 4 to 20 mA as c_3 varies from 3 to 14 lb solute/ft^3.

(iii) The zero of the composition transmitter is adjusted so that the transmitter output varies from 4 to 20 mA as c_3 varies from 4 to 10 lb solute/ft^3.

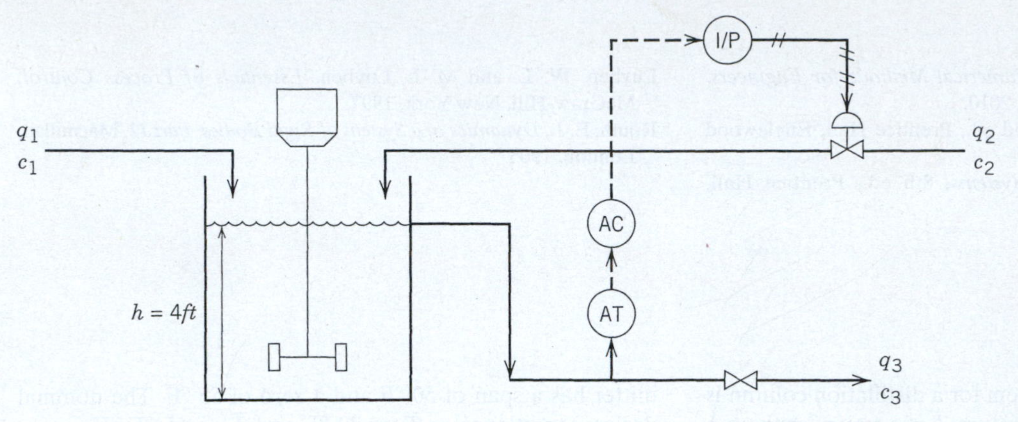

Figure E11.4

Available Information

1. The tank is perfectly mixed.

2. An overflow pipe is used to keep the mixture height at 4 ft.

3. The volumetric flow rate and solute concentration of stream 2, q_2 and c_2, vary with time, whereas those of stream 1 are constant.

4. The density of all three streams are identical and do not vary with time.

5. A 2-min time delay is associated with the composition measurement. The transmission output signal varies linearly from 4 to 20 mA as c_3 varies from 3 to 9 lb solute/ft^3.

6. The pneumatic control valve has negligible dynamics. Its steady-state behavior is summarized below where p_t is the air pressure signal to the control valve from the I/P transducer.

p_t (psi)	q_2 (gal/min)
6	20
9	15
12	10

7. An electronic, direct-acting PI controller is used.

8. The current-to-pressure transducer has negligible dynamics and a gain of 0.3 psi/mA.

9. The nominal operating conditions are:

$\rho = 75$ lb/ft^3 $\quad \bar{c}_3 = 5$ lb solute/ft^3

$\bar{q}_1 = 75$ lb/ft^3 $\quad \bar{c}_2 = 7$ lb solute/ft^3

$\bar{q}_2 = 75$ lb/ft^3 $\quad \bar{c}_1 = 2$ lb solute/ft^3

D = tank diameter = 4 ft.

11.5 A control system has the following transfer functions in its block diagram (see Fig. 11.8): $G_c = 1$, $G_v = 2$, $G_d = G_p = \dfrac{2}{s(s+1)}$, $G_m = 1$. For a unit step change in Y_{sp}, determine

(a) $Y(s)/Y_{sp}(s)$

(b) $y(\infty)$

(c) Offset (note proportional control is being used)

(d) $y(0.5)$

(e) if the closed-loop response is oscillatory

11.6 For a liquid-level control system similar to that in Fig. 11.22, Appelpolscher has argued that integral control action is not required because the process acts as an integrator (cf. Eq. 11-77). To evaluate his assertion, determine whether proportional-only control will eliminate offset for step changes in (a) set point and (b) disturbance variable.

11.7 A block diagram for internal model control, a control technique that is considered in Chapter 12, is shown in Fig. E11.7. Transfer function $\tilde{G}_p$ denotes the process model, while G_p denotes the actual process transfer function. It has been assumed that $G_v = G_m = 1$ for simplicity. Derive closed-loop transfer functions for both the servo and regulator problems.

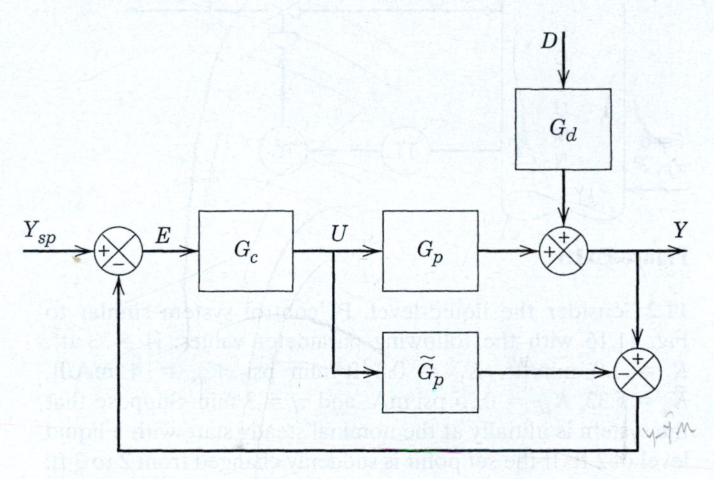

Figure E11.7

11.8 An electrically heated, stirred-tank system is shown in Fig. E11.8. Using the given information, do the following:

(a) Draw a block diagram for the case where T_3 is the controlled variable and voltage signal V_2 is the manipulated variable. Derive an expression for each transfer function.

(b) Repeat part (a) using V_1 as the manipulated variable.

(c) Which of these two control configurations would provide better control? Justify your answer.

Available Information

1. The volume of liquid in each tank is kept constant using an overflow line.

2. Temperature T_0 is constant.

3. A 0.75-gal/min decrease in q_0 ultimately makes T_1 increase by 3 °F. Two-thirds of this total temperature change

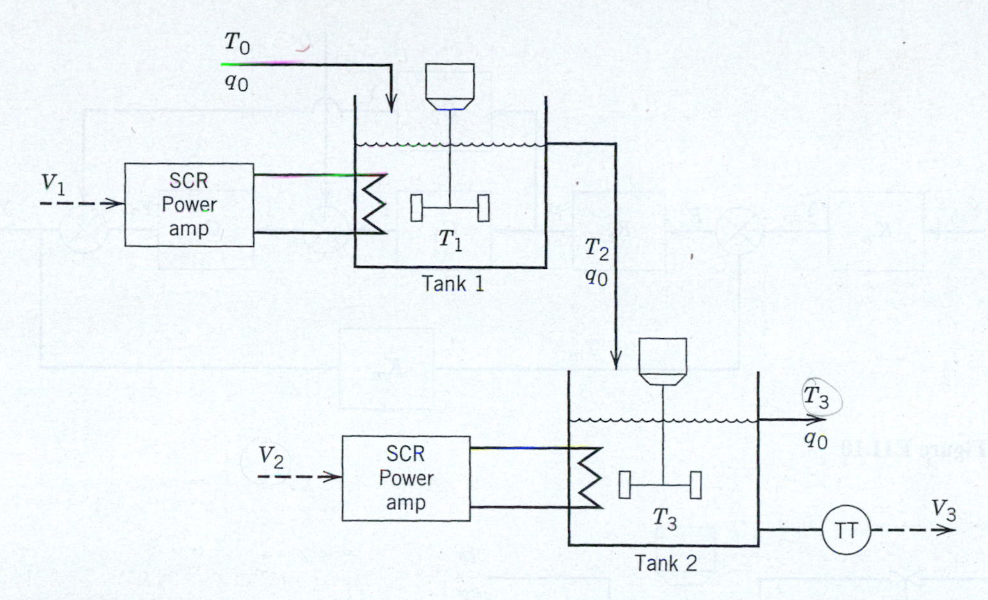

Figure E11.8

occurs in 12 min. This change in q_0 ultimately results in a 5 °F increase in T_3.

4. A change in V_1 from 10 to 12 volts ultimately causes T_1 to change from 70 to 78 °F. A similar test for V_2 causes T_3 to change from 85 to 90 °F. The apparent time constant for these tests is 10 min.

5. A step change in T_2 produces a transient response in T_3 that is essentially complete in 50 (= 5 τ min).

6. The thermocouple output is amplified to give $V_3 = 0.15T_3 + 5$, where V_3 [=] volts and T_3 [=] °F.

7. The pipe connecting the two tanks has a mean residence time of 30 s.

11.9 The block diagram of a special feedback control system is shown in Fig. E11.9. Derive an expression for the closed-loop transfer function, $Y(s)/D(s)$.

11.10 A block diagram of a closed-loop system is shown in Fig. E11.10.

(a) Derive a closed-loop transfer function for disturbance changes, $Y(s)/D(s)$.

(b) For the following transfer functions, what values of K_c will result in a stable closed-loop system?

$$G_1(s) = 5 \qquad G_2(s) = \frac{4}{2s+1}$$

$$K_m = 1 \qquad G_3(s) = \frac{1}{s-1}$$

11.11 A mixing process consists of a single stirred-tank instrumented as shown in Fig. E11.11. The concentration of a single species A in the feed stream varies. The controller attempts to compensate for this by varying the flow rate of pure A through the control valve. The transmitter dynamics are negligible.

(a) Draw a block diagram for the controlled process.

(b) Derive a transfer function for each block in your block diagram.

Process

(i) The volume is constant (5 m³).

(ii) The feed flow rate is constant ($\overline{q}_F = 7$ m³/min).

(iii) The flow rate of the A stream varies but is small compared to $\overline{q}_F$ ($\overline{q}_A = 0.5$ m³/min).

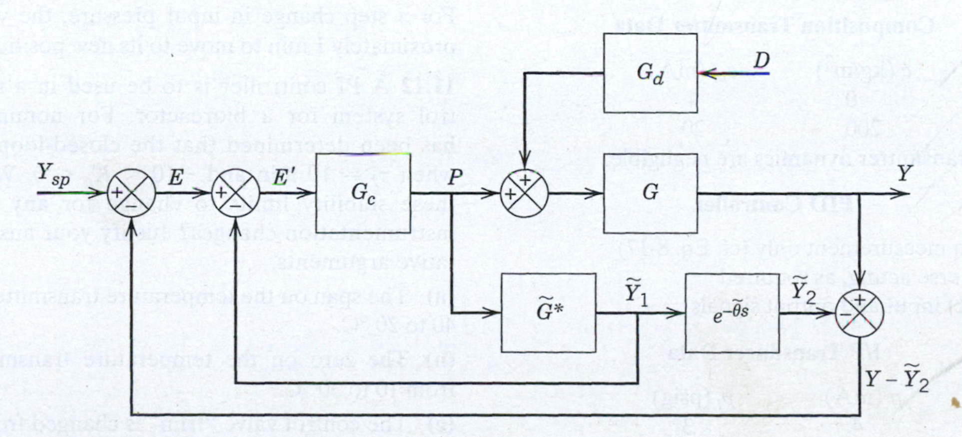

Figure E11.9

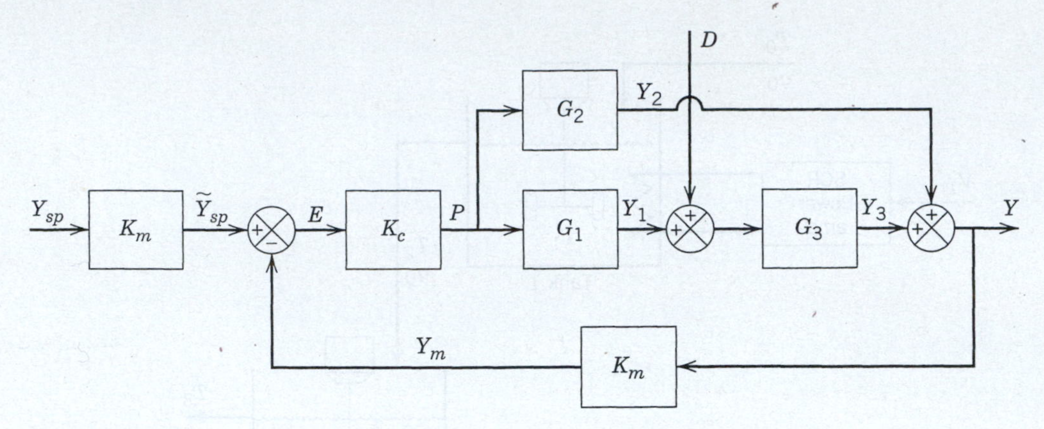

Figure E11.10

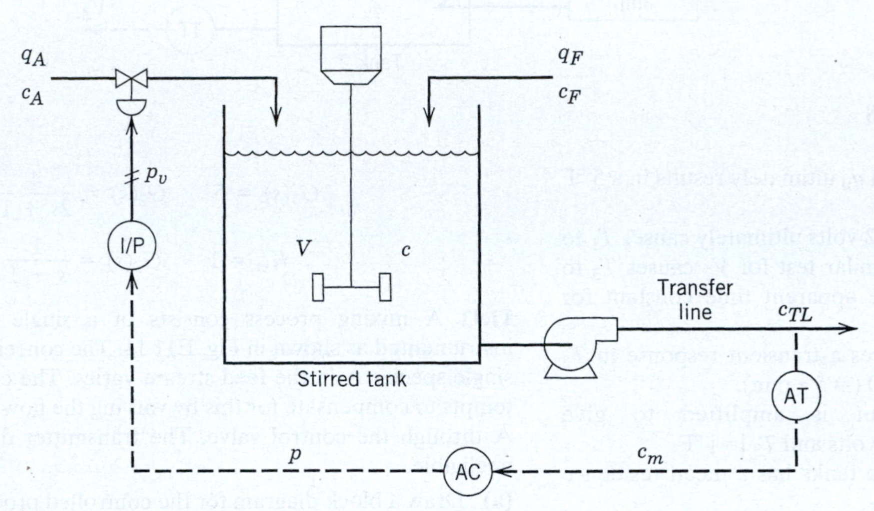

Figure E11.11

(iv) $\bar{c}_F = 50$ kg/m^3 and $\bar{c}_A = 800$ kg/m^3.
(v) All densities are constant and equal.

Transfer Line

(i) The transfer line is 20 m long and has 0.5 m inside diameter.
(ii) Pump volume can be neglected.

Composition Transmitter Data

c (kg/m^3)	c_m (mA)
0	4
200	20

Transmitter dynamics are negligible.

PID Controller

(i) Derivative on measurement only (cf. Eq. 8-17)
(ii) Direct or reverse acting, as required
(iii) Current (mA) input and output signals

I/P Transducer Data

p (mA)	p_t (psig)
4	3
20	15

Control Valve

An equal percentage valve is used, which has the following relation:

$$q_A = 0.17 + 0.03\,(20)^{\frac{p_V - 3}{12}}$$

For a step change in input pressure, the valve requires approximately 1 min to move to its new position.

11.12 A PI controller is to be used in a temperature control system for a bioreactor. For nominal conditions, it has been determined that the closed-loop system is stable when $\tau_I = 10$ min and $-10 < K_c < 0$. Would you expect these stability limits to change for any of the following instrumentation changes? Justify your answers using qualitative arguments.

(a) The span on the temperature transmitter is reduced from 40 to 20 °C.

(b) The zero on the temperature transmitter is increased from 10 to 30 °C.

(c) The control valve "trim" is changed from linear to equal percentage.

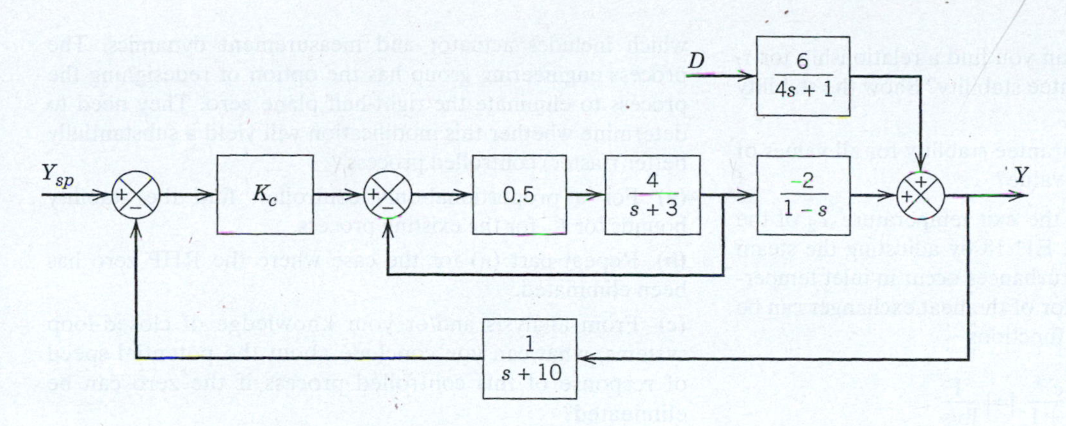

Figure E11.14

11.13 A process is described by the transfer function

$$G(s) = \frac{K}{(\tau s + 1)(s + 1)}$$

Find the range of controller settings that yield *stable* closed-loop systems for:

(a) A proportional-only controller.

(b) A proportional-integral controller.

(c) What can you say about the effect of adding the integral mode on the stability of the controlled system; that is, does it tend to stabilize or destabilize the system relative to proportional-only control? Justify your answer.

11.14 The block diagram of a feedback control system is shown in Fig. E11.14. Determine the values of K_c that result in a stable closed-loop system.

11.15 The question has been raised whether an open-loop unstable process can be stabilized with a proportional-only controller.

(a) For the process and controller shown in Fig. E11.15*a*, find the range of K_c values that yield a stable response. (Note that τ is positive.)

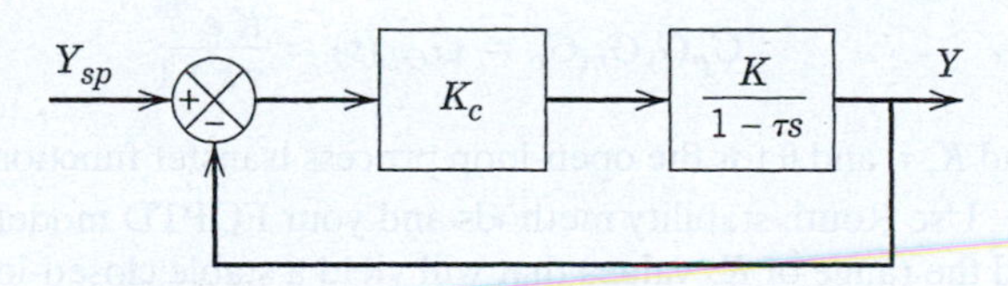

Figure E11.15a

(b) Check the gain of $Y(s)/Y_{sp}(s)$ to make sure that the process responds in the correct direction if K_c is within the range of part (a).

(c) For $K = 10$ and $\tau = 20$, find the value of K_c that yields a pole at $s = -0.1$. What is the offset for these conditions?

(d) Suppose that you had designed the controller neglecting a second smaller time constant. Would the controller still yield a stable closed-loop response? To check the "robustness" of your design, find the general conditions on K_c and τ_m for stability if the system is as shown in Fig. E11.15*b*. Are these conditions generally easy (or difficult) to meet? Why? Show for $\tau_m = 5$ that the value of K_c from part (c) does or does not still yield a stable system.

11.16 For the liquid-level control system in Fig. 11.22, determine the numerical values of K_c and τ_I that result in a stable closed-loop system. The level transmitter has negligible dynamics, while the control valve has a time constant of 10 s. The following numerical values are available:

$$A = 3 \text{ ft}^2$$
$$\bar{q}_3 = 10 \text{ gal/min}$$
$$K_v = -1.3 \text{ gal/min/mA}$$
$$K_m = 4 \text{ mA/ft}$$

11.17 As a newly hired engineer of the Ideal Gas Company, you are trying to make a reputation in the Process Control Group. However, this objective turns out to be a real challenge with I. M. Appelpolscher as your supervisor. At lunch one day, I.M.A. declares that a simple second-order process with a PI controller will always have a stability upper limit on K_c; that is, K_c is limited for all values of $\tau_I > 0$. His best argument is that the open-loop process with the controller is third order. Furthermore, he claims that any critically damped second-order process will show he is right.

Muttering "au contraire," you leave the table and quickly investigate the properties of

$$G_v G_p G_m = \frac{5}{(10s + 1)^2}$$

(a) What are the necessary and sufficient conditions for closed-loop stability for a PI controller?

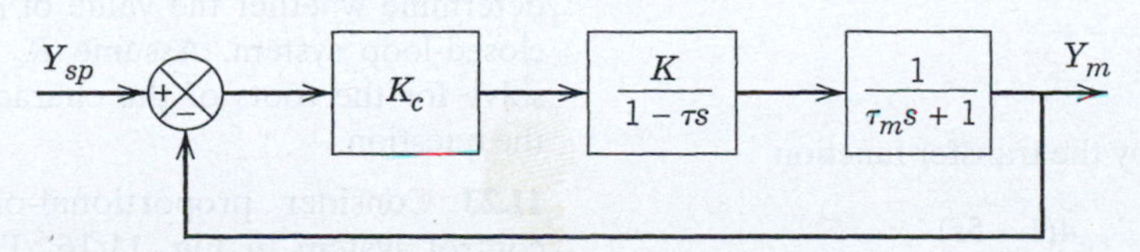

Figure E11.15b

(b) From these conditions, can you find a relationship for τ_I in terms of K_c that will guarantee stability? Show the stability region in a plot of τ_I versus K_c.

(c) Do some values of τ_I guarantee stability for all values of K_c? If so, what is the smallest value?

11.18 It is desired to control the exit temperature T_2 of the heat exchanger shown in Fig. E11.18 by adjusting the steam flow rate w_s. Unmeasured disturbances occur in inlet temperature T_1. The dynamic behavior of the heat exchanger can be approximated by the transfer functions

$$\frac{T_2'(s)}{W_s'(s)} = \frac{2.5e^{-s}}{10s+1} \; [=] \; \frac{°\text{F}}{\text{lb/s}}$$

$$\frac{T_2'(s)}{T_1'(s)} = \frac{0.9e^{-2s}}{5s+1} \; [=] \; \text{dimensionless}$$

where the time constants and time delays have units of seconds. The control valve has the following steady-state characteristic:

$$w_s = 0.6\sqrt{p-4}$$

where p is the controller output expressed in mA. At the nominal operating condition, p = 12 mA. After a sudden change in the controller output, w_s reaches a new steady-state value in 20 s (assumed to take five time constants). The temperature transmitter has negligible dynamics and is designed so that its output signal varies linearly from 4 to 20 mA as T_2 varies from 120 to 160 °F.

(a) If a proportional-only feedback controller is used, what is K_{cm}? What is the frequency of the resulting oscillation when $K_c = K_{cm}$? (*Hint:* Use the direct-substitution method and Euler's identity.)

(b) Estimate K_{cm} using the Routh criterion and a 1/1 Padé approximation for the time-delay term. Does this analysis provide a satisfactory approximation?

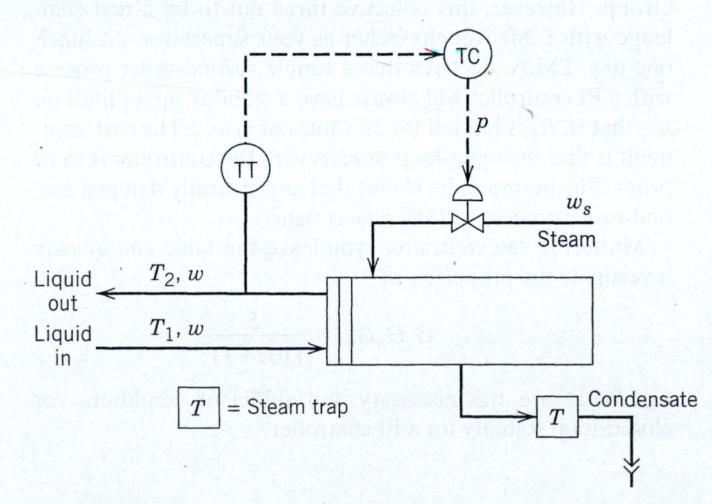

Figure E11.18

11.19 A process is described by the transfer function

$$G(s) = \frac{4(1-5s)}{(25s+1)(4s+1)(2s+1)}$$

which includes actuator and measurement dynamics. The process engineering group has the option of redesigning the process to eliminate the right-half plane zero. They need to determine whether this modification will yield a substantially better (faster) controlled process.

(a) For a proportional-only controller, find the stability bounds for K_c for the existing process.

(b) Repeat part (a) for the case where the RHP zero has been eliminated.

(c) From analysis and/or your knowledge of closed-loop systems, what can you conclude about the potential speed of response of this controlled process if the zero can be eliminated?

11.20 A feedback control system has the open-loop transfer function, $G_{OL}(s) = 0.5K_c e^{-3s}/(10s + 1)$. Determine the values of K_c for which the closed-loop system is stable using two approaches:

(a) An approximate analysis using the Routh stability criterion and a 1/1 Padé approximation for e^{-3s}.

(b) An exact analysis based on substitution of $s = j\omega$. (*Hint:* Recall Euler's identity.)

11.21 A process control system contains the following transfer functions:

$$G_p(s) = \frac{2\,e^{-1.5s}}{(60\,s+1)(5\,s+1)}$$

$$G_v(s) = \frac{0.5\,e^{-0.3s}}{3\,s+1}$$

$$G_m(s) = \frac{3\,e^{-0.2s}}{2\,s+1}$$

$$G_c(s) = K_c$$

(a) Show how $G_{OL}(s)$ can be approximated by a FOPTD model;

$$G_pG_vG_mG_c = G_{OL}(s) \approx \frac{K\,e^{-\theta s}}{\tau\,s+1}$$

Find K, τ, and θ for the open-loop process transfer function.

(b) Use Routh stability methods and your FOPTD model to find the range of K_c values that will yield a stable closed-loop system. Repeat for the full-order model using simulation.

(c) Determine K_{cm} and the corresponding value of ω.

11.22 For the control system based on the standard feedback control configuration in Fig. 10.8

$$G_c = K_c \quad G_p = \frac{5 - as}{(s+1)^3} \quad G_v = G_m = 1 \quad a > 0$$

determine whether the value of a affects the stability of the closed-loop system. Assume $K_c > 0$. You do not need to solve for the roots of the characteristic equation to answer the question.

11.23 Consider proportional-only control of the level control system in Fig. 11.16. The level transmitter has a span of 10 ft. and a zero of 15 ft. Recall that the standard instrument ranges are 4 to 20 mA and 3 to 15 psia. The nominal design conditions are $h = 20$ ft. The controller has a gain of 5 while the gain for the control valve is $K_v = 0.4$ cfm/psi, respectively. The time constant for the tank is $\tau = 5$ min. After the set-point is changed from 20 to 22 ft, the tank level eventually reaches a new steady-state value of 21.92 ft.

(a) What is the offset?

(b) What are the gains K_m and K_p in Fig. 11.16? (Give their units also)

(c) How could the controller be modified to obtain zero offset?

11.24 A control system has $G_v = G_m = 1$ and a second-order process G_p with $K_p = 2$, $\tau_1 = 4$ min, and $\tau_2 = 1$ min, which is to be controlled by a PI controller with $K_c = 2$ and $\tau_I = \tau_1 = 4$ min (i.e., the integral time of the controller is set equal to the dominant time constant). For a set-point change

(a) Determine the closed-loop transfer function.

(b) Derive the characteristic equation, which is a quadratic polynomial in s. Is it overdamped or underdamped?

(c) Can a large value of K_c make the closed-loop process unstable?

11.25 The set-point of the control system under proportional control ($K_c = 2.0$) undergoes a step change of magnitude 2. For $G_vG_p = \dfrac{5}{(s+1)(2s+1)}$ and $G_m = 1$,

(a) Determine when the maximum value of y occurs.

(b) Determine the offset.

(c) Determine the period of oscillation.

(d) Draw a sketch of $y(t)$ as a function of time, showing key characteristics as determined in (a), (b), and (c).

11.26 A batch process has a process gain of E Å/min (but no dynamics). The maniulated variable is the etch time, so the controlled variable is the film thickness. There are no time constants that need to be included. Assume $G_v = G_m = 1$. Derive the closed-loop transfer function for a set-point change for two different controllers:

(a) $G_c = K_c$

(b) $G_c = \dfrac{1}{\tau_I s}$

In both cases, analyze the effect of a unit step set-point change. Sketch the response and show whether there is offset or not.

11.27 Determine whether the following closed-loop transfer functions for (Y/Y_{sp}) are stable or unstable or undetermined (requires further analysis):

(a) $\dfrac{8K_c}{5s+1}$

(b) $\dfrac{8K_c}{s^2+3s+2}$

(c) $\dfrac{8K_c}{s^3+6s^2+12s+8+8K_c}$

Give a reason for each answer, i.e., does the value of K_c affect stability of the controlled system?

11.28 Derive the characteristic equation and construct the Routh array for a control system with the following transfer functions: $G_c = K_c$,

$$G_vG_p = \frac{1}{(s+1)(0.5s+1)},\quad G_m = \frac{3}{s+3}.$$

Is the system stable for (a) $K_c = 9$, (b) $K_c = 11$, (c) $K_c = 13$? Check your answers using simulation.

11.29 Suppose a control system is modeled by $G_vG_p = \dfrac{1}{(s+1)^3}$, $G_m = 1$, and $G_c = K_c$. Find the highest value of K_c for a proportional controller for which the system is stable, using the Routh array, and verify your result using simulation. Replace the controller with a PD controller ($K_c = 10$). Determine the range of τ_D for which the system is stable. Check your results using simulation.

Chapter 12

PID Controller Design, Tuning, and Troubleshooting

CHAPTER CONTENTS

Several examples in Chapter 11 demonstrated that the controller settings have a major effect on closed-loop stability. For most control problems, the closed-loop system is stable for a wide range of controller settings. Consequently, there is an opportunity to specify controller settings so that the desired control system performance is achieved.

To further illustrate the influence of controller settings, we consider a simple closed-loop system that consists of a first-order-plus-time-delay model and a PI controller. The simulation results in Fig. 12.1 show the disturbance responses for nine combinations of the controller gain K_c and integral time τ_I. As K_c increases or τ_I decreases, the response to the step disturbance becomes more aggressive. Controller 1 produces an unstable response, while Controller 5 arguably provides the best response. This example demonstrates that controller settings can be adjusted to achieve the desired closed-loop system performance, a procedure referred to as *controller tuning*.

In this chapter, we consider general controller design methods and tuning relations for PID controllers based on transfer function models and transient response criteria. Controller settings based on frequency response criteria will be presented in Chapter 14. Advanced process control strategies are considered later, beginning with Chapter 15.

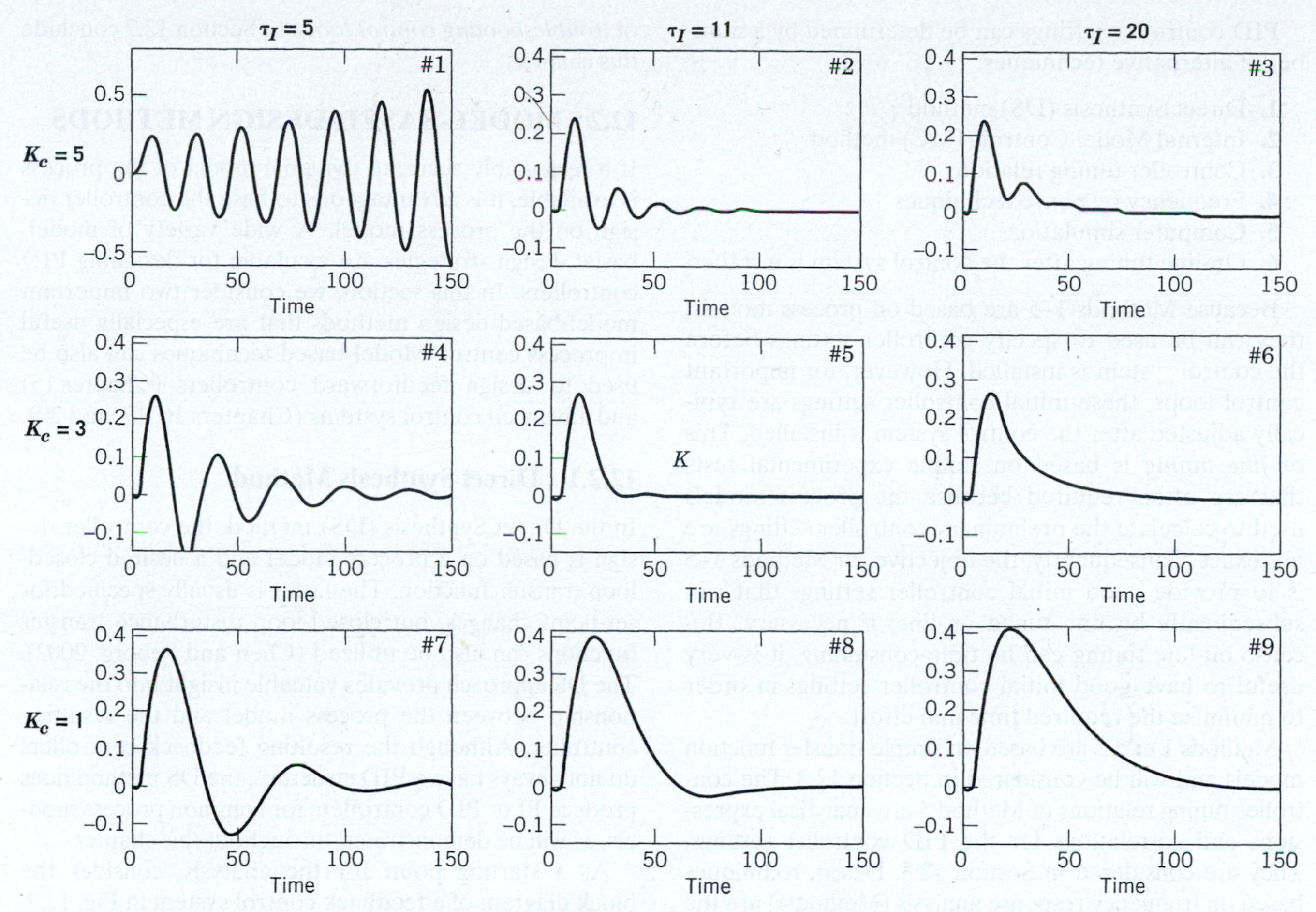

Figure 12.1 Unit-step disturbance responses for the candidate controllers (FOPTD Model: $K = 1$, $\theta = 4$, $\tau = 20$) and $G = G_d$.

12.1 PERFORMANCE CRITERIA FOR CLOSED-LOOP SYSTEMS

The function of a feedback control system is to ensure that the closed-loop system has desirable dynamic and steady-state response characteristics. Ideally, we would like the closed-loop system to satisfy the following performance criteria:

1. The closed-loop system must be stable
2. The effects of disturbances are minimized, providing good *disturbance rejection*
3. Rapid, smooth responses to set-point changes are obtained, that is, good *set-point tracking*
4. Steady-state error (offset) is eliminated
5. Excessive control action is avoided
6. The control system is robust, that is, insensitive to changes in process conditions and to inaccuracies in the process model

In typical control applications, it is not possible to achieve all of these goals simultaneously, because they involve inherent conflicts and tradeoffs. The tradeoffs must balance two important objectives, *performance* and *robustness*. A control system exhibits a high degree of performance if it provides rapid and smooth responses to disturbances and set-point changes with little, if any, oscillation. A control system is *robust* if it provides satisfactory performance for a wide range of process conditions and for a reasonable degree of model inaccuracy. Robustness can be achieved by choosing conservative controller settings (typically, small values of K_c and large values of τ_I), but this choice tends to result in poor performance. Thus, conservative controller settings sacrifice performance in order to achieve robustness. Robustness analysis is considered in Appendix J.

A second type of tradeoff occurs because PID controller settings that provide excellent disturbance rejection can produce large overshoots for set-point changes. On the other hand, if the controller settings are specified to provide excellent set-point tracking, the disturbance responses can be very sluggish. Thus, a tradeoff between set-point tracking and disturbance rejection occurs for standard PID controllers. Fortunately, this tradeoff can be avoided by using a *controller with two degrees of freedom*, as shown in Section 12.4.

PID controller settings can be determined by a number of alternative techniques:

1. Direct Synthesis (DS) method
2. Internal Model Control (IMC) method
3. Controller tuning relations
4. Frequency response techniques
5. Computer simulation
6. On-line tuning after the control system is installed

Because Methods 1–5 are based on process models, they can be used to specify controller settings before the control system is installed. However, for important control loops, these initial controller settings are typically adjusted after the control system is installed. This *on-line tuning* is based on simple experimental tests that are often required because the process models used to calculate the preliminary controller settings are not exact. Consequently, the objective for Methods 1–5 is to provide good initial controller settings that can subsequently be fine tuned on-line, if necessary. Because on-line tuning can be time-consuming, it is very useful to have good initial controller settings in order to minimize the required time and effort.

Methods 1 and 2 are based on simple transfer function models and will be considered in Section 12.2. The controller tuning relations of Method 3 are analytical expressions and correlations for the PID controller settings. They are considered in Section 12.3. Design techniques based on frequency response analysis (Method 4) are the subject of Chapter 14. Computer simulation of the controlled process (Method 5) can provide considerable insight into dynamic behavior and control system performance. In particular, software such as MATLAB and LabVIEW facilitates the comparison of alternative control strategies and different controller settings. (See Appendices C and E of Doyle (2000).) Method 6, on-line tuning, is considered in Section 12.5.

A comparison of PID tuning relations in Section 12.6 and an introduction to the important practical problem of *troubleshooting control loops* in Section 12.7 conclude this chapter.

12.2 MODEL-BASED DESIGN METHODS

If a reasonably accurate dynamic model of the process is available, it is advantageous to base the controller design on the process model. A wide variety of model-based design strategies are available for designing PID controllers. In this section, we consider two important model-based design methods that are especially useful in process control. Model-based techniques can also be used to design feedforward controllers (Chapter 15) and advanced control systems (Chapters 16, 17, and 20).

12.2.1 Direct Synthesis Method

In the Direct Synthesis (DS) method, the controller design is based on a process model and a desired closed-loop transfer function. The latter is usually specified for set-point changes, but closed-loop disturbance transfer functions can also be utilized (Chen and Seborg, 2002). The DS approach provides valuable insight into the relationship between the process model and the resulting controller. Although the resulting feedback controllers do not always have a PID structure, the DS method does produce PI or PID controllers for common process models, as will be demonstrated throughout this chapter.

As a starting point for the analysis, consider the block diagram of a feedback control system in Fig. 12.2. The closed-loop transfer function for set-point changes was derived in Section 11.2:

$$\frac{Y}{Y_{sp}} = \frac{K_m G_c G_v G_p}{1 + G_c G_v G_p G_m} \tag{12-1}$$

For simplicity, let $G \triangleq G_v G_p G_m$ and assume that $G_m = K_m$. Then Eq. 12-1 reduces to[1]

$$\frac{Y}{Y_{sp}} = \frac{G_c G}{1 + G_c G} \tag{12-2}$$

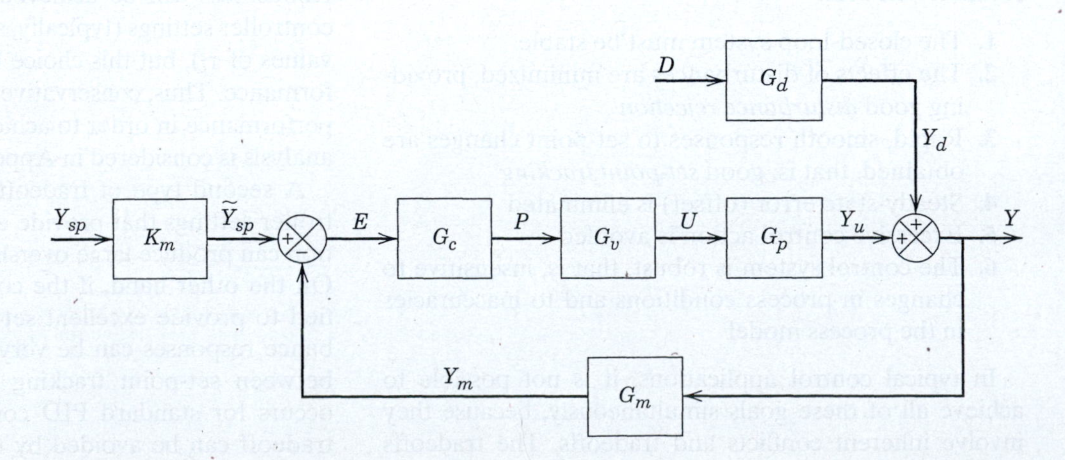

Figure 12.2 Block diagram for a standard feedback control system.

[1]We use the symbols G and G_c to denote $G(s)$ and $G_c(s)$, for the sake of simplicity.

Rearranging and solving for G_c gives an expression for the ideal feedback controller:

$$G_c = \frac{1}{G}\left(\frac{Y/Y_{sp}}{1 - Y/Y_{sp}}\right) \quad (12\text{-}3a)$$

Equation 12-3a cannot be used for controller design, because the closed-loop transfer function Y/Y_{sp} is not known *a priori*. Also, it is useful to distinguish between the actual process G and the model, $\tilde{G}$, that provides an approximation of the process behavior. A practical design equation can be derived by replacing the unknown G by $\tilde{G}$, and Y/Y_{sp} by a *desired closed-loop transfer function*, $(Y/Y_{sp})_d$:

$$G_c = \frac{1}{\tilde{G}}\left[\frac{(Y/Y_{sp})_d}{1 - (Y/Y_{sp})_d}\right] \quad (12\text{-}3b)$$

The specification of $(Y/Y_{sp})_d$ is the key design decision and will be considered later in this section. Note that the controller transfer function in (12-3b) contains the inverse of the process model due to the $1/\tilde{G}$ term. This feature is a distinguishing characteristic of model-based control.

Desired Closed-Loop Transfer Function

The performance of the DS controller in Eq. 12-3b strongly depends on the specification of the desired closed-loop transfer function, $(Y/Y_{sp})_d$. Ideally, we would like to have $(Y/Y_{sp})_d = 1$ so that the controlled variable tracks set-point changes instantaneously without any error. However, this ideal situation, called *perfect control*, cannot be achieved by feedback control because the controller does not respond until $e \neq 0$. For processes without time delays, the first-order model in Eq. 12-4 is a more reasonable choice

$$\left(\frac{Y}{Y_{sp}}\right)_d = \frac{1}{\tau_c s + 1} \quad (12\text{-}4)$$

where τ_c is the *desired closed-loop time constant*. This model has a settling time of $\sim 5\tau_c$, as shown in Section 5.2. Because the steady-state gain is one, no offset occurs for set-point changes. By substituting (12-4) into (12-3b) and solving for G_c, the controller design equation becomes

$$G_c = \frac{1}{\tilde{G}}\frac{1}{\tau_c s} \quad (12\text{-}5)$$

The $1/\tau_c s$ term provides integral control action and thus eliminates offset. Design parameter τ_c provides a convenient controller tuning parameter that can be used to make the controller more aggressive (small τ_c) or less aggressive (large τ_c).

If the process transfer function contains a known time delay θ, a reasonable choice for the desired closed-loop transfer function is

$$\left(\frac{Y}{Y_{sp}}\right)_d = \frac{e^{-\theta s}}{\tau_c s + 1} \quad (12\text{-}6)$$

The time-delay term in (12-6) is essential, because it is physically impossible for the controlled variable to respond to a set-point change at $t = 0$, before $t = \theta$. If the time delay is unknown, θ must be replaced by an estimate. Combining Eqs. 12-6 and 12-3b gives

$$G_c = \frac{1}{\tilde{G}}\frac{e^{-\theta s}}{\tau_c s + 1 - e^{-\theta s}} \quad (12\text{-}7)$$

Although this controller is not in a standard PID form, it is physically realizable. Sometimes, the symbol λ is used instead of τ_c in Eq. 12-6, and the Direct Synthesis method is referred to as the *lambda-tuning method* (McMillan, 2006).

Next, we show that the design equation in Eq. 12-7 can be used to derive PID controllers for simple process models. The following derivation is based on approximating the time-delay term in the denominator of (12-7) with a truncated Taylor series expansion:

$$e^{-\theta s} \approx 1 - \theta s \quad (12\text{-}8)$$

Substituting (12-8) into the denominator of Eq. 12-7 and rearranging gives

$$G_c = \frac{1}{\tilde{G}}\frac{e^{-\theta s}}{(\tau_c + \theta)s} \quad (12\text{-}9)$$

Note that this controller also contains integral control action.

Time-delay approximations are less accurate when the time delay is relatively large compared to the dominant time constant of the process. Note that it is not necessary to approximate the time-delay term in the numerator, because it is canceled by the identical term in $\tilde{G}$, when the time delay is known exactly.

Next, we derive controllers for two important process models. For each derivation, we assume that the model is perfect ($\tilde{G} = G$).

First-Order-Plus-Time-Delay (FOPTD) Model

Consider the standard first-order-plus-time-delay model,

$$\tilde{G}(s) = \frac{Ke^{-\theta s}}{\tau s + 1} \quad (12\text{-}10)$$

Substituting Eq. 12-10 into Eq. 12-9 and rearranging gives a PI controller, $G_c = K_c(1 + 1/\tau_I s)$, with the following controller settings:

$$K_c = \frac{1}{K}\frac{\tau}{\theta + \tau_c}, \quad \tau_I = \tau \quad (12\text{-}11)$$

The expressions for the PI controller settings in (12-11) provide considerable insight. Controller gain K_c depends inversely on model gain K, which is reasonable based on the stability analysis in Chapter 11. In particular, if the product $K_c K$ is constant, the characteristic equation and stability characteristics of the closed-loop system do not change. It is also reasonable that $\tau_I = \tau$, because slow processes have large values of τ, and thus τ_I should also be large for satisfactory control. As τ_c decreases, K_c increases, because a faster set-point response requires more strenuous control action and thus a larger value of K_c. The time delay θ imposes an upper limit on K_c, even for the limiting case where $\tau_c \to 0$. By contrast, K_c becomes unbounded when $\theta = 0$ and $\tau_c \to 0$.

Second-Order-Plus-Time-Delay (SOPTD) Model Consider a second-order-plus-time delay model,

$$\widetilde{G}(s) = \frac{Ke^{-\theta s}}{(\tau_1 s + 1)(\tau_2 s + 1)} \tag{12-12}$$

Substitution into Eq. 12-9 and rearrangement gives a PID controller in parallel form,

$$G_c = K_c\left(1 + \frac{1}{\tau_I s} + \tau_D s\right) \tag{12-13}$$

where

$$K_c = \frac{1}{K}\frac{\tau_1 + \tau_2}{\tau_c + \theta}, \quad \tau_I = \tau_1 + \tau_2, \quad \tau_D = \frac{\tau_1 \tau_2}{\tau_1 + \tau_2} \tag{12-14}$$

The tuning relations in Eq. 12-14 indicate that for large values of θ, K_c decreases, but τ_I and τ_D do not. Again, the time delay imposes an upper limit on K_c as $\tau_c \to 0$.

The controller settings in Eqs. 12-11 and 12-14 become more conservative (smaller K_c) as τ_c increases. If θ is relatively large (for example, $\theta/\tau_1 > 0.5$), a conservative choice of τ_c is prudent, because the controller design equations are based on the time-delay approximation in Eq. 12-8.

A number of guidelines for choosing τ_c that are applicable to both the Direct Synthesis method and the Internal Model Control method are presented in Section 12.2.2.

EXAMPLE 12.1

Use the DS design method to calculate PID controller settings for the process:

$$G = \frac{2e^{-s}}{(10s + 1)(5s + 1)}$$

Consider three values of the desired closed-loop time constant: τ_c = 1, 3, and 10. Evaluate the controllers for unit step changes in both the set point and the disturbance, assuming that $G_d = G$. Repeat the evaluation for two cases:

(a) The process model is perfect ($\widetilde{G} = G$).

(b) The model gain is incorrect, $\widetilde{K} = 0.9$, instead of the actual value, $K = 2$. Thus,

$$\widetilde{G} = \frac{0.9e^{-s}}{(10s + 1)(5s + 1)}$$

SOLUTION

The controller settings for this example are

	$\tau_c = 1$	$\tau_c = 3$	$\tau_c = 10$
K_c ($\widetilde{K}=2$)	3.75	1.88	0.682
K_c ($\widetilde{K}=0.9$)	8.33	4.17	1.51
τ_I	15	15	15
τ_D	3.33	3.33	3.33

The values of K_c decrease as τ_c increases, but the values of τ_I and τ_D do not change, as indicated by Eq. 12-14. Although some of these controller settings have been reported with three significant figures for purposes of comparison, calculating a third digit is not necessary in practice. For example, controllers with K_c values of 8.33 and 8.3 would produce essentially the same closed-loop responses.

Figures 12.3 and 12.4 compare the closed-loop responses for the three DS controllers. As τ_c increases, the responses

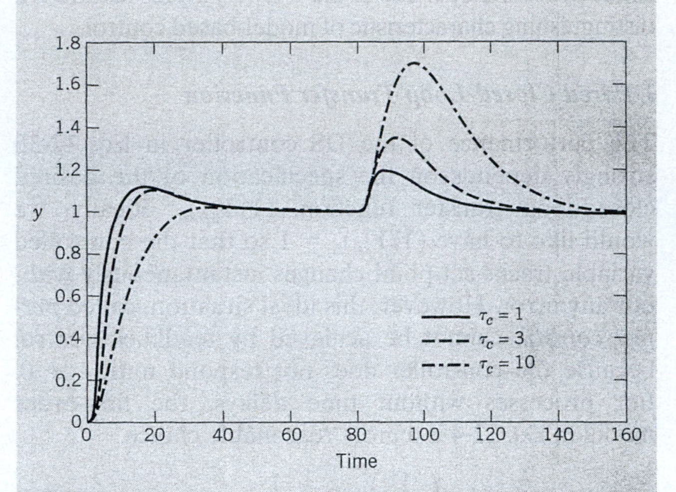

Figure 12.3 Simulation results for Example 12.1 (a): correct model gain.

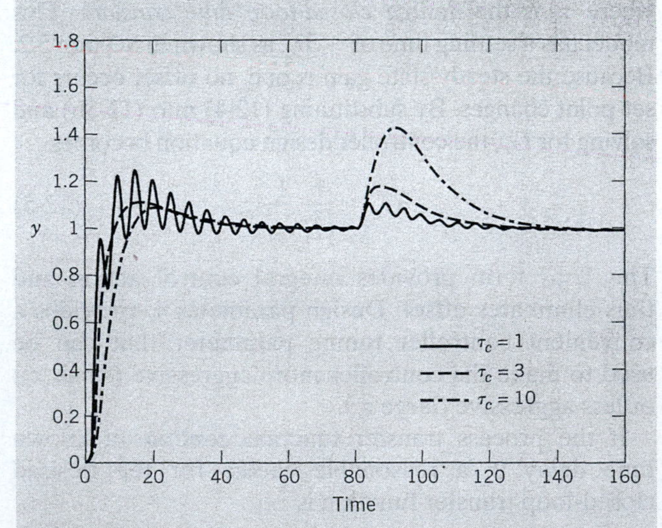

Figure 12.4 Simulation results for Example 12.1 (b): incorrect model gain.

become more sluggish, and the maximum deviation is larger after the disturbance occurs at $\tau = 80$. For case (b), when the model gain is 0.9, about 50% too low, the closed-loop response for $\tau_c = 1$ in Fig. 12.4 is excessively oscillatory and would even become unstable if $\widetilde{K} = 0.8$ had been considered. The disturbance responses for $\tau_c = 3$ and $\tau_c = 10$ in Fig. 12.4 are actually better than the corresponding responses in Fig. 12.3 because the former have shorter settling times and smaller maximum deviations. This improvement is due to the larger values of K_c for case (b).

The Simulink diagram for this example is quite simple, as shown in Fig. 12.5. (See Appendix C.) However, the simulation results for Figs. 12.3 and 12.4 were generated using a modified controller that eliminated derivative kick (see Chapter 8).

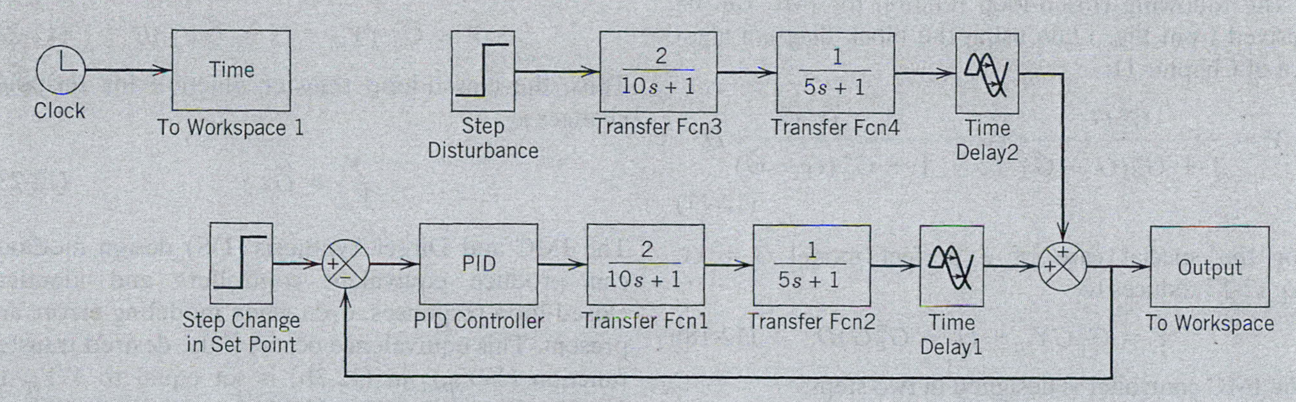

Figure 12.5 Simulink diagram for Example 12.1.

The specification of the desired closed-loop transfer function, $(Y/Y_{sp})_d$, should be based on the assumed process model, as well as the desired set-point response. The FOPTD model in Eq. 12-6 is a reasonable choice for many processes but not all. For example, suppose that the process model contains a right-half plane zero term denoted by $(1 - \tau_a s)$ where $\tau_a > 0$. Then if Eq. 12-6 is selected, the DS controller will have the $(1 - \tau_a s)$ term in its denominator and thus be unstable, a very undesirable feature. This problem can be avoided by replacing (12-6) with (12-15):

$$\left(\frac{Y}{Y_{sp}}\right)_d = \frac{(1 - \tau_a s)e^{-\theta s}}{\tau_c s + 1} \tag{12-15}$$

The DS approach should not be used directly for process models with unstable poles. However, it can be applied if the model is first stabilized by an additional feedback control loop.

12.2.2 Internal Model Control (IMC)

A more comprehensive model-based design method, *Internal Model Control* (*IMC*), was developed by Morari and coworkers (Garcia and Morari, 1982; Rivera et al., 1986). The IMC method, like the DS method, is based on an assumed process model and leads to analytical expressions for the controller settings. These two design methods are closely related and produce identical controllers if the design parameters are specified in a consistent manner. However, the IMC approach has the advantage that it allows model uncertainty and tradeoffs between performance and robustness to be considered in a more systematic fashion.

The IMC method is based on the simplified block diagram shown in Fig. 12.6*b*. A process model $\widetilde{G}$ and the controller output P are used to calculate the model response, $\widetilde{Y}$. The model response is subtracted from the actual response Y, and the difference, $Y - \widetilde{Y}$, is used as the input signal to the IMC controller, G_c^*. In general, $Y \neq \widetilde{Y}$ due to modeling errors ($\widetilde{G} \neq G$) and unknown disturbances ($D \neq 0$) that are not accounted for in the model.

The block diagrams for conventional feedback control and IMC are compared in Fig. 12.6. It can be

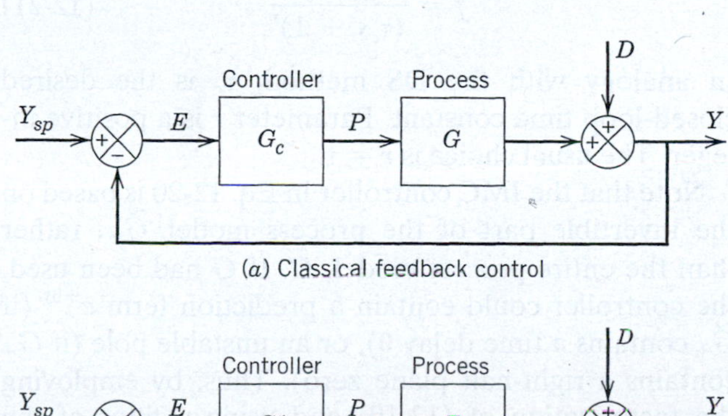

Figure 12.6 Feedback control strategies.

shown that the two block diagrams are equivalent if controllers G_c and G_c^* satisfy the relation

$$G_c = \frac{G_c^*}{1 - G_c^* \widetilde{G}} \tag{12-16}$$

Thus, any IMC controller G_c^* is equivalent to a standard feedback controller G_c, and vice versa.

The following closed-loop relation for IMC can be derived from Fig. 12.6*b* using the block diagram algebra of Chapter 11:

$$Y = \frac{G_c^* G}{1 + G_c^*(G - \widetilde{G})} Y_{sp} + \frac{1 - G_c^* \widetilde{G}}{1 + G_c^*(G - \widetilde{G})} D \tag{12-17}$$

For the special case of a perfect model, $\widetilde{G} = G$, Eq. 12-17 reduces to

$$Y = G_c^* G Y_{sp} + (1 - G_c^* G) D \tag{12-18}$$

The IMC controller is designed in two steps:

Step 1. The process model is factored as

$$\widetilde{G} = \widetilde{G}_+ \widetilde{G}_- \tag{12-19}$$

where $\widetilde{G}_+$ contains any time delays and right-half plane zeros. In addition, $\widetilde{G}_+$ is required to have a steady-state gain equal to one in order to ensure that the two factors in Eq. 12-19 are unique.

Step 2. The IMC controller is specified as

$$G_c^* = \frac{1}{\widetilde{G}_-} f \tag{12-20}$$

where f is a *low-pass filter* with a steady-state gain of one.[2] It typically has the form

$$f = \frac{1}{(\tau_c s + 1)^r} \tag{12-21}$$

In analogy with the DS method, τ_c is the desired closed-loop time constant. Parameter r is a positive integer. The usual choice is $r = 1$.

Note that the IMC controller in Eq. 12-20 is based on the invertible part of the process model, $\widetilde{G}_-$, rather than the entire process model, $\widetilde{G}$. If $\widetilde{G}$ had been used, the controller could contain a prediction term $e^{+\theta s}$ (if $\widetilde{G}_+$ contains a time delay θ), or an unstable pole (if $\widetilde{G}_+$ contains a right-half plane zero). Thus, by employing the factorization of (12-19) and using a filter of the form of (12-21), the resulting IMC controller G_c^* is guaranteed to be physically realizable and stable. In general, the noninvertible part of the model, $\widetilde{G}_+$, places limitations on the performance that can be achieved by any control system (Goodwin et al., 2001). Because the standard IMC design method is based on pole-zero cancellation, the IMC approach must be modified for processes that are open-loop unstable.

For the ideal situation where the process model is perfect ($\widetilde{G} = G$), substituting Eq. 12-20 into (12-18) gives the closed-loop expression

$$Y = \widetilde{G}_+ f Y_{sp} + (1 - \widetilde{G}_+ f) D \tag{12-22}$$

Thus, the closed-loop transfer function for set-point changes is

$$\frac{Y}{Y_{sp}} = \widetilde{G}_+ f \tag{12-23}$$

The IMC and Direct Synthesis (DS) design methods can produce equivalent controllers and identical closed-loop responses, even when modeling errors are present. This equivalence occurs if the desired transfer function $(Y/Y_{sp})_d$ in (12-3b) is set equal to Y/Y_{sp} in (12-23). Recall that Eq. 12-16 shows how to convert G_c^* to the equivalent G_c.

The IMC design method is illustrated in the following example.

EXAMPLE 12.2

Use the IMC design method to design two controllers for the FOPTD model in (12-10). Assume that f is specified by (12-21) with $r = 1$, and consider two approximations for the time-delay term:

(a) 1/1 Padé approximation:

$$e^{-\theta s} \cong \frac{1 - \frac{\theta}{2} s}{1 + \frac{\theta}{2} s} \tag{12-24a}$$

(b) First-order Taylor series approximation:

$$e^{-\theta s} \cong 1 - \theta s \tag{12-24b}$$

SOLUTION

(a) Substituting Eq. 12-24a into (12-10) gives

$$\widetilde{G}(s) = \frac{K\left(1 - \frac{\theta}{2} s\right)}{\left(1 + \frac{\theta}{2} s\right)(\tau s + 1)} \tag{12-25}$$

Factor this model as $\widetilde{G} = \widetilde{G}_+ \widetilde{G}_-$ where

$$\widetilde{G}_+ = 1 - \frac{\theta}{2} s \tag{12-26}$$

---

[2] The term *low-pass filter* is a frequency response concept that will be explained in Chapter 14.

and

$$\tilde{G}_- = \frac{K}{\left(1 + \frac{\theta}{2}s\right)(\tau s + 1)} \quad (12\text{-}27)$$

Note that $\tilde{G}_+$ has a steady-state gain of one, as required in the IMC design procedure.

Substituting Eqs. 12-27 and 12-21 into Eq. 12-20 and setting $r = 1$ gives

$$G_c^* = \frac{\left(1 + \frac{\theta}{2}s\right)(\tau s + 1)}{K(\tau_c s + 1)} \quad (12\text{-}28)$$

The equivalent controller G_c can be obtained from Eq. 12-16,

$$G_c = \frac{\left(1 + \frac{\theta}{2}s\right)(\tau s + 1)}{K\left(\tau_c + \frac{\theta}{2}\right)s} \quad (12\text{-}29)$$

and rearranged into the PID controller of (12-13) with:

$$K_c = \frac{1}{K}\frac{2\left(\frac{\tau}{\theta}\right)+1}{2\left(\frac{\tau_c}{\theta}\right)+1}, \quad \tau_I = \frac{\theta}{2} + \tau, \quad \tau_D = \frac{\tau}{2\left(\frac{\tau}{\theta}\right)+1} \quad (12\text{-}30)$$

(b) Repeating this derivation for the Taylor series approximation gives a standard PI controller for G_c:

$$K_c = \frac{1}{K}\frac{\tau}{\tau_c + \theta}, \quad \tau_I = \tau \quad (12\text{-}31)$$

A comparison of (12-30) and (12-31) indicates that the type of controller that is designed depends on the time-delay approximation. Furthermore, the IMC controller in (12-31) is identical to the DS controller for a first-order-plus-time-delay model. This equivalence can be confirmed by noting that the DS controller settings in (12-14) reduce to the IMC settings in (12-31) for $\tau_1 = \tau$ and $\tau_2 = 0$.

Selection of τ_c

The choice of design parameter τ_c is a key decision in both the DS and IMC design methods. In general, increasing τ_c produces a more conservative controller because K_c decreases while τ_I increases. Several IMC guidelines for τ_c have been published for the FOPTD model in Eq. 12-10:

1. $\tau_c/\theta > 0.8$ and $\tau_c > 0.1\tau$ (Rivera et al., 1986)
2. $\tau > \tau_c > \theta$ (Chien and Fruehauf, 1990)
3. $\tau_c = \theta$ (Skogestad, 2003)

For more general process models with a dominant time constant, τ_{dom}, guideline (2) can be generalized to: $\tau_{dom} > \tau_c > \theta$. For example, setting $\tau_c = \tau_{dom}/3$ means that the desired closed-loop response is three times faster than the open-loop response.

12.3 CONTROLLER TUNING RELATIONS

In the last section, we have seen that model-based design methods such as DS and IMC produce PI or PID controllers for certain classes of process models. Analytical expressions for PID controller settings have been derived from other perspectives as well. These expressions are referred to as *controller tuning relations*, or just *tuning relations*. In this section we present some of the most widely used tuning relations as well as some promising new ones.

12.3.1 IMC Tuning Relations

The IMC method can be used to derive PID controller settings for a variety of transfer function models. Different tuning relations can be derived depending on the type of lowpass filter f and time-delay approximation (e.g., Eq. 12-24) that are selected (Rivera et al., 1986; Chien and Fruehauf, 1990; Skogestad, 2003).

Table 12.1 presents the PID controller tuning relations for the parallel form that were derived by Chien and Fruehauf (1990) for common types of process models. The IMC filter f was selected according to Eq. 12-21 with $r = 1$ for first-order and second-order models. For models with integrating elements, the following expression was employed:

$$f = \frac{(2\tau_c - C)s + 1}{(\tau_c s + 1)^2} \quad \text{where} \quad C = \left.\frac{d\tilde{G}_+}{ds}\right|_{s=0} \quad (12\text{-}32)$$

In Table 12.1 two controllers are listed for some process models (cf. controllers G and H, and also M and N). For these models, the PI controller in the first row was derived based on the time-delay approximation in (12-24b), while the PID controller in the next row was derived based on (12-24a). The tuning relations in Table 12.1 were derived for the parallel form of the PID controller in Eq. 12-13. The derivations are analogous to those for Example 12.2. Chien and Fruehauf (1990) have reported the equivalent tuning relations for the series form of the PID controller in Chapter 8. The controller settings for the parallel form can easily be converted to the corresponding settings for the series form, and *vice versa*, as shown in Table 12.2.

The following example illustrates the use of the tuning relations in Table 12.1.

Table 12.1 IMC-Based PID Controller Settings for $G_c(s)$ (Chien and Fruehauf, 1990)

Case	Model	K_cK	τ_I	τ_D
A	$\frac{K}{\tau s + 1}$	$\frac{\tau}{\tau_c}$	τ	—
B	$\frac{K}{(\tau_1 s + 1)(\tau_2 s + 1)}$	$\frac{\tau_1 + \tau_2}{\tau_c}$	$\tau_1 + \tau_2$	$\frac{\tau_1\tau_2}{\tau_1 + \tau_2}$
C	$\frac{K}{\tau^2 s^2 + 2\zeta\tau s + 1}$	$\frac{2\zeta\tau}{\tau_c}$	$2\zeta\tau$	$\frac{\tau}{2\zeta}$
D	$\frac{K(-\beta s + 1)}{\tau^2 s^2 + 2\zeta\tau s + 1}, \beta > 0$	$\frac{2\zeta\tau}{\tau_c + \beta}$	$2\zeta\tau$	$\frac{\tau}{2\zeta}$
E	$\frac{K}{s}$	$\frac{2}{\tau_c}$	$2\tau_c$	—
F	$\frac{K}{s(\tau s + 1)}$	$\frac{2\tau_c + \tau}{\tau_c^2}$	$2\tau_c + \tau$	$\frac{2\tau_c\tau}{2\tau_c + \tau}$
G	$\frac{Ke^{-\theta s}}{\tau s + 1}$	$\frac{\tau}{\tau_c + \theta}$	τ	—
H	$\frac{Ke^{-\theta s}}{\tau s + 1}$	$\frac{\tau + \frac{\theta}{2}}{\tau_c + \frac{\theta}{2}}$	$\tau + \frac{\theta}{2}$	$\frac{\tau\theta}{2\tau + \theta}$
I	$\frac{K(\tau_3 s + 1)e^{-\theta s}}{(\tau_1 s + 1)(\tau_2 s + 1)}$	$\frac{\tau_1 + \tau_2 - \tau_3}{\tau_c + \theta}$	$\tau_1 + \tau_2 - \tau_3$	$\frac{\tau_1\tau_2 - (\tau_1 + \tau_2 - \tau_3)\tau_3}{\tau_1 + \tau_2 - \tau_3}$
J	$\frac{K(\tau_3 s + 1)e^{-\theta s}}{\tau^2 s^2 + 2\zeta\tau s + 1}$	$\frac{2\zeta\tau - \tau_3}{\tau_c + \theta}$	$2\zeta\tau - \tau_3$	$\frac{\tau^2 - (2\zeta\tau - \tau_3)\tau_3}{2\zeta\tau - \tau_3}$
K	$\frac{K(-\tau_3 s + 1)e^{-\theta s}}{(\tau_1 s + 1)(\tau_2 s + 1)}$	$\frac{\tau_1 + \tau_2 + \frac{\tau_3\theta}{\tau_c + \tau_3 + \theta}}{\tau_c + \tau_3 + \theta}$	$\tau_1 + \tau_2 + \frac{\tau_3\theta}{\tau_c + \tau_3 + \theta}$	$\frac{\tau_3\theta}{\tau_c + \tau_3 + \theta} + \frac{\tau_1\tau_2}{\tau_1 + \tau_2 + \frac{\tau_3\theta}{\tau_c + \tau_3 + \theta}}$
L	$\frac{K(-\tau_3 s + 1)e^{-\theta s}}{\tau^2 s^2 + 2\zeta\tau s + 1}$	$\frac{2\zeta\tau + \frac{\tau_3\theta}{\tau_c + \tau_3 + \theta}}{\tau_c + \tau_3 + \theta}$	$2\zeta\tau + \frac{\tau_3\theta}{\tau_c + \tau_3 + \theta}$	$\frac{\tau_3\theta}{\tau_c + \tau_3 + \theta} + \frac{\tau^2}{2\zeta\tau + \frac{\tau_3\theta}{\tau_c + \tau_3 + \theta}}$
M	$\frac{Ke^{-\theta s}}{s}$	$\frac{2\tau_c + \theta}{(\tau_c + \theta)^2}$	$2\tau_c + \theta$	—
N	$\frac{Ke^{-\theta s}}{s}$	$\frac{2\tau_c + \theta}{\left(\tau_c + \frac{\theta}{2}\right)^2}$	$2\tau_c + \theta$	$\frac{\tau_c\theta + \frac{\theta^2}{4}}{2\tau_c + \theta}$
O	$\frac{Ke^{-\theta s}}{s(\tau s + 1)}$	$\frac{2\tau_c + \tau + \theta}{(\tau_c + \theta)^2}$	$2\tau_c + \tau + \theta$	$\frac{(2\tau_c + \theta)\tau}{2\tau_c + \tau + \theta}$

Table 12.2 Equivalent PID Controller Settings for the Parallel and Series Forms

Parallel Form	Series Form
$G_c(s) = K_c\left(1 + \frac{1}{\tau_I s} + \tau_D s\right)$	$G_c(s) = K'_c\left(1 + \frac{1}{\tau'_I s}\right)(1 + \tau'_D s)^\dagger$
$K_c = K'_c\left(1 + \frac{\tau'_D}{\tau'_I}\right)$	$K'_c = \frac{K_c}{2}\left(1 + \sqrt{1 - 4\tau_D/\tau_I}\right)$
$\tau_I = \tau'_I + \tau'_D$	$\tau'_I = \frac{\tau_I}{2}\left(1 + \sqrt{1 - 4\tau_D/\tau_I}\right)$
$\tau_D = \frac{\tau'_D\tau'_I}{\tau'_I + \tau'_D}$	$\tau'_D = \frac{\tau_I}{2}\left(1 - \sqrt{1 - 4\tau_D/\tau_I}\right)$

†These conversion equations are only valid if $\tau_D/\tau_I \leq 0.25$.

EXAMPLE 12.3

A process model for a liquid storage system is given by Chien and Fruehauf (1990):

$$\widetilde{G}(s) = \frac{Ke^{-7.4s}}{s}$$

Use Table 12.1 to calculate PI and PID controller settings for $K = 0.2$ and $\tau_c = 8$. Repeat for $\tau_c = 15$ and do the following:

(a) Compare the four controllers for unit step changes in the set point and disturbance, assuming that $G_d = \widetilde{G}$.

(b) In order to characterize the robustness of each controller of part (a), determine K_{max}, the largest value of K that results in a stable closed-loop system for each controller.

SOLUTION

(a) For this integrating process, $\tilde{G}_+ = e^{-\theta s}$, and thus $C = -\theta$ in (12-32). The IMC controller settings for controllers M and N in Table 12.1 are

	K_c	τ_I	τ_D
PI ($\tau_c = 8$)	0.493	23.4	—
PI ($\tau_c = 15$)	0.373	37.4	—
PID ($\tau_c = 8$)	0.857	23.4	3.12
PID ($\tau_c = 15$)	0.535	37.4	3.33

The closed-loop responses in Fig. 12.7 are more sluggish and less oscillatory for $\tau_c = 15$ than they are for $\tau_c = 8$. Also, for $\tau_c = 15$ the overshoot is smaller for the set-point change, and the maximum deviation is larger after the disturbance. The PID controller provides a better disturbance response than the PI controller with a smaller maximum deviation. In addition, the PID controller has a very short settling time for $\tau_c = 8$, which gives it the best performance of the four controllers considered.

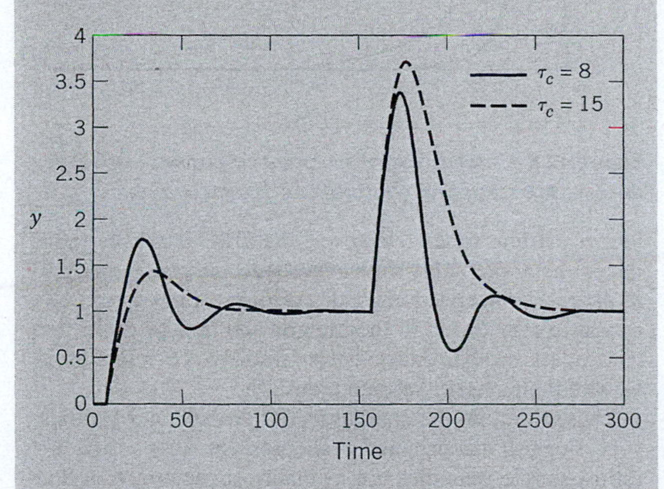

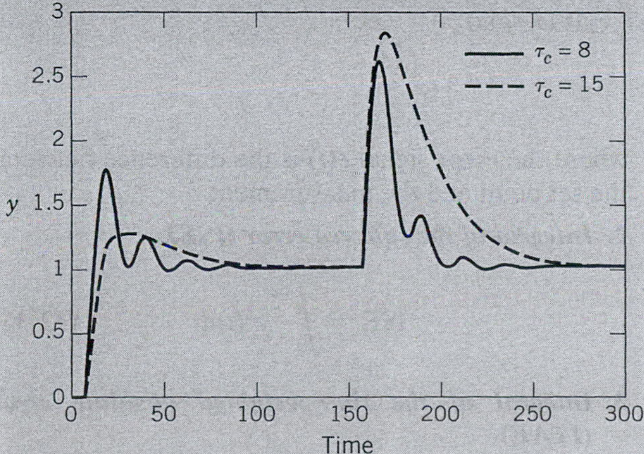

Figure 12.7 Simulation results for Example 12.3: PI control (top) and PID control (bottom).

(b) Let K_{max} denote the maximum value of K that results in a stable closed-loop system. The numerical value of K_{max} can be obtained from a stability analysis. For example, the Routh Stability Criterion of Chapter 11 can be used with a time-delay approximation (e.g., Eq. 12-24) to calculate an approximate value of K_{max}. The exact values can be obtained by applying frequency response stability criteria that will be introduced in Chapter 14.

The numerical results shown in the following table indicate that K can increase significantly from its nominal value of 0.2 before the closed-loop system becomes unstable. Thus, these IMC controllers are quite robust and become even more so as τ_c increases. The approximate values of K_{max} were obtained by using the time-delay approximation in Eq. 12-24b.

		K_{max}	
Controller	τ_c	Approximate	Exact
PI	8	0.274	0.356
PI	15	0.363	0.515
PID	8	0.376	0.277
PID	15	0.561	0.425

Lag-Dominant Models ($\theta/\tau << 1$)

First- or second-order models with relatively small time delays ($\theta/\tau << 1$) are referred to as *lag-dominant models*. The IMC and DS methods provide satisfactory set-point responses, but very slow disturbance responses, because the value of τ_I is very large. Fortunately, this problem can be solved in two different ways.

1. ***Approximate the lag-dominant model by an integrator-plus-time-delay model*** (Chien and Fruehauf, 1990). As indicated in Section 7.2.3, the integrator-plus-time-delay model in Eq. 12-33 provides an accurate approximation to the FOPTD model in Eq. 12-10 for the initial portion of the step response:

$$G(s) = \frac{K^* e^{-\theta s}}{s} \tag{12-33}$$

In Eq. 12-33, $K^* \triangleq K/\tau$. Then the IMC tuning relations in Table 12.1 for either controller M or N can be applied.

2. ***Limit the value of*** τ_I. For lag-dominant models, the standard IMC controllers for first- and second-order models provide sluggish disturbance responses because τ_I is very large. For example, controller G in Table 12.1 has $\tau_I = \tau$ where τ is very large. As a remedy, Skogestad (2003) has proposed limiting the value of τ_I:

$$\tau_I = \min\{\tau, 4(\tau_c + \theta)\} \tag{12-34}$$

EXAMPLE 12.4

Consider a lag-dominant model with $\theta/\tau = 0.01$:

$$\widetilde{G}(s) = \frac{100}{100s + 1} e^{-s}$$

Design three PI controllers:

(a) IMC ($\tau_c = 1$)

(b) IMC ($\tau_c = 2$) based on the integrator approximation of Eq. 12-33

(c) IMC ($\tau_c = 1$) with Skogestad's modification (Eq. 12-34)

Evaluate the three controllers by comparing their performance for unit step changes in both set point and disturbance. Assume that the model is perfect and that $G_d(s) = G(s)$.

SOLUTION

The PI controller settings are

Controller	K_c	τ_I
(a) IMC	0.5	100
(b) Integrator approximation	0.556	5
(c) Skogestad	0.5	8

The simulation results in Fig. 12.8 indicate that the IMC controller provides an excellent set-point response, while the other two controllers have significant overshoots and longer settling times. However, the IMC controller produces an unacceptably slow disturbance response owing to its large τ_I value, although the response does eventually return to zero owing to the integral action. The other two controllers provide much better disturbance rejection in view of their smaller settling times.

Thus, although the standard IMC tuning rules produce very sluggish disturbance responses for very small θ/τ ratios, simple remedies are available, as demonstrated in cases (b) and (c).

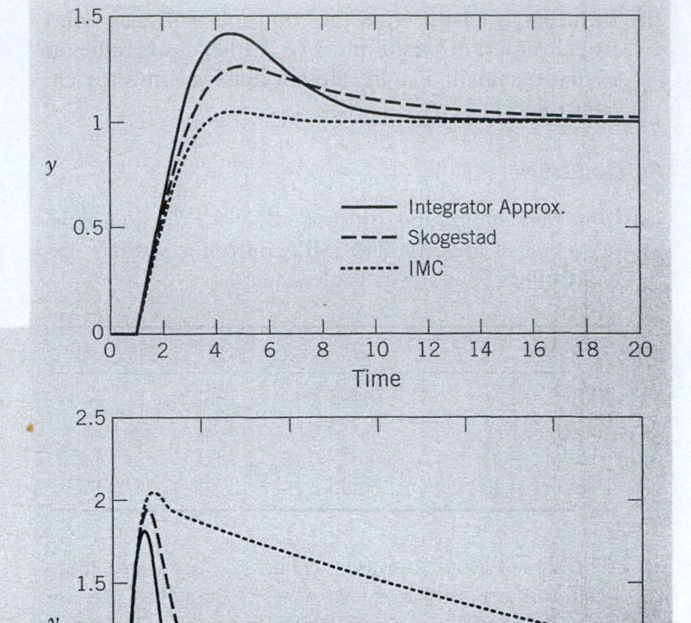

Figure 12.8 Comparison of set-point responses (top) and disturbance responses (bottom) for Example 12.4.

12.3.2 Tuning Relations Based on Integral Error Criteria

Controller tuning relations have been developed that optimize the closed-loop response for a simple process model and a specified disturbance or set-point change. The optimum settings minimize an *integral error criterion*. Three popular integral error criteria are

1. ***Integral of the absolute value of the error (IAE)***

$$\text{IAE} = \int_0^\infty |e(t)|dt \tag{12-35}$$

where the error signal $e(t)$ is the difference between the set point and the measurement.

2. ***Integral of the squared error (ISE)***

$$\text{ISE} = \int_0^\infty e^2(t)dt \tag{12-36}$$

3. ***Integral of the time-weighted absolute error (ITAE)***

$$\text{ITAE} = \int_0^\infty t|e(t)|dt \tag{12-37}$$

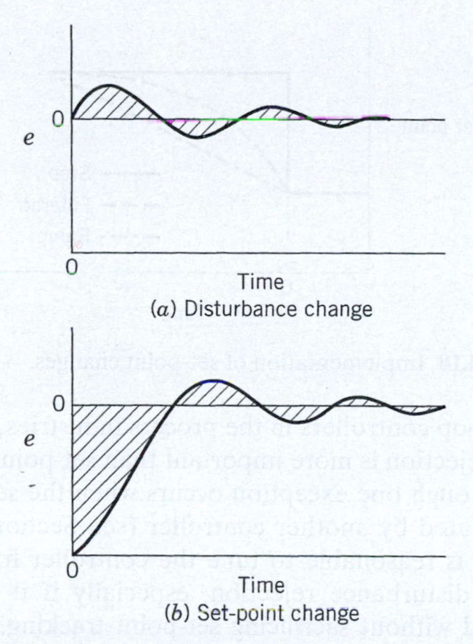

(a) Disturbance change

(b) Set-point change

Figure 12.9 Graphical interpretation of IAE. The shaded area is the IAE value.

The ISE criterion penalizes large errors, while the ITAE criterion penalizes errors that persist for long periods of time. In general, the ITAE is the preferred criterion, because it usually results in the most conservative controller settings. By contrast, the ISE criterion provides the most aggressive settings, while the IAE criterion tends to produce controller settings that are between those for the ITAE and ISE criteria. A graphical interpretation of the IAE performance index is shown in Fig. 12.9.

Controller tuning relations for the ITAE performance index are shown in Table 12.3. These relations were developed for the FOPTD model of Eq. 12-10 and the parallel form of the PID controller in Eq. 12-13. It was also assumed that the disturbance and process transfer functions in Fig. 12.2 are identical (that is, $G_d = G$). Note that the optimal controller settings are different for set-point changes and step disturbances. In general, the controller settings for set-point changes are more conservative.

12.3.3 Miscellaneous Tuning Relations

Two early controller tuning relations were published by Ziegler and Nichols (1942) and Cohen and Coon (1953). These well-known tuning relations were developed to provide closed-loop responses that have a 1/4 decay ratio (see Section 5.4). Because a response with a 1/4 decay ratio is considered to be excessively oscillatory for most process control applications, these tuning relations are not recommended.

Other PID design methods and tuning relations are available in books (Åström and Hägglund, 2006; McMillan, 2006; Visioli, 2006) and in an extensive compilation (O'Dwyer, 2003).

12.3.4 Comparison of Controller Design and Tuning Relations

Although the design and tuning relations of the previous sections are based on different performance criteria, several general conclusions can be drawn:

1. The controller gain K_c should be inversely proportional to the product of the other gains in the feedback loop (i.e., $K_c \propto 1/K$ where $K = K_v K_p K_m$).
2. K_c should decrease as θ/τ, the ratio of the time delay to the dominant time constant increases. In general, the quality of control decreases as θ/τ

Table 12.3 Controller Design Relations Based on the ITAE Performance Index and a First-Order-plus-Time-Delay Model (Smith and Corripio, 1997)[a]

Type of Input	Type of Controller	Mode	A	B
Disturbance	PI	P	0.859	−0.977
		I	0.674	−0.680
Disturbance	PID	P	1.357	−0.947
		I	0.842	−0.738
		D	0.381	0.995
Set point	PI	P	0.586	−0.916
		I	1.03[b]	−0.165[b]
Set point	PID	P	0.965	−0.85
		I	0.796[b]	−0.1465[b]
		D	0.308	0.929

[a] Design relation: $Y = A(\theta/\tau)^B$ where $Y = KK_c$ for the proportional mode, τ/τ_I for the integral mode, and τ_D/τ for the derivative mode.

[b] For set-point changes, the design relation for the integral mode is $\tau/\tau_I = A + B(\theta/\tau)$.

increases due to longer settling times and larger maximum deviations from the set point.

3. Both τ_I and τ_D should increase as θ/τ increases. For many controller tuning relations, the ratio, τ_D/τ_I, is between 0.1 and 0.3. As a rule of thumb, use $\tau_D/\tau_I = 0.25$ as a first guess.
4. When integral control action is added to a proportional-only controller, K_c should be reduced. The further addition of derivative action allows K_c to be increased to a value greater than that for proportional-only control.

Similar trends occur for the control system design methods based on frequency response criteria that will be considered in Chapter 14.

Although the tuning relations in the previous sections were developed for the parallel form of PID control, they can be converted to the series form by using Table 12.2.

EXAMPLE 12.5

A blending system with a measurement time delay can be modeled as

$$G(s) = \frac{1.54e^{-1.07s}}{5.93s + 1}$$

Calculate PI controller settings using the following tuning relations:

(a) IMC ($\tau_c = \tau/3$)
(b) IMC ($\tau_c = \theta$)
(c) ITAE (disturbance)
(d) ITAE (set point)

SOLUTION

The calculated PI controller settings are

	K_c	τ_I
IMC ($\tau_c = \tau/3 = 1.97$)	1.27	5.93
IMC ($\tau_c = \theta = 1.07$)	1.80†	5.93†
ITAE (disturbance)	2.97	2.75
ITAE (set point)	1.83	5.93

The ITAE (disturbance) settings are the most aggressive, and the IMC settings ($\tau_c = \tau/3$) are the least aggressive. The settings for IMC ($\tau_c = \theta$) and ITAE (set point) are almost identical for this example, but this is not true in general.

12.4 CONTROLLERS WITH TWO DEGREES OF FREEDOM

The specification of controller settings for a standard PID controller typically requires a tradeoff between set-point tracking and disturbance rejection. For most single-loop controllers in the process industries, disturbance rejection is more important than set-point tracking, although one exception occurs when the set point is calculated by another controller (see Section 16.1). Thus, it is reasonable to tune the controller for satisfactory disturbance rejection, especially if it can be achieved without sacrificing set-point tracking. Fortunately, two simple strategies can be used to adjust the set-point and disturbance responses independently. These strategies are referred to as *controllers with two degrees of freedom* (Goodwin et al., 2001).

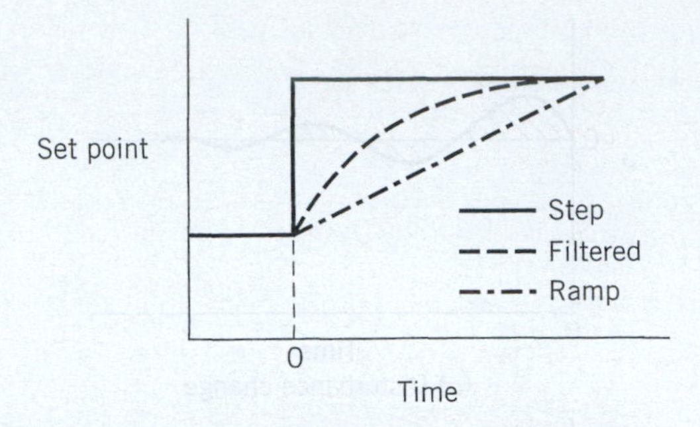

Figure 12.10 Implementation of set-point changes.

The first strategy is very simple. Set-point changes are introduced gradually rather than as abrupt step changes. For example, the set point can be ramped to the new value as shown in Fig. 12.10 or "filtered" by passing it through a first-order transfer function,

$$\frac{Y^*_{sp}}{Y_{sp}} = \frac{1}{\tau_f s + 1} \tag{12-38}$$

where Y^*_{sp} denotes the *filtered set point* that is used in the control calculations. The filter time constant, τ_f, determines how quickly the filtered set point will attain the new value, as shown in Fig. 12.10. This strategy can significantly reduce, or even eliminate, overshoot for set-point changes.

A second strategy for independently adjusting the set-point response is based on a simple modification of the PID control law in Chapter 8,

$$p(t) = \bar{p} + K_c\left[e(t) + \frac{1}{\tau_I}\int_0^t e(t^*)dt^* - \tau_D\frac{dy_m}{dt}\right] \tag{8-17}$$

where y_m is the measured value of y and e is the error signal, $e \triangleq y_{sp} - y_m$. The control law modification consists of multiplying the set point in the proportional term by a *set-point weighting factor*, β:

$$p(t) = \bar{p} + K_c[\beta y_{sp}(t) - y_m(t)] + K_c\left[\frac{1}{\tau_I}\int_0^t e(t^*)dt^* - \tau_D\frac{dy_m}{dt}\right] \tag{12-39}$$

The set-point weighting factor is bounded, $0 < \beta < 1$, and serves as a convenient tuning factor (Åström and Hägglund, 2006). Note that the integral and derivative control terms in (8-17) and (12-39) are the same. Consequently, offset is eliminated for all values of β.

In general, as β increases, the set-point response becomes faster but exhibits more overshoot. When $\beta = 1$, the modified PID control law in Eq. 12-39 reduces to the standard PID control law in Eq. 8-17.

EXAMPLE 12.6

For the first-order-plus-time-delay model of Example 12.4, the PI controller for case (b) provided the best disturbance response. However, its set-point response had a significant overshoot. Can set-point weighting significantly reduce the overshoot without adversely affecting the settling time?

SOLUTION

Figure 12.11 compares the set-point responses for a PI controller with and without set-point weighting. Set-point weighting with $\beta = 0.5$ provides a significant improvement, because the overshoot is greatly reduced and the settling time is significantly decreased. Because the disturbance response in Fig. 12.8 is independent of the value of β, the stated goal is achieved.

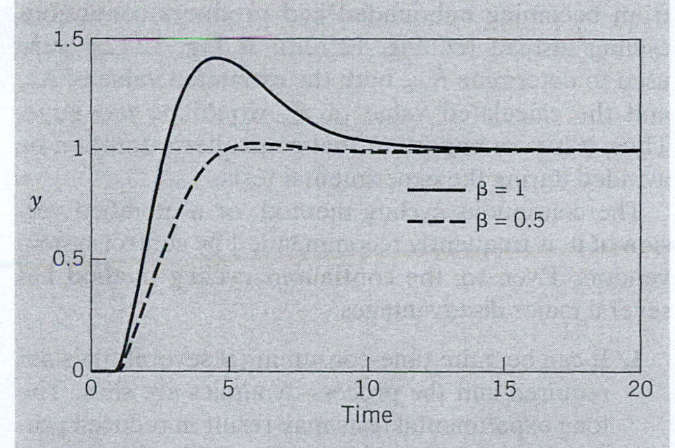

Figure 12.11 Influence of set-point weighting on closed-loop responses for Example 12.6.

We have considered two simple but effective strategies for adjusting the set-point response without affecting the disturbance response: set-point filtering (or ramping) and the use of a set-point weighting factor β.

12.5 ON-LINE CONTROLLER TUNING

The control systems for modern industrial plants typically include thousands of individual control loops. During control system design, preliminary controller settings are specified based on process knowledge, control objectives, and prior experience. After a controller is installed, the preliminary settings often prove to be satisfactory. But for critical control loops, the preliminary settings may have to be adjusted in order to achieve satisfactory control. This on-site adjustment is referred to by a variety of names: *on-line tuning, field tuning*, or *controller tuning*.

Because on-line controller tuning involves plant testing, often on a trial-and-error basis, the tuning can be quite tedious and time-consuming. Consequently, good initial controller settings are very desirable to reduce the required time and effort. Ideally, the preliminary settings from the control system design can be used as the initial field settings. If the preliminary settings are not satisfactory, alternative settings can be obtained from simple experimental tests. If necessary, the settings can be fine-tuned by a modest amount of trial and error.

In this chapter, we present three important methods for on-line controller tuning. More detailed analysis and a wealth of practical experience are available in books by industrial practitioners (McMillan, 2006; Shinskey, 1994) and university researchers (Åström and Hägglund, 2006; Yu, 1999). Software for controller tuning is also widely available.

Next, we make a few general observations:

1. ***Controller tuning inevitably involves a tradeoff between performance and robustness.*** The performance goals of excellent set-point tracking and disturbance rejection should be balanced against the robustness goal of stable operation over a wide range of conditions.
2. ***Controller settings do not have to be precisely determined.*** In general, a small change in a controller setting from its best value (for example, ±10%) has little effect on closed-loop responses.
3. ***For most plants, it is not feasible to manually tune each controller.*** Tuning is usually done by a control specialist (engineer or technician) or by a plant operator. Because each person is typically responsible for 300 to 1,000 control loops, it is not feasible to tune every controller. Instead, only the control loops that are perceived to be the most important or the most troublesome receive detailed attention. The other controllers typically operate using the preliminary settings from the control system design.
4. ***Diagnostic techniques for monitoring control system performance are available.*** This topic is introduced in Chapter 21.

Next we consider three important on-line tuning methods.

12.5.1 Continuous Cycling Method

Over 60 years ago, Ziegler and Nichols (1942) published a classic paper that introduced the *continuous cycling method* for controller tuning. It is based on the following trial-and-error procedure:

Step 1. After the process has reached steady state (at least approximately), eliminate the integral and derivative control action by setting τ_D to zero and τ_I to the largest possible value.

Step 2. Set K_c equal to a small value (e.g., 0.5) and place the controller in the automatic mode.

Step 3. Introduce a small, momentary set-point change so that the controlled variable moves away from the set point. Gradually increase K_c in small increments until continuous cycling occurs. The term *continuous cycling* refers to a sustained oscillation with a constant amplitude. The numerical value of K_c that produces continuous cycling (for proportional-only control) is called the *ultimate gain*, K_{cu}. The period of the corresponding sustained oscillation is referred to as the *ultimate period*, P_u.

Step 4. Calculate the PID controller settings using the Ziegler-Nichols (Z-N) tuning relations in Table 12.4 or the more conservative Tyreus-Luyben settings.

Step 5. Evaluate the Z-N controller settings by introducing a small set-point change and observing the closed-loop response. Fine-tune the settings, if necessary.

The tuning relations reported by Ziegler and Nichols (1942) were determined empirically to provide closed-loop responses that have a 1/4 decay ratio. For proportional-only control, the Z-N settings in Table 12.4 provide a safety margin of two for K_c, because it is equal to one-half of the stability limit, K_{cu}. When integral action is added to form a PI controller, K_c is reduced from $0.5K_{cu}$ to $0.45K_{cu}$. The stabilizing effect of derivative action allows K_c to be increased to 0.6 K_{cu} for PID control.

Typical results for the trial-and-error determination of K_{cu} are shown in Fig. 12.12. For $K_c < K_{cu}$, the closed-loop response $y(t)$ is usually overdamped or slightly oscillatory. For the ideal case where $K_c = K_{cu}$, continuous cycling occurs (Fig. 12.12*b*). For $K_c > K_{cu}$, the closed-loop system is unstable and will theoretically have an unbounded response (Fig. 12.12*c*). But in practice, controller saturation prevents the response from becoming unbounded and produces continuous cycling instead (cf. Fig. 12.12*d*). If Fig. 12.12*d* were used to determine K_{cu}, both the estimated value of K_{cu} and the calculated value of K_c would be too large. Thus, it is very important that controller saturation be avoided during the experimental tests.

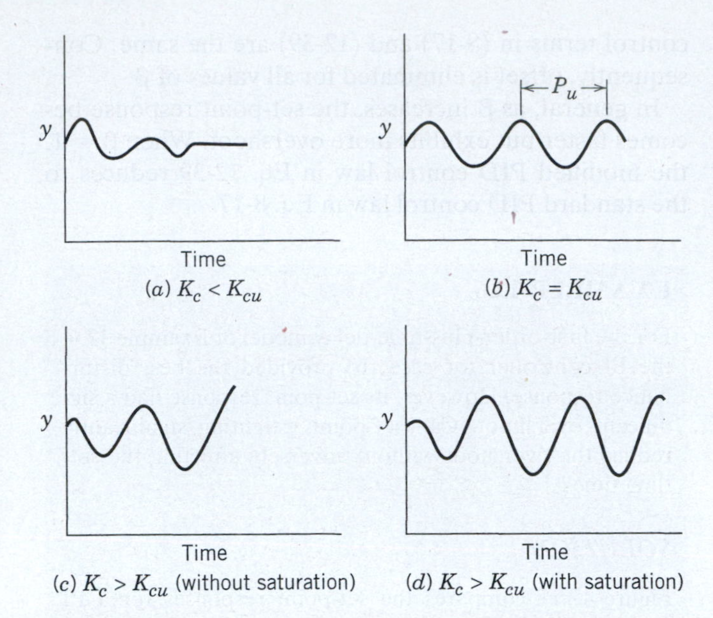

Figure 12.12 Experimental determination of the ultimate gain K_{cu}.

Table 12.4 Controller Settings based on the Continuous Cycling Method

Ziegler-Nichols	K_c	τ_I	τ_D
P	$0.5K_{cu}$	—	—
PI	$0.45K_{cu}$	$P_u/1.2$	—
PID	$0.6K_{cu}$	$P_u/2$	$P_u/8$
Tyreus-Luyben[†]	K_c	τ_I	τ_D
PI	$0.31K_{cu}$	$2.2P_u$	—
PID	$0.45K_{cu}$	$2.2P_u$	$P_u/6.3$

[†] Luyben and Luyben (1997).

The continuous cycling method, or a modified version of it, is frequently recommended by control system vendors. Even so, the continuous cycling method has several major disadvantages:

1. It can be quite time-consuming if several trials are required and the process dynamics are slow. The long experimental tests may result in reduced production or poor product quality.
2. In many applications, continuous cycling is objectionable, because the process is pushed to the stability limits. Consequently, if external disturbances or process changes occur during the test, unstable operation or a hazardous situation could result (e.g., a "runaway" chemical reaction).
3. This tuning procedure is not applicable to integrating or open-loop unstable processes, because their control loops typically are unstable at both high and low values of K_c, while being stable for intermediate values.
4. For first-order and second-order models without time delays, the ultimate gain does not exist, because the closed-loop system is stable for all values of K_c, providing that its sign is correct. However, in practice, it is unusual for a control loop not to have an ultimate gain.

$P_u = \frac{2\pi}{\omega_{cu}}$

Fortunately, the first two disadvantages can be avoided by using either the *relay auto-tuning method* or the *step test method* described later in this section. Alternatively, if a process model is available, K_{cu} and P_u can be determined from a frequency response analysis, as described in Chapter 14.

The Z-N controller settings have been widely used as a benchmark for evaluating different tuning methods and control strategies. Because they are based on a 1/4 decay ratio, the Z-N settings tend to produce oscillatory responses and large overshoots for set-point changes. Consequently, more conservative controller settings are preferable such as the Tyreus-Luyben settings in Table 12.4.

Despite their prominence in the process control literature, it is not certain whether the famous Z-N tuning relations for PID control were developed for the series or parallel form of the controller (Åström et al., 2001). Although the PID equations were developed for a Taylor Instruments pneumatic PID controller that had a series structure, simulation studies were conducted with a differential analyzer (Blickley, 1990) that may have facilitated simulation of the parallel structure. Furthermore, applying the Z-N settings in the parallel form produces more conservative control (Skogestad, 2003). Consequently, it is reasonable to apply the Z-N settings to the parallel form and then convert the settings to series form, if necessary, using Table 12.2.

EXAMPLE 12.7

For the process model of Example 12.1,

$$G = \frac{2e^{-s}}{(10s + 1)(5s + 1)}$$

compare PID controllers with the following settings:

(a) Ziegler-Nichols (Z-N) settings (Table 12.4)

(b) Tyreus-Luyben (T-L) settings (Table 12.4)

(c) Direct Synthesis (DS) method with $\tau_c = 3$ (see Eq. 12-14)

Evaluate these controllers for unit step changes in both the set point and the disturbance, assuming that $G_d = G$.

SOLUTION

The ultimate gain and ultimate period were determined by trial and error to be $K_{cu} = 7.88$ and $P_u = 11.66$. The calculated PID controller settings are

Method	K_c	τ_I	τ_D
Z-N	4.73	5.8	1.45
T-L	3.55	25.8	1.84
DS	1.88	15.0	3.33

These controller settings and the closed-loop responses in Fig. 12.13 indicate that the Z-N settings are the most aggressive and produce oscillatory responses. The Z-N controller provides the best control for the disturbance and the worst for the set-point change. The T-L and DS controllers result in satisfactory set-point responses but sluggish disturbance responses.

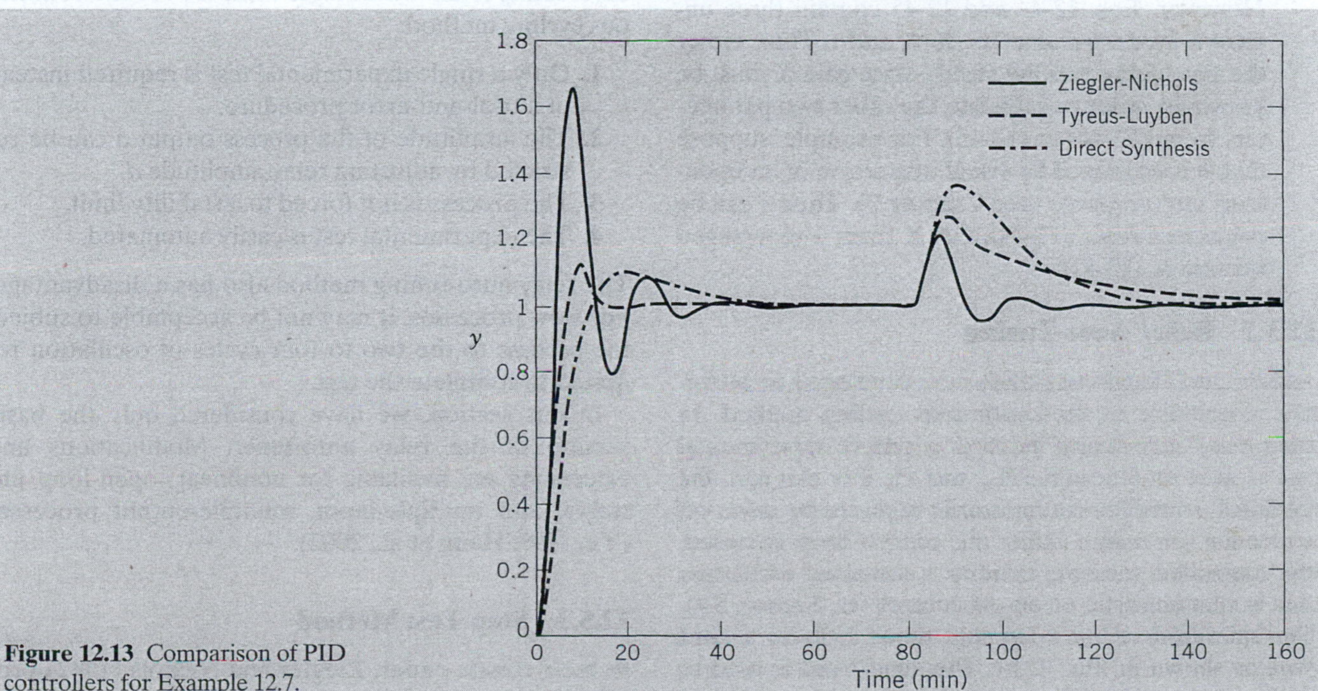

Figure 12.13 Comparison of PID controllers for Example 12.7.

Calculation of Model Parameters from K_{cu} and P_u

Model parameters for simple transfer function models can be calculated from K_{cu} and P_u. Then the model-based design methods in Sections 12.2 and 12.3 can be used to calculate controller settings from the model parameters. We illustrate this strategy for two important process models.

1. *Integrator-plus-time-delay model:*

$$G(s) = \frac{Ke^{-\theta s}}{s} \tag{12-40}$$

The model parameters can be calculated from the following equations (Yu, 1999):

$$K = \frac{2\pi}{K_{cu}P_u} \tag{12-41}$$

$$\theta = \frac{P_u}{4} \tag{12-42}$$

2. *First-order-plus-time-delay model:*

$$G(s) = \frac{Ke^{-\theta s}}{\tau s + 1} \tag{12-43}$$

The time constant can be calculated from either one of two equations (Yu, 1999):

$$\tau = \frac{P_u}{2\pi} \tan\left[\frac{\pi(P_u - 2\theta)}{P_u}\right] \tag{12-44}$$

or

$$\tau = \frac{P_u}{2\pi}\sqrt{(KK_{cu})^2 - 1} \tag{12-45}$$

However, Eqs. 12-44 and 12-45 contain three unknown model parameters, K, θ, and τ. Thus, either the time delay θ or the steady-state gain K must be known in order to calculate the other two parameters from (12-44) or (12-45). For example, suppose that θ is estimated by visual inspection of an open-loop step response (see Chapter 7). Then τ can be calculated from (12-44) and K from a rearranged version of (12-45).

12.5.2 Relay Auto-Tuning

Åström and Hägglund (1984) have developed an attractive alternative to the continuous cycling method. In their *relay auto-tuning* method, a simple experimental test is used to determine K_{cu} and P_u. For this test, the feedback controller is temporarily replaced by an on-off controller (or *relay*). After the control loop is closed, the controlled variable exhibits a sustained oscillation that is characteristic of on-off control (cf. Section 8.4). The operation of the relay auto-tuner includes a *dead zone* as shown in Fig. 12.14. The dead band is used to avoid frequent switching caused by measurement noise.

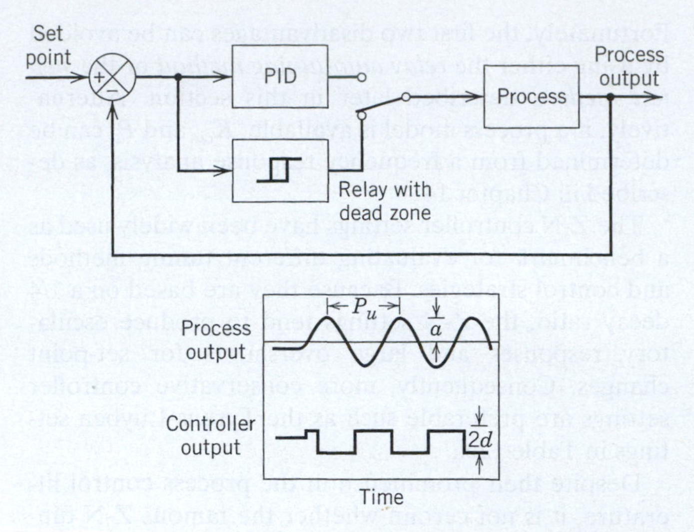

Figure 12.14 Auto-tuning using a relay controller.

The ultimate gain and the ultimate period can easily be obtained from Fig. 12.14. The ultimate period P_u is equal to the period of oscillation for the process output. Åström and Hägglund (1984) derived an approximate expression for the ultimate gain,

$$K_{cu} = \frac{4\,d}{\pi\,a} \tag{12-46}$$

where d is the relay amplitude (set by the user) and a is the measured amplitude of the process oscillation. PID controller settings can then be calculated from the Z-N settings in Table 12.6 or from the model parameters in Eqs. 12-40 to 12-45. The relay auto-tuning method has several important advantages compared to the continuous cycling method:

1. Only a single experimental test is required instead of a trial-and-error procedure.
2. The amplitude of the process output a can be restricted by adjusting relay amplitude d.
3. The process is not forced to a stability limit.
4. The experimental test is easily automated.

The relay auto-tuning method also has a disadvantage. For slow processes, it may not be acceptable to subject the process to the two to four cycles of oscillation required to complete the test.

In this section, we have considered only the basic version of the relay auto-tuner. Modifications and extensions are available for nonlinear, open-loop unstable, and multiple-input, multiple-output processes (Yu, 1999; Hang et al., 2002).

12.5.3 Step Test Method

In their classic paper, Ziegler and Nichols (1942) proposed a second on-line tuning technique based on a single step test. The experimental procedure is quite simple. After the process has reached steady state (at least approximately), the controller is placed in the manual mode. Then a small step change in the controller output (e.g., 3 to 5%) is introduced. The controller settings are based on the *process reaction curve* (Section 7.2), the open-loop step response. Consequently, this on-line tuning technique is referred to as *the step test method* or *the process reaction curve method.*

Two types of process reaction curves are shown in Fig. 12.15 for step changes occurring at $t = 0$. After an initial transient, the measured response y_m for Case (a) increases at a constant rate, indicating that the process appears to act as an integrating element and thus is *not* self-regulating. In contrast, the hypothetical process considered in Case (b) is self-regulating, because the step response reaches a new steady state. Both step responses are characterized by two parameters: S, the slope of the tangent through the inflection point, and θ, the apparent time delay. The graphical determination of S and θ was described in Section 7.2. For Case (b), the slope S is equal to K/τ for a first-order-plus-time-delay model.

An appropriate transfer function model can be obtained from the step response by using the parameter estimation methods of Chapter 7. For processes that have monotonically increasing step responses, such as the responses in Fig. 12.15, the models in Eqs. 12-40 and 12-43 are appropriate. Then, any of the model-based tuning relations in Sections 12.2 and 12.3 can be employed.

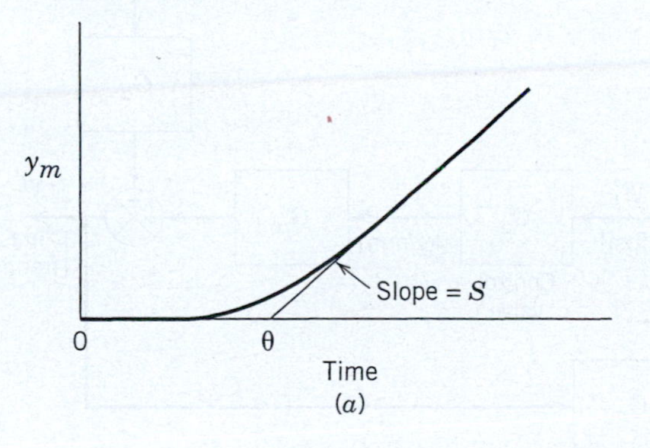

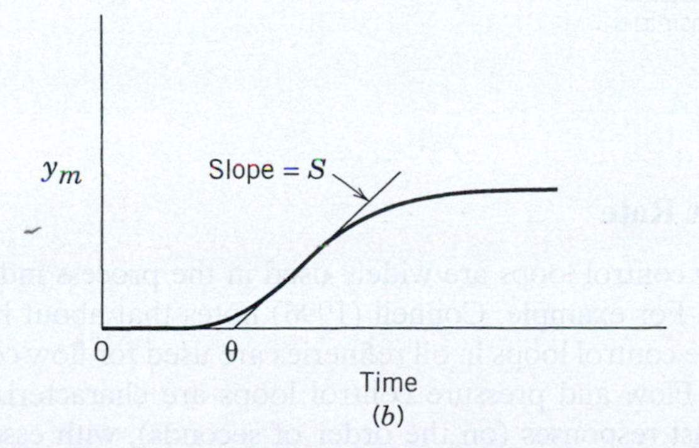

Figure 12.15 Typical process reaction curves: (a) non-self-regulating process, (b) self-regulating process.

The chief advantage of the step test method is that only a single experimental test is necessary. But the method does have four disadvantages:

1. The experimental test is performed under open-loop conditions. Thus, if a significant disturbance occurs during the test, no corrective action is taken. Consequently, the process can be upset and the test results may be misleading.
2. For a nonlinear process, the test results can be sensitive to the magnitude and direction of the step change. If the magnitude of the step change is too large, process nonlinearities can influence the result. But if the step magnitude is too small, the step response may be difficult to distinguish from the usual fluctuations due to noise and disturbances. The direction of the step change (positive or negative) should be chosen so that the controlled variable will not violate a constraint.
3. The method is not applicable to open-loop unstable processes.
4. For continuous controllers, the method tends to be sensitive to controller calibration errors. By contrast, the continuous cycling method is less sensitive to calibration errors in K_c, because it is adjusted during the experimental test.

Closed-loop versions of the step test method have been proposed as a partial remedy for the first disadvantage (Yuwana and Seborg, 1982; Lee et al., 1990). Typically, a step change in the set point is introduced while the process is controlled using proportional-only control. Then the parameters in simple transfer function models can be estimated from the closed-loop step response.

The second disadvantage can be avoided by making multiple step changes instead of a single step. For example, if a series of both positive and negative changes is made, the effects of disturbances, nonlinearities, and control valve hysteresis will become apparent.

EXAMPLE 12.8

Consider the feedback control system for the stirred-tank blending process shown in Fig. 11.1 and the following step test. The controller was initially in the manual mode, and then its output was suddenly changed from 30% to 43%. The resulting process reaction curve is shown in Fig. 12.16. Thus, after the step change occurred at $t = 0$, the measured exit composition changed from 35% to 55% (expressed as a percentage of the measurement span), which is equivalent to the mole fraction changing from 0.10 to 0.30. Determine an appropriate process model for $G \triangleq G_{IP}G_vG_pG_m$.

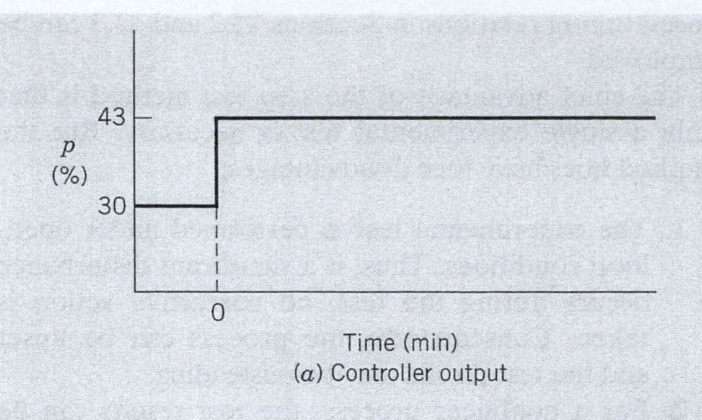

(a) Controller output

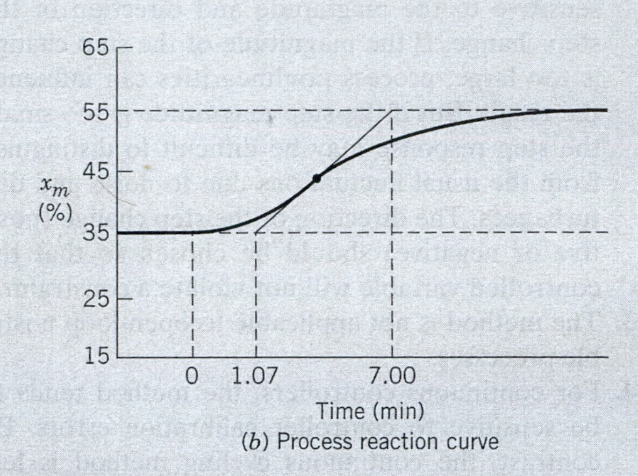

(b) Process reaction curve

Figure 12.16 Process reaction curve for Example 12.8.

SOLUTION

A block diagram for the closed-loop system is shown in Fig. 12.17. This block diagram is similar to Fig. 11.7, but the feedback loop has been opened between the controller and the current-to-pressure (I/P) transducer. A first-order-plus-time-delay model can be developed from the process reaction curve in Fig. 12.16 using the graphical method of Section 7.2. The tangent line through the inflection point intersects the horizontal lines for the initial and final composition values at 1.07 min and 7.00 min, respectively. The slope of the line is

$$S = \left(\frac{55\% - 35\%}{7.00 - 1.07\ \text{min}}\right) = 3.37\%/\text{min}$$

The model parameters can be calculated as

$$K = \frac{\Delta x_m}{\Delta p} = \frac{55\% - 35\%}{43\% - 30\%} = 1.54\ \text{(dimensionless)}$$

$$\theta = 1.07\ \text{min}$$

$$\tau = 7.00 - 1.07\ \text{min} = 5.93\ \text{min}$$

The apparent time delay of 1.07 min is subtracted from the intercept value of 7.00 min for the τ calculation.

The resulting empirical process model can be expressed as

$$\frac{X'_m(s)}{P'(s)} = G(s) = \frac{1.54e^{-1.07s}}{5.93s + 1}$$

Example 12.5 in Section 12.3 provided a comparison of PI controller settings for this model that were calculated using different tuning relations.

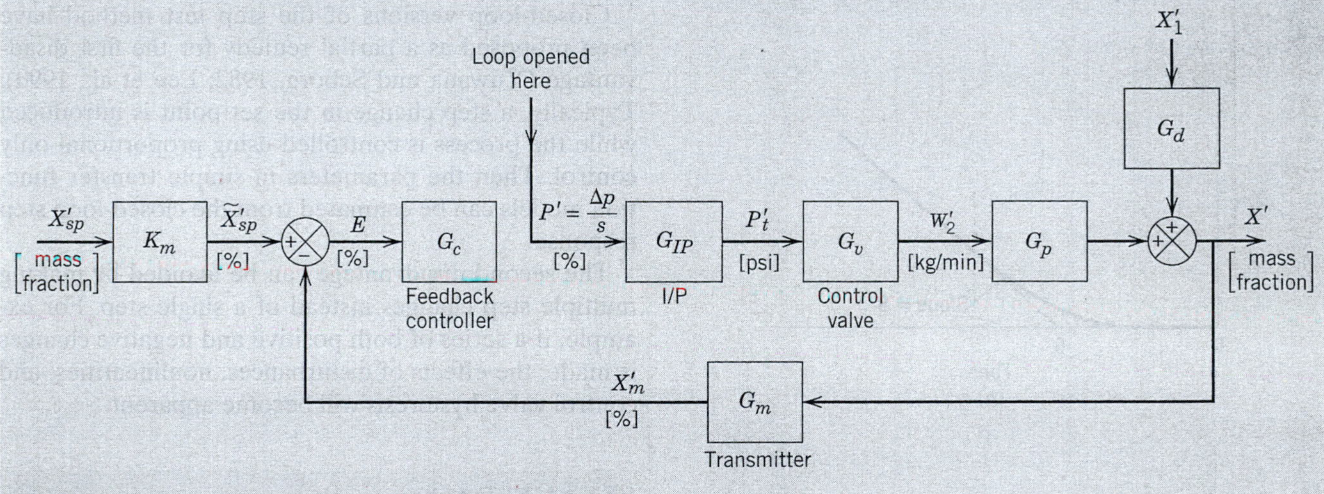

Figure 12.17 Block diagram for Example 12.8.

12.6 GUIDELINES FOR COMMON CONTROL LOOPS

General guidelines for selection of controller type (P, PI, etc.) and controller settings are available for common process variables such as flow rate, liquid level, gas pressure, temperature, and composition. The general guidelines presented below are useful but they should be used with caution, because exceptions do occur.

Flow Rate

Flow control loops are widely used in the process industries. For example, Connell (1996) notes that about half of the control loops in oil refineries are used for flow control. Flow and pressure control loops are characterized by fast responses (on the order of seconds), with essentially no time delay. The process dynamics result from compressibility (in a gas stream) or inertial effects (in a liquid) plus control valve dynamics for large-diameter pipelines. Disturbances in flow control systems tend to be frequent but generally small. Most of the disturbances are high-frequency noise (periodic or random) due to upstream turbulence, valve changes, and pump vibration.

For flow control loops, PI control is generally used with intermediate values of the controller gain. Fruehauf et al. (1994) recommend the following controller settings: $0.5 < K_c < 0.7$ and $0.2 < \tau_I < 0.3$ min. The presence of recurring high-frequency noise discourages the use of derivative action, because it amplifies the noise. Furthermore, because flow control loops usually have relatively small settling times (compared to other control loops), there is little incentive to use derivative action to make the loop respond even faster.

Liquid Level

A liquid storage vessel with a pump on its exit line can act as an integrating process, as has been discussed in Chapters 2 and 11. Standard P or PI controllers are widely used for level control. However, as shown in Section 11.3, these level control problems have an unusual characteristic: increasing the gain of a PI controller can increase stability, while reducing the gain can increase the degree of oscillation and thus reduce stability. Of course, if K_c becomes too large, oscillations or even instability can result. Integral control action is often used but can be omitted if small offsets in the liquid level ($\pm$ 5%) can be tolerated. Derivative action is not normally used for level control, because the level measurements are often noisy as a result of the splashing and turbulence of the liquid entering the tank.

It is common industrial practice to use a liquid storage tank as a surge tank in order to damp out fluctuations in the inlet streams. The control objectives are that (1) the exit flow rate from the tank change should change gradually, rather than abruptly, in order to avoid upsetting downstream process units, (2) the liquid level should be maintained within specified upper and lower limits, and (3) the steady-state mass balance must be satisfied so that the inlet and outlet flows are equal. These three goals can be achieved by allowing the liquid level to rise or fall in response to inlet flow disturbances. This strategy is referred to as *averaging level control.*

Because offset is not important in averaging level control, it is reasonable to use a proportional-only controller. But if integral control action is desired, St. Clair (1993) recommends the following PI controller settings for averaging level control:

$$K_c = \frac{100\%}{\Delta h} \quad (12\text{-}47)$$

$$\tau_I = \frac{4V}{K_c Q_{\max}} \quad (12\text{-}48)$$

where

$$\Delta h \triangleq \min\,(h_{\max} - h_{sp}, h_{sp} - h_{\min}) \quad (12\text{-}49)$$

In Eq. 12-49 $h_{\max}$ and $h_{\min}$ are the maximum and minimum allowable liquid levels, and h_{sp} is the set point. Each is expressed as a percentage of the level transmitter range. In Eq. 12-47, V is the tank volume, and $Q_{\max}$ is the maximum flow rate through the control valve. Equations 12-47 and 12-49 ensure that the controller output will be at a saturation limit (0% or 100%) when the absolute value of the controller error is larger than Δh.

Nonlinear versions of PI control are sometimes used for averaging level control, especially *error-squared controllers* where the controller gain is proportional to the error signal (see Chapter 16). Error-squared controllers offer the advantage of a large control action when the controlled variable is far from the set point, and a small control action when it is near. However, they must be tuned carefully in order to guarantee stable responses over the entire operating range (Shinskey, 1994).

For some applications, tight level control is desirable. For example, a constant liquid level is desirable for some chemical reactors or bioreactors in order to keep the residence time constant. In these situations, the level controller settings can be specified using standard tuning methods. If level control also involves heat transfer, such as for a vaporizer or an evaporator, the controller design becomes much more complicated. In such situations special control methods can be advantageous (Shinskey, 1994).

Gas Pressure

The control of gas pressure is very analogous to the control of liquid level in the sense that some applications use averaging control while others require tight control around a set point. However, high and low limits are usually a more serious concern for pressure control than for level control, because of safety and operational issues. For self-regulating processes, pressure is relatively easy to control, except when the gas is in equilibrium with a liquid. Gas pressure is self-regulating when the vessel (or pipeline) admits more feed when the pressure is low, and reduces the intake when the pressure becomes high. Integrating processes occur when the exit pressure is determined by a compressor, in analogy to liquid level when there is a pump for the exit stream. For pressure control, PI controllers are normally used with only a small amount of integral control action (i.e., τ_I is large). Usually the pressure vessel is not large, leading to relatively small residence times and time constants. Derivative action is normally not needed because the process response times are usually quite small compared to those of other process operations.

Temperature

General guidelines for temperature control loops are difficult to state because of the wide variety of processes and equipment involving heat transfer and their different time scales. For example, the temperature control problems are quite different for heat exchangers, distillation columns, chemical reactors, and evaporators. The presence of time delays and/or multiple thermal capacitances will usually place a stability limit on the controller gain. PID controllers are commonly employed to provide quicker responses than can be obtained with PI controllers.

Composition

Composition control loops generally have characteristics similar to temperature loops but with certain differences:

1. Measurement (instrument) noise is a more significant problem in composition loops.
2. The time delay associated with the analyzer and its sampling system may be a significant factor.

These two factors can limit the effectiveness of derivative action. Because of their importance and the difficulty of control, composition and temperature loops often are prime candidates for the advanced control strategies discussed in Chapters 16, 18, and 20.

12.7 TROUBLESHOOTING CONTROL LOOPS

If a control loop is not performing satisfactorily, then troubleshooting is necessary to identify the source of the problem. Ideally, it would be desirable to evaluate the control loop over the full range of process operating conditions during the commissioning of the plant. In practice, this is seldom feasible. Furthermore, the process characteristics can vary with time for a variety of reasons, including changes in equipment and instrumentation, different operating conditions, new feedstocks or products, and large disturbances. Surveys of thousands of control loops have confirmed that a large fraction of industrial control loops perform poorly. For example, surveys have reported that about one-third of the industrial control loops were in the manual mode, and another one-third actually increased process variability over manual control, a result of poor controller tuning (Ender, 1993; Desborough and Miller, 2002). Clearly, controller tuning and control loop troubleshooting are important activities.

This section provides a brief introduction to the basic principles and strategies that are useful in troubleshooting control loops. More detailed analyses that provide useful insights are available elsewhere (Buckley, 1973; Ender, 1992; Riggs and Karim, 2006; Lieberman, 2008).

An important consideration for troubleshooting activities is to be aware that the control loop consists of a number of individual components: sensor/transmitter, controller, final control element, instrument lines, computer-process interface (for digital control), as well as the process itself. Serious control problems can result from a malfunction of any single component. On the other hand, even if each individual component is functioning properly, there is no guarantee that the overall system will perform properly. Thus, a *systems approach* is required.

As Buckley (1973) has noted, operating and maintenance personnel unfortunately tend to use controller retuning as a cure-all for control loop problems. Based on experience in the chemical industry, he has observed that a control loop that once operated satisfactorily can become either unstable or excessively sluggish for a variety of reasons that include

a. Changing process conditions, usually changes in throughput rate
b. Sticking control valve stem
c. Plugged line in a pressure or differential pressure transmitter
d. Fouled heat exchangers, especially reboilers for distillation columns
e. Cavitating pumps (usually caused by a suction pressure that is too low)

Note that only Items (a) and (d) provide valid reasons for re-tuning the controller.

The starting point for troubleshooting is to obtain enough background information to clearly define the problem. Many questions need to be answered:

1. What is the process being controlled?
2. What is the controlled variable?
3. What are the control objectives?
4. Are closed-loop response data available?
5. Is the controller in the manual or automatic mode? Is it reverse- or direct-acting?
6. If the process is cycling, what is the cycling frequency?
7. What control algorithm is used? What are the controller settings?
8. Is the process open-loop stable?
9. What additional documentation is available, such as control loop summary sheets, piping and instrumentation diagrams, etc.?

After acquiring this background information, the next step is to check out each component in the control loop. In particular, one should determine that the process, measurement device (sensor), and control valve are all in proper working condition. Typically, sensors and control valves that are located in the field require more maintenance than control equipment located in the central control room. Any recent change to the equipment or instrumentation could very well be the source of the problem. For example, cleaning heat exchanger tubes, using a new shipment of catalyst, or changing a transmitter span could cause control-loop performance to change.

Sensor problems are often associated with the small-diameter lines that transport process fluids to the sensors. For example, the presence of solid material, ice, or bubbles in a line can result in erroneous measurements. Simple diagnostic checks can be performed to detect certain types of sensor problems. An abnormally small amount of variability in a set of consecutive measurements can indicate a "dead" sensor (see Section 10.3), while a break in an instrument line (e.g., a thermocouple) can be detected by a *rate-of-change* or *noise-spike* filter (see Chapter 17).

Control valve problems can also be detected by performing a simple diagnostic test. The controller is placed in manual, and a small step change is made in the controller output. If the control valve is working properly, the flow rate through the control valve should change accordingly. But if the valve is stuck or sticking badly, the flow rate will not change. Then, larger step changes should be made to estimate the size of the valve dead band and to ensure that the control valve is functioning properly. Finally, controller re-tuning may be necessary if the control loop exhibits undesirable oscillations or excessively sluggish responses.

EXAMPLE 12.9

A control loop exhibits excessively oscillatory behavior, even though there have been no recent equipment, instrument, or personnel changes. Suggest a general troubleshooting strategy to diagnose the problem.

SOLUTION

Oscillatory control loops are often the result of either (i) a cyclic process disturbance, (ii) a sticking control valve, or (iii) a poorly tuned controller. A simple test can be used to distinguish between Case (i), and Cases (ii) and (iii). If the controller is placed in the manual mode for a short period of time and the oscillations die out, then the oscillation is caused by the control loop, rather than an external disturbance. In order to distinguish between Cases (ii) and (iii), the controller should be placed in the manual mode, and one or more small step changes made in the controller output. The test results can be used to characterize the deadband and hysteresis of the control valve. If the control valve appears to be functioning properly, then the controller should be re-tuned.

EXAMPLE 12.10

The bottom composition of a pilot-scale, methanol-water distillation column is measured using an on-line gas chromatograph (GC). The composition measurement is sent to a digital PI controller that adjusts the steam flow rate to the reboiler. Recently the control-loop performance has deteriorated: the closed-loop composition response is more oscillatory and the period of oscillation is much larger than usual. The troubleshooting strategy employed in Example 12.9 has indicated that neither a cyclic disturbance nor a sticking control valve is the source of the problem. According to the maintenance records, the filter in the sample line has been replaced recently. Suggest additional diagnostic tests that should be considered before controller re-tuning is performed as a last resort.

SOLUTION

The combination of a more oscillatory response and a larger period of oscillation could occur if the time delay associated with the composition measurement had increased. For example, the transport delay associated with the sampling line to the GC would increase if the flow rate in the sampling line decreased. A decrease could occur due to a partial blockage in the line, or perhaps due to the new filter in the sample line. Thus, the filter and the sample line should be inspected.

If the old filter had inadvertently been replaced with a new filter that had a smaller sieve size, a larger pressure drop would occur, and the downstream pressure would decrease. Consequently, the liquid velocity in the sample line would decrease, and the transport delay would increase. As a result, the composition control loop would become more oscillatory and exhibit a longer period of oscillation.

SUMMARY

In this chapter we have considered three important issues associated with feedback control systems: design, on-line tuning, and control-loop troubleshooting. Control system design should consider the inevitable trade-offs between control system performance and robustness to modeling errors and process changes. Model-based design and tuning methods are recommended, because they provide considerable insight and usually have one (or zero) adjustable parameters. However, a reasonably accurate process model must be available. After a control system is installed, on-line tuning is commonly used to improve the performance of key control loops.

This chapter has presented a variety of controller design and tuning methods. Consequently, it is appropriate to summarize specific conclusions and recommendations:

1. Model-based techniques are recommended for control system design, especially the Internal Model Control and Direct Synthesis methods.

However, if the process is "lag dominant" (very small θ/τ ratio), the standard design methods should be modified, as discussed in Section 12.3.1.

2. For most process control applications, the controller should be tuned for disturbances rather than set-point changes. The set-point tracking can be adjusted independently by using a set-point filter or the set-point weighting factor β in Eq. 12-39.
3. Controller tuning should be based on a process model, if a model is available. The IMC tuning rules in Table 12.1 are applicable to common model forms.
4. Many controller tuning relations exhibit the same general features, as noted in Section 12.3.4.
5. Tuning relations based on a one-quarter decay ratio, such as the Ziegler-Nichols and Cohen-Coon methods, are not recommended.
6. The tuning relations based on integral error criteria provide useful benchmarks, but the resulting controllers are typically not very robust.
7. For on-line controller tuning, the relay auto-tuning and step response methods are recommended.

If a control loop is not performing satisfactorily, then troubleshooting is necessary to identify the source of the problem. This diagnostic activity should be based on a "systems approach" that considers the overall performance of the control loop as well as the performance of individual components. Retuning the controller is not a cure-all, especially if a sensor or a control valve is the source of the problem. Automated monitoring techniques for control loops are currently being developed and are commercially available (see Chapter 21).

REFERENCES

Åström, K. J., D. B. Ender, S. Skogestad, and R. C. Sorensen, personal communications, 2001.

Åström, K. J. and T. Hägglund, Automatic Tuning of Simple Regulators with Specification on the Gain and Phase Margins, *Automatica*, **20,** 645 (1984).

Åström, K. J. and T. Hägglund, *Advanced PID Control*, ISA, Research Triangle Park, NC, 2006.

Blickley, G. J., Modern Control Started with Ziegler-Nichols Tuning, *Control Eng.*, **38** (10), 11 (1990).

Buckley, P., A Modern Perspective on Controller Tuning, *Proc., Texas A&M Instrument. Sympos.*, pp. 80–88 (January 1973).

Chen, D., and D. E. Seborg, PI/PID Controller Design Based on Direct Synthesis and Disturbance Rejection, *Ind. Eng. Chem. Res.*, **41,** 4807 (2002).

Chien, I-L. and P. S. Fruehauf, Consider IMC Tuning to Improve Controller Performance, *Chem. Eng. Progress*, **86** (10), 33 (1990).

Cohen, G. H., and G. A. Coon, Theoretical Considerations of Retarded Control, *Trans. ASME*, **75,** 827 (1953).

Connell, B., *Process Instrumentation Applications Manual*, McGraw-Hill, New York, 1996.

Desborough, L., and R. Miller, Increasing Customer Value of Industrial Control Performance Monitoring, *Proc. of the 6th Internat. Conf. on Chemical Process Control*, CPC-VII, AICHE Sympos. Series, **98,** 169 (2002).

Doyle, F. J., III, *Process Control Modules: A Software Laboratory for Control System Design*, Prentice Hall PTR, Upper Saddle River, NJ, 2000.

Ender, D. B., Troubleshooting Your PID Control Loop, *InTech*, **39** (5), 35 (1992).

Ender, D. B., Process Control Performance: Not as Good as You Think, *Control Eng.*, **40** (10), 189 (1993).

Fruehauf, P. S., I-L. Chien, and M. D. Lauritsen, Simplified IMCPID Tuning Rules, *ISA Trans.*, **33,** 43 (1994).

Garcia, C. E., and M. Morari, Internal Model Control I. A Unifying Review and Some New Results, *Ind. Eng. Chem. Process Des. Dev.*, **21,** 308 (1982).

Goodwin, G. C., S. F. Graebe, and M. E. Salgado, *Control System Design*, Prentice Hall, Upper Saddle River, NJ, 2001.

Hang, C. C., K. J. Åström, and Q. G. Wang, Relay Feedback Auto-Tuning of Process Controllers—A Tutorial Review, *J. Process Control*, **12,** 143 (2002).

Lee, J., W. Cho, and T. F. Edgar, An Improved Technique for PID Controller Tuning from Closed-Loop Tests, *AIChE J.*, **36,** 1891 (1990).

Lieberman, N., *Troubleshooting Process Plant Control*, Wiley, New York, 2008.

Luyben, W. L. and M. L. Luyben, *Essentials of Process Control*, McGraw-Hill, New York (1997).

McMillan, G. K., *Tuning and Control-Loop Performance*, 3d ed., ISA, Research Triangle Park, NC, 1994.

McMillan, G. M., *Good Tuning: A Pocket Guide, 2d ed.*, ISA, Research Triangle Park, NC, 2006.

O'Dwyer, A., *Handbook of Controller Tuning Rules*, Imperial College Press, London, 2003.

Riggs, J. B., and M. N. Karim, *Chemical and Bio-Process Control*, 3rd ed., Ferret Pub., Lubbock, TX, 2006.

Rivera, D. E., M. Morari, and S. Skogestad, Internal Model Control, 4. PID Controller Design, *Ind. Eng. Process Design Dev.*, **25,** 252 (1986).

Shinskey, F. G., *Feedback Controllers for the Process Industries*, McGraw-Hill, New York, 1994.

Skogestad, S., Simple Analytic Rules for Model Reduction and PID Controller Tuning, *J. Process Control* **13,** 291 (2003).

Smith, C. A., and A. B. Corripio, *Principles and Practice of Automatic Control*, 2d ed., John Wiley, New York, 1997.

St. Clair, D. W., *Controller Tuning and Control Loop Performance: "PID without the Math,"* 2d ed., Straight-Line Control Co. Inc., Newark, DE, 1993.

Visioli, A., *Practical PID Control*, Springer, London, 2006.

Yu, C.-C., *Autotuning of PID Controllers*, Springer-Verlag, New York, 1999.

Yuwana, M., and D. E. Seborg, A New Method for On-Line Controller Tuning, *AIChE J.*, **28,** 434 (1982).

Ziegler, J. G., and N. B. Nichols, Optimum Settings for Automatic Controllers, *Trans. ASME*, **64,** 759 (1942).

EXERCISES

12.1 A process has the transfer function,

$$G(s) = \frac{K}{(10s + 1)(5s + 1)}$$

where K has a nominal value of $K = 1$. PID controller settings are to be calculated using the Direct Synthesis approach with $\tau_c = 5$ min. Suppose that these controller constants are employed and that K changes unexpectedly from 1 to $1 + \alpha$.

(a) For what values of α will the closed-loop system be stable?

(b) Suppose that the PID controller constants are calculated using the nominal value of $K = 1$ but it is desired that the resulting closed-loop system be stable for $|\alpha| \leq 0.2$. What is the smallest value of τ_c that can be used?

(c) What conclusions can be made concerning the effect that the choice of τ_c has on the robustness of the closed-loop system to changes in steady-state gain K?

12.2 Consider the two feedback control strategies shown in Fig. 12.6 (with $G = G_v G_p G_m$) and the following transfer functions:

$$G_p(s) = G_d(s) = \frac{2(1 - 3s)}{s}, \quad G_v(s) = 4, \quad G_m(s) = 1$$

(a) Design an IMC controller, G_c^*, using a filter, $f = 1/(\tau_c s + 1)$.

(b) Suppose that the IMC controller will be implemented as a controller G_c in the classical feedback control configuration of Fig. 12.6(a). Derive an expression for G_c and report it in a standard form (e.g., P, PI, or PID).

12.3 A process has the transfer function, $G(s) = 2e^{-0.2s}/(s + 1)$. Compare the PI controller settings for the following design approaches:

(a) Direct Synthesis method ($\tau_c = 0.2$)

(b) Direct Synthesis method ($\tau_c = 1.0$)

(c) ITAE performance index (disturbance)

(d) ITAE performance index (set point)

(e) Which controller has the most conservative settings? Which has the least conservative?

(f) For the two controllers of part (e), simulate the closed-loop responses to a unit step disturbance, assuming that $G_d(s) = G(s)$.

12.4 A process, including the sensor and control valve, can be modeled by the transfer function:

$$G(s) = \frac{4e^{-3s}}{s}$$

(a) If a proportional-only controller is used, what is the maximum value of controller gain K_c that will result in a stable closed-loop system? (Determine the *exact* value of K_c, not an approximate value.)

(b) Specify a PI controller using the Tyreus-Luyben controller settings.

12.5 A process stream is heated using a shell and tube heat exchanger. The exit temperature is controlled by adjusting the steam control valve shown in Fig. E12.5. During an open-loop experimental test, the steam pressure P_s was suddenly changed from 18 to 20 psig and the temperature data shown below were obtained. At the nominal conditions, the control valve and current-to-pressure transducers have gains of $K_v = 0.9$ psi/psi and $K_{IP} = 0.75$ psi/mA, respectively. Determine appropriate PID controller settings using the following approaches:

(a) Internal Model Control (select a reasonable value of τ_c)

(b) ITAE (set point)

(c) ITAE (disturbance)

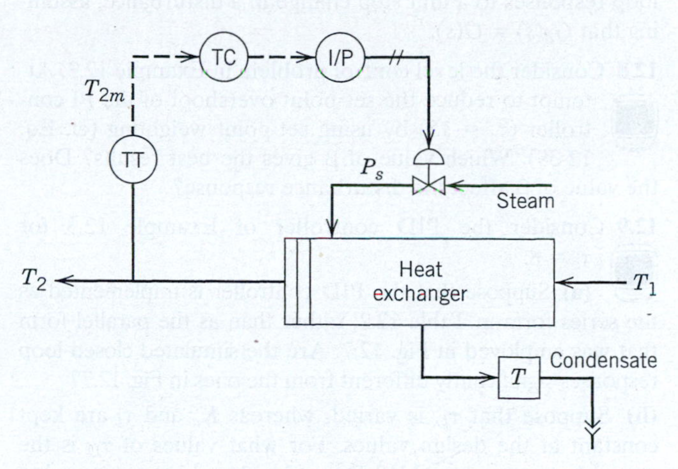

Figure E12.5

t(min)	T_{2m} (mA)
0	12.0
1	12.0
2	12.5
3	13.1
4	14.0
5	14.8
6	15.4
7	16.1
8	16.4
9	16.8
10	16.9
11	17.0
12	16.9

12.6 Suggest a modification of the Direct Synthesis approach that will allow it to be applied to open-loop unstable processes. (*Hint:* First stabilize the process using a proportional-only feedback controller.) Draw a block diagram for your proposed control scheme.

12.7 A process including sensor and control valve can be modeled by a fourth-order transfer function:

$$G(s) = \frac{1}{(s + 1)(0.2s + 1)(0.04s + 1)(0.008s + 1)}$$

(a) Design PID controllers using two design methods:

(i) A second-order-plus-time-delay model using the model reduction approach proposed by Skogestad (Section 6.3) and the IMC tuning relation in Table 12.1.

(ii) The Tyreus-Luyben settings in Table 12.4.

(b) Evaluate the two controllers by simulating the closed-loop responses to a unit step change in a disturbance, assuming that $G_d(s) = G(s)$.

12.8 Consider the level control problem in Example 12.3. Attempt to reduce the set-point overshoot of the PI controller ($\tau_c = 15$) by using set-point weighting (cf. Eq. 12-39). Which value of β gives the best results? Does the value of β affect the disturbance response?

12.9 Consider the PID controller of Example 12.3 for $\tau_c = 8$.

(a) Suppose that the PID controller is implemented as the series form in Table 12.2, rather than as the parallel form that was employed in Fig. 12.7. Are the simulated closed-loop responses significantly different from the ones in Fig. 12.7?

(b) Suppose that τ_D is varied, whereas K_c and τ_I are kept constant at the design values. For what values of τ_D is the closed-loop system stable? Does the closed-loop system become more or less oscillatory as τ_D increases?

12.10 Consider the blending system shown in Fig. E12.10. A feedback control system is used to reduce the effect of disturbances in feed composition x_1 on the controlled variable, product composition, x. Inlet flow rate w_2 can be manipulated. Do the following:

(a) Draw a block diagram of the feedback control system.

(b) Using the information shown below, derive a transfer function for each block.

(c) Simulate the closed-loop response for the PI controller settings given below and a step disturbance of +0.2 in x_1.

(d) Repeat part (c) for a set-point change of −0.1. Attempt to obtain better closed-loop responses by tuning the PI controller. Which controller settings give the best results?

(e) Attempt to obtain improved control by adding derivative action to your best PI controller of part (d). Try several values of derivative time, τ_D. Which one gives the best results?

(f) Suppose that the sampling line to the composition analyzer becomes partially plugged so that the measurement time delay is now three minutes. Using your best controller settings of part (d), simulate the closed-loop response for the same set-point change and the new time-delay value. Explain your new simulation results. Does the larger time delay have a major effect on control system performance?

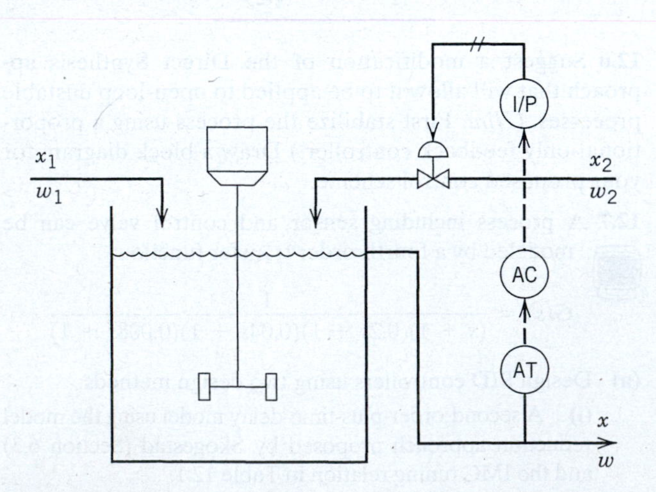

Figure E12.10

Process Information

The pilot-scale blending tank has an internal diameter of 2 m and a height of 3 m. Inlet flow rate w_1 and inlet composition x_2 are constant. The nominal steady-state operating conditions are as follows:

$$\overline{w}_1 = 650 \text{ kg/min} \quad \overline{x}_1 = 0.2 \quad \overline{h} = 1.5 \text{ m}$$
$$\overline{w}_2 = 350 \text{ kg/min} \quad \overline{x}_2 = 0.6$$
$$\rho = 1 \text{ g/cm}^3 \quad \overline{x} = 0.34$$

The overflow line maintains a constant liquid volume in the tank.

Instrumentation: The range for all of the electronic signals is 4 to 20 mA.

Current-to-pressure transducer: The I/P transducer acts as a linear device with negligible dynamics. The output signal changes from 3 to 15 psi when the input signal changes full-scale from 4 to 20 mA.

Control valve: The behavior of the control valve can be approximated by a first-order transfer function with a time constant of 5 s (i.e., 0.0833 min). A 1.2-psi change in the signal to the control valve produces a 300 kg/min change in w_2.

Composition measurement: The zero and span of the composition transmitter for the exit composition are 0 and 0.50 (mass fraction), respectively. A one-minute time delay is associated with the measurement.

Feedback controller: Initially, consider a standard PI controller tuned using the IMC relations in Table 12.1. Justify your choice of τ_c.

12.11 A PID controller is used to control the temperature of a jacketed batch reactor by adjusting the flow rate of coolant to the jacket. The temperature controller has been tuned to provide satisfactory control at the nominal operating conditions. Would you anticipate that the temperature controller may have to be retuned for any of the following instrumentation changes? Justify your answers.

(a) The span of the temperature transmitter is reduced from 30 to 15 °C.

(b) The zero of the temperature transmitter is increased from 50 to 60 °C.

(c) The control valve "trim" is changed from linear to equal percentage.

(d) The temperature of the coolant leaving the jacket is used as the controlled variable instead of the temperature in the reactor.

12.12 Suppose that a process can be adequately modeled by the first-order-plus-time-delay model in Eq. 12-10.

(a) Calculate PI controller settings using the IMC tuning relations in Table 12.1.

(b) Cohen and Coon (1953) reported the following tuning relations to PI controllers:

$$K_c = \frac{1}{K}\frac{\tau}{\theta}[0.9 + \theta/12\tau]$$

$$\tau_I = \frac{\theta[30 + 3(\theta/\tau)]}{9 + 20(\theta/\tau)}$$

These tuning relations were developed to provide closed-loop responses with a quarter decay ratio. Compare the PI settings calculated from these equations to the controller settings of part (a). Which would you expect to be more conservative?

(c) Simulate the controllers of parts (a) and (b) for a unit step change in set point, followed by a unit step disturbance at $t = 40$. Assume that the $G_d(s) = G(s)$ and that the model parameters are $K = 2$, $\tau = 3$, and $\theta = 1$. Which controller provides better control?

12.13 Consider the experimental step response data for the heat exchanger of Exercise 12.5. Determine the PI controller settings using the step response method and two controller tuning relations:

(a) Direct Synthesis method with $\tau_c = \tau/3$

(b) Ziegler-Nichols settings in Table 12.4.

Which controller provides the more conservative controller settings?

12.14 Consider the transfer function model in Eq. 12-10 with $K = 2$, $\tau = 5$, and $\theta = 1$. Compare the PID controller settings obtained from the Ziegler-Nichols and Tyreus-Luyben tuning relations in Table 12.4. Simulate the closed-loop responses for a unit step change in the set point.

12.15 IGC's operations area personnel are experiencing problems with a particular feedback control loop on an interstage cooler. Appelpolscher has asked you to assess the situation and report back what remedies, if any, are available. The control loop is exhibiting an undesirable sustained oscillation that the operations people are sure is caused by the feedback loop itself (e.g., poor controller tuning). They want assistance in retuning the loop. Appelpolscher thinks that the oscillations are caused by external disturbances (e.g., cyclic disturbances such as cycling of the cooling water temperature); he wants the operations people to deal with the problem themselves. Suggest a simple procedure that will allow you to determine quickly what is causing the oscillations. How will you explain your logic to Appelpolscher?

12.16 A problem has arisen in the level control loop for the flash separation unit shown in Fig. E12.16. The level control loop had functioned in a satisfactory manner for a long period of time. However, the liquid level is gradually increasing with time, even though the PI level controller output has saturated. Furthermore, the liquid flow rate is well above the nominal value, while the feed flow rate is at the nominal value, according to the recorded measurements from the two flow transmitters. The accuracy of the level transmitter measurement has been confirmed by comparison with sight glass readings for the separator. The two flow measurements are obtained via orifice plates and differential pressure transmitters, as described in Chapter 9. Suggest possible causes for this problem and describe how you would troubleshoot this situation.

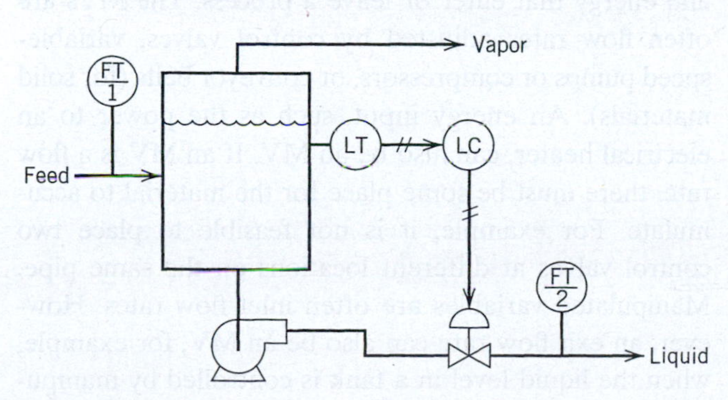

Figure E12.16

12.17 Consider the PCM furnace module of Appendix E. Assume that hydrocarbon temperature T_{HC} is the CV, that fuel gas flow rate F_{FG} is the MV, and that they are related by the following transfer function model:

$$\frac{T_{HC}}{F_{FG}} = G_p = \frac{220e^{-2s}}{6.5s + 1}, \qquad G_v = G_m = 1$$

(a) Design a PID controller based on the IMC tuning relations and a reasonable choice for τ_c.

(b) Design a PID controller based on the relay auto-tuning feature of the PCM, and the Ziegler-Nichols settings.

(c) Simulate each controller for a sudden change in an unmeasured disturbance, the hydrocarbon flow rate F_{HC}, at $t = 10$ min, from 0.035 to 0.040 m^3/min. Which controller is superior? Justify your answer.

(d) Attempt to obtain improved control system performance by fine tuning your better controller for part (c). For the performance criteria, consider maximum deviation from set point and settling time.

12.18 Consider the PCM distillation column module of Appendix E. Assume that distillate MeOH composition x_D is the CV, that reflux ratio R is the MV, and that they are related by the following transfer function model:

$$\frac{X_D(s)}{R(s)} = G_p(s) = \frac{0.126e^{-138s}}{762s + 1}, \qquad G_v = G_m = 1$$

(a) Design a PID controller based on the IMC tuning relations and a reasonable choice for τ_c.

(b) Design a PID controller based on the relay auto-tuning feature of the PCM and the Ziegler-Nichols settings.

(c) Simulate each controller for a sudden change in an unmeasured disturbance, feed composition x_F, at $t = 10$ min, from 0.50 to 0.55 (mole fraction). Which controller is superior? Justify your answer.

(d) Attempt to obtain improved control system performance by fine-tuning your better controller for part (c). For the performance criteria, consider maximum deviation from set point and settling time.

Chapter 13

Control Strategies at the Process Unit Level

CHAPTER CONTENTS

Previous chapters have emphasized process control problems with a single controlled variable and single manipulated variable. In this chapter, we show that these concepts and analysis methods are also applicable to control problems at the process unit level that have multiple controlled variables (CVs) and multiple manipulated variables (MVs). These types of control problems are considered further in Chapters 18, 20, 25, and 26.

For control system design and analysis, it is convenient to classify process variables as being either output variables or input variables. The output variables (or outputs) are dependent variables that typically are associated with exit streams or conditions within a process vessel (e.g., compositions, temperatures, levels, and flow rates). Some outputs must be controlled in order to operate a process in a satisfactory manner. They are called controlled variables (CVs). Input variables are process variables that affect one or more output variables.

Input variables are classified as either manipulated variables (MVs) or disturbance variables (DVs). Manipulated variables are used to adjust the rates of material and energy that enter or leave a process. The MVs are often flow rates adjusted by control valves, variable-speed pumps or compressors, or conveyor belts (for solid materials). An energy input, such as the power to an electrical heater, can also be an MV. If an MV is a flow rate, there must be some place for the material to accumulate. For example, it is not feasible to place two control valves at different locations on the same pipe. Manipulated variables are often inlet flow rates. However, an exit flow rate can also be an MV, for example, when the liquid level in a tank is controlled by manipulating an exit flow rate.

By definition, disturbance variables are input variables that cannot be manipulated. Common DVs include ambient conditions and feed streams from upstream process units.

13.1 DEGREES OF FREEDOM ANALYSIS FOR PROCESS CONTROL

The important concept of degrees of freedom, N_F, was introduced in Section 2.3 in connection with process modeling. It is the number of process variables that must be specified in order to be able to determine the remaining process variables. If a dynamic model is available, N_F can be determined from a relation in Chapter 2,

$$N_F = N_V - N_E \tag{13-1}$$

where N_V is the number of process variables and N_E is the number of independent equations.

For process control applications, it is very important to determine the maximum number of process variables that can be independently controlled, that is, to determine the *control degrees of freedom*, N_{FC}:

Definition. *The control degrees of freedom,* N_{FC}, *is the number of process variables that can be controlled independently.*

In order to make a clear distinction between N_F and N_{FC}, we refer to N_F as the model degrees of freedom and to N_{FC} as the control degrees of freedom. They are related by,

$$N_F = N_{FC} + N_D \tag{13-2}$$

where N_D is the number of DVs.

13.1.1 Control Degrees of Freedom

The control degrees of freedom N_{FC} is closely related to the number of independent MVs that are available:

General Rule. *For most practical control problems, the control degrees of freedom* N_{FC} *is equal to the number of independent input variables that can be manipulated.*

It is important that the manipulated inputs be independent. For example, if a process stream splits, or if two process streams merge to form a third stream, it is not possible to adjust all three flow rates independently. These situations are shown in Fig. 13.1.

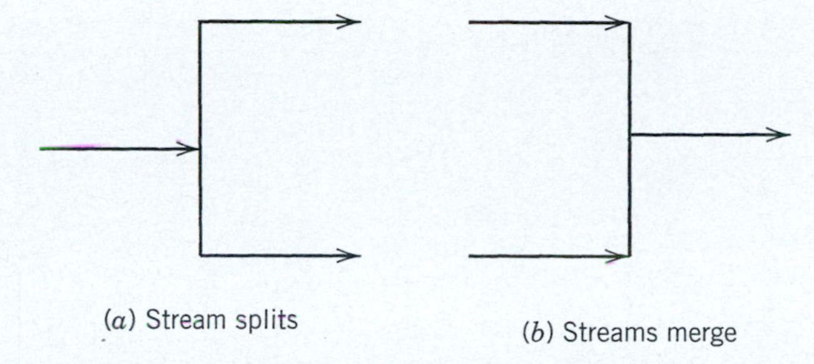

Figure 13.1 Two examples where all three streams cannot be manipulated independently.

Two examples illustrate the General Rule.

EXAMPLE 13.1

Determine N_F and N_{FC} for the steam-heated, stirred-tank system model in Eqs. 2-50 through 2-52 in Chapter 2. Assume that only steam pressure P_s can be manipulated.

SOLUTION

To calculate N_F from Eq. 13-1, we need to determine N_V and N_E. The dynamic model contains three equations ($N_E = 3$) and six process variables ($N_V = 6$): T_s, P_s, w, T_i, T, and T_w. Thus, $N_F = 6 - 3 = 3$. If feed temperature T_i and mass flow rate w are considered to be disturbance variables, $N_D = 2$ and thus $N_{FC} = 1$ from Eq. 13-2. This single degree of freedom could be used to control temperature T by manipulating steam pressure, P_s.

EXAMPLE 13.2

A conventional distillation column with a single feed stream and two product streams is shown in Fig. 13.2. The feed conditions are disturbance variables. Determine the control degrees of freedom N_{FC} and identify potential MVs and CVs.

SOLUTION

For a typical distillation column, five input variables can be manipulated: product flow rates, B and D, reflux flow rate R, coolant flow rate q_c, and heating medium flow rate q_h. Thus, according to the General Rule, $N_{FC} = 5$. This result can also be obtained from Eqs. 13-1 and 13-2, but considerable effort is required to develop the required dynamic model. Although five output variables could be selected as CVs, x_D, x_B, h_B, h_D, and P, for many distillation control problems, it is not necessary to control all five. Also, if it not feasible to measure the product compositions on-line, tray temperatures near the top and bottom of the column are often controlled instead, as discussed in the next section.

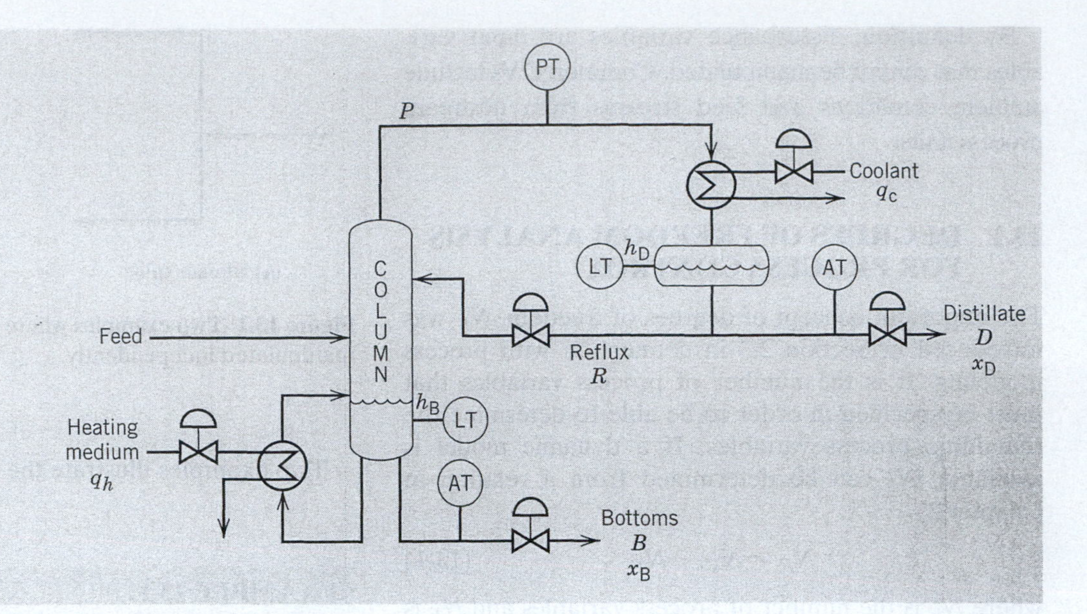

Figure 13.2 Schematic diagram of a distillation column.

Although the General Rule is simple and widely applicable, there are exceptions where it is not valid. For example, N_{FC} should be reduced by 1 when a MV does not have a significant steady-state effect on any of the CVs, that is, when these steady-state gains are very small. This situation is illustrated in Example 13.3.

EXAMPLE 13.3

The blending system in Fig. 13.3 has a bypass stream that allows a fraction f of inlet stream w_2 to bypass the stirred tank. It is proposed that product composition x be controlled by adjusting f via the control valve. Analyze the feasibility of this control scheme by considering its steady-state and dynamic characteristics. In your analysis, assume that x_1 is the principal disturbance variable and that x_2, w_1, and w_2 are constant. Variations in liquid volume V can be neglected because $w_2 << w_1$.

SOLUTION

The dynamic characteristics of the proposed control scheme are quite favorable because product composition x responds rapidly to changes in the bypass flow rate. In order to evaluate the steady-state characteristics, consider a component balance over the entire system:

$$w_1\bar{x}_1 + w_2x_2 = w\bar{x} \tag{13-3}$$

Solving for the controlled variable gives,

$$\bar{x} = \frac{w_1\bar{x}_1 + w_2x_2}{w} \tag{13-4}$$

Thus $\bar{x}$ depends on disturbance variable $\bar{x}_1$ and four constants (w_1, w_2, x_2 and w), but it does not depend on bypass fraction, f. Thus, it is not possible to compensate for sustained disturbances in x_1 by adjusting f. For this reason, the proposed control scheme is not feasible.

Because f does not appear in (13-4), the steady-state gain between x and f is zero. Thus the bypass flow rate can be adjusted, but it does not provide a control degree of freedom. However, if w_2 could also be adjusted, manipulating both f and w_2 could produce excellent control of the product composition.

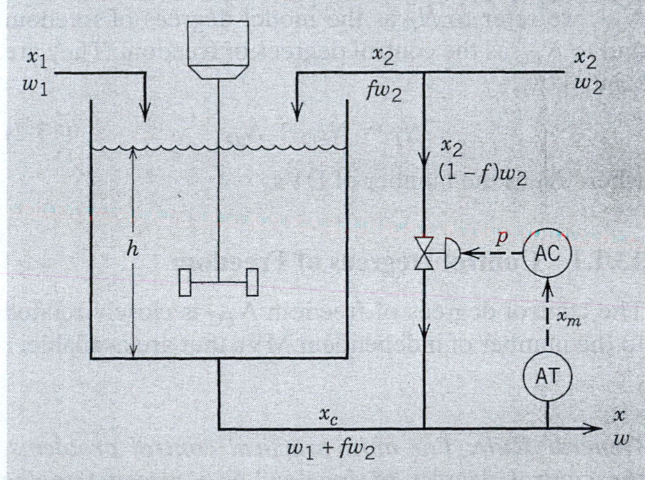

Figure 13.3 Blending system with bypass stream.

13.1.2 Effect of Feedback Control

The addition of a feedback controller can change the control degrees of freedom, N_{FC}. In general, adding a feedback controller utilizes a control degree of freedom, because an MV is now adjusted by the controller. However, if the controller set-point is adjusted by a higher-level (or supervisory) control system, neither N_F nor N_{FC} changes. The reason is as follows. Adding a controller introduces a new equation, the control law, and a new variable, the set point. Thus both N_E and N_V increase by one. But Eqs. 13-1 and 13-2 indicate that N_F and N_{FC} do not change.

13.2 SELECTION OF CONTROLLED, MANIPULATED, AND MEASURED VARIABLES

A general representation of a control problem is shown in Fig. 13.4. In general, it is desirable to have at least as many MVs as CVs. But this is not always possible, and special types of control systems sometimes need to be utilized (see Chapter 16). It may not be feasible to control all of the output variables for several reasons:

1. It may not be possible or economical to measure all of the outputs, especially chemical compositions.
2. There may not be enough MVs.
3. Potential control loops may be impractical because of slow dynamics, low sensitivity to the MVs, or interactions with other control loops.

In general, CVs are measured on-line, and the measurements are used for feedback control. But sometimes it is possible to control an unmeasured CV by using a process model (a *soft sensor*) to estimate it from measurements of other process variables. This strategy is referred to as inferential control (see Chapter 16).

13.2.1 Controlled Variables

Consideration of plant and control objectives has produced guidelines for the selection of CVs from the available output variables (Newell and Lee, 1989).

Guideline 1 *All variables that are not self-regulating must be controlled.* In Chapter 5, a non-self-regulating variable was defined to be an output variable that exhibits an unbounded response after a sustained input change such as a step change. A common example is liquid level in a tank that has a pump on an exit line (see Chapter 11). Non-self-regulating variables must be controlled in order for the controlled process to be stable.

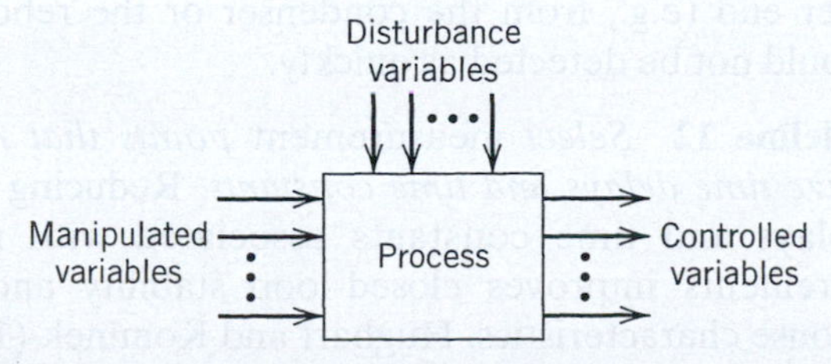

Figure 13.4 Process with multiple inputs and multiple outputs.

Guideline 2 *Choose output variables that must be kept within equipment and operating constraints (e.g., temperatures, pressures, and compositions).* The constraints are due to safety, environmental, and operational requirements.

Guideline 3 *Select output variables that are a direct measure of product quality (e.g., composition, refractive index) or that strongly affect it (e.g., temperature or pressure).*

Guideline 4 *Choose output variables that seriously interact with other controlled variables.* The pressure in a steam header that supplies steam to downstream units is a good example. If this supply pressure is not well regulated, it will act as a significant disturbance to downstream units.

Guideline 5 *Choose output variables that have favorable dynamic and static characteristics.* Output variables that have large measurement time delays, large time constants, or are insensitive to the MVs are poor choices.

Except for Guideline 1, these guidelines are not strict rules. For specific situations, the guidelines may be inconsistent or conflicting. For example, suppose that one output variable must be kept within specified limits for safety reasons (Guideline 2), whereas a second interacts strongly with other output variables (Guideline 4). Guideline 2 would prevail because of safety considerations. Thus, the first output variable should be controlled if there is only a single MV.

13.2.2 Manipulated Variables

Based on the process and control objectives, a number of guidelines have been proposed for the selection of MVs from among the input variables (Newell and Lee, 1989). Inlet or exit flow rates can be manipulated in order to adjust mass balances and thus control CVs such as liquid level and pressure. Temperatures and vapor pressures are controlled by adjusting the energy balance.

Guideline 6 *Select inputs that have large effects on controlled variables.* Ideally, an MV should have a significant, rapid effect on only one controlled variable. In other words, the corresponding steady-state gain should be large. Furthermore, it is desirable that the effects of this MV on the other CVs should be negligible (that is, the other steady-state gains should be small or zero). It is also important that each manipulated variable be able to accommodate a wide range of conditions. For example, if a distillation column has a reflux ratio of 5, it will be much more effective to control the reflux drum level by manipulating the large reflux flow rate rather than the small distillate flow rate, because larger disturbances in the vapor flow rate could be handled. However, the effect of this choice on the control of product compositions must also be considered in making the final decision.

Guideline 7 *Choose inputs that rapidly affect the controlled variables.* For multiloop control, it is desirable that each manipulated variable have a rapid effect on its corresponding controlled variable.

Guideline 8 *The manipulated variables should affect the controlled variables directly, rather than indirectly.* Compliance with this guideline usually results in a control loop with favorable static and dynamic characteristics. For example, consider the problem of controlling the exit temperature of a process stream that is heated by steam in a shell and tube heat exchanger. It is preferable to throttle the steam flow to the heat exchanger rather than the condensate flow from the shell, because the steam flow rate has a more direct effect on the steam pressure and on the rate of heat transfer.

Guideline 9 *Avoid recycling of disturbances.* As Newell and Lee (1989) have noted, it is preferable not to manipulate an inlet stream or a recycle stream, because disturbances tend to be propagated forward, or recycled back, to the process. This problem can be avoided by manipulating a utility stream to absorb disturbances or an exit stream that allows the disturbances to be passed downstream, provided that the exit stream changes do not unduly upset downstream process units.

Note that these guidelines for MVs may be in conflict. For example, a comparison of the effects of two inputs on a single controlled variable could indicate that one has a larger steady-state gain (Guideline 6) but slower dynamics (Guideline 7). In this situation, a trade-off between static and dynamic considerations must be made in selecting the appropriate manipulated variable from the two candidates.

13.2.3 Measured Variables

Safe, efficient operation of processing plants requires on-line measurement of key process variables. Clearly, the CVs should be measured. Other output variables can be measured to provide additional information or for use in model-based control schemes such as inferential control. It is also desirable to measure MVs because they provide useful information for tuning controllers and troubleshooting control loops (see Chapter 12). Measurements of DVs provide the basis for feedforward control (see Chapter 15).

In choosing sensor locations, both static and dynamic considerations are important, as discussed in Chapter 9.

Guideline 10 *Reliable, accurate measurements are essential for good control.* Inadequate measurements are a key factor in poor process control performance. Hughart and Kominek (1977) cite common measurement problems that they observed in distillation-column control applications: orifice runs without enough straight piping, analyzer sample lines with large time delays, temperature probes located in insensitive regions, and flow rate measurement of liquids that are at, or near, their boiling points, which can lead to liquid flashing at the orifice plate. They note that these types of measurement problems can be readily resolved during the process design stage, but changing a measurement location after the process is operating can be both difficult and costly.

Guideline 11 *Select measurement points that have an adequate degree of sensitivity.* As an example, consider product composition control in a tray-distillation column. If the product composition cannot be measured on-line, it is often controlled indirectly by regulating a tray temperature near that end of the column. But for high-purity separations, the location of the temperature measurement point can be quite important. If a tray near an end of the column is selected, the tray temperature tends to be insensitive, because the tray composition can vary significantly, even though the tray temperature changes very little. For example, suppose that an impurity in the vapor leaving the top tray has a nominal value of 20 ppm. A feed composition change could cause the impurity level to change significantly (for example, from 20 to 40 ppm) but produce only a negligible change in the tray temperature. By contrast, suppose that the temperature measurement point were moved to a tray that is closer to the feed tray. Then the temperature sensitivity is improved because the impurity level is higher, but disturbances entering the column at either end (e.g., from the condenser or the reboiler) would not be detected as quickly.

Guideline 12 *Select* measurement *points that minimize time delays and time constants.* Reducing time delays and time constants associated with measurements improves closed-loop stability and response characteristics. Hughart and Kominek (1977) have observed distillation columns with the sample connection for the bottom analyzer located 200 ft downstream from the column. This large distance introduced a significant time delay and made the column difficult to control, particularly because the time delay varied with the bottom flow rate.

An evaporator control problem will now be used to illustrate the Guidelines.

EXAMPLE 13.4

The evaporator shown in Fig. 13.5 is used to concentrate a dilute solution of a single, nonvolatile solute in a volatile solvent by evaporating solvent using heat supplied by a steam coil. Three process variables can be manipulated: steam pressure, P_s, product flow rate, B, and vapor flow rate of solvent, D. The chief DVs are feed composition, x_F, and feed flow rate, F. The compositions are expressed as mole fractions of solute, and the flow rates are in molar units.

Propose multiloop control strategies for two situations:

(a) The product composition x_B can be measured on-line

(b) x_B cannot be measured on-line

SOLUTION

Case (a): Product Composition x_B Is Measured On-Line

First, we select the CVs. Because the chief objective for an evaporator is to produce a product stream with a specified composition, mole fraction x_B is the primary CV (Guideline 3). Liquid level h must be controlled because of operating constraints and safety considerations (Guideline 2). If the level is too high, liquid could be entrained in the solvent stream; if the level is too low, the tubes of the steam chest would be exposed to vapor, rather than liquid, a potentially dangerous situation. In this latter situation, the heat transfer rate from the steam to the evaporator liquid would be significantly lower, and thus overheating and damage to the steam chest could result. Pressure P should also be controlled, because it has a major influence on the evaporator operation (Guideline 2). Large pressure variations affect the

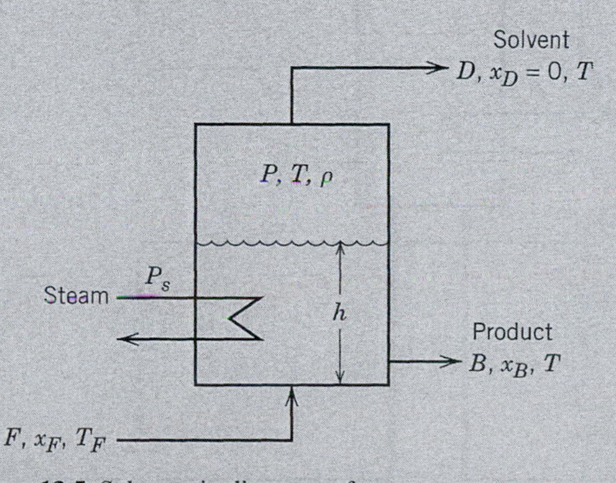

Figure 13.5 Schematic diagram of an evaporator.

temperature T and could shift the boiling regime from film boiling to nucleate boiling, or vice versa. This type of regime shift could produce a major process upset. For these reasons, three CVs are selected: x_B, h, and P.

Next we select the MVs. Because the feed conditions cannot be adjusted, the obvious MVs are B, D, and P_s. Product flow rate B has a significant effect on h, but relatively small effects on P and x_B. Therefore, it is reasonable to control h by manipulating B (Guideline 6) unless B is only a small fraction of F (for example, less than 10%). In this latter case, it would be desirable to have F become an MV. Vapor flow rate D has a direct and rapid effect on P but has less direct effects on h and x_B. Thus, P should be paired with D (Guideline 6). This leaves the P_s-x_B pairing for the third control loop. This pairing is physically reasonable, because the most direct way of regulating x_B is by adjusting the evaporation of solvent via the steam pressure (Guideline 8).

Finally, we consider which process variables to measure. Clearly, the three CVs should be measured. It is also desirable to measure the three MVs because this information is useful for controller tuning and troubleshooting. If large and frequent feed disturbances occur, measurements of disturbance variables F and x_F could be used in a feedforward control strategy that would complement the feedback control scheme. It is not necessary to measure T_F, because sensible heat changes in the feed stream are typically small compared to the heat fluxes in the evaporator.

A schematic diagram of the controlled evaporator for Case (a) is shown in Figure 13.6.

Case (b): Product Composition Cannot Be Measured On-Line

The CVs are the same as in Case (a), but, because the third controlled variable x_B cannot be measured, standard feedback control is not possible. A simple feedforward control strategy can be developed based on a steady-state component balance for the solute,

$$0 = \overline{F}\overline{x}_F - \overline{B}\overline{x}_B \tag{13-5}$$

where the bar denotes the nominal steady-state value. Rearranging gives,

$$\overline{B} = \overline{F}\frac{\overline{x}_F}{\overline{x}_B} \tag{13-6}$$

Equation 13-6 provides the basis for the feedforward control law. Replacing $\overline{B}$ and $\overline{F}$ by the actual flow rates, $B(t)$ and $F(t)$, and replacing $\overline{x}_B$ by the set-point value, x_{Bsp}, gives

$$B(t) = F(t)\frac{\overline{x}_F}{x_{Bsp}} \tag{13-7}$$

Thus B is adjusted based on the measured value of F, the set point x_{Bsp}, and the nominal value of the feed composition, $\overline{x}_F$.

The MVs are the same as for Case (a): D, B, and P_s. Bottom flow rate B has already been used in the feedforward control strategy of (13-7). Clearly, the P-D

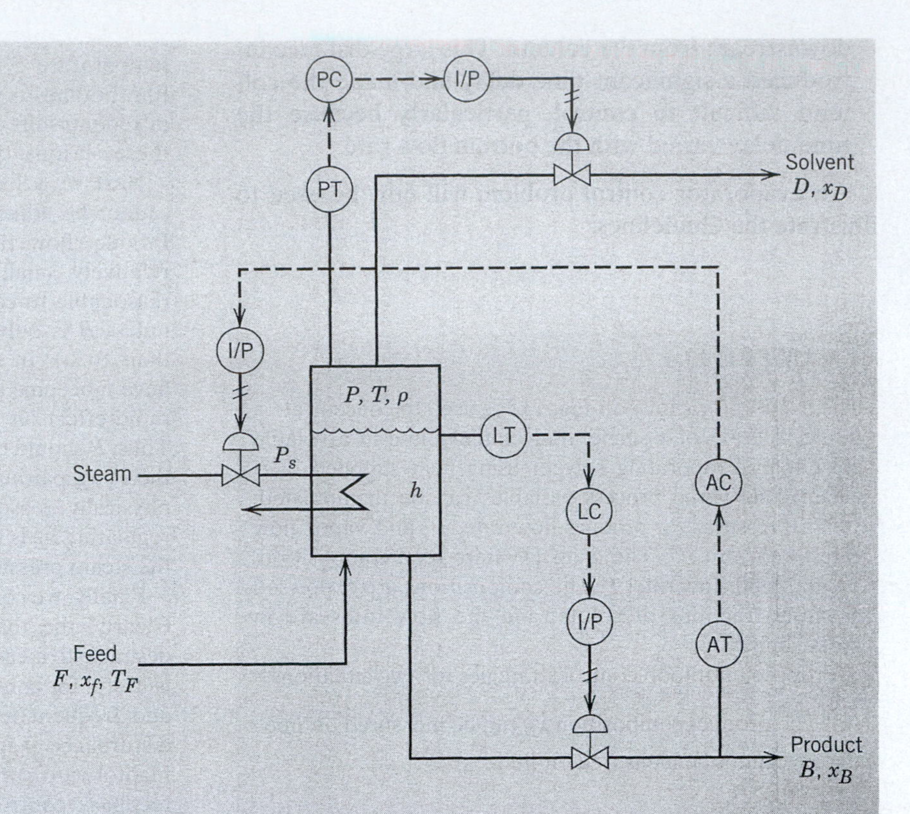

Figure 13.6 Evaporator control strategy for Case (a).

pairing is still desirable for the reasons given for Case (a). This leaves h to be controlled by adjusting the rate of evaporation via P_s. A schematic diagram of the controlled evaporator is shown in Fig. 13.7.

This control strategy has two disadvantages. First, it is based on the assumption that the unmeasured feed composition is constant at a known steady-state value. Second, because the feedforward control technique was based on a steady-state analysis, it may not perform well during transient conditions. Nevertheless, this scheme provides a simple, indirect method for controlling a product composition when it cannot be measured.

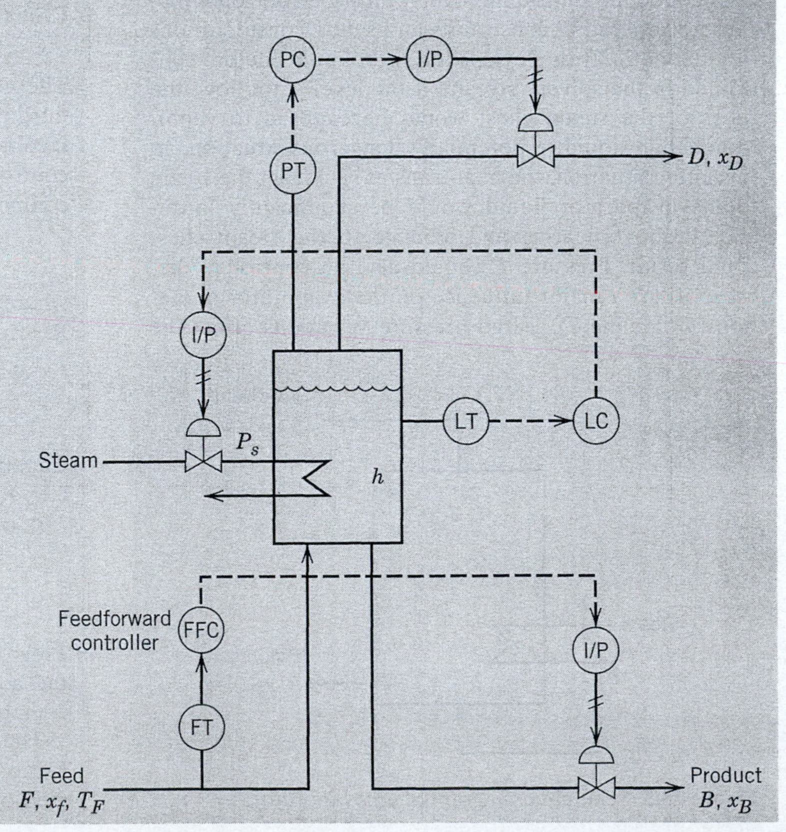

Figure 13.7 Evaporator control strategy for Case (b).

13.3 APPLICATIONS

In this section, we describe four representative examples of control problems at the process unit level, rather than at the individual control loop level, in order to provide an introduction to more complex control problems. For each of these case studies, key aspects of their control system design are considered:

(a) *Process objectives.* For control system design, it is essential to know the process objectives. For simple processes, the process objectives are fairly obvious. But for others, they may not be. For example, a chemical reactor can be operated to maximize the yield, selectivity, or throughput subject to satisfying process constraints (e.g., safety and the environmental constraints.) Similarly, a distillation column can be operated to maximize throughput or minimize energy consumption, while satisfying product specifications and other constraints.

(b) *Control objectives.* The control objectives should be carefully formulated based on a number of considerations that include process objectives, process constraints, and economic data. Even for simple control problems, there can be alternative control objectives. For example, consider a very common control application, liquid-level control. In some applications, the control objective is to achieve tight level control at a specified set point. This situation might occur for a continuous bioreactor, when maintaining a constant residence time is important. On the other hand, many process vessels are used as intermediate storage tanks for surge control. Here, the process objective is to reduce the effect of upstream disturbances on downstream units, by having the exit stream from an intermediate storage tank change gradually in response to large, rapid changes in its inlet steams. In this situation, tight level control would be undesirable, because inlet flow rate disturbances would be propagated to the outlet stream. The more appropriate control objective would be averaging control, in which the liquid level is allowed to vary between specified upper and lower limits, thus providing surge capacity and more gradual changes in the exit flow rate.

(c) *Choice of control configuration.* A key decision in control system design is to decide whether a conventional multiloop control strategy, consisting of individual feedback control loops, will provide satisfactory control. If not, an advanced process control strategy is required. This decision should be based primarily on the control objectives and knowledge of the static and dynamic behavior of the process. Advanced control strategies are considered in Chapters 16, 18, and 20.

(d) *Pairing of MVs and CVs.* If a multiloop control strategy is selected, the next step is to determine how the CVs and MVs should be paired. A systematic method for making these decisions, the Relative Gain Method, is considered in Chapter 18.

13.3.1 Distillation Column

For continuous distillation, the primary process objective is to separate a feed mixture into two (or more) product streams with specified compositions. Thus the product compositions are the most important CVs. For some distillation columns, one product composition is much more important than the other. For example, consider a series of columns where the bottom stream of a column serves as the feed stream to the next column. The distillate composition for each column is more important because it is a product stream (Guideline 3). By contrast, the bottom stream for each column (except the last column in the series) undergoes further separation, so the bottom compositions are of less concern. Consequently, in these situations, the bottom compositions may not have to be controlled.

In addition to one or more product compositions, other process variables need to be controlled. Consider the separation of a binary mixture and the conventional tray-distillation column shown in Fig. 13.2. Assume that the chief control objective is to control both product compositions, x_D and x_B. However, the liquid levels in the reflux drum, h_D, and the column base (or sump), h_B, must be kept between upper and lower limits (Guideline 2). The column pressure, P, must also be controlled in order to avoid weeping or flooding in the column and to control the vapor inventory in the column. Thus, this column has a total of five CVs.

The MVs are selected from among six input variables: feed flow rate F, product flow rates, D and B, reflux flow rate R, and the heat duties for the condenser and reboiler, q_D and q_B. If the feed stream comes from an upstream process, instead of from a storage tank, it cannot be manipulated, and thus F is considered to be a DV. In this situation, there are five MVs.

Distillation column control can be difficult for the following reasons.

1. **There can be significant interaction between process variables.** One important example is that changing a single MV can have significant effects

on many CVs. For example, increasing heat duty q_B by increasing the steam flow rate causes more liquid to be boiled and thus increases the vapor flow in the column. Consequently, the q_B increase causes sump level h_B, pressure P, and bottom composition x_B to change rather quickly (Guideline 7). However, the increase in q_B also affects reflux drum level h_D and distillate composition x_D more slowly, after the increased vapor flow reaches the top of the column (Guideline 8).

Similarly, changes in R or D affect h_D and x_D rather quickly and h_B and x_B more slowly. Other interactions arise when the hot bottom stream from the column is used to heat the cold feed stream in a bottom-feed heat exchanger, in order to reduce energy consumption.

2. **The column behavior can be very nonlinear, especially for high-purity separations.** For example, the amount of effort required to reduce an impurity level in a product composition from 5% to 4% is typically much less than the effort required to reduce it from 1.5 % to 0.5%.
3. **Distillation columns often have very slow dynamics.** The dominant time constants can be several hours, or even longer, and long time delays are also common. Because slow dynamics result in long response times with feedback control, the addition of feedforward control can be very advantageous.
4. **Process constraints are important.** The most profitable operating conditions typically occur when some MVs and CVs are at upper or lower limits. For example, maximum separation, or maximum recovery of a valuable feed component, often occurs for maximum reboiler heating or condenser cooling.
5. **Product compositions are often not measured.** Although product compositions are the primary CVs, their on-line measurement is often difficult, expensive, and costly to maintain. Consequently, tray or product stream temperatures are commonly measured and controlled as proxies for product compositions. This strategy is easier to implement but makes tight composition control more difficult.

Another major complication is that there are many different column configurations, especially for reboilers and condensers, and many alternative process and control objectives. Consequently, each column control application tends to be different and to require individual analysis. Fortunately, there is an extensive literature available on both the practical (Shinskey, 1984; Luyben., 1992) and theoretical (Skogestad, 1997) aspects of distillation column control.

13.3.2 Fired-Tube Furnace

Fired-tube furnaces (or heaters) are widely used in the process industries to heat process streams and to "crack" high-molecular-weight hydrocarbon feeds, in order to produce more valuable lower-molecular-weight compounds. In this case study, we consider a fired-tube furnace used to heat a liquid hydrocarbon feed steam that passes through the furnace in a set of tubes. A simplified schematic diagram is shown in Fig. 13.8. The combustion of the fuel gas (FG) generates heat, which is transferred to the hydrocarbon (HC). The major gaseous combustion reactions in the furnace are

$$CH_4 + \frac{3}{2}O_2 \rightarrow CO + 2H_2O$$

$$CO + \frac{1}{2}O_2 \rightarrow CO_2$$

A fired-tube furnace is one of the case studies in the Process Control Modules (PCM) in Appendix E. The PCM furnace model is a nonlinear state-space model that consists of 26 nonlinear ordinary differential equations based on conservation equations and reaction rate expressions for combustion (Doyle et al., 1998). The key process variables for the furnace model are listed in Table 13.1.

Important dynamic characteristics of the furnace model include the different time scales associated with mass and energy transfer, the nonlinear behavior of the model, time delays, and the process interactions between the input and output variables. The term process interaction means that changes in an input variable affect more than one output variable. For example, the step responses in Fig. 13.9 illustrate that a step change greater than 20% in the inlet air temperature affects three of the four output variables, and their corresponding response times are quite different.

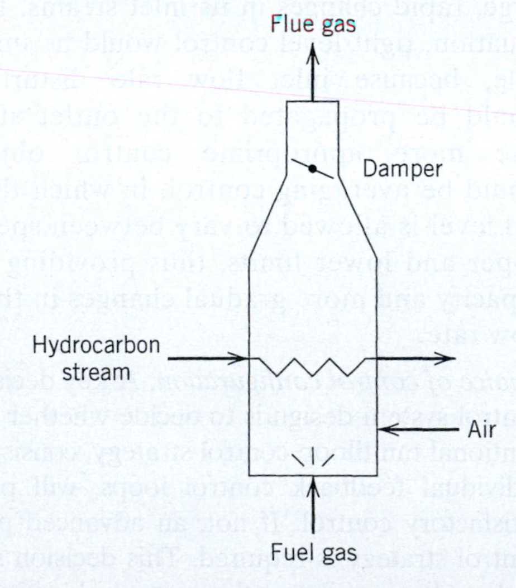

Figure 13.8 Schematic diagram of a tube-fired furnace.

Table 13.1 key process variables for the PCM furnace module

Measured Output Variables	Disturbance Variables (DVs)	Manipulated Variables (MVs)
HC outlet temperature	HC inlet temperature	Air flow rate
Furnace temperature	HC flow rate	FG flow rate
Flue gas (or exhaust gas) flow rate	Inlet air temperature	
O_2 exit concentration	FG temperature	
	FG purity (CH_4 concentration)	

The control objectives for the furnace are the following:

1. To heat the hydrocarbon stream to a desired exit temperature
2. To avoid unsafe conditions resulting from the interruption of fuel gas or hydrocarbon feed
3. To operate the furnace economically by maintaining an optimum air-fuel ratio

Because the furnace model has two MVs, two CVs should be specified. Of the four measured outputs in Table 13.1, the most important are the primary CV, HC outlet temperature, and the O_2 exit concentration in the flue gas. The latter provides a good indication of the combustion efficiency of the furnace. A very low O_2 measurement indicates that the FG combustion is incomplete; a very high measurement indicates excess air and thus low furnace efficiency. A high furnace efficiency strongly depends on maintaining the optimum air-fuel ratio. For these reasons, the HC outlet temperature and the O_2 exit concentration in the flue gas are selected to be the CVs (Guideline 3).

The chief DV is the FG composition, which can vary significantly depending on the source of the fuel gas. Large composition changes affect the FG heating value and thus upset the combustion process and heat generation.

Conventional furnace control strategies involve both feedforward and feedback aspects (Lipták, 2003, Shinskey, 1996). The HC exit temperature can be controlled by adjusting either the FG flow rate or FG pressure. The O_2 exit concentration is controlled by adjusting the furnace draft, i.e., the difference between air inlet and outlet pressures, by changing either the inlet air flow rate or the damper in the furnace stack. The air-fuel ratio can be controlled using a special type of feedforward control referred to as ratio control (see Chapter 15). The measured HC inlet flow rate can also be used as a measured disturbance for feedforward control. The PCM include advanced model-based control strategies for the furnace, including decoupling (Chapter 18) and model predictive control (Chapter 20).

Safety considerations are a primary concern for furnace operation because of the large amount of combustible material that is present. In particular, it is important to ensure that unreacted FG is not allowed to accumulate. This unsafe condition can result if the air-fuel ratio is too low or if burner flameout occurs because of a FG interruption. Safety interlocks (see Chapter 10) are used to shut off the fuel gas in these

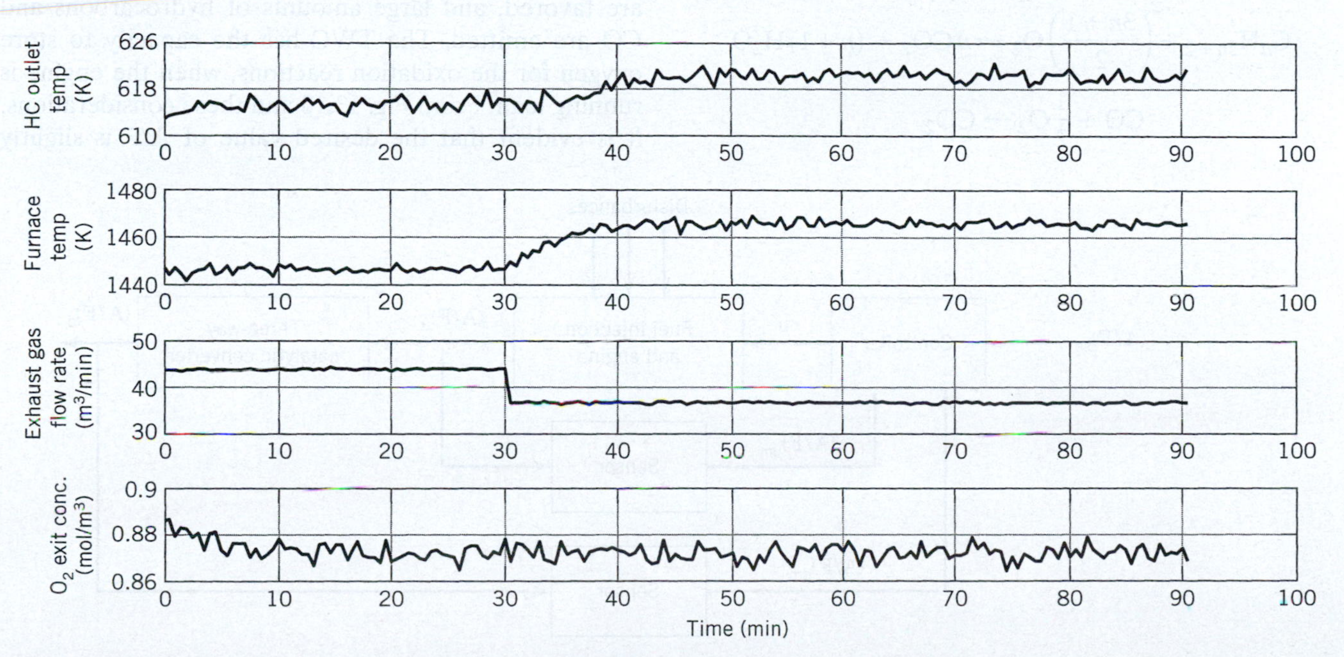

Figure 13.9 Step response to a + 20% step change in inlet air temperature at t = 30 min.

situations. Interruption of the feed stream also poses a serious hazard, because it can result in overheating and possible rupture of tubes in the furnace (Lipták, 2003). Thus, an interlock shuts off the fuel gas if a low flow alarm occurs for the feed flow rate.

For additional information on furnace and heater control problems, see the books by Shinskey (1996) and the Lipták (2003).

13.3.3 Catalytic Converters for Automobiles

In many urbanized areas of the world, automobiles are the single greatest producer of harmful vehicle exhaust emissions. For over 30 years, catalytic converters have been used to significantly reduce harmful exhaust emissions from internal combustion engines, especially automobiles. In North America, automobiles with standard gasoline engines manufactured since 2004 are required to have three-way catalytic (TWC) converters. This section presents an overview of control issues and control strategies associated with TWC. More detailed information is available elsewhere (Fiengo et al., 2005; Balenović et al., 2006; Guzzella, 2008).

Three-way catalytic converters (TWC) are designed to reduce three types of harmful automobile emissions: carbon monoxide (CO), unburned hydrocarbons in the fuel (HC), and nitrogen oxides (NO_x). The term *nitrogen oxides* refers to both NO_2 and NO. The TWC accomplishes these tasks using precious metal catalysts (platinum, palladium, and rhodium) for chemical reactions that take place at high temperatures (e.g., 1000–1600°F) and short residence times (~0.05 s).

The desired TWC oxidation reactions are (Schmidt, 2005),

$$C_nH_{2n+2} + \left(\frac{3n+1}{2}\right)O_2 \rightarrow nCO_2 + (n+1)H_2O$$

$$CO + \frac{1}{2}O_2 \rightarrow CO_2$$

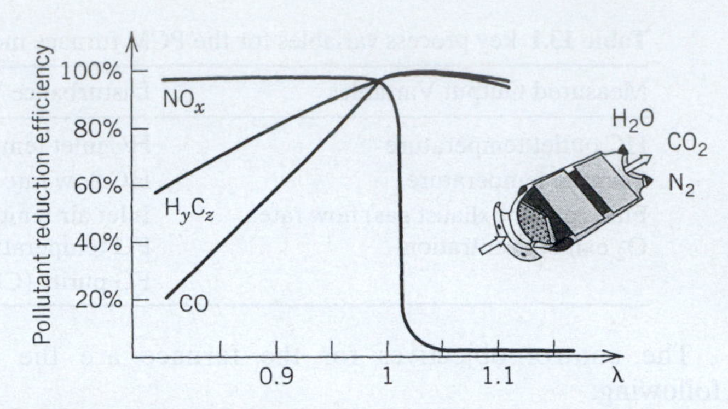

Figure 13.10 TWC efficiency as a function of air-to-fuel ratio (Guzzella, 2008).

and the desired reduction reaction is:

$$2NO_x \rightarrow xO_2 + N_2$$

The catalytic converter is most effective when the automobile engine operates with an air-to-fuel ratio (A/F) that is slightly above the stoichiometric ratio of 1/14.7 for gasoline. A normalized air-to-fuel ratio λ is defined as:

$$\lambda = \frac{w_{air}}{w_{fuel}}\frac{1}{14.7} \tag{13-8}$$

where w_{air} is the mass flow rate of air, and w_{fuel} is the mass flow rate of fuel. The pollutant removal efficiencies for the three pollutants vary strongly with air-to-fuel ratio, as shown in Fig. 13.10. For $\lambda \approx 1$, the three pollutants are essentially eliminated, with over 98% removal. When $\lambda > 1$, there is excess O_2, and the engine is said to be running lean. For these conditions, the oxidation reactions are favored, and excessive amounts of NO_x are emitted. Conversely, when $\lambda < 1$, the engine is said to be running rich. Here, the reduction reactions are favored, and large amounts of hydrocarbons and CO are emitted. The TWC has the capacity to store oxygen for the oxidation reactions, when the engine is running lean. From Fig. 13.10 and these considerations, it is evident that the desired value of A/F is slightly

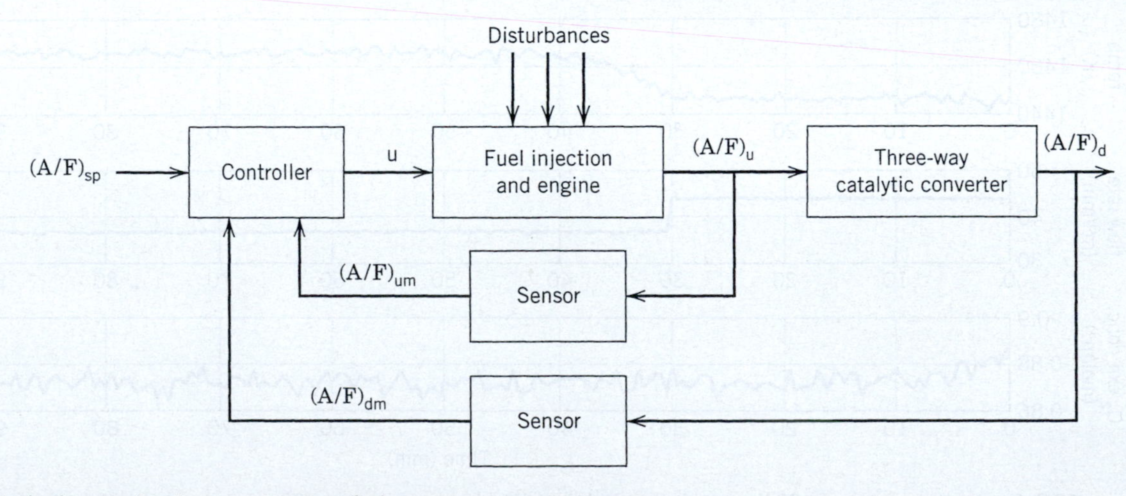

Figure 13.11 Block diagram for the three way catalytic converter control system.

greater than stoichiometric, or, equivalently, λ should be slightly greater than 1.

A general representation of a TWC control strategy is shown in the block diagram of Fig. 13.11. The A/F is measured both upstream, $(A/F)_u$, and downstream $(A/F)_d$, of the TWC, using O_2 sensors (also called lambda sensors). Based on the A/F measurements and the desired value, $(A/F)_{sp}$, the feedback controller calculates an appropriate output signal u that adjusts the fuel injection system for the engine (Guidelines 6–8). Sometimes, the two A/F measurements are used in a cascade control configuration, a topic that is considered in Chapter 16. The static and dynamic behavior of the TWC varies with the operating conditions (e.g., engine load) and aging of the components, including the O_2 sensors. Thus automatic adjustment of the controller settings (adaptive control) is a promising approach (Guzzella, 2008).

Many TWC control systems are operated so that A/F rapidly alternates between being slightly rich and slightly lean ($\lambda = 1 \pm 0.05$) to ensure that the reduction catalyst (rhodium) does not become overloaded and that the oxidation catalysts (platinum and palladium) do not become oxygen-starved. The switching time between the two modes is very small, less than a second. The switching strategy can be implemented by cycling the set-point, $(A/F)_{sp}$.

The performance of a TWC is strongly affected by its temperature, as well as the A/F value. The TWC does not begin to operate properly until it heats up to approximately 550°F; efficient operation does not occur until the temperature reaches about 750°F. Consequently, a significant amount of emissions occur during cold starts of the engine. This problem can be alleviated by the addition of an electrical heater that can heat the TWC prior to cold starts. The TWC can operate properly up to sustained temperatures of 1500°F.

If the A/F ratio is rich, unburned fuel from the engine undergoes combustion in the TWC, which can raise the TWC exit temperature to several hundred degrees above the inlet temperature. Consequently, temperature sensors located before and after the TWC can provide useful diagnostic information (Guideline 11). If the difference in temperature measurements is unusually large, it indicates that rich conditions occur. On the other hand, if the difference is essentially zero, the TWC has stopped functioning (e.g., as a result of catalyst poisoning).

As emission standards for automobiles become tighter, improved closed-loop TWC control strategies become even more critical. The development of advanced TWC control strategies includes custom model-based methods (Fiengo et al., 2005; Balenović et al., 2006; Guzzella, 2008).

13.3.4 Plasma Etching in Semiconductor Processing

Solid-state devices are manufactured on circular disks of semiconducting material called wafers (Edgar et al., 2000). These devices are three-dimensional structures made up of stacked layers. Each layer is typically manufactured in batch operations, such as deposition and etching. The purpose of deposition is to grow a thin layer of a specific material on the wafer surface. In etching, part of the layer is removed chemically, using gases such as CF_4 and HF. Etching can remove silicon dioxide, silicon nitride, polysilicon, aluminum, photoresist, and other thin film materials. It creates the final layer definition by transforming a single layer of semiconductor material into the patterns, features, lines, and interconnects that make up an integrated circuit.

The polysilicon (poly) gate etch process is shown schematically in Fig. 13.12. Photoresist (PR) etching and polysilicon etching are the most critical batch steps for

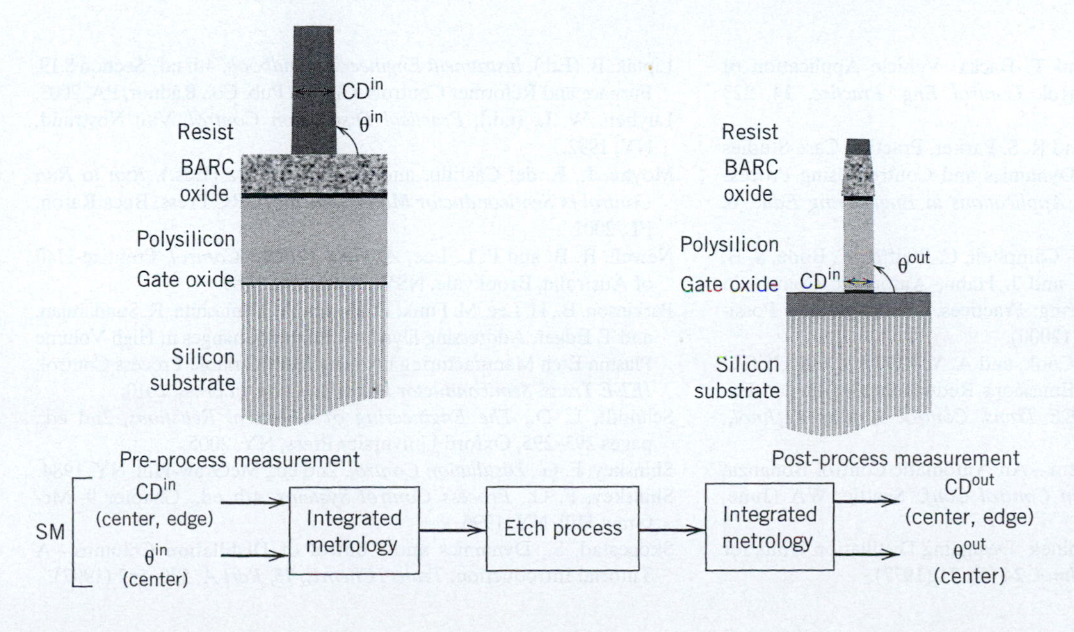

Figure 13.12 Inputs and outputs for polysilicon gate etch process in semiconductor manufacturing. The measured inputs (CD^{in} and θ^{in}) in the incoming wafer can be used in feedforward control, while the measured outputs (CD^{out} and θ^{out}) are used in feedback control. BARC is bottom anti-reflective coating.

creating the profile of polysilicon (side views are shown in Fig. 13.12). Photoresist etching entails isotropic etching of the top layer of photoresist, which determines the critical dimension (CD), or width, of polysilicon. This step is followed by polysilicon etching, which is anisotropic (etches in a single downward direction); the final profile of polysilicon is determined in this step.

The etching process can be used to illustrate the application of Guidelines 3, 6, 8, and 10 from Section 13.2. For Guideline 3, the key CVs in plasma etching, CD and θ (sidewall angle), are shown in Fig. 13.12. The CD affects transistor speed, which is the most important electrical property of a logic chip (a product quality variable). Ideally, θ should be 90°, but a target of 87° represents a trade-off between θ and CD because of the interactions between the variables. Attempting to control θ closer to 90° causes the CD to move further from the target. The uniformity of the CD over the wafer is a third CV affected by the inputs. Excessive nonuniformity makes the wafer lower quality because of chip inconsistency.

A plasma etcher has a number of MVs that can be adjusted in order to achieve the desired chip geometry. By applying Guidelines 6 and 8, several input variables can be selected from the four possible MVs: etch time, pressure, plasma power, and flow rates of gases such as N_2, O_2, and Cl_2. Steady-state nonlinear models can be obtained from experimental test wafers by specifying the inputs and measuring each CV; the data can then be fitted using polynomial models as the input-output relationships. These models also allow the process gain to be calculated for each input-output pair.

A controller determines the set of input variables (known as the etch recipe) that keep CD, θ, and uniformity as close as possible to their targets while satisfying MV constraints. Consistent with Guideline 10, integrated metrology (IM), shown in Fig. 13.12, uses optical techniques such as ellipsometry or scatterometry to measure the incoming wafer profile (CD^{in} and θ^{in}) at multiple sites. IM then sends the measurements to a computer that calculates the values of the MVs for the batch. At the end of the etch process, the output CVs for the batch are measured using IM. The errors for the CVs are calculated and used to adjust the control strategy (Parkinson et al., 2010).

Advanced control strategies for microelectronics applications such as including plasma etching are discussed elsewhere (Edgar et al., 2000; Moyne et al., 2001). Batch process control is discussed in Chapter 22.

SUMMARY

This chapter has considered two important issues in control system design. The first issue was that the control system design is strongly influenced by the control degrees of freedom that are available, N_{FC}. In most situations, N_{FC} is simply the number of process variables that can be manipulated. In general, $N_{FC} < N_F$, where N_F is the model degrees of freedom that was introduced in Chapter 2. The second issue concerned the selection of the controlled, manipulated, and measured variables, a key step in the control system design. These choices should be based on the guidelines presented in Section 13.2.

The chapter concluded with four case studies that illustrated control problems at the process-unit level, rather than at the individual control-loop level.

REFERENCES

Balenović, M., J. Edwards, and T. Backx, Vehicle Application of Model-Based Catalyst Control, *Control Eng. Practice,* **14**, 223 (2006).

Doyle, F. J. III, E. P. Gatzke, and R. S. Parker, Practical Case Studies for Undergraduate Process Dynamics and Control Using Process Control Modules, *Computer Applications in Engineering Educ.* **6**, 181–191 (1998).

Edgar, T. F., S. W. Butler, W. J. Campbell, C. Pfeiffer, C. Bode, S. B. Hwang, K. S. Balakrishnan, and J. Hahn, Automatic Control in Microelectronics Manufacturing: Practices, Challenges and Possibilities, *Automatica*, **36**, 1567 (2000).

Fiengo, G., J. W. Grizzle, J. A. Cook, and A. Y. Karnik, Dual-UEGO Active Catalyst Control for Emissions Reduction: Design and Experimental Verification, *IEEE Trans. Control Systems Technol.*, **13**,722 (2005).

Guzzella, L., Automotive System—An Automatic Control Bonanza, Plenary Talk, *2008 American Control Conf.,* Seattle, WA (June, 2008).

Hughart, C. L. and K. W. Kominek, Designing Distillation Units for Controllability, *Instrum. Technol.* **24** (5), 71 (1977).

Lipták, B. (Ed.), *Instrument Engineers' Handbook,* 4th ed., Section 8.19, Furnace and Reformer Controls, Chilton Pub. Co., Radnor, PA, 2003.

Luyben, W. L. (ed.), *Practical Distillation Control*, Van Nostrand, NY, 1992.

Moyne, J., E. del Castillo, and A. M. Hurwitz (Eds.), *Run to Run Control in Semiconductor Manufacturing*, CRC Press, Boca Raton, FL, 2001.

Newell, R. B. and P. L. Lee, *Applied Process Control,* Prentice-Hall of Australia, Brookvale, NSW, Australia, 1989.

Parkinson, B., H. Lee, M. Funk, D. Prager, A. Yamashita, R. Sundarajan, and T. Edgar, Addressing Dynamic Process Changes in High Volume Plasma Etch Manufacturing by using Multivariable Process Control, *IEEE Trans. Semiconductor Manufacturing*, in press, 2010.

Schmidt, L. D., *The Engineering of Chemical Reactions,* 2nd ed., pages 293–295, Oxford University Press, NY, 2005.

Shinskey, F. G., *Distillation Control,* 2nd ed., McGraw-Hill, NY, 1984.

Shinskey, F. G., *Process Control Systems,* 4th ed., Chapter 9, McGraw-Hill, NY, 1996.

Skogestad, S., Dynamics and Control of Distillation Columns—A Tutorial Introduction, *Trans IChemE,* **75**, *Part A*, 539–562 (1997).

EXERCISES

13.1 Consider the distillation column shown in Fig. 13.2. It would be reasonable to control the liquid level in the reflux drum, h_D, by manipulating either reflux flow rate, R, or distillate flow rate, D. How would the nominal value of the reflux ratio (R/D) influence your choice? As a specific example, assume that $R/D = 4$.

13.2 A stirred-tank blending system with a bypass stream is shown in Fig. E13.2. The control objective is to control the composition of a key component in the exit stream, x_4. The chief disturbance variables are the mass fractions of the key component in the inlet streams, x_1 and x_2. Using the following information, discuss which flow rate should be selected as the manipulated variable: (i) inlet flow rate w_2, (ii) the bypass fraction f, or (iii) exit flow rate, w_4. Your choice should reflect both steady-state and dynamic considerations.

Available Information:

(a) The tank is perfectly mixed.

(b) Constant physical properties can be assumed because the composition changes are quite small.

(c) Because the variations in liquid level are small, h does not have to be controlled.

(d) The bypass piping results in a negligible time delay.

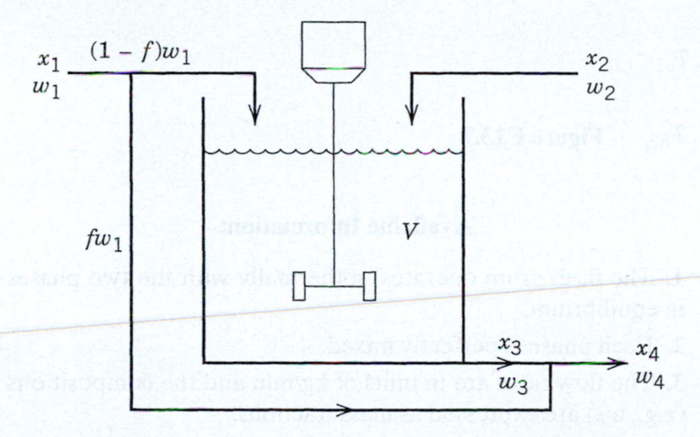

Figure E13.2

13.3 Suppose that the distillation column shown in Fig. 13.2 has been designed to separate a methanol-water mixture that is 50% methanol (MeOH). This high-purity column has a large number of trays and a nominal distillate composition of $x_D = 5$ ppm of MeOH. Because a composition analyzer is not available, it is proposed to control x_D indirectly, by measuring and controlling the liquid temperature at one of the following locations:

(a) The reflux stream

(b) The top tray in the rectifying section

(c) An intermediate tray in the rectifying section, midway between the feed tray and the top tray

Discuss the relative advantages and disadvantages of each choice, based on both steady-state and dynamic considerations.

13.4 It has been suggested that the capital cost for the distillation column in Fig. 13.2 can be reduced by using a "flooded condenser." In the proposed design, the reflux drum would be eliminated, and the condensed vapor in the condenser would provide the liquid inventory for the reflux and distillate streams, as shown in Fig. E13.4. As a result, the coolant tubes in the condenser would be partially covered (or flooded), and the area available for heat transfer would change as the liquid level changes.

Discuss the dynamic and control implications of this proposed process change for both pressure control and liquid-level control. You may assume that the conventional control configuration for this column is to control column pressure P by manipulating coolant flow rate, q_C, and liquid level h_D by manipulating distillate flow rate, D.

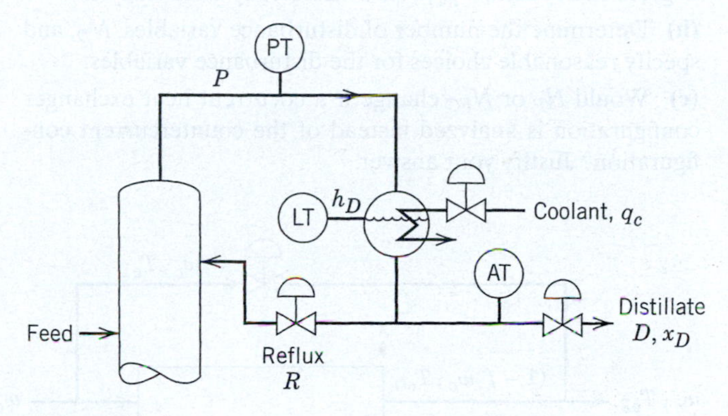

Figure E13.4

13.5 The exit stream from a chemical reactor is sent to a storage tank, as shown in Fig. E13.5. The exit stream from the storage tank serves as the feed stream to a separation process. The function of the intermediate storage tank is to "damp" feed disturbances and to allow the separation process to continue to operate when the reactor is shut down for short periods of time.

(a) Discuss the design vs. control trade-offs that are inherent in specifying the capacity of the storage tank.

(b) Suppose that the chemical reactor must produce a variety of products and, consequently, the set-point for the exit composition changes frequently. How would this consideration influence your specification of the tank capacity?

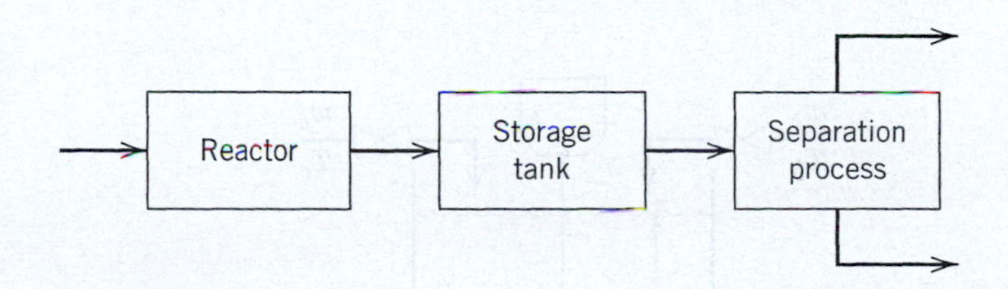

Figure E13.5

13.6 Consider the liquid storage system shown in Fig. E13.6. Only volumetric flow rates, q_1 and q_2, can be manipulated. Determine the model degrees of freedom, N_F, and the control degrees of freedom, N_{FC}.

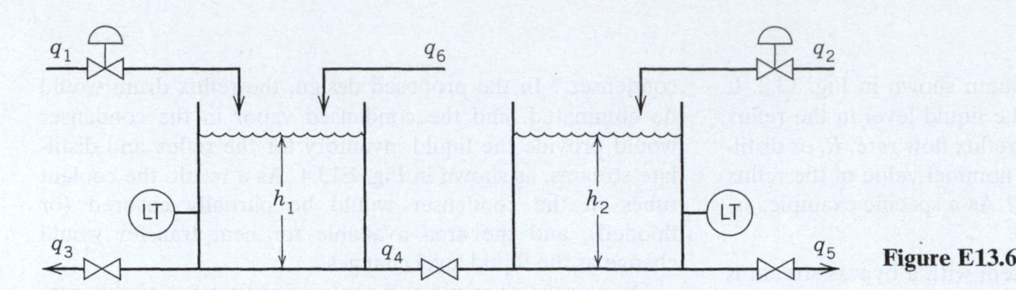

Figure E13.6

13.7 A double pipe heat exchanger with a partial bypass for the cold stream is shown in Fig. E13.7. The mass flow rate of the hot stream, w_h, and the bypass fraction, f, can be manipulated. Heat losses can be neglected.

(a) Determine the model degrees of freedom, N_F, and control degrees of freedom, N_{FC}, based on a steady-state analysis.

(b) Determine the number of disturbance variables, N_D, and specify reasonable choices for the disturbance variables.

(c) Would N_F or N_{FC} change if a cocurrent heat exchanger configuration is analyzed instead of the countercurrent configuration? Justify your answer.

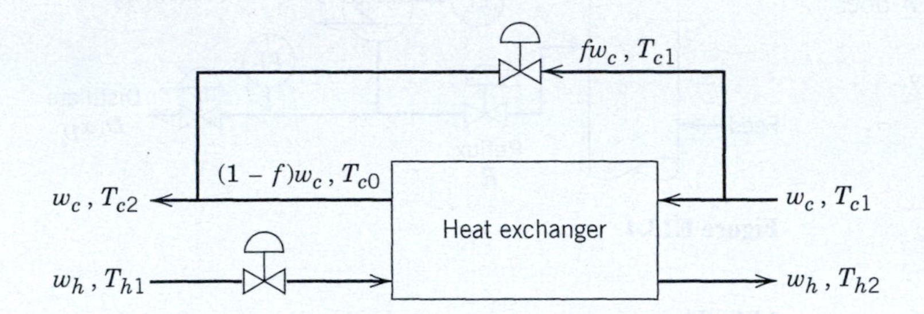

Figure E13.7

13.8 Consider the blending system of Exercise 13.2. Inlet flow rate, w_2, and the bypass fraction, f, can be manipulated. Determine the model degrees of freedom, N_F, and the control degrees of freedom, N_{FC}.

13.9 A stirred-tank heating system is shown in Fig. E13.9. Briefly critique these two control strategies.

(a) It is proposed that h and T be controlled by manipulating w_h and w_c using two PI controllers.

(b) Suppose that two PI controllers are to be used, with h controlled by manipulating w_h and T controlled by manipulating w.

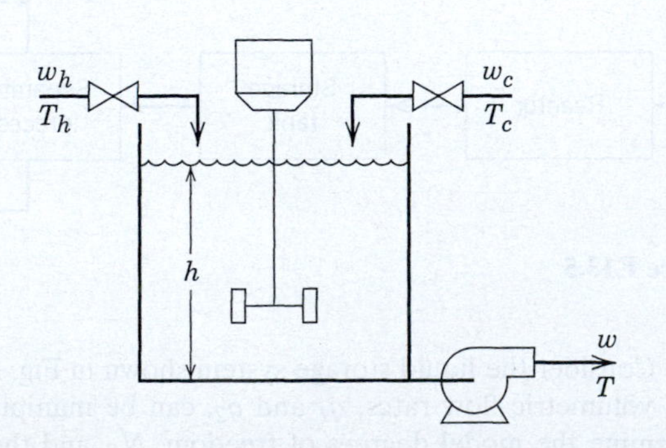

Figure E13.9

13.10 A two-phase feed to the gas-liquid separator (or flash drum), shown in Fig. E13.10, consists of a mixture of two hydrocarbons. Because the vessel pressure P is lower than the feed pressure, the feed flashes as it enters the separator. Using the following information and a degrees of freedom analysis, do the following:

(a) Determine the model degrees of freedom, N_F.

(b) Determine the control degrees of freedom, N_{FC}.

(c) Specify reasonable CVs and MVs. Justify your answers.

Available Information:

1. The flash drum operates isothermally with the two phases in equilibrium.
2. Each phase is perfectly mixed.
3. The flow rates are in units of kg/min and the compositions (e.g., w_F) are expressed as mass fractions.
4. Each flow rate can be adjusted by a control valve (not shown).
5. For the uncontrolled process, the exit flow rates are related to vessel conditions by empirical equations that have the following forms: $G = f_1(P)$ and $L = f_2(h)$.

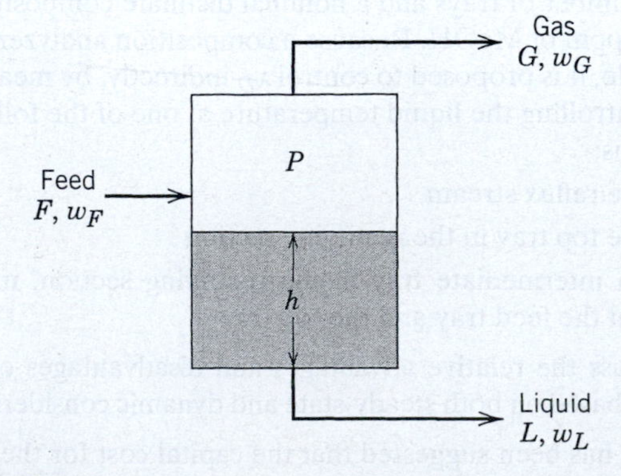

Figure E13.10

Chapter 14

Frequency Response Analysis and Control System Design

CHAPTER CONTENTS

In previous chapters, Laplace transform techniques were used to calculate transient responses from transfer functions. This chapter focuses on an alternative way to analyze dynamic systems by using frequency response analysis. Frequency response concepts and techniques play an important role in stability analysis, control system design, and robustness analysis. Historically, frequency response techniques provided the conceptual framework for early control theory and important applications in the field of communications (MacFarlane, 1979). We introduce a simplified procedure to calculate the frequency response characteristics from the transfer function of any linear process. Two concepts, the Bode and Nyquist stability criteria, are generally applicable for feedback control systems and stability analysis. Next we introduce two useful metrics for relative stability, namely gain and phase margins. These metrics indicate how close to instability a control system is. A related issue is robustness, which addresses the sensitivity of control system performance to process variations and to uncertainty in the process model.

14.1 SINUSOIDAL FORCING OF A FIRST-ORDER PROCESS

We start with the response properties of a first-order process when forced by a sinusoidal input and show how the output response characteristics depend on the frequency of the input signal. This is the origin of the term *frequency response*. The responses for first- and second-order processes forced by a sinusoidal input were presented in Chapter 5. Recall that these responses consisted of sine, cosine, and exponential terms. Specifically, for a first-order transfer function with gain K and time constant τ, the response to a general sinusoidal input, $x(t) = A \sin \omega t$, is

$$y(t) = \frac{KA}{\omega^2\tau^2 + 1}\left(\omega\tau e^{-t/\tau} - \omega\tau \cos \omega t + \sin \omega t\right) \quad (5\text{-}25)$$

Note that in (5-25) y is in deviation form.

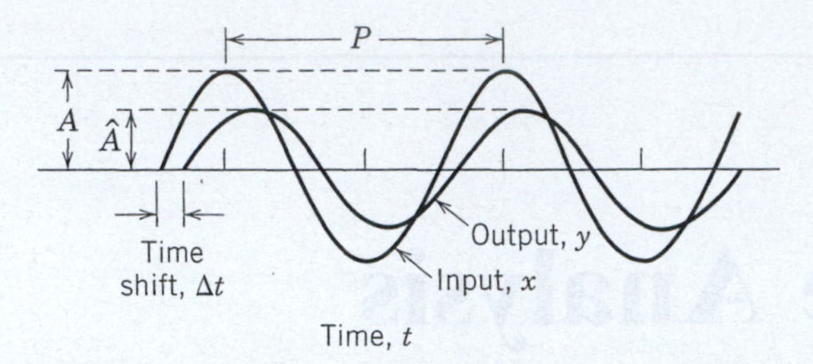

Figure 14.1 Attenuation and time shift between input and output sine waves. The phase angle ϕ of the output signal is given by $\phi = \Delta t/P \times 360°$, where Δt is the time shift and P is the period of oscillation.

If the sinusoidal input is continued for a long time, the exponential term $(\omega\tau e^{-t/\tau})$ becomes negligible. The remaining sine and cosine terms can be combined via a trigonometric identity to yield

$$y_\ell(t) = \frac{KA}{\sqrt{\omega^2\tau^2+1}} \sin(\omega t + \phi) \tag{14-1}$$

where $\phi = -\tan^{-1}(\omega\tau)$. The long-time response $y_\ell(t)$ is called the frequency response of the first-order system and has two distinctive features (see Figure 14.1).

1. The output signal is a sine wave that has the same frequency, but its phase is shifted relative to the input sine wave by the angle ϕ (referred to as the *phase shift* or the *phase angle*); the amount of phase shift depends on the forcing frequency ω.
2. The output signal is a sine wave that has an amplitude A that also is a function of the forcing frequency:

$$\hat{A} = \frac{KA}{\sqrt{\omega^2\tau^2+1}} \tag{14-2}$$

Dividing both sides of (14-2) by the input signal amplitude A yields the *amplitude ratio* (AR)

$$AR = \frac{\hat{A}}{A} = \frac{K}{\sqrt{\omega^2\tau^2+1}} \tag{14-3a}$$

which can, in turn, be divided by the process gain to yield the *normalized amplitude ratio* (AR_N):

$$AR_N = \frac{AR}{K} = \frac{1}{\sqrt{\omega^2\tau^2+1}} \tag{14-3b}$$

Because the steady-state gain K is constant, the normalized amplitude ratio often is used for frequency response analysis.

Next we examine the physical significance of the preceding equations, with specific reference to the blending process example discussed earlier. In Chapter 4, the transfer function model for the stirred-tank blending system was derived as

$$X'(s) = \frac{K_1}{\tau s+1}X_1'(s) + \frac{K_2}{\tau s+1}W_2'(s) + \frac{K_3}{\tau s+1}W_1'(s) \tag{4-69}$$

Suppose flow rate w_2 is varied sinusoidally about a constant value, while the other inlet conditions are kept constant at their nominal values; that is, $w_1'(t) = x_1'(t) = 0$. Because $w_2(t)$ is sinusoidal, the output composition deviation $x'(t)$ eventually becomes sinusoidal according to Eq. 5-26. However, there is a phase shift in the output relative to the input, as shown in Fig. 14.1, owing to the material holdup of the. tank. If the flow rate w_2 oscillates very slowly relative to the residence time $\tau(\omega << 1/\tau)$, the phase shift is very small, approaching 0°, whereas the normalized amplitude ratio$(\hat{A}/KA)$ is very nearly unity. For the case of a low-frequency input, the output is in phase with the input, tracking the sinusoidal input as if the process model were $G(s) = K$.

On the other hand, suppose that the flow rate is varied rapidly by increasing the input signal frequency. For $\omega >> 1/\tau$, Eq. 14-1 indicates that the phase shift approaches a value of $-\pi/2$ radians (−90°). The presence of the negative sign indicates that the output lags behind the input by 90°; in other words, the phase lag is 90°. The amplitude ratio approaches zero as the frequency becomes large, indicating that the input signal is almost completely attenuated; namely, the sinusoidal deviation in the output signal is very small.

These results indicate that positive and negative deviations in w_2 are essentially canceled by the capacitance of the liquid in the blending system if the frequency is high enough. In this case, high frequency implies $\omega >> 1/\tau$. Most processes behave qualitatively, similar to the stirred-tank blending system, when subjected to a sinusoidal input. For high-frequency input changes, the process output deviations are so completely attenuated that the corresponding periodic variation in the output is difficult (perhaps impossible) to detect or measure.

Input-output phase shift and attenuation (or amplification) occur for any stable transfer function, regardless of its complexity. In all cases, the phase shift and amplitude ratio are related to the frequency ω of the sinusoidal input signal. In developments up to this point, the expressions for the amplitude ratio and phase shift were derived using the process transfer function. However, the frequency response of a process can also be obtained experimentally. By performing a series of tests in which a sinusoidal input is applied to the process, the resulting amplitude ratio and phase shift can be measured for different frequencies. In this case, the frequency response is expressed as a table of measured amplitude ratios and phase shifts for selected values of ω. However, the method is very time-consuming because of the repeated experiments for different values of ω. Thus, other methods, such as pulse testing (Ogunnaike and Ray, 1994), are utilized, because only a single test is required.

In this chapter, the focus is on developing a powerful analytical method to calculate the frequency response for any process transfer function, as shown in the following. Later in this chapter, we show how this information can be used to design controllers and analyze the properties of the controlled system responses.

14.2 SINUSOIDAL FORCING OF AN *n*TH-ORDER PROCESS

This section presents a general approach for deriving the frequency response of any stable transfer function. We show that a rather simple procedure can be employed to find the sinusoidal response.

Setting $s = j\omega$ in $G(s)$, by algebraic manipulation we can separate the expression into real (R) and imaginary (I) expressions.

$$G(j\omega) = R(\omega) + jI(\omega) \tag{14-4}$$

Similar to Eq. (14-1), we can express the time-domain response for any linear system as

$$y_\ell(t) = \hat{A}\sin(\omega t + \phi) \tag{14-5}$$

$\hat{A}$ and ϕ are related to $I(\omega)$ and $R(\omega)$ by the following relations (Seborg et al., 2004):

$$\hat{A} = A\sqrt{R^2 + I^2} \tag{14-6a}$$

$$\phi = \tan^{-1}(I/R) \tag{14-6b}$$

Both $\hat{A}$ and ϕ are functions of frequency ω. A simple but elegant relation for the frequency response can be derived, where the amplitude ratio is given by

$$AR = \frac{\hat{A}}{A} = |G(j\omega)| = \sqrt{R^2 + I^2} \tag{14-7}$$

and the phase shift between the sinusoidal output and input is given by

$$\phi = \angle G = \tan^{-1}(I/R) \tag{14-8}$$

Because $R(\omega)$ and $I(\omega)$ (and hence AR and ϕ) can be obtained without calculating the complete transient response $y(t)$, these characteristics provide a shortcut method to determine the frequency response of the first-order transfer function.

More important, Eqs. 14-7 and 14-8 provide a convenient shortcut technique for calculating the frequency response characteristics of any stable $G(s)$, including those with time-delay terms. However, the physical interpretation of frequency response is not valid for unstable systems, because a sinusoidal input produces an unbounded output response instead of a sinusoidal response.

The shortcut method can be summarized as follows:

Step 1. Substitute $s = j\omega$ in $G(s)$ to obtain $G(j\omega)$.

Step 2. Rationalize $G(j\omega)$. Express $G(j\omega)$ as $R + jI$, where R and I are functions of ω and possibly model parameters, using complex conjugate multiplication. Find the complex conjugate of the denominator of $G(j\omega)$ and multiply both numerator and denominator of $G(j\omega)$ by this quantity.

Step 3. The output sine wave has amplitude $\hat{A} = A\sqrt{R^2 + I^2}$ and phase angle $\phi = \tan^{-1}(I/R)$. The amplitude ratio is $AR = \sqrt{R^2 + I^2}$ and is independent of the value of A.

EXAMPLE 14.1

Find the frequency response of a first-order system, with

$$G(s) = \frac{1}{\tau s + 1} \tag{14-9}$$

SOLUTION

First substitute $s = j\omega$ in the transfer function

$$G(j\omega) = \frac{1}{\tau j\omega + 1} = \frac{1}{j\omega\tau + 1} \tag{14-10}$$

Then multiply both numerator and denominator by the complex conjugate of the denominator, that is, $-j\omega\tau + 1$

$$G(j\omega) = \frac{-j\omega\tau + 1}{(j\omega\tau + 1)(-j\omega\tau + 1)} = \frac{-j\omega\tau + 1}{\omega^2\tau^2 + 1}$$

$$= \frac{1}{\omega^2\tau^2 + 1} + j\frac{(-\omega\tau)}{\omega^2\tau^2 + 1} = R + jI \tag{14-11}$$

where

$$R = \frac{1}{\omega^2\tau^2 + 1} \tag{14-12a}$$

and

$$I = \frac{-\omega\tau}{\omega^2\tau^2 + 1} \tag{14-12b}$$

From (14-7)

$$AR = |G(j\omega)| = \sqrt{\left(\frac{1}{\omega^2\tau^2 + 1}\right)^2 + \left(\frac{-\omega\tau}{\omega^2\tau^2 + 1}\right)^2}$$

Simplifying,

$$AR = \sqrt{\frac{(1 + \omega^2\tau^2)}{(\omega^2\tau^2 + 1)^2}} = \frac{1}{\sqrt{\omega^2\tau^2 + 1}} \tag{14-13a}$$

$$\phi = \angle G(j\omega) = \tan^{-1}(-\omega\tau) = -\tan^{-1}(\omega\tau) \tag{14-13b}$$

If the process gain had been a positive value K instead of 1,

$$AR = \frac{K}{\sqrt{\omega^2\tau^2 + 1}} \tag{14-14}$$

and the phase angle would be unchanged (Eq. 14-13b). Both the amplitude ratio and phase angle are identical to those values calculated in Section 14.1 using Eq. 5-25.

From this example, we conclude that direct analysis of the complex transfer function $G(j\omega)$ is computationally easier than solving for the actual long-time output response. The computational advantages are even greater when dealing with more complicated processes, as shown in the following. Start with a general transfer function in factored form

$$G(s) = \frac{G_a(s)G_b(s)G_c(s)\cdots}{G_1(s)G_2(s)G_3(s)\cdots} \tag{14-15}$$

$G(s)$ is converted to the complex form $G(j\omega)$ by the substitution $s = j\omega$:

$$G(j\omega) = \frac{G_a(j\omega)G_b(j\omega)G_c(j\omega)\cdots}{G_1(j\omega)G_2(j\omega)G_3(j\omega)\cdots} \tag{14-16}$$

We can express the magnitude and angle of $G(j\omega)$ as follows:

$$|G(j\omega)| = \frac{|G_a(j\omega)\| G_b(j\omega)\| G_c(j\omega)\cdots}{|G_1(j\omega)\| G_2(j\omega)\|| G_3(j\omega)\cdots} \tag{14-17a}$$

$$\angle G(j\omega) = \angle G_a(j\omega) + \angle G_b(j\omega) + \angle G_c(j\omega) + \cdots$$
$$- [\angle G_1(j\omega) + \angle G_2(j\omega) + \angle G_3(j\omega) + \cdots] \tag{14-17b}$$

Eqs. 14-17a and 14-17b greatly simplify the computation of $|G(j\omega)|$ and $\angle G(j\omega)$ and, consequently, AR and ϕ. These expressions eliminate much of the complex arithmetic associated with the rationalization of complicated transfer functions. Hence, the factored form (Eq. 14-15) may be preferred for frequency response analysis. On the other hand, if the frequency response curves are generated using MATLAB, there is no need to factor the numerator or denominator, as discussed in Section 14.3.

EXAMPLE 14.2

Calculate the amplitude ratio and phase angle for the overdamped second-order transfer function

$$G(s) = \frac{K}{(\tau_1 s + 1)(\tau_2 s + 1)}$$

SOLUTION

Using Eq. 14-15, let

$$G_a = K$$
$$G_1 = \tau_1 s + 1$$
$$G_2 = \tau_2 s + 1$$

Substituting $s = j\omega$

$$G_a(j\omega) = K$$
$$G_1(j\omega) = j\omega\tau_1 + 1$$
$$G_2(j\omega) = j\omega\tau_2 + 1$$

The magnitudes and angles of each component of the complex transfer function are:

$$|G_a| = K \qquad \angle G_a = 0$$
$$|G_1| = \sqrt{\omega^2\tau_1^2 + 1} \qquad \angle G_1 = \tan^{-1}(\omega\tau_1)$$
$$|G_2| = \sqrt{\omega^2\tau_2^2 + 1} \qquad \angle G_2 = \tan^{-1}(\omega\tau_2)$$

Combining these expressions via Eqs. 14-17a and 14-17b yields

$$AR = |G(j\omega)| = \frac{|G_a|}{|G_1||G_2|}$$
$$= \frac{K}{\sqrt{\omega^2\tau_1^2 + 1}\sqrt{\omega^2\tau_2^2 + 1}} \tag{14-18a}$$

$$\phi = \angle G(j\omega) = \angle G_a - (\angle G_1 + \angle G_2)$$
$$= -\tan^{-1}(\omega\tau_1) - \tan^{-1}(\omega\tau_2) \tag{14-18b}$$

14.3 BODE DIAGRAMS

The Bode diagram or Bode plot, provides a convenient display of the frequency response characteristics of a transfer function model in which AR and ϕ are each plotted as a function of ω. Ordinarily, ω is expressed in units of radians/time to simplify inverse tangent calculations (e.g., Eq. 14-18b) where the arguments must be dimensionless, that is, in radians. Occasionally, a cyclic frequency, $\omega/2\pi$, with units of cycles/time, is used. Phase angle ϕ is normally expressed in degrees rather than radians. For reasons that will become apparent in the following development, the Bode diagram consists of: (1) a log-log plot of AR versus ω and (2) a semilog plot of ϕ versus ω. These plots are particularly useful for rapid analysis of the response characteristics and stability of closed-loop systems.

14.3.1 First-Order Process

In the past, when frequency response plots had to be generated by hand, they were of limited utility. A much more practical approach now utilizes spreadsheets or control-oriented software such as MATLAB to simplify calculations and generate Bode plots. Although spreadsheet software can be used to generate Bode plots, it is much more convenient to use software designed specifically for control system analysis. Thus, after describing the qualitative features of Bode plots of simple transfer functions, we illustrate how the AR and ϕ components of such a plot are generated by a MATLAB program in Example 14.3.

For a first-order process, $K/(\tau s + 1)$, Fig. 14.2 shows a log-log plot of the normalized amplitude ratio versus $\omega\tau$, so that the figure applies for all values of K and τ. A semilog plot of ϕ versus $\omega\tau$ is also shown. In Figure 14.2, the abscissa $\omega\tau$ has units of radians. If K and τ are known,

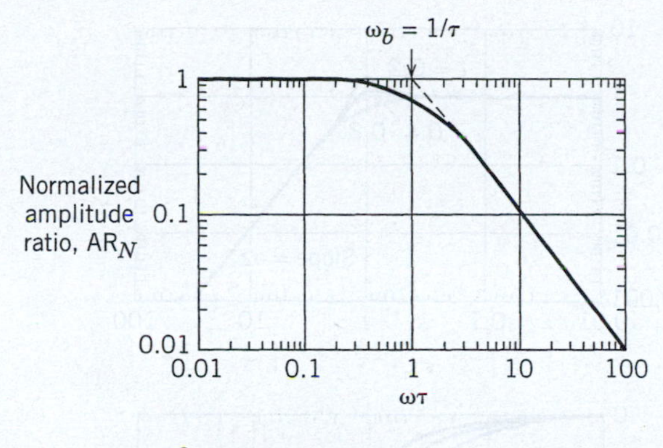

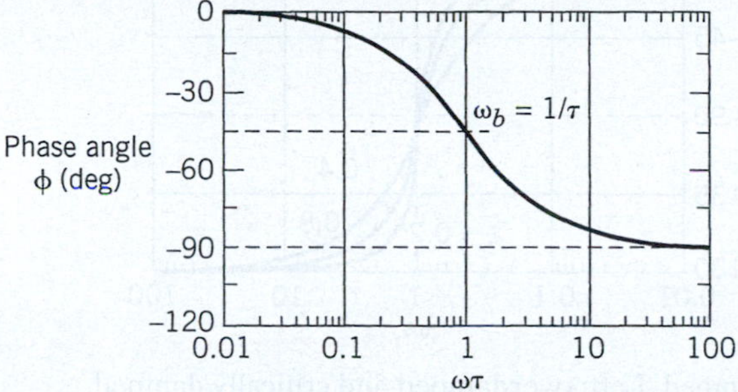

Figure 14.2 Bode diagram for a first-order process

AR_N (or AR) and ϕ can be plotted as a function of ω. Note that, at high frequencies, the amplitude ratio drops to an infinitesimal level, and the phase lag (the phase shift expressed as a positive value) approaches a maximum value of 90°.

Some books and software define AR differently, in terms of decibels. The amplitude ratio in decibels AR_{db} is defined as

$$AR_{db} = 20 \log \text{AR} \tag{14-19}$$

The use of decibels merely results in a rescaling of the Bode plot AR axis. The decibel unit is employed in electrical communication and acoustic theory and is seldom used today in the process control field. Note that the MATLAB bode routine uses decibels as the default option; however, it can be modified to plot AR results, as done in Fig 14.2. In the rest of this chapter, we only derive frequency responses for simple transfer function elements (integrator, first-order, second-order, zeros, time delay). Software should be used for calculating frequency responses of combinations of these elements.

14.3.2 Integrating Process

The transfer function for an integrating process was given in Chapter 5.

$$G(s) = \frac{Y(s)}{U(s)} = \frac{K}{s} \tag{5-34}$$

Because of the single pole located at the origin, this transfer function represents a marginally stable process. The shortcut method of determining frequency response outlined in the preceding section was developed for stable processes, that is, those that converge to a bounded oscillatory response. Because the output of an integrating process is bounded when forced by a sinusoidal input, the shortcut method does apply for this marginally stable process:

$$AR = |G(j\omega)| = \left|\frac{K}{j\omega}\right| = \frac{K}{\omega} \tag{14-20}$$

$$\phi = \angle G(j\omega) = \angle K - \tan^{-1}(j\ \infty\ ) = -90^\circ \tag{14-21}$$

The effect of an integrator multiplied by a stable transfer function G_1 is to change the overall phase angle of G_1 by −90°.

14.3.3 Second-Order Process

A general transfer function for a second-order system without numerator dynamics is

$$G(s) = \frac{K}{\tau^2 s^2 + 2\zeta\tau s + 1} \tag{14-22}$$

Substituting $s = j\omega$ and rearranging into real and imaginary parts (see Example 14.1) yields

$$AR = \frac{K}{\sqrt{(1-\omega^2\tau^2)^2 + (2\zeta\omega\tau)^2}} \tag{14-23a}$$

$$\phi = \tan^{-1}\left[\frac{-2\zeta\omega\tau}{1-\omega^2\tau^2}\right] \tag{14-23b}$$

Note that, in evaluating ϕ, multiple results are obtained because Eq. 14-23b has infinitely many solutions, each differing by $n180°$, where n is a positive integer. The appropriate solution of (14-23b) for the second-order system yields $-180° < \phi < 0$.

Figure 14.3 shows the Bode plots for overdamped ($\xi > 1$), critically damped ($\xi = 1$), and underdamped ($0 < \xi = 1$) processes as a function of $\omega\tau$. The low-frequency limits of the second-order system are identical to those of the first-order system. However, the limits are different at high frequencies, $\omega\tau >> 1$.

$$AR_N \approx 1/(\omega\tau)^2 \tag{14-24a}$$

$$\phi \approx -180^\circ \tag{14-24b}$$

For overdamped systems, the normalized amplitude ratio is attenuated ($\hat{A}/KA < 1$) for all ω. For underdamped systems, the amplitude ratio plot exhibits a maximum (for values of $0 < \zeta < \sqrt{2}/2$) at the resonant frequency

$$\omega_r = \frac{\sqrt{1-2\zeta^2}}{\tau} \tag{14-25}$$

$$(AR_N)_{\max} = \frac{1}{2\zeta\sqrt{1-\zeta^2}} \tag{14-26}$$

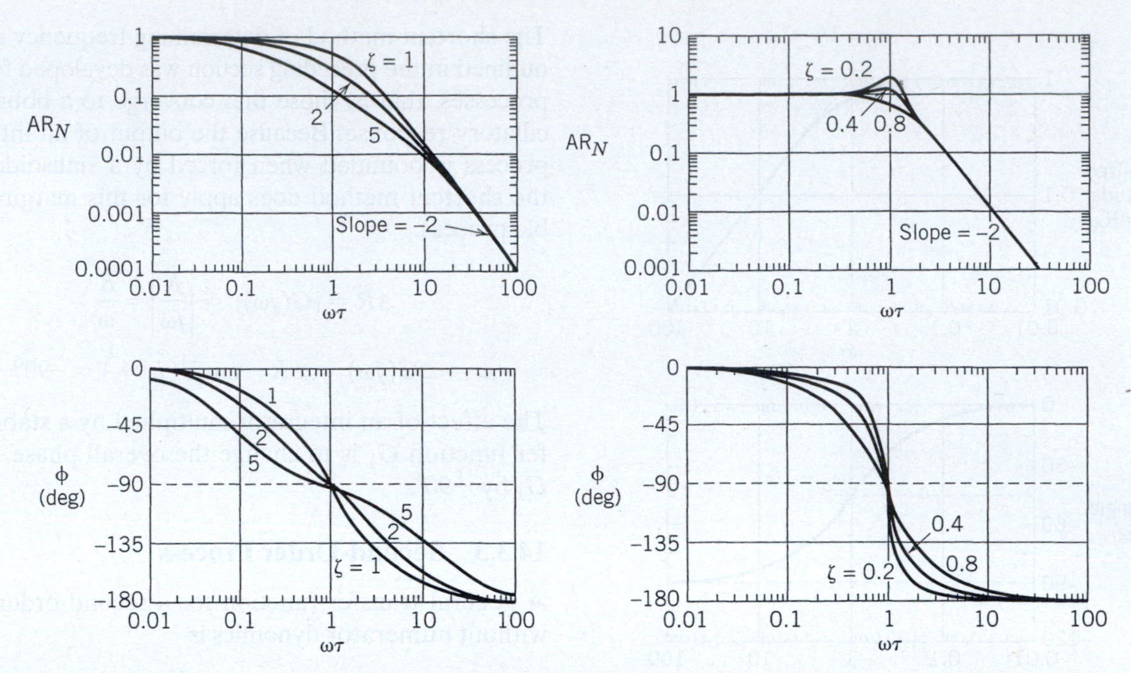

Figure 14.3 Bode diagrams for second-order processes. Right: underdamped. Left: overdamped and critically damped.

These expressions can be derived by the interested reader. The resonant frequency ω_r is that frequency for which the sinusoidal output response has the maximum amplitude for a given sinusoidal input. Eqs. (14-25) and (14-26) indicate how ω_r and $(AR_N)_{\max}$ depend on ξ. This behavior is used in designing organ pipes to create sounds at specific frequencies. However, excessive resonance is undesirable, for example, in automobiles, where a particular vibration is noticeable only at a certain speed. For industrial processes operated without feedback control, resonance is seldom encountered, although some measurement devices are designed to exhibit a limited amount of resonant behavior. On the other hand, feedback controllers can be tuned to give the controlled process a slight amount of oscillatory or underdamped behavior in order to speed up the controlled system response (see Chapter 12).

14.3.4 Process Zero

A term of the form $\tau s + 1$ in the denominator of a transfer function is sometimes referred to as a process lag, because it causes the process output to lag the input (the phase angle contribution is negative). Similarly, a process zero of the form $\tau s + 1$ $(\tau > 0)$ in the numerator (see Section 6.1) causes the sinusoidal output of the process to lead the input $(\phi > 0)$; hence, a left-half plane (LHP) zero often is referred to as process lead. Next we consider the amplitude ratio and phase angle for such a term.

Substituting $s = j\omega$ into $G(s) = \tau s + 1$ gives

$$G(j\omega) = j\omega\tau + 1 \tag{14-27}$$

from which

$$AR = |G(j\omega)| = \sqrt{\omega^2\tau^2 + 1} \tag{14-28a}$$

$$\phi = \angle G(j\omega) = \tan^{-1}(\omega\tau) \tag{14-28b}$$

Therefore, a process zero contributes a positive phase angle that varies between 0 and +90°. The output signal amplitude becomes very large at high frequencies (i.e., $AR \to \infty$ *as* $\omega \to \infty$), which is a physical impossibility. Consequently, a process zero is always found in combination with one or more poles. The order of the numerator of the process transfer function must be less than or equal to the order of the denominator, as noted in Section 6.1.

Suppose that the numerator of a transfer function contains the term $1 - \tau s$, with $\tau > 0$. As shown in Section 6.1, a right-half plane (RHP) zero is associated with an inverse step response. The frequency response characteristics of $G(s) = 1 - \tau s$ are

$$AR = \sqrt{\omega^2\tau^2 + 1} \tag{14-29a}$$

$$\phi = -\tan^{-1}(\omega\tau) \tag{14-29b}$$

Hence, the amplitude ratios of LHP and RHP zeros are identical. However, an RHP zero contributes phase lag to the overall frequency response because of the negative sign. Processes that contain an RHP zero or time delay are sometimes referred to as nonminimum phase systems because they exhibit more phase lag than another transfer function that has the same AR characteristics (Franklin et al., 2005). Exercise 14.11 illustrates the importance of zero location on the phase angle.

14.3.5 Time Delay

The time delay $e^{-\theta s}$ is the remaining important process element to be analyzed. Its frequency response characteristics can be obtained by substituting $s = j\omega$:

$$G(j\omega) = e^{-j\omega\theta} \tag{14-30}$$

which can be written in rational form by substitution of the Euler identity

$$G(j\omega) = \cos \omega\theta - j \sin \omega\theta \tag{14-31}$$

From (14-6)

$$AR = |G(j\omega)| = \sqrt{\cos^2\omega\theta + \sin^2\omega\theta} = 1 \tag{14-32}$$

$$\phi = \angle G(j\omega) = \tan^{-1}\left(-\frac{\sin \omega\theta}{\cos \omega\theta}\right)$$

or

$$\phi = -\omega\theta \tag{14-33}$$

Because ω is expressed in radians/time, the phase angle in degrees is $-180\omega\theta/\pi$. Figure 14.4 illustrates the Bode plot for a time delay. The phase angle is unbounded, that is, it approaches $-\infty$ as ω becomes large. By contrast, the phase angle of all other process elements is smaller in magnitude than some multiple of 90°. This unbounded phase lag is an important attribute of a time delay and is detrimental to closed-loop system stability, as is discussed in Section 14.6.

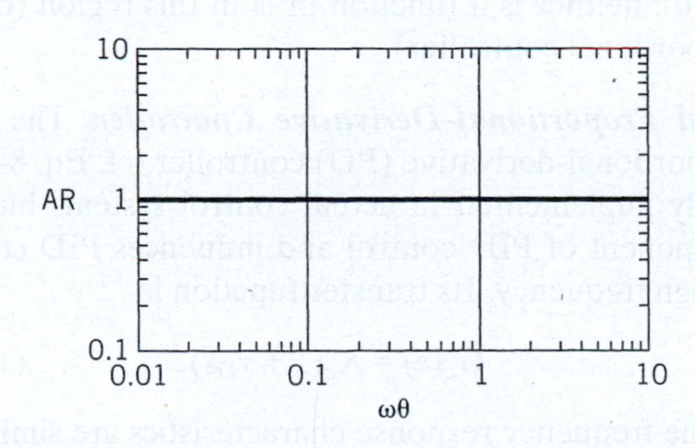

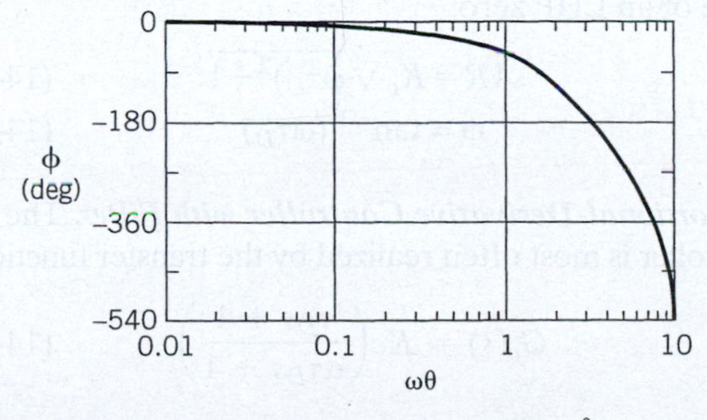

Figure 14.4 Bode diagram for a time delay, $e^{-\theta s}$.

EXAMPLE 14.3

Generate the Bode plot for the transfer function

$$G(s) = \frac{5(0.5s + 1)e^{-0.5s}}{(20s + 1)(4s + 1)}$$

where the time constants and time delay have units of minutes.

SOLUTION

The Bode plot is shown in Fig. 14.5. The steady-state gain (K = 5) is the value of AR when $\omega \to 0$. The phase angle at high frequencies is dominated by the time delay. The MATLAB listing for generating a Bode plot of the transfer function is shown in Table 14.1. In this listing the normalized AR is used (AR_N).

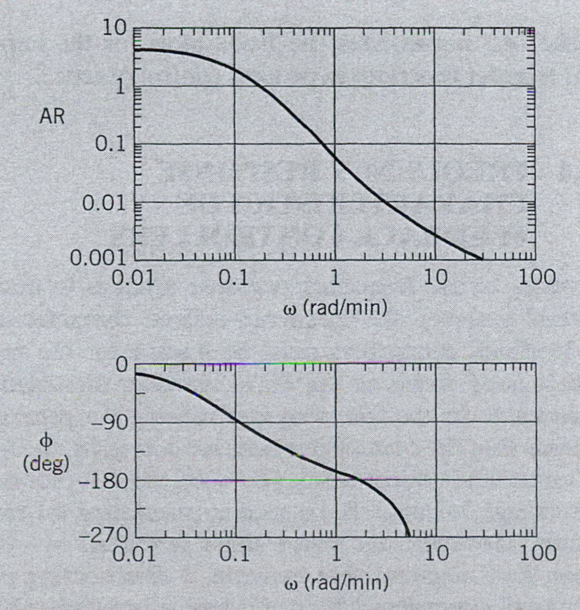

Figure 14.5 Bode plot of the transfer function in Example 14.3

Table 14.1 MATLAB Program to Calculate and Plot the Frequency Response of a Complex Process in Example 14.3

```
%Making a Bode plot for G=5 (0.5s+1)e^-0.5s/(20s+1)
%(4s+1)
close all
gain=5;
tdead=0. 5;
num=[0. 5 1];
den=[80 24 1];
G=tf (gain*num, den) %Define the system as a transfer
%function
points=500;           %Define the number of points
ww=logspace (-2, 2, points); %Frequencies to be evaluated
[mag, phase, ww]=bode (G,ww); % Generate numerical
%values for Bode plot
AR=zeros (points, 1); % Preallocate vectors for Amplitude
%Ratio and Phase Angle
PA=zeros (points, 1);
for i=1 : points
    AR(i)=mag (1,1,i)/gain; %Normalized AR
    PA(i)= phase (1,1,i) - ((180/pi) *tdead*ww(i));
end
figure
subplot (2,1,1)
loglog(ww, AR)
axis ([0.01 100 0.001 1])
title ('Frequency Response of a SOPTD with Zero')
ylabel('AR/K')
subplot (2,1,2)
semilogx(ww,PA)
axis ([0.01 100 -270 0])
ylabel('Phase Angle (degrees)')
xlabel('Frequency (rad/time)')
```

Table 14.2 summarizes the Bode plots for the important transfer functions in process control practice.

14.4 FREQUENCY RESPONSE CHARACTERISTICS OF FEEDBACK CONTROLLERS

In order to use frequency response analysis to design control systems, the frequency-related characteristics of feedback controllers must be known for the most widely used forms of the PID controller discussed in Chapter 8. In the following derivations, we generally assume that the controller is reverse-acting ($K_c > 0$). If a Controller is direct-acting ($K_c < 0$), the AR plot does not change, because $|K_c|$ is used in calculating the magnitude. However, the phase angle is shifted by –180° when K_c is negative. For example, a direct-acting proportional controller ($K_c < 0$) has a constant phase angle of –180°. Table 14.3 provides a summary of the frequency response characteristics of the most important industrial controllers.

Proportional Controller. Consider a proportional controller with positive gain

$$G_c(s) = K_c \tag{14-34}$$

In this case, $|G_c(j\omega)| = K_c$, which is independent of ω. Therefore,

$$AR = K_c \tag{14-35}$$

and

$$\phi = 0° \tag{14-36}$$

Proportional-Integral Controller. A proportional-integral (PI) controller has the transfer function,

$$G_c(s) = K_c\left(1 + \frac{1}{\tau_I s}\right) = K_c\left(\frac{\tau_I s + 1}{\tau_I s}\right) \tag{14-37}$$

Substituting $s = j\omega$ gives

$$G_c(j\omega) = K_c\left(1 + \frac{1}{\tau_I j\omega}\right) = K_c\left(1 - \frac{j}{\omega\tau_I}\right) \tag{14-38}$$

Thus, the amplitude ratio and phase angle are

$$AR = |G_c(j\omega| = K_c\sqrt{1 + \frac{1}{(\omega\tau_I)^2}} = K_c\frac{\sqrt{(\omega\tau_I)^2 + 1}}{\omega\tau_I} \tag{14-39}$$

$$\phi = -G_c(j\omega) = \tan^{-1}(-1/\omega\tau_I) = \tan^{-1}(\omega\tau_I) - 90° \tag{14-40}$$

Based on Eqs. 14-39 and 14-40, at low frequencies, the integral action dominates. As $\omega \to 0$, AR $\to \infty$, and $\phi \to -90°$. At high frequencies, AR $= K_c$ and $\phi = 0°$; neither is a function of ω in this region (cf. the proportional controller).

Ideal Proportional-Derivative Controller. The ideal proportional-derivative (PD) controller (cf. Eq. 8-11) is rarely implemented in actual control systems but is a component of PID control and influences PID control at high frequency. Its transfer function is

$$G_c(s) = K_c(1 + \tau_D s) \tag{14-41}$$

The frequency response characteristics are similar to those of an LHP zero:

$$AR = K_c\sqrt{\omega\tau_D)^2 + 1} \tag{14-42}$$

$$\phi = \tan^{-1}(\omega\tau_D) \tag{14-43}$$

Proportional-Derivative Controller with Filter. The PD controller is most often realized by the transfer function

$$G_c(s) = Kc\left(\frac{\tau_D s + 1}{\alpha\tau_D s + 1}\right) \tag{14-44}$$

where α has a value in the range 0.05 – 0.2. The frequency response for this controller is given by

$$AR = K_c\sqrt{\frac{(\omega\tau_D)^2+1}{(\alpha\omega\tau_D)^2+1}} \quad (14\text{-}45)$$

$$\phi = \tan^{-1}(\omega\tau_D) - \tan^{-1}(\alpha\omega\tau_D) \quad (14\text{-}46)$$

The pole in Eq. 14-44 bounds the high-frequency asymptote of the AR

$$\lim_{\omega\to\infty} \text{AR} = \lim_{\omega\to\infty} |G_c(j\omega)| = K_c/\alpha = 2/0.1 = 20 \quad (14\text{-}47)$$

Note that this feature actually is an advantage, because the ideal derivative action in (14-41) would amplify high-frequency input noise, due to its large value of AR in that region. In contrast, the PD controller with derivative filter exhibits a bounded AR in the high-frequency region. Because its numerator and denominator orders are both one, the high-frequency phase angle returns to zero.

Parallel PID Controller. The PID controller can be developed in both parallel and series forms, as discussed in Chapter 8. Either version exhibits features of both the PI and PD controllers. The simpler version is the following parallel form (cf. Eq. 8-14):

$$G_c(s) = K_c\left(1 + \frac{1}{\tau_I s} + \tau_D s\right) = K_c\left(\frac{1 + \tau_I s + \tau_I\tau_D s^2}{\tau_I s}\right) \quad (14\text{-}48)$$

Substituting $s = j\omega$ and rearranging gives

$$G_c(s) = K_c\left(1 + \frac{1}{j\omega\tau_I} + j\omega\tau_D\right) = K_c\left[1 + j\left(\omega\tau_D - \frac{1}{\omega\tau_I}\right)\right] \quad (14\text{-}49)$$

Figure 14.6 shows a Bode plot for a PID controller, with and without a derivative filter (see Table 8.1). The controller settings are $K_c = 2$, $\tau_I = 10$ min, $\tau_D = 4$ min, and $\alpha = 0.1$. The phase angle varies from –90° ($\omega \to 0$) to + 90° ($\omega \to \infty$).

By adjusting the values of τ_I and τ_D, one can prescribe the shape and location of the notch in the AR curve. Decreasing τ_I and increasing τ_D narrows the notch, whereas the opposite changes broaden it. Figure 14.6 indicates that the center of the notch is located at $\omega = 1/\sqrt{\tau_I\tau_D}$ where $\phi = 0°$ and $AR = K_c$. Varying K_c merely moves the amplitude ratio curve up or down, without affecting the width of the notch. Generally, the integral time τ_I should be larger than τ_D, typically $\tau_I \approx 4\tau_D$.

Series PID Controller. The simplest version of the series PID controller is

$$G_c(s) = K_c\left(\frac{\tau_1 s + 1}{\tau_1 s}\right)(\tau_D s + 1) \quad (14\text{-}50)$$

This controller transfer function can be interpreted as the product of the transfer functions for PI and PD controllers. Because the transfer function in (14-50) is physically unrealizable and amplifies high-frequency noise, we consider a more practical version that includes a derivative filter.

Series PID Controller with a Derivative Filter. The series controller with a derivative filter was described in Chapter 8.

$$G_c(s) = K_c\left(\frac{\tau_I s + 1}{\tau_I s}\right)\left(\frac{\tau_D s + 1}{\alpha\tau_D s + 1}\right) \quad (14\text{-}51)$$

where $0.05 < \alpha << 1.0$. A comparison of the amplitude ratios in Fig. 14.6 indicates that the AR for the controller without the derivative filter in (14-50) is unbounded at high frequencies, in contrast to the controller with the derivative filter (Eq. 14-51), which has a bounded AR at all frequencies. Consequently, the addition of the derivative filter makes the series PID controller less sensitive to high-frequency noise. For the typical value of $\alpha = 0.05$, Eq. 14-51 yields at high frequencies:

$$AR_{w\to\infty} = \lim_{\omega\to\infty} |G_c(j\omega)| = K_c/\alpha = 20\,K_c \quad (14\text{-}52)$$

When $\tau_D = 0$, the series PID controller with filter is the same as the PI controller of Eq. 14-52.

As a practical matter, it is possible to use the absolute value of K_c to calculate ϕ when designing closed-loop control systems, because stability considerations (see Chapter 11) require that $K_c < 0$ only when $K_v K_p K_m < 0$. This choice guarantees that the open-loop gain ($K_{OL} = K_c K_v K_p K_m$) will always be positive. Use of this convention conveniently yields $\phi = 0°$ for any proportional controller and, in general, eliminates the need to consider the –180° phase shift contribution of the controller gain.

14.5 NYQUIST DIAGRAMS

The Nyquist diagram is an alternative representation of frequency response information, a polar plot of $G(j\omega)$ in which frequency ω appears as an implicit parameter. The Nyquist diagram for a transfer function $G(s)$ can be constructed directly from $|G(j\omega)|$ and $\angle G(j\omega)$ for different values of ω. Alternatively, the Nyquist diagram can be constructed from the Bode diagram, because $AR = |G(j\omega)|$ and $\phi = \angle G(j\omega)$. Advantages of Bode plots are that frequency is plotted explicitly as the abscissa, and the log-log and semilog coordinate systems facilitate block multiplication. The Nyquist diagram, on the other hand, is more compact and is sufficient for many important analyses, for example, determining system stability (see Appendix J). Most of the recent interest in Nyquist diagrams has been in connection with designing multiloop controllers and for robustness (sensitivity) studies (Maciejowski, 1989; Skogestad and Postlethwaite, 2005). For single-loop controllers, Bode plots are used more often.

Table 14.2 Frequency Response Characteristics of Important Process Transfer Functions

Transfer Function	$G(s)$	$AR = \lvert G(j\omega)\rvert$	Plot of log AR_N vs. log ω	$\phi = \angle G(j\omega)$	Plot of ϕ vs. log ω
1. First-order	$\dfrac{K}{\tau s + 1}$	$\dfrac{K}{\sqrt{(\omega\tau)^2+1}}$	1; $\omega_b = \frac{1}{\tau}$; slope 1/1	$-\tan^{-1}(\omega\tau)$	0°; −45°; −90°; $\omega_b = \frac{1}{\tau}$
2. Integrator	$\dfrac{K}{s}$	$\dfrac{K}{\omega}$	slope 1/1	$-90°$	0°; −90°
3. Derivative	Ks	$K\omega$	slope 1/1	$+90°$	90°; 0°
4. Overdamped second-order	$\dfrac{K}{(\tau_1 s + 1)(\tau_2 s + 1)}$	$\dfrac{K}{\sqrt{(\omega\tau_1)^2+1}\sqrt{(\omega\tau_2)^2+1}}$	1; $\omega_{b1} = \frac{1}{\tau_1}$; $\omega_{b2} = \frac{1}{\tau_2}$; slope 1/2	$-\tan^{-1}(\omega\tau_1) - \tan^{-1}(\omega\tau_2)$	0°; −90°; −180°; $\omega_b = \frac{1}{\sqrt{\tau_1\tau_2}}$
5. Critically damped second-order	$\dfrac{K}{(\tau s + 1)^2}$	$\dfrac{K}{(\omega\tau)^2 + 1}$	1; $\omega_b = \frac{1}{\tau}$; slope 1/2	$-2\tan^{-1}(\omega\tau)$	0°; −90°; −180°; $\omega_b = \frac{1}{\tau}$

6. Underdamped second-order	$\dfrac{K}{\tau^2 s^2 + 2\zeta\tau s + 1}$	$\dfrac{K}{\sqrt{(1-(\omega\tau_1)^2)^2+(2\zeta\omega\tau)^2}}$	1; $\omega_b = \dfrac{1}{\tau}$; 1, 2	$-\tan^{-1}\left[\dfrac{2\zeta\omega\tau}{1-(\omega\tau)^2}\right]$	0°, −90°, −180°; $\omega_b = \dfrac{1}{\tau}$
7. Left-half plane (positive) zero	$K(\tau_a s + 1)$	$K\sqrt{(\omega\tau_a)^2 + 1}$	1; 1, 1; $\omega_b = \dfrac{1}{\tau_a}$	$+\tan^{-1}(\omega\tau_a)$	90°, 45°, 0°; $\omega_b = \dfrac{1}{\tau_a}$
8. Right-half plane (negative) zero	$-\tau_a s + 1$	$K\sqrt{(\omega\tau_a)^2 + 1}$	1; 1, 1; $\omega_b = \dfrac{1}{\tau_a}$	$-\tan^{-1}(\omega\tau_a)$	0°, −45°, −90°; $\omega_b = \dfrac{1}{\tau_a}$
9. Lead–lag unit ($\tau_a < \tau_1$)	$K\dfrac{\tau_a s + 1}{\tau_1 s + 1}$	$K\dfrac{\sqrt{(\omega\tau_a)^2 + 1}}{\sqrt{(\omega\tau_1)^2 + 1}}$	1; $\omega_{b1} = \dfrac{1}{\tau_1}$; 1, 1; $\omega_{ba} = \dfrac{1}{\tau_a}$	$+\tan^{-1}(\omega\tau_a) - \tan^{-1}(\omega\tau_1)$	0°, −90°
10. Lead–lag unit ($\tau_a > \tau_1$)	$K\dfrac{\tau_a s + 1}{\tau_1 s + 1}$	$K\dfrac{\sqrt{(\omega\tau_a)^2 + 1}}{\sqrt{(\omega\tau_1)^2 + 1}}$	1; $\omega_{b1} = \dfrac{1}{\tau_1}$; 1, 1; $\omega_{ba} = \dfrac{1}{\tau_a}$	$+\tan^{-1}(\omega\tau_a) - \tan^{-1}(\omega\tau_2)$	90°, 0°
11. Time delay	$Ke^{-\theta s}$	K	1	$-\omega\theta$	0°

Table 14.3 Frequency Response Characteristics of Important Controller Transfer Functions

Controller	$G_c(s)$	AR = $\lvert G_c(j\omega)\rvert$	Plot of log AR_N vs. log ω	$\phi = \angle G_c(j\omega)$	Plot of ϕ vs. log ω
1. PI	$K_c\left(1+\frac{1}{\tau_I s}\right)=K_c\left(\frac{\tau_I s+1}{\tau_I s}\right)$	$K_c\left(\frac{\sqrt{(\omega\tau_I)^2+1}}{\omega\tau_I}\right)$	K_c $\omega_b=\frac{1}{\tau_I}$	$\tan^{-1}(\omega\tau_I)-90°$	0°, −45°, −90°; $\omega_b=\frac{1}{\tau_I}$
2. Ideal PD	$K_c(\tau_D s+1)$	$K_c\sqrt{(\omega\tau_D)^2+1}$	K_c $\omega_b=\frac{1}{\tau_D}$	$\tan^{-1}(\omega\tau_D)$	90°, 45°, 0°; $\omega_b=\frac{1}{\tau_D}$
3. PD with Derivative Filter	$K_c\left(\frac{\tau_D s+1}{\alpha\tau_D s+1}\right)$	$K_c\sqrt{\frac{(\omega\tau_D)^2+1}{(\alpha\omega\tau_D)^2+1}}$	$\omega_b=\frac{1}{\alpha\tau_D}$ K_c $\omega_b=\frac{1}{\tau_D}$	$\tan^{-1}(\omega\tau_D)$ $-\tan^{-1}(\alpha\omega\tau_D)$	0°
4. Parallel PID	$K_c\left(1+\frac{1}{\tau_I s}+\tau_D s\right)$	$K_c\sqrt{\left(\omega\tau_D-\frac{1}{\omega\tau_I}\right)^2+1}$	$\frac{1}{\sqrt{\tau_I\tau_D}}$	$\tan^{-1}\left(\omega\tau_D-\frac{1}{\omega\tau_I}\right)$	90°, 0°, −90°; $\frac{1}{\sqrt{\tau_I\tau_D}}$
5. Series PID	$K_c\left(\frac{\tau_I s+1}{\tau_I s}\right)(\tau_D s+1)$	$K_c\frac{\sqrt{(\omega\tau_I)^2+1}}{\omega\tau_I}\sqrt{(\omega\tau_D)^2+1}$	$\frac{1}{\sqrt{\tau_I\tau_D}}$	$\tan^{-1}(\omega\tau_I)+\tan^{-1}(\omega\tau_D)-90°$	90°, 0°, −90°; $\frac{1}{\sqrt{\tau_I\tau_D}}$
6. Series PID with Filter	$K_c\left(\frac{\tau_I s+1}{\tau_I s}\right)\left(\frac{\tau_D s+1}{\alpha\tau_D s+1}\right)$	$K_c\left(\frac{\sqrt{(\omega\tau_I)^2+1}}{\omega\tau_I}\right)\sqrt{\frac{(\omega\tau_D)^2+1}{(\alpha\omega\tau_D)^2+1}}$	$\frac{1}{\sqrt{\tau_I\tau_D}}$	$\tan^{-1}(\omega\tau_I)+\tan^{-1}(\omega\tau_D)$ $-\tan^{-1}(\alpha\omega\tau_D)-90°$	0°, −90°; $\frac{1}{\sqrt{\tau_I\tau_D}}$

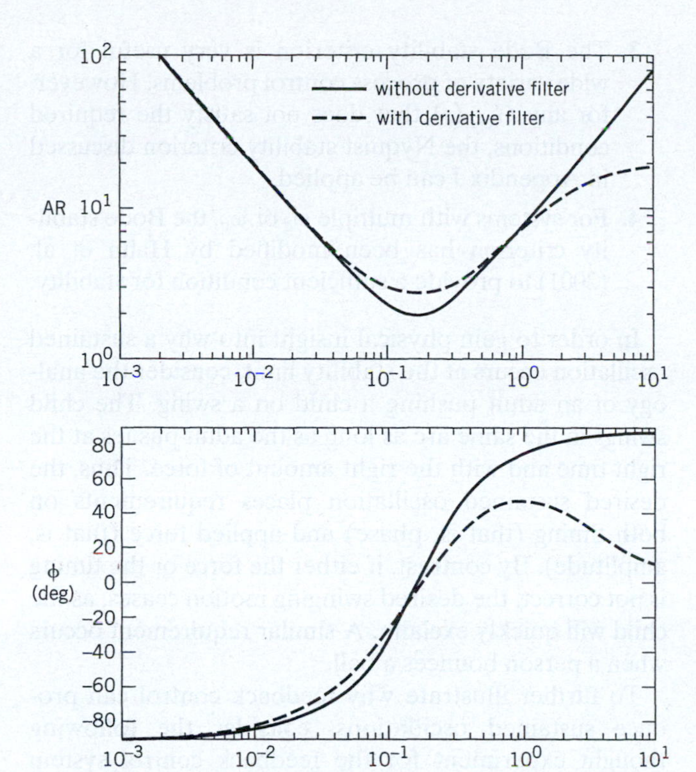

Figure 14.6. Bode plots of ideal parallel PID controller and series PID controller with derivative filter ($\alpha = 1$)

Ideal parallel: $G_c(s) = 2\left(1 + \dfrac{1}{10s} + 4s\right)$

Series with derivative filter:

$$G_c(s) = 2\left(\frac{10s+1}{10s}\right)\left(\frac{4s+1}{0.4s+1}\right)$$

14.6 BODE STABILITY CRITERION

The Bode stability criterion has two important advantages in comparison with the Routh stability criterion of Chapter 11:

1. It provides exact results for processes with time delays, while the Routh stability criterion provides only approximate results because of the polynomial approximation that must be substituted for the time delay.
2. The Bode stability criterion provides a measure of the relative stability rather than merely a yes or no answer to the question "Is the closed-loop system stable?"

Before considering the basis for the Bode stability criterion, it is useful to review the General Stability Criterion of Section 11.1: A feedback control system is stable if and only if all roots of the characteristic equation lie to the left of the imaginary axis in the complex plane.

Thus, the imaginary axis divides the complex plane into stable and unstable regions. Recall that the characteristic equation was defined in Chapter 11 as

$$1 + G_{OL}(s) = 0 \tag{14-53}$$

where the open-loop transfer function in (14-53) is $G_{OL}(s) = G_c(s)G_v(s)G_p(s)G_m(s)$.

The root locus diagrams of Section 11.5 (e.g., Fig. 11.27) show how the roots of the characteristic equation change as controller gain K_c changes. By definition, the roots of the characteristic equation are the numerical values of the complex variables that satisfy Eq. 14-53. Thus, each point on the root locus also satisfies (14-54), which is a rearrangement of (14-53):

$$G_{OL}(s) = -1 \tag{14-54}$$

The corresponding magnitude and argument are:

$$|G_{OL}(j\omega)| = 1 \text{ and } \angle G_{OL}(j\omega) = -180^\circ \tag{14-55}$$

In general, the ith root of the characteristic equation can be expressed as a complex number, $r_i = a_i \pm b_i j$. Note that complex roots occur as complex conjugate pairs. When a pair is located on the imaginary axis, the real part is zero ($a_i = 0$) and the closed-loop system is at the stability limit. As indicated in Chapter 11, this condition is referred to as marginal stability or conditional stability. When the closed-loop system is marginally stable and $b_i \neq 0$, the closed-loop response exhibits a sustained oscillation after a set-point change or a disturbance. Thus, the amplitude neither increases nor decreases. However, if K_c is increased slightly, the closed-loop system becomes unstable, because the complex roots on the imaginary axis move into the unstable region.

For a marginally stable system, with $b_i \neq 0$, the frequency of the sustained oscillation, ω_c, is given by $\omega_c = b_i$. This oscillatory behavior is caused by the pair of roots on the imaginary axis at $s = \pm\,\omega_c j$ (see Chapter 3). Substituting this expression for s into Eq. 14-55 gives the following expressions for a conditionally stable system:

$$AR_{OL}(\omega_c) = |G_{OL}(j\omega_c)| = 1 \tag{14-56}$$

$$\phi_{OL}(\omega_c) = \angle G_{OL}(j\omega_c) = -180^\circ \tag{14-57}$$

for some specific value of $\omega_c > 0$. Equations (14-56) and (14-57) provide the basis for both the Bode stability criterion discussed as follows.

Before stating the Bode stability criterion, we need to introduce two important definitions:

1. A critical frequency ω_c is defined to be a value of ω for which $\phi_{OL}(\omega) = -180^\circ$. This frequency is also referred to as a phase crossover frequency.
2. A gain crossover frequency ω_g is defined to be a value of ω for which $AR_{OL}(\omega) = 1$.

For a marginally stable system, $\omega_c = \omega_g$.

For many control problems, there is only a single ω_c and a single ω_g. But multiple values for ω_c can occur, as shown in Fig. 14.7. In this somewhat unusual situation, the closed-loop system is stable for two different ranges

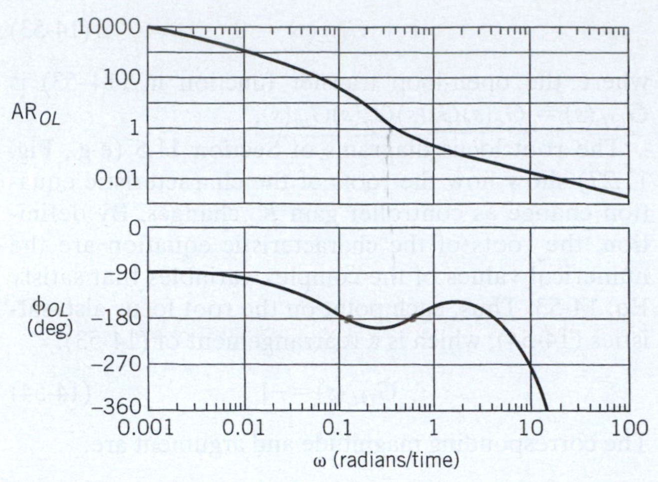

Figure 14.7 Bode plot exhibiting multiple critical frequencies

of the controller gain (Luyben and Luyben, 1997). Consequently, increasing the absolute value of K_c can actually improve the stability of the closed-loop system for certain ranges of K_c.

Next, we state one of the most important results of frequency response analysis, the Bode stability criterion. It allows the stability of a closed-loop system to be determined from the open-loop transfer function.

Bode Stability Criterion. *Consider an open-loop transfer function $G_{OL} = G_cG_vG_pG_m$ that is strictly proper (more poles than zeros) and has no poles located on or to the right of the imaginary axis, with the possible exception of a single pole at the origin. Assume that the open-loop frequency response has only a single critical frequency ω_c and a single gain crossover frequency ω_g. Then the closed-loop system is stable if $AR_{OL}(\omega_c) < 1$. Otherwise, it is unstable.*

Some of the important properties of the Bode stability criterion are:

1. It provides a necessary and sufficient condition for closed-loop stability, based on the properties of the open-loop transfer function.
2. Unlike the Routh stability criterion of Chapter 11, the Bode stability criterion is applicable to systems that contain time delays.
3. The Bode stability criterion is very useful for a wide variety of process control problems. However, for any $G_{OL}(s)$ that does not satisfy the required conditions, the Nyquist stability criterion discussed in Appendix J can be applied.
4. For systems with multiple ω_c or ω_g, the Bode stability criterion has been modified by Hahn et al. (2001) to provide a sufficient condition for stability.

In order to gain physical insight into why a sustained oscillation occurs at the stability limit, consider the analogy of an adult pushing a child on a swing. The child swings in the same arc as long as the adult pushes at the right time and with the right amount of force. Thus, the desired sustained oscillation places requirements on both timing (that is, phase) and applied force (that is, amplitude). By contrast, if either the force or the timing is not correct, the desired swinging motion ceases, as the child will quickly exclaim. A similar requirement occurs when a person bounces a ball.

To further illustrate why feedback control can produce sustained oscillations, consider the following thought experiment for the feedback control system shown in Fig. 14.8. Assume that the open-loop system is stable and that no disturbances occur ($D = 0$). Suppose that the set-point is varied sinusoidally at the critical frequency, $y_{sp}(t) = A \sin(\omega_c t)$, for a long period of time. Assume that, during this period, the measured output, y_m, is disconnected, so that the feedback loop is broken before the comparator. After the initial transient dies out, y_m will oscillate at the excitation frequency ω_c because the response of a linear system to a sinusoidal input is a sinusoidal output at the same frequency (see Section 14.2). Suppose that two events occur simultaneously: (i) the set-point is set to zero, and (ii) y_m is reconnected. If the feedback control system is marginally stable, the controlled variable y will then exhibit a sustained sinusoidal oscillation with amplitude A and frequency ω_c.

To analyze why this special type of oscillation occurs only when $\omega = \omega_c$, note that the sinusoidal signal E in Fig. 14.8 passes through transfer functions G_c, G_v, G_p, and G_m before returning to the comparator. In order

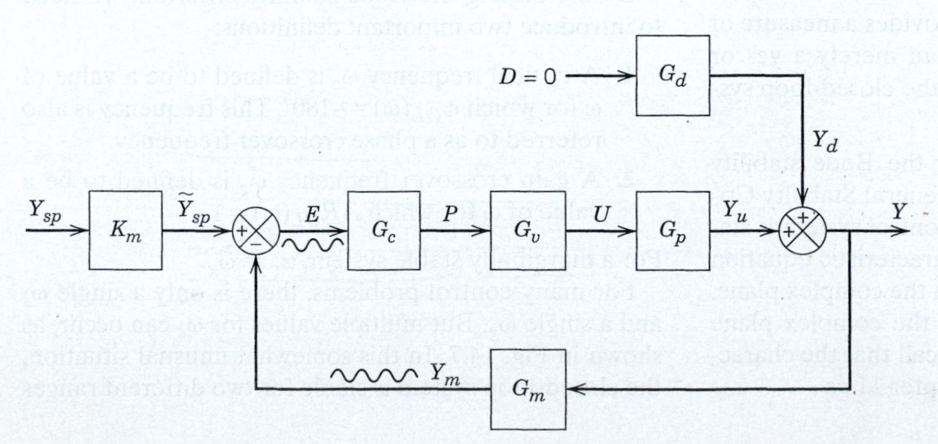

Figure 14.8 Sustained oscillation in a feedback control system

to have a sustained oscillation after the feedback loop is reconnected, signal Y_m must have the same amplitude as E and a 180° phase shift relative to E. Note that the comparator also provides a –180° phase shift because of its negative sign. Consequently, after Y_m passes through the comparator, it is in phase with E and has the same amplitude, A. Thus, the closed-loop system oscillates indefinitely after the feedback loop is closed because the conditions in Eqs. 14-56 and 14-57 are satisfied. But what happens if K_c is increased by a small amount? Then, $AR_{OL}(\omega_c)$ is greater than one, the oscillations grow, and the closed-loop system becomes unstable. In contrast, if K_c is reduced by a small amount, the oscillation is damped and eventually dies out.

EXAMPLE 14.4

A process has the third-order transfer function (time constant in minutes),

$$G_p(s) = \frac{2}{(0.5s+1)^3}$$

Also, $G_v = 0.1$ and $G_m = 10$. For a proportional controller, evaluate the stability of the closed-loop control system using the Bode stability criterion and three values of K_c: 1, 4, and 20.

SOLUTION

For this example,

$$G_{OL} = G_c G_v G_p G_m = (K_c)(0.1)\frac{2}{(0.5s+1)^3}(10) = \frac{2K_c}{(0.5s+1)^3}$$

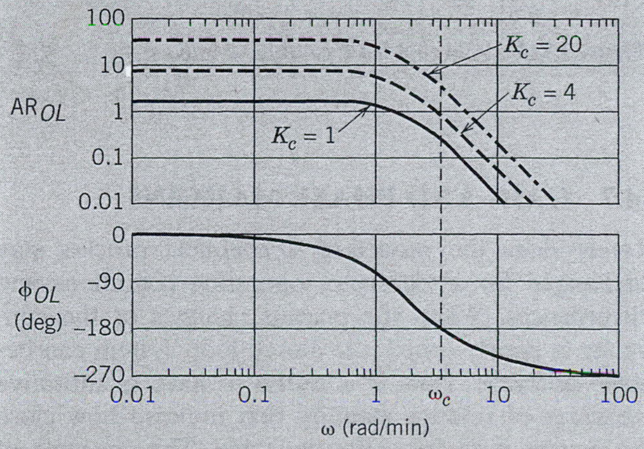

Figure 14.9 Bode plots for $G_{OL} = 2K_c/(0.5s + 1)^3$

Figure 14.9 shows a Bode plot of G_{OL} for three values of K_c. Note that all three cases have the same phase angle plot, because the phase lag of a proportional controller is zero for $K_c > 0$.

From the phase angle plot, we observe that $\omega_c = 3.46$ rad/min. This is the frequency of the sustained oscillation that occurs at the stability limit, as discussed previously. Next, we consider the amplitude ratio AR_{OL} for each value of K_c. Based on Figure 14.9, we make the following classifications:

K_c	AR_{OL} (for $\omega = \omega_c$)	Classification
1	0.25	Stable
4	1	Marginally stable
20	5	Unstable

In Section 12.5.1, the concept of the ultimate gain was introduced. For proportional-only control, the ultimate gain K_{cu} was defined to be the largest value of K_c that results in a stable closed-loop system. The value of K_{cu} can be determined graphically from a Bode plot for transfer function $G = G_v G_p G_m$. For proportional-only control, $G_{OL} = K_c G$. Because a proportional controller has zero phase lag, ω_c is determined solely by G. Also,

$$AR_{OL}(\omega) = K_c AR_G(\omega) \quad (14\text{-}58)$$

where AR_G denotes the amplitude ratio of G. At the stability limit, $\omega = \omega_c$, $AR_{OL}(\omega_c) = 1$ and $K_c = K_{cu}$. Substituting these expressions into (14-58) and solving for K_{cu} gives an important result:

$$K_{cu} = \frac{1}{AR_G(\omega_c)} \quad (14\text{-}59)$$

The stability limit for K_c can also be calculated for PI and PID controllers, as demonstrated by Example 14.5.

EXAMPLE 14.5

Consider PI control of an overdamped second-order process (time constants in minutes),

$$G_p(s) = \frac{5}{(s+1)(0.5s+1)}$$

$$G_m = G_v = 1$$

Determine the value of K_{cu}. Use a Bode plot to show that controller settings of $K_c = 0.4$ and $\tau_I = 0.2$ min produce an unstable closed-loop system. Find K_{cm}, the maximum value of K_c that can be used with $\tau_I = 0.2$ min and still have closed-loop stability. Show that $\tau_I = 1$ min results in a stable closed-loop system for all positive values of K_c.

SOLUTION

In order to determine K_{cu}, we let $G_c = K_c$. The open-loop transfer function is $G_{OL} = K_c G$ where $G = G_v G_p G_m$. Because a proportional controller does not introduce any phase lag, G and G_{OL} have identical phase angles. Consequently, the critical frequency can be determined graphically from the phase angle plot for G. However, curve a in Fig. 14.10 indicates that ω_c does not exist, because ϕ_{OL} is always greater than $-180°$. As a result, K_{cu} does not exist, and thus K_c does not have a stability limit.

Conversely, the addition of integral control action can produce closed-loop instability. Curve b in Fig. 14.10 indicates that an unstable closed-loop system occurs for $G_c(s) = 0.4\,(1 + 1/0.2s)$, because $AR_{OL} > 1$ when $\phi_{OL} = -180°$. To find K_{cm} for $\tau_I = 0.2$ min, we note that ω_c depends on τ_I but not on K_c, because K_c has no effect on ϕ_{OL}. For curve b in Fig. 14.10, $\omega_c = 2.2$ rad/min, and the corresponding amplitude ratio is $AR_{OL} = 1.38$. To find K_{cm}, multiply the current value of K_c by a factor, 1/1.38. Thus, $K_{cm} = 0.4/1.38 = 0.29$.

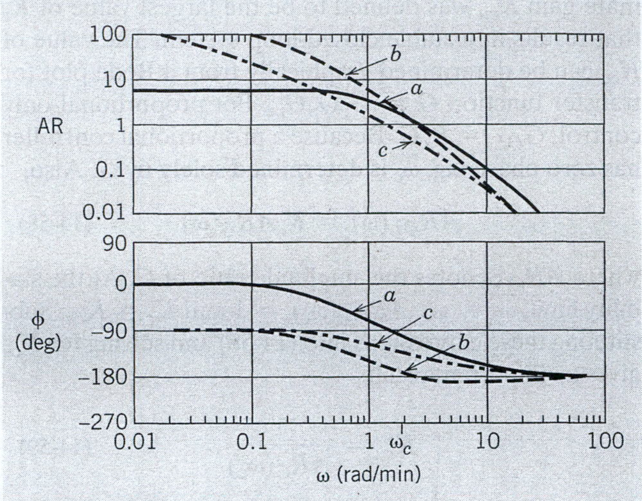

Figure 14.10 Bode plots for Example 14.5

Curve a: $G_p(s)$

Curve b: $G_{OL}(s) : G_c(s) = 0.4\left(1 + \frac{1}{0.2s}\right)$

Curve c: $G_{OL}(s)$: $G_c(s) = 0.4\left(1 + \frac{1}{s}\right)$

When τ_I is increased to 1 min, curve c in Fig. 14.10 results. Because curve c does not have a critical frequency, the closed-loop system is stable for all positive values of K_c.

EXAMPLE 14.6

Find the critical frequency for the following process and PID controller, assuming $G_v = G_m = 1$:

$$G_p(s) = \frac{e^{-0.3s}}{(9s + 1)(11s + 1)} \qquad G_c(s) = 20\left(1 + \frac{1}{2.5s} + s\right)$$

SOLUTION

Figure 14.7 shows the open-loop amplitude ratio and phase angle plots for G_{OL}. Note that the phase angle crosses $-180°$ at three points. Because there is more than one value of ω_c, the Bode stability criterion cannot be applied.

EXAMPLE 14.7

Evaluate the stability of the closed-loop system for:

$$G_p(s) = \frac{4e^{-s}}{5s + 1}$$

The time constant and time delay have units of minutes and,

$$G_v = 2, \; G_m = 0.25, \; G_c = K_c$$

Obtain ω_c and K_{cu} from a Bode plot.

SOLUTION

The Bode plot for G_{OL} and $K_c = 1$ is shown in Fig. 14.7. For $\omega_c = 1.69$ rad/min, $\phi_{OL} = -180°$, and $AR_{OL} = 0.235$. For $K_c = 1$, $AR_{OL} = AR_G$ and K_{cu} can be calculated from Eq. 14-59. Thus, $K_{cu} = 1/0.235 = 4.25$. Setting $K_c = 1.5\,K_{cu}$ gives $K_c = 6.38$. A larger value of K_c causes the closed-loop system to become unstable. Only values of K_c less than K_{cu} result in a stable closed-loop system.

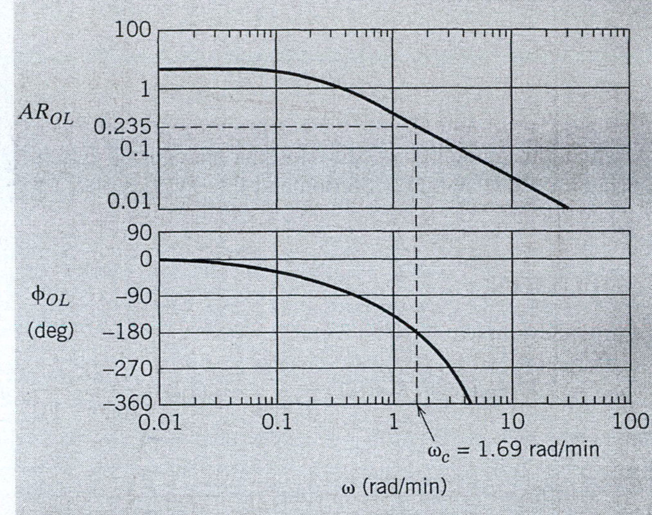

Figure 14.11 Bode plot for Example 14.6, $K_c = 1$

14.7 GAIN AND PHASE MARGINS

Rarely does the model of a chemical process stay unchanged for a variety of operating conditions and disturbances. When the process changes or the controller is poorly tuned, the closed-loop system can become unstable. Thus, it is useful to have quantitative measures of relative stability that indicate how close the system is to becoming unstable. The concepts of gain margin (GM) and phase margin (PM) provide useful metrics for relative stability.

Let AR_c be the value of the open-loop amplitude ratio at the critical frequency ω_c. Gain margin GM is defined as:

$$GM \triangleq \frac{1}{AR_c} \tag{14-60}$$

According to the Bode stability criterion, AR_c must be less than one for closed-loop stability. An equivalent stability requirement is that $GM > 1$. The gain margin provides a measure of relative stability, because it indicates how much any gain in the feedback loop component can increase before instability occurs. For example, if $GM = 2.1$, either process gain K_p or controller gain K_c could be doubled, and the closed-loop system would still be stable, although probably very oscillatory.

Next, we consider the phase margin. In Fig. 14.12, ϕ_g denotes the phase angle at the gain-crossover frequency ω_g where $AR_{OL} = 1$. Phase margin PM is defined as

$$PM \triangleq 180 + \phi_g \tag{14-61}$$

The phase margin also provides a measure of relative stability. In particular, it indicates how much additional time delay can be included in the feedback loop before instability will occur. Denote the additional time delay as $\Delta\theta_{max}$. For a time delay of $\Delta\theta_{max}$, the phase angle is $-\Delta\theta_{max}\,\omega$ (see Section 14.3.5). Thus, $\Delta\theta_{max}$ can be calculated from the following expression,

$$PM = \Delta\theta_{max}\,\omega_g\left(\frac{180^o}{\pi}\right) \tag{14-62}$$

or

$$\Delta\theta_{max} = \left(\frac{PM}{\omega_g}\right)\left(\frac{\pi}{180^o}\right) \tag{14-63}$$

where the $(\pi/180^o)$ factor converts PM from degrees to radians. Graphical representations of the gain and phase margins in a Bode plot are shown in Fig. 14.12.

The specification of phase and gain margins requires a compromise between performance and robustness. In general, large values of GM and PM correspond to sluggish closed-loop responses, whereas smaller values result in less sluggish, more oscillatory responses. The

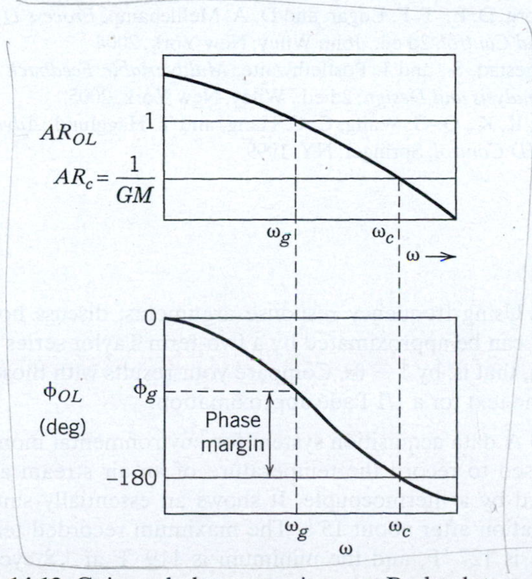

Figure 14.12 Gain and phase margins on a Bode plot

choices for GM and PM should also reflect model accuracy and the expected process variability.

Guideline. *In general, a well-tuned controller should have a gain margin between 1.7 and 4.0 and a phase margin between 30° and 45°.*

Recognize that these ranges are approximate and that it may not be possible to choose PI or PID controller settings that result in specified GM and PM values. Tan et al. (1999) have developed graphical procedures for designing PI and PID controllers that satisfy GM and PM specifications. The GM and PM concepts are easily evaluated when the open-loop system does not have multiple values of ω_c or ω_g. However, for systems with multiple ω_g, gain margins can be determined from Nyquist plots (Doyle et al., 1992).

EXAMPLE 14.8

For the FOPTD model of Example 14.7, calculate PID controller settings for the two tuning relations in Table 12.4:

(a) Ziegler-Nichols

(b) Tyreus-Luyben

Assume that the two PID controllers are implemented in the parallel form with a derivative filter ($\alpha = 0.1$) in Table 8.1. Plot the open-loop Bode diagram and determine the gain and phase margins for each controller.

For the Tyreus-Luyben settings, determine the maximum increase in the time delay $\Delta\theta_{max}$ that can occur while still maintaining closed-loop stability.

SOLUTION

From Example 14.7, the ultimate gain is $K_{cu} = 4.25$, and the ultimate period is $P_u = 2\pi/1.69 = 3.72$ min. Therefore, the PID controller settings are:

Controller Settings	K_c	τ_I (min)	τ_D (min)
Ziegler-Nichols	2.55	1.86	0.46
Tyreus-Luyben	1.91	8.18	0.59

The open-loop transfer function is:

$$G_{OL} = G_c G_v G_p G_m = G_c \frac{2e^{-s}}{5s+1}$$

Figure 14.13 shows the frequency response of G_{OL} for the two controllers. The gain and phase margins can be determined by inspection of the Bode diagram or by using the MATLAB command: margin.

Controller	GM	PM	ω_c (rad/min)
Ziegler-Nichols	1.6	40°	2.29
Tyreus-Luyben	1.8	76°	2.51

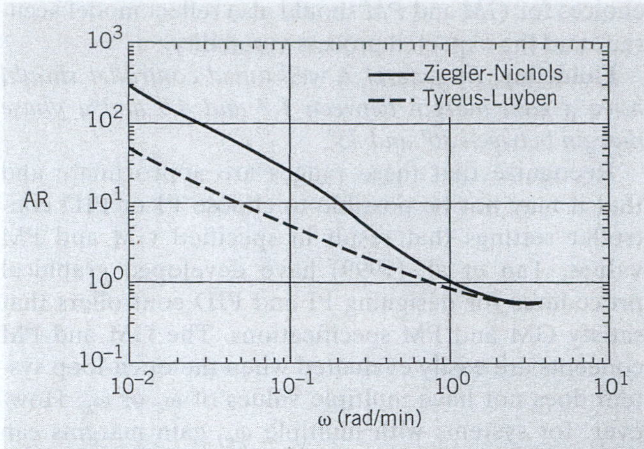

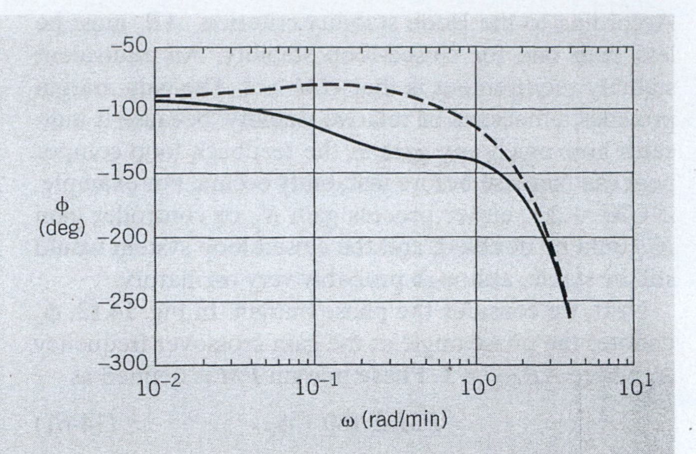

Figure 14.13 Comparison of G_{OL} Bode plots for Example 14.7

The Tyreus-Luyben controller settings are more conservative, due to the larger gain and phase margins. The value of $\Delta\theta_{max}$ is calculated from Eq. 14-63, and the information in the preceding table:

$$\Delta\theta_{max} = \frac{(76^\circ)(\pi \text{ rad})}{(0.79 \text{ rad/ min})(180^\circ)} = 1.7 \text{ min}$$

Thus, time delay θ can increase by as much as 70% and still maintain closed-loop stability.

SUMMARY

Frequency response techniques are powerful tools for the design and analysis of feedback control systems. The frequency response characteristics of a process, its amplitude ratio AR and phase angle, characterize the dynamic behavior of the process and can be plotted as functions of frequency in Bode diagrams. The Bode stability criterion provides exact stability results for a wide variety of control problems, including processes with time delays. It also provides a convenient measure of relative stability, such as gain and phase margins. Control system design involves trade-offs between control system performance and robustness. Modern control systems are typically designed using a model-based technique, such as those described in Chapter 12.

REFERENCES

Doyle, J. C., B. A. Francis, and A. R. Tannenbaum, *Feedback Control Theory*, Macmillan, NY, 1992.

Franklin, G. F., J. D. Powell, and A. Emami-Naeini, *Feedback Control of Dynamic Systems*, 5th ed., Prentice Hall, Upper Saddle River, NJ, 2005.

Hahn, J., T. Edison, and T. F. Edgar, A Note on Stability Analysis Using Bode Plots, *Chem. Eng. Educ.* **35**(3), 208 (2001).

Luyben, W. L., and M. L. Luyben, *Essentials of Process Control*, McGraw-Hill, NY, 1997, Chapter 11.

MacFarlane, A. G. J., The Development of Frequency Response Methods in Automatic Control, *IEEE Trans. Auto. Control*, **AC-24**, 250 (1979).

Maciejowski, J. M., *Multivariable Feedback Design*, Addison-Wesley, NY, 1989.

Ogunnaike, B. A., and W. H. Ray, *Process Dynamics, Modeling, and Control*, Oxford University Press, New York, 1993.

Seborg, D. E., T. F. Edgar, and D. A, Mellichamp, *Process Dynamics and Cantrol*, 2d ed., John Wiley, New York, 2004

Skogestad, S., and I. Postlethwaite, *Multivariable Feedback Design: Analysis and Design*, 2d ed., Wiley, New York, 2005.

Tan, K. K., Q.-G. Wang, C. C. Hang, and T. Hägglund, *Advances in PID Control*, Springer, NY, 1999.

EXERCISES

14.1 A heat transfer process has the following transfer function between a temperature T and an inlet flow rate q where the time constants have units of minutes:

$$\frac{T'(s)}{Q'(s)} = \frac{3(1-s)}{s(2s+1)}$$

If the flow rate varies sinusoidally with an amplitude of 2 L/min and a period of 0.5 min, what is the amplitude of the temperature signal after the transients have died out?

14.2 Using frequency response arguments, discuss how well $e^{-\theta s}$ can be approximated by a two-term Taylor series expansion, that is, by $1 - \theta s$. Compare your results with those given in the text for a 1/1 Padé approximation.

14.3 A data acquisition system for environmental monitoring is used to record the temperature of an air stream as measured by a thermocouple. It shows an essentially sinusoidal variation after about 15 s. The maximum recorded temperature is 127 °F, and the minimum is 119 °F at 1.8 cycles per min. It is estimated that the thermocouple has a time constant of 4.5 s. Estimate the actual maximum and minimum air temperatures.

14.4 A perfectly stirred tank is used to heat a flowing liquid. The dynamics of the system have been determined to be as shown in Fig. E14.4.

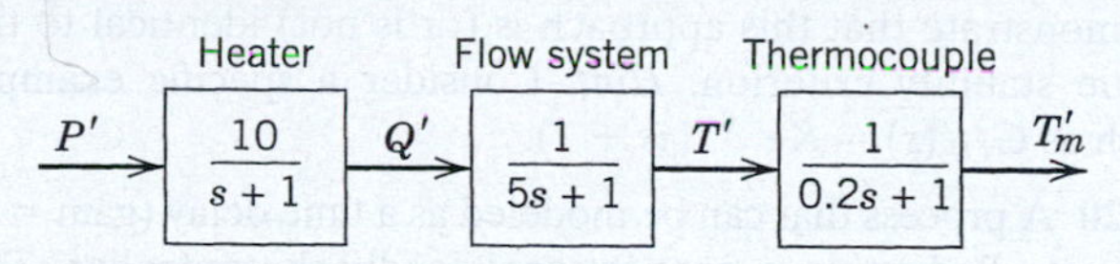

Figure E14.4

where:

P is the power applied to the heater
Q is the heating rate of the system
T is the actual temperature in the tank
T_m is the measured temperature

A test has been made with P' varied sinusoidally as

$$P' = 0.5 \sin 0.2t$$

For these conditions, the measured temperature is

$$T'_m = 3.464 \sin (0.2t + \phi)$$

Find a value for the maximum error bound between T' and T'_m if the sinusoidal input has been applied for a long time.

14.5 For each of the following transfer functions, develop both the amplitude ratio and phase angle of the Bode plot. Find AR and ϕ for each transfer function at values of $\omega = 0.1$, 1, and 10.

(a) $\dfrac{5}{(5s+1)(s+1)}$

(b) $\dfrac{5}{(5s+1)(s+1)^2}$

(c) $\dfrac{5(s+1)}{(5s+1)(0.2s+1)}$

(d) $\dfrac{5(-s+1)}{(5s+1)(0.2s+1)}$

(e) $\dfrac{5}{s(5s+1)}$

(f) $\dfrac{5(s+1)}{s(5s+1)(0.2s+1)}$

14.6 A second-order process transfer function is given by

$$G(s) = \frac{K(\tau_a s+1)}{\tau^2 s^2 + 2\zeta\tau s + 1}$$

(a) Find $|G|$ and $\angle G$ when $\zeta = 0.2$.

(b) Plot $|G(j\omega)|$ vs. $\omega\tau$ and $\angle G(j\omega)$ vs. $\omega\tau$ for the range $0.01 \le \omega\tau \le 100$ and values of $\tau_a/\tau = (0, 0.1, 1, 10)$.

Note that you can obtain the $\omega\tau$ plots by setting $\tau = 1$ and reparameterizing τ_a/τ accordingly.

14.7 Plot the Bode diagram ($0.1 \le \omega \le 100$) of the third-order transfer function,

$$G(s) = \frac{4}{(10s+1)(2s+1)(s+1)}$$

Find both the value of ω that yields a $-180°$ phase angle and the value of AR at that frequency.

14.8 Using MATLAB, plot the Bode diagram of the following transfer function:

$$G(s) = \frac{6(s+1)e^{-2s}}{(4s+1)(2s+1)}$$

Repeat for the situation where the time-delay term is replaced by a 1/1 Padé approximation.

14.9 Two thermocouples, one of them a known standard, are placed in an air stream whose temperature is varying sinusoidally. The temperature responses of the two thermocouples are recorded at a number of frequencies, with the phase angle between the two measured as shown below. The standard is known to have first-order dynamics and to have a time constant of 0.15 min when operating in the air stream. From the data, show that the unknown thermocouple also is first order and determine its time constant.

Frequency (cycles/min)	Phase Difference (deg)
0.05	4.5
0.1	8.7
0.2	16.0
0.4	24.5
0.8	26.5
1.0	25.0
2.0	16.7
4.0	9.2

14.10 Exercise 5.19 considered whether a two-tank liquid surge system provided better damping of step disturbances than a single-tank system with the same total volume. Reconsider this situation, this time with respect to sinusoidal disturbances; that is, determine which system better damps sinusoidal inputs of frequency ω. Does your answer depend on the value of ω?

14.11 For the process described in Exercise 6.5, plot the composite amplitude ratio and phase angle curves on a single Bode plot for each of the four cases of numerator dynamics. What can you conclude concerning the importance of the zero location for the amplitude and phase characteristics of this second-order system?

14.12 Develop expressions for the amplitude ratio as a function of ω of each of the two forms of the PID controller:

(a) The parallel controller of Eq. 8-14.

(b) The series controller of Eq. 8-15.

Display the results on a single plot along with asymptotic representations of each AR curve. You may assume that $\tau_I = 4\tau_D$ and $\alpha = 0.1$.

For what region(s) of ω are the differences significant? By how much?

14.13 You are using proportional control ($G_c = K_c$) for a process with $G_v = \dfrac{4}{2s+1}$ and $G_p = \dfrac{0.6}{50s+1}$ (time constants in secs). You have a choice of two measurements, both of which exhibit first-order dynamic behavior, $G_{m1} = \dfrac{2}{s+1}$ or

$G_{m2} = \dfrac{2}{20s+1}$. Can G_c be made unstable for either process? Which measurement is preferred for the best stability and performance properties? Why?

14.14 For the following statements, discuss whether they are always true, sometimes true, always false, or sometimes false. Cite evidence from this chapter.

(a) Increasing the controller gain speeds up the response for a set-point change.

(b) Increasing the controller gain always causes oscillation in the response to a setpoint change.

(c) Increasing the controller gain too much can cause instability in the control system.

(d) Selecting a large controller gain is a good idea in order to minimize offset.

14.15 Use arguments based on the phase angle in frequency response to determine if the following combinations of $G = G_v G_p G_m$ and G_c can become unstable for some value of K_c.

(a) $G = \dfrac{1}{(4s+1)(2s+1)}$ $\quad G_c = K_c$

(b) $G = \dfrac{1}{(4s+1)(2s+1)}$ $\quad G_c = K_c(1 + \dfrac{1}{5s})$

(c) $G = \dfrac{s+1}{(4s+1)(2s+1)}$ $\quad G_c = K_c\dfrac{(2s+1)}{s}$

(d) $G = \dfrac{1-s}{(4s+1)(2s+1)}$ $\quad G_c = K_c$

(e) $G = \dfrac{e^{-s}}{(4s+1)}$ $\quad G_c = K_c$

14.16 Plot the Bode diagram for a composite transfer function consisting of $G(s)$ in Exercise 14.8 multiplied by that of a parallel-form PID controller with $K_c = 0.21$, $\tau_I = 5$, and $\tau_D = 0.42$.

Repeat for a series PID controller with filter that employs the same settings. How different are these two diagrams? In particular, by how much do the two amplitude ratios differ when $\omega = \omega_c$?

14.17 For the process described by the transfer function

$$G(s) = \frac{12}{(8s+1)(2s+1)(0.4s+1)(0.1s+1)}$$

(a) Find two second-order-plus-time-delay models that approximate $G(s)$ and are of the form

$$\hat{G}(s) = \frac{Ke^{-\theta s}}{(\tau_1 s+1)(\tau_2 s+1)}$$

One of the approximate models can be found by using the method discussed in Section 6.3; the other, by visual inspection of the frequency responses of G and $\hat{G}$.

(b) Compare all three models (exact and approximate) in the frequency domain and also by plotting their impulse responses.

14.18 Obtain Bode plots for both the transfer function:

$$G(s) = \frac{10(2s+1)e^{-2s}}{(20s+1)(4s+1)(s+1)}$$

and a FOPTD approximation obtained using the method discussed in Section 6.3. What do you conclude about the accuracy of the approximation relative to the original transfer function?

14.19 In Chapter 11 we presented a stability analysis based on substituting $s = j\omega$ into the characteristic equation. Demonstrate that this approach is (or is not) identical to the Bode stability criterion. *Hint:* Consider a specific example such as $G_{OL}(s) = Ke^{-\theta s}/(\tau s + 1)$.

14.20 A process that can be modeled as a time delay (gain = 1) is controlled using a proportional feedback controller. The control valve and measurement device have negligible dynamics and steady-state gains of $K_v = 0.5$ and $K_m = 1$, respectively. After a small set-point change is made, a sustained oscillation occurs, which has period of 10 min.

(a) What controller gain is being used? Explain.

(b) How large is the time delay?

14.21 A Bode diagram for a process, valve, and sensor is shown in Fig. E14.21.

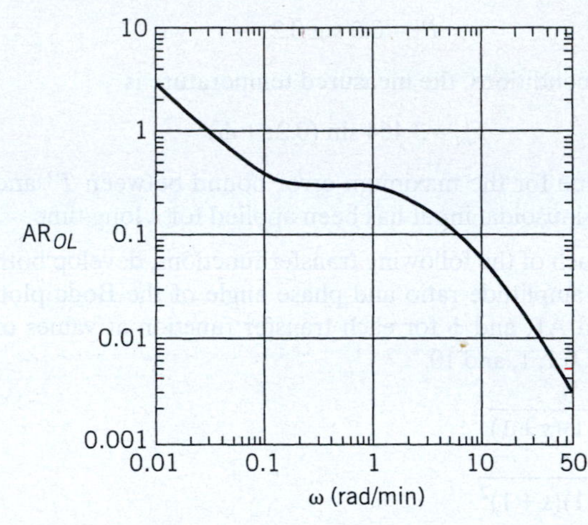

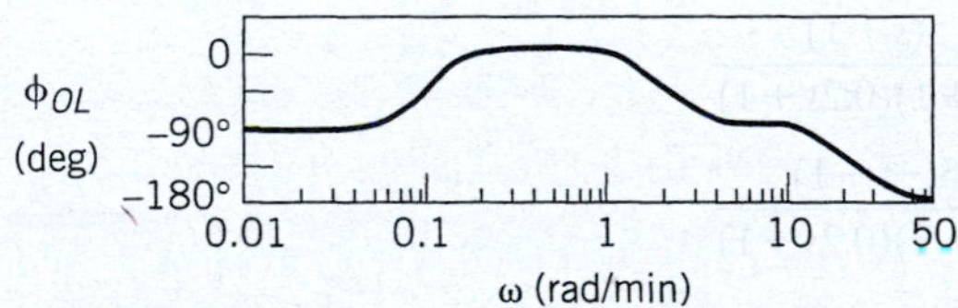

Figure E14.21

(a) Determine an approximate transfer function for this system.

(b) Suppose that a proportional controller is used and that a value of K_c is selected so as to provide a phase margin of 30°. What is the gain margin? What is the phase margin?

14.22 Consider the storage tank with sightglass in Fig. E14.22. The parameter values are $R_1 = 0.5$ min/ft², $R_2 = 2$ min/ft²,

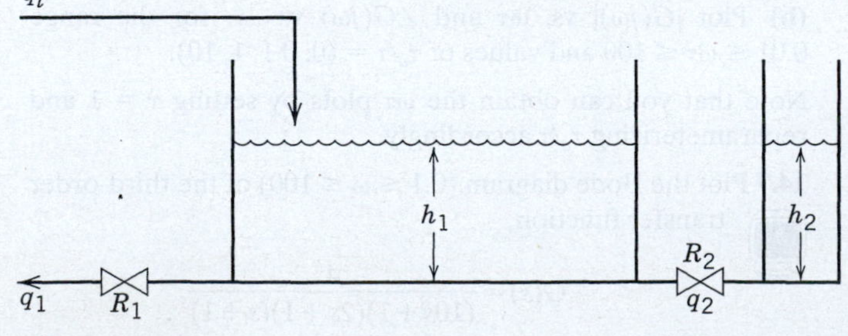

Figure E14.22

$A_1 = 10\text{ ft}^2$, $K_v = 2.5$ cfm/mA, $A_2 = 0.8\text{ ft}^2$, $K_m = 1.5$ mA/ft, and $\tau_m = 0.5$ min.

(a) Suppose that R_2 is decreased to 0.5 min/ft². Compare the old and new values of the ultimate gain and the critical frequency. Would you expect the control system performance to become better or worse? Justify your answer.

(b) If PI controller settings are calculated using the Ziegler-Nichols rules, what are the gain and phase margins? Assume $R_2 = 2$ min/ft.

14.23 A process (including valve and sensor-transmitter) has the approximate transfer function, $G(s) = 2e^{-0.2s}/(s+1)$ with time constant and time delay in minutes. Determine PI controller settings and the corresponding gain margins by two methods:

(a) Direct synthesis ($\tau_c = 0.3$ min).

(b) Phase margin = 40° (assume $\tau_I = 0.5$ min).

(c) Simulate these two control systems for a unit step change in set point. Which controller provides the better performance?

14.24 Consider the feedback control system in Fig. 14.8, and the following transfer functions:

$$G_c = K_c\left(\frac{2s+1}{0.1s+1}\right) \qquad G_v = \frac{2}{0.5s+1}$$

$$G_p = \frac{0.4}{s(5s+1)} \qquad G_d = \frac{3}{5s+1}$$

$$G_m = 1$$

(a) Plot a Bode diagram for the open-loop transfer function.

(b) Calculate the value of K_c that provides a phase margin of 30°.

(c) What is the gain margin when $K_c = 10$?

14.25 Frequency response data for a process are tabulated below. These results were obtained by introducing a sinusoidal change in the controller output (under manual control) and recording the measured response of the controlled variable. This procedure was repeated for various frequencies.

(a) If the PI controller is adjusted so that $\tau_I = 0.4$ min, what value of K_c will result in a phase margin of 45°?

(b) If the controller settings in part (a) are used, what is the gain margin?

ω (rad/min)	AR	ϕ (deg)
0.1	2.40	−3
0.10	1.25	−12
0.20	0.90	−22
0.5	0.50	−41
1.0	0.29	−60
2.0	0.15	−82
5.0	0.05	−122
10.0	0.017	−173
15.0	0.008	−230

14.26 For the process in Exercise 14.23, the measurement is to be filtered using a noise filter with transfer function $G_F(s) = 1/(0.1s+1)$. Would you expect this change to result in better or worse control system performance? Compare the ultimate gains and critical frequencies with and without the filter. Justify your answer.

14.27 The block diagram of a conventional feedback control system contains the following transfer functions:

$$G_c = K_c\left(1 + \frac{1}{5s}\right) \qquad G_v = 1$$

$$G_m = \frac{1}{s+1}$$

$$G_p = G_d = \frac{5e^{-2s}}{10s+1}$$

(a) Plot the Bode diagram for the open-loop transfer function.

(b) For what values of K_c is the system stable?

(c) If $K_c = 0.2$, what is the phase margin?

(d) What value of K_c will result in a gain margin of 1.7?

14.28 The dynamic behavior of the heat exchanger shown in Fig. E14.28 can be described by the following transfer functions (H. S. Wilson and L. M. Zoss, *ISA J.*, 9, 59 (1962)):

Process:

$$\frac{T'}{W_s'} = \frac{2\ ^\circ\text{F/lb min}}{(0.432s+1)(0.017s+1)}$$

Control valve:

$$\frac{X'}{P'} = \frac{0.047\text{ in/psi}}{0.083s+1} \qquad \frac{W_s'}{X'} = 112\ \frac{\text{lb}}{\text{min in}}$$

Temperature sensor-transmitter:

$$\frac{P'}{T'} = \frac{0.12\text{ psi/}^\circ\text{F}}{0.024s+1}$$

The valve lift x is measured in inches. Other symbols are defined in Fig. E14.27.

(a) Find the Ziegler-Nichols settings for a PI controller.

(b) Calculate the corresponding gain and phase margins.

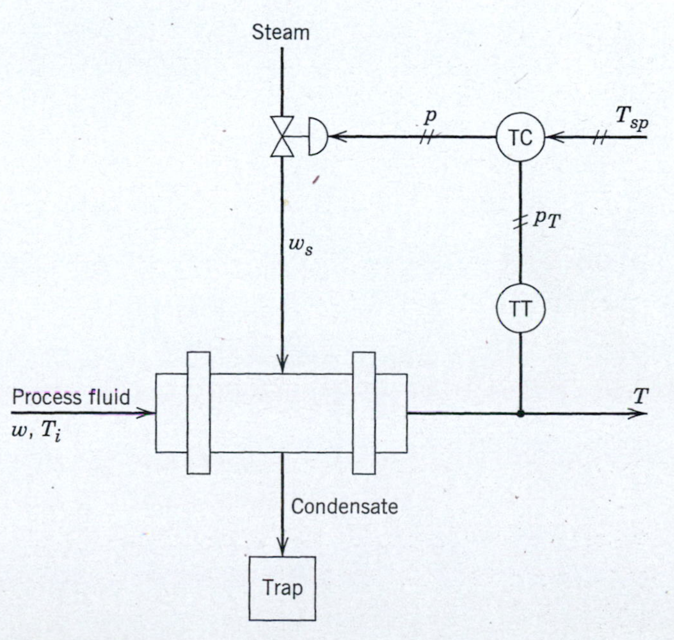

Figure E14.28

14.29 Consider a standard feedback control system with the following transfer functions:

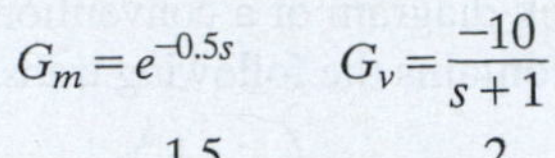

$$G_m = e^{-0.5s} \qquad G_v = \frac{-10}{s+1}$$

$$G_p = \frac{1.5}{10s+1} \qquad G_d = \frac{2}{6s+1}$$

(a) Plot the Bode diagram for the transfer function, $G = G_v G_p G_m$.

(b) Design a PI controller for this process and sketch the asymptotic Bode diagram for the open-loop transfer function, $G_{OL} = G_c G$.

(c) Analyze the stability of the resulting feedback control system.

(d) Suppose that under *open-loop* conditions, a sinusoidal set-point change, $y_{sp}(t) = 1.5 \sin(0.5t)$, is introduced. What is the amplitude of the measured output signal $y_m(t)$ that is also sinusoidal in nature?

(e) Repeat the same analysis for closed-loop conditions.

(f) Compare and discuss your results of parts (d) and (e).

14.30 Hot and cold liquids are mixed at the junction of two pipes. The temperature of the resulting mixture is to be controlled using a control valve on the hot stream. The dynamics of the mixing process, control valve, and temperature sensor/transmitter are negligible and the sensor-transmitter gain is 6 mA/mA. Because the temperature sensor is located well downstream of the junction, an 8 s time delay occurs. There are no heat losses/gains for the downstream pipe.

(a) Draw a block diagram for the closed-loop system.

(b) Determine the Ziegler-Nichols settings (continuous cycling method) for both PI and PID controllers.

(c) For each controller, simulate the closed-loop responses for a unit step change in set point.

(d) Does the addition of derivative control action provide a significant improvement? Justify your answer.

14.31 Consider the control problem of Exercise 14.28 and a PI controller with $K_c = 5$ and $\tau_I = 0.3$ min.

(a) Plot the Bode diagram for the open-loop system.

(b) Determine the gain margin from the Bode plot.

14.32 Determine if the following processes can be made unstable by increasing the gain of a proportional controller to a sufficiently large value:

(a) $G_p G_v G_m = \dfrac{2}{s+1} \qquad G_c = K_c$

(b) $G_p G_v G_m = \dfrac{3}{(s+1)(2s+1)} \qquad G_c = K_c$

(c) $G_p G_v G_m = \dfrac{4}{(s+1)(2s+1)(3s+1)} \qquad G_c = K_c$

(d) $G_p G_v G_m = \dfrac{5e^{-s}}{2s+1} \qquad G_c = K_c$

14.33 (a) Using the process, sensor, and valve transfer functions in Exercise 11.21, find the ultimate controller gain K_{cu} using a Bode plot. Using simulation, verify that values of $K_c > K_{cu}$ cause instability.

(b) Next fit a FOPTD model to G and tune a PI controller for a set-point change. What is the gain margin for the controller?

14.34 Two engineers are analyzing step-test data from a bioreactor. Engineer A says that the data indicate a second-order overdamped process, with time constants of 2 and 6 min but no time delay. Engineer B insists that the best fit is a FOPTD model, with $\tau = 7$ min and $\theta = 1$ min. Both engineers claim a proportional controller can be set at a large value for K_c to control the process and that stability is no problem. Based on their models, who is right, who is wrong, and why? Use a frequency-response argument.

Chapter 15

Feedforward and Ratio Control

CHAPTER CONTENTS

In Chapter 8 it was emphasized that feedback control is an important technique that is widely used in the process industries. Its main advantages are as follows.

1. Corrective action occurs as soon as the controlled variable deviates from the set point, regardless of the source and type of disturbance.
2. Feedback control requires minimal knowledge about the process to be controlled; in particular, a mathematical model of the process is *not* required, although it can be very useful for control system design.
3. The ubiquitous PID controller is both versatile and robust. If process conditions change, re-tuning the controller usually produces satisfactory control.

However, feedback control also has certain inherent disadvantages:

1. No corrective action is taken until after a deviation in the controlled variable occurs. Thus, *perfect control*, where the controlled variable does not deviate from the set point during disturbance or set-point changes, is theoretically impossible.
2. It does not provide predictive control action to compensate for the effects of known or measurable disturbances.
3. It may not be satisfactory for processes with large time constants and/or long time delays. If large and frequent disturbances occur, the process may operate continuously in a transient state and never attain the desired steady state.
4. In some situations, the controlled variable cannot be measured on-line, so feedback control is not feasible.

For situations in which feedback control by itself is not satisfactory, significant improvement can be achieved by adding feedforward control. But feedforward control requires that the disturbances be measured (or estimated) on-line.

In this chapter, we consider the design and analysis of feedforward control systems. We begin with an overview of feedforward control. Then ratio control, a special type of feedforward control, is introduced. Next, design techniques for feedforward controllers are developed based on either steady-state or dynamic models. Then alternative configurations for combined feedforward–feedback control systems are considered. This chapter concludes with a section on tuning feedforward controllers.

15.1 INTRODUCTION TO FEEDFORWARD CONTROL

The basic concept of feedforward control is to measure important disturbance variables and take corrective action before they upset the process. In contrast, a feedback controller does not take corrective action until after the disturbance has upset the process and generated a nonzero error signal. Simplified block diagrams for feedforward and feedback control are shown in Fig. 15.1.

Feedforward control has several disadvantages:

1. The disturbance variables must be measured on-line. In many applications, this is not feasible.
2. To make effective use of feedforward control, at least an approximate process model should be available. In particular, we need to know how the controlled variable responds to changes in both the disturbance variable and the manipulated variable. The quality of feedforward control depends on the accuracy of the process model.
3. Ideal feedforward controllers that are theoretically capable of achieving perfect control may not be physically realizable. Fortunately, practical approximations of these ideal controllers can provide very effective control.

Feedforward control was not widely used in the process industries until the 1960s (Shinskey, 1996). Since then, it has been applied to a wide variety of processes that include boilers, evaporators, solids dryers, direct-fired heaters, and waste neutralization plants (Shinskey et al., 1995). However, the basic concept is much older and was applied as early as 1925 in the three-element level control system for boiler drums. We will use this control application to illustrate the use of feedforward control.

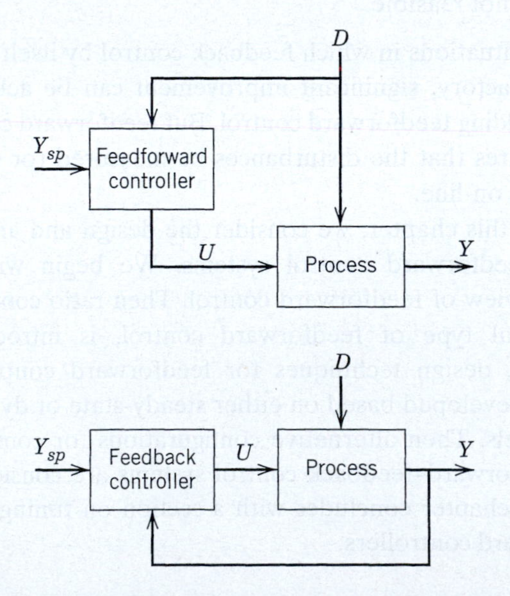

Figure 15.1 Simplified block diagrams for feedforward and feedback control.

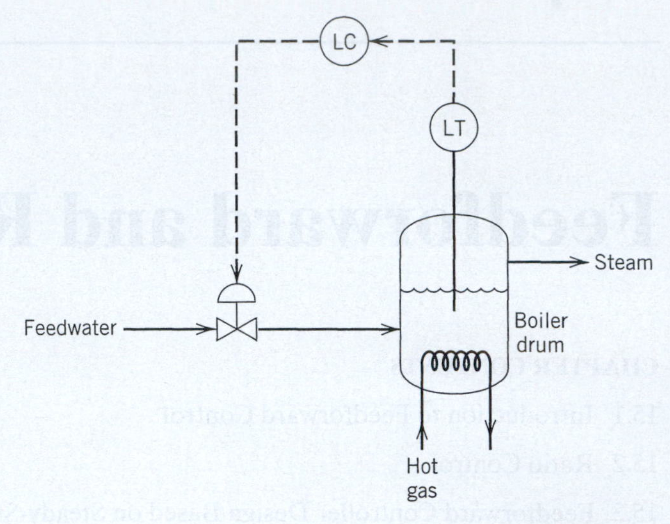

Figure 15.2 Feedback control of the liquid level in a boiler drum.

A boiler drum with a conventional feedback control system is shown in Fig. 15.2. The level of the boiling liquid is measured and used to adjust the feedwater flow rate. This control system tends to be quite sensitive to rapid changes in the disturbance variable, steam flow rate, as a result of the small liquid capacity of the boiler drum. Rapid disturbance changes are produced by steam demands made by downstream processing units. Another difficulty is that large controller gains cannot be used because level measurements exhibit rapid fluctuations for boiling liquids. Thus a high controller gain would tend to amplify the measurement noise and produce unacceptable variations in the feedwater flow rate.

The feedforward control scheme in Fig. 15.3 can provide better control of the liquid level. The steam flow rate is measured, and the feedforward controller

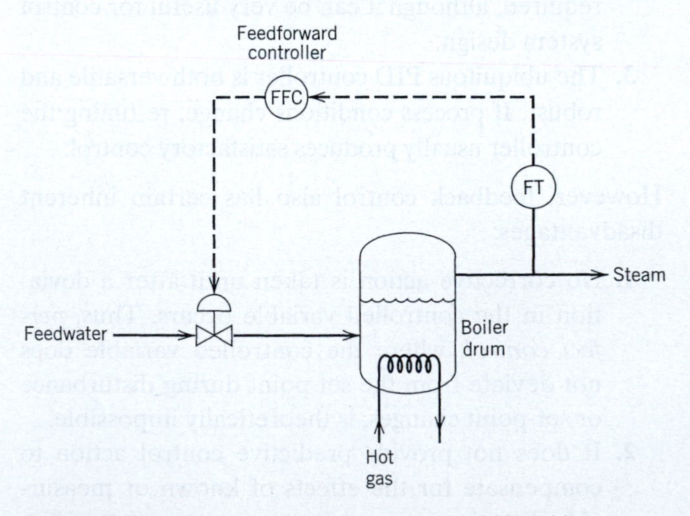

Figure 15.3 Feedforward control of the liquid level in a boiler drum.

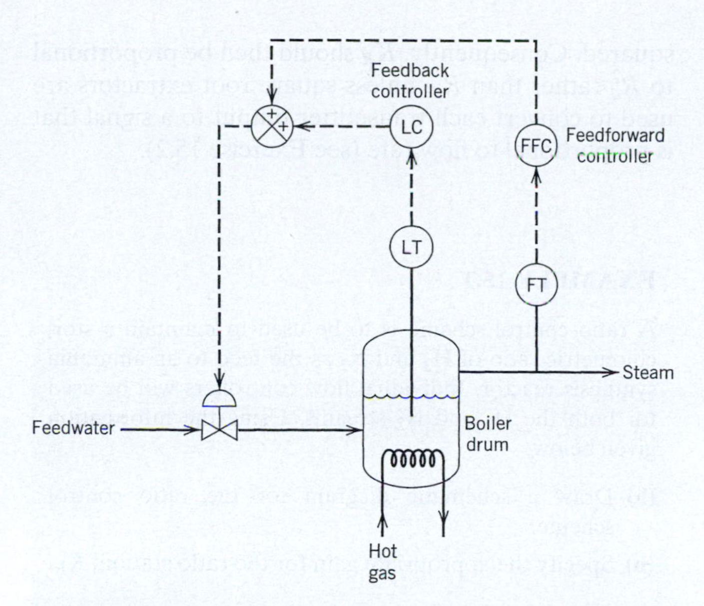

Figure 15.4 Feedforward–feedback control of the boiler drum level.

adjusts the feedwater flow rate so as to balance the steam demand. Note that the controlled variable, liquid level, is not measured. As an alternative, steam pressure could be measured instead of steam flow rate.

Feedforward control can also be used advantageously for level control problems where the objective is surge control (or averaging control), rather than tight level control. For example, the input streams to a surge tank will be intermittent if they are effluent streams from batch operations, but the tank exit stream can be continuous. Special feedforward control methods have been developed for these batch-to-continuous transitions to balance the surge capacity requirement for the measured inlet flow rates with the surge control objective of gradual changes in the tank exit stream (Blevins et al., 2003).

In practical applications, feedforward control is normally used in combination with feedback control. Feedforward control is used to reduce the effects of measurable disturbances, while *feedback trim* compensates for inaccuracies in the process model, measurement errors, and unmeasured disturbances. The feedforward and feedback controllers can be combined in several different ways, as will be discussed in Section 15.6. A typical configuration is shown in Fig. 15.4, where the outputs of the feedforward and feedback controllers are added together and the combined signal is sent to the control valve.

15.2 RATIO CONTROL

Ratio control is a special type of feedforward control that has had widespread application in the process industries. Its objective is to maintain the ratio of two process variables at a specified value. The two variables

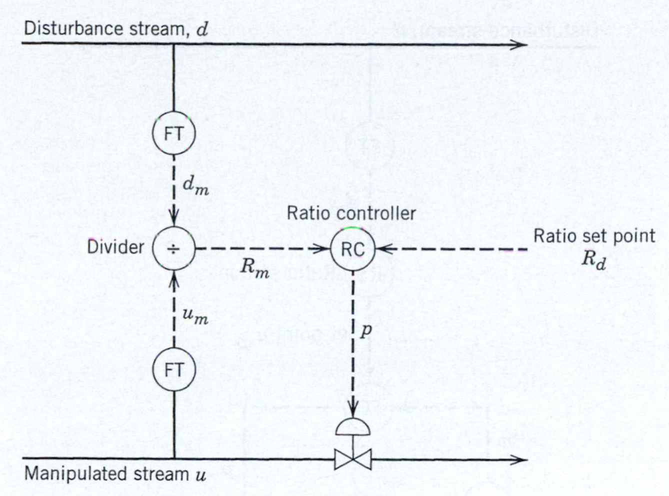

Figure 15.5 Ratio control, Method I.

are usually flow rates, a manipulated variable u and a disturbance variable d. Thus, the ratio

$$R = \frac{u}{d} \tag{15-1}$$

is controlled rather than the individual variables. In Eq. 15-1, u and d are physical variables, not deviation variables.

Typical applications of ratio control include (1) specifying the relative amounts of components in blending operations, (2) maintaining a stoichiometric ratio of reactants to a reactor, (3) keeping a specified reflux ratio for a distillation column, and (4) holding the fuel-air ratio to a furnace at the optimum value.

Ratio control can be implemented in two basic schemes. For Method I in Fig. 15.5, the flow rates for both the disturbance stream and the manipulated stream are measured, and the measured ratio, $R_m = u_m/d_m$, is calculated. The output of the divider element is sent to a ratio controller (RC) that compares the calculated ratio R_m to the desired ratio R_d and adjusts the manipulated flow rate u accordingly. The ratio controller is typically a PI controller with the desired ratio as its set point.

The main advantage of Method I is that the measured ratio R_m is calculated. A key disadvantage is that a divider element must be included in the loop, and this element makes the process gain vary in a nonlinear fashion. From Eq. 15-1, the process gain

$$K_p = \left(\frac{\partial R}{\partial u}\right)_d = \frac{1}{d} \tag{15-2}$$

is inversely related to the disturbance flow rate d. Because of this significant disadvantage, the preferred scheme for implementing ratio control is Method II, which is shown in Fig. 15.6.

In Method II the flow rate of the disturbance stream is measured and transmitted to the ratio station (RS),

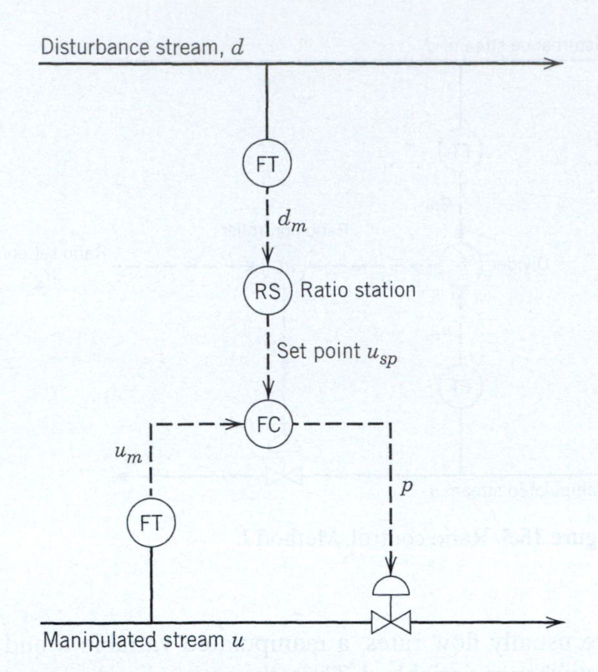

Figure 15.6 Ratio control, Method II.

which multiplies this signal by an adjustable gain, K_R, whose value is the desired ratio. The output signal from the ratio station is then used as the set point u_{sp} for the flow controller, which adjusts the flow rate of the manipulated stream, u. The chief advantage of Method II is that the process gain remains constant. Note that disturbance variable d is measured in both Methods I and II. Thus, ratio control is, in essence, a simple type of feedforward control.

A disadvantage of both Methods I and II is that the desired ratio may not be achieved during transient conditions as a result of the dynamics associated with the flow control loop for u. Thus, after a step change in disturbance d, the manipulated variable will require some time to reach its new set point, u_{sp}. Fortunately, flow control loops tend to have short settling times and this transient mismatch between u and d is usually acceptable. For situations where it is not, a modified version of Method II proposed by Hägglund (2001) can be applied.

Regardless of how ratio control is implemented, the process variables must be scaled appropriately. For example, in Method II the gain setting for the ratio station K_d must take into account the spans of the two flow transmitters. Thus, the correct gain for the ratio station is

$$K_R = R_d \frac{S_d}{S_u} \tag{15-3}$$

where R_d is the desired ratio, and S_u and S_d are the spans of the flow transmitters for the manipulated and disturbance streams, respectively. If orifice plates are used with differential-pressure transmitters, then the transmitter output is proportional to the flow rate squared. Consequently, K_R should then be proportional to R_d^2 rather than R_d, unless square root extractors are used to convert each transmitter output to a signal that is proportional to flow rate (see Exercise 15.2).

EXAMPLE 15.1

A ratio control scheme is to be used to maintain a stoichiometric ratio of H_2 and N_2 as the feed to an ammonia synthesis reactor. Individual flow controllers will be used for both the H_2 and N_2 streams. Using the information given below,

(a) Draw a schematic diagram for the ratio control scheme.

(b) Specify the appropriate gain for the ratio station, K_R.

Available information:

(i) The electronic flow transmitters have built-in square root extractors. The spans of the flow transmitters are 30 L/min for H_2 and 15 L/min for N_2.

(ii) The control valves have pneumatic actuators.

(iii) Each required current-to-pressure (I/P) transducer has a gain of 0.75 psi/mA.

(iv) The ratio station is an electronic instrument with 4–20 mA input and output signals.

SOLUTION

The stoichiometric equation for the ammonia synthesis reaction is

$$3H_2 + N_2 \rightleftarrows 2NH_3$$

In order to introduce a feed mixture in stoichiometric proportions, the ratio of the molar flow rates (H_2/N_2) should be 3 : 1. For the sake of simplicity, we assume that the ratio of the molar flow rates is equal to the ratio of the volumetric flow rates. But in general, the volumetric flow rates also depend on the temperature and pressure of each stream (cf. the ideal gas law).

(a) The schematic diagram for the ammonia synthesis reaction is shown in Fig. 15.7. The H_2 flow rate is considered to be the disturbance variable, although this choice is arbitrary, because both the H_2 and N_2 flow rates are controlled. Note that the ratio station is merely a device with an adjustable gain. The input signal to the ratio station is d_m, the measured H_2 flow rate. Its output signal u_{sp} serves as the set point for the N_2 flow control loop. It is calculated as $u_{sp} = K_R d_m$.

(b) From the stoichiometric equation, it follows that the desired ratio is $R_d = u/d = 1/3$. Substitution into Eq. 15-3 gives

$$K_R = \left(\frac{1}{3}\right)\left(\frac{30 \text{ L/min}}{15 \text{ L/min}}\right) = \frac{2}{3}$$

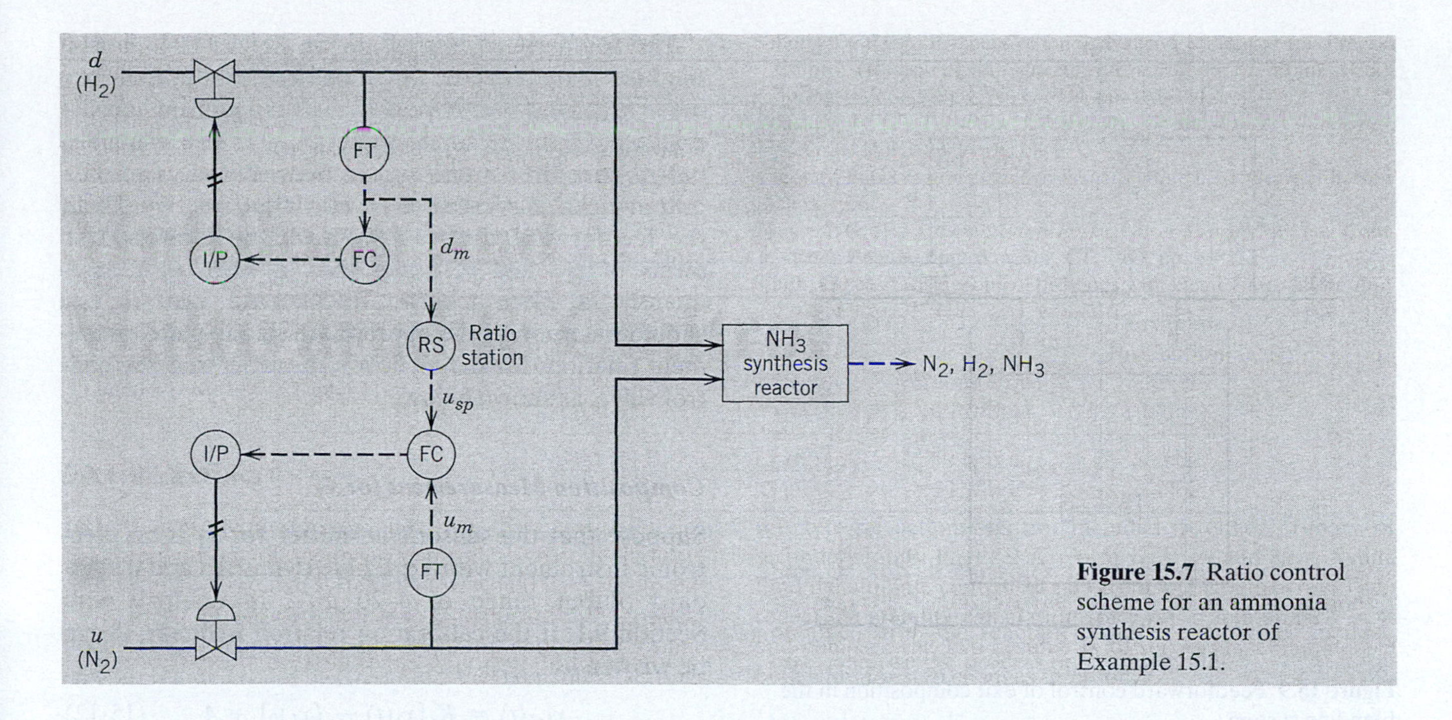

Figure 15.7 Ratio control scheme for an ammonia synthesis reactor of Example 15.1.

15.3 FEEDFORWARD CONTROLLER DESIGN BASED ON STEADY-STATE MODELS

A useful interpretation of feedforward control is that it continually attempts to balance the material or energy that must be delivered to the process against the demands of the disturbance (Shinskey, 1996). For example, the level control system in Fig. 15.3 adjusts the feedwater flow so that it balances the steam demand. Thus, it is natural to base the feedforward control calculations on material and energy balances. For simplicity, we will first consider designs based on steady-state balances using physical variables rather than deviation variables. Design methods based on dynamic models are considered in Section 15.4.

To illustrate the design procedure, consider the distillation column shown in Fig. 15.8, which is used to separate a binary mixture. Feedforward control has gained widespread acceptance for distillation column control owing to the slow responses that typically occur with feedback control. In Fig. 15.8, the symbols B, D, and F denote molar flow rates, while x, y, and z are the mole fractions of the more volatile component. The objective is to control the distillate composition y despite measurable disturbances in feed flow rate F and feed composition z, by adjusting distillate flow rate D. It is assumed that measurements of x and y are not available.

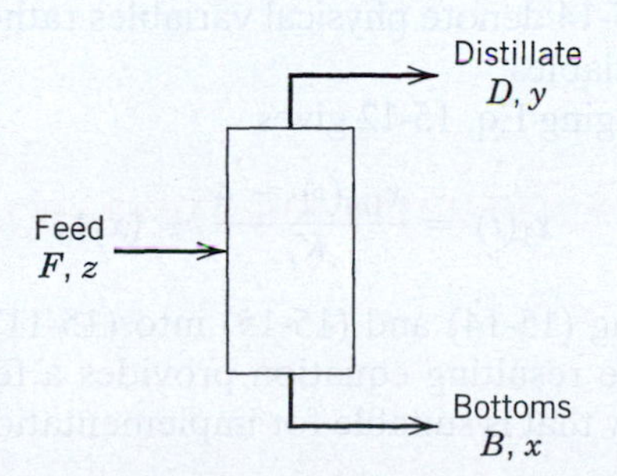

Figure 15.8 A simplified schematic diagram of a distillation column.

The steady-state mass balances for the distillation column can be written as

$$\overline{F} = \overline{D} + \overline{B} \tag{15-4}$$

$$\overline{F}\overline{z} = \overline{D}\overline{y} + \overline{B}\overline{x} \tag{15-5}$$

Solving (15-4) for $\overline{B}$ and substituting into (15-5) gives

$$\overline{D} = \frac{\overline{F}(\overline{z} - \overline{x})}{\overline{y} - \overline{x}} \tag{15-6}$$

Because x and y are not measured, we replace these variables by their set points to yield the feedforward control law:

$$D(t) = \frac{F(t)[z(t) - x_{sp}]}{y_{sp} - x_{sp}} \tag{15-7}$$

Thus, the feedforward controller calculates the required value of the manipulated variable D from measurements of the disturbance variables, F and z, and knowledge of the composition set points x_{sp} and y_{sp}. Note that the control law is nonlinear owing to the product of F and z.

15.3.1 Blending System

To further illustrate the design method, consider the blending system and feedforward controller shown in

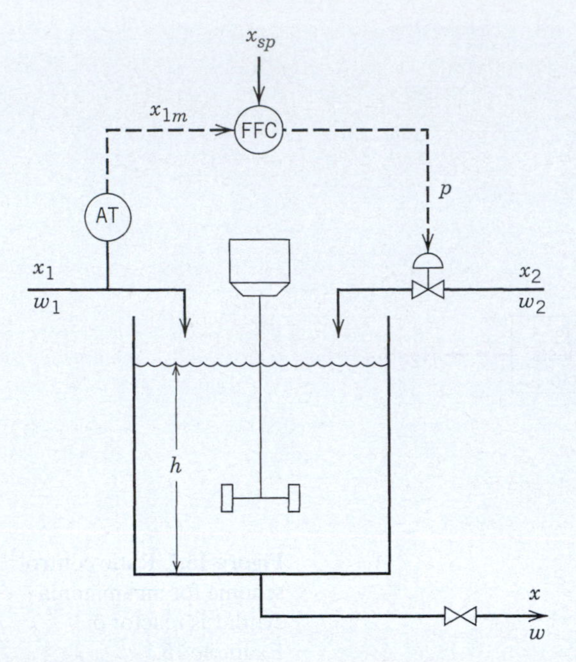

Figure 15.9 Feedforward control of exit composition in the blending system.

Fig. 15.9. We wish to design a feedforward control scheme to maintain exit composition x at a constant set point x_{sp}, despite disturbances in inlet composition, x_1. Suppose that inlet flow rate w_1 and the composition of the other inlet stream x_2 are constant. It is assumed that x_1 is measured but that x is not. If x were measured, then feedback control would also be possible. The manipulated variable is inlet flow rate w_2. The flow-head relation for the valve on the exit line is given by $w = C_v\sqrt{h}$. Note that the feedforward controller has two input signals: the x_1 measurement x_{1m}, and the set point for the exit composition x_{xp}.

The starting point for the feedforward controller design is the steady-state mass and component balances that were considered in Chapter 1,

$$\overline{w} = \overline{w}_1 + \overline{w}_2 \quad (15\text{-}8)$$

$$\overline{w}\overline{x} = \overline{w}_1\overline{x}_1 + \overline{w}_2\overline{x}_2 \quad (15\text{-}9)$$

where the bar over the variable denotes a steady-state value. These equations are the steady-state version of the dynamic model in Eqs. 2-12 and 2-13. Substituting Eq. 15-8 into Eq. 15-9 and solving for $\overline{w}_2$ gives:

$$\overline{w}_2 = \frac{\overline{w}_1(\overline{x} - \overline{x}_1)}{\overline{x}_2 - \overline{x}} \quad (15\text{-}10)$$

In order to derive a feedforward control law, we replace $\overline{x}$ by x_{sp}, and $\overline{w}_2$ and $\overline{x}_1$ by $w_2(t)$ and $x_1(t)$, respectively:

$$w_2(t) = \frac{\overline{w}_1[x_{sp} - x_1(t)]}{\overline{x}_2 - x_{sp}} \quad (15\text{-}11)$$

Note that this feedforward control law is based on physical variables rather than deviation variables.

The feedforward control law in Eq. 15-11 is not in the final form required for actual implementation, because it ignores two important instrumentation considerations: First, the actual value of x_1 is not available, but its measured value x_{1m} is. Second, the controller output signal is p rather than inlet flow rate, w_2. Thus, the feedforward control law should be expressed in terms of x_{1m} and p, rather than x_1 and w_2. Consequently, a more realistic feedforward control law should incorporate the appropriate steady-state instrument relations for the w_2 flow transmitter and the control valve, as shown below.

Composition Measurement for x_1

Suppose that the sensor/transmitter for x_1 is an electronic instrument with negligible dynamics and a standard output range of 4–20 mA. In analogy with Section 9.1, if the calibration relation is linear, it can be written as

$$x_{1m}(t) = K_t[x_1(t) - (x_1)_0] + 4 \quad (15\text{-}12)$$

where $(x_1)_0$ is the zero of this instrument and K_t is its gain. From Eq. 9.1,

$$K_t = \frac{\text{output range}}{\text{input range}} = \frac{20 - 4 \text{ mA}}{S_t} \quad (15\text{-}13)$$

where S_t is the span of the instrument.

Control Valve and Current-to-Pressure Transducer

Suppose that the current-to-pressure transducer and the control valve operate as linear devices with negligible dynamics. Then its input-output relationship can be written as

$$w_2(t) = K_v K_{IP}[p(t) - 4] + (w_2)_0 \quad (15\text{-}14)$$

where K_v and K_{IP} are the steady-state gains for the control valve and I/P transducer, respectively, while $(w_2)_0$ is the w_2 flow rate that corresponds to the minimum controller output signal of 4 mA. This value also corresponds to the minimum signal of 3 psi from the I/P transducer. Note that all of the symbols in Eqs. 15-8 through 15-14 denote physical variables rather than deviation variables.

Rearranging Eq. 15-12 gives

$$x_1(t) = \frac{x_{1m}(t) - 4}{K_t} + (x_1)_0 \quad (15\text{-}15)$$

Substituting (15-14) and (15-15) into (15-11) and rearranging the resulting equation provides a feedforward control law that is suitable for implementation:

$$p(t) = C_1 + C_2\left[\frac{K_t x_{sp} - x_{1m}(t) + C_3}{\overline{x}_2 - x_{sp}}\right] \quad (15\text{-}16)$$

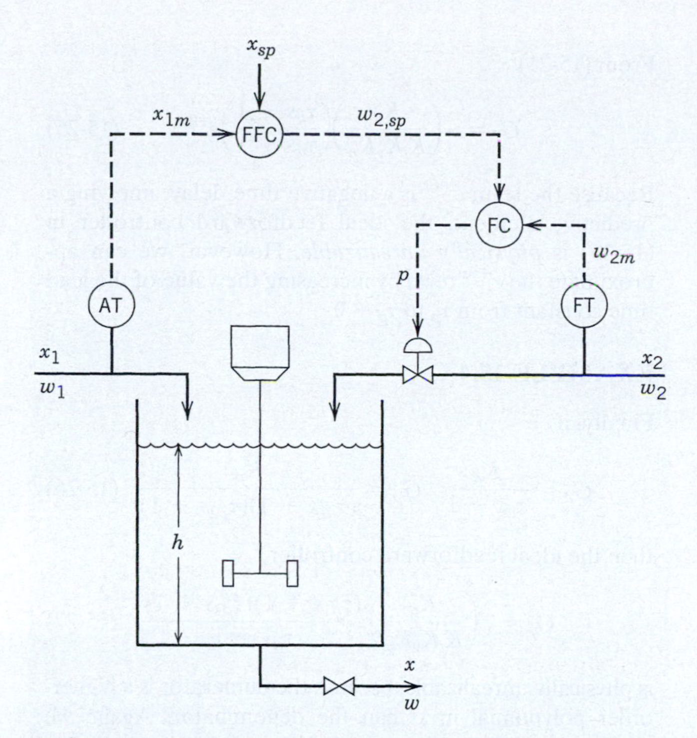

Figure 15.10 Feedforward control of exit composition using a flow control loop.

where

$$C_1 \triangleq 4 - \frac{(w_2)_0}{K_v K_{IP}} \tag{15-17}$$

$$C_2 \triangleq \frac{\overline{w}_1}{K_v K_{IP} K_t} \tag{15-18}$$

$$C_3 \triangleq 4 + K_t(x_1)_0 \tag{15-19}$$

An alternative feedforward control scheme for the blending system is shown in Fig. 15.10. Here the feedforward controller output signal serves as a set point to a feedback controller for flow rate w_2. The advantage of this configuration is that it is less sensitive to valve sticking and upstream pressure fluctuations. Because the feedforward controller calculates the w_2 set point rather than the signal to the control valve p, it would not be necessary to incorporate Eq. 15-14 into the feedforward control law.

The blending and distillation column examples illustrate that feedforward controllers can be designed using steady-state mass and energy balances. The advantages of this approach are that the required calculations are quite simple, and a detailed process model is not required. However, a disadvantage is that process dynamics are neglected, and consequently the control system may not perform well during transient conditions. The feedforward controllers can be improved by adding dynamic compensation, usually in the form of a lead–lag unit. This topic is discussed in Section 15.7. An alternative approach is to base the controller design on a dynamic model of the process, as discussed in the next section.

15.4 FEEDFORWARD CONTROLLER DESIGN BASED ON DYNAMIC MODELS

In this section, we consider the design of feedforward control systems based on dynamic, rather than steady-state, process models. We will restrict our attention to design techniques based on linear dynamic models. But nonlinear process models can also be used (Smith and Corripio, 2006).

As a starting point for our discussion, consider the block diagram shown in Fig. 15.11. This diagram is similar to Fig. 11.8 for feedback control, but an additional signal path through G_t and G_f has been added. The disturbance transmitter with transfer function G_t sends a measurement of the disturbance variable to the feedforward controller G_f. The outputs of the feedforward

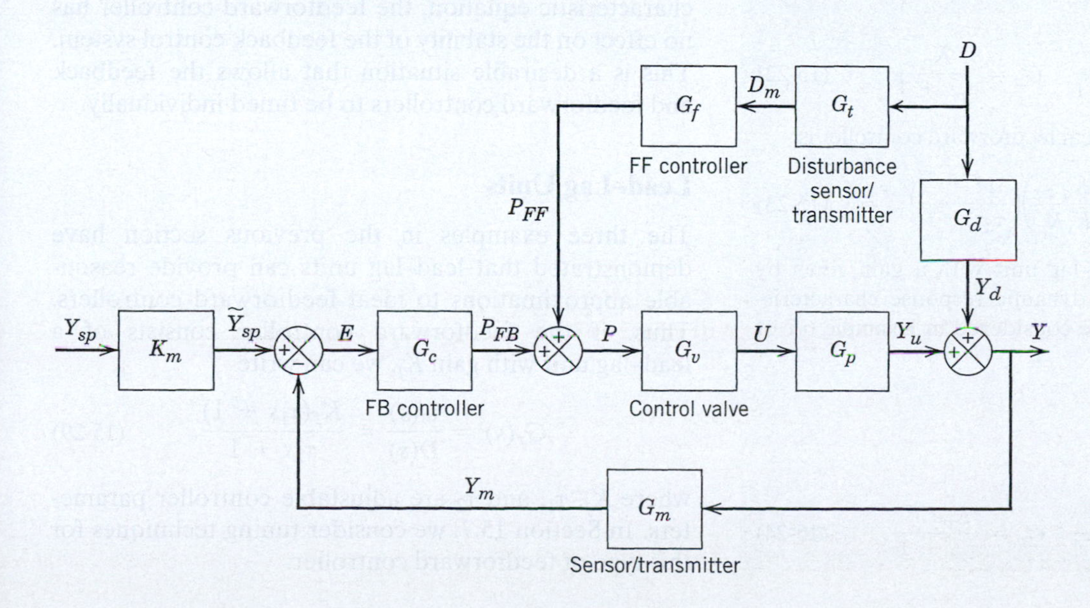

Figure 15.11 A block diagram of a feedforward-feedback control system.

and feedback controllers are then added together, and the sum is sent to the control valve. In contrast to steady-state feedforward control, the block diagram in Fig. 15.11 is based on deviation variables.

The closed-loop transfer function for disturbance changes in Eq. 15-20 can be derived using the block diagram algebra that was introduced in Chapter 11:

$$\frac{Y(s)}{D(s)} = \frac{G_d + G_t G_f G_v G_p}{1 + G_c G_v G_p G_m} \tag{15-20}$$

Ideally, we would like the control system to produce *perfect control*, where the controlled variable remains exactly at the set point despite arbitrary changes in the disturbance variable, D. Thus, if the set point is constant ($Y_{sp}(s) = 0$), we want $Y(s) = 0$, even though $D(s) \neq 0$. This condition can be satisfied by setting the numerator of (15-20) equal to zero and solving for G_f:

$$G_f = -\frac{G_d}{G_t G_v G_p} \tag{15-21}$$

Figure 15.11 and Eq. 15-21 provide a useful interpretation of the ideal feedforward controller. Figure 15.11 indicates that a disturbance has two effects: it upsets the process via the disturbance transfer function G_d; however, a corrective action is generated via the path through $G_t G_f G_v G_p$. Ideally, the corrective action compensates exactly for the upset so that signals Y_d and Y_u cancel each other and $Y(s) = 0$.

Next, we consider three examples in which we derive feedforward controllers for various types of process models. For simplicity, it is assumed that the disturbance transmitters and control valves have negligible dynamics, that is, $G_t(s) = K_t$ and $G_v(s) = K_v$, where K_t and K_v denote steady-state gains.

EXAMPLE 15.2

Suppose that

$$G_d = \frac{K_d}{\tau_d s + 1} \quad G_p = \frac{K_p}{\tau_p s + 1} \tag{15-22}$$

Then, from (15-21), the ideal feedforward controller is

$$G_f = -\left(\frac{K_d}{K_t K_v K_p}\right)\left(\frac{\tau_p s + 1}{\tau_d s + 1}\right) \tag{15-23}$$

This controller is a lead–lag unit with a gain given by $K_f = -K_d/K_t K_v K_p$. The dynamic response characteristics of lead–lag units were considered in Example 6.1 of Chapter 6.

EXAMPLE 15.3

Now consider

$$G_d = \frac{K_d}{\tau_d s + 1} \quad G_p = \frac{K_p e^{-\theta s}}{\tau_p s + 1} \tag{15-24}$$

From (15-21)

$$G_f = -\left(\frac{K_d}{K_t K_v K_p}\right)\left(\frac{\tau_p s + 1}{\tau_d s + 1}\right)e^{+\theta s} \tag{15-25}$$

Because the term $e^{+\theta s}$ is a negative time delay, implying a predictive element, the ideal feedforward controller in (15-25) is *physically unrealizable*. However, we can approximate the $e^{+\theta s}$ term by increasing the value of the lead time constant from τ_p to $\tau_p + \theta$.

EXAMPLE 15.4

Finally, if

$$G_d = \frac{K_d}{\tau_d s + 1} \quad G_p = \frac{K_p}{(\tau_{p1} s + 1)(\tau_{p2} s + 1)} \tag{15-26}$$

then the ideal feedforward controller,

$$G_f = -\left(\frac{K_d}{K_t K_v K_p}\right)\frac{(\tau_{p1} s + 1)(\tau_{p2} s + 1)}{\tau_d s + 1} \tag{15-27}$$

is physically unrealizable, because the numerator is a higher-order polynomial in s than the denominator. Again, we could approximate this controller by a physically realizable one such as a lead–lag unit, where the lead time constant is the sum of the two time constants, $\tau_{p1} + \tau_{p2}$.

Stability Considerations

To analyze the stability of the closed-loop system in Fig. 15.11, we consider the closed-loop transfer function in Eq. 15-20. Setting the denominator equal to zero gives the characteristic equation,

$$1 + G_c G_v G_p G_m = 0 \tag{15-28}$$

In Chapter 11 it was shown that the roots of the characteristic equation completely determine the stability of the closed-loop system. Because G_f does not appear in the characteristic equation, the feedforward controller has no effect on the stability of the feedback control system. This is a desirable situation that allows the feedback and feedforward controllers to be tuned individually.

Lead–Lag Units

The three examples in the previous section have demonstrated that lead–lag units can provide reasonable approximations to ideal feedforward controllers. Thus, if the feedforward controller consists of a lead–lag unit with gain K_f, we can write

$$G_f(s) = \frac{U(s)}{D(s)} = \frac{K_f(\tau_1 s + 1)}{\tau_2 s + 1} \tag{15-29}$$

where K_f, τ_1, and τ_2 are adjustable controller parameters. In Section 15.7, we consider tuning techniques for this type of feedforward controller.

EXAMPLE 15.5

Consider the blending system of Section 15.3, but now assume that a pneumatic control valve and an I/P transducer are used. A feedforward-feedback control system is to be designed to reduce the effect of disturbances in feed composition x_1 on the controlled variable, product composition x. Inlet flow rate w_2 can be manipulated. Using the information given below, design the following control systems and compare the closed-loop responses for a +0.2 step change in x_1.

(a) A feedforward controller based on a steady-state model of the process.

(b) Static and dynamic feedforward controllers based on a linearized, dynamic model.

(c) A PI feedback controller based on the Ziegler-Nichols settings for the continuous cycling method.

(d) The combined feedback-feedforward control system that consists of the feedforward controller of part (a) and the PI controller of part (c). Use the configuration in Fig. 15.11.

Process Information

The pilot-scale blending tank has an internal diameter of 2 m and a height of 3 m. Inlet flow rate w_1 and inlet composition x_2 are constant. The nominal steady-state operating conditions are as follows:

$$\overline{w}_1 = 650 \text{ kg/min} \quad \overline{x}_1 = 0.2 \quad \overline{h} = 1.5 \text{ m}$$

$$\overline{w}_2 = 350 \text{ kg/min} \quad \overline{x}_2 = 0.6 \quad \rho = 1 \text{ g/cm}^3 \quad \overline{x} = 0.34$$

The flow-head relation for the valve on the exit line is given by $w = C_v\sqrt{h}$.

Instrumentation (The range for each electronic signal is 4 to 20 mA.)

Current-to-pressure transducer: The I/P transducer acts as a linear device with negligible dynamics. The output signal changes from 3 to 15 psi when the input signal changes full-scale from 4 to 20 mA.

Control valve: The behavior of the control valve can be approximated by a first-order transfer function with a time constant of 5 s (0.0833 min). A 3–15 psi change in the signal to the control valve produces a 300-kg/min change in w_2.

Composition measurement: The zero and span of each composition transmitter are 0 and 0.50 (mass fraction), respectively. The output range is 4–20 mA. A one-minute time delay is associated with each measurement.

SOLUTION

A block diagram for the feedforward-feedback control system is shown in Fig. 15.12.

(a) Using the given information, we can calculate the following steady-state gains:

$$K_{IP} = (15 - 3)/(20 - 4) = 0.75 \text{ psi/mA}$$

$$K_v = 300/12 = 25 \text{ kg/min psi}$$

$$K_t = (20 - 4)/0.5 = 32 \text{ mA}$$

Substitution into Eqs. 15-16 to 15-19 with $(w_2)_0 = 0$ and $(x_1)_0 = 0$ gives the following feedforward control law:

$$p(t) = 4 + 1.083\left[\frac{32x_{sp} - x_{1m}(t) + 4}{0.6 - x_{sp}}\right] \tag{15-30}$$

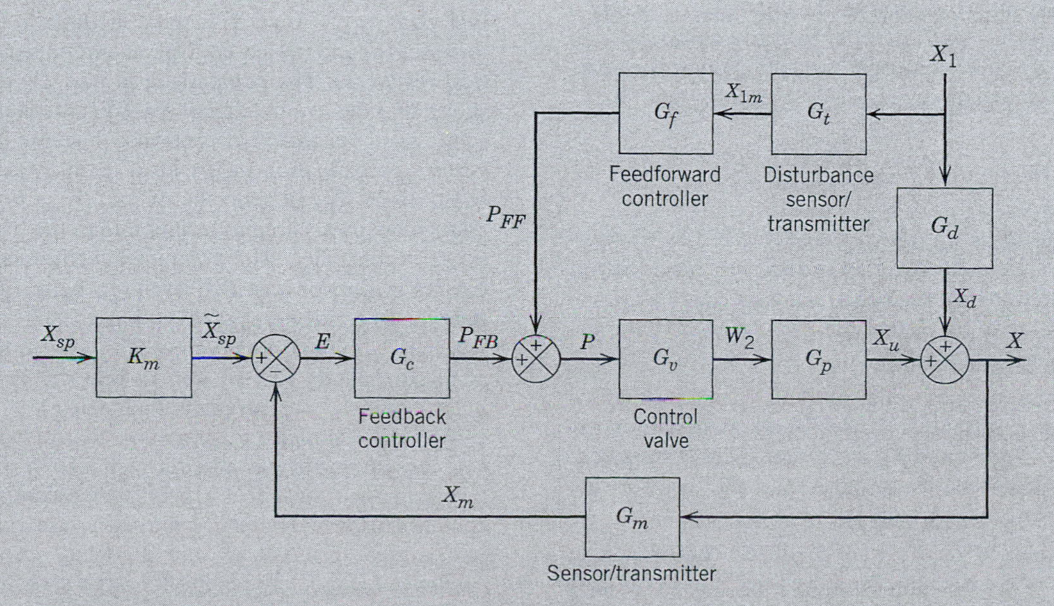

Figure 15.12 Block diagram for feedforward-feedback control of the blending system.

(b) The following expression for the ideal feedforward controller can be derived in analogy with the derivation of Eq. 15-21:

$$G_f = -\frac{G_d}{K_{IP} G_t G_v G_p} \tag{15-31}$$

The process and disturbance transfer functions are similar to the ones derived in Example 4.1:

$$\frac{X'(s)}{W'_2(s)} = G_p(s) = \frac{K_p}{\tau s + 1}$$

$$\frac{X'(s)}{X'_1(s)} = G_d(s) = \frac{K_d}{\tau s + 1}$$

where

$$K_p = \frac{\overline{x}_2 - \overline{x}}{\overline{w}}, \quad K_d = \frac{\overline{w}_1}{\overline{w}}, \quad \tau = V\rho/\overline{w}, \quad V = \pi R^2 h$$

Substituting numerical values gives

$$\frac{X'(s)}{W'_2(s)} = \frac{2.6 x 10^{-4}}{4.71s + 1} \qquad \frac{X'(s)}{X'_1(s)} = \frac{0.65}{4.71s + 1} \tag{15-32}$$

The transfer functions for the instrumentation can be determined from the given information:

$$G_{IP} = K_{IP} = 0.75 \text{ psi/mA},$$

$$G_t(s) = G_m(s) = K_t e^{-\theta s} = 32e^{-s}$$

$$G_v(s) = \frac{K_v}{\tau_v s + 1} = \frac{25}{0.0833s + 1}$$

Substituting the individual transfer functions into Eq. 15-21 gives the ideal dynamic feedforward controller:

$$G_f(s) = -4.17(0.0833s + 1)e^{+s} \tag{15-33}$$

Note that $G_f(s)$ is physically unrealizable. The static (or steady-state) version of the controller is simply a gain, $G_f(s) = -4.17$. In order to derive a physically realizable dynamic controller, we approximate the unrealizable controller in (15-33) by a lead–lag unit:

$$G_f(s) = -4.17\frac{1.0833s + 1}{\alpha(1.0833)s + 1} \tag{15-34}$$

Equation 15-34 was derived from (15-33) by (i) omitting the time-delay term, (ii) adding the time delay of one minute to the lead time constant, and (iii) introducing a small time constant of $\alpha \times 1.0833$ in the denominator, with $\alpha = 0.1$.

(c) The ultimate gain and ultimate period obtained from the continuous cycling method (Chapter 12) are $K_{cu} = 48.7$, and $P_u = 4.0$ min. The corresponding Ziegler-Nichols settings for PI control are $K_c = 0.45K_{cu} = 21.9$, and $\tau_I = P_u/1.2 = 3.33$ min.

(d) The combined feedforward-feedback control system consists of the dynamic feedforward controller of part (b) and the PI controller of part (c).

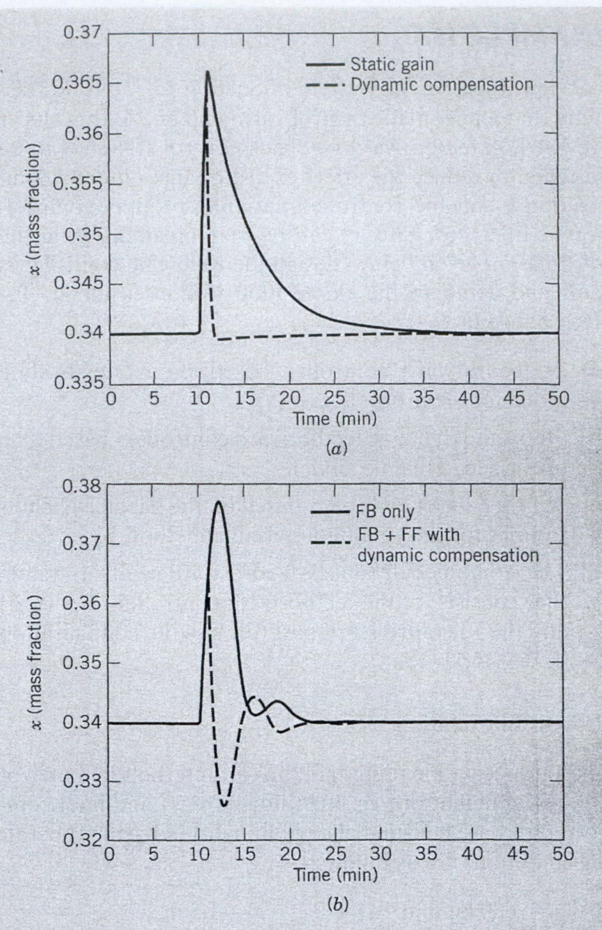

Figure 15.13 Comparison of closed-loop responses: (a) feedforward controllers with and without dynamic compensation; (b) feedback control and feedforward-feedback control.

The closed-loop responses to a step change in x_1 from 0.2 to 0.4 are shown in Fig. 15.13. The set point is the nominal value, $x_{sp} = 0.34$. The static feedforward controllers for cases (a) and (b) are equivalent and thus produce identical responses. The comparison in part (a) of Fig. 15.13 shows that the dynamic feedforward controller is superior to the static feedforward controller, because it provides a better approximation to the ideal feedforward controller of Eq. 15-33. The PI controller in part (b) of Fig. 15.13 produces a larger maximum deviation than the dynamic feedforward controller. The combined feedforward-feedback control system of part (d) results in better performance than the PI controller, because it has a much smaller maximum deviation and IAE value. The peak in the response at approximately $t = 13$ min in Fig. 15.13b is a consequence of the x_1 measurement time delay.

For this example, feedforward control with dynamic compensation provides a better response to the measured x_1 disturbance than does combined feedforward-feedback control. However, feedback control is essential to cope with unmeasured disturbances and modeling errors. Thus, a combined feedforward-feedback control system is preferred in practice.

15.5 THE RELATIONSHIP BETWEEN THE STEADY-STATE AND DYNAMIC DESIGN METHODS

In the previous two sections, we considered two design methods for feedforward control. The design method of Section 15.3 was based on a nonlinear steady-state process model, while the design method of Section 15.4 was based on a transfer function model and block diagram analysis. Next, we show how the two design methods are related.

The block diagram of Fig. 15.11 indicates that the manipulated variable is related to the disturbance variable by

$$\frac{U(s)}{D(s)} = G_v(s)G_f(s)G_t(s) \tag{15-35}$$

Let the steady-state gain for this transfer function be denoted by K. Thus,

$$K = \lim_{s \to 0} G_v(s)G_f(s)G_t(s) \tag{15-36}$$

Suppose that the disturbance changes from a nominal value, $\bar{d}$, to a new value, d_1. Denote the change as $\Delta d = d_1 - \bar{d}$. Let the corresponding steady-state change in the manipulated variable be denoted by $\Delta u = u_1 - \bar{u}$. Then, from Eqs. (15-35) and (15-36) and the definition of a steady-state gain in Chapter 4, we have

$$K = \frac{\Delta u}{\Delta d} \tag{15-37}$$

The steady-state design method of Section 15.3 produces a feedforward control law that has the general nonlinear form:

$$u = f(d, y_{sp}) \tag{15-38}$$

Let K_{loc} denote the local derivative of u with respect to d at the nominal value $\bar{d}$:

$$K_{loc} = \left(\frac{\partial u}{\partial d}\right)_{\bar{d}} \tag{15-39}$$

A comparison of Eqs. 15-37 and 15-39 indicates that if Δd is small, $K_{loc} \approx K$. If the steady-state feedforward control law of Eq. 15-38 is indeed linear, then $K_{loc} = K$ and the gains for the two design methods are equivalent.

15.6 CONFIGURATIONS FOR FEEDFORWARD-FEEDBACK CONTROL

As mentioned in Section 15.1 and illustrated in Example 15.5, *feedback trim* is normally used in conjunction with feedforward control to compensate for modeling errors and unmeasured disturbances. Feedforward and feedback controllers can be combined in several different ways. In a typical control configuration, the outputs

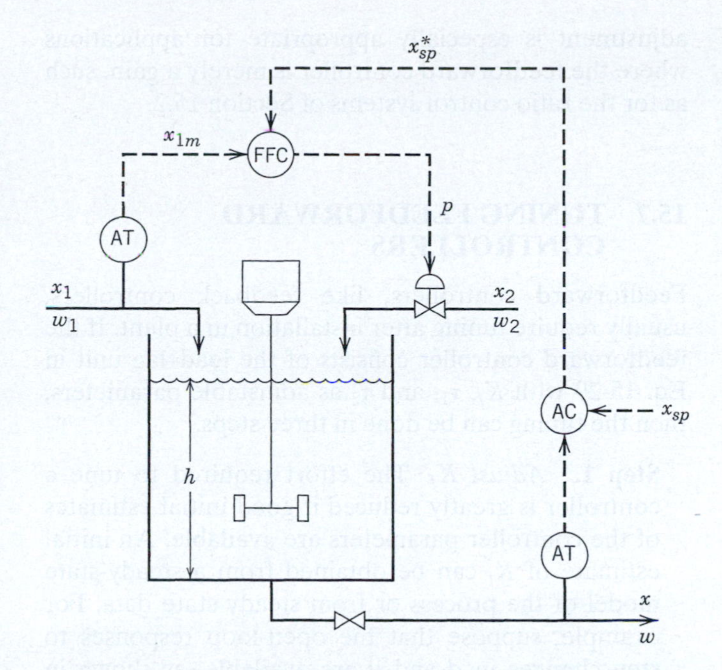

Figure 15.14 Feedforward-feedback control of exit composition in the blending system.

of the feedforward and feedback controllers are added together, and the sum is sent to the final control element. This configuration was introduced in Figs. 15.4 and 15.11. Its chief advantage is that the feedforward controller theoretically does not affect the stability of the feedback control loop. Recall that the feedforward controller transfer function $G_f(s)$ does not appear in the characteristic equation of Eq. 15-28.

An alternative configuration for feedforward-feedback control is to have the feedback controller output serve as the set point for the feedforward controller. It is especially convenient when the feedforward control law is designed using steady-state material and energy balances. For example, a feedforward-feedback control system for the blending system is shown in Fig. 15.14. Note that this control system is similar to the feedforward scheme in Fig. 15.9 except that the feedforward controller set point is now denoted as x_{sp}^*. It is generated as the output signal from the feedback controller. The actual set point x_{sp} is used as the set point for the feedback controller. In this configuration, the feedforward controller can affect the stability of the feedback control system, because it is now an element in the feedback loop. If dynamic compensation is included, it should be introduced outside of the feedback loop. Otherwise, it will interfere with the operation of the feedback loop, especially when the controller is placed in the manual model (Shinskey, 1996).

Alternative ways of incorporating feedback trim into a feedforward control system include having the feedback controller output signal adjust either the feedforward controller gain or an additive bias term. The gain adjustment is especially appropriate for applications where the feedforward controller is merely a gain, such as for the ratio control systems of Section 15.2.

15.7 TUNING FEEDFORWARD CONTROLLERS

Feedforward controllers, like feedback controllers, usually require tuning after installation in a plant. If the feedforward controller consists of the lead–lag unit in Eq. 15-29 with K_f, τ_1, and τ_2 as adjustable parameters, then the tuning can be done in three steps.

Step 1. *Adjust K_f.* The effort required to tune a controller is greatly reduced if good initial estimates of the controller parameters are available. An initial estimate of K_f can be obtained from a steady-state model of the process or from steady-state data. For example, suppose that the open-loop responses to step changes in d and u are available, as shown in Fig. 15.15. After K_p and K_d have been determined, the feedforward controller gain can be calculated from the steady-state version of Eq. 15-21:

$$K_f = -\frac{K_d}{K_t K_v K_p} \tag{15-40}$$

Gains K_t and K_v are available from the steady-state characteristics of the transmitter and control valve.

To tune the controller gain, K_f is set equal to an initial value and a small step change (3 to 5%) in the disturbance variable d is introduced, if this is feasible.

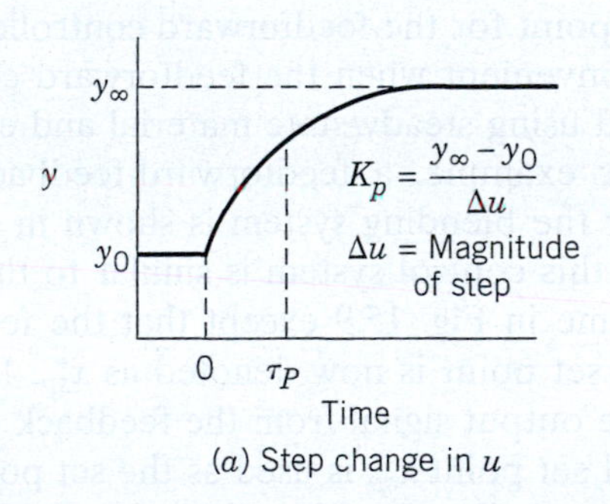

(*a*) Step change in u

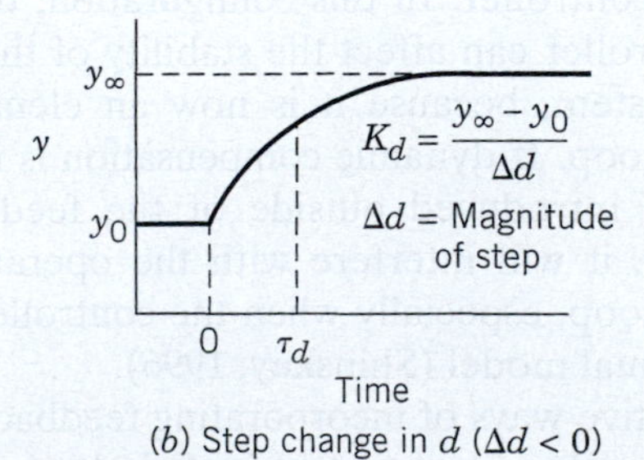

(*b*) Step change in d ($\Delta d < 0$)

Figure 15.15 The open-loop responses to step changes in u and d.

If an offset results, then K_f is adjusted until the offset is eliminated. While K_f is being tuned, τ_1 and τ_2 should be set equal to their minimum values, ideally zero.

Step 2. *Determine initial values for τ_1 and τ_2.* Theoretical values for τ_1 and τ_2 can be calculated if a dynamic model of the process is available. Alternatively, initial estimates can be determined from open-loop response data. For example, if the step responses have the shapes shown in Fig. 15.15, a reasonable process model is

$$G_p(s) = \frac{K_p}{\tau_p s + 1} \quad G_d(s) = \frac{K_d}{\tau_d s + 1} \tag{15-41}$$

where τ_p and τ_d can be calculated using one of the methods of Chapter 7. A comparison of Eqs. 15-23 and 15-29 leads to the following expressions for τ_1 and τ_2:

$$\tau_1 = \tau_p \tag{15-42}$$

$$\tau_2 = \tau_d \tag{15-43}$$

These values can then be used as initial estimates for the fine tuning of τ_1 and τ_2 in Step 3.

If neither a process model nor experimental data are available, the relations $\tau_1/\tau_2 = 2$ or $\tau_1/\tau_2 = 0.5$ may be used, depending on whether the controlled variable responds faster to the disturbance variable or to the manipulated variable.

Step 3. *Fine-tune τ_1 and τ_2.* The final step is a trial-and-error procedure to fine-tune τ_1 and τ_2 by making small step changes in d, if feasible. The desired step response consists of small deviations in the controlled variable with equal areas above and below the set point (Shinskey, 1996), as shown in Fig. 15.16. For simple process models, it can be shown theoretically that equal areas above and below the set point imply that the difference, $\tau_1 - \tau_2$, is correct. In subsequent tuning to reduce the size of the areas, τ_1 and τ_2 should be adjusted so that $\tau_1 - \tau_2$ remains constant.

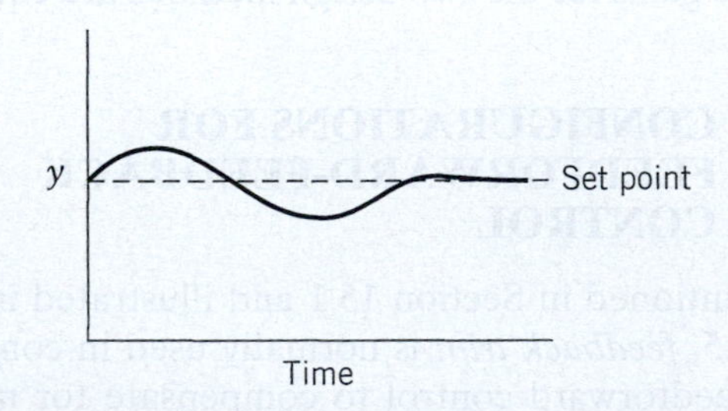

Figure 15.16 The desired response for a well-tuned feedforward controller. Note approximately equal areas above and below the set point.

As a hypothetical illustration of this trial-and-error tuning procedure, consider the set of responses shown in Fig. 15.17 for positive step changes in disturbance variable d. It is assumed that $K_p > 0$, that $K_d < 0$, and that controller gain K_f has already been adjusted so that offset is eliminated. For the initial values of τ_1 and τ_2 in Fig. 15.17*a*, the controlled variable is below the set point, which implies that τ_1 should be increased to speed up the corrective action. (Recall that $K_p > 0$, that $K_d < 0$, and that positive step changes in d are introduced.) Increasing τ_1 from 1 to 2 gives the response in Fig. 15.17*b*, which has equal areas above and below the set point. Thus, in subsequent tuning to reduce the size of each area, $\tau_1 - \tau_2$ should be kept constant. Increasing both τ_1 and τ_2 by 0.5 reduces the size of each area, as shown in Fig. 15.17*c*. Because this response is considered to be satisfactory, no further controller tuning is required.

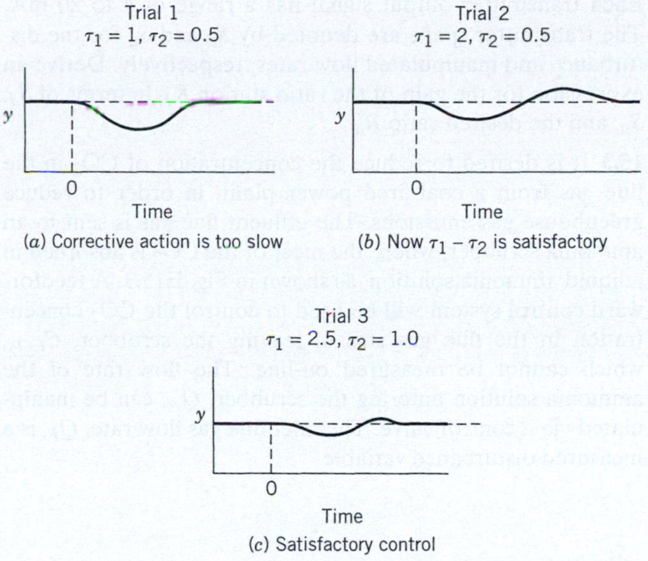

Trial 1 $\tau_1 = 1, \tau_2 = 0.5$ — (*a*) Corrective action is too slow

Trial 2 $\tau_1 = 2, \tau_2 = 0.5$ — (*b*) Now $\tau_1 - \tau_2$ is satisfactory

Trial 3 $\tau_1 = 2.5, \tau_2 = 1.0$ — (*c*) Satisfactory control

Figure 15.17 An example of feedforward controller tuning.

SUMMARY

Feedforward control is a powerful strategy for control problems wherein important disturbance variable(s) can be measured on-line. By measuring disturbances and taking corrective action before the controlled variable is upset, feedforward control can provide dramatic improvements for regulatory control. Its chief disadvantage is that the disturbance variable(s) must be measured (or estimated) on-line, which is not always possible. Ratio control is a special type of feedforward control that is useful for applications such as blending operations where the ratio of two process variables is to be controlled.

Feedforward controllers tend to be custom-designed for specific applications, although a lead–lag unit is often used as a generic feedforward controller. The design of a feedforward controller requires knowledge of how the controlled variable responds to changes in the manipulated variable and the disturbance variable(s). This knowledge is usually represented as a process model. Steady-state models can be used for controller design; however, it may then be necessary to add a lead–lag unit to provide dynamic compensation. Feedforward controllers can also be designed using dynamic models.

Feedfoward control is normally implemented in conjunction with feedback control. Tuning procedures for combined feedforward-feedback control schemes have been described in Section 15.7. For these control configurations, the feedforward controller is usually tuned before the feedback controller.

REFERENCES

Blevins, T. L., G. K. McMillan, W. K. Wojsznis, and M. W. Brown, *Advanced Control Unleashed: Plant Performance Management for Optimum Benefit,* Appendix B, ISA, Research Triangle Park, NC, 2003.

Hägglund, T., The Blend Station—A New Ratio Control Structure, *Control Eng. Practice*, **9,** 1215 (2001).

Shinskey, F. G., *Process Control Systems: Application, Design, and Tuning,* 4th ed. McGraw-Hill, New York, 1996, Chapter 7.

Shinskey, F. G., M. F. Hordeski, and B. G. Lipták, Feedback and Feedforward Control, in *Instrument Engineer's Handbook: Vol. 2, Process Control*, 3d ed., B. G. Lipták (Ed.), Chilton Book Co., Radnor, PA, 1995, Section 1.8.

Smith, C. A., and A. B. Corripio, *Principles and Practice of Automatic Process Control*, 3rd ed., Wiley, New York, 2006.

EXERCISES

15.1 In ratio control, would the control loop gain for Method I (Fig. 15.5) be less variable if the ratio were defined as $R = d/u$ instead of $R = u/d$? Justify your answer.

15.2 Consider the ratio control scheme shown in Fig. 15.6. Each flow rate is measured using an orifice plate and a differential pressure (D/P) transmitter. The electrical output signals from the D/P transmitters are related to the flow rates by the expressions

$$d_m = d_{m0} + K_1 d^2$$
$$u_m = u_{m0} + K_2 u^2$$

Each transmitter output signal has a range of 4 to 20 mA. The transmitter spans are denoted by S_d and S_u for the disturbance and manipulated flow rates, respectively. Derive an expression for the gain of the ratio station K_R in terms of S_d, S_u, and the desired ratio R_d.

15.3 It is desired to reduce the concentration of CO_2 in the flue gas from a coal-fired power plant, in order to reduce greenhouse gas emissions. The effluent flue gas is sent to an ammonia scrubber, where the most of the CO_2 is absorbed in a liquid ammonia solution, as shown in Fig. E15.3. A feedforward control system will be used to control the CO_2 concentration in the flue gas stream leaving the scrubber, C_{CO_2}, which cannot be measured on-line. The flow rate of the ammonia solution entering the scrubber, Q_A, can be manipulated via a control valve. The inlet flue gas flow rate, Q_F, is a measured disturbance variable.

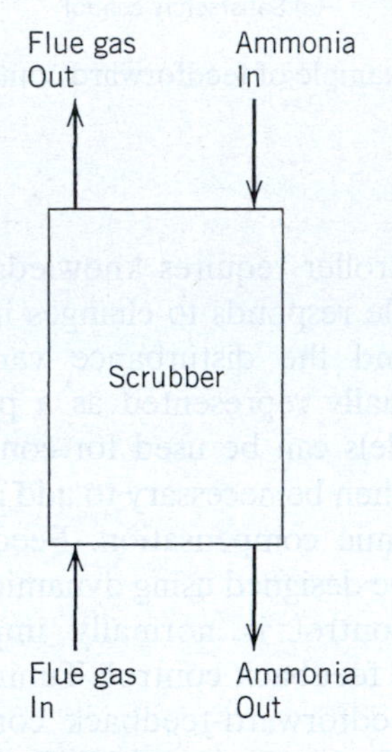

Figure E15.3

Using the available information, do the following:

(a) Draw a block diagram of the feedforward control system. (It is *not* necessary to derive transfer functions.)

(b) Design a feedforward control system to reduce CO_2 emissions based on a *steady state analysis.*

Available Information:

(i) The flow sensor-transmitter and the control valve have negligible dynamics.

(ii) The flow sensor-transmitter has a steady-state gain of 0.08 mA/(L/min).

(iii) The control valve has a steady-state gain of 4 (gal/min)/mA.

(iv) The following steady-state data are available for a series of changes in Q_A:

Q_A (gal/min)	C_{CO_2} (ppm)
30	125
60	90
90	62

(v) The following steady-state data are available for a series of changes in Q_F,

Q_F (L/min)	C_{CO_2} (ppm)
200	75
300	96
400	122

15.4 For the liquid storage system shown in Fig. E15.4, the control objective is to regulate liquid level h_2 despite disturbances in flow rates, q_1 and q_4. Flow rate q_2 can be manipulated. The two hand valves have the following flow-head relations:

$$q_3 = C_1\sqrt{h_1} \qquad q_5 = C_2\sqrt{h_2}$$

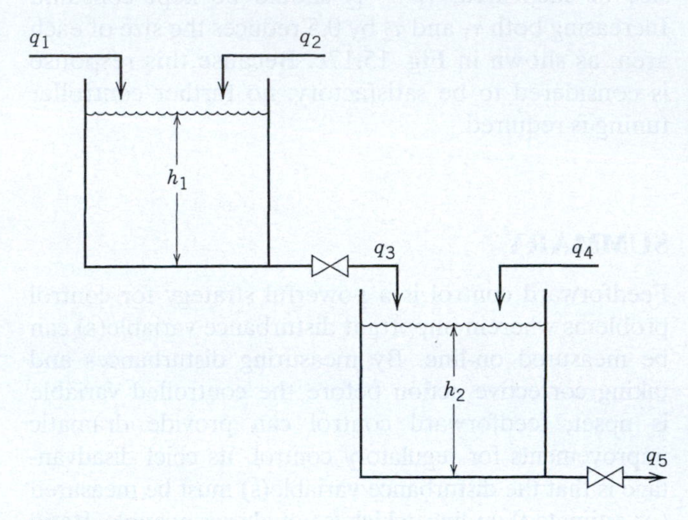

Figure E15.4

Do the following, assuming that the flow transmitters and the control valve have negligible dynamics. Also assume that the objective is tight level control.

(a) Draw a block diagram for a feedforward control system for the case where q_1 can be measured and variations in q_4 are neglected.

(b) Design a feedforward control law for case (a) based on a steady-state analysis.

(c) Repeat part (b), but consider dynamic behavior.

(d) Repeat parts (a) through (c) for the situation where q_4 can be measured and variations in q_1 are neglected.

15.5 The closed-loop system in Fig. 15.11 has the following transfer functions:

$$G_p(s) = \frac{1}{s+1} \qquad G_d(s) = \frac{2}{(s+1)(5s+1)}$$

$$G_v = G_m = G_t = 1$$

(a) Design a feedforward controller based on a steady-state analysis.

(b) Design a feedforward controller based on a dynamic analysis.

(c) Design a feedback controller based on the IMC approach of Chapter 12 and $\tau_c = 2$.

(d) Simulate the closed-loop response to a unit step change in the disturbance variable using feedforward control only and the controllers of parts (a) and (b).

(e) Repeat part (d) for the feedforward-feedback control scheme of Fig. 15.11 and the controllers of parts (a) and (c) as well as (b) and (c).

15.6 A feedforward control system is to be designed for the two-tank heating system shown in Fig. E15.6. The design objective is to regulate temperature T_4, despite variations in disturbance variables T_1 and w. The voltage signal to the heater p is the manipulated variable. Only T_1 and w are measured. Also, it can be assumed that the heater and transmitter dynamics are negligible and that the heat duty is linearly related to voltage signal p.

(a) Design a feedforward controller based on a steady-state analysis. This control law should relate p to T_{1m} and w_m.

(b) Is dynamic compensation desirable? Justify your answer.

15.7 Consider the liquid storage system of Exercise 15.4 but suppose that the hand valve for q_5 is replaced by a pump and a control valve (cf. Fig. 11.22). Repeat parts (a) through (c) of Exercise 15.4 for the situation where q_5 is the manipulated variable and q_2 is constant.

15.8 A liquid-phase reversible reaction, A $\rightleftarrows$ B, takes place isothermally in the continuous stirred-tank reactor shown in Fig. E15.8. The inlet stream does not contain any B. An overflow line maintains constant holdup in the reactor. The reaction rate for the disappearance of A is given by

$$-r_A = k_1 c_A - k_2 c_B, \qquad r_A [=] \left[\frac{\text{moles of A reacting}}{\text{(time) (volume)}} \right]$$

The control objective is to control exit concentration c_B by manipulating volumetric flow rate, q. The chief disturbance variable is feed concentration c_{Ai}. It can be measured on-line, but the exit stream compositions cannot. The control valve and sensor-transmitter have negligible dynamics and positive steady-state gains.

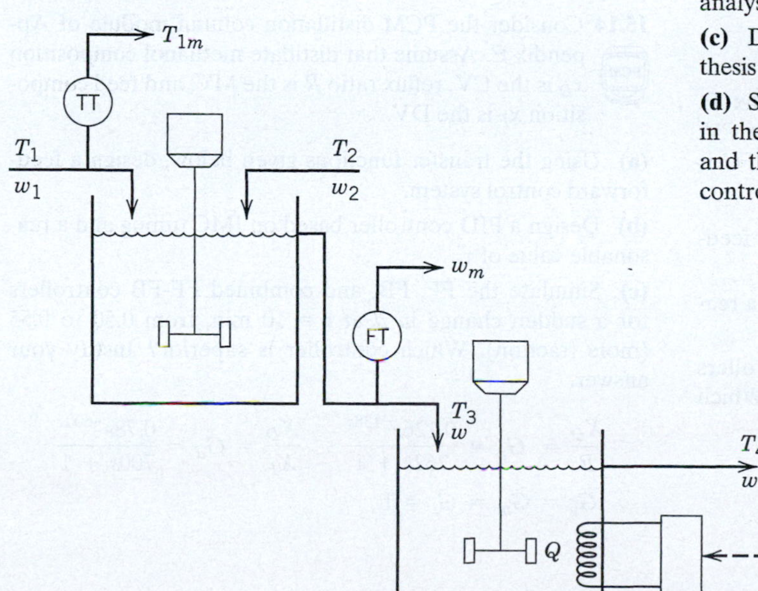

Figure E15.6

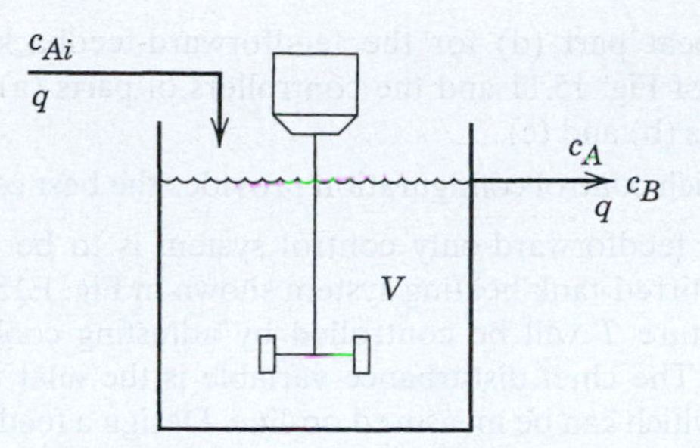

Figure E15.8

(a) Design a feedforward controller based on an *unsteady-state analysis*.

(b) If the exit concentration c_B could be measured and used for feedback control, should this feedback controller be reverse- or direct-acting? Justify your answer.

15.9 Design a feedforward-feedback control system for the blending system in Example 15.5, for a situation in which an improved sensor is available that has a smaller time delay of 0.1 min. Repeat parts (b), (c), and (d) of Example 15.5. For part (c), approximate $G_v G_p G_m$ with a first-order plus time-delay transfer function, and then use a PI controller with ITAE controller tuning for disturbances (see Table 12.3). For the feedforward controller in (15-34), use $\alpha = 0.1$.

Develop a Simulink diagram for feedforward-feedback control and generate two graphs similar to those in Fig. 15.13.

15.10 The distillation column in Fig. 15.8 has the following transfer function model:

$$\frac{Y'(s)}{D'(s)} = \frac{2e^{-20s}}{95s + 1} \qquad \frac{Y'(s)}{F'(s)} = \frac{0.5e^{-30s}}{60s + 1}$$

with $G_v = G_m = G_t = 1$.

(a) Design a feedforward controller based on a steady-state analysis.

(b) Design a feedforward controller based on a dynamic analysis.

(c) Design a PI feedback controller based on the Direct Synthesis approach of Chapter 12 with $\tau_c = 30$.

(d) Simulate the closed-loop response to a unit step change in the disturbance variable using feedforward control only and the controllers of parts (a) and (b). Does the dynamic controller of part (b) provide a significant improvement?

(e) Repeat part (d) for the feedforward-feedback control scheme of Fig. 15.11 and the controllers of parts (a) and (c), as well as (b) and (c).

(f) Which control configuration provides the best control?

15.11 A feedforward-only control system is to be designed for the stirred-tank heating system shown in Fig. E15.11. Exit temperature T will be controlled by adjusting coolant flow rate, q_c. The chief disturbance variable is the inlet temperature T_i which can be measured on-line. Design a feedforward-only control system based on a dynamic model of this process and the following assumptions:

1. The rate of heat transfer, Q, between the coolant and the liquid in the tank can be approximated by

$$Q = U(1 + q_c)A(T - T_c)$$

where U, A, and the coolant temperature T_c are constant.

2. The tank is well mixed, and the physical properties of the liquid remain constant.

3. Heat losses to the ambient air can be approximated by the expression $Q_L = U_L A_L(T - T_a)$, where T_a is the ambient temperature.

4. The control valve on the coolant line and the T_i sensor/transmitter (not shown in Fig. E15.11) exhibit linear behavior. The dynamics of both devices can be neglected, but there is a time delay θ associated with the T_i measurement due to the sensor location.

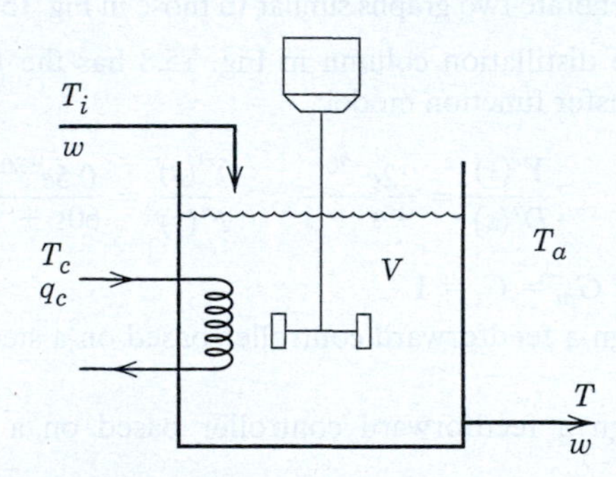

Figure E15.11

15.12 Consider the PCM furnace module of Appendix E. Assume that oxygen exit concentration c_{O_2} is the CV, air flow rate AF is the MV, and fuel gas purity FG is the DV.

(a) Using the transfer functions given below, design a feedforward control system.

(b) Design a PID controller based on IMC tuning and a reasonable value of τ_c.

(c) Simulate the FF, FB, and combined FF-FB controllers for a sudden change in d at $t = 10$ min, from 1 to 0.9. Which controller is superior? Justify your answer.

$$\frac{c_{O_2}}{AF} = G_p = \frac{0.14e^{-4s}}{4.2s + 1},$$

$$\frac{c_{O_2}}{FG} = G_d = -\frac{2.82e^{-4s}}{4.3s + 1}, \quad G_v = G_m = G_t = 1$$

15.13 It is desired to design a feedforward control scheme in order to control the exit composition x_4 of the two-tank blending system shown in Fig. E15.13. Flow rate q_2 can be manipulated, while disturbance variables, q_5 and x_5, can be measured. Assume that controlled variable x_4 cannot be measured and that each process stream has the same density. Also, assume that the volume of liquid in each tank is kept constant by using an overflow line. The transmitters and control valve have negligible dynamics.

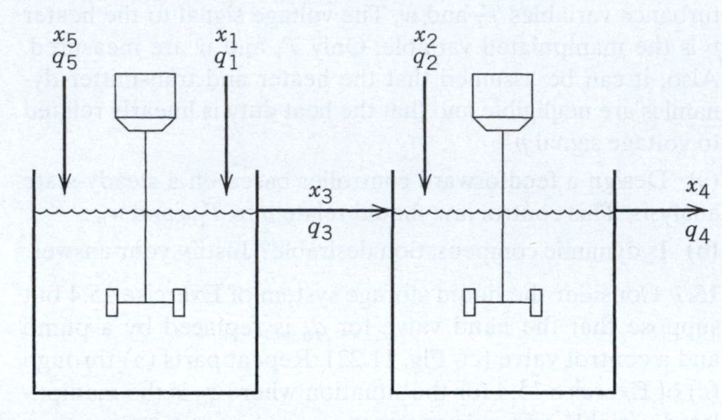

Figure E15.13

(a) Using the steady-state data given below, design an ideal feedforward control law based on steady-state considerations. State any additional assumptions that you make.

(b) Do you recommend that dynamic compensation be used in conjunction with this feedforward controller? Justify your answer.

Steady-State Data

Stream	*Flow (gpm)*	*Mass Fraction*
1	1900	0.000
2	1000	0.990
3	2400	0.167
4	3400	0.409
5	500	0.800

15.14 Consider the PCM distillation column module of Appendix E. Assume that distillate methanol composition x_D is the CV, reflux ratio R is the MV, and feed composition x_F is the DV.

(a) Using the transfer functions given below, design a feedforward control system.

(b) Design a PID controller based on IMC tuning and a reasonable value of τ_c.

(c) Simulate the FF, FB, and combined FF-FB controllers for a sudden change in d at $t = 10$ min, from 0.50 to 0.55 (mole fraction). Which controller is superior? Justify your answer.

$$\frac{X_D}{R} = G_p = \frac{0.126e^{-138s}}{762s + 1} \qquad \frac{X_D}{X_F} = G_d = \frac{0.78e^{-600s}}{700s + 1},$$

$$G_v = G_m = G_t = 1$$

Chapter 16

Enhanced Single-Loop Control Strategies

CHAPTER CONTENTS

In this chapter, we introduce several specialized strategies that provide enhanced process control beyond what can be obtained with conventional single-loop PID controllers. As processing plants become more and more complex in order to increase efficiency or reduce costs, there are incentives for using such enhancements, which also fall under the general classification of advanced control. Although new methods are continually evolving and being field-tested (Henson and Badgwell, 2006; Seborg, 1999; Rawlings et al., 2002), this chapter emphasizes six different strategies that have been proven commercially:

1. Cascade control
2. Time-delay compensation
3. Inferential control
4. Selective and override control
5. Nonlinear control
6. Adaptive control

These techniques have gained increased industrial acceptance over the past 20 years, and in many cases they utilize the principles of single-loop PID feedback controller design. These strategies can incorporate additional measurements, controlled variables, or manipulated variables, and they can also incorporate alternative block diagram structures.

16.1 CASCADE CONTROL

A disadvantage of conventional feedback control is that corrective action for disturbances does not begin until after the controlled variable deviates from the set point. As discussed in Chapter 15, feedforward control offers large improvements over feedback control for processes that have large time constants or time delays. However, feedforward control requires that the disturbances be measured explicitly, and that a model be available to calculate the controller output. An alternative approach, and one that can significantly improve the dynamic response to disturbances, employs a secondary measurement point and a secondary feedback controller. The secondary measurement point is located so that it recognizes the upset condition sooner than the controlled variable, but the disturbance is not necessarily measured. This approach, called *cascade control*, is

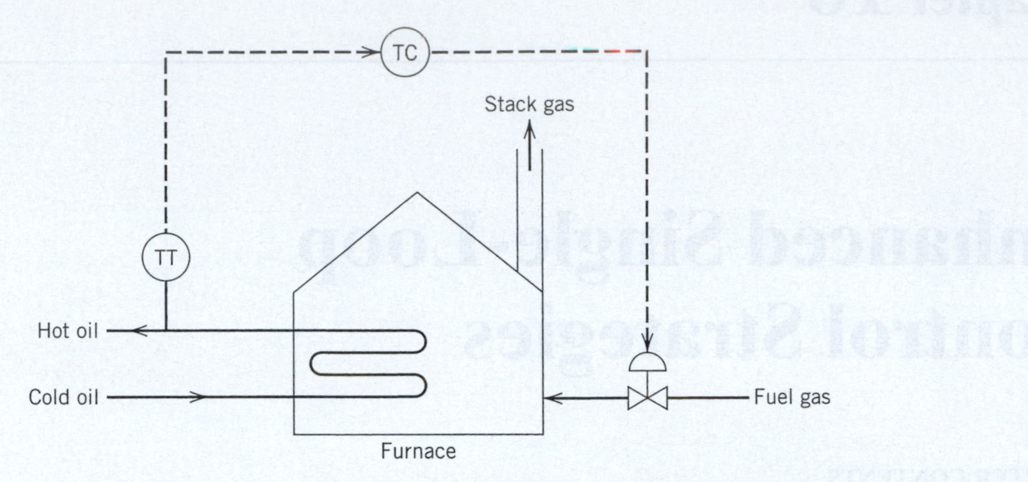

Figure 16.1 A furnace temperature control scheme that uses conventional feedback control.

widely used in the process industries and is particularly useful when the disturbances are associated with the manipulated variable or when the final control element exhibits nonlinear behavior (Shinskey, 1996).

As an example of where cascade control may be advantageous, consider the natural draft furnace temperature control problem shown in Fig. 16.1. The conventional feedback control system in Fig. 16.1 may keep the hot oil temperature close to the set point despite disturbances in oil flow rate or cold oil temperature. However, if a disturbance occurs in the fuel gas supply pressure, the fuel gas flow will change, which upsets the furnace operation and changes the hot oil temperature. Only then will the temperature controller (TC) begin to take corrective action by adjusting the fuel gas flow. Thus, we anticipate that conventional feedback control may result in very sluggish responses to changes in fuel gas supply pressure. This disturbance is clearly associated with the manipulated variable.

Figure 16.2 shows a cascade control configuration for the furnace, which consists of a primary control loop (utilizing TT and TC) and a secondary control loop that controls the pressure via PT and PC. The primary measurement is the hot oil temperature that is used by the *primary* (*master*) controller (TC) to establish the set point for the *secondary* (*slave*) loop controller. The secondary measurement is the fuel gas pressure, which is transmitted to the slave controller (PC). If a disturbance in supply pressure occurs, the pressure controller will act very quickly to hold the fuel gas pressure at its set point. The cascade control scheme provides improved performance, because the control valve will be adjusted as soon as the change in supply pressure is detected.

Because the pressure control loop responds rapidly, the supply pressure disturbance will have little effect on furnace operation and exit oil temperature. Some engineers prefer that flow control, rather than pressure control, be employed in the slave loop to deal with discharge pressure variations. If the performance improvements for disturbances in oil flow rate or inlet temperature are not large enough, then feedforward control could be utilized for those disturbances (see Chapter 15).

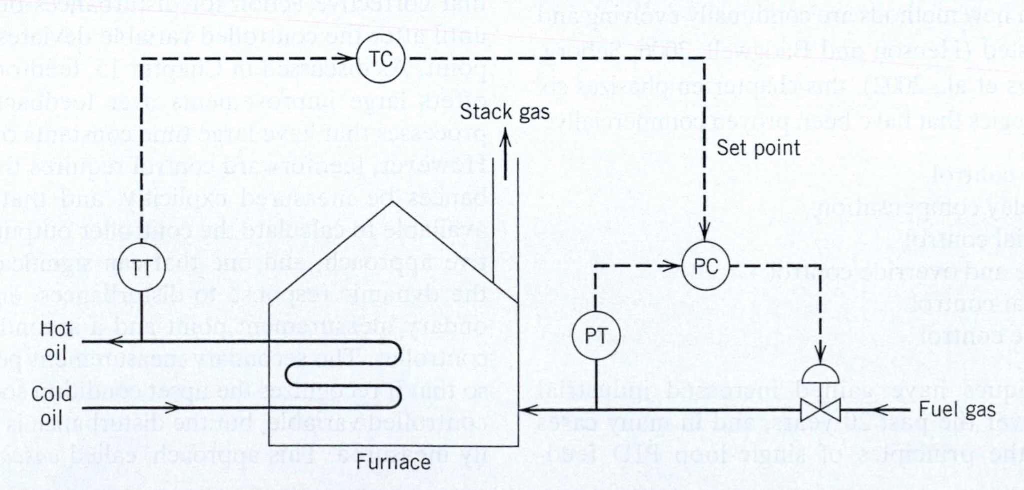

Figure 16.2 A furnace temperature control scheme using cascade control.

The cascade control loop structure has two distinguishing features:

1. The output signal of the master controller serves as the set point for the slave controller.
2. The two feedback control loops are nested, with the secondary control loop (for the slave controller) located inside the primary control loop (for the master controller).

Thus there are two controlled variables, two sensors, and one manipulated variable, while the conventional control structure has one controlled variable, one sensor, and one manipulated variable.

The primary control loop can change the set point of the pressure control loop based on deviations of the hot oil temperature from its set point. Note that all variables in this configuration can be viewed as deviation variables. If the hot oil temperature is at its set point, the deviation variable for the pressure set point is also zero, which keeps the pressure at its desired steady-state value.

Figure 16.3 shows a second example of cascade control, a stirred chemical reactor where cooling water flows through the reactor jacket to regulate the reactor temperature. The reactor temperature is affected by changes in disturbance variables such as reactant feed temperature or feed composition. The simplest control strategy would handle such disturbances by adjusting a control valve on the cooling water inlet stream. However, an increase in the inlet cooling water temperature, an unmeasured disturbance, can cause unsatisfactory performance. The resulting increase in the reactor temperature, due to a reduction in heat removal rate, may occur slowly. If appreciable dynamic lags occur in the jacket as well as in the reactor, the corrective action taken by the controller will be delayed. To avoid this disadvantage, a feedback controller for the jacket temperature, whose set point is determined by the reactor temperature controller, can be added to provide cascade control, as shown in Fig. 16.3. The control system measures the jacket temperature, compares it to a set point, and adjusts the cooling water makeup. The reactor temperature set point and both measurements are used to adjust a single manipulated variable, the cooling water makeup rate. The principal advantage of the cascade control strategy is that a second measured variable is located close to a potential disturbance and its associated feedback loop can react quickly, thus improving the closed-loop response. However, if cascade control does not improve the response, feedforward control should be the next strategy considered, with cooling water temperature as the measured disturbance variable.

The block diagram for a general cascade control system is shown in Fig. 16.4. Subscript 1 refers to the primary control loop, whereas subscript 2 refers to the secondary control loop. Thus, for the furnace temperature control example,

Y_1 = hot oil temperature
Y_2 = fuel gas pressure
D_1 = cold oil temperature (or cold oil flow rate)
D_2 = supply pressure of fuel gas
Y_{m1} = measured value of hot oil temperature
Y_{m2} = measured value of fuel gas pressure
Y_{sp1} = set point for Y_1
$\tilde{Y}_{sp2}$ = set point for Y_2

All of these variables represent deviations from the nominal steady state. Because disturbances can affect both the primary and secondary control loops, two

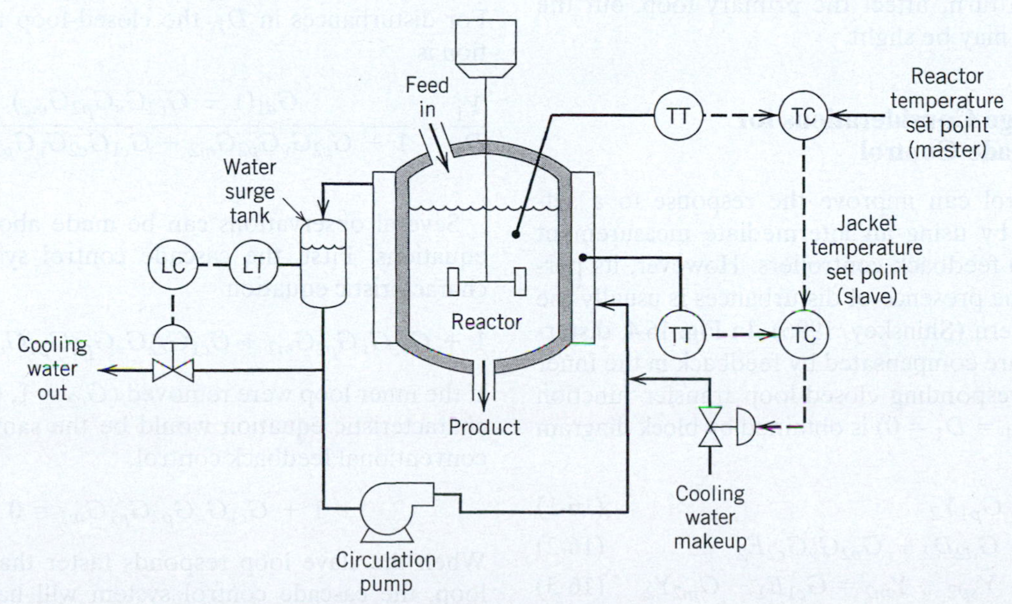

Figure 16.3 Cascade control of an exothermic chemical reactor.

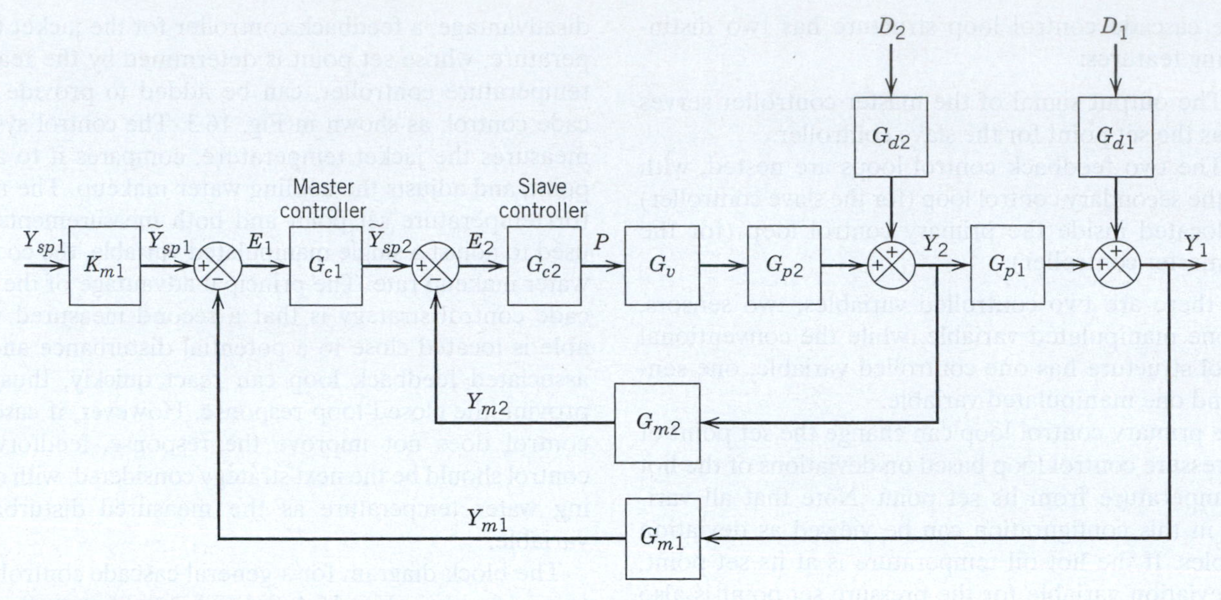

Figure 16.4 Block diagram of the cascade control system.

disturbance variables (D_1 and D_2) and two disturbance transfer functions (G_{d1} and G_{d2}) are shown in Fig. 16.4. Note that Y_2 serves as both the controlled variable for the secondary loop and the manipulated variable for the primary loop.

Figures 16.2 and 16.4 clearly show that cascade control will effectively reduce the effects of pressure disturbances entering the secondary loop (i.e., D_2 in Fig. 16.4). But what about the effects of disturbances such as D_1, which enter the primary loop? Cascade control can provide an improvement over conventional feedback control when both controllers are well tuned. The cascade arrangement will reduce the response times of the elements in the secondary loop, which will, in turn, affect the primary loop, but the improvement may be slight.

16.1.1 Design Considerations for Cascade Control

Cascade control can improve the response to a set-point change by using an intermediate measurement point and two feedback controllers. However, its performance in the presence of disturbances is usually the principal concern (Shinskey, 1996). In Fig. 16.4, disturbances in D_2 are compensated by feedback in the inner loop; the corresponding closed-loop transfer function (assuming $Y_{sp1} = D_1 = 0$) is obtained by block diagram algebra:

$$Y_1 = G_{p1} Y_2 \tag{16-1}$$

$$Y_2 = G_{d2} D_2 + G_{p2} G_v G_{c2} E_2 \tag{16-2}$$

$$E_2 = \widetilde{Y}_{sp2} - Y_{m2} = G_{c1} E_1 - G_{m2} Y_2 \tag{16-3}$$

$$E_1 = -G_{m1} Y_1 \tag{16-4}$$

Eliminating all variables except Y_1 and D_2 gives

$$\frac{Y_1}{D_2} = \frac{G_{p1} G_{d2}}{1 + G_{c2} G_v G_{p2} G_{m2} + G_{c1} G_{c2} G_v G_{p2} G_{p1} G_{m1}} \tag{16-5}$$

By similar analysis, the set-point transfer functions for the outer and inner loops are

$$\frac{Y_1}{Y_{sp1}} = \frac{G_{c1} G_{c2} G_v G_{p1} G_{p2} K_{m1}}{1 + G_{c2} G_v G_{p2} G_{m2} + G_{c1} G_{c2} G_v G_{p2} G_{p1} G_{m1}} \tag{16-6}$$

$$\frac{Y_2}{\widetilde{Y}_{sp2}} = \frac{G_{c2} G_v G_{p2}}{1 + G_{c2} G_v G_{p2} G_{m2}} \tag{16-7}$$

For disturbances in D_1, the closed-loop transfer function is

$$\frac{Y_1}{D_1} = \frac{G_{d1}(1 + G_{c2} G_v G_{p2} G_{m2})}{1 + G_{c2} G_v G_{p2} G_{m2} + G_{c1} G_{c2} G_v G_{p2} G_{p1} G_{m1}} \tag{16-8}$$

Several observations can be made about the above equations. First, the cascade control system has the characteristic equation

$$1 + G_{c2} G_v G_{p2} G_{m2} + G_{c1} G_{c2} G_v G_{p2} G_{p1} G_{m1} = 0 \tag{16-9}$$

If the inner loop were removed ($G_{c2} = 1$, $G_{m2} = 0$), the characteristic equation would be the same as that for conventional feedback control,

$$1 + G_{c1} G_v G_{p2} G_{p1} G_{m1} = 0 \tag{16-10}$$

When the slave loop responds faster than the master loop, the cascade control system will have improved stability characteristics and thus should allow larger values of K_{c1} to be used in the primary control loop. Cascade control also makes the closed-loop process less sensitive to errors in the process model used to design the controller.

EXAMPLE 16.1

Consider the block diagram in Fig. 16.4 with the following transfer functions:

$$G_v = \frac{5}{s+1} \quad G_{p1} = \frac{4}{(4s+1)(2s+1)} \quad G_{p2} = 1$$

$$G_{d2} = 1 \quad G_{m1} = 0.05 \quad G_{m2} = 0.2 \quad G_{d1} = \frac{1}{3s+1}$$

where the time constants have units of minutes and the gains have consistent units. Determine the stability limits for a conventional proportional controller as well as for a cascade control system consisting of two proportional controllers. Assume $K_{c2} = 4$ for the secondary controller. Calculate the resulting offset for a unit step change in the secondary disturbance variable D_2.

SOLUTION

For the cascade arrangement, first analyze the inner loop. Substituting into Eq. 16-7 gives

$$\frac{Y_2}{\tilde{Y}_{sp2}} = \frac{4\left(\frac{5}{s+1}\right)}{1+4\left(\frac{5}{s+1}\right)(0.2)} = \frac{20}{s+5} = \frac{4}{0.2s+1} \quad (16\text{-}11)$$

From Eq. 16-11 the closed-loop time constant for the inner loop is 0.2 min. In contrast, the conventional feedback control system has a time constant of 1 min because in this case, $Y_2(s)/\tilde{Y}_{sp2}(s) = G_v = 5/(s+1)$. Thus, cascade control significantly speeds up the response of Y_2. Using a proportional controller in the primary loop ($G_{c1} = K_{c1}$), the characteristic equation becomes

$$1 + (K_{c1})(4)\left(\frac{5}{s+1}\right)\left(\frac{4}{(4s+1)(2s+1)}\right)(0.05) + 4\left(\frac{5}{s+1}\right)(0.2) = 0 \quad (16\text{-}12)$$

which reduces to

$$8s^3 + 46s^2 + 31s + 5 + 4K_{c1} = 0 \quad (16\text{-}13)$$

By use of the Routh array (Chapter 11), the ultimate gain for marginal stability is $K_{c1,u} = 43.3$.

For the conventional feedback system with proportional-only control, the characteristic equation in (16-10) reduces to

$$8s^3 + 14s^2 + 7s + 1 + K_{c1} = 0 \quad (16\text{-}14)$$

The Routh array gives $K_{c1,u} = 11.25$. Therefore, the cascade configuration has increased the ultimate gain by nearly a factor of four. Increasing K_{c2} will result in even larger values for $K_{c1,u}$. For this example, there is no theoretical upper limit for K_{c2}, except that large values will cause the valve to saturate for small set-point changes or disturbances.

The offset of Y_1 for a unit step change in D_2 can be obtained by setting $s = 0$ in the right side of (16-5); equivalently, the Final Value Theorem of Chapter 3 can be applied for a unit step change in D_2 ($Y_{sp1} = 0$):

$$e_1(t \to \infty) = y_{sp1} - y_1(t \to \infty) = \frac{-4}{5 + 4K_{c1}} \quad (16\text{-}15)$$

By comparison, the offset for conventional control ($G_{m2} = 0$, $G_{c2} = 1$) is

$$e_1(t \to \infty) = \frac{-4}{1 + K_{c1}} \quad (16\text{-}16)$$

By comparing (16-15) and (16-16), it is clear that for the same value of K_{c1}, the offset is much smaller (in absolute value) for cascade control.

For a cascade control system to function properly, the secondary control loop must respond faster than the primary loop. The secondary controller is normally a P or PI controller, depending on the amount of offset that would occur with proportional-only control. Note that small offsets in the secondary loop can be tolerated, because the primary loop will compensate for them. Derivative action is rarely used in the secondary loop. The primary controller is usually PI or PID.

For processes with higher-order dynamics and/or time delay, the model can first be approximated by a low-order model, or the frequency response methods described in Chapter 14 can be employed to design controllers. First, the inner loop frequency response for a set-point change is calculated from (16-7), and a suitable value of K_{c2} is determined. The offset is checked to determine whether PI control is required. After K_{c2} is specified, the outer loop frequency response can be calculated, as in conventional feedback controller design. The open-loop transfer function used in this part of the calculation is

$$G_{OL} = \frac{G_{c1}K_{c2}G_vG_{p2}}{1 + K_{c2}G_vG_{p2}G_{m2}} G_{p1}G_{m1} \quad (16\text{-}17)$$

For the design of G_{c1}, we should consider the closed-loop transfer functions for set-point changes Y_1/Y_{sp1} and for disturbances, Y_1/D_2 and Y_1/D_1. Generally, cascade control is superior to conventional control in this regard and provides superior time-domain responses. Figure 16.5 shows the closed-loop response for Example 16.1 and disturbance variable D_2. The cascade configuration has a PI controller in the primary loop and a proportional controller in the secondary loop. Each controller was tuned using frequency response analysis (see Section 14.7). Figure 16.5 demonstrates that the cascade control system is superior to a conventional PI controller for a secondary loop disturbance. Figure 16.6

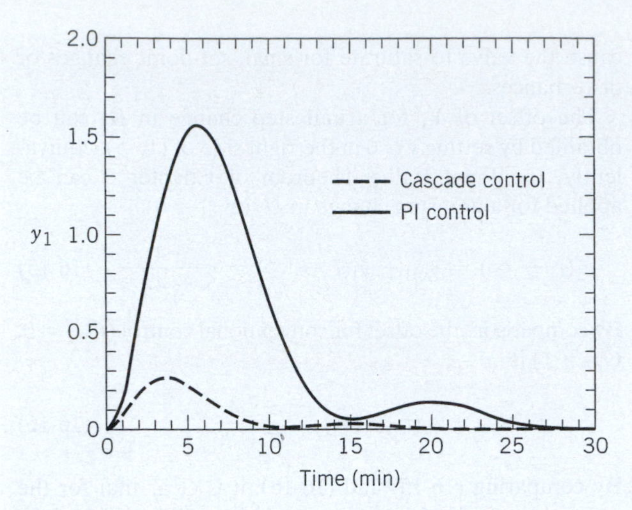

Figure 16.5 A comparison of D_2 unit step responses with and without cascade control.

shows a similar comparison for a step change in the primary loop disturbance D_1.

When a cascade control system is tuned after installation, the secondary controller should be tuned first with the primary controller in the manual mode. Then the primary controller is transferred to automatic, and it is tuned. The relay auto-tuning technique presented in Chapter 12 can be used for each control loop. If the secondary controller is retuned for some reason, usually the primary controller must also be retuned. Alternatively, Lee et al. (1998) have developed a tuning method based on Direct Synthesis where both loops are tuned simultaneously. When there are limits on either controller (saturation constraints), Brosilow and Joseph (2002) have recommended design modifications based on the Internal Model Control (IMC) approach.

A commonly used form of cascade control involves a *valve positioner*, which addresses the problem of valve nonidealities. Valves may stick or exhibit dead zones or hysteresis, and so they may not achieve the same percentage stem position required for a given controller output. The valve positioner senses the valve stem position and uses an internal proportional controller in the inner loop of a cascade control system

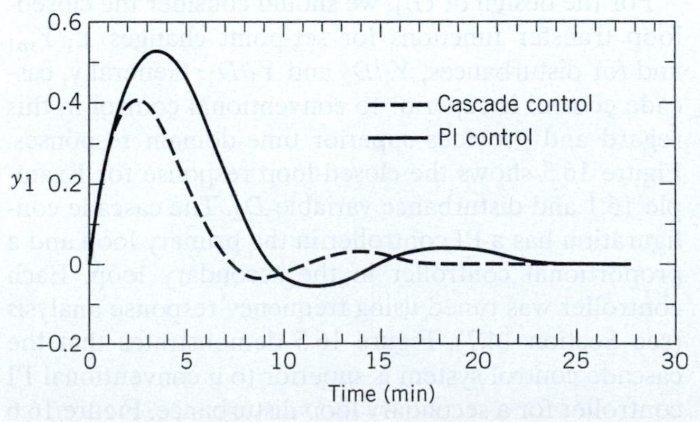

Figure 16.6 A comparison of D_1 step responses.

to attain the desired stem position. The set point is provided from the flow controller. Valve positioners are almost always beneficial when used in control loops. See Chapter 9 for more details on these devices.

16.2 TIME-DELAY COMPENSATION

In this section we present an advanced control technique, *time-delay compensation*, which deals with a problematic area in process control—namely, the occurrence of significant time delays. Time delays commonly occur in the process industries because of the presence of distance velocity lags, recycle loops, and the analysis time associated with composition measurement. As discussed in Chapters 12 and 14, the presence of time delays in a process limits the performance of a conventional feedback control system. From a frequency response perspective, a time delay adds phase lag to the feedback loop, which adversely affects closed-loop stability. Consequently, the controller gain must be reduced below the value that could be used if no time delay were present, and the response of the closed-loop system will be sluggish compared to that of the control loop with no time delay.

EXAMPLE 16.2

Compare the set-point responses for a second-order process with a time delay (θ = 2 min) and without the delay. The transfer function for the delay case is

$$G_p(s) = \frac{e^{-\theta s}}{(5s+1)(3s+1)} \tag{16-18}$$

Assume that $G_m = G_v = 1$, with time constants in minutes. Use the following PI controllers. For $\theta = 0$, $K_c = 3.02$ and $\tau_I = 6.5$ min, while for $\theta = 2$ min the controller gain must be reduced to meet stability requirements ($K_c = 1.23$, $\tau_I = 7.0$ min).

SOLUTION

The closed-loop responses are shown in Fig. 16.7. For $\theta = 2$, the resulting response is more sluggish. Clearly the closed-loop response for the time-delay case has deteriorated, with a 50% increase in response time (30 vs. 20 min). This response is much longer than might be expected from the size of the time delay.

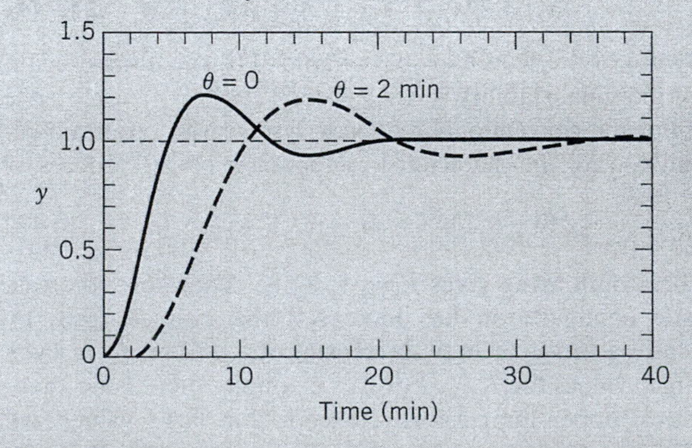

Figure 16.7 A comparison of closed-loop set-point changes.

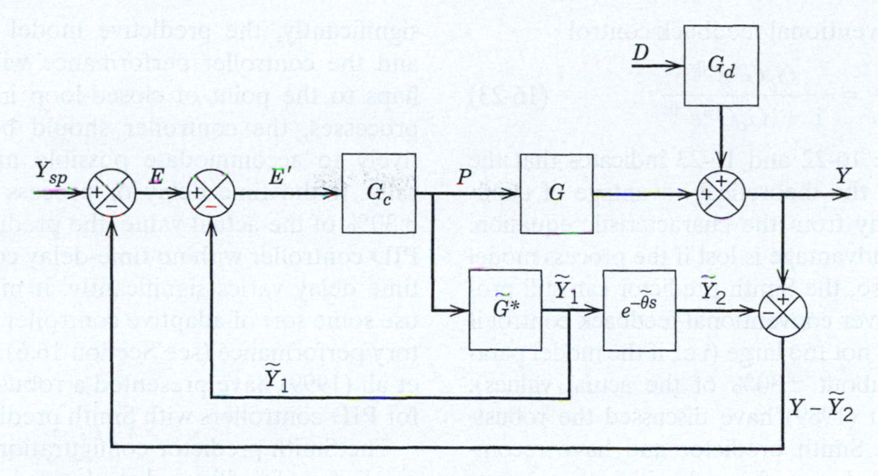

Figure 16.8 Block diagram of the Smith predictor.

In order to improve the performance of system containing time delays, special control strategies have been developed that provide effective time-delay compensation. The *Smith predictor* technique is the best known strategy (Smith, 1957). A related method, the *analytical predictor*, has been developed specifically for digital control applications, and it is discussed in Chapter 17. Various investigators have found that the performance of a controller incorporating the Smith predictor for set-point changes is better than a conventional PI controller based on the integral-squared-error criterion. However, the Smith predictor performance may not be superior for all types of disturbances.

A block diagram of the Smith predictor controller is shown in Fig. 16.8, where $G_v = G_m = 1$ for simplicity. Here the process model $\widetilde{G}(s)$ is divided into two parts: the part without a time delay, $\widetilde{G}^*(s)$, and the time-delay term, $e^{-\widetilde{\theta}s}$. Thus, the total transfer function model is $\widetilde{G}(s) = \widetilde{G}^*(s)e^{-\widetilde{\theta}s}$. The model of the process without the time delay, $\widetilde{G}^*(s)$, is used to predict the effect of control actions on the undelayed output. The controller then uses the predicted response $\widetilde{Y}_1$ to calculate its output signal. The predicted process output is also delayed by the amount of the time delay $\widetilde{\theta}$, for comparison with the actual undelayed output Y. This delayed model output is denoted by $\widetilde{Y}_2$ in Fig. 16.8. From the block diagram,

$$E' = E - \widetilde{Y}_1 = Y_{sp} - \widetilde{Y}_1 - (Y - \widetilde{Y}_2) \tag{16-19}$$

if the process model is perfect and the disturbance is zero, then $\widetilde{Y}_2 = Y$ and

$$E' = Y_{sp} - \widetilde{Y}_1 \tag{16-20}$$

For this ideal case, the controller responds to the error signal that would occur if no time delay were present.

Figure 16.9 shows an alternative (equivalent) configuration for the Smith predictor that includes an inner feedback loop, somewhat similar to that in cascade control. Assuming there is no model error ($\widetilde{G} = G$), the inner loop has the effective transfer function

$$G' = \frac{P}{E} = \frac{G_c}{1 + G_c G^*(1 - e^{-\theta s})} \tag{16-21}$$

where G^* is defined analogously to $\widetilde{G}^*$, that is, $G = G^* e^{-\theta s}$. After some rearrangement, the closed-loop set-point transfer function is obtained:

$$\frac{Y}{Y_{sp}} = \frac{G_c G^* e^{-\theta s}}{1 + G_c G^*} \tag{16-22}$$

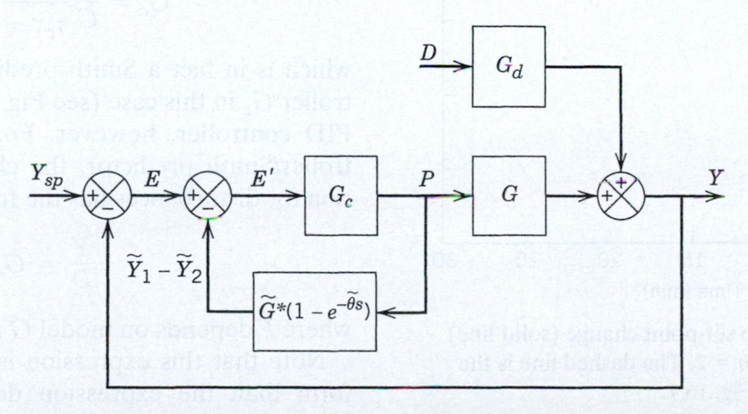

Figure 16.9 An alternative block diagram of a Smith predictor.

By contrast, for conventional feedback control

$$\frac{Y}{Y_{sp}} = \frac{G_c G^* e^{-\theta s}}{1 + G_c G^* e^{-\theta s}} \tag{16-23}$$

Comparison of Eqs. 16-22 and 16-23 indicates that the Smith predictor has the theoretical advantage of eliminating the time delay from the characteristic equation. Unfortunately, this advantage is lost if the process model is inaccurate. Even so, the Smith predictor can still provide improvement over conventional feedback control if the model errors are not too large (i.e., if the model parameters are within about ±30% of the actual values). Morari and Zafiriou (1989) have discussed the robustness aspects of the Smith predictor and have recommended that tuning be performed with other inputs besides step inputs.

Figure 16.10 shows the closed-loop responses for the Smith predictor ($\theta = 2$) and PI control ($\theta = 0$). The controller settings are the same as those developed in Example 16.2 for $\theta = 0$. A comparison of Fig. 16.7 (dashed line) and Fig. 16.10 shows the improvement in performance that can be obtained with the Smith predictor. Note that the responses in Fig. 16.10 for $\theta = 2$ and $\theta = 0$ are identical, except for the initial time delay. This closed-loop time delay results from the numerator delay term in (16-22). The process model $G^*(s)$ is second-order and thus readily yields a stable closed-loop system for P-only control, but not necessarily for PI control (cf. Example 14.4).

One disadvantage of the Smith predictor approach is that it is model-based; that is, a dynamic model of the process is required. If the process dynamics change significantly, the predictive model will be inaccurate and the controller performance will deteriorate, perhaps to the point of closed-loop instability. For such processes, the controller should be tuned conservatively to accommodate possible model errors. Typically, if the time delay or process gain is not within ±30% of the actual value, the predictor is inferior to a PID controller with no time-delay compensation. If the time delay varies significantly, it may be necessary to use some sort of adaptive controller to achieve satisfactory performance (see Section 16.6). Alternatively, Lee et al. (1999) have presented a robust tuning procedure for PID controllers with Smith predictors.

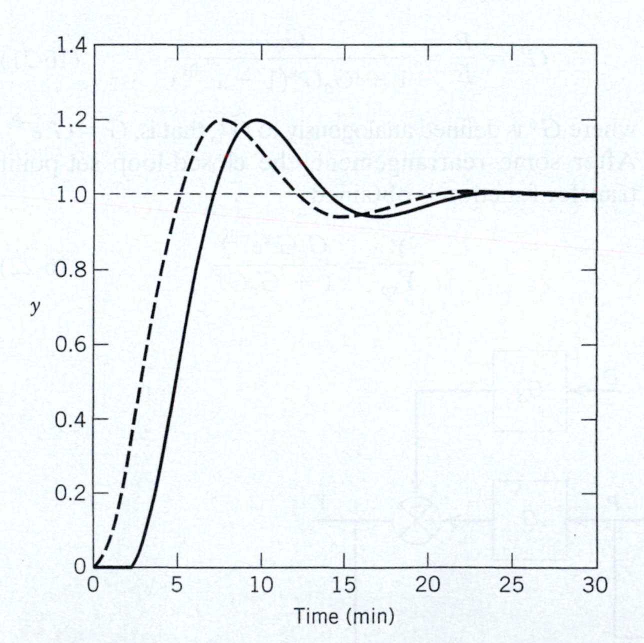

Figure 16.10 Closed-loop set-point change (solid line) for Smith predictor with $\theta = 2$. The dashed line is the response for $\theta = 0$ from Fig. 16.7.

The Smith predictor configuration generally is beneficial for handling disturbances. However, under certain conditions, a conventional PI controller can provide better regulatory control than the Smith predictor. This somewhat anomalous behavior can be attributed to the closed-loop transfer function for a disturbance and a perfect model:

$$\frac{Y}{D} = \frac{G_d[1 + G_c G^*(1 - e^{-\theta s})]}{1 + G_c G^*} \tag{16-24}$$

The denominators of Y/D in (16-24) and Y/Y_{sp} in (16-22) are the same, but the numerator terms are quite different in form. Figure 16.11 shows disturbance responses for Example 16.2 ($\theta = 2$) for PI controllers with and without the Smith predictor. By using the two degree-of-freedom controllers discussed in Chapter 12, it is possible to improve the response for disturbances and avoid this undesirable behavior. In fact, Ingimundarson and Hägglund (2002) have shown that for step disturbances and a FOPTD process, the performance of a properly tuned PID controller is comparable to or better than a PI controller with time-delay compensation.

The Direct Synthesis approach can be used to derive a controller with time-delay compensation, as discussed in Chapter 12. If a FOPTD model is stipulated for the closed-loop transfer function $(Y/Y_{sp})_d$ (see Eq. 12-6), then the resulting design equation for G_c is

$$G_c = \frac{1}{G} \frac{e^{-\theta s}}{\tau_c s + 1 - e^{-\theta s}} \tag{12-7}$$

which is in fact a Smith predictor. The feedback controller G_c in this case (see Fig. 16.9) is not necessarily a PID controller, however. For this model-based controller/Smith predictor, the closed-loop transfer function for disturbances has the following form,

$$\frac{Y}{D} = G_d F \tag{16-25}$$

where F depends on model G and design parameter τ_c.

Note that this expression is considerably simpler in form than the expression derived for a PI or PID

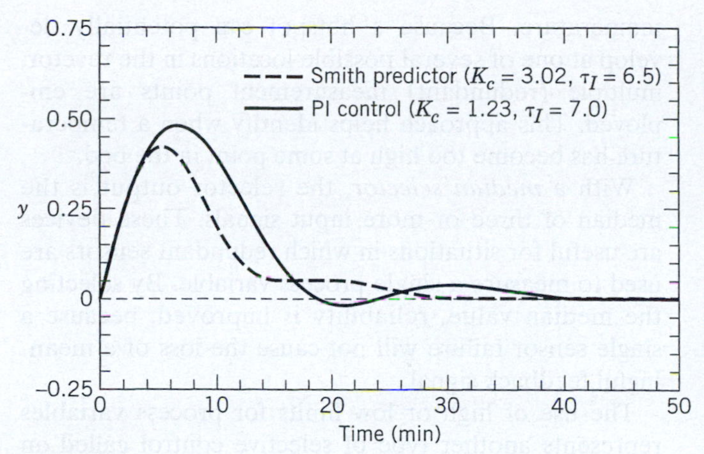

Figure 16.11 A comparison of disturbance changes for the Smith predictor and a conventional PI controller.

controller shown in (16-24), and that the model-based controller designed to obtain a desired set-point change does not influence the disturbance dynamics embedded in G_d. In fact, if the dominant time constant in G_d is relatively large, a controller designed for set-point changes by either Direct Synthesis or IMC will not be able to speed up the disturbance response, as demonstrated in Example 12.4.

Another control strategy for treating both disturbances and set-point changes is the analytical predictor, which utilizes a prediction of the process behavior in the future based on the process and disturbance transfer functions, G and G_d. In the context of Eq. 16-23, if G_c included a term $e^{+\theta s}$ (a perfect prediction θ units of time ahead), then the time delay would effectively be eliminated from the characteristic equation. However, this is an idealized view, and further details are given in Chapter 17.

16.3 INFERENTIAL CONTROL

The previous discussion of time-delay compensation assumed that measurements of the controlled variable were available. In some control applications, the process variable that is to be controlled cannot be conveniently measured on-line. For example, product composition measurement may require that a sample be sent to the plant analytical laboratory from time to time. In this situation, measurements of the controlled variable may not be available frequently enough or quickly enough to be used for feedback control.

One solution to this problem is to employ *inferential control*, where process measurements that can be obtained more rapidly are used with a mathematical model, sometimes called a *soft sensor*, to infer the value of the controlled variable. For example, if the overhead product stream in a distillation column cannot be analyzed on-line, sometimes measurement of a selected tray temperature can be used to infer the actual composition. For a binary mixture, the Gibbs phase rule indicates that a unique relation exists between composition and temperature if pressure is constant and there is vapor-liquid equilibrium. In this case, a thermodynamic equation of state can be employed to infer the composition from a tray temperature.

On the other hand, for the separation of multicomponent mixtures, approximate methods to estimate compositions must be used. Based on process models and plant data, simple algebraic correlations can be developed that relate the mole fraction of the heavy key component to several different tray temperatures (usually in the top half of the column above the feed tray). The overhead composition can then be inferred from the available temperature measurements and used in the control algorithm. The parameters in the correlation may be updated, if necessary, as composition measurements become available. For example, if samples are sent to the plant's analytical laboratory once per hour, the correlation parameters can be adjusted so that the predicted values agree with the measured values. Figure 16.12 shows the general structure of an inferential controller. X is the secondary measurement, which is available on a nearly continuous basis (fast sampling), while Y is the primary measurement, which is obtained intermittently and less frequently (e.g., off-line laboratory sample analysis). Note that X and/or Y can be used for control. One type of nonlinear model that could be used as a soft sensor is a neural network (see Chapter 7). The inferential model is obtained by analyzing and fitting accumulated X and Y data. Dynamic linear or nonlinear models (called *observers*) can also be used for inferential control, as reviewed by Doyle (1998).

The concept of inferential control can be employed for other process operations, such as chemical reactors, where composition is normally the controlled variable. Selected temperature measurements can be used to estimate the outlet composition if it cannot be measured on-line. However, when inferential control

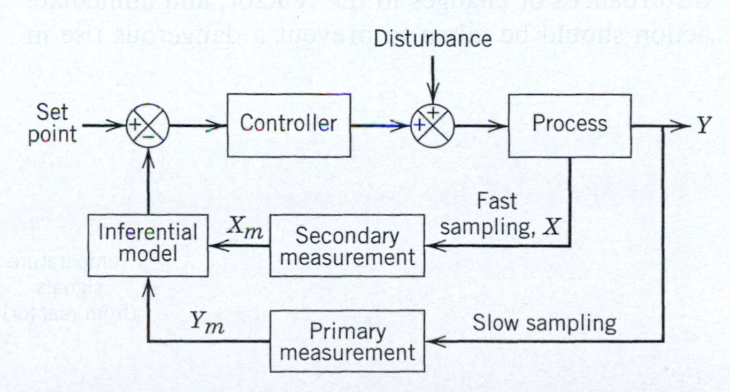

Figure 16.12 Block diagram of a soft sensor used in inferential control.

does not perform satisfactorily, an incentive exists to introduce other on-line measurements for feedback control. Consequently, there is ongoing interest in the development of new sensors, such as novel process analyzers, which can be used on-line and whose response times are very short.

16.4 SELECTIVE CONTROL/OVERRIDE SYSTEMS

Most process control applications have an equal number of controlled variables and manipulated variables. However, if fewer manipulated variables than controlled variables are available, it is not possible to eliminate offset in all the controller variables for arbitrary disturbances or set-point changes. This assertion is evident from a degrees-of-freedom analysis of a steady-state model. For control problems with fewer manipulated variables than controlled variables, selectors are employed for sharing the manipulated variables among the controlled variables.

16.4.1 Selectors

A *selector* is a practical solution for choosing the appropriate signal from among a number of available measurements. Selectors can be based on multiple measurement points, multiple final control elements, or multiple controllers, as discussed below. Selectors are used to improve the control system performance as well as to protect equipment from unsafe operating conditions. On instrumentation diagrams, the symbol (>) denotes a high selector and the symbol (<) a low selector.

For one type of selector, the output signal is the highest (or lowest) of two or more input signals. This approach is often referred to as *auctioneering* (Shinskey, 1996). For example, a high selector can be used to determine the *hotspot* temperature in a fixed-bed chemical reactor as shown in Fig. 16.13. In this reactor application, the output from the high selector is the input to the temperature controller. In an exothermic catalytic reaction, the process may "run away" due to disturbances or changes in the reactor, and immediate action should be taken to prevent a dangerous rise in temperature. Because a hotspot can potentially develop at one of several possible locations in the reactor, multiple (redundant) measurement points are employed. This approach helps identify when a temperature has become too high at some point in the bed.

With a *median selector*, the selector output is the median of three or more input signals. These devices are useful for situations in which redundant sensors are used to measure a single process variable. By selecting the median value, reliability is improved, because a single sensor failure will not cause the loss of a meaningful feedback signal.

The use of high or low limits for process variables represents another type of selective control called an *override*, where a second controller can "override" or take over from the first controller. This is a less extreme action than an interlock, which is used for emergency shutdown of the process (see Chapter 10). The anti-reset windup feature in feedback controllers (cf. Chapter 8) is a type of override. Another example is a distillation column that has lower and upper limits on the heat input to the column reboiler. The minimum level ensures adequate liquid inventory on the trays, while the upper limit exists to prevent the onset of flooding (Buckley et al., 1985; Shinskey, 1996). Overrides are also often used in forced draft combustion control systems to prevent an imbalance between air flow and fuel flow, which could result in unsafe operating conditions (Singer, 1981).

Other types of selective systems employ multiple final control elements or multiple controllers. For example, in *split-range control* several manipulated variables are used to control a single controlled variable. Typical examples include the adjustment of both inflow and outflow from a chemical reactor in order to control reactor pressure or the use of both acid and base to control pH in wastewater treatment. Another example is in reactor control, where both heating and cooling are used to maintain precise regulation of the reactor temperature. Figure 16.14 shows how the control loop in Fig. 16.3 can be modified to accommodate both heating and cooling using a single controller and two control valves. This split-range control is achieved using the controller input-output relationship shown in Fig. 16.14*b*.

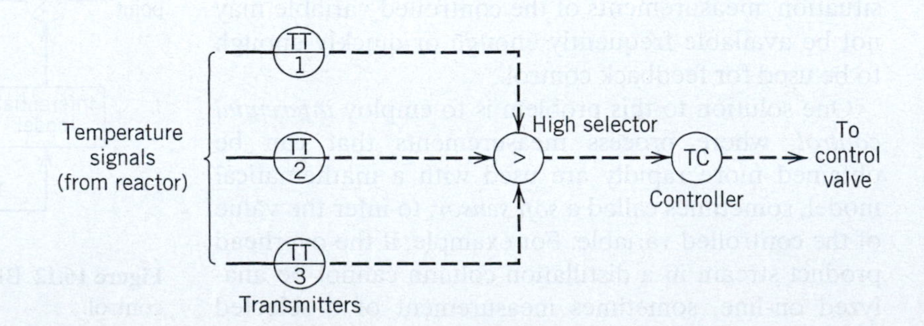

Figure 16.13 Control of a reactor hotspot temperature by using a high selector.

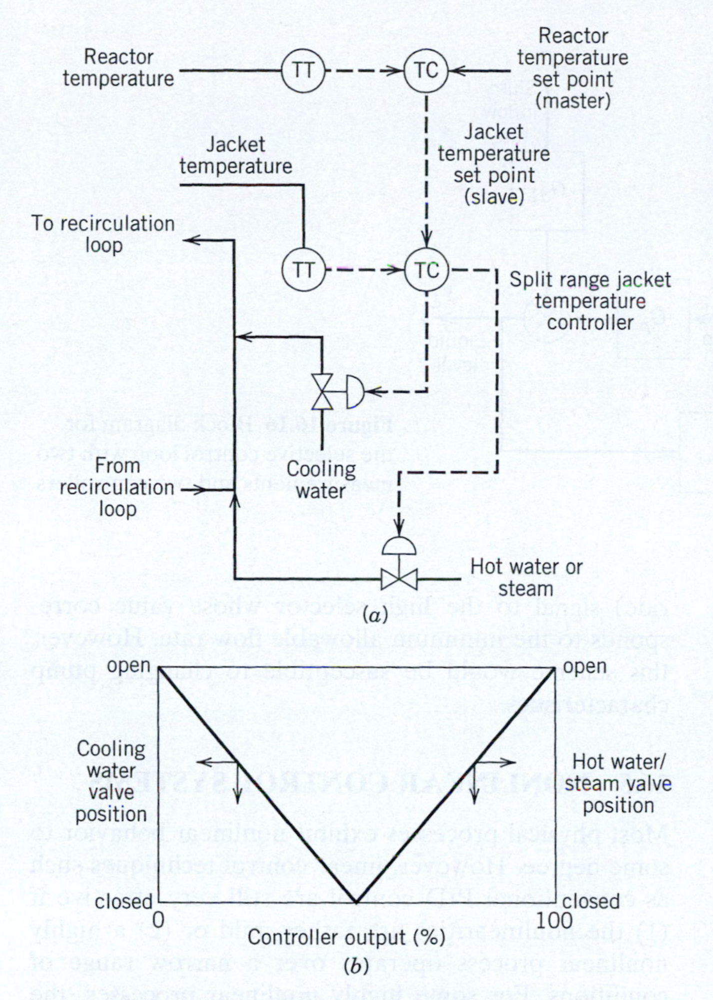

Figure 16.14 Split range control: (*a*) control loop configuration, (*b*) valve position–controller output relationship.

An alternative solution for reactor temperature control can be applied when both heating and cooling are necessary, or when a median temperature is regulated. Although the physical processes are configured in the same way as in Fig. 16.14, the controller output is mapped to the valves so that both valves are always active; that is, cooling/heating medium is always flowing through the heat exchange equipment. However, it is not a requirement that both valves must be of the same size and characteristics. Note also that in this scheme, the heat removal capability is more or less linear with respect to the controller output, regardless of valve selection.

Constraint control is another type of selector or override that is intended to keep the controlled variable near a constraining or limiting value. Chapter 19 discusses how constraints influence the selection of operating conditions and why it is necessary in many cases to operate near a constraint boundary. Riggs (1998) has described a constraint control application for distillation columns with dual composition control, where reboiler duty Q_R controls bottoms composition of x_B and reflux flow R controls overhead composition x_D. The reboiler becomes constrained at its upper limit when the steam flow control valve is completely open. Several abnormal situations can result: (1) the column pressure increases, (2) heat transfer surfaces become fouled, or (3) the column feed rate increases. When the reboiler duty reaches the upper limit, it is no longer able to control bottoms composition, so constraint control forces one composition (the more valuable product) to be controlled with the reflux ratio while the other product composition is left uncontrolled (allowed to "float"). Computer control logic must be added to determine when the column has returned to normal operation, and thus the constraint control should be made inactive.

The selective control system shown in Fig. 16.15 is used to regulate the level and exit flow rate in a pumping system for a sand/water slurry. During normal operation, the level controller (LC) adjusts the slurry exit flow by changing the pump speed. A variable-speed pump is used rather than a control valve owing to the abrasive nature of the slurry. The slurry velocity in the

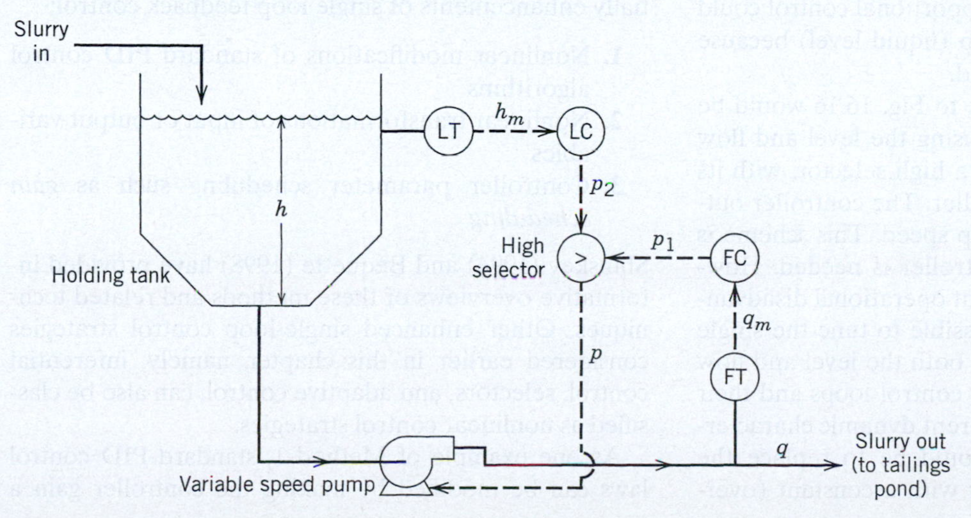

Figure 16.15 A selective control system to handle a sand/water slurry.

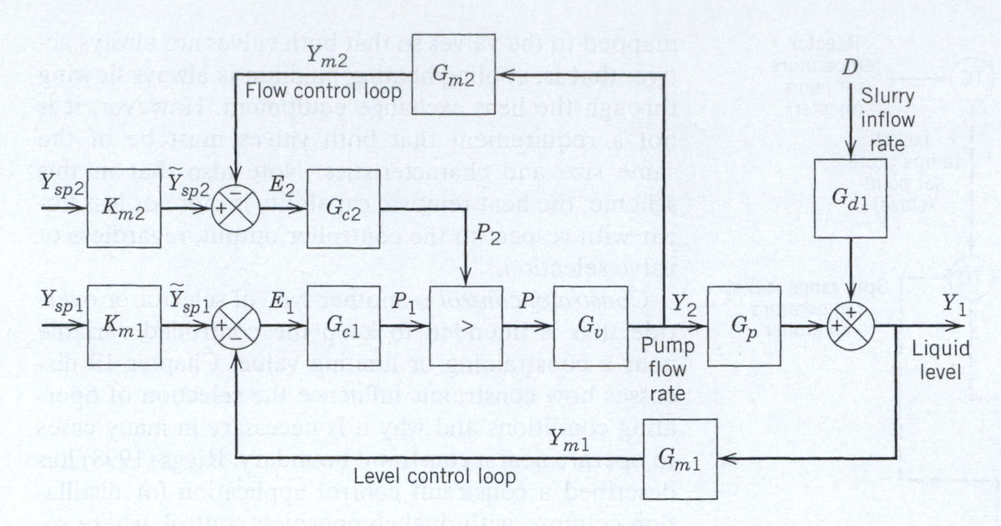

Figure 16.16 Block diagram for the selective control loop with two measurements and two controllers.

exit line must be kept above a minimum value at all times to prevent the line from sanding up. Consequently, the selective control system is designed so that as the flow rate approaches the lower limit, the flow controller takes over from the level controller and speeds up the pump. The strategy is implemented in Fig. 16.15 using a high selector and a reverse-acting flow controller with a high gain. The set point and gain of the flow controller are chosen so that the controller output is at the maximum value when the measured flow is near the constraint.

The block diagram for the selector control loop used in the slurry example is shown in Fig. 16.16. The selector compares signals P_1 and P_2, both of which have the same units (e.g., mA or %). There are two parallel feedback loops. Note that G_v is the transfer function for the final control element, the variable-speed drive pump. A stability analysis of Fig. 16.16 would be rather complicated because the high selector introduces a nonlinear element into the control system. Typically, the second loop (pump flow) will be faster than the first loop (level) and uses PI control (although reset windup protection will be required). Proportional control could be employed in the slower loop (liquid level) because tight level control is not required.

One alternative arrangement to Fig. 16.16 would be to employ a single controller, using the level and flow transmitter signals as inputs to a high selector, with its output signal sent to the controller. The controller output would then adjust the pump speed. This scheme is simpler, because only one controller is needed. However, it suffers from an important operational disadvantage; namely, it may not be possible to tune the single controller to meet the needs of both the level and flow control loops. In general, these control loops and their transmitters will have very different dynamic characteristics. A second alternative would be to replace the flow transmitter and controller with a constant (override) signal to the high selector whose value corresponds to the minimum allowable flow rate. However, this scheme would be susceptible to changing pump characteristics.

16.5 NONLINEAR CONTROL SYSTEMS

Most physical processes exhibit nonlinear behavior to some degree. However, linear control techniques such as conventional PID control are still very effective if (1) the nonlinearities are rather mild or (2) a highly nonlinear process operates over a narrow range of conditions. For some highly nonlinear processes, the second condition is not satisfied and as a result, linear control strategies may not be adequate. For these situations, nonlinear control strategies can provide significant improvements over PID control. In this section, we consider several traditional nonlinear control strategies that have been applied in industry. Newer model-based techniques are described by Henson and Seborg (1997).

Three types of nonlinear control strategies are essentially enhancements of single loop feedback control:

1. Nonlinear modifications of standard PID control algorithms
2. Nonlinear transformations of input or output variables
3. Controller parameter scheduling such as *gain scheduling*

Shinskey (1994) and Bequette (1998) have provided informative overviews of these methods and related techniques. Other enhanced single-loop control strategies considered earlier in this chapter, namely, inferential control, selectors, and adaptive control, can also be classified as nonlinear control strategies.

As one example of Method 1, standard PID control laws can be modified by making the controller gain a function of the control error. For example, the controller gain can be higher for larger errors and smaller for small errors by making the controller gain vary linearly with the absolute value of the error signal

$$K_c = K_{c0}(1 + a|e(t)|) \tag{16-26}$$

where K_{c0} and a are constants. The resulting controller is sometimes referred to as an *error-squared controller*, because the controller output is proportional to $|e(t)|e(t)$. Error-squared controllers have been used for level control in surge vessels where it is desirable to take stronger action as the level approaches high or low limits. However, care should be exercised when the error signal is noisy (Shinskey, 1994).

The design objective for Method 2 is to make the closed-loop operation as linear as possible. If successful, this general approach allows the process to be controlled over a wider range of operating conditions and in a more predictable manner. One approach uses simple linear transformations of input or output variables. Common applications include using the logarithm of a product composition as the controlled variable for high-purity distillation columns or adjusting the ratio of feed flow rates in blending problems. The major limitation of this approach is that it is difficult to generalize, because the appropriate variable transformations are application-specific.

In distillation column control, some success has been found in using logarithmic transformations to linearize the error signal so that the controller makes adjustments that are better scaled. For example, a transformed composition variable x_D^* has been used in commercial applications (Shinskey, 1996):

$$x_D^* = \log \frac{1 - x_D}{1 - x_{Dsp}} \tag{16-27}$$

where x_{Dsp} is the desired value of x_D. However, this approach may not work in all cases. Another linearizing function can be used to treat the nonlinear behavior observed in flow systems. When there are pipe resistances in series with a control valve, a nonlinear gain results between stem position l and flow rate. In this case a nonlinear function, called a *valve characterizer*, can be used to transform the controller output p:

$$f(p) = \frac{p}{L + (1 - l)p} \tag{16-28}$$

where L is a parameter used to fit the shape of the nonlinearity. Shinskey (1994) has discussed this approach and related control strategies.

In Method 3, *controller parameter scheduling*, one or more controller settings are adjusted automatically based on the measured value of a scheduling variable. Adjustment of the controller gain, *gain scheduling*, is the most common method. The scheduling variable is usually the controlled variable or set point, but it could be the manipulated variable or some other measured variable. Usually, only the controller gain is adjusted, because many industrial processes exhibit variable steady-state gains but relatively constant dynamics (for example, pH neutralization).

The scheduling variable is usually a process variable that changes slowly, such as a controlled variable, rather than one that changes rapidly, such as a manipulated variable. To develop a parameter-scheduled controller, it is necessary to decide how the controller settings should be adjusted as the scheduling variable(s) change. Three general strategies are:

a. The controller parameters vary continuously with the scheduling variable.

b. One or more scheduling variables are divided into regions where the process characteristics are quite different. Different controller settings can be assigned to each region.

c. The current controller settings are based on the value of the scheduling variable and interpolation of the settings for the different regions. Thus Method (c) is a combination of methods (a) and (b). It is similar to fuzzy logic control, the topic of Section 16.5.2.

Approach (a) is illustrated in the next section. Approaches (b) and (c) can be implemented in several different ways. For example, different values of the control settings can be stored for each region (that is, a *table look-up* approach). Then the controller settings are switched whenever the scheduling variable enters a new region. Alternatively, a dynamic model can be developed for each region and a different controller designed for each model (Bequette, 1998).

16.5.1 Gain Scheduling

The most widely-used type of controller parameter scheduling is *gain scheduling*. A simple version has a piecewise constant controller gain that varies with a single scheduling variable, the error signal e:

$$\begin{aligned} K_c &= K_{c1} \quad \text{for} \quad e_1 \le e < e_2 \\ K_c &= K_{c2} \quad \text{for} \quad e_2 \le e < e_3 \\ K_c &= K_{c3} \quad \text{for} \quad e_3 \le e \le e_4 \end{aligned} \tag{16-29}$$

This gain-scheduling approach is shown in Fig. 16.17 and can easily be extended to more than three regions. A special case, the *error gap controller*, includes a dead band around $e = 0$. In this case, $K_c = 0$ for $e_2 \le e < e_3$, while $K_c \neq 0$ outside this region. Note that the nonlinear gain expression in Eq. 16-26 is another example of gain scheduling.

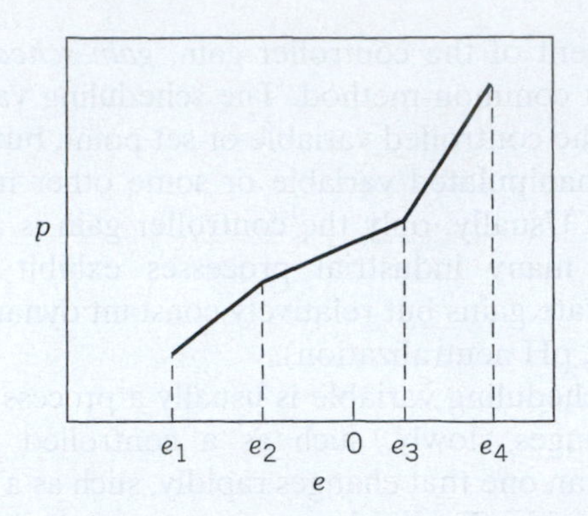

Figure 16.17 A gain-scheduled proportional controller with a controller gain that is piecewise constant.

Next, we consider an example of Method 1 from the previous section. A relationship can be developed between the controller settings and the scheduling variable(s). The resulting strategy is sometimes called *programmed adaptation* (Lipták, 2005; Shinskey, 1996). Programmed adaptation is limited to applications where the process dynamics depend on known, measurable variables and the necessary controller adjustments are not too complicated. As an example of programmed adaptation, consider a once-through boiler (Lipták, 2005). Here, feedwater passes through a series of heated tube sections before emerging as superheated steam. The steam temperature must be accurately controlled by adjusting the flow rate of the hot gas that is used to heat the water. The feedwater flow rate has a significant effect on both the steady-state and dynamic behavior of the boiler. For example, Fig. 16.18 shows typical open-loop responses to a step change in controller output at two different feedwater flow rates, 50% and 100% of the maximum flow. Suppose that an empirical FOPTD model is chosen to approximate the process. The steady-state gain, time delay, and dominant time constant are all twice as large at 50% flow as the corresponding values are at 100% flow. Lipták's proposed solution to this control problem is to have the PID controller settings vary

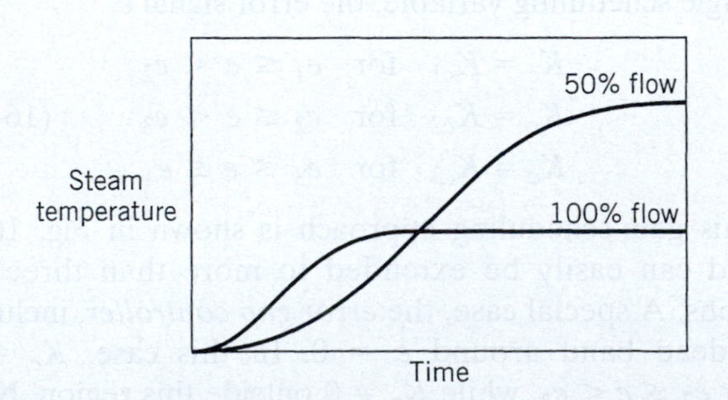

Figure 16.18 Open-loop step responses for a once-through boiler.

with w, the fraction of full-scale flow ($0 \leq w \leq 1$), in the following manner:

$$K_c = w\overline{K}_c \tag{16-30}$$

$$\tau_I = \overline{\tau}_I / w$$

$$\tau_D = \overline{\tau}_D / w$$

where $\overline{K}_c$, $\overline{\tau}_I$, and $\overline{\tau}_D$ are the controller settings for 100% flow. Note that this recommendation for programmed adaptation is qualitatively consistent with the controller tuning rules of Chapter 12. The recommended settings in Eq. 16-30 are based on the assumption that the effects of flow changes are linearly related to flow rate over the full range of operation.

In this example, step responses were available to categorize the process behavior for two different conditions. In other applications, such test data may not be available but there may be some knowledge of process nonlinearities. For pH control problems involving a strong acid and/or a strong base, the pH curve can be very nonlinear, with gain variations over several orders of magnitude. If the process gain changes significantly with the operating conditions, an appropriate gain scheduling strategy is to keep the product of the controller and process gains constant, $K_cK_p = C$, where C is a specified constant. This strategy helps maintain the desired gain margin (see Chapter 14). This gain scheduling strategy could be implemented as follows. Suppose that pH is used as the scheduling variable and an empirical equation is available that relates K_p to pH. Then the current value of K_p could be calculated from the pH measurement and K_c could be determined as, $K_c = C/K_p$.

Another representative nonlinearity is illustrated by the step responses in Fig. 16.19, where the process

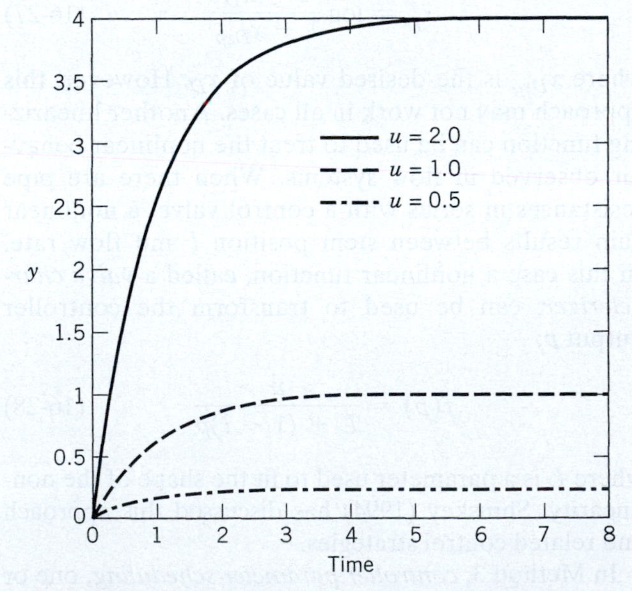

Figure 16.19 Step responses for a nonlinear model for different input magnitudes.

gain K_p depends on the input signal u, $K_p(u)$. Note that the process gain double when the input size is increased from $u = 0.5$ to $u = 1.0$. Because the process dynamics are independent of the magnitude of u, the dynamic behavior can be approximated by a FOPTD model with a gain that varies with u:

$$\tau \frac{dy}{dt} = -y + K_p(u)u(t - \theta) \qquad (16\text{-}31)$$

For example, $K_p(u)$ could be described by a second-order polynomial,

$$K_p(u) = a_0 + a_1 u + a_2 u^2 \qquad (16\text{-}32)$$

The model parameters in (16-32) can be obtained by regression of K_p values for steady-state values of u and y, using the parameter estimation methods of Chapter 7. Note that a transfer function with constant parameters, $Y(s)/U(s)$, cannot be derived for the dynamic model in Eqs. 16-31 and 16-32, because it is not a linear system. However, it is fairly straightforward to design a PID controller (with gain-scheduling) for this process. Suppose that the nominal gain is $K_{p0} = a_0$ for $u = 0$. Then the values of K_c, τ_I, and τ_D can be determined using any of the tuning rules in Chapter 12 for given values of K_{p0}, τ, and θ. As u varies, we would like to keep the product of $K_p(u)$ and $K_c(u)$ constant, in order to maintain a satisfactory gain margin. Thus, we specify that $K_c(u)K_p(u) = K_{c0}K_{p0}$, where K_{c0} is the nominal controller gain for K_{p0}. The controller gain for the current u can be calculated as, $K_c(u) = K_{c0}K_{p0}/K_p(u)$. A similar relationship can be developed for the case where K_p is a function of y, rather than u.

16.5.2 Fuzzy Logic Control

Engineers normally consider physical variables in a quantitative manner, such as specifying a temperature of 78 °C or reporting a flow rate as 10 L/min. However, qualitative information can also be very useful both in engineering and everyday life. For example, a person in a shower is aware of whether the water temperature is too hot, too cold, or just right. An accurate temperature measurement is not necessary. Also, such qualitative information can be used to good advantage for feedback control. For example, if the shower temperature is too cold and the flow rate is too low, the person would increase the hot water flow rate. In the process industries, experienced plant operators sometimes take control actions based on qualitative information, such as the observed color or uniformity of a solid material.

Fuzzy logic control (FLC) is a feedback control technique that utilizes qualitative information through using verbal or linguistic rules of the if–then form (Babuška and Verbruggen, 1996; Rhinehart et al., 1996; Passino and Yurkovich, 1998). To derive the control law, the FLC uses fuzzy sets theory, the set of rules, and a fuzzy inference system. FLC has been used in consumer products such as washing machines, vacuum cleaners, automobiles, battery chargers, air conditioning systems, and camera autofocusing. It has also been applied to such industrial control problems as furnace temperature control, wind energy, power system stability, biological processes, a jet engine fuel system, and control of robots.

Fuzzy Sets

The *Fuzzy Set A* in *U* is defined as a set of ordered pairs: A $\triangleq \{\langle x, \mu_A(x)\rangle x \in U\}$ (Jantzen, 2007), where U is called the universe of discourse, $\mu_A(x)$ is called the membership function for the set of all objects x in U. This definition means that the membership function relates to each x a membership grade $\mu_A(x)$, a real number in the closed interval [0,1]. This concept is illustrated in the following example. Consider a room temperature T to be a qualitative variable with three possible classifications: Hot, OK, or Cold. One possible classification scheme is

Hot: if $T > 24$ °C
OK: if $18\ °C < T < 24$ °C
Cold: if $T < 18$ °C

However, these class boundaries are arbitrary and somewhat inappropriate. For example, do we really want to classify temperatures of 23.5°C as *OK* and 24.5°C as *Hot*? A more appropriate classification scheme is based on the concept of a *fuzzy set*.

As was mentioned in the definition in fuzzy set theory, a physical variable such as a temperature is converted into a qualitative category such as *Hot* by use of a *membership function*, μ. Figure 16.20 shows membership functions for three categories of room temperature. Each μ is bounded between zero (no membership) and one (complete membership). Also, the universe of discourse U is defined in the range from 15 to 27°C. A distinguishing feature of fuzzy set theory is that a physical variable can simultaneously have membership in

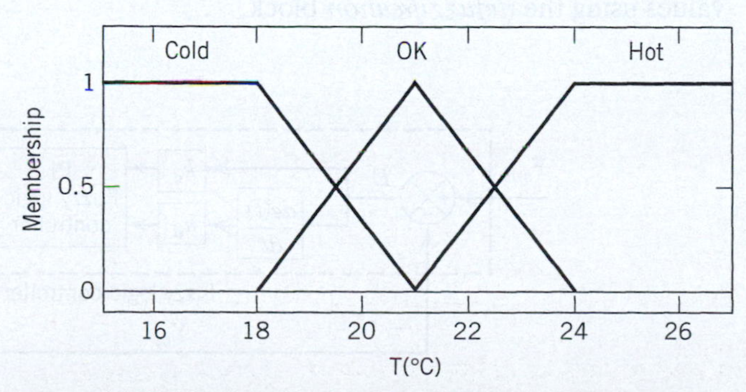

Figure 16.20 Membership functions for room temperature.

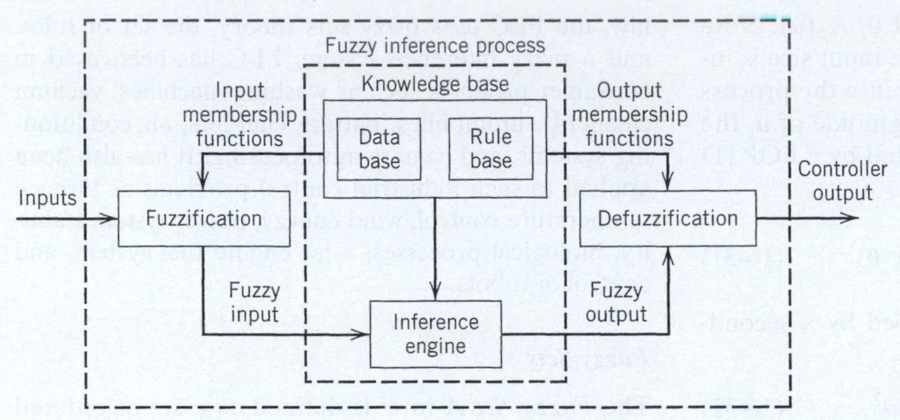

Figure 16.21 Basic configuration of a fuzzy logic controller (FLC).

more than one category. For example, Figure 16.20 indicates that when $T = 22°C$, the room temperature is considered to be both *Hot* and *OK*, with $\mu_{Hot} = 0.33$ and $\mu_{OK} = 0.67$, respectively. The value of μ is called the *membership grade*. For each value of T, the membership grades sum to one, a universal requirement for membership functions. Although the membership functions in Fig. 16.20 consist of linear segments, curved membership functions, such as Gaussian functions, can also be used (Passino and Yurkovich, 1998; Jantzen, 2007).

Each membership function defines a *fuzzy set*, also referred to as a *linguistic variable*. For example, the fuzzy set *Hot* consists of the values of T and the membership function μ_{Hot}. Thus, it can be expressed as $Hot = \{T, \mu_{Hot}(T)\}$.

Fuzzy Inference Systems

Figure 16.21 shows the main blocks of the fuzzy system which is the main part of the fuzzy logic controller (Lee, 1990; Passino and Yurkovich, 1998). The *fuzzification* block converts the inputs or physical variables, for instance the error signal, *e(t)*, into suitable fuzzy sets, as was shown in the example of Figure 16.20. The *fuzzy inference process* combines membership functions with the control rules to derive the fuzzy output, for example, the fuzzy controller output, *u(t)*. This process is also often called *fuzzy reasoning*. Finally, these outputs of the fuzzy computations are translated into terms of real values using the *defuzzification* block.

The most classical inference engine models used in FLC systems are the Mamdani and Takagi-Sugeno (TS) models. Equation 16-33 shows the form of each rule for the Mamdani type inference model.

$$\text{Rule } k\text{: If } x_1 \text{ is } A_{1k}, \text{ and } x_2 \text{ is } A_{2k}, \ldots, \text{ and } x_N \text{ is } A_{Nk}, \text{ Then } y_k \text{ is } B_k \tag{16-33}$$

where A_{Nk} and B_k represent membership functions, and x_N and y_k are the inputs/outputs of the fuzzy system.

The rules in the Takagi-Sugeno type inference model have the form shown in Eq. 16-34.

$$\text{Rule } k\text{: If } x_1 \text{ is } A_{1k}, \text{ and } x_2 \text{ is } A_{2k}, \ldots, \text{ and } x_N \text{ is } A_{Nk}, \text{ Then } y_k = f_k(x_1, x_2, \ldots, x_N) \tag{16-34}$$

where $f_k(x_1, x_2, \ldots, x_N) = b_{0k} + b_{1k}x_1 + \cdots + b_{Nk}x_N$

The rules of both inference systems have antecedents that are comprised of fuzzy sets, but the consequents for the Mamdani type are fuzzy sets. This chapter considers controllers based on the Mamdani inference system.

Fuzzy Control Architecture and Calculations

There are many ways to set up a fuzzy logic controller through using the Mamdani inference system (Babuška and Verbruggen, 1996; Passino and Yurkovich, 1998; Jantzen, 2007). Figure 16.22 shows a block diagram of a PI fuzzy controller, inspired by the PI classical control law, but including a fuzzy inference system.

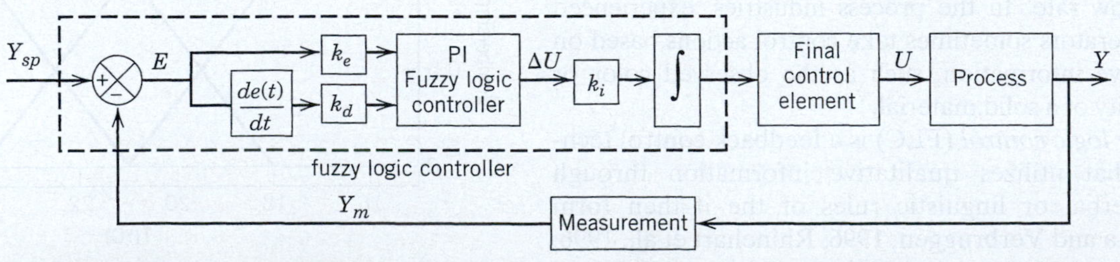

Figure 16.22 PI fuzzy controller.

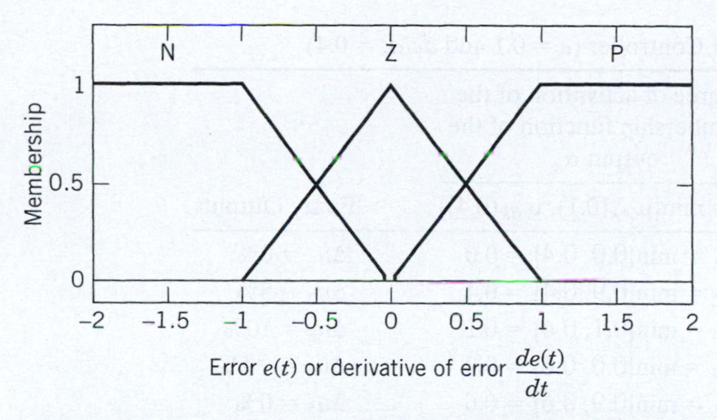

Figure 16.23 Membership functions for the inputs of the PI fuzzy controller (N is negative, P is positive, and Z is zero).

Equation 16-35 shows the control law for a PI fuzzy control. The inputs in Eq. 16-35 are the error $e(t)$ and the derivative of the error de/dt and the output is the change of u, $\Delta u(t)$, which results from evaluating the function $f(\cdot)$ that is the fuzzy system shown in Fig. 16.21. Thus, to get the output $u(t)$, an integrator is added at the output of the FLC as is shown in Fig. 16.22. The constants k_e, k_d, and k_i are used as scaling factors.

$$\Delta u(t) = k_i f\left(k_e\, e(t), k_d \frac{de(t)}{dt} \right) \tag{16-35}$$

Fuzzy logic control calculations are executed by using both membership functions of the inputs and outputs and a set of rules called a *rule base,* as shown in Fig. 16.21. Typical membership functions for the inputs, e and de/dt, are shown in Fig. 16.23, where it is assumed that these inputs have identical membership functions with the following characteristics: three linguistic variables which are negative (N), positive (P), and zero (Z) with trapezoidal, triangular and trapezoidal membership function forms respectively. Input variables e and de/dt have been scaled so that the membership functions overlap for the range from -1 to $+1$. Furthermore, Fig. 16.24 shows the membership functions of the output $\Delta u(t)$, which are defined, by both five linguistic variables and singleton functions, as follows:

$$\begin{aligned} \text{LI} &= \text{large increase} = +10\% \\ \text{MI} &= \text{medium increase} = +5\% \\ \text{Z} &= \text{zero} = 0\% \\ \text{MD} &= \text{medium decrease} = -5\% \\ \text{LD} &= \text{large decrease} = -10\% \end{aligned}$$

where singleton membership functions are defined as in Eq. 16-36 and the universe of discourse is determined in the range from -10 to $+10$.

$$\mu_{\Delta u_i}(x) = \begin{cases} 1 & x = u_i \\ 0 & otherwise \end{cases} \tag{16-36}$$

On the other hand, the rules are specified based on process understanding and past experience. Table 16.1 shows a typical PI-FLC rule base which consists of the nine user-selected rules. For each pair of e and de/dt values, the corresponding entry in the table is a fuzzy output for that rule. For example, Eq. 16-37 shows how Rule 2 for Δu_2 can be expressed in the if-then form.

Rule 2: If *e is zero* AND $\frac{de}{dt}$ is positive,

$$\text{then } \Delta u_2 = \text{MI} \tag{16-37}$$

where Δu_2 denotes the fuzzy output for Rule 2. For this set of rules, there are nine fuzzy controller outputs, one for each rule. These rules also have the form of Eq. 16-33, indicating that this fuzzy controller has a fuzzy-type Mamdani inference system. Additionally, these membership functions and rule-based sets can be configured by using the MATLAB Fuzzy Toolbox.

Once the membership functions of inputs/outputs are determined and the rule base is set, it becomes easier to understand how the block in Fig. 16.21 works. First, each rule k causes a fuzzy membership value μ_{Ak} (*error*) and μ_{Bk} (*derivative of error*), defined as a fuzzification, determined by the input error and derivative of error, which correspond to the inputs of the PI-FLC system (Fig. 16.22). Second, the degree of fulfillment of

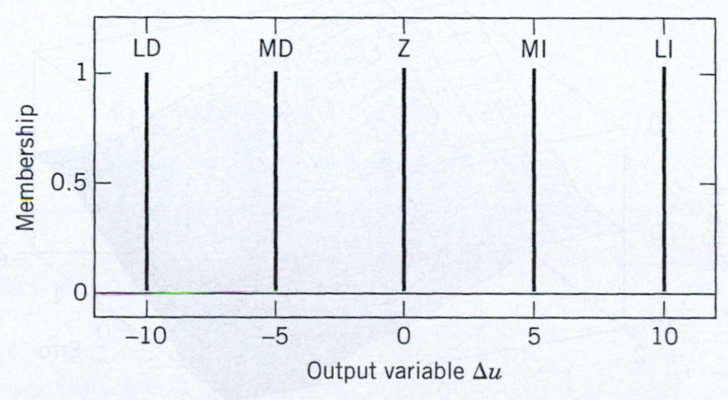

Figure 16.24 Membership functions for the output of the PI fuzzy controller (LD is large decrease, MD is medium decrease, Z is zero, MI is medium increase, and LI is large increase)

Table 16.1 Rule Base for the Fuzzy PI Controller

		$e(t)$		
		N	Z	P
$\frac{de(t)}{dt}$	P	**1** Z	**2** MI	**3** LI
	Z	**4** MD	**5** Z	**6** MI
	N	**7** LD	**8** MD	**9** Z

Table 16.2 Control Calculations for the Fuzzy PI Controller (e = 0.1 and de/dt = 0.4)

	Membership grades associated to the antecedents		Degree of activation of the membership function of the output α_k	
Rule	$\mu_{Ak}(0.1)$	$\mu_{Bk}(0.4)$	$\alpha_k = \min\{\mu_{Ak}(0.1), \mu_{Bk}(0.4)\}$	Fuzzy Outputs
1	$\mu_{N1} = 0.0$	$\mu_{P1} = 0.4$	$\alpha_1 = \min\{0.0, 0.4\} = 0.0$	$\Delta u_1 = 0\%$
2	$\mu_{Z2} = 0.9$	$\mu_{P2} = 0.4$	$\alpha_2 = \min\{0.9, 0.4\} = 0.4$	$\Delta u_2 = 5\%$
3	$\mu_{P3} = 0.1$	$\mu_{P3} = 0.4$	$\alpha_3 = \min\{0.1, 0.4\} = 0.1$	$\Delta u_3 = 10\%$
4	$\mu_{N4} = 0.0$	$\mu_{Z4} = 0.6$	$\alpha_4 = \min\{0.0, 0.6\} = 0.0$	$\Delta u_4 = -5\%$
5	$\mu_{Z5} = 0.9$	$\mu_{Z5} = 0.6$	$\alpha_5 = \min\{0.9, 0.6\} = 0.6$	$\Delta u_5 = 0\%$
6	$\mu_{P6} = 0.1$	$\mu_{Z6} = 0.6$	$\alpha_6 = \min\{0.1, 0.6\} = 0.1$	$\Delta u_6 = 5\%$
7	$\mu_{N7} = 0.0$	$\mu_{N7} = 0.0$	$\alpha_7 = \min\{0.0, 0.0\} = 0.0$	$\Delta u_7 = -10\%$
8	$\mu_{Z8} = 0.9$	$\mu_{N8} = 0.0$	$\alpha_8 = \min\{0.0, 0.0\} = 0.0$	$\Delta u_8 = -5\%$
9	$\mu_{P9} = 0.1$	$\mu_{N9} = 0.0$	$\alpha_9 = \min\{0.0, 0.0\} = 0.0$	$\Delta u_9 = 0\%$

each rule, as a result of the combination of these membership values, is obtained through the firing strength α_k for every rule k; this process is called aggregation and is defined by Eq. 16-38.

$$\alpha_k = \min\{\mu_{Ak}(error), \mu_{Bk}(derivative\ of\ error)\} \quad (16\text{-}38)$$

In (16-38), α_k represents the degree of activation of the membership function of the output for rule k. Third, all activated conclusions, for the whole set of rules are accumulated using the set union operation; this result corresponds to the fuzzy output (see Fig. 16.21). The procedure described above is often called fuzzy reasoning. Finally, the control output, which is $\Delta u(t)$ (see Fig. 16.22), is obtained by applying a defuzzication method.

The defuzzification method can be defined for an arbitrary number of R rules. Equation 16-39 shows how $\Delta u(t)$ is calculated as a weighted sum of the fuzzy controller outputs (Passino and Yurkovich, 1998; Jantzen, 2007).

$$\Delta u(t) = \frac{\sum_{k=1}^{R} \alpha_k \Delta u_k}{\sum_{k=1}^{R} \alpha_k} \quad (16\text{-}39)$$

where Δu_k is the change in the fuzzy controller output for Rule k and α_k is the degree of activation for Rule k.

As an example, to illustrate the fuzzy control calculations, suppose that the error $e = +0.1$ and the derivative of error $de/dt = +0.4$. The membership grades determined from Fig. 16.23, associated to each rule, are summarized in Table 16.2. The rules 1, 4, 7, 8, and 9 are inactive. This situation occurs because the firing strength for each of these five rules is $\alpha_k = 0$; thus these rules do not contribute to the defuzzification calculation of Eq. 16-39. These α_k values are zero because two conditions occur: (1) α_k is defined as the minimum of the two membership grades associated with Rule k, and (2) the membership grades are zero for N; thus, for the five rules that involve N, $\alpha_k = 0$.

Next, we illustrate the calculation of controller output Δu in Eq. 16-40. The firing strength values and fuzzy controller outputs for the nine rules are summarized in Table 16.2. Substitution into (16-39) gives $\Delta u = 2.92\%$.

$$\Delta u(t) = \frac{\alpha_2 \cdot \Delta u_2 + \alpha_3 \cdot \Delta u_3 + \alpha_5 \cdot \Delta u_5 + \alpha_6 \cdot \Delta u_6}{\alpha_2 + \alpha_3 + \alpha_5 + \alpha_6}$$

$$\Delta u(t) = \frac{0.4 * 5 + 0.1 * 10 + 0.6 * 0 + 0.1 * 5}{0.4 + 0.1 + 0.6 + 0.1} = 2.92\% \quad (16\text{-}40)$$

Finally, Fig. 16.25 shows the control surface that represents the mapping between inputs and outputs of the PI-FLC. This is a typical nonlinear surface obtained by using the membership functions, controller rules, the inference engine, and defuzzification. This surface represents the nonlinear behavior of the controller.

In summary, FLC can be viewed as a formal methodology for incorporating process knowledge and experience,

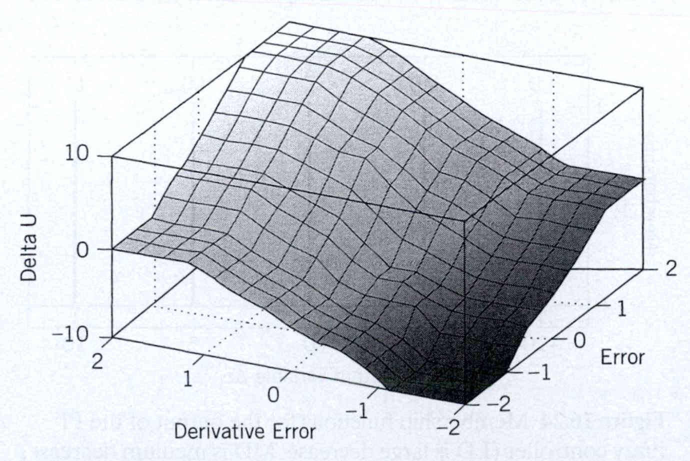

Figure 16.25 Control surface of the PI fuzzy controller.

expressed through defining both membership functions and the fuzzy rules into control system design. It can also be interpreted as a heuristic design method for nonlinear controllers. A fuzzy-type Mamdani controller inspired by PID classical strategies was presented in this section. In order to design this type of controller, it is necessary to incorporate knowledge of the process and the experience of operators. Although the fuzzy PI controller example employed elementary fuzzification and defuzzification steps, more complicated alternatives are available (Passino and Yurkovich, 1998; Jantzen, 2007).

16.6 ADAPTIVE CONTROL SYSTEMS

Process control problems inevitably require on-line tuning of the controller settings to achieve a satisfactory degree of control. If the process operating conditions or the environment changes significantly, the controller may then have to be retuned. If these changes occur frequently, then adaptive control techniques should be considered. An *adaptive control system* is one in which the controller parameters are adjusted automatically to compensate for changing process conditions. Many adaptive control techniques have been proposed for situations where the process changes are largely unknown or unpredictable, as contrasted with situations amenable to the gain-scheduling approach discussed in the previous section. In this section, we are concerned principally with automatic adjustment of feedback controller settings.

Examples of changing process conditions that may require controller retuning or adaptive control are

1. Changes in equipment characteristics (e.g., heat exchanger fouling, catalyst deactivation)
2. Unusual operational status, such as failures, start-up, and shutdown, or batch operations
3. Large, frequent disturbances (feed composition, fuel quality, etc.)
4. Ambient variations (rain storms, daily cycles, etc.)
5. Changes in product specifications (grade changes) or product flow rates
6. Inherent nonlinear behavior (e.g., the dependence of chemical reaction rates on temperature)

In situations where the process changes can be anticipated or measured directly and the process is reasonably well understood, then the gain-scheduling approach (or programmed adaptation) discussed in the previous section can be employed. When the process changes cannot be measured or predicted, the adaptive control strategy must be implemented in a feedback manner, because there is little opportunity for a feedforward type of strategy such as programmed adaptation. Many such controllers are referred to as *self-tuning controllers* or *self-adaptive controllers* (Åström and Wittenmark, 1995; Lipták, 2005).

In self-tuning control, the parameters in the process model are updated as new data are acquired (using on-line estimation methods), and the control calculations are based on the updated model. For example, the controller settings could be expressed as a function of the model parameters and the estimates of these parameters updated on-line as process input/output data are received. Self-tuning controllers generally are implemented as shown in Fig. 16.26 (Åström and Wittenmark, 1995).

In Fig. 16.26, three sets of computations are employed: estimation of the model parameters, calculation of the controller settings, and implementation of the controller output in a feedback loop. Most real-time parameter estimation techniques require that an external forcing signal occasionally be introduced to allow accurate estimation of model parameters (Hang et al., 1993). Such a pertubation signal can be deliberately introduced through the set point or added to the controller output.

During each disturbance or set-point change, the process response is compared to the predicted model response, and then the model can be updated based on

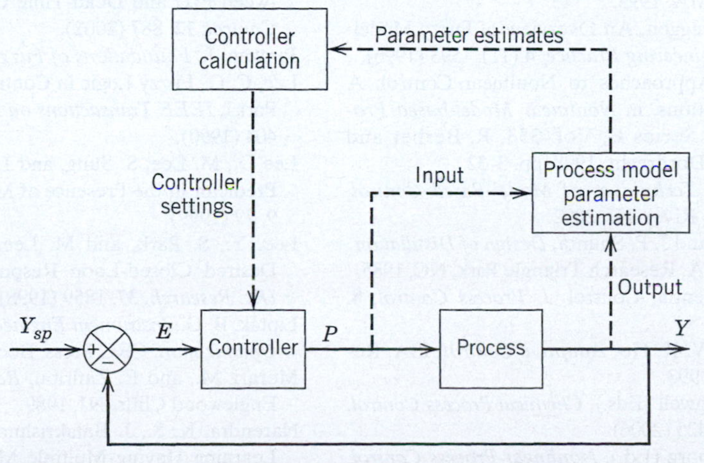

Figure 16.26 A block diagram for self-tuning control.

the prediction error. On-line parameter estimation can be problematic when there is a high level of signal noise or unmeasured disturbances (that are not included in the model). The plant–model mismatch also present difficulties. However, it is possible to successfully implement adaptive control through diagnostics that assess when the estimator is not behaving correctly. In addition, limits can be placed on control parameter changes to make the controller more robust. One approach that deals with models changing with varying operating conditions is *multiple model adaptive control* (Narendra et al., 1995; Schott and Bequette, 1997), where a set of models and corresponding controllers is employed. A weighting function is used to choose the combination of models that best matches the process input-output behavior. This technique has been used in a variety of applications, including drug infusion control (Schott and Bequette, 1997).

Two advantages of the self-tuning control approach is that the model in Fig. 16.26 is not restricted to low-order linear transfer functions or difference equations and the controller is not required to have a PID controller structure. The model structure can be multivariable in nature and even incorporate nonlinear modeling elements, such as artificial neural nets. If disturbances are measured explicitly, it is possible to update a disturbance model and implement adaptive feedforward control.

The subject of adaptive control continues to be of considerable interest. Many new and improved algorithms have been proposed since the early 1970s, and a number of commercial products were introduced during the 1980s. However, adaptive control has tended to be a niche application rather than being used pervasively in industrial applications. In addition, there is a concern that use of an adaptive controller can lead to unstable operations or unsafe operating conditions. In the future, it is expected that control products that use some form of adaptation will become more common as algorithms improve and process knowledge increases.

SUMMARY

In this chapter, we have presented a number of control strategies that offer the potential of enhanced performance over what can be achieved with conventional single-loop PID controllers. These techniques are especially attractive for difficult control problems, such as those characterized by unmeasured process variables and disturbances, long time delays, process constraints, changing operating conditions, and process nonlinearities and uncertainties. All of these characteristics can be treated by one or more of the advanced feedback control techniques discussed here: cascade control, time-delay compensation, inferential control, selective control, nonlinear control, and adaptive control. Note that feedforward control (Chapter 15), multivariable control (Chapter 18), and model predictive control (Chapter 20) augment the more specialized methods treated here.

REFERENCES

Åström, K. J., and B. Wittenmark, *Adaptive Control Systems*, 2d ed., Addison-Wesley, Reading, MA, 1995.

Babuška, R., and H. B. Verbruggen, An Overview of Fuzzy Modeling for Control, *Control Engineering Practice,* **4** (11), 1593 (1996).

Bequette, B. W., Practical Approaches to Nonlinear Control: A Review of Process Applications, in *Nonlinear Model-based Process Control*, NATO ASI Series E, Vol. 353, R. Berber and C. Kravaris (Eds.), Kluwer, Dordrecht, 1998, pp. 3–32.

Brosilow, C., and B. Joseph, *Techniques of Model-Based Control*, Prentice Hall, Upper Saddle River, NJ, 2002.

Buckley, P. S., W. L. Luyben, and J. P. Shunta, *Design of Distillation Column Control Systems*, ISA, Research Triangle Park, NC, 1985.

Doyle, F. J., Nonlinear Inferential Control, *J. Process Control*, **8**, 339 (1998).

Hang, C. C., T. H. Lee, and W. K. Ho, *Adaptive Control*, ISA, Research Triangle Park, NC, 1993.

Henson, M. A., and T. A. Badgwell (Eds.), *Chemical Process Control*, **7**, *Comput. Chem. Eng*., **30**, 425 (2006).

Henson, M. A., and D. E. Seborg (Ed.), *Nonlinear Process Control*, Prentice-Hall, Upper Saddle River, NJ, 1997.

Ingimundarson, I., and T. Hägglund, Performance Comparison Between PID and Dead-Time Compensating Controllers, *J. Process Control*, **12**, 887 (2002).

Jantzen, J., *Foundations of Fuzzy Control,* Wiley, New York, 2007.

Lee, C. C., Fuzzy Logic in Control Systems: Fuzzy Logic Controller—Part I, *IEEE Transactions on Systems, Man and Cybernetics,* **20** (2), 404 (1990).

Lee D., M. Lee, S. Sung, and I. Lee, Robust PID Tuning for Smith Predictor in the Presence of Model Uncertainty, *J. Process Control*, **9**, 79 (1999).

Lee, Y., S. Park, and M. Lee, PID Controller Tuning to Obtain Desired Closed-Loop Responses for Cascade Control Systems, *IEC Research*, **37**, 1859 (1998).

Lipták, B. G., *Instrument Engineers Handbook*, 4th ed., *Process Control Optimization*, CRC Press, Boca Raton, FL, 2005.

Morari, M., and E. Zafiriou, *Robust Process Control*, Prentice Hall, Englewood Cliffs, NJ, 1989.

Narendra, K. S., J. Balakrishnan, and M. K. Ciliz, Adaptation and Learning Having Multiple Models, Switching and Tuning, *IEEE Control Systems*, **15** (3), 37 (1995).

Palm, R., D. Driankov, and H. Hellendoorn, *Model-Based Fuzzy Control*, Springer-Verlag, New York, 1997.

Passino, K. M., and S. Yurkovich, *Fuzzy Control*, Addison-Wesley, Reading, MA, 1998.

Rawlings, J. B., B. A. Ogunnaike, and J. W. Eaton (Eds.), *Chemical Process Control VI, AIChE Symp. Ser.*, **97**, No. 326, 2002.

Rhinehart, R. R., H. H. Li, and P. Murugan, Improve Process Control Using Fuzzy Logic, *Chem. Eng. Progress*, **92** (11), 60 (1996).

Riggs, J. B., Improve Distillation Column Control, *Chem. Eng. Prog.*, **94** (10), 31 (1998).

Schott, K. D., and B. W. Bequette, Multiple Model Adaptive Control, in *Multiple Model Approaches to Modeling and Control*, R. Murray-Smith and T. Johansen, (Eds.), Taylor and Francis, London, U.K., 1997, Chapter 11.

Seborg, D. E., A Perspective on Advanced Strategies for Process Control (Revisited), *Advances in Control*, P. M. Frank (Ed.), Springer-Verlag, New York, 1999, pp. 103–134.

Shinskey, F. G., *Feedback Controllers for the Process Industries*, McGraw-Hill, New York, 1994.

Shinskey, F. G., *Process Control Systems*, 4th ed., McGraw-Hill, New York 1996.

Singer, J. G. (Ed.), *Combustion-Fossil Power Systems*, 4th ed., Combustion Engineering, Windsor, CT, 1993, pp. 14–27.

Smith, O.J.M., Closer Control of Loops with Dead Time, *Chem. Eng. Prog.*, **53** (5), 217 (1957).

EXERCISES

16.1 Measurement devices and their dynamics influence the design of feedback controllers. Briefly indicate which of the two systems below would have its closed-loop performance enhanced significantly by application of cascade control (see Fig. 16.4 for notation and assume $G_{p2} = 1$ and $G_{d1} = G_{p1}$). Using the controller settings shown below, evaluate the effect of a unit step disturbance in D_1 on both systems A and B.

System A	*System B*
$G_v = 5$	$G_v = 5$
$G_{p1} = \dfrac{2}{10s+1}$	$G_{p1} = \dfrac{2}{10s+1}$
$G_{m2} = \dfrac{0.5}{0.5s+1}$	$G_{m2} = \dfrac{2}{5s+1}$
$G_{m1} = \dfrac{1}{5s+1}$	$G_{m1} = \dfrac{0.2}{5s+1}$
$K_{c1} = 0.5$	$K_{c1} = 2.5$
$\tau_{I1} = 15$	$\tau_{I1} = 15$
$K_{c2} = 1.0$	$K_{c2} = 0.25$

All time constants are in minutes.

16.2 In Example 16.1, the ultimate gain for the primary controller was found to be 43.3 when $K_{c2} = 5$.

(a) Derive the closed-loop transfer functions for Y_1/D_1 and Y_1/D_2 as a function of K_{c1} and K_{c2}.

(b) Examine the effect of K_{c2} on the critical gain of K_{c1} by varying K_{c2} from 1 to 20. For what values of K_{c2} do the benefits of cascade control seem to be less important? Is there a stability limit on K_{c2}?

(c) Integral action was not included in either primary or secondary loops. First set $K_{c2} = 5$, $\tau_{I1} = \infty$, and $\tau_{I2} = 5$ min. Find the ultimate controller gain using the Routh array. Then repeat the stability calculation for $\tau_{I1} = 5$ min and $\tau_{I2} = \infty$ and compare the two results. Is offset for Y_1 eliminated in both cases for step changes in D_1 or D_2?

16.3 Consider the cascade control system in Fig. E16.3.

(a) Specify K_{c2} so that the gain margin ≥ 1.7 and phase margin $\geq 30°$ for the slave loop.

(b) Then specify K_{c1} and τ_I for the master loop using the Ziegler-Nichols tuning relation and simulate the closed-loop response for a set-point change.

16.4 Solve Exercise 16.3 using MATLAB, but use IMC tuning rules for both the master and slave controllers. Design K_{c2} first, and then use that value to design G_{c1} (PI controller). The higher-order transfer function will need to be approximated first by a FOPTD model using a step test. Plot closed-loop responses for different values of the IMC closed-loop time constant for both outer loop and inner loop for a set point change.

16.5 Consider the stirred-tank heating system shown in Fig. E16.5. It is desired to control temperature T_2 by adjusting the heating rate Q_1 (Btu/h) via voltage signal V_1 to the SCR. It has been suggested that measurements of T_1 and T_0, as well as of T_2, could provide improved control of T_2.

(a) Briefly describe how such a control system might operate, and sketch a schematic diagram. State any assumptions that you make.

(b) Indicate how you would classify your control scheme—for example, feedback, cascade, or feedforward. Briefly justify your answer.

(c) Draw a block diagram for the control system.

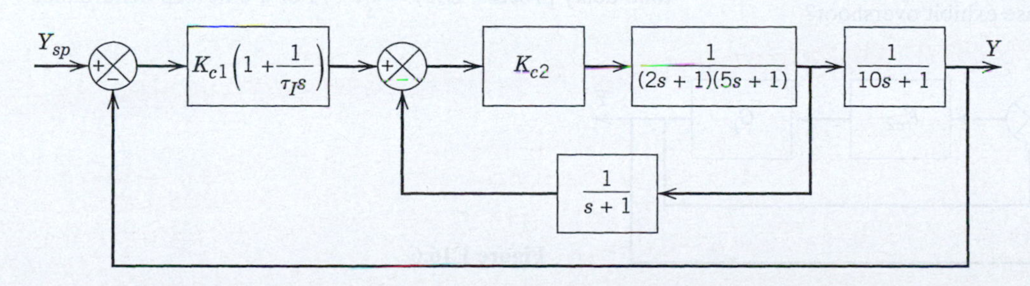

Figure E16.3

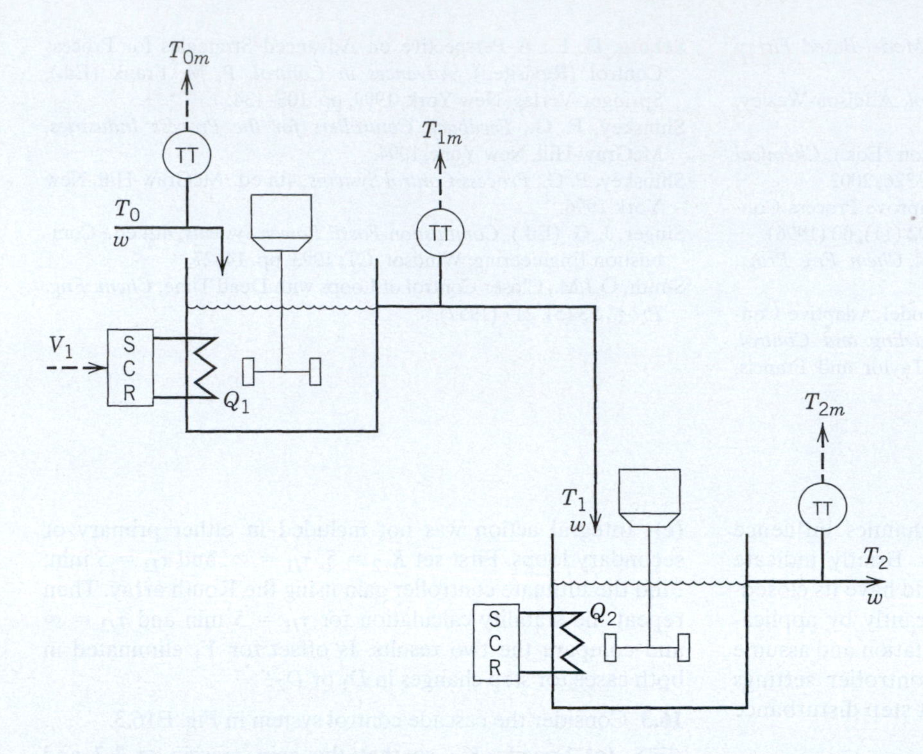

Figure E16.5

16.6 Consider Figs. 16.3 and 16.4 illustrating cascade control.

(a) Suppose you were to apply feedforward control, instead of cascade control, to handle disturbances D_1 and D_2. Where do you expect feedforward control to be more beneficial: for D_1, or D_2? Explain why.

(b) Draw a block diagram that is a modification of Figure 16.4 that uses in feedforward control of D_1.

(c) What additional sensors would be required for feedforward control of D_1?

16.7 Design a time-delay compensator (Smith predictor) for

$$G_p = \frac{e^{-\theta s}}{5s + 1}$$

when $G_v = G_m = 1$. Show closed-loop responses for unit step set-point and disturbance changes ($G_d = G_p$), $G_c = K_c = 1$, and $\theta = 1$.

16.8 Shinskey (1994) has proposed a delay-time compensator of the form

$$G_c = K_c\left(\frac{1 + \tau_1 s}{1 + \tau_1 s - e^{-\theta s}}\right)$$

for a FOPTD process, with $K_c = \dfrac{1}{K_p}$ and $\tau_I = \tau$.

(a) Derive the closed-loop transfer function and show that the time delay is eliminated from the characteristic equation.

(b) Will the closed-loop response exhibit overshoot?

16.9 Applepolscher has designed a Smith predictor with proportional control for a control loop that regulates blood glucose concentration with insulin flow. Based on simulation results for a FOPTD model, he tuned the controller so that it will not oscillate. However, when the controller was implemented, severe oscillations occurred. He has verified through numerous step tests that the process model is linear. What explanations can be offered for this anomalous behavior?

16.10 The closed-loop transfer function for the Smith predictor in Eq. 16-22 was derived assuming no model error.

(a) Derive a formula for Y/Y_{sp} when $G_p \neq \widetilde{G}_p$. What is the characteristic equation?

(b) Let $G_p = 2e^{-2s}/(5s + 1)$. A proportional controller with $K_c = 15$ and a Smith predictor are used to control this process. Simulate set-point changes for $\pm 20\%$ errors in process gain (K_p), time constant (τ), and time delay (six different cases). Discuss the relative importance of each type of error.

(c) What controller gain would be satisfactory for $\pm 50\%$ changes in all three model parameters?

(d) For $K_c = 15$, how large a change in either K_p, τ, or θ can be tolerated before the loop goes unstable?

16.11 A Smith predictor is to be used with an integrator-plus-time-delay process, $G(s) = \dfrac{2}{s}e^{-3s}$. For a unit step disturbance

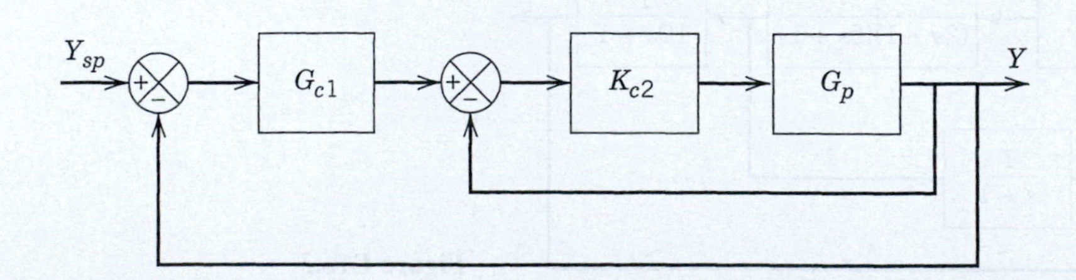

Figure E16.6

and $G_d = G$, show that PI control will not eliminate offset even when the model is known perfectly. Use Eq. 16-24 as the starting point for your analysis.

16.12 In Chapter 12, we introduced the Direct Synthesis design method, in which the closed-loop servo response is specified and the controller transfer functions are calculated algebraically. For an IMC controller (see Chapter 12), show that setting $G_+ = e^{-\theta s}$ leads to a Smith predictor controller structure when $G = \tilde{G}$ for a FOPTD process.

16.13 A CSTR is used to produce a specialty chemical. The reaction is exothermic and exhibits first-order kinetics. Laboratory analyses for the product quality are time-consuming, requiring several hours to complete. No on-line composition measurement has been found satisfactory. It has been suggested that composition can be inferred from the exit temperature of the CSTR. Using the linearized CSTR model in Example 4.8, determine whether this inferential control approach would be feasible. Assume that measurements of feed flow rate, feed temperature, and coolant temperature are available.

16.14 The pressure of a reactor vessel can be adjusted by changing either the inlet or outlet gaseous flow rate. The outlet flow is kept fixed as long as the tank pressure remains between 100 and 120 psi, and pressure changes are treated by manipulating the inlet flow control valve. However, if the pressure goes higher than these limits, the exit gas flow is then changed. Finally, if the pressure exceeds 200 psi, a vent valve on the vessel is opened and transfers the gas to a storage vessel. Design a control scheme that meets the performance objectives. Draw a process instrumentation diagram for the resulting control system.

16.15 Selectors are normally used in combustion control systems to prevent unsafe situations from occurring. Figure E16.15 shows the typical configuration for high and low selectors are applied to air and fuel flow rates. The energy demand signal comes from the steam pressure controller. Discuss how the selectors operate in this control scheme when the furnace pressure drops suddenly.

16.16 Buckley et al. (1985) discuss using a selector to control condensate temperature at 100 °C in a reflux drum, where the manipulated variable is the cooling water flow rate. If the condensate temperature becomes too low, the temperature controller reduces the cooling water flow rate, causing the cooling water exit temperature to rise. However, if the water temperature exceeds 50 to 60 °C, excessive fouling and corrosion can result. Draw a schematic diagram that uses a selector to keep the exit temperature below 50 °C. Determine the valve action (A–O or A–C) for the flow control valve, and whether the level controller should be reverse- or direct-acting.

16.17 For many chemical and biological processes, the steady-state gain changes when a process operating condition such as throughput changes. Consider a biomedical application where a drug flow rate is used to control blood pressure. The steady-state gain K_p varies with the manipulated variable u according to the relation

$$K_p = a + \frac{b}{u}$$

where $u > 0$ and a and b are constants that have been determined by fitting steady-state data. Suggest a modification for the standard PID controller to account for this variation in the process gain. Justify your answer. (In the above equation, u is *not* a deviation variable.)

16.18 The product quality from a catalytic tubular reactor is controlled by the flow rate of the entering stream, utilizing composition measurements from a process gas chromatograph. The

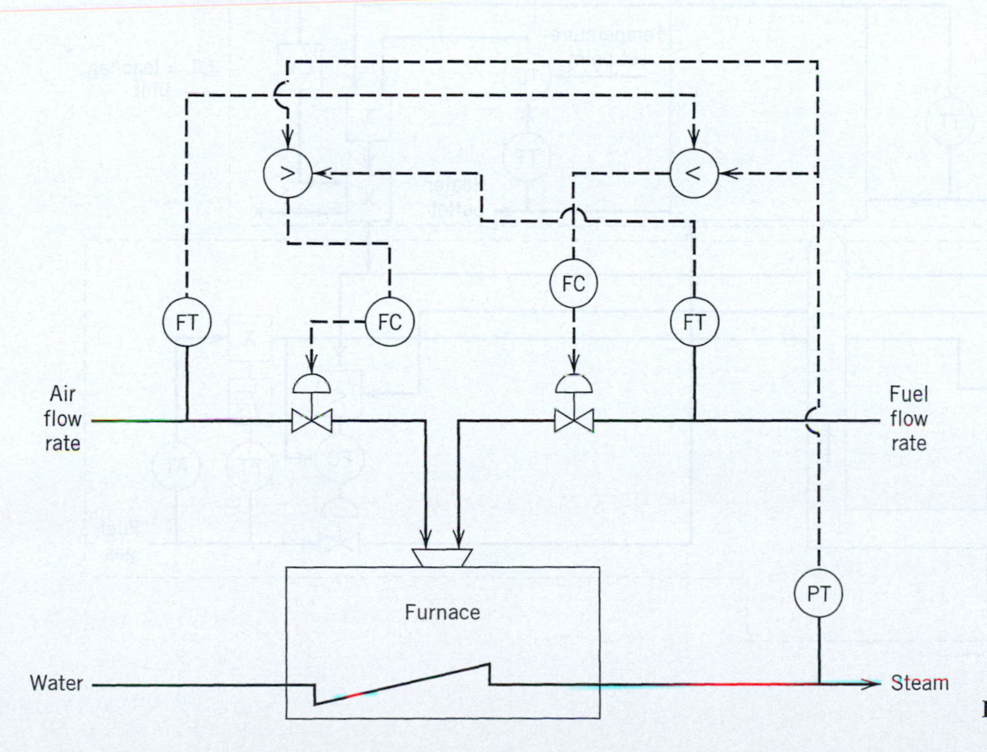

Figure E16.15

catalyst decays over time and once its overall activity drops below 50%, it must be recharged. Deactivation usually takes two to three months to occur. One measure of catalyst activity is the average of three temperature measurements that are used to estimate the peak temperature. Discuss how you would employ an adaptive control scheme to maintain product quality at acceptable levels. What transfer functions would need to be determined, and why?

16.19 A second-order process is controlled by a PID controller. The desired closed-loop servo transfer function is

$$\frac{Y}{Y_{sp}} = \frac{1}{\tau_c s + 1} = \frac{G_p G_c}{1 + G_p G_c} \qquad G_v = G_m = 1$$

and the process model is

$$G_p = \frac{K_p}{(\tau_1 s + 1)(\tau_2 s + 1)}$$

(a) Derive a control law that shows how to adjust K_c, τ_I, and τ_D based on variations in K_p, τ_1, and τ_2 and the desired closed-loop time constant τ_c.

(b) Suppose $\tau_1 = 3$, $\tau_2 = 5$, and $K_p = 1$. Calculate values of K_c, τ_I, and τ_D to achieve $\tau_c = 1.5$. Show how the response deteriorates for changes in the following model parameters when the controller remains unchanged:

i. $K_p = 2$
ii. $K_p = 0.5$
iii. $\tau_2 = 10$
iv. $\tau_2 = 1$

16.20 The Ideal Gas Company has a process that requires an adaptive PI controller, but the company capital budget has been frozen. Appelpolscher has been given the job to develop a homegrown, cheap adaptive controller. It has been suggested that the closed-loop response after a disturbance can be studied to determine how to adjust K_c and τ_I incrementally up or down, using measures such as settling time, peak error, and decay ratio. Appelpolscher has proposed the following algorithm: If decay ratio > 0.25, reduce K_c. If decay ratio < 0.25, increase K_c. He is not sure how to adjust τ_I. Critique his rule for K_c, and propose a rule for changing τ_I.

16.21 An instrumentation diagram for a fired heater control system is shown in Fig. E16.21. Identify advanced control

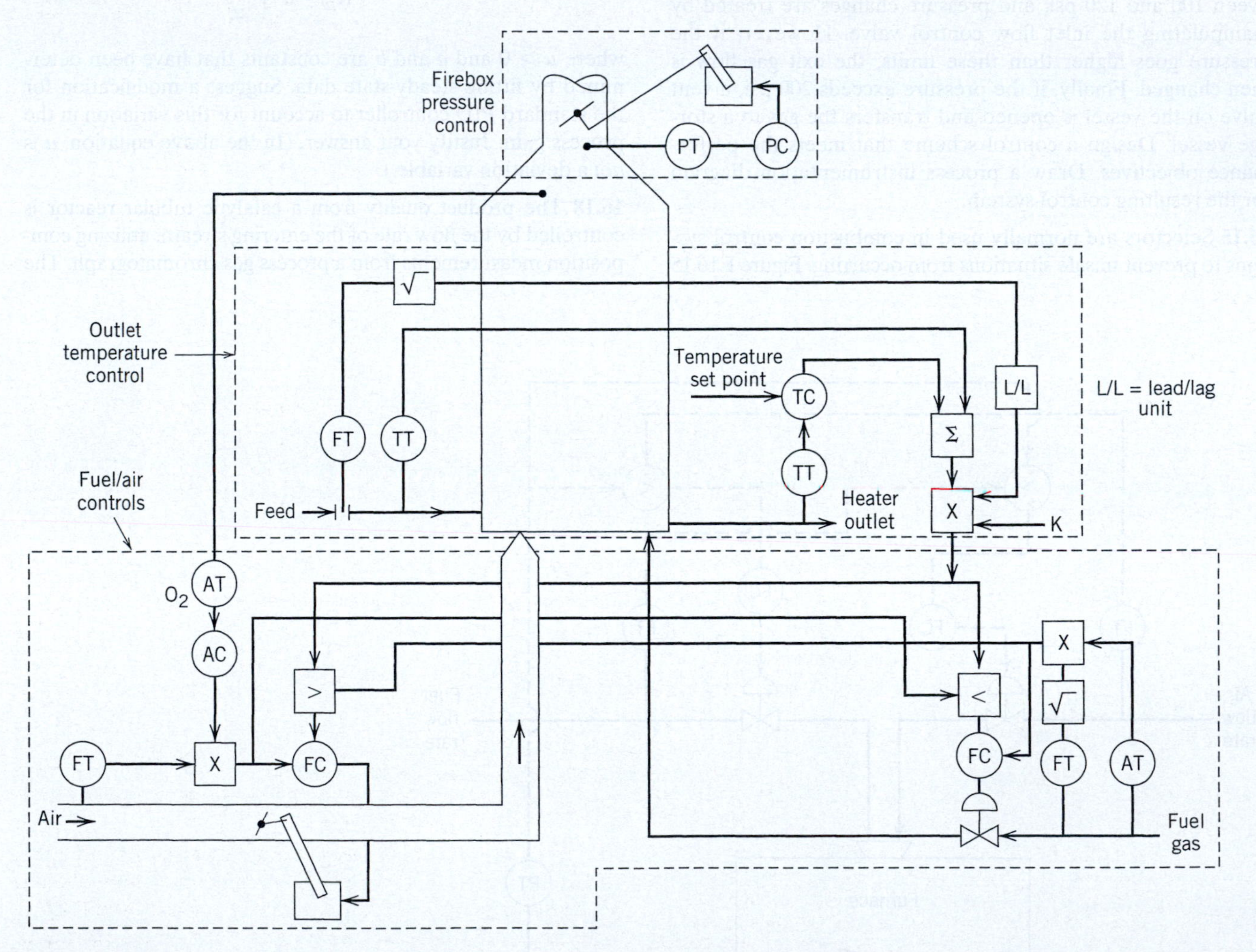

Figure E16.21

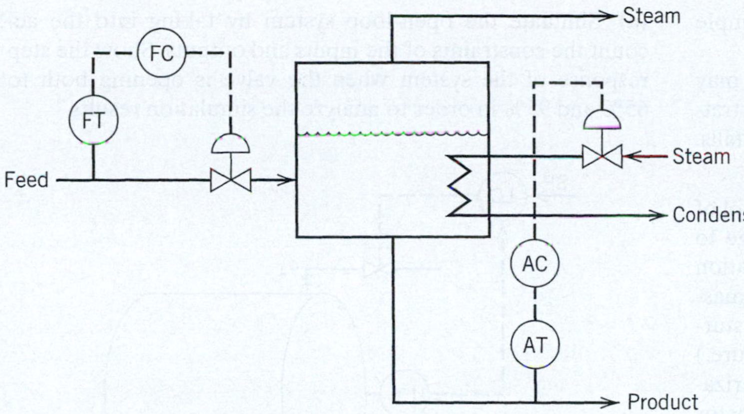

Figure E16.22

strategies based on material from Chapters 15 and 16. Discuss the rationale for each advanced method.

16.22 A liquid is concentrated by evaporating water in an evaporator. The available measurements and control valves are shown in Fig. E16.22. During normal operation, the concentration controller output P_{ac} is less than or equal to 80%. Also, the product concentration is to be controlled by adjusting the steam control valve, while the feed flow rate is regulated via a flow controller. However, if the feed concentration is low for a sustained time period, the concentration controller tends to saturate, and consequently the product concentration can be significantly below its set point. Both control valves are fail close, whereas each transmitter is direct-acting. In order to cope with this undesirable situation, it is proposed to temporarily (and automatically) reduce the feed flow rate when the concentration controller output signal exceeds 80%. Propose a control strategy that will accomplish this goal and draw the corresponding schematic diagram. Justify your choice. (*Note:* The feed composition cannot be measured.)

16.23 A waste stream (dilute nitric acid) is neutralized by adding a base stream (sodium hydroxide) of known concentration to a stirred neutralization tank, as shown in Fig. E16.23.

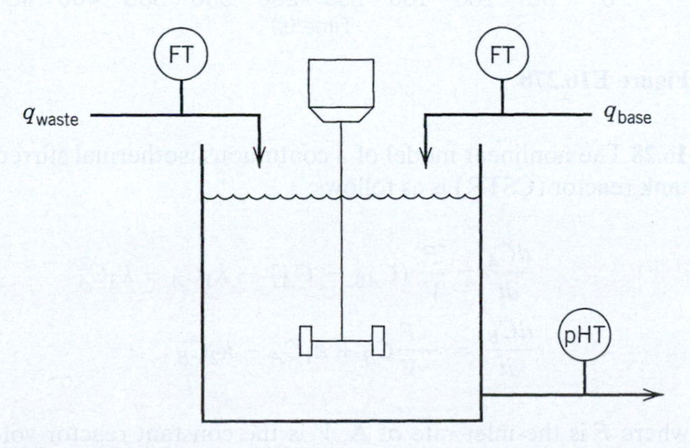

Figure E16.23

The concentration and the flow rate of the waste acid stream vary unpredictably. The flow rates of the waste stream and base stream can be measured. The effluent stream pH can be measured, but a significant time delay occurs due to the downstream location of the pH probe.

Past experience has indicated that it is not possible to tune a standard PID controller so that satisfactory control occurs over the full range of operating conditions. As a process control specialist, you have been requested to recommend an advanced control strategy that has the potential of greatly improved control. Justify your proposed method (be as specific as possible). Also cite any additional information that you will need.

16.24 Flow control loops are usually fast compared to other loops, so they can be considered to be at steady state (essentially). In this case, integral control is recommended. Show that for $G_d = G_p = K_p$, integral control provides satisfactory control for both set-point changes and disturbances. Assume $G_v = G_m = 1$.

16.25 Diabetes mellitus is characterized by insufficiency of the pancreas to produce enough insulin to regulate the blood sugar level. In type I diabetes, the pancreas produces no insulin, and the patient is totally dependent on insulin from an external source to be infused at a rate to maintain blood sugar levels at normal levels. Hyperglycemia occurs when blood glucose level rises much higher than the norm (>8 mmol/L) for prolonged periods of time; hypoglycemia occurs when the blood sugar level falls below values of 3 mmol/L. Both situations can be deleterious to the individual's health. The normal range of blood sugar falls between 3.8 and 5.6 mmol/L, the target range for a controller regulating blood sugar.

A patient with type I diabetes needs your help to maintain her blood sugar within an acceptable range (3 mmol/L < glucose < 8 mmol/L). She has just eaten a large meal (a disturbance) that you estimate will release glucose according to $D(t) = 0.5\, e^{-0.05t}$, where t is in minutes and $D(t)$ is in mmol/L − min. She has a subcutaneous insulin pump that can release insulin up to 115 mU/min (mU = 10^{-3} Unit of Insulin). The flow rate of insulin is the manipulated variable. Assume that the blood glucose level can be measured by taking a blood sample periodically.

Discuss control strategies from Chapters 15 and 16 that may be useful for solving this problem. Explain why a given strategy might be appropriate but also indicate possible pitfalls. Chapter 23 discusses a diabetes simulation.

16.26 The figure below shows cascade temperature control of a polymerization reactor, which uses feed heat exchange to adjust the reactor temperature. Using the instrumentation diagram, explain how this cascade control system (both master and slave components) handles the following disturbances. (describe what happens to the reactor temperature.) Assume normal temperatures of coolant (70°F), polymerization feed (200°F), exchanger effluent (100°F), and reactor outlet (800°F).

(a) Feed temperature becomes too high.

(b) Feed flow rate becomes too high.

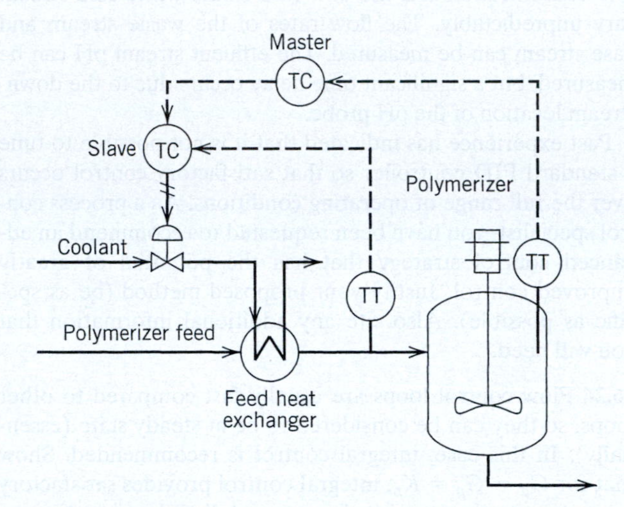

Figure E16.26

16.27 Consider the horizontal cylindrical tank shown in Fig. E16.27a, which is based on the model presented in Example 4.7. The output of the system, the controlled variable, is the height of the tank, $h(t)$, and the input of the system, the manipulated variable, is the opening of the valve, x, which is proportional to the input flow, q_i. The nonlinear dynamic model of the system is represented by the following equation:

$$\frac{dh(t)}{dt} = \frac{1}{2L\sqrt{(D - h(t))h(t)}}(q_i - q)$$

where

$$q_i = 0.2\,x$$

$$q = 15\sqrt{h(t)}$$

For this model, assume $L = 1\,\text{m}$ and $D = 1\,\text{m}$ as values for the parameters of the model, with the following restrictions for the input and output variables: $0 < h(t) \leq 1$, $0 < x \leq 100$ [%]. For this process,

(a) Simulate the open-loop system by taking into the account the constraints of the inputs and outputs. Show the step response of the system when the valve is opening both to 65% and 95% in order to analyze the simulation results.

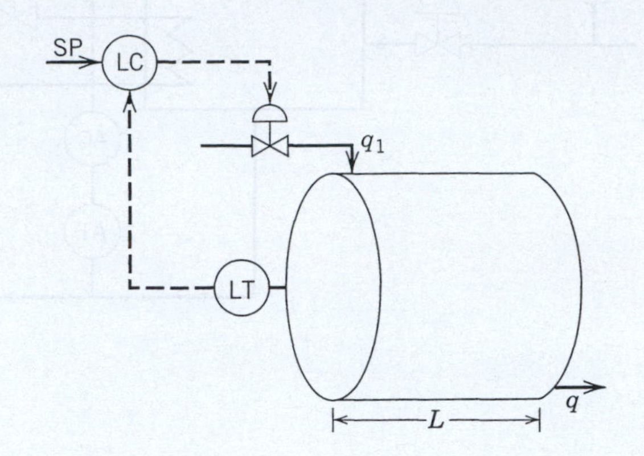

Figure E16.27a

(b) Design a fuzzy-type Mamdani controller based on the PI fuzzy controller shown in Fig. 16.22. Utilize the same membership functions for inputs, $e(t)$ and de/dt, and the output, $\Delta u(t)$, which are shown in Figs. 16.23 and 16.24, respectively. Utilize the rules defined in Table 16.2. Assume the scaling factors, k_e, k_d, and k_i, are equal to 1. Finally, evaluate the performance of the controller by applying the set point trajectory shown in Fig. E16.27b.

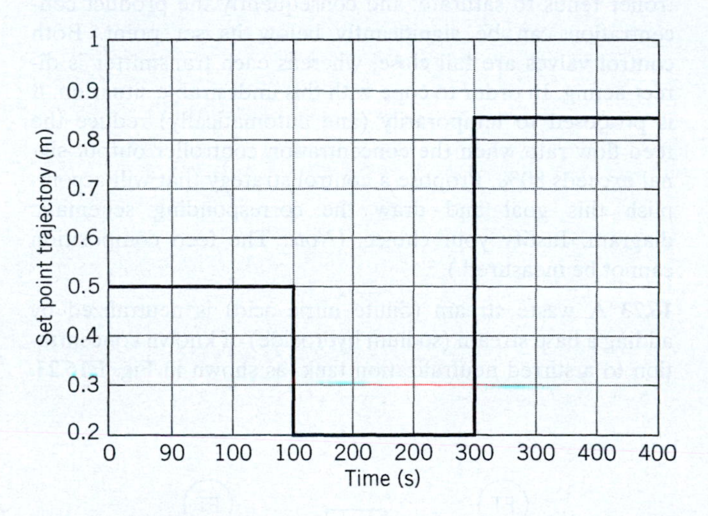

Figure E16.27b

16.28 The nonlinear model of a continuous isothermal stirred tank reactor (CSTR) is as follows:

$$\frac{dC_A}{dt} = \frac{F}{V}(C_{A0} - C_A) - k_1 C_A - k_3 C_A^2$$

$$\frac{dC_B}{dt} = -\frac{F}{V}C_B + k_1 C_A - k_2 C_B$$

where F is the inlet rate of A, V is the constant reactor volume, C_A and C_B are the concentrations of species A and B, respectively, and $C_{A0} = 10$ gmol/L is the concentration of A in the feed stream. The values of the reaction rate constants are $k_1 = 50\text{s}^{-1}$, $k_2 = 100\text{s}^{-1}$ and $k_3 = 10$ L/gmol · s

(a) Design a PI fuzzy controller for the case when the controlled variable is C_B and the manipulated variable is the dilution rate F/V. Assume the scaling factors, k_e, k_d, and k_i, are equal to 1. Evaluate the performance of the controller by using the set point trajectory equal to 1.2 gmol/L. Improve the performance of this controller by changing the universe of discourse of the fuzzy output. Evaluate the effects of making this change in the closed-loop performance.

(b) Improve the performance of the closed-loop system by manipulating the scaling factors. Simulate by using the set point trajectory of 1.2 gmol/L. Compare the results with the controller obtained in part (a).

Chapter 17

Digital Sampling, Filtering, and Control

CHAPTER CONTENTS

The specifications for a computer-based system to perform data acquisition and control must address several questions:

1. How often should data be acquired from each sensor? That is, what sampling rate should be employed?
2. Do the measurements contain a significant amount of noise? If so, can the data be conditioned (*filtered*) to reduce the effects of noise?
3. What digital control algorithm should be employed?

17.1 SAMPLING AND SIGNAL RECONSTRUCTION

When a digital computer is used for control, continuous measurements are converted into digital form by an analog-to-digital converter (ADC) (see Appendix A). This operation is necessary because the digital computer cannot directly process a continuous (analog) signal; first, the signal must be sampled, and then each analog value must be assigned its corresponding digital value. The time interval between successive samples is

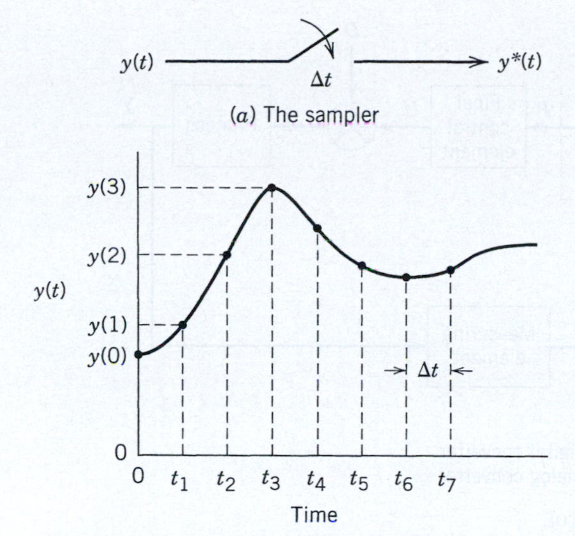

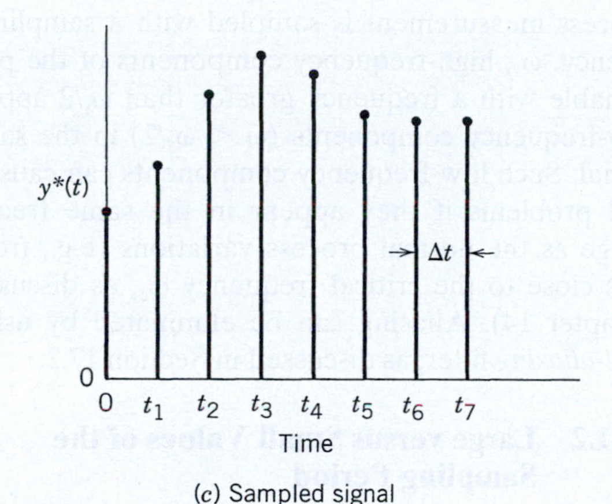

Figure 17.1 Idealized, periodic sampling. For a uniform sampling period Δt, the sampling instants $t_1, t_2, \ldots, t_k$ correspond to times $\Delta t, 2\Delta t, \ldots, k\Delta t$.

referred to as the sampling period Δt, which corresponds to the sampling or *scan* rate, $f_s = 1/\Delta t$ (cycles/time), or, equivalently, the sampling frequency, $\omega_s = 2\pi/\Delta t$ (radians/time).

Figure 17.1 shows an idealized periodic sampling operation in which the sampled signal $y^*(t)$ is a series of impulses that represents the measurements $y(0)$, $y(1)$, $y(2)$. . . . at the sampling instants $t = 0, \Delta t, 2\Delta t, \ldots$ The representation in Fig. 17.1, also referred to as *impulse modulation* (Franklin et al., 1997), is based on the assumption that the sampling operation occurs instantaneously.

Most computer control systems require a device called a DAC (*digital-to-analog converter*), which changes a series of pulses (from the digital computer or controller) into a continuous signal. This signal is then transferred to a final control element such as a control valve. In process control, the final control element normally requires a continuous input signal rather than a pulsed input (although a stepping motor is one exception). The DAC usually contains a *zero-order hold* (ZOH) to convert

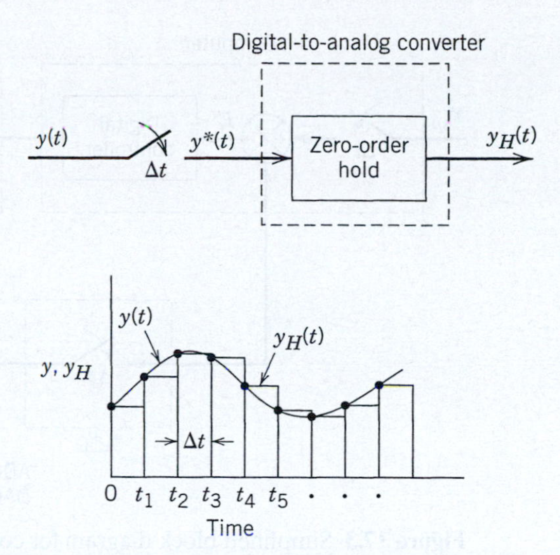

Figure 17.2 Digital-to-analog conversion using a zero-order hold.

the digital (pulsed) signal from the controller into a staircase function $y_H(t)$, as shown in Fig. 17.2. Note that the output signal from the zero-order hold $y_H(t)$ is held constant for one sampling period until the next sample is received, which can be expressed as

$$y_H(t) = y(k-1) \quad \text{for } t_{k-1} \le t < t_k \tag{17-1}$$

Other types of hold devices can be employed for *signal reconstruction*; for example, a *first-order-hold* extrapolates the digital signal linearly during the time interval from t_{k-1} to t_k based on the change during the previous interval:

$$y_H(t) = y(k-1) + \left(\frac{t - t_{k-1}}{\Delta t}\right)[y(k-1) - y(k-2)]$$
$$\text{for} \quad t_{k-1} \le t < t_k \tag{17-2}$$

Although second-order and other higher-order holds can be designed and implemented as special-purpose DACs (Ogata, 1994; Åström and Wittenmark, 1997; Franklin et al., 1997), these more complicated approaches do not offer significant advantages for most process control problems. Consequently, we will emphasize the zero-order hold, because it is the most widely used hold device for process control.

Figure 17.3 shows the block diagram for a typical feedback control loop with a digital controller. Note that both continuous (analog) and sampled (digital) signals appear in the block diagram. The two samplers typically have the same sampling period and operate synchronously, which means that they acquire sampled signals at exactly the same time. However, *multirate sampling* is sometimes used, in which one sampler operates at a faster rate than the other. For example, we may wish to sample a process variable and filter the measurements quite frequently while performing the control calculations less often in order to avoid excessive wear in the actuator or control valve. The block diagram in Fig. 17.3 is symbolic in that

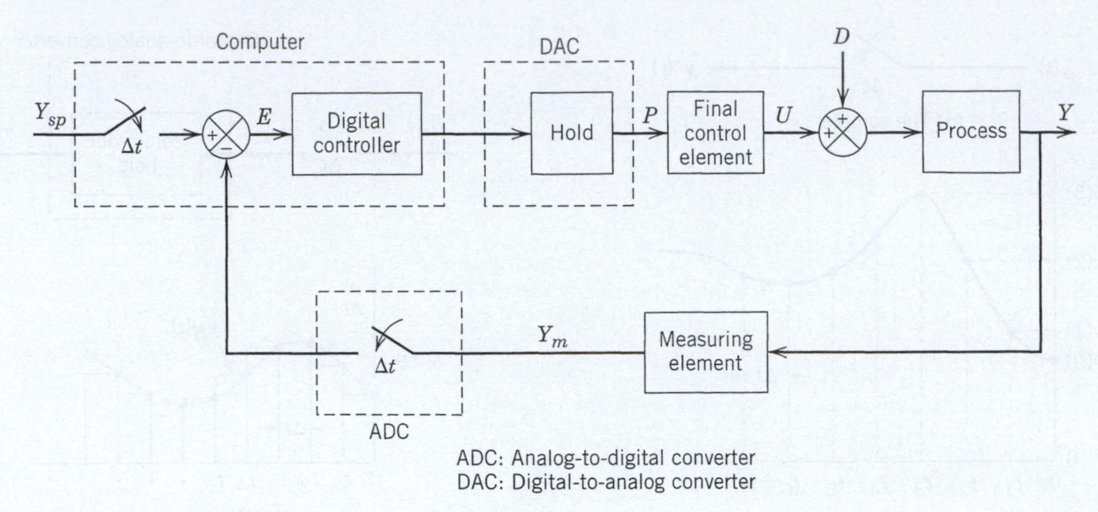

Figure 17.3 Simplified block diagram for computer control.

the mathematical relations between the various signals (e.g., transfer functions) are not shown.

17.1.1 Aliasing

The sampling rate must be large enough that significant process information is not lost, as illustrated in Fig. 17.4. Suppose that a sinusoidal signal is sampled at a rate of 4/3 samples per cycle (i.e., 4/3 samples per period). This sampling rate causes the reconstructed signal to appear as a sinusoid with a much longer period than the original signal, as shown in Fig. 17.4*a*. This phenomenon is known as *aliasing*. Note that if the original sinusoidal signal were sampled only twice per period, then a constant sampled signal would result, as shown in Fig. 17.4*d*. According to Shannon's sampling theorem (Franklin et al., 1997), a sinusoidal signal must be sampled *more* than twice each period to recover the original signal; that is, the sampling frequency must be at least twice the frequency of the sine wave.

Aliasing also occurs when a process variable that is *not* varying sinusoidally is sampled. In general, if a process measurement is sampled with a sampling frequency, ω_s, high-frequency components of the process variable with a frequency greater than $\omega_s/2$ appear as low-frequency components ($\omega < \omega_s/2$) in the sampled signal. Such low-frequency components can cause control problems if they appear in the same frequency range as the normal process variations (e.g., frequencies close to the critical frequency ω_c, as discussed in Chapter 14). Aliasing can be eliminated by using an *anti-aliasing* filter, as discussed in Section 17.2.

17.1.2 Large versus Small Values of the Sampling Period

Sampling too slowly can reduce the effectiveness of a feedback control system, especially its ability to cope with disturbances. In an extreme case, if the sampling period is longer than the process response time, then a disturbance can affect the process, but the influence of the disturbance will disappear before the controller takes corrective action. In this situation, the control system cannot handle transient disturbances

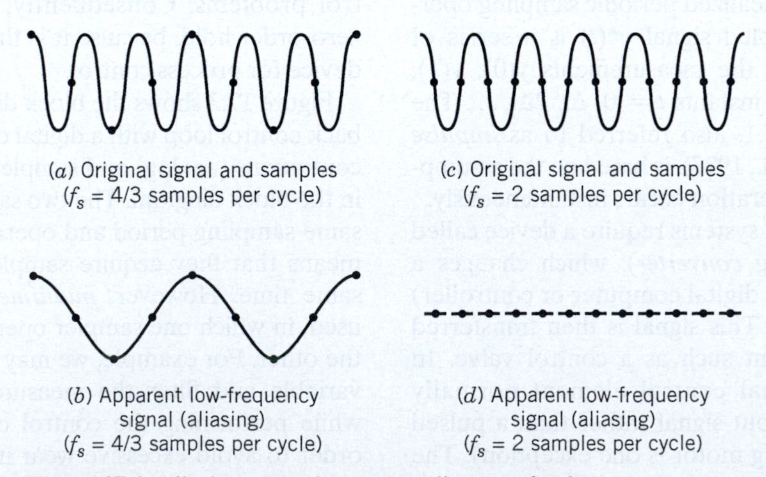

Figure 17.4 Aliasing error due to sampling too slowly.

and is capable only of steady-state control. Thus, it is important to consider the process dynamics (including disturbance characteristics) in selecting the sampling period. Commercial digital controllers, which handle a specified number of control loops (e.g., 8 to 16), typically employ a fixed scan rate less than or equal to 1 s but can vary the sampling period for control calculations. For $\Delta t \leq 1$ s, the performance of a digital controller closely approximates that for continuous (analog) control in normal process control applications.

17.1.3 Guidelines for Selecting the Sampling Period

Selection of the sampling period remains more of an art than a science. A number of guidelines and rules of thumb have been reported for both PID controllers and model-based controllers such as the Internal Model Control (IMC) approach of Chapter 12 (Åström and Wittenmark, 1997; Isermann, 1989). There is a difference between the sampling period used by the computer control hardware (typically 1 s or less) for data acquisition and the sampling period used for controller output changes. For the sampling period Δt in the control algorithm, Åström and Wittenmark (1997) have proposed several guidelines in terms of dominant time constant τ_{dom} or settling time t_s.

$$0.01 \leq \frac{\Delta t}{\tau_{dom}} \leq 0.05 \tag{17-3a}$$

$$\frac{t_s}{15} \leq \Delta t \leq \frac{t_s}{6} \tag{17-3b}$$

In some cases, the process time delay can become a factor, and the sampling period must be reduced to speed up the response time for disturbances. Simulation using different sampling periods can be carried out to make the final selection. Again, we should mention that process data may be acquired at a higher rate than that indicated above.

17.2 SIGNAL PROCESSING AND DATA FILTERING

In process control applications, noise associated with measurements can arise from a number of sources: the measurement device, electrical equipment, or the process itself. The effects of electrically generated noise can be minimized by following established procedures concerning shielding of cables, grounding, and so forth (McConnell and Jernigan, 2005). Process-induced noise can arise from variations resulting from incomplete mixing, turbulence, and nonuniform multiphase flows. The effects of both process noise and measurement noise can be reduced by signal conditioning or filtering. In signal processing parlance, the term *filter* is synonymous with transfer function, because a filter transforms input signals to yield output signals. A filter effectively increases valve life, because valve movements are reduced when the controller receives filtered measurements.

17.2.1 Analog Filters

Analog filters are used to smooth noisy experimental data. For example, an *exponential filter* can be used to damp out high-frequency fluctuations due to electrical noise; hence, it is called a low-pass filter. Its operation is described by a first-order transfer function, or, equivalently, a first-order differential equation.

$$\tau_F \frac{dy_F(t)}{dt} + y_F(t) = y_m(t) \tag{17-4}$$

where y_m is the measured value (the filter input), y_F is the filtered value (the filter output), and τ_F is the time constant of the filter. Note that the filter has a steady-state gain of one. The exponential filter is also called an *RC filter,* because it can be constructed from a simple RC electrical circuit.

Figure 17.4 showed that relatively slow sampling of a high-frequency analog signal can produce an artificial low-frequency signal. Therefore, it is desirable to use an analog filter to *pre-filter* process data before sampling in order to remove high-frequency noise as much as possible. For these applications, the analog filter is often referred to as an *anti-aliasing filter* in which the sampling period can be selected independently, with τ_F set to approximately $0.5\Delta t$. However, to treat slowly varying signals, digital filtering can also be used, as described in Section 17.2.2 (McConnell and Jernigan, 2005).

The filter time constant τ_F in (17-4) should be much smaller than the dominant time constant of the process τ_{dom} to avoid introducing a significant dynamic lag in the feedback control loop. For example, choosing $\tau_F < 0.1\ \tau_{dom}$ generally satisfies this requirement. On the other hand, if the noise amplitude is high, then a larger value of τ_F may be required to smooth the noisy measurements. The frequency range of the noise is another important consideration. Suppose that the lowest noise frequency expected is denoted by ω_N. Then τ_F should be selected so that $\omega_F < \omega_N$, where $\omega_F = 1/\tau_F$. For example, suppose we specify $\omega_F = 0.1\omega_N$, which corresponds to $\tau_F = 10/\omega_N$. Then noise at frequency ω_N will be attenuated by a factor of 10, according to Eq. 14-13 and the Bode diagram of Fig. 14.2. In summary, τ_F should be selected so that $1/\omega_N \leq \tau_F \leq 0.1/\tau_{dom}$.

17.2.2 Digital Filters

In this section, we consider several widely used digital filters. A more comprehensive treatment of digital filtering and signal processing techniques is available elsewhere (Oppenheim and Shafer, 1999).

Exponential Filter

First we consider a digital version of the exponential filter in Eq. 17-4. Denote the samples of the measured variable as $y_m(k-1), y_m(k), \cdots$ and the corresponding filtered values as $y_F(k-1), y_F(k), \cdots$ where k refers to the current sampling instant. The derivative in (17-4) at time step k can be approximated by a first-order backward difference:

$$\frac{dy_F}{dt} \cong \frac{y_F(k) - y_F(k-1)}{\Delta t} \tag{17-5}$$

Substituting in (17-4) and replacing $y_F(t)$ by $y_F(k)$ and $y_m(t)$ by $y_m(k)$ yields

$$\tau_F \frac{y_F(k) - y_F(k-1)}{\Delta t} + y_F(k) = y_m(k) \tag{17-6}$$

Rearranging gives

$$y_F(k) = \frac{\Delta t}{\tau_F + \Delta t} y_m(k) + \frac{\tau_F}{\tau_F + \Delta t} y_F(k-1) \tag{17-7}$$

We define the dimensionless parameter

$$\alpha \triangleq = \frac{\Delta t}{\tau_F + \Delta t} \tag{17-8a}$$

where $0 < \alpha \leq 1$. Then

$$1 - \alpha = 1 - \frac{\Delta t}{\tau_F + \Delta t} = \frac{\tau_F}{\tau_F + \Delta t} \tag{17-8b}$$

so that (17-7) can be written as

$$y_F(k) = \alpha y_m(k) + (1-\alpha) y_F(k-1) \tag{17-9}$$

Equation 17-9 indicates that the filtered measurement is a weighted sum of the current measurement $y_m(k)$ and the filtered value at the previous sampling instant $y_F(k-1)$. This operation is also called *single exponential smoothing* or the *EWMA filter*, for *exponentially weighted moving average*. Limiting cases for α are

$\alpha = 1$: No filtering (the filter output is the raw measurement $y_m(k)$).

$\alpha \rightarrow 0$: The measurement is ignored.

Equation 17-8a indicates that $\alpha = 1$ corresponds to a filter time constant of zero (no filtering).

Alternative expressions for α in (17-9) can be derived if the forward difference or other integration schemes for dy/dt are utilized (Franklin et al., 1997). Analytical integration of (17-4) to yield a difference equation can be performed for a piecewise constant input, leading to the result previously obtained in Chapter 7 (Eq. 7-34).

Double Exponential Filter

Another useful digital filter is the double exponential or second-order filter, which offers some advantages for dealing with signal drift: the second-order filter is equivalent to two first-order filters in series where the second filter input is the output signal $y_F(k)$ from the exponential filter in Eq. 17-9. The second filter (with output $\bar{y}_F(k)$ and filter constant γ) can be expressed as

$$\bar{y}_F(k) = \gamma y_F(k) + (1-\gamma)\bar{y}_F(k-1) \tag{17-10}$$

or

$$\bar{y}_F(k) = \gamma\alpha y_m(k) + \gamma(1-\alpha) y_F(k-1) + (1-\gamma)\bar{y}_F(k-1) \tag{17-11}$$

Writing Eq. 17-10 for the previous sampling instant gives

$$\bar{y}_F(k-1) = \gamma y_F(k-1) + (1-\gamma)\bar{y}_F(k-2) \tag{17-12}$$

Solving for $y_F(k-1)$,

$$y_F(k-1) = \frac{1}{\gamma}\bar{y}_F(k-1) - \frac{1-\gamma}{\gamma} y_F(k-2) \tag{17-13}$$

Substituting (17-13) into (17-11) and rearranging gives the following expression for the double exponential filter:

$$\bar{y}_F(k) = \gamma\alpha y_m(k) + (2-\gamma-\alpha)\bar{y}_F(k-1) - (1-\alpha)(1-\gamma)\bar{y}_F(k-2) \tag{17-14}$$

A common simplification is to select $\gamma = \alpha$, yielding

$$\bar{y}_F(k) = \alpha^2 y_m(k) + 2(1-\alpha)\bar{y}_F(k-1) - (1-\alpha)^2 \bar{y}_F(k-2) \tag{17-15}$$

The advantage of the double exponential filter over the exponential filter of Eq. 17-9 is that it provides better filtering of high-frequency noise, especially if $\gamma = \alpha$. On the other hand, it is sometimes difficult to tune γ and α properly for a given application or data set. It is also hard to tune a controller in series with a double exponential filter.

Although the double exponential filter is beneficial in some cases, the single exponential filter is more widely used in process control applications.

Moving-Average Filter

A moving-average filter averages a specified number of past data points, giving equal weight to each data point. It is usually less effective than the exponential filter, which gives more weight to the most recent data. The moving-average filter can be expressed mathematically as

$$y_F(k) = \frac{1}{N^*} \sum_{i=k-N^*+1}^{k} y_{mi} \qquad (17\text{-}16)$$

where N^* is the number of past data points that are being averaged. Equation 17-16 also can be expressed in terms of the $k-1$ filtered value, $y_F(k-1)$:

$$y_F(k-1) = \frac{1}{N^*} \sum_{i=k-N^*}^{k-1} y_m(i) \qquad (17\text{-}17)$$

Subtracting (17-17) from (17-16) gives the recursive form of the moving-average filter:

$$y_F(k) = y_F(k-1) + \frac{1}{N^*}(y_m(k) - y_m(k-N^*)) \qquad (17\text{-}18)$$

The moving-average filter is a low-pass filter that eliminates high-frequency noise.

Noise-Spike Filter

If a noisy measurement changes suddenly by a large amount and then returns to the original value (or close to it) at the next sampling instant, a *noise spike* is said to occur. Figure 17.5 shows two noise spikes appearing in the experimental temperature data for a fluidized sand bath. In general, noise spikes can be caused by spurious electrical signals in the environment of the sensor. If noise spikes are not removed by filtering before the noisy measurement is sent to the controller, the controller will produce large, sudden changes in the manipulated variable.

Noise-spike filters (or *rate-of-change* filters) are used to limit how much the filtered output is permitted to change from one sampling instant to the next. If Δy denotes the maximum allowable change, the noise-spike filter can be written as

$$y_F(k) = \begin{cases} y_m(k) & \text{if } [y_m(k) - y(k-1)] \leq \Delta y \\ y_F(k-1) - \Delta y & \text{if } y_F(k-1) - y_m(k) > \Delta y \\ y_F(k-1) + \Delta y & \text{if } y_m(k) - y_F(k-1) > \Delta y \end{cases} \qquad (17\text{-}19)$$

If a large change in the measurement occurs, the filter replaces the measurement by the previous filter output plus (or minus) the maximum allowable change. This filter can also be used to detect instrument malfunctions such as a power failure, a break in a thermocouple or instrument line, or an ADC "glitch."

More complicated digital filters are available but have not been commonly used in process control applications. These include high-pass filters and band-pass filters (Isermann, 1989; Oppenheim and Shafer, 1999).

EXAMPLE 17.1

To compare the performance of alternative filters, consider a square-wave signal with a frequency of $f = 0.33$ cycles/min and an amplitude 0.5 corrupted by

(i) High-frequency sinusoidal noise (amplitude = 0.25, $f_N = 9$ cycles/min)

(ii) Random (Gaussian) noise with zero mean and a variance of 0.01

Evaluate both analog and digital exponential filters, as well as a moving-average filter, and assess the effect of sampling interval Δt.

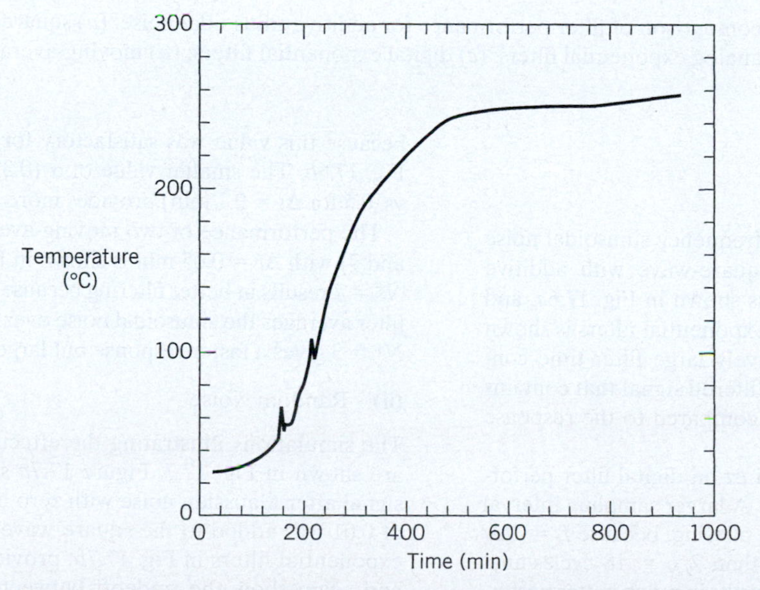

Figure 17.5 Temperature response data from a fluidized sand bath contains two noise spikes (Phillips and Seborg, 1987).

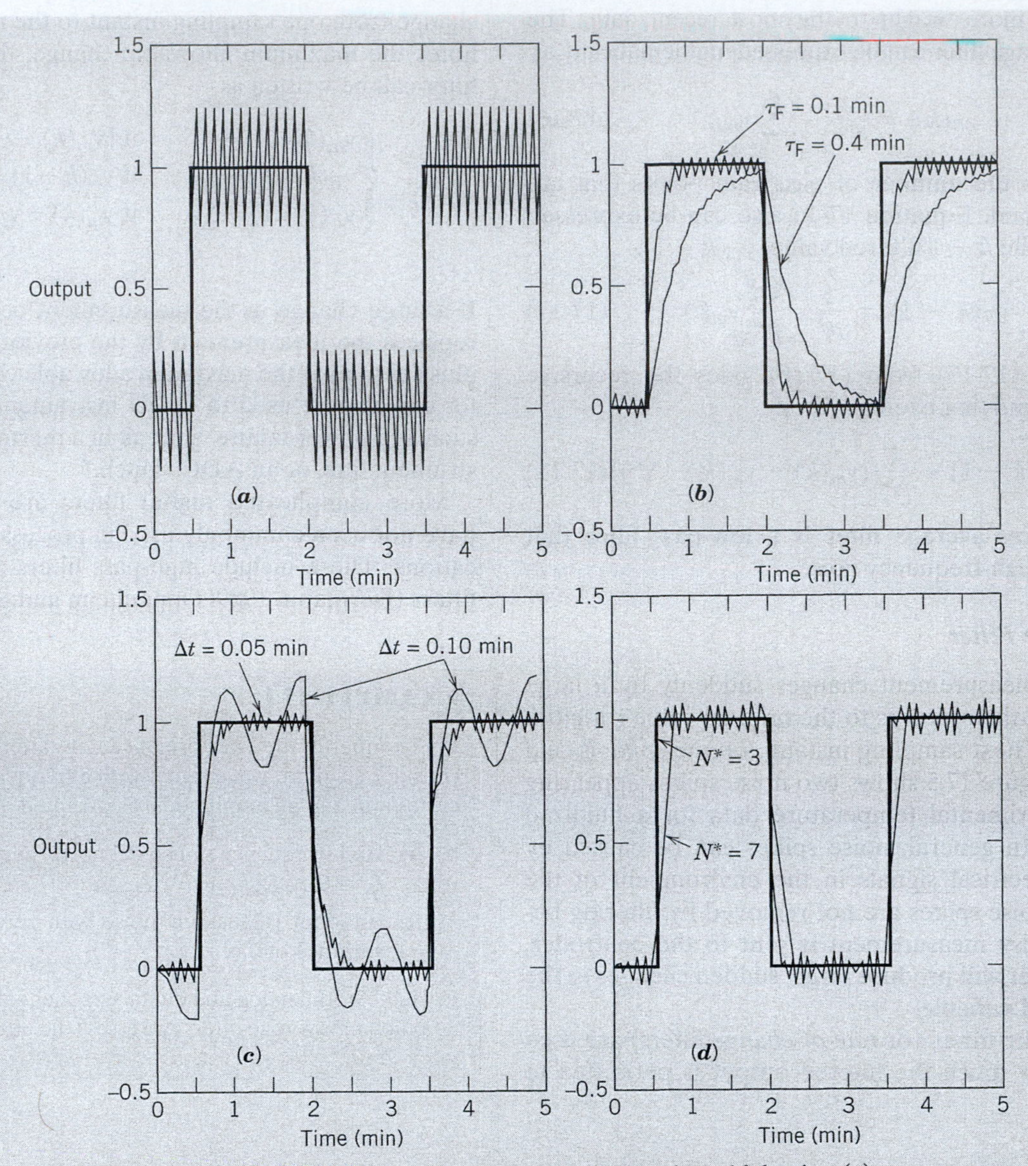

Figure 17.6 A comparison of filter performance for additive sinusoidal noise: (*a*) square-wave plus noise; (*b*) analog exponential filters; (*c*) digital exponential filters; (*d*) moving-average filters.

SOLUTION

(i) Sinusoidal Noise

Representative results for high frequency sinusoidal noise are shown in Fig. 17.6. The square-wave with additive noise, the signal to be filtered, is shown in Fig. 17.6*a*, and the performance of two analog exponential filters is shown in Fig. 17.6*b*. Choosing a relatively large filter time constant ($\tau_F = 0.4$ min) results in a filtered signal that contains less noise but is more sluggish, compared to the response for $\tau_F = 0.1$ min.

The effect of sampling period Δt on digital filter performance is illustrated in Fig. 17.6*c*. A larger sampling interval ($\Delta t = 0.1$ min) results in serious aliasing, because $f_s = 1/\Delta t = 10$ cycles/min, which is less than $2f_N = 18$ cycles/min. Reducing Δt by a factor of two results in much better performance. For each filter, a value of $\tau_F = 0.1$ min was chosen, because this value was satisfactory for the analog filter of Fig. 17.6*b*. The smaller value of α (0.33 for $\Delta t = 0.05$ min vs. 0.5 for $\Delta t = 0.1$ min) provides more filtering.

The performance of two moving-average filters ($N^* = 3$ and 7) with $\Delta t = 0.05$ min is shown in Fig. 17.6*d*. Choosing $N^* = 7$ results in better filtering because this moving-average filter averages the sinusoidal noise over several cycles, while $N^* = 3$ gives a faster response but larger fluctuations.

(ii) Random Noise

The simulations illustrating the effects of this noise level are shown in Fig. 17.7. Figure 17.7*a* shows the unfiltered signal after Gaussian noise with zero mean and a variance of 0.01 was added to the square-wave signal. The analog, exponential filters in Fig. 17.7*b*, provide effective filtering and again show the tradeoff between degree of filtering and sluggish response that is inherent in the choice of τ_F.

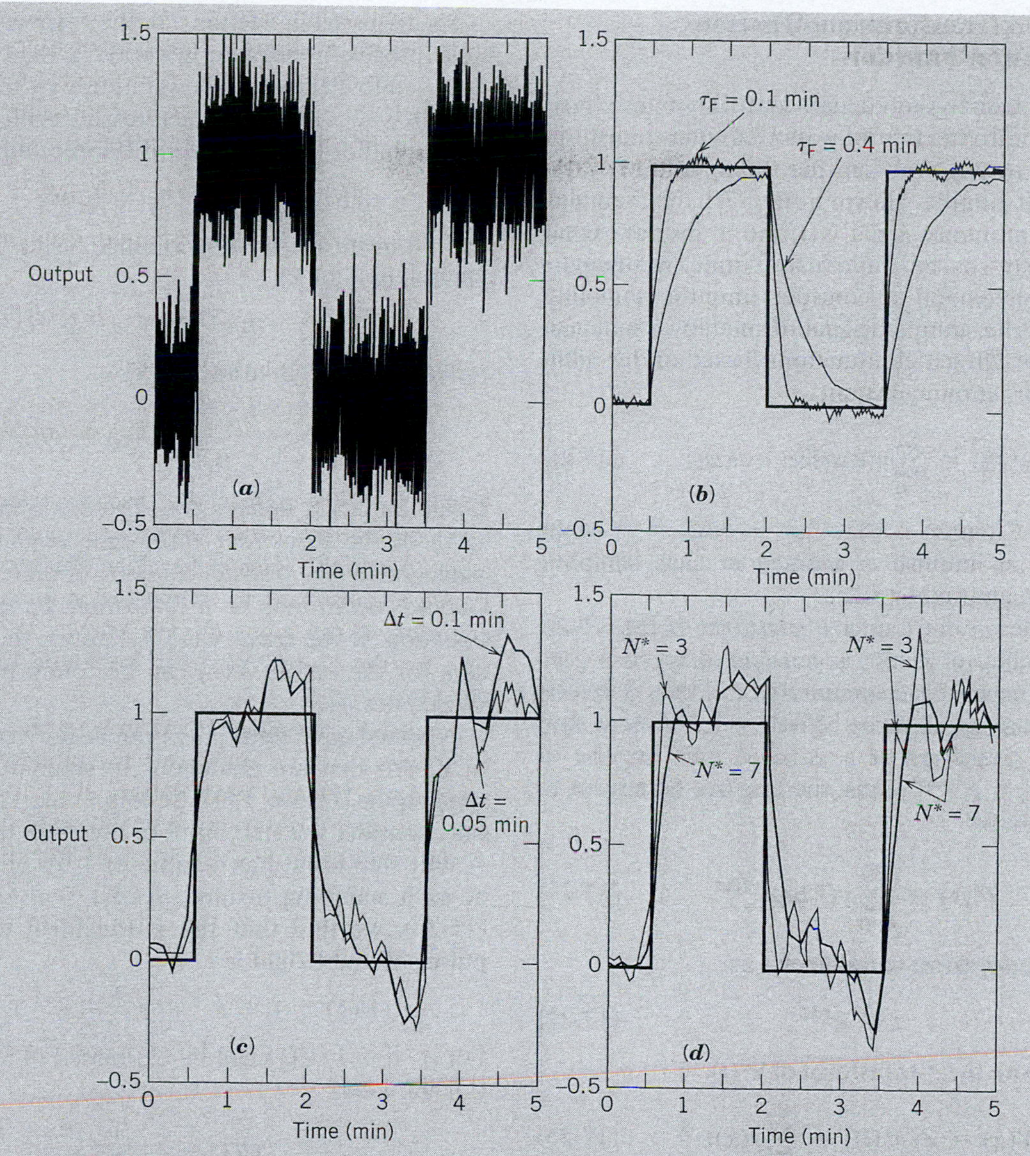

Figure 17.7 Comparison of filter performance for additive Gaussian noise: (*a*) Square-wave plus noise; (*b*) analog exponential filters; (*c*) digital exponential filters; (*d*) moving-average filters.

The digital filters in Fig. 17.7*c* and *d* are less effective, even though different values of Δt and N^* were considered. Some aliasing occurs owing to the high-frequency components of the random noise, which prevents the digital filter from performing as well as the analog filter.

In conclusion, both analog and digital filters can smooth noisy signals, providing that the filter design parameters (including sampling period) are carefully selected.

17.3 z-TRANSFORM ANALYSIS FOR DIGITAL CONTROL

In this section, we introduce the z-transform in order to analyze discrete-time systems. Once a continuous system is interfaced with a discrete system, such as shown in Fig. 17.3, it is necessary to analyze the behavior of the closed-loop system in discrete time. It is possible to simulate the discrete and continuous elements of the closed-loop control system using software such as Simulink; however, a simulation-based approach does not provide a rigorous basis to interpret or analyze discrete-time behavior. This analysis includes such items as process and controller discrete models; effect of poles, zeros, and system order on dynamic behavior; physical realizability; and stability of closed-loop systems. Key concepts for these topics are discussed below. More extensive presentations are available in Franklin et al. (1997) and the first edition of this book (Seborg et al., 1989).

17.3.1 The *z*-Transform and Discrete Transfer Functions

The design and analysis of digital control systems is facilitated by the introduction of a discrete-time transform, namely, the *z-transform*. Consider the operation of the ideal, periodic sampler shown in Fig. 17.1. The sampler converts a continuous signal $y(t)$ into a discrete signal $y^*(t)$ at equally spaced intervals of time. Mathematically, it is convenient to consider impulse sampling, where $y^*(t)$ is the sampled signal formed by a sequence of impulses or Dirac delta functions based on the value of $y(t)$ at each sampling instant:

$$y^*(t) = \sum_{k=0}^{\infty} y(k\Delta t)\delta(t - k\Delta t) \qquad (17\text{-}20)$$

Recall from Chapter 3 that $\delta(t - k\Delta t) = 1$ when $t = k\Delta t$, so an impulse is formed at each sampling instant with magnitude $y(k\Delta t)$.

Next, we derive the Laplace transform of Eq. 17-20, $Y^*(s)$. The value of $y(k\Delta t)$ is considered to be a constant in each term of the summation and thus is invariant when transformed. Since $\mathcal{L}[\delta(t)] = 1$, it follows that the Laplace transform of a delayed unit impulse is $\mathcal{L}[\delta(t - k\Delta t)] = e^{-k\Delta ts}$. Thus, the Laplace transform of (17-20) is given by

$$Y^*(s) = \sum_{n=0}^{\infty} y(k\Delta t)e^{-k\Delta ts} \qquad (17\text{-}21)$$

Define the *z*-transform variable of *z* as

$$z \triangleq e^{s\Delta t} \qquad (17\text{-}22)$$

Let $Y(z)$ denote the *z*-transform of $y^*(t)$,

$$Y(z) = Z[y^*(t)] = \sum_{k=0}^{\infty} y(k)z^{-k} \qquad (17\text{-}23)$$

where the notation is simplified by using $y(k)$ to denote $y(k\Delta t)$.

We can use *z*-transforms in a similar way to Laplace transforms and ultimately express a transfer function for discrete time that corresponds to a difference equation. First we need to derive some properties of *z*-transforms. Using (17-23), we develop the *real translation theorem* as follows:

$$Z(y(t - i\Delta t)) = \sum_{k=0}^{\infty} y(k\Delta t - i\Delta t)z^{-k} \qquad (17\text{-}24)$$

Substituting $j = k - i$ and because $y(j\Delta t) = 0$ for $j < 0$, then

$$Z(y(t - i\Delta t)) = z^{-i}\sum_{j=0}^{\infty} y(j\Delta t)z^{-j} = z^{-i}Y(z) \qquad (17\text{-}25)$$

The translation theorem therefore states that $Z(y(k - i)) = z^{-i}Y(z)$; hence, $Z(y(k - 1)) = z^{-1}Y(z)$.

As discussed in Section 7.4, the response of a continuous process at discrete intervals of time ($y(k)$, $k = 0, 1, 2\ldots$) to changes in the input at past intervals ($u(k)$, $k = 0, 1, 2\ldots$) can be expressed using a difference equation. For the first-order difference equation,

$$y(k) + a_1y(k - 1) = b_1u(k - 1) \qquad (17\text{-}26)$$

the *z*-transform can be obtained using (17-25) for a general input $u(k)$:

$$Y(z) + a_1z^{-1}Y(z) = b_1z^{-1}U(z) \qquad (17\text{-}27)$$

Solving for $Y(z)$ in terms of $U(z)$,

$$Y(z) = \frac{b_1z^{-1}}{1 + a_1z^{-1}}U(z) = G(z)U(z) \qquad (17\text{-}28)$$

Equation 17-28 defines the discrete transfer function $G(z)$ of the first-order difference equation, which is analogous to the transfer function obtained by applying Laplace transforms to a first-order linear differential equation. If the input $U(z)$ is known, then an expression for the output $Y(z)$ can be found by multiplying $G(z)$ times $U(z)$.

A pulsed input signal $U(z)$ can be derived for a variety of signals that are analogous to standard continuous-time inputs (Ogata, 1994; Seborg et al., 1989). Here we only consider the step input to illustrate the procedure. A unit step input has a value of 1 for all time; hence, at each sampling instant, $u(k\Delta t) = u(k) = 1$. Using (17-23), we find that the z-transform of a series of pulses of unit height is

$$U(z) = 1 + z^{-1} + z^{-2} + z^{-3} + \ldots. \qquad (17\text{-}29)$$

For $|z^{-1}| < 1$, $U(z)$ can be expressed in closed form as (Ogata, 1994)

$$U(z) = \frac{1}{1 - z^{-1}} \qquad (17\text{-}30)$$

To calculate the response of a discrete transfer function, which corresponds to the response of the equivalent difference equation, we can use direct simulation of the difference equation based on the specified input. Alternatively, the output *z*-transform can be calculated using long division, which is a power series expansion in terms of z^{-k}. We will illustrate this calculation in Examples 17.2 and 17.3.

EXAMPLE 17.2

Calculate the response of the first-order difference equation (17-26) for $a_1 = -0.368$, $b_1 = 1.264$, and $y(0) = 0$ using z-transforms and long division for $k = 0, 1, \ldots 5$. Compare the result with the unit step response for a first-order continuous-time system ($K = 20$, $\tau = 1$), where $a_1 = -e^{-\Delta t/\tau}$, $b_1 = K(1 - e^{-\Delta t/\tau})$, and $\Delta t = 1$, as discussed in Section 7.4.

SOLUTION

Using (17-28), we find that the response for a step input ($U(z) = 1/(1 - z^{-1})$) is

$$Y(z) = \frac{1.264}{1 - 0.368z^{-1}} \cdot \frac{1}{1 - z^{-1}} = \frac{1.264z^{-1}}{1 - 1.368z^{-1} + 0.368z^{-2}} \tag{17-31}$$

Next long division is used to divide the denominator into the numerator. The order of the numerator and denominator polynomials starts with the lowest powers of z^{-k} for the division operation.

$$\begin{array}{r|l} & 1.264z^{-1} + 1.729z^{-2} + 1.900z^{-3} + \ldots \\ 1 - 1.368z^{-1} + 0.368z^{-2} & 1.264z^{-1} \\ & \underline{1.264z^{-1} - 1.729z^{-2} + 0.465z^{-3}} \\ & 1.729z^{-2} - 0.465z^{-3} \\ & \underline{1.729z^{-2} - 2.365z^{-3} + 0.636z^{-4}} \\ & 1.900z^{-3} - 0.636z^{-4} \\ & \text{(etc.)} \end{array}$$

Because of space limitations, only the first three terms are shown above: $y(1) = 1.264$, $y(2) = 1.729$, and $y(3) = 1.900$. Continuing on, we calculate $y(4) = 1.963$ and $y(5) = 1.986$. Ultimately, $y(k)$ reaches its steady-state value of 2.0 (k large), which agrees with the fact that the process gain K is 2 and the input is a unit step change. The step response in continuous time is $y(t) = 2(1 - e^{-t})$, and the sampled values of the discrete-time response for $\Delta t = 1$ are the same ($k = 0, 1, 2, 3 \ldots$). Thus, the discretization is exact; that is, it is based on the analytical solution for a piecewise constant input.

The same answer could be obtained from simulating the first-order difference equation (17-26), with $u(k) = 1$ for $k \geq 0$; that is,

$$y(k) = 0.368\, y(k - 1) + 1.264\, (1)$$

Starting with $y(0) = 0$, it is easy to generate recursively the values of $y(1) = 1.264$, $y(2) = 1.729$, and so on. Note that the steady-state value can be obtained in the above equation by setting $y(k) = y(k - 1) = y_{ss}$ and solving for y_{ss}. In this case $y_{ss} = 2.0$, as expected.

EXAMPLE 17.3

For the difference equation,

$$\begin{aligned} y(k) = {} & 0.9744\, y(k - 1) - 0.2231\, y(k - 2) \\ & - 0.3225\, u(k - 2) + 0.5712\, u(k - 3) \end{aligned} \tag{17-32}$$

derive its discrete transfer function and step response for $U(z) = \dfrac{1}{1 - z^{-1}}$. Use long division to obtain $Y(z)$ for $k = 0$ to $k = 9$. Compare this result with simulating the original difference equation, (17-32), and a unit step change in $u(k)$ at $k = 0$. Assume $y(0) = 0$.

SOLUTION

Taking the z-transform of (17-32),

$$\begin{aligned} Y(z) = {} & 0.9744\, z^{-1}\, Y(z) - 0.2231\, z^{-2}\, Y(z) \\ & - 0.3225\, z^{-2}\, U(z) + 0.5712\, z^{-3}\, U(z) \end{aligned} \tag{17-33}$$

Rearranging gives the discrete transfer function

$$G(z) = \frac{Y(z)}{U(z)} = \frac{-0.3225z^{-2} + 0.5712z^{-3}}{1 - 0.9744z^{-1} + 0.2231z^{-2}} \tag{17-34}$$

Note that the numerator of $G(z)$ has a common factor of z^{-2}, which indicates the presence of an apparent time delay of two sampling periods.

To determine the step response, set $U(z) = \dfrac{1}{1 - z^{-1}}$ and multiply it by the transfer function to find the power series for $Y(z)$. Long division as done in Example 17.2 yields

$$\begin{aligned} Y(z) = {} & -0.3225z^{-2} - 0.0655z^{-3} + 0.2568z^{-4} \\ & + 0.5136z^{-5} + 0.6918z^{-6} + 0.8082z^{-7} \\ & + 0.8820z^{-8} + 0.9277z^{-9} + \ldots \end{aligned} \tag{17-35}$$

Note that $y(k) = 0$ for $k = 0$ and $k = 1$, and $y(2) = -0.3225$, which indicates a two-unit time delay in $G(z)$. After an initial transient period, it appears that $y(k)$ is steadily increasing and may approach a steady-state value. For a unit step change in $U(z)$, the steady-state value of the response $Y(z)$ can be found by determining the steady-state gain of $G(z)$. In analogy to continuous-time transforms, the steady-state gain can be found by setting $s = 0$ in $z = e^{s\Delta t}$, or $z = 1$. In this case, $G(z = 1) = 1$, so $y(k) = 1$ at steady state for a unit step change.

The same result can be obtained using the original difference equation. A table (or spreadsheet) could be constructed to track the various terms of the difference equation for $k = 0, 1, 2 \ldots 9$. The top row of the table is structured using the same terms as in the difference equation, and the step response is generated using spreadsheet software. We assume that the original system is at steady state at $k = 0$, so y and u terms corresponding to $k < 0$ are equal to 0. Using the difference equation, Eq. 17-32, the value of $y(k)$ can be obtained from the entries (cells) in the same row by performing the appropriate multiplications.

Next, consider a general higher-order difference equation given by

$$\begin{aligned} & a_0 y(k) + a_1 y(k - 1) + \cdots + a_m y(k - m) = \\ & b_0 u(k) + b_1 u(k - 1) + \cdots + b_n u(k - n) \end{aligned} \tag{17-36}$$

where $\{a_i\}$ and $\{b_i\}$ are sets of constant coefficients, n and m are positive integers, $u(k)$ is the input, and $y(k)$ is the output. Taking the z-transform of both sides of Eq. 17-36 gives

$$\begin{aligned} & a_0 Y(z) + a_1 z^{-1} Y(z) + \cdots + a_m z^{-m} Y(z) \\ & = b_0 U(z) + b_1 z^{-1} U(z) + \cdots + b_n z^{-n} U(z) \end{aligned} \tag{17-37}$$

Rearranging (17-37) gives the transfer function form,

$$Y(z) = \frac{b_0 + b_1 z^{-1} + \cdots + b_n z^{-n}}{a_0 + a_1 z^{-1} + \cdots + a_m z^{-m}}\, U(z) \tag{17-38}$$

The ratio of polynomials in the discrete transfer function, $G(z)$, can be derived by algebraic manipulations for any difference equation (Åström and Wittenmark, 1997). In this case

$$G(z) = \frac{B(z)}{A(z)} = \frac{b_0 + b_1z^{-1} + \cdots + b_nz^{-n}}{a_0 + a_1z^{-1} + \cdots + a_mz^{-m}} \quad (17\text{-}39)$$

where $B(z)$ and $A(z)$ are polynomials in z^{-1}. For most processes, b_0 is zero, indicating that the input does not instantaneously affect the output; in Eq. 17-26, if $b_0 \neq 0$, then the input term would include $u(k)$ in the difference equation. In addition, the leading coefficient in the denominator can be set equal to unity by dividing both numerator and denominator by a_0. The steady-state gain of G in (17-39) can be found by setting $z = 1$.

The dynamic behavior of (17-39) can be characterized by its poles and zeros in analogy to continuous-time systems. To do this, we must first convert $G(z)$ to positive powers of z by multiplying (17-39) by z^n/z^m, leading to modified polynomials, $B'(z)/A'(z)$. The stability of $G(z)$ is determined by its poles, the roots of the characteristic equation, $A'(z) = 0$; $A'(z)$ is called the characteristic polynomial. Note that z is a complex variable, because it is related to complex variable s by the definition in (17-22). The *unit circle* in the complex z-plane is defined as a circle with unit radius where $|z| = 1$. The unit circle is the dividing line between the stable and unstable regions. Any pole that lies inside the unit circle is stable and thus provides a stable response to a bounded input. In contrast, a pole lying outside the unit circle is unstable (Åström and Wittenmark, 1997). The zeros of $G(z)$ are the roots of $B'(z) = 0$. In discrete time analysis, time delays are usually assumed to be an integer multiple N of the sampling period Δt. This time delay produces N roots of $A'(z)$ that are located at the origin and thus can be expressed as a factor, z^N. By definition, the order of $G(z)$ is $P + N$, where P is the number of poles.

Figure 17.8 shows the effect of pole location on the possible responses for a simple first-order transfer function, $G(z) = b_0/(1 - az^{-1})$, forced by an impulse at $k = 0$. The corresponding continuous-time model responses are also shown. Poles 3 and 4 are inside the unit circle and thus are stable, while poles 1 and 6 are outside the unit circle and cause an unstable response. Poles 2 and 5 lie on the unit circle and are marginally stable. Negative poles such as 4–6 produce oscillatory responses, even for a first-order discrete-time system, in contrast to continuous-time first-order systems.

Some important properties of sampled-data systems can be obtained from long division of their z-transforms. For example, for the first-order z-transform,

$$Y(z) = \frac{b_0}{1 - a_1z^{-1}} \quad (17\text{-}40)$$

an equivalent sampled signal can be found by long division, resulting in the infinite series

$$\overline{Y}(z) = b_0(1 + a_1z^{-1} + a_1^2z^{-2} + \cdots + a_1^nz^{-n} + \cdots) \quad (17\text{-}41)$$

If $a_1 = b_0 = 1$, then $Y(z) = \dfrac{1}{1 - z^{-1}}$ and $\overline{Y}(z) = \sum_{i=0}^{\infty} z^{-i}$.

Hence, the equivalent difference equation would be a summation of all previous values of $u(k)$.

17.3.2 Convolution Model Form

For the transfer function in (17-38), dividing the numerator by the denominator (starting with a_0 and determining the remainders) leads to a model equivalent to

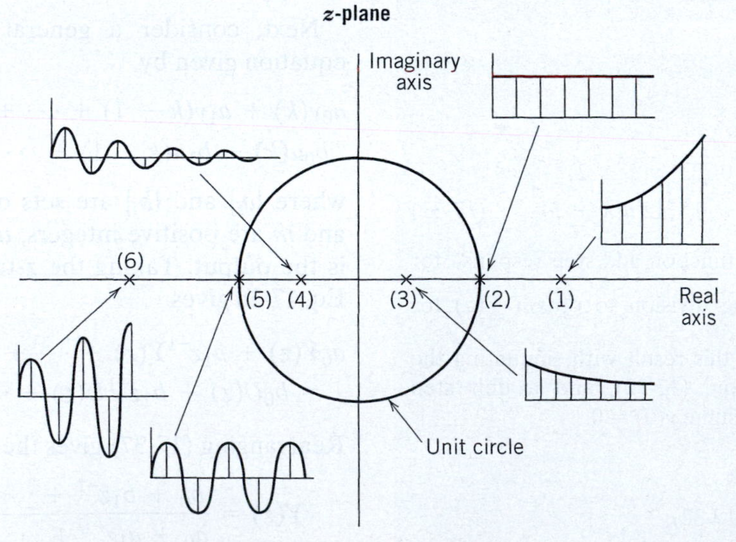

Figure 17.8 Time-domain responses for different locations of the pole a, indicated by an ×, of a first-order discrete transfer function and a pulse input at $k = 0$.

a discrete convolution model (see Section 7.5.1). Long division of (17-38) yields for the first three terms of the series $\sum_{k=0}^{\infty} c_k z^{-k}$

$$c_0 = \frac{b_0}{a_0} \tag{17-42}$$

$$c_1 = \frac{b_1}{a_0} - \frac{b_0 a_1}{a_0^2} \tag{17-43}$$

$$c_2 = b_2 - \frac{b_0 a_2}{a_0} - \frac{b_1 a_1}{a_0} + \frac{b_0 a_1^2}{a_0^2} \tag{17-44}$$

The transfer function is therefore

$$G(z) = c_0 + c_1 z^{-1} + c_2 z^{-2} + \cdots \tag{17-45}$$

which corresponds to the convolution model form,

$$y(k) = c_0 + c_1 u(k-1) + c_2 u(k-2) + \cdots \tag{17-46}$$

17.3.3 Physical Realizability

Chapter 4 addressed the notion of physical realizability for continuous-time transfer functions. An analogous condition can be stated for a difference equation or its transfer function—namely, that a discrete-time model cannot have an output signal that depends on future inputs. Otherwise, the model is not physically realizable. Consider the ratio of polynomials given in Eq. 17-39. The discrete transfer function will be physically realizable as long as $a_0 \neq 0$, assuming that $G(z)$ has been reduced so that common factors in the numerator and denominator have been canceled. To show this property, examine Eq. 17-36. If $a_0 = 0$, the difference equation is

$$a_1 y(k-1) = b_0 u(k) + b_1 u(k-1) + \cdots + b_k u(k-n) - a_2 y(k-2) - \cdots - a_m y(k-m) \tag{17-47}$$

or, by shifting the index from k to $k + 1$,

$$a_1 y(k) = b_0 u(k+1) + b_1 u(k) + \ldots \tag{17-48}$$

This equation requires a future input $u(k + 1)$ to influence the present value of the output $y(k)$, which is physically impossible (unrealizable). Physical realizability of models (process or controller) should be checked prior to their use for simulation or control.

Discrete Transfer Function of an Integral Controller

Now we derive a transfer function for a digital integral controller, where the output is $p(k)$ and the input is the error signal $e(k)$ (cf. Eq. 8-7). The integral of $e(t)$ in continuous time can be approximated by a summation in discrete time. By using a finite difference approximation to the integral

$$p(t) = \overline{p} + \frac{1}{\tau_I}\int_0^t e(t')dt' \approx \frac{\Delta t}{\tau_I}\sum_{k=0}^{n} e(k) \tag{17-49}$$

Then

$$p(k) - \overline{p} = \frac{\Delta t}{\tau_I}\sum_{k=0}^{n} e(k) \tag{17-50}$$

Taking the z-transform,

$$P(z) = \frac{\Delta t}{\tau_I}\left(\sum_{k=0}^{n} z^{-k}\right)E(z) \tag{17-51}$$

When n is large, the summation can be expressed in closed form as

$$P(z) = \frac{\Delta t}{(1 - z^{-1})\tau_I}E(z) \tag{17-52}$$

The continuous-time analog to (17-43) for an integral controller is

$$P(s) = \frac{E(s)}{\tau_I s}$$

Comparing the above two expressions, we can observe a relationship between z^{-1} and s, which is

$$s \cong \frac{1 - z^{-1}}{\Delta t} \tag{17-53}$$

This expression is known as the *backward-difference* (BD) approximation of s (equivalent to a first-order Taylor series), and it can be used to convert a continuous-time expression (in s) into an approximate discrete-time expression in z^{-1} simply by direct substitution. The same relationship can be obtained by recognizing that $\mathscr{L}\left(\frac{de}{dt}\right) = sE(s)$. This expression can be compared with $\frac{de}{dt} \approx \frac{e(k) - e(k-1)}{\Delta t}$ for which the z-transform is $\frac{1 - z^{-1}}{\Delta t}E(z)$. Therefore, we can approximate the Laplace transform of a continuous-time function such as a PID controller in discrete time by substituting the BD approximation for s given in (17-53), as shown in the next example.

EXAMPLE 17.4

Derive the discrete transfer function for the parallel form of a PID controller.

$$G_c(s) = K_c\left(1 + \frac{1}{\tau_I s} + \tau_D s\right) \tag{17-54}$$

using the backward-difference substitution for s. Compare the result with the velocity form of the PID algorithm given in Eq. 8-28.

SOLUTION

Substituting $s \cong (1 - z^{-1})/\Delta t$ into (17-54) gives

$$G_c(z) = \frac{K_c(a_0 + a_1 z^{-1} + a_2 z^{-2})}{1 - z^{-1}} \tag{17-55}$$

where $a_0 = 1 + \frac{\Delta t}{\tau_I} + \frac{\tau_D}{\Delta t}$, $a_1 = -\left(1 + \frac{2\tau_D}{\Delta t}\right)$, $a_2 = \frac{\tau_D}{\Delta t}$

Because $E(z)$ is the error signal (input) and $P(z)$ is the controller output, $P(z) = G_c(z)E(z)$. Multiplying both sides of (17-46) by $(1 - z^{-1})$ yields

$$(1 - z^{-1})P(z) = K_c(a_0 + a_1 z^{-1} + a_2 z^{-2})E(z) \quad (17\text{-}56)$$

Converting the controller transfer function into difference equation form gives

$$p(k) - p(k-1) = K_c a_0 e(k) + K_c a_1 e(k-1) + K_c a_2 e(k-2) \quad (17\text{-}57)$$

Substituting for a_0, a_1, and a_2 and collecting terms with respect to the controller settings K_c, τ_I, and τ_D gives

$$p(k) - p(k-1) = K_c\left[(e(k) - e(k-1)) + \frac{\Delta t}{\tau_I}e(k) + \frac{\tau_D}{\Delta t}(e(k) - 2e(k-1) + e(k-2))\right] \quad (17\text{-}58)$$

Note that this equation is identical to Eq. 8-28, which was derived using a finite-difference approximation in the time domain.

Ogata (1994) has listed more accurate formulas for algebraic substitution into a transfer function $G(s)$. Approximate substitution is a procedure that should always be used with care. When feasible, it is preferable to use exact conversion of the continuous process model into discrete time rather than finite difference approximations such as Eq. 17-53. Table 17.1 presents a conversion table of commonly used transfer functions $G(s)$ based on a zero-order hold (which yields a piecewise constant input u). This table is also consistent with the exact (analytical) conversion formulas for first- and second-order process models in Eqs. 7-33 through 7-40.

Block Diagram Algebra

It is important to realize that in the block diagram of the closed-loop system, Fig. 17.3, the open-loop transfer function of the continuous components includes the product of the final control element (or valve), the process, and the measurement, or $G(s) = G_v(s)G_p(s)G_m(s)$. It is mathematically incorrect to find separate discrete-time versions $G_v(z)$, $G_p(z)$,

Table 17.1 Discrete Transfer Functions Obtained Using a Zero-Order Hold

Transfer Function $G(s)$	$G(z)$	
$\frac{K}{s}$	$\frac{b_1 z^{-1}}{1 + a_1 z^{-1}}$	$a_1 = -1$ $b_1 = K\Delta t$
$\frac{K}{s + r}$	$\frac{b_1 z^{-1}}{1 + a_1 z^{-1}}$	$a_1 = -\exp(-r\Delta t)$ $b_1 = \frac{K}{r}[1 - \exp(-r\Delta t)]$
$\frac{K}{(s + r)(s + p)}$	$\frac{b_1 z^{-1} + b_2 z^{-2}}{1 + a_1 z^{-1} + a_2 z^{-2}}$	$a_1 = -\exp(-r\Delta t) - \exp(-p\Delta t)$ $a_2 = \exp[-(r + p)\Delta t]$ $b_1 = [K/rp(r - p)][(r - p) - r\exp(-p\Delta t) + p\exp(-r\Delta t)]$ $b_2 = [K/rp(r - p)]\{(r - p)\exp[-(r + p)\Delta t] + p\exp(-p\Delta t) - r\exp(-r\Delta t)\}$
$\frac{K}{s(s + r)}$	$\frac{b_1 z^{-1} + b_2 z^{-2}}{1 + a_1 z^{-1} + a_2 z^{-2}}$	$a_1 = -\{1 + \exp(-r\Delta t)\}$ $a_2 = \exp(-r\Delta t)$ $b_1 = -(K/r^2)[1 - r\Delta t - \exp(-r\Delta t)]$ $b_2 = (K/r^2)[1 - \exp(-r\Delta t) - r\Delta t\exp(-r\Delta t)]$
$\frac{K(s + v)}{(s + r)(s + p)}$	$\frac{b_1 z^{-1} + b_2 z^{-2}}{1 + a_1 z^{-1} + a_2 z^{-2}}$	$a_1 = -\{\exp(-p\Delta t) + \exp(-r\Delta t)\}$ $a_2 = \exp[-(r + p)\Delta t]$ $b_1 = \frac{K}{p - r}\{\exp(-p\Delta t) - \exp(-r\Delta t) + (v/p)[1 - \exp(-p\Delta t)] - (v/r)[1 - \exp(-r\Delta t)]\}$ $b_2 = K\{(v/rp)\exp[-(r + p)\Delta t] + [(p - v)/p(r - p)]\exp(-r\Delta t) + [(v - r)/r(r - p)]\exp(-p\Delta t)\}$

and $G_m(z)$ and then multiply them together (Ogata, 1994; Seborg et al., 1989). Instead, the z-transform of $G(s)$ should be based on exact discretization of the Laplace transform (Åström and Wittenmark, 1997). In the case where $G_m = K_m$, this leads to the closed-loop expression (analogous to Chapter 12),

$$\frac{Y(z)}{Y_{sp}(z)} = \frac{G_c(z)G(z)}{1 + G_c(z)G(z)} \quad (17\text{-}59)$$

In (17-59), $G_c(z)$ is the discrete transfer function for the digital controller. A digital controller is inherently a discrete-time device, but with the zero-order hold, the discrete-time controller output is converted to a continuous signal that is sent to the final control element. So the individual elements of G are inherently continuous, but by conversion to discrete-time we compute their values at each sampling instant. The discrete closed-loop transfer function in (17-59) provides a framework to perform closed-loop analysis and controller design, as discussed in the next section. Additional material on closed-loop analysis for discrete-time systems is available elsewhere (Ogata, 1994; Seborg et al., 1989).

17.3.4 Stability Analysis

For sampled-data systems the stability of a transfer function can be tested by determining whether any roots of its characteristic polynomial lie outside the unit circle (see Fig. 17.8). To apply this stability test, write the denominator of the transfer function in (17-39) in terms of positive powers of z:

$$\begin{aligned} A'(z) &= z^m A(z) \\ &= a_m z^m + a_{m-1} z^{m-1} + \cdots + a_1 z + a_0 = 0 \quad (17\text{-}60) \end{aligned}$$

Note that if the time delay is an integer multiple of Δt, a polynomial in z will always result. Any roots of characteristic polynomial $A'(z)$ can be found using a root-finding subroutine. If any root lies outside the unit circle, then the discrete transfer function is unstable (see Fig. 17.8). This conclusion can be verified by simulating the transfer function response for a pulse input to see if the output $y(k)$ grows with respect to k.

Because of space limitation, we have deliberately not included rigorous coverage of stability of z-transform models, such as the bilinear transform or Jury's test. See Franklin et al. (1997) and Ogata (1994) for more details on these techniques.

17.4 TUNING OF DIGITAL PID CONTROLLERS

In this section, the main emphasis is on the digital PID controller and how it should be tuned. Digital versions of the PID controller in the form of difference equations were previously presented in Eqs. 8-26 and 8-28 (position and velocity forms) by using the backward difference approximation. For small values of Δt (relative to the process response time), the finite difference approximations for integral and derivative control action discussed in Section 8.3 are reasonably accurate. Hence, suitable controller settings obtained for a continuous controller can also be utilized in a digital PID controller. As noted by Isermann (1989), the continuous and the discrete PID controllers will have essentially the same behavior as long as $\Delta t/\tau_I \leq 0.1$. Åström and Wittenmark (1997) have discussed the effect of sampling period for designing a wide range of digital controllers.

If Δt is not small, use of the zero-order hold in digital control systems requires a modification in the controller design procedure, because the sampler plus a ZOH introduces an effective time delay in the signal to the final control element. In this case, the dynamic behavior of the sampler plus ZOH should be approximated by a time delay equal to one-half the sampling period (Franklin et al., 1997). Thus, it is a common practice in tuning digital PID controllers to add $\Delta t/2$ to the process time delay θ before using the controller tuning relations in Chapter 12. Using the backward-difference (BD) version of a PID controller (Eq. 17-55) will result in greater stability margins in discrete time (vs. other finite difference schemes); see Franklin et al. (1997). As shown in Example 17.5 in the next section, using s-domain Direct Synthesis of a PID controller followed by BD conversion to discrete time yields satisfactory results.

In earlier chapters, Simulink was used to simulate linear continuous-time control systems described by transfer function models. For digital control systems, Simulink can also be used to simulate open- and closed-loop responses of discrete-time systems. As shown in Fig. 17.3, a computer control system includes both continuous and discrete components. In order to carry out detailed analysis of such a hybrid system, it is necessary to convert all transfer functions to discrete time and then carry out analysis using z-transforms (Åström and Wittenmark, 1997; Franklin et al., 1997). On the other hand, simulation can be carried out with Simulink using the control system components in their native forms, either discrete or continuous. This approach is beneficial for tuning digital controllers.

In this section, we show how to perform closed-loop simulations for various digital controllers. Although the controller is represented by a discrete transfer function, all other components of the control loop (models for the final control element, process, sensor, and disturbance) will normally be available as continuous transfer functions, which can be directly entered into a Simulink block diagram as functions of s. To

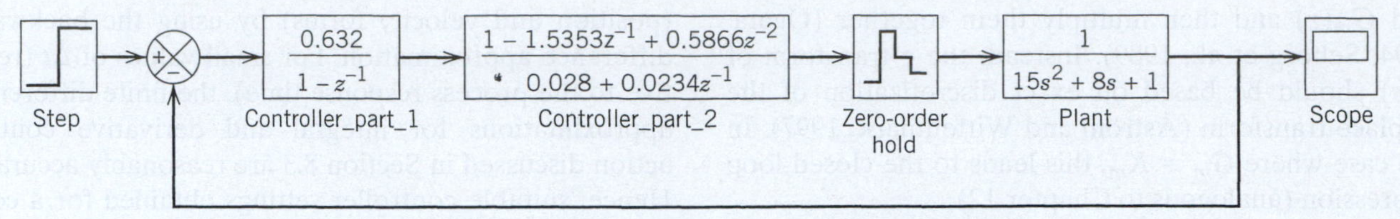

Figure 17.9 Simulink diagram for a discrete-time controller, continuous-time process, and a step change in set point.

introduce a step change in set point requires only selecting the icon in Simulink for the step input (no s or z transform is needed). A step change in the disturbance can be entered in a similar fashion. The zero-order hold (ZOH) icon is placed after the digital controller, because it is necessary to convert the series of controller pulses into a continuous signal to drive a final control element such as a control valve. However, in Simulink the user does not have to include the ZOH, because the software performs the calculations as if the ZOH were there.

Figure 17.9 shows the Simulink block diagram for a digital control system with

1. Two controller blocks that are multiplied ($G_c = G_{c1}G_{c2}$)
2. Zero-order hold
3. Continuous-time process G (time constants of 5 and 3, gain of 1)
4. Unit step in set point and graphical output ("scope") for the process output, which is a continuous-time response

Other blocks can be added from the Simulink menu to explicitly include transfer functions for the final control element, disturbance, and sensor (as many as desired). Simulink can also be employed for open-loop simulation.

Simulink will be used in subsequent examples to demonstrate the performance of various types of digital feedback controllers, as was done in Chapter 12. The stability of closed-loop systems can also be checked using trial-and-error simulations of the block diagram to determine the maximum controller gain for stability.

EXAMPLE 17.5

A digital controller is used to control the pressure in a tank by adjusting a purge stream. The control valve is air-to-open, and the process model has been identified as

$$G = G_v G_p G_m = \frac{-20e^{-s}}{5s + 1}$$

The gain is dimensionless, and the time constant and time delay are in minutes. The sampling period is $\Delta t = 1$ min.

Compare the closed-loop performance of a discrete PI controller using the ITAE (disturbance) tuning rules in Table 12.3. Approximate the sampler and ZOH by a time delay equal to $\Delta t/2$. Use Simulink to check the effect of sampling period for different controllers, with $\Delta t = 0.05$, 0.25, 0.5, and 1.0 min.

SOLUTION

First adjust the process time delay for each controller calculation by adding $\Delta t/2$, which accounts for the time delay due to the sampler plus the ZOH. The controller settings calculated from Table 12.2 for different sampling periods are (τ_I in min):

	PI	
Δt	K_c	τ_I
0.05	−0.21	2.48
0.25	−0.19	2.68
0.5	−0.17	2.89
1.0	−0.14	3.27

These continuous controller settings are then substituted into Eq. 17-55 to obtain the corresponding settings for the digital controller. Figure 17.10 shows that smaller sampling periods result in faster closed-loop responses for PI control. There is no change in performance for $\Delta t \leq 0.05$ min. It is interesting to compare the results of this example with guidelines for choosing the sampling periods given in (17-3).

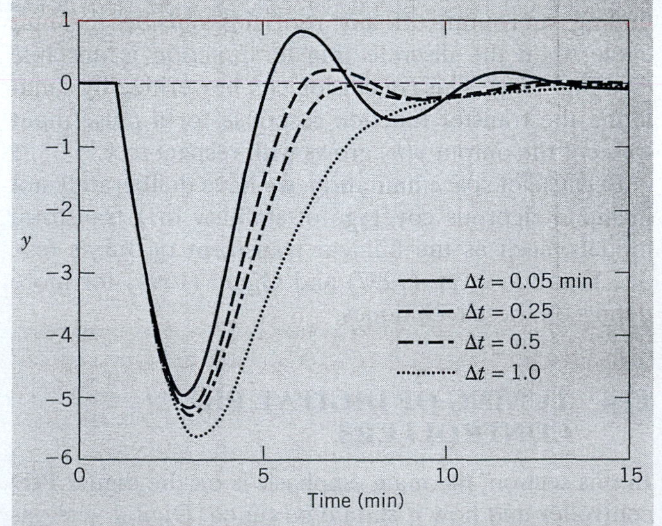

Figure 17.10 Closed-loop responses for PI controllers with different sampling periods and a step disturbance.

Effect of Filter Selection on PID Controller Performance

Digital and analog filters are valuable for smoothing data and eliminating high-frequency noise, but they also affect control system performance. In particular, a filter is an additional dynamic element in the feedback loop that introduces additional phase lag. Consequently, it reduces the stability margin for a feedback controller, compared to the situation where no filter is present. Therefore, the controller may have to be retuned if the filter constant is changed. When derivative action is used, it is important to filter noisy signals before the derivative control calculations are performed. Because derivative action tends to amplify noise in the process measurement, filtering helps prevent controller saturation and wear in the final control elements. Some PID controllers include a filter in the controller equation; see Chapter 8 for more details.

17.5 DIRECT SYNTHESIS FOR DESIGN OF DIGITAL CONTROLLERS

In this section, the Direct Synthesis (DS) method presented in Section 12.2 is extended to the design of digital controllers. We begin with special cases that lead to a PID controller, and then show how other types of digital feedback controllers can be derived using the Direct Synthesis technique. Both G_c and G must be expressed as discrete-time in the closed-loop transfer function (17-59). In Direct Synthesis, the designer specifies the desired closed-loop transfer function $(Y/Y_{sp})_d$. The controller G_c that yields the desired performance is obtained from (17-59)

$$G_c = \frac{1}{G}\frac{(Y/Y_{sp})_d}{1 - (Y/Y_{sp})_d} \tag{17-61}$$

Equation 17-61 is the model-based control law used for the Direct Synthesis design method in discrete time, analogous to the procedure discussed in Section 12.2 for continuous-time controllers.

Next two Direct Synthesis algorithms for discrete-time application are considered: Dahlin's method and the Vogel-Edgar method. The discrete-time version of a related method considered in Chapter 12, Internal Model Control, is also presented.

17.5.1 Dahlin's Method (Lambda Tuning)

Dahlin's method (Dahlin, 1968) specifies that the desired closed-loop transfer function is a FOPTD model:

$$\left(\frac{Y}{Y_{sp}}\right)_d = \frac{e^{-hs}}{\lambda s + 1} \tag{17-62}$$

where λ is the desired closed-loop time constant (cf. Chapter 12) and h is the specified time delay. After selecting $h = \theta$ (the process time delay), Table 17.1 indicates that the discrete transfer function corresponding to Eq. 17-62 is

$$\left(\frac{Y}{Y_{sp}}\right)_d = \frac{(1 - A)z^{-N-1}}{1 - Az^{-1}} \tag{17-63}$$

where $A = e^{-\Delta t/\lambda}$, $K = 1/\lambda$, $r = 1/\lambda$, $N = \theta/\Delta t$.

Substituting (17-63) into Eq. 17-61 yields the general form of G_c for Dahlin's method, which is denoted by G_{DC}:

$$G_{DC} = \frac{1}{G}\frac{(1 - A)z^{-N-1}}{1 - Az^{-1} - (1 - A)z^{-N-1}} \tag{17-64}$$

As a special case, consider G to be the discrete version of a FOPTD transfer function with gain K and time constant τ.

$$G = \frac{K(1 - a_1)}{1 - a_1 z^{-1}} z^{-N-1} \tag{17-65}$$

where $a_1 = e^{-\Delta t/\tau}$. Dahlin's controller is

$$G_{DC} = \frac{(1 - A)}{1 - Az^{-1} - (1 - A)z^{-N-1}} \frac{1 - a_1 z^{-1}}{K(1 - a_1)} \tag{17-66}$$

For all values of N, $(1 - z^{-1})$ is a factor of the denominator, indicating the presence of integral action. The result is consistent with steady-state gain calculations in (17-62) with $s = 0$ and in (17-63) with $z = 1$, which specify zero steady-state error for set-point changes.

Because the desired time constant λ for the closed-loop system serves as a convenient tuning parameter, this approach is often referred to as *lambda-tuning*. Small values of λ produce faster responses, while large values of λ give more sluggish control. This flexibility is especially useful in situations where the model parameters, especially the time delay, are subject to error or are time-varying because of changes in the process. In an aggressively controlled process, an inaccurate time delay can cause poor control and an unstable response. By choosing a larger λ and having more conservative control action, the controller can better accommodate the inaccurate model. As $\lambda \rightarrow 0$ (i.e., $A \rightarrow 0$), Dahlin's algorithm is equivalent to *minimal prototype control*, but such aggressive tuning is usually not desirable for process control applications, because it is quite sensitive to parameter changes (Seborg et al., 1989).

One important feature of a Direct Synthesis method such as Dahlin's method is that the resulting controller contains the reciprocal of the process transfer function. This feature causes the poles of G to become zeros of G_c, while the zeros of G become poles of the controller, unless the poles and zeros of G are canceled by terms in $(Y/Y_{sp})_d$. The inversion of G in (17-61) can lead to operational difficulties, just as in the continuous-time case. If G contains a zero that lies outside the unit circle, then G_c will contain an unstable pole lying outside the unit circle. In this case, G_c is an *unstable* controller and produces an unbounded output sequence for a step change in set point. Although the product G_cG in (17-59) indicates that the unstable pole and zero will cancel, in practice there will always be some model error that prevents exact cancellation. Nevertheless, problems associated with unstable zeros can be successfully treated by judicious selection of $(Y/Y_{sp})_d$, as discussed below.

Digital controllers of the Direct Synthesis type share yet one more characteristic: namely, they contain time-delay compensation in the form of a Smith predictor (see Chapter 16). In Eq. 17-61, for G_c to be physically realizable, $(Y/Y_{sp})_d$ must also contain a term equivalent to $e^{-\theta s}$, which is z^{-N}, where $N = \theta/\Delta t$. In other words, if there is a term z^{-N} in the open-loop discrete transfer function, the closed-loop process cannot respond before $N\Delta t$ or θ units of time have passed. Using $(Y/Y_{sp})_d$ of this form in Eq. 17-63 yields a G_c containing the mathematical equivalence of time-delay compensation, because the time delay has been eliminated from the characteristic equation.

EXAMPLE 17.6

A process is modeled in continuous time by a second-order-plus-time-delay transfer function with $K = 1$, $\tau_1 = 5$, and $\tau_2 = 3$. For $\Delta t = 1$, the discrete-time equivalent (with zero-order hold) is

$$G = \frac{(b_1 + b_2 z^{-1})z^{-N-1}}{1 + a_1 z^{-1} + a_2 z^{-2}} \tag{17-67}$$

where $a_1 = -1.5353$, $a_2 = 0.5866$, $b_1 = 0.0280$, $b_2 = 0.0234$, and $N = 0$ (cf. Eq. 7-36 to 7-40). For Dahlin's controller with $\lambda = \Delta t = 1$, plot the response for a unit change in set point at $t = 5$ for $0 \leq t \leq 10$ using Simulink.

SOLUTION

The Simulink diagram for the example was shown earlier in Fig. 17.9. Using Eq. 17-63, the desired closed-loop transfer function for $\theta = 0$ $(N = 0)$ and $\lambda = \Delta t$ is $(Y/Y_{sp})_d = 0.632z^{-1}/(1 - 0.368z^{-1})$. By applying (17-64), the formula for the controller is

$$G_{DC} = \frac{1 + a_1 z^{-1} + a_2 z^{-2}}{b_1 z^{-1} + b_2 z^{-2}} \frac{0.632z^{-1}}{1 - z^{-1}} \tag{17-68}$$

Substituting the numerical values for a_1, a_2, b_1, and b_2, the controller is

$$G_{DC} = \frac{1 - 1.5353z^{-1} + 0.5866z^{-2}}{0.0280 + 0.0234z^{-1}} \frac{0.632}{1 - z^{-1}} \tag{17-69}$$

When this controller is implemented, an undesirable characteristic appears, namely, *intersample ripple*. Figure 17.11*a* shows the response y and ZOH output p to a unit step change in set point at $t = 5$. Although the response does satisfy $y(k) = 1$ at each sampling instant $(\Delta t = 1)$ for $k \geq 6$, the response is quite oscillatory; that is, intersample ripple occurs. This result is caused by the controller output cycling back and forth between positive and negative deviations from the steady-state value. The behavior, called *ringing*, of course is unacceptable for a control system.

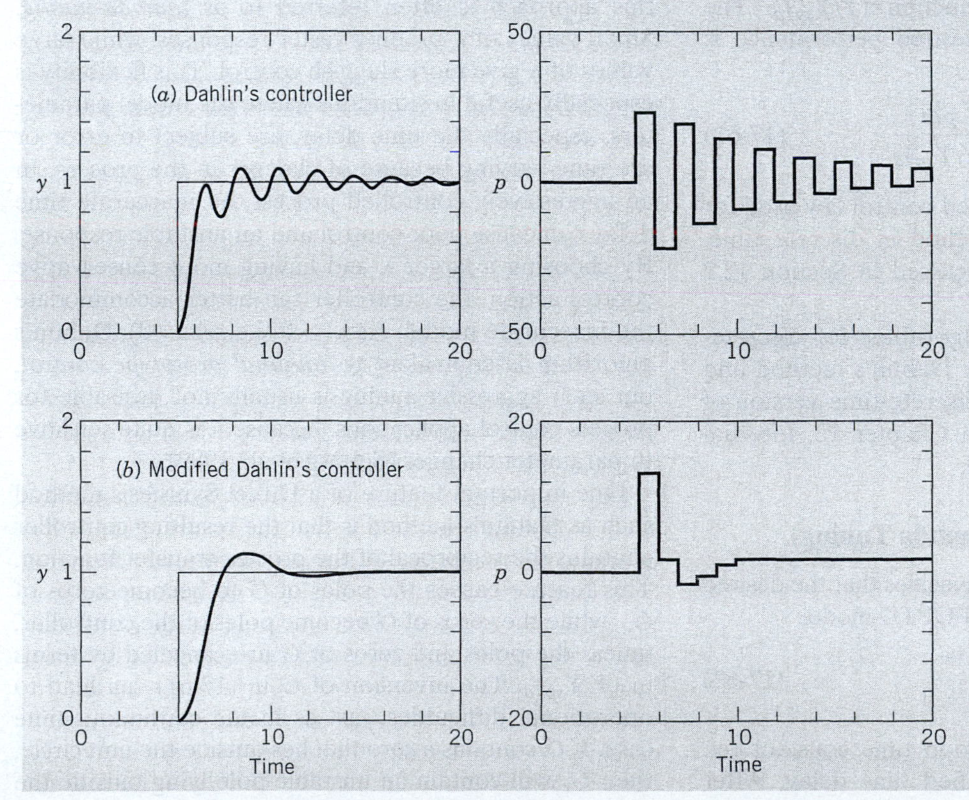

Figure 17.11 Comparison of (*a*) ringing and (*b*) nonringing Dahlin's controllers for a second-order process ($\lambda = 1$), Example 17.6 (y = controlled variable, p = controller output after zero-order hold).

The controller ringing results from the presence of the term $(0.0280 + 0.0234z^{-1})$ in the denominator of (17-56), which corresponds to a controller pole at -0.836, quite close to the unit circle. As can be shown using long division, this term, when transformed to the time domain, causes a change in sign at each sampling instant in the manipulated variable. Dahlin (1968) suggested that ringing can be eliminated by setting $z^{-1} = 1$ in the ringing term, in this case replacing $(0.0280 + 0.0234z^{-1})$ by a constant $(0.0280 + 0.0234 = 0.0514)$. Let the nonringing version of Dahlin's controller be denoted by $\overline{G}_{DC}$. Figure 17.11*b* shows $y(t)$ and $p(t)$ for this case, indicating that the ringing behavior has disappeared. Interestingly, the closed-loop response now exhibits an overshoot, which contradicts the original design criterion of a first-order approach to the set point (Eq. 17-62). Therefore, the closed-loop performance of Dahlin's controller modified for ringing is not always predictable. This lack of predictability represents a major disadvantage of this technique.

17.5.2 Vogel-Edgar Algorithm

For processes that can be described by a second-order-plus-time-delay model (Eq. 17-67), Vogel and Edgar (1988) have developed a controller that eliminates the ringing pole caused by inverting G. The desired closed-loop transfer function is similar to that for Dahlin's controller (cf. (17-63) and includes time-delay compensation (z^{-N-1} in the closed-loop transfer function):

$$\left(\frac{Y}{Y_{sp}}\right)_d = \frac{(1-A)}{1-Az^{-1}}\,\frac{b_1 + b_2 z^{-1}}{b_1 + b_2}z^{-N-1} \quad (17\text{-}70)$$

However, the zeros of model G are also included as zeros of the closed-loop transfer function (in this case, divided by $b_1 + b_2$ to ensure that the closed-loop steady-state gain equals one). Although this choice may slow down the response somewhat, it makes the controller less sensitive to model errors and also eliminates the possibility of ringing. The Vogel-Edgar controller corresponding to (17-70) is

$$G_{VE} = \frac{(1 + a_1 z^{-1} + a_2 z^{-2})(1-A)}{(b_1 + b_2)(1 - Az^{-1}) - (1-A)(b_1 + b_2 z^{-1})z^{-N-1}} \quad (17\text{-}71)$$

Because of the form of $(Y/Y_{sp})_d$ in (17-70), this controller does not attempt to cancel the numerator terms of the process transfer function and thus does not include the potential ringing pole. Note that for $a_2 = b_2 = 0$ (a first-order process), Eq. 17-71 reverts to Dahlin's controller, Eq. 17-64. This is acceptable, because a first-order process will not lead to a ringing controller.

EXAMPLE 17.7

For the same process model used in Example 17.6, plot the response and the controller output for a unit set-point change ($\lambda = \Delta t = 1.0$) for the nonringing Dahlin's controller and compare it with the Direct Synthesis PID (BD conversion) and Vogel-Edgar approaches.

SOLUTION

Applying Eq. 17-63 for $N = 0$, the controller transfer function for the nonringing version of Dahlin's algorithm is

$$\overline{G}_{DC} = \left(\frac{0.6321}{1 - z^{-1}}\right)\left(\frac{1 - 1.5353z^{-1} + 0.5866z^{-2}}{0.0514}\right) \quad (17\text{-}72)$$

For this controller the response $y(t)$ and the controller output $p(t)$ are shown in Fig. 17.12*a*. Next we derive a PID controller using the IMC approach presented in Chapter 12. Starting with the continuous-time second-order transfer function ($\tau_1 = 5$, $\tau_2 = 3$, $K = 1$, $\theta = 0$), τ_c in Eq. 12-27 is set equal to one to provide the best response for a set-point change. When the BD approximation $s = (1 - z^{-1})/\Delta t$ is substituted into the PID controller transfer function,

$$G_{BD} = 4.1111\frac{3.1486 - 5.0541z^{-1} + 2.0270z^{-2}}{1.7272 - 2.4444z^{-1} + 0.7222z^{-2}} \quad (17\text{-}73)$$

The closed-loop response is shown in Fig. 17.12*b*. When the BD-PID digital controller is designed for this system, there is no need to correct for ringing such as is required for Dahlin's controller. This is true for wide ranges of λ (i.e., τ_c) and Δt that have been investigated.

Figure 17.12 shows the closed-loop response for the Vogel-Edgar controller (G_{VE}) for the same second-order model. This controller is

$$G_{VE} = 0.6321\frac{1 - 1.5353z^{-1} + 0.5866z^{-2}}{0.0514 - 0.0366z^{-1} - 0.0148z^{-2}} \quad (17\text{-}74)$$

The tuning parameter A is selected to be 0.368 ($\lambda = 1$). For this second-order system, the controlled variable response for G_{VE} is superior to $\overline{G}_{DC}$. If a time delay is added to the model, the comparative performance of G_{VE} and $\overline{G}_{DC}$ is still the same, because both controllers utilize the same form of the Smith predictor.

Note that the controller parameters in (17-72) to (17-74) have been reported with four decimal points in order to avoid roundoff errors in the control calculations.

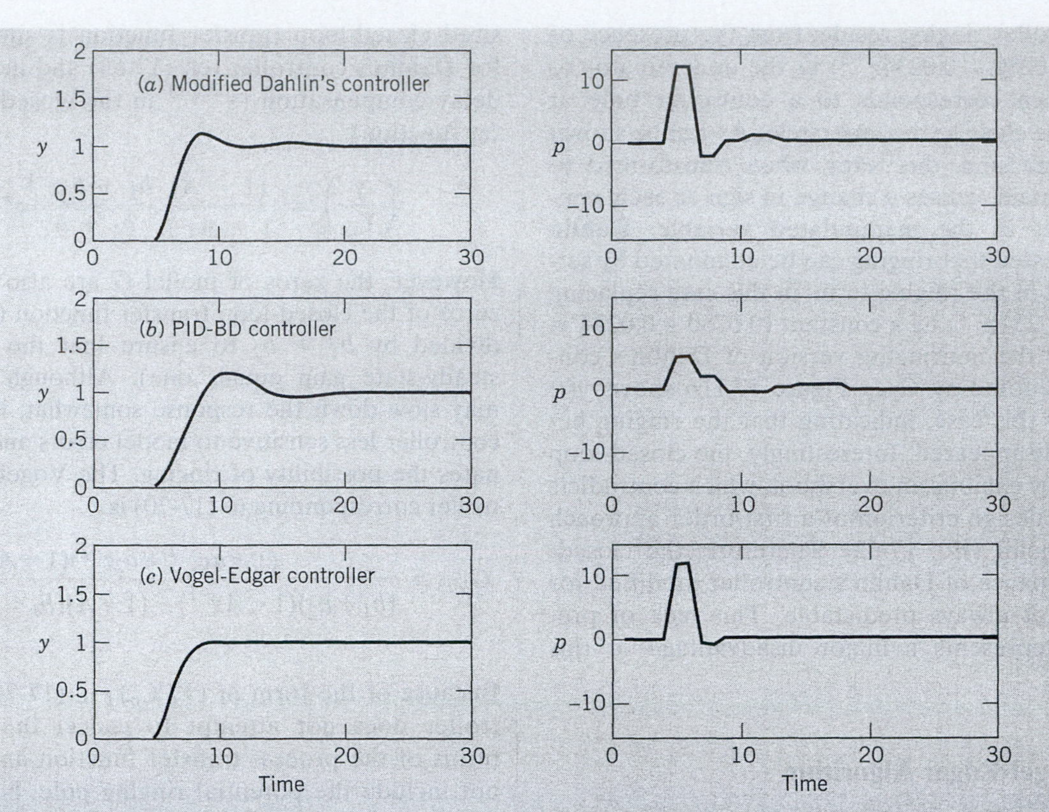

Figure 17.12 Comparison of closed-loop response for a second-order process in Example 17.7 using (*a*) nonringing Dahlin's controller ($\lambda = 1$), (*b*) backwards difference PID controller and (*c*) Vogel-Edgar controller (y = controlled variable, p = controller output after zero-order hold).

Vogel and Edgar (1988) have shown that their controller satisfactorily handles first-order or second-order process models with positive zeros (inverse response) or negative zeros as well as simulated process and measurement noise. Many higher-order process models can be successfully controlled with G_{VE}. Neither G_{VE} nor $\overline{G}_{DC}$ are suitable for unstable process models, however. The robustness of the Vogel-Edgar controller is generally better than Dahlin's controller when model errors are present. G_{VE} can be used as an adaptive controller when the discrete-time model is updated on-line. For processes with zeros outside the unit circle, Dahlin's controller can become unstable, while the stability of the Vogel-Edgar controller is unaffected.

17.5.3 Internal Model Control (IMC)

The general design methodology of Internal Model Control presented in Section 12.2.3 for continuous-time systems can be extended to sampled-data systems (Garcia and Morari, 1982; Zafiriou and Morari, 1985). Figure 12.5 shows the block diagram used for IMC contrasted with that for conventional feedback control. Here the notation G_c^* is used instead of G_c for the controller transfer function because of the different block diagram structure and controller design methodology used with IMC. The perfect IMC controller is simply the inverse of the process model.

$$G_c^*(z) = 1/\widetilde{G} \tag{17-75}$$

However, a perfect controller is usually not physically realizable, or it may be impractical because of model error. The two key steps involved in digital controller design are as follows (cf. Section 12.2.2).

1. The process model is factored as

$$\widetilde{G}(z) = \widetilde{G}_+(z)\,\widetilde{G}_-(z) \tag{17-76}$$

where $\widetilde{G}_+$ contains the time-delay term z^{-N-1}, zeroes that lie outside the unit circle, and zeroes that lie inside the unit circle near $(-1, 0)$. Also, $\widetilde{G}_+$ has a steady-state gain of unity.

2. The controller is obtained by inverting $\widetilde{G}_-$ (the invertible part of $\widetilde{G}$) and then multiplying by a first-order filter F to improve the robustness of the controller as well as to ensure the physical realizability of G_c^*:

$$G_c^*(z) = \frac{F(z)}{\widetilde{G}_-(z)} \tag{17-77}$$

The filter F usually contains one or more tuning parameters. Zeros of $\widetilde{G}$ that lie outside the unit circle (the so-called nonminimum phase zeroes) would yield unstable controller poles if such terms were included in $\widetilde{G}_-$, instead of $\widetilde{G}_+$. Negative zeroes on the real axis near $z = -1$ will result in a ringing controller if they are inverted; hence, they are also included in $\widetilde{G}_+$. The closed-loop transfer function using the above design rules, assuming the process model is perfect, is

$$\frac{Y}{Y_{sp}} = \tilde{G}_+(z)F(z) \qquad (17\text{-}78)$$

The IMC design framework can yield Dahlin's controller and the Vogel-Edgar controller for appropriate choices of $F(z)$ and $\tilde{G}_+(z)$. It can also readily be applied to higher-order systems, where Direct Synthesis is not as reliable. For details on treatment of process model zeroes and selection of the filter, see Zafiriou and Morari (1985).

The IMC block diagram in Fig. 12.5 can be expanded to include a block A^* in the feedback path as well as a disturbance transfer function G_d. The block A^* can be used to predict the effect of the disturbance on the error signal to the controller, and it can also provide time-delay compensation. This two-degree-of-freedom controller (see Chapter 12) is known as an *analytical predictor* (Doss and Moore, 1982; Wellons and Edgar, 1987).

17.6 MINIMUM VARIANCE CONTROL

In this design method, the objective is to reduce the variability of the controlled variable y when the set point is constant and the process is subject to unknown, random disturbances. In statistical terms, the objective is to minimize the variance of y. This approach is especially relevant for processes where the disturbances are *stochastic* (that is, random) rather than *deterministic* (for example, steps or drifts). Sheet-making processes for producing paper and plastic film or sheets are common examples (Featherstone et al., 2000).

The Minimum Variance Control (MVC) design method generates the form of the feedback control law, as well as the values of the controller parameters. Like the Direct Synthesis and Internal Model Control design methods, the MVC method results in PI or PID controllers for simple transfer function models (MacGregor, 1988; Ogunnaike and Ray, 1994; Box and Luceño, 1997). Although MVC tends to be quite aggressive, the design method can be modified to be less aggressive (Bergh and MacGregor, 1987). Because Minimum Variance Control is a limiting case on actual controller performance, it provides a useful benchmark for monitoring control–loop performance (Harris and Seppala, 2002); see Chapter 21.

The starting point for the MVC design method is the following discrete transfer function model:

$$Y(z) = G(z)U(z) + D(z) \qquad (17\text{-}79)$$

The disturbance $D(z)$ can be written as a zero, mean white (e.g., Gaussian) noise signal, $a(z)$, and a disturbance transfer function $G_d(z)$:

$$D(z) = G_d(z)a(z) \qquad (17\text{-}80)$$

Previous discussions on accommodating disturbances focused on deterministic changes in the disturbance, such

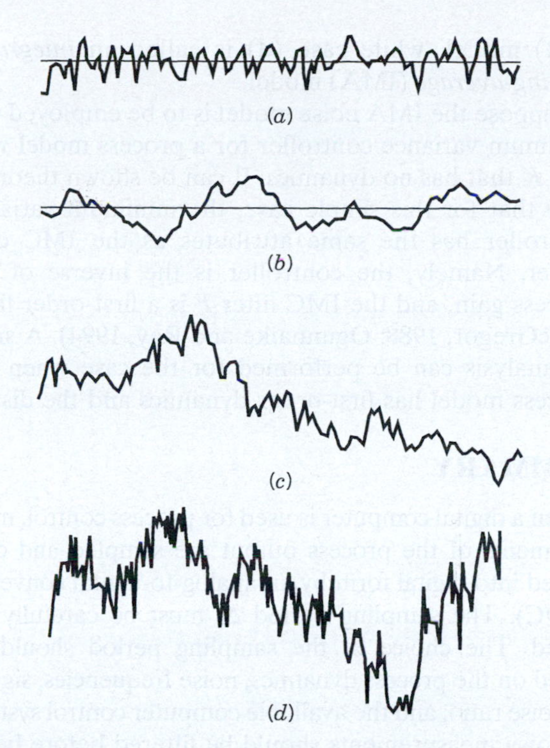

Figure 17.13 Four models for $d(k)$: (*a*) stationary white noise disturbance; (*b*) stationary autoregressive disturbance; (*c*) nonstationary disturbance (random walk); (*d*) integrated (nonstationary) moving-average disturbance (adapted from Box and Luceño, 1997).

as step changes. Four alternative disturbance models are shown graphically in Fig. 17.13. These disturbances are persistent (as a result of the random component) but may also exhibit features such as dynamics, drift, or trending. A typical process disturbance seldom will be random but will depend on past values of the disturbance. These models can be constructed by starting with an input $a(z)$ that is a white noise sequence. This input passes through a dynamic model such as a first-order transfer function or an integrating transfer function. The output $D(z)$ is an auto-correlated disturbance to the process.

Table 17.2 gives important time-series models that are commonly encountered in industrial process control, including statistical process control applications (see Chapter 21). Stationary disturbance models (a) and (b) have a fixed mean; that is, the sums of deviations above and below the line are equal to zero, but case (a) rarely occurs in industrial processes. Nonstationary disturbance models (c) and (d) do not have a fixed mean but are drifting in nature. Case (c), so-called *random walk behavior*, is often used to describe stock market index patterns. Case (b) is called an *autoregressive*

Table 17.2 Disturbance Models for Figure 17.13

	Model	
a.	$d(k) = a(k)$	(white noise)
b.	$d(k) = \phi d(k-1) + a(k)$	$(\phi \le 1)$
c.	$d(k) = d(k-1) + a(k)$	
d.	$d(k) = d(k-1) + a(k) + \psi a(k-1)$	$(\psi \le 1)$

(AR) model, while case (d) is called an *integrated moving-average* (IMA) model.

Suppose the IMA noise model is to be employed in a minimum variance controller for a process model with gain K that has no dynamics. It can be shown theoretically that for this simple case, the minimum variance controller has the same attributes as the IMC controller. Namely, the controller is the inverse of the process gain, and the IMC filter F is a first-order filter (MacGregor, 1988; Ogunnaike and Ray, 1994). A similar analysis can be performed for the case when the process model has first-order dynamics and the disturbance model is described by the IMA noise model. As shown by Ogunnaike and Ray (1994), these model assumptions lead to a PI controller. A Minimum Variance Controller can be obtained for higher-order process models, models with time delay, and different disturbance model structures (Bergh and MacGregor, 1987; Middleton and Goodwin, 1990). MVC can also be extended to problems where there is a cost associated with the control effort, as well as upper and lower bounds on the controller output. This approach requires that an objective function be defined, as in model predictive control (see Chapter 20).

SUMMARY

When a digital computer is used for process control, measurements of the process output are sampled and converted into digital form by an analog-to-digital converter (ADC). The sampling period Δt must be carefully selected. The choice of the sampling period should be based on the process dynamics, noise frequencies, signal-to-noise ratio, and the available computer control system.

Noisy measurements should be filtered before being sent to the controller. Analog filters are effective in removing high-frequency noise and avoiding aliasing. Digital filters are also widely used both for low-pass filters and other purposes such as the elimination of noise spikes. The choice of a filter and the filter parameters (e.g., τ_F) should be based on the process dynamics, the noise characteristics, and the sampling period. If a filter parameter is changed, it may be necessary to retune the controller, because the filter is a dynamic element in the feedback control loop.

We have presented a number of different approaches for designing digital feedback controllers. Digital controllers that emulate continuous-time PID controllers can include a number of special features to improve operability. Controllers based on Direct Synthesis or IMC can be tuned in continuous or discrete time, avoid ringing, eliminate offset, and provide a high level of performance for set-point changes. Minimum variance control can be very effective if a disturbance model is available.

REFERENCES

Åström, K. J., and B. Wittenmark, *Computer Controlled Systems: Theory and Design*, 3d ed., Prentice Hall, Upper Saddle River, NJ, 1997.

Bergh, L. G., and J. F. MacGregor, Constrained Minimum Variance Controllers: Internal Model Structure and Robustness Properties, *Ind. Eng. Chem. Res.*, **26**, 1558 (1987).

Box, G. E. P., G. M. Jenkins, and G. C. Reinsel, *Time Series Analysis, Forecasting, and Control*, 3d ed., Prentice Hall, Englewood Cliffs, NJ, 1994.

Box, G. E. P., and A. Luceño, *Statistical Control by Monitoring and Feedback Control*, Wiley, New York, 1997.

Corripio, A. B., C. L. Smith, and P. W. Murrill, Filter Design for Digital Control Loops, *Instrum. Tech.*, **20** (1), 33 (1973).

Dahlin, E. B., Designing and Tuning Digital Controllers, *Instrum. Control Systems*, **41** (6), 77 (1968).

Doss, J. E., and C. F. Moore, The Discrete Analytical Predictor—A Generalized Dead-Time Compensation Technique, *ISA Trans.*, **20** (4), 77 (1982).

Featherstone, A. P., J. G. VanAntwerp, and R. D. Braatz, *Identification and Control of Sheet and Film Processes*, Springer-Verlag, London, 2000.

Franklin, G. F., D. J. Powell, M. L. Workman, and J. D. Powell, *Digital Control of Dynamic Systems*, 3d ed., Addison-Wesley, Reading, MA, 1997.

Garcia, C. E., and M. Morari, Internal Model Control, 1. A Unifying Review and Some New Results. *IEC Proc. Des. Dev.*, **21**, 308 (1982).

Harris, T. J., and C. T. Seppala, Recent Developments in Controller Performance Monitoring and Assessment Techniques, Chemical Process Control VI, *AIChE Symp. Ser.*, **98**, No. 326, 208 (2002).

Isermann, R., *Digital Control Systems*, 2d ed., Springer-Verlag, New York, 1989.

Ljung, L., *System Identification: Theory for the User*, 2d ed., Prentice Hall, Upper Saddle River, NJ, 1999.

MacGregor, J. F., On-Line Statistical Process Control, *Chem. Engr. Prog.*, **84**, 21 (October 1988).

McConnell, E., and D. Jernigan, Data Acquisition, Sect. 18.3, p. 1938, in *The Electronics Handbook*, 2nd ed., J. C. Whitaker (Ed.), CRC Press, Boca Raton, FL, 2005.

Middleton, R. H., and G. C. Goodwin, *Digital Control and Identification—A Unified Approach*, Prentice Hall, Englewood Cliffs, NJ, 1990.

Ogata, K., *Discrete-Time Control Systems*, 2d ed., Prentice Hall, Englewood Cliffs, NJ, 1994.

Ogunnaike, B. A., and W. H. Ray, *Process Dynamics, Modeling, and Control*, Oxford University Press, New York, 1994.

Oppenheim, A. V., and R. W. Shafer, *Discrete Time Signal Processing*, 2d ed., Prentice Hall, Englewood Cliffs, NJ, 1999.

Phillips, S. F., and D. E. Seborg, Adaptive Control Strategies for Achieving Desired Temperature Control Profiles during Process Startup, *Ind. Eng. Chem. Res.*, **27**, 1434 (1987).

Seborg, D. E., T. F. Edgar, and D. A. Mellichamp, *Process Dynamics and Control*, 1st ed., Wiley, New York, 1989.

Vogel, E. G., and T. F. Edgar, An Adaptive Pole Placement Controller for Chemical Processes and Variable Dead Time, *Comp. Chem. Engr.*, **12**, 15 (1988).

Wellons, M. C., and T. F. Edgar, The Generalized Analytical Predictor, *Ind. Eng. Chem. Res.*, **26**, 1523 (1987).

Zafiriou, E., and M. Morari, Digital Controllers for SISO Systems: A Review and a New Algorithm, *Int. J. Control*, **42**, 885 (1985).

EXERCISES

17.1 The mean arterial pressure P in a patient is subjected to a unit step change in feed flow rate F of a drug. Normalized response data are shown below. Previous experience has indicated that the transfer function,

$$\frac{P(s)}{F(s)} = \frac{5}{10s + 1}$$

provides an accurate dynamic model. Filter these data using an exponential filter with two different values of α, 0.5 and 0.8. Graphically compare the noisy data, the filtered data, and the analytical solution for the transfer function model for a unit step input.

Time (min)	P	Time (min)	P
0	0	11	3.336
1	0.495	12	3.564
2	0.815	13	3.419
3	1.374	14	3.917
4	1.681	15	3.884
5	1.889	16	3.871
6	2.078	17	3.924
7	2.668	18	4.300
8	2.533	19	4.252
9	2.908	20	4.409
10	3.351		

17.2 Show that the digital exponential filter output can be written as a function of previous measurements $y_m(k)$ and the initial filter output $y_F(0)$.

17.3 A signal given by

$$y_m(t) = t + 0.5\sin(t^2)$$

is to be filtered with an exponential digital filter over the interval $0 \le t \le 20$. Using three different values of α (0.8, 0.5, 0.2), determine the output of the filter at each sampling time. Do this for sampling periods of 1.0 and 0.1. Compare the three filters for each value of Δt.

17.4 The following product quality data y_m were obtained from a bioreactor, based on a photometric measurement evaluation of the product:

Time (min)	y_m (absorbance)
0	0
1	1.5
2	0.3
3	1.6
4	0.4
5	1.7
6	1.5
7	2.0
8	1.5

(a) Filter the data using an exponential filter with $\Delta t = 1$ min. Consider α = 0.2 and α = 0.5.

(b) Use a moving-average filter with $N^* = 4$.

(c) Implement a noise-spike filter with $\Delta y = 0.5$.

(d) Plot the filtered data and the raw data for purposes of comparison.

17.5 The analog exponential filter in Eq. 17-4 is used to filter a measurement before it is sent to a proportional controller with $K_c = 1$. The other transfer functions for the closed-loop system are $G_v = G_m = 1$, and $G_p = G_d = 1/(5s + 1)$. Compare the closed-loop responses to a sinusoidal disturbance, $d(t) = \sin t$, for no filtering ($\tau_F = 0$), and for an exponential filter ($\tau_F = 3$ min).

17.6 Consider the first-order transfer function $Y(s)/U(s) = 1/(s + 1)$. Generate a set of data ($t = 1, 2, \ldots 20$) by integrating this equation for $u = 1$ and randomly adding binary noise to the output, ±0.05 units at each integer value of t. Design a digital filter for this system and compare the filtered and noise-free step responses for $t = 1$. Justify your choice of τ_F. Repeat for other noise levels, for example, ±0.01 and ±0.1.

17.7 Find the response $y(k)$ for the difference equation

$$y(k) - y(k-1) + 0.21y(k-2) = u(k-2)$$

Let $y(0) = y(1) = 0$, $u(0) = 1$, $u(k) = 0$ for $k \ge 1$. Perform direct integration using a spreadsheet. What is the steady-state value of y?

17.8 The dynamic behavior of a temperature sensor and transmitter can be described by the FOPTD transfer function

$$\frac{T_m'(s)}{T'(s)} = \frac{e^{-2s}}{8s + 1}$$

where the time constant and time delay are in seconds and:

T' = actual temperature (deviation)

T_m' = measured temperature (deviation)

The actual temperature changes as follows (t in s):

$$T = \begin{cases} 70\ °\text{C} & \text{for } t < 0 \\ 85\ °\text{C} & \text{for } 0 \le t < 10 \\ 70\ °\text{C} & \text{for } t \ge 10 \end{cases}$$

If samples of the measured temperature are automatically logged in a digital computer every two minutes beginning at $t = 0$, what is the maximum value of the logged temperature? Use simulation with a zero-order hold to find the answer.

17.9 For a process given by

$$\frac{Y(z)}{U(z)} = \frac{2.7z^{-2} + 8.1z^{-3}}{1 - 0.5z^{-1} + 0.06z^{-2}}$$

(a) Calculate the response $y(k\Delta t)$ to a unit step change in u using simulation of the difference equation.

(b) Check your answer in (a) by using simulation.

(c) What is the steady-state value of y?

17.10 A dissolved oxygen analyzer in a bioreactor is used to provide composition measurements at each sampling time in a feedback control loop. The open-loop transfer function is given by

$$G_{OL} = G_c G$$

$$G_c = 2\left(1 + \frac{1}{8s}\right) \qquad G = \left(\frac{10}{12s + 1}\right)e^{-2s}$$

(a) Suppose that a sampling period of $\Delta t = 1$ min is selected. Derive $G_c(z)$ using a backward-difference approximation of $G_c(s)$.

(b) If a unit step change in the controller error signal $e(t)$ is made, calculate the sampled open-loop response $y_m(k\Delta t)$ using simulation with a zero-order hold after the controller G_c and before the process G in the block diagram constructed via Simulink.

17.11 The discrete-time transfer function of a process is given by

$$\frac{Y(z)}{U(z)} = \frac{5z^{-1} + 3z^{-2}}{1 + z^{-1} + 0.41z^{-2}}$$

(a) Convert this transfer function to an equivalent difference equation.

(b) Calculate the response $y(k)$ to a unit step change in u using simulation of the difference equation.

(c) Check your answer in (a) by using simulation.

(d) What is the steady-state value of y?

17.12 To determine the effects of pole and zero locations, simulate the unit step responses of the discrete transfer functions shown below for the first six sampling instants, $k = 0$ to $k = 5$. What conclusions can you make concerning the effect of pole and zero locations?

(a) $\dfrac{1}{1 - z^{-1}}$

(b) $\dfrac{1}{1 + 0.7z^{-1}}$

(c) $\dfrac{1}{1 - 0.7z^{-1}}$

(d) $\dfrac{1}{(1 + 0.7z^{-1})(1 - 0.3z^{-1})}$

(e) $\dfrac{1 - 0.5z^{-1}}{(1 + 0.7z^{-1})(1 - 0.3z^{-1})}$

(f) $\dfrac{1 - 0.2z^{-1}}{(1 + 0.6z^{-1})(1 - 0.3z^{-1})}$

17.13 A process operation under proportional-only digital control with $\Delta t = 1$ has

$$G_p(s) = \frac{3}{(5s + 1)(2s + 1)}, \quad K_c = 1, \quad G_m = 0.25$$

Determine whether the controlled system is stable by calculating the response to a set-point change using simulation.

17.14 Determine how the maximum allowable digital controller gain for stability varies as a function of Δt for the following system:

$$G_p(s) = \frac{1}{(5s + 1)(s + 1)} \qquad G_c = K_c \qquad G_m = 1$$

Use $\Delta t = 0.01, 0.1, 0.5$, and closed-loop simulation to find the maximum K_c for each Δt; K_c can range between 10 and 1200. What do you conclude about how sampling period affects the allowable controller gain?

17.15 A temperature control loop includes a second-order overdamped process described by the discrete transfer function.

$$G(z) = \frac{(0.0826 + 0.0368z^{-1})z^{-1}}{(1 - 0.894z^{-1})(1 - 0.295z^{-1})}$$

and a digital PI controller

$$G_c(z) = K_c\left(1 + \frac{1}{8(1 - z^{-1})}\right)$$

Find the maximum controller gain K_{cm} for stability by trial-and-error.

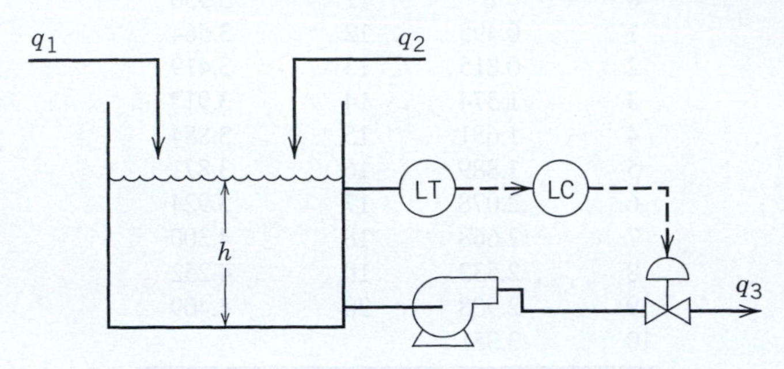

Figure E17.16

17.16 A digital controller is used to control the liquid level of the storage tank shown in Fig. E17.16. The control valve has negligible dynamics and a steady-state gain, $K_v = 0.1$ ft^3/(min)(mA). The level transmitter has a time constant of 30 s and a steady-state gain of 4 mA/ft. The tank is 4 ft in diameter. The exit flow rate is not directly influenced by the liquid level; that is, if the control valve stem position is kept constant, $q_3 \neq f(h)$. Suppose that a proportional digital controller and a digital-to-analog converter with 4 to 20 mA output are used. If the sampling period for the analog-to-digital converter is $\Delta t = 1$ min, for what values of controller gain K_c is the closed-loop system stable? Use simulation and trial values for K_c of −10, −50, and −90. Will offset occur for the proportional controller after a change in set point?

17.17 The block diagram of a digital control system is shown in Fig. E17.17. The sampling period is $\Delta t = 1$ min.

(a) Design the digital controller $G_c(z)$ so that the closed-loop system exhibits a first-order response to a unit step change in the set point (after an appropriate time delay).

(b) Will this controller eliminate offset after a step change in the set point? Justify your answer.

(c) Is the controller physically realizable? Justify your answer.

(d) Design a digital PID controller based on the ITAE (set-point) method in Chapter 12 and examine its performance for a step change in set point. Approximate the sampler and zero-order-hold by a time delay of $\theta = \Delta t/2$.

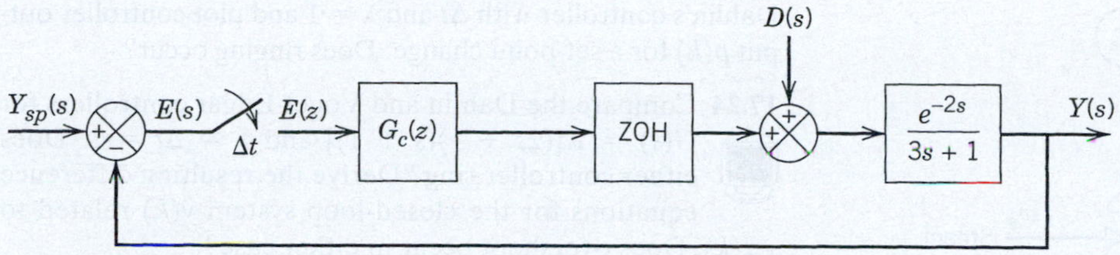

Figure E17.17

17.18 The exit composition c_3 of the blending system in Fig. E17.18 is controlled using a digital feedback controller. The exit stream is automatically sampled every minute, and the composition measurement is sent from the composition transmitter (AT) to the digital controller. The controller output is sent to the ZOH device before being transmitted to the control valve.

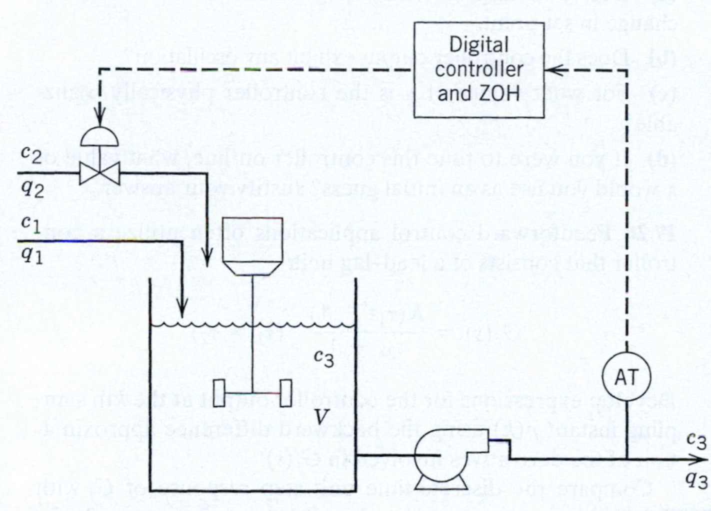

Figure E17.18

(a) Derive an expression for the discrete open-loop transfer function C_3/Q_2, where C_3 and Q_2 are deviation variables, by deriving the continuous transfer function and then deriving the equivalent discrete time model.

(b) The closed-loop system exhibits a first-order response to a unit step change in the disturbance C_2. Specify the form of the desired response $(C_3/C_{3sp})_d$. It is not necessary to derive an expression for $G_c(z)$, but you should justify your choice for $(C_3/C_{3sp})_d$.

Available Information

(i) Because flow rate q_2 is quite small, the liquid volume in the tank V remains essentially constant at 30 ft^3. The tank is perfectly mixed.

(ii) The primary disturbance variable is inlet composition c_2.

(iii) The control valve has negligible dynamics and a steady-state gain of 0.1 ft^3/min mA.

(iv) The composition transmitter (AT) has a steady gain of 2.5 mA/(lb-mole solute/ft^3). Composition samples are analyzed every minute; that is, the sampling period is $\Delta t = 1$ min. There is also a 1-min time delay associated with the composition analysis.

(v) Nominal steady-state values are

$$\bar{q}_2 = 0.1 \text{ ft}^3/\text{min} \quad \bar{c}_2 = 1.5 \text{ lb-mol solute/ft}^3$$

$$\bar{q}_3 = 3 \text{ ft}^3/\text{min} \quad \bar{c}_3 = 0.21 \text{ lb-mol solute/ft}^3$$

17.19 The block diagram of a sampled-data control system is shown in Fig. E17.19. Design a Dahlin controller $G_c(z)$ that is physically realizable and based on a change in set point. The sampling period is $\Delta t = 1$ min. Calculate the closed-loop response when this controller is used and a unit step change in disturbance occurs.

17.20 It is desired to control the exit temperature T_2 of the heat exchanger shown in Fig. E17.20 by adjusting the steam flow rate w_s. Unmeasured disturbances occur in inlet temperature T_1. The dynamic behavior of the heat exchanger can be approximated by the transfer function

$$\frac{T_2'(s)}{W_s'(s)} = \frac{2.5}{10s + 1} \; [=] \; \frac{°\text{F}}{\text{lb/s}}$$

where the time constant has units of seconds and the primes denote deviation variables. The control valve and temperature transmitter have negligible dynamics and steady-state gains of $K_v = 0.2$ lb/s/mA and $K_m = 0.25$ mA/°F. Design a minimal

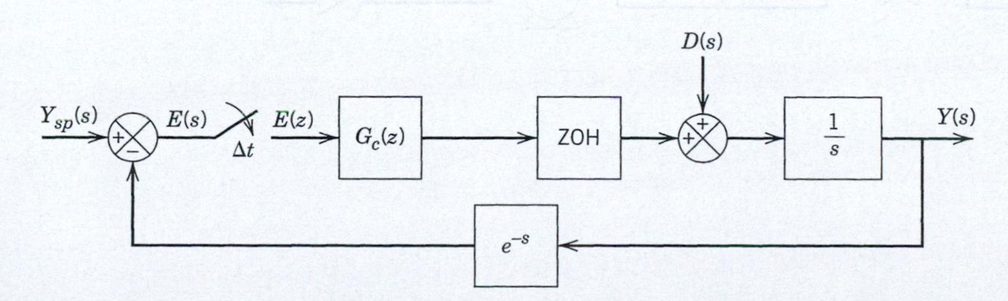

Figure E17.19

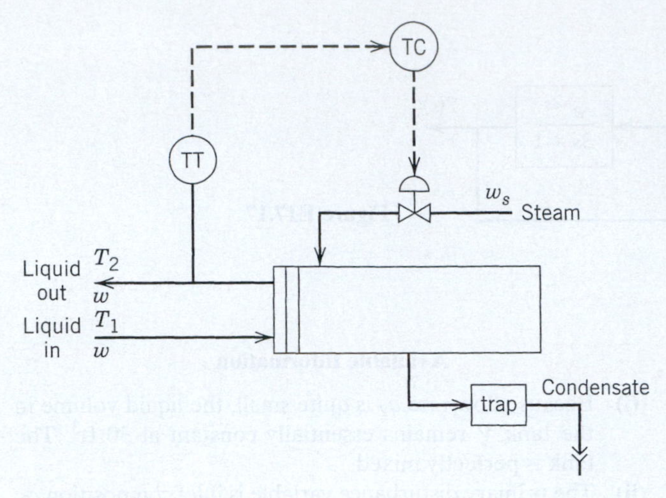

Figure E17.20

prototype controller (i.e., Dahlin's controller with $\lambda = 0$) that is physically realizable and based on a unit step change in the set point. Assume that a zero-order hold is used and that the sampling period is $\Delta t = 2$ s.

17.21 A second-order system G with $K = 1$, $\tau_1 = 6$, and $\tau_2 = 4$ is to be controlled using the Vogel-Edgar controller with $\lambda = 5$ and $\Delta t = 1$. Assuming a step change in y_{sp}, calculate the controlled variable $y(k)$ for $k = 0, 1, \ldots, 25$, and plot $y(k)$ and the controller output $p(k)$.

17.22 Compare PID (ITAE for set-point changes) and Dahlin controllers for $\Delta t = 1$, $\lambda = 1$, and $G(s) = 2e^{-s}/(10s + 1)$. For the ITAE controller, approximate the sampler and ZOH by a time delay equal to $\Delta t/2$. Adjust for ringing, if necessary. Plot the closed-loop responses for a set-point change as well as the controller output for each case.

17.23 For a process including control valve and sensor, $G(s) = 1.25e^{-5s}/(5s + 1)$, derive the equation for Dahlin's controller with Δt and $\lambda = 1$ and plot controller output $p(k)$ for a set-point change. Does ringing occur?

17.24 Compare the Dahlin and Vogel-Edgar controllers for $G(s) = 1/[(2s + 1)(s + 1)]$ and $\lambda = \Delta t = 1$. Does either controller ring? Derive the resulting difference equations for the closed-loop system $y(k)$ related to $y_{sp}(k)$. Does overshoot occur in either case?

17.25 Design a digital controller for the liquid level in the storage system shown in Fig. E17.25. Each tank is 2.5 ft in diameter. The piping between the tanks acts as a linear resistance to flow with $R = 2$ min/ft^2. The liquid level is sampled every 30 s. The digital controller also acts as a zero-order hold device for the signal sent to the control valve. The control valve and level transmitter have negligible dynamics. Their gains are $K_v = 0.25$ ft^3/min/mA and $K_m = 8$ mA/ft, respectively. The nominal value of q_1 is 0.5 ft^3/min.

(a) Derive Dahlin's control algorithm based on a step change in set point.

(b) Does the controller output exhibit any oscillation?

(c) For what values of λ is the controller physically realizable?

(d) If you were to tune this controller on-line, what value of λ would you use as an initial guess? Justify your answer.

17.26 Feedforward control applications often utilize a controller that consists of a lead–lag unit:

$$G_f(s) = \frac{K(\tau_1 s + 1)}{\tau_2 s + 1} \quad (\tau_1 \neq \tau_2)$$

Develop expressions for the controller output at the kth sampling instant $p(k)$ using the backward difference approximation of the derivatives involved in $G_f(s)$.

Compare the discrete-time unit step response of G_f with the continuous-time response when $K = 1$, $\tau_1 = 5$ min, $\tau_2 = 2$ min, and $\Delta t = 1$ min.

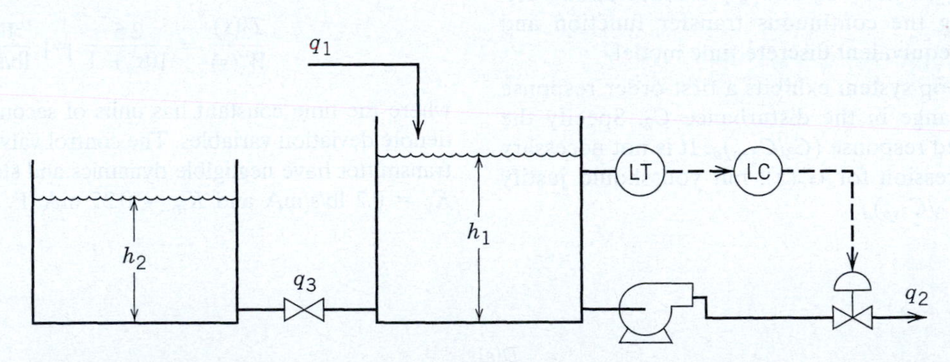

Figure E17.25

Chapter 18

Multiloop and Multivariable Control

CHAPTER CONTENTS

In previous chapters, we have emphasized control problems that have only one controlled variable and one manipulated variable. These problems are referred to as single-input, single-output (SISO), or *single-loop*, control problems. But in many practical control problems, typically a number of variables must be controlled, and a number of variables can be manipulated. These problems are referred to as multiple-input, multiple-output (MIMO) control problems. For almost all important processes, at least two variables must be controlled: product quality and throughput.

Several examples of processes with two controlled variables and two manipulated variables are shown in Fig. 18.1. These examples illustrate a characteristic feature of MIMO control problems, namely, the presence of *process interactions*; that is, each manipulated variable can affect both controlled variables. Consider the in-line blending system shown in Fig. 18.1*a*. Two streams containing species A and B, respectively, are to be blended to produce a product stream with mass flow rate w and composition x, the mass fraction of A. Adjusting either manipulated flow rate, w_A or w_B, affects both w and x.

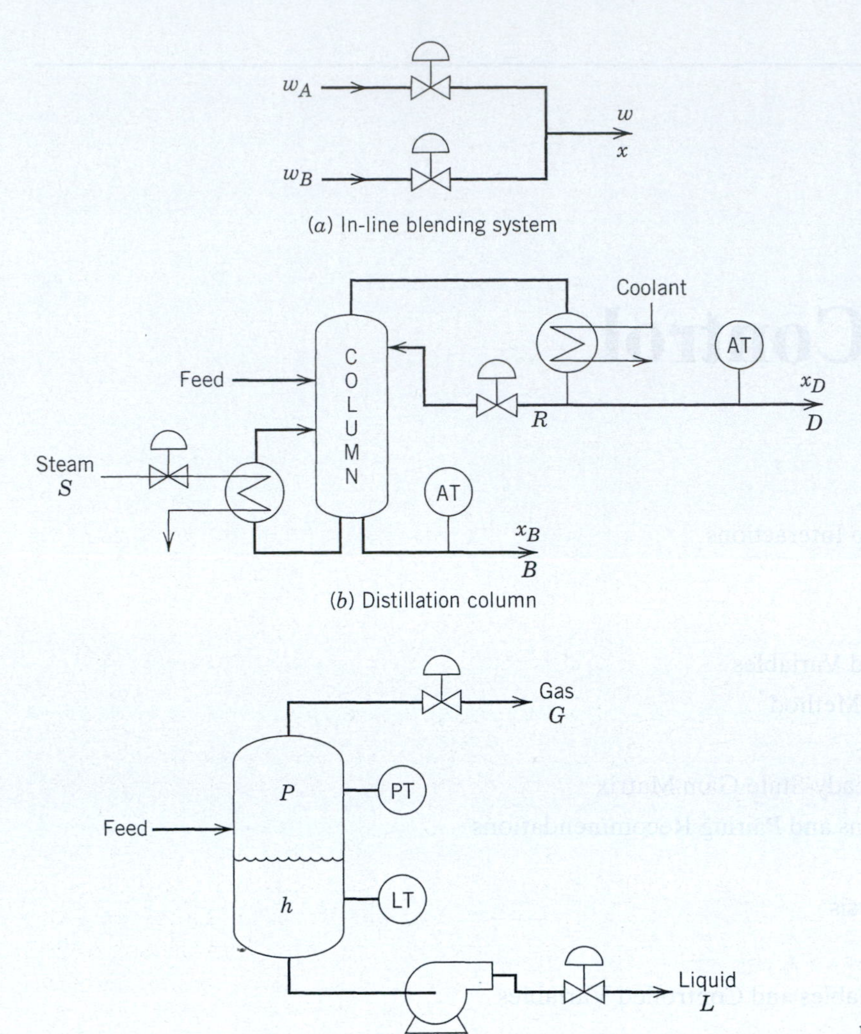

Figure 18.1 Physical examples of multivariable control problems.

Similarly, for the distillation column in Fig. 18.1*b*, adjusting either reflux flow rate R or steam flow rate S will affect both distillate composition x_D and bottoms composition x_B. For the gas-liquid separator in Fig. 18.1*c*, adjusting gas flow rate G will have a direct effect on pressure P and a slower, indirect effect on liquid level h, because changing the pressure in the vessel will tend to change the liquid flow rate L and thus affect h. In contrast, adjusting the other manipulated variable L directly affects h but has only a relatively small and indirect effect on P.

When significant process interactions are present, the selection of the most effective control configuration may not be obvious. For example, in the blending problem, suppose that a conventional feedback control strategy, consisting of two PI controllers, is to be used. This control system, referred to as a *multiloop control system* because it employs two single-loop feedback controllers, raises several questions. Should the composition controller adjust w_A and the flow controller adjust w_B, or *vice versa*? How can we determine which of these two multiloop control configurations will be more effective? Will control loop interactions generated by the process interactions cause problems?

In the next section, we consider techniques for selecting an appropriate multiloop control configuration. If the process interactions are significant, even the best multiloop control system may not provide satisfactory control. In these situations there are incentives for considering *multivariable control strategies* such as decoupling control (Section 18.5) and model predictive control (Chapter 20). But first we examine the phenomenon of control loop interactions.

18.1 PROCESS INTERACTIONS AND CONTROL LOOP INTERACTIONS

A schematic representation of several SISO and MIMO control applications is shown in Fig. 18.2. For convenience, it is assumed that the number of manipulated variables is equal to the number of controlled variables.

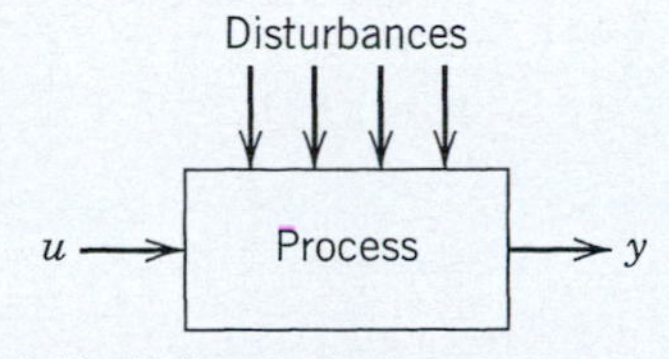

(a) Single-input, single-output process with multiple disturbances

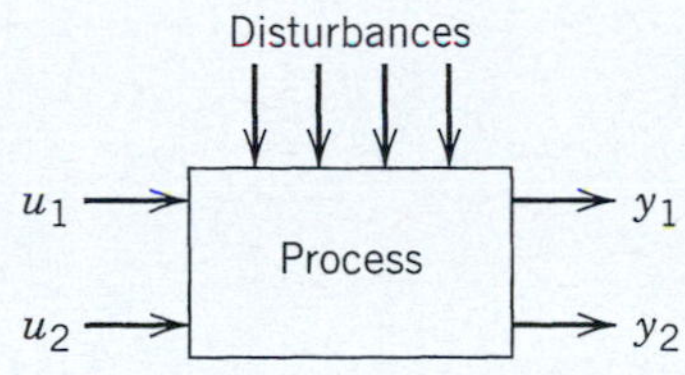

(b) Multiple-input, multiple-output process (2 × 2)

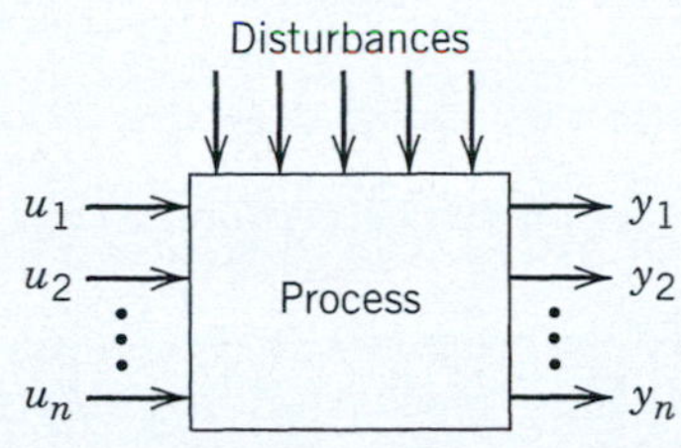

(c) Multiple-input, multiple-output process ($n \times n$)

Figure 18.2 SISO and MIMO control problems.

This allows pairing of a single controlled variable and a single manipulated variable via a feedback controller. On the other hand, more general multivariable control strategies do not make such restrictions (see Chapter 20). MIMO control problems are inherently more complex than SISO control problems because process interactions occur between controlled and manipulated variables. In general, a change in a manipulated variable, say u_1, will affect all of the controlled variables $y_1, y_2, \ldots y_n$. Because of the process interactions, the selection of the best pairing of controlled and manipulated variables for a multiloop control scheme can be a difficult task. In particular, for a control problem with n controlled variables and n manipulated variables, there are $n!$ possible multiloop control configurations.

18.1.1 Block Diagram Analysis

Consider the 2 × 2 control problem shown in Fig. 18.2*b*. Because there are two controlled variables and two manipulated variables, four process transfer functions are necessary to completely characterize the process dynamics:

$$\frac{Y_1(s)}{U_1(s)} = G_{p11}(s) \qquad \frac{Y_1(s)}{U_2(s)} = G_{p12}(s)$$
$$\frac{Y_2(s)}{U_1(s)} = G_{p21}(s) \qquad \frac{Y_2(s)}{U_2(s)} = G_{p22}(s) \tag{18-1}$$

The transfer functions in Eq. 18-1 can be used to determine the effect of a change in either U_1 or U_2 on Y_1 and Y_2. From the Principle of Superposition (Section 3.1), it follows that *simultaneous* changes in U_1 and U_2 have an additive effect on each controlled variable:

$$Y_1(s) = G_{p11}(s)U_1(s) + G_{p12}(s)U_2(s) \tag{18-2}$$

$$Y_2(s) = G_{p21}(s)U_1(s) + G_{p22}(s)U_2(s) \tag{18-3}$$

These input-output relations can also be expressed in vector-matrix notation as

$$\boldsymbol{Y}(s) = \boldsymbol{G}_p(s)\boldsymbol{U}(s) \tag{18-4}$$

where $\boldsymbol{Y}$(s) and $\boldsymbol{U}$(s) are vectors with two elements,

$$\boldsymbol{Y}(s) = \begin{bmatrix} Y_1(s) \\ Y_2(s) \end{bmatrix} \qquad \boldsymbol{U}(s) = \begin{bmatrix} U_1(s) \\ U_2(s) \end{bmatrix} \tag{18-5}$$

and $\boldsymbol{G}_p(s)$ is the process transfer function matrix,

$$\boldsymbol{G}_p(s) = \begin{bmatrix} G_{p11}(s) & G_{p12}(s) \\ G_{p21}(s) & G_{p22}(s) \end{bmatrix} \tag{18-6}$$

The matrix notation in Eq. 18-4 provides a compact representation for problems larger than 2 × 2. Recall that a transfer function matrix for an MIMO system, a stirred-tank blending system, was derived in Section 6.5. The steady-state process transfer function matrix ($s = 0$) is called the process *gain matrix* and is denoted by $\boldsymbol{K}$.

Suppose that a conventional multiloop control scheme consisting of two feedback controllers is to be used. The two possible control configurations are shown in Fig. 18.3. In scheme (a), Y_1 is controlled by adjusting U_1, while Y_2 is controlled by adjusting U_2. Consequently, this configuration will be referred to as the 1-1/2-2 control scheme. The alternative strategy is to pair Y_1 with U_2 and Y_2 with U_1, the 1-2/2-1 control scheme shown in Fig. 18.3*b*. Note that these block diagrams have been simplified by omitting the transfer functions for the final control elements and the sensor-transmitters. Also, the disturbance variables have been omitted.

Figure 18.3 indicates that the process interactions can induce undesirable interactions between the control loops. For example, suppose that the 1-1/2-2 control scheme is used and a disturbance moves Y_1 away from its set point, Y_{sp1}. Then the following events occur:

1. The controller for loop 1 (G_{c1}) adjusts U_1 so as to force Y_1 back to the set point. However, U_1 also affects Y_2 via transfer function G_{p21}.
2. Since Y_2 has changed, the loop 2 controller (G_{c2}) adjusts U_2 so as to bring Y_2 back to its set point, Y_{2sp}. However, changing U_2 also affects Y_1 via transfer function G_{p12}.

These controller actions proceed simultaneously until a new steady state is reached. Note that the initial change in U_1 has two effects on Y_1: (1) a *direct* effect and (2) an

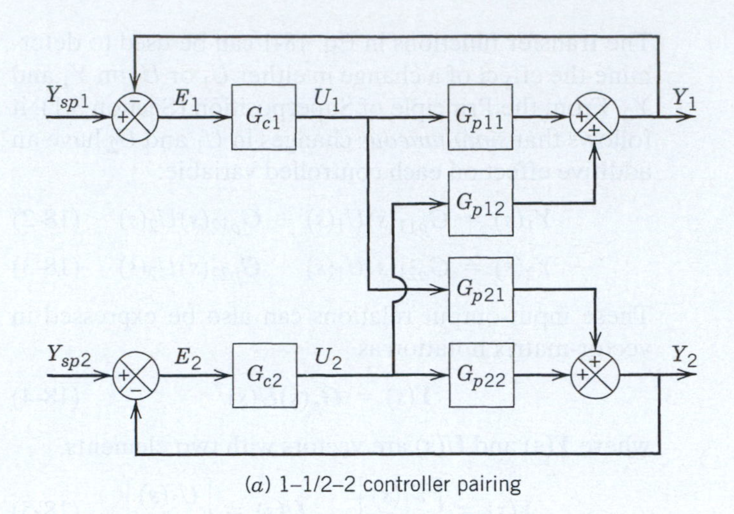

(a) 1–1/2–2 controller pairing

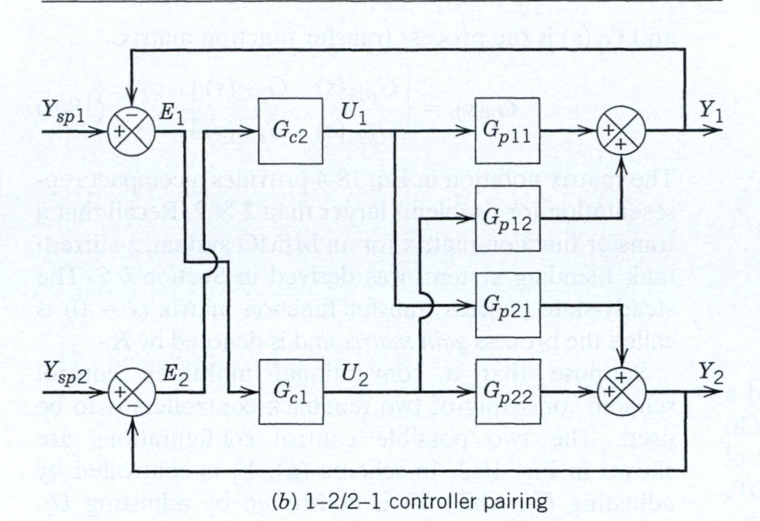

(b) 1–2/2–1 controller pairing

Figure 18.3 Block diagrams for 2×2 multiloop control schemes.

indirect effect via the control loop interactions. Although it is instructive to view this dynamic behavior as a sequence of events, in practice, the process variables would change continuously and simultaneously.

The control loop interactions in a 2×2 control problem result from the presence of a third feedback loop that contains the two controllers and two of the four process transfer functions (Shinskey, 1996). Thus, for the 1-1/2-2 configuration, *this hidden feedback loop* contains G_{c1}, G_{c2}, G_{p12}, and G_{p21}, as shown in Fig. 18.4. A similar hidden feedback loop is also present in the 1-2/2-1 control scheme of Fig. 18.3*b*. The third feedback loop causes two potential problems:

1. It tends to destabilize the closed-loop system.
2. It makes controller tuning more difficult.

Next we show that the transfer function between a controlled variable and a manipulated variable depends on whether the other feedback control loops are open or closed. Consider the control system in Fig. 18.3*a*. If the controller for the second loop G_{c2} is out of service or is placed in the manual mode with the controller output constant at its nominal value, then $U_2 = 0$. For this situation, the transfer function between Y_1 and U_1 is merely G_{p11}:

$$\frac{Y_1}{U_1} = G_{p11} \quad (Y_2 - U_2 \text{ loop open}) \tag{18-7}$$

If both loops are closed, then the contributions to Y_1 from the two loops are added together:

$$Y_1 = G_{p11}U_1 + G_{p12}U_2 \tag{18-8}$$

However, if the second feedback controller is in the automatic mode with $Y_{2sp} = 0$, then, using block diagram algebra,

$$Y_2 = \frac{G_{p21}U_1}{1 + G_{c2}G_{p22}} \tag{18-9}$$

The signal to the first loop from the second loop is

$$G_{p12}U_2 = -G_{p12}G_{c2}Y_2 \tag{18-10}$$

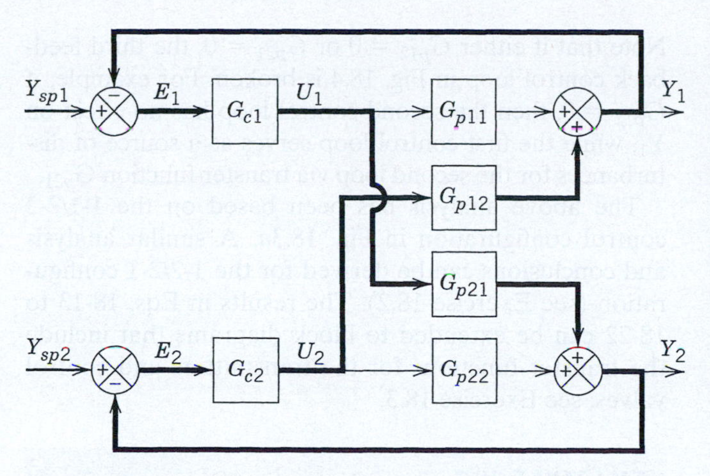

Figure 18.4 The hidden feedback control loop (in dark lines) for a 1-1/2-2 controller pairing.

If we substitute for $G_{p12}U_2$ in (18-8) using (18-10) and then substitute for Y_2 using (18-9), the overall closed-loop transfer function between Y_1 and U_1 is

$$\frac{Y_1}{U_1} = G_{p11} - \frac{G_{p12}G_{p21}G_{c2}}{1 + G_{c2}G_{p22}} \quad (Y_2 - U_2 \text{ loop closed}) \tag{18-11}$$

Thus, the transfer function between Y_1 and U_1 depends on the controller for the second loop G_{c2} via the interaction term. Similarly, transfer function Y_2/U_2 depends on G_{c1} when the first loop is closed (see Exercise 18.1). These results have important implications for controller tuning because they indicate that the two controllers should not be tuned independently. For general $n \times n$ processes, Balchen and Mummé (1988) have derived analogous results that illustrate the effect of closing all but one of the n feedback loops.

EXAMPLE 18.1

Consider the following empirical model of a pilot-scale distillation column (Wood and Berry, 1973)

$$\begin{bmatrix} X_D(s) \\ X_B(s) \end{bmatrix} = \begin{bmatrix} \dfrac{12.8e^{-s}}{16.7s+1} & \dfrac{-18.9e^{-3s}}{21s+1} \\ \dfrac{6.6e^{-7s}}{10.9s+1} & \dfrac{-19.4e^{-3s}}{14.4s+1} \end{bmatrix} \begin{bmatrix} R(s) \\ S(s) \end{bmatrix} \tag{18-12}$$

where the notation is defined in Fig. 18.1*b*. Suppose that a multiloop control system consisting of two PI controllers is used. Compare the closed-loop set-point changes that result if the $X_D - R/X_B - S$ pairing is selected and

(a) A set-point change is made in each loop with the other loop in manual

(b) The set-point changes are made with both controllers in automatic

Assume that the controller settings are based on the ITAE tuning method for set-point changes in Chapter 12.

Table 18.1 Controller Settings for Example 18.1

Controller Pairing	K_c	τ_I (min)
$x_D - R$	0.604	16.37
$x_B - S$	−0.127	14.46

SOLUTION

Table 18.1 shows the single-loop ITAE settings, and Fig. 18.5 shows simulation results for set-point changes for each controlled variable. The ITAE settings provide satisfactory set-point responses for either control loop when the other controller is in manual (solid line). However, when both controllers are in automatic, the control loop interactions produce very oscillatory responses especially in x_B (dashed line). McAvoy (1981) has discussed various approaches for improving the performance of the multiloop controllers. See Exercise 18.1 for a similar MIMO control problem where the loops also exhibit oscillations.

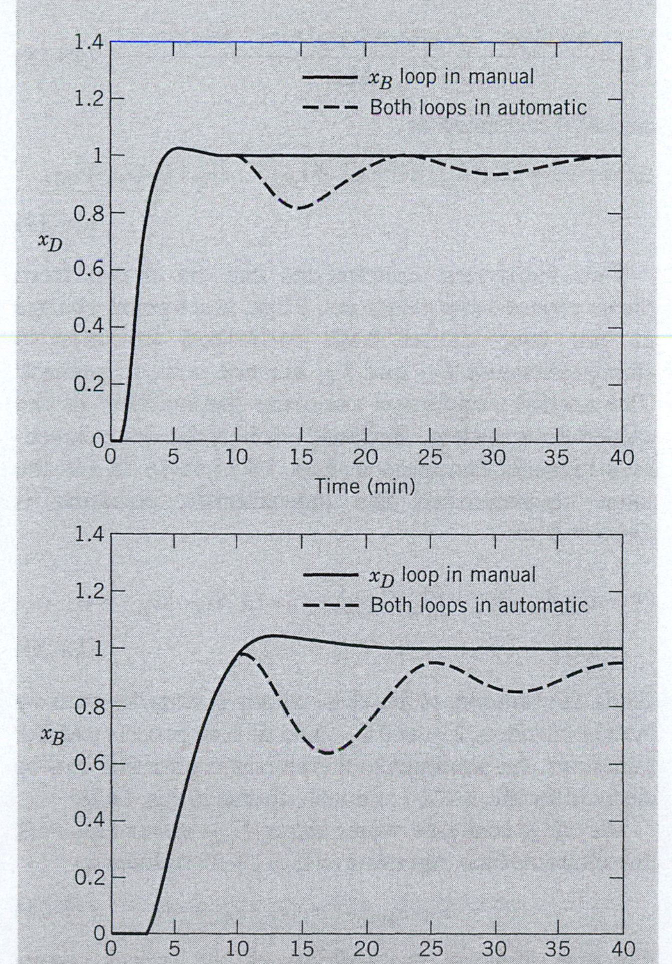

Figure 18.5 Set-point responses for Example 18.1 using ITAE tuning.

18.1.2 Closed-Loop Stability

To evaluate the effects of control loop interactions further, again consider the block diagram for the 1-1/2-2 control scheme in Fig. 18.3*a*. Using block diagram algebra (see Chapter 11), we can derive the following expressions relating controlled variables and set points:

$$Y_1 = \Gamma_{11} Y_{sp1} + \Gamma_{12} Y_{sp2} \tag{18-13}$$

$$Y_2 = \Gamma_{21} Y_{sp1} + \Gamma_{22} Y_{sp2} \tag{18-14}$$

where the closed-loop transfer functions are

$$\Gamma_{11} = \frac{G_{c1}G_{p11} + G_{c1}G_{c2}(G_{p11}G_{p22} - G_{p12}G_{p21})}{\Delta(s)} \tag{18-15}$$

$$\Gamma_{12} = \frac{G_{c2}G_{p12}}{\Delta(s)} \tag{18-16}$$

$$\Gamma_{21} = \frac{G_{c1}G_{p21}}{\Delta(s)} \tag{18-17}$$

$$\Gamma_{22} = \frac{G_{c2}G_{p22} + G_{c1}G_{c2}(G_{p11}G_{p22} - G_{p12}G_{p21})}{\Delta(s)} \tag{18-18}$$

and $\Delta(s)$ is defined as

$$\Delta(s) = (1 + G_{c1}G_{p11})(1 + G_{c2}G_{p22}) - G_{c1}G_{c2}G_{p12}G_{p21} \tag{18-19}$$

Two important conclusions can be drawn from these closed-loop relations. First, a set-point change in one loop causes both controlled variables to change because Γ_{12} and Γ_{21} are not zero, in general. The second conclusion concerns the stability of the closed-loop system. Because each of the four closed-loop transfer functions in Eqs. 18-15 to 18-18 has the same denominator, the characteristic equation is $D(s) = 0$, or

$$(1 + G_{c1}G_{p11})(1 + G_{c2}G_{p22}) - G_{c1}G_{c2}G_{p12}G_{p21} = 0 \tag{18-20}$$

Thus, the stability of the closed-loop system depends on both controllers, Γ_{12} and Γ_{21}, and all four process transfer functions. An analogous characteristic equation can be derived for the 1-2/2-1 control scheme in Fig. 18.3*b*.

For the special case where either $G_{p12} = 0$ or $G_{p21} = 0$, the characteristic equation in Eq. 18-20 reduces to

$$(1 + G_{c1}G_{p11})(1 + G_{c2}G_{p22}) = 0 \tag{18-21}$$

For this situation, the stability of the overall system merely depends on the stability of the two individual feedback control loops and their characteristic equations.

$$1 + G_{c1}G_{p11} = 0 \quad \text{and} \quad 1 + G_{c2}G_{p22} = 0 \tag{18-22}$$

Note that if either $G_{p12} = 0$ or $G_{p21} = 0$, the third feedback control loop in Fig. 18.4 is broken. For example, if $G_{p12} = 0$, then the second control loop has no effect on Y_1, while the first control loop serves as a source of disturbances for the second loop via transfer function G_{p21}.

The above analysis has been based on the 1-1/2-2 control configuration in Fig. 18.3*a*. A similar analysis and conclusions can be derived for the 1-2/2-1 configuration (see Exercise 18.2). The results in Eqs. 18-13 to 18-22 can be extended to block diagrams that include the transfer functions for the transmitters and control valves; see Exercise 18.3.

EXAMPLE 18.2

Consider a process that can be described by the transfer function matrix (Gagnepain and Seborg, 1982):

$$\mathbf{G}_p(s) = \begin{bmatrix} \dfrac{2}{10s+1} & \dfrac{1.5}{s+1} \\ \dfrac{1.5}{s+1} & \dfrac{2}{10s+1} \end{bmatrix}$$

Assume that two proportional feedback controllers are to be used so that $G_{c1} = K_{c1}$ and $G_{c2} = K_{c2}$. Determine the values of K_{c1} and K_{c2} that result in closed-loop stability for both the 1-1/2-2 and 1-2/2-1 configurations.

SOLUTION

The characteristic equation for the closed-loop system (1-1/2-2 pairing) is obtained by substitution into Eq. 18-20 and collecting powers of s as follows:

$$a_4 s^4 + a_3 s^3 + a_2 s^2 + a_1 s + a_0 = 0 \tag{18-23}$$

where
$$a_4 = 100$$
$$a_3 = 20K_{c1} + 20K_{c2} + 220$$
$$a_2 = 42K_{c1} + 42K_{c2} - 221\,K_{c1}K_{c2} + 141$$
$$a_1 = 24K_{c1} + 24K_{c2} - 37K_{c1}K_{c2} + 22$$
$$a_0 = 2K_{c1} + 2K_{c2} + 1.75\,K_{c1}K_{c2} + 1$$

Note that the characteristic equation in (18-23) is fourth-order, even though each individual transfer function in $\mathbf{G}_p(s)$ is first order.

The controller gains that result in a stable closed-loop system can be determined by applying the Routh stability criterion (Chapter 11) for specified values of K_{c1} and K_{c2}. The resulting stability regions are shown in Fig. 18.6. If either K_{c1} or K_{c2} is close to zero, the other controller gain can be an arbitrarily large, positive value and still have a stable closed-loop system. This result is a consequence of having process transfer functions that are first order without time delay, which is an idealistic case. MIMO control systems normally have an upper bound for stability for both controller gains for all values of K_{ci}.

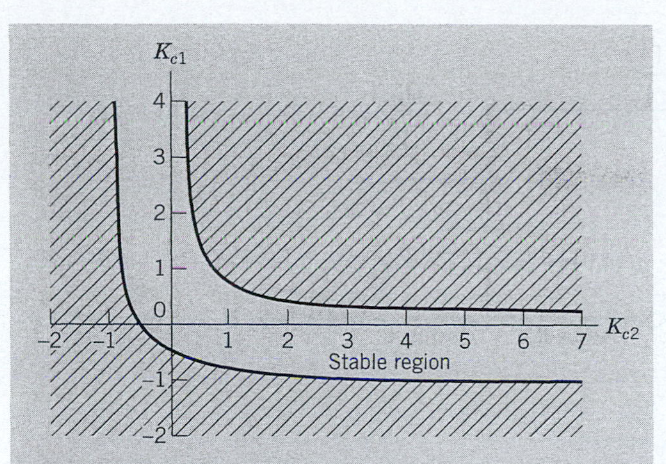

Figure 18.6 Stability region for Example 18.2 with 1-1/2-2 controller pairing.

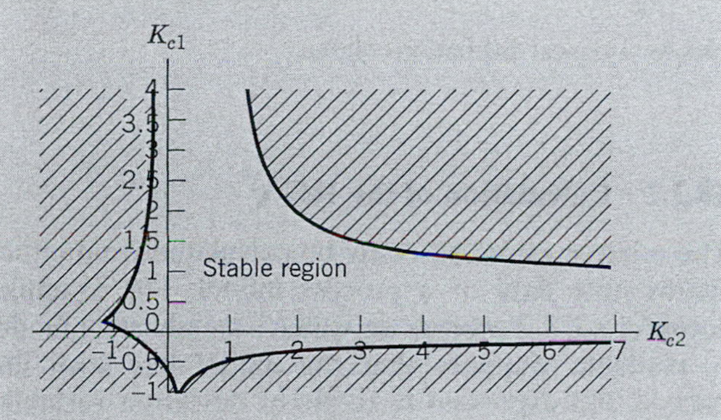

Figure 18.7 Stability region for Example 18.2 with 1-2/2-1 controller pairing.

A similar stability analysis can be performed for the 1-2/2-1 control configuration. The calculated stability regions are shown in Fig. 18.7. A comparison of Figs. 18.6 and 18.7 indicates that the 1-2/2-1 control scheme results in a larger stability region because a wider range of controller gains can be used. For example, suppose that $K_{c1} = 2$. Then Fig. 18.6 indicates that the 1-1/2-2 configuration will be stable if $-0.8 < K_{c2} < 0.5$. By contrast, Fig. 18.7 shows that the corresponding stability limits for the 1-2/2-1 configuration are $-0.3 < K_{c2} < 2.0$.

This example illustrates that closed-loop stability depends on the control configuration as well as the numerical values of the controller settings. If PI control had been considered instead of proportional-only control, the stability analysis would have been much more complicated due to the larger number of controller settings and the higher-order characteristic equation.

18.2 PAIRING OF CONTROLLED AND MANIPULATED VARIABLES

In this section, we consider the important general problem of how the controlled variables and the manipulated variables should be paired in a multiloop control scheme. An incorrect pairing can result in poor control system performance and reduced stability margins, as was the case for the 1-1/2-2 pairing in Example 18.2. As an illustrative example, consider the distillation column shown in Fig. 18.8. A typical distillation column has five possible controlled variables and five manipulated variables (Shinskey, 1996). The controlled variables in Fig. 18.8 are product composition x_D and x_B, column pressure P, and the liquid levels in the reflux drum h_D and column base h_B. The five manipulated variables are product flows D and B, reflux flow R, and the heat duties for the condenser and reboiler, Q_D and Q_B. The heat duties are adjusted via the control valves on the steam and coolant lines. If a multiloop control scheme consisting of five feedback controllers is used, there are 5! = 120 different ways of pairing the controlled and manipulated variables. Some of these control configurations would be immediately rejected as being impractical or unworkable, for example, any scheme that attempts to control base level h_B by adjusting distillate flow D or condenser heat duty Q_D. However, there may be a number of alternative pairings that seem promising; the question then facing the control system designer is how to determine the most effective pairing.

Next, we consider a systematic approach for determining the best pairing of controlled and manipulated variables, the *relative gain array* method. An alternative approach based on singular value analysis is described later in this chapter.

18.2.1 Bristol's Relative Gain Array Method

Bristol (1966) developed a systematic approach for the analysis of multivariable process control problems. His approach requires only steady-state information (the process gain matrix $\boldsymbol{K}$) and provides two important items of information:

1. A measure of process interactions.
2. A recommendation concerning the most effective pairing of controlled and manipulated variables.

Bristol's approach is based on the concept of a *relative gain.* Consider a process with n controlled variables and n manipulated variables. The relative gain λ_{ij} between a controlled variable y_i and a manipulated variable u_j is defined to be the dimensionless ratio of two steady-state gains:

$$\lambda_{ij} \triangleq \frac{(\partial y_i/\partial u_j)_u}{(\partial y_i/\partial u_j)_y} = \frac{\text{open-loop gain}}{\text{closed-loop gain}} \tag{18-24}$$

for $i = 1, 2, \ldots, n$ and $j = 1, 2, \ldots, n$.

In Eq. 18-24 the symbol $(\partial y_i/\partial u_j)_u$ denotes a partial derivative that is evaluated with all of the manipulated variables except u_j held constant. Thus, this term

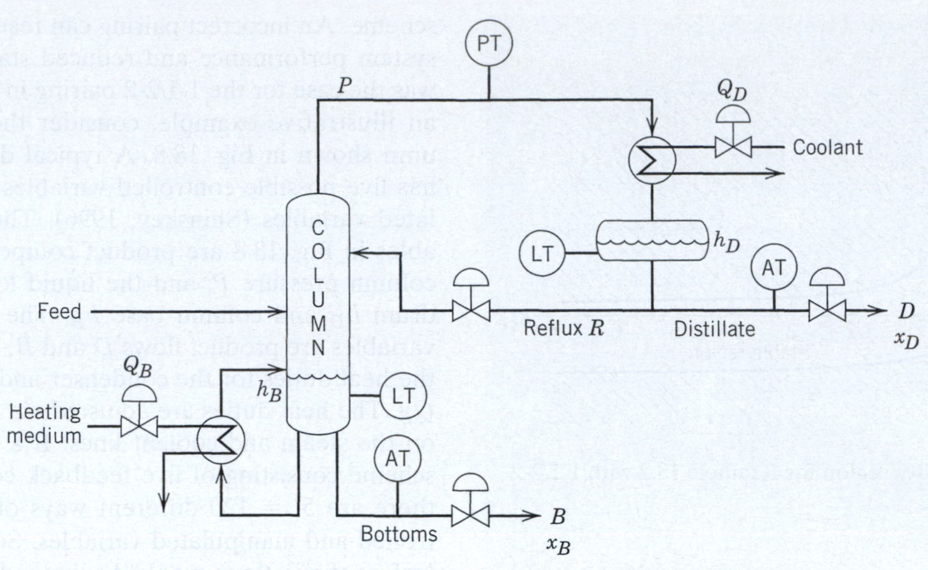

Figure 18.8 Controlled and manipulated variables for a typical distillation column.

is the *open-loop gain* (or steady-state gain) between y_i and u_j, which corresponds to the gain matrix element K_{ij}. Similarly, $(\partial y_i/\partial u_j)_y$ is evaluated with all of the controlled variables except y_i held constant. This situation could be achieved in practice by adjusting the other manipulated variables using controllers with integral action. Thus, $(\partial y_i/\partial u_j)_y$ can be interpreted as a *closed*-loop gain that indicates the effect of u_j on y_j when all of the other controlled variables ($y_i \neq y_j$) are held constant.

It is convenient to arrange the relative gains in a *relative gain array* (RGA), denoted by $\boldsymbol{\Lambda}$:

$$\boldsymbol{\lambda} = \begin{array}{c} \\ y_1 \\ y_2 \\ \cdots \\ y_n \end{array} \begin{array}{c} \begin{array}{cccc} u_1 & u_2 & \cdots & u_n \end{array} \\ \begin{bmatrix} \lambda_{11} & \lambda_{12} & \cdots & \lambda_{1n} \\ \lambda_{21} & \lambda_{22} & \cdots & \lambda_{2n} \\ \cdots & \cdots & \cdots & \cdots \\ \lambda_{n1} & \lambda_{n2} & & \lambda_{nm} \end{bmatrix} \end{array} \tag{18-25}$$

The RGA has several important properties for steady-state process models (Bristol, 1966; McAvoy, 1983):

1. It is normalized because the sum of the elements in each row or column is equal to one.
2. The relative gains are dimensionless and thus not affected by choice of units or scaling of variables.
3. The RGA is a measure of sensitivity to element uncertainty in the gain matrix $\boldsymbol{K}$. The gain matrix can become singular if a single element K_{ij} is changed to $K_{ij}(1 - 1/\lambda_{ij})$. Thus a large RGA element indicates that small changes in K_{ij} can markedly change the process control characteristics.

18.2.2 Calculation of the RGA

The relative gains can easily be calculated from either steady-state data or a process model. For example, consider a 2×2 process for which a steady-state model is available. Suppose that the model has been linearized and expressed in terms of deviation variables as follows:

$$y_1 = K_{11}u_1 + K_{12}u_2 \tag{18-26}$$

$$y_2 = K_{21}u_1 + K_{22}u_2 \tag{18-27}$$

where K_{ij} denotes the steady-state gain between y_i and u_j. This model can be expressed more compactly in matrix notation as

$$\boldsymbol{y} = \boldsymbol{K}\boldsymbol{u} \tag{18-28}$$

For stable processes, the steady-state (gain) model in Eq. 18-28 is related to the dynamic model in Eq. 18-4 by

$$\boldsymbol{K} = \boldsymbol{G}_p(0) = \lim_{s \to 0} \boldsymbol{G}_p(s) \tag{18-29}$$

Next, we consider how to calculate λ_{11}. It follows from Eq. 18-26 that

$$\left(\frac{\partial y_1}{\partial u_1}\right)_{u_2} = K_{11} \tag{18-30}$$

Before calculating $(\partial y_1/\partial u_1)_{y_2}$ from Eq. 18-26, we first must eliminate u_2. This is done by solving Eq. 18-27 for u_2 and holding y_2 constant at its nominal value, $y_2 = 0$:

$$u_2 = -\frac{K_{21}}{K_{22}}u_1 \tag{18-31}$$

Then substituting into Eq. 18-26 gives

$$y_1 = K_{11}\left(1 - \frac{K_{12}K_{21}}{K_{11}K_{22}}\right)u_1 \quad (18\text{-}32)$$

It follows that

$$\left(\frac{\partial y_1}{\partial u_1}\right)_{y_2} = K_{11}\left(1 - \frac{K_{12}K_{21}}{K_{11}K_{22}}\right) \quad (18\text{-}33)$$

Substituting Eqs. 18-30 and 18-33 into Eq. 18-24 gives an expression for relative gain λ_{11}:

$$\lambda_{11} = \frac{1}{1 - \dfrac{K_{12}K_{21}}{K_{11}K_{22}}} \quad (18\text{-}34)$$

Because each row and each column of $\boldsymbol{\Lambda}$ in (18-25) sums to one, the other relative gains are easily calculated from λ_{11} for the 2×2 case:

$$\lambda_{12} = \lambda_{21} = 1 - \lambda_{11} \quad \text{and} \quad \lambda_{22} = \lambda_{11} \quad (18\text{-}35)$$

Thus, the RGA for a 2×2 system can be expressed as

$$\boldsymbol{\Lambda} = \begin{bmatrix} \lambda & 1-\lambda \\ 1-\lambda & \lambda \end{bmatrix}$$

where the symbol λ is now used to denote λ_{11}. Note that the RGA for a 2×2 process is always symmetric. However, this will not necessarily be the case for a higher-dimension process ($n > 2$).

For higher-dimension processes, the RGA can be calculated from the expression

$$\boldsymbol{\Lambda} = \boldsymbol{K} \otimes \mathbf{H} \quad (18\text{-}36)$$

where $\otimes$ denotes the Schur product (element by element multiplication):

$$\lambda_{ij} = K_{ij}H_{ij} \quad (18\text{-}37)$$

K_{ij} is the (i, j) element of $\boldsymbol{K}$ in Eq. 18-28, and H_{ij} is the (i, j) element of $\boldsymbol{H} = (\boldsymbol{K}^{-1})^T$; that is, H_{ij} is an element of the transpose of the matrix inverse of $\boldsymbol{K}$. Because computer software is readily available to perform matrix algebra, Eq. 18-37 can be easily evaluated. Note that Eq. 18-36 does not imply that $\boldsymbol{\Lambda} = \boldsymbol{K}(\boldsymbol{K}^{-1})^T$.

18.2.3 Methods for Obtaining the Steady-State Gain Matrix

Equation 18-37 shows how the RGA can be calculated from a linearized steady-state process model with gains K_{ij}. The open-loop process gains can be obtained numerically from a simulation model or directly from experimental data. For a multivariable process, one input (u_1) can be changed in a stepwise fashion (Δu_1) while holding all other inputs constant. The responses for y_1, y_2, and so on are then observed. All loops are kept open during this test; that is, no feedback controllers are operational. Then step changes can be made in the other inputs, one at a time, and open-loop response data can be obtained for all the controlled variables. The steady-state gain depends only on the final value of each y_i, from which the change in y, Δy_i, can be calculated. Thus, the individual process gains are given by the formula (see Chapter 7):

$$K_{ij} = \frac{\Delta y_i}{\Delta u_j} \quad (18\text{-}38)$$

For example, K_{11} can be evaluated as $K_{11} = \Delta y_1/\Delta u_1$. This approach can be used whether the gains are obtained from a mathematical model (simulator) or from actual process data. Of course, in the latter case, usually more effort and cost are required to obtain the necessary information; hence, it is advantageous to use a model when one is available.

For experimental determination of K_{ij}, it is also desirable to perform several step tests for the same input, using different magnitudes and directions for the input change and then average the results. When using a simulator, it is easier to control the conditions of the simulated step change, and the results are less prone to error. However, it is not mandatory to have a dynamic simulator in order to perform the gain calculation. Alternatively, a steady-state simulator can be used by starting from a base case and then changing a single input to two new values, one higher ($+\Delta u_j$) and one lower ($-\Delta u_j$) than the base case, and then finding the corresponding changes in y for each input change. The perturbations in u_j should be chosen carefully so that the calculated gains (and hence the RGA) do not change significantly with the size of the perturbation (as it is increased or decreased). See McAvoy (1983) for more details.

18.2.4 Measure of Process Interactions and Pairing Recommendations

Equation 18-32 can be used to interpret further the relative gain array of a 2×2 process. Assuming that y_2 is kept at its set point of zero under closed-loop control, rearranging (18-32) gives

$$\frac{y_1}{u_1} = K_{11} - \frac{K_{12}K_{21}}{K_{22}} = K_{11}\left[1 - \frac{K_{12}K_{21}}{K_{11}K_{22}}\right] \quad (18\text{-}39)$$

The term, $-K_{12}K_{21}/K_{11}K_{22}$, can be thought of as an interaction term that modifies the open-loop process gain K_{11} due to the effect of the controller in the other loop (G_{c2}). This effect can be positive or negative, depending on the ratio of the gains. Using (18-34) as the definition of $\lambda_{11} = \lambda$, we can then write

$$\frac{y_1}{u_1} = K_{11}\left[\frac{1}{\lambda}\right] = K_{11}\left[1 + \frac{1-\lambda}{\lambda}\right] \quad (18\text{-}40)$$

Thus λ can be interpreted as a divisor of the open-loop gain K_{11}, or the term $(1 - \lambda)/\lambda$ is a correction to K_{11}. Another way to view (18-40) is that closed-loop gain = open-loop gain/λ, which is a restatement of the definition in (18-24). For example, if $\lambda = 1$, there is no correction to the gain, and the open-loop gain is the same as the closed-loop gain. If λ is a large positive value, the closed-loop gain is much smaller than the open-loop gain. The practical implication of a large value of λ is that u_1 no longer has much influence on y_1, which could have important operational implications. Finally, if λ is negative, the closed-loop gain changes sign from K_{11}, which indicates serious difficulties in controller design if the first loop is closed.

RGA analysis of cases larger than 2 × 2 may not lead to clear conclusions, especially if one is evaluating an operating process that already has some mixture of open and closed loops. It is important to understand in a multivariable process how other control loops affect the process gain of a given open loop. Sometimes it is not obvious that the pathological behavior of a particular input-output combination may be due to other controllers. Further troubleshooting analysis may lead to the conclusion that a different control strategy is required before a critical output variable can be successfully controlled (see Appendices G and H for a discussion of plantwide control).

For the 2 × 2 process in Eqs. 18-26 and 18-27, five pairing cases can be considered (recall that λ defines the entire matrix Λ because the sum of the RGA elements is unity for each row and for each column):

1. **$\lambda = 1$.** In this situation, it follows from (18-24) that the open-loop and closed-loop gains between y_1 and u_1 are identical. In this ideal situation, opening or closing loop 2 has no effect on loop 1. It follows that y_1 should be paired with u_1 (i.e., a 1-1/2-2 configuration should be employed).
2. **$\lambda = 0$.** Equation 18-24 indicates that the open-loop gain between y_1 and u_1 is zero, and thus u_1 has no direct effect on y_1. Consequently, u_1 should be paired with y_2 rather than y_1 (i.e., the 1-2/2-1 configuration should be utilized).
3. **$0 < \lambda < 1$.** From Eq. 18-24, the closed-loop gain between y_1 and u_1 is larger than the open-loop gain. Within this range, the interaction between the two loops is largest when $\lambda = 0.5$, which indicates that the second term in Eq. 18-33 is equal to −1 (McAvoy, 1981; Shinskey, 1996).
4. **$\lambda > 1$.** For this situation, closing the second loop reduces the gain between y_1 and u_1. Thus, the control loops interact. As λ increases, the degree of interaction increases and becomes most severe as $\lambda \to \infty$. When λ is very large, it is impossible to control both outputs independently (Skogestad and Postlethwaite, 2005).
5. **$\lambda < 0$.** When λ is negative, the open-loop and closed-loop gains between y_1 and u_1 have different signs. Thus, opening or closing loop 2 has an adverse effect on the behavior of loop 1 such as oscillation. It follows that y_1 should not be paired with u_1. For $\lambda < 0$ the control loops interact by trying to "fight each other" (McAvoy, 1983; Shinskey, 1996), and the closed-loop system may become unstable.

Based on these considerations, the RGA analysis for a 2 × 2 process leads to the conclusion that y_1 should be paired with u_1 only if $\lambda \geq 0.5$. Otherwise, y_1 should be paired with u_2, the reverse pairing. This reasoning can be extended to $n \times n$ processes and leads to Bristol's original recommendation for controller pairing:

Recommendation: *Pair the controlled and manipulated variables so that corresponding relative gains are positive and as close to one as possible.*

At this point, it is appropriate to make several remarks about the RGA approach.

1. The above recommendation is based solely on steady-state information. However, dynamic behavior should also be considered in choosing a controller pairing. In particular, closed-loop stability should be checked using a theorem that is presented in the next section.
2. If $\lambda = 0$ or $\lambda = 1$, the two control loops for a 2 × 2 process either do not interact at all or exhibit only a one-way interaction, based on this steady-state analysis. Furthermore, at least one of the four process gains must be zero, according to Eq. 18-34.
3. If a pairing of inputs and outputs in a 2 × 2 process corresponds to a negative relative gain, then the closed-loop system will exhibit instability either in the overall closed-loop system or in the loop with the negative relative gain by itself.

One property of interest in control loop design is called *loop integrity*. Skogestad and Postlethwaite (2005) have considered the case when the RGA element is negative for a given pairing, the process is stable, and the feedback controller contains integral action. If the control loop with the negative pairing is disabled owing to failure (or being taken out of service) or because of saturation of the controller output, the multiloop control system will become unstable. A related property is called *decentralized integral controllability* (DIC). When each loop contains integral action, the property of DIC means that the gain of each controller can be reduced to zero without the closed-loop system becoming unstable. Campo and Morari (1994) have developed conditions of λ_{ij} that ensure DIC; for the 2×2 case, for example, $\lambda_{11} > 0$ guarantees DIC.

A one-way interaction occurs when one loop affects the other loop but not vice versa. Suppose that $\boldsymbol{K}$ has the structure

$$\boldsymbol{K} = \begin{bmatrix} K_{11} & K_{12} \\ 0 & K_{22} \end{bmatrix}$$

Then loop 1 does not affect loop 2, because $K_{21} = 0$, and thus u_1 has no effect on y_2. However, loop 2 does affect loop 1 via u_2 if $K_{12} \neq 0$. This one-way interaction does not affect the closed-loop stability because $K_{21} = 0$; consequently, the characteristic equation in Eq. 18-20 reduces to the two equations in (18-22). Thus, for this one way interaction, loop 2 tends to act as a source of disturbances for loop 1.

To illustrate how the RGA can be used to determine controller pairing, we next consider several examples. Additional examples have been presented by McAvoy (1983).

EXAMPLE 18.3

Consider the in-line blending system of Fig. 18.1*a*. It is proposed that w and x be controlled using a conventional multiloop control scheme, with w_A and w_B as the manipulated variables. Derive an expression for the RGA and recommend the best controller pairing for the following conditions: $w = 4$ lb/min and $x = 0.4$.

SOLUTION

Assuming perfect mixing, a process model can be derived from the following steady-state mass balances:

$$\text{Total mass:} \quad w = w_A + w_B \tag{18-41}$$

$$\text{Component A:} \quad xw = w_A \tag{18-42}$$

Substituting (18-41) into (18-42) and rearranging gives

$$x = \frac{w_A}{w_A + w_B} \tag{18-43}$$

The RGA for the blending system can be expressed as

$$\Lambda = \begin{array}{c} \\ w \\ x \end{array} \begin{array}{c} \begin{array}{cc} w_A & w_B \end{array} \\ \begin{bmatrix} \lambda & 1-\lambda \\ 1-\lambda & \lambda \end{bmatrix} \end{array}$$

Relative gain λ can be calculated from Eq. 18-34 after the four steady-state gains are calculated:

$$K_{11} = \left(\frac{\partial w}{\partial w_A}\right)_{w_B} = 1 \tag{18-44a}$$

$$K_{12} = \left(\frac{\partial w}{\partial w_B}\right)_{w_A} = 1 \tag{18-44b}$$

$$K_{21} = \left(\frac{\partial x}{\partial w_A}\right)_{w_B} = \frac{w_B}{(w_A + w_B)^2} = \frac{1-x}{w} \tag{18-44c}$$

$$K_{22} = \left(\frac{\partial x}{\partial w_B}\right)_{w_A} = \frac{-w_A}{(w_A + w_B)^2} = -\frac{x}{w} \tag{18-44d}$$

Substituting into Eq. 18-34 gives $\lambda = x$. Thus, the RGA is

$$\Lambda = \begin{array}{c} \\ w \\ x \end{array} \begin{array}{c} \begin{array}{cc} w_A & w_B \end{array} \\ \begin{bmatrix} x & 1-x \\ 1-x & x \end{bmatrix} \end{array}$$

Note that the recommended pairing depends on the desired product composition x. For $x = 0.4$, w should be paired with w_B and x with w_A. Because all four relative gains are close to 0.5, control loop interactions will be a serious problem. On the other hand, if $x = 0.9$, w should be paired with w_A and x with w_B. In this case, the control loop interactions will be small. Note that for both cases, total flow rate w is controlled by the larger component flow rate, w_A or w_B.

EXAMPLE 18.4

The relative gain array for a refinery distillation column associated with a hydrocracker discussed by Nisenfeld and Schultz (1971) is given by

$$\Lambda = \begin{array}{c} \\ y_1 \\ y_2 \\ y_3 \\ y_4 \end{array} \begin{array}{c} \begin{array}{cccc} u_1 & u_2 & u_3 & u_4 \end{array} \\ \begin{bmatrix} 0.931 & 0.150 & 0.080 & -0.164 \\ -0.011 & -0.429 & 0.286 & 1.154 \\ -0.135 & 3.314 & -0.270 & -1.910 \\ 0.215 & -2.030 & 0.900 & 1.919 \end{bmatrix} \end{array} \tag{18-45}$$

The four controlled variables are the compositions of the top and bottom product streams (y_1, y_2) and the two side streams (y_3, y_4). The manipulated variables are the four flow rates numbered from the top of the column; for example, the top flow rate is u_1. Find the recommended pairing using the RGA.

SOLUTION

To determine the recommended controller pairs, we identify the positive relative gains that are closest to one in each row and column. From the rows, it is apparent that the recommended pairings are y_1-u_1, y_2-u_4, y_3-u_2, and y_4-u_3. Note that this pairing assigns u_2 to y_3 rather than to y_1, even though its relative gain of 3.314 is farther from one. This choice is required because pairing any other manipulated variable with y_3 corresponds to a negative relative gain, which is undesirable. The y_1-u_1 and y_4-u_3 relative gains are close to one, so these selections are straightforward.

The two previous examples have shown how the RGA can be calculated from steady-state gain information. For integrating processes such as the liquid storage system considered in Section 5.3, one or more steady-state gains do not exist. Consequently, the standard RGA analysis must be modified for such systems (Woolverton, 1980; Arkun and Downs, 1990). The RGA analysis proceeds in the usual manner, except that any controlled variable that is the output of an integrating element should be replaced by its rate of change. Thus, if a liquid level h is both a controlled variable and the output of an integrating element, then h will be replaced by dh/dt in the RGA analysis. This procedure is illustrated in Exercise 18.8.

Useful information about the stability of a proposed multiloop control system can be obtained using a theorem originally reported by Niederlinski (1971) and later corrected by Grosdidier et al. (1985). Like the RGA analysis, the theorem is based solely on steady-state information. It is assumed that the steady-state gain matrix $\boldsymbol{K}$ has been arranged so that the diagonal elements correspond to the proposed pairing; that is, it is assumed that y_1 is paired with u_1, y_2 and u_2, and so on. This arrangement can always be obtained by reordering the elements of the $\boldsymbol{y}$ and $\boldsymbol{u}$ vectors if necessary.

The following theorem is based on three assumptions similar to those stated by Grosdidier et al. (1985):

1. Let $G_{pij}(s)$ denote the (i, j) element or process transfer function matrix, $\boldsymbol{G}_p(s)$. Each $G_{pij}(s)$ must be stable, rational, and proper; that is, the order of the denominator must be at least as great as the order of the numerator.
2. Each of the n feedback controllers in the multiloop control system contains integral action.
3. Each individual control loop is stable when any of the other $n - 1$ loops are opened.

Stability Theorem. Suppose that a multiloop control system is used with the pairing $y_1 - u_1, y_2 - u_2, \ldots, y_n - u_n$. If the closed-loop system satisfies Assumptions 1–3, then the closed-loop system is unstable if

$$\frac{|\boldsymbol{K}|}{\prod_{i=1}^{n} K_{ii}} < 0 \tag{18-46}$$

where $|\boldsymbol{K}|$ denotes the determinant of $\boldsymbol{K}$.

Note that this theorem provides a sufficient (but not necessary) condition for instability. Thus, if the inequality is satisfied, the closed-loop system will be unstable. However, if the inequality is not satisfied, the closed-loop system may or may not be unstable, depending on the numerical values of the controller settings. The inequality is also satisfied if the proposed pairing for a 2 × 2 system corresponds to a negative value of a relative gain. McAvoy (1983, p. 84) reports several examples where apparently reasonable RGA pairings result in unstable closed-loop systems. Thus, it is important to consider the process dynamics and also check to ensure that a proposed pairing does not satisfy the inequality in Eq. 18-46.

Assumption 1 requires that each $G_{pij}(s)$ be a rational function; hence, the theorem does not strictly apply to processes that contain time delays. Because time delays do not affect the steady-state matrix $\boldsymbol{K}$, the theorem still provides useful insight into the stability of such systems, even though the analysis is no longer rigorous (Grosdidier et al., 1985).

Processes with poorly conditioned $\boldsymbol{K}$ matrices tend to require large changes in the manipulated variables in order to influence the controlled variables. This assertion can be justified as follows. Solving Eq. 18-28 for $\boldsymbol{u}$,

$$\boldsymbol{u} = \boldsymbol{K}^{-1}\,\boldsymbol{y} \tag{18-47}$$

and substituting set point $\boldsymbol{y}_{sp}$ for $\boldsymbol{y}$ gives

$$\boldsymbol{u} = \boldsymbol{K}^{-1}\boldsymbol{y}_{sp} \tag{18-48}$$

The inverse of $\boldsymbol{K}$ in (18-48) can be calculated from the standard formula,

$$\boldsymbol{K}^{-1} = \frac{\text{adjoint of } \boldsymbol{K}}{|\boldsymbol{K}|} \tag{18-49}$$

The adjoint of $\boldsymbol{K}$ is formed from its cofactors (Strang, 1988).

If $|\boldsymbol{K}|$ is small ($<< 1$), we conclude from (18-48) and (18-49) that the required adjustments in $\boldsymbol{u}$ will be very large, resulting in excessive control actions. Small values of $|\boldsymbol{K}|$ also lead to large values of the relative gain array (cf. Section 18.2). For a 2 × 2 process, the relative gain array is characterized by a single parameter λ. The following expression for λ can be obtained by rearranging Eq. 18-34:

$$\lambda = \frac{K_{11}K_{22}}{K_{11}K_{22} - K_{12}K_{21}} = \frac{K_{11}K_{22}}{|\boldsymbol{K}|} \tag{18-50}$$

Thus, if $|\boldsymbol{K}|$ is small, λ becomes very large, and process interactions are extremely strong, leading to control difficulties.

18.2.5 Dynamic Considerations

An important disadvantage of the standard RGA approach is that it ignores process dynamics, which can be an important factor in the pairing decision. For example, if the transfer function between y_1 and u_1 contains a very large time delay or time constant (relative to the other transfer functions), y_1 will respond very slowly to changes in u_1. Thus, in this situation, a y_1-u_1 pairing is not desirable from a dynamic perspective (see

Example 18.2). McAvoy (1983, p. 214) has noted that dynamic interactions tend to be more important for 2×2 processes when $\lambda > 1$ than when $0 < \lambda < 1$. However, dynamic considerations can still affect the pairing decision even when $0 < \lambda < 1$, as illustrated in the following example.

EXAMPLE 18.5

Consider the transfer function model of Example 18.2 but with a gain of −2 in $G_{p11}(s)$ and a time delay of unity in each transfer function:

$$G_p(s) = \begin{bmatrix} \dfrac{-2e^{-s}}{10s+1} & \dfrac{1.5e^{-s}}{s+1} \\ \dfrac{1.5e^{-s}}{s+1} & \dfrac{2e^{-s}}{10s+1} \end{bmatrix} \tag{18-51}$$

Use the RGA approach to determine the recommended controller pairing based on steady-state considerations. Do dynamic considerations suggest the same pairing?

SOLUTION

The corresponding steady-state gain matrix is

$$\boldsymbol{K} = \begin{bmatrix} -2 & 1.5 \\ 1.5 & 2 \end{bmatrix} \tag{18-52}$$

Using the formula in Eq. 18-34, we obtain $\lambda_{11} = 0.64$. Thus, the RGA analysis indicates that the 1-1/2-2 pairing should be used. However, the off-diagonal time constants in Eq. 18-51 are only one-tenth of the diagonal time constants. Thus, y_1 responds 10 times faster to u_2 than to u_1; similarly, y_2 responds 10 times faster to u_1 than to u_2. Consequently, the 1-2/2-1 pairing is favored based on dynamic considerations, and a conflict exists between steady-state and dynamic considerations. A computer simulation of the two alternative control configurations for this example has shown that the 1-2/2-1 configuration provides better control (Gagnepain and Seborg, 1982). Here the RGA analysis provides an incorrect recommendation concerning the more effective controller pairing. Extensions of the RGA to include process dynamics are discussed next.

18.2.6 Extensions of the RGA Analysis

Several researchers have suggested interaction measures that consider the process dynamics or frequency response as well as steady-state gains (Witcher and McAvoy, 1977; Tung and Edgar, 1981; Grosdidier and Morari, 1986; Skogestad and Postlethwaite, 2005). Although these newer methods are more complicated than the standard RGA approach, they offer additional insights concerning the closed-loop behavior of the system. The frequency domain interpretation of the RGA indicates how dynamics should be considered in the pairing of inputs and outputs. The frequency-dependent RGA, analogous to Eq. 18-36, is the Schur product (each variable is a function of $s = j\omega$):

$$\Lambda = \boldsymbol{G} \otimes (\boldsymbol{G}^{-1})^T \qquad \Lambda(j\omega) = \boldsymbol{G}(j\omega) \otimes (\boldsymbol{G}^{-1}(j\omega))^T \tag{18-53}$$

Skogestad and Postlethwaite (2005) recommend pairings for which the relative gains at the gain crossover and critical frequencies are close to one (see Appendix J). For this analysis, the input and output variables are reordered based on the recommended pairing, which yields an RGA that is diagonally dominant (diagonal terms have larger magnitudes than off-diagonal terms) and close in magnitude to the identity matrix (Grosdidier and Morari, 1986). Plants with large RGA elements around the critical frequency are inherently difficult to control because of sensitivity to errors in the model parameters or model mismatch, which makes design approaches such as decoupling unattractive (see Section 18.5).

Other papers have extended the RGA approach to consider the effect of model uncertainty and disturbances on multiloop control systems (Stanley et al., 1985; Chen and Seborg, 2002). The *relative disturbance gain* (RDG) provides a measure of the change in the effect of a given disturbance caused by multiloop (decentralized) control. For the 2×2 case, the steady-state RDG is

$$\beta_1 = \lambda_{11}\left(1 - \frac{K_{d2}K_{12}}{K_{d1}K_{22}}\right) \tag{18-54}$$

where K_{di} is the gain of the disturbance variable d_i on y_i and β_1 is a dimensionless parameter. It is desirable to keep β_1 small, because small values indicate that the loop interactions actually reduce the effect of the disturbance. Skogestad and Postlethwaite (2005) have discussed the frequency dependence of RDG.

18.3 SINGULAR VALUE ANALYSIS

Singular value analysis (SVA) is a powerful analytical technique that can be used to solve several important control problems:

1. Selection of controlled, measured, and manipulated variables
2. Evaluation of the robustness of a proposed control strategy
3. Determination of the best multiloop control configuration

Singular value analysis and extensions such as singular value decomposition (SVD) also have many uses in numerical analysis and the design of multivariable control systems, which is beyond the scope of this book (Bjorck, 1996; Skogestad and Postlethwaite, 2005). In this section, we provide a brief introduction to SVA that is based on an analysis of steady-state gains from the process models.

Again, we consider the linear steady-state process model in Eq. 18-28.

$$y = Ku \tag{18-55}$$

One desirable property of K is that the n linear equations in n unknowns represented by (18-55) be linearly independent. In contrast, if the equations are dependent, then not all of the n controlled variables can be independently regulated. This characteristic property of linear independence can be checked by several methods (Bjorck, 1996). For example, if the determinant of K is zero, the matrix is singular and the n equations in (18-55) are not linearly independent.

Another way to check for linear independence is to calculate one of the most important properties of a matrix: its eigenvalues. The eigenvalues of matrix K are the roots of the equation

$$|K - \alpha I| = 0 \tag{18-56}$$

where $|K - \alpha I|$ denotes the determinant of matrix $K - \alpha I$, and I is the $n \times n$ identity matrix. The n eigenvalues of K will be denoted by $\alpha_1, \alpha_2, \ldots \alpha_n$. If any of the eigenvalues are zero, K is a singular matrix, and difficulties will be encountered in controlling the process, as noted above. If one eigenvalue is very small compared to the others, then very large changes in one or more manipulated variables will be required to control the process, as will be shown at the end of this section.

Another important property of K is its *singular values*, $\sigma_1, \sigma_2, \ldots \sigma_n$ (Roat et al., 1986; Bjorck, 1996). The singular values are nonnegative numbers that are defined as the positive square roots of the eigenvalues of the matrix product K^TK. The first r singular values are positive numbers, where r is the rank of K^TK. The remaining $n - r$ singular values are zero. Usually, the nonzero singular values are ordered, with σ_1 denoting the largest and σ_r the smallest.

The singular values arise from the decomposition of K (Laub, 2004):

$$K = W\Sigma V^T \tag{18-57}$$

where Σ is the diagonal matrix of singular values. W and V are unitary matrices such that

$$WW^T = I \tag{18-58}$$

$$VV^T = I \tag{18-59}$$

Note that for a unitary matrix, the transpose of W (or V) is also its inverse. The columns of W are referred to as the input singular vectors (and are orthonormal), and the columns of V are the output singular vectors (also orthonormal). W, V, and Σ can be easily calculated from computer software for matrix analysis.

The final matrix property of interest here is the *condition number* (CN). Assume that K is nonsingular. Then the condition number of K is a positive number defined as the ratio of the largest and smallest nonzero singular values:

$$\text{CN} = \frac{\sigma_1}{\sigma_r} \tag{18-60}$$

If K is singular, then it is ill-conditioned, and by convention, CN $= \infty$. The concept of a condition number can also be extended to nonsquare matrices (Bjorck, 1996).

One significant difference between the RGA and SVA is that the elements of the RGA are independent of scaling, whereas the singular values (and CN) depend on scaling or normalization of inputs and outputs. The usual SVA convention is to divide each u_i and y_i by its corresponding range. Thus, input u_i is scaled as

$$u_i^* = \frac{u_i}{u_i^{\max} - u_i^{\min}} \tag{18-61}$$

where u_i^* is the scaled input. Skogestad and Postlethwaite (2005) discuss the notion of the *minimum condition number*, where all possible scalings are evaluated in order to find the minimum CN.

The condition number also provides useful information about the sensitivity of the matrix properties to variations in the elements of the matrices. This important topic, which is related to control system robustness, will be considered later in this section. But first we consider a simple example.

EXAMPLE 18.6

A 2×2 process has the following steady-state gain matrix:

$$K = \begin{bmatrix} 1 & K_{12} \\ 10 & 1 \end{bmatrix} \tag{18-62}$$

Calculate the determinant, RGA, eigenvalues, and singular values of K. Use $K_{12} = 0$ as the base case; then recalculate the matrix properties for a small change, $K_{12} = 0.1$.

SOLUTION

By inspection, the determinant for the base case is $|K| = 1$, and the RGA is $\Lambda = I$, so $\lambda_{11} = 1.0$ and pairing is straightforward ($y_1 - u_1$, $y_2 - u_2$). The eigenvalues can be calculated as follows:

$$|\boldsymbol{K} - \alpha \boldsymbol{I}| = \begin{vmatrix} 1-\alpha & 0 \\ 10 & 1-\alpha \end{vmatrix} = 0 \qquad (18\text{-}63)$$

Thus, $(1 - \alpha)^2 = 0$ and the eigenvalues are $\alpha_1 = \alpha_2 = 1$.

Now calculate the singular values, which arise from

$$\boldsymbol{K}^T\boldsymbol{K} = \begin{bmatrix} 1 & 10 \\ 0 & 1 \end{bmatrix}\begin{bmatrix} 1 & 0 \\ 10 & 1 \end{bmatrix} = \begin{bmatrix} 101 & 10 \\ 10 & 1 \end{bmatrix} \qquad (18\text{-}64)$$

The eigenvalues of $\boldsymbol{K}^T\boldsymbol{K}$, denoted by α', can be calculated from $|\boldsymbol{K}^T\boldsymbol{K} - \alpha'\boldsymbol{I}| = 0$, which again yields a second-order polynomial:

$$(101 - \alpha')(1 - \alpha') - 100 = 0 \qquad (18\text{-}65)$$

Solving (18-65) gives $\alpha'_1 = 101.99$, and $\alpha'_2 = 0.01$. The singular values of $\boldsymbol{K}$ are then

$$\sigma_1 = \sqrt{101.99} = 10.1 \qquad (18\text{-}66)$$

$$\sigma_2 = \sqrt{0.01} = 0.1 \qquad (18\text{-}67)$$

and the condition number is

$$\text{CN} = \frac{\sigma_1}{\sigma_2} = \frac{10.1}{0.1} = 101 \qquad (18\text{-}68)$$

Thus, $\boldsymbol{K}$ is considered to be poorly conditioned because of the large CN value.

Now consider the case where $K_{12} = 0.1$, a small change from the base case. The determinant of $\boldsymbol{K}$ is zero, which indicates that $\boldsymbol{K}$ is singular and the RGA does not exist for this perturbation. The eigenvalues of $\boldsymbol{K}$ calculated from (18-63) are $\alpha_1 = 2$ and $\alpha_2 = 0$. The singular values of $\boldsymbol{K}$ are $\sigma_1 = 10.1$, $\sigma_2 = 0$, and the condition number is $\text{CN} = \infty$, because $\boldsymbol{K}$ is singular.

This example shows that the original $\boldsymbol{K}$ matrix (with $K_{12} = 0$) is poorly conditioned and very sensitive to small variations in the K_{12} element. The large condition number (CN = 101) indicates the poor conditioning. In contrast, the value for the determinant ($|\boldsymbol{K}| = 1$) and the RGA give no indication of poor conditioning. The value of $\lambda = 1$ for $K_{12} = 0$ is quite misleading, because it suggests that the process model in Example 18.6 has no interactions and that a 1-1/2-2 controller pairing will be suitable. However, the large condition number of 101 for this case implies that the process is poorly conditioned and thus will be difficult to control with any controller pairing. The example demonstrates that the condition number is superior to the determinant in providing a more reliable measure of ill-conditioning and potential sensitivity problems.

18.3.1 Selection of Manipulated Variables and Controlled Variables

The SVA and RGA methods can be used as a way to screen subsets of the possible manipulated variables (MVs) and controlled variables (CVs) for a MIMO control system. Because these analyses are based on the steady-state gain matrix, it is recommended that promising combinations of MVs and CVs be identified and then investigated in more detail using simulation and dynamic analysis. The two steps shown below can be used to identify promising subsets of MVs and CVs, recognizing that for multiloop control the number of MVs should equal the number of CVs (a square system).

1. Arrange the singular values from largest to smallest (σ_n, σ_{n-1}, ... σ_1); if $\sigma_i/\sigma_{i-1} > 10$ for some $i \geq 2$, then these singular values can be neglected, and at least one MV and one CV should be omitted, as discussed in step 2.
2. Generate alternative gain matrices by deleting one row and one column at a time and calculating the singular values and condition numbers. Elements of $\boldsymbol{W}$ and $\boldsymbol{V}$ can be used in some cases to guide the choice of which MV and CV should be removed (Skogestad, 1992). The most promising gain matrices have the smallest condition numbers. Then perform dynamic simulation to choose the best MV/CV set out of the remaining alternatives.

Skogestad and Postlethwaite (2005) have indicated that for nonsquare plants (more inputs than outputs or vice versa), the RGA can be used to eliminate some inputs or outputs. For this case the RGA is also nonsquare, and elements in each row (or each column) do not necessarily sum to one. For more inputs than outputs, if all the elements in a column in the RGA are small ($<< 1$), then the corresponding input can be deleted without much loss in performance. Similarly, for more outputs than inputs, if all elements in a row of the RGA are small, then that output cannot be controlled easily and other outputs should be selected. Chang and Yu (1990) have developed an RGA-based methodology for non-square multivariable systems.

Roat et al. (1986) analyzed the choice of manipulated variables for a complex, four-component distillation column. The four components were propane, isobutane, n-butane, and isopentane. There were six possible manipulated variables, and ratios of these variables were also permissible. Table 18.2 shows the condition numbers for six control configurations that were evaluated for the column. Note that the last three strategies have approximately the same low CN. Subsequently, these

Table 18.2 Condition Numbers for the Gain Matrices Relating Controlled Variables to Various Sets of Manipulated Variables for a Distillation Column (Roat et al., 1986)

Controlled Variables

x_D = Mole fraction of propane in distillate D
x_{64} = Mole fraction of isobutane in tray 64 sidedraw
x_{15} = Mole fraction of n-butane in tray 15 sidedraw
x_B = Mole fraction of isopentane in bottoms B

Possible Manipulated Variables

L = Reflux flow rate — B = Bottoms flow rate
D = Distillate flow rate — S_{64} = Sidedraw flow rate at tray 64
V = Steam flow rate — S_{15} = Sidedraw flow rate at tray 15

Strategy Number[a]	Manipulated Variables	Condition Number
1	$L/D, S_{64}, S_{15}, V$	9,030
2	$V/L, S_{64}, S_{15}, V$	60,100
3	$D/V, S_{64}, S_{15}, V$	116,000
4	D, S_{64}, S_{15}, V	51.5
5	L, S_{64}, S_{15}, B	57.4
6	L, S_{64}, S_{15}, V	53.8

[a] In each control strategy, the first controlled variable is paired with the first manipulated variable, and so on. Thus, for Strategy 1, x_D is paired with L/D, and x_B is paired with V.

three strategies were selected for further evaluation using dynamic simulation. Based on simulation results, the best control strategy in Table 18.2 was number 4.

EXAMPLE 18.7

Determine the preferred multiloop control strategy for a process with the following steady-state gain matrix, which has been scaled by dividing the process variables by their maximum values:

$$\begin{bmatrix} y_1 \\ y_2 \\ y_3 \end{bmatrix} = \begin{bmatrix} 0.48 & 0.90 & -0.006 \\ 0.52 & 0.95 & 0.008 \\ 0.90 & -0.95 & 0.020 \end{bmatrix} \begin{bmatrix} u_1 \\ u_2 \\ u_3 \end{bmatrix} \tag{18-69}$$

SOLUTION

The singular value analysis in Eqs. (18-57) through (18-60) yields

$$\mathbf{W} = \begin{bmatrix} 0.5714 & 0.3766 & 0.7292 \\ 0.6035 & 0.4093 & -0.6843 \\ -0.5561 & 0.8311 & 0.0066 \end{bmatrix} \tag{18-70}$$

$$\Sigma = \begin{bmatrix} 1.618 & 0 & 0 \\ 0 & 1.143 & 0 \\ 0 & 0 & 0.0097 \end{bmatrix} \tag{18-71}$$

$$\mathbf{V} = \begin{bmatrix} 0.0541 & 0.9984 & 0.0151 \\ 0.9985 & -0.0540 & -0.0068 \\ -0.0060 & 0.0154 & -0.9999 \end{bmatrix} \tag{18-72}$$

$$\text{CN} = \frac{\sigma_1}{\sigma_3} = \frac{1.618}{0.0097} = 166.5$$

The RGA is as follows:

$$\Lambda = \begin{bmatrix} -2.4376 & 3.0241 & 0.4135 \\ 1.2211 & -0.7617 & 0.5407 \\ 2.2165 & -1.2623 & 0.0458 \end{bmatrix} \tag{18-73}$$

Note that a preliminary pairing based on the RGA would be y_1-u_2, y_2-u_3, y_3-u_1. However, two of the singular values (σ_1, σ_2) are of the same magnitude, but σ_3 is much smaller. The CN value suggests that only two output variables can be controlled effectively. If we eliminate one input variable and one output variable, the condition number, σ_1/σ_2, can be recalculated, as shown in Table 18.3.

In order to assess which two inputs should be used and which measured variables should be controlled, Table 18.3 shows nine pairings, along with CN and λ. Based on their having small condition numbers and acceptable values of λ, pairings 4 (y_1-u_2, y_3-u_1) and 7 (y_2-u_2, y_3-u_1) appear to be the most promising ones. In both cases, u_1 and u_2 are the preferred set of inputs, probably because the gain matrix has small entries in the column corresponding to u_3. Note also that λ is acceptable for pairings 5 and 9, but the CN is very high for each case, thus ruling them out. Pairing 4 is consistent that the original 3 × 3 RGA in (18-73), but pairing 7 is not. The final choice of either pairing 4 or 7 should be based on dynamic simulation of the closed-loop systems.

Table 18.3 CN and λ for Different 2×2 Pairings, Example 18.7

Pairing Number	Controlled Variables	Manipulated Variables	CN	λ
1	y_1, y_2	u_1, u_2	184	39.0
2	y_1, y_2	u_1, u_3	72.0	0.552
3	y_1, y_2	u_2, u_3	133	0.558
4	y_1, y_3	u_2, u_1	1.51	0.640
5	y_1, y_3	u_1, u_3	69.4	0.640
6	y_1, y_3	u_2, u_3	139	1.463
7	y_2, y_3	u_2, u_1	1.45	0.634
8	y_2, y_3	u_1, u_3	338	3.25
9	y_2, y_3	u_2, u_3	67.9	0.714

There may be considerable value in using the various measures discussed in this section (RGA and SVA) for plantwide control analysis, where the number of process variables can be very large. Screening approaches can identify possible control configurations, which reduces the number of dynamic simulation cases to a manageable number (McAvoy and Braatz, 2003).

This topic is currently an open research area; more details on plantwide control are provided in Appendices G and H.

18.4 TUNING OF MULTILOOP PID CONTROL SYSTEMS

Multiloop (decentralized) PID control systems are often used to control interacting multiple-input, multiple-output processes because they are easy to understand and require fewer parameters to tune than more general multivariable controllers. Another advantage of multiloop controllers is that *loop failure tolerance* of the resulting control system can be easily checked. Loop failure tolerance is important in practical applications, because some loops may be placed in manual mode, or the manipulated variables of some loops can be saturated at their limits so they cannot be changed to avoid instability.

We consider four types of tuning methods for multiloop PID control systems:

1. Detuning method (Luyben, 1986)
2. Sequential loop tuning method (Hovd and Skogestad, 1994)
3. Independent loop method (Grosdidier and Morari, 1987; Skogestad and Morari, 1989)
4. Relay auto-tuning (Shen and Yu, 1994)

In the detuning method, each controller of the multiloop control system is first designed, ignoring process interactions from the other loops. Then interactions are taken into account by detuning each controller until a performance criterion is met. Typically, controller settings are made more conservative; that is, the gains are decreased, and the integral times are increased in one or more loops. For example, in a 2×2 control problem, one could choose to detune the control loop for the less important controlled variable. The *biggest log-modulus tuning* (BLT) method proposed by Luyben (1986) is a well-known detuning method. Initially, the Ziegler-Nichols settings (Section 12.5) are determined for each control loop (K_{ZN}, τ_{ZN}). For PI controllers, the detuning is performed by adjusting a single parameter F that adjusts the controller gain and the integral time as follows.

$$K_c = \frac{K_{ZN}}{F} \qquad \tau_I = F\tau_{ZN} \tag{18-74}$$

The detuning parameter F is increased from one until the biggest log-modulus reaches a specified value. The biggest log modulus is a measure of how far the closed-loop system is from being unstable (Luyben, 1986).

In the sequential loop tuning method (Hovd and Skogestad, 1994), the controller for a selected input-output pair is tuned and this loop is closed. Then a second controller is tuned for a second pair while the first control loop remains closed, and so on. Because each controller can be tuned using SISO methods, it is simpler than the detuning method. A disadvantage is that the controller settings depend strongly on which loop is tuned first. Usually, the fastest loops are tuned first. In the independent loop method, each controller is designed based on the corresponding open-loop and closed-loop transfer functions, while satisfying inequality constraints on the process interactions (Grosdidier and Morari, 1987; Skogestad and Morari, 1989). Then the IMC approach is used to obtain PID controller settings for each loop, usually with a single tuning parameter for each loop.

Relay auto-tuning can also be used to tune multiloop control systems. The loops can be tuned in a sequential manner or simultaneously. Shen and Yu (1994) use relay auto-tuning of each single loop in succession. For a 2 × 2 system, they first put one loop in manual while tuning the second loop. Then with the first loop in automatic, they auto-tune the second loop. Then the first loop is tuned again with the second controller in automatic. This procedure is repeated until convergence occurs.

18.5 DECOUPLING AND MULTIVARIABLE CONTROL STRATEGIES

In this section, we discuss several strategies for reducing control-loop interactions.

18.5.1 Decoupling Control

One of the early approaches to multivariable control is *decoupling control*. By adding additional controllers called *decouplers* to a conventional multiloop configuration, the design objective to reducing control loop interactions can be realized. In principle, decoupling control schemes can reduce control loop interactions, and a set-point change for one controlled variable has little or no effect on the other controlled variables. In practice, these benefits may not be fully realized due to imperfect process models. A typical decoupler is based on a simple process model that can be either a steady-state or dynamic model.

One type of decoupling control system for a 2 × 2 process and a 1-1/2-2 control configuration is shown in Fig. 18.9. Note that four controllers are used: two conventional feedback controllers, G_{c1} and G_{c2}, plus two decouplers, T_{12} and T_{21}. The input signal to each decoupler is the output signal from a feedback controller. In Fig. 18.9, the transfer functions for the transmitters, disturbances, and final control elements have been omitted for the sake of simplicity. Skogestad and Postlethwaite (2005) have discussed the more general case where these transfer functions are included. The decouplers are designed to compensate for undesirable process interactions. For example, decoupler T_{21} can be designed so as to cancel Y_{21}, which arises from the undesirable process interaction between U_1 and Y_2.

This cancellation will occur at the Y_2 summer if the decoupler output U_{21} satisfies

$$G_{p21}\,U_{11} + G_{p22}\,U_{21} = 0 \tag{18-75}$$

Substituting for $U_{21} = T_{21}\,U_{11}$ and factoring gives

$$(G_{p21} + G_{p22}\,T_{21})\,U_{11} = 0 \tag{18-76}$$

Note that $U_{11}(s) \neq 0$, because U_{11} is a controller output that is time dependent. Thus, to satisfy Eq. 18-76, it follows that

$$G_{p21} + G_{p22}\,T_{21} = 0 \tag{18-77}$$

Solving for T_{21} gives an expression for the ideal decoupler,

$$T_{21} = -\frac{G_{p21}}{G_{p22}} \tag{18-78}$$

Similarly, a design equation for T_{12} can be derived by imposing the requirement that U_{22} have no net effect on Y_1. Thus, the compensating signal U_{12} and the process interaction due to G_{p12} should cancel at the Y_1 summer. Similar to 18-78, the ideal decoupler is given by

$$T_{12} = -\frac{G_{p12}}{G_{p11}} \tag{18-79}$$

The ideal decouplers in Eqs. 18-78 and 18-79 are very similar to the ideal feedforward controller in Eq. 15-21 with $G_t = G_v = 1$. In fact, one can interpret a decoupler as a type of feedforward controller with an input signal

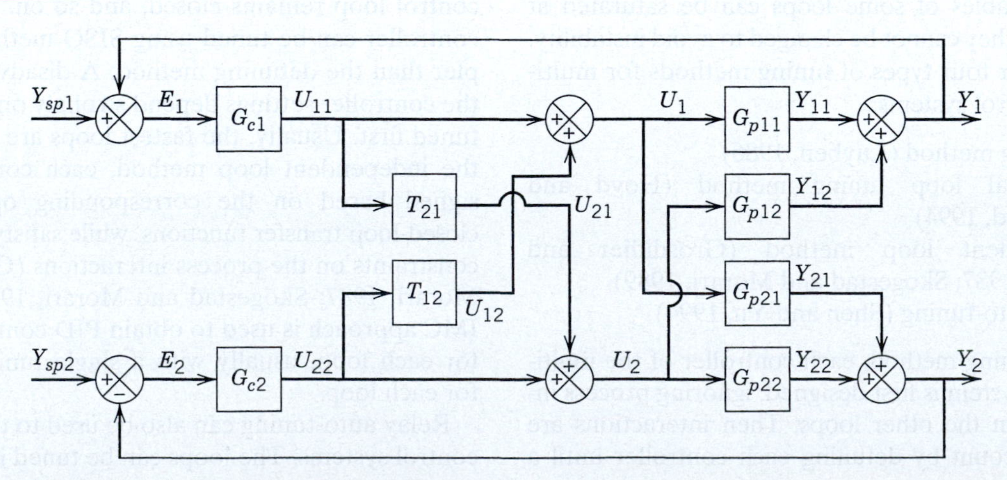

Figure 18.9 A decoupling control system.

that is a manipulated variable rather than a disturbance variable. Recall from Chapter 15 that the ideal feedforward controller may not be physically realizable. Similarly, ideal decouplers are not always physically realizable and they may suffer from model error; hence a steady state model may be assumed for simplicity—e.g.,

$$T_{21} = -\frac{K_{p21}}{K_{p22}} \quad \text{and} \quad T_{12} = -\frac{K_{p12}}{K_{p11}} \tag{18-80}$$

This is called static decoupling (McAvoy, 1979). Another simplification is called partial decoupling, where only one decoupler is implemented. A disadvantage of static decoupling is that control loop interactions still exist during transient conditions; e.g., a set-point change for Y_1 will tend to upset Y_2. However, if the dynamics of the two loops are similar, static decoupling can produce excellent transient responses. For most multivariable control problems, model predictive control (Chapter 20) is the preferred technique, and decoupling is rarely implemented in new control applications in industry.

18.5.2 General Multivariable Control Techniques

The term *multivariable control* refers generically to the class of control strategies in which each manipulated variable is adjusted on the basis of the errors in all of the controlled variables, rather than the error in a single variable, as is the case for multiloop control. For example, a simple multivariable proportional control strategy for a 2 × 2 process could have four controller gains and the following form:

$$u_1(t) = K_{c11}e_1(t) + K_{c12}e_2(t) \tag{18-81}$$

$$u_2(t) = K_{c21}e_1(t) + K_{c22}e_2(t) \tag{18-82}$$

If $K_{c12} = K_{c21} = 0$, the multivariable control system reduces to a 1-1/2-2 multiloop control system, because each manipulated variable is adjusted based on a single error signal. Similarly, a 1-2/2-1 multiloop control system results if $K_{c11} = K_{c22} = 0$. Thus, multiloop control is a special case of the more general multivariable control. Note that the decoupling control scheme shown in Fig. 18.9 is also a multivariable control strategy, because each manipulated variable depends on both error signals.

Equations 18-81 and 18-82 illustrate multivariable proportional control for a 2 × 2 process. Multivariable control strategies can also be developed that include integral, derivative, and feedforward control action. The books by Goodwin et al. (2001) and Skogestad and Postlethwaite (2005) provide additional information. In this text, we emphasize the use of model predictive control as the method of choice for designing multivariable controllers, as discussed in Chapter 20.

18.6 STRATEGIES FOR REDUCING CONTROL LOOP INTERACTIONS

In Section 18.1 we described how process interactions between manipulated and controlled variables can result in undesirable control loop interactions. When control loop interactions are a problem, a number of alternative strategies are available:

1. Select different manipulated or controlled variables.
2. Re-tune one or more multiloop PID controllers, taking process interactions into account.
3. Consider a more general multivariable control method, such as model predictive control.

Strategy 1 is illustrated in the next section, while strategy 2 was considered in previous sections. Strategy 3 is considered in Chapter 20.

18.6.1 Selection of Different Manipulated or Controlled Variables

For some control problems, loop interactions can be significantly reduced by choosing alternative controlled and manipulated variables. For example, the new controlled or manipulated variable could be a simple function of the original variables such as a sum, difference, or ratio (Weber and Gaitonde, 1982; McAvoy, 1983; Waller and Finnerman, 1987). Industrial distillation columns have been controlled using simple, nonlinear functions of x_D and x_B as the controlled variables, rather than x_D and x_B (e.g., Weber and Gaitonde, 1982). The selection of appropriate manipulated and controlled variables that reduce control loop interactions tends to be an art rather than a science.

EXAMPLE 18.8

For the blending system of Example 18.3, choose a new set of manipulated variables that will reduce control loop interactions by making $\lambda = 1$.

SOLUTION

From the expression for the relative gain in Eq. 18-34, it is clear that $\lambda = 1$ if K_{12} and/or $K_{21} = 0$. Thus, we want to choose manipulated variables so that the steady-state gain matrix has a zero for at least one of the off-diagonal elements. Inspection of the process model in Eqs. 18-41 and 18-42 suggests that suitable choices for the manipulated variables are $u_1 = w_A + w_B$ and $u_2 = w_A$. Substitution into the process model gives an equivalent model in terms of the new variables:

$$w = u_1 \tag{18-83}$$

$$x = \frac{u_2}{u_1} \tag{18-84}$$

Linearizing (18-84) gives the gain matrix $\boldsymbol{K}$ in Eq. (18-28) where

$$\boldsymbol{K} = \begin{bmatrix} 1 & 0 \\ -\dfrac{u_2}{u_1^2} & \dfrac{1}{u_1} \end{bmatrix} \tag{18-85}$$

vectors $\boldsymbol{y}$ and $\boldsymbol{u}$ are defined as $\boldsymbol{y} = [w, x]^T$ and $\boldsymbol{u} = [u_1, u_2]^T$. Because w depends on u_1 but not u_2, the only feasible controller pairing is w-u_1 and x-u_2. From Eqs. 18-85 and 18-34, it follows that $K_{12} = 0$ and $\lambda = 1$. Because $K_{21} \neq 0$, there will be a one-way interaction, with the w-u_1 control loop generating disturbances that affect the x-u_2 loop but not vice versa.

McAvoy (1983, p. 136) suggests alternative manipulated variables, namely, $u_1 = w_A + w_B$ and $u_2 = w_A/(w_A + w_B)$. This choice is motivated by the process model in Eqs. 18-41 and 18-42 and means that the controlled variables are identical to the manipulated variables! Thus, K is the identity matrix, $\lambda = 1$, and the two control loops do not interact at all. This situation is fortuitous, and also unusual, because it is seldom possible to choose manipulated variables that are, in fact, the controlled variables.

SUMMARY

In this chapter we have considered control problems with multiple inputs (manipulated variables) and multiple outputs (controlled variables), with the main focus on using a set of single-loop controllers (multiloop control). Such MIMO control problems are more difficult than SISO control problems because of the presence of process interactions. Process interactions can produce undesirable control loop interactions for multiloop control. If these control loop interactions are unacceptable, then one can either detune one or more of the control loops, choose a different set of manipulated or controlled variables, or employ multivariable control. Model-based multivariable control strategies such as model predictive control can provide significant improvements over conventional multiloop control, as discussed in Chapter 20.

REFERENCES

Arkun, Y., and J. J. Downs, A General Method to Calculate Input-Output Gains and the Relative Gain Array for Integrating Processes, *Comput. Chem. Eng.*, **14**, 1101 (1990).

Arkun, Y., V. Manousiouthakis, and A. Palazoglu, Robustness Analysis of Process Control Systems: A Case Study of Decoupling Control in Distillation, *IEC Process Des. Dev.*, **23**, 93 (1984).

Balchen, J. G., and K. I. Mummé, *Process Control: Structures and Applications*, Van Nostrand Reinhold, New York, 1988.

Bjorck, A., *Numerical Methods for Least Squares Problems*, SIAM, Philadelphia, PA, 1996.

Bristol, E. H., On a New Measure of Interactions for Multivariable Process Control, *IEEE Trans. Auto. Control,* **AC-11**, 133 (1966).

Campo, P. J. and M. Morari, Achievable Closed-loop Properties of Systems under Decentralized Control: Conditions Involving the Steady State Gain, *IEEE Trans. Auto. Control*, **AC-39**, 932 (1994).

Chang, J., and C. C. Yu, The Relative Gain for Non-Square Multivariable Systems, *Chem. Engr. Sci.*, **45**, 1309 (1990).

Chen, D. and D. E. Seborg, Relative Gain Array Analysis for Uncertain Process Models, *AIChE J.*, **48**, 302 (2002).

Fagervik, K. C., K. V. Waller, and L. G. Hammarström, Two-Way or One-Way Decoupling in Distillation?, *Chem. Eng. Commun.*, **21**, 235 (1983).

Gagnepain, J. P., and D. E. Seborg, Analysis of Process Interactions with Applications to Multiloop Control System Design, *IEC Process Des. Dev.*, **21**, 5 (1982).

Goodwin, G. C., S. F. Graebe, and M. E. Salgado, *Control System Design*, Prentice Hall, Upper Saddle River, NJ, 2001.

Grosdidier, P., M. Morari, and B. R. Holt, Closed-Loop Properties from Steady-State Information, *IEC Fund.*, **24**, 221 (1985).

Grosdidier, P., and M. Morari, Interaction Measures for Systems under Decentralized Control, *Automatica*, **22**, 309 (1986).

Grosdidier, P., and M. Morari, A Computer-Aided Methodology for the Design of Decentralized Controllers, *Comput. Chem. Eng.*, **11**, 423 (1987).

Hovd, M., and S. Skogestad, Sequential Design of Decentralized Controllers, *Automatica*, **30**, 1601 (1994).

Laub, A. J., *Matrix Analysis for Scientists and Engineers*, SIAM, Philadelphia, PA, 2004.

Luyben, W. L., Simple Method for Tuning SISO Controllers in Multivariable Systems, *IEC Process Des. Dev.*, **25**, 654 (1986).

McAvoy, T. J., Steady-State Decoupling of Distillation Columns, *IEC Fund*, **18**, 269 (1979).

McAvoy, T. J., Connection between Relative Gain and Control Loop Stability and Design, *AIChE J.*, **27**, 613 (1981).

McAvoy, T. J., *Interaction Analysis*, ISA, Research Triangle Park, NC, 1983.

McAvoy, T. J., and R. D. Braatz, Contollability of Processes with Large Singular Values, *Ind. Eng. Chem. Res.*, **42**, 6155 (2003).

Niederlinski, A., A Heuristic Approach to the Design of Linear Multivariable Interacting Control Systems, *Automatica*, **7**, 691 (1971).

Nisenfeld, A. E., and H. M. Schultz, Interaction Analysis in Control System Design, Paper 70-562, *Advances in Instrum.* **25**, Pt. 1, ISA, Pittsburgh, PA, 1971.

Roat, S. D., J. J. Downs, E. F. Vogel, and J. E. Doss, The Integration of Rigorous Dynamic Modeling and Control System Synthesis for Distillation Columns: An Industrial Approach, *Chemical Process Control CPC-III*, M. Morari and T. J. McAvoy (Eds.), CACHE-Elsevier, New York, 1986, p. 99.

Shen, S. H., and C.-C. Yu, Use of Relay-feedback Test for Automatic Tuning of Multivariable Systems, *AIChE J.*, **40**, 627 (1994).

Shinskey, F. G., *Process Control Systems*, 4th ed., McGraw-Hill, New York, 1996, Chapter 8.

Skogestad, S., Robust Control, *Practical Distillation Control*, W. L. Luyben (Ed.), Van Nostrand-Reinhold, New York, 1992, Chapter 14.

Skogestad, S., and M. Morari, Robust Performance of Decentralized Control Systems by Independent Design, *Automatica*, **25**, 119 (1989).

Skogestad, S., and I. Postlethwaite, *Multivariable Feedback Control*, 2d ed., Wiley, New York, 2005.

Stanley, G., M. Marino-Galarraga, and T. J. McAvoy. Shortcut Operability Analysis 1. The Relative Disturbance Gain, *IEC Process Des. Dev.*, **24**, 1181 (1985).

Strang, G., *Linear Algebra and Its Applications*, 4th ed., Brooks-Cole, Independence, KY, 2005.

Tung, L. S., and T. F. Edgar, Analysis of Control-Output Interactions in Dynamic Systems, *AIChE J.*, **27**, 690 (1981).

Waller, M., J. B. Waller, and K. V. Waller, Decoupling Revisited, *Ind. Eng. Chem. Res.* **42**, 4575 (2003).

Waller, K. V., and D. H. Finnerman, On Using Sums and Differences to Control Distillation, *Chem. Eng. Commun.*, **56**, 253 (1987).

Weber, R., and N. Y. Gaitonde, Non-Interactive Distillation Tower Analyzer Control, *Proc. Amer. Control Conf.*, Arlington, VA, 1982, p. 87.

Weischedel, K., and T. J. McAvoy, Feasibility of Decoupling in Conventionally Controlled Distillation Columns, *IEC Fund.*, **19**, 379 (1980).

Witcher, M. E., and T. J. McAvoy, Interacting Control Systems: Steady-State and Dynamic Measurement of Interaction, *ISA Trans.*, **16** (3), 35 (1977).

Wood, R. K., and M. W. Berry, Terminal Composition Control of a Binary Distillation Column, *Chem. Eng. Sci.*, **28**, 1707 (1973).

Woolverton, P. E., How to Use Relative Gain Analysis in Systems with Integrating Variables, *InTech*, **27** (9), 63 (1980).

EXERCISES

18.1 Luyben and Vinante (*Kem. Teollisuus*, **29**, 499 (1972)) developed a distillation column model relating temperatures on the 4th and 17th trays from the bottom of the column (T_4, T_{17}) to the reflux ratio R and the steam flow rate to the reboiler S:

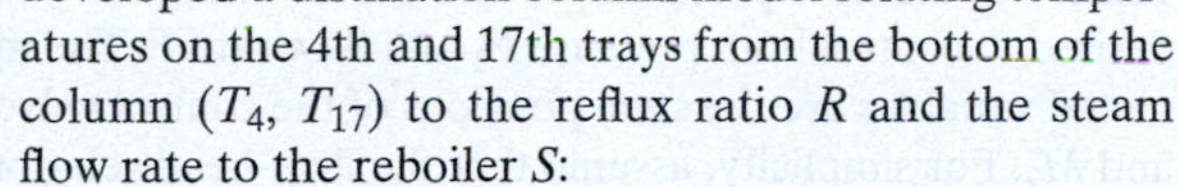

$$\begin{bmatrix} T_{17}(s) \\ T_4(s) \end{bmatrix} = \begin{bmatrix} \dfrac{-2.16e^{-s}}{8.25s+1} & \dfrac{1.26e^{-0.3s}}{7.05s+1} \\ \dfrac{-2.75e^{-1.8s}}{8.25s+1} & \dfrac{4.28e^{-0.35s}}{9.0s+1} \end{bmatrix} \begin{bmatrix} R(s) \\ S(s) \end{bmatrix}$$

Compare the closed-loop set-point changes that result from the T_{17}–R/T_4–S pairing and the Ziegler-Nichols continuous cycling method in Chapter 12. Consider two cases:

(a) A set-point change is made in each loop with the other loop in manual.

(b) The set-point changes are made with both controllers in automatic.

18.2 Derive an expression for the characteristic equation for the 1-2/2-1 configuration in Fig. 18.3*b*. Simplify and interpret this equation for the special situation where either G_{p11} or G_{p22} is zero.

18.3 Derive equivalent closed-loop formulas to (18-9) through (18-11) for the case where there are sensor transfer functions (G_{m1}, G_{m2}) for the outputs (y_1, y_2).

18.4 Consider the stirred-tank heating system of Fig. 6.14 and assume that the manipulated inputs are w_h and w. Suggest a reasonable pairing for a multiloop control scheme and justify your answer.

18.5 For the in-line blending system of Example 18.8, draw block diagrams for two multiloop control schemes:

(a) The standard scheme for $x = 0.4$.

(b) The less interacting scheme where $u_1 = w_A + w_B$ and $u_2 = w_A$.

You may assume that each transmitter and control valve can be represented by a first-order transfer function, and that PI controllers are utilized.

18.6 A conventional multiloop control scheme consisting of two PI controllers is to be used to control the product compositions x_D and x_B of the distillation column shown in Fig. 18.1*b*. The manipulated variables are the reflux flow rate R and the steam flow rate to the reboiler S. Experimental data for a number of steady-state conditions are summarized below. Use this information to do the following:

(a) Calculate the RGA and determine the recommended pairing between controlled and manipulated variables.

(b) Does this pairing seem appropriate from dynamic considerations? Justify your answer.

Table E18.6

Run	R (lb/min)	S (lb/min)	x_D	x_B
1	125	22	0.97	0.04
2	150	22	0.95	0.05
3	175	22	0.93	0.06
4	150	20	0.94	0.06
5	150	24	0.96	0.04

18.7 For the Wood-Berry distillation column model in Example 18.1:

(a) Which pairing of controlled and manipulated variables would you recommend based on steady-state considerations?

(b) Which pairing based on dynamic considerations? Justify your answers.

18.8 A dynamic model of the stirred-tank heating system in Fig. 6.14 was derived in Chapter 6. Use this model to do the following:

(a) Derive an expression for the relative gain array.

(b) Design an ideal decoupling control system, assuming that the transmitters and control valves have negligible dynamics.

(c) Are these decouplers physically realizable? If not, suggest appropriate modifications.

18.9 A binary distillation column has three tray temperature measurements (17th, 24th, 30th trays) that can be used as possible controlled variables. Controlling temperature is equivalent to controlling composition. Step testing gives the following steady-state input-output relationships (u_1 = steam pressure in reboiler; u_2 = reflux ratio):

$$T'_{17} = 1.5u_1 + 0.5u_2 \quad (1)$$

$$T'_{24} = 2.0u_1 + 1.7u_2 \quad (2)$$

$$T'_{30} = 3.4u_1 + 2.9u_2 \quad (3)$$

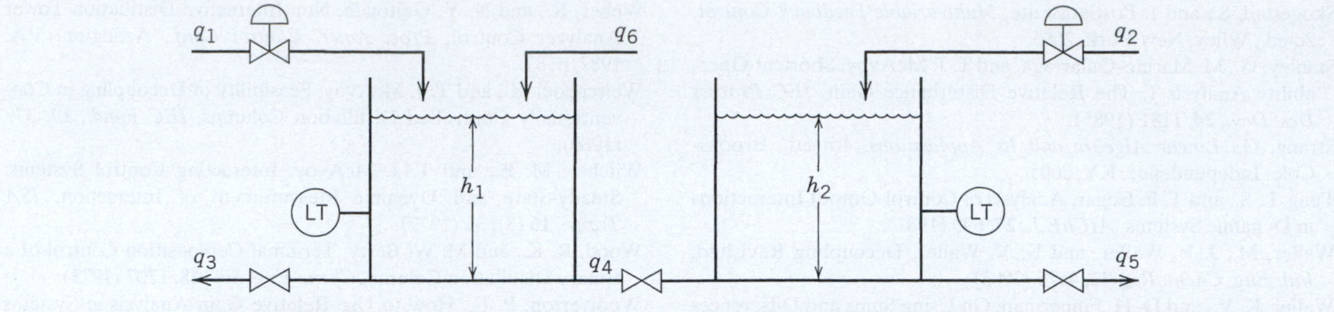

Figure E18.10

All variables are deviation variables. Select the 2×2 control system that has the most desirable interactions, as determined by the RGA (note that there are three possible 2×2 control configurations). Explain why the combination of T_{24} and T_{30} is the least desirable controlled variable set, based on analyzing Eqs. (2) and (3) and the resulting determinant.

18.10 For the liquid storage system shown in Fig. E18.10, it is desired to control liquid levels h_1 and h_2 by adjusting volumetric flow rates q_1 and q_2. Flow rate q_6 is the major disturbance variable. The flow-head relations are given by

$$q_3 = C_{v1}\sqrt{h_1} \quad q_5 = C_{v2}\sqrt{h_2} \quad q_4 = K(h_1 - h_2)$$

where C_{v1}, C_{v2}, and K are constants.

(a) Derive an expression for the relative gain array for this system.

(b) Use the RGA to determine the recommended pairing of controlled and manipulated variables for the following conditions:

Parameter Values

$K = 3$ gal/min ft

$C_{v1} = 3$ gal/min ft$^{0.5}$

$C_{v2} = 3.46$ gal/min ft$^{0.5}$

$D_1 = D_2 = 3.5$ ft (tank diameters)

Nominal Steady-State Values

$\bar{h}_1 = 4$ ft, $\bar{h}_2 = 3$ ft

18.11 For the liquid-level storage system in Exercise 18.10:

(a) Derive a transfer function model of the form,

$$Y(s) = G_p(s)U(s) + G_d(s)\,D(s)$$

where D is the disturbance variable and G_d is a 2×1 matrix of disturbance transfer functions.

(b) Draw a block diagram for a multiloop control system based on the following pairing: h_1–q_1/h_2–q_2. Do not attempt to derive transfer functions for the transmitters, control valves, or controllers.

18.12 For the flow-pressure process shown in Fig. E18.12, it is desired to control both pressure P_1 and flow rate F. The manipulated variables are the stem positions of the control valves, M_1 and M_2. For simplicity, assume that the flow-head relations for the two valves are given by

$$F = 20M_1(P_0 - P_1)$$
$$F = 30M_2(P_1 - P_2)$$

The nominal steady-state conditions are $F = 100$ gal/min, $P_0 = 20$ psi, $P_1 = 10$ psi, and $P_2 = 5$ psi. Use the RGA approach to determine the best controller pairing.

18.13 A blending system is shown in Fig. E18.13. Liquid level h and exit composition c_3 are to be controlled by adjusting flow rates q_1 and q_3. Based on the information below, do the following:

(a) Derive the process transfer function matrix, $\boldsymbol{G_p(s)}$.

(b) If a conventional multiloop control system is used, which controller pairing should be used? Justify your answer.

(c) Obtain expressions for the ideal decouplers $T_{21}(s)$ and $T_{12}(s)$ in the configuration of Fig. 18.9.

Available Information

(i) The tank is 3 ft in diameter and is perfectly mixed.

(ii) Nominal steady-state values are

$\bar{h} = 3$ ft $\quad \bar{q}_3 = 20$ ft^3/min

$\bar{c}_1 = 0.4$ mole/ft^3 $\quad \bar{c}_2 = 0.1$ mole/ft^3

$\bar{q}_1 = 10$ ft^3/min

(iii) The density of each process stream remains constant at $\rho = 60$ lb/ft^3.

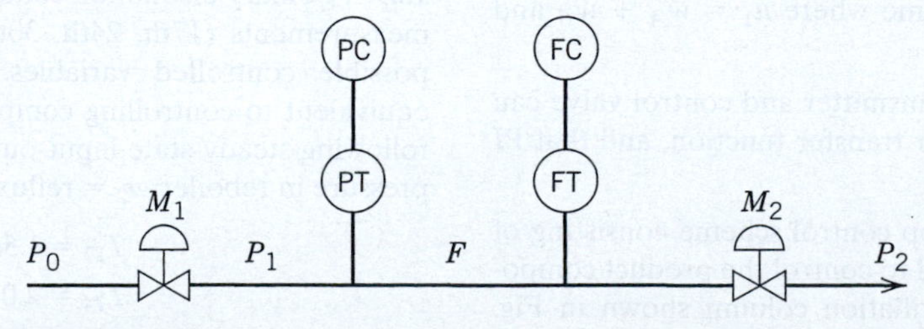

Figure E18.12

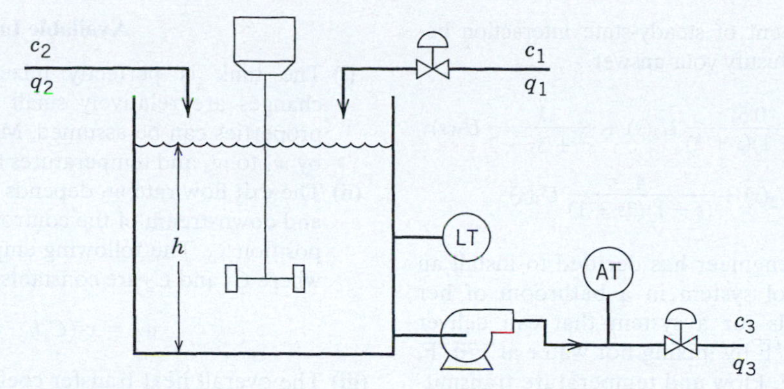

Figure E18.13

(iv) The primary disturbance variable is flow rate q_2.

(v) Inlet compositions c_1 and c_2 are constant.

(vi) The transmitter characteristics are approximated by the following transfer functions with time constants in minutes:

$$G_{m11}(s) = \frac{4}{0.1s + 1} \quad \text{(mA/ft)}$$

$$G_{m22}(s) = \frac{100}{0.2s + 1} \quad \text{(mA ft}^3\text{/mole)}$$

(vii) Each control valve has a gain of 0.15 ft^3/min mA and a time constant of 10 s.

8.14 (Modified from McAvoy, 1983). A decanter shown in Fig. E18.14 is used to separate a feed that consists of two completely immiscible liquids, a light component and a heavy component. Because of the large difference in their densities, the two components form separate liquid phases very rapidly after the feed enters the decanter. The decanter is always full of liquid. The level of the interface I between the two liquid phases is measured by a dp cell. Each liquid flow rate can be adjusted by using a control valve, which is connected to a standard PI controller. The control valve equations relate flow rates, pressures, and controller output signals (m_1, m_2, m_3):

$$F_1 = m_1(P_0 - P_1)$$
$$F_2 = m_2(P_1 - P_2)$$
$$F_3 = m_3(P_1 - P_3)$$

Using the following information, propose a pairing of controlled and manipulated variables for a conventional multiloop control configuration based on *physical* arguments. It is *not* necessary to calculate a RGA.

Available Information

(a) Pressures P_0 and P_2 are constant:

$$P_0 = 250 \text{ psi} \quad P_2 = 30 \text{ psi}$$

(b) The feed composition can vary. The nominal value is $w_H = 0.99$, where w_H is the weight fraction of the heavy component.

(c) The densities of the pure components are

$$\rho_H = 9 \text{ lb/gal} \quad \rho_L = 3 \text{ lb/gal}$$

(d) At the nominal steady state,

$$F_1 = 2093 \text{ gal/min}, F_2 = 60 \text{ gal/min}, P_1 = 180 \text{ psi}$$

(e) The transmitters and control valves have negligible dynamics compared to the process dynamics.

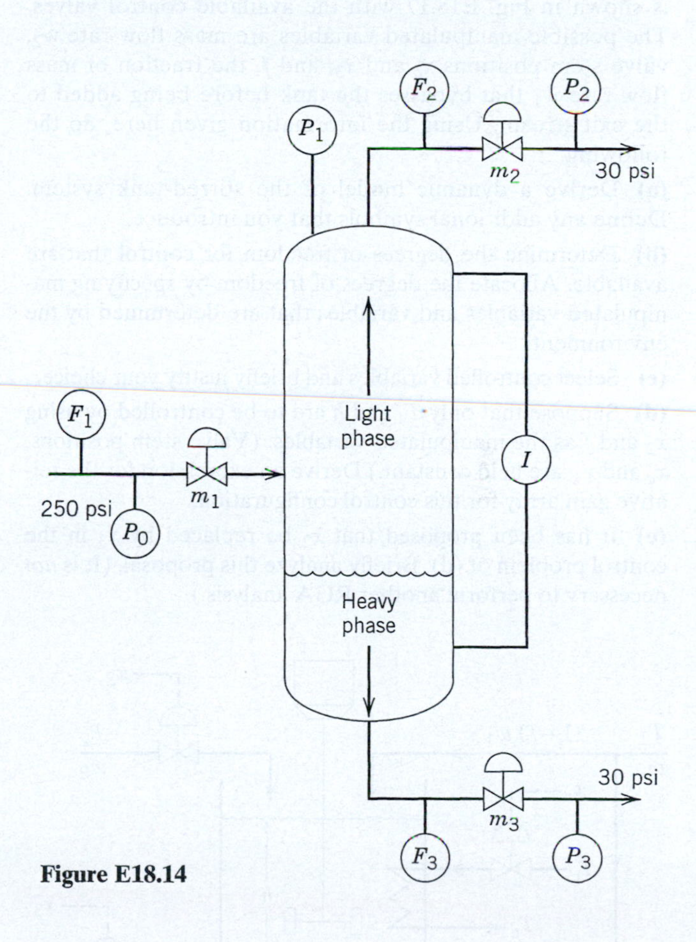

Figure E18.14

18.15 A process to be controlled has two controlled variables Y_1 and Y_2, and three inputs that can be used as manipulated variables, U_1, U_2, and U_3. However, it is desired to use only two of these three manipulated variables in a conventional multiloop feedback control system. Transfer functions for the process are shown below. Which multiloop control configuration will result in the smallest amount of steady-state interaction between inputs and outputs? Justify your answer.

$$Y_1(s) = \frac{3}{2s+1}U_1(s) - \frac{0.5}{(s+1)(s+3)}U_2(s) + \frac{1}{s^2+3s+2}U_3(s)$$

$$Y_2(s) = -10U_1(s) + \frac{2}{s+1}U_2(s) + \frac{4}{(s+1)(3s+1)}U_3(s)$$

18.16 A process control engineer has decided to install an automated shower control system in a bathroom of her mansion. The design calls for a system that can deliver 3 gal/min of water at 110 °F by mixing hot water at 170 °F, with colder water at 80 °F. Flow and temperature transmitters are available along with control valves for adjusting the hot and cold water flow rates.

(a) Calculate the required flow rates of hot and cold water, assuming that the density and heat capacity of water are constant.

(b) Calculate the relative gain array for the system, and recommend a pairing of controlled and manipulated variables.

18.17 A stirred-tank heat exchanger with a bypass stream is shown in Fig. E18.17 with the available control valves. The possible manipulated variables are mass flow rate w_2, valve stem positions x_c and x_3, and f, the fraction of mass flow rate w_1 that bypasses the tank before being added to the exit stream. Using the information given here, do the following:

(a) Derive a dynamic model of the stirred-tank system. Define any additional symbols that you introduce.

(b) Determine the degrees of freedom for control that are available. Allocate the degrees of freedom by specifying manipulated variables and variables that are determined by the environment.

(c) Select controlled variables and briefly justify your choice.

(d) Suppose that only T_4 and h are to be controlled by using x_2 and f as the manipulated variables. (Valve stem positions, x_c and x_3, are held constant.) Derive an expression for the relative gain array for this control configuration.

(e) It has been proposed that x_2 be replaced by x_3 in the control problem of (d). Briefly analyze this proposal. (It is *not* necessary to perform another RGA analysis.)

Available Information

(i) The tank is perfectly mixed, and the temperature changes are relatively small so that constant physical properties can be assumed. Mass flow rates are denoted by w_1 to w_4 and temperatures by T_1 to T_4.

(ii) The exit flow rate w_3 depends on the pressures upstream and downstream of the control valve, and the valve stem position x_3. The following empirical relation is available where C_1 and C_2 are constants:

$$w_3 = x_3(C_1 h - C_2 f w_1)$$

(iii) The overall heat transfer coefficient for the cooling coil U depends on the velocity of the coolant in the line and hence on the valve stem position x_c, according to the relation below where C_3 is a constant:

$$U = C_3 x_c$$

(iv) The pump on the bypass line operates in the "flat part" of the pump curve so that the mass flow rate, fw_1, depends only on the control valve.

(v) The following process variables remain constant: T_1, w_1, T_2, and T_c.

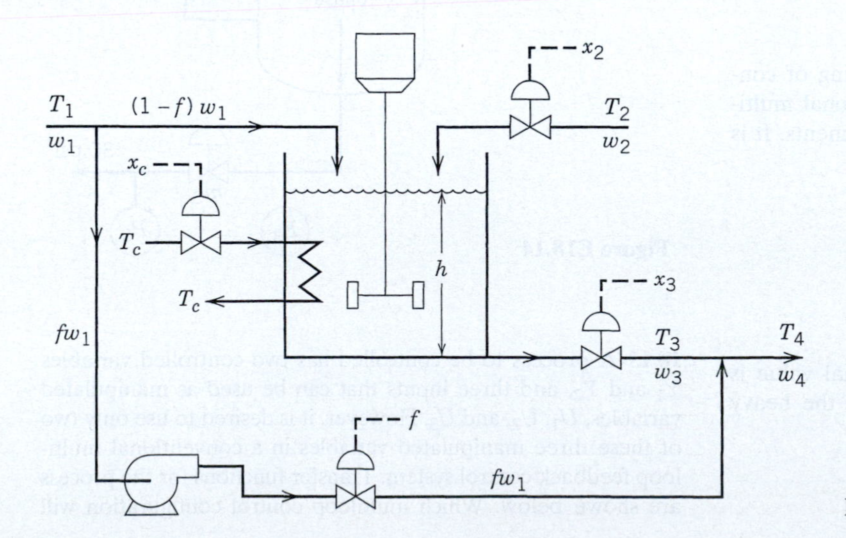

Figure E18.17

18.18 Water (F_1) is blended with a stream F_2 with 40% ethanol to make a whiskey product that is 30% ethanol. Assume F_1 = 6 gal/min and F_2 = 2 gal/min.

(a) Develop a steady-state material balance model for the blending operation. Find the linearized gains for the 2 × 2 transfer function model using Example 18.3 as a guide. F_1 and F_2 are manipulated variables, and the controlled variables are the outlet composition z and the total flow rate F.

(b) Determine the RGA and the preferred pairing for the controllers.

18.19 A schematic diagram for a pH neutralization process is shown in Fig. E18.19. The transfer function matrix and relative gain array are also shown.

(a) Suppose that a multiloop control system consisting of four PID controllers is to be designed. Recommend a pairing of controlled and manipulated variables. Briefly justify your recommendation based on steady-state, dynamic, and physical considerations.

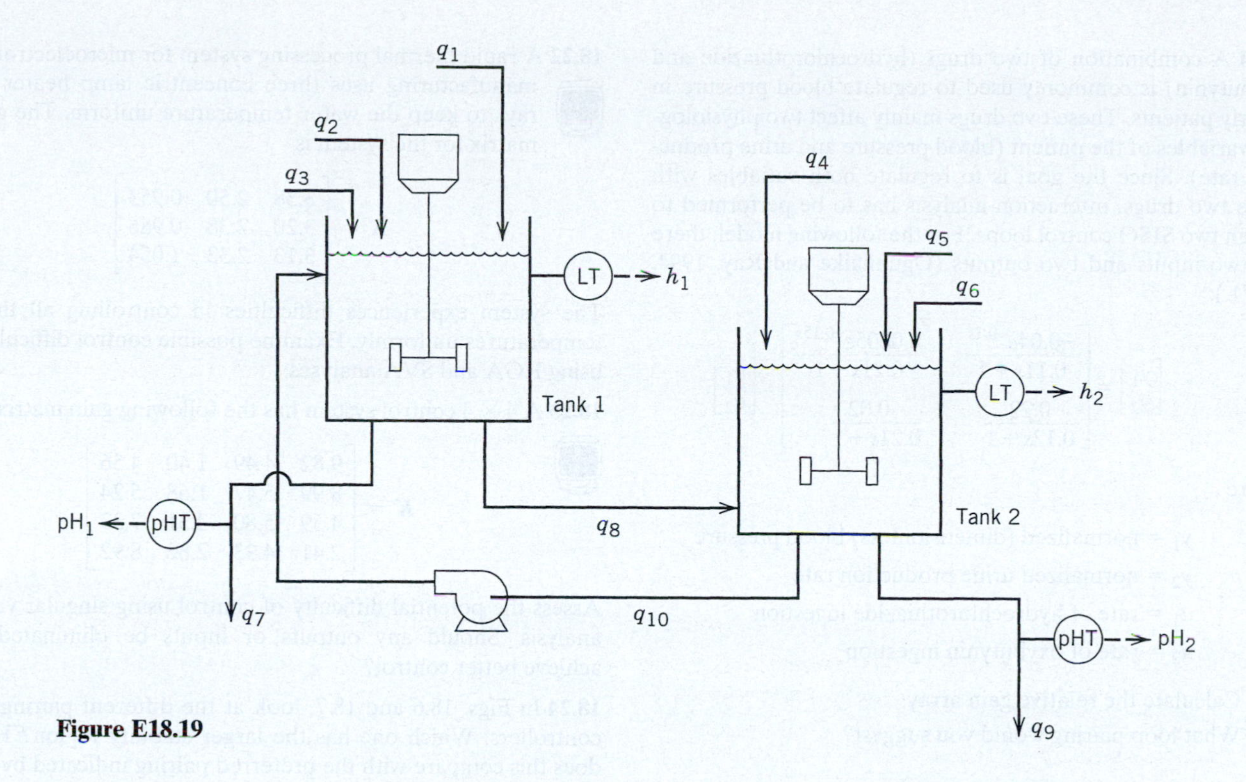

Figure E18.19

(b) Suppose that only pH_2 and h_2 are to be controlled using Q_4 and Q_6 as the manipulated variables (Q_1 and Q_3 are held constant).

(i) What is the RGA for this 2 × 2 control problem?

(ii) What pairing of controlled and manipulated variables do you recommend? (Justify your answer.)

$$\begin{bmatrix} h_1 \\ pH_1 \\ h_2 \\ pH_2 \end{bmatrix} = \begin{bmatrix} \dfrac{0.43\,e^{-0.8s}}{4.32\,s+1} & \dfrac{0.43\,e^{-0.1s}}{3.10\,s+1} & \dfrac{0.23\,e^{-1.0s}}{5.24\,s+1} & \dfrac{0.22\,e^{-0.5s}}{4.42\,s+1} \\ \dfrac{-0.33\,e^{-1.0s}}{2.56\,s+1} & \dfrac{0.32\,e^{-0.5s}}{2.58\,s+1} & \dfrac{-0.20\,e^{-1.8s}}{2.82\,s+1} & \dfrac{0.20\,e^{-0.8s}}{3.30\,s+1} \\ \dfrac{0.22\,e^{-1.1s}}{5.52\,s+1} & \dfrac{0.23\,e^{-0.3s}}{4.49\,s+1} & \dfrac{0.42\,e^{-0.4s}}{3.32\,s+1} & \dfrac{0.41\,e^{-0.1s}}{2.07\,s+1} \\ \dfrac{-0.22\,e^{-1.5s}}{3.24\,s+1} & \dfrac{0.22\,e^{-1.2s}}{2.65\,s+1} & \dfrac{-0.32\,e^{-0.8s}}{2.36\,s+1} & \dfrac{0.32\,e^{-0.4s}}{2.03\,s+1} \end{bmatrix} \begin{bmatrix} Q_1 \\ Q_3 \\ Q_4 \\ Q_6 \end{bmatrix}$$

$$\text{RGA} = \begin{bmatrix} 0.64 & 0.72 & -0.20 & -0.20 \\ 0.87 & 0.85 & -0.35 & -0.35 \\ -0.18 & -0.21 & 0.70 & 0.70 \\ -0.36 & -0.37 & 0.85 & 0.88 \end{bmatrix}$$

18.20 A control scheme is to be developed for the evaporator shown in Fig. E18.20. The feed and product streams are mixtures of a solute and a solvent, while the vapor stream is pure solvent. The liquid level is tightly controlled by manipulating the feed flow rate, w_F. The product composition, x_p, and the feed flow rate, w_F, are to be controlled by manipulating the product flow, w_p, and the steam flow rate, w_s. The evaporator "economy" is approximately constant, because E kg of solvent are evaporated for each kg of steam. The flow rates have units of kg/min, while the compositions are expressed in weight fraction of solute.

Derive an expression for the relative gain array for this system.

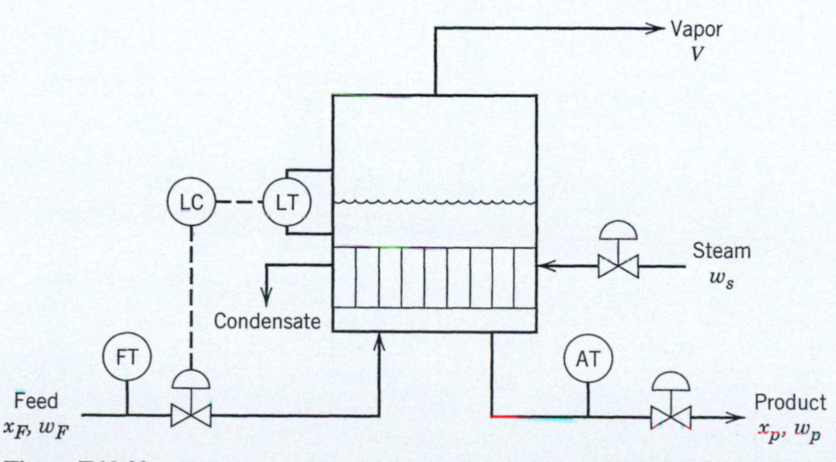

Figure E18.20

18.21 A combination of two drugs (hydrochlorothiazide and oxybutynin) is commonly used to regulate blood pressure in elderly patients. These two drugs mainly affect two physiological variables of the patient (blood pressure and urine production rate). Since the goal is to regulate both variables with these two drugs, interaction analysis has to be performed to design two SISO control loops. For the following model, there are two inputs and two outputs (Ogunnaike and Ray, 1994, p. 771.):

$$\begin{bmatrix} y_1 \\ y_2 \end{bmatrix} = \begin{bmatrix} \dfrac{-0.04e^{-0.1s}}{0.11s+1} & \dfrac{0.0005e^{-0.15s}}{0.21s+1} \\ \dfrac{0.22}{0.12s+1} & \dfrac{-0.02}{0.21s+1} \end{bmatrix} \begin{bmatrix} u_1 \\ u_2 \end{bmatrix}$$

where

y_1 = normalized (dimensionless) blood pressure

y_2 = normalized urine production rate

u_1 = rate of hydrochlorothiazide ingestion

u_2 = rate of oxybutynin ingestion

(a) Calculate the relative gain array.

(b) What loop pairing would you suggest?

18.22 A rapid thermal processing system for microelectronics manufacturing uses three concentric lamp heater arrays to keep the wafer temperature uniform. The gain matrix for the system is

$$\boldsymbol{K} = \begin{bmatrix} 3.38 & 2.50 & 0.953 \\ 3.20 & 2.38 & 0.986 \\ 3.13 & 2.33 & 1.054 \end{bmatrix}$$

The system experiences difficulties in controlling all three temperatures uniformly. Examine possible control difficulties using RGA and SVA analyses.

18.23 A 4 × 4 control system has the following gain matrix:

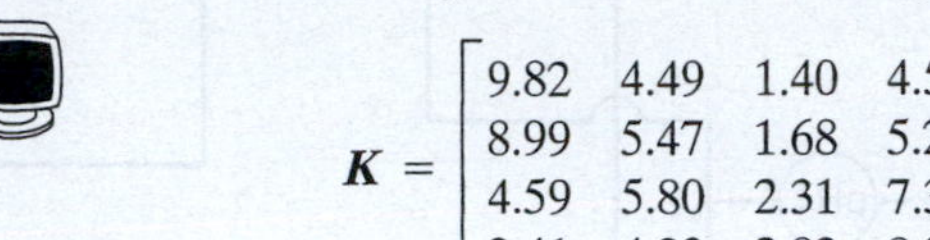

$$\boldsymbol{K} = \begin{bmatrix} 9.82 & 4.49 & 1.40 & 4.56 \\ 8.99 & 5.47 & 1.68 & 5.24 \\ 4.59 & 5.80 & 2.31 & 7.33 \\ 2.41 & 4.33 & 2.82 & 8.92 \end{bmatrix}$$

Assess the potential difficulty of control using singular value analysis. Should any outputs or inputs be eliminated to achieve better control?

18.24 In Figs. 18.6 and 18.7, look at the different pairings of controllers. Which one has the larger stability region? How does this compare with the preferred pairing indicated by the RGA (is it the same or is it different)? Can you suggest a reason for the results (dynamic vs. steady-state effects)?

Chapter 19

Real-Time Optimization

CHAPTER CONTENTS

Previous chapters have considered the development of process models and the design of controllers from an unsteady-state point of view. Such an approach focuses on obtaining reasonable closed-loop responses for set-point changes and disturbances. Up to this point, we have only peripherally mentioned how set points should be specified for the process. The on-line calculation of optimal set points, also called *real-time optimization* (RTO), allows the profits from the process to be maximized (or costs to be minimized) while satisfying operating constraints. The appropriate optimization techniques are implemented in the computer control system. Steady-state models are normally used, rather than dynamic models, because the process is intended to be operated at steady state except when the set point is changed.

This chapter first discusses basic RTO concepts and then describe typical applications to process control. Guidelines for determining when RTO can be advantageous are also presented. Subsequently, set-point selection is formulated as an optimization problem, involving economic information and a steady-state process model. Optimization techniques that are used in the process industries are briefly described. For more information, see textbooks on optimization methodology (Ravindran et al., 2006; Griva et al., 2008; Edgar et al., 2001).

Figure 19.1 is a detailed version of Fig. 1.7, which shows the five levels in the process control hierarchy where various optimization, control, monitoring, and data acquisition activities are employed. The relative position of each block in Fig. 19.1 is intended to be conceptual, because there can be overlap in the functions carried out, and often several levels may utilize the same computing platform. The relative time scale for each level's activity is also shown. Process data (flows, temperatures, pressures, compositions, etc.) as well as *enterprise data*, consisting of commercial and financial information, are used with the methodologies shown to make decisions in a timely fashion. The highest level (planning and scheduling) sets production goals to meet supply and logistics constraints and addresses time-varying capacity and manpower utilization decisions. This *enterprise resource planning* (ERP) and the *supply*

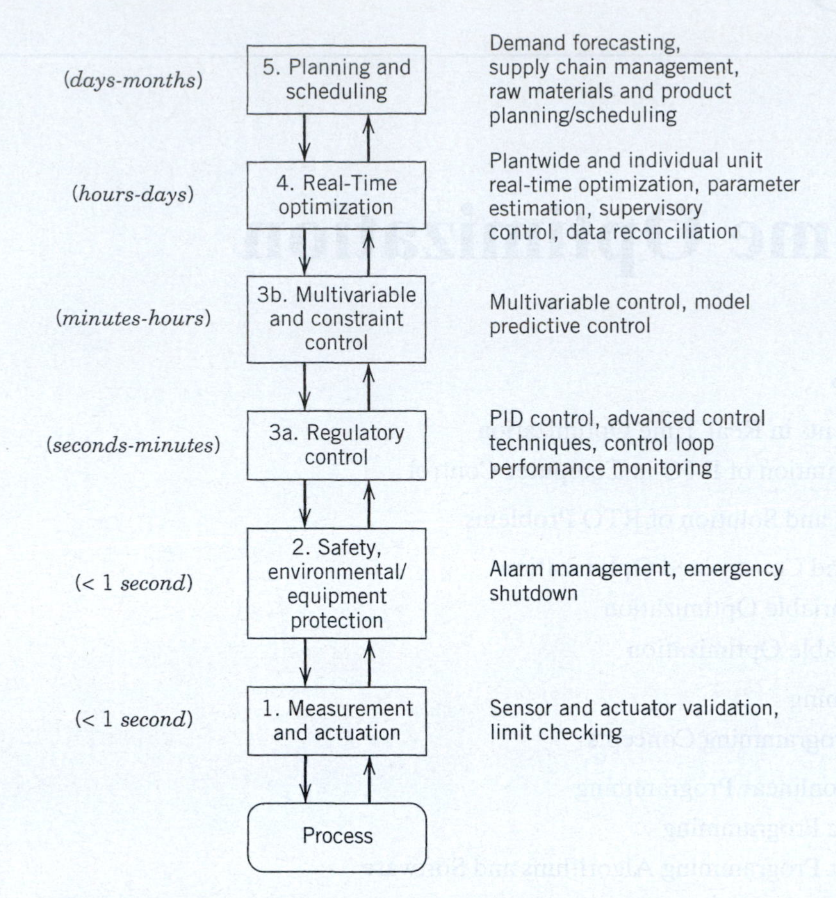

Figure 19.1 The five levels of process control and optimization in manufacturing. Time scales are shown for each level.

chain management in level 5 refer to the links in a web of relationships involving retailing (sales), distribution, transportation, and manufacturing (Bryant, 1993). Planning and scheduling usually operate over relatively long time scales and tend to be decoupled from the rest of the activities in lower levels (Geddes and Kubera, 2000). For example, Baker (1993) and Shobrys and White (2002) indicate that all of the refineries owned by an oil company are usually included in a comprehensive planning and scheduling model. This model can be optimized to obtain target levels and prices for inter-refinery transfers, crude oil and product allocations to each refinery, production targets, inventory targets, optimal operating conditions, stream allocations, and blends for each refinery.

In Level 4, RTO is utilized to coordinate the network of process units and to provide optimal set points for each unit, which is called *supervisory control.* For multivariable control or processes with active constraints, set-point changes are performed in Level 3b (e.g., model predictive control discussed in Chapter 20). For single-loop or multiloop control the regulatory control is performed at Level 3a. Level 2 (safety and environmental/equipment protection) includes activities such as alarm management and emergency shutdowns. Although software implements the tasks shown, there is also a separate hardwired safety system for the plant, as discussed in Chapter 10. Level 1 (process measurement and actuation) provides data acquisition and on-line analysis and actuation functions, including some sensor validation. Ideally, there is bidirectional communication between levels, with higher levels setting goals for lower levels and the lower levels communicating constraints and performance information to the higher levels. The time scale for decision-making at the highest level (planning and scheduling) may be of the order of months, while at lower levels (for example, regulatory control), decisions affecting the process can be made frequently (e.g., in fractions of a second). The main focus of this chapter is on Level 4.

Historically, the focus of optimization in chemical plants has been during the design phase, but since the 1990s this has changed because plant profitability can be enhanced by performing optimization of operating conditions on a repetitive basis. In a large plant, the improved profits attained with RTO can be substantial (Bailey et al., 1993; White, 2010). Optimal operating points can sometimes change markedly from day to day, or even during the course of one day. For example, the price of delivered electrical power can vary by a factor of five from highest to lowest price (due to time-of-day pricing by electrical utilities). Other changes that require periodic optimization of operating conditions include variations in the quality and cost of feedstocks, processing and storage limits, and product demands. With recent advances in digital hardware and optimization software, RTO can be easily incorporated into computer control systems. The scale at which industrial RTO can be implemented is impressive. Problems with over 100,000 variables and equality/inequality constraints are routinely solved (Georgiou et al., 1997).

19.1 BASIC REQUIREMENTS IN REAL-TIME OPTIMIZATION

The steady-state model used in RTO typically is obtained either from fundamental knowledge of the plant or from experimental data. It utilizes the plant operating conditions for each unit such as temperature, pressure, and feed flow rates to predict properties such as product yields (or distributions), production rates, and measurable product characteristics (e.g., purity, viscosity, and molecular weight). The economic model involves the costs of raw materials, values of products, and costs of production as functions of operating conditions, projected sales figures, and so on. An objective function is specified in terms of these quantities; in particular, *operating profit* over some specific period of time can be expressed as

$$P = \sum_s F_s V_s - \sum_r F_r C_r - OC \qquad (19\text{-}1)$$

where

P = operating profit/time

$\sum_s F_s V_s$ = sum of product flow rates times respective product values

$\sum_r F_r C_r$ = sum of feed flow rate times respective unit cost

OC = operating costs/time

Both the operating and economic models typically will include constraints on

1. ***Operating conditions:*** Process variables must be within certain limits due to valve ranges (0% to 100% open) and environmental restrictions (e.g., furnace firing constraints).
2. ***Feed and production rates:*** A feed pump has a maximum capacity; sales are limited by market projections.
3. ***Storage and warehousing capacities:*** Storage tank capacity cannot be exceeded during periods of low demand.
4. ***Product impurities:*** A salable product cannot contain more than the maximum amount of a specified contaminant or impurity.

Process operating situations that are relevant to maximizing operating profits include

1. ***Sales limited by production.*** In this type of market, sales can be increased by increasing production. This can be achieved by optimizing operating conditions and production schedules.
2. ***Sales limited by market.*** This situation is susceptible to optimization only if improvements in efficiency at current production rates can be obtained. An increase in thermal efficiency, for example, usually leads to a reduction in manufacturing costs (e.g., utilities or feedstocks).
3. ***Large throughput.*** Units with large production rates (or throughputs) offer great potential for increased profits. Small savings in product costs per unit throughput or incremental improvements in yield, plus large production rates, can result in major increases in profits.
4. ***High raw material or energy consumption.*** These are major cost factors in a typical plant and thus offer potential savings. For example, the optimal allocation of fuel supplies and steam in a plant can reduce costs by minimizing fuel consumption.
5. ***Product quality better than specification.*** If the product quality is significantly better than the customer requirements, it can cause excessive production costs and wasted capacity. By operating closer to the customer requirement (e.g., impurity level), cost savings can be obtained, but this strategy also requires lower process variability (see Fig. 1.9).
6. ***Losses of valuable or hazardous components through waste streams.*** The chemical analysis of plant waste streams, both to air and water, will indicate whether valuable materials are being lost. Adjustment of air/fuel ratios in furnaces to minimize unburned hydrocarbon losses and to reduce nitrogen-oxide emissions is one such example.

Timmons et al. (2000) and Latour (1979) have discussed opportunities for the application of on-line optimization or supervisory control in refinery operations. Three general types of optimization problems commonly encountered in industrial process operations are discussed next.

Operating Conditions

Common examples include optimizing distillation column reflux ratio and reactor temperature. Consider the

RTO of a fluidized catalytic cracker (FCC) (Latour, 1979). The FCC reaction temperature largely determines the conversion of a light gas oil feedstock to lighter (i.e., more volatile) components. The product distribution (gasoline, middle distillate, fuel oil, light gases) changes as the degree of conversion is increased. Accurate process models of the product distribution as a function of FCC operating conditions and catalyst type are required for real-time optimization. Feedstock composition, downstream unit capacities (e.g., distillation columns), individual product prices, product demand, feed preheat, gas oil recycle, and utilities requirements must be considered in optimizing an FCC unit. The large throughput of the FCC implies that a small improvement in yield translates to a significant increase in profits. Biegler et al. (1997) have discussed an RTO case study on a hydrocracker and fractionation plant originally formulated by Bailey et al. (1993).

Olefins plants in which ethylene is the main product are another application where RTO has had a significant impact (Darby and White, 1998; Starks and Arrieta, 2007). A full plant model can have as many as 1,500 submodels, based on the development of fundamental chemical engineering relations for all unit operations involved, that is, furnaces, distillation columns, mixers, compressors, and heat exchangers (Georgiou et al., 1997). In the ExxonMobil olefins plant (Beaumont, TX), the detailed model contained about 200,000 variables and equations, and optimization is used to obtain the values of about 50 targets or set points. Although standard approaches are used for developing separation and heat exchange models, the furnace models are quite elaborate and are typically usually proprietary. In the ExxonMobil application, 12 furnaces are operated in parallel with up to eight possible gas feeds and five liquid feeds (different hydrocarbons) to be cracked, along with three different coil geometries. The key optimization variables are conversion, feed rate, and steam/oil ratio, subject to feedstock availability and equipment constraints. This particular application has led to benefits in the range of millions of dollars per year.

Allocation

Allocation problems involve the optimal distribution of a limited resource among several parallel (alternative) process units. Typical examples include (Latour, 1979; Marlin and Hrymak, 1997):

Steam Generators. Optimum load distribution among several boilers of varying size and efficiency.

Refrigeration Units. Optimum distribution of a fixed refrigeration capacity among several low-temperature condensers associated with distillation columns.

Parallel Distillation Columns. Minimization of "off-spec" products and utilities consumption while maximizing overall capacity.

Planning and Scheduling

Examples of scheduling problems encountered in continuous plants include catalyst regeneration, furnace decoking, and heat exchanger cleaning, which deal with the tradeoff between operating efficiency and lost production due to maintenance. Planning problems normally entail optimization of continuous plant operations over a period of months. This approach is commonly used in refinery optimization. In batch processing, optimal scheduling is crucial to match equipment to product demands and to minimize cycle times. In a batch campaign, several batches of product may be produced using the same recipe. In order to optimize the production process, the engineer needs to determine the recipe that satisfies product quality requirements; the production rates to fulfill the product demand; the availability of raw material inventories; product storage availability; and the run schedule. Recent examples of optimal batch scheduling include specialty polymer products by McDonald (1998) and pharmaceuticals by Schulz and Rudof (1998). See Chapter 22 for more details on batch processing.

19.1.1 Implementation of RTO in Computer Control

In RTO the computer control system performs all data transfer and optimization calculations and sends set-point information to the controllers. The RTO system should perform all tasks without unduly upsetting plant operations. Several steps are necessary for implementation of RTO, including data gathering and validation (or reconciliation), determination of the plant steady state, updating of model parameters (if necessary) to match current operations, calculation of the new (optimized) set points, and implementation of these set points.

To determine whether a process unit is at steady state, software in the computer control system monitors key plant measurements (e.g., compositions, product rates, flow rates, etc.) and determines whether the plant operating conditions are close enough to steady state to start the RTO sequence. Only when all of the key measurements are within the allowable tolerances is the plant considered to be at steady state and the optimization calculations started; see Cao and Rhinehart (1995) for a statistical technique that determines the existence of steady-state conditions. The optimization software screens the measurements for unreasonable data (gross error detection). Data validity checking automatically adjusts the model updating procedure to reflect the presence of bad data or equipment that has been taken out of service. Data reconciliation based on satisfying material and energy balances can be carried out using separate optimization software (Narasimhan and Jordache, 2000). Data validation and reconciliation is an extremely

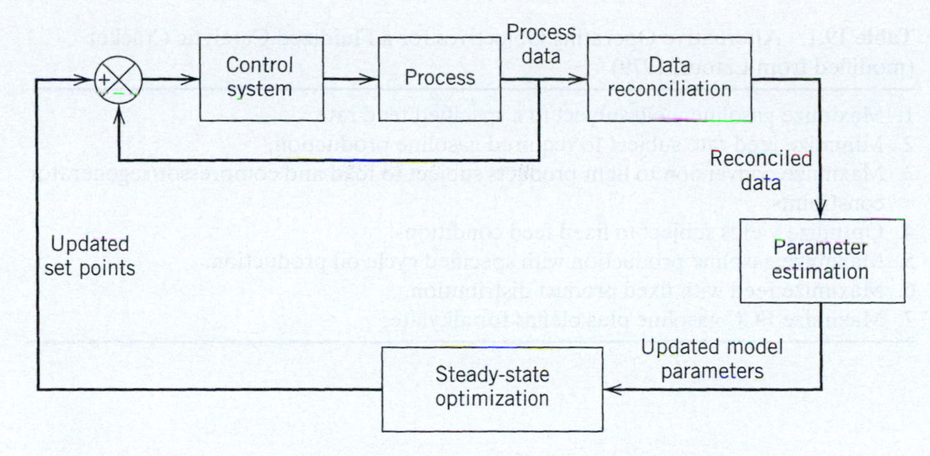

Figure 19.2 A block diagram for RTO and regulatory feedback control.

critical part of any optimization activity. If measurement errors resulting from poor instrument calibration are not considered, the data reconciliation step or subsequent parameter estimation step will not provide meaningful answers (Soderstrom et al., 2000).

The optimization software can update model parameters to match current plant data, using regression techniques. Typical model parameters include exchanger heat transfer coefficients, reactor performance parameters, and furnace efficiencies. The parameters appear in material and energy balances for each unit in the plant as well as constitutive equations for physical properties. Parameter updating compensates for plant changes and degradation of process equipment, although there is a loss of performance when the model parameters are uncertain or the plant data contain noise (Perkins, 1998). Considerable plant knowledge and experience is required in deciding which parameters to update and which data to use for the updates. After completion of the parameter estimation, the information regarding the current plant constraints, the control status data, and the economic values for feeds, products, utilities, and other operating costs are collected. The department in charge of planning and scheduling updates the economic values on a regular basis. The optimization software then calculates the optimum set points. The steady-state condition of the plant is rechecked after the optimization calculation. If the individual processes are confirmed to still be at the same steady state, then the new set points are transferred to the computer control system for implementation. Subsequently, the process control computer repeats the steady-state detection calculations, restarting the cycle. If the new optimum set points are not statistically different from the previous ones, no changes are made (Marlin and Hrymak, 1997).

The combination of RTO and regulatory control can be viewed as analogous to cascade control. As shown in Fig. 19.2, the outer RTO loop will operate more slowly than the inner loop, and a poor design of this interaction results in poor performance. The dynamic controller (or layer 3) handles the transformation between the steady-state model used in RTO and the actual dynamic operation of the process. If the RTO model and dynamic model have very different gains, the resulting combination can perform poorly. As in cascade control (cf. Chapter 16), the inner loop should be faster than the outer loop; otherwise poor closed-loop performance may result (Marlin and Hrymak, 1997).

19.2 THE FORMULATION AND SOLUTION OF RTO PROBLEMS

Once a process has been selected for RTO, an appropriate problem statement must be formulated and then solved. As mentioned earlier, the optimization of set points requires

1. The economic model, an objective function to be maximized or minimized, that includes costs and product values
2. The operating model, which includes a steady-state process model and all constraints on the process variables

Edgar et al. (2001) have listed six steps that should be used in solving any practical optimization problem. A summary of the procedure with comments relevant to RTO is given below.

Step 1. ***Identify the process variables.*** The important input and output variables for the process must be identified. These variables are employed in the objective function and the process model (see Steps 2 and 3).

Step 2. ***Select the objective function.*** Converting a verbal statement of the RTO goals into a meaningful

Table 19.1 Alternative Operating Objectives for a Fluidized Catalytic Cracker (modified from Latour, 1979)

1. Maximize gasoline yield subject to a specified feed rate.
2. Minimize feed rate subject to required gasoline production.
3. Maximize conversion to light products subject to load and compressor/regenerator constraints.
4. Optimize yields subject to fixed feed conditions.
5. Maximize gasoline production with specified cycle oil production.
6. Maximize feed with fixed product distribution.
7. Maximize FCC gasoline plus olefins for alkylate.

objective function can be difficult. The verbal statement often contains multiple objectives and implied constraints. To arrive at a single objective function based on operating profit, the quantity and quality of each product must be related to the consumption of utilities and the feedstock composition. The specific objective function selected may vary depending on plant configuration as well as the supply/demand situation. Table 19.1 shows different operating objectives that may arise for a fluidized catalytic cracker.

Step 3. ***Develop the process model and constraints.*** Steady-state process models are formulated, and operating limits for the process variables are identified. The process model can be based on the physics and chemistry of the process (see Chapter 2), or it can be based on empirical relations obtained from experimental process data (see Chapter 7). Inequality constraints arise because many physical variables, such as composition or pressure, can only have positive values, or there may be maximum temperature or maximum pressure restrictions. These inequality constraints are a key part of the optimization problem statement and can have a profound effect on the optimum operating point. In most cases, the optimum lies on a constraint.

Step 4. ***Simplify the model and objective function.*** Before undertaking any computation, the mathematical statement developed in steps 1–3 may be simplified to be compatible with the most effective solution techniques. A nonlinear objective function and nonlinear constraints can be linearized in order to use a fast, reliable optimization method such as linear programming.

Step 5. ***Compute the optimum.*** This step involves choosing an optimization technique and calculating the optimum set points. Most of the literature on the subject of optimization is concerned with this step. Over the past 20 years, much progress has been made in developing efficient and robust numerical methods for optimization calculations (Edgar et al., 2001; Griva et al., 2008; Nocedal and Wright, 2006). Virtually all optimization methods are iterative; thus a good initial estimate of the optimum can reduce the required computer time.

Step 6. ***Perform sensitivity studies.*** It is useful to know which parameters in an optimization problem are the most important in determining the optimum. By varying model and cost parameters individually and recalculating the optimum, the most sensitive parameters can be identified.

Example 19.1 illustrates the six steps.

EXAMPLE 19.1

A section of a chemical plant makes two specialty products (E, F) from two raw materials (A, B) that are in limited supply. Each product is formed in a separate process as shown in Fig. 19.3. Raw materials A and B do not have to be totally consumed. The reactions involving A and B are as follows:

$$\text{Process 1:} \quad A + B \rightarrow E$$
$$\text{Process 2:} \quad A + 2B \rightarrow F$$

The processing cost includes the costs of utilities and supplies. Labor and other costs are \$200/day for process 1 and \$350/day for process 2. These costs occur even if the production of E or F is zero. Formulate the objective function as the total operating profit per day. List the equality and inequality constraints (Steps 1, 2, and 3).

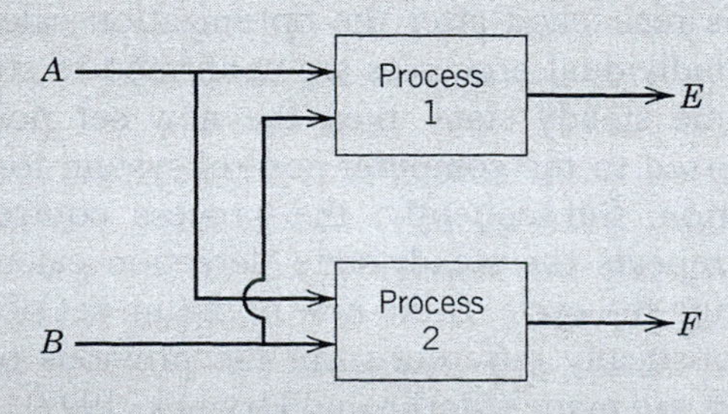

Figure 19.3 A flow diagram of a chemical plant (Example 19.1).

Available Information

Raw Material	Maximum Available (lb/day)	Cost (¢/lb)
A	40,000	15
B	30,000	20

Process	Product	Reactant Requirements (lb) per lb Product	Processing Cost	Selling Price of Product	Maximum Production Level (lb/day)
1	E	2/3 A, 1/3 B	15 ¢/lb E	40 ¢/lb E	30,000
2	F	1/2 A, 1/2 B	5 ¢/lb F	33 ¢/lb F	30,000

SOLUTION

The optimization problem is formulated using the first three steps delineated above.

Step 1. The relevant process variables are the mass flow rates of reactants and products (see Fig. 19.3):

$$x_1 = \text{lb/day A consumed}$$
$$x_2 = \text{lb/day B consumed}$$
$$x_3 = \text{lb/day E produced}$$
$$x_4 = \text{lb/day F produced}$$

Step 2. In order to use Eq. 19-1 to compute the operating product per day, we need to specify product sales income, feedstock costs, and operating costs:

$$\text{Sales income (\$/day)} = \sum_s F_s V_s = 0.4x_3 + 0.33x_4 \quad (19\text{-}2)$$

$$\text{Feedstock costs (\$/day)} = \sum_r F_r C_r = 0.15x_1 + 0.2x_2 \quad (19\text{-}3)$$

$$\text{Operating costs (\$/day)} = \text{OC} = 0.15x_3 + 0.05x_4 + 350 + 200 \quad (19\text{-}4)$$

Substituting into (19-1) yields the daily profit:

$$\begin{aligned} P &= 0.4x_3 + 0.33x_4 - 0.15x_1 - 0.2x_2 \\ &\quad - 0.15x_3 - 0.05x_4 - 350 - 200 \\ &= 0.25x_3 + 0.28x_4 - 0.15x_1 - 0.2x_2 - 550 \end{aligned} \quad (19\text{-}5)$$

Step 3. Not all variables in this problem are unconstrained. First consider the material balance equations, obtained from the reactant requirements, which in this case comprise the process operating model:

$$x_1 = 0.667x_3 + 0.5x_4 \quad (19\text{-}6a)$$

$$x_2 = 0.333x_3 + 0.5x_4 \quad (19\text{-}6b)$$

The limits on the feedstocks and production levels are:

$$0 \le x_1 \le 40{,}000 \quad (19\text{-}7a)$$

$$0 \le x_2 \le 30{,}000 \quad (19\text{-}7b)$$

$$0 \le x_3 \le 30{,}000 \quad (19\text{-}7c)$$

$$0 \le x_4 \le 30{,}000 \quad (19\text{-}7d)$$

Equations (19-5) through (19-7) constitute the optimization problem to be solved. Because the variables appear linearly in both the objective function and constraints, this formulation is referred to as a *linear programming problem*, which is discussed in Section 19.4.

19.3 UNCONSTRAINED AND CONSTRAINED OPTIMIZATION

Unconstrained optimization refers to the situation where there are no inequality constraints and all equality constraints can be eliminated by variable substitution in the objective function. First we consider single-variable optimization, followed by optimization problems with multiple variables. Because optimization techniques are iterative in nature, we focus mainly on efficient methods that can be applied on-line. Most RTO applications are multivariable problems, which are considerably more challenging than single-variable problems.

19.3.1 Single-Variable Optimization

Some RTO problems involve determining the value of a single independent variable that maximizes (or minimizes) an objective function. Examples of single-variable optimization problems include optimizing the reflux ratio in a distillation column or the air/fuel ratio in a furnace. Optimization methods for single-variable problems are typically based on the assumption that the objective function $f(x)$ is *unimodal* with respect to x over the region of the search. In other words, a single maximum (or minimum) occurs in this region. To use these methods, it is necessary to specify upper and lower bounds for x^{opt}, the optimum value of x, by evaluating $f(x)$ for trial values of x within these bounds and observing where $f(x)$ is a maximum (or minimum). The values of x nearest this apparent optimum are specified to be the region of the search. This region is also referred to as the *interval of uncertainty* or *bracket*, and is used to initiate the formal optimization procedure.

Efficient single-variable (or *one-dimensional*) optimization methods include Newton and quasi-Newton methods and polynomial approximation (Edgar et al., 2001). The second category includes quadratic interpolation, which utilizes three points in the interval of uncertainty to fit a quadratic polynomial to $f(x)$ over this interval. Let x_a, x_b, and x_c denote three values of x in the interval of uncertainty and f_a, f_b and f_c denote the corresponding values of $f(x)$. Then a quadratic polynomial, $\hat{f}(x) = a_0 + a_1 x + a_2 x^2$, can be fit to these data to provide a local approximation to $f(x)$. The resulting equation for $\hat{f}(x)$ can be differentiated, set equal to zero, and solved for its optimum value, which is denoted by x^*. The expression for x^* is

$$x^* = \frac{1}{2}\frac{(x_b^2 - x_c^2) f_a + (x_c^2 - x_a^2) f_b + (x_a^2 - x_b^2) f_c}{(x_b - x_c) f_a + (x_c - x_a) f_b + (x_a - x_b) f_c} \tag{19-8}$$

After one iteration, x^* usually is not equal to x^{opt}, because the true function $f(x)$ is not necessarily quadratic. However, x^* is expected to be an improvement over x_a, x_b, and x_c. By saving the best two of the three previous points and finding the actual objective function at x^*, the search can be continued until convergence is indicated.

EXAMPLE 19.2

A free radical reaction involving nitration of decane is carried out in two sequential reactor stages, each of which operates like a continuous stirred-tank reactor (CSTR). Decane and nitrate (as nitric acid) in varying amounts are added to each reactor stage, as shown in Fig. 19.4. The reaction of nitrate with decane is very fast and forms the following products by successive nitration: DNO_3, $D(NO_3)_2$, $D(NO_3)_3$, $D(NO_3)_4$, and so on. The desired product is DNO_3, whereas dinitrate, trinitate, etc., are undesirable products.

The flow rates of D_1 and D_2 are chosen to satisfy temperature requirements in the reactors, while N_1 and N_2 are optimized to maximize the amount of DNO_3 produced from stage 2, subject to satisfying an overall level of nitration. In this case, we stipulate that $(N_1 + N_2)/(D_1 + D_2) = 0.4$. There is an excess of D in each stage, and $D_1 = D_2 = 0.5$ mol/s. A steady-state reactor model has been developed to maximize selectivity. Define $r_1 \triangleq N_1/D_1$ and $r_2 \triangleq N_2/(D_1 + D_2)$. The amount of DNO_3 leaving stage 2 (as mol/s in F_2) is given by

$$f_{DNO3} = \frac{r_1 D_1}{(1 + r_1)^2(1 + r_2)} + \frac{r_2 D_2}{(1 + r_1)(1 + r_2)^2} \tag{19-9}$$

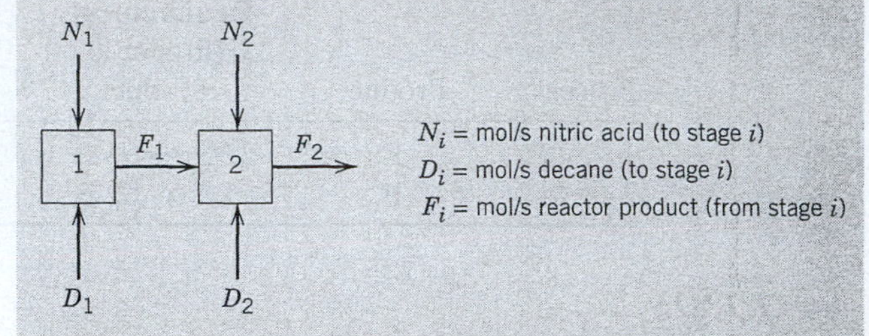

Figure 19.4 A schematic diagram of a two-stage nitration reactor.

This equation can be derived from the steady-state equations for a continuous stirred reactor with the assumption that all reaction rate constants are equal.

Formulate a one-dimensional search problem in r_1 that will permit the optimum values of r_1 and r_2 to be found. Employ quadratic interpolation using an initial interval of $0 \le r_1 \le 0.8$. Use enough iterations so that the final value of f_{DNO3} is within $\pm$ 0.0001 of the maximum.

SOLUTION

The six steps described earlier are used to formulate the optimization problem.

Step 1. ***Identify the process variables.*** The process variables to be optimized are N_1 and N_2, the nitric acid molar flow rates for each stage. Because D_1 and D_2 are specified, we can just as well use r_1 and r_2, because the conversion model is stated in terms of r_1 and r_2.

Step 2. ***Select the objective function.*** The objective is to maximize production of DNO_3 which can be made into useful products, while other nitrates cannot. We assume that the unwanted byproducts have a value of zero. The objective function f is given in (19-9). We do not need to state it explicitly as a profit function, as in Eq. 19-1, because the economic value (selling price) of DNO_3 is merely a multiplicative constant.

Step 3. ***Develop models for the process and constraints.*** The values of N_1 and N_2 are constrained by the overall nitration level:

$$\frac{N_1 + N_2}{D_1 + D_2} = 0.4 \tag{19-10}$$

which can be expressed in terms of r_1 and r_2 as

$$\frac{r_1 D_1 + r_2 D_1 + r_2 D_2}{D_1 + D_2} = 0.4 \tag{19-11}$$

Inequality constraints on r_1 and r_2 do exist, namely, $r_1 \geq 0$ and $r_2 \geq 0$—because all N_i and D_i are positive. These constraints can be ignored except when the search method incorrectly leads to negative values of r_1 or r_2.

Step 4. ***Simplify the model.*** Because $D_1 = D_2 = 0.5$, then, from (19-11),

$$r_2 = 0.4 - 0.5r_1 \tag{19-12}$$

We select r_1 to be the independent variable for the one-dimensional search in Eq. (19-9), and then r_2 is a dependent variable. Because r_1 and r_2 are nonnegative, Eq. 19-12 implies that $r_1 \leq 0.8$ and $r_2 \leq 0.4$. After variable substitution, there is only one independent variable (r_1) in the objective function.

Step 5. ***Compute the optimum.*** Because r_1 lies between 0 and 0.8 (the interval of uncertainty), select the three interior points for the search to be $r_1 = 0.2$, 0.4, and 0.6. The corresponding values of r_2 are 0.3, 0.2, and 0.1. Table 19.2 shows the numerical results for three iterations, along with objective function values. After the first iteration, the worst point ($r_1 = 0.2$) is discarded and the new point ($r = 0.4536$) is added. After the second iteration, the point with the lowest value of $f(r_1 = 0.6)$ is discarded. The tolerance on the objective function change is satisfied after only three iterations, with the value of r_1 that maximizes f_{DNO3} computed to be $r_1^{opt} = 0.4439$. The converted mononitrate is 0.1348 mol/s from stage 2; the remainder of the nitrate is consumed to make higher molecular weight byproducts.

Step 6. ***Perform sensitivity studies.*** Based on the results in Table 19.2, the yield is not significantly different from the optimum as long as $0.4 \leq r_1 \leq 0.6$. Practically speaking, this situation is beneficial, because it allows a reasonable range of decane flows to achieve temperature control. If either D_1 or D_2 changes by more than 10%, we should recalculate the optimum. There also might be a need to reoptimize r_1 and r_2 if ambient conditions change (e.g., summer vs. winter operation). Even a 1% change in yield can be economically significant if production rates and the selling price of the product are sufficiently high.

Table 19.2 Search Iterations for Example 19.2 (Quadratic Interpolation)

Iteration	x_a	f_a	x_b	f_b	x_c	f_c	x^*
1	0.2	0.1273	0.4	0.1346	0.6	0.1324	0.4536
2	0.4	0.1346	0.6	0.1324	0.4536	0.1348	0.4439
3	0.4	0.1346	0.4536	0.1348	0.4439	0.1348	(not needed)
	$r_1^{opt} = 0.4439$						

If the function to be optimized is not unimodal, then some care should be taken in applying the quadratic interpolation method. Selecting multiple starting points for the initial scanning before quadratic interpolation is initiated ensures that an appropriate search region has been selected. For a single variable search, scanning the region of search is a fairly simple and fast procedure, but evaluating the presence of multiple optima can become problematic for multivariable optimization problems.

19.3.2 Multivariable Optimization

In multivariable optimization problems, there is no guarantee that a given optimization technique will find the optimum point in a reasonable amount of computer time. The optimization of a general nonlinear multivariable objective function, $f(\boldsymbol{x}) = f(x_1, x_2, \ldots, x_{N_V})$, requires that efficient and robust numerical techniques be employed. Efficiency is important, because the solution requires an iterative approach. Trial-and-error solutions are usually out of the question for problems with more than two or three variables. For example, consider a four-variable grid search, where an equally spaced grid for each variable is prescribed. For 10 values of each of the 4 variables, there are 10^4 total function evaluations required to find the best answer out of the 10^4 grid intersections. Even then, this computational effort may not yield a result sufficiently close to the true optimum. Grid search is a very inefficient method for multivariable optimization.

The difficulty of optimizing multivariable functions often is resolved by treating the problem as a series of single-variable (or one-dimensional) searches. From a given starting point, a search direction is specified, and then the optimum point along that direction is determined by a one-dimensional search. Then a new search direction is determined, followed by another one-dimensional search in that direction. In choosing an algorithm to determine the search direction, we can draw upon extensive numerical experience with various optimization methods (Griva et al, 2008; Nocedal and Wright, 2006; Edgar et al., 2001).

Multivariable RTO of nonlinear objective functions using function derivatives is recommended with more than two variables. In particular, the conjugate gradient

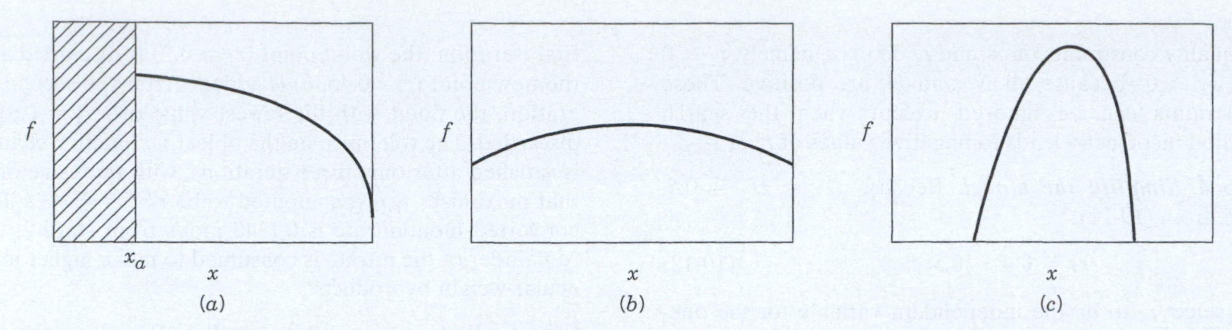

Figure 19.5 Three types of optimal operating conditions.

and quasi-Newton methods (Griva et al, 2008; Edgar et al., 2001) are extremely effective in solving such problems. Applications of multivariable RTO have experienced rapid growth as a result of advances in computer hardware and software. We consider such methods in more detail in Section 19.5.

An important application of unconstrained optimization algorithms is to update parameters in steady-state models from the available data. Usually, only a few model parameters are estimated on-line, and then RTO is based on the updated model. Guidelines for parameter estimation have been provided by Marlin and Hrymak (1997) and Forbes et al. (1994).

Most practical multivariable problems include constraints, which must be treated using enhancements of unconstrained optimization algorithms. The next two sections describe two classes of constrained optimization techniques that are used extensively in the process industries. When constraints are an important part of an optimization problem, constrained techniques must be employed, because an unconstrained method might produce an optimum that violates the constraints, leading to unrealistic values of the process variables. The general form of an optimization problem includes a nonlinear objective function (profit) and nonlinear constraints and is called a *nonlinear programming* problem.

$$\text{maximize} \quad f(x_1, x_2, \ldots, x_{N_V}) \tag{19-13}$$

$$\text{subject to:} \quad h_i(x_1, x_2, \ldots, x_{N_V}) = 0 \; (i = 1, \ldots, N_E) \tag{19-14}$$

$$g_i(x_1, x_2, \ldots, x_{N_V}) \leq 0 \; (i = 1, \ldots, N_I) \tag{19-15}$$

In this case, there are N_V process variables, N_E equality constraints and N_I inequality constraints.

Skogestad (2000) and Perkins (1998) have discussed the interplay of constraints, and the selection of the optimal operating conditions. Skogestad identified three different cases for RTO that are illustrated in Fig. 19.5. In each case, a single variable x is used to maximize a profit function, $f(x)$.

(a) ***Constrained optimum:*** The optimum value of the profit is obtained when $x = x_a$. Implementation of an active constraint is straightforward; for example, it is easy to keep a valve closed.

(b) ***Unconstrained flat optimum:*** In this case, the profit is insensitive to the value of x, and small process changes or disturbances do not affect profitability very much.

(c) ***Unconstrained sharp optimum:*** A more difficult problem for implementation occurs when the profit is sensitive to the value of x. If possible, we may want to select a different input variable for which the corresponding optimum is flatter, so that the operating range can be wider without reducing the profit very much.

In some cases, an actual process variable (such as yield) can be the objective function, and no process model is required. Instead, the process variables are varied systematically to find the best value of the objective function from the specific data set, sometimes involving design of experiments as discussed by Myers and Montgomery (2002). In this way, improvements in the objective function can be obtained gradually. Usually, only a few variables can be optimized in this way, and it is limited to batch operations. Methods used in industrial batch process applications include EVOP (evolutionary operation) and response surface analysis (Edwards and Jutan, 1997; Box and Draper, 1998; Myers and Montgomery, 2002).

19.4 LINEAR PROGRAMMING

An important class of constrained optimization problems has a linear objective function and linear constraints. The solution of these problems is highly structured and can be obtained rapidly via *linear programming* (LP). This powerful approach is widely used in RTO applications.

For processing plants, different types of *linear* inequality and equality constraints often arise that make the LP method of great interest. The constraints can change on a daily or even an hourly basis.

1. *Production constraints.* Equipment throughput restrictions, storage limits, or market constraints (no additional product can be sold) are frequently encountered in manufacturing. These constraints have the form of $x_i \leq c_i$ or $g_i = x_i - c_i \leq 0$ (cf. Eq. 19-15).
2. *Raw material limitations.* Feedstock supplies are frequently limited owing to supplier capability or production levels of other plants within the same company.
3. *Safety restrictions.* Common examples are limitations on operating temperature and pressure.
4. *Product specifications.* Constraints placed on the physical properties or composition of the final product fall into this category. For blends of various liquid products in a refinery, it is commonly assumed that a blend property can be calculated by averaging pure component properties. Thus, a blend of N_c components with physical property values ψ_k and volume fractions y_k (based on volumetric flow rates) has a calculated blend property of

$$\overline{\psi} = \sum_{k=1}^{N_c} \psi_k y_k \tag{19-16}$$

If there is an upper limit α on $\overline{\psi}$, the resulting constraint is

$$\sum_{k=1}^{N_c} \psi_k y_k \leq \alpha \tag{19-17}$$

5. *Material and energy balances.* Although items 1–4 generally are considered to be inequality constraints, the steady-state material and energy balances are equality constraints.

19.4.1 Linear Programming Concepts

For simplicity, consider a multivariable process with two inputs (u_1, u_2) and two outputs (y_1, y_2). The set of inequality constraints for $\boldsymbol{u}$ and $\boldsymbol{y}$ define an *operating window* for the process. A simple example of an operating window for a process with two inputs (to be optimized) is shown in Fig. 19.6. The upper and lower limits for u_1 and u_2 define a rectangular region. There are also upper limits for y_1 and y_2 and a lower limit for y_2. For a linear process model,

$$\boldsymbol{y} = \boldsymbol{K}\boldsymbol{u} \tag{19-18}$$

the inequality constraints on $\boldsymbol{y}$ can be converted to constraints in $\boldsymbol{u}$, which reduces the size of the operating

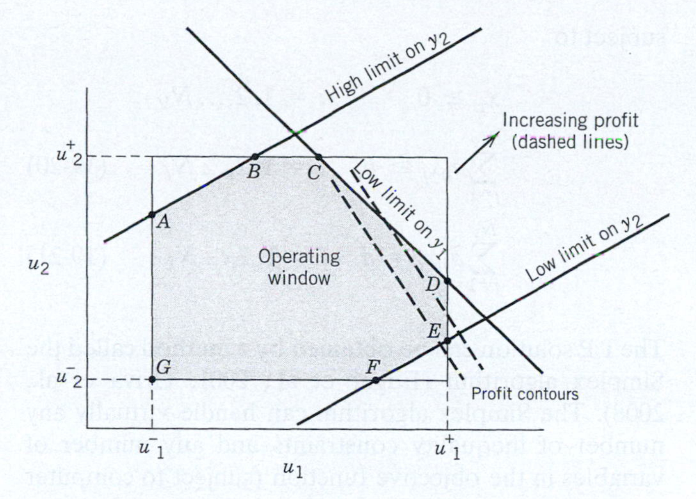

Figure 19.6 Operating window for a 2 × 2 optimization problem. The dashed lines are objective function contours, increasing from left to right. The maximum profit occurs where the profit line intersects the constraints at vertex *D*.

window to the shaded region in Fig. 19.6. If a linear cost function is selected, the optimum operating condition occurs on the boundary of the operating window at a point where constraints intersect (Griva et al, 2008; Edgar et al., 2001). These points of intersections are called vertices. Thus, in Fig. 19.6 the optimum operating point, $\boldsymbol{u}^{\text{opt}}$ occurs at one of the seven vertices, points *A–G*. For the indicated linear profit function (dashed lines), the maximum occurs at vertex *D*. This graphical concept can be extended to problems with more than two inputs because the operating window is a closed convex region, providing that the process model, cost function, and inequality constraints are all linear. Using Eq. 19-18, we can calculate the optimal set points $\boldsymbol{y}_{\text{sp}}$ from the value of $\boldsymbol{u}^{\text{opt}}$.

The number of independent variables in a constrained optimization problem can be found by a procedure analogous to the degrees of freedom analysis in Chapter 2. For simplicity, suppose that there are no constraints. If there are N_V process variables (which includes process inputs and outputs) and the process model consists of N_E independent equations, then the number of independent variables is $N_F = N_V - N_E$. This means N_F set points can be specified independently to maximize (or minimize) the objective function. The corresponding values of the remaining $(N_V - N_F)$ variables can be calculated from the process model. However, the presence of inequality constraints that can become active changes the situation, because the N_F set points cannot be selected arbitrarily. They must satisfy all of the equality and inequality constraints.

The standard linear programming (LP) problem can be stated as follows:

$$\text{minimize } f = \sum_{r=1}^{N_V} c_i x_i \tag{19-19}$$

subject to

$$x_i \geq 0 \qquad i = 1, 2, \ldots N_V$$

$$\sum_{j=1}^{N_V} a_{ij} x_j \geq b_i \quad i = 1, 2, \ldots N_I \qquad (19\text{-}20)$$

$$\sum_{j=1}^{N_V} \tilde{a}_{ij} x_j = d_i \quad i = 1, 2, \ldots N_E \qquad (19\text{-}21)$$

The LP solution can be obtained by a method called the Simplex algorithm (Edgar et al., 2001; Griva et al., 2008). The Simplex algorithm can handle virtually any number of inequality constraints and any number of variables in the objective function (subject to computer time limitations, of course). Maximization problems can be converted to the form of (19-19) by multiplying the objective function by −1. Inequality constraints are handled by the introduction of artificial variables called *slack variables*, which convert the inequality constraints (19-20) to equality constraints by subtracting a non-negative slack variable from the left-hand side of each inequality. The slack variable then provides a measure of the distance from the constraint for a given set of variables, and these artificial variables are introduced for computational purposes. When a slack variable is zero, the constraint is active. Because there are a limited number of intersections of constraint boundaries where the optimum must occur, the amount of computer time required to search for the optimum is reduced considerably compared to more general nonlinear optimization problems. Hence, many nonlinear optimization problems (even those with nonlinear constraints) are often linearized so that the LP algorithm can be employed. This procedure allows optimization problems with over 100,000 variables to be solved.

In the 1980s, a major change in optimization software occurred when linear programming solvers and then nonlinear programming solvers were interfaced to spreadsheet software for desktop computers. The spreadsheet has become a popular user interface for entering and manipulating numeric data. Spreadsheet software increasingly incorporates analytic tools that are accessible from the spreadsheet interface and permit access to external databases. For example, Microsoft Excel incorporates an optimization-based routine called Solver that operates on the values and formulas of a spreadsheet model. Current versions (4.0 and later) include LP and NLP solvers and mixed integer programming (MIP) capability for both linear and nonlinear problems. The user specifies a set of cell addresses to be independently adjusted (the decision variables), a set of formula cells whose values are to be constrained (the constraints), and a formula cell designated as the optimization objective, as shown in the following example.

EXAMPLE 19.3

Consider a simple version of a refinery blending and production problem. This example is more illustrative of a scheduling application (Level 5 in Fig. 19.1) that has been used extensively since the 1960s in the chemical process industries. Figure 19.7 is a schematic diagram of feedstocks and products for the refinery (costs and selling prices are given in parentheses). Table 19.3 lists the information pertaining to the expected yields of the two types of crude oils when processed by the refinery. Note that the product distribution from the refinery is quite different for the two crude oils. Table 19.3 also lists the limitations on the established markets for the various products in terms of the allowed maximum daily production. In addition, processing costs are given.

To set up the linear programming problem, formulate an objective function and constraints for the refinery operation. From Fig. 19.7, six variables are involved, namely, the flow rates of the two raw materials and the four products. Solve the LP using the Excel Solver.

SOLUTION

Let the variables be

$$x_1 = \text{bbl/day of crude \#1}$$
$$x_2 = \text{bbl/day of crude \#2}$$
$$x_3 = \text{bbl/day of gasoline}$$
$$x_4 = \text{bbl/day of kerosene}$$
$$x_5 = \text{bbl/day of fuel oil}$$
$$x_6 = \text{bbl/day of residual}$$

The linear objective function f (to be maximized) is the profit, the difference between income and costs:

$$f = \text{income} - \text{raw material cost} - \text{processing cost}$$

where the following items are expressed as dollars per day:

$$\left.\begin{cases} \text{Income} = 36x_3 + 24x_4 + 21x_5 + 10x_6 \\ \text{Raw material cost} = 24x_1 + 15x_2 \\ \text{Processing cost} = 0.5x_1 + x_2 \end{cases}\right\} \qquad (19\text{-}22)$$

$$f = 36x_3 + 24x_4 + 21x_5 + 10x_6 - 24.5x_1 - 16x_2 \qquad (19\text{-}23)$$

The yield data provide four linear equality constraints (material balances) relating x_1 through x_6:

$$\text{Gasoline:} \quad x_3 = 0.80x_1 + 0.44x_2 \qquad (19\text{-}24)$$
$$\text{Kerosene:} \quad x_4 = 0.05x_1 + 0.10x_2 \qquad (19\text{-}25)$$
$$\text{Fuel oil:} \quad x_5 = 0.10x_1 + 0.36x_2 \qquad (19\text{-}26)$$
$$\text{Residual:} \quad x_6 = 0.05x_1 + 0.10x_2 \qquad (19\text{-}27)$$

Other constraints that exist or are implied in this problem are given in Table 19.3, which lists certain restrictions on the $\{x_i\}$ in terms of production limits. These can be formulated as inequality constraints:

$$\text{Gasoline:} \quad x_3 \leq 24{,}000 \qquad (19\text{-}28)$$
$$\text{Kerosene:} \quad x_4 \leq 2{,}000 \qquad (19\text{-}29)$$
$$\text{Fuel oil:} \quad x_5 \leq 6{,}000 \qquad (19\text{-}30)$$

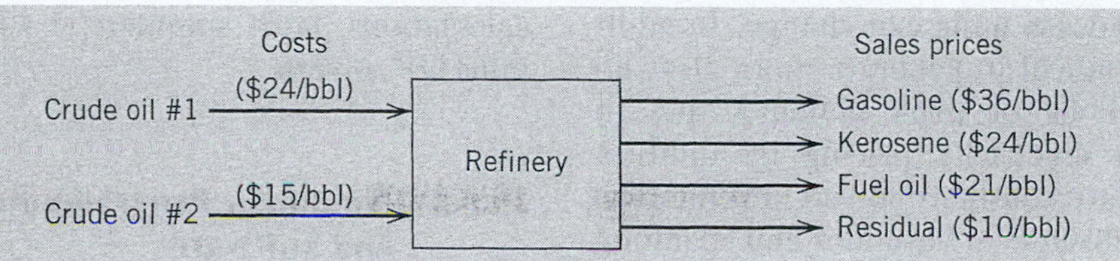

Figure 19.7 Refinery input and output schematic.

Table 19.3 Data for the Refinery Feeds and Products

	Volume percent yield		Maximum allowable production
	Crude #1	Crude #2	(bbl/day)
Gasoline	80	44	24,000
Kerosene	5	10	2,000
Fuel oil	10	36	6,000
Processing cost ($/bbl)	0.50	1.00	

One other set of constraints, although not explicitly stated in the formulation of the problem, is composed of the non-negativity restrictions, namely, $x_i \geq 0$. All process variables must be zero or positive, because it is meaningless to have negative production rates.

The formal statement of the linear programming problem is now complete, consisting of Eqs. 19-23 to 19-30. We can now proceed to solve the LP problem using the Excel Solver option. The problem statement can be introduced into the spreadsheet as illustrated in the Solver Parameter dialog box in Fig. 19.8. There are four equality constraints and three inequality constraints; the first three equality constraints are shown in the dialog box in Fig. 19.8. The objective function is in the target cell A10, and the six variable cells are in cells A4–F4.

In the refinery blending problem, the optimum $\boldsymbol{x}$ obtained by Excel occurs at the intersection of the gasoline and kerosene constraints. For these active constraints, the optimum is therefore

$$x_1 = 26{,}207$$
$$x_2 = 6{,}897$$
$$x_3 = 24{,}000 \text{ (gasoline constraint)}$$
$$x_4 = 2{,}000 \text{ (kerosene constraint)}$$
$$x_5 = 5{,}103$$
$$x_6 = 2{,}000$$
$$f = \$286{,}758/\text{day}$$

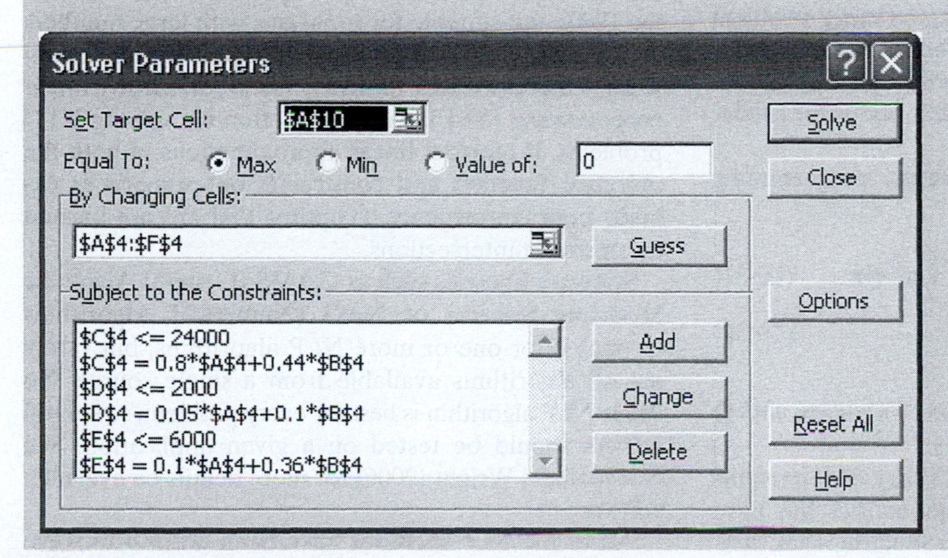

Figure 19.8 Solver parameter dialog box for Example 19.3 (Refinery LP).

In the process industries, the Simplex algorithm has been applied to a wide range of problems, such as the optimization of a total plant utility system. A general steam utility configuration, typically involving as many as 100 variables and 100 constraints, can be easily optimized using linear programming (Bouilloud, 1969; Edgar et al., 2001; Marlin, 2000). The process variables can be updated on an hourly basis because steam demands in process units can change. In addition, it may be economical to generate more electricity locally during times of peak demand, due to variable time-of-day electricity pricing by utilities. Larger LP problems are routinely solved in refineries, numbering in the thousands of variables and spanning several months of operations (Pike, 1986).

19.5 QUADRATIC AND NONLINEAR PROGRAMMING

The most general optimization problem occurs when both the objective function and constraints are nonlinear, a case referred to as *nonlinear programming* (NLP), which is stated mathematically in Eqs. 19-13 to 19-15. The leading constrained optimization methods include (Nocedal and Wright, 2006; Griva et al., 2008; Edgar et al., 2001)

1. Quadratic programming
2. Generalized reduced gradient
3. Successive quadratic programming (SQP)
4. Successive linear programming (SLP)

19.5.1 Quadratic Programming

In quadratic programming (QP), the objective function is quadratic and the constraints are linear. Although the solution is iterative, it can be obtained quickly as in linear programming.

A quadratic programming problem minimizes a quadratic function of n variables subject to m linear inequality or equality constraints. A convex QP is the simplest form of a nonlinear programming problem with inequality constraints. A number of practical optimization problems are naturally posed as a QP problem, such as constrained least squares and some model predictive control problems.

In compact notation, the quadratic programming problem is

$$\text{Minimize} \quad f(\boldsymbol{x}) = \boldsymbol{c}^T\boldsymbol{x} + \frac{1}{2}\boldsymbol{x}^T\boldsymbol{Q}\boldsymbol{x} \tag{19-31}$$

$$\text{Subject to} \quad \boldsymbol{A}\boldsymbol{x} = \boldsymbol{b} \tag{19-32}$$
$$\boldsymbol{x} \geq \boldsymbol{0}$$

where $\boldsymbol{c}$ is a vector $(n \times 1)$, $\boldsymbol{A}$ is an $m \times n$ matrix, and $\boldsymbol{Q}$ is a symmetric $n \times n$ matrix.

The equality constraint of (19-32) may contain some constraints that were originally inequalities but have been converted to equalities by introducting slack variables, as is done for LP problems. Computer codes for quadratic programming allow arbitrary upper and lower bounds on $\boldsymbol{x}$; here we assume $\boldsymbol{x} \geq \boldsymbol{0}$ for simplicity. QP software finds a solution by using LP operations to minimize the sum of constraint violations. Because LP algorithms are employed as part of the QP calculations, most commercial LP software also contains QP solvers.

19.5.2 Nonlinear Programming Algorithms and Software

One of the older and most accessible NLP algorithms uses iterative linearization and is called the *generalized reduced gradient (GRG)* algorithm. The GRG algorithm employs linear or linearized constraints and uses slack variables to convert all constraints to equality constraints. It then develops a reduced basis by eliminating a subset of the variables, which is removed by inversion of the equalities. The gradient or search direction is then expressed in terms of this reduced basis. The GRG algorithm is used in the Excel Solver. CONOPT is a reduced gradient algorithm that works well for large-scale problems and nonlinear constraints. CONOPT and GRG work best for problems where the number of degrees of freedom is small (the number of constraints is nearly equal to the number of variables).

Successive quadratic programming (SQP) solves a sequence of quadratic programs that approach the solution of the original NLP by linearizing the constraints and using a quadratic approximation to the objective function. Lagrange multipliers are introduced to handle constraints, and the search procedure generally employs some variation of Newton's method, a second-order method that approximates the Hessian matrix using first derivatives (Biegler et al., 1997; Edgar et al., 2001). MINOS and NPSOL, software packages developed in the 1980s, are suitable for programs with large numbers of variables (more variables than equations) and constraints that are linear or nearly linear. *Successive linear programming (SLP)* is used less often for solving RTO problems. It requires linear approximations of both the objective function and constraints but sometimes exhibits poor convergence to optima that are not located at constraint intersections.

Software libraries such as GAMS (General Algebraic Modeling System) or NAG (Numerical Algorithms Group) offer one or more NLP algorithms, but rarely are all algorithms available from a single source. No single NLP algorithm is best for every problem, so several solvers should be tested on a given application. See Nocedal and Wright (2006) for more details on available software.

All of the NLP methods have been utilized to solve nonlinear programming problems in the field of chemical engineering design and operations. Although in the following example we illustrate the use of GRG in the Excel Solver, large-scale NLP problems in RTO are more frequently solved using SQP owing to its superior ability in handling a large number of active constraints.

EXAMPLE 19.4

Consider the problem of minimizing fuel costs in a boilerhouse. The boilerhouse contains two turbine generators, each of which can be simultaneously operated with two fuels: fuel oil and medium Btu gas (MBG); see Fig. 19.9. The MBG is produced as a waste off-gas from another part of the plant, and it must be flared if it cannot be used on site. The goal of the RTO scheme is to find the optimum flow rates of fuel oil and MBG and provide 50 MW of power at all times, so that steady-state operations can be maintained while minimizing costs. It is desirable to use as much of the MBG as possible (which has zero cost) while minimizing consumption of expensive fuel oil. The two turbine generators (G1, G2) have different operating characteristics; the efficiency of G1 is higher than that of G2.

Data collected on the fuel requirements for the two generators yield the following empirical relations:

$$P_1 = 4.5x_1 + 0.1x_1^2 + 4.0x_2 + 0.06x_2^2 \quad (19\text{-}33)$$

$$P_2 = 4.0x_3 + 0.05x_3^2 + 3.5x_4 + 0.02x_4^2 \quad (19\text{-}34)$$

where

P_1 = power output (MW) from G1
P_2 = power output (MW) from G2
x_1 = fuel oil to G1 (tons/h)
x_2 = MBG to G1 (fuel units/h)
x_3 = fuel oil to G2 (tons/h)
x_4 = MBG to G2 (fuel units/h)

The total amount of MBG available is 5 fuel units/h. Each generator is also constrained by minimum and maximum power outputs: generator 1 output must lie between 18 and 30 MW, while generator 2 can operate between 14 and 25 MW.

Formulate the optimization problem by applying the methodology described in Section 19.2. Then solve for the optimum operating conditions (x_1, x_2, x_3, x_4, P_1, P_2) using the Excel Solver.

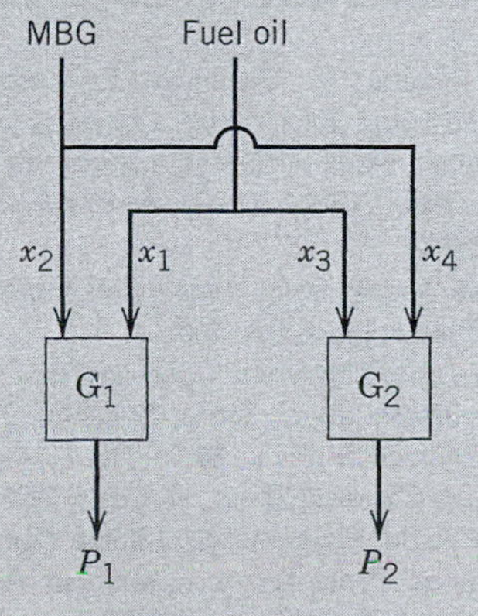

Figure 19.9 The allocation of two fuels in a boilerhouse with two turbine generators (G_1, G_2).

SOLUTION

Step 1. ***Identify the variables.*** Use x_1 through x_4 as the four process variables. Variables P_1 and P_2 are dependent because of the equality constraints (see Steps 3 and 4).

Step 2. ***Select the objective function.*** The way to minimize the cost of operation is to minimize the amount of fuel oil consumed. This implies that we should use as much MBG as possible, because it has zero cost. The objective function can be stated in terms of variables defined above; that is, we wish to minimize

$$f = x_1 + x_3 \quad (19\text{-}35)$$

Step 3. ***Specify process model and constraints.*** The constraints given in the problem statement are as follows:

(1) Power relations

$$P_1 = 4.5x_1 + 0.1x_1^2 + 4.0x_2 + 0.06x_2^2 \quad (19\text{-}33)$$

$$P_2 = 4.0x_3 + 0.05x_3^2 + 3.5x_4 + 0.2x_4^2 \quad (19\text{-}34)$$

(2) Power range $\quad 18 \le P_1 \le 30 \quad$ (19-36)

$\quad 14 \le P_2 \le 25 \quad$ (19-37)

(3) Total power $\quad 50 = P_1 + P_2 \quad$ (19-38)

(4) MBG supply $\quad 5 = x_2 + x_4 \quad$ (19-39)

Note that all variables defined above are nonnegative.

Step 4. ***Simplify the model and objective function.*** Although there are two independent variables in this problem (six variables and four equality constraints), there is no need to carry out variable substitution or further simplification, because the Excel Solver can easily handle the solution of this fairly small NLP problem.

Step 5. ***Compute the optimum.*** The Solver dialog box is shown in Fig. 19.10. The objective function value is in the target cell of the spreadsheet, written as a function of $x_1 - x_4$ (Eq. 19-35). These four variables are changed in the series of cells A4–D4. The constraints shown above are expressed in cells B12, B9, E12, and E9.

At the optimum $f = 6.54$, $x_1 = 1.82$, and $x_3 = 4.72$, meaning that 1.82 tons/h of fuel oil are delivered to generator G1, while 4.72 tons/h are used in G2. G1 utilizes all of the MBG (x_2), while G2 uses none ($x_4 = 0$), due to its lower efficiency with MBG.

Step 6. ***Perform a sensitivity analysis.*** Many operating strategies may be satisfactory, though not optimal, for the above problem. The procedure discussed above can also be repeated if parameters in the original constraint equations are changed as plant operating conditions vary. For example, suppose the total power requirement is changed to 55 MW; as an exercise, determine whether any of the active constraints change for the increased power requirement.

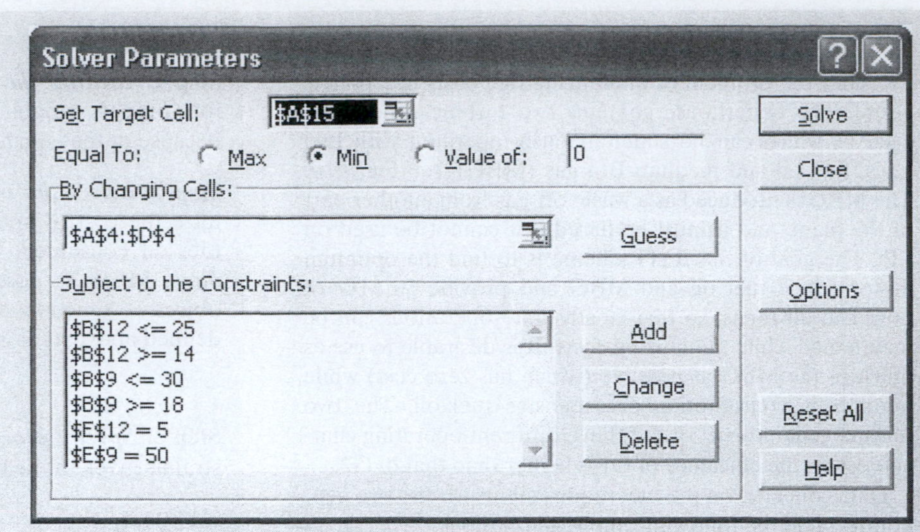

Figure 19.10 Excel Solver parameter dialog box.

SUMMARY

Although the economic benefits from feedback control are not always readily quantifiable, RTO offers a direct method of maximizing the steady-state profitability of a process or group of processes. The optimization of the set points is performed as frequently as necessary, depending on changes in operating conditions or constraints. It is important to formulate the optimization problem carefully; a methodology for formulation and solution of optimization problems is presented in this chapter. A wide range of optimization techniques can be used, depending on (1) the number of variables, (2) the nature of the equality and inequality constraints, and (3) the nature of the objective function. Because we have presented only introductory concepts in optimization here, the reader is advised to consult other comprehensive references on optimization such as Edgar et al. (2001) before choosing a particular method for RTO.

REFERENCES

Bailey, J. K., A. N. Hrymak, S. S. Treiba, and R. B. Hawkins, Nonlinear Optimization of a Hydrocracker Fractionation Plant, *Comput. Chem. Eng.*, **17**, 123 (1993).

Baker, T. E., An Integrated Approach to Planning and Scheduling, *Foundations of Computer Aided Process Operations (FOCAPO)*, D. W. T. Rippin, J. C. Hale, and J. F. Davis (eds.), CACHE Corporation, Austin, TX, 1993, p. 237.

Biegler, L. T., I. E. Grossmann, and A. W. Westerberg, *Systematic Methods of Chemical Process Design*, Prentice-Hall, Upper Saddle River, NJ, 1997.

Bouilloud, P., Compute Steam Balance by LP, *Hydrocarb. Proc.* **48**(8), 127 (1969).

Box, G. E. P., and N. R. Draper, *Evolutionary Operation: A Statistical Method for Process Improvement*, Wiley, New York, 1998.

Bryant, G. F., Developments in Supply Chain Management Control Systems Design, *Foundations of Computer Aided Process Operations (FOCAPO)*, D. W. T. Rippin, J. C. Hale, and J. F. Davis (Eds.), CACHE Corporation, Austin, TX, 1993, p. 317.

Cao, S., and R. R. Rhinehart, An Efficient Method for On-line Identification of Steady State, *J. Process Control*, **5**, 363 (1995).

Darby, M. L., and D. C. White, On-line Optimization of Complex Process Units, *Chem. Engr. Prog.*, **84**(10), 51 (1998).

Edgar, T. F., D. M. Himmelblau, and L. S. Lasdon, *Optimization of Chemical Processes*, 2d ed., McGraw-Hill, New York, 2001.

Edwards, I. M., and A. Jutan, Optimization and Control Using Response Surface Methods, *Comput. Chem. Eng.*, **21**, 441 (1997).

Forbes, F., T. Marlin, and J. F. MacGregor, Model Selection Criteria for Economics-Based Optimizing Control. *Comput. Chem. Eng.*, **18**, 497 (1994).

Geddes, D., and T. Kubera, Integration of Planning and Real-Time Optimization in Olefins Productions, *Comput. Chem. Eng.*, **24**, 1645 (2000).

Georgiou, A., P. Taylor, R. Galloway, L. Casey, and A. Sapre. Plantwide Closed-Loop Real Time Optimization and Advanced Control of Ethylene Plant—(CLRTO) Improves Plant Profitability and Operability, *Proc. NPRA Computer Conference*, New Orleans, LA, November 1997.

Griva, I., S. G. Nash, and A. Sofer, *Linear and Nonlinear Optimization*, 2nd ed., SIAM, Philadelphia, PA, 2008

Latour, P. R. On Line Computer Optimization, 1. What It Is and Where to Do It, *Hydro. Proc.*, **58**(6), 73 (1979); 2. Benefits and Implementation, *Hydrocarb. Proc.*, **58**(7), 219 (1979).

Marlin, T. E., *Process Control*, 2d ed., McGraw-Hill, New York, 2000.

Marlin, T. E., and A. N. Hrymak, Real-Time Operations Optimization of Continuous Processes, in *Chemical Process Control V, AIChE Symp. Ser.* **93**, No. 316, 156 (1997).

McDonald, C. M., Synthesizing Enterprise-Wide Optimization with Global Information Technologies, *Foundations of Computer Aided Process Operations (FOCAPO), AIChE Symp. Ser.*, **94**, No. 320, 62 (1998).

Myers, R. H., and D. C. Montgomery, *Response Surface Methodology: Process and Product Optimization Using Designed Experiments*, 2d ed., Wiley, New York, 2002.

Narasimhan, S., and C. Jordache, *Data Reconciliation and Gross Error Detection*, Gulf Publishing, Houston, TX, 2000.

Nocedal, J., and S. J. Wright, *Numerical Optimization*, 2d ed., Springer, New York, 2006.

Perkins, J. D., Plant-wide Optimization: Opportunities and Challenges, *Foundations of Computer-Aided Process Operations*, J. F. Pekny and G. E. Blau (Eds.), *AIChE Symp. Ser.*, **94**, No. 320, 15 (1998).

Pike, R. W., *Optimization for Engineering Systems*, Van-Nostrand Reinhold, New York, 1986.

Schulz, C., and R. Rudof, Scheduling of a Multiproduct Polymer Plant, *Foundations of Computer-Aided Process Operations, AIChE Symp. Ser.*, **94**, No. 320, 224 (1998).

Shobrys, D. E., and D. C. White, Planning, Scheduling, and Control Systems: Why They Cannot Work Together," *Comput. Chem. Eng.*, **26**, 149 (2002).

Skogestad, S., Self-optimizing Control: The Missing Link Between Steady-State Optimization and Control. *Compute Chem. Eng.*, **24**, 569 (2000).

Soderstrom, T. A., T. F. Edgar, L. P. Russo, and R. E. Young, Industrial Application of a Large-Scale Dynamic Data Reconciliation Strategy, *Ind. Eng. Chem. Res.*, **39**, 1683 (2000).

Starks, D. M., and E. Arrieta, Maintaining AC & O Applications, Sustaining the Gain, AIChE Spring Meeting Houston, TX, March, 2007.

Timmons, C., J. Jackson, and D. C. White, Distinguishing On-line Optimization Benefits from Those of Advanced Controls, *Hydrocarb Proc.*, **79**(6), 69 (2000).

White, D. C., Save Energy Through Automation, *Chem. Eng. Prog.*, **106**, 26 (January, 2010).

EXERCISES

19.1 A laboratory filtration study has been carried out at constant rate. The filtration time (t_f in hours) required to build up a specific cake thickness has been correlated as

$$t_f = 5.3\, x_i e^{-3.6x_i + 2.7}$$

where x_i = mass fraction solids in the cake. Find the value of x_i that maximizes t_f using quadratic interpolation.

19.2 The thermal efficiency of a natural gas boiler versus air/fuel ratio is plotted in Fig. E19.2. Using physical arguments, explain why a maximum occurs.

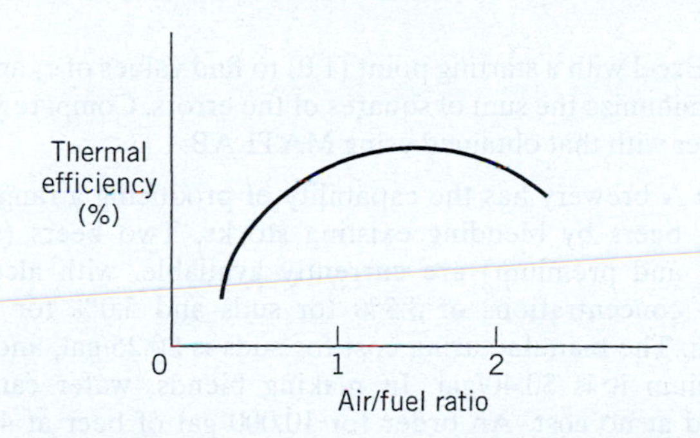

Figure E19.2

19.3 A plasma etcher has a yield of good chips that is influenced by pressure (X_1) and gas flow rate (X_2). Both X_1 and X_2 are scaled variables ($0 \le X_i \le 2$). A model has been developed based on operating data as follows:

$$Y = -0.1X_1^4 + 0.2X_2X_1^2 - 0.09X_2^2 - 0.11X_1^2 + 0.15X_1 + 0.5$$

Use Excel to maximize yield Y, using starting points of (1,1) and (0,0).

19.4 A specialty chemical is produced in a batch reactor. The time required to successfully complete one batch of product depends on the amount charged to (and produced from) the reactor. Using reactor data, a correlation is $t = 2.0P^{0.4}$, where P is the amount of product in pounds per batch and t is given in hours. A certain amount of nonproduction time is associated with each batch for charging, discharging, and minor maintenance, namely, 14 h/batch. The operating cost for the batch system is \$50/h. Other costs, including storage, depend on the size of each batch and have been estimated to be C_1 = \$800 $P^{0.7}$(\$/yr). The required annual production is 300,000 lb/yr, and the process can be operated 320 days/yr (24 h/day). Total raw material cost at this production level is \$400,000/yr.

(a) Formulate an objective function using P as the only variable. (Show algebraic substitution.)

(b) What are the constraints on P?

(c) Solve for the optimum value of P analytically. Check that it is a minimum. Also check applicable constraints.

19.5 A refinery processes two crude oils that have the yields shown in the following table. Because of equipment and storage limitations, production of gasoline, kerosene, and fuel oil must be limited as shown below. There are no plant limitations on the production of other products such as gas oils. The profit on processing crude No. 1 is \$2.00/bbl, and on crude No. 2 it is \$1.40/bbl. Find the optimum daily feed rates of the two crudes to this plant via linear programming using the Excel Solver.

Yields (Volume %)

	Crude No. 1	Crude No. 2	Maximum Allowable Production Rate (bbl/day)
Gasoline	70	31	6,000
Kerosene	6	9	2,400
Fuel oil	24	60	12,000

19.6 Linear programming is to be used to optimize the operation of the solvent splitter column shown in Fig. E19.6. The feed is naphtha, which has a value of \$40/bbl in its alternate use as a gasoline blending stock. The light ends sell at \$50/bbl, while the bottoms are passed through a second distillation column to yield two solvents. A medium solvent comprising 50 to 70% of the bottoms can be sold for \$70/bbl., while the remaining heavy solvent (30 to 50% of the bottoms) can be sold for \$40/bbl.

Another part of the plant requires 200 bbl/day of medium solvent; an additional 200 bbl/day can be sold to an external market. The maximum feed that can be processed in column 1 is 2,000 bbl/day. The operational cost (i.e., utilities) associated

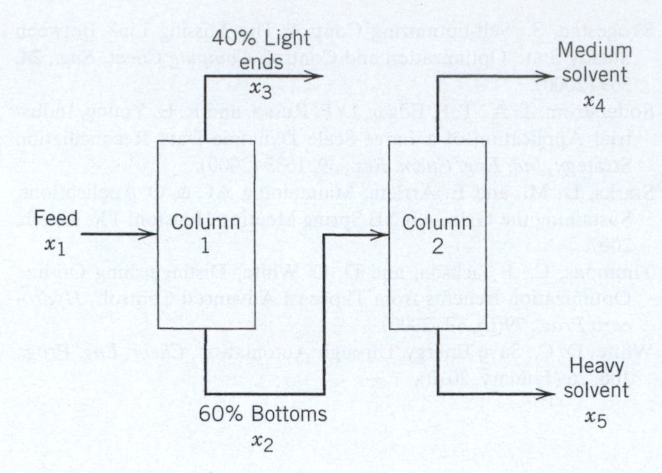

Figure E19.6

with each distillation column is $2.00/bbl feed. The operating range for column 2 is given as the percentage split of medium and heavy solvent. Solve the linear programming problem to determine the maximum revenue and percentages of output streams in column 2.

19.7 Reconciliation of inaccurate process measurements is an important problem in process control that can be solved using optimization techniques. The flow rates of streams B and C have been measured three times during the current shift (shown in Fig. E19.7). Some errors in the measurement devices exist. Assuming steady-state operation (w_A = constant), find the optimal value of w_A (flow rate in kg/h) that minimizes the sum of the squares of the errors for the material balance, $w_A + w_C = w_B$.

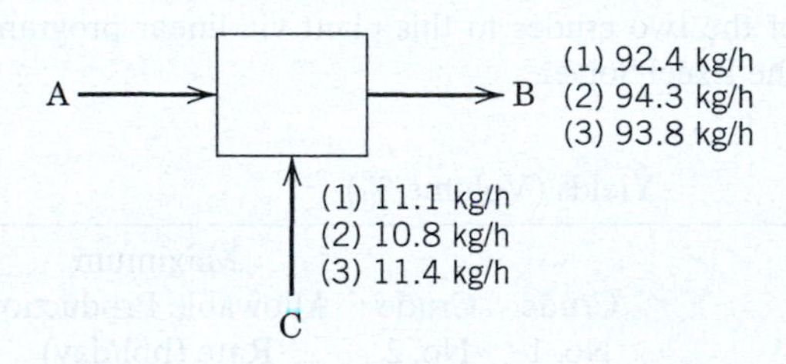

Figure E19.7

19.8 A reactor converts reactant BC to product CB by heating the material in the presence of an additive A (mole fraction = x_A). The additive can be injected into the reactor, while steam can be injected into a heating coil inside the reactor to provide heat. Some conversion can be obtained by heating without addition of A, and vice versa. The product CB can be sold for $50 per lb-mol. For 1 lb-mol of feed, the cost of the additive (in dollars per lb-mol feed) as a function of x_A is given by the formula $2.0 + 10x_A + 20x_A^2$. The cost of the steam (in dollars per lb-mol feed) as a function of S is $1.0 + 0.003S + 2.0 \times 10^{-6}S^2$ (S = lb steam/lb-mol feed). The yield equation is $y_{CB} = 0.1 + 0.3x_A + 0.0001S - 0.0001x_AS$.

$$y_{CB} = \frac{\text{lb-mol product CB}}{\text{lb-mol feed}}$$

(a) Formulate the profit function (basis of 1.0 lb-mol feed) in terms of x_A and S.

$$f = \text{income} - \text{costs}$$

(b) Maximize f subject to the constraints

$$0 \le x_A \le 1 \qquad S \ge 0$$

19.9 Optimization methods can be used to fit equations to data. Parameter estimation involves the computation of unknown parameters that minimize the squared error between data and the proposed mathematical model. The step response of an overdamped second-order dynamic process can be described using the equation

$$\frac{y(t)}{K} = \left(1 - \frac{\tau_1 e^{-t/\tau_1} - \tau_2 e^{-t/\tau_2}}{\tau_1 - \tau_2}\right)$$

where τ_1 and τ_2 are process time constants and K is the process gain.

The following normalized data have been obtained from a unit step test (K is equal to $y(\infty)$):

time, t	0	1	2	3	4	5
y_i/K	0.0	0.0583	0.2167	0.360	0.488	0.600
t	6	7	8	9	10	
y_i/K	0.692	0.772	0.833	0.888	0.925	

Use Excel with a starting point (1,0) to find values of τ_1 and τ_2 that minimize the sum of squares of the errors. Compare your answer with that obtained using MATLAB.

19.10 A brewery has the capability of producing a range of beers by blending existing stocks. Two beers (suds and premium) are currently available, with alcohol concentrations of 3.5% for suds and 5.0% for premium. The manufacturing cost for suds is $0.25/gal, and for premium it is $0.40/gal. In making blends, water can be added at no cost. An order for 10,000 gal of beer at 4.0% has been received for this week. There is a limited amount of suds available (9000 gal), and, because of aging problems, the brewery must use at least 2,000 gal of suds this week. What amounts of suds, premium, and water must be blended to fill the order at minimum cost?

19.11 A specialty chemicals facility manufactures two products A and B in barrels. Products A and B utilize the same raw material; A uses 120 kg/bbl, while B requires 100 kg/bbl. There is an upper limit on the raw material supply of 9,000 kg/day. Another constraint is warehouse storage space (40 m^2 total; both A and B require 0.5 m^2/bbl). In addition, production time is limited to 7 h per day. A and B can be produced at 20 bbl/h and 10 bbl/h, respectively. If the profit per bbl is $10 for A and $14 for B, find the production levels that maximize profit.

19.12 Supervisory control often involves the optimization of set points in order to maximize profit. Can the same results be achieved by optimizing PID controller tuning (K_c, τ_I, τ_D), in order to maximize profits? Are regulatory (feedback) control and supervisory control complementary?

19.13 A dynamic model of a continuous-flow, biological chemostat has the form

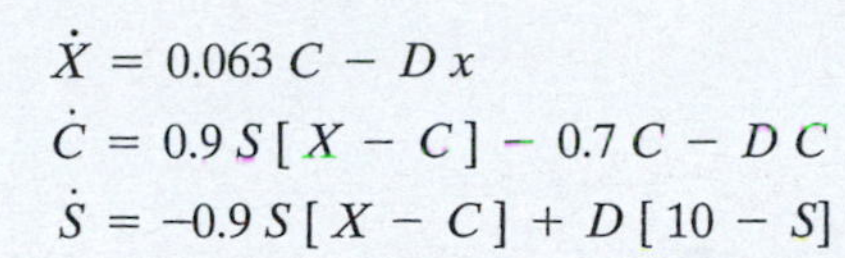

$$\dot{X} = 0.063\,C - D\,x$$
$$\dot{C} = 0.9\,S\,[\,X - C\,] - 0.7\,C - D\,C$$
$$\dot{S} = -0.9\,S\,[\,X - C\,] + D\,[\,10 - S\,]$$

where X is the biomass concentration, S is the substrate concentration, and C is a metabolic intermediate concentration. The dilution rate, D, is an independent variable, which is defined to be the flow rate divided by the chemostat volume.

Determine the value of D, which maximizes the steady-state production rate of biomass, f, given by

$$f = DX$$

19.14 A reversible chemical reaction, A ⇄ B, occurs in the isothermal continuous stirred-tank reactor shown in Fig. E19.14. The rate expressions for the forward and reverse reactions are

$$r_1 = k_1 C_A$$
$$r_2 = k_2 C_B$$

Using the information given below, use a numerical search procedure to determine the value of F_B (L/h) that maximizes the production rate of C_B (i.e., the amount of C_B that leaves the reactor, mol B/h). The allowable values of F_B are $0 \leq F_B \leq 200$ L/h.

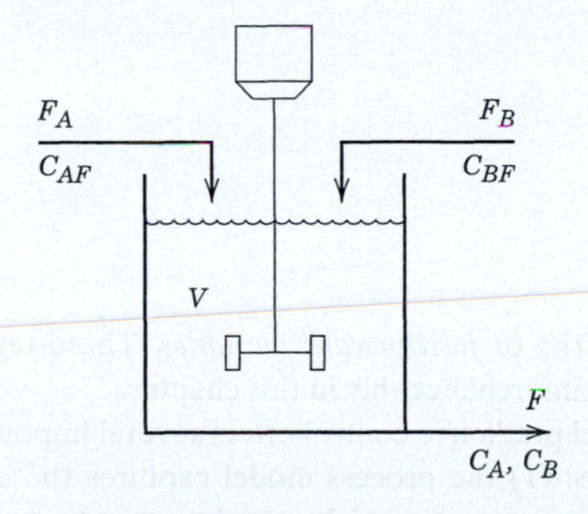

Figure E19.14

Available Information

(i) The reactor is perfectly mixed.

(ii) The volume of liquid, V, is maintained constant using an overflow line (not shown in the diagram).

(iii) The following parameters are kept constant at the indicated numerical values:

$$V = 200 \text{ L} \qquad F_A = 150 \text{ L/h}$$
$$C_{AF} = 0.3 \text{ mol A/L} \qquad C_{BF} = 0.3 \text{ mol B/L}$$
$$k_1 = 2\text{h}^{-1} \qquad k_2 = 1.5\text{ h}^{-1}$$

19.15 A reversible chemical reaction, A ⇄ B, occurs in the isothermal continuous stirred-tank reactor shown in Fig. E19.14. The rate expressions for the forward and reverse reactions are

$$r_1 = k_1 C_A \qquad r_2 = k_2 C_B$$

where the rate constants have the following temperature dependence:

$$k_1 = 3.0 \times 10^6 \exp(-5000/T)$$
$$k_2 = 6.0 \times 10^6 \exp(-5500/T)$$

Each rate constant has units of h^{-1}, and T is in K.

Use the MATLAB Optimization Toolbox or Excel to determine the optimum values of temperature $T(K)$ and flow rate F_B (L/h) that maximize the steady-state production rate of component B. The allowable values are $0 \leq F_B \leq 200$ and $300 \leq T \leq 500$.

Available Information

(i) The reactor is perfectly mixed.

(ii) The volume of liquid, V, is maintained constant using an overflow line (not shown in the diagram).

(iii) The following parameters are kept constant at the indicated numerical values:

$$V = 200 \text{ L} \qquad F_A = 150 \text{ L/h}$$
$$C_{AF} = 0.3 \text{ mol A/L} \qquad C_{BF} = 0.3 \text{ mol B/L}$$

Chapter 20

Model Predictive Control

CHAPTER CONTENTS

In this chapter we consider *model predictive control (MPC)*, an important advanced control technique for difficult multivariable control problems. The basic MPC concept can be summarized as follows. Suppose that we wish to control a multiple-input, multiple-output process while satisfying inequality constraints on the input and output variables. If a reasonably accurate dynamic model of the process is available, model and current measurements can be used to predict future values of the outputs. Then the appropriate changes in the input variables can be calculated based on both predictions and measurements. In essence, the changes in the individual input variables are coordinated after considering the input-output relationships represented by the process model. In MPC applications, the output variables are also referred to as *controlled variables* or *CVs*, while the input variables are also called *manipulated variables* or *MVs*. Measured disturbance variables are called *DVs* or *feedforward variables*. These terms will be used interchangeably in this chapter.

Model predictive control offers several important advantages: (1) the process model captures the dynamic and static interactions between input, output, and disturbance variables, (2) constraints on inputs and outputs are considered in a systematic manner, (3) the control calculations can be coordinated with the calculation of optimum set points, and (4) accurate model predictions can provide early warnings of potential problems. Clearly, the success of MPC (or any other model-based approach) depends on the accuracy of the process model. Inaccurate predictions can make matters worse, instead of better.

First-generation MPC systems were developed independently in the 1970s by two pioneering industrial research groups. Dynamic Matrix Control *(DMC)*, devised by Shell Oil (Cutler and Ramaker, 1980), and a related approach developed by ADERSA (Richalet et al., 1978) have quite similar capabilities. An adaptive MPC technique, Generalized Predictive Control (GPC), developed by Clarke et al. (1987) has also received considerable attention. Model predictive control has had a major impact on industrial practice. For example, an MPC survey by Qin and Badgwell (2003) reported that there were over 4,500 applications worldwide by the end of 1999, primarily in oil refineries and petrochemical plants. In these industries, MPC has become the method of choice for difficult multivariable control problems that include inequality constraints.

In view of its remarkable success, MPC has been a popular subject for academic and industrial research. Major extensions of the early MPC methodology have been developed, and theoretical analysis has provided insight into the strengths and weaknesses of MPC. Informative reviews of MPC theory and practice are available in books (Camacho and Bordons, 2003; Maciejowski, 2002; Rossiter, 2003; Richalet and O'Donovan, 2009); tutorials (Hokanson and Gerstle, 1992; Rawlings, 2000), and survey papers (Morari and Lee, 1999; Qin and Badgwell, 2003; Canney, 2003; Kano and Ogawa, 2009).

20.1 OVERVIEW OF MODEL PREDICTIVE CONTROL

The overall objectives of an MPC controller have been summarized by Qin and Badgwell (2003):

1. Prevent violations of input and output constraints.
2. Drive some output variables to their optimal set points, while maintaining other outputs within specified ranges (see Section 20.4.2).
3. Prevent excessive movement of the input variables.
4. Control as many process variables as possible when a sensor or actuator is not available.

A block diagram of a model predictive control system is shown in Fig. 20.1. A process model is used to predict the current values of the output variables. The residuals, the differences between the actual and predicted outputs, serve as the feedback signal to a *Prediction* block. The predictions are used in two types of MPC calculations that are performed at each sampling instant: set-point calculations and control calculations. Inequality constraints on the input and output variables, such as upper and lower limits, can be included in either type of calculation. Note that the MPC configuration is similar to both the internal model control configuration in Chapter 12 and the Smith predictor configuration of Chapter 16, because the model acts in parallel with the process and the residual serves as a feedback signal. However, the coordination of the control and set-point calculations is a unique feature of MPC. Furthermore, MPC has had a much greater impact on industrial practice than IMC or Smith predictor, because it is more suitable for constrained MIMO control problems.

The set points for the control calculations, also called *targets*, are calculated from an economic optimization based on a steady-state model of the process, traditionally, a linear model. Typical optimization objectives include maximizing a profit function, minimizing a cost function, or maximizing a production rate. The optimum values of set points change frequently due to varying process conditions, especially changes in the inequality constraints (see Chapter 19). The constraint changes are due to variations in process conditions, equipment, and instrumentation, as well as economic data such as prices and costs. In MPC the set points are typically calculated each time the control calculations are performed, as discussed in Section 20.5.

The MPC calculations are based on current measurements and predictions of the future values of the outputs. The objective of the MPC control calculations is to determine a sequence of *control moves* (that is, manipulated input changes) so that the predicted response moves to the set point in an optimal manner. The actual output y, predicted output $\hat{y}$, and manipulated input u for SISO control are shown in Fig. 20.2. At the current sampling instant, denoted by k, the MPC strategy calculates a set of M values of the input $\{u(k + i - 1),$

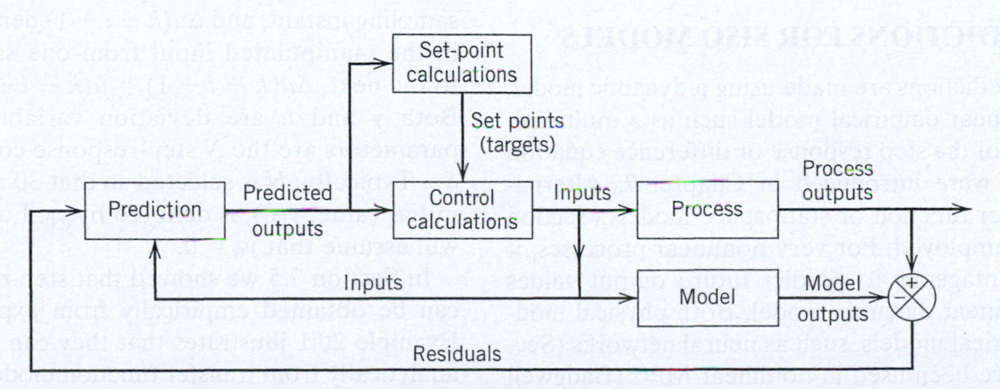

Figure 20.1 Block diagram for model predictive control.

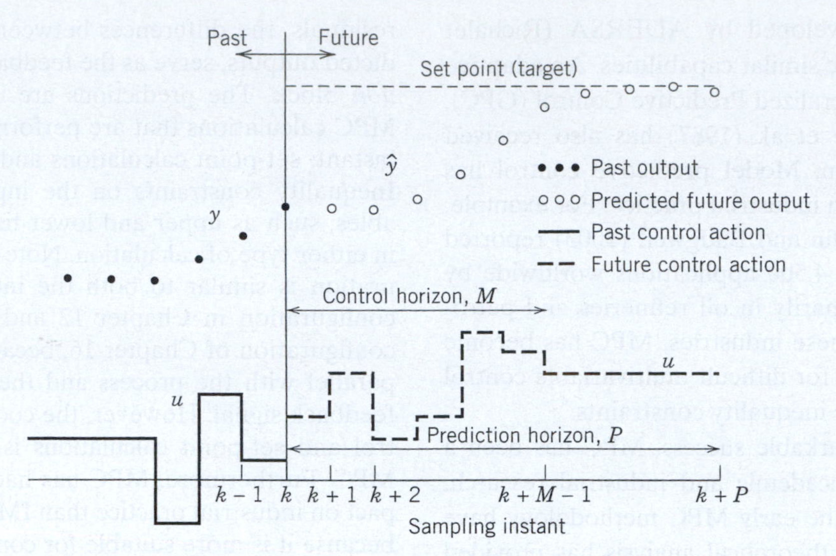

Figure 20.2 Basic concept for model predictive control.

$i = 1, 2, \ldots, M\}$. The set consists of the current input $u(k)$ and $M - 1$ future inputs. The input is held constant after the M control moves. The inputs are calculated so that a set of P predicted outputs $\hat{y}\,(k+i), i = 1, 2, \ldots, P\}$ reaches the set point in an optimal manner. The control calculations are based on optimizing an objective function (cf. Section 20.4). The number of predictions P is referred to as the *prediction horizon* while the number of control moves M is called the *control horizon*.

A distinguishing feature of MPC is its *receding horizon approach*. Although a sequence of M control moves is calculated at each sampling instant, only the first move is actually implemented. Then a new sequence is calculated at the next sampling instant, after new measurements become available; again only the first input move is implemented. This procedure is repeated at each sampling instant. But why is an M-step control strategy calculated if only the first step is implemented? We will answer this question in Section 20.4.

20.2 PREDICTIONS FOR SISO MODELS

The MPC predictions are made using a dynamic model, typically a linear empirical model such as a multivariable version of the step response or difference equation models that were introduced in Chapter 7. Alternatively, transfer function or state-space models (Section 6.5) can be employed. For very nonlinear processes, it can be advantageous to predict future output values using a nonlinear dynamic model. Both physical models and empirical models, such as neural networks (Section 7.3), have been used in nonlinear MPC (Badgwell and Qin, 2001; White, 2008). Step-response models offer the advantage that they can represent stable processes with unusual dynamic behavior that cannot be accurately described by simple transfer function models (cf. Example 7.6). Their main disadvantage is the large number of model parameters. Although step-response models are not suitable for unstable processes, they can be modified to represent integrating processes, as shown in Section 20.2.2.

Next, we demonstrate how step-response models can be used to predict future outputs. Similar predictions can be made using other types of linear models such as transfer function or state-space models.

The step-response model of a stable, single-input, single-output process can be written as

$$y(k+1) = y_0 + \sum_{i=1}^{N-1} S_i \Delta u(k-i+1) + S_N u(k-N+1) \qquad (20\text{-}1)$$

where $y(k+1)$ is the output variable at the $(k+1)$-sampling instant, and $\Delta u(k-i+1)$ denotes the change in the manipulated input from one sampling instant to the next, $\Delta u(k-i+1) = u(k-i+1) - u(k-i)$. Both y and u are deviation variables. The model parameters are the N step-response coefficients, S_1 to S_N. Typically, N is selected so that $30 \leq N \leq 120$. The initial value, $y(0)$, is denoted by y_0. For simplicity, we will assume that $y_0 = 0$.

In Section 7.5 we showed that step-response models can be obtained empirically from experimental data. Example 20.1 illustrates that they can also be derived analytically from transfer function models.

EXAMPLE 20.1

Consider a first-order-plus-time-delay model:

$$\frac{Y(s)}{U(s)} = \frac{Ke^{-\theta s}}{\tau s + 1} \tag{20-2}$$

(a) Derive the equivalent step-response model by considering the analytical solution to a unit step change in the input.

(b) Calculate the step-response coefficients, $\{S_i\}$, for the following parameter values: $K = 5$, $\tau = 15$ min, $\theta = 3$ min, and a sampling period of $\Delta t = 1$ min. Also, calculate and plot the response $y(k)$ for $0 \leq k \leq 80$ after a step change in u from 0 to 3 occurs at $t = 2$ min.

SOLUTION

(a) The step response for a first-order model without a time delay ($\theta = 0$) was derived in Chapter 5

$$y(t) = KM(1 - e^{-t/\tau}) \tag{5-18}$$

where M is the magnitude of the step change. The corresponding response for the model with a time delay is

$$\begin{aligned} y(t) &= 0 && \text{for } t \leq \theta \\ y(t) &= KM\,(1 - e^{-(t-\theta)/\tau}) && \text{for } t > \theta \end{aligned} \tag{20-3}$$

The sampling instants are denoted by $t = i\Delta t$ where Δt is the sampling period and $i = 1, 2, \ldots$. Substituting $t = i\Delta t$ into (20-3) gives the response for $0 \leq i \leq 80$:

$$\left.\begin{aligned} y(i\Delta t) &= 0 && \text{for } i\Delta t \leq \theta \\ y(i\Delta t) &= KM(1 - e^{-(i\Delta t-\theta)/\tau}) && \text{for } \Delta t > \theta \end{aligned}\right\} \tag{20-4}$$

where $i = 1, 2, \ldots, 80$

The number of step-response coefficients, N in (20-1), is specified to be $N = 80$ so that $N\Delta t$ is slightly larger than the process settling time of approximately $5\tau + \theta$. As indicated in Section 7.5, the ith step-response coefficient is the value of the unit step response at the ith sampling instant. Thus, the step-response coefficients can be determined from (20-4) after setting $M = 1$:

$$\left.\begin{aligned} S_i &= 0 && \text{for } i\Delta t \leq \theta \\ S_i &= K(1 - e^{-(i\Delta t-\theta)/\tau}) && \text{for } i\Delta t > \theta \end{aligned}\right\} \tag{20-5}$$

where $i = 1, 2, \ldots, 80$

(b) Substituting numerical values into (20-5) gives the step-response coefficients in Table 20.1. The step response $y(k)$ in Fig. 20.3 can be calculated either from (20-1) and (20-5), or from (20-4). For $\theta = 2$ min, $S_1 = S_2 = 0$, and the step response is zero until $t > 2$ min. The new steady-state value is $y = 15$ because the steady-state gain in (20-2) is $K = 5$ and the magnitude of the step change is $M = 3$. Because $\Delta t = 1$ min and the step change occurs at $t = 3$ min, $\Delta u(k) = M = 3$ for $k = 3$ and $\Delta u(k) = 0$ for all other values of k. Recall that $\Delta u(k)$ is defined as $\Delta u(k) \triangleq u(k) - u(k-1)$.

For this example, the response $y(k)$ could be calculated analytically from (20-4) because a transfer function model was assumed. However, in many MPC applications, the transfer function model is not known, and thus the response must be calculated from the step-response model in (20-1).

Table 20.1 Step-Response Coefficients for Example 20.1

Sampling Instant	S_i	Sampling Instant	S_i	Sampling Instant	S_i
1	0	28	4.06	55	4.84
2	0	29	4.12	56	4.85
3	0	30	4.17	57	4.86
4	0.32	31	4.23	58	4.87
5	0.62	32	4.28	59	4.88
6	0.91	33	4.32	60	4.89
7	1.17	34	4.37	61	4.90
8	1.42	35	4.41	62	4.90
9	1.65	36	4.45	63	4.91
10	1.86	37	4.48	64	4.91
11	2.07	38	4.52	65	4.92
12	2.26	39	4.55	66	4.93
13	2.43	40	4.58	67	4.93
14	2.60	41	4.60	68	4.93
15	2.75	42	4.63	69	4.94
16	2.90	43	4.65	70	4.94
17	3.03	44	4.68	71	4.95
18	3.16	45	4.70	72	4.95
19	3.28	46	4.72	73	4.95
20	3.39	47	4.73	74	4.96
21	3.49	48	4.75	75	4.96
22	3.59	49	4.77	76	4.96
23	3.68	50	4.78	77	4.96
24	3.77	51	4.80	78	4.97
25	3.85	52	4.81	79	4.97
26	3.92	53	4.82	80	4.97
27	3.99	54	4.83		

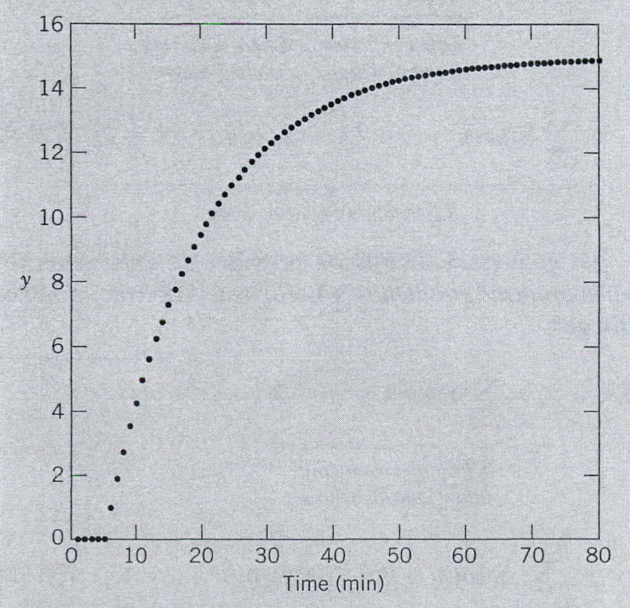

Figure 20.3 Step response for Example 20.1.

Model predictive control is based on predictions of future outputs over a prediction horizon, P. We now consider the calculation of these predictions. Let k denote the current sampling instant and $\hat{y}(k+1)$ denote the prediction of $y(k+1)$ that is made at time k. If $y_0 = 0$, this one-step-ahead prediction can be obtained from Eq. (20-1) by replacing $y(k+1)$ with $\hat{y}(k+1)$:

$$\hat{y}(k+1) = \sum_{i=1}^{N-1} S_i \Delta u(k-i+1) + S_N u(k-N+1) \tag{20-6}$$

Equation 20-6 can be expanded as

$$\hat{y}(k+1) = \underbrace{S_1 \Delta u(k)}_{\substack{\text{Effect of current}\\ \text{control action}}} + \underbrace{\sum_{i=2}^{N-1} S_i \Delta u(k-i+1) + S_N u(k-N+1)}_{\text{Effect of past control actions}} \tag{20-7}$$

The first term on the right-hand side indicates the effect of the current manipulated input $u(k)$ because $\Delta u(k) = u(k) - u(k-1)$. The second term represents the effects of past inputs, $\{u(i), i < k\}$. An analogous expression for a two-step-ahead prediction can be derived in a similar manner. Substitute $k = k' + 1$ into Eq. 20-6:

$$\hat{y}(k'+2) = \sum_{i=1}^{N-1} S_i \Delta u(k'-i+2) + S_N u(k'-N+2) \tag{20-8}$$

Because Eq. 20-8 is valid for all positive values of k', without loss of generality, we can replace k' with k and then expand the right-hand side to identify the contributions relative to the current sampling instant, k:

$$\hat{y}(k+2) = \underbrace{S_1 \Delta u(k+1)}_{\substack{\text{Effect of future}\\ \text{control action}}} + \underbrace{S_2 \Delta u(k)}_{\substack{\text{Effect of current}\\ \text{control action}}} + \underbrace{\sum_{i=3}^{N-1} S_i \Delta u(k-i+2) + S_N u(k-N+2)}_{\text{Effect of past control actions}} \tag{20-9}$$

An analogous derivation provides an expression for a *j-step-ahead prediction* where j is an arbitrary positive integer:

$$\hat{y}(k+j) = \underbrace{\sum_{i=1}^{j} S_i \Delta u(k+j-i)}_{\substack{\text{Effect of current and}\\ \text{future control actions}}} + \underbrace{\sum_{i=j+1}^{N-1} S_i \Delta u(k+j-i) + S_N u(k+j-N)}_{\text{Effect of past control actions}} \tag{20-10}$$

The second and third terms on the right-hand side of Eq. 20-10 represent the predicted response when there are no current or future control actions, that is, the predicted response when $u(k+i) = u(k-1)$ for $i \geq 0$, or equivalently, $\Delta u(k+i) = 0$ for $i \geq 0$. Because this term accounts for past control actions, it is referred to as the *predicted unforced response* and is denoted by the symbol, $\hat{y}^o(k+j)$. Thus, we define $\hat{y}^o(k+j)$ as

$$\hat{y}^o(k+j) \triangleq \sum_{i=j+1}^{N-1} S_i \Delta u(k+j-i) + S_N u(k+j-N) \tag{20-11}$$

and write Eq. 20-10 as

$$\hat{y}(k+j) = \sum_{i=1}^{j} S_i \Delta u(k+j-i) + \hat{y}^o(k+j) \tag{20-12}$$

Examples 20.2 and 20.3 demonstrate that Eq. 20-12 can be used to derive a simple predictive control law based on a *single* prediction.

EXAMPLE 20.2

Derive a predictive control law that is based on the following concept. A single control move, $\Delta u(k)$, is calculated so that the J-step-ahead prediction is equal to the set point, that is, $\hat{y}(k+J) = y_{sp}$ where integer J is a tuning parameter. This sampling instant, $k+J$, is referred to as a *coincidence point*. Assume that u is held constant after the single control move, so that $\Delta u(k+i) = 0$ for $i > 0$.

SOLUTION

In the proposed predictive control strategy, only a single prediction for J steps ahead is considered. Thus, we let $j = J$ in Eq. 20-12. Similarly, because we are only interested in calculating the current control move, $\Delta u(k)$, the future control moves in Eq. 20-12 are set equal to zero: $\Delta u(k+J-i) = 0$ for $i = 1, 2, \ldots, J-1$. Thus, (20-12) reduces to

$$\hat{y}(k+J) = S_J \Delta u(k) + \hat{y}^o(k+J) \tag{20-13}$$

Setting $\hat{y}(k+J) = y_{sp}$ and rearranging gives the desired predictive controller:

$$\Delta u(k) = \frac{y_{sp} - \hat{y}^o(k+J)}{S_J} \tag{20-14}$$

The predicted unforced response $\hat{y}^o(k+J)$ can be calculated from Eq. 20-11 with $j = J$.

The control law in (20-14) is based on a single prediction that is made for J steps in the future. Note that the control law can be interpreted as the inverse of the predictive model in (20-13).

EXAMPLE 20.3

Apply the predictive control law of Example 20.2 to a fifth-order process:

$$\frac{Y(s)}{U(s)} = \frac{1}{(5s+1)^5} \tag{20-15}$$

Evaluate the effect of tuning parameter J on the set-point responses for values of $J = 3, 4, 6,$ and 8 and $\Delta t = 5$ min.

SOLUTION

The y and u responses for a unit set-point change at $t = 0$ are shown in Figs. 20.4 and 20.5, respectively. As J increases, the y responses become more sluggish while the u responses become smoother. These trends occur because larger values of J allow the predictive controller more time before the J-step ahead prediction $\hat{y}(k+J)$ must equal the set point. Consequently, less strenuous control action is required. The Jth step-response coefficient S_J increases monotonically as J increases. Consequently, the input moves calculated from (20-14) tend to become smaller as S_J increases. (The u responses for $J = 4$ and 8 are omitted from Fig. 20.5.)

The previous two examples have considered a simple predictive controller based on single prediction made J steps ahead. Now, we consider the more typical situation in which the MPC calculations are based on multiple predictions rather than on a single prediction. The notation is greatly simplified if vector-matrix notation is employed. Consequently, we define a vector of predicted responses for the next P sample instants as

$$\hat{\mathbf{Y}}(k+1) \triangleq col\,[\hat{y}(k+1), \hat{y}(k+2), \ldots, \hat{y}(k+P)] \tag{20-16}$$

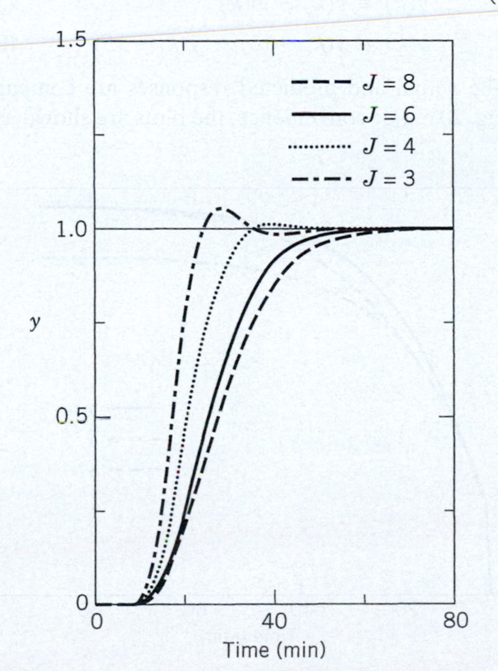

Figure 20.4 Set-point responses for Example 20.3 and different values of J.

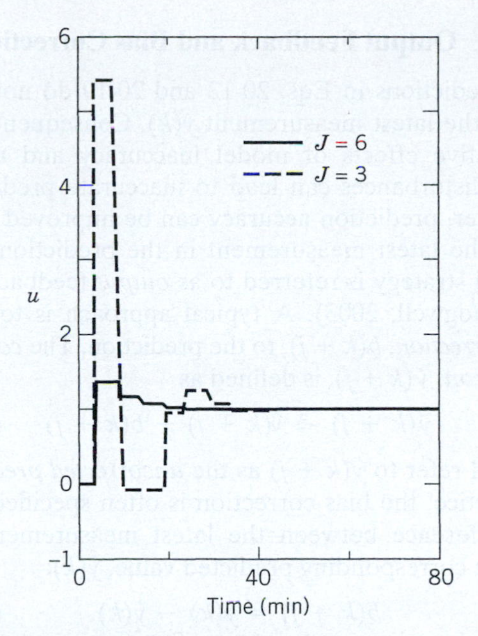

Figure 20.5 Input responses for Fig. 20.4.

where col denotes a column vector. Similarly, a vector of predicted unforced responses from Eq. 20-11 is defined as

$$\hat{\mathbf{Y}}^o(k+1) \triangleq col\,[\hat{y}^o(k+1), \hat{y}^o(k+2), \ldots, \hat{y}^o(k+P)] \tag{20-17}$$

Define $\Delta \mathbf{U}(k)$ to be a vector of control actions for the next M sampling instants:

$$\Delta \mathbf{U}(k) \triangleq col\,[\Delta u(k), \Delta u(k+1), \ldots, \Delta u(k+M-1)] \tag{20-18}$$

The control horizon M and prediction horizon P are key design parameters, as discussed in Section 20.6. In general, $M \le P$ and $P \le N + M$.

The MPC control calculations are based on calculating $\Delta \mathbf{U}(k)$ so that the predicted outputs move optimally to the new set points. For the control calculations, the model predictions in Eq. 20-12 are conveniently written in vector-matrix notation as

$$\hat{\mathbf{Y}}(k+1) = \mathbf{S}\Delta\mathbf{U}(k) + \hat{\mathbf{Y}}^o(k+1) \tag{20-19}$$

where $\mathbf{S}$ is the $P \times M$ *dynamic matrix*:

$$\mathbf{S} \triangleq \begin{bmatrix} S_1 & 0 & \cdots & 0 \\ S_2 & S_1 & 0 & \vdots \\ \vdots & \vdots & \ddots & 0 \\ S_M & S_{M-1} & \cdots & S_1 \\ S_{M+1} & S_M & \cdots & S_2 \\ \vdots & \vdots & & \vdots \\ S_P & S_{P-1} & \cdots & S_{P-M+1} \end{bmatrix} \tag{20-20}$$

Equations 20-19 and 20-20 can be derived from (20-12) and (20-16) to (20-18).

20.2.1 Output Feedback and Bias Correction

The predictions in Eqs. 20-12 and 20-19 do not make use of the latest measurement $y(k)$. Consequently, the cumulative effects of model inaccuracy and unmeasured disturbances can lead to inaccurate predictions. However, prediction accuracy can be improved by utilizing the latest measurement in the predictions. This general strategy is referred to as *output* feedback (Qin and Badgwell, 2003). A typical approach is to add a *bias correction*, $b(k + j)$, to the prediction. The *corrected prediction*, $\tilde{y}(k + j)$, is defined as

$$\tilde{y}(k + j) \triangleq \hat{y}(k + j) + b(k + j) \quad (20\text{-}21)$$

We will refer to $\hat{y}(k + j)$ as the *uncorrected prediction*. In practice, the bias correction is often specified to be the difference between the latest measurement $y(k)$ and the corresponding predicted value, $\hat{y}(k)$:

$$b(k + j) = y(k) - \hat{y}(k) \quad (20\text{-}22)$$

The difference, $y(k) - \tilde{y}(k)$, is also referred to as a *residual* or an *estimated disturbance*. The block diagram for MPC in Fig. 20.1 includes the bias correction.

In (20-22) $\hat{y}(k)$ is a one-step ahead prediction made at the previous sampling instant, $k - 1$. Using Eq. 20-22 is equivalent to assuming that a process disturbance is added to the output and is constant for $j = 1, 2, \ldots, P$. Furthermore, the assumed value of the additive disturbance is the residual, $y(k) - \tilde{y}(k)$.

Substituting Eq. 20-22 into 20-21 gives

$$\tilde{y}(k + j) \triangleq \hat{y}(k + j) + [y(k) - \hat{y}(k)] \quad (20\text{-}23)$$

In a similar fashion, adding the bias correction to the right side of Eq. 20-19 provides a vector of corrected predictions,

$$\tilde{\mathbf{Y}}(k+1) = \mathbf{S}\Delta\mathbf{U}(k) + \hat{\mathbf{Y}}^o(k+1) + [y(k) - \hat{y}k]\mathbf{1} \quad (20\text{-}24)$$

where $\mathbf{1}$ is a P-dimensional column vector with each element having a value of one. Thus the same correction is made for all P predictions. Vector $\tilde{\mathbf{Y}}(k + 1)$ is defined as

$$\tilde{\mathbf{Y}}(k + 1) \triangleq col\,[\tilde{y}(k + 1), \tilde{y}(k + 2), \ldots, \tilde{y}(k + P)] \quad (20\text{-}25)$$

Incorporating output feedback as a bias correction has been widely applied, but it can result in excessively sluggish responses for certain classes of disturbances. Consequently, other types of output feedback and disturbance estimation methods have been proposed (Maciejowski, 2002; Qin and Badgwell, 2003).

EXAMPLE 20.4

The benefits of using corrected predictions will be illustrated by a simple example, the first-order plus time-delay model:

$$\frac{Y(s)}{U(s)} = \frac{5e^{-2s}}{15s + 1} \quad (20\text{-}26)$$

Assume that the disturbance transfer function is identical to the process transfer function, $G_d(s) = G_p(s)$. A unit change in u occurs at time $t = 2$ min, and a step disturbance, $d = 0.15$, occurs at $t = 8$ min. The sampling period is $\Delta t = 1$ min.

(a) Compare the process response $y(k)$ with the predictions that were made 15 steps earlier based on a step-response model with $N = 80$. Consider both the corrected prediction $\tilde{y}(k)$ and the uncorrected prediction $\hat{y}(k)$ over a time period, $0 \le k \le 90$.

(b) Repeat (a) for the situation where the step-response coefficients are calculated using an incorrect model:

$$\frac{Y(s)}{U(s)} = \frac{4e^{-2s}}{20s + 1} \quad (20\text{-}27)$$

SOLUTION

The output response to the step changes in u and d can be derived from (20-26) using the analytical techniques developed in Chapters 3 and 4. Because $\theta = 2$ min and the step in u begins at $t = 2$ min, $y(t)$ first starts to respond at $t = 5$ min. The disturbance at $t = 8$ min begins to affect y at $t = 11$ min. Thus, the response can be written as

$$\begin{aligned} y(t) &= 0 && \text{for } t \le 4 \text{ min} \\ y(t) &= 5(1)(1 - e^{-(t-4)/15}) && \text{for } 4 < t \le 10 \text{ min} \\ y(t) &= 5(1)(1 - e^{-(t-4)/15}) && \text{for } t > 10 \text{ min} \\ &\quad + 5(0.15)(1 - e^{-(t-10)/15}) \end{aligned} \quad (20\text{-}28)$$

The 15-step-ahead prediction, $\hat{y}(k + 15)$, can be obtained using Eq. 20-12 with $j = 15$ and $N = 80$. The corrected prediction, $\hat{y}(k + 15)$, can be calculated from Eqs. 20-21 and 20-22 with $j = 15$. But in order to compare actual and predicted responses, it is more convenient to write these equations in an equivalent form:

$$\tilde{y}(k) \triangleq \hat{y}(k) + b(k) \quad (20\text{-}29)$$

$$b(k) = y(k - 15) - \hat{y}(k - 15) \quad (20\text{-}30)$$

(a) The actual and predicted responses are compared in Fig. 20.6. For convenience, the plots are shown as lines

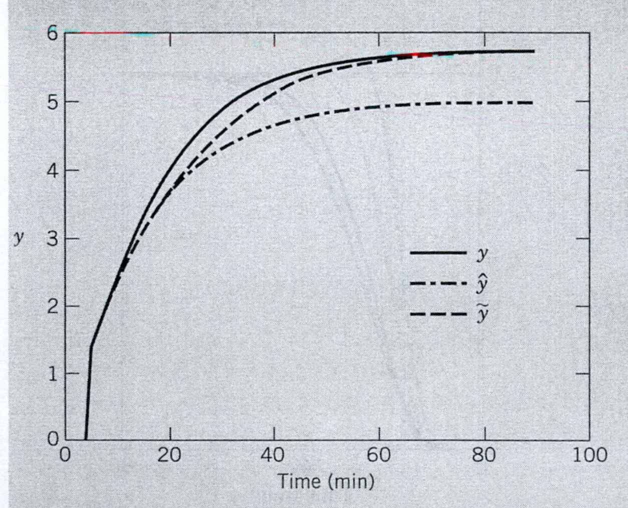

Figure 20.6 Comparison of actual (y), predicted ($\hat{y}$), and corrected ($\tilde{y}$) responses when the model is perfect.

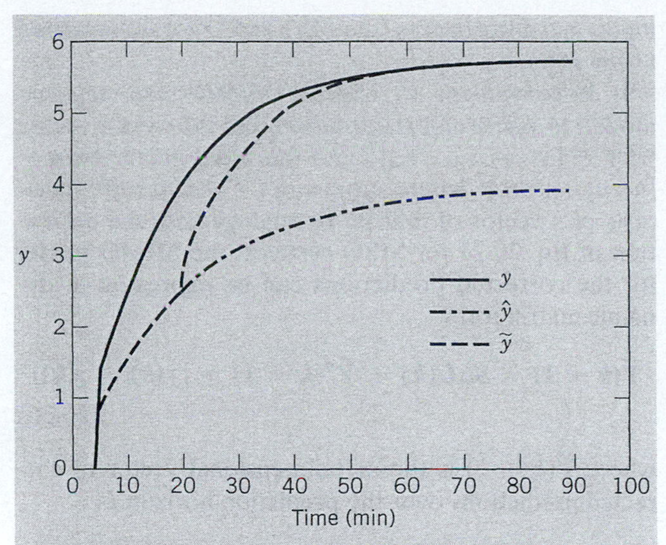

Figure 20.7 Comparison of actual and predicted responses for plant-model mismatch.

rather than as discrete points. After the step change in u at $t = 2$ min (or equivalently, at $k = 4$), the 15-step-ahead predictions are identical to the actual response until the step disturbance begins to affect $y(k)$ starting at $k = 11$. For $k > 10$, $\hat{y}(k) < y(k)$ because $\hat{y}(k)$ does not include the effect of the unknown disturbance. Note that $\tilde{y}(k) = \hat{y}(k)$ for $10 < k < 25$ because $b(k) = 0$ during this period. For $k > 25$, $b(k) \neq 0$ and $\tilde{y}(k)$ converges to $y(k)$. Thus, the corrected prediction $\tilde{y}(k)$ is more accurate than the uncorrected prediction, $\hat{y}(k)$.

(b) Figure 20.7 compares the actual and predicted responses for the case of the plant-model mismatch in Eqs. 20-26 and 20-27. The responses in Fig. 20.7 are similar to those in Fig. 20.6, but there are a few significant differences. Both of the predicted responses in Fig. 20.7 differ from the actual response for $t > 4$, as a result of the model inaccuracy. Figure 20.7 demonstrates that the corrected predictions are much more accurate than the uncorrected predictions even when a significant plant-model mismatch occurs. This improvement occurs because new information is used as soon as it becomes available.

20.2.2 Extensions of the Basic MPC Model Formulation

We will now consider several extensions of the basic MPC problem formulation that are important for practical applications.

Integrating Processes

The standard step-response model in Eq. 20-6 is not appropriate for an integrating process because its step response is unbounded. However, because the output rate of change, $\Delta y(k) = y(k+1) - y(k)$, is bounded, a simple modification eliminates this problem. Replacing $\hat{y}(k+1)$ in Eq. 20-6 by $\Delta\hat{y}(k + 1 = \hat{y}(k+1) - \hat{y}(k)$ provides an appropriate step-response model for integrating processes (Hokanson and Gerstle, 1992):

$$\Delta\hat{y}(k+1) = \sum_{i=1}^{N-1} S_i \Delta u(k-i+1) + S_N u(k-N+1) \tag{20-31}$$

or, equivalently,

$$\hat{y}(k+1) = \hat{y}(k) + \sum_{i=1}^{N-1} S_i \Delta u(k-i+1) + S_N u(k-N+1) \tag{20-32}$$

Although the bias correction approach of Eq. 20-22 is not valid for integrating processes, several modifications are available (Qin and Badgwell, 2003).

Known Disturbances

If a disturbance variable is known or can be measured, it can be included in the step-response model. Let d denote a measured disturbance and $\{S_i^d\}$ its step-response coefficients. Then the standard step-response model in Eq. 20-6 can be modified by adding a disturbance term,

$$\hat{y}(k+1) = \sum_{i=1}^{N-1} S_i \Delta u(k-i+1) + S_N u(k-N+1) + \sum_{i=1}^{N_d-1} S_i^d \Delta d(k-i+1) + S_N^d d(k-N_d+1) \tag{20-33}$$

where N_d is the number of step-response coefficients for the disturbance variable (in general, $N_d \neq N$). This same type of modification can be made to other step-response models such as Eq. 20-19 or 20-24. However, predictions made more than one step ahead require an assumption about future disturbances. If no other information is available, the usual assumption is that the future disturbances will be equal to the current disturbance: $d(k+j) = d(k)$ for $j = 1, 2, \ldots, P$. However, if a disturbance model is available, the prediction accuracy can improve.

20.3 PREDICTIONS FOR MIMO MODELS

The previous analysis for SISO systems can be generalized to MIMO systems by using the Principle of Superposition. For simplicity, we first consider a process control problem with two outputs, y_1 and y_2, and two inputs, u_1 and u_2. The predictive model consists of two equations and four individual step-response models, one for each input-output pair:

$$\hat{y}_1(k+1) = \sum_{i=1}^{N-1} S_{11,i} \Delta u_1(k-i+1) + S_{11,N} u_1(k-N+1) + \sum_{i=1}^{N-1} S_{12,i} \Delta u_2(k-i+1) + S_{12,N} u_2(k-N+1) \tag{20-34}$$

$$\hat{y}_2(k+1) = \sum_{i=1}^{N-1} S_{21,i}\Delta u_1(k-i+1) + S_{21,N}\, u_1(k-N+1)$$

$$+ \sum_{i=1}^{N-1} S_{22,i}\Delta u_2(k-i+1) + S_{22,N}\, u_2(k-N+1) \quad (20\text{-}35)$$

where $S_{12,i}$ denotes the ith step-response coefficient for the model that relates y_1 and u_2. The other step-response coefficients are defined in an analogous manner. This MIMO model is a straightforward generalization of the SISO model in Eq. 20-6. In general, a different model horizon can be specified for each input-output pair. For example, the upper limits for the summations in Eq. 20-35 can be specified as N_{21} and N_{22}, if y_2 has very different settling times for changes in u_1 and u_2.

Next, the analysis is generalized to MIMO models with arbitrary numbers of inputs and outputs. Suppose that there are r inputs and m outputs. In a typical MPC application, $r < 20$ and $m < 40$, but applications with much larger numbers of inputs and outputs have also been reported (Qin and Badgwell, 2003; Canney, 2003). It is useful to display the individual step-response models graphically as shown in Fig. 20.8 (Hokanson and Gerstle, 1992), where the output variables (or CVs) are arranged as the columns and the inputs and disturbances (the MVs and DVs) are arranged as the rows.

It is convenient to express MIMO step-response models in vector-matrix notation. Let the output vector be $\mathbf{y} = [y_1, y_2, \ldots, y_m]^T$ and the input vector be $\mathbf{u} = [u_1, u_2, \ldots, u_r]^T$ where superscript T denotes the transpose of a vector of matrix. In analogy with the derivation of Eq. 20-24 for SISO systems, the MIMO model for the corrected predictions can be expressed in dynamic matrix form:

$$\tilde{\mathbf{Y}}(\mathbf{k}+\mathbf{1}) = \mathbf{S}\Delta\mathbf{U}(k) + \hat{\mathbf{Y}}^{o}(k+1) + [y(k) - \hat{y}(k)] \quad (20\text{-}36)$$

where $\tilde{\mathbf{Y}}(k+1)$ is the mP-dimensional vector of corrected predictions over the prediction horizon P,

$$\tilde{\mathbf{Y}}(k+1) \triangleq col\,[\tilde{\mathbf{y}}(k+1), \tilde{\mathbf{y}}(k+2), \ldots, \tilde{\mathbf{y}}(k+P)] \quad (20\text{-}37)$$

$\hat{\mathbf{Y}}^{o}(k+1)$ is the mP-dimensional vector of predicted unforced responses,

$$\hat{\mathbf{Y}}^{o}(k+1) \triangleq col\,[\hat{\mathbf{y}}^{o}(k+1), \hat{\mathbf{y}}^{o}(k+2), \ldots, \hat{\mathbf{y}}^{o}(k+P)] \quad (20\text{-}38)$$

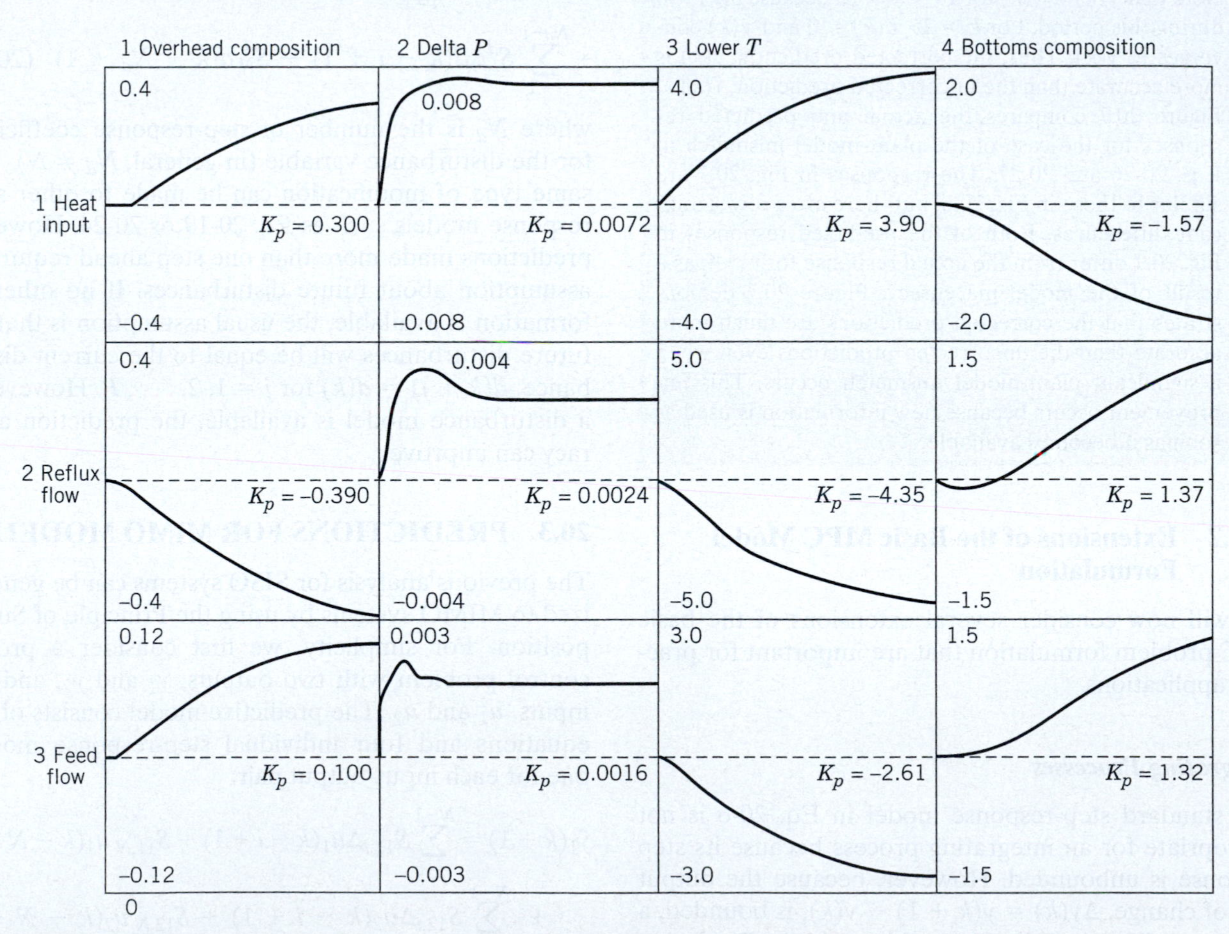

Figure 20.8 Individual step-response models for a distillation column with three inputs and four outputs. Each model represents the step response for 120 minutes (Hokanson and Gerstle, 1992).

and $\Delta \boldsymbol{U}(k)$ is the rM-dimensional vector of the next M control moves,

$$\Delta \boldsymbol{U}(k) \triangleq col\,[\Delta \boldsymbol{u}(k), \Delta \boldsymbol{u}(k+1), \ldots, \Delta \boldsymbol{u}(k+M-1)] \tag{20-39}$$

The $mP \times m$ matrix (in Eq. 20-36) is defined as

$$S \triangleq [\underbrace{\boldsymbol{I}_m \boldsymbol{I}_m \cdots \boldsymbol{I}_m}_{P\ times}]^T \tag{20-40}$$

where $\boldsymbol{I}_m$ denotes the $m \times m$ identity matrix.

The dynamic matrix $\boldsymbol{S}$ is defined as

$$\boldsymbol{S} \triangleq \begin{bmatrix} \boldsymbol{S}_1 & \boldsymbol{0} & \cdots & \boldsymbol{0} \\ \boldsymbol{S}_2 & \boldsymbol{S}_1 & \boldsymbol{0} & \vdots \\ \vdots & \vdots & \cdots & \boldsymbol{0} \\ \boldsymbol{S}_M & \boldsymbol{S}_{M-1} & \cdots & \boldsymbol{S}_1 \\ \boldsymbol{S}_{M+1} & \boldsymbol{S}_M & \cdots & \boldsymbol{S}_2 \\ \vdots & \vdots & \ddots & \vdots \\ \boldsymbol{S}_P & \boldsymbol{S}_{P-1} & \cdots & \boldsymbol{S}_{P-M+1} \end{bmatrix} \tag{20-41}$$

where $\boldsymbol{S}_i$ is the $m \times r$ matrix of step-response coefficients for the ith time step.

$$\boldsymbol{S}_i \triangleq \begin{bmatrix} S_{11,i} & S_{12,i} & \cdots & S_{1r,i} \\ S_{21,i} & & & S_{2r,i} \\ \vdots & \vdots & \vdots & \vdots \\ S_{m1,i} & \cdots & \cdots & S_{mr,i} \end{bmatrix} \tag{20-42}$$

Note that the dynamic matrix in Eq. 20-41 for MIMO systems has the same structure as the one for SISO systems in Eq. 20-20.

The dimensions of the vectors and matrices in Eq. 20-36 are as follows. Both $\tilde{\boldsymbol{Y}}(k+1)$ and $\tilde{\boldsymbol{Y}}^o(k+1)$ are mP-dimensional vectors where m is the number of outputs and P is the prediction horizon. Also, $\Delta \boldsymbol{U}(k)$ is an rM-dimensional vector where r is the number of manipulated inputs and M is the control horizon. Consequently, the dimensions of step-response matrix $\boldsymbol{S}$ are $mP \times rM$. The MIMO model in (20-36) through (20-42) is the MIMO generalization of the SISO model in (20-24). It is also possible to write MIMO models in an alternative form, a generalization of Eqs. 20-34 and 20-35. An advantage of this alternative formulation is that the new dynamic matrix is partitioned into the individual SISO models, a convenient form for real-time predictions.

For stable models, the predicted unforced response, $\hat{\boldsymbol{Y}}^o(k+1)$ in Eq. 20-38, can be calculated from a recursive relation (Lundström et al., 1995) that is in the form of a discrete-time version of a state-space model:

$$\hat{\boldsymbol{Y}}^o(k+1) = \boldsymbol{M}\,\hat{\boldsymbol{Y}}^o(k) + \boldsymbol{S}^* \Delta \boldsymbol{u}(k) \tag{20-43}$$

where:

$$\hat{\boldsymbol{Y}}^o(k) = col\,[\hat{\boldsymbol{y}}^o(k), \hat{\boldsymbol{y}}^o(k+1), \ldots, \hat{\boldsymbol{y}}^o(k+P-1)] \tag{20-44}$$

$$\boldsymbol{M} \triangleq \begin{bmatrix} \boldsymbol{0} & \boldsymbol{I}_m & \boldsymbol{0} & \cdots & \boldsymbol{0} \\ \boldsymbol{0} & \boldsymbol{0} & \boldsymbol{I}_m & \ddots & \boldsymbol{0} \\ \vdots & \vdots & \vdots & \ddots & \boldsymbol{0} \\ \boldsymbol{0} & \boldsymbol{0} & \cdots & \boldsymbol{0} & \boldsymbol{I}_m \\ \boldsymbol{0} & \boldsymbol{0} & \cdots & \boldsymbol{0} & \boldsymbol{I}_m \end{bmatrix} \tag{20-45}$$

$$\boldsymbol{S}^* \triangleq \begin{bmatrix} \boldsymbol{S}_1 \\ \boldsymbol{S}_2 \\ \vdots \\ \boldsymbol{S}_{P-1} \\ \boldsymbol{S}_P \end{bmatrix} \tag{20-46}$$

where $\boldsymbol{M}$ is an $mP \times mP$ matrix and $\boldsymbol{S}^*$ is an $mP \times r$ matrix. The MIMO models in Eqs. 20-36 through 20-46 can be extended to include measured disturbances and integrating variables, in analogy to the SISO case in the previous section.

Most of the current MPC research is based on state-space models, because they provide an important theoretical advantage, namely, a unified framework for both linear and nonlinear control problems. State-space models are also more convenient for theoretical analysis and facilitate a wider range of output feedback strategies (Rawlings, 2000, Maciejowski, 2002; Qin and Badgwell, 2003).

20.4 MODEL PREDICTIVE CONTROL CALCULATIONS

The flowchart in Fig. 20.9 provides an overview of the MPC calculations. The seven steps are shown in the order they are performed at each control execution time. For simplicity, we assume that the control execution times coincide with the measurement sampling instants.

In MPC applications, the calculated MV moves are usually implemented as set points for regulatory control loops at the Distributed Control System (DCS) level, such as flow control loops. If a DCS control loop has been disabled or placed in manual, the MV is no longer available for control. In this situation, the control degrees of freedom are reduced by one. Even though an MV is unavailable for control, it can serve as a disturbance variable if it is measured.

In Step 1 of the MPC calculations, new process data are acquired via the regulatory control system (DCS) that is interfaced to the process. Then new output predictions are calculated in Step 2 using the process model and the new data (see Eqs. 20-21 and 20-22, for example).

Before each control execution, it is necessary to determine which outputs (CVs), inputs (MVs), and disturbance variables (DVs) are currently available for the MPC calculations. This Step 3 activity is referred to

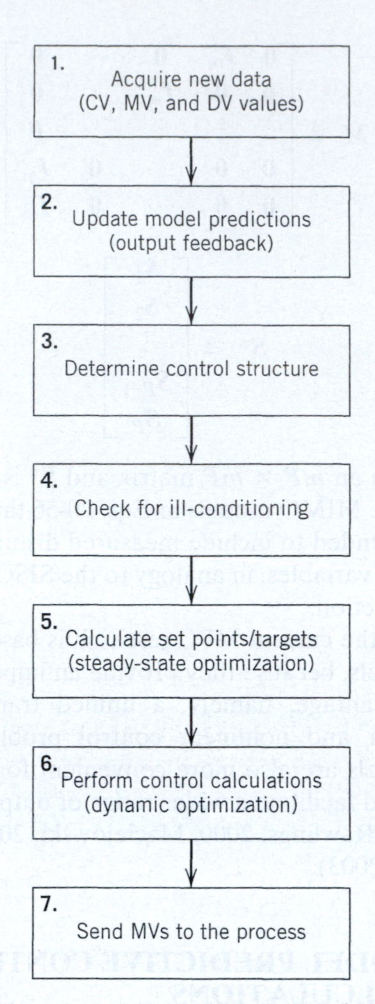

Figure 20.9 Flow chart for MPC calculations (modified from Qin and Badgwell (2003)).

as determining the current control structure. The variables available for the control calculations can change from one control execution time to the next, for a variety of reasons. For example, a sensor may not be available due to routine maintenance or recalibration. Output variables are often classified as being either *critical* or *noncritical*. If the sensor for a critical output is not available, the MPC calculations can be stopped immediately or after a specified number of control execution steps. For a noncritical output, missing measurements could be replaced by model predictions or the output could be removed from the control structure (Qin and Badgwell, 2003).

If the control structure changes from one control execution time to another, the subsequent control calculations can become *ill-conditioned*. It is important to identify and correct these situations before executing the MPC calculations in Steps 5 and 6. Ill-conditioning occurs when the available MVs have very similar effects on two or more outputs. For example, consider a high-purity distillation column where the product compositions are controlled by manipulating the reflux flow rate and the reboiler heat duty. Ill-conditioning occurs because each input has approximately the same effect on both outputs, but in different directions. As a result, the process gain matrix is nearly singular, and large input movements are required to control these outputs independently. Consequently, it is important to check for ill-conditioning (Step 4) by calculating the *condition number* of the process gain matrix for the current control structure (see Chapter 18). If ill-conditioning is detected, effective strategies are available for its removal (Maciejowski, 2002; Qin and Badgwell, 2003).

In MPC applications, the major benefits result from determining the optimal operating conditions (set-point calculations) and from moving the process to these set points in an optimal manner based on the control calculations. Both types of calculations optimize a specified objective function while satisfying inequality constraints, such as upper and lower limits on the inputs or outputs. Set-point calculations are the subject of Section 20.5, while control calculations are considered in the next section.

The final step, Step 7 of Fig. 20.9, is to implement the calculated control actions, usually as set points to regulatory PID control loops at the DCS level.

20.4.1 Unconstrained MPC

This section considers the control calculations of Step 6 for the special case in which inequality constraints are not included in the problem formulation. In Section 20.4.2, the analysis is extended to the more typical situation where there are inequality constraints on $\boldsymbol{u}$, $\Delta \boldsymbol{u}$, and $\boldsymbol{y}$.

As noted earlier, the MPC control calculations are based on both current measurements and model predictions. The control objective is to calculate a set of control moves (MV changes) that make the corrected predictions as close to a reference trajectory as possible. Thus, an optimization approach is employed. For unconstrained linear control problems, an analytical expression for the MPC control law is available.

Reference Trajectory

In MPC applications, a reference trajectory can be used to make a gradual transition to the desired set point. The *reference trajectory* $\boldsymbol{y}_r$ can be specified in several different ways (Maciejowski, 2002; Qin and Badgwell, 2003; Rossiter, 2003). We briefly introduce several typical approaches.

Let the reference trajectory over the prediction horizon P be denoted as

$$\boldsymbol{Y}_r(k+1) \triangleq col[\boldsymbol{y}_r(k+1), \boldsymbol{y}_r(k+2), \ldots, \boldsymbol{y}_r(k+P)] \tag{20-47}$$

where $\boldsymbol{Y}_r$ is an mP-dimensional vector. A reasonable approach is to specify the reference trajectory to be the *filtered set point*,

$$y_{i,r}(k+j) = (\alpha_i)^j y_{i,r}(k) + [1 - (\alpha_i)^j]\, y_{i,sp}(k)$$
$$\text{for } i = 1, 2, \ldots, m \quad \text{and} \quad j = 1, 2, \ldots, P \qquad (20\text{-}48)$$

where $y_{i,r}$ is the ith element of $\boldsymbol{y}_r$, $\boldsymbol{y}_{sp}$ denotes the set point, and α_i is a filter constant, $0 < \alpha_i < 1$. For $j = 1$, Eq. 20-48 reduces to the set-point filtering expression for PID controllers that was considered in Chapter 12. It is also equivalent to the exponential filter introduced in Chapter 17. Note that $\boldsymbol{y}_r = \boldsymbol{y}_{sp}$ for the limiting case of $\alpha_i = 0$. An alternative approach is to specify the reference trajectory for the ith output as an exponential trajectory from the current measurement $y_i(k)$ to the set point, $y_{i,sp}(k)$:

$$y_{i,r}(k+j) = (\alpha_i)^j y_i(k) + [1 - (\alpha_i)^j]\, y_{i,sp}(k)$$
$$\text{for } i = 1, 2, \ldots, m \quad \text{and} \quad j = 1, 2, \ldots, P \qquad (20\text{-}49)$$

In some commercial MPC products, the desired reference trajectory for each output is specified indirectly by a *performance ratio* for the output. The performance ratio is defined to be the ratio of the desired closed-loop settling time to the open-loop settling time. Thus, small values of the performance ratios correspond to small values of α_i in (20-48) or (20-49).

Model Predictive Control Law

The control calculations are based on minimizing the predicted deviations from the reference trajectory. Let k denote the current sampling instant. The *predicted error* vector, $\hat{\boldsymbol{E}}(k+1)$, is defined as

$$\hat{\boldsymbol{E}}(k+1) \triangleq \boldsymbol{Y}_r(k+1) - \tilde{\boldsymbol{Y}}(k+1) \qquad (20\text{-}50)$$

where $\tilde{\boldsymbol{Y}}(k+1)$ was defined in (20-37). Similarly, $\hat{\boldsymbol{E}}^o(k+1)$ denotes the *predicted unforced error vector,*

$$\hat{\boldsymbol{E}}^o(k+1) \triangleq \boldsymbol{Y}_r(k+1) - \tilde{\boldsymbol{Y}}^o(k+1) \qquad (20\text{-}51)$$

where the *corrected prediction for the unforced case,* $\tilde{\boldsymbol{Y}}^o(k+1)$, is defined as

$$\tilde{\boldsymbol{Y}}^o(k+1) \triangleq \hat{\boldsymbol{Y}}^o(k+1) + \mathbf{I}[\boldsymbol{y}(k) - \hat{\boldsymbol{y}}(k)] \quad (20\text{-}52)$$

Thus, $\hat{\boldsymbol{E}}^o(k+1)$ represents the predicted deviations from the reference trajectory when no further control action is taken, that is, the predicted deviations when $\Delta\boldsymbol{u}(k+j) = \boldsymbol{0}$ for $j = 0, 1, \ldots, M-1$. Note that $\hat{\boldsymbol{E}}(k+1)$ and $\hat{\boldsymbol{E}}^o(k+1)$ are mP-dimensional vectors.

The general objective of the MPC control calculations is to determine $\Delta\boldsymbol{U}(k)$, the control moves for the next M time intervals,

$$\Delta\boldsymbol{U}(k) = col\,[\Delta\boldsymbol{u}(k), \Delta\boldsymbol{u}(k+1), \ldots, \Delta\boldsymbol{u}(k+M-1)] \qquad (20\text{-}53)$$

The rM-dimensional vector $\Delta\boldsymbol{U}(k)$ is calculated so that an objective function (also called a *performance* index) is minimized. Typically, either a linear or a quadratic objective function is employed. For unconstrained MPC, the objective function is based on minimizing some (or all) of three types of deviations or errors (Qin and Badgwell, 2003):

1. The predicted errors over the predicted horizon, $\hat{\boldsymbol{E}}(k+1)$
2. The next M control moves, $\Delta\boldsymbol{U}(k)$
3. The deviations of $\boldsymbol{u}(k+i)$ from its desired steady-state value $\boldsymbol{u}_{sp}$ over the control horizon

For MPC based on linear process models, both linear and quadratic objective functions can be used (Maciejowski, 2002; Qin and Badgwell, 2003). To demonstrate the MPC control calculations, consider a quadratic objective function J based on the first two types of deviations:

$$\min_{\Delta\boldsymbol{U}(k)} \boldsymbol{J} = \hat{\boldsymbol{E}}(k+1)^T \boldsymbol{Q} \hat{\boldsymbol{E}}(k+1) + \Delta\boldsymbol{U}(k)^T \boldsymbol{R} \Delta\boldsymbol{U}(k) \qquad (20\text{-}54)$$

where $\boldsymbol{Q}$ is a positive-definite weighting matrix and $\boldsymbol{R}$ is a positive semi-definite matrix. Both are usually diagonal matrices with positive diagonal elements. The weighting matrices are used to weight the most important elements of $\hat{\boldsymbol{E}}(k+1)$ or $\Delta\boldsymbol{U}(k)$, as described in Section 20.6. If diagonal weighting matrices are specified, these elements are weighted individually.

The MPC control law that minimizes the objective function in Eq. (20-54) can be calculated analytically.

$$\Delta\boldsymbol{U}(k) = (\boldsymbol{S}^T\boldsymbol{Q}\,\boldsymbol{S} + \boldsymbol{R})^{-1}\boldsymbol{S}^T\boldsymbol{Q}\,\hat{\boldsymbol{E}}^o(k+1) \qquad (20\text{-}55)$$

This control law can be written in a more compact form,

$$\Delta\boldsymbol{U}(k) = \boldsymbol{K}_c\hat{\boldsymbol{E}}^o(k+1) \qquad (20\text{-}56)$$

where the controller gain matrix $\boldsymbol{K}_c$ is defined to be

$$\boldsymbol{K}_c \triangleq (\boldsymbol{S}^T\,\boldsymbol{Q}\,\boldsymbol{S} + \boldsymbol{R})^{-1}\,\boldsymbol{S}^T\,\boldsymbol{Q} \qquad (20\text{-}57)$$

Note that $\boldsymbol{K}_c$ is an $rM \times mP$ matrix that can be evaluated off-line rather than on-line provided that the dynamic matrix $\boldsymbol{S}$ and weighting matrices, $\boldsymbol{Q}$ and $\boldsymbol{R}$, are constant.

The MPC control law in Eq. 20-56 can be interpreted as a multivariable, proportional control law based on the predicted error rather than the conventional control error (set point–measurement). The control law utilizes the latest measurement $\boldsymbol{y}(k)$ because it appears in the expressions for the corrected prediction $\tilde{\boldsymbol{y}}(k)$, and thus also in the predicted unforced error, $\tilde{\boldsymbol{E}}^o(k+1)$. Furthermore, the MPC control law in Eq. (20-56) implicitly contains integral control action because $\boldsymbol{u}$ tends to change until the unforced error $\hat{\boldsymbol{E}}^o$ becomes zero. Thus, offset is eliminated for set-point changes or sustained disturbances.

Although the MPC control law calculates a set of M input moves, $\Delta\boldsymbol{U}(k)$, only the first control move,

$\Delta u(k)$, is actually implemented. Then at the next sampling instant, new data are acquired and a new set of control moves is calculated. Once again, only the first control move is implemented. These activities are repeated at each sampling instant, and the strategy is referred to as a *receding horizon approach*. The first control move, $\Delta u(k)$, can be calculated from Eqs. 20-53 and 20–56,

$$\Delta u(k) = K_{c1}\hat{E}^o(k+1) \tag{20-58}$$

where matrix K_{c1} is defined to be the first r rows of K_c. Thus, K_{c1} has dimensions of $r \times mP$.

It may seem strange to calculate an M-step control policy and then only implement the first move. The important advantage of this receding horizon approach is that new information in the form of the most recent measurement $y(k)$ is utilized immediately instead of being ignored for the next M sampling instants. Otherwise, the multistep predictions and control moves would be based on old information and thus be adversely affected by unmeasured disturbances, as demonstrated in Example 20.4.

The calculation of K_c requires the inversion of an $rM \times rM$ matrix where r is the number of input variables and M is the control horizon. For large problems with many inputs, the required computational effort can be reduced by using *input blocking* (Maciejowski, 2002; Qin and Badgwell, 2003). In this approach, the inputs are not changed at every sampling instant. Instead, $\Delta u = 0$ for "blocks" of sampling instants. Input blocking is illustrated in Fig. 20.10 where a single input is changed at each sampling instant for the first four sampling instants (k through $k+3$). Starting at $k+4$, u is blocked so that it changes every three sampling instants until the steady-state value is reached at $k+13$. The design parameters are the block length and the time at which blocking begins.

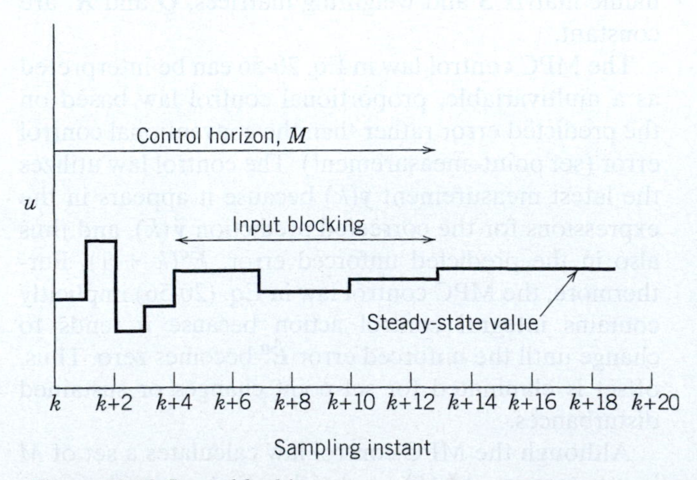

Figure 20.10 Input blocking.

20.4.2 MPC with Inequality Constraints

Inequality constraints on input and output variables are important characteristics for MPC applications. In fact, inequality constraints were a primary motivation for the early development of MPC. Input constraints occur as a result of physical limitations on plant equipment such as pumps, control valves, and heat exchangers. For example, a manipulated flow rate might have a lower limit of zero and an upper limit determined by the pump, control valve, and piping characteristics. The dynamics associated with large control valves impose rate-of-change limits on manipulated flow rates.

Constraints on output variables are a key component of the plant operating strategy. For example, a common distillation column control objective is to maximize the production rate while satisfying constraints on product quality and avoiding undesirable operating regimes such as flooding or weeping. Additional examples of inequality constraints were given in Chapter 19. The set of inequality constraints for u and y define an operating window for the process, as shown in Fig. 19.6.

Inequality constraints can be included in the control calculations in many different ways (Maciejowski, 2002; Qin and Badgwell, 2003). It is convenient to make a distinction between *hard constraints* and *soft constraints*. As the name implies, a hard constraint cannot be violated at any time. By contrast, a soft constraint can be violated, but the amount of violation is penalized by a modification of the objective function, as described below. This approach allows small constraint violations to be tolerated for short periods of time.

For MPC the inequality constraints for u and Δu are typically hard constraints specified as upper and lower limits:

$$u^-(k) \le u(k+j) \le u^+(k) \qquad j = 0, 1, \ldots, M-1 \tag{20-59}$$

$$\Delta u^-(k) \le \Delta u(k+j) \le \Delta u^+(k) \quad j = 0, 1, \ldots, M-1 \tag{20-60}$$

The analogous hard constraints for the predicted outputs are:

$$y^-(k+j) \le \tilde{y}(k+j) \le y^+(k+j) \quad j = 1, 2, \ldots, P \tag{20-61}$$

Unfortunately, hard output constraints can result in infeasible solutions for the optimization problem, especially for large disturbances. Consequently, output constraints are usually expressed as soft constraints involving *slack variables* s_j (Qin and Badgwell, 2003):

$$y^-(k+j) - s_j \le \tilde{y}(k+j) \le y^+(k+j) + s_j \qquad j = 1, 2, \ldots, P \tag{20-62}$$

The numerical values of the slack variables can be determined during constrained optimization if the performance index in Eq. 20-54 is modified by adding a penalty term for the slack variables. Thus, an mP-dimensional vector of slack variables is defined as $\bar{S} \triangleq col\,[s_1, s_2, \ldots, s_P]$. The modified performance index is

$$\min_{\Delta U(k)} J = \hat{E}(k+1)^T Q\, \hat{E}(k+1) + \Delta U(k)^T R\, \Delta U(k) + \bar{S}^T T \bar{S} \quad (20\text{-}63)$$

where T is an $mP \times mP$ weighting matrix for the slack variables. Note that inequality constraints in (20-61) and (20-62) are imposed on the corrected prediction $\tilde{y}$, rather than the actual output y, because future values of y are not available. Consequently, y may violate a constraint even though $\tilde{y}$ does not, as a result of modeling errors. Slack variables can also be used to weight positive and negative errors, differently.

Range Control

An unusual feature of MPC applications is that many output variables do not have set points. For these outputs, the control objective is to keep them between upper and lower limits, an approach called *range control* (or *zone control*). The limits can vary with time, as shown in Eq. 20-61. The advantage of range control is that it creates additional degrees of freedom for the control calculations. Furthermore, many output variables such as the liquid level in a surge tank do not have to be regulated at a set point. Consequently, in many MPC applications, range control is the rule rather than the exception. Set points are only specified for output variables that must be kept close to a specified value (for example, pH or a quality variable). Note that control to a set point can be considered to be a special case of range control that occurs when the upper and lower limits in (20-61) are equal.

The constraint limits in Eqs. 20-59 to 20-62 can vary with time as a result of changes in process equipment or instrumentation. However, it can also be beneficial to allow the limits to change in a specified manner over the control or prediction horizons. For example, in the *limit funnel* technique, the output limits in (20-61) or (20-62) gradually become closer together over the prediction horizon (Maciejowski, 2002; Qin and Badgwell, 2003).

The introduction of inequality constraints results in a constrained optimization problem that can be solved numerically using linear or quadratic programming techniques (Edgar et al., 2001). As an example, consider the addition of inequality constraints to the MPC design problem in the previous section. Suppose that it is desired to calculate the M-step control policy $\Delta U(k)$ that minimizes the quadratic objective function J in Eq. 20-54, while satisfying the constraints in Eqs. 20-59, 20-60, and 20-61. The output predictions are made using the step-response model in Eq. 20-36. This MPC design problem can be solved numerically using the quadratic programming technique in Chapter 19.

20.5 SET-POINT CALCULATIONS

As indicated in Section 20.1 and Fig. 20.9, the MPC calculations at each control execution time are typically performed in two steps. First, the optimum set points (or *targets*) for the control calculations are determined. Then, a set of M control moves are generated by the control calculations, and the first move is implemented. In practical applications, significant economic benefits result from both types of calculations, but the steady-state optimization is usually more important. In this section, the set-point calculations are described in more detail.

The MPC set points are calculated so that they maximize or minimize an economic objective function. The calculations are usually based on linear steady-state models and a simple objective function, typically a linear or quadratic function of the MVs and CVs. The linear model can be a linearized version of a complex nonlinear model or the steady-state version of the dynamic model that is used in the control calculations. Linear inequality constraints for the MVs and CVs are also included in the steady-state optimization. The set-point calculations are repeated at each sampling instant because the active constraints can change frequently due to disturbances, instrumentation, equipment availability, or varying process conditions.

Because the set-point calculations are repeated as often as every minute, the steady-state optimization problem must be solved quickly and reliably. If the optimization problem is based on a linear process model, linear inequality constraints, and either a linear or a quadratic cost function, the linear and quadratic programming techniques discussed in Chapter 19 can be employed.

20.5.1 Formulation of the Set-Point Optimization Problem

Next, we provide an overview of the set-point calculation problem. More detailed descriptions are available elsewhere (Sorensen and Cutler, 1998; Kassman et al., 2000; Rawlings, 2000; Maciejowski, 2002).

Consider an MIMO process with r MVs and m CVs. Denote the current values of u and y as $u(k)$ and $y(k)$. The objective is to calculate the optimum set point y_{sp} for the next control calculation (at $k+1$) and also to determine the corresponding steady-state value of u, u_{sp}. This value is used as the set point for u for the next control calculation.

A general, linear steady-state process model can be written as

$$\Delta y = K \Delta u \quad (20\text{-}64)$$

where $\boldsymbol{K}$ is the steady-state gain matrix and $\Delta \boldsymbol{y}$ and $\Delta \boldsymbol{u}$ denote steady-state changes in $\boldsymbol{y}$ and $\boldsymbol{u}$. It is convenient to define $\Delta \boldsymbol{y}$ and $\Delta \boldsymbol{u}$ as

$$\Delta \boldsymbol{y} \triangleq \boldsymbol{y}_{sp} - \boldsymbol{y}_{OL}(k) \tag{20-65}$$

$$\Delta \boldsymbol{u} \triangleq \boldsymbol{u}_{sp} - \boldsymbol{u}(k) \tag{20-66}$$

In Eq. 20-65 $\boldsymbol{y}_{OL}(k)$ represents the steady-state value of $\boldsymbol{y}$ that would result if $\boldsymbol{u}$ were held constant at its current value, $\boldsymbol{u}(k)$, until steady state was achieved. In general, $\boldsymbol{y}_{OL}(k) \neq \boldsymbol{y}(k)$ except for the ideal situation where the process is at steady state at time k. In order to incorporate output feedback, the steady-state model in Eq. 20-64 is modified as

$$\Delta \boldsymbol{y} = \boldsymbol{K}\Delta \boldsymbol{u} + [\boldsymbol{y}(k) - \hat{\boldsymbol{y}}(k)] \tag{20-67}$$

A representative formulation for the set-point optimization is to determine the optimum values, $\boldsymbol{u}_{sp}$ and $\boldsymbol{y}_{sp}$, that minimize a quadratic objective function,

$$\min_{\boldsymbol{u}_{sp}, \boldsymbol{y}_{sp}} J_s = \boldsymbol{c}^T \boldsymbol{u}_{sp} + \boldsymbol{d}^T \boldsymbol{y}_{sp} + \boldsymbol{e}_y^T \boldsymbol{Q}_{sp} \boldsymbol{e}_y + \boldsymbol{e}_u^T \boldsymbol{R}_{sp} \boldsymbol{e}_u + \boldsymbol{S}^T \boldsymbol{T}_{sp} \boldsymbol{S} \tag{20-68}$$

subject to satisfying Eq. 20-64 and inequality constraints on the MVs and CVs:

$$\boldsymbol{u}^- \leq \boldsymbol{u}_{sp} \leq \boldsymbol{u}^+ \tag{20-69}$$

$$\Delta \boldsymbol{u}^- \leq \Delta \boldsymbol{u}_{sp} \leq \Delta \boldsymbol{u}^+ \tag{20-70}$$

$$\boldsymbol{y}^- - \boldsymbol{s} \leq \boldsymbol{y}_{sp} \leq \boldsymbol{y}^+ + \boldsymbol{s} \tag{20-71}$$

where

$$\boldsymbol{e}_y \triangleq \boldsymbol{y}_{sp} - \boldsymbol{y}_{\text{ref}} \tag{20-72}$$

$$\boldsymbol{e}_u \triangleq \boldsymbol{u}_{sp} - \boldsymbol{u}_{\text{ref}} \tag{20-73}$$

The $\boldsymbol{s}$ vector in (20-71) denotes the slack elements. In (20-72) and (20-73), $\boldsymbol{y}_{\text{ref}}$ and $\boldsymbol{u}_{\text{ref}}$ are the desired steady-state values of $\boldsymbol{y}$ and $\boldsymbol{u}$ that are often determined by a higher-level optimization (for example, Level 4 in Fig. 19.1). The weighting factors in (20-68), $\boldsymbol{c}$, $\boldsymbol{d}$, $\boldsymbol{Q}_{sp}$, $\boldsymbol{R}_{sp}$, and $\boldsymbol{T}_{sp}$, are selected based on economic considerations. Although the weighting factors are constants in Eq. 20-68, in MPC applications they can vary with time to accommodate process changes or changes in economic conditions such as product prices or raw material costs. Similarly, it can be advantageous to allow the limits in Eqs. 20-69 to 20-71 ($\boldsymbol{u}^-$, $\boldsymbol{u}^+$, etc.) to vary from one execution time to the next, as discussed in Section 20.4. Fortunately, new values of weighting factors and constraint limits are easily accommodated, because the optimum set points are recalculated at each execution time.

It is important to make a distinction between $\boldsymbol{y}_{\text{ref}}$ and $\boldsymbol{u}_{\text{ref}}$, and $\boldsymbol{y}_{sp}$ and $\boldsymbol{u}_{sp}$. Both pairs represent desired values of $\boldsymbol{y}$ and $\boldsymbol{u}$, but they have different origins and are used in different ways. Reference values, $\boldsymbol{y}_{\text{ref}}$ and $\boldsymbol{u}_{\text{ref}}$, are often determined infrequently by a higher-level optimization. They are used as the desired values for the steady-state optimization of Step 5 of Fig. 20.9. By contrast, $\boldsymbol{y}_{sp}$ and $\boldsymbol{u}_{sp}$ are calculated at each MPC control execution time and serve as set points for the control calculations of Step 6.

We have emphasized that the goal of this steady-state optimization is to determine $\boldsymbol{y}_{sp}$ and $\boldsymbol{u}_{sp}$, the set points for the control calculations in Step 6 of Fig. 20.9. But why not use $\boldsymbol{y}_{\text{ref}}$ and $\boldsymbol{u}_{\text{ref}}$ for this purpose? The reason is that $\boldsymbol{y}_{\text{ref}}$ and $\boldsymbol{u}_{\text{ref}}$ are ideal values that may not be attainable for the current plant conditions and constraints, which could have changed since $\boldsymbol{y}_{\text{ref}}$ and $\boldsymbol{u}_{\text{ref}}$ were calculated. Thus, steady-state optimization (Step 5) is necessary to calculate $\boldsymbol{y}_{sp}$ and $\boldsymbol{u}_{sp}$, target values that more accurately reflect current conditions. In Eq. 20-68, $\boldsymbol{y}_{sp}$ and $\boldsymbol{u}_{sp}$ are shown as the independent values for the optimization. However, $\boldsymbol{y}_{sp}$ can be eliminated by substituting the steady-state model, $\boldsymbol{y}_{sp} = \boldsymbol{K}\boldsymbol{u}_{sp}$.

Next, we demonstrate that the objective function J_s is quite flexible, by showing how it is defined for three different types of applications.

Application 1: *Maximize operating profit.*
In Chapter 19, real-time optimization was considered problems where the operating profit was expressed in terms of product values and feedstock and utility costs. If the product, feedstock, and utility flow rates are manipulated or disturbance variables in the MPC control structure, they can be included in objective function J_s. In order to maximize the operating profit (OP), the objective function is specified to be $J_s = -OP$, because minimizing J_s is equivalent to maximizing $-J_s$. The weighting matrices for two quadratic terms, $\boldsymbol{Q}_{sp}$ and $\boldsymbol{R}_{sp}$, are set equal to zero.

Application 2: *Minimize deviations from the reference values.*
Suppose that the objective of the steady-state optimization is to calculate $\boldsymbol{y}_{sp}$ and $\boldsymbol{u}_{sp}$ so that they are as close as possible to the reference values, $\boldsymbol{y}_{\text{ref}}$ and $\boldsymbol{u}_{\text{ref}}$. This goal can be achieved by setting $\boldsymbol{c} = \boldsymbol{0}$ and $\boldsymbol{d} = \boldsymbol{0}$ in (20-68). Weighting matrices $\boldsymbol{Q}_{sp}$, $\boldsymbol{R}_{sp}$, and $\boldsymbol{T}_{sp}$ should be chosen according to the relative importance of the MVs, CVs, and constraint violations.

Application 3: *Maximize the production rate.*
Suppose that the chief control objective is to maximize a production rate while satisfying inequality constraints on the inputs and the outputs. Assume that the production rate can be adjusted via a flow control loop whose set point is denoted as u_{1sp} in the MPC control structure. Thus, the optimization objective is to maximize u_{1sp}, or equivalently, to minimize $-u_{1sp}$. Consequently, the performance index in (20-68) becomes $J_s = -u_{1sp}$. This expression can be derived by setting all of the weighting factors equal to zero except for c_1, the first element of $\boldsymbol{c}$. It is chosen to be $c_1 = -1$.

The set-point optimization problem can be summarized as follows. At each sampling instant, the optimum values of $\boldsymbol{u}$ and $\boldsymbol{y}$ for the next sampling instant ($\boldsymbol{u}_{sp}$ and $\boldsymbol{y}_{sp}$) are calculated by minimizing the cost function in Eq. 20-68, subject to satisfying the model equation 20-64 and the constraints in Eqs. 20-69 to 20-71. This optimization problem can be solved efficiently using the standard LP or QP techniques of Chapter 19.

Infeasible calculations can occur if the calculations of Steps 5 and 6 are based on constrained optimization, because feasible solutions do not always exist (Edgar et al., 2001). Infeasible problems can result when the control degrees of freedom are reduced (e.g., control valve maintenance), large disturbances occur, or the inequality constraints are inappropriate for current conditions. For example, the allowable operating window in Fig. 19.6 could disappear for inappropriate choices of the y_1 and y_2 limits. Other modifications can be made to ensure that the optimization problem always has a feasible solution (Kassmann et al., 2000).

In view of the dramatic decreases in the ratio of computer cost to performance in recent years, it can be argued that physically based, nonlinear process models should be used in the set-point calculations, instead of approximate linear models. However, linear models are still widely used in MPC applications for three reasons: First, linear models are reasonably accurate for small changes in $\boldsymbol{u}$ and $\boldsymbol{d}$ and can easily be updated based on current data or a physically based model. Second, some model inaccuracy can be tolerated, because the calculations are repeated on a frequent basis and they include output feedback from the measurements. Third, the computational effort required for constrained, nonlinear optimization is still relatively large, but is decreasing.

20.6 SELECTION OF DESIGN AND TUNING PARAMETERS

A number of design parameters must be specified in order to design an MPC system. In this section, we consider key design issues and recommended values for the parameters. Several design parameters can also be used to tune the MPC controller. The effects of the MPC design parameters will be illustrated in two examples.

Sampling period Δt and model horizon N. The sampling period Δt and model horizon N (in Eq. 20-6) should be chosen so that $N\Delta t = t_s$ where t_s is the settling time for the open-loop response. This choice ensures that the model reflects the full effect of a change in an input variable over the time required to reach steady state. Typically, $30 \leq N \leq 120$. If the output variables respond on different time scales, a different value of N can be used for each output, as noted earlier. Also, different model horizons can be used for the MVs and DVs, as illustrated in Eq. 20-33.

Control M and prediction P horizons. As control horizon M increases, the MPC controller tends to become more aggressive and the required computational effort increases. However, the computational effort can be reduced by input blocking, as shown in Fig. 20.10. Some typical rules of thumb are $5 \leq M \leq 20$ and $N/3 < M < N/2$. A different value of M can be specified for each input.

The prediction horizon P is often selected to be $P = N + M$ so that the full effect of the last MV move is taken into account. Decreasing the value of P tends to make the controller more aggressive. A different value of P can be selected for each output if their settling times are quite different. An infinite prediction horizon can also be used and has significant theoretical advantages (Maciejowski, 2002; Rawlings, 2000).

Weighting Matrices, $\boldsymbol{Q}$ and $\boldsymbol{R}$

The output weighting matrix $\boldsymbol{Q}$ in Eq. 20-54 allows the output variables to be weighted according to their relative importance. Thus, an $mP \times mP$ diagonal $\boldsymbol{Q}$ matrix allows the output variables to be weighted individually, with the most important variables having the largest weights. For example, if a reactor temperature is considered more important than a liquid level, the temperature will be assigned a larger weighting factor. The inverse of a diagonal weighting factor is sometimes referred to as an *equal concern factor* (Qin and Badgwell, 2003).

It can be advantageous to adjust the output weighting over the prediction horizon. For example, consider an SISO model with a time delay θ. Suppose that an input change Δu occurs at $k = 0$. Then $y(k) = 0$ until $k\Delta t > \theta$ due to the time delay. Consequently, it would be reasonable to set the corresponding elements of the $\boldsymbol{Q}$ matrix equal to zero, or, equivalently, to make the corresponding predictions zero. These approaches tend to make the control calculations better conditioned (see Section 20.4).

As a second example, the elements of $\boldsymbol{Q}$ that correspond to predicted errors early in the prediction horizon (for example, at time $k+1$) can be weighted more heavily than the predicted errors at the end of the horizon, $k + P$, or vice versa. The use of coincidence points is a special case of this strategy. Here, the corrected errors only have nonzero weights for a subset of the P sampling instants called coincidence points. The corrected errors at other times are given zero weighting. In Example 20.2 a simple predictive control strategy was derived based on a single coincidence point.

A time-varying $\boldsymbol{Q}$ matrix can also be used to implement soft constraints by real-time adjustment of $\boldsymbol{Q}$. For example, if an output variable approaches an upper or lower limit, the corresponding elements of $\boldsymbol{Q}$ would be temporarily increased.

In a similar fashion, $\boldsymbol{R}$ in Eq. 20-54 allows input MVs to be weighted according to their relative importance. This $rM \times rM$ matrix is referred to as the *input weighting matrix* or the *move suppression matrix*. It is usually chosen to be a diagonal matrix with the diagonal elements r_{ii}, referred to as *move suppression factors*. They provide convenient tuning parameters, because increasing the value of r_{ii} tends to make the MPC controller more conservative by reducing the magnitudes of the MV moves.

If a reference trajectory is employed, move suppression is not required, and thus $\boldsymbol{R}$ can be set equal to zero.

Reference Trajectory α_i

In MPC applications, the desired future output behavior can be specified in several different ways: as a set point, high and low limits, a reference trajectory, or a funnel (Qin and Badgwell, 2003). Both the reference trajectory and the funnel approaches have a tuning factor that can be used to adjust the desired speed of response for each output. Consider Eq. 20-48 or 20-49, for example. As α_i increases from zero to one, the desired reference trajectory becomes slower. Alternatively, the performance ratio concept can be used to specify the reference trajectories. As mentioned earlier, the performance ratio is defined to be the ratio of the desired closed-loop settling time to the open-loop settling time.

The influence of MPC design parameters is illustrated by a simple example.

EXAMPLE 20.5

A process has the transfer function,

$$\frac{Y(s)}{U(s)} = \frac{e^{-s}}{(10s+1)(5s+1)}$$

(a) Use Eq. 20-57 to calculate the controller gain matrix, $\mathbf{K}_c$, for $\mathbf{Q} = \mathbf{I}$, $\mathbf{R} = \mathbf{0}$ two cases:

(i) $P = 3$, $M = 1$

(ii) $P = 4$, $M = 2$

Assume that $N = 70$, $\Delta t = 1$, and that u is unconstrained for each case.

(b) Compare the set-point responses of two MPC controllers and a digital PID controller with $\Delta t = 0.5$ and ITAE set-point tuning (Chapter 12): $K_c = 2.27$, $\tau_I = 16.6$, and $\tau_D = 1.49$. Compare both y and u responses.

(c) Repeat **(b)** for a unit step disturbance and a PID controller with ITAE disturbance tuning: $K_c = 3.52$, $\tau_I = 6.98$, and $\tau_D = 1.73$.

SOLUTION

(a) The step-response coefficients are obtained by evaluating the step response at the sampling instants, $t = i\Delta t = i$ (because $\Delta t = 1$):

$$S_1 = 0$$

$$S_i = 1 - 2e^{-0.1(i-1)} + e^{-0.2(i-1)} \quad \text{for } i = 2, 3, \ldots, 70$$

The controller matrix $\mathbf{K}_c$ for each case is shown in Table 20.2. Note that the dimensions of $\boldsymbol{K}$ are different for the two cases, because $\mathbf{K}_c$ has dimensions of $rM \times mP$, as noted earlier. For this SISO example, $r = m = 1$, and the values of M and P differ for the two cases.

Table 20.2 Feedback Matrices $\mathrm{K_c}$ for Example 20.5

Case	K_c
For $P = 3$ and $M = 1$:	$K_c = [0 \quad 7.79 \quad 28.3]$
For $P = 4$ and $M = 2$:	$K_c = \begin{bmatrix} 0 & 33.1 & 48.8 & -13.4 \\ 0 & -71.4 & -97.4 & 57.3 \end{bmatrix}$

(b) The unit step response can be derived analytically using Lapace transforms:

$$y(t) = 0 \qquad \text{for } t \le 1$$

$$y(t) = 1 - 2e^{-0.1(t-1)} + e^{-0.2(t-1)} \quad \text{for } t > 1$$

Figure 20.11 compares the y and u responses for a unit set-point change. The two MPC controllers provide superior output responses with very small settling times, but their initial MV changes are larger than those for the PID controller. (Note the expanded time scale for u.)

(c) For the step disturbance, the output responses for the MPC controllers in Fig. 20.12 have relatively small maximum deviations and are nonoscillatory. By comparison, the PID controller results in the largest maximum deviation and an oscillatory response. Of the two MPC controllers, the one designed using $P = 3$ and $M = 1$ provides a slightly more conservative response.

20.6.1 MPC Application: Distillation Column Model

In order to illustrate the effects of the MPC design parameters (M, P, $\boldsymbol{Q}$, and $\boldsymbol{R}$) for an MIMO problem, consider the Wood-Berry model that was introduced in Example 18.1:

$$\begin{bmatrix} X_D(s) \\ X_B(s) \end{bmatrix} = \begin{bmatrix} \dfrac{12.8e^{-s}}{16.7s+1} & \dfrac{-18.9e^{-3s}}{21s+1} \\ \dfrac{6.6e^{-7s}}{10.9s+1} & \dfrac{-19.4e^{-3s}}{14.4s+1} \end{bmatrix} \begin{bmatrix} R(s) \\ S(s) \end{bmatrix} + \begin{bmatrix} \dfrac{3.8e^{-8.1s}}{14.9s+1} \\ \dfrac{4.9e^{-3.4s}}{13.2s+1} \end{bmatrix} F(s) \qquad (20\text{-}74)$$

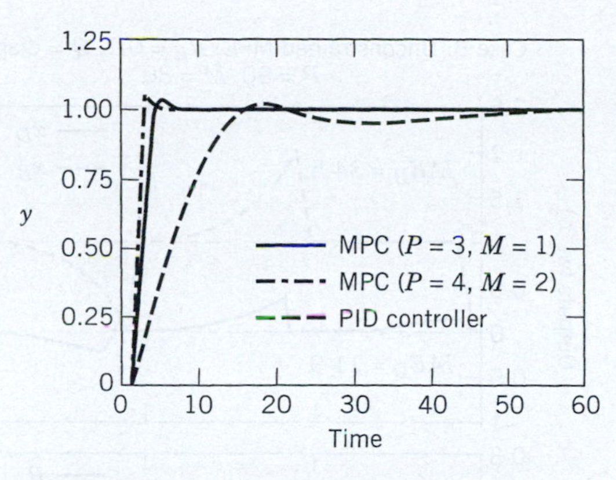

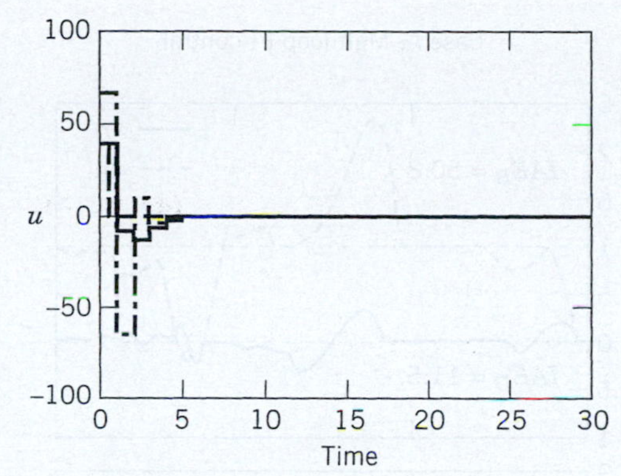

Figure 20.11 Set-point responses for Example 20.5.

The controlled variables (outputs) are the distillate and bottoms compositions (X_D and X_B); the manipulated variables (inputs) are the reflux flow rate and the steam flow rate to the reboiler (R and S); and feed flow rate F is an unmeasured disturbance variable.

Next, we compare a variety of MPC controllers and a multiloop control system, based on simulations performed using the MATLAB Model Predictive Control Toolbox (Bemporad et al., 2009).[2] For each simulation the sampling period was $\Delta t = 1$ min, and saturation limits of $\pm$ 0.15 were imposed on each input. Unconstrained MPC controllers were designed using Eq. 20-55, while the constrained MPC controllers were based on the input constraints in Eq. 20–59. Some constrained MPC controllers were designed using an additional hard-output constraint of $|y_i| \leq 1.8$. In order to compare MPC and a standard multiloop control system, two PI controllers were simulated using the $X_D - R/X_B - S$ control configuration from Example 18.1 and the controller settings in Table 20.3 reported by Lee et al. (1998).

Figures 20.13 and 20.14 compare the performance of the MPC and multiloop control systems for a +1% set-point change in X_B at $t = 0$, followed by two feed flow rate disturbances: a +30% increase at $t = 50$ min and a return to the original value at $t = 100$ min. The input and output variables are displayed as deviation variables. The numerical values of the integral of the absolute error (IAE) performance index (Chapter 12) are included for each output.

A comparison of Cases A and B in Fig. 20.13 indicates that unconstrained MPC is superior to the multiloop control system, because its output variables exhibit faster set-point responses, less oscillation, and smaller IAE values. In addition, the changes in the input variables are

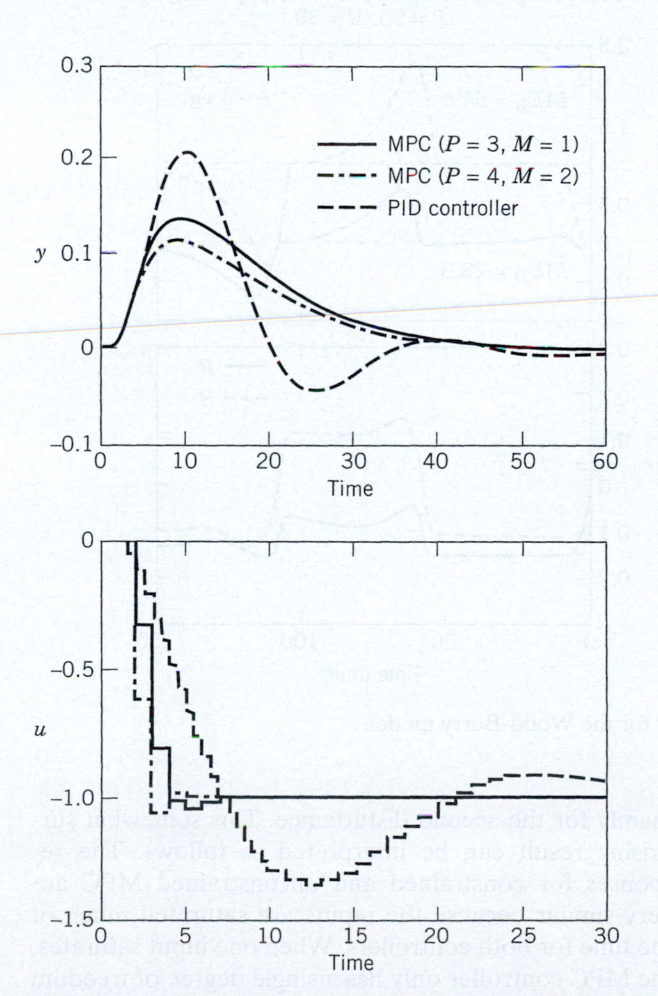

Figure 20.12 Disturbance responses for Example 20.5.

Table 20.3 PI Controller Settings for the Wood-Berry Model

Control Loop	K_c	τ_I (min)
$X_D - R$	0.85	7.21
$X_B - S$	−0.089	8.86

[2]The code for the Wood-Berry example is available in this MATLAB Toolbox. A modified version of the code is included with Exercise 20.9. A newer version of the MPC Toolbox is also available, but without this example (Bemporad et al., 2009).

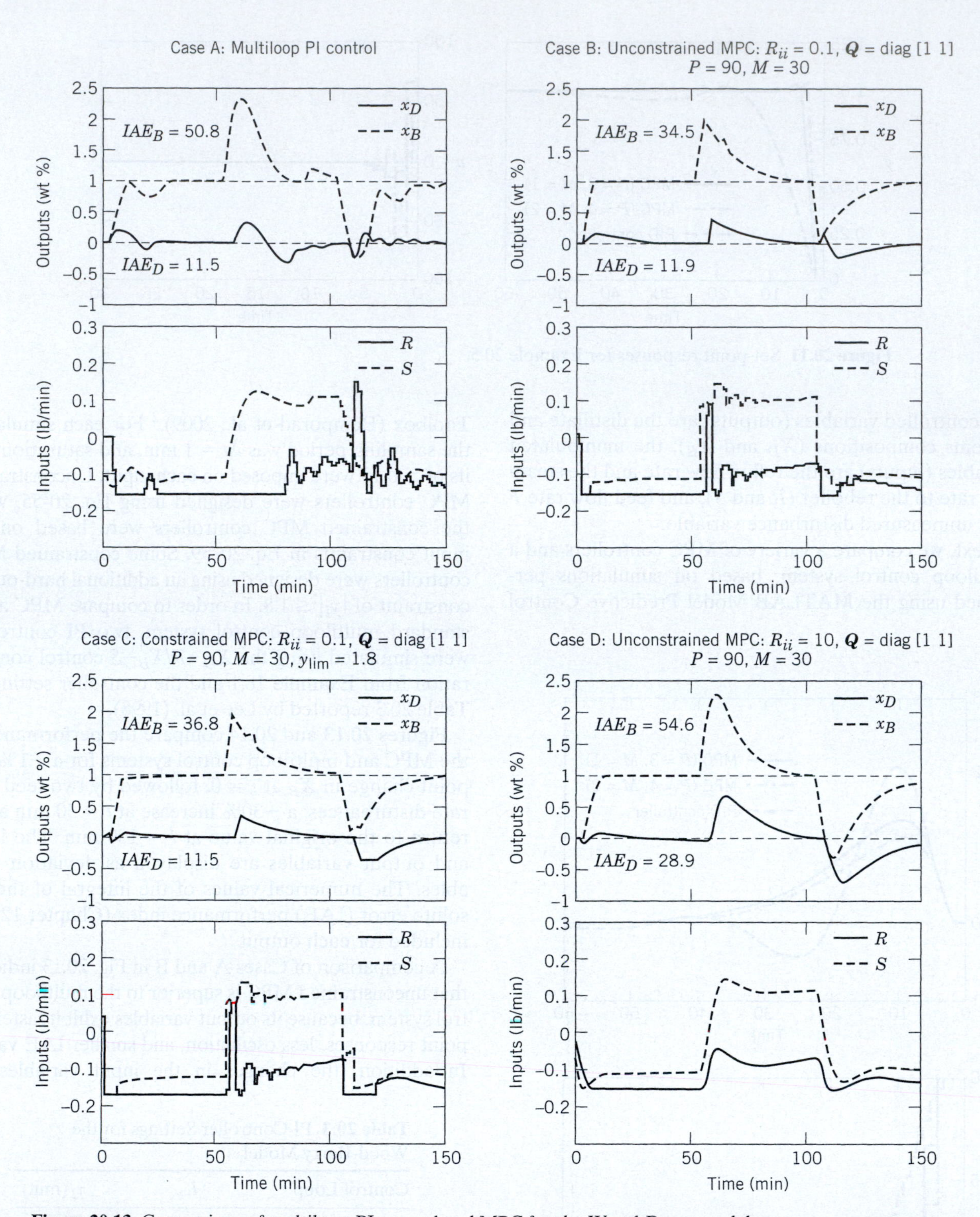

Figure 20.13 Comparison of multiloop PI control and MPC for the Wood-Berry model.

smoother for MPC. Case B is used as a "base case" for the comparisons in Figs. 20.13 and 20.14. Its MPC design parameters are shown in Fig. 20.13 and were selected according to the guidelines presented earlier.

Cases B and C in Fig. 20.13 provide a comparison of constrained and unconstrained MPC. These responses are very similar, with only small differences occurring, mainly for the second disturbance. This somewhat surprising result can be interpreted as follows. The responses for constrained and unconstrained MPC are very similar because the inputs are saturated much of the time for both controllers. When one input saturates, the MPC controller only has a single degree of freedom left, the other input. By contrast, for larger control

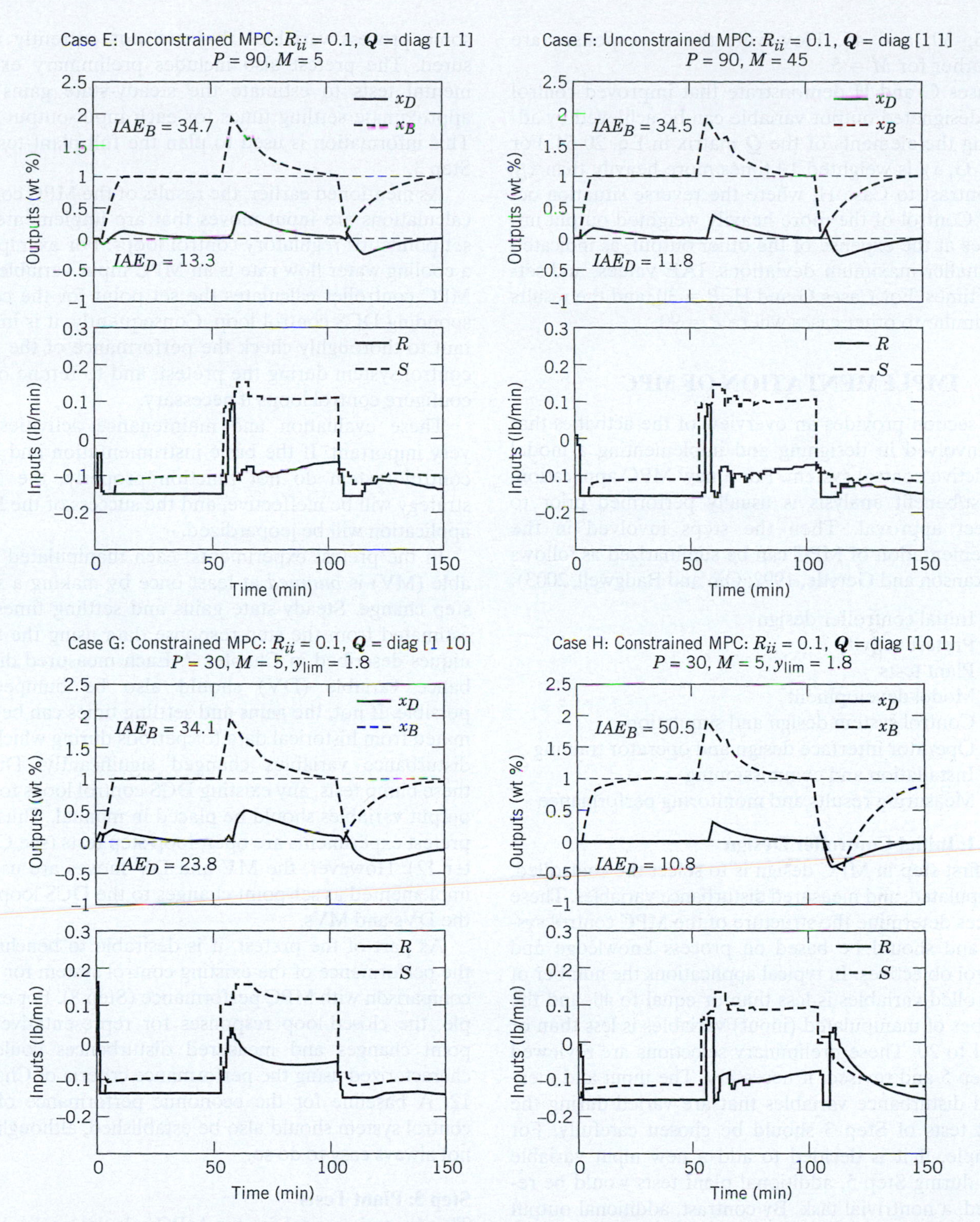

Figure 20.14 Effects of MPC design parameters for the Wood-Berry model.

problems (for example, 10 × 10), constrained MPC will have many more degrees of freedom. For these larger problems, constrained MPC tends to provide improved control due to the extra degrees of freedom and its awareness of the constraints and process interactions.

The effect of a diagonal move suppression matrix $\boldsymbol{R}$ is apparent from a comparison of Cases B and D. When the diagonal elements, R_{ii}, are increased from 0.1 to 10, the MPC inputs become smoother and the output responses have larger deviations, higher IAE values, and longer settling times.

The effect of changing control horizon, M, is shown in Cases B, E, and F. The y responses and IAE values are quite similar for all three values of

M: 5, 30, and 45. However, the u responses are smoother for $M = 5$.

Cases G and H demonstrate that improved control of a designated output variable can be achieved by adjusting the elements of the $\mathbf{Q}$ matrix in Eq. 20-54. For Case G, x_B is weighted 10 times more heavily than x_D, in contrast to Case H, where the reverse situation occurs. Control of the more heavily weighted output improves at the expense of the other output, as indicated by smaller maximum deviations, IAE values, and settling times. For Cases G and H, $P = 30$, and the results are similar to other cases where $P = 90$.

20.7 IMPLEMENTATION OF MPC

This section provides an overview of the activities that are involved in designing and implementing a model predictive control system. For a new MPC application, a cost/benefit analysis is usually performed prior to project approval. Then the steps involved in the implementation of MPC can be summarized as follows (Hokanson and Gerstle, 1992; Qin and Badgwell, 2003):

1. Initial controller design
2. Pretest activity
3. Plant tests
4. Model development
5. Control system design and simulation
6. Operator interface design and operator training
7. Installation and commissioning
8. Measuring results and monitoring performance

Step 1: Initial Controller Design

The first step in MPC design is to select the controlled, manipulated, and measured disturbance variables. These choices determine the structure of the MPC control system and should be based on process knowledge and control objectives. In typical applications the number of controlled variables is less than or equal to 40, and the number of manipulated (input) variables is less than or equal to 20. These preliminary selections are reviewed in Step 5 and revised, if necessary. The input and measured disturbance variables that are varied during the plant tests of Step 3 should be chosen carefully. For example, if it is decided to add a new input variable later during Step 5, additional plant tests would be required, a nontrivial task. By contrast, additional output variables can be added to the MPC control structure later, if necessary, provided that these measurements were recorded during the plant tests.

Step 2: Pretest Activity

During the pretest activity (or *pretest*, for short), the plant instrumentation is checked to ensure that it is working properly. Remedial action may be required for faulty sensors, sticking control valves, and the like. Also, a decision may be made to install sensors for some process variables that are not currently measured. The pretest also includes preliminary experimental tests to estimate the steady-state gains and approximate settling times for each input-output pair. This information is used to plan the full plant tests of Step 3.

As mentioned earlier, the results of the MPC control calculations are input moves that are implemented as set points for regulatory control loops. For example, if a cooling water flow rate is an MPC input variable, the MPC controller calculates the set point for the corresponding DCS control loop. Consequently, it is important to thoroughly check the performance of the DCS control system during the pretest, and to retune or reconfigure control loops if necessary.

These evaluation and maintenance activities are very important. If the basic instrumentation and DCS control system do not function properly, the MPC strategy will be ineffective, and the success of the MPC application will be jeopardized.

In the pretest experiments, each manipulated variable (MV) is *bumped* at least once by making a small step change. Steady-state gains and settling times are estimated from the step-response data using the techniques described in Chapter 7. Each measured disturbance variable (DV) should also be bumped, if possible. If not, the gains and settling times can be estimated from historical data for periods during which the disturbance variables changed significantly. During these bump tests, any existing DCS control loops for the output variables should be placed in manual. Thus, the pretest experiments are open-loop step tests (see Chapter 12). However, the MV and DV moves are usually implemented as set-point changes to the DCS loops for the DVs and MVs.

As part of the pretest, it is desirable to benchmark the performance of the existing control system for later comparison with MPC performance (Step 8). For example, the closed-loop responses for representative set-point changes and measured disturbances could be characterized using the performance criteria of Chapter 12. A baseline for the economic performance of the control system should also be established, although it is not always easy to do so.

Step 3: Plant Tests

The dynamic model for the MPC calculations is developed from data collected during special plant tests. The plant testing can be very time-consuming, typically requiring days, or even weeks, of around-the-clock experiments. The required test duration depends on the settling times of the outputs and the numbers of MVs and DVs. The excitation for the plant tests usually consists of changing an input variable or a disturbance variable (if possible) from one value to another, using either a series of step changes with different durations or the pseudorandom-binary sequence (PRBS) that was introduced in Chapter 7. The plant test experiments are implemented in the same manner as the pretest experiments of Step 2.

It is traditional industrial practice to move each MV and DV individually. The magnitudes of the moves should be carefully chosen, because too small a move may result in the step responses being obscured by normal process fluctuations and measurement noise. On the other hand, too large a change may result in an output constraint violation or nonlinear process behavior that cannot be accurately described by a linear model.

The magnitude of the maximum allowable input changes can be estimated from knowledge of the output constraints and the estimated steady-state gains from the pretest. For example, suppose that $(\Delta u_j)_{\max}$ denotes the maximum change that can be made in u_j without violating a constraint for y_i. It can be estimated from the expression,

$$(\Delta u_j)_{\max} = \frac{(\Delta y_i)_{\max}}{\hat{K}_{ij}} \tag{20-75}$$

where $(\Delta y_i)_{\max}$ is the maximum allowable change in y_i and $\hat{K}_{ij}$ is the estimated steady-state gain between y_i and u_j. However, this steady-state analysis does not guarantee that each y_i satisfies its constraints during transient responses.

The duration of the longest step test is equal to $t_{\max}$, the longest settling time that was observed during the pretest. Shorter step changes are also made, with the durations typically varying from $t_{\max}/8$ to $t_{\max}/2$. In order to ensure that sufficient data are obtained for model identification, each input variable is typically moved 8–15 times (Qin and Badgwell, 2003).

Some MPC vendors recommend a total plant testing period of $t_{\text{test}} = 6(r + p)t_{\max}$ where r is the number of input variables and p is the number of measured disturbance variables. In principle, t_{test} can be reduced by making simultaneous changes to several input (or disturbance) variables rather than the traditional sequential ("one-at-a-time") approach. Also, it can be very difficult to identify poorly conditioned process models using the sequential approach. However, because of a number of practical considerations, input moves are traditionally made sequentially. In particular, simultaneous input moves tend to complicate the test management and make it more difficult to identify periods of abnormal operation by visual inspection of the test data. It is also more difficult to ensure that output constraints will not be violated. Because of similar practical considerations, step changes have been traditionally preferred over the pseudorandom binary sequence (PRBS) of Chapter 7.

Step 4: Model Development

The dynamic model is developed from the plant test data by selecting a model form (for example, a step-response model) and then estimating the model parameters. However, first it is important to eliminate periods of test data during which plant upsets or other abnormal situations have occurred, such as control valve saturation or a DCS control loop having been placed in manual. Decisions to omit portions of the test data are based on visual inspection of the data, knowledge of the process, and experience. Parameter estimation is usually based on least-squares estimation (Chapter 7).

As part of the model development step, the model accuracy should be characterized, because this information is useful for subsequent system design and tuning. The characterization can include confidence intervals for the model predictions and/or model parameters. The confidence intervals can be calculated using standard statistical techniques (Ljung, 1999).

Step 5: Control System Design and Simulation

The MPC design is based on the control and optimization objectives, process constraints, and the dynamic model of the process. The preliminary control system design from Step 1 is critically evaluated and modified, if necessary. Then the MPC design parameters in Section 20.6 are selected, including the sampling periods, weighting factors, and control and prediction horizons. Next, the closed-loop system is simulated using the identified process model and a wide variety of process conditions to evaluate control system performance. The MPC design parameters are adjusted, if necessary, to obtain satisfactory control system performance and robustness over the specified range of operating conditions.

Step 6: Operator Interface Design and Operator Training

Because plant operators play a key role in manufacturing plants, it is important that the MPC operator interface meet their needs. Operator training is also important, because MPC concepts such as predictive control, multivariable interactions, and constraint handling are very different from conventional regulatory control concepts. For a standard multiloop control system, each input is adjusted based on measurements of a single output. By contrast, in MPC each input depends on all of the outputs. Thus, understanding why the MPC system responds the way that it does, especially in unusual operating conditions, can be very challenging for both operators and engineers.

Step 7: Installation and Commissioning

After a MPC control system is installed, it is first evaluated in a "prediction mode." Model predictions are compared with measurements, but the process continues to be controlled by the existing control system (e.g., DCS). After the output predictions are judged to be satisfactory, the calculated MPC control moves are evaluated to see if they are reasonable. Finally, the MPC software is evaluated during closed-loop operation with the calculated control moves implemented as set points to the DCS control loops. The MPC design parameters are tuned, if necessary. The commissioning period typically requires some troubleshooting and can take as long as, or even longer than, the plant tests of Step 3.

Step 8: Measuring Results and Monitoring Performance The evaluation of MPC system performance is not easy, and widely accepted metrics and monitoring strategies are not available. However, useful diagnostic information is provided by basic statistics, such as the means and standard deviations for both measured variables, and calculated quantities, such as control errors and model residuals. Another useful statistic is the relative amount of time that an input is saturated or a constraint is violated, expressed as a percentage of the total time the MPC system is in service. These types of routine monitoring activities are considered in more detail in Chapter 21.

In Chapter 12, we considered a number of classical metrics for characterizing control system performance, such as the IAE index, overshoot, settling time, and degree of oscillation. Though helpful, these metrics provide an incomplete picture of overall MPC performance. An important motivation for MPC is that it facilitates process operation closer to targets and limiting constraints. Thus, an evaluation of MPC performance should include measures of whether these objectives have been realized. If so, a noticeable improvement in process operation should be reflected in economically meaningful measures such as product quality, throughput, or energy costs. The various performance metrics should be calculated before and after the MPC system is installed.

MPC system performance should be monitored on a regular basis to ensure that performance does not degrade owing to changes in the process, instrumentation, or process conditions, including disturbances. If performance becomes significantly worse, retuning the controller or reidentifying all (or part of) the process model may be required. The development of MPC monitoring strategies is an active research area (Kozub, 2002, Jelali, 2006; McIntosh and Canney, 2008; Badwe et al., 2009).

SUMMARY

Model predictive control is an important model-based control strategy devised for large multiple-input, multiple-output control problems with inequality constraints on the inputs and/or outputs. This chapter has considered both the theoretical and practical aspects of MPC. Applications typically involve two types of calculations: (1) a steady-state optimization to determine the optimum set points for the control calculations, and (2) control calculations to determine the MV changes that will drive the process to the set points. The success of model-based control strategies such as MPC depends strongly on the availability of a reasonably accurate process model. Consequently, model development is the most critical step in applying MPC. As Rawlings (2000) has noted, "feedback can overcome some effects of poor models, but starting with a poor process model is akin to driving a car at night without headlights." Finally, the MPC design parameters should be chosen carefully, as illustrated by two simulation examples.

Model predictive control has had a major impact on industrial practice, with thousands of applications worldwide. MPC has become the method of choice for difficult control problems in the oil refining and petrochemical industries. However, it is not a panacea for all difficult control problems (Shinskey, 1994; Hugo, 2000; McIntosh and Canney, 2008). Furthermore, MPC has had much less impact in the other process industries. Performance monitoring of MPC systems is an important topic of current research interest.

REFERENCES

Badgwell, T. A., and S. J. Qin, A Review of Nonlinear Model Predictive Control Applications, in *Nonlinear Predictive Control: Theory and Practice*, B. Kouvaritakis and M. Cannon (Eds.), Inst. Electrical Eng., London, 2001, Chapter 1.

Badwe, A. S., R. D. Gudi, R. S. Patwardhan, S. L. Shah, and S. C. Patwardhan, Detection of Model-Plant Mismatch in MPC Applications, *J. Process Control*, **19**, 1305 (2009).

Bemporad, A., M. Morari, and N. L. Ricker, *Model Predictive Control Toolbox 3*, User's Guide, Mathworks, Inc., Natick, MA, 2009.

Camacho, E. F., and C. Bordons, *Model Predictive Control* 2nd ed., Springer-Verlag, New York, 2003.

Canney, W. M., The Future of Advanced Process Control Promises More Benefits and Sustained Value, *Oil & Gas Journal*, **101** (16), 48 (2003).

Clarke, D. W., C. Mohtadi, and P. S. Tufts, Generalized Predictive Control—Part I. The Basic Algorithm, *Automatica*, **23**, 137 (1987).

Cutler, C. R., and B. L. Ramaker, Dynamic Matrix Control—A Computer Control Algorithm, *Proc. Joint Auto. Control Conf., Paper WP5-B*, San Francisco (1980).

Edgar, T. F., D. M. Himmelblau, and L. Lasdon, *Optimization of Chemical Processes*, 2d ed., McGraw-Hill, New York, 2001.

Hokanson, D. A., and J. G. Gerstle, Dynamic Matrix Control Multivariable Controllers, in *Practical Distillation Control*, W. L. Luyben (Ed.) Van Nostrand Reinhold, New York, 1992, p. 248.

Hugo, A., Limitations of Model Predictive Controllers, *Hydrocarbon Process.*, **79** (1), 83 (2000).

Jelali M., An Overview of Control Performance Assessment Technology and Industrial Applications, *Control Eng. Practice*, **14**, 441 (2006).

Kano, M., and M. Ogawa, The State of the Art in Advanced Chemical Process Control in Japan, *Proc. IFAC Internat. Sympos on Advanced Control of Chemical Processes (ADCHEM 2009)*, Paper 240, Istanbul, Turkey (July 2009).

Kassman, D. E., T. A. Badgwell, and R. B. Hawkins, Robust Steady-State Target Calculations for Model Predictive Control, *AIChE J.*, **46**, 1007 (2000).

Kozub, D. J. Controller Performance Monitoring and Diagnosis. Industrial Perspective, *Preprints of the 15th Triennial World IFAC Congress*, Barcelona, Spain (July 2002).

Lee, J., W. Cho, and T. F. Edgar, Multiloop PI Controller Tuning for Interacting Multivariable Processes, *Computers and Chem. Engng.*, **22**, 1711 (1998).

Ljung, L., *System Identification*, 2d ed., Prentice Hall, Upper Saddle River, NJ, 1999.

Lundström, P., J. H. Lee, M. Morari, and S. Skogestad, Limitations of Dynamic Matrix Control, *Computers and Chem. Engng.*, **19**, 409 (1995).

Maciejowski, J. M., *Predictive Control with Constraints*, Prentice Hall, Upper Saddle River, NJ, 2002.

Maurath, P. R. D. A. Mellichamp, and D. E. Seborg, Predictive Controller Design for SISO Systems, *IEC Research*, **27**, 956 (1988).

McIntosh, A. R. and W. M. Canney, The Dirty Secrets of Model Predictive Controller Sustained Value, *Proc. Internat. Sympos. on Advanced Control of Industrial Processes (ADCONIP 2008)*, Paper MoB1.4, Jasper, Alberta, Canada (May 2008).

Morari, M., and J. H. Lee, Model Predictive Control: Past, Present, and Future, *Comput. and Chem. Eng.*, **23**, 667 (1999).

Morari, M., and N. L. Ricker, *Model Predictive Control Toolbox*, The Mathworks, Inc., Natick, MA, 1994.

Qin, S. J., and T. A. Badgwell, A Survey of Industrial Model Predictive Control Technology, *Control Eng. Practice*, **11**, 733 (2003).

Rawlings, J. B., Tutorial Overview of Model Predictive Control, *IEEE Control Systems*, **20**(3), 38 (2000).

Richalet, J. and D. O'Donovan, *Predictive Functional Control: Principles and Industrial Applications*, Springer-Verlag Ltd., London, 2009.

Richalet, J., A. Rault, J. L. Testud, and J. Papon, Model Predictive Heuristic Control: Applications to Industrial Processes, *Automatica*, **14**, 413 (1978).

Rossiter, J. A., *Model-Based Predictive Control: A Practical Approach*, CRC Press, Boca Raton, FL, 2003.

Shinskey, F. G., *Feedback Controllers for the Process Industries*, McGraw-Hill, New York, 1994.

Sorensen, R. C., and C. R. Cutler, LP Integrates Economics into Dynamic Matrix Control, *Hydrocarbon Process.*, **77**(9), 57 (1998).

White, D. C., Multivariable Control of a Chemical Plant Incinerator: An Industrial Case Study of Co-Linearity and Non-Linearity, *Proc. Internat. Sympos. on Advanced Control of Industrial Processes (ADCONIP 2008)*, Paper MoB1.1, Jasper, Alberta, Canada (May 2008).

EXERCISES

20.1 For the transfer functions

$$G_P(s) = \frac{2e^{-s}}{(10s+1)(5s+1)} \qquad G_v = G_m = 1$$

(a) Derive an analytical expression for the step response to a unit step change. Evaluate the step-response coefficients, $\{S_i\}$, for a sampling period of $\Delta t = 1$.

(b) What value of model horizon N should be specified in order to ensure that the step-response model covers a period of at least 99% of the open-loop settling time? (That is, we require that $N\Delta t \geq t_{99}$ where t_{99} is the 99% settling time.)

20.2 A process (including sensor and control valve) can be modeled by the transfer function,

$$G(s) = \frac{2(1-9s)}{(15s+1)(3s+1)}$$

(a) Derive an analytical expression for the response to a unit step change in the input.

(b) Suppose that the maximum allowable value for the model horizon is $N = 30$. What value of the sampling period Δt should be specified to ensure that the step-response model covers a period of at least 99% of the open-loop settling time? (That is, we require that $N\Delta t \geq t_{99}$ where t_{99} is the 99% settling time.)

Use the analytical solution and this value of Δt to obtain a step-response model in the form of Eq. 20-1.

20.3 Control calculations for a control horizon of $M = 1$ can be performed either analytically or numerically. For the process model in Exercise 20.1, derive $\boldsymbol{K}_{c1}$ for $\Delta t = 1, N = 50$, and $P = 5$, $\mathbf{Q} = \mathbf{I}$ and $\mathbf{R} = \mathbf{0}$, using Eq. 20-57. Compare your answer with the analytical result reported by Maurath et al. (1988).

$$\boldsymbol{K}_{c1} = \frac{1}{\sum_{i=1}^{P} S_i^2} [S_1\, S_2\, S_3 \ldots S_P]$$

20.4 Consider the transfer function model of Exercise 20.1. For each of the four sets of design parameters shown below, design a model predictive controller. Then do the following:

(a) Compare the controllers for a unit step change in set point. Consider both the y and u responses.

(b) Repeat the comparison of (a) for a unit step change in disturbance, assuming that $G_d(s) = G(s)$.

(c) Which controller provides the best performance? Justify your answer.

Set No.	Δt	N	M	P	$\boldsymbol{R}$
(i)	2	40	1	5	0
(ii)	2	40	20	20	0
(iii)	2	40	3	10	0.01
(iv)	2	40	3	10	0.1

20.5 For Exercise 20.1, suppose that a constraint is placed on the manipulated variable, $u_{k+j} \leq 0.2$ for $j = 1, 2, \ldots, M-1$. Let $\Delta t = 2$ and $N = 40$. Select values of M, P, and $\boldsymbol{R}$ so that these constraints are not violated after a unit step disturbance occurs.

20.6 For Exercise 20.1, consider two sets of design parameters and simulate unit step changes in both the disturbance and the set point. Assume that the disturbance model is identical to the process model. The design parameters are

(a) $M = 7 \quad P = 10 \quad \boldsymbol{R} = \mathbf{0}$

(b) $M = 3 \quad P = 10 \quad \boldsymbol{R} = \mathbf{0}$

Which controller provides the best control? Justify your answer.

20.7 Consider the unconstrained, SISO version of MPC in Eq. 20-57. Suppose that the controller is designed so that the control horizon is $M = 1$ and the weighting matrices are $\boldsymbol{Q} = \boldsymbol{I}$ and $\boldsymbol{R} = 1$. The prediction horizon P can be chosen arbitrarily.

Demonstrate that the resulting MPC controller has a simple analytical form.

20.8 A theoretical advantage of MPC for ideal conditions is that it guarantees that both controlled and manipulated variables satisfy specified inequality constraints. Briefly discuss why this theoretical advantage may not be realized in practical applications.

20.9 In Section 20.6.1, MPC was applied to the Wood-Berry distillation column model. A MATLAB program for this example and constrained MPC is shown in Table E20.9. The design parameters have the base case values (Case B in Fig. 20.13) except for $P = 10$ and $M = 5$. The input constraints are the saturation limits for each input (−0.15 and +0.15). Evaluate the effects of control horizon M and input weighting matrix $\boldsymbol{R}$ by simulating the set-point change and the first disturbance of Section 20.6.1 for the following parameter values:

(a) Control horizon, $M = 2$ and $M = 5$

(b) Input weighting matrix, $\boldsymbol{R} = 0.1\boldsymbol{I}$ and $\boldsymbol{R} = \boldsymbol{I}$

Table E20.9 MATLAB Program (Based on a program by Morari and Ricker (1994))

```
g11=poly2tfd(12.8,[16.7 1],0,1);   % model
g21=poly2tfd(6.6,[10.9 1],0,7);
g12=poly2tfd(−18.9,[21.0 1],0,3);
g22=poly2tfd(−19.4,[14.4 1],0,3);
gd1=poly2tfd(3.8,[14.9 1],0,8.1);
gd2=poly2tfd(4.9,[13.2 1],0,3.4);
tfinal=120;     % Model horizon, N
delt=1;         % Sampling period
ny=2;           % Number of outputs
model=tfd2step(tfinal,delt,ny,g11,g21,g12,g22)
plant=model;    % No plant/model mismatch
dmodel=[]       % Default disturbance model
dplant=tfd2step(tfinal,delt,ny,gd1,gd2)
P=10;   M=5;   % Horizons
ywt=[1 1];   uwt=[0.1 0.1];   % Q and R
tend=120;       % Final time for simulation
r=[0 1];   % Set-point change in XB
a=zeros([1,tend]);
for i=51:tend
   a(i)=0.3*2.45; % 30 % step in F at t=50 min.
end
dstep=[a'];
ulim=[−.15 −.15 .15 .15 1000 1000]; % u limits
ylim=[];        % No y limits
tfilter=[ ];
[y1,u1]=cmpc(plant,model,ywt,uwt,M,P,tend,r,
ulim,ylim, tfilter,dplant,dmodel,dstep);
figure(1)
subplot(211)
plot(y1)
legend('XD','XB')
xlabel('Time (min)')
subplot(212)
stairs(u1) % Plot inputs as staircase functions
legend('R','S')
xlabel('Time (min)')
```

Consider plots of both inputs and outputs. Which choices of M and $\boldsymbol{R}$ provide the best control? Do any of these MPC controllers provide significantly better control than the controllers shown in Figs. 20.13 and 20.14? Justify your answer.

20.10 Design a model predictive controller for the process

$$G_p(s) = \frac{e^{-6s}}{10s+1} \qquad G_v = G_m = 1$$

Select a value of N based on 95% completion of the step response and $\Delta t = 2$. Simulate the closed-loop response for a set-point change using the following design parameters:

(a) $M = 1$ $P = 7$ $\boldsymbol{R} = \boldsymbol{0}$

(b) $M = 1$ $P = 5$ $\boldsymbol{R} = \boldsymbol{0}$

(c) $M = 4$ $P = 30$ $\boldsymbol{R} = \boldsymbol{0}$

20.11 Repeat Exercise 20.9 for the situation where the input constraints have been changed to −0.3 and +0.3.

20.12 Consider the PCM furnace module of Appendix E with the following variables (HC denotes hydrocarbon):

CVs: HC exit temperature T_{HC} and oxygen exit concentration c_{O_2}

MVs: fuel gas flow rate F_{FG} and air flow rate F_A

DV: HC flow rate F_{HC}

Do the following, using the transfer function models given below:

(a) Design an MPC system using the following design parameters: $\Delta t = 1$ min, $\boldsymbol{Q}$ = diagonal [0.1, 1], $\boldsymbol{R}$ = diagonal [0.1, 0.1], $P = 20$, and M = 1.

(b) Repeat part (a) for the same design parameters, but where $\boldsymbol{R}$ = diagonal [0.5, 0.5].

(c) Simulate the two MPC controllers for a step change in the c_{O_2} set point to 1.0143 mol/m^3 at $t = 10$ min.

(d) Repeat part (c) for a step change in F_{HC} at $t = 10$ min to 0.035 m^3/min.

(e) Based on your results for parts (c) and (d), which MPC controller is superior? Justify your answer.

Process transfer function matrix:

$$\begin{array}{c} \\ T_{HC} \\ \\ c_{O_2} \end{array} \begin{array}{c} \quad F_{FG} \qquad\qquad F_A \\ \begin{bmatrix} \dfrac{220\,e^{-2s}}{6.5\,s+1} & \dfrac{-13\,e^{-2s}}{6.2\,s+1} \\[2ex] \dfrac{-2.0\,e^{-4s}}{3.8\,s+1} & \dfrac{0.14\,e^{-4s}}{4.2\,s+1} \end{bmatrix} \end{array}$$

20.13 Repeat Exercise 20.12 for $\boldsymbol{R}$ = diagonal [0.1, 0.1] and: (i) $M = 1$ and (ii) $M = 4$.

Chapter 21

Process Monitoring

CHAPTER CONTENTS

In industrial plants, large numbers of process variables must be maintained within specified limits in order for the plant to operate properly. Excursions of key variables beyond these limits can have significant consequences for plant safety, the environment, product quality, and plant profitability. Earlier chapters have indicated that industrial plants rely on feedback and feedforward control to keep process variables at or near their set points. A related activity, *process monitoring*, also plays a key role in ensuring that the plant performance satisfies the operating objectives. In this chapter, we introduce standard monitoring techniques as well as newer strategies that have gained industrial acceptance in recent years. In addition to process monitoring, the related problem of monitoring the performance of the control system itself is also considered.

The general objectives of process monitoring are:

1. *Routine Monitoring.* Ensure that process variables are within specified limits.
2. *Detection and Diagnosis.* Detect abnormal process operation and diagnose the root cause.
3. *Preventive Monitoring.* Detect abnormal situations early enough that corrective action can be taken before the process is seriously upset.

Abnormal process operation can occur for a variety of reasons, including equipment problems (heat exchanger fouling), instrumentation malfunctions (sticking control valves, inaccurate sensors), and unusual disturbances (reduced catalyst activity, slowly drifting feed composition). Severe abnormal situations can have serious consequences, even forcing a plant shutdown. It has been estimated that improved handling of abnormal situations could result in savings of $10 billion each year to the U.S. petrochemical industry (ASM, 2009). Thus, process monitoring and abnormal situation management are important activities.

The traditional approach for process monitoring is to compare measurements against specified limits. This *limit checking* technique is a standard feature of computer control systems and is widely used to validate measurements of process variables such as flow rate, temperature, pressure, and liquid level. Process variables are measured quite frequently with sampling periods that typically are much smaller than the process settling time (see Chapter 17). However, for most industrial plants, many important *quality variables* cannot be measured on-line. Instead, samples of the product are taken on an infrequent basis (e.g., hourly or daily) and sent to the quality control laboratory for analysis. Due to the infrequent measurements, standard feedback control methods like PID control cannot be applied, Consequently, statistical process control techniques are implemented to ensure that the product quality meets the specifications.

The terms *statistical process control* (*SPC*) and *statistical quality control* (*SQC*) refer to a collection of statistically–based techniques that rely on *quality control charts* to monitor product quality. These terms tend to be used on an interchangeable basis. However, the term SPC is sometimes used to refer to a broader set of statistical techniques that are employed to improve process performance as well as product quality (MacGregor, 1988). In this chapter, we emphasize the classical SPC techniques that are based on quality control charts (also called control charts). The simplest control chart, a Shewhart chart, merely consists of measurements plotted vs. sample number, and control limits that indicate the upper and lower limits for normal process operation.

The major objective in SPC is to use process data and statistical techniques to determine whether the process operation is normal or abnormal. The SPC methodology is based on the fundamental assumption that normal process operation can be characterized by random variations about a mean value. If this situation exists, the process is said to be *in a state of statistical control* (or *in control*), and the control chart measurements tend to be normally distributed about the mean value. By contrast, frequent control chart violations would indicate abnormal process behavior or an *out-of-control* situation. Then, a search would be initiated to attempt to identify the root cause of the abnormal behavior. The root cause is referred to as the *assignable cause* or the *special cause* in the SPC literature, while the normal process variability is referred to as *common cause* or *chance cause*. From an engineering perspective, SPC is more of a monitoring technique than a control technique because no automatic corrective action is taken after an abnormal situation is detected. A brief comparison of conventional feedback control and SPC is presented in Section 21.2.4. More detailed comparisons are available elsewhere (MacGregor, 1988; Box and Luceño, 1997).

The basic SPC concepts and control chart methodology were introduced by Shewhart (1931). The current widespread interest in SPC techniques began in the 1950s when they were successfully applied first in Japan and then in North America, Europe, and the rest of the world. Control chart methodologies are now widely used in discrete-parts manufacturing and in some sectors of the process industries, especially for the production of semiconductors, synthetic fibers, polymers, and specialty chemicals. SPC techniques are also widely used for product quality control and for monitoring control system performance (Shunta, 1995). The basic SPC methodology is described in introductory statistics texts (Montgomery and Runger, 2007) and books on SPC (Ryan 2000; Montgomery, 2009).

SPC techniques played a key role in the renewed industrial emphasis on product quality that is sometimes referred to as the *Quality Revolution*. During the 1980s, Deming (1986) had a major impact on industrial management in North America by convincing corporations that quality should be a top corporate priority. He argued that the failure of a company to produce quality products was largely a failure in management rather than a shortcoming of the plant equipment or employees. His success led to the establishment of many process and quality improvement programs, including the *Six Sigma* methodology that is considered in Section 21.3.

In this chapter, we first introduce traditional process monitoring techniques (Section 21.1) that are based on limit checking of measurements and process performance calculations. In Section 21.2, the theoretical basis of SPC monitoring techniques and the most widely used control charts are considered. We also introduce *process capability indices* and compare SPC with standard automatic feedback control. Traditional SPC monitoring techniques consider only a single measured variable at a time, a *univariate* approach. But when the measured variables are highly correlated, improved monitoring can be achieved by applying the *multivariate* techniques that are introduced in Section 21.4. In addition to monitoring process performance, it can be very beneficial to assess control system performance. This topic is considered in Section 21.5.

Monitoring strategies have been proposed based on process models, neural networks, and expert systems (Davis et al., 2000; Chiang et al., 2001). However, these topics are beyond the scope of this book.

21.1 TRADITIONAL MONITORING TECHNIQUES

In this section, we consider two relatively simple but very effective process monitoring techniques: limit checking and performance calculations.

21.1.1 Limit Checking

Process measurements should be checked to ensure that they are between specified limits, a procedure referred to as *limit checking*. The most common types of measurement limits are (see Chapter 10):

1. High and low limits
2. High limit for the absolute value of the rate of change
3. Low limit for the sample variance

The limits are specified based on safety and environmental considerations, operating objectives, and equipment limitations. For example, the high limit on a reactor temperature could be set based on metallurgical limits or the onset of undesirable side reactions. The low limit for a slurry flow rate could be selected to avoid having solid material settle and plug the line. Sometimes a second set of limits serves as *warning limits*. For example, in a liquid storage system, when the level drops to 15% (the low limit), a low-priority alarm signal could be sent to the operator. But when the level decreases to 5% (the low-low limit), a high-priority alarm would be generated for this more serious situation. Similarly, in order to avoid having the tank overflow, a high limit of 85% and a high-high limit of 95% level could be specified. The high-high and low-low limits are also referred to as *action limits*.

In practice, there are physical limitations on how much a measurement can change between consecutive sampling instants. For example, we might conclude that a temperature in a process vessel cannot change by more than 2 °C from one sampling instant to the next, based on knowledge of the energy balance and the process dynamics. This *rate-of-change limit* can be used to detect an abnormal situation such as a noise spike or a sensor failure. (Noise-spike filters were considered in Chapter 17.)

A set of process measurements inevitably exhibits some variability, even for "steady-state operation." This variability occurs as a result of measurement noise, turbulent flow near a sensor, and other process disturbances. However, if the amount of variability becomes unusually low, it could indicate an abnormal situation such as a "dead sensor" or a sticking control valve. Consequently, it is common practice to monitor a measure of variability such as the variance or standard deviation of a set of measurements. For example, the variability of a set of n measurements can be characterized by the *sample standard deviation*, s, or the *sample variance*, s^2,

$$s^2 \triangleq \frac{1}{n-1}\sum_{i=1}^{n}(x_i - \bar{x})^2 \tag{21-1}$$

where x_i denotes the ith measurement and $\bar{x}$ is the sample mean:

$$\bar{x} \triangleq \frac{1}{n}\sum_{i=1}^{n} x_i \tag{21-2}$$

For a set of data, $\bar{x}$ indicates the average value, while s and s^2 provide measures of the spread of the data. Either s or s^2 can be monitored to ensure that it is above a threshold that is specified based on process operating experience.

The flow rate data in Fig. 21.1 includes three noise spikes and a sensor failure. The rate of change limit would detect the noise spikes, while an abnormally low sample variance would identify the failed sensor. After a limit check violation occurs, an alarm signal can be sent to the plant operator in a number of different ways. A relatively minor alarm might merely be "logged" in a computer file. A more important alarm could be displayed as a flashing message on a computer terminal and require operator acknowledgment. A critical alarm could result in an audible sound or a flashing warning light in the control room. Other alarm options are available, as discussed in Chapter 10.

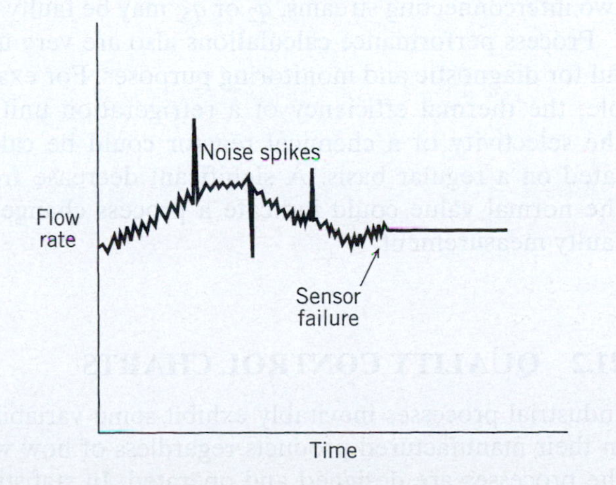

Figure 21.1 Flow rate measurement.

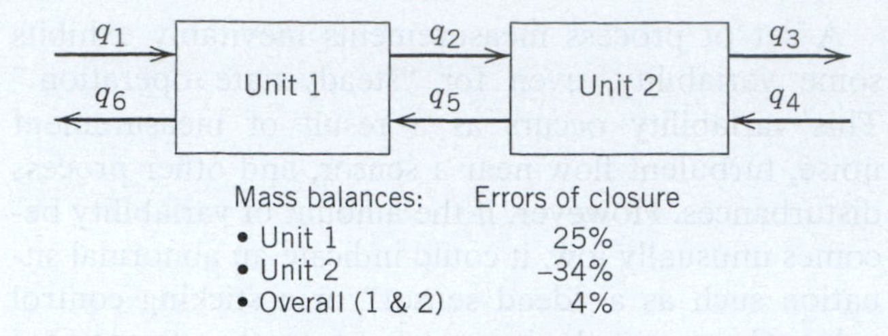

Figure 21.2 Countercurrent flow process.

21.1.2 Performance Calculations

A variety of performance calculations can be made to determine whether the process and instrumentation are working properly. In particular, steady-state mass and energy balances are calculated using data that are averaged over a period of time (for example, one hour). The percent error of closure for a total mass balance can be defined as

$$\%\ \textit{error of closure} \triangleq \frac{\textit{rate in} - \textit{rate out}}{\textit{rate in}} \times 100\% \quad (21\text{-}3)$$

A large error of closure may be caused by an equipment problem (e.g., a pipeline leak) or a sensor problem. *Data reconciliation* based on a statistical analysis of the errors of closure provides a systematic approach for deciding which measurements are suspect (Romagnoli and Sanchez, 2000).

Both redundant measurements and conservation equations can be used to good advantage. A process consisting of two units in a countercurrent flow configuration is shown in Fig. 21.2. Three steady-state mass balances can be written, one for each unit plus an overall balance around both units. Although the three balances are not independent, they provide useful information for monitoring purposes. Figure 21.2 indicates that the error of closure is small for the overall balance but large for each individual balance. This situation suggests that the flow rate sensor for one of the two interconnecting streams, q_2 or q_5, may be faulty.

Process performance calculations also are very useful for diagnostic and monitoring purposes. For example, the thermal efficiency of a refrigeration unit or the selectivity of a chemical reactor could be calculated on a regular basis. A significant decrease from the normal value could indicate a process change or faulty measurement.

21.2 QUALITY CONTROL CHARTS

Industrial processes inevitably exhibit some variability in their manufactured products regardless of how well the processes are designed and operated. In statistical process control, an important distinction is made between normal (random) variability and abnormal (nonrandom) variability. Random variability is caused by the cumulative effects of a number of largely unavoidable phenomena such as electrical measurement noise, turbulence, and random fluctuations in feedstock or catalyst preparation. The random variability can be interpreted as a type of "background noise" for the manufacturing operation. Nonrandom variability can result from process changes (e.g., heat exchanger fouling, loss of catalyst activity), faulty instrumentation, or human error. As mentioned earlier, the source of this abnormal variability is referred to as a *special cause* or an *assignable cause*.

21.2.1 Normal Distribution

Because the *normal distribution* plays a central role in SPC, we briefly review its important characteristics. The normal distribution is also known as the *Gaussian distribution*.

Suppose that a random variable x has a normal distribution with a mean μ and a variance σ^2 denoted by $N(\mu, \sigma^2)$. The probability that x has a value between two arbitrary constants, a and b, is given by:

$$P(a < x < b) = \int_a^b f(x)dx \quad (21\text{-}4)$$

where $P(\cdot)$ denotes the probability that x lies within the indicated range and $f(x)$ is the probability density function for the normal distribution:

$$f(x) = \frac{1}{\sigma\sqrt{2\pi}} \exp\left[-\frac{(x-\mu)^2}{2\sigma^2}\right] \quad (21\text{-}5)$$

The following probability statements are valid for the normal distribution (Montgomery and Runger, 2007):

$$\begin{aligned} P(\mu - \sigma < x < \mu + \sigma) &= 0.6827 \\ P(\mu - 2\sigma < x < \mu + 2\sigma) &= 0.9545 \quad (21\text{-}6) \\ P(\mu - 3\sigma < x < \mu + 3\sigma) &= 0.9973 \end{aligned}$$

A graphical interpretation of these expressions is shown in Fig. 21.3 where each probability corresponds to an area under the $f(x)$ curve. Equation 21-6 and Fig. 21.3 demonstrate that if a random variable x is normally distributed, there is a very high probability (0.9973) that a measurement lies within 3σ of the mean μ. This important result provides the theoretical basis for widely used SPC techniques. Similar probability statements can be formulated based on statistical tables for the normal distribution. For the sake of generality, the tables are expressed in terms of the *standard normal distribution*, $N(0, 1)$, and the *standard normal variable*, $z \triangleq (x - \mu)/\sigma$.

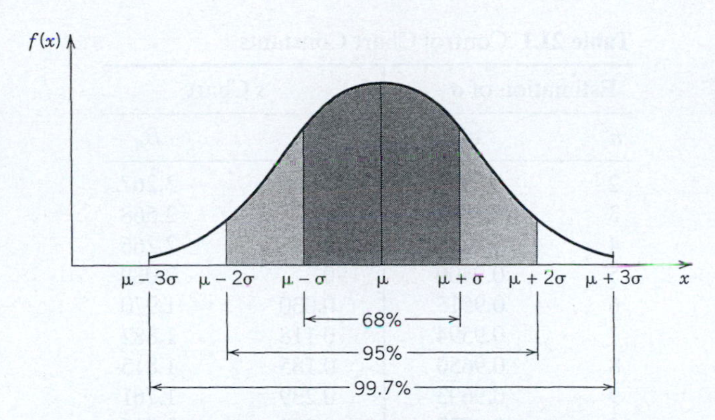

Figure 21.3 Probabilities associated with the normal distribution. From Montgomery and Runger (2007).

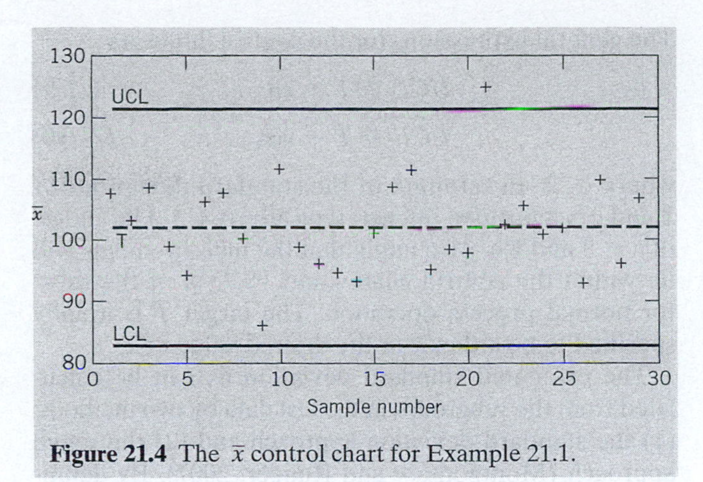

Figure 21.4 The $\bar{x}$ control chart for Example 21.1.

It is important to distinguish between the theoretical mean μ and the sample mean $\bar{x}$. If measurements of a process variable are normally distributed, $N(\mu, \sigma^2)$, the sample mean is also normally distributed. Of course, for any particular sample, $\bar{x}$ is not necessarily equal to μ.

21.2.2 The $\bar{x}$ Control Chart

In statistical process control, Control Charts (or Quality Control Charts) are used to determine whether the process operation is normal or abnormal. The widely used $\bar{x}$ control chart is introduced in the following example. This type of control chart is often referred to as a Shewhart Chart, in honor of the pioneering statistician, Walter Shewhart, who first developed it in the 1920s.

EXAMPLE 21.1

A manufacturing plant produces 10,000 plastic bottles per day. Because the product is inexpensive and the plant operation is normally satisfactory, it is not economically feasible to inspect every bottle. Instead, a sample of n bottles is randomly selected and inspected each day. These n items are called a *subgroup*, and n is referred to as the *subgroup size*. The inspection includes measuring the toughness x of each bottle in the subgroup and calculating the sample mean $\bar{x}$.

The $\bar{x}$ control chart in Fig. 21.4 displays data for a 30-day period. The control chart has a *target* (T), an *upper control limit* (UCL), and a *lower control limit* (LCL). The target (or *centerline*) is the desired (or *expected*) value for $\bar{x}$, while the region between UCL and LCL defines the range of typical variability, as discussed below. If all of the $\bar{x}$ data are within the control limits, the process operation is considered to be normal, or "in a state of control." Data points outside the control limits are considered to be abnormal, indicating that the process operation is out of control. This situation occurs for the twenty-first sample. A single measurement located slightly beyond a control limit is not necessarily a cause for concern. But frequent or large chart violations should be investigated to determine a special cause.

The concept of a *rational subgroup* plays a key role in the development of quality control charts. The basic idea is that a subgroup should be specified so that it reflects typical process variability but not assignable causes. Thus, it is desirable to select a subgroup so that a special cause can be detected by a comparison of subgroups, but it will have little effect within a subgroup (Montgomery, 2009). For example, suppose that a small chemical plant includes six batch reactors and that a product quality measurement for each reactor is made every hour. If the monitoring objective is to determine whether overall production is satisfactory, then the individual reactor measurements could be pooled to provide a subgroup size of $n = 6$ and a sampling period of $\Delta t = 1$ h. On the other hand, if the objective is to monitor the performance of individual reactors, the product quality data for each reactor could be plotted on an hourly basis ($n = 1$) or averaged over an eight-hour shift ($n = 8$ and $\Delta t = 8$ h). When only a single measurement is made at each sampling instant, the subgroup size is $n = 1$ and the control chart is referred to as an *individuals chart*.

The first step in devising a control chart is to select a set of representative data for a period of time when the process operation is believed to be normal, rather than abnormal. Suppose that these test data consist of N subgroups that have been collected on a regular basis (for example, hourly or daily) and that each subgroup consists of n randomly selected items. Let x_{ij} denote the jth measurement in the ith subgroup. Then, the subgroup sample means can be calculated:

$$\bar{x}_i \triangleq \frac{1}{n} \sum_{j=1}^{n} x_{ij} \quad (i = 1, 2, \ldots, N) \tag{21-7}$$

The *grand mean* $\bar{\bar{x}}$ is defined to be the average of the subgroup means:

$$\bar{\bar{x}} \triangleq \frac{1}{N} \sum_{i=1}^{N} \bar{x}_i \tag{21-8}$$

The general expressions for the control limits are

$$UCL \triangleq T + c\hat{\sigma}_{\bar{x}} \tag{21-9}$$

$$LCL \triangleq T + c\hat{\sigma}_{\bar{x}} \tag{21-10}$$

where $\hat{\sigma}_{\bar{x}}$ is an estimate of the standard deviation for $\bar{x}$ and c is a positive integer; typically, $c = 3$. The choice of $c = 3$ and Eq. 21-6 imply that the measurements will lie within the control chart limits 99.73% of the time, for normal process operation. The target T is usually specified to be either $\bar{\bar{x}}$ or the desired value of $\bar{x}$.

The estimated standard deviation $\hat{\sigma}_{\bar{x}}$ can be calculated from the subgroups in the test data by two methods: (1) the standard deviation approach and (2) the range approach (Montgomery and Runger, 2007). By definition, the *range* R is the difference between the maximum and minimum values. Historically, the R approach has been emphasized, because R is easier to calculate than s, an advantage for hand calculations. However, the standard deviation approach is now preferred because it uses all of the data, instead of only two points in each subgroup. It also has the advantage of being less sensitive to *outliers* (i.e., bad data points). However, for small values of n, the two approaches tend to produce similar control limits (Ryan, 2000). Consequently, we will only consider the standard deviation approach.

The average sample standard deviation $\bar{s}$ for the N subgroups is

$$\bar{s} \triangleq \frac{1}{N}\sum_{i=1}^{N} s_i \tag{21-11}$$

where the standard deviation for the ith subgroup is

$$s_i \triangleq \sqrt{\frac{1}{n-1}\sum_{j=1}^{n}(x_{ij} - \bar{x}_i)^2} \tag{21-12}$$

If the x data are normally distributed, then $\hat{\sigma}_{\bar{x}}$ is related to $\bar{s}$ by

$$\hat{\sigma}_{\bar{x}} = \frac{1}{c_4\sqrt{n}}\bar{s} \tag{21-13}$$

where c_4 is a constant that depends on n (Montgomery and Runger, 2007) and is tabulated in Table 21.1.

21.2.3 The *s* Control Chart

In addition to monitoring average process performance, it is also advantageous to monitor process variability. The variability within a subgroup can be characterized by its range, standard deviation, or sample variance. Control charts can be developed for all three statistics, but our discussion will be limited to the control chart for the standard deviation, the *s control chart*.

Table 21.1 Control Chart Constants

Estimation of σ		*s* Chart	
n	c_4	B_3	B_4
2	0.7979	0	3.267
3	0.8862	0	2.568
4	0.9213	0	2.266
5	0.9400	0	2.089
6	0.9515	0.030	1.970
7	0.9594	0.118	1.882
8	0.9650	0.185	1.815
9	0.9693	0.239	1.761
10	0.9727	0.284	1.716
15	0.9823	0.428	1.572
20	0.9869	0.510	1.490
25	0.9896	0.565	1.435

Source: Adapted from Ryan (2000).

The centerline for the s chart is $\bar{s}$, which is the average standard deviation for the test set of data. The control limits are

$$UCL = B_4\bar{s} \tag{21-14}$$

$$LCL = B_3\bar{s} \tag{21-15}$$

Constants B_3 and B_4 depend on the subgroup size n, as shown in Table 21.1.

The control chart limits for the $\bar{x}$ and s charts in Eqs. 21-9 to 21-15 have been based on the assumption that the x data are normally distributed.

When individual measurements are plotted ($n = 1$), the standard deviation for the subgroup does not exist. In this situation, the *moving range* (*MR*) of two successive measurements can be employed to provide a measure of variability. The moving range is defined as the absolute value of the difference between successive measurements. Thus, for the kth sampling instant, $MR(k) = |x(k) - x(k-1)|$. The $\bar{x}$ and s control charts are also applicable when the sample size n varies from one sample to the next.

Example 21.2 illustrates the construction of $\bar{x}$ and s control charts.

EXAMPLE 21.2

In semiconductor processing, the photolithography process is used to transfer the circuit design to silicon wafers. In the first step of the process, a specified amount of a polymer solution, *photoresist*, is applied to a wafer as it spins at high speed on a turntable. The resulting photoresist thickness x is a key process variable. Thickness data for 25 subgroups are shown in Table 21.2. Each subgroup consists of three randomly selected wafers. Construct $\bar{x}$ and s control charts for these test data and critically evaluate the results.

Table 21.2 Thickness Data (in Å) for Example 21.2

No.		*x* Data		$\bar{x}$	*s*
1	209.6	207.6	211.1	209.4	1.8
2	183.5	193.1	202.4	193.0	9.5
3	190.1	206.8	201.6	199.5	8.6
4	206.9	189.3	204.1	200.1	9.4
5	260.0	209.0	212.2	227.1	28.6
6	193.9	178.8	214.5	195.7	17.9
7	206.9	202.8	189.7	199.8	9.0
8	200.2	192.7	202.1	198.3	5.0
9	210.6	192.3	205.9	202.9	9.5
10	186.6	201.5	197.4	195.2	7.7
11	204.8	196.6	225.0	208.8	14.6
12	183.7	209.7	208.6	200.6	14.7
13	185.6	198.9	191.5	192.0	6.7
14	202.9	210.1	208.1	207.1	3.7
15	198.6	195.2	150.0	181.3	27.1
16	188.7	200.7	207.6	199.0	9.6
17	197.1	204.0	182.9	194.6	10.8
18	194.2	211.2	215.4	206.9	11.2
19	191.0	206.2	183.9	193.7	11.4
20	202.5	197.1	211.1	203.6	7.0
21	185.1	186.3	188.9	186.8	1.9
22	203.1	193.1	203.9	200.0	6.0
23	179.7	203.3	209.7	197.6	15.8
24	205.3	190.0	208.2	201.2	9.8
25	203.4	202.9	200.4	202.2	1.61

SOLUTION

The following sample statistics can be calculated from the data in Table 21.2: $\bar{\bar{x}} = 199.8$ Å, $\bar{s} = 10.4$ Å. For $n = 3$ the required constants from Table 21.1 are $c_4 = 0.8862$, $B_3 = 0$, and $B_4 = 2.568$. Then the $\bar{x}$ and *s* control limits can be calculated from Eqs. 21-9 to 21-15. The traditional value of $c = 3$ is selected for Eqs. (21-9) and (21-10). The resulting control limits are labeled as the "original limits" in Fig. 21.5.

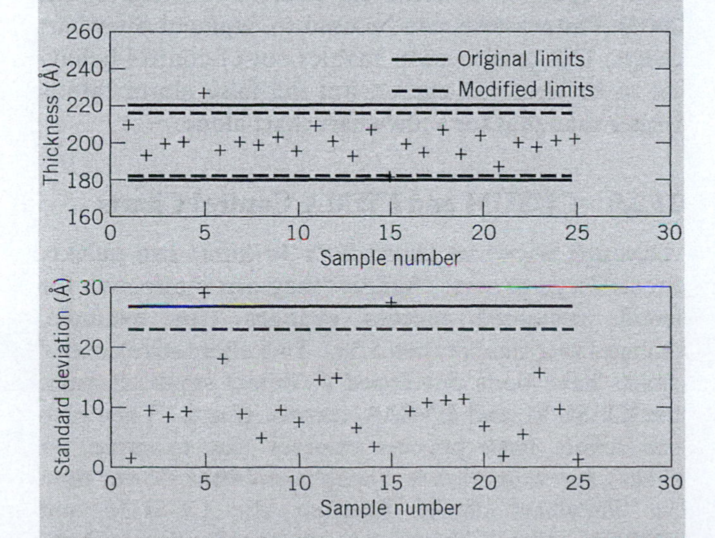

Figure 21.5 The $\bar{x}$ and *s* control charts for Example 21.2.

Figure 21.5 indicates that sample #5 lies beyond the UCL for both the $\bar{x}$ and *s* control charts, while sample #15 is very close to a control limit on each chart. Thus, the question arises whether these two samples are "outliers" that should be omitted from the analysis. Table 21.2 indicates that sample #5 includes a very large value (260.0), while sample #15 includes a very small value (150.0). However, unusually large or small numerical values by themselves do not justify discarding samples; further investigation is required.

Suppose that a more detailed evaluation has discovered a specific reason as to why measurements #5 and #15 should be discarded (e.g., faulty sensor, data misreported, etc.). In this situation, these two samples should be removed and the control limits should be recalculated based on the remaining 23 samples. These modified control limits are tabulated below as well as in Fig. 21.5.

	Original Limits	Modified Limits (omit samples #5 and #15)
$\bar{x}$ Chart Control Limits		
UCL	220.1	216.7
LCL	179.6	182.2
***s* Chart Control Limits**		
UCL	26.6	22.7
LCL	0	0

21.2.4 Theoretical Basis for Quality Control Charts

The traditional SPC methodology is based on the assumption that the natural variability for "in control" conditions can be characterized by random variations around a constant average value,

$$x(k) = x^* + e(k) \tag{21-16}$$

where $x(k)$ is the measurement at time k, x^* is the true (but unknown) value, and $e(k)$ is an additive random error. Traditional control charts are based on the following assumptions:

1. Each additive error, $\{e(k), k = 1, 2, \ldots\}$, is a zero-mean, random variable that has the same normal distribution, $N(0, \sigma^2)$.
2. The additive errors are statistically independent and thus uncorrelated. Consequently, $e(k)$ does not depend on $e(j)$ for $j \neq k$.
3. The true value x^* is constant.
4. The subgroup size n is the same for all of the subgroups.

The second assumption is referred to as *independent and identically distributed (IID)*.

Consider an ideal individuals control chart for x with x^* as its target and "3σ control limits":

$$UCL \triangleq x^* + 3\sigma \tag{21-17}$$
$$LCL \triangleq x^* - 3\sigma \tag{21-18}$$

These control limits are a special case of Eqs. 21-9 and 21-10 for the idealized situation where σ is known, $c = 3$, and the subgroup size is $n = 1$. The typical choice of $c = 3$ can be justified as follows. Because x is $N(0, \sigma^2)$, the probability p that a measurement lies outside the 3σ control limits can be calculated from Eq. 21-6: $p = 1 - 0.9973 = 0.0027$. Thus on average, approximately three out of every 1,000 measurements will be outside of the 3σ limits. The average number of samples before a chart violation occurs is referred to as the *average run length* (ARL). For the normal ("in control") process operation,

$$\text{ARL} \triangleq \frac{1}{p} = \frac{1}{0.0027} = 370 \qquad (21\text{-}19)$$

Thus, a Shewhart chart with 3σ control limits will have an average of one control chart violation every 370 samples, even when the process is *in a state of control.*

This theoretical analysis justifies the use of 3σ limits for $\bar{x}$ and other control charts. However, other values of c are sometimes used. For example, 2σ warning limits can be displayed on the control chart in addition to the 3σ control limits. Although the 2σ warning limits provide an early indication of a process change, they have a very low average run length value of ARL = 22. In general, larger values of c result in wider chart limits and larger ARL values. Wider chart limits mean that process changes will not be detected as quickly as they would be for smaller c values. Thus, the choice of c involves a classical engineering compromise between early detection of process changes (low value of c) and reducing the frequency of false alarms (high value of c).

Standard SPC techniques are based on the four assumptions listed above. However, because these assumptions are not always valid for industrial processes, standard techniques can give misleading results. In particular, the implications of violating the normally distributed and IID assumptions have received considerable theoretical analysis (Ryan, 2000). Although modified SPC techniques have been developed for these nonideal situations, commercial SPC software is usually based on these assumptions.

Industrial plant measurements are not normally distributed. However, for large subgroup sizes ($n > 25$), $\bar{x}$ is approximately normally distributed even if x is not, according to the famous *Central Limit Theorem* of statistics (Montgomery and Runger, 2007). Fortunately, modest deviations from "normality" can be tolerated. In addition, the standard SPC techniques can be modified so that they are applicable to certain classes of nonnormal data (Jacobs, 1990).

In industrial applications, the control chart data are often *serially correlated*, because the current measurement is related to previous measurements. For example, the flow rate data in Fig. 21.1 are serially correlated. Standard control charts such as the $\bar{x}$ and s charts can provide misleading results if the data are serially correlated. But if the degree of correlation is known, the control limits can be adjusted accordingly (Montgomery, 2009). Serially correlated data also can be modeled using time-series analysis, as described in Section 17.6.

21.2.5 Pattern Tests and the Western Electric Rules

We have considered how abnormal process behavior can be detected by comparing individual measurements with the $\bar{x}$ and s control chart limits. However, the pattern of measurements can also provide useful information. For example, if ten consecutive measurements are all increasing, then it is very unlikely that the process is *in a state of control.*

A wide variety of pattern tests (also called *zone rules*) can be developed based on the IID and normal distribution assumptions and the properties of the normal distribution. For example, the following excerpts from the *Western Electric Rules* (Western Electric Company, 1956; Montgomery and Runger, 2007) indicate that the process is *out of control* if one or more of the following conditions occur:

1. One data point is outside the 3σ control limits.
2. Two out of three consecutive data points are beyond a 2σ limit.
3. Four out of five consecutive data points are beyond a 1σ limit and on one side of the centerline.
4. Eight consecutive points are on one side of the centerline.

Note that the first condition corresponds to the familiar Shewhart chart limits of Eqs. 21-9 and 21-10 with $c = 3$. Additional pattern tests are concerned with other types of nonrandom behavior (Montgomery, 2009). Pattern tests can be used to augment Shewhart charts. This combination enables out-of-control behavior to be detected earlier, but the false alarm rate is higher than that for a Shewhart chart alone.

21.2.6 CUSUM and EWMA Control Charts

Although Shewhart charts with 3σ limits can quickly detect large process changes, they are ineffective for small, sustained process changes (for example, changes in μ smaller than 1.5σ). Two alternative control charts have been developed to detect small changes: the CUSUM and EWMA control charts. They also can detect large process changes (for example, 3σ shifts), but detection is usually somewhat slower than for Shewhart charts. Because the CUSUM and EWMA control charts can effectively detect both large and small process shifts, they provide viable alternatives to the widely used Shewhart charts. Consequently, they will now be considered. The *cumulative sum (CUSUM)* is defined to be a running summation of the deviations of the plotted variable from its target. If the sample mean is plotted, the cumulative sum, $C(k)$, is

$$C(k) = \sum_{j=1}^{k} (\bar{x}(j) - T) \tag{21-20}$$

where T is the target for $\bar{x}$. During normal process operation, $C(k)$ fluctuates around zero. But if a process change causes a small shift in $\bar{x}$, $C(k)$ will drift either upward or downward.

The CUSUM control chart was originally developed using a graphical approach based on *V-masks* (Montgomery, 2009). However, for computer calculations, it is more convenient to use an equivalent algebraic version that consists of two recursive equations,

$$C^+(k) = \max[0, \bar{x}(k) - (T + K) + C^+(k-1)] \tag{21-21}$$

$$C^-(k) = \max[0, (T - K) - \bar{x}(k) + C^-(k-1)] \tag{21-22}$$

where C^+ and C^- denote the sums for the high and low directions and K is a constant, the *slack parameter*. The CUSUM calculations are initialized by setting $C^+(0) = C^-(0) = 0$. A deviation from the target that is larger than K increases either C^+ or C^-. A control limit violation occurs when either C^+ or C^- exceeds a specified control limit (or *threshold*), H. After a limit violation occurs, that sum is reset to zero or to a specified value.

The selection of the threshold H can be based on considerations of average run length. Suppose that we want to detect whether the sample mean $\bar{x}$ has shifted from the target by a small amount, δ. The slack parameter K is usually specified as $K = 0.5\,\delta$. For the ideal situation where the normally distributed and IID assumptions are valid, ARL values have been tabulated for specified values of δ, K, and H (Ryan, 2000; Montgomery, 2009).

Table 21.3 summarizes ARL values for two values of H and different values of δ. (The values of δ are usually expressed as multiples of $\hat{\sigma}_{\bar{x}}$.) The ARL values indicate the average number of samples before a change of δ is detected. Thus, the ARL values for $\delta = 0$ indicate the average time between "false alarms," that is, the average time between successive CUSUM alarms when no shift in $\bar{x}$ has occurred. Ideally, we would like the ARL value to be very large for $\delta = 0$, and small for $\delta \neq 0$. Table 21.3 shows that as the magnitude of the shift δ increases, ARL decreases, and thus the CUSUM control chart detects the change faster. Increasing the value of H from 4σ to 5σ increases all of the ARL values and thus provides a more conservative approach.

Table 21.3 Average Run Lengths for CUSUM Control Charts

Shift from Target (in multiples of $\hat{\sigma}_{\bar{x}}$)	ARL for $H = 4\hat{\sigma}_{\bar{x}}$	ARL for $H = 5\hat{\sigma}_{\bar{x}}$
0	168.0	465.0
0.25	74.2	139.0
0.50	26.6	38.0
0.75	13.3	17.0
1.00	8.38	10.4
2.00	3.34	4.01
3.00	2.19	2.57

Source: Adapted from Ryan (2000).

CUSUM control charts also are constructed for measures of variability such as the range or standard deviation (Ryan, 2000; Montgomery, 2009).

EWMA Control Chart

Information about past measurements can also be included in the control chart calculations by exponentially weighting the data. This strategy provides the basis for the *exponentially weighted moving-average (EWMA) control chart.* Let $\bar{x}$ denote the sample mean of the measured variable and z denote the EWMA of $\bar{x}$. A recursive equation is used to calculate $z(k)$,

$$z(k) = \lambda\bar{x}(k) + (1 - \lambda)z(k-1) \tag{21-23}$$

where λ is a constant, $0 \le \lambda \le 1$. Note that Eq. 21-23 has the same form as the first-order (or exponential) filter that was introduced in Chapter 17. The EWMA control chart consists of a plot of $z(k)$ vs. k, as well as a target and upper and lower control limits. Note that the EWMA control chart reduces to the Shewhart chart for $\lambda = 1$. The EWMA calculations are initialized by setting $z(0) = T$.

If the $\bar{x}$ measurements satisfy the IID condition, the EWMA control limits can be derived. The theoretical 3σ limits are given by

$$T \pm 3\hat{\sigma}_{\bar{x}} \sqrt{\frac{\lambda}{2 - \lambda}} \tag{21-24}$$

where $\hat{\sigma}_{\bar{x}}$ is determined from a set of test data taken when the process is *in a state of control* (Montgomery, 2009). The target T is selected to be either the desired value of $\bar{x}$ or the grand mean for the test data, $\bar{\bar{x}}$. Time-varying control limits can also be derived that provide narrower limits for the first few samples, for applications where early detection is important (Montgomery, 2009; Ryan, 2000). Tables of ARL values have been developed for the EWMA method, similar to Table 21.3 for the CUSUM method (Ryan, 2000).

The EWMA performance can be adjusted by specifying λ. For example, $\lambda = 0.25$ is a reasonable choice, because it results in an ARL of 493 for no mean shift ($\delta = 0$) and an ARL of 11 for a mean shift of $\sigma_{\bar{x}}$ ($\delta = 1$). EWMA control charts can also be constructed for measures of variability such as the range and standard deviation.

EXAMPLE 21.3

In order to compare Shewhart, CUSUM, and EWMA control charts, consider simulated data for the tensile strength of a phenolic resin. It is assumed that the tensile strength x is normally distributed with a mean of $\mu = 70$ MPa and a standard deviation of $\sigma = 3$ MPa. A single measurement is available at each sampling instant. A constant ($\delta = 0.5\sigma = 1.5$) was added to $x(k)$ for $k \geq 10$ in order to evaluate each chart's ability to detect a small process shift. The CUSUM chart was designed using $K = 0.5\sigma$ and $H = 5\sigma$, while the EWMA parameter was specified as $\lambda = 0.25$.

The relative performance of the Shewhart, CUSUM, and EWMA control charts is compared in Fig. 21.6. The Shewhart chart fails to detect the 0.5σ shift in x. However, both the CUSUM and EWMA charts quickly detect this change, because limit violations occur about ten samples after the shift occurs (at $k = 20$ and $k = 21$, respectively). The mean shift can also be detected by applying the Western Electric Rules in the previous section.

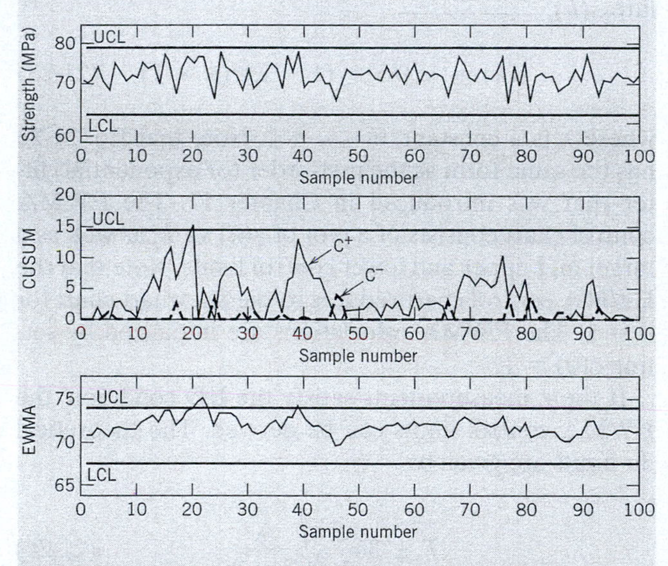

Figure 21.6 Comparison of Shewhart (top), CUSUM (middle), and EWMA (bottom) control charts for Example 21.3.

21.3 EXTENSIONS OF STATISTICAL PROCESS CONTROL

Now that the basic quality control charts have been presented, we consider several other important topics in statistical process control.

21.3.1 Process Capability Indices

Process capability indices (or *process capability ratios*) provide a measure of whether an "in control" process is meeting its product specifications. Suppose that a quality variable x must have a volume between an *upper specification limit* (USL) and a *lower specification limit* (LSL) in order for product to satisfy customer requirements. The C_p *capability index* is defined as

$$C_p \triangleq \frac{USL - LSL}{6\sigma} \tag{21-25}$$

where σ is the standard deviation of x. Suppose that $C_p = 1$ and x is normally distributed. Based on Eq. 21-6, we would expect that 99.73% of the measurements satisfy the specification limits. If $C_p > 1$, the product specifications are satisfied; for $C_p < 1$, they are not.

A second capability index C_{pk} is based on average process performance ($\bar{x}$), as well as process variability (σ). It is defined as

$$C_{pk} \triangleq \frac{\min[\bar{x} - LSL, USL - \bar{x}]}{3\sigma} \tag{21-26}$$

Although both C_p and C_{pk} are used, we consider C_{pk} to be superior to C_p for the following reason. If $\bar{x} = T$, the process is said to be "centered" and $C_{pk} = C_p$. But for $\bar{x} \neq T$, C_p does not change, even though the process performance is worse, while C_{pk} decreases. For this reason, C_{pk} is preferred.

If the standard deviation σ is not known, it is replaced by an estimate $\hat{\sigma}$ in Eqs. 21-25 and 21-26. For situations where there is only a single specification limit, either USL or LSL, the definitions of C_p and C_{pk} can be modified accordingly (Ryan, 2000).

In practical applications, a common objective is to have a capability index of 2.0, while a value greater than 1.5 is considered to be acceptable (Shunta, 1995). If the C_{pk} value is too low, it can be improved by making a change that either reduces process variability or causes $\bar{x}$ to move closer to the target. These improvements can be achieved in a number of ways, including better process control, better process maintenance, reduced variability in raw materials, improved operator training, and changes in process operating conditions.

Three important points should be noted concerning the C_p and C_{pk} capability indices:

1. The data used in the calculations do *not* have to be normally distributed.
2. The specification limits, USL and LSL, and the control limits, UCL and LCL, are *not* related. The specification limits denote the desired process performance, while the control limits represent actual performance during normal operation when the process is *in control*.

3. The numerical values of the C_p and C_{pk} capability indices in (21-25) and (21-26) are only meaningful when the process is *in a state of control*. However, other *process performance indices* are available to characterize process performance when the process is not *in a state of control*. They can be used to evaluate the incentives for improved process control (Shunta, 1995).

EXAMPLE 21.4

Calculate the average values of the C_p and C_{pk} capability indices for the photolithography thickness data in Example 21.2. Omit the two outliers (samples #5 and #15), and assume that the upper and lower specification limits for the photoresist thickness are USL = 235 Å and LSL = 185 Å.

SOLUTION

After samples #5 and #15 are omitted, the grand mean is $\bar{\bar{x}} = 199$ Å, and the standard deviation of $\bar{x}$ (estimated from Eq. 21-13 with $c_4 = 0.8862$) is

$$\hat{\sigma}_{\bar{x}} = \frac{\bar{s}}{c_4\sqrt{n}} = \frac{8.83}{0.8862\sqrt{3}} = 5.75 \text{ Å}$$

From Eqs. 21-25 and 21-26,

$$C_p = \frac{235 - 185}{6(5.75)} = 1.45$$

$$C_{pk} = \frac{\min[199.5 - 185, 235 - 199.5]}{3(5.75)} = 0.84$$

Note that C_{pk} is much smaller than the C_p, because $\bar{\bar{x}}$ is closer to the LSL than the USL.

21.3.2 Six Sigma Approach

Product quality specifications continue to become more stringent as a result of market demands and intense worldwide competition. Meeting quality requirements is especially difficult for products that consist of a very large number of components and for manufacturing processes that consist of hundreds of individual steps. For example, the production of a microelectronics device typically requires 100 to 300 batch processing steps. Suppose that there are 200 steps, and that each one must meet a quality specification in order for the final product to function properly. If each step, is independent of the others and has a 99% success rate, the overall yield of satisfactory product is $(0.99)^{200} = 0.134$, or only 13.4%. This low yield is clearly unsatisfactory. Similarly, even when a processing step meets 3σ specifications (99.73% success rate), it will still result in an average of 2,700 "defects" for every million produced. Furthermore, the overall yield for this 200-step process is still only 58.2%.

These examples demonstrate that for complicated products or processes, 3σ quality is no longer adequate, and there is no place for failure. These considerations and economic pressures have motivated the development of the *six sigma approach* (Pande et al., 2000). The statistical motivation for this approach is based on the properties of the normal distribution. Suppose that a product quality variable x is normally distributed, $N(\mu, \sigma^2)$. As indicated on the left portion of Fig. 21.7, if the product specifications are $\mu \pm 6\sigma$, the product will meet the specifications 99.999998% of the time. Thus, on average, there will only be two defective products for every billion produced. Now suppose that the process operation changes so that the mean value is shifted from $\bar{x} = \mu$ to either $\bar{x} = \mu + 1.5\sigma$ or $\bar{x} = \mu - 1.5\sigma$, as shown on the right side of Fig. 21.7. Then the product specifications will still be satisfied 99.99966% of the time, which corresponds to 3.4 defective products per million produced.

In summary, if the variability of a manufacturing operation is so small that the product specification limits are equal to $\mu \pm 6\sigma$, then the limits can be satisfied

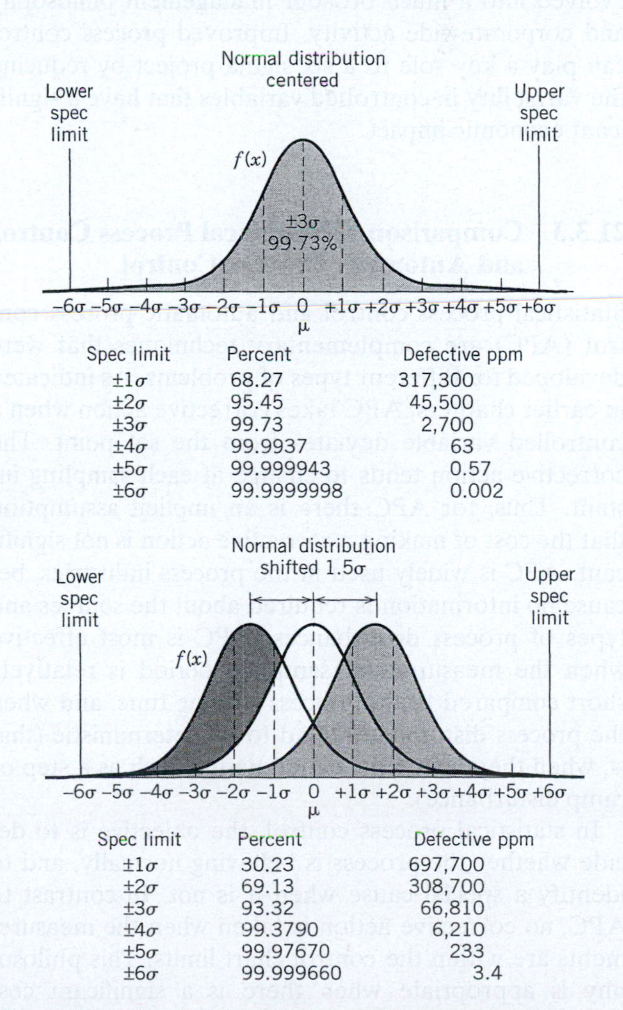

Figure 21.7 The Six Sigma Concept (Montgomery and Runger, 2007). Top: No shift in the mean. Bottom: 1.5σ shift.

even if the mean value of x shifts by as much as 1.5σ. This very desirable situation of near perfect product quality is referred to as *six sigma quality*.

The six sigma approach was pioneered by the Motorola and General Electric companies in the early 1980s as a strategy for achieving both six sigma quality and continuous improvement. Since then, other large corporations have adopted companywide programs that apply the six sigma approach to all of their business operations, both manufacturing and nonmanufacturing. Thus, although the six sigma approach is "data-driven" and based on statistical techniques, it has evolved into a broader management philosophy that has been implemented successfully by many large corporations. Six sigma programs have also had a significant financial impact. Large corporations have reported savings of billions of dollars that were attributed to successful six sigma programs.

In summary, the six sigma approach based on statistical monitoring techniques has had a major impact on both manufacturing and business practice during the past two decades. It is based on SPC concepts but has evolved into a much broader management philosophy and corporatewide activity. Improved process control can play a key role in a six sigma project by reducing the variability in controlled variables that have a significant economic impact.

21.3.3 Comparison of Statistical Process Control and Automatic Process Control

Statistical process control and automatic process control (APC) are complementary techniques that were developed for different types of problems. As indicated in earlier chapters, APC takes corrective action when a controlled variable deviates from the set point. The corrective action tends to change at each sampling instant. Thus, for APC there is an implicit assumption that the cost of making a corrective action is not significant. APC is widely used in the process industries, because no information is required about the sources and types of process disturbances. APC is most effective when the measurement sampling period is relatively short compared to the process settling time, and when the process disturbances tend to be deterministic (that is, when they have a sustained nature such as a step or ramp disturbance).

In statistical process control, the objective is to decide whether the process is behaving normally, and to identify a special cause when it is not. In contrast to APC, no corrective action is taken when the measurements are within the control chart limits. This philosophy is appropriate when there is a significant cost associated with taking a corrective action, such as when shutting down a process unit or taking an instrument out of service for maintenance. From an engineering perspective, SPC is viewed as a *monitoring*, rather than a *control*, strategy. It is very effective when the normal process operation can be characterized by random fluctuations around a mean value. SPC is an appropriate choice for monitoring problems where the sampling period is long compared to the process settling time and the process disturbances tend to be random rather than deterministic. SPC has been widely used for quality control in both discrete-parts manufacturing and the process industries.

In summary, SPC and APC should be regarded as complementary rather than competitive techniques. They were developed for different types of situations and have been successfully used in the process industries. Furthermore, a combination of the two methods can be very effective. For example, in model-based control such as model predictive control (Chapter 20), APC can be used for feedback control, while SPC is used to monitor the model residuals, the differences between the model predictions and the actual values.

21.4 MULTIVARIATE STATISTICAL TECHNIQUES

In Chapters 13 and 18, we have emphasized that many important control problems are multivariable in nature because more than one process variable must be controlled and more than one variable can be manipulated. Similarly, for common SPC monitoring problems, two or more quality variables are important, and they can be highly correlated. For example, ten or more quality variables are typically measured for synthetic fibers (MacGregor, 1996). For these situations, multivariable SPC techniques can offer significant advantages over the single-variable methods discussed in Section 21.2. In the statistics literature, these techniques are referred to as *multivariate methods*, while the standard Shewhart and CUSUM control charts are examples of *univariate methods*. The advantage of a multivariate monitoring approach is illustrated in Example 21.5.

EXAMPLE 21.5

The effluent stream from a wastewater treatment process is monitored to make sure that two process variables, the biological oxidation demand (BOD) and the solids content, meet specifications. Representative data are shown in Table 21.4. Shewhart charts for the sample means are shown in parts (*a*) and (*b*) of Fig. 21.8. These univariate control charts indicate that the process appears to be in-control because no chart violations occur for either variable. However, the bivariate control chart in Fig. 21.8*c* indicates that the two variables are highly correlated, because the solids content tends to be large when the BOD is large, and vice versa. When the two variables are

Table 21.4 Wastewater Treatment Data

Sample Number	BOD (mg/L)	Solids (mg/L)
1	17.7	1380
2	23.6	1458
3	13.2	1322
4	25.2	1448
5	13.1	1334
6	27.8	1485
7	29.8	1503
8	9.0	1540
9	14.3	1341
10	26.0	1448
11	23.2	1426
12	22.8	1417
13	20.4	1384
14	17.5	1380
15	18.4	1396
16	16.8	1345
17	13.8	1349
18	19.4	1398
19	24.7	1426
20	16.8	1361
21	14.9	1347
22	27.6	1476
23	26.1	1454
24	20.0	1393
25	22.9	1427
26	22.4	1431
27	19.6	1405
28	31.5	1521
29	19.9	1409
30	20.3	1392

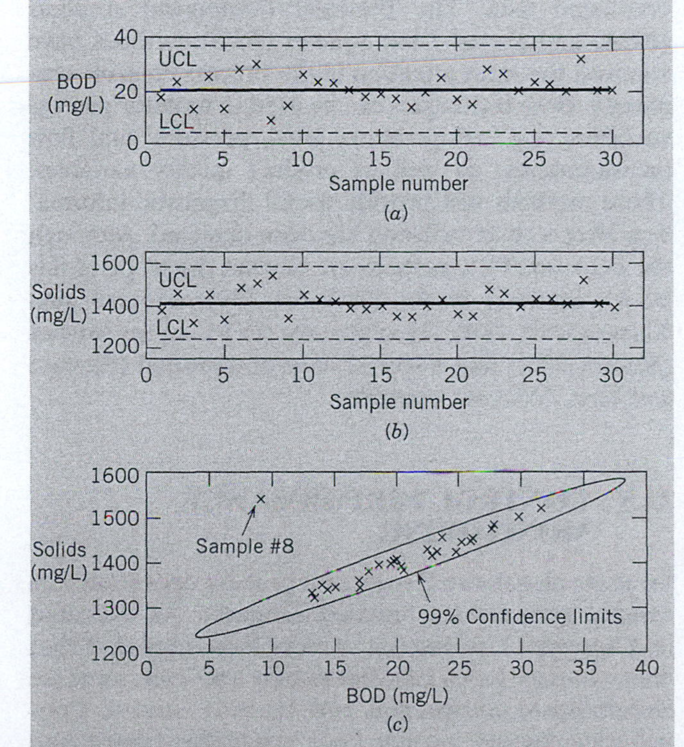

Figure 21.8 Confidence regions for Example 21.5. Univariate in (a) and (b), bivariate in (c).

considered together, their joint confidence limit (e.g., at the 99% confidence level) is an ellipse, as shown in Fig. 21.8c.[1] Sample #8 lies well beyond the 99% limit, indicating an out-of-control condition. By contrast, this sample lies within the Shewhart control chart limits for both individual variables.

This example has demonstrated that univariate SPC techniques such as Shewhart charts can fail to detect abnormal process behavior when the process variables are highly correlated. By contrast, the abnormal situation was readily apparent from the multivariate analysis.

Figure 21.9 provides a general comparison of univariate and multivariate SPC techniques (Alt et al., 1998). When two variables, x_1 and x_2, are monitored individually, the two sets of control limits define a rectangular region, as shown in Fig. 21.9. In analogy with Example 21.5, the multivariate control limits define the dark, ellipsoidal region that represents *in-control* behavior. Figure 21.9 demonstrates that the application of univariate SPC techniques to correlated multivariate data can result in two types of misclassification: false alarms and out-of-control conditions that are not detected. The latter type of misclassification occurred at sample #8 for the two Shewhart charts in Fig. 21.8.

In the next section, we consider some well-known multivariate monitoring techniques.

21.4.1 Hotelling's T^2 Statistic

Suppose that it is desired to use SPC techniques to monitor p variables, which are correlated and normally distributed. Let $\boldsymbol{x}$ denote the column vector of these p variables, $\boldsymbol{x} = col[x_1, x_2, \ldots, x_p]$. At each sampling instant, a subgroup of n measurements is made for each variable. The subgroup sample means for the kth sampling instant can be expressed as a column vector: $\bar{\boldsymbol{x}}(k) = col[\bar{x}_1(k), \ \bar{x}_2(k), \ldots, \ \bar{x}_p(k)]$. Multivariate control charts are traditionally based on *Hotelling's T^2 statistic* (Montgomery, 2009).

$$T^2(k) \triangleq n[\bar{\boldsymbol{x}}(k) - \bar{\bar{\boldsymbol{x}}}]^T \boldsymbol{S}^{-1} \, [\bar{\boldsymbol{x}}(k) - \bar{\bar{\boldsymbol{x}}}] \qquad (21\text{-}27)$$

where $T^2(k)$ denotes the value of the T^2 statistic at the kth sampling instant. The vector of grand means $\bar{\bar{\boldsymbol{x}}}$

[1] If two random variables are correlated and normally distributed, the confidence limit is in the form of an ellipse and can be calculated from the well-known F distribution (Montgomery and Runger, 2007).

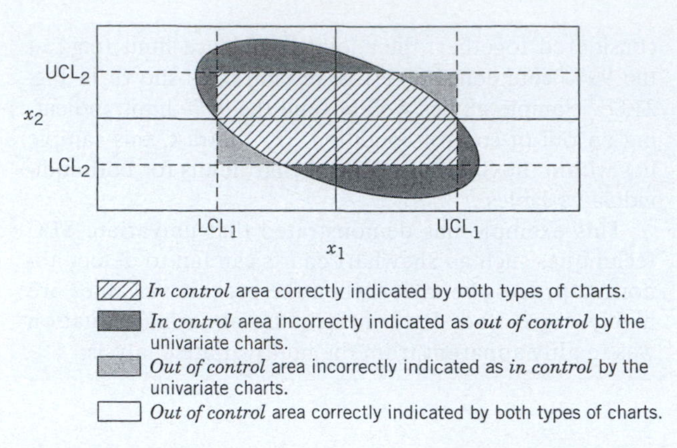

Figure 21.9 Univariate and bivariate confidence regions for two random variables, x_1 and x_2 (modified from Alt et al., 1998).

and the covariance matrix $\mathbf{S}$ are calculated for a test set of data for *in-control* conditions. By definition S_{ij}, the (i, j)-element of matrix $\mathbf{S}$, is the sample covariance of x_i and x_j:

$$S_{ij} \triangleq \frac{1}{N} \sum_{k=1}^{N} [\bar{x}_i(k) - \bar{\bar{x}}_i]\,[\bar{x}_j(k) - \bar{\bar{x}}_j] \quad (21\text{-}28)$$

In Eq. 21-28 N is the number of subgroups, and $\bar{\bar{x}}_i$ denotes the mean for $\bar{x}_i$.

Note that T^2 is a scalar, even though the other quantities in Eq. 21-27 are vectors and matrices. The inverse of the sample covariance matrix, $\mathbf{S}^{-1}$, scales the p variables and accounts for correlation among them.

A multivariate process is considered to be out-of-control at the kth sampling instant if $T^2(k)$ exceeds an upper control limit (UCL). (There is no target or lower control limit.) The UCL values are tabulated in statistics books and depend on the number of variables p and the subgroup size n. The T^2 control chart consists of a plot of $T^2(k)$ vs. k and an UCL. Thus, the T^2 control chart is the multivariate generalization of the $\bar{x}$ chart introduced in Section 21.2.2. Multivariate generalizations of the CUSUM and EWMA charts are also available (Montgomery, 2009).

EXAMPLE 21.6

Construct a T^2 control chart for the wastewater treatment problem of Example 21.5. The 99% control chart limit is $T^2 = 11.63$. Is the number of T^2 control chart violations consistent with the results of Example 21.5?

SOLUTION

The T^2 control chart is shown in Fig. 21.10. All of the T^2 values lie below the 99% confidence limit except for sample #8. This result is consistent with the bivariate control chart in Fig. 21.8c.

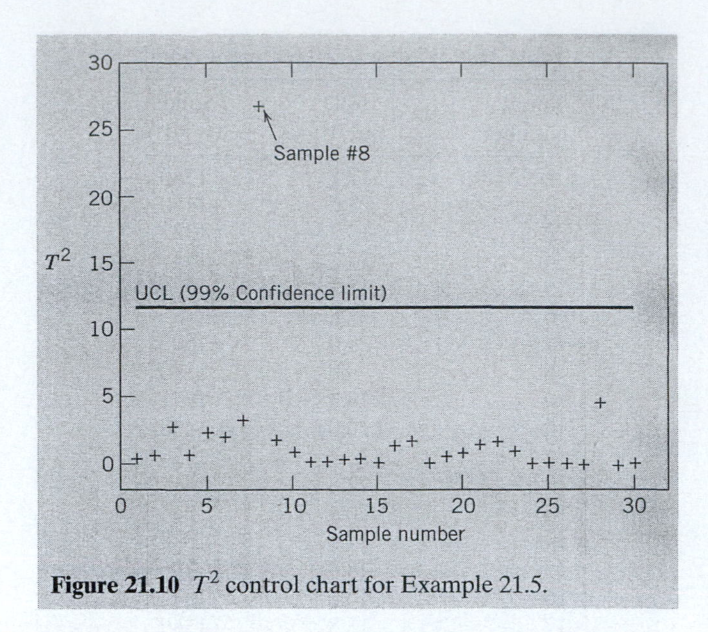

Figure 21.10 T^2 control chart for Example 21.5.

21.4.2 Principal Component Analysis and Partial Least Squares

Multivariate monitoring based on Hotelling's T^2 statistic can be effective if the data are not highly correlated and the number of variables p is not large (for example, $p < 10$). For highly correlated data, the $\mathbf{S}$ matrix is poorly conditioned and the T^2 approach becomes problematic. Fortunately, alternative multivariate monitoring techniques have been developed that are very effective for monitoring problems with large numbers of variables and highly correlated data. The *Principal Component Analysis* (*PCA*) and *Partial Least Squares* (*PLS*) methods have received the most attention in the process control community. Both techniques can be used to monitor process variables (e.g., temperature, level, pressure, and flow measurements) as well as product quality variables. These methods can provide useful diagnostic information after a chart violation has been detected. Although the PCA and PLS methods are beyond the scope of this book, excellent books (Jackson, 1991; Piovoso and Khosanovich, 1996; Montgomery, 2009), survey articles (Kourti, 2002) and a special issue of a journal (Piovoso and Hoo, 2002) are available.

21.5 CONTROL PERFORMANCE MONITORING

In order to achieve the desired process operation, the control system must function properly. As indicated in Chapter 12, industrial surveys have reported that many control loops perform poorly and even increase variability in comparison with manual control. Contributing factors include poor controller tuning and control valves that are incorrectly sized or tend to

Table 21.5 Basic Data for Control Loop Monitoring

- Service factors (time in use/total time period)
- Mean and standard deviation for the control error (set point − measurement)
- Mean and standard deviation for the controller output
- Alarm summaries
- Operator logbooks and maintenance records

stick due to excessive frictional forces. In large processing plants, each plant operator is typically responsible for 200 to 1,000 loops. Thus, there are strong incentives for automated *control* (or *controller*) *performance monitoring* (*CPM*). The overall objectives of CPM are (1) to determine whether the control system is performing in a satisfactory manner and (2) to diagnose the cause of any unsatisfactory performance.

21.5.1 Basic Information for Control Performance Monitoring

In order to monitor the performance of a single standard PI or PID control loop, the basic information in Table 21.5 should be available.

Service factors should be calculated for key components of the control loop such as the sensor and final control element. Low service factors and/or frequent maintenance suggest chronic problems that require attention. The fraction of time that the controller is in the automatic mode is a key metric. A low value indicates that the loop is frequently in the manual mode and thus requires attention. Service factors for computer hardware and software should also be recorded.

Simple statistical measures such as the sample mean and standard deviation can indicate whether the controlled variable is achieving its target and how much control effort is required. An unusually small standard deviation for a measurement could result from a faulty sensor with a constant output signal, as noted in Section 21.1. By contrast, an unusually large standard deviation could be caused by equipment degradation or even failure, for example, inadequate mixing caused by a faulty vessel agitator.

A high alarm rate can be indicative of poor control system performance (see Section 10.2). Operator logbooks and maintenance records are valuable sources of information, especially if this information has been captured in a computer database.

21.5.2 Control Performance Monitoring Techniques

Chapters 6 and 12 introduced traditional control loop performance criteria such as rise time, settling time, overshoot, offset, degree of oscillation, and integral error criteria. CPM methods have been developed based on these and other criteria, and commercial CPM software is available. A comprehensive review of CPM techniques and industrial applications has been reported by Jelali (2006).

If a process model is available, then process monitoring techniques based on monitoring the model residuals can be employed (Chiang et al., 2001; Davis et al., 2000; Cinar et al., 2007). Simple CPM methods have also been developed that do not require a process model. Control loops that are excessively oscillatory or very sluggish can be identified using correlation or frequency response techniques (Hägglund, 1999; Miao and Seborg, 1999, Tangirala et al., 2005), or by evaluating standard deviations (Rhinehart, 1995; Shunta, 1995). A common problem, control valve stiction, can be detected from routine operating data (Shoukat Choudhury et al., 2008).

Control system performance can be assessed by comparison with a *benchmark*. For example, historical data representing periods of satisfactory control can be used as a benchmark. Alternatively, the benchmark could be an ideal control system performance, such as *minimum variance control*. As the name implies, a minimum variance controller minimizes the variance of the controlled variable when unmeasured, random disturbances occur. This ideal performance limit can be estimated from closed-loop operating data; then the ratio of minimum variance to the actual variance is used as the measure of control system performance. This statistically based approach has been commercialized, and many successful industrial applications have been reported (Kozub, 1997; Desborough and Miller, 2002; Harris and Seppala, 2002; Hoo et al., 2003; Paulonis and Cox, 2003).

Additional information on statistically-based CPM is available in a tutorial (MacGregor, 1988), survey articles (Piovoso and Hoo, 2002; Kourti, 2005), and books (Box and Luceño, 1997; Huang and Shah, 1999; Cinar et al., 2007). Extensions to MIMO control problems, including MPC, have also been reported (Huang et al., 2000; Qin and Yu, 2007; Cinar et al., 2007).

SUMMARY

Process monitoring is essential to ensure that plants operate safely and economically while meeting environmental standards. In recent years, control system performance monitoring has also been recognized as a key component of the overall monitoring activity. Process variables are monitored by making simple limit and performance calculations. Statistical process control (SPC) techniques based on control charts are monitoring techniques widely used for product quality control and other applications where the sampling periods are long relative to process settling times. In particular, Shewhart control charts are used to detect large shifts in mean process behavior, while CUSUM and EWMA control charts are better at detecting small, sustained changes. Multivariate monitoring techniques such as PCA and PLS can offer significant improvements over these traditional univariate methods when the measured variables are highly correlated. SPC and APC are complementary techniques that can be used together to good advantage. Control performance monitoring techniques have been developed and commercialized, especially methods based on on-line, statistical analysis of operating data.

REFERENCES

Abnormal Situation Management Consortium (ASM), http://www.asmconsortium.com (2009).

Alt, F. B., N. D. Smith, and K. Jain, Multivariate Quality Control, in *Handbook of Statistical Methods for Scientists and Engineers*, 2d ed., H. M. Wadsworth (Ed.), McGraw-Hill, New York, 1998, Chapter 21.

Box, G., and A. Luceño, *Statistical Control by Monitoring and Feedback*, Wiley, New York, 1997.

Chiang, L. H., E. L. Russell, and R. D. Braatz, *Fault Detection and Diagnosis in Industrial Systems*, Springer, New York, 2001.

Cinar, A. A. Palazoglu, and F. Kayihan, *Chemical Process Performance Evaluation*, CRC Press, Boca Raton, FL, 2007.

Davis, J. F., M. J. Piovoso, K. A. Hoo, and B. R. Bakshi, Process Data Analysis and Interpretation, *Advances in Chem. Eng.*, **25**, Academic Press, New York (2000).

Deming, W. E., *Out of the Crisis*, MIT Center for Advanced Engineering Study, Cambridge, MA, 1986.

Desborough, L., and R. Miller, Increasing Customer Value of Industrial Control Performance Monitoring—Honeywell's Experience, *Chemical Process Control, CPC-VI*, J. B. Rawlings, B. A. Ogunnaike, and J. Eaton (Eds.), *AIChE Symposium Series*, **98**, 169 (2002).

Hägglund, T., Automatic Detection of Sluggish Control Loops, *Control Eng. Practice*, **7**, 1505 (1999).

Harris, T. J., and C. T. Seppala, Recent Developments in Controller Performance Monitoring and Assessment Techniques, *Chemical Process Control, CPC-VI*, J. B. Rawlings, B. A. Ogunnaike, and J. Eaton (Eds.), *AIChE Symposium Series*, **98**, 208 (2002).

Hoo, K. A., M. J. Piovoso, P. D. Schnelle, and D. A. Rowan, Process and Controller Performance Monitoring: Overview with Industrial Applications, *Int. J. Adaptive Control and Signal Processing*, **17**, 635 (2003).

Huang, B., R. Kadali, X. Zhao, E. C. Tamayo, and A Hanafi, An Investigation into the Poor Performance of a Model Predictive Control System on An Industrial CGO Coker, *Control Eng. Practice*, **8**, 619 (2000).

Huang, B., and S. L. Shah, *Performance Assessment of Control Loops: Theory and Applications*, Springer-Verlag, New York (1999).

Jackson, J. E. *A User's Guide to Principal Components*, Wiley-Interscience, New York, 1991.

Jacobs, D. C., Watch Out for Nonnormal Distributions, *Chem. Eng. Progress*, **86**, (11), 19 (1990).

Jelali, M., An Overview of Performance Assessment Technology and Industrial Applications, *Control Eng. Practice*, **14**, 441 (2006).

Kourti, T., Process Analysis and Abnormal Situation Detection: From Theory to Practice, *IEEE Control Systems*, **22** (5), 10 (2002).

Kourti, T., Application of Latent Variable Methods to Process Control and Statistical Process Control in Industry, *Int. J. Adaptive Control and Signal Processing*, **19**, 213 (2005).

Kozub, D. J., Monitoring and diagnosis of chemical processes with automated process control in *Chemical Process Control, CPC-V*, J. C. Kantor, C. E. Garcia, and B. Carnahan, *AIChE Symposium Series*, **93** (No. 316), 83 (1997).

MacGregor, J. F., On-line Statistical Process Control, *Chem. Eng. Progress*, **84**, (10), 21 (1988).

MacGregor, J. F., Using On-line Process Data to Improve Quality, *ASQC Statistics Division Newsletter*, **16** (2), 6 (1996).

Miao, T., and D. E. Seborg, Automatic Detection of Excessively Oscillatory Feedback Control Loops, *Proc. IEEE Internat. Conf. on Control Applications*, 359, Kohala Coast, HI, USA (1999).

Montgomery, D. C., *Introduction to Statistical Quality Control*, 6th ed., Wiley, New York, 2009.

Montgomery, D. C., and G. C. Runger, *Applied Statistics and Probability for Engineers*, 4th ed., Wiley, New York, 2007.

Pande, P. S., R. P. Neuman, and R. R. Cavanagh, *The Six Sigma Way*, McGraw-Hill, New York, 2000.

Paulonis, M. A., and J. W. Cox, A Practical Approach for Large-Scale Controller Performance Assessment, Diagnosis, and Improvement, *J. Process Control*, **13**, 155 (2003).

Piovoso, M. J., and K. A. Hoo, Multivariate Statistics for Process Control, *IEEE Control Systems*, **22** (5), 8 (2002).

Piovoso, M. J., and K. A. (Hoo) Kosanovich, The Use of Multivariate Statistics in Process Control, in *The Control Handbook*, W. S. Levine and R. C. Dorf (Eds.), CRC Press, Boca Raton, FL, 1996, Chapter 33.

Qin, S. J. and J. Yu, Recent Developments in Multivariable Controller Performance Monitoring, *J. Process Control*, **17**, 221 (2007).

Rhinehart, R. R., A Watchdog for Controller Performance Monitoring, *Proc. Amer. Control Conf.*, 2239 (1995).

Romagnoli, J., and M. C. Sanchez, *Data Processing and Reconciliation in Chemical Process Operations*, Academic Press, San Diego, CA, 2000.

Ryan, T. P., *Statistical Methods for Quality Improvement*, 2d ed., Wiley, New York, 2000.

Shewhart, W. A., *Economic Control of Quality*, Van Nostrand, New York, 1931.

Shoukat Choudhury, M. A. A., M. Jain, S. L., and Shah, Stiction—Definition, Modelling, Detection and Quantification, *J. Process Control*, **18**, 232 (2008).

Shunta, J. P., *Achieving World Class Manufacturing Through Process Control*, Prentice Hall PTR, Englewood Cliffs, NJ, 1995.

Tangirala, A. K., S. L. Shah, and N. F. Thornhill, PSCMAP: A New Tool for Plant-Wide Oscillation Detection, *J. Process Control*, **15**, 931 (2005).

Western Electric Company, *Statistical Quality Control Handbook*, Delmar Printing Company, Charlotte, NC, 1956.

EXERCISES

21.1 A standard signal range for electronic instrumentation is 4–20 mA. For purposes of monitoring instruments using limit checks, would it be preferable to have an instrument range of 0–20 mA? Justify your answer.

21.2 An analyzer measures the pH of a process stream every 15 minutes. During normal process operation, the mean and standard deviation for the pH measurement are $\bar{x} = 5.75$ and $s = 0.05$, respectively. When the process is operating normally, what is the probability that a pH measurement will exceed 5.9?

21.3 In a computer control system, the high and low warning limits for a critical temperature measurement are set at the "2-sigma limits," $\overline{T} \pm 2\hat{\sigma}_T$, where $\overline{T}$ is the nominal temperature and $\hat{\sigma}_T$ is the estimated standard deviation. If the process operation is normal and the temperature is measured every minute, how many "false alarms" (that is, measurements that exceed the warning limits) would you expect to occur during an eight-hour period?

21.4 In order to improve the reliability of a critical control loop, it is proposed that redundant sensors be used. Suppose that three independent sensors are employed and each sensor works properly 95% of the time.

(a) What is the probability that all three sensors are functioning properly?

(b) What is the probability that none of the sensors are functioning properly?

(c) It is proposed that the average of the three measurements be used for feedback control. Briefly critique this strategy.

Hint: See Appendix F for a review of basic probability concepts.

21.5 In a manufacturing process, the impurity level of the product is measured on a daily basis. When the process is operating normally, the impurity level is approximately normally distributed with a mean value of 0.800% and a standard deviation of 0.021%. The laboratory measurements for a period of eight consecutive days are shown below. From an SPC perspective, is there strong evidence to believe that the mean value of the impurity has shifted? Justify your answer.

Day	*Impurity (%)*	*Day*	*Impurity (%)*
1	0.812	5	0.799
2	0.791	6	0.833
3	0.841	7	0.815
4	0.814	8	0.807

21.6 A drought in southern California resulted in water rationing and extensive discussion of alternative water supplies. Some people believed that this drought was the worst one ever experienced in Santa Barbara County. But was this really true? Rainfall data for a 120-year period are shown in Table E21.6. In order to distinguish between normal and abnormal drought periods, do the following.

(a) Consider the data before the year 1920 to be a set of "normal operating data." Use these data to develop the target and control limits for a Shewhart chart. Determine if any of the data for subsequent years are outside the chart limits.

(b) Use the data prior to 1940 to construct an s chart that is based on a subgroup of 10 data points for each decade. How many chart violations occur for subsequent decades?

21.7 Develop CUSUM and EWMA charts for the rainfall data of Exercise 21.6 considering the data for 1900 to 1930 to be the "normal operating data." Use the following design parameters: $K = 0.5$, $H = 5$, $\lambda = 0.25$. Based on these charts, do any of the next three decades appear to be abnormally dry or wet?

21.8 An SPC chart is to be designed for a key process variable, a chemical composition, which is also a controlled variable. Because the measurements are very noisy, they must be filtered before being sent to a PI controller. The question arises whether the variable plotted on the SPC chart should be the filtered value or the raw measurement. Are both alternatives viable? If so, which one do you recommend? (Briefly justify your answers.)

21.9 For the BOD data of Example 21.5, develop CUSUM and EWMA charts. Do these charts indicate an "abnormal situation"? Justify your answer. For the CUSUM chart, use $K = 0.5s$ and $H = 5s$ where s is the sample standard deviation. For the EWMA chart, use $\lambda = 0.25$.

21.10 Calculate the average values of the C_p and C_{pk} capability indices for the BOD data of Example 21.5, assuming that LSL = 5 mg/L and USL = 35 mg/L. Do these values of the indices indicate that the process performance is satisfactory?

21.11 Repeat Exercise 21.10 for the solids data of Example 21.5, assuming that USL = 1,600 mg/L and LSL = 1,200 mg/L.

21.12 Consider the wastewater treatment problem of Examples 21.5 and 21.6 and five new pairs of measurements shown below. Calculate the value of Hotelling's T^2 statistic for each pair using the information for Example 21.6, and plot the data on a T^2 chart. Based on the number of chart violations for the new data, does it appear that the current process behavior is normal or abnormal?

Sample Number	*BOD (mg/L)*	*Solids (mg/L)*
1	18.1	1281
2	36.8	1430
3	16.0	1510
4	28.2	1343
5	31.0	1550

Note: The required covariance matrix $\boldsymbol{S}$ in Eq. 21-27 can be calculated using either the *cov* command in MATLAB or the *covar* command in EXCEL.

Table E21.6 Rainfall Data, 1870–1990

Year	Rain (in)	Year	Rain (in)	Year	Rain (in)
1870	10.47	1911	31.94	1951	11.29
1871	8.84	1912	16.35	1952	31.20
1872	14.94	1913	12.78	1953	12.98
1873	10.52	1914	31.57	1954	15.37
1874	14.44	1915	21.46	1955	17.07
1875	18.71	1916	25.88	1956	19.58
1876	23.07	1917	21.84	1957	13.89
1877	4.49	1918	21.66	1958	31.94
1878	28.51	1919	12.16	1959	9.06
1879	13.61	1920	14.68	1960	10.82
1880	25.64	1921	14.31	1961	9.99
1881	15.23	1922	19.25	1962	28.22
1882	14.27	1923	17.24	1963	15.73
1883	13.41	1924	6.36	1964	10.19
1884	34.47	1925	12.26	1965	18.48
1885	13.79	1926	15.83	1966	14.39
1886	24.24	1927	22.73	1967	24.96
1887	12.96	1928	13.48	1968	13.67
1888	21.73	1929	14.54	1969	30.47
1889	21.04	1930	13.91	1970	12.03
1890	32.47	1931	14.99	1971	14.02
1891	17.31	1932	22.13	1972	8.64
1892	10.75	1933	6.64	1973	23.33
1893	27.02	1934	13.43	1974	17.33
1894	7.02	1935	21.12	1975	18.87
1895	16.34	1936	18.21	1976	8.83
1896	13.37	1937	25.51	1977	16.49
1897	18.50	1938	26.10	1978	41.71
1898	4.57	1939	13.35	1979	21.74
1899	12.35	1940	14.94	1980	24.59
1900	12.65	1941	45.71	1981	15.04
1901	15.40	1942	12.87	1982	15.11
1902	14.21	1943	24.37	1983	38.25
1903	20.74	1944	17.95	1984	14.70
1904	11.58	1945	15.23	1985	14.00
1905	29.64	1946	11.33	1986	22.12
1906	22.68	1947	13.35	1987	11.45
1907	27.74	1948	9.34	1988	15.45
1908	19.00	1949	10.43	1989	8.90
1909	35.82	1950	13.15	1990	6.57
1910	19.61				

Chapter 22

Batch Process Control

CHAPTER CONTENTS

Batch processing is an alternative to continuous processing. In batch processing, a sequence of one or more steps, either in a single vessel or in multiple vessels, is performed in a defined order, yielding a specific quantity of a finished product. Because the volume of product is normally small, large production runs are achieved by repeating the process steps on a predetermined schedule. In batch processing, the production amounts are usually smaller than for continuous processing; hence, it is usually not economically feasible to dedicate processing equipment to the manufacture of a single product. Instead, batch processing units are organized so that a range of products (from a few to possibly hundreds) can be manufactured with a given set of process equipment. Batch processing can be complicated by having multiple stages, multiple products made in the same equipment, or parallel processing lines. The key challenge for batch plants is to consistently manufacture each product in accordance with its specifications while maximizing the utilization of available equipment. Benefits include reduced inventories and shortened response times to make a specialty product compared to continuous processing plants. Typically, it is not possible to use blending of multiple batches in order to obtain the desired product quality, so product quality specifications must be satisfied by each batch.

Batch processing is widely used to manufacture specialty chemicals, metals, electronic materials, ceramics, polymers, food, biochemicals and pharmaceuticals, multiphase materials/blends, coatings, and composites, an extremely broad range of processes and products. The unit operations in batch processing are also quite diverse, and some are analogous to operations for continuous processing.

As one example of batch processing, batch distillation is used in the production of many chemicals and pharmaceuticals. A batch column or still can be used to separate products with different purity specifications. Compared to continuous distillation, it is easier to tailor product specification on a batch-to-batch basis, giving a flexible, easily operated separation unit with low capital cost (Muhrer, 1992; Diwekar, 1995). The general arrangement of a typical batch still and the important controlled and manipulated variables are shown in Fig. 22.1. As is typical of batch processes, a sequence of steps must be carried out; each step involves the

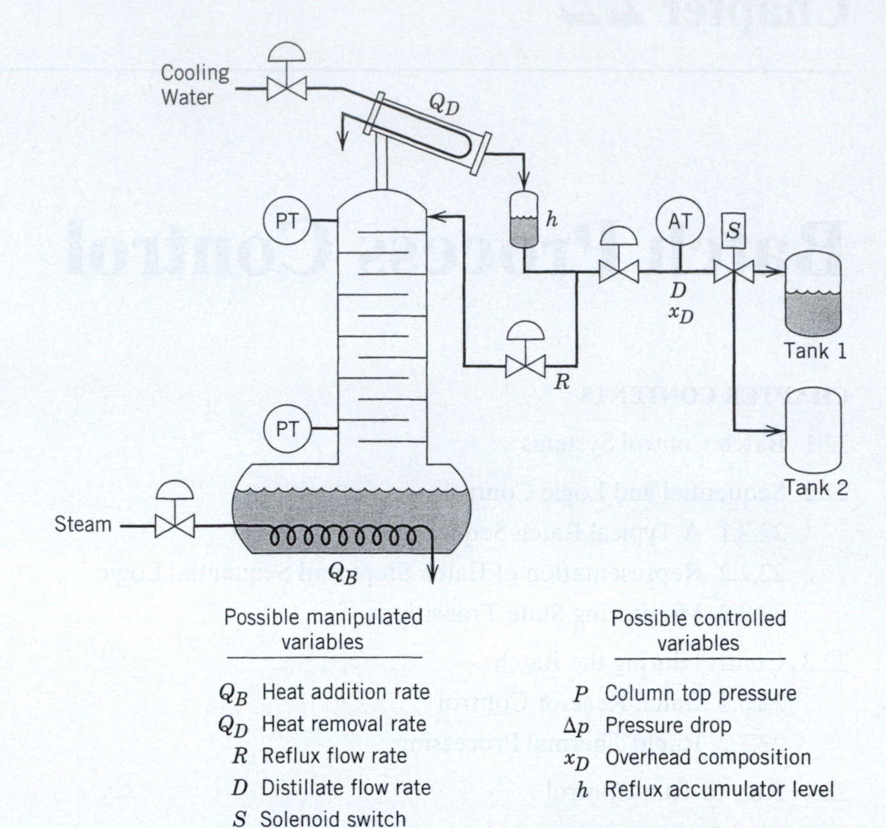

Figure 22.1 Batch distillation schematic.

opening and closing of different valves at specified times and in a specific order. After charging the kettle at the base of the column, the feed flow is stopped and heat is applied at the reboiler. Using cooling at the condenser, the reflux flow rate can be manipulated so that the column reaches a certain overhead distillate composition x_{Dsp} (the set point) prior to product withdrawal. At this point, distillate is withdrawn into a product receiver at a flow rate D in order to meet a product specification. At selected times, the product receiver may be switched to make multiple products with different purity specifications in various tanks. At the end of the last product withdrawal, the column is shut down, the remaining bottoms residue and receiver holdup(s) are pumped to storage, and the column is readied for the next batch. Fig. 22.2 shows the time profile of ethanol composition for an ethanol–water batch fractionation at a constant distillate rate. Notice that the overhead ethanol mole fraction (the main product) remains nearly constant for the first 45 minutes of the run, while the bottoms composition undergoes a gradual decline.

In an effort to increase the safety, efficiency, and affordability of medicines, the FDA has recently proposed a new framework for the regulation of pharmaceutical development, manufacturing, and quality assurance. The primary focus of the initiative is to reduce variability through a better understanding of processes than can be obtained by the traditional approach. PAT (Process Analytical Technology) has become an acronym in the pharma industry for designing, analyzing, and controlling manufacturing through measurements (i.e., during processing) of critical quality and performance attributes of raw and in-process materials and processes, with the goal of ensuring final product quality. Process variations that could possibly contribute to patient risk are determined through modeling and timely measurements of

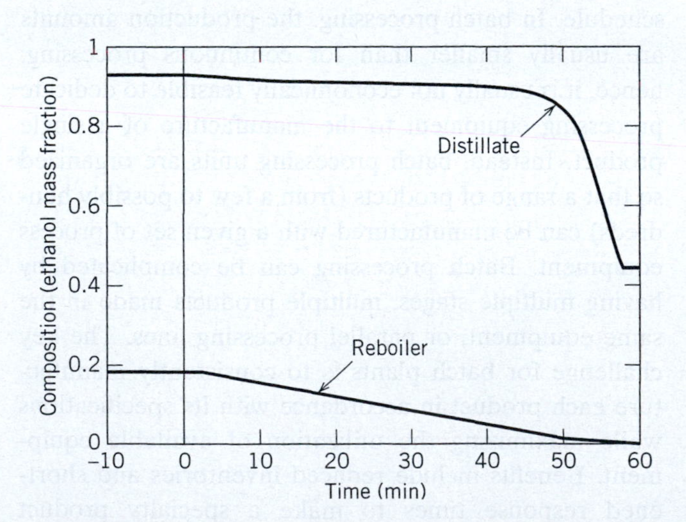

Figure 22.2 The time variation of distillate and bottom ethanol compositions during batch fractionation of ethanol and water for constant distillate flow.

critical quality attributes, which are then addressed by process control.

This chapter provides an introduction to batch process control. First we introduce the operational practices and control system design for batch plants, which differ markedly from continuous plants. In batch processing, there is a much greater emphasis on production scheduling of batch equipment; this procedure is critical to matching available production equipment and raw materials with the demands for a range of specialty products, each having different specifications. Batch control systems, in contrast to continuous process control, involve *binary logic* and *discrete event analysis* applied to the sequencing of different process steps in the same vessel, usually requiring the application of programmable logic controllers (PLCs). Feedback controllers are utilized in order to handle set-point changes and disturbances, but they may require certain enhancements to treat the wide operating ranges, because there is no steady-state operating point. In several sections, we highlight the use of batch process control in semiconductor manufacturing, where individual wafers or groups of wafers are repetitively processed through a variety of unit operations such as etching and lithography. The practice of *run-to-run control*, a form of supervisory control in which operating conditions or trajectories are changed only between runs (batches) and not during a batch, is also described.

22.1 BATCH CONTROL SYSTEMS

In analogy with the different levels of plant control depicted in Fig. 19.1, batch control systems operate at various levels:

- Batch sequencing and logic control (Levels 1 and 2)
- Control during the batch (Level 3)
- Run-to-run control (Levels 4 and 5)
- Batch production management (Level 5)

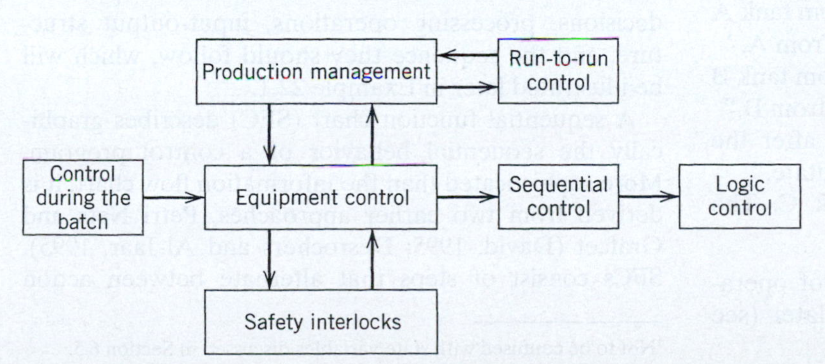

Figure 22.3 shows the interconnections of the different types of control used in a typical batch process. Run-to-run control is a type of supervisory control that is part of the production management block. In contrast to continuous processing, the focus of control shifts from regulation to set-point changes, and sequencing of batches and equipment takes on a more important role.

Batch control systems must be very versatile to be able to handle pulse inputs and discrete I/O as well as analog signals for sensors and actuators. Functional control activities are summarized as follows.

1. ***Batch sequencing and logic control:*** The sequence of control steps that follow a recipe involves, for example: mixing of ingredients, heating, waiting for a reaction to complete, cooling, and discharging the resulting product. Transfer of materials to and from batch tanks or reactors includes metering of materials as they are charged (as specified by the recipe), as well as transfer of materials at the completion of the process operation. In addition to discrete logic for the control steps, logic is needed for safety interlocks to protect personnel, equipment, and the environment from unsafe conditions (see Chapter 10). Process interlocks ensure that process operations can only occur in the correct time sequence.
2. ***Control during the batch:*** Feedback control of flow rate, temperature, pressure, composition, and level, including advanced control strategies, falls in this category, which is also called "within-the-batch" control (Bonvin, 1998). In complex applications, this requires specification of an operating trajectory for the batch (that is, temperature or flow rate as a function of time). In simpler cases, it involves tracking of set points of the controlled variables, which includes ramping the controlled variables up and down and/or holding them constant for a prescribed period of time. Detection of when the batch operations should be terminated (*end point*) may be performed by inferential measurements of product quality, if direct measurement is not feasible (see Chapter 16).

Figure 22.3 Overview of a batch control system.

3. ***Run-to-run control:*** Also called batch-to-batch control, this supervisory function is based on off-line product quality measurements at the end of a run. Operating conditions and profiles for the batch are adjusted between runs to improve the product quality using tools such as optimization.
4. ***Batch production management:*** This activity entails advising the plant operator of process status and how to interact with the recipes and the sequential, regulatory, and discrete controls. Complete information (recipes) is maintained for manufacturing each product grade, including the names and amounts of ingredients, process variable set points, ramp rates, processing times, and sampling procedures. Other database information includes batches produced on a shift, daily, or weekly basis, as well as material and energy balances. Scheduling of process units is based on availability of raw materials and equipment and customer demand.

22.2 SEQUENTIAL AND LOGIC CONTROL

Sequential logic is used to ensure that the batch process undergoes the proper sequence of events, because the time order of steps is important. Sequential logic must not be confused with combinational logic, which depends only on instantaneous values for the variables. This type of logic is especially suitable for interlocks or for permissive actions; for example, the reactor discharge valve must be closed, or the vent must be open in order for the feed valve to be opened. Both sequential and combinational logic can be implemented with a digital device, a microprocessor, or a computer. Digital devices can be intrinsically discrete (producing only discrete outputs, such as integers) or can mimic continuous devices such as a PID controller, as discussed in Chapter 17.

22.2.1 A Typical Batch Sequence

Batch processing requires that the process proceed through the proper sequence of steps. For example, a simple blending sequence might consist of the following steps:

1. Transfer specified amount of material from tank A to tank R. The process step is "Transfer from A."
2. Transfer specified amount of material from tank B to tank R. The process step is "Transfer from B."
3. Agitate for a specified period of time after the feeds are added. The process step is "Agitate."
4. Discharge the product to storage tank C. The process step is "Transfer from R."

A more detailed example of the sequence of operations for a batch mixing tank is described later (see Example 22.1).

For each process step, the various discrete-acting devices are expected to be in a specified *device state*[1], usually a binary value (0 to 1). Then, for process step "Transfer from A," the device states might be as follows:

1. Tank A discharge valve: open
2. Tank R inlet valve: open
3. Tank A transfer pump: running
4. Tank R agitator: off
5. Tank R cooling valve: closed

Sequential logic is coupled to device states. For example, device state 0 could be a valve closed, agitator off, and so on, while device state 1 could be the valve open or the agitator on. Basically, the sequential logic determines when the process should proceed from the current set of operating conditions to the next. Sequential logic must encompass both normal and abnormal process operations, such as equipment failures.

When failure occurs, the simplest action is to stop or hold at the current operating state in response to any abnormal condition, and let the process operator determine the cause of the problem. However, some failures lead to hazardous conditions that require immediate action; waiting for the operator to decide what to do is not acceptable. The appropriate response to such situations is best determined in conjunction with process hazards and operability (HAZOP) studies. For example, guidelines for safe operation of batch reaction systems have been published (Center for Chemical Process Safety, 1999).

22.2.2 Representation of Batch Steps and Sequential Logic

There are several ways to depict the sequential logic in batch operation, which is a prerequisite to incorporating binary logic into the computer control system. Two process-oriented representations are considered: the *information flow diagram* and the *sequential function chart*. These can be used to develop digital logic diagrams including *ladder logic diagrams* and *binary logic diagrams*.

To create an information flow chart, a complete list of steps for a batch process must be documented and displayed. From this representation, it is straightforward to prepare a sequential function chart. Figure 22.4 shows the flow-chart symbols that indicate the points of decisions, processing operations, input-output structure, and the sequence they should follow, which will be illustrated later in Example 22.1.

A sequential function chart (SFC) describes graphically the sequential behavior of a control program. More sophisticated than the information flow chart, it is derived from two earlier approaches, Petri Nets and Grafcet (David, 1995; Desrochers and Al-Jaar, 1995). SFCs consist of steps that alternate between action

[1]Not to be confused with state variables discussed in Section 6.5.

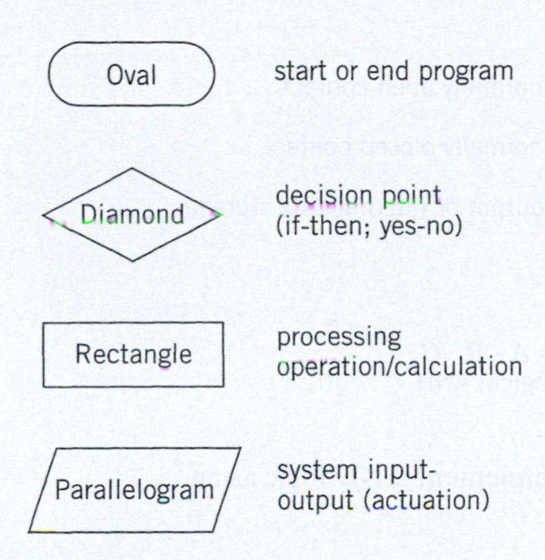

Figure 22.4 Flowchart symbols and their definitions.

blocks and transitions. Each step corresponds to a state of the batch process. A transition is associated with a condition that, when true, activates the next step and deactivates the previous step. Steps are linked with action blocks that perform a specific control action. SFC and Grafcet are standard languages established by the International Electrotechnical Commission (IEC) and are supported by an association of vendors and users called PLCopen; see *www.plcopen.org*. Figure 22.5 gives a simple illustration of the SFC notation. The steps are denoted as rectangles (a double rectangle is the initial step), and the transition symbol is a small horizontal bar on the line linking control steps. A double bar is used for branching, and it can precede a transition when two or more paths can be followed. Similarly, a double bar indicates where two or more parallel paths

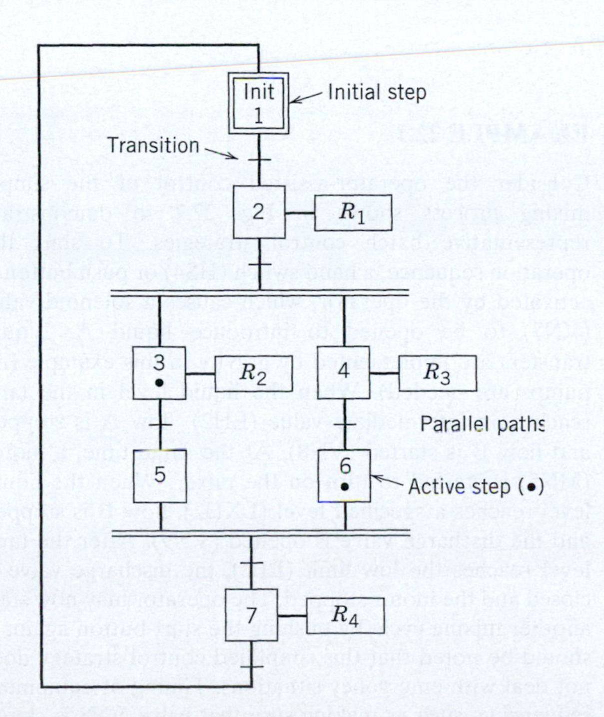

Figure 22.5 A generic sequential function chart (active steps are 3 and 6). R_1 through R_4 denote specified actions.

Table 22.1 Truth Table for AND, OR, NOT Binary Logic Operations

A B	$A \cdot B$ (AND)	$A + B$ (OR)	$\overline{A}$ (NOT)	$\overline{B}$ (NOT)
0 0	0	0	1	1
0 1	0	1	1	0
1 0	0	1	0	1
1 1	1	1	0	0

join together into a single path. In Fig. 22.5, both steps 5 and 6 must be completed before moving to step 7. The active steps are shown with a black dot in the box.

Ladder logic and binary logic diagrams provide alternative graphical formats for representing logical functions and can be analyzed using *truth tables* (Platt and Gilbert, 1995). In binary logic, the main operations are AND, OR, NAND (not AND), and NOR (not OR). When two input variables A and B are "ANDed" together ($= A \cdot B$), the output is 1 if and only if both inputs A and B are 1. When two inputs A and B are ORed together, the output ($A + B$) is 1 if either A or B is 1. The NOT operation changes the input A to the complementary binary value $\overline{A}$. Table 22.1 gives the *truth table* for several standard operations with four combinations of binary variables A and B.

A ladder diagram contains two vertical uprights, which are the power source (on the left) and neutral (on the right). A number of horizontal rungs indicate various paths between the two uprights, which can contain logical switches (normally open or closed) and an output. Very few symbols are required to construct a ladder diagram (Erickson and Hedrick, 1999; Johnson, 2005). Two or more switches (also called contacts) on the same rung form an AND gate. Contacts on two or more parallel branches of a rung form an OR gate, as discussed below. Two vertical bars are used to depict a normally open contact, while a slash across the bars indicates a normally closed contact.

By clever construction of parallel and series relay circuits, designers can implement sophisticated logical statements for a sequence of logical steps (Platt and Gilbert, 2005). As an example, a set of three relays wired in series can be used to implement the three-input AND condition shown in Fig. 22.6*a*. The output is actuated only when all three input relays (A, B, and C) are actuated.

Similarly, a set of relays wired in parallel can be used to implement the OR condition shown in Fig. 22.6*b*. Here actuation of any one or more of these input relays will cause the output to be actuated. To illustrate NOR or NAND gates in ladder logic, slashed contacts are drawn. In binary logic diagrams, a small circle is appended to the OR or AND symbol. These contacts are then normally closed rather than normally open, as is the case shown in Fig. 22.6. Sequencing operations typically require the use of *latching relays* (which hold a

A B C D

|| = normally open contact

|/| = normally closed contact

○ = output or internal data storage

A
B
C
D

Out = $A \cdot B \cdot C$

$\cdot$ = Logical AND

Figure 22.6*a* Use of contacts connected in series to implement AND logic as an "AND gate" in ladder and binary logic.

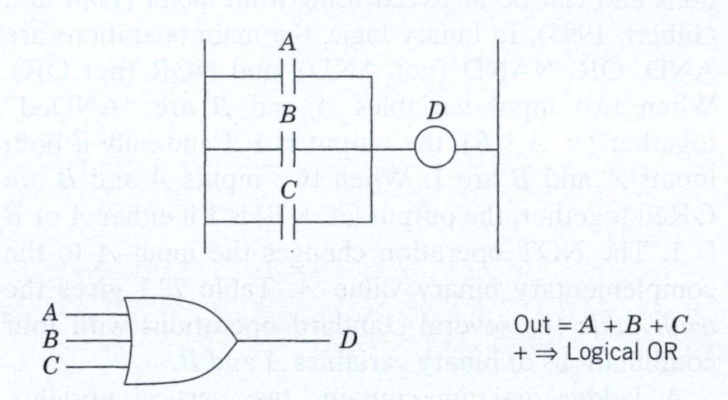

Figure 22.6*b* Use of contacts connected in parallel to implement OR logic as an "OR GATE" in ladder and binary logic.

state indefinitely once actuated, much like a solid-state flip-flop) and *delay relays* (which delay a preprogrammed time interval before operating, after actuation). Latching relays are shown as parallel connected rungs, as shown in Example 22.1.

From a process point of view, the sequential function chart is preferred to relay ladder logic and binary logic because it clearly shows the sequence of steps and also indicates concurrency, that is, when some subsystems are partially independent. The input-output structure and behavior are more clearly delineated for these subsystems by SFC. SFCs are also useful for interfacing with expert system software for supervisory control, monitoring, and diagnosis (Arzen, 1994). It is possible to convert binary ladder logic diagrams into SFCs using computer algorithms (Falcione and Krogh, 1993).

Programmable Logic Controllers

Programmable logic controllers (*PLCs*) are widely used in batch process control to execute the desired binary logic operations and to implement the desired sequencing. The inputs to the PLC are a set of relay contacts representing various device states (for example, limit switches indicate whether a valve is fully open or fully closed). Various operator inputs (for example, start/stop buttons) are also provided. The PLC output signals energize (actuate) a set of relays that turn pumps on or off, actuate lights on a display panel, operate solenoid or motor-driven valves, and so on. PLCs are discussed in more detail in Section A.3.1; see also Hughes (2005) and Webb and Reis (2002). PLCs can easily implement all variations of PID control. Consequently, it is relatively easy to program a single PLC to deal with most requirements of batch processing. However, it would be difficult for a PLC to optimize batch cycle operations or to implement inferential control of a reactor or separator product composition, functions that are easier to implement by integrating the PLC with a general-purpose computer or a distributed control system.

EXAMPLE 22.1

Consider the operator-assisted control of the simple mixing process shown in Fig. 22.7 to demonstrate representative batch control strategies. To start the operation sequence, a hand switch (HS4) or push button is activated by the operator, which causes a solenoid valve (VN7) to be opened to introduce liquid A. Liquid transfers are implemented by gravity in this example (no pumps are needed). When the liquid level in the tank reaches an intermediate value (LH2), flow A is stopped and flow B is started (VN8). At the same time, a motor (MN5) is started to turn on the mixer. When the liquid level reaches a specified level (LXH2), flow B is stopped and the discharge valve is opened (VN9). After the tank level reaches the low limit (LL2), the discharge valve is closed and the motor stopped. The operator may now start another mixing cycle by pushing the start button again. It should be noted that this simplified control strategy does not deal with emergency situations. Timing of equipment sequencing, such as making sure that valve VN8 is closed before opening discharge valve VN9, is also not considered.

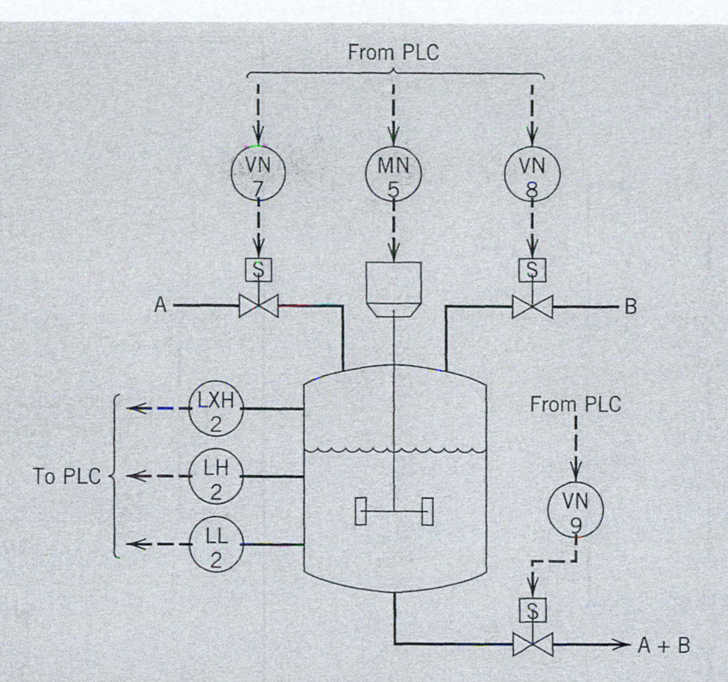

Figure 22.7 Schematic diagram for a batch mixing tank.

Develop an information flow diagram, a sequential function chart, and binary and ladder logic diagrams for this batch operation. Assume the batch proceeds uninterrupted (i.e., do not consider the case where the operator could accidentally activate the hand switch after the batch sequence starts).

For the binary logic diagram, denote the tank level by L and use the following binary values for the different device states:

operator push button HS4:	on	$\Rightarrow 1$	off	$\Rightarrow 0$
LH2 indicator:	$L \geq LH2$	$\Rightarrow 1$	$L < LH2$	$\Rightarrow 0$
LXH2 indicator:	$L \geq LXH2$	$\Rightarrow 1$	$L < LXH2$	$\Rightarrow 0$
LL2 indicator:	$L \geq LL2$	$\Rightarrow 0$	$L < LL2$	$\Rightarrow 1$
Valves VN7, VN8, VN9:	open	$\Rightarrow 1$	closed	$\Rightarrow 0$
Mixer MN5:	motor on	$\Rightarrow 1$	off	$\Rightarrow 0$

SOLUTION

Figures 22.8 and 22.9 show the series of events on the information flow diagram and the sequential function chart. For implementation via hardware (or software) interlocks, the binary logic diagram is shown in Fig. 22.10. Figure 22.6 defines the symbols used in this logic diagram. Gate 1 (an AND gate) ensures that the process will not start, if requested, when the tank level is not low. Gate 3 opens valve VN7 for flow A only if valve VN8 is not open (small circle on gate 3 denotes NOT). Gate 2 (an OR gate) latches the operator request once valve VN7 is opened so that the operator may release the push button. Gates 4 and 7 start flow B and the mixer motor when the intermediate liquid level is reached. This start signal (AND gate 3) terminates flow A. At the high tank liquid level, gate 6 opens the discharge valve. Gate 7 is used to prevent VN8 from opening during the discharge cycle. Gate 8 starts the mixer MN5 at the same time that valve VN8 is opened and ensures that the mixer continues to operate when VN8 is closed and VN9 is opened. The high-level signal LH2 is fed into gate 4 to stop flow B and the mixer motor. Gate 5 holds the discharge signal until the tank is drained.

Figure 22.11 presents the ladder logic diagram for the same mixing process. An example of a normally open contact is C1 (no slash), and an example of a normally closed contact is CR8. The operator-actuated push button HS4 and contact C1 form an AND gate equivalent to gate 1 in Fig. 22.10. The junction connecting rungs 1 and 1a is equivalent to the output of the OR gate 2 in Fig. 22.10. Contact C1 on rung 1 is normally open unless the tank level is low (condition LL2). Contact relay CR8 is usually closed unless relay CR8 on rung 2 is energized. When the level is at (or below) LL2, contact C1 is closed. Then if push button HS4 is engaged, rung 1 becomes energized. Because contact CR8 is normally closed, this action energizes CR7, which results in starting flow A. At the same time, because CR7 is energized, it closes the normally open CR7 on rung 1a. As soon as CR7 on rung 1a is closed, the operator can release HS4, but rung 1/1a will stay energized (or latched) to continue introducing flow A. As long as the level is not at or above LH2, C2 on rung 2 is open. Although CR9 and C3 on rung 2 are closed, rung 2 is not energized because of C2. Thus, contact CR8 remains "not energized," and the mixer motor and flow B valve are not active.

This all changes when the level reaches LH2. First, C2 on rung 2 closes. Because all three contacts on rung 2 are

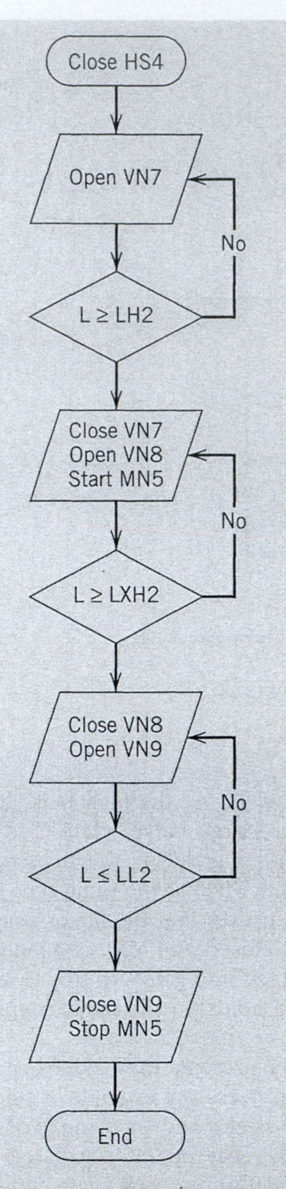

Figure 22.8 Information flow diagram for control of the mixing tank.

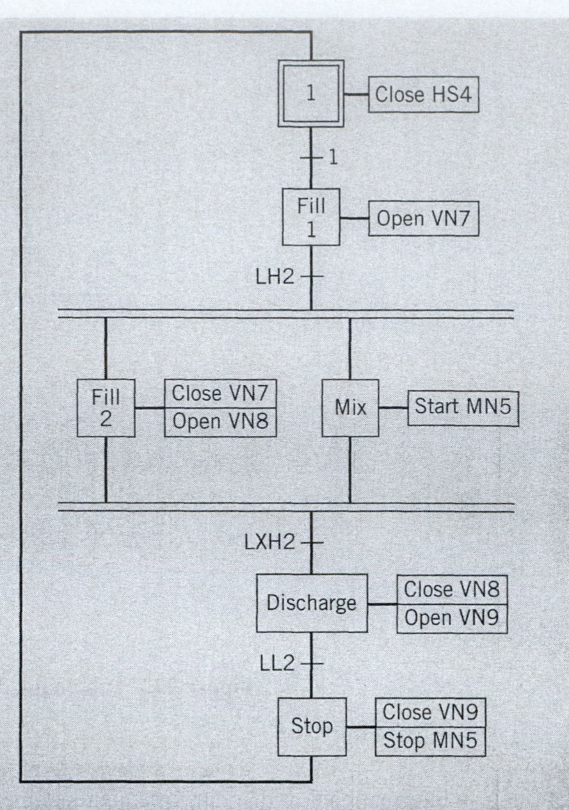

Figure 22.9 Sequential function chart (SFC) for control of the mixing tank.

now closed, rung 2 and CR8 are energized. This activates solenoid valve control VN8. At the same time, as soon as CR8 on rung 2 is energized, it opens CR8 on rung 1 and closes CR8 on rung 3. This deenergizes rungs 1 and 1a and CR7, causing two things to happen. First, CR7 on rung 1a is opened; that is, rungs 1 and 1a are "tripped." More importantly, because CR7 is deenergized, the solenoid valve for flow A is also deenergized, closing the valve and stopping flow A.

In rung 3, CR5 is energized, turning on the mixer motor control MN5. Rung 3a assures that CR5 remains energized after the contact CR8 is deenergized. The junction

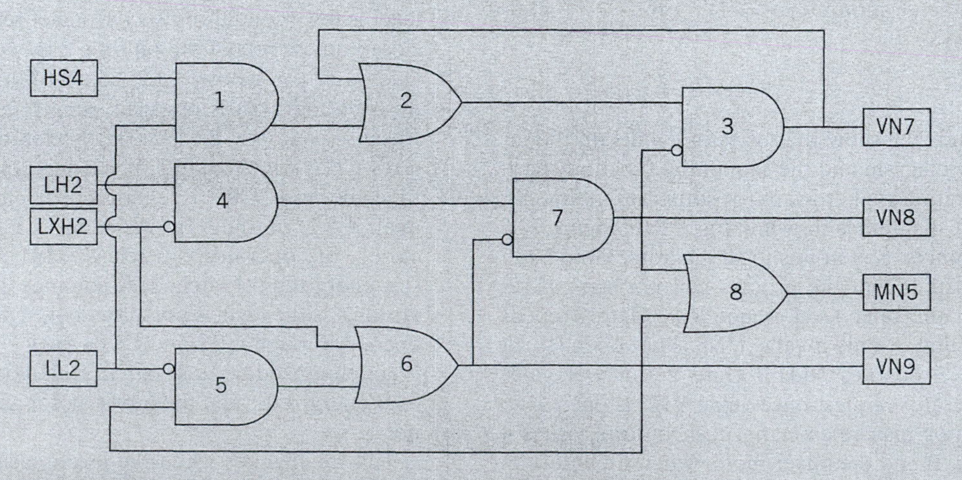

Figure 22.10 Binary logic diagram for control of the mixing tank.

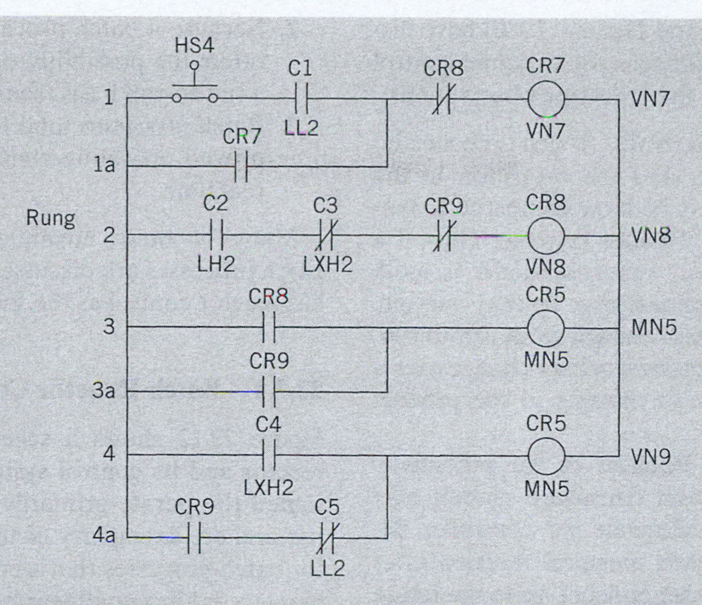

Figure 22.11 Ladder logic diagram for control of the mixing tank.

connecting rungs 3 and 3a is equivalent to the output of the OR gate 8 in Fig. 22.10. For rung 4, as soon as the liquid level reaches LXH2, C4 closes energizing the discharge solenoid valve VN9. The discharge should not stop until the level drops below LXH2, so rung 4a is designed to latch LXH2 (CR9). This strategy allows discharge to continue until the level reaches LL2 (which is tripped by normally closed contact C5).

22.2.3 Monitoring State Transitions

An automated batch facility utilizes sequential logic and discrete devices to change the process state subject to the interlocks. Discrete devices such as two-position valves can be driven to one of two possible states (open/closed). Such devices can be equipped with limit switches that indicate the state of the device. The discrete-device driver is the software routine that generates the output to a discrete device such as a valve and that also monitors process measurements to ascertain that the discrete device actually attains the desired state.

Valves do not instantly change states; rather, they require *travel times* associated with the change from one state to another. Thus, the processing logic within the discrete device driver must provide for a user-specified transition time for each field device. The transition between states can be implemented as:

1. ***Drive and wait.*** Further actions are delayed until the device attains its assigned state.
2. ***Drive and proceed.*** Further actions are initiated while the device is in the transition state.

Although two-state devices are most common, the need occasionally arises for devices with three or more states. For example, an agitator may be on high speed, low speed, or off.

Batch control software packages permit the control computer to:

1. Generate the necessary commands to drive each device to its proper state.
2. Monitor the status of each device to determine when all devices have attained their proper states.
3. Continue to monitor the state of each device to ensure that the devices are in their proper states.

If any discrete device does not reach its target state, failure logic is initiated.

22.3 CONTROL DURING THE BATCH

Control during a batch or "within-the-batch" control (Bonvin, 1998) is different from the sequential and logic control discussed above because it is concerned with an *operating trajectory*, that is, how the manipulated and controlled variables change as a function of time (vs. a sequence of on-off device states). Tracking of the set point (which may be a function of time) is challenging for this type of control, because there is no steady-state operating point and wide operating ranges may be encountered due to frequent start-up and shut-down.

Bonvin (1998) and Juba and Hamer (1986) have discussed the operational challenges for dynamic control during a batch and provide the following observations:

1. ***Time-varying characteristics.*** There is no steady-state operating point, and the transition in the controlled variable may be large compared to typical excursions for continuous systems. Thus, if a standard linear transfer function model is used, the gain and time constants may be time-varying. Batch characteristics can change from run to run, and even the process chemistry may change over a period of months due to changes in the product specifications.
2. ***Nonlinear behavior.*** Because of the potentially wide range of operation, linearized models may be inaccurate and inadequate for controller design. For example, batch chemical reaction rates may have a nonlinear dependence on temperature and concentration, and a nonlinear relationship may exist between heat transferred from a reactor and the flow rate of the cooling medium.
3. ***Model inaccuracies.*** Often, mechanistic or fundamental models are not available for batch processes, thus limiting the ability to design and tune controllers *a priori*.
4. ***Sensors.*** Often on-line sensors are not available or are inaccurate due to the wide operating ranges; hence, infrequent samples are analyzed by the plant laboratory. The inability to measure a process variable in real time reduces the safety margin for a process, potentially leading to an undesirable operating condition, for example, a runaway reaction (Center for Chemical Process Safety, 1999).
5. ***Constrained operation.*** This is a consequence of the wide operating ranges, which makes operating against constraints more likely.
6. ***Unmeasured disturbances.*** Operator error (e.g., wrong feed tank chosen), fouling of vessel walls and heat transfer surfaces, and raw material impurities are sources of major disturbances.
7. ***Irreversible behavior.*** It is often impossible to reverse the effects of history-dependent evolution in product properties such as molecular weight distribution in a polymer or crystal size distribution in a pharmaceutical product. In the semiconductor industry, once a semiconductor wafer is made, it is difficult to modify its electrical properties by further processing or rework.

On the other hand, a batch process has several advantages over a continuous process in meeting product quality requirements.

1. The batch duration can be adjusted in order to meet quality specifications.
2. Because a batch process is repetitive in nature, it offers the possibility of making improvements on a run-to-run basis (see Section 22.4).
3. Batch processes tend to be fairly slow so that improved operating conditions can be computed in real time.

Many of these advantages and disadvantages of batch processes are discussed in the next section, which has reactor control as the focus.

22.3.1 Batch Reactor Control

Figure 22.12 shows a schematic diagram of a batch reactor and its control system. Batch reactors are designed to operate primarily in an unsteady-state manner and are exemplary of the seven control challenges for batch processes that were cited earlier. Many batch reactors exhibit nonlinear behavior due to the coupling of reaction kinetics and reactor temperature while operating over a wide temperature range. Exothermic reactions produce heat that must be removed by a cooling system. Figure 22.12 shows the recommended control system for an exothermic batch reactor (cf. Fig. 16.1). The circulating pump for the coolant loop is essential to minimize the time delay and keep it constant; without it, the time delay varies inversely with cooling load. Because heating is also required to raise the temperature to reaction conditions, the valves are operated in split range. The heating valve opens when the controller output is between 50 and 100%, and the cooling valve opens for the 0 to 50% range. Sometimes cascade control of the cooling water temperature is utilized (see Chapter 16).

Figure 22.13 shows a typical *batch reactor cycle* consisting of (1) charging each of three reactants sequentially, (2) a heat-up operation, (3) reaction, (4) a cool-down sequence, and (5) discharge of the final product mix for separation and subsequent processing. In implementing batch process control, there are several important differences compared with control of continuous processes. The start-up of a batch process can be carried out by operators with all controllers placed in the manual mode. In charging materials to the unit, *totalizers* (see Fig. 22.12) are often used to determine the end point of a charge, that is, the total amount of material that has been transferred to the reactor. Thus, the ability to control flow rate accurately is not as important as the ability to measure and integrate flow rate accurately. If a weight mechanism is used, such as a load cell for the batch reactor, flow rate measurement is not particularly important. For the reaction period, it is important to determine when to terminate the batch reaction (the end-point composition). If the reactor conversion cannot be measured directly in real time, end-point composition can be predicted by some type

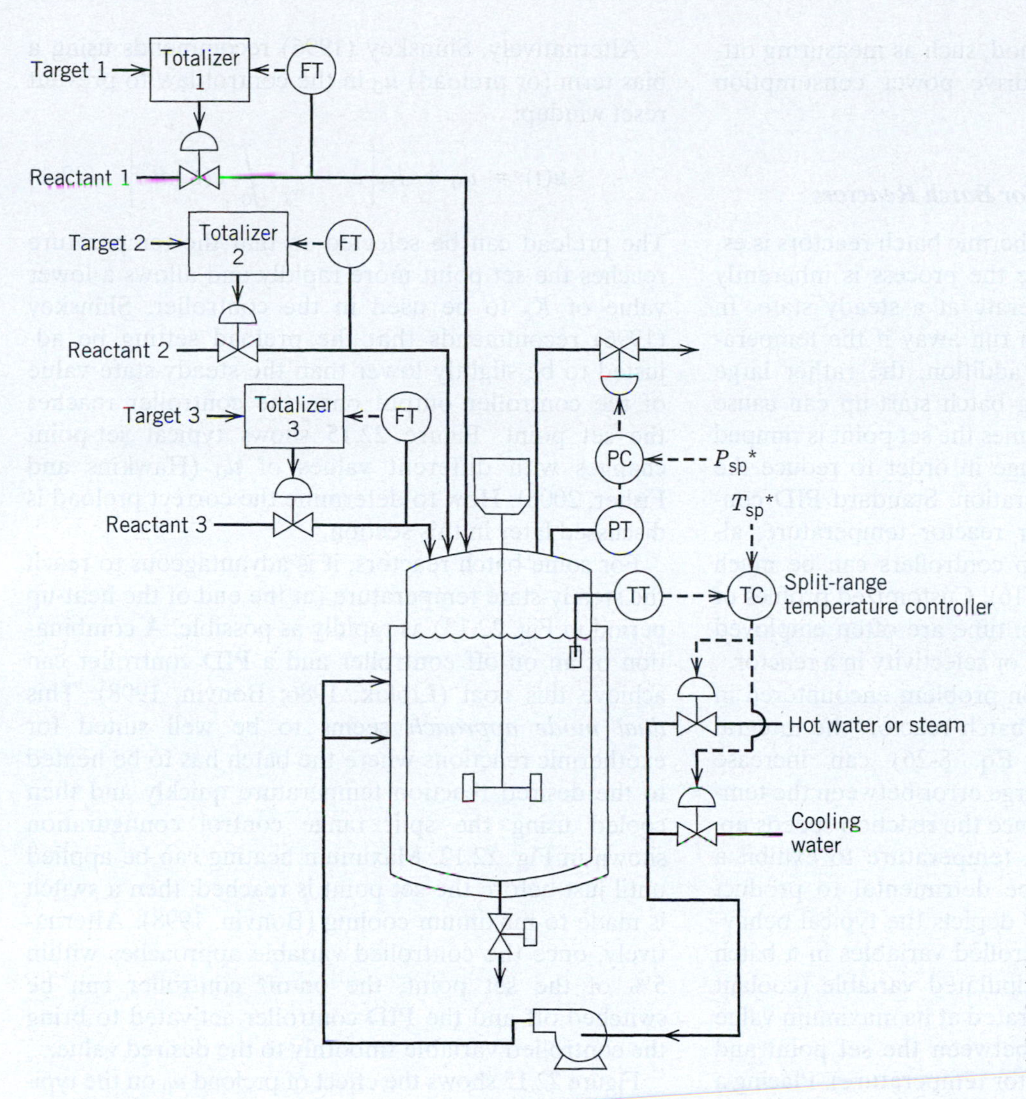

Figure 22.12 Schematic diagram of a batch reactor (* denotes a set point that is a function of time).

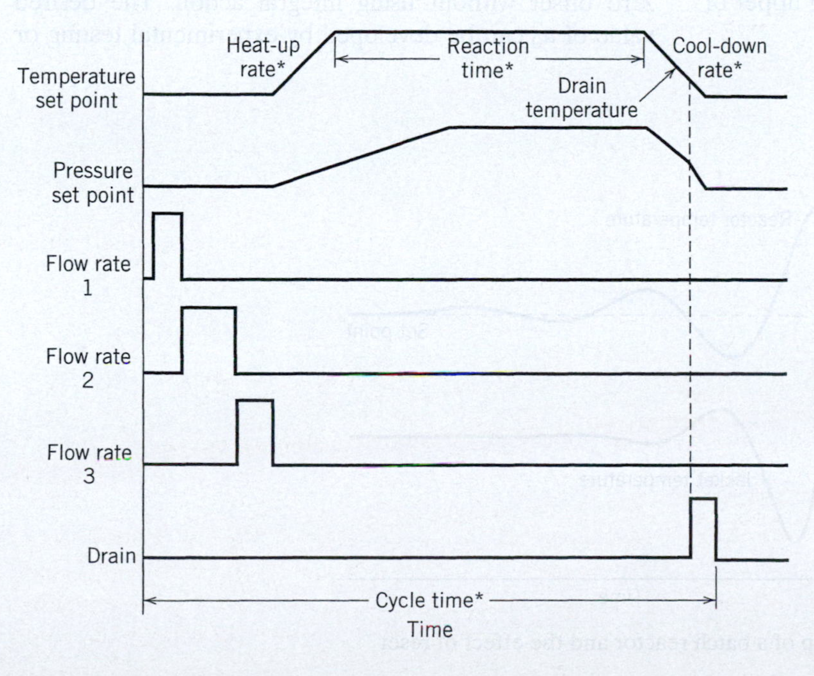

Figure 22.13 Set points and flow sequences of a batch reactor cycle (* denotes externally set function of time).

of indirect (inferential) method, such as measuring off-gas evolution or agitator drive power consumption (torque requirements).

Design of PID Controllers for Batch Reactors

Temperature control of exothermic batch reactors is especially challenging, because the process is inherently nonlinear and does not operate at a steady state. In some cases, the reaction can run away if the temperature becomes too high. In addition, the rather large changes in set points during batch start-up can cause controller saturation. Sometimes the set point is ramped instead of using a step change in order to reduce the possibility of controller saturation. Standard PID control may be satisfactory for reactor temperature, although enhanced single-loop controllers can be much more effective (see Chapter 16). Customized profiles of the temperature set point vs. time are often employed to obtain the maximum yield or selectivity in a reactor.

Reset windup is a common problem encountered in batch process control. In a batch reactor, the integral term (the summation in Eq. 8-26) can increase (windup) as a result of the large error between the temperature and its set point. Once the reaction speeds up, reset windup can cause the temperature to exhibit a large overshoot that may be detrimental to product quality control. Figure 22.14 depicts the typical behavior of manipulated and controlled variables in a batch reactor. Note that the manipulated variable (coolant temperature) is initially saturated at its maximum value because of the large error between the set point and the controlled variable (reactor temperature). Placing a limit on the integral term is a common way of implementing anti-reset windup (see Chapter 8). It acts to keep the controller from saturating by placing upper or lower limits on the summation term.

Alternatively, Shinskey (1996) recommends using a bias term (or preload) u_0 in the control law to prevent reset windup:

$$u(t) = u_0 + K_c\left[e + \frac{1}{\tau_I}\int_0^t e(t^*)dt^*\right] \quad (22\text{-}1)$$

The preload can be selected so that the temperature reaches the set point more rapidly and allows a lower value of K_c to be used in the controller. Shinskey (1996) recommends that the preload setting be adjusted to be slightly lower than the steady-state value of the controller output once the controller reaches the set point. Figure 22.15 shows typical set-point changes with different values of u_0 (Hawkins and Fisher, 2006). How to determine the correct preload is discussed later in this section.

For some batch reactors, it is advantageous to reach the steady-state temperature (at the end of the heat-up period in Fig. 22.13) as rapidly as possible. A combination of an on-off controller and a PID controller can achieve this goal (Lipták, 1986; Bonvin, 1998). This *dual mode approach* seems to be well suited for exothermic reactions where the batch has to be heated to the desired reaction temperature quickly and then cooled using the split range control configuration shown in Fig. 22.12. Maximum heating can be applied until just before the set point is reached; then a switch is made to maximum cooling (Bonvin, 1998). Alternatively, once the controlled variable approaches within 5% of the set point, the on-off controller can be switched off and the PID controller activated to bring the controlled variable smoothly to the desired value.

Figure 22.15 shows the effect of preload u_0 on the typical responses of batch reactor temperature, which is a nonlinear process. u_0 should be ideally selected to achieve zero offset without using integral action. The desired value of u_0 can be developed by experimental testing or

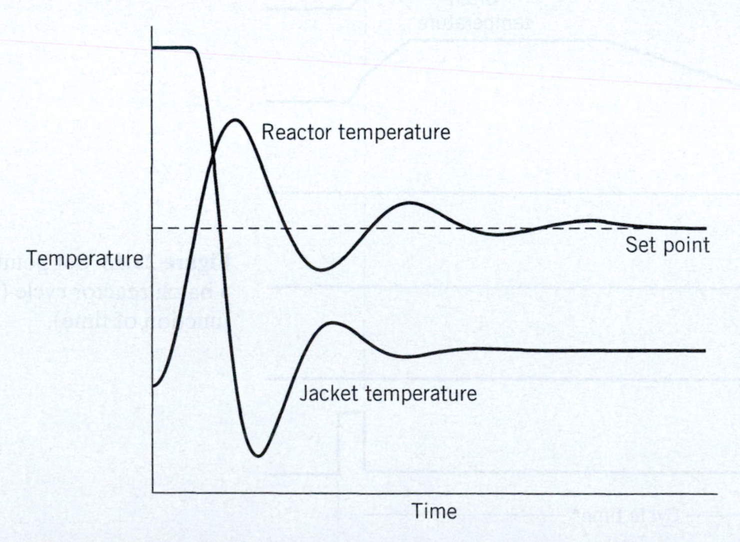

Figure 22.14 Start-up of a batch reactor and the effect of reset windup.

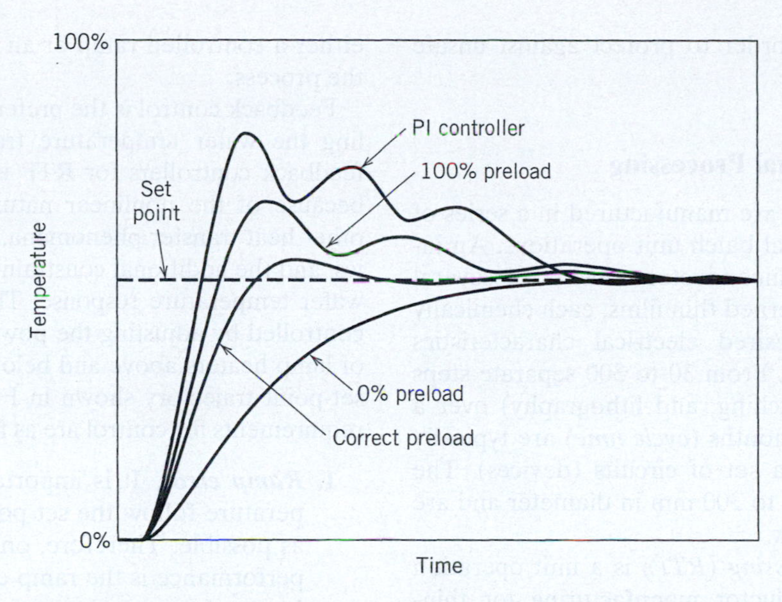

Figure 22.15 Transient responses for batch switch with preload.

from the process model. It is instructive to analyze a linear dynamic model to see how preload eliminates offset with a proportional controller. For a first-order model with gain $K(=K_vK_pK_m)$ and time constant τ,

$$\tau\frac{dy}{dt}+y=Ku \tag{22-2}$$

Assume the proportional controller has preload u_0 and gain K_c:

$$u=u_0+K_c(y_{sp}-y) \tag{22-3}$$

the closed-loop model is

$$\tau\frac{dy}{dt}+y=K\left[(u_0+K_c(y_{sp}-y))\right] \tag{22-4}$$

If $y_{sp}=Ku_0$ at the final steady state and the gain K is known, the appropriate preload expression is $u_0=y_{sp}/K$. Then the closed-loop equation becomes

$$\tau\frac{dy}{dt}+y=K\left[\frac{y_{sp}}{K}+K_c(y_{sp}-y)\right] \tag{22-5}$$

The steady-state solution of (22-5) yields $y_{sp}=y$. Thus, there is no steady-state error with proportional control (K_c can be any value) as long as the process gain is known. K_c can be adjusted in order to tune the speed of the closed-loop response. In theory it is possible to eliminate offset without using integral action, but in practice integral action should be included to deal with model inaccurary and unanticipated disturbances. For nonlinear processes, the preload will not match the required controller output ("100% preload" in Fig. 22.15), thus an adjustment in u_0 must be made ("correct preload" in Fig. 22.15). This controller will normally be superior to the standard PI controller for a batch reactor shown in Fig. 22.15.

Advanced Batch Reactor Control

When the process nonlinearities are significant during the batch transition or start-up, a standard PID controller may not be satisfactory. As discussed in Section 16.6, a gain scheduling or multimodel control approach can be used to deal with excessive nonlinearity. If transfer function models are available for the starting and ending points of the batch trajectory, and model-based PID controllers are available (cf. Chapter 12), then the controller settings can be switched at some point during the set-point change. On the other hand, an adaptive control strategy can be employed, as discussed in Chapter 16. Huzmezan et al. (2002) applied an adaptive control strategy to both a PVC reactor and an ethoxylated fatty acid reactor. In both cases, the variability of the reactor temperatures was reduced by 60% or more.

Juba and Hamer (1986) have described the advantages of using model-based controllers to address the challenges of control of composition or yield when there are highly exothermic reactions (and the potential for a runaway reaction). Typically, three process characteristics must be determined:

1. How steam and/or cooling water affects the reactor temperature
2. How reactor temperature affects reaction chemistry and reaction rates
3. How reaction rate affects heat generation

Simplified nonlinear relationships can be developed based on operating data. When the development of a detailed reaction kinetics model (model 2) is not feasible, a *heat-release-rate estimator* using material balance information can be employed, in effect combining models 2 and 3. It is especially important to understand the sensitivity of the resulting controller to variations in reaction chemistry in order to protect against unsafe conditions.

22.3.2 Rapid Thermal Processing

Semiconductor devices are manufactured in a series of physical and/or chemical batch unit operations. An integrated circuit or semiconductor consists of several layers of carefully patterned thin films, each chemically altered to achieve desired electrical characteristics (Badgwell et al., 1995). From 30 to 300 separate steps (such as deposition, etching, and lithography) over a total duration of two months (*cycle time*) are typically required to construct a set of circuits (devices). The wafers range in size up to 300 mm in diameter and are 400 to 700 microns thick.

Rapid thermal processing (*RTP*) is a unit operation employed in semiconductor manufacturing for thin-film deposition, such as nitridation or oxidation, and for annealing. RTP provides high ramp rates in wafer temperature that lead to short thermal processing times, thus increasing wafer throughput (Edgar et al., 2000). In RTP the wafer temperatures have specified ramp rates and steady-state values. It is imperative that the wafer temperature be controlled precisely to the specified ramp rate and steady-state temperature in order to meet process specifications. The temperature trajectory can be divided into three regions: the ramp, steady state, and cool-down periods shown in Fig. 22.16. The overall duration of the three steps is less than several minutes. In the ramp region, the ramp rates can vary between 25 and 200 °C/s. Steady-state temperatures depend on the RTP process (e.g., oxidation, nitridation, annealing, etc.). Cool-down can be

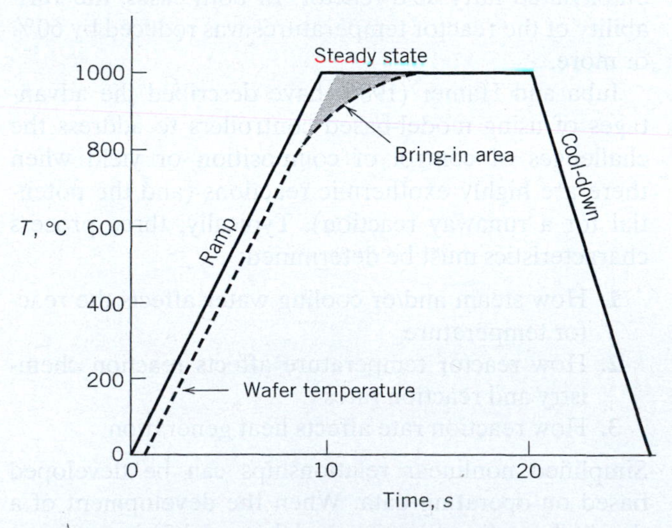

Figure 22.16 RTP set-point change. The solid line is the temperature set point, and the dashed line is the wafer temperature. The wafer temperature has been shifted to the right, for clarity. Consequently, the ramp error exceeds 10 °C.

either a controlled ramp or an uncontrolled cooling of the process.

Feedback control is the preferred method for controlling the wafer temperature trajectory. The design of feedback controllers for RTP is a challenging problem because of the nonlinear nature of the radiative and other heat transfer phenomena, the fast process dynamics, and the additional constraints that are placed on the wafer temperature response. The wafer temperature is controlled by adjusting the power output from an array of lamp heaters above and below the wafer to track the set-point trajectory shown in Fig. 22.16. The three key requirements for control are as follows:

1. ***Ramp error.*** It is important that the wafer temperature follow the set-point trajectory as closely as possible. Therefore, one measure of controller performance is the ramp error, defined as the difference between the set point and wafer temperature at any given time during the ramp. It should be less than 10 °C during the ramp.
2. ***Bring-in.*** As mentioned earlier, it is important that the wafer temperature reach steady state as rapidly as possible. Bring-in is a criterion that indicates controller performance during the transition from ramp to steady state (that is, where the trajectory "turns the corner"). Bring-in is defined as the enclosed area between the desired set-point trajectory (shifted in time, as shown in Fig. 22.16) and the set-point temperature at steady state; bring-in should be minimized. The shifting of the set-point signal takes into account the dynamic error that normally occurs.
3. ***Overshoot.*** The corner of the set-point trajectory should be turned without overshoot.

Balakrishnan and Edgar (2000) evaluated gain-scheduled control of a commercial RTP reactor. They determined that a PID controller based on a semi-empirical model of the heating process provided effective temperature control of the reactor. Derivation of a fundamental heat transfer model based on an unsteady-state energy balance yielded an approximate second-order transfer function with wafer heating time constant τ_w and heating lamp time constant $\tau_L (\tau_L << \tau_w)$.

$$G(s) = \frac{K}{(\tau_w s + 1)(\tau_L s + 1)} \tag{22-6}$$

The controlled variable is the measured wafer temperature and the manipulated variable is the percent power to the heating lamps above and below the wafer. For a series of step tests, the time constants varied with temperature between 650 °C and 1000 °C ($3 \leq \tau_w \leq 8$ s; $0.5 \leq \tau_L \leq 0.7$ s), and the gain K varied from 4 to 12 °C/%.

Application of the Direct Synthesis method (Chapter 12) and Eq. (12-14) yielded the following PID

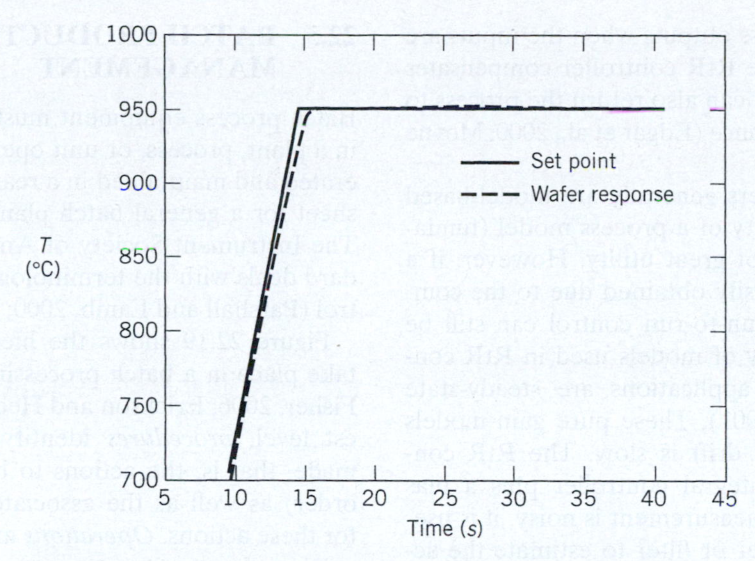

Figure 22.17 Gain-scheduled PID controller results for RTP temperature cycle and a steady-state temperature of 950 °C.

controller parameters, where τ_c is the desired closed-loop time constant:

$$K_c = \frac{1}{K}\frac{\tau_w + \tau_L}{\tau_c} \cong \frac{\tau_w}{\tau_c K} \qquad (22\text{-}7)$$

$$\tau_I = \tau_w + \tau_L \cong \tau_w$$

$$\tau_D = \frac{\tau_w \tau_L}{\tau_w + \tau_L} \cong \tau_L$$

A bias term for the controller was calculated from a physical heat transfer model. It was necessary to include this preload in order to avoid overshoot and to satisfy the ramp error and bring-in requirements. The control algorithm was programmed with gain-scheduled PID parameters determined from the linear models. Because the model gain and time constants varied with temperature, seven gain-scheduled regions were used between room temperature and 1100 °C. Experiments were performed for final steady-state temperatures ranging from 750 to 1050 °C and a ramp rate of 50 °C/s. The closed-loop experimental response for 950 °C is shown in Fig. 22.17.

22.4 RUN-TO-RUN CONTROL

Recipe modifications from one run to the next are common in many batch processes. Typical examples are modifying the reaction time, feed stoichiometry, or reactor temperature. When such modifications are done at the beginning of a run (rather than during a run), the control strategy is called *run-to-run* (*RtR*) *control*. Run-to-run control is frequently motivated by the lack of on-line measurements of the product quality during a batch run. In batch chemical production, on-line measurements are often not available during the run, but the product can be analyzed by laboratory samples at the end of the run (Bonvin, 1998). The process engineer must specify a recipe that contains the values of the inputs (which may be time-varying) that will meet the product requirements. The task of the run-to-run controller is to adjust the recipe after each run to reduce variability in the output product from the stated specifications.

In semiconductor manufacturing, the goal is to control qualities such as film thickness or electrical properties that are difficult, if not impossible, to measure in real-time in the process environment. Most semiconductor products must be transferred from the processing chamber to a metrology tool (measuring device) before an accurate measurement of the controlled variable can be taken. The scope of run-to-run control applications in the semiconductor industry is significant.

Batch run-to-run control can be viewed as implementing a series of set-point changes to the underlying batch process controllers at the end of each run. By analyzing the results of previous batches, the run-to-run controller adjusts the batch recipe in order to reduce quality variations. Thus, run-to-run control is equivalent to controlling a sequence of the controlled variable at times k, $k + 1$, $k + 2, \ldots$, analogous to a standard control problem.

Run-to-run control is particularly useful to compensate for processes where the controlled variable drifts over time. For example, in a chemical vapor deposition process or in a batch chemical reactor, the reactor walls may become fouled due to byproduct deposition. This slow drift in the reactor chamber condition requires occasional changes to the batch recipe in order to ensure that the controlled variables remain on-target. Eventually, the reactor chamber must be cleaned to remove the wall deposits, effectively causing a step disturbance to the process outputs when the inputs are held constant. Just as the RtR controller compensates for the drifting process, it can also return the process to target after a step disturbance (Edgar et al., 2000; Moyne et al., 2001).

Because RtR controllers generally are model-based controllers, the availability of a process model (fundamental or empirical) is of great utility. However, if a dynamic model is not easily obtained due to the complexity of the process, run-to-run control can still be carried out. The majority of models used in RtR control for semiconductor applications are steady-state models (Moyne et al., 2001). These pure gain models assume that the process drift is slow. The RtR controller is typically an integral controller plus a bias term. When the output measurement is noisy, it is useful to employ an observer or filter to estimate the actual process output. In this case, controllers can be designed using the techniques presented in Sections 17.5 and 17.6.

Use of Optimization in Batch Control

A batch trajectory may be changed on a run-to-run basis in order to optimize product yield or selectivity while satisfying process constraints. The best set-point profile can be obtained theoretically using optimal control techniques (Bonvin et al., 2002). An alternative approach uses parameterization of the manipulated variable as a function of the batch time t, for example, $u(t) = a_0 + a_1 t + a_2 t^2$. This type of control law is not based on feedback from the available sensors; parameters a_0, a_1, and a_2 would be adjusted after each batch based on the product quality measurements at the end of the run. This approach is beneficial when unmeasured slow disturbances are encountered, namely, those that do not change much from run to run. Faster-acting disturbances would need to be managed with a feedback control system, as discussed in Section 22.3.

The minimal information needed in carrying out this type of RtR control is a static model relating the manipulated variable to the quality variables at the end of a batch. It can be as simple as a steady-state (constant) gain relationship, or as complicated as a nonlinear model that includes the effects of different initial conditions and the batch time. In contrast, a time-dependent profile for the manipulated variable during the batch can be adjusted from run-to-run to meet the end-of-the-batch quality requirements as well as operating constraints, for example, upper and lower bounds on the manipulated variables (Bonvin et al., 2002). Other variations of RtR controllers in different applications have been reported by Zafiriou et al. (1995), Clarke-Pringle and MacGregor (1998), and Zisser et al., (2005).

22.5 BATCH PRODUCTION MANAGEMENT

Batch process equipment must be properly configured in a plant, process, or unit operation in order to be operated and maintained in a reasonable manner. A flowsheet for a general batch plant is shown in Fig. 22.18. The Instrument Society of America (ISA) SP-88 standard deals with the terminology involved in batch control (Parshall and Lamb, 2000; Strothman, 1995).

Figure 22.19 shows the hierarchy of activities that take place in a batch processing system (Hawkins and Fisher, 2006; Erickson and Hedrick, 1999). At the highest level, *procedures* identify how the products are made, that is, the actions to be performed (and their order) as well as the associated control requirements for these actions. *Operations* are equivalent to unit operations in continuous processing and include such steps as charging, reacting, separating, and discharging. Within each operation are logical points called *phases*, where processing can be interrupted by operator or computer interaction. Examples of different phases include the sequential addition of ingredients, heating a batch to a prescribed temperature, mixing, and so. *Control steps* involve direct commands to final control elements, specified by individual control instructions in software. As an example, for {*operation* = *charge reactant*} and {*phase* = *add ingredient B*}, the control steps would be (1) open the B supply valve, (2) total the flow of B over a period of time until the prescribed amount has been added, and (3) close the B supply valve. Such sequential control operations were discussed in Section 22.2.

The term *recipe* has a range of definitions in batch processing, but in general a recipe is a procedure with the set of data, operations, and control steps required to manufacture a particular grade of product. A *formula* is the list of recipe parameters, which includes the raw materials, processing parameters, and product outputs. A recipe *procedure* has operations for both normal and abnormal conditions. Each *operation* contains resource requests for certain ingredients (and their amounts). The operations in the recipe can adjust set points and turn equipment on and off. The complete production run for a specific recipe is called a *campaign* (multiple batches).

In *multigrade batch processing*, the instructions remain the same from batch to batch, but the formula can be changed to yield modest variations in the product. For example, in emulsion polymerization, different grades of polymers are manufactured by changing the formula. In *flexible batch processing*, both the formula (recipe parameters) and the processing instructions can change from batch to batch. The recipe for each product must specify both the raw materials required and

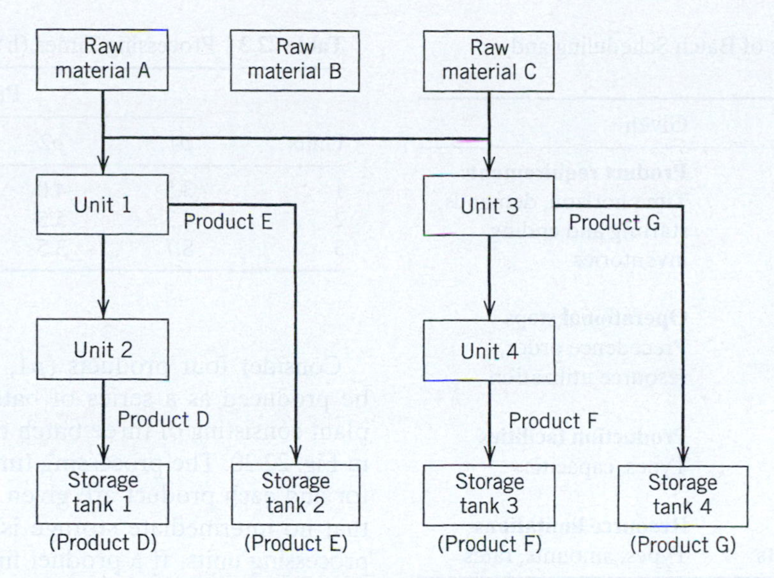

Figure 22.18 Flowsheet for a multiproduct batch plant.

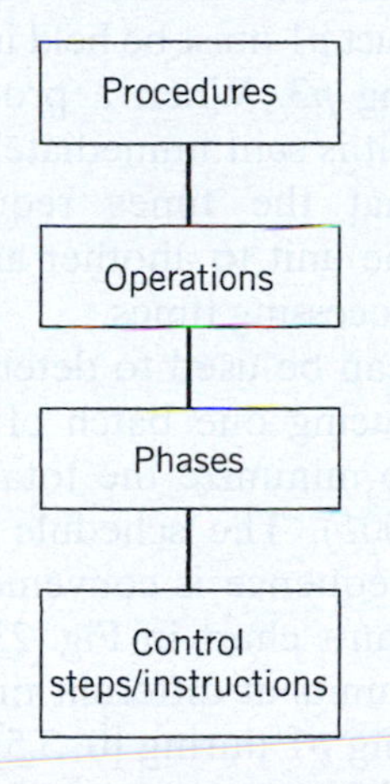

Figure 22.19 Hierarchy of activities in implementing a batch recipe.

how conditions within the reactor are to be sequenced in order to make the desired product.

Many batch plants, especially those used to manufacture pharmaceuticals, are certified by the International Standards Organization (ISO). ISO 9000 (and related ISO standards 9001–9004) state that every manufactured product should have an established, documented procedure, and the manufacturer should be able to document that the procedure was followed. Companies must pass periodic audits to main ISO 9000 status. Both ISO 9000 and the U.S. Food and Drug Administration (FDA) require that only a certified recipe be used. Thus, if the operation of a batch becomes "abnormal," performing any unusual corrective action to bring it back within the normal limits is not an option. In addition, if a slight change in the recipe apparently produces superior batches, the improvement cannot be implemented unless the entire recipe is recertified. The FDA typically requires product and raw materials tracking, so that product abnormalities can be traced back to their sources.

Batch Scheduling and Planning

For recipe management, each batch is tracked as it moves through the production stages, which may involve sequential processing operations on various pieces of equipment. As the batch proceeds from one piece of equipment to the next, recipe management is responsible for ensuring that the proper type of process equipment is used, the specific equipment is not currently being used by another batch, and materials are charged to the correct batch. The complexity in such operations demands that a computer control system be utilized to minimize operator errors and off-specification batches.

A production run typically consists of a sequence of a specified number of batches using the same raw materials and making the same product to satisfy customer demand; the accumulated batches are called a *lot*. When a production run is scheduled, the necessary equipment items are assigned and the necessary raw materials are allocated to the production run. As the individual batches proceed, the consumption of raw materials must be monitored for consistency with the original allocation of raw materials to the production run, because parallel trains of equipment may be involved. A typical scheduling and planning scenario is shown in Table 22.2. Various optimization techniques can be employed to solve the problem, ranging from linear programming to mixed-integer nonlinear programming (Pekny and Reklaitis, 1998; Edgar et al., 2001).

Table 22.2 Characteristics of Batch Scheduling and Planning

Determine	Given
What? Product amounts: lot sizes, batch sizes	**Product requirements** Time horizon, demands, starting and ending inventories
When? Timing of specific operations, run lengths	**Operational steps** Precedence order, resource utilization
Where? Sites, units, equipment	**Production facilities** Types, capacities
How? Resource types and amounts	**Resource limitations** Types, amounts, rates

Source: Pekny and Reklaitis (1998).

When several products are similar in nature, they require the same processing steps and hence pass through the same series of processing units; often the batches are produced sequentially. Because of different processing time requirements, the total time required to produce a set of batches (also called the *makespan* or cycle time) depends on the sequence in which they are produced. To maximize plant productivity, that is, the maximum amounts of each product for the fixed capital investment, the batches should be produced in a sequence that minimizes the makespan. The plant schedule corresponding to such a sequence can be represented graphically in the form of a Gantt chart. The Gantt chart provides a timetable of plant operations showing which products are produced by which units and at what times.

Table 22.3 Processing Times (h) of Products

	Products			
Units	$p1$	$p2$	$p3$	$p4$
1	3.5	4.0	3.5	12.0
2	4.3	5.5	7.5	3.5
3	8.7	3.5	6.0	8.0

Consider four products ($p1$, $p2$, $p3$, $p4$) that are to be produced as a series of batches in a multiproduct plant consisting of three batch reactors in series shown in Fig. 22.20. The processing times for each batch reactor and each product are given in Table 22.3. Suppose that no intermediate storage is available between the processing units. If a product finishes its processing on unit k and unit $k + 1$ is not free because it is still processing a previous product, then the completed product must be kept in unit k until unit $k + 1$ becomes free. As an example, product $p1$ must be held in unit 1 until unit 2 finishes processing $p3$. When a product finishes processing in unit 3, it is sent immediately to product storage. Assume that the times required to transfer products from one unit to another are negligible compared with the processing times.

Optimization can be used to determine the time sequence for producing one batch of each of the four products so as to minimize the total production time (Edgar et al., 2001). The schedule corresponding to this production sequence is conveniently displayed in the form of a Gantt chart in Fig. 22.21, which shows the status of the units at different times. For instance, unit 1 is processing $p1$ during [0, 3.5]. When $p1$ leaves unit 1 at $t = 3.5$ h, it starts processing $p3$. It processes $p3$ during [3.5, 7]. However, it is unable to discharge $p3$

Raw materials → Unit 1 → Unit 2 → Unit 3 → Products $p1, p2, p3, p4$

Figure 22.20 Multiproduct plant.

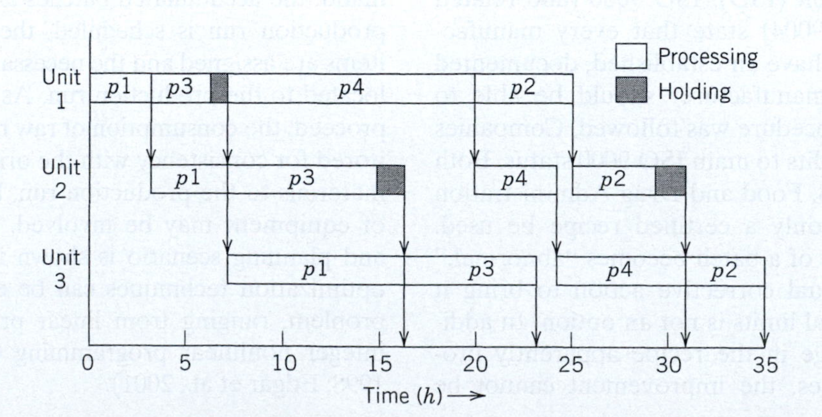

Figure 22.21 Gantt chart for the optimal multiproduct plant schedule.

to unit 2, because unit 2 is still processing $p1$. So unit 1 holds $p3$ during [7, 7.8]. When unit 2 discharges $p3$ to unit 3 at 16.5 h, unit 1 is still processing $p4$; therefore, unit 2 remains idle during [16.5, 19.8]. It is common in batch plants to have units blocked due to busy downstream units, or units waiting for upstream units to finish. This happens because the processing times vary from unit to unit and from product to product, reducing the time utilization of units in a batch plant. The finished batches of $p1$, $p3$, $p4$, and $p2$ are completed at times 16.5, 23.3, 31.3, and 34.8 h. The minimum makespan is 34.8 h.

Many different kinds of planning and scheduling software systems are used in batch processing. Figure 22.22 gives an expanded view of batch scheduling and recipe management, along with the different types of control involved. In the top half of Fig. 22.22, Enterprise Resource Planning (ERP) software provides the following information to the operator console: production planning, equipment scheduling, recipe management, selection of resources and rates, and lot sizing (Erickson and Hedrick, 1999). Typically, the activities are structured hierarchically, with higher-level tasks carried out infrequently to determine the operating conditions and set points that must be addressed by regulatory control as well as by interlocks at the equipment level. Within the scope of a scheduling area (defined as a few units or machines that make a group of products), and a time horizon of hours to day, much greater detail is needed to sequence batches and calculate the exact schedule for operations.

The ability to handle recipe changes after a recipe has started is a challenging aspect of batch control systems. Many times it is desirable to change the grade of the batch to meet product demand, or to change the resources used by the batch after the batch has started. In other cases, the grade is unknown until near the end of the batch, or off-spec laboratory analysis necessitates a change in grade. Because every batch of product is not always good, special-purpose control recipes are needed to fix, rework, blend, or dispose of bad batches, if that is allowable. It is important to be able to respond to unusual situations by creating special-purpose recipes and still meet the demand. This procedure is referred to as *reactive scheduling*.

When ample storage capacity is available, the normal practice has been to build up large inventories of raw materials and ignore the inventory carrying cost. However, improved scheduling can be employed to minimize inventory costs, which implies that *supply chain management* techniques may be necessary to implement the schedule (Pekny and Reklaitis, 1998; Grossmann, 2008).

Movable storage tanks and processing equipment are used in a number of plants. In *flexible manufacturing plants*, flexhoses, manned vehicles, and automated guided vehicles (AGVs) are used to move material between the different groups of equipment (Realff et al., 1996), instead of having permanently installed connections. Even most nonflexible chemical plants have a great number of flexhoses; with an array of pipe headers, cleaned out flexhoses can connect virtually any

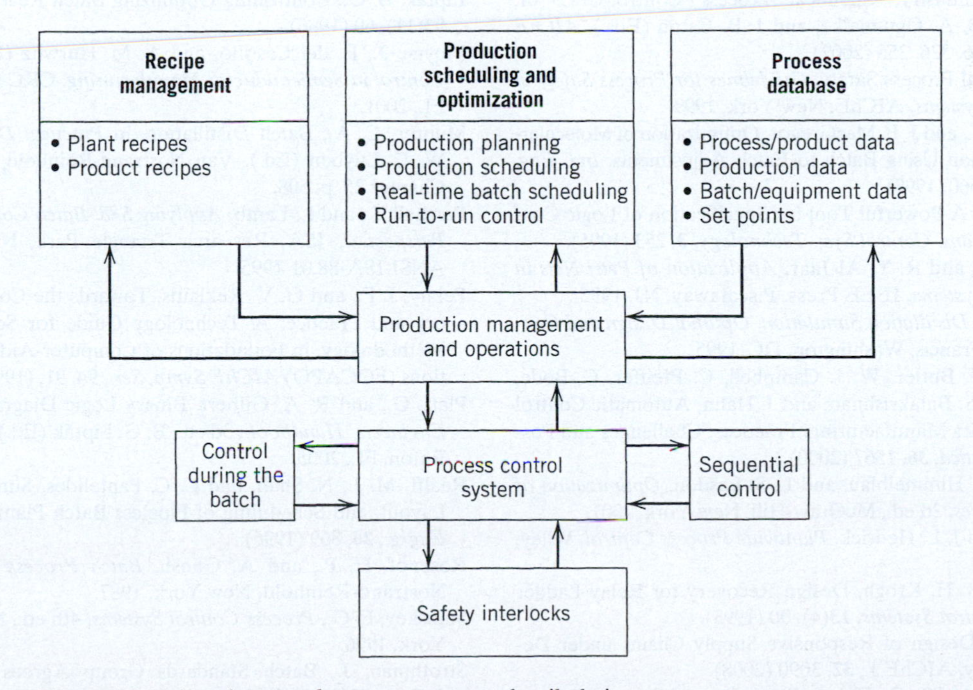

Figure 22.22 Batch control system—a more detailed view.

two pieces of equipment. Some pharmaceutical plants have taken the flexhose concept much further by eliminating all permanent piping. Junctions are made and broken as needed. Although this setup is not economical for large plants with long distances to traverse, it is practical for small plants that make a large number of products. A third approach is to have fixed piping and movable units. The batch reactors and separators are moved when transfers need to be made. If a batch reactor needs to operate for a few hours with no transfers occurring, it can be moved off to the side and left unconnected.

SUMMARY

This chapter has surveyed the broad field of batch process control and emphasized topics and techniques that are unique to batch processing, for example, sequential logic and batch scheduling. Binary logic and ladder logic diagrams, sequential function charts, and Gantt charts are specialized tools that are introduced and applied in this chapter. Batch processes present significant challenges for the design of feedback control systems, especially because of the process nonlinearity that is normally encountered. Various methods to design controllers for batch proccess have been presented. In the future, the chemical industry will rely more heavily on specialty products via batch processing. Many electronic materials and pharmaceutical products are already manufactured by batch processing. Thus, it is important to understand the key concepts in this developing area of process control.

REFERENCES

Arzen, K. E., Grafcet for Intelligent Supervisory Control Applications, *Automatica*, 1513 (1994).

Badgwell, T. A., T. Breedijk, S. G. Bushman, S. W. Butler, S. Chatterjee, T. F. Edgar, A. J. Toprac, and I. Trachtenberg, Modeling and Control of Microelectronics Materials Processing, *Comput. Chem. Engr.*, **19**, 1 (1995).

Balakrishnan, K. S., and T. F. Edgar, Model-Based Control of Rapid Thermal Processing, *J. Thin Solid Films*, **365**, 322 (2000).

Bonvin, D., Optimal Operation of Batch Reactors—A Personal View, *J. Process Control*, **8**, 355 (1998).

Bonvin, D., B. Srinivasan, and D. Ruppen, Dynamic Optimization in the Chemical Industry, Chemical Process Control-CPC VI, J. B. Rawlings, B. A. Ogunnaike, and J. B. Eaton (Eds.), *AIChE Symp. Ser.*, **97**, No. 326, 255 (2002).

Center for Chemical Process Safety, *Guidelines for Process Safety in Batch Reaction Systems*, AIChE, New York, 1999.

Clarke-Pringle, T. L., and J. F. MacGregor, Optimization of Molecular-Weight Distribution Using Batch to Batch Adjustments, *Ind. Eng. Chem. Res.*, **37**, 3660 (1998).

David, R., Grafcet: A Powerful Tool for Specification of Logic Controllers, *IEEE Trans. Control Syst. Technology*, **3**, 253 (1995).

Desrochers, A. A., and R. Y. Al-Jaar, *Application of Petri Nets in Manufacturing Systems*, IEEE Press, Piscataway, NJ, 1995.

Diwekar, U., *Batch Distillation Simulation, Optimal Design and Control*, Taylor and Francis, Washington, DC, 1995.

Edgar, T. F., S. W. Butler, W. J. Campbell, C. Pfeiffer, C. Bode, S. B. Hwang, K. S. Balakrishnan, and J. Hahn, Automatic Control in Microelectronics Manufacturing: Practices, Challenges and Possibilities, *Automatica*, **36**, 1567 (2000).

Edgar, T. F., D. M. Himmelblau, and L. S. Lasdon, *Optimization of Chemical Processes*, 2d ed., McGraw-Hill, New York, 2001.

Erickson, K. T., and J. L. Hedrick, *Plantwide Process Control*, Wiley, New York, 1999.

Falcione, A., and B. H. Krogh, Design Recovery for Relay Ladder Logic, *IEEE Control Systems*, **13**(4), 90 (1993).

Grossmann, I. E., Design of Responsive Supply Chains under Demand Uncertainty, AIChE J., **32**, 3090 (2008).

Hawkins, W. M. J., and T. G., Fisher, *Batch Control Systems: Design, Application, and Implementation* 2d ed., ISA, Research Triangle Park, NC, 2006.

Hughes, T. A., *Programmable Controllers*, 4th ed., ISA, Research Triangle Park, NC, 2005.

Huzmezan, M., B. Gough, and S. Kovac, Advanced Control of Batch Reactor Temperature, *Proc. Amer. Control Conf.*, 1156 (2002).

Johnson, C. D., *Process Control Instrumentation Technology*, 8th ed., Prentice Hall, Upper Saddle River, NJ, 2005.

Juba, M. R., and J. W. Hamer, Progress and Challenges in Batch Process Control, *Chemical Process Control—CPC IV*, M. Morari, and T. J. McAvoy (Eds.) CACHE-Elsevier, Amsterdam, 1986, p. 139.

Lipták, B. G., Controlling Optimizing Batch Reactors, *Chem. Engr.*, **93**(11), 69 (1986).

Moyne, J., E. del Castillo, and A. M. Hurwitz (Eds.), *Run to Run Control in Semiconductor Manufacturing*, CRC Press, Boca Raton, FL, 2001.

Muhrer, C. A., Batch Distillation, in *Practical Distillation Control*, W. L. Luyben (Ed.), Van Nostrand-Reinhold, New York, 1992, Chapter 25, p. 508.

Parshall, J., and L. Lamb, *Applying S88: Batch Control from a User's Perspective,* ISA, Research Triangle Park, NC, 2000; see also ANSI-ISA-88.01-1995.

Pekny, J. F., and G. V. Reklaitis, Towards the Convergence of Theory and Practice: A Technology Guide for Scheduling/Planning Methodology, in Foundations of Computer-Aided Process Operations (FOCAPO) *AIChE Symp. Ser.*, **94**, 91, (1998).

Platt, G., and R. A. Gilbert, Binary Logic Diagrams, in *Instrument Engineers' Handbook*, 3d ed., B. G. Lipták (Ed.), CRC Press, Boca Raton, FL, 2005.

Realff, M. J., N. Shah, and C. C. Pantelides, Simultaneous Design, Layout, and Scheduling of Pipeless Batch Plants, *Comput. Chem. Engrg.*, **20**, 869 (1996).

Rosenof, H. P., and A. Ghosh, *Batch Process Automation*, Van Nostrand-Reinhold, New York, 1987.

Shinskey, F. G., *Process Control Systems*, 4th ed., McGraw-Hill, New York, 1996.

Strothman, J., Batch Standards Group Agrees on Terminology, *InTech*, **42**(8), 31 (1995).

Webb, J. W., and R. A. Reis, *Programmable Logic Controllers*, 5th ed., Prentice Hall, Upper Saddle River, NJ, 2002.

Zafiriou, E., R. Adomaitis, and G. Gattu, An Approach to Run-to-run Control for Rapid Thermal Processing, *Proc. Amer. Cont. Conf.*, 1286 (1995).

Zisser, H., L. Joranorich, F. Doyle III, P. Ogata, and C. Owens, Run-to-Run Control of Meal Related Insulin Dosing, *Diabetes Technol. Ther.*, **7**, 48 (2005).

EXERCISES

22.1 Consider the microwave oven as an example of a discrete state process. The process variables that can be on or off include fan, light, timer, rotating base, microwave generator, and door switch. The process steps include opening the oven door/placing food inside, closing the door, setting the timer, heating up food, and cooking completed. Prepare a table for each process step that shows the process variable status (on or off). What interlocks between variables are important safety issues?

22.2 A pump motor operates by a push button that once actuated keeps the pump on until the operator pushes a stop button. Also there is an emergency stop if the pump overheats ($T > T_H$). Draw a logic diagram using AND/OR symbols. Also draw the equivalent ladder logic diagram.

22.3 A truth table for a set of inputs is shown below where A and B are inputs and Y is an output:

A	B	Y
0	0	1
1	0	1
0	1	0
1	1	1

Construct binary operations (AND, OR, NOT) in series or parallel that yield Y from A and B.

22.4 A batch operation is used to heat a liquid to a specified temperature. There is a start button, a stop button, inlet/outlet valves, and limit sensors for low tank level (LL) and high tank level (LH). Flow is performed by gravity transfer. The process steps are

(a) Push the start button to start the process.

(b) Fill the tank up to LH by opening an inlet valve with the exit line closed.

(c) Heat the liquid to the temperature set point while stirring.

(d) Turn off the stirrer and empty the tank down to LL by opening the exit valve (but the inlet valve must be closed).

Draw an information flow diagram, sequential function chart, and ladder logic diagram.

22.5 Consider a process that consists of a liquid chemical tank with two level indicators, a heater, inlet and outlet pumps, and two valves. Assume that the following sequence of operations are to be performed:

(a) Start the sequence by pressing button S.

(b) Fill the tank with liquid by opening valve V_1 and turning on pump P_1 until the upper level L_1 is reached.

(c) Heat the liquid until the temperature is greater than T_H. The heating can start as soon as the liquid is above level L_0.

(d) Empty the liquid by opening the valve V_2 and turning on pump P_2 until the lower level L_0 is reached.

(e) Close the valves and go to step (a) and wait for a new sequence to start.

Draw an information flow diagram, sequence function chart, and ladder logic diagram.

22.6 A two-tank filling system is shown in Fig. E22.6. Both tanks are used in a similar way. Tank 1 is considered to be empty when the level is less than L_1 and is considered to be full when the level is greater than L_2. Initially, both tanks are empty. If push button S is pressed, both tanks are filled by opening valves V_1 and V_2. When a tank is full (e.g., tank 1), filling stops by closing valve V_1, and its contents start to be used (by opening valve W_1). When tank 1 is empty, valve W_1 is closed. Filling may only start up again when both tanks are empty, and if button S is pressed. Draw an information flow diagram, a sequential function chart, and ladder logic diagram for the system. Use the notation $V_1 = 1$ to denote that valve V_1 is open.

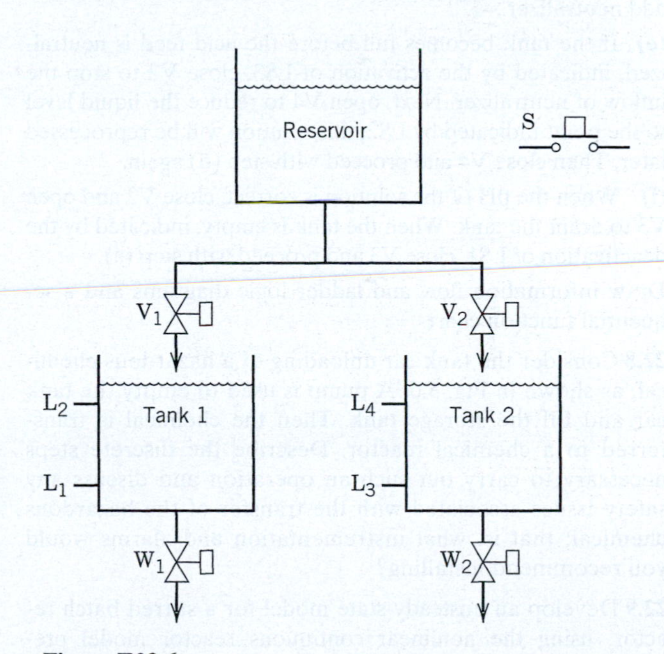

Figure E22.6

22.7 Consider a neutralization system shown in Fig. E22.7, where a certain amount of acid feed is added to a tank, chemically treated, and then sent to the next tank. Sensor pHS indicates whether or not the solution has the correct pH. When pHS is activated, the neutralization is complete. Level switches LS1, LS2, and LS3 are activated when the level in the tank is at or above a given level. The neutralization process proceeds with the following steps:

(a) Initially, all the valves are closed, the mixer is off, and the neutralization tank is empty.

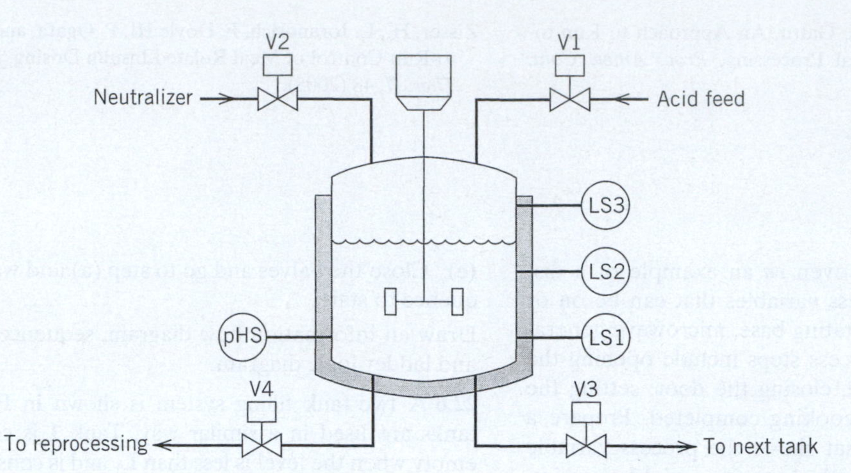

Figure E22.7

(b) When the start button (not shown) is pressed, V1 opens and LS2 is activated. These actions fill the tank with the solution to be neutralized.

(c) When the solution level rises above LS2, start mixer M. When the level drops below LS1, stop the mixer.

(d) Whenever the pH of the solution is too low, open V2 to add neutralizer.

(e) If the tank becomes full before the acid feed is neutralized, indicated by the activation of LS3, close V2 to stop the inflow of neutralizer. Next, open V4 to reduce the liquid level to the point indicated by LS2; this solution will be reprocessed later. Then close V4 and proceed with step (d) again.

(f) When the pH of the solution is correct, close V2 and open V3 to drain the tank. When the tank is empty, indicated by the deactivation of LS1, close V3 and proceed with step (a).

Draw information flow and ladder logic diagrams and a sequential function chart.

22.8 Consider the tank car unloading of a hazardous chemical, as shown in Fig. 5.6. A pump is used to empty the tank car and fill the storage tank. Then the chemical is transferred to a chemical reactor. Describe the discrete steps necessary to carry out such an operation and discuss any safety issues associated with the transfer of the hazardous chemical; that is, what instrumentation and alarms would you recommend installing?

22.9 Develop an unsteady-state model for a stirred batch reactor, using the nonlinear continuous reactor model presented in Example 4.8 as a starting point. For the parameter values given below, compare the dynamics of the linearized models of the batch reactor and the continuous reactor, specifically the time constants of the open-loop transfer function between c'_A and T'_c, the concentration of A, and the jacket temperature, respectively. Assume constant physical properties and the following data:

Initial steady-state conditions and parameter values for the continuous case are

$$\overline{T} = 150\ °\text{F},\quad c_{Ai} = 0.8\ \text{mol/ft}^3,\quad \overline{q} = 26\ \text{ft}^3/\text{min},$$

$$UA = 142.03\ \frac{kJ}{\text{min °F}},\quad V = 1336\ \text{ft}^3,\quad T_c = 77\ °\text{F}$$

The physical property data are

$$C = 0.843\ \text{Btu/lb °F},\quad \rho = 52\ \text{lb/ft}^3,\quad -\Delta H_R = 500\ \text{kJ/mol}.$$

The reaction rate is first order with a rate constant (in min^{-1})

$$k = 2.4 \times 10^{15}\, e^{-20{,}000/T}\ (T \text{ in °R}).$$

For the batch case, linearize the model around $T = \overline{T}$.

22.10 A batch reactor converts component A into B, which in turn decomposes into C:

$$A \xrightarrow{k_1} B \xrightarrow{k_2} C$$

where $k_1 = k_{10}e^{-E_1/RT}$ and $k_2 = k_{20}e^{-E_2/RT}$.

The concentrations of A and B are denoted by x_1 and x_2, respectively. The reactor model is

$$\frac{dx_1}{dt} = -k_{10}x_1e^{-E_1/RT}$$

$$\frac{dx_2}{dt} = k_{10}x_1e^{-E_1/RT} - k_{20}x_2e^{-E_2/RT}$$

Thus, the ultimate values of x_1 and x_2 depend on the reactor temperature as a function of time. For

$$k_{10} = 1.335 \times 10^{10}\ \text{min}^{-1},\quad k_{20} = 1.149 \times 10^{17}\ \text{min}^{-1}$$

$$E_1 = 75{,}000\ \text{J/g mol},\quad E_2 = 125{,}000\ \text{J/gmol}$$

$$R = 8.31\ \text{J/(gmol } K),\quad x_{10} = 0.7\ \text{mol/L},\quad x_{20} = 0$$

Find the constant temperature that maximizes the amount of B, for $0 \le t \le 8$ min. Next allow the temperature to change as a cubic function of time

$$T(t) = a_0 + a_1t + a_2t^2 + a_3t^3$$

Find the values of a_0, a_1, a_2, a_3 that maximize x_2 by integrating the model and using a suitable optimization method.

22.11 Suppose a batch reactor such as the one in Fig. 22.12 has a gas ingredient added to the liquid feed. As long as the reaction is proceeding normally, the gas is absorbed in the liquid (where it reacts), keeping the pressure low. However, if the reaction slows or the gas feed is greater than can be absorbed, the pressure will start to rise. The pressure rise can be compensated by an increase in liquid feed, but this may cause the cooling capacity to be exceeded. Describe a solution to this problem using overrides (see Chapter 16).

22.12 Fogler[2] describes a safety accident in which a batch reactor was used to produce nitroanaline from ammonia and o-nitro chlorobenzene. On the day of the accident, the feed composition was changed from the normal operating value. Using the material/energy balances and data provided by Fogler, show that the maximum cooling rate will not be sufficient to prevent a temperature runaway under conditions of the new feed composition. Use a simulator to solve the model equations.

22.13 Consider the batch reactor system simulated by Aziz et al.[3] The two reactions, $A + B \rightarrow C$ and $A + C \rightarrow D$, are carried out in a jacketed batch reactor, where C is the desired product and D is a waste product. The manipulated variable is the temperature of the coolant in the cooling jacket. There are two inequality constraints: input bounds on the coolant temperature and an upper limit on the maximum reactor temperature. Using the model parameters specified by Aziz et al., evaluate the following control strategies for a set-point change from 20 °C to 92 °C.

(a) PID controller

(b) Batch unit

(c) Batch unit with preload

(d) Dual-mode controller

---

[2] *Elements of Chemical Reaction Engineering*, 4th ed., Prentice Hall, Upper Saddle River, NJ, 2005, Chapter 9.

[3] N. Aziz, M. A. Hussain, and I. M. Mujtaba, Performance of Different Types of Controllers in Tracking Optimal Temperature Profiles in Batch Reactors, *Comput. Chem. Eng*, **24**, 1069 (2000).

Chapter 23

Biosystems Control Design

CHAPTER CONTENTS

Previous chapters have introduced the concepts of process dynamics and strategies for process control, emphasizing traditional applications from the petrochemical industries, such as chemical reactors and distillation columns. In this chapter, we introduce the application fields of bioprocessing and biomedical devices, and illustrate the characteristics that these processes share with traditional chemical processes. Differences will also be highlighted, including the nature of uncertainty in biological processes, as well as the safety considerations in medical closed-loop systems. Control system design for three bioprocessing operations is described: crystallization, fermentation, and granulation. Finally, a number of problems in controlled drug delivery are reviewed, and control strategies are demonstrated in the areas of diabetes and blood pressure regulation. Biological applications are expanded in Chapter 24, with a discussion of control systems opportunities, including applications to systems biology.

23.1 PROCESS MODELING AND CONTROL IN PHARMACEUTICAL OPERATIONS

A typical flowsheet in the pharmaceutical industry contains many of the same categories of operations as occur in a traditional petrochemical processing plant: reactors to generate products from raw materials, purification steps to extract desired products from the by-products and unreacted feed materials, and downstream processing associated with the final formulation of the product. Pharmaceutical processes are unique in several respects: (1) the main reactions involve biological materials, such as cells and tissues from more complex organisms, and (2) most of the products are formulated in solid form, which requires a unique set of bulk solids processing steps to purify and formulate the desired end product (e.g., a medicinal tablet). Consequently, the upstream processing involves sterilization and fermentation, and the manipulated inputs for the reactor often include "inducers" to activate the expression of particular genes in microbes in the reactor (gene expression is covered in more detail in Chapter 24). The downstream section of the flowsheet includes crystallization or chromatographic purification, to extract a high-purity product with desirable properties (e.g., chirality). Subsequent steps may involve solids handling and processing to produce final particulates with desirable properties, including dissolution attributes and tableting capability. These processes include mixing, classification, milling, grinding, crushing, granulation (agglomeration), tableting, and coating. Each of these operations has its own challenges and unique dynamic characteristics.

In the following sections, we consider three of the main processing steps in the pharmaceutical flowsheet: fermentation, crystallization, and granulation. Several of these processes appear in other industries as well (e.g., food, semiconductor, and specialty chemical), so the process control methods described find broad application in industry. It is important to note that the industry has a new emphasis on process systems engineering methods, driven by changes in FDA regulations (see PAT discussion in chapter 22).

23.1.1 Bioreactors

Fermentation reactors are widely used in the pharmaceutical industries to make an array of important compounds, including penicillin, insulin, and human growth hormone. In recent years, genetic engineering has further expanded the portfolio of useful products that can be synthesized using fermentation methods (Buckland, 1984; Lim 1991; Schügerl, 2001). Despite the importance of this unit operation, the state of pharmaceutical fermentation operations is often characterized as more art than science, as with the winemaking industry (Fleet, 1993; Alford, 2006). Since 1990, there has been a focused effort to develop more sophisticated control architectures for fermentation operation, driven by the availability of new technologies for monitoring the quality of the contents of the fermentor (see, for example, Boudreau and McMillan, 2007). A schematic of a fermentation process is given in Figure 23.1.

In a general sense, fermentation involves the generation of cell mass (product) from a substrate according to a simple reaction:

$$aC_\alpha H_\beta O_\gamma + bO_2 + cNH_3 \rightarrow C_\delta H_\varepsilon O_\zeta N_\eta + dCO_2 + eH_2O \tag{23-1}$$

where $C_\alpha H_\beta O_\gamma$ is the substrate (reactant) and $C_\delta H_\varepsilon O_\zeta N_\eta$ is the cell mass(product). For example in beer making, the substrate is glucose (derived in the wort from grains), and the products are the alcohol and carbon dioxide gas, both of which contribute to the quality of the final product. This apparently straightforward reaction is complicated by the fact that it does not obey simple mass action kinetics; instead, the complex biochemistry underlying the reaction gives rise to unusual nonlinear rate expressions that characterize the enzymatic processes.

A simple dynamic model of a fed-batch bioreactor was given in Chapter 2, Eqs. 2-98 to 2-101. This model can be converted into mass balances on individual components as follows:

$$\frac{dX}{dt} = \mu(S)X - \frac{F}{V}X \tag{23-2}$$

$$\frac{dP}{dt} = Y_{P/X}\,\mu(S)X - \frac{F}{V}P \tag{23-3}$$

$$\frac{dS}{dt} = \frac{F}{V}(S_f - S) - \mu(S)X/Y_{X/S} \tag{23-4}$$

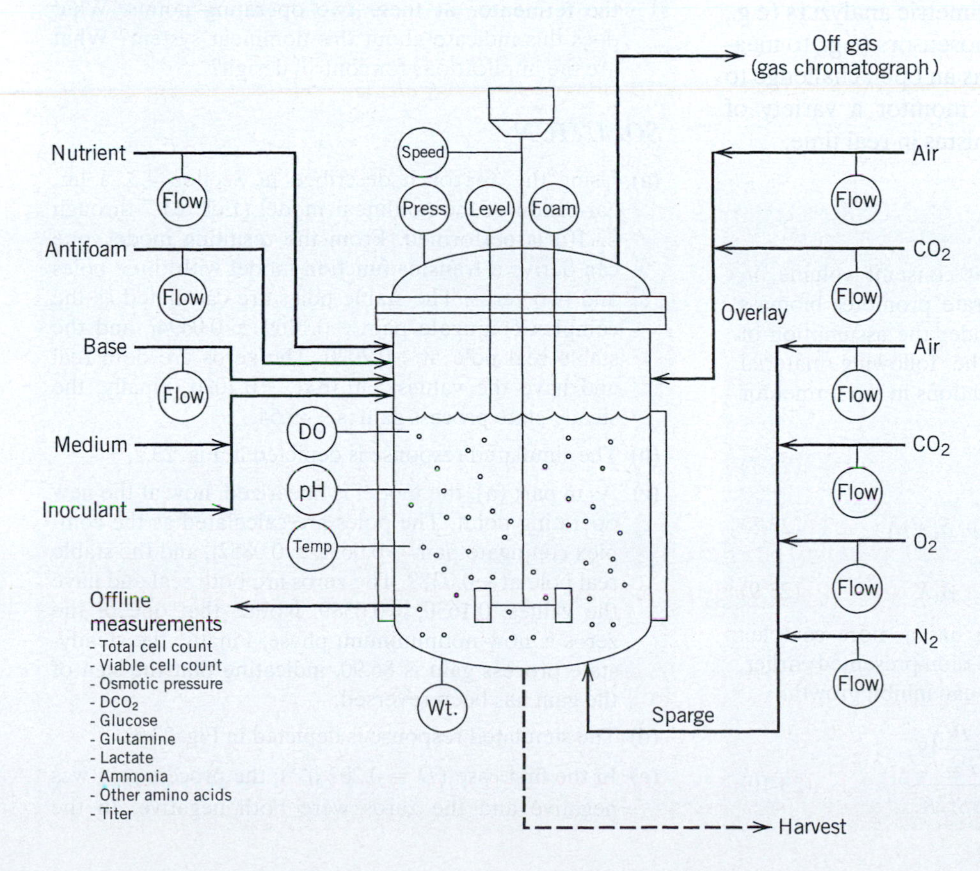

Figure 23.1 Schematic of a typical industrial fermentor. (Figure from Jon Gunther, PhD Thesis, Dept of Chemical Eng., UCSB, 2008).

$$\frac{dV}{dt} = F \qquad (23\text{-}5)$$

where the material balance in (23-2) details the conservation of biomass (X), Eq. (23-3) describes the production of metabolites by the cells (biomass), Eq. (23-4) details the conservation of substrate (S), and Eq. (23-5) is the overall material balance. The ratio F/V is often denoted as the dilution rate, D. The constants that appear in this equation include the feed concentration of the substrate (S_f), the yield of cell mass from substrate ($Y_{X/S}$), and the product yield coefficient ($Y_{P/X}$). The rate of the biochemical reaction, $\mu(S)$, typically utilizes Monod kinetics given by the saturating function:

$$\mu(S) = \frac{\mu_m S}{K_m + S} \qquad (23\text{-}6)$$

where μ_m is the maximum specific growth rate (limiting value of the rate), and K_m is the substrate saturation constant. Control of this simple reactor involves manipulating the influx of substrate (via the dilution rate, D) to achieve an optimal level of production. More sophisticated control inputs are also possible, including inducers that stimulate the transcription of key genes in the microorganisms, leading to the synthesis of enzymes that maximize product yield.

One of the primary challenges to controlling these reactors in industry is the difficulty in measuring the status of the microorganisms in the fermentor. Specialized sensors (Mandenius, 1994) include enzyme electrodes (e.g., to measure glucose, lactate), calorimetric analyzers (e.g., to measure penicillin), and immunosensors (e.g., to measure antigens). However, it remains an open challenge to develop *in situ* sensors that can monitor a variety of metabolites within the microorganisms in real time.

EXAMPLE 23.1

Consider a fermentor, operated at constant volume, in which a single, rate-limiting substrate promotes biomass growth and product formation. Under the assumption of constant yield, one can derive the following material balances that describe the concentrations in the fermentor (Henson and Seborg, 1991):

$$\dot{X} = -DX + \mu(S, P)X \qquad (23\text{-}7)$$

$$\dot{S} = D(S_f - S) - \frac{1}{Y_{X/S}}\mu(S, P)X \qquad (23\text{-}8)$$

$$\dot{P} = -DP + [\alpha\mu(S, P) + \beta]X \qquad (23\text{-}9)$$

For this reactor, the growth term has a more complex shape than the simple Monod expression presented earlier, because both substrate and product can inhibit growth:

$$\mu(S, P) = \frac{\mu_m\left(1 - \dfrac{P}{P_m}\right)S}{K_m + S + S^2/K_i} \qquad (23\text{-}10)$$

The variables X, S, and P are the biomass, substrate, and product concentrations, respectively; D is the manipulated variable (dilution rate); S_f is the feed substrate concentration, and the remaining variables are fixed constants (yield parameters). Take the following values for the fixed parameters:

Table 23.1 Parameter values and units for fermentor in Example 23.1

$Y_{X/S}$	0.4 g/g	α	2.2 g/g
β	0.2 h^{-1}	μ_m	0.48 h^{-1}
P_m	50 g/L	K_m	1.2 g/L
K_i	22 g/L	S_f	20 g/L

(a) Assume a nominal operating point is $D = 0.202$ h^{-1} (dilution rate). The corresponding steady-state or equilibrium values of X, S, and P are [6.0 g/L; 5.0 g/L; 19.14 g/L]. Calculate the linearized model at this operating point, and determine the poles, zeros, and steady-state gain.

(b) Simulate the biomass X response to ±10% relative changes in dilution rate.

(c) Next, change the nominal dilution rate to $D = 0.0389$ h^{-1}. The corresponding equilibrium values of X, S, and P are [6.0 g/L; 5.0 g/L; 44.05 g/L]. (Does anything look unusual here?). Recalculate the linearized model at this operating point, as well as the poles, zeros, and steady-state gain.

(d) Simulate the biomass response to ±10% relative changes in dilution rate.

(e) Comment on the extreme differences in behavior of the fermentor at these two operating points. What does this indicate about this nonlinear system? What are the implications for control design?

SOLUTION

(a) Using the approach described in Section 4.3, a linearization of the nonlinear model (Eqs. 23-7 through 23-10) is performed. From the resulting model, one can derive a transfer function model with three poles and two zeros. The stable poles are calculated as the complex conjugate pair, −0.1469 ± 0.0694j, and the stable real pole at −0.2020. The zeros are both real and have the values: −0.1631, −0.2020. Finally, the steady-state process gain is −39.54.

(b) The simulated response is depicted in Fig. 23.2.

(c) As in part (a), the model is linearized, now at the new operating point. The poles are calculated as the complex conjugate pair, −0.0632 ± 0.0852j, and the stable real pole at −0.0389. The zeros are both real and have the values: 0.1630, −0.0389. Notice that one of the zeros is now nonminimum phase. Finally, the steady-state process gain is 86.90, indicating that the sign of the gain has been reversed.

(d) The simulated response is depicted in Fig. 23.3.

(e) In the first case ($D = 0.202$ h^{-1}), the process gain was negative and the zeros were both negative. In the

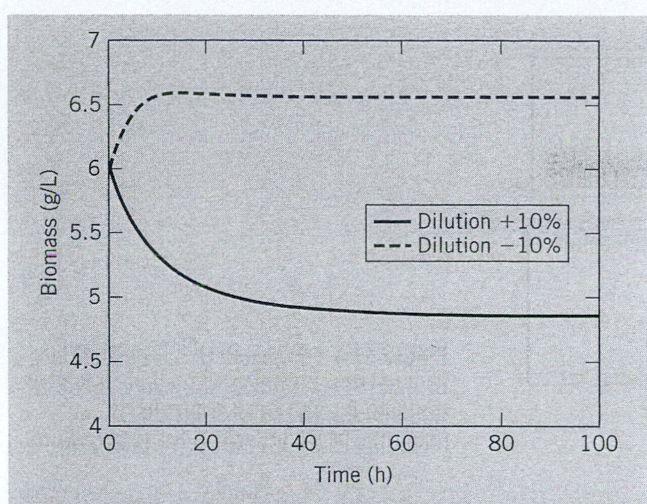

Figure 23.2 Step response of fermentor model to symmetric changes in D of magnitude 10% from the nominal value of $D = 0.202\ h^{-1}$.

second case ($D = 0.0389\ h^{-1}$), the process gain was positive and one of the zeros becomes nonminimum phase, exhibiting inverse response. This suggests that the fermentor exhibits a dramatic nonlinearity, in which the gain can change sign and process zeros can change from negative to positive (indeed, a plot of the steady-state relationship between the dilution rate and the biomass for this fermentor reveals a parabolic shape; also see Exercise 2.15). This suggests that operation across this gain change requires a nonlinear controller or an adaptive control scheme.

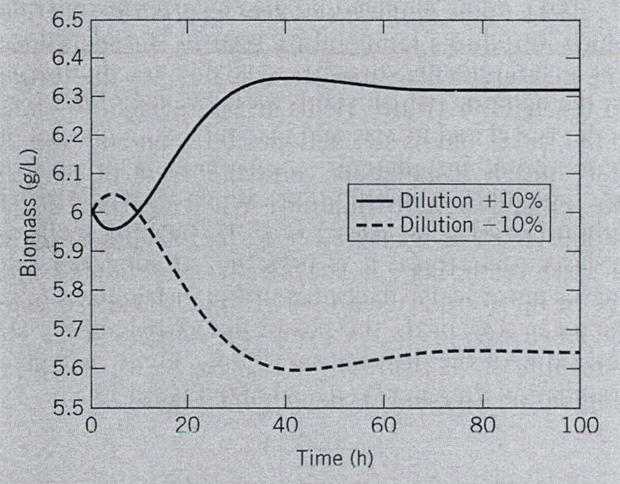

Figure 23.3 Step response of fermentor model to symmetric changes in dilution of magnitude 10% from the nominal value of $D = 0.0389\ h^{-1}$.

23.1.2 Crystallizers

The operation of crystallization allows the separation of one phase (in this case, the solid from a solution mixture) so that the product has desirable properties. The solid product that results from crystallization is a highly ordered solid structure, which may have other desirable attributes, including morphology (e.g., shape), that are of direct benefit to the value of the final product. In the pharmaceutical industry, crystal size and shape may facilitate downstream solids processing and/or may be directly related to the final drug formulation, such as bioavailability, shelf life, toxicity, and drug dissolution (Fujiwara et al., 2005). Crystallization also finds application in the food industry to improve taste, as well as shelf life, for a diverse range of products (Larsen et al., 2006).

In order to explain the process control strategies employed in the operation of an industrial crystallizer, it is important to review briefly the concept of supersaturation and its relevance to crystallization (Larsen et al., 2006). Saturation refers to the property of phase equilibrium, in this case the equilibrium between the liquid and the dissolved solid (i.e., the solubility of the solid in the liquid). The state of supersaturation refers to the condition in which the liquid solution contains more solid than the amount that corresponds to the solubility (equilibrium), and the system exists in a so-called metastable state. Crystal formation can be induced by changing the operating conditions, such as temperature, so that the supersaturation state cannot be sustained, and a crystal is nucleated, or created, from the solution. As the dissolved component moves from the solution to the solid crystal phase, the concentration is, of course, lowered. Once a crystal is formed, it continues to grow as a function of the operating temperature and the concentration in the solution. In effect, the operation of crystallization involves the manipulation of this supersaturation state, trading off the formation of new crystals against the growth of existing ones.

An industrial crystallizer is operated typically in a batch mode, so that the management of the supersaturation state is accomplished over the course of the batch cycle time. The available manipulated inputs are the cooling jacket and steam flow rate, for temperature management, and the inflow of antisolvent (a component that lowers the solvation capability of the liquid) and solvent, to regulate the concentration of the solution (Zhou et al., 2006). A typical crystallization flowsheet is depicted in Fig. 23.4. Measurement of temperature is straightforward, and there are an increasing number of sophisticated instruments available for measurement of the crystal properties. These include turbidity sensors (to detect presence of solids), laser scattering instruments (to extract the distribution of crystal sizes in the unit), and spectroscopic instruments, e.g., attenuated total reflectance-Fourier transformed infrared (ATR-FTIR), for measuring solution concentrations (Fujiwara et al., 2005; Larsen et al., 2006). More recently, a variety of imaging techniques have been used to measure crystal-shape properties (morphology), such as width and length. As mentioned earlier, these size and shape

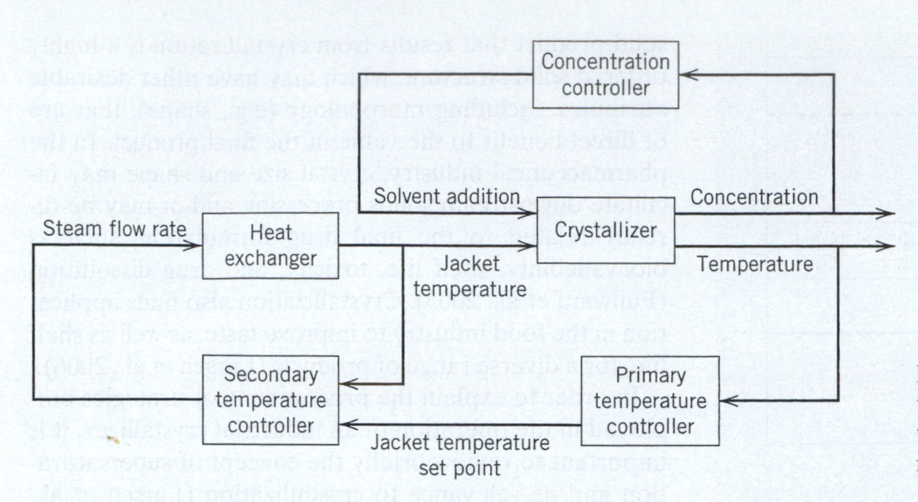

Figure 23.4 Flowsheet of a typical industrial batch crystallizer, showing concentration and temperature controllers, including cascade control for temperature.

properties are major determinants for the resulting utility of the product (e.g., drug solubility), and it may be desirable to produce crystals with very uniform properties (i.e., narrow size distribution).

23.1.3 Granulation

Granulation is a widely used process in which small particles agglomerate into larger granules. In wet granulation processes, the coagulation of particles is improved by the addition of a binder liquid, sprayed over an agitated powder in a tumbling drum or pan. The particles are wetted by the binder and a nucleate. The resulting binder-coated granules then collide and stick to form larger granules. These granules can also compact and consolidate as the binder liquid is brought to the surface of the aggregates by stirring in the granulator. Particles can also break because of collisions with the other particles or the granulator walls during mixing. Thus, the main phenomena in granulation processes are granule wetting and nucleation, consolidation and growth, and aggregation and breakage (Mort et al., 2001).

Granulation plays a key role in producing particles with special characteristics, such as time-release attributes (e.g., fertilizer, pharmaceutical tablet). However, in practice, inefficient operation, with very small yields and large recycle ratios (typically 4:1, recycle:product), often occurs. This inefficiency is due to the difficulty in designing and controlling granulation circuits that allow maintenance of specified size ranges for the granules.

As for crystallization, the key challenges for controlling a granulator are to produce particles with desirable attributes, to simplify downstream processing, and to realize end-product properties. In the pharmaceutical industry, granulation is usually accomplished in batch reactors, owing to the relatively small amount of material throughput. The key particle process that must be regulated is the agglomeration of smaller particles into larger particles. Manipulated inputs include binder spray addition (and/or viscosity), particle flow rate, recycle of oversize (crushed) and undersize (fines) particles, and changing the rate of agitation (mixing) in the vessel (Pottmann et al., 2000; Mort et al., 2001). Some applications also incorporate heating, which introduces temperature control considerations. The measurements currently available are the torque on the agitator (which yields an inference of the load in the vessel and its size and moisture content) and, in more recent installations, measurements of particle size (possibly as a distribution). When a particle distribution (PSD) is measured (e.g., by imaging methods or laser scattering), it is typically consolidated into one or more scalar measures of the distribution (e.g., the mean size, or d_x, the size of the particle in the xth percentile of the distribution (d_5, d_{90}, etc.)). A typical granulation flowsheet is depicted in Figure 23.5.

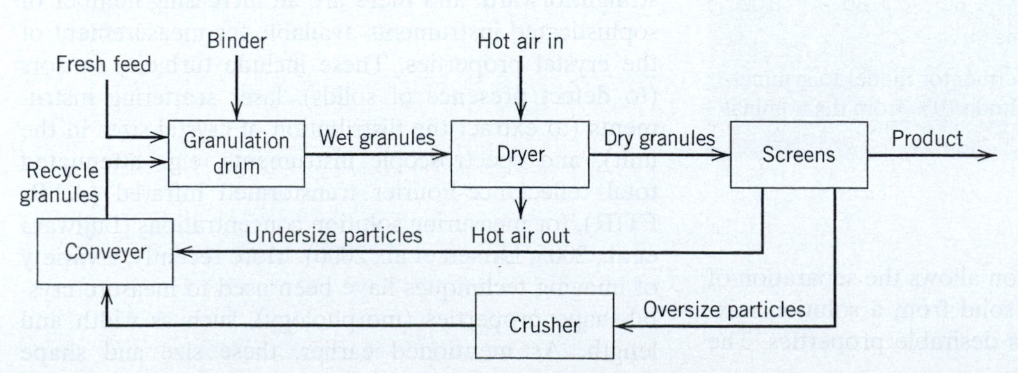

Figure 23.5 Process flowsheet for granulation circuit with recycle.

EXAMPLE 23.2

A simplified granulation flowsheet (Pottmann et al., 2000) is shown in Figure 23.6. The manipulated inputs are the liquid flow rates of binder introduced in three different nozzles, and the measured controlled variables are the bulk density and the 5th and 90th percentiles of the particle size distribution (d_5 and d_{90}, respectively). Pottmann et al. (2000) identified a first-order-plus time-delay model for each combination of inputs and outputs (3 × 3 problem) with the following parameters (time units are dimensionless):

$$G_{ij}(s) = \frac{K_{ij}}{\tau_{ij}s + 1} e^{-\theta_{ij}s} \tag{23-11}$$

$$K_{ij} = \begin{bmatrix} 0.20 & 0.58 & 0.35 \\ 0.25 & 1.10 & 1.30 \\ 0.30 & 0.70 & 1.20 \end{bmatrix} \tag{23-12}$$

$$\tau_{ij} = \begin{bmatrix} 2 & 2 & 2 \\ 3 & 3 & 3 \\ 4 & 4 & 4 \end{bmatrix} \tag{23-13}$$

$$\theta_{ij} = \begin{bmatrix} 3 & 3 & 3 \\ 3 & 3 & 3 \\ 3 & 3 & 3 \end{bmatrix} \tag{23-14}$$

The units for both the manipulated inputs and the measurements are dimensionless, and the nominal conditions (on which deviation variables are based) are: 180 for all three nozzles, 40 for bulk density, 400 for d_5, and 1600 for d_{90}.

(a) Using the relative gain array (RGA), determine the most effective pairings between the inputs and the outputs.

(b) Design three PI + Smith predictor controllers using the IMC design method (see Table 12.1, and assume that a value of $\tau_c = 5$ is employed). Keep in mind that the IMC/PI tuning is for the delay-free part of the plant with a Smith predictor.

(c) Simulate the system response for the three PI + Smith predictor controllers using a step set-point change of [10 0 0]. Be sure to enforce the constraints on the manipulated variables (lower bound of 105; upper bound of 345). Repeat the simulation for a step set-point change of [50 0 0].

SOLUTION

(a) The relative gain array is calculated as detailed in Section 18.2 from the process gains:

$$RGA = K \otimes (K^{-1})^T$$

$$= \begin{bmatrix} 1.0256 & 0.6529 & -0.6785 \\ -1.4103 & 1.8574 & 0.5528 \\ 1.3846 & -1.5103 & 1.1257 \end{bmatrix} \tag{23-15}$$

Hence, the diagonal pairing of the controllers is recommended (1-1/2-2/3-3), because there are no negative values, and two of the three loops are paired on *RGA* values very close to 1.

(b) The individual controllers are given calculated from Table 12.1, row A:

Loop 1: $K_c = (2/5)/.2 = 2.0$; $\tau_I = 3$

Loop 2: $K_c = (3/5)/1.1 = 0.545$; $\tau_I = 4$

Loop 3: $K_c = (4/5)/1.2 = 0.667$; $\tau_I = 5$

(c) The simulation results are shown in Fig. 23.7 and 23.8. Note that enforcing the constraints for the second case leads to an unattainable set point for the first output.

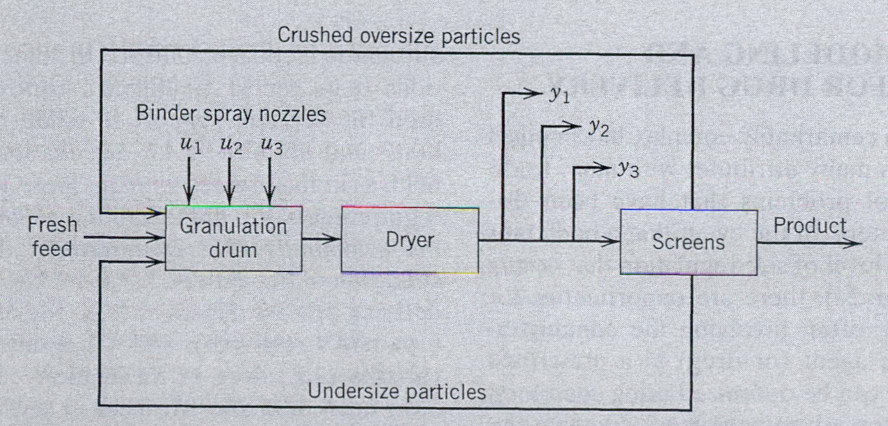

Figure 23.6 Simplified process flowsheet for granulator example. Here u_1, u_2, and u_3 are, respectively, nozzles 1, 2, and 3, and y_1, y_2, and y_3 are, respectively, bulk density, d_5, and d_{90}.

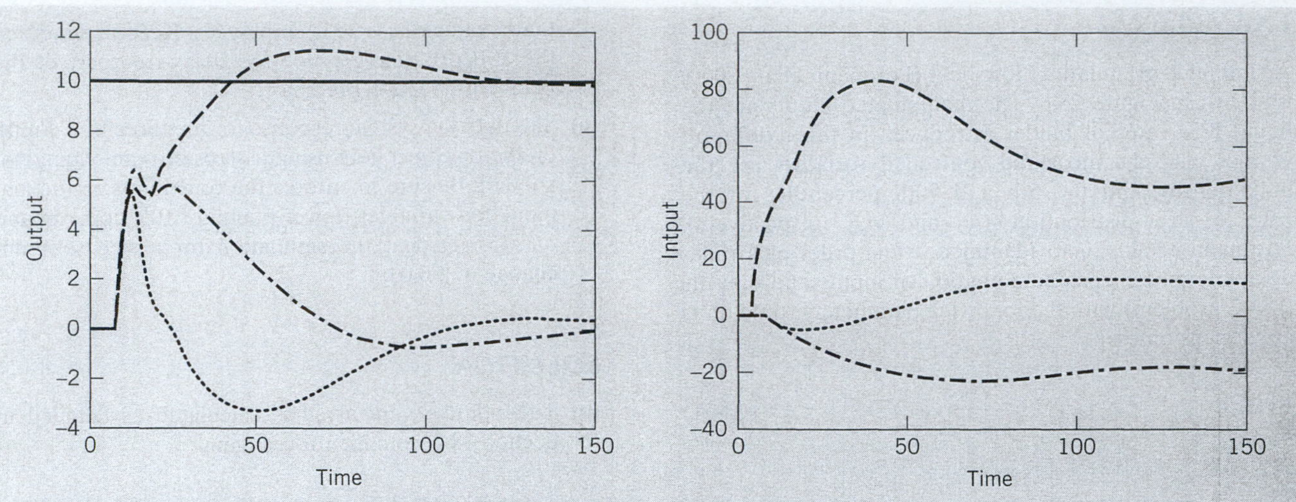

Figure 23.7 Closed-loop response of granulator to +10 step change in set point for y_1: left plot is CVs, right plot is MVs (----, y_1 and u_1; -.-.-.-., y_3 and u_3; ...., y_2 and u_2).

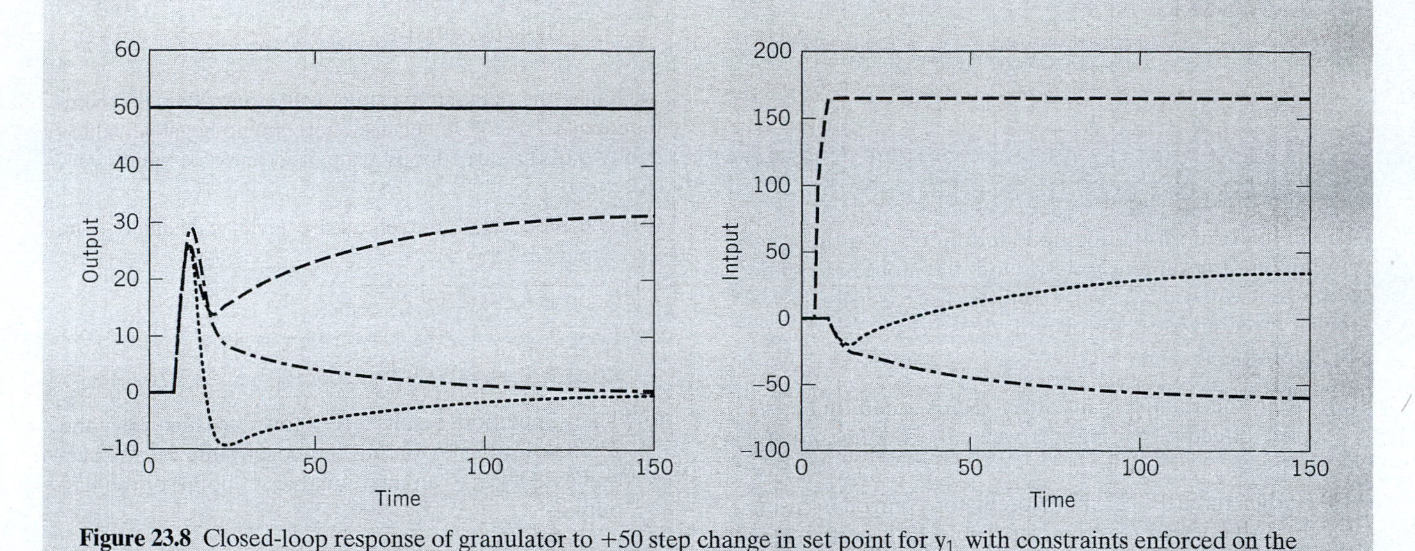

Figure 23.8 Closed-loop response of granulator to +50 step change in set point for y_1, with constraints enforced on the inputs. The left plot is CVs, right plot is MVs (----, y_1 and u_1; -.-.-.-., y_3 and u_3; ...., y_2 and u_2).

23.2 PROCESS MODELING AND CONTROL FOR DRUG DELIVERY

The human body is a remarkably complex biochemical process, and it shares many attributes with more traditional process control problems that have been discussed in earlier chapters. In the event that a body fails to achieve the robust level of self-regulation that occurs naturally (cf. Chapter 24), there are opportunities for medical intervention, often involving the administration of a therapeutic agent (or drug) in a prescribed manner. The therapy can be optimized using open-loop methods, but it is often advantageous to automate the process, thus removing the human from the feedback loop (much as a chemical plant removes the operator from the loop in the transition from manual control to automatic feedback control). In some medical applications (e.g., cancer treatment), control design can be used for decision support to guide medical interventions, and not strictly for automation. In the medical field, as in the process domain, there are three essential requirements for implementing feedback control: (1) the availability of a measurement that indicates the condition of the patient, (2) some knowledge of the underlying process dynamics (e.g., the effect of a drug on a patient's response), and (3) a suitable manipulated variable (e.g., drug or medication). Since 1990, there have been dramatic advances in sensor technology, as well as modeling and control strategies, for a variety of medical problems (see, for example, Hahn et al., 2002; Heller, 2005; Doyle et al., 2007).

In the following sections, a diverse range of biomedical applications that motivate the application of process control are described.

23.2.1 Type 1 Diabetes

In a healthy individual, the concentration of blood sugar (glucose), the body's primary energy source, is regulated primarily by the pancreas, using a combination of manipulated inputs that are analogous to the brake and gas pedal system used to control the speed of an automobile. As the blood sugar falls, the pancreas responds with the release of the hormone glucagon from the α-cells, which stimulates the breakdown of glycogen in the liver to create glucose, thus leading to an increase in glucose (i.e., the gas pedal). On the other hand, as blood glucose rises, the pancreatic β-cells release the hormone insulin that stimulates the uptake of glucose by muscle and fat tissue (Ashcroft and Ashcroft, 1992), and, consequently, the blood glucose level is decreased (i.e., the brake).

Type 1 diabetes mellitus is a disease characterized by failure of the pancreatic β-cells. In contrast, the primary manifestation of Type 2 diabetes is an inability, or resistance, of the cells to respond to insulin. The only treatment for Type 1 diabetes consists of exogenous insulin injections, traditionally administered in an open-loop manner by the patient. The insufficient secretion of insulin by the pancreas results in large excursions of blood glucose outside of the target range of approximately 80–120 mg/dL, leading to brief, or often sustained, periods of hyperglycemia (elevated glucose levels). Intensive insulin therapy can often have the unintended consequence of overdosing, which can then lead to hypoglycemia (low glucose levels). The consequences of such inadequate glucose regulation include an increased risk for retinopathy, nephropathy, and peripheral vascular disease (DCCT, 1993; Jovanovič, 2000; Zisser at al., 2005).

As illustrated in Fig. 23.9, a feedback controller can be used to regulate blood glucose using an insulin pump (widely available on the market today). There are preliminary clinical trials testing the efficacy of PID controllers for this delivery (Steil et al., 2006). The ADA has published guidelines (American Diabetes Association, 2006) recommending the following target zones for a blood sample drawn from a vein (a whole-blood sample):

- 80 mg/dL to 120 mg/dL before meals
- Less than 160 mg/dL 1 to 2 hours after meals

As indicated in Chapter 22, batch processes can benefit from recipe modifications in between consecutive batches or cycles, using a run-to-run (RtR) strategy. Run-to-run control strategies have also been developed for diabetes control, by considering glucose data for a meal response or an entire day to be the batch of interest. The similarities between the diabetic patient and the batch reactor recipe that motivate the application of this technique are the following:

1. The recipe (24-h cycle) for a human patient consists of a repeated meal protocol (typically three meals), with some variation on meal type, timing, and duration.
2. There is not an accurate dynamic model available to describe the detailed glucose response of each individual to the meal profile.
3. There are selected measurements available that might be used to characterize the quality of the glucose response for a 24-h day, including maximum and minimum glucose values.

Using currently available glucose meters, the blood sampling is very sparse, typically about 6–8 measurements per day; hence, the overall quality (i.e., glycemic regulation) has to be inferred from these infrequent samples. The results of a subsequent clinical trial (Zisser et al., 2005) demonstrated that a large fraction of the patients responded favorably to this type of control.

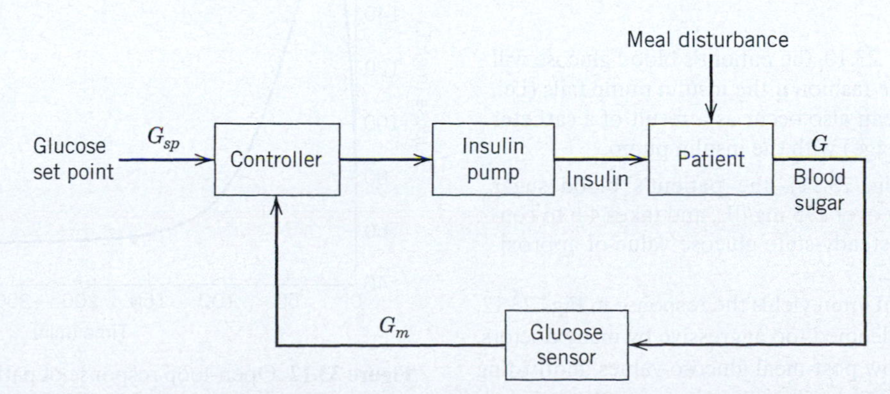

Figure 23.9 Block diagram for artificial β-cell, illustrating the meal as the most common disturbance. G denotes the blood sugar of the patient, G_m is the output of the glucose sensor, and G_{sp} is the glucose set point.

EXAMPLE 23.3

A patient with Type I diabetes needs an automated scheme to maintain her glucose within an acceptable range, widened here to allow less conservative control (54 mg/dL < G < 144 mg/dL). She has just eaten a large meal (a disturbance) that you estimate will introduce glucose into her bloodstream according to $D(t) = 9.0\, e^{-0.05t}$, where t is in minutes and $D(t)$ is in mg/dL-min. She has a subcutaneous insulin pump that can release insulin up to 115 mU/min (mU = 10^{-3} Units of insulin). The "U" is a standard convention used to denote the strength of an insulin solution. The flow rate of insulin is the manipulated variable.

A simple model of her blood glucose level is given by Bequette (2002):

$$\frac{dG}{dt} = -p_1 G - X(G + G_{Basal}) + D \qquad (23\text{-}16)$$

$$\frac{dX}{dt} = -p_2 X + p_3 I \qquad (23\text{-}17)$$

$$\frac{dI}{dt} = -n(I + I_{Basal}) + \frac{U}{V_1} \qquad (23\text{-}18)$$

where the constants are defined as follows: p_1 = 0.028735 [min^{-1}], p_2 = 0.028344 [min^{-1}], p_3 = 5.035E-5 [min^{-1}], V_1 = 12 [L], and n = .0926 [min^{-1}]. G, X, and I are values for glucose concentration (deviation) in the blood (mg/dL), insulin concentration (deviation) at the active site (mU/L), and blood insulin concentration, expressed in deviation variables. Basal values refer to the initial or baseline values for G and I (G_{basal} = 81 mg/dL and I_{basal} = 15 mU/L). D is the rate of glucose release into the blood (mg/dL-min) as the disturbance. U is the flow rate of insulin (mU/min) as the manipulated variable.

(a) What will happen to her blood glucose level if the pump is shut off initially?

(b) What will happen to her blood glucose level if the pump injects insulin at a constant rate of 15 mU/min?

(c) Is there a constant value of U that will help her stay within an acceptable glucose range (54 mg/dL < G < 170 mg/dL) for the next 400 min?

SOLUTION

(a) As shown in Fig. 23.10, the patient's blood glucose will rise in a ramplike fashion if the insulin pump fails (i.e., shuts off). This can also occur as a result of a catheter occlusion (blockage) with the insulin pump.

(b) In this case (Fig. 23.11), the patient's blood sugar peaks, at slightly over 175 mg/dL, and takes 4 h to converge back to a steady-state glucose value of approximately 90 mg/dL.

(c) A setting of 25 mU/min yields the response in Fig. 23.12, which might be deemed too aggressive by many doctors, because of the low post-meal glucose values, motivating a more advanced (i.e., closed-loop) approach to glucose management.

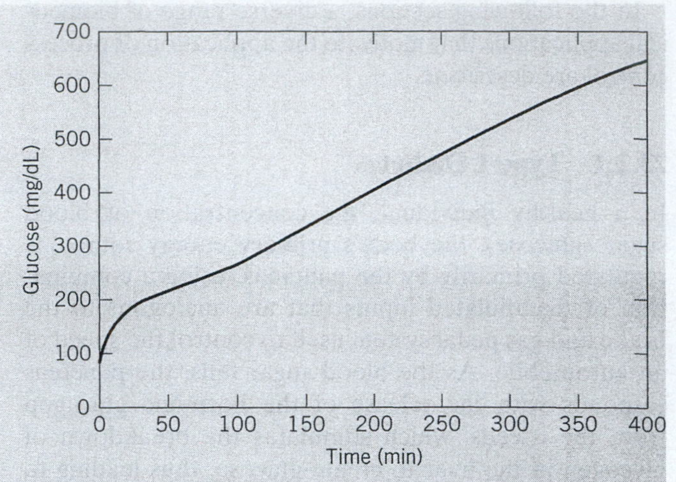

Figure 23.10 Open-loop response of patient's blood glucose when the insulin pump is turned off.

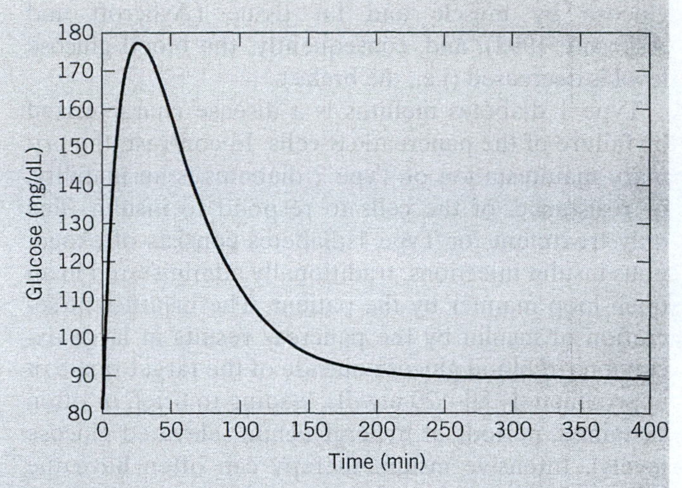

Figure 23.11 Open-loop response of the patient's blood glucose to a constant infusion rate of 15 mU/min from her insulin pump.

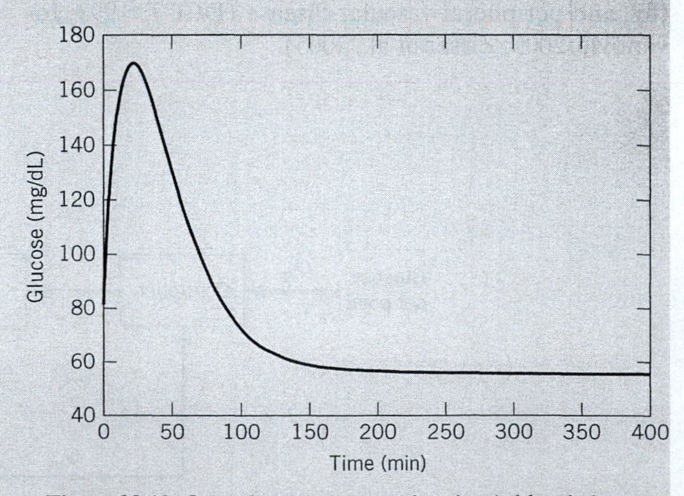

Figure 23.12 Open-loop response of patient's blood glucose to a constant infusion rate of 25 mU/min from her insulin pump.

23.2.2 Blood Pressure Regulation

In both the operating room and postoperative care contexts, closed-loop control of blood pressure and related variables (such as cardiac output and depth of anesthesia) have been studied for a number of years (e.g., Rao et al., 1999), and human clinical trials have proved the efficacy of the approach (Bailey and Haddad, 2005; Araki and Furutani, 2005). The post-operative application was handled typically by the administration of sodium nitroprusside (SNP) by a nurse via a continuous intravenous (IV) pump. SNP is a vasodilator that achieves blood pressure reduction by relaxing the muscles controlling the vascular resistance to flow through blood vessels. The current technology for both sensors and infusion pumps is facilitating the design of completely automated control strategies.

The context of the operating room is more complicated, with many critical variables that must be monitored. But an advantage of this setting is that nonportable sensors can be employed that would be too cumbersome or impractical for ambulatory applications. The measured variables include mean arterial pressure (MAP), cardiac output (CO), and depth of anesthesia (DOA). The DOA has been the subject of intense research activity over the last decade, and sensors are available to determine the depth of anesthesia through correlations. These sensors are inferential (see Chapter 16), in that they do not directly measure the medical state of anesthesia, which is characterized by such patient responses as hypnosis, amnesia, analgesia, and muscle relaxation (Araki and Furutani, 2005); rather, they measure the state of electrical activity in the patient's brain. One of the more promising methods is the bispectral index, derived from signal analysis of an electroencephalograph (EEG) (Bailey and Haddad, 2005). A variety of manipulated inputs are also available, resulting in an intrinsically multivariable control problem. Some candidate manipulated variables include vasoactive drugs, such as dopamine and SNP, as well as anesthetics (isoflurane, propofol, etc.).

EXAMPLE 23.4

Consider the following model for predicting the influence of two drugs: SNP, [μg/kg-min]) and dopamine (DPM, [μg/kg-min]), on two medical variables (MAP, [mmHg]) and CO, [L/(kg-min)]), where time is measured in minutes (Bequette, 2007):

$$\begin{bmatrix} MAP \\ CO \end{bmatrix} = \begin{bmatrix} \dfrac{-6e^{-0.75s}}{0.67s+1} & \dfrac{3e^{-s}}{2.0s+1} \\ \dfrac{12e^{-0.75s}}{0.67s+1} & \dfrac{5e^{-s}}{5.0s+1} \end{bmatrix} \begin{bmatrix} SNP \\ DPM \end{bmatrix} \tag{23-19}$$

(a) Calculate the RGA for this problem and propose the appropriate control-loop pairing.

(b) Consider the pairing, SNP-MAP and DPM-CO, as is typically used in practice. Design a pair of PI controllers for this process, using the IMC tuning rules (Table 12.1) and choosing a value of τ_c for each controller that is equal to the corresponding open-loop time constant for that subsystem.

(c) Simulate the closed-loop response to a –10 mmHg change in the MAP set point, while holding CO constant. Discuss the extent of control-loop interactions.

SOLUTION

(a) Using the RGA calculation in Eq. 18-34, $\lambda_{11} = 0.4545$; therefore, the loop pairings apparently should be the 1-2/2-1 pairing, SNP-CO and DPM-MAP.

(b) From Table 12.1, the following values for the PI controller settings are calculated:

Loop 1: $K_c = -(0.67)/(6*(0.67 + .75)) = -0.0786$

$\tau_I = 0.67$

Loop 2: $K_c = (5)/(5*(5 + 1)) = 0.1667$

$\tau_I = 5$

(c) The simulated response for the MAP set point change is depicted in Fig. 23.13, where there is a modest undershoot in the MAP response; however, the interacting nature of the process leads to a large excursion in CO. The control tuning (τ_c) could be refined to trade-off speed versus overshoot and interaction, or by designing a multivariable controller, such as MPC.

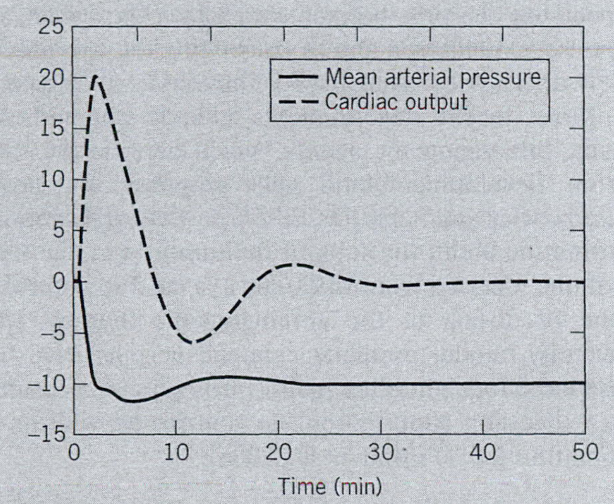

Figure 23.13 Closed-loop response of patient's mean arterial blood pressure and cardiac output to a –10 mmHg change in the MAP set point.

23.2.3 Cancer Treatment

Cancer treatment has changed dramatically over the past decade, in large part enabled by advances in imaging technology. Surgery has been the classical method for attacking cancerous tumors, and more recently

X-ray radiation has been employed. An unfortunate side effect in both cases is that healthy tissue can be compromised by inappropriate surgery or delivery of radiation, respectively. Chemotherapy is often be used, alone or in conjunction with surgery or X-ray treatment, and has the advantage that undetected metastases (cancer cells that have circulated through the bloodstream) can be attacked with this method. Thermal therapies (radiofrequency, microwave, or laser techniques) have also been demonstrated to be effective, with similar requirements on targeting the energy to the localized region of the tumor (Dodd et al., 2000). In thermal and radiation treatment, feedback control is finding application to the optimized delivery of the treatment (radiation, heat) to the targeted area (Salomir et al., 2000; Davison and Hwang, 2003; Ledzewicz and Schättler, 2007; Moonen, 2007). In one feedback-based therapy (Salomir et al., 2000), the heat source power was adjusted based on the deviation of temperature from a target at a particular location in the body, including an integral term, very similar to a PID controller. The desired response was that the temperature should rise quickly to the target without overshoot or oscillations.

Parker (2007) describes a strategy for "model-informed" treatment design for delivery of a chemotherapeutic agent. Using a combination of pharmacokinetic and pharmacodynamic models, predictions can be made about the patient's response (e.g., tumor volume) to the manipulated variables, which in this case could include the drug dosage level and schedule for drug administration. This strategy can be implemented by specifying the time horizon over which the patient's response is monitored and by calculating the optimal drug delivery protocol using the RTO methods of Chapter 19.

More recent developments include chemotherapy using antiangiogenic agents, which deprive the tumor from developing blood cells required for growth (Ledzewicz and Schättler, 2007). In this application, information about the state of the tumor (e.g., the tumor volume, derived from MRI data) is used to control the rate of dosing of the antiangiogenic therapy. More recently, model predictive control designs have been proposed for chemotherapeutic protocols, as an example of a decision support tool, as contrasted with an automation tool (Florian et al., 2008).

23.2.4 Controlled Treatment for HIV/AIDS

To address the global problem of HIV/AIDS, a number of mathematical models and control algorithms have been proposed to help design better treatments for the disease. The drug categories that have been considered include reverse transcriptase inhibitors and protease inhibitors, which affect reproduction of the virus via transcription and production of the virus from infected cells, respectively. The most effective strategies to date have involved a so-called cocktail of multiple drugs, thus attacking the disease in a vector direction (i.e., multiple, simultaneous targets). Measurements are problematic, consisting of relatively slow techniques based on off-line sampling of blood. However, the slow progression of the disease does not warrant real-time measurements, and thus feedback can still be accomplished on this slow time scale.

In their simplest form, mathematical models have been developed that describe the interactions of healthy CD4+ T cells, infected CD4+ T cells, and free viruses in the form of three coupled ordinary differential equations (Craig and Xia, 2005). Such a model can be the basis of simple model-based feedback strategies for control and can also be extended to generate more complex models suitable for a model predictive control strategy (Zurakowski et al., 2004).

23.2.5 Cardiac-Assist Devices

Cardiac-assist devices are mechanical pumps that provide cardiac output at an appropriate pressure, to allow normal circulation of blood through the patient's body, subject to the changing demands for cardiac output as a function of the patient's state (e.g., level of exercise, emotion, posture, etc.). The ideal device would mimic the body's own mechanisms for maintaining cardiac output at target levels; however, currently available devices are rather primitive in terms of automation, requiring the patient to adjust the set point (Boston et al., 2000). The first such implantable device received approval by the FDA over a decade ago.

One of the more interesting aspects of the control design problem for ventricular-assist devices is the placement of the sensors and actuators: there are the issues of susceptibility to infection, as well as anatomical placement (Paden et al., 2000).

23.2.6 Additional Medical Opportunities for Process Control

There are many other challenges in drug therapy, in which an optimized delivery regimen could be calculated using principles of process control and process optimization, e.g., the modeling and control of the anticoagulant drug, heparin (McAvoy, 2007). Another medical application is the treatment of acute neuropatients with brain hypothermia, to lower the intracranial pressure (ICP). A mathematical model can be developed to relate temperature effects with blood flow. The model can then be used to create an automated closed-loop controller (Gaohua and Kimura, 2006) to adjust the coolant temperature (e.g., by using cold-water circulating blankets) in an effort to regulate the ICP.

SUMMARY

Biological and biomedical processes share a great deal in common with the process applications considered in preceding chapters. The latter applications have a characteristic time constant, often exhibit time delays associated with measurements, and typically are multivariable in nature. In contrast, the types of uncertainties in bioprocesses are quite different, owing to the complex nature of biological regulation (see Chapter 24). In addition, there are multiple safety and regulatory issues that are unique to medical closed-loop systems. Process control strategies for several key unit operations in the bioprocess industries (fermentation, crystallization, and granulation) have been described, along with biomedical applications, such as drug delivery for diabetes and blood pressure control and other examples. In Chapter 24, we focus on the molecular scale of biological systems and consider the feedback mechanisms inherent in naturally occurring biophysical networks.

REFERENCES

Alford, J., Bioprocess Control: Advantages and Challenges, Proc. Conf. Chemical Process Control, Alberta (2006).

American Diabetes Association, Standards of Medical Care in Diabetes—2006, *Diabetes Care*, **29**, S4 (2006).

Araki, M., and E. Furutani, Computer Control of Physiological States of Patients under and after Surgical Operation Annu. Rev. Control, 29, 229 (2005).

Ashcroft, F. M., and S. J. H. Ashcroft, *Insulin: Molecular Biology to Pathology*, Oxford University Press, NY, 1992.

Bequette, B. W., *Process Control*, Prentice-Hall, Upper Saddle River, NJ, 2002.

Bequette, B. W., Modeling and Control of Drug Infusion in Critical Care, *J. Process Control*, **17**, 582 (2007).

Boston, J., M. Simaan, J. Antaki, and Y. Yu, Control Issues in Rotary Heart Assist Devices, *Proc. American Control Conference*, 3473 (2000).

Boudreau, M. A., and G. K. McMillan, New Directions in Bioprocess Modeling and Control, ISA, NC, 2007.

Buckland, B. C., The Translation of Scale in Fermentation Processes: The Impact of Computer Process Control, *Nature Biotechnology*, **2**, 875 (1984).

Craig, I., and X. Xia, Can HIV/AIDS Be Controlled?, *IEEE Control Sys.*, **25**, 80 (2005).

Davison, D. E., and E.-S. Hwang, Automating Radiotherapy Cancer Treatment: Use of Multirate Observer-Based Control, *Proc. American Control Conference*, 1194 (2003).

DCCT—The Diabetes Control and Complications Trial Research Group, The Effect of Intensive Treatment of Diabetes on the Development and Progression of Long-Term Complications in Insulin-Dependent Diabetes Mellitus, *New Engl. J. Med.*, **329**, 977 (1993).

Dodd, G. D., M. C. Soulen, R. A. Kane, T. Livraghi, W. R. Lees, Y. Yamashita, A. R. Gillams, O. I. Karahan, and H. Rhim, Minimally Invasive Treatment of Malignant Hepatic Tumors: At the Threshold of a Major Breakthrough, *Radiographics*, **20**, 9 (2000).

Doyle, F., L. Jovanovič, D. Seborg, R. S. Parker, B. W. Bequette, A. M. Jeffrey, X. Xia, I. K. Craig, and T. J. McAvoy, A Tutorial of Biomedical Process Control, *J. Process Control*, **17**, 571 (2007).

Doyle, F., L. Jovanovič, and D. E. Seborg, Glucose Control Strategies for Treating Type 1 Diabetes Mellitus, *J. Process Control*, **17**, 572 (2007).

Fleet, G. H. (Ed.), *Wine Microbiology and Biotechnology*, Taylor & Francis, London, U.K., 1993.

Florian, J. A., J. L. Eiseman, and R. S. Parker. Nonlinear Model Predictive Control for Dosing Daily Anticancer Agents: A Tamoxifen Treatment of Breast Cancer Case Study, *Comput. Biol. Med.*, **38**, 339 (2008).

Fujiwara, M., Z. K. Nagy, J. W. Chew, and R. D. Braatz, First-Principles and Direct Design Approaches for the Control of Pharmaceutical Crystallization, *J. Process Control*, **15**, 493 (2005).

Gaohua, L., and H. Kimura, A Mathematical Model of Intracranial Pressure Dynamics for Brain Hypothermia Treatment, *J. Theor. Biol.*, **238**, 882 (2006).

Gatzke, E. P., and F. J. Doyle III, Model Predictive Control of a Granulation System Using Soft Output Constraints and Prioritized Control Objectives, *Powder Tech.*, **121**, 149 (2001).

Hahn, J., T. Edison, and T. F. Edgar, Adaptive IMC Control for Drug Infusion for Biological Systems, *Control Eng. Practice*, **10**, 45 (2002).

Heller, A., Integrated Medical Feedback Systems for Drug Delivery, *AIChE J.*, **51**, 1054 (2005).

Henson, M. A., and D. E. Seborg, An Internal Model Control Strategy for Nonlinear Systems, *AIChE J.*, **37**, 1065 (1991).

Jovanovič, L., The Role of Continuous Glucose Monitoring in Gestational Diabetes Mellitus, *Diabetes Technol. Ther.*, **2**, S67 (2000).

Larsen, P. A., D. B. Patience, and J. B. Rawlings, Industrial Crystallization Process Control, *IEEE Control Systems*, **26**, 70 (2006).

Ledzewicz, U., and H. Schättler, Antiangiogenic Therapy in Cancer Treatment as an Optimal Control Problem, *SIAM J. Control Optim.*, **46**, 1052 (2007).

Mahadevan, R., S. K. Agrawal, and F. J. Doyle III, Differential Flatness Based Nonlinear Predictive Control of Fed-Batch Bioreactors, *Control Eng. Prac.*, **9**, 889 (2001).

Mandenius, C. -F., Process Control in the Bioindustry, *Chemistry Today*, 19 (1994).

McAvoy, T. J., Modeling and Control of the Anticoagulant Drug Heparin, *J. Process Control*, **17**, 590 (2007).

Moonen, C. T. W., Spatio-Temporal Control of Gene Expression and Cancer Treatment Using Magnetic Resonance Imaging-Guided Focused Ultrasound, *Clin. Cancer Res.*, **13**, 3482 (2007).

Mort, P. R., S. W. Capeci, and J. W. Holder, Control of Agglomeration Attributes in a Continuous Binder-Agglomeration Process, *Powder Tech.*, **117**, 173 (2001).

Paden, B., J. Ghosh, and J. Antaki, Control System Architectures for Mechanical Cardiac Assist Devices, *Proc. American Control Conference*, 3478 (2000).

Parker, R. S., Modeling for Anti-Cancer Chemotherapy Design, *J. Process Control*, **17**, 576 (2007).

Pottmann, M., B. A. Ogunnaike, A. A. Adetayo, and B. J. Ennis, Model-Based Control of a Granulation System, *Powder Tech.*, **108**, 192 (2000).

Rao, R. R., J. W. Huang, B. W. Bequette, H. Kaufman, and R. J. Roy, Control of a Nonsquare Drug Infusion System: A Simulation Study, *Biotechnol. Prog.*, **15**, 556 (1999).

Rohani, S., M. Haeri, and H. C. Wood, Modeling and Control of a Continuous Crystallization Process Part 1. Linear and Non-Linear Modeling, *Comput. Chem. Eng.*, **23**, 263 (1999).

Salomir, R., F. C. Vimeux, J. A. de Zwart, N. Grenier, and C. T. W. Moonen, Hyperthermia by MR-Guided Focused Ultrasound: Accurate Temperature Control Based on Fast MRI and a Physical Model of Local Energy Deposition and Heat Conduction, *Magnetic Resonance in Medicine*, **43**, 342 (2000).

Schügerl, K., Progress in Monitoring, Modeling and Control of Bioprocesses during the Last 20 Years, *J. Biotechnol.*, **85**, 149 (2001).

Steil, G. M., K. Rebrin, C. Darwin, F. Hariri, and M. F. Saad, Feasibility of Automating Insulin Delivery for the Treatment of Type 1 Diabetes, *Diabetes*, **55**, 3344 (2006).

Zisser, H., L. Jovanovič, F. Doyle III, P. Ospina, and C. Owens, Run-to-Run Control of Meal-Related Insulin Dosing, *Diabetes Technol. Ther.*, **7**, 48 (2005).

Zhou, G. X., M. Fujiwara, X. Y. Woo, E. Rusli, H. -H. Tung, C. Starbuck, O. Davidson, Z. Ge, and R. D. Braatz, Direct Design of Pharmaceutical Antisolvent Crystallization through Concentration Control, *Crystal Growth & Design*, **6**, 892 (2006).

Zurakowski, R., M. J. Messina, S. E. Tuna, and A. R. Teel, HIV Treatment Scheduling Via Robust Nonlinear Model Predictive Control, *Proc. Asian Control Conference*, 25 (2004).

EXERCISES

23.1 Consider the fermentor problem in Example 23.1.

(a) Design an IMC controller for the first operating point (dilution = .202 h^{-1}), and simulate the response to both a +0.5 [g/L] and a −0.5 [g/L] change in the biomass concentration set point. Then, simulate the response to both a +1 [g/L] and a −1 [g/L] change in the biomass concentration set point.

(b) Simulate the response of a −12.5% step change in the maximum growth rate (μ_m). How well does the controller perform?

(c) Comment on the observed nonlinearity in the system.

(d) Discuss how the controller design would change if there were a requirement to operate at the lower dilution operating point. What do you need to consider in this case?

23.2 Consider the granulation model that was given in Example 23.2.

(a) Design an MPC controller, using the nominal process model. Initially consider a control horizon of $M = 2$ and a prediction horizon of $P = 40$ (with a sampling period of $\Delta = 1$). Use equal weights on the manipulated inputs and penalize the two percentile outputs equally, but use a larger weight on the bulk density (y_1).

(b) Consider the effect of a plant-model mismatch. Use the problem statement for control design, but assume that the actual process is characterized by the following parameters:

$$\widehat{K}_{i,j} = \begin{bmatrix} 0.10 & 0.90 & 0.15 \\ 0.25 & 1.10 & 1.30 \\ 0.50 & 0.80 & 1.00 \end{bmatrix}$$

$$\widehat{\tau}_{i,j} = \begin{bmatrix} 1 & 2 & 1.5 \\ 3 & 3 & 3 \\ 3 & 3 & 3 \end{bmatrix}$$

$$\widehat{\theta}_{i,j} = \begin{bmatrix} 2 & 2 & 4 \\ 2 & 3 & 4 \\ 2 & 3 & 4 \end{bmatrix}$$

(c) These models are in deviation variables, but the actual steady-state flow rates for the nozzles are 175, 175, and 245, respectively. The steady-state outputs are 40, 400, and 1620, respectively. Nozzle flow rates are limited to values between 100 and 340, and it is desired to keep the 5th percentile (y_2) above 350 and the 90th percentile (y_3) below 1650. Simulate the response of the controller to the following changes:

(i) Step change in bulk density from 40 to 90

(ii) Simultaneous change in the 5th percentile from 400 to 375 and 90th percentile from 1620 to 1630

Comment on the performance of your controller (and retune as necessary).

23.3 Gaohua and Kimura (2006) derived an empirical patient model for the manipulation of ambient temperature u (°C) to influence the patient's brain intracranial pressure (ICP) y (mm Hg). The medical data support the following empirical values for a first-order plus time-delay model to describe the effect of cooling temperature (°C) on the ICP (mmHg) in time units of hours:

$$G(s) = \frac{4.7}{9.6s + 1} e^{-s}$$

The nominal values for the process variables are: ICP = 20 mmHg; ambient temperature = 30°C.

(a) Using the IMC tuning rules, derive an appropriate PI controller for this medical experiment. (*Hint:* begin with a value τ_c =1.0 [h]). What does that value of τ_c mean?

(b) Simulate the response of a 10-mmHg reduction in ICP. What is the overshoot? What is the minimum value of the temperature? What is the settling time?

(c) Comment on whether this is a reasonable controller design for a biomedical application. How might you improve the design?

23.4 In a rehabilitation training experiment for a neurological patient, a step change in treadmill speed of +2.5 km/h was made. The patient heart rate response HR is given in Figure E23.4.

(a) Derive an appropriate first-order plus time-delay model for the patient dynamics.

(b) The doctors wish to control the patient's heart rate to a nearly constant value by adjusting treadmill speed. Using the IMC tuning rules, design a suitable PI controller for this patient. Simulate the response of the controller for a step change in the HR of +10 bpm. Calculate the settling time, overshoot, and rise time for the controller. Do these values seem reasonable for a medical application?

(c) How would you improve the procedure for fitting the patient's initial dynamics?

23.5 A crystallizer is used to separate a pharmaceutical product from the fermentation extract. The three manipulated variables are the fines dissolution rate (u_1), the crystallizer temperature (u_2), and the flow rate in the overflow (u_3). The nominal values of these three inputs are 2.25 × 10^{-6} m^3/s, 310 K, and 1.5 × 10^{-6} m^3/s, respectively. The three variables to be controlled are the crystal size distribution, as calculated by the fines suspension density (y_1); the crystal purity, as

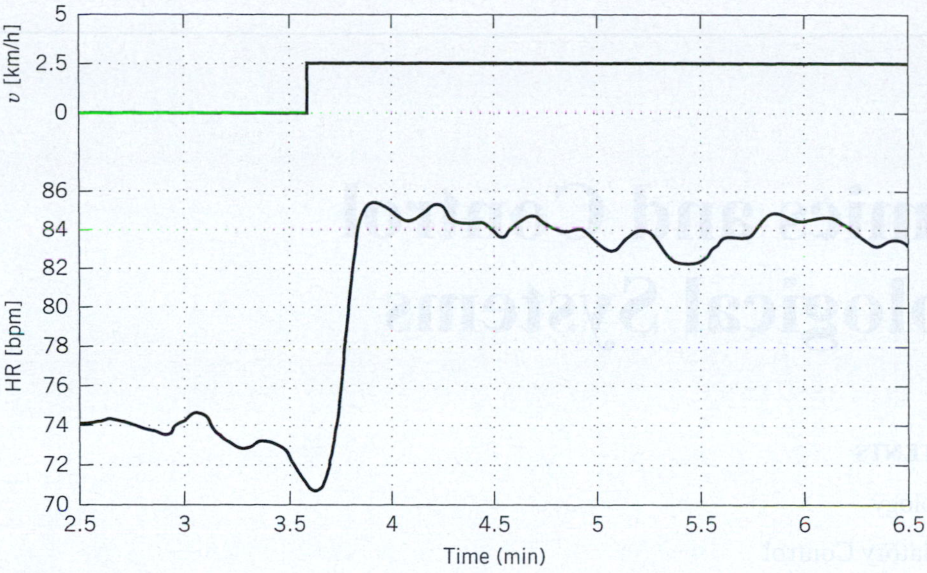

Figure E23.4

calculated by supersaturation conditions (y_2); and the product rate (y_3). The nominal values of these three inputs are 0.55 K, 11.23 K, and 0.12 kg/kg H_2O, respectively. These variables have multiple interactions, and the following model has been identified from experimental data for a continuous crystallizer, where time is measured in s (Rohani et al., 1999):

$$\begin{bmatrix} y_1 \\ y_2 \\ y_3 \end{bmatrix} = \begin{bmatrix} \dfrac{72{,}600}{s+0.2692} & \dfrac{0.025082(s-20.0)(s-10.4)}{s^2+10.11s+96.57} & \dfrac{125{,}000(s-1.25)}{s+0.39} \\ \dfrac{568{,}000}{s+2.11} & \dfrac{-0.15095}{s+0.1338} & \dfrac{-1{,}830{,}000(s+0.089)}{s+0.43} \\ \dfrac{-1{,}870}{s+0.21} & \dfrac{-0.0071}{s+0.235} & \dfrac{16{,}875}{s+0.2696} \end{bmatrix} \times \begin{bmatrix} u_1 \\ u_2 \\ u_3 \end{bmatrix}$$

(a) Calculate the RGA, and determine the appropriate pairings for SISO feedback control. Comment on the role of dynamics in your decision.

(b) Using the IMC tuning rules, design three PI controllers for this process.

(c) Simulate the process response to a step set point, separately, in each of the controlled outputs [use a magnitude of +10% (relative) change]. Next, simulate the response of the system to a simultaneous pair of step changes (again, use a magnitude of +10% relative) in each of the second (purity) and third (product rate) controlled outputs. Try to tune the controller to improve the transient response to the simultaneous step changes.

23.6 Consider the diabetic patient in Example 23.3. Your goal is to design an automated device to administer insulin infusion in response to meal disturbances.

(a) Considering only the insulin-glucose dynamics, calculate an approximate second-order patient model by fitting the responses (changes in insulin) obtained from simulations of the equations given in the example.

(b) Using the IMC tuning rules, design a PID controller for this process.

(c) Simulate the closed-loop system response to a step set point change in blood glucose of –20 mg/dl. Try to tune the controller to improve the transient response.

(d) Simulate the closed-loop system response to the meal disturbance described in Example 23.4. Is the controller able to maintain the safety boundaries for blood glucose (54 mg/dL $< G <$ 144 mg/dL)?

(e) In practice, the sensors are available for measuring blood glucose sample from the subcutaneous tissue (the layer of fat under the skin, as opposed to directly from the blood stream). Assuming that such a procedure introduces a pure delay, repeat the simulation from part (d) with a 10-min sensor delay. How has the performance changed? What is the maximum time delay that the closed-loop design will tolerate before it becomes unstable?

Chapter 24

Dynamics and Control of Biological Systems

CHAPTER CONTENTS

Previous chapters have emphasized the design of controllers for chemical process systems, as well as for biomedical systems (Chapter 23). In this chapter, we consider the analysis of intrinsically closed-loop systems that exist in biological circuits, from gene level through cellular level. There is no external controller to be synthesized; rather, the tools that were developed in the first half of this textbook are applied to the analysis of networks that exploit principles of feedback and feedforward control. These biophysical networks display the same rich character as those encountered in process systems engineering: multivariable interactions, complex dynamics, and nonlinear behavior. Examples are drawn from gene regulatory networks, as well as from protein signal transduction networks, with an emphasis on the role of feedback. A glossary of key technical terms is provided at the end of the Chapter.

24.1 SYSTEMS BIOLOGY

Biophysical networks are remarkably diverse, cover a wide spectrum of scales, and are characterized by a range of complex behaviors. These networks have attracted a great deal of attention at the level of gene regulation, where dozens of input connections may characterize the regulatory domain of a single gene in a eukaryote, as well as at the protein level, where hundreds to thousands of interactions have been mapped in protein interactome diagrams that illustrate the potential coupling of pairs of proteins (Campbell and Heyer, 2007; Barabasi, 2004). However, these networks also exist at higher levels, including the coupling of individual cells via signaling molecules, the coupling of organs via endocrine signaling, and, ultimately, the coupling of organisms in ecosystems. The biochemical notion of signaling is discussed in Section 24.3. To elucidate the mechanisms employed by these networks, biological experimentation and intuition by themselves are insufficient. Instead, investigators characterize dynamics via mathematical models and apply control principles, with the goal of guiding further experimentation to better understand the biological network (Kitano, 2002). Increased understanding can facilitate drug discovery and therapeutic treatments.

A simple example that illustrates the roles of feedback and feedforward control in nature is the heat shock response exhibited by simple bacteria (El-Samad et al., 2006), as illustrated in Fig. 24.1. When the organism experiences an increase in temperature, it leads to the misfolding of protein, which disrupts a number of

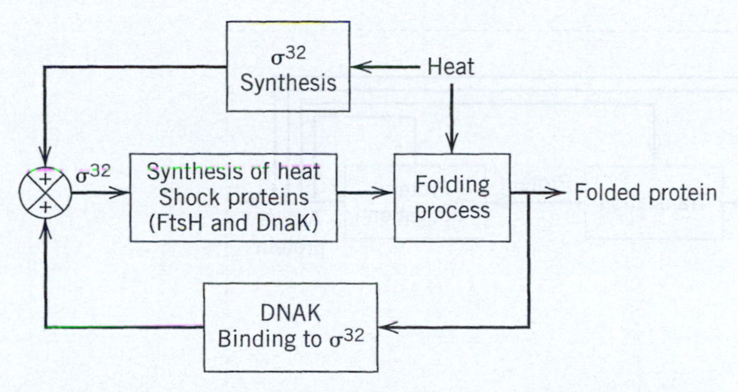

Figure 24.1 Feedback and feedforward control loops that regulate heat shock in bacteria (modified from El-Samad, et al., 2006) (positive feedback is common in biological systems).

metabolic processes. One of the immediate effects of a heat disturbance is the feedforward activation of a component, σ^{32}, which turns on the transcription process for a pair of genes (FtsH and DnaK) that facilitates the repair mechanism for a misfolded protein. In particular, the FtsH gene is a protease, which catalyzes the destruction of the improperly folded protein. In parallel, and independently, the protein product of one of those genes (DNAK) monitors the state of protein misfolding. It binds to σ^{32} and releases σ^{32} when misfolded protein is detected, leading to feedback activation of DnaK transcription.

A second example of networked biological control is the circadian clock, which coordinates daily physiological behaviors of most organisms. The word *circadian* comes from the Latin for "approximately one day," and the circadian clock is vital to regulation of metabolic processes in everything from simple fungi to humans. The mammalian circadian master clock resides in the hypothalamus region of the brain (Reppert and Weaver, 2002). It is a network of multiple autonomous noisy oscillators, which communicate via neuropeptides to synchronize and form a coherent oscillator (Herzog et al., 2004; Liu et al., 2007). At the core of the clock is a gene regulatory network, in which approximately six classes of genes are regulated through an elegant array of time-delayed negative feedback circuits (see Figure 24.2, which illustrates two of those six gene classes). The activity states of the proteins in this network are modulated (activated/inactivated) through a series of chemical reactions, including phosphorylation and dimerization. These networks exist at the subcellular level. Above this layer is the signaling that leads to a synchronized response from the population of thousands of clock neurons in the brain. Ultimately, this coherent oscillator then coordinates the timing of daily behaviors, such as the sleep/wake cycle. An interesting property of the clock is that, under conditions of constant darkness, the clock free-runs, with a period of approximately 24 h (i.e., "circa"), such that its internal time, or phase, drifts away from that of its environment. However, in the presence of an entraining cue (i.e., forcing signal, such as the rising and setting of the sun), the clock locks on to the period of that cue (Boulos et al., 2002; Dunlap et al., 2004; Daan and Pittendrigh, 1976). This gives rise to a precise 24-h period for the oscillations in protein concentrations for the feedback circuit in Fig. 24.2.

The Central Dogma tenet that most students learn in high school biology is a good starting point to understand these complex networks. Information in the cell is encoded in the DNA, and that information is expressed by the gene to produce messenger RNA. The mRNA is translated into a protein, which is one of the key building blocks of cells and which plays a critical role in cellular regulation. This form of the Central Dogma suggests a serial process, or a feedforward process, in which the genetic code influences the outcome (protein level and protein function). In some of the early publicity surrounding the Human Genome project, this type of logic was pervasive, and there was an understanding in some circles that the "parts list" (genetic code) would illuminate the cause of diseases. An engineer immediately recognizes the flaw in this logic: by analogy, if one were provided with the raw materials list for an aircraft (sheet metal, nuts, bolts, rivets, etc.), it would be an impossible leap to conclude

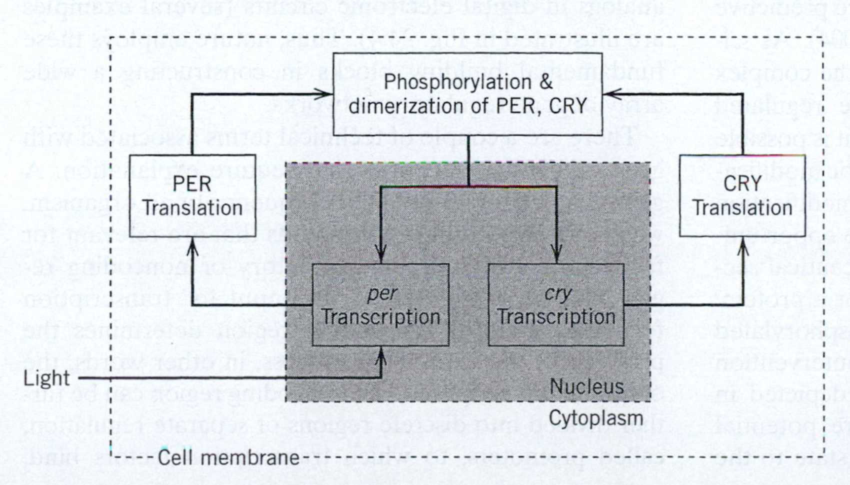

Figure 24.2 The gene regulatory circuit responsible for mammalian circadian rhythms (by convention, italics and lowercase refer to genes, uppercase refers to proteins).

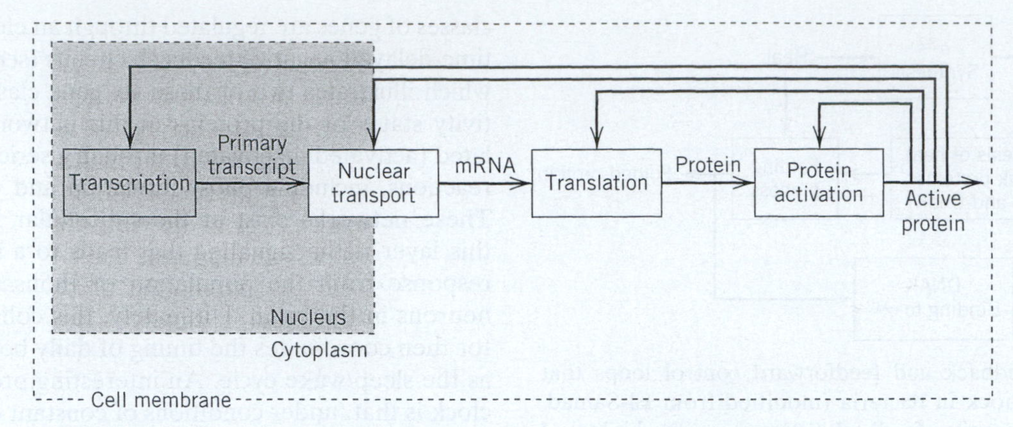

Figure 24.3 The layers of feedback control in the Central Dogma (modified from Alberts et al., 1998)

anything about the principles of aerodynamics. Critical missing elements are the manner in which the parts are arranged into a network and, more important, how the components are controlled (or regulated). The same reasoning applies equally to biological networks as well, and this notion of the systems perspective has driven current research in systems biology.

According to the Central Dogma tenet, the additional layers of control and regulation that are mentioned in the preceding paragraph can be incorporated schematically, as shown in Fig. 24.3. Feedback control plays a key role in (i) regulation of the transcription event; (ii) processing of the RNA, including its stability and potential silencing via RNA interference; (iii) regulation of the ribosomal machinery that accomplishes translation; and (iv) modulation of the activity state of protein, through, for example, degradation, conformation changes, and phosphorylation. Recalling the circadian clock schematic in Fig. 24.2, the process of controlling the concentration of a phosphorylated form of the PER protein can be broken down into each of the elementary steps indicated by the Central Dogma schematic in Fig. 24.3.

Systems biology holds great promise to revolutionize the practice of medicine, enabling a far more predictive and preventative capability (Hood et al., 2004). As scientists and engineers begin to understand the complex networks of genes and proteins that are regulated through feedback and feedforward control, it is possible to develop novel therapies through systematic modification of these closed-loop systems. These modification sites are referred to as targets, and they are opportunities for the design of drugs by the pharmaceutical sector. A drug may target a particular gene, or a protein, or an activity state of a protein (e.g., phosphorylated form), suggesting that there are multiple intervention points in the Central Dogma process, as depicted in Fig. 24.3. In control terminology, they are potential manipulated variables to restore a healthy state to the network. Likewise, medical scientists and engineers are looking for markers that reveal the pattern of a disease in the signature of the network response. Again, they are understood in control terms as novel sensors that form the basis of an inferential strategy to monitor the status of an unmeasurable disease state. Just as process control engineers test the efficacy of their control system designs through simulation, systems biologists evaluate these new drug targets through extensive simulations of patient populations.

24.2 GENE REGULATORY CONTROL

As described in the previous section, genes are regulated through complex feedback control networks. These networks exhibit a remarkable degree of robustness, because the transcription of critical genes is reliable and consistent, even in the face of disturbances from both within the cell and external to the organism. One of the very compelling features of gene regulatory networks is the recurring use of circuit elements that occur in engineering networks. It has been shown that groups of two to four genes exhibit recurring connection topologies, so-called motifs, which have direct analogs in digital electronic circuits (several examples are illustrated in Fig. 24.4). Thus, nature employs these fundamental building blocks in constructing a wide array of gene regulatory networks.

There are a couple of technical terms associated with gene regulatory networks that require explanation. A gene is a portion of the DNA sequence of an organism, which has two primary subregions that are relevant for feedback control: (i) the regulatory or noncoding region can be considered as the input for transcription feedback, and (ii) the coding region determines the products of the expression process, in other words, the output of transcription. The noncoding region can be further divided into discrete regions of separate regulation, called promoters, to which transcription factors bind,

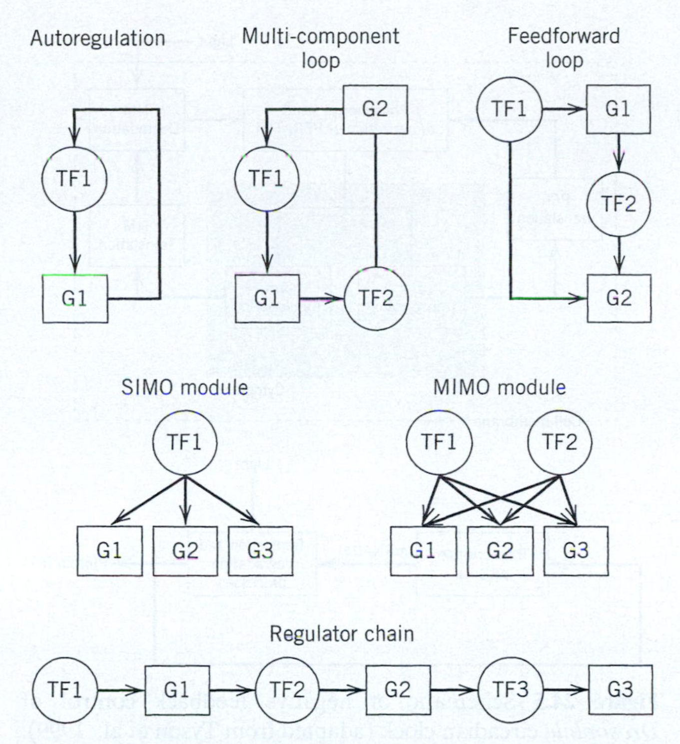

Figure 24.4 Examples of circuit motifs in yeast (adapted from Lee et al., 2002). The rectangles denote promoter regions on a gene (G1, G2, etc.), and the circles are transcription factors (TF1, TF2, etc.).

leading to activation or inhibition of the expression of the gene (the transcription process). The combination of transcription factors and promoter regions are the controller for the gene transcription process.

There are three dominant network motifs found in *E. coli* (Shen-Orr et al., 2002): (i) a feedforward loop, in which one transcription factor regulates another factor, and, in turn, the pair jointly regulates a third transcript factor; (ii) a single-input multiple-output (SIMO) block architecture; and (iii) a multiple-input multiple-output (MIMO) block architecture, referred to as a densely overlapping regulon by biologists.

A completely different organism, *S. cerevisiae*, has six closely related network motifs (Lee et al., 2002): (i) an autoregulatory motif, in which a regulator binds to the promoter region of its own gene; (ii) a feedforward loop; (iii) a multicomponent loop, consisting of a feedback closed-loop with two or more transcription factors; (iv) a regulator chain, consisting of a cascade of serial transcription factor interactions; (v) a single-input multiple-output (SIMO) module; and (vi) a multiple-input multiple-output (MIMO) module. These motifs are illustrated in Fig. 24.4.

In effect, these studies prove that, in both eukaryotic and prokaryotic systems, cell function is controlled by complex networks of control loops, which are cascading and interconnected with other (transcriptional) control

Table 24.1 Analogies between process control concepts and gene transcription control concepts

Process Control Concept	Biological Control Analog
Sensor	Concentration of a protein
Set point	Implicit: equilibrium concentration of protein
Controller	Transcription factors
Final control element	Transcription apparatus; ribosomal machinery for protein translation
Process	Cellular homeostasis

loops. The complex networks that underlie biological regulation appear to be constructed of elementary systems components, not unlike a digital circuit. This lends credibility to the notion that analysis tools from process control are relevant in systems biology.

Some of the analogies between process control concepts and biological control concepts are summarized in Table 24.1, at the level of gene transcription. Keep in mind that there are many levels of analysis in biological circuits, and one can draw comparisons to engineering circuits at each of these levels.

EXAMPLE 24.1

The control strategy of gene regulatory circuits can often be approximated using simple logic functions, much like the functions employed in Chapter 22 for batch recipe control. Consider the logic underlying the regulation of the *lacZ* gene, which is involved in sugar metabolism (Ptashne and Gann, 2002). This gene codes for the enzyme β-galactosidase, which is responsible for cleaving lactose, a less efficient source of energy for a bacterium than the preferred glucose supply. The state of the gene (activated or inhibited) is determined by the transcription factors that bind to the regulatory domain of the gene. One of those transcription factors, catabolite activator protein (CAP), binds to the appropriate promoter domain when glucose is absent and lactose is present, leading to the activation of *lacZ*. The other transcription factor, rep (short for Lac repressor), binds to the appropriate promoter domain in the absence of lactose. Once bound, rep inhibits the expression of the gene. If neither rep nor CAP is present, you may assume that only a very small (basal) rate of gene expression occurs.

(a) Develop a logic table for the permutations in outcome (transcription of gene *lacZ*) as a function of the two input signals, CAP and rep.

(b) Write a simple logic rule for the expression of the *lacZ* gene as a function of the presence of lactose and glucose (ignore the basal state).

SOLUTION

(a) The logic table is given in Table 24.2.

Table 24.2 Logic table for activity state of gene *lacZ* as a function of input signals CAP and rep

CAP	rep	*lacZ* state
+	−	off
+	+	basal
−	+	activated
−	−	off

(b) A simple rule for the expression logic is given as:

lacZ = lactose AND (NOT(glucose))

because the gene (and its enzyme product) are only required when the primary sugar source (glucose) is not present and the secondary source (lactose) is present.

24.2.1 Circadian Clock Network

Recall from the previous section that the circadian clock orchestrates a number of important metabolic processes in an organism. It does this by regulating the concentration of key proteins in a cycle manner, with a period of (approximately) 24 h. Consider a simplified model of the *Drosophila melanogaster* circadian clock involving two key genes: the period gene (denoted *per*) and the timeless gene (denoted *tim*). Those genes are transcribed into mRNA, exported from the nucleus, and translated into their respective proteins (denoted in Fig. 24.5 by the uppercase convention as PER and TIM). The protein monomers form a dimer, and the dimers of both PER and TIM combine to form a heteromeric complex that reenters the nucleus and suppresses the rate of transcription of the two genes via negative feedback. The kinetic mechanisms for the phosphorylation events are assumed to be Michaelis-Menten form, and the kinetic mechanism for gene regulation (inhibition) follows a Hill mechanism (with a Hill coefficient of 2).

For the assumptions made by Tyson et al. (1999), the two genes can be lumped together, as well as their corresponding proteins and the nuclear and cytoplasmic forms of the dimer. Finally, assuming rapid equilibrium between the monomer and dimer, a second-order set of balances can be developed for the mRNA state M and the protein state P. The resulting pair of differential equations captures the dynamics of the feedback-controlled circuit:

$$\frac{dM}{dt} = \frac{v_m}{1 + (P(1-q)/2P_{crit})^2} - k_m M \qquad (24\text{-}1)$$

$$\frac{dP}{dt} = v_p M \frac{k_{p1}Pq + k_{p2}P}{J_p + P} - k_{p3}P \qquad (24\text{-}2)$$

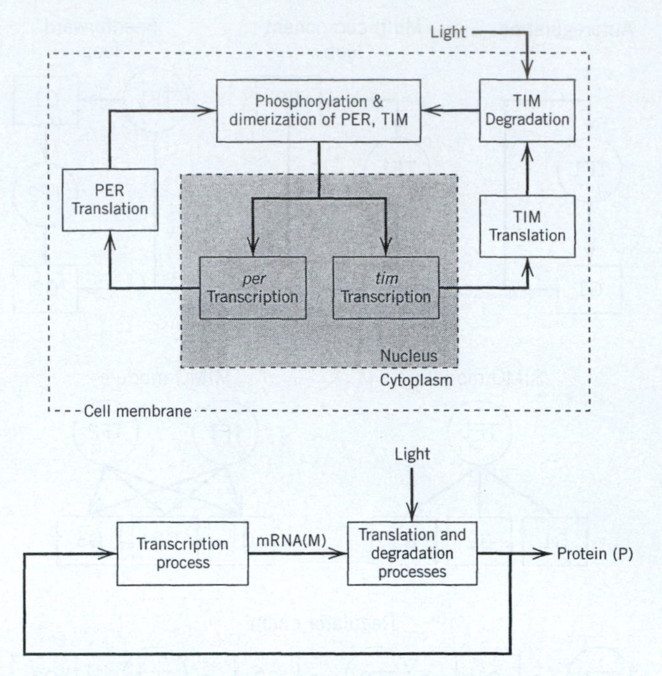

Figure 24.5 Schematic of negative feedback control of *Drosophila* circadian clock (adapted from Tyson et al., 1999): detailed system (top), and simplified model (bottom).

An additional algebraic relationship introduces a more complex dependence of the transcription rate on the protein concentration P:

$$q = \frac{2}{1 + \sqrt{1 + 8K_{eq}P}} \qquad (24\text{-}3)$$

The model parameters and their definitions are a result of the work of Tyson et al. (1999) and are summarized in Table 24.3.

Using a computer package, such as Simulink/MATLAB, the gene regulatory circuit using these defined parameters can be simulated with initial values of M and P equal to [2.0; 2.0]. A 100-h simulation is shown in Fig. 24.6; the period can be calculated from either the mRNA (M) or the Protein (P) trajectory (e.g., time between peaks) and is 24.2 h (i.e., approximately 24 h or "circadian").

A common property of biological closed-loop circuits is that they exhibit remarkable robustness to disturbances and fluctuations in operating conditions. For example, the clock should maintain a nearly 24-hr period, even though the organism is exposed to temperature changes, which affect the rates of biochemical reactions. The model circadian clock can be simulated by perturbing values of the kinetic constants. The same clock simulation is evaluated for the following values of the parameter μ_m: [1.0; 1.1; 1.5; 2.0; 4.0]. The period of the

Table 24.3 Parameter values for circadian clock circuit in Figure 24.5 (C_m denotes transcript concentration and C_p denotes protein concentration).

Parameter	Value	Units	Description
v_m	1	$C_m h^{-1}$	Maximum rate of mRNA synthesis
k_m	0.1	h^{-1}	First-order constant for mRNA degradation
v_p	0.5	$C_p C_m h^{-1}$	Rate constant for translation of mRNA
k_{p1}	10	$C_p h^{-1}$	V_{max} for monomer phosphorylation
k_{p2}	0.03	$C_p h^{-1}$	V_{max} for dimer phosphorylation
k_{p3}	0.1	h^{-1}	First-order rate constant for proteolysis
k_{eq}	200	C_p^{-1}	Equilibrium constant for dimerization
P_{crit}	0.1	C_p	Dimer concentration at half-maximum transcription rate
J_p	0.05	C_p	Michaelis constant for protein kinase

clock lengthens as μ_m is increased, as shown in Fig. 24.7. The period increases as follows: [24.2; 24.5 25.5; 26.4] corresponding to the first four values of μ_m. At the extreme value of 4.0, oscillations are no longer observed, and the system settles to a stable equilibrium. The stability of the oscillations is quite remarkable for such large perturbations in μ_m (over 100%).

Another important feature of the circadian clock is its ability to entrain (i.e., track) an external signal (sunlight), so that the period of the oscillations of mRNA and Protein match exactly the period of the external signal. In this manner, the organism's clock is reset to a period of precisely 24 h. Tyson et al. (1999) show that this can be simulated in the present model by switching the value of K_{eq} to emulate dark-light cycles (i.e., using a square wave with even intervals of light and dark and a 24-h period). In the fly, sunlight appears to modulate the rate of degradation of one of the key proteins in the circuit. This can be achieved in the same simulation model by altering K_{eq}, between 100 and 200, and observing the period of the driven system. Fig. 24.8 illustrates that the oscillations in mRNA and Protein do indeed exhibit a period equal to the forcing signal (in this case, 20 h).

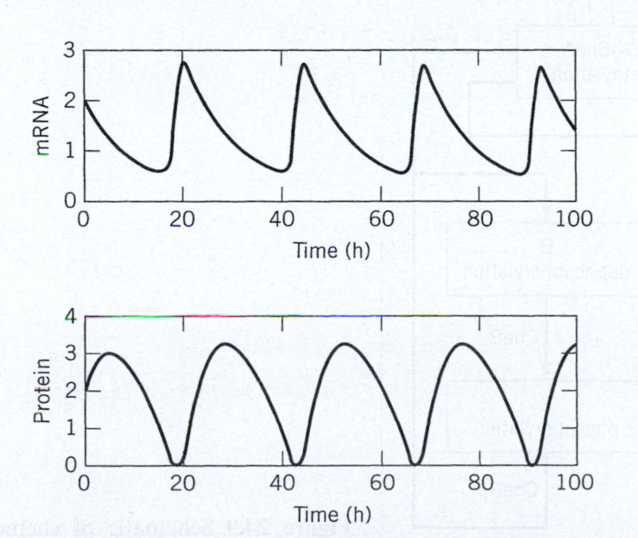

Figure 24.6 Simulation of the circadian clock model.

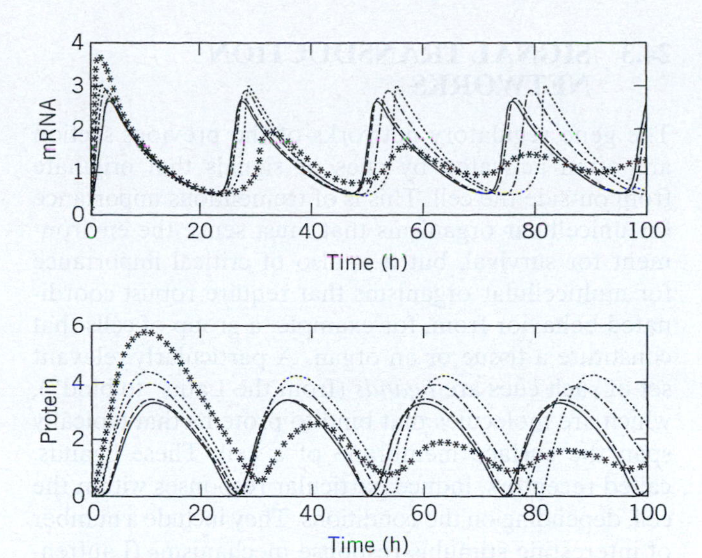

Figure 24.7 Simulation of circadian clock model for varying values of v_m [1.0 (solid), 1.1 (dashed), 1.5 (dash-dot), 2.0 (dotted), 4.0 (asterisk)].

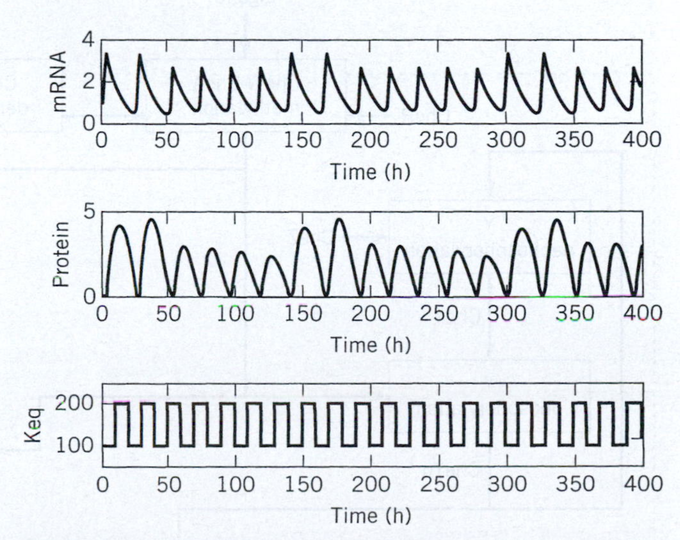

Figure 24.8 Simulation of circadian clock model for entraining signal with period of 20 h.

24.3 SIGNAL TRANSDUCTION NETWORKS

The gene regulatory networks of the previous section are often activated by cues or signals that originate from outside the cell. This is of tremendous importance for unicellular organisms that must sense the environment for survival, but it is also of critical importance for multicellular organisms that require robust coordinated behavior from, for example, a group of cells that constitute a tissue or an organ. A particularly relevant set of such cues are *ligands* (from the Latin "to bind"), which are molecules that bind to proteins that typically span the surface membrane of a cell. These ligands, called receptors, induce particular responses within the cell, depending on the conditions. They include a number of interesting stimulus-response mechanisms (Lauffenburger and Linderman, 1993):

- Growth factors → cell division
- Necrosis factor → programmed cell death (apoptosis)
- Chemoattractant → chemotaxis
- Insulin → glucose uptake
- Neurotransmitter → secretion by nerve cell
- Extracellular matrix (ECM) protein → adhesion

Once the ligand binds to the receptor, it initiates a series of biochemical reactions that induce a short-term response (e.g., phosphorylation state of an intermediate protein) and/or a longer-term response as a result of a regulated gene response. These networks respond relatively rapidly, exhibiting dynamics with characteristic time scales of seconds to minutes. A cell is often presented with multiple, competing cues, and it processes that information in rich signal transduction networks, to result in the appropriate cellular fate, depending on the context.

In this section, we highlight several signal transduction cascades, to illustrate the rich processing dynamics manifested by these networks.

24.3.1 Chemotaxis

The process of chemotaxis is the directed motion of a cell or cellular organism toward a chemical source, typically a food molecule. This mechanism is also invoked in the response to a detected toxin (i.e., motion away from that source) and is involved in more complex processes, such as development. The process is initiated by the detection of a ligand (e.g., a food molecule) at the cell surface, which invokes a signal transduction cascade and results in the alteration of the motor apparatus responsible for moving the cell.

A simplified version of the biochemical pathway that underlies chemotaxis in *E. coli* is shown in Fig. 24.9. The binding of an attractant molecule (ligand) to the receptor complex CheW-CheA (denoted as W-A) induces the phosphorylation of protein CheY (Y), and the phosphorylated form (Yp) invokes a tumbling motion from the bacteria's flagella. This tumbling motion allows the organism to reorient and search the surrounding space; otherwise, the organism proceeds in a straight run. The ability of CheW-CheA to phosphorylate CheY depends on the methylation state of that complex, which is fine-tuned by the proteins CheR (R) and the phosphorylated form of CheB (Bp), as illustrated in the figure. Feedback is evident in Fig. 24.9, because CheB phosphorylation is mediated by the CheW-CheA complex.

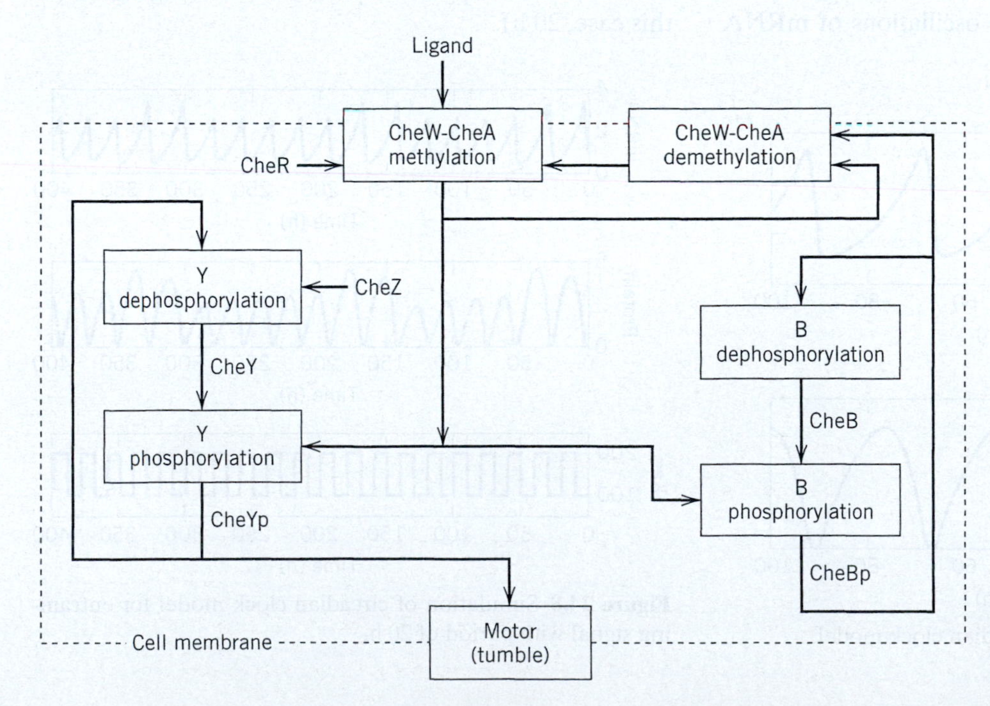

Figure 24.9 Schematic of chemotaxis signaling pathway in *E. coli* (adapted from Rao et al., 2004).

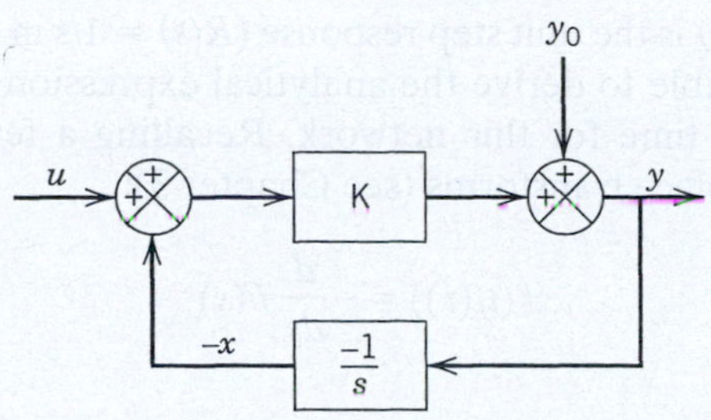

Figure 24.10 Integral control feedback circuit representation of chemotaxis (adapted from Yi et al., 2000).

The signal transduction system that mediates chemotaxis exhibits a type of adaptation in which the response to a persistent stimulus is reset to the pre-stimulus value, thereby enabling an enhanced sensitivity. Several mechanistic explanations can be postulated for this robust behavior, including the following: (i) precise fine-tuning of several parameters to yield a consistent (robust) response under varied conditions, or (ii) inherent regulation that yielded this robust behavior. Utilizing process control principles, it has been demonstrated that the regulatory system exploits integral feedback control to achieve the robust level of adaptation exhibited in chemotaxis (Yi et al., 2000). The chemotaxis network can be reduced to the simple block diagram in Fig. 24.10, in which u denotes the chemoattractant, y denotes the receptor activity, and $-x$ denotes the methylation level of the receptors. It is left as an exercise to show that this circuit ensures that perfect adaptation is achieved (i.e., the receptor activity always resets to zero asymptotically).

This understanding suggests that many seemingly complex biological networks may employ redundancy and other structural motifs or modules to achieve relatively simple overall system behavior.

24.3.2 Insulin-Mediated Glucose Uptake

Muscle, liver, and fat cells in the human body take up glucose as an energy source in response to, among other signals, the hormone insulin, which is secreted by the pancreas. As discussed in Chapter 23, the release of insulin is regulated in a feedback manner by the blood glucose level. In Type 2 diabetes, the insulin signal transduction network is impaired such that insulin does not lead to glucose uptake in these cells. A simplified model of the insulin signaling network can be decomposed into three submodules, as shown in Fig. 24.11. The first submodule describes insulin receptor dynamics: insulin binds to insulin receptor, causing subsequent receptor autophosphorylation. The receptor can also be recycled, introducing additional dynamics in the network. The second submodule describes the phosphorylation cascade downstream from the insulin receptor. The final submodule describes the activation of movement and fusion of specialized glucose transporter

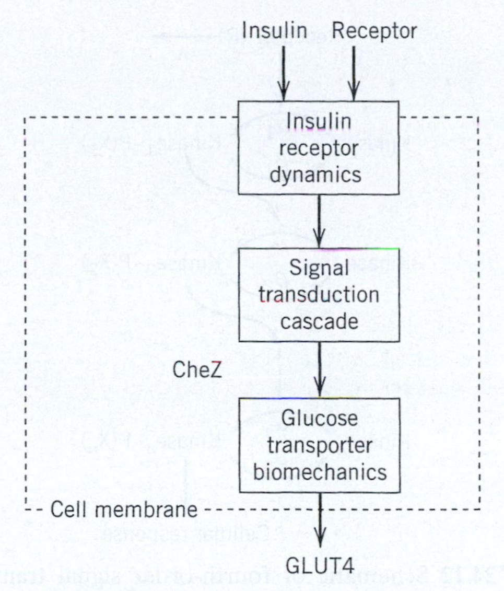

Figure 24.11 Simplified insulin signaling pathway for glucose uptake.

(GLUT4) storage vesicles with the plasma membrane by the intermediate proteins from the second module. These GLUT4 transporters allow glucose molecules to enter the cell. Each of the three modules contains submodules that consist of layers of feedback.

24.3.3 Simple Phosphorylation Transduction Cascade

In signal transduction, a receptor signal is processed in a cascaded pathway, to yield a cellular response. For the example considered here, the processing consists of a sequence of kinase- and phosphatase-catalyzed reaction steps, consisting of phosphorylation and dephosphorylation, respectively. The key performance attributes of such a system are (i) the speed at which a signal arrives to the destination, (ii) the duration of the signal, and (iii) the strength of the signal. Under conditions of weak activation (low degree of phosphorylation), the individual steps in the signal transduction cascade can be modeled as a set of linear ODEs (Heinrich et al., 2002):

$$\frac{dX_i}{dt} = \alpha_i X_{i-1} - \beta_i X_i \tag{24-4}$$

where α_i is a pseudo first-order rate constant for phosphorylation, β_i is the rate constant for dephosphorylation, and X_i is the phosphorylated form of the kinase (i). Assume that the cascade consists of four stages (levels of phosphorylation), that the corresponding rate constants are equal for all stages ($\alpha_i = \alpha$; $\beta_i = \beta$), and that the receptor inactivation is approximated as an exponential decay with time constant $1/\lambda$ (see Fig. 24.12). The resulting cellular response can be written in the Laplace domain as,

$$Y(s) = \left(\frac{s}{\frac{1}{\lambda}s + 1}\right)\left(\frac{\alpha^4}{(s+\beta)^4}\right)R(s) \tag{24-5}$$

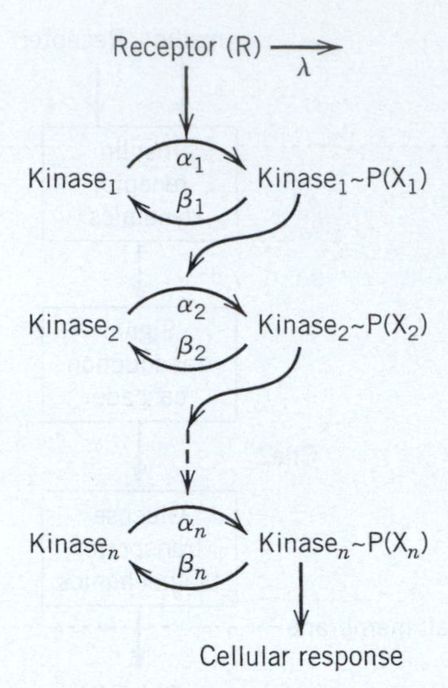

Figure 24.12 Schematic of fourth-order signal transduction cascade for Example 24.3, combined with first-order receptor activation (adapted from Heinrich et al., 2002).

where $R(s)$ is the receptor input and $Y(s)$ is the cellular response.

If the signaling T_{sig} time is defined as the average time to activate a kinase, a suitable expression in the time domain for this quantity is:

$$T_{sig} = \frac{\int_0^\infty ty(t)dt}{\int_0^\infty y(t)dt} \tag{24-6}$$

where $y(t)$ is the unit step response ($R(s) = 1/s$ in (24-5)). It is possible to derive the analytical expression for the signaling time for this network. Recalling a few rules from Laplace transforms (see Chapter 3):

$$\mathcal{L}(tf(t)) = -\frac{d}{ds}F(s) \tag{24-7}$$

and

$$\mathcal{L}\left(\int_0^\infty f(t)dt\right) = F(s = 0) \tag{24-8}$$

Then the following expression for the signal time can be derived:

$$T_{sig} = \frac{\left(-\frac{d}{ds}Y(s)\right)_{s=0}}{Y(0)} = \frac{[\lambda\alpha^4(s+\lambda)^{-2}(s+\beta)^{-4} + 4\lambda\alpha^4(s+\lambda)^{-1}(s+\beta)^{-5}]_{s=0}}{\alpha^4/\beta^4} \tag{24-9}$$

which simplifies to:

$$T_{sig} = \frac{1}{\lambda} + \frac{4}{\beta} \tag{24-10}$$

Notice that the average time through the network (i.e., the signaling time) is not dependent on the rate of phosphorylation (α).

SUMMARY

In this chapter, a number of biological circuit diagrams have been introduced that illustrate the rich array of dynamics and feedback control that exist in all living organisms. Two particular biological processes were considered: the regulation of gene transcription and the protein signal transduction that characterizes cellular stimulus-response mechanisms. The recurring motifs of feedback and feedforward control motivated the application of process control analysis to these problems, to shed light on both the healthy functioning state as well as to promote the investigation of therapies for cases where the natural circuit is impaired (i.e., a disease state).

The rapidly developing field of systems biology continues to make great advancements in the area of medical problems, and the increased understanding of the biological circuits underlying diseases will likely lead to novel therapeutic strategies, as well to the discovery of new drugs. More information is available in more specialized books, including those of Klipp et al. (2005), Palsson (2006), and Alon (2007).

GLOSSARY

Eukaryote: an organism that is comprised of cells (or possibly a single cell, as in yeast) that are divided into substructures by membranes, notably containing a nucleus. Examples include animals, plants, and fungi.

Kinase: an enzyme that catalyzes the transfer of phosphate group to a substrate, leading to phosphorylation of that substrate.

Prokaryote: an organism that is comprised of a single cell that does not contain a separate nucleus. Examples include bacteria and archae.

Promoter: a region of a DNA involved in the regulation of transcription of the corresponding gene.

REFERENCES

Alon, U., *An Introduction to Systems Biology: Design Principles of Biological Circuits*, Chapman & Hall/CRC, New York, 2007.

Alberts, B., D. Bray, A. Johnson, J. Lewis, M. Raff, K. Roberts, and P. Walters. *Essential Cell Biology*, Garland Pub., Inc., New York, 1998.

Barabasi, A. L., Network Biology: Understanding the Cell's Functional Organization, *Nature Rev. Genetics*, **5**, 101 (2004).

Boulos, Z., M. M. Macchi, M. P. Sturchler, K. T. Stewart, G. C. Brainard, A. Suhner, G. Wallace, and R. Steffen, Light Visor Treatment for Jet Lag after Westward Travel across Six Time Zones, *Aviat. Space Environ. Med.*, **73**, 953 (2002).

Campbell, A. M., and L. J. Heyer. *Discovering Genomics, Proteomics, & Bioinformatics*, Benjamin Cummings, San Francisco, CA, 2007.

Daan, S., and C. S. Pittendrigh, A Functional Analysis of Circadian Pacemakers in Nocturnal Rodents. II. The Variability of Phase Response Curves, *J. Comp. Physiol.*, **106**, 253 (1976).

Dunlap, J. C., J. J. Loros, and P. J. DeCoursey (Eds.). *Chronobiology: Biological Timekeeping*, Sinauer Associates, Inc., Sunderland, MA, 2004.

El-Samad, H., S. Prajna, A. Papachristodoulou, J. Doyle, and M. Khammash, Advanced Methods and Algorithms for Biological Networks Analysis, *Proc. IEEE*, **94**, 832 (2006).

Heinrich, R., B. G. Neel, and T. A. Rapoport, Mathematical Models of Protein Kinase Signal Transduction, *Molecular Cell, 9, 957 (2002).*

Herzog, E. D., S. J. Aton, R. Numano, Y. Sakaki, and H. Tei, Temporal Precision in the Mammalian Circadian System: A Reliable Clock from Less Reliable Neurons, *J. Biol. Rhythms*, **19**, 35 (2004).

Hood, L., J. R. Heath, M. E. Phelps, and B. Lin, Systems Biology and New Technologies Enable Predictive and Preventative Medicine, *Science*, **306**, 640 (2004).

Kitano, H., Systems Biology: A Brief Overview, *Science*, **295**, 1662 (2002).

Klipp, E., R. Herwig, A. Kowald, C. Wierling, and H. Lehrach. *Systems Biology in Practice*, Wiley-VCH, Weinheim, 2005.

Lauffenburger, D. A., and J. J. Linderman. *Receptors: Models for Binding, Trafficking, and Signaling*, Oxford University Press, New York, 1993.

Lee, T. I., N. J. Rinaldi, F. Robert, D. T. Odom, Z. Bar-Joseph, G. K. Gerber, N. M. Hannett, C. T. Harbison, C. M. Thompson, I. Simon, J. Zeitlinger, E. G. Jennings, H. L. Murray, D. B. Gordon, B. Ren, J. J. Wyrick, J. B. Tagne, T. L. Volkert, E. Fraenkel, D. K. Gifford and R. A. Young, Transcriptional Regulatory Networks in *Saccharomyces cerevisiae*. *Science*, **298**, 799 (2002).

Liu, A. C., D. K. Welsh, C. H. Ko, H. G. Tran, E. E. Zhang, A. A. Priest, E. D. Buhr, O. Singer, K. Meeker, I. M. Verma, F. J. Doyle III, J. S. Takahashi, and S.A. Kay, Intercellular Coupling Confers Robustness against Mutations in the SCN Circadian Clock Network, *Cell*, **129**, 605 (2007).

National Research Council, *Network Science*, National Academies Press, Washington DC, 2005.

Palsson, B., *Systems Biology: Properties of Reconstructed Networks*, Cambridge University Press, New York, 2006

Ptashne, M. and A. Gann. *Genes and Signals*, Cold Spring Harbor Laboratory Press, Cold Spring Harbor, New York, 2002.

Rao, C.V., J. R. Kirby, and A. P. Arkin, Design and Diversity in Bacterial Chemotaxis: A Comparative Study in *Escherichia coli* and *Bacillus subtilis*, *PLoS Biology*, **2**, 239 (2004).

Reppert, S. M. and D. R. Weaver, Coordination of Circadian Timing in Mammals, *Nature*, **418**, 935 (2002).

Sedaghat, A. R., A. Sherman, and M. J. Quon, A Mathematical Model of Metabolic Insulin Signaling Pathways, *Amer. J. Physio-End. Met.*, **283**, E1084 (2002).

Shen-Orr, S. S., R. Milo, S. Mangan and U. Alon, Network Motifs in the Transcriptional Regulation Network of *Escherichia coli*, *Nature Genetics*, **31**, 64 (2002).

Tyson, J. J., C. I. Hong, C. D. Thron, and B. Novak, A Simple Model of Circadian Rhythms Based on Dimerization and Proteolysis of PER and TIM, *Biophys. J.*, **77**, 2411 (1999).

Yi, T. M., Y. Huang, M. I. Simon, and J. Doyle, Robust Perfect Adaptation in Bacterial Chemotaxis through Integral Feedback Control, *Proc. Nat. Acad. Sci. USA*, **97**, 4649 (2000).

EXERCISES

24.1 In this exercise, treat the components as simple (reactive) chemical species and perform the appropriate (dynamic) material balance. Assume that a messenger RNA (*mRNA*) is produced by a constant (basal) expression rate from a particular gene. In addition, assume that the *mRNA* degrades according to a first-order decay rate.

(a) Write the equation for the dynamics of the mRNA concentration as a function of the expression rate (G_0) and the decay rate constant (k_d^{mRNA}).

(b) Assume that each *mRNA* molecule is translated to form *p* copies of a protein product, *P*. Furthermore, the protein is subject to first-order degradation, with a decay rate constant (k_P^{mRNA}). Write the equation for the dynamics of the protein concentration.

(c) Assume that the system has been operating for some time at a constant gene expression rate (G_0), and then the expression rate changes instantaneously to a value G_1. Derive an analytical expression for the transient responses for *mRNA* and *P*.

24.2 Consider the block diagram in Fig. E24.2 of the multiple feedback loops involved in the Central Dogma schematic from Fig. 24.3, namely genetic regulation (C_1), translational regulation (C_2), and enzyme inhibition (C_3). Assume that the processes P_1, P_2, and P_3 obey first-order dynamics, with corresponding gains and time constants (K_i, τ_i).

(i) Derive the transfer function from the external input (u) to the output (y) for each of the three cases shown in Figure E24.2 (a), (b), (c).

(ii) Assume that the feedback mechanisms operate via proportional control with corresponding controller gains (K_{ci}). Derive the closed-loop transfer function from the external input (u) to the output (y) in block diagram (b).

(iii) Consider a simplified biological circuit in which only genetic regulation is active (C_1). Derive the closed-loop transfer function and comment on the key differences between this transfer function and the one from part (b).

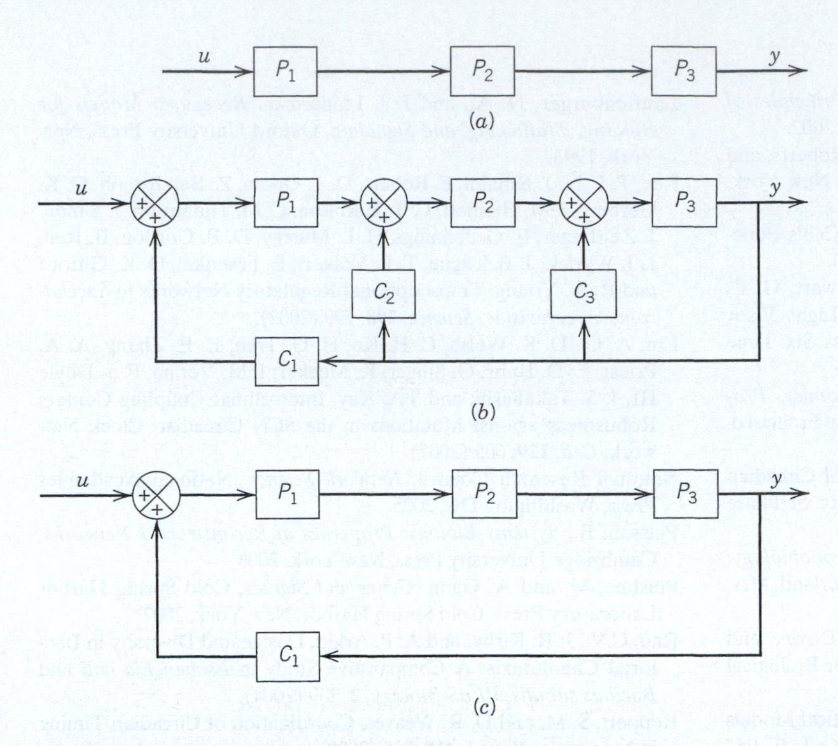

Figure E24.2

(iv) Give several reasons why the natural feedback architecture with all three controllers operating is more effective than the control architecture in part (c).

24.3 As a specific biological example for Exercise 24.2 and Figure E24.2(b),* the synthesis of tryptophan can be described by the following set of material balances:

$$\frac{d}{dt}[O_R] = k_1[O_t]C_1[T] - k_{d1}[O_R] - \mu[O_R]$$

$$\frac{d}{dt}[mRNA] = k_2[O_R]C_2[T] - k_{d2}[mRNA] - \mu[mRNA]$$

$$\frac{d}{dt}[E] = k_3\,[mRNA] - \mu[E]$$

$$\frac{d}{dt}[T] = k_4C_3[T][E] - g\,\frac{[T]}{[T]+K_g} - \mu[T]$$

where k_1, k_2, k_3, and k_4 represent kinetic rate constants for the synthesis of free operator, mRNA transcription, translation, and tryptophan synthesis, respectively. Parameters O_t, μ, k_{d1}, and k_{d2} refer to total operator site concentration, specific growth rate of *E. coli*, degradation rate constants of free operator O_R, and mRNA, respectively. E and T represent concentrations of enzyme anthranilate synthase and tryptophan, respectively, in the cell. K_g and g are the half saturation constant and kinetic constant for the uptake of tryptophan for protein synthesis in the cell. Model parameter values are as follows: $k_1 = 50$ min^{-1}; $k_2 = 15$ min^{-1}; $k_3 = 90$ min^{-1}; $k_4 = 59$ min^{-1}; $O_t = 3.32$ nM; $k_{d1} = 0.5$ min^{-1}; $k_{d2} = 15$ min^{-1}, $\mu = 0.01$ min^{-1}; $g = 25$ μM. min^{-1}; $K_g = 0.2$ μM. Here, controllers $C_1(T)$, $C_2(T)$, and $C_3(T)$ represent repression, attenuation, and inhibition, respectively, by tryptophan and are modeled by a particular form of Michaelis-Menten kinetics (the Hill equation) as follows:

$$C_1(T) = \frac{K_{i,1}^{\eta_H}}{K_{i,1}^{\eta_H} + T^{\eta_H}},\; C_2(T) = \frac{K_{i,2}^{1.72}}{K_{i,2}^{1.72} + T^{1.72}},\; C_3(T) = \frac{K_{i,3}^{1.2}}{K_{i,3}^{1.2} + T^{1.2}}$$

$K_{i,1}$, $K_{i,2}$, and $K_{i,3}$ represent the half-saturation constants, with values $K_{i,1} = 3.53$ μM; $K_{i,2} = 0.04$ μM; $K_{i,3} = 810$ μM, whereas sensitivity of genetic regulation to tryptophan concentration, $\eta_H = 1.92$.

(a) Draw a block diagram, using one block for each of the four states. Comment on the similarities between this diagram and schematic (b) in Fig. E24.2.

(b) Simulate the response of the system to a step change in the concentration of the medium (change g from 25 to 0 μM).

(c) Calculate the rise time, overshoot, decay ratio, and settling time for the closed-loop response.

(d) Omit the inner two feedback loops (by setting C_2 and C_3 to 0) and change the following rate constants: $K_{i,1} = 8 \times 10^{-8}$ μM; $\eta_H = 0.5$. Repeat the simulation described in part (b), and obtain the new closed-loop properties for this network (compared to part (c)).

24.4 Consider Section 24.3.3, where the dynamic properties of a signal transduction were analyzed. Two properties of interest are the signal duration and the amplitude of the signal.

(a) The following definition is used for signal duration:

$$T_{dur} = \sqrt{\frac{\int_{t=0}^{\infty} t^2 y(t)dt}{\int_{t=0}^{\infty} y(t)dt} - T_{sig}^2}$$

where T_{sig} was defined in Section 24.3.3. Use Laplace transforms to derive an expression for the signal duration as a function of the parameters in the phosphorylation cascade.

*The authors acknowledge Profs. Bhartiya, Venkatesh, and Gayen for their help with formulating this problem.

(b) Define the signal amplitude as:

$$A = \frac{\int_{t=0}^{\infty} y(t)dt}{2T_{dur}}$$

Use Laplace transforms to derive an expression for the signal amplitude as a function of the parameters in the phosphorylation cascade.

24.5 Consider the simplified version of the chemotaxis circuit in Fig. 24.10.

(a) Derive the conditions for the process gain K that ensure that the receptor activity is always reset to zero and even for the case of a persistent ligand signal.
(b) Show that the closed-loop transfer function from the ligand to the receptor activity is equivalent to a first-order transfer function with numerator dynamics.
(c) Comment on the biological relevance of the result in part (b), particularly for a ligand signal that is fluctuating.

24.6 An interesting motif in biological circuits is a switch, in which the system can change from (effectively) one binary state to another. An analysis of a continuous reaction network reveals a rise to a switchlike response (also referred to as ultrasensitivity). Consider interconversion of a protein from its native state P to an activated form P^*, catalyzed by the enzymes E_1 and E_2:

$$P + E_1 \leftrightarrow PE_1 \leftrightarrow P^* + E_1$$

$$P^* + E_2 \rightarrow P^*E_2 \rightarrow P + E_2$$

(a) Assume that all reaction steps obey mass-action kinetics. What is the steady-state dependence of P^* as a function of the concentration of E_1? (Assume that total amount of E_1 E_2, and P are all constant and that P is in excess compared to E_1 and E_2)
(b) Alternate starting point for problem: you should be able to rearrange the solution as follows

$$\frac{V_1}{V_2} = \frac{P^*/P_T(1-P^*/P_T+K_1)}{(1-P^*/P_T)(P^*/P_T+K_2)}$$

where V_1 is proportional to the total E_1 in the system, E_2 is proportional to the total V_2 in the system, P_T is the total protein concentration (in all forms), and K_1 and K_2 are suitable combinations of the rate constants for the reactions previously described.

For $K_1 = 1.0$, $K_2 = 1.0$, plot the steady-state locus of solutions for P^*/P_T versus V_1/V_2.
(c) Assume that the two enzymes operate in a saturated regime, i.e., the reactions follow zero-order kinetics with respect to the enzymes. Use the expression from part (b) to plot the steady-state locus for this extreme situation (i.e., $K_1 = 0$, $K_2 = 0$).
(d) Comment on the difference in shape of the gain functions in parts (b) and (c). Based on the initial problem description, explain how biology can produce switchlike behavior in this system.

Appendix A

Digital Process Control Systems: Hardware and Software

APPENDIX CONTENTS

Process control implemented by computers has undergone extensive changes in both concepts and equipment during the past 50 years. The feasibility of digital computer control in the chemical process industries was first investigated in the mid-1950s. During that period, studies were performed to identify chemical processes that were suitable for process monitoring and control by computers. These efforts culminated in several successful applications, the first ones being a Texaco refinery and a Monsanto chemical plant (both on the Gulf Coast) using mainframe computers. The first commercial systems were slow in execution and massive in size compared with the computers available today. They also had very limited capacity. For example, a typical first-generation process control computer had 32K RAM and disk storage of 1MB.

The functionalities of these early control systems were limited by capabilities of the existing computers rather than the process characteristics. These limitations, coupled with inadequate operator training and an unfriendly user interface, led to designs that were difficult to operate, maintain, and expand. In addition, many systems had customized specifications, making them extremely expensive. Although valuable experience was gained in systems design and implementation, the lack of financial success hindered the infusion of digital system applications into the process industries until about 1970, when inexpensive microprocessors became available commercially (Lipták, 2005).

During the past 40 years, developments in microelectronics and software technologies have led to the widespread application of computer control systems. Digital control systems have largely replaced traditional analog instrument panels, allowing computers to control process equipment while monitoring process conditions. Technological advancements, such as VLSI (very large-scale integrated) circuitry, object-oriented programming techniques, and distributed configurations have improved system reliability and maintainability while reducing manufacturing and implementation cost. This cost reduction has allowed small-scale applications in new areas, for example, microprocessors in single-loop controllers and smart instruments (Herb, 1999). Programmable logic controllers have also gained a strong foothold in the process industries.

Increased demand for digital control systems created a new industry, consisting of systems engineering and service organizations. Manufacturing companies moved toward enterprise-wide computer networks by interfacing process control computers with business computer networks. These networks permit all computers to use the same databases in planning and scheduling (see Chapter 19), and they also allow access to operator station information from locations outside the plant.

In the following sections, we provide an overview of the hardware and software used for process control. The distributed control system configuration is described first, followed by data acquisition for different signal types. Digital hardware is then considered, and concluding with a description of control system software organization and architectures.

A.1 DISTRIBUTED DIGITAL CONTROL SYSTEMS

The revolutionary development in microelectronics and telecommunications hastened the evolution of distributed computer networks. In the 1970s, first-generation *distributed control systems* (DCS) replaced the single-mainframe design used previously in process control with a number of identical minicomputers that operated independently of each other. Removable media such as magnetic tapes were used for information transmission. Networking allowed these computers to share resources and/or information.

Computers physically located in different plant areas, which control nearby processes, are said to be *geographically distributed*. More than one computer may share the control of one or more processes. When the control functions are distributed over more than one computer or device, the system is said to be *logically distributed*. Process control networks tend to be distributed both geographically and logically, the extent of which depends on execution priority and complexity. Applications often utilize a variety of digital devices, such as workstations in a distributed control system, personal computers (PC), single-loop controllers (SLC), and programmable logic controllers (PLC).

During the 1980s, the standard distributed digital control network topology was the star configuration, where individual satellite nodes communicated with each other via an arbitrator node. The arbitrator was often the main computer of that system, located in or near the central control room. This computer supported the operator interface and a number of other functions not normally implemented in the satellite computers, which were located in processing areas. One inherent flaw of this scheme is that the operator supervisory and control capability was lost when the main computer failed, even though the satellites continued to function (Lipták, 2005).

Currently open system designs with global bus architecture and local area networks (LANs) are being used for computer control, as shown in Fig. A.1. Unlike earlier networks, which were normally isolated, the LANs are often connected to other networks via gateway devices. The traditional host computer functions are divided functionally and are implemented in separate autonomous computers, which share the same data bus. When more than one operator interface node is installed, the operator interface to the process can be maintained even when several operator stations fail. A DCS for process control is fundamentally the same as for other real-time distributed systems used in business data centers or server facilities (Herb, 1999; Lewis et al., 2006), although specialized hardware such as data acquisition equipment is required.

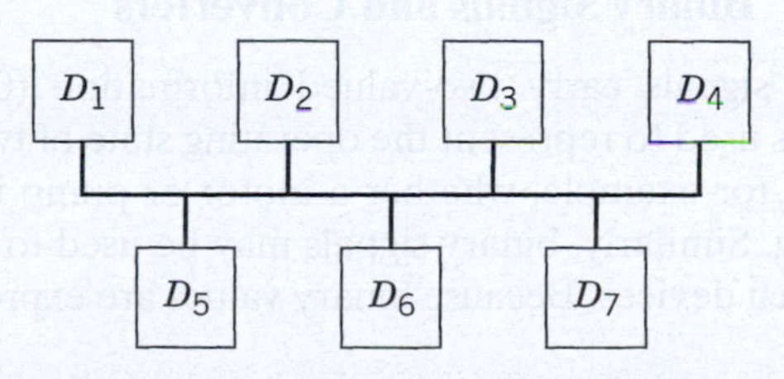

Figure A.1 Global bus architecture for digital process control with different devices D_i.

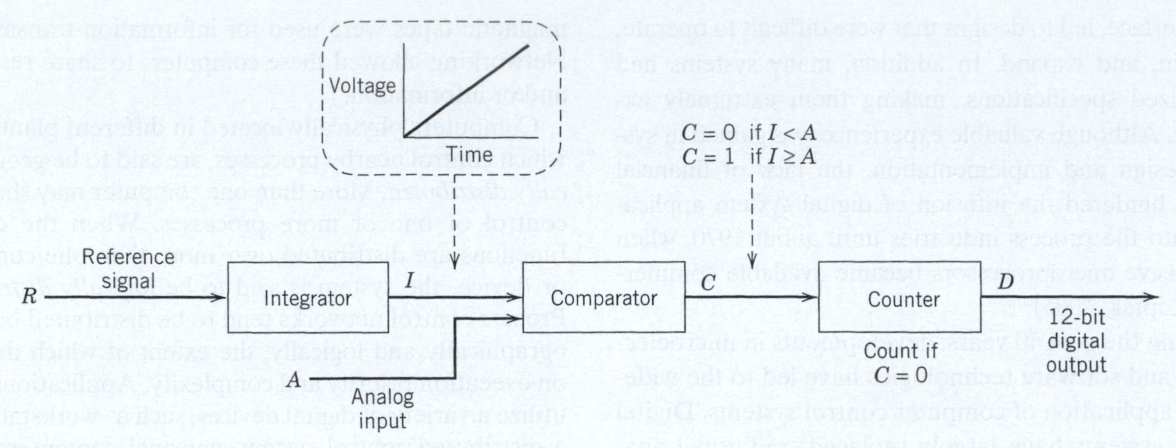

Figure A.2 A conceptual diagram of a voltage A/D converter.

A.2 ANALOG AND DIGITAL SIGNALS AND DATA TRANSFER

Field instrumentation is essential for process control and monitoring. For digital computers to monitor and control processes, they must be able to acquire data from these instruments and implement control based on the sensed information. Special devices are required to convert analog signals to and from digital form. Although analog signals have traditionally been used to transmit data within the plant, the availability of digital architectures such as Fieldbus and Profibus (Berge, 2002) is changing this situation.

A.2.1 Analog Signal Representation

Analog signals occur in the physical world as continuous time-varying signals that can have any value in a specified range. In contrast, discrete signals are limited to a defined set of values. To interface digital computers with measurements from field instruments, electrical signals must be converted to a form acceptable to digital computers, and vice versa. Analog-to-digital and digital-to-analog conversion is performed by simple devices called A/D and D/A converters[1] (ADCs and DACs). Analog electrical signals are in either voltage or current form. As a result of the transmitter standards discussed in Chapter 9, standard interfaces are available for every measured and manipulated variable.

A.2.2 Binary Signals and Converters

Binary signals carry two-valued information (0 or 1), which is used to represent the operating state of two-state devices, for example, whether a motor or pump is off or running. Similarly, binary signals may be used to start or stop such devices. Because binary values are expressed in bits (elements that are 0 or 1), they are packed in clusters of a certain length (bytes = 8 bits) or according to computer word sizes. Often, a light-emitting diode (LED) is attached to an instrument for state indication.

Practically all binary signals use a zero voltage (AC or DC) to represent *logical zero*. Different voltage levels, for example, 5, 15, and 24 VDC and 24 and 110 VAC, are used to represent *logical one*. Other commonly used types of binary signals include pulse trains, which are described below. More details are available elsewhere (Khambata, 1987; Johnson, 2005).

A.2.3 Analog Signals and Converters

Because digital computers are not capable of storing data with infinite precision, measurements must be *quantized*. Similarly, the control actions calculated from these process measurements are quantized according to the computer precision. Thus, a fixed number of bits is used to represent the digitized version of an analog measurement. Most process control-oriented ADCs and DACs utilize a 12-bit unsigned integer representation. Thus, there are 2^{12} or 4096 quantization levels for each process variable. This resolution is better than 0.025%, which is lower than typical noise levels in electrical signals. For high-precision applications, up to 24-bit representation is used.

To digitize an analog input, the unknown process signal is compared with a known signal (Johnson, 2005). Figure A.2 illustrates a simple voltage ADC, which includes an integrator, a comparator, and a counter. The unknown signal is used as an input of the comparator, which compares it with a trial signal generated by the integrator. In this example, the trial signal is a ramp voltage. Assuming 12-bit representation, the ramp voltage is increased by 1/4096 of the nominal voltage span of the

[1]Pronounced "A-to-D" and "D-to-A."

unknown signal, at a preset frequency determined by a quartz crystal. The counter is incremented at the same frequency. The comparator signals the counter to freeze its content when the ramp voltage equals or exceeds the unknown voltage. The time it takes the ramp voltage to equal or exceed the unknown signal is proportional to the magnitude of the unknown signal. At the beginning of the next sampling interval, the value stored in the counter is transferred to another register for processing. The counter is zeroed, and the comparator output is reset. The ramp voltage is then returned to the lower bound of the nominal voltage range, and the process described above is repeated. Other A/D conversion methods exist, but these methods use different techniques to generate the trial signals.

D/A converters are based on a different principle that involves arrays of resistors. To convert a digital value, the bits of its digital representation are fed into the resistor array simultaneously. The array performs the electrical equivalence of a weighted sum of each bit. The voltage level of the array output is proportional to the analog value and amplified to the desired signal level.

Current signal converters, for both inputs and outputs, operate using the same principles as their voltage counterparts. Because of the trial-and-match type operations, ADCs for analog signals are slow in execution compared with other types of signal converters. It should be noted that among the various types of signal converters discussed, only ADCs require the explicit use of microprocessors.

A.2.4 Pulse Trains

A *pulse train* is a special type of binary signal that is used to convey analog information. This can be accomplished by measuring the frequency of the pulses (usually for inputs), while on-time ratio over a period (the fraction that the period of the pulse is equal to 1) is used for outputs. This is also called *duty cycle* or absolute-on-time (Johnson, 2005). Although process control computers are quite capable of handling low-frequency pulse signals, this is rarely done unless only a few signals are involved. To process high-frequency pulses for a large number of signals, special pulse-counting ADCs and pulse-generating DACs are used. A value indicating the pulse frequency is required to process a pulse output. The low and high instrument limits represent 0% and 100%, respectively, on-time of the pulses.

A pulse input consisting of a train of pulses can be digitized by using a pulse counter, which measures the pulse frequency and converts pulse frequency to a digital representation. The computer maintains an accumulator for the pulse counts; its output after a period of time is proportional to the pulse frequency (Johnson, 2005). For example, a turbine flow meter utilizes a pulse counter to measure the rate of fluid flow. In one full revolution, a fixed amount of fluid flows through the meter, and a single pulse is generated. By determining the pulse frequency, the fluid flow rate can be calculated. Pulse outputs are normally used to manipulate two-state devices to control process variables. For example, suppose a heater is equipped with a constant wattage power supply. Temperature can be controlled by limiting power consumed by the heater, which can be accomplished by turning the heater on and off periodically while regulating the percent on-time. The higher the on-off frequency, the smoother the maintained temperature. For pulse duration outputs (PDOs), the duration of a pulse is proportional to the incremental control applied to an analog device. For example, the pulse duration corresponds to the magnitude of change in valve opening via a stepping motor.

Some very important measurement devices require a programming interface. For example, on-line gas chromatographs are extensively used to measure the compositions of multicomponent streams. The digital output signal indicates the composition, as well as related information, such as the time that the sample was analyzed.

A.2.5 Multiplexers and Signal Multiplexing

A typical DCS monitors a large number of inputs and generates a much smaller number of outputs. Instead of using an ADC for each input signal, a *multiplexer* (MUX) is employed so that a group of signals can share an ADC. The multiplexing and data retrieval are synchronized by a computer and are applicable to high-level signals that are measured in volts. For low-level input signals, such as millivolts from thermocouples and strain gauges, low-level multiplexing must be performed. These MUXs are electromechanical in nature. Alternatively, amplifiers can be used to boost low-level signals in order to employ high-level MUXs directly (Johnson, 2005). Although hardware costs have dropped, the use of MUXs to reduce the number of ADCs still merits consideration in certain cases.

A.3 MICROPROCESSORS AND DIGITAL HARDWARE IN PROCESS CONTROL

Digital systems employed for process control increase in size, scope, and cost according to the following hierarchy:

1. Single-loop controllers
2. Programmable logic controllers
3. Personal computer controllers
4. Distributed control system

These categories are discussed in four subsections below (A.3.1 through A.3.4). Even at the lowest level (SLCs), miniaturization of the integrated circuits permits up to 16 control loops to be incorporated into special-purpose microprocessors. All four types of control hardware systems include redundant hardware for failure protection. They can operate under extreme environmental conditions such as high temperature and can withstand vibrations and shocks. They are often enclosed in special cabinets when sited in explosive or corrosive atmospheres. Nitrogen purge gas is used to maintain a slight positive pressure inside the cabinets and isolate the systems from the hazardous environment or airborne contaminants.

A.3.1 Single-Loop Controllers

The single-loop controller (SLC) is the digital equivalent of analog single-loop controllers. It is a self-contained microprocessor-based unit that can be rack-mounted. Although the basic three-mode (PID) controller function is the same as its analog counterpart, the processor-based SLC allows the operator to select a control strategy from a predefined set of control functions, such as PID, on/off, lead/lag, adder/subtractor, multiply/divider, filter functions, signal selector, peak detector, and analog track. SLCs feature auto/manual transfer switching, multi-set point, self-diagnosis, gain scheduling, and perhaps also time sequencing. Many manufacturers produce single processor units that handle cascade control or multiple loops, typically 4, 8, or 16 loops per unit, and incorporate self-tuning or auto-tuning PID control algorithms. Although designed to operate independently, single-loop controllers have digital communications capability similar to that for a distributed control system (DCS), as discussed in Section A.3.4.

A.3.2 Programmable Logic Controllers

Programmable logic controllers (PLCs) are simple digital devices that are widely used to control sequential and batch processes (see Chapter 22). Although PLCs were originally designated to replace electromechanical relays, they now have additional functions that are usually associated with microprocessors. For example, PLCs can implement PID control and other mathematical operations via specialized software (Hughes, 2005; Webb and Reis, 2002).

PLCs can be utilized as standalone devices or in conjunction with digital computer control systems. Hughes (2005) and Lipták (2005) have summarized the general characteristics of PLCs:

1. ***Inputs/Outputs (I/O).*** Up to several thousand discrete (binary) inputs and outputs can be accommodated. Large PLCs have several hundred analog inputs and outputs for data logging and/or continuous PID control.
2. ***Logic handling capability.*** All PLCs are designed to handle binary logic operations efficiently. Because the logical functions are stored in main memory, one measure of a PLC's capability is its memory scan rate. Another measure is the average time required to scan each step in a logic or ladder diagram (see Chapter 22). Thousands of steps can be processed by a single unit. Most PLCs also handle sequential logic and are equipped with an internal timing capability to delay an action by a prescribed amount of time, to execute an action at a prescribed time, and so on.
3. ***Continuous control capability.*** PLCs with analog I/O capability usually include PID control algorithms to handle up to several hundred control loops. More elaborate PLCs incorporate virtually all of the commonly used control functions covered in Chapters 12, 15, and 16, including PID, on/off, integral action only, ratio and cascade control, low- or high-signal select, lead–lag elements, and so forth. Such PLCs are quite efficient, because internal logic signals are available to switch controller functions.
4. ***Operator communication.*** Older PLCs provide virtually no operator interface other than simple signal lamps to indicate the states of discrete inputs and outputs. Newer models often are networked to serve as one component of a DCS control system, with operator I/O provided by a separate component in the network.
5. ***PLC programming.*** A distinction is made between *configurable* and *programmable* PLCs. The term *configurable* implies that logical operations (performed on inputs to yield a desired output) are located in PLC memory, perhaps in the form of *ladder diagrams* by selecting from a PLC menu or by direct interrogation of the PLC. Usually, the logical operations are put into PLC memory in the form of a higher-level programming language. Most control engineers prefer the simplicity of configuring the PLC to the alternative of programming it. However, some batch applications, particularly those involving complex sequencing, are best handled by a programmable approach, perhaps through a higher-level, computer control system.

A.3.3 Personal Computer Controllers

Because of their high performance, low cost, and ease of use, personal computers (PCs) are a popular platform for process control. When configured to perform scan, control, alarm, and data acquisition (*SCADA*) functions, and when combined with a spreadsheet or database management application, the PC controller can be a low-cost, basic alternative to the DCS.

In order to use a PC for real-time control, it must be interfaced to the process instrumentation. The I/O interface can be located on a board in an expansion slot, or the PC can be connected to an external I/O module using a standard communication port on the PC (e.g., RS-232, RS-422, or IEEE-488). The controller card/module supports 16- or 32-bit microprocessors. Standardization and the high-volume PC market has resulted in a large selection of hardware and software tools for PC controllers (McConnell and Jernigan, 1996; Auslander and Ridgely, 2002).

In comparison with PLCs, PCs have the advantages of lower purchase cost, graphics output, large memory, large selection of software products (including databases and development tools), more programming options (use of C or Java vs. ladder logic), richer operating systems, and open networking. PLCs have the following advantages: lower maintenance cost, operating system and hardware optimized for control, fast boot times, ruggedness, low mean time between failures, longer support for product models, and self-contained units. PC-based control systems are predicted to continue to grow at a much faster rate than PLCs and DCSs during the next decade.

Process control systems should also be *scalable*, which means that the size of the control and instrumentation system is easily expanded by simply adding more devices. This feature is possible because of the availability of open systems (i.e., "plug-and-play" between devices), smaller size, lower cost, greater flexibility, and more off-the-shelf hardware and software in digital control systems. A typical system includes personal computers, an operating system, object-oriented database technology, modular field-mounted controllers, and plug-and-play integration of both system and intelligent field devices. New devices are automatically recognized and configured with the system. Advanced control algorithms can be executed at the PC level (Lipták, 2005).

A.3.4 Distributed Control System

Figure A.3 depicts a representative distributed control system. The DCS system consists of many commonly used DCS components, including MUXs, single-loop and multiple-loop controllers, PLCs, and smart devices. A system includes some or all of the following components (Lipták, 2005):

1. ***Control Network.*** The control network is the communication link between the individual components of a network. Coaxial cable and, more recently, fiber-optic cable have often been used, in competition with ethernet protocols. A redundant pair of cables (dual redundant highway) is normally supplied to reduce the possibility of link failure.
2. ***Workstations.*** Workstations are the most powerful computers in the system, capable of performing functions not normally available in other units. A workstation acts both as an arbitrator unit to

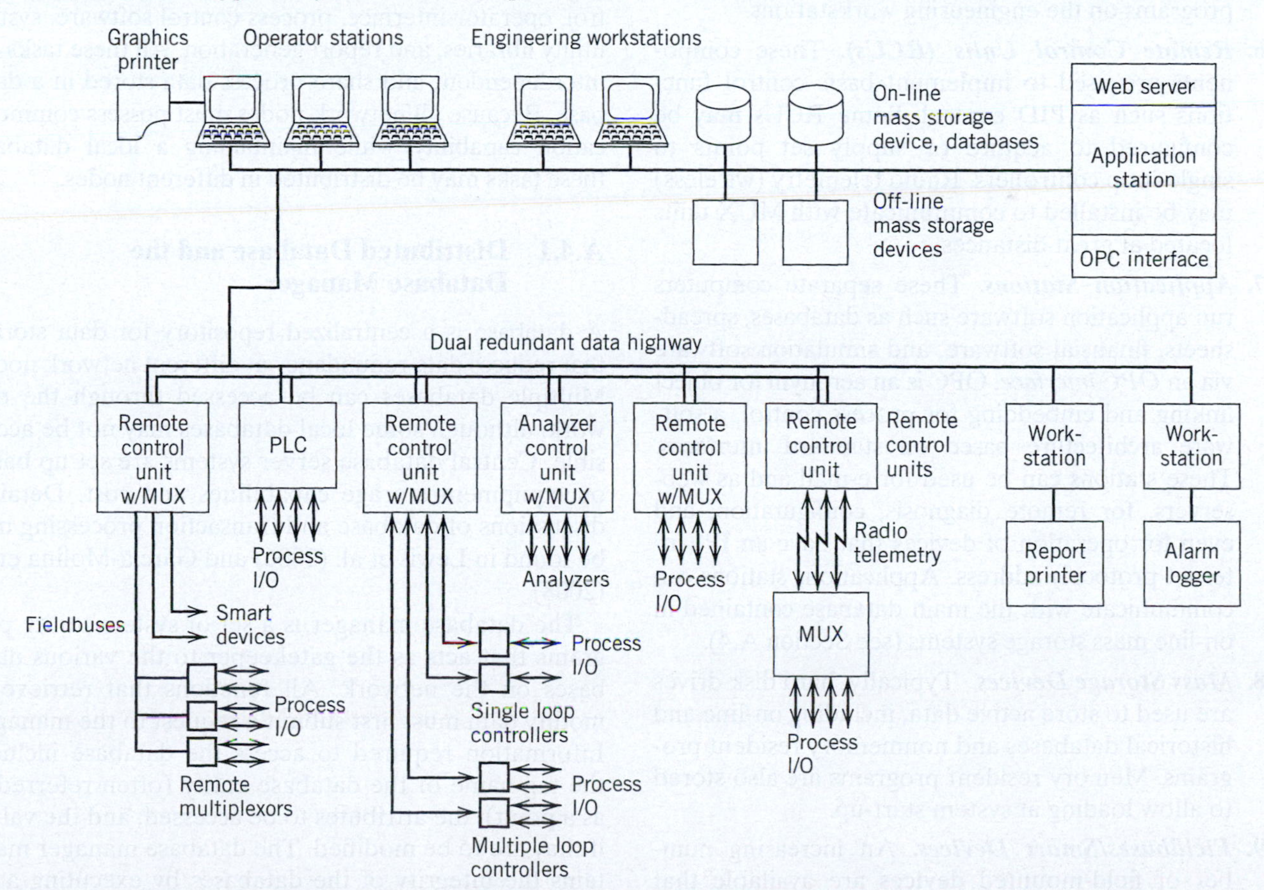

Figure A.3 A typical distributed control system (DCS).

route internodal communications and the database server. An operator interface is supported, and various peripheral devices are coordinated through the workstations. Computationally intensive tasks, such as real-time optimization (Chapter 19) or model predictive control (Chapter 20), are implemented in a workstation.

3. ***Real-Time Clocks.*** Process control systems must respond to events in a timely manner and should have the capability of real-time control. Some DCSs are connected to atomic clock signals to maintain accuracy.
4. ***Operator Stations.*** Operator stations typically consist of color graphics monitors with special keyboards to perform dedicated functions. Operators supervise and control processes from these workstations. Operator stations may be connected directly to printers for alarm logging, printing reports, or process graphics.
5. ***Engineering Workstations.*** These are similar to operator stations but can also be used as programming terminals, that is, used to develop system software. This arrangement reduces compatibility problems between the development and application environments for the system software. Typically, users may also develop their own application programs on the engineering workstations.
6. ***Remote Control Units (RCUs).*** These components are used to implement basic control functions such as PID control. Some RCUs may be configured to acquire or supply set points to single-loop controllers. Radio telemetry (wireless) may be installed to communicate with MUX units located at great distances.
7. ***Application Stations.*** These separate computers run application software such as databases, spreadsheets, financial software, and simulation software via an *OPC interface*. OPC is an acronym for object linking and embedding for process control, a software architecture based on standard interfaces. These stations can be used for e-mail and as webservers, for remote diagnosis, configuration, and even for operation of devices that have an IP (Internet protocol) address. Applications stations can communicate with the main database contained in on-line mass storage systems (see Section A.4).
8. ***Mass Storage Devices.*** Typically, hard disk drives are used to store active data, including on-line and historical databases and nonmemory resident programs. Memory resident programs are also stored to allow loading at system start-up.
9. ***Fieldbuses/Smart Devices.*** An increasing number of field-mounted devices are available that support digital communication of the process I/O in addition to, or in place of, the traditional 4–20 mA current signal. These devices have greater functionality, resulting in reduced setup time, improved control, combined functionality of separate devices, and control-valve diagnostic capabilities. Digital communication also allows the control system to become completely distributed where, for example, a PID control algorithm could reside in a valve positioner or in a sensor/transmitter. See Section A.4.3 for more details.

A.4 SOFTWARE ORGANIZATION

In distributed control systems, computers and other components from a number of vendors may be part of the network. Consequently, software compatibility and portability is a major concern. Portable software is used to ensure consistent computer performance and to avoid duplicating development efforts. Object-oriented programming techniques are employed to minimize customization for different computers and applications.

For a DCS to function properly, a concerted effort of many software tasks is required (Miklovic, 1993; Lipták, 2005). The core of each network node must be a reliable real-time multitasking operating system that is divided functionally into different tasks—that is, communication between DCS nodes, data acquisition and control, operator interface, process control software, system utility libraries, and report generation. All these tasks are interdependent and share process data stored in a database. Because all network nodes must possess communication capability while maintaining a local database, these tasks may be distributed in different nodes.

A.4.1 Distributed Database and the Database Manager

A database is a centralized repository for data storage that reduces data redundancy at different network nodes. Multiple databases can be accessed through the network, although some local databases may not be accessible. Central database server systems are set up based on equipment storage capabilities and cost. Detailed discussions of database and transaction processing may be found in Lewis et al. (2006) and Garcia-Molina et al. (2008).

The database manager is a set of system utility programs that acts as the gatekeeper to the various databases on the network. All functions that retrieve or modify data must first submit a request to the manager. Information required to access the database includes the tag name of the database entity (often referred to as a point), the attributes to be accessed, and the values if they are to be modified. The database manager maintains the integrity of the databases by executing a request only when it is not processing other conflicting

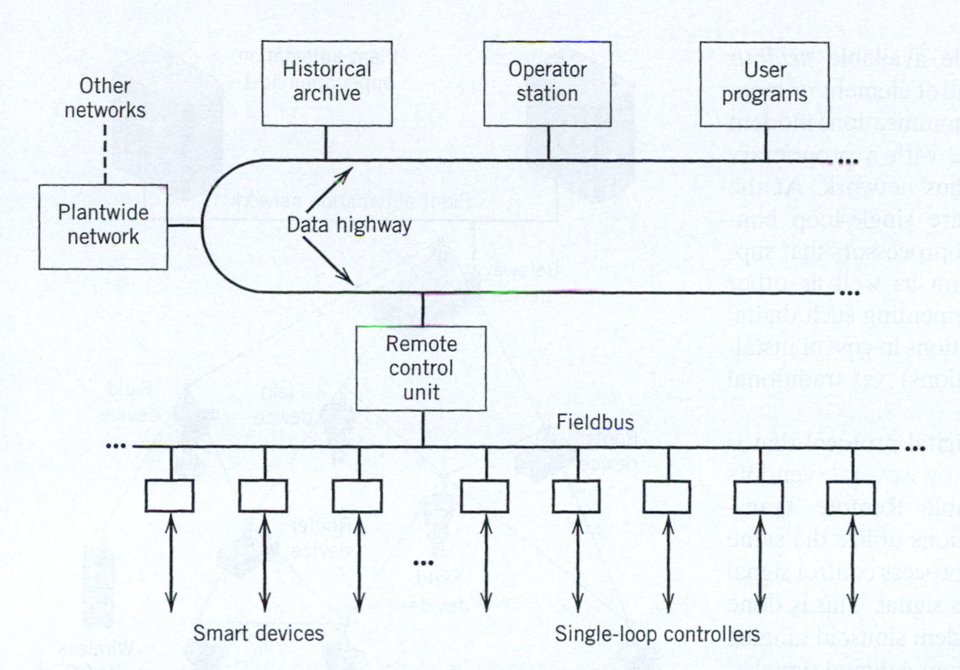

Figure A.4 A DCS using a broadband (high-bandwidth) data highway and fieldbus connected to a single remote control unit that operates smart devices and single-loop controllers.

requests. Although a number of tasks may simultaneously read the same data, simultaneous read/write of the same data item is not permitted.

A.4.2 Internodal Communications

In order for a group of computers to become a network, intercomputer communication is required. Prior to the 1980s, each system vendor used a proprietary protocol to network its computers. Ad hoc approaches were sometimes used to connect third-party equipment but were not cost-effective with regard to system maintenance, upgrade, and expansion. The introduction of standardized communication protocols has decreased capital cost. Most current DCS network protocol designs are based on the ISO-OSI[2] seven-layer model with physical, data link, network, transport, session, presentation, and application layers (Herb, 1999).

An effort in standardizing communication protocols for plant automation was initiated by General Motors in the early 1980s. This work culminated in the Manufacturing Automation Protocol (MAP), which adopted the ISO-OSI standards as its basis. MAP specifies a broadband backbone local area network (LAN) that incorporates a selection of existing standard protocols suitable for discrete component manufacturing. MAP was intended to address the integration of DCSs used in process control. Subsequently, TCP/IP (transmission control protocol/Internet protocol) was adopted for communication between nodes that have different operating systems.

Communication programs also act as links to the database manager. When data are requested from a remote node, the database manager transfers the request to the remote node database manager via the communication programs. The remote node communication programs then relay the request to the resident database manager and return the requested data. The remote database access and the existence of communications equipment and software are transparent to the user.

A.4.3 Digital Field Communications and Fieldbus

Microprocessor-based equipment, such as smart instruments and single-loop controllers with digital communications capability, are now used extensively in process plants. A *fieldbus*, which is a low-cost protocol, is necessary to perform efficient communication between the DCS and devices that may be obtained from different vendors. Figure A.4 illustrates a LAN-based DCS with fieldbuses and smart devices connected to a data highway.

Presently, there are several regional and industry-based fieldbus standards, including the French standard (FIP), the German standard (Profibus), and proprietary standards by DCS vendors, generally in the United States, led by the Fieldbus Foundation, a not-for-profit corporation (Berge, 2002; Thomesse, 1999). International standards organizations have adopted all of these fieldbus standards rather than a single unifying standard. However, there will likely be further developments in fieldbus standards in the future. A benefit of standardizing the fieldbus is that it has encouraged third-party traditional equipment manufacturers to enter the smart equipment market, resulting in increased competition and improved equipment quality.

[2]Abbreviated from *International Organization for Standardization-Open System Interconnection.*

Several manufacturers have made available *fieldbus controllers* that reside in the final control element or measurement transmitter. A suitable communications modem is present in the device to interface with a proprietary PC-based, or hybrid analog/digital bus network. At the present time, fieldbus controllers are single-loop controllers containing 8- and 16-bit microprocessors that support the basic PID control algorithm as well as other functionalities. Case studies in implementing such digital systems have shown significant reductions in cost of installation (mostly cabling and connections) vs. traditional analog field communication.

An example of a hybrid analog/digital protocol that is open (not proprietary) and used by several vendors is the *HART* (Highway Addressable Remote Transducer) protocol. Digital communications utilize the same two wires to provide the 4 to 20 mA process control signal without disrupting the actual process signal. This is done by superimposing a frequency-dependent sinusoid ranging from −0.5 mA to +0.5 mA to represent a digital signal.

A general movement has also begun in the direction of using the high-speed ethernet standard (100 Mbit/s or higher), allowing data transfer by TCP/IP that is used pervasively in computer networking. This allows any smart device to communicate directly with others in the network or to be queried by the operator regarding its status and settings. However, considerable changes in the ethernet standard will be required to make it suitable for process control applications, which provides a more challenging environment than corporate data networks.

Because the HART protocol is widely used due to its similarity to the traditional 4- to 20-mA field signaling, it represents a safe, controlled transition to wireless field communications as an alternative to fieldbus. The HART protocol is principally a master/slave protocol, which means that a field device (slave) speaks only when requested by a master device. An optional communication mode, "burst mode," allows a HART slave device to continuously broadcast updates without stimulus requests from the master device, which is an important attribute for wireless data transmission.

Wireless digital communication to and from the final control element is now commercially available. The advantage of a wireless field network is the potentially reduced cost vs. a wired installation. Hurdles for wireless transmissions include security from non-network sources, transmission reliability in the plant environment, limited bus speed, battery life, and the resistance of the process industry to change. Both point to point and mesh architectures are being commercialized at the device level. *Mesh architectures* utilize the other transmitting devices in the area to receive and then pass on any data transmission, thus re-routing communications around sources of interference. Multiple frequencies within the radio band are utilized to transmit data.

The most recent version of the HART standard, version 7, included a major new communication protocol,

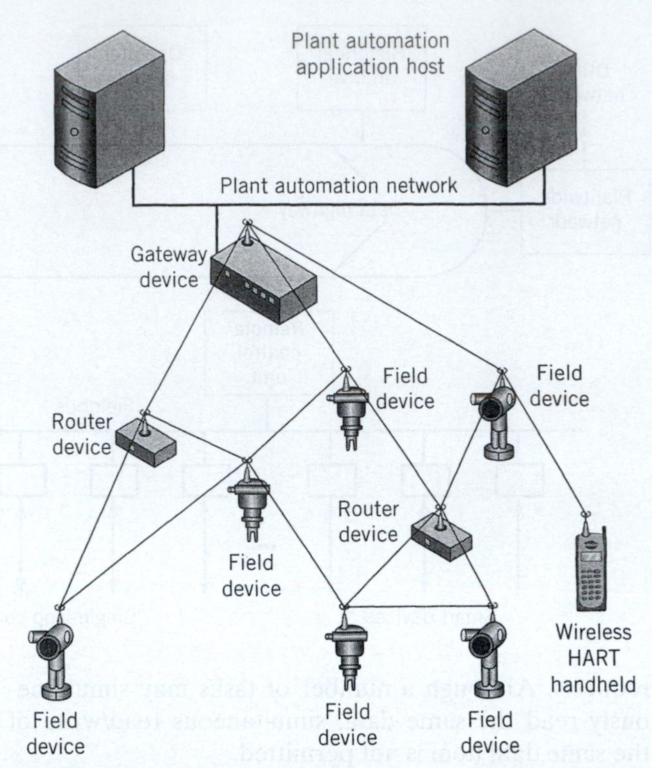

Figure A.5 Example of a WirelessHART network (Mesh Architecture).

WirelessHART™, supporting wireless applications operating in the 2.4 GHz ISM radio band. WirelessHART utilizes IEEE 802.15.4 compatible radios with channel hopping on a packet by packet basis. A WirelessHART network supports a wide variety of devices from many manufacturers. Figure A.5 illustrates the basic network device types for a mesh architecture.

WirelessHART communications are scheduled with precise time synchronization, using an approach referred to as *Time Division Multiple Access* (TDMA). Scheduling is performed by a centralized network manager that uses overall network routing information, in combination with communication requirements of each device. The vast majority of communications are directed along graph routes. The network manager continuously adapts the overall network graph and network schedule to changes in network topology and communication demand.

A.4.4 Data Acquisition

The data acquisition software is utilized to coordinate signal converters and MUXs discussed in Section A.2. Process data are preprocessed before being transferred to databases for storage and retrieval. Alarm condition screening is performed on process data on a periodic basis. A number of data fields and parameters are required for data acquisition and utilization in process control. A *tag name* is an alphanumeric string that uniquely identifies a process I/O point. Most commercial systems use some numeric sequences to associate database points to signal converters and MUXs. Process system and smart devices can frequently monitor the quality of each point and direct it to appropriate operator and control strategies. Lists of tag names and parameters are stored in EEPROM or Flash ROM to prevent loss due to system failure.

Most DCSs provide a pair of alarm bits associated with the instrument limits. For an instrument output signal, the limits prevent transmitting a value that is outside of the specified ranges. If an input value is outside the limits, an alarm action is taken (see Chapter 10).

A.4.5 Process Control Languages

Originally, software for process control utilized high-level programming languages such as FORTRAN and BASIC. Some companies have incorporated libraries of software routines for these languages, but others have developed specialty languages characterized by natural language statements. The most widely adopted user-friendly approach is the fill-in-the-forms or table-driven *process control languages* (PCLs). Typical PCLs include function block diagrams, ladder logic, and programmable logic. The core of these languages is a number of basic function blocks or software modules, such as analog in, digital in, analog out, digital out, PID, summer and splitter. Using a module is analogous to calling a subroutine in conventional Fortran or C programs.

In general, each module contains one or more inputs and an output. The programming involves connecting outputs of function blocks to inputs of other blocks via the graphical user interface. Some modules may require additional parameters to direct module execution. Users are required to fill in templates to indicate the sources of input values, the destinations of output values, and the parameters for forms/tables prepared for the modules. The source and destination blanks may specify process I/O channels and tag names when appropriate. To connect modules, some systems require filling in the tag names of modules originating or receiving data. A completed control strategy resembles a data flow diagram such as the one shown in Fig. A.6.

Many DCSs allow users to write custom code (much as with BASIC) and attach it to data points, so that the code is executed each time the point is scanned. The use of custom code allows many tasks to be performed that cannot be carried out by standard blocks.

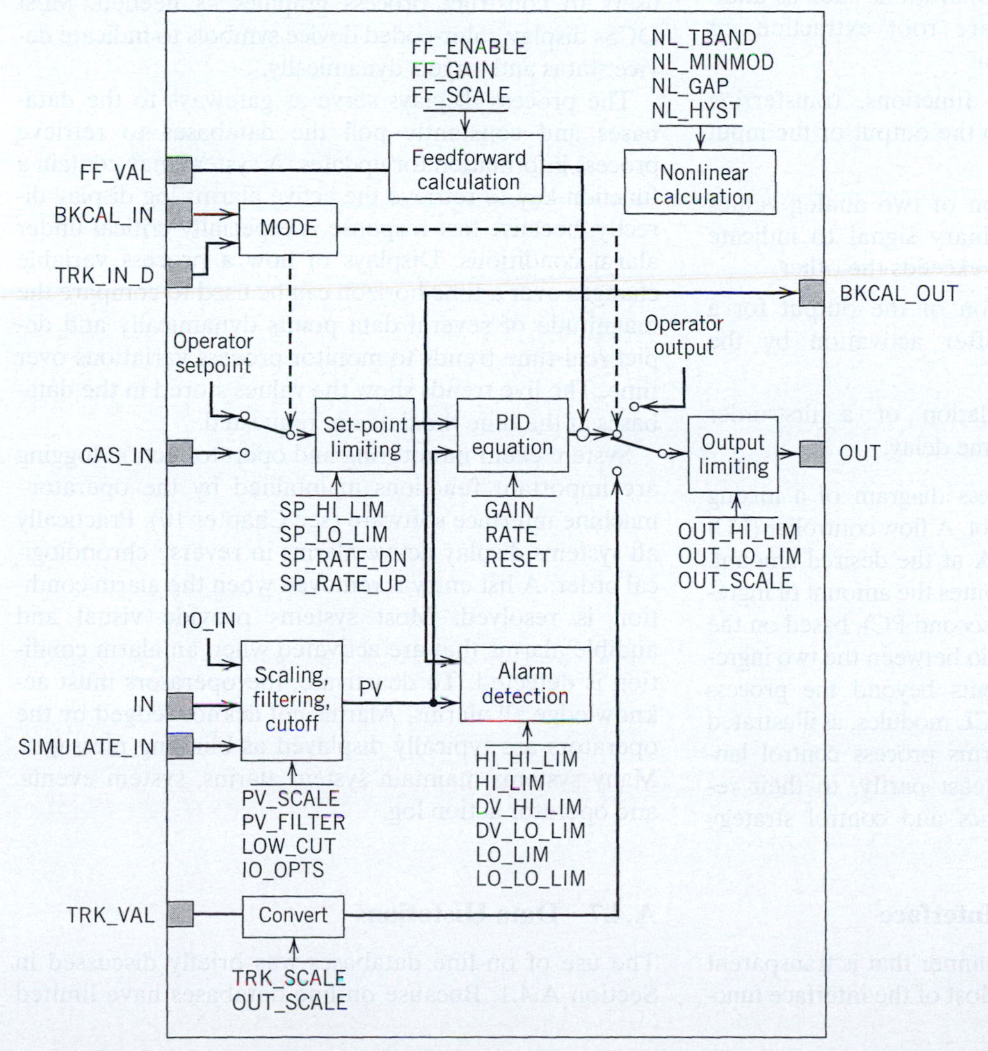

Figure A.6 Function block representation (Courtesy Fisher-Rosemount Systems).

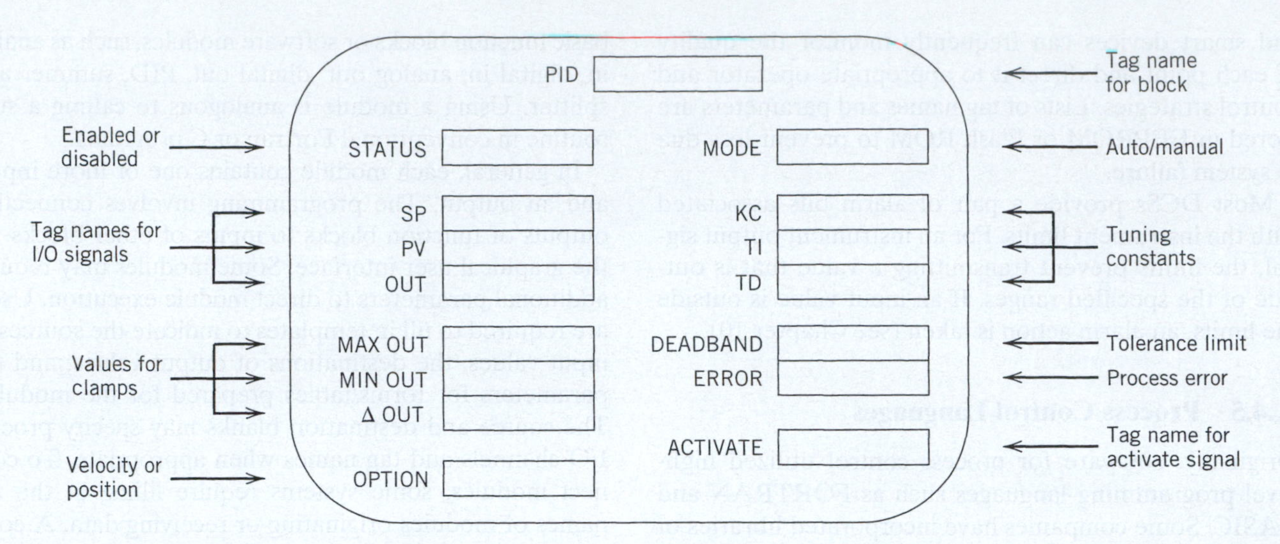

Figure A.7 A display template for PID blocks.

All process control languages contain PID control blocks of different forms (Fig. A.7, also see Chapter 8). Other categories of function blocks include

1. ***Logical operators.*** AND, OR, and exclusive OR (XOR) functions.
2. ***Calculations.*** Algebraic operations such as addition, multiplication, square root extraction, or special function evaluation.
3. ***Selectors.*** Min and max functions, transferring data in a selected input to the output or the input to a selected output.
4. ***Comparators.*** Comparison of two analog values and transmission of a binary signal to indicate whether one analog value exceeds the other.
5. ***Timers.*** Delayed activation of the output for a programmed duration after activation by the input signal.
6. ***Process Dynamics.*** Emulation of a first-order process lag (or lead) and time delay.

Figure A.8*a* shows the process diagram of a mixing process under analog ratio control. A flow controller (FC) is used to maintain ingredient A at the desired amount. An analog calculator (FY) computes the amount of ingredient B to be maintained (by a second FC), based on the desired amount of A and the ratio between the two ingredients. All hardware components beyond the process equipment can be replaced by PCL modules, as illustrated in Fig. A.8*b*. The fill-in-the-forms process control languages owe their success, at least partly, to their resemblance to process schematics and control strategy diagrams.

A.4.6 Operator–Machine Interface

Most DCS tasks execute in a manner that is transparent to the operators or engineers. Most of the interface functions are integrated in the operator control stations equipped with color graphics monitors. Through monitor displays, the operators observe the process operations and their status and issue commands via associated peripheral devices. Operator stations support some graphics building/generation capability, allowing system users to construct process graphics as needed. Most DCSs display color-coded device symbols to indicate device status and targets dynamically.

The process displays serve as gateways to the databases and constantly poll the databases to retrieve process information for updates. A system may contain a function key to retrieve the active alarms log display directly, because fast response is especially critical under alarm conditions. Displays of how a process variable changes over a time horizon can be used to compare the magnitude of several data points dynamically and depict real-time trends to monitor process variations over time. The live trends show the values stored in the databases at the time the data are requested.

System event monitoring and operator action logging are important functions maintained by the operator–machine interface software (see Chapter 10). Practically all systems display active alarms in reverse chronological order. A list entry is removed when the alarm condition is resolved. Most systems provide visual and audible alarms that are activated when an alarm condition is detected. To deactivate, the operators must acknowledge all alarms. Alarms not acknowledged by the operators are typically displayed as blinking messages. Many systems maintain system alarms, system events, and operator action log.

A.4.7 Data Historians

The use of on-line databases was briefly discussed in Section A.4.1. Because on-line databases have limited

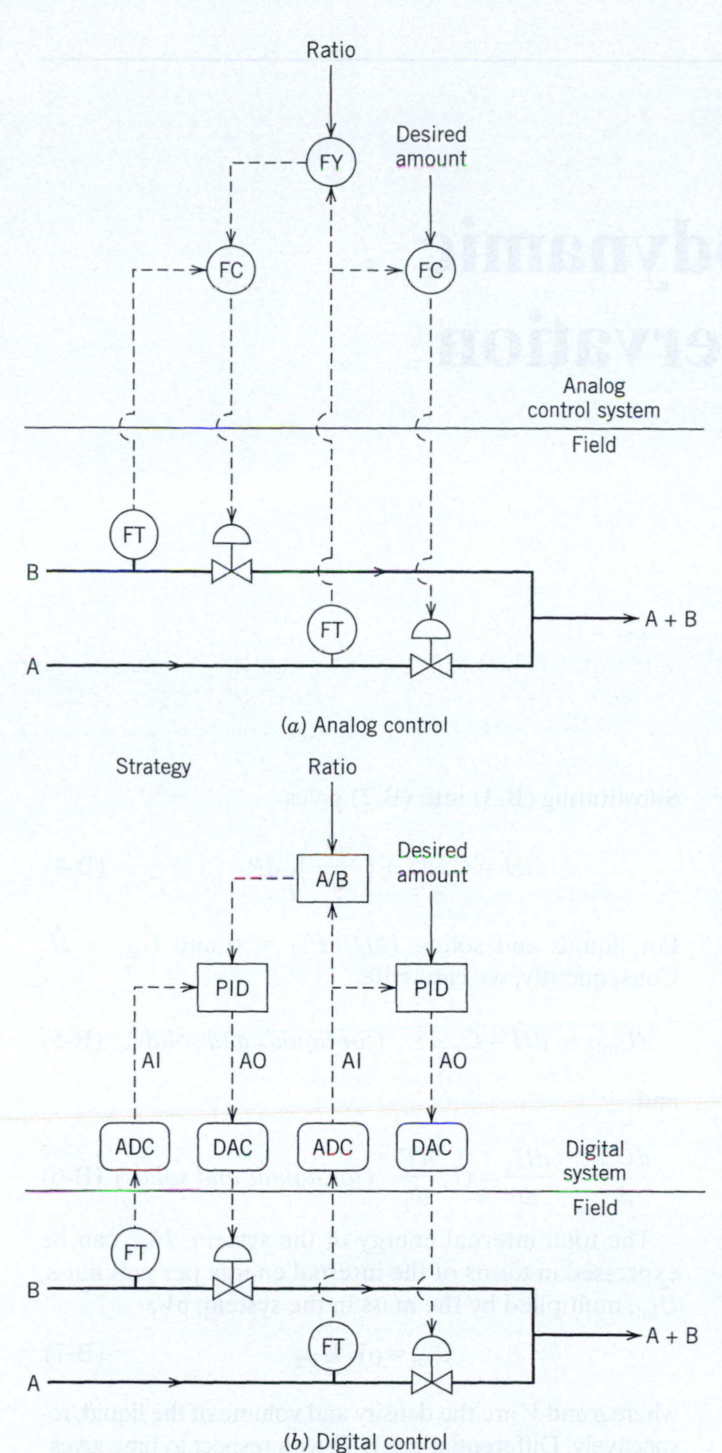

Figure A.8 Analog and digital control of a mixing process.

capacity, the oldest data points are periodically transferred to a historical database. The data stored in a historical database are not normally accessed directly by other subsystems for process control and monitoring. These databases tend to be set up as relational databases, similar to corporate databases (Garcia-Molina et al., 2008; Silberschatz et al., 2005). Periodic reports and long-term trends are generated based on the historical (or archived) data. The reports are often used for long-term planning and system performance evaluations such as statistical process control. The trends may be used to detect process drifts or to compare process variations at different times.

Large industrial plants can have as many as 50,000 measured variables. Sampling periods for many process variables range from seconds to a few minutes. All the acquired data can be stored for relatively short periods of time (for example, weeks to months). However, it is not feasible to store years of historical data as individual data points. Consequently, *data compression* techniques are widely employed. A simple approach is to average data over a specified period of time such as an hour or a day. Other data compression methods only store a new measurement when the process variable has changed significantly from the last stored value (Singhal and Seborg, 2003). Data compression methods based on *wavelet analysis* allow accurate reconstruction of the original data (Walnut, 2002).

REFERENCES

Auslander, D. M., and J. R. Ridgely, *Design and Implementation of Real-Time Software for the Control of Mechanical Systems*, Prentice Hall, Upper Saddle River, NJ, 2002.

Berge, J., *Process Fieldbuses—Engineering Operation and Maintenance*, ISA, Research Triangle Park, NC, 2002.

Garcia-Molina, H., J. D. Ullman, and J. D. Widom, *Database Systems: The Complete Book*, 2d ed., Prentice Hall, Upper Saddle River, NJ, 2008.

Herb, S. M., *Understanding Distributed Processor Systems for Control*, ISA, Research Triangle Park, NC, 1999.

Hughes, T. A., *Programmable Controllers*, 2d ed., ISA, Research Triangle Park, NC, 2005.

Johnson, C. D., *Process Control Instrumentation Technology*, 8th ed., Prentice Hall, Upper Saddle River, NJ, 2005.

Khambata, A. J., *Microprocessors/Microcomputers: Architecture, Software, and Systems*, 2d ed., Wiley, New York, 1987.

Lewis, P. M., A. Bernstein, and M. Kifer, *Databases and Transaction Processing: An Application-Oriented Approach*, 2d ed., Addison-Wesley, New York, 2006.

Lipták, B. G., *Instrument Engineers Handbook*, 4th ed., CRC Press, Boca Raton, FL, 2005.

McConnell, E., and D. Jernigan, Data Acquisition, *The Electronics Handbook*, 2d ed., J. C. Whitaker (Ed.), CRC Press, Boca Raton, FL, 1938, 2005.

Miklovic, D. T., *Real-Time Control Networks*, ISA, Research Triangle Park, NC, 1993.

Silberschatz, A., H. Korth, and S. Sudarshan, *Database Systems Concepts*, 5th ed., McGraw-Hill, New York, 2005.

Singhal, A., and D. E. Seborg, Data Compression Issues with Pattern Matching in Historical Data, *Proc. Amer. Control Conf.*, 3696 (2003).

Thomesse, J. P., Fieldbuses and Interoperability, *Control Engr. Practice*, **7**, 81 (1999).

Walnut, D. F., *An Introduction to Wavelet Analysis*, Birkhanser, Boston, MA, 2002.

Webb, J. W., and R. A. Reis, *Programmable Logic Controllers*, 4th ed., Prentice Hall, Upper Saddle River, NJ, 2002.

Appendix B

Review of Thermodynamic Concepts for Conservation Equations

APPENDIX CONTENTS

The general energy balances in Eqs. 2-10 and 2-11 provide a useful starting point for the development of dynamic models. However, expressions for U_{int} and $\hat{H}$ (or $\tilde{H}$) are required and can be derived from thermodynamic principles. In this appendix, we review fundamental thermodynamic concepts, first for single components and then for multicomponent mixtures. Additional background information is available in thermodynamics textbooks such as Sandler (2006).

B.1 SINGLE-COMPONENT SYSTEMS

Consider a fluid or a solid that consists of a single component such as water or silicon. The enthalpy per unit mass, $\hat{H}$, depends on temperature and pressure. With a slight abuse of standard mathematical notation, we can write

$$\hat{H} = \hat{H}(T, P) \tag{B-1}$$

For differential changes in T and P,

$$d\hat{H} = \left(\frac{\partial \hat{H}}{\partial T}\right)_P dT + \left(\frac{\partial \hat{H}}{\partial P}\right)_T dP \tag{B-2}$$

By definition, the heat capacity at constant pressure, C_p, is defined as

$$C_p \triangleq \left(\frac{\partial \hat{H}}{\partial T}\right)_P \tag{B-3}$$

Substituting (B-3) into (B-2) gives

$$d\hat{H} = C_p\, dT + \left(\frac{\partial \hat{H}}{\partial P}\right)_T dP \tag{B-4}$$

For liquids and solids, $(\partial \hat{H}/\partial P)_T \approx 0$ and $\hat{U}_{int} \approx \hat{H}$. Consequently, we can write

$$d\hat{U}_{int} \approx d\hat{H} = C_p\, dT \quad \textit{(for liquids and solids)} \tag{B-5}$$

and

$$\frac{d\hat{U}_{int}}{dt} \approx \frac{d\hat{H}}{dt} = C_p \frac{dT}{dt} \quad \textit{(for liquids and solids)} \tag{B-6}$$

The total internal energy of the system, U_{int}, can be expressed in terms of the internal energy per unit mass, $\hat{U}_{int}$, multiplied by the mass in the system, ρV,

$$U_{int} = \rho V\, \hat{U}_{int} \tag{B-7}$$

where ρ and V are the density and volume of the liquid, respectively. Differentiating (B-7) with respect to time gives

$$\frac{dU_{int}}{dt} = \frac{d(\rho V \hat{U}_{int})}{dt} \tag{B-8}$$

Suppose that ρ and V are constant. Then substituting (B-6) into (B-8) gives

$$\frac{dU_{int}}{dt} = \rho V \frac{\hat{U}_{int}}{dt} = \rho V C_p \frac{dT}{dt} \tag{B-9}$$

(for liquids with constant ρ and V)

For some modeling activities, it is more convenient to express U_{int} in terms of molar quantities,

$$U_{int} = n\tilde{U}_{int} \tag{B-10}$$

where n is the total number of moles. Then equations analogous to (B-8) and (B-9) can be derived. Equations B-8 and B-9 provide general expressions for the accumulation term in the energy balance of (2-10).

For ideal gases, $\hat{H}$ and $\hat{U}_{int}$ are functions only of temperature, and the following relationships hold:

$$d\hat{H} = C_p dT \quad \text{(for ideal gases)} \tag{B-11}$$

$$\hat{H} = \hat{U}_{int} + RT \quad \text{(for ideal gases)} \tag{B-12}$$

For nonideal (*real*) gases, $\hat{H}$ and $\hat{U}_{int}$ depend on pressure, as well as temperature, as shown in Eq. B-4. Numerical values of $\hat{H}$ and $\hat{U}_{int}$ can be obtained from tables of thermodynamic data or relations.

Consider a liquid or ideal gas at a temperature T. Integrating Eq. B-5 or B-11 from a reference temperature T_{ref} to T provides an expression for the difference between $\hat{H}$ and $\hat{H}_{ref}$, the value of $\hat{H}$ at T_{ref}:

$$\hat{H} - \hat{H}_{ref} = C(T - T_{ref}) \tag{B-13}$$

In (B-13) C is the mean heat capacity over the temperature range from T_{ref} to T. Without loss of generality, we assume that $\hat{H}_{ref} = 0$.

The value of T_{ref} for enthalpy calculations can be selected arbitrarily. For example, the triple point of water is used as the reference point for the steam tables, while 25 °C is a typical choice for physical property tables. For process control calculations, it is often convenient to set $T_{ref} = 0$ or to choose T_{ref} to be an inlet temperature or an initial temperature.

B.2 MULTICOMPONENT SYSTEMS

A key issue for multicomponent systems is: how are the properties of the mixture related to pure component properties? Consider a system that consists of k components. Because the enthalpy depends on composition as well as temperature and pressure, the enthalpy per unit mole of the system, $\tilde{H}$, can be expressed as

$$\tilde{H} = \tilde{H}(T, p, \tilde{x}) \tag{B-14}$$

where $\tilde{x}$ denotes chemical composition. In general,

$$\tilde{H}(T, P, \tilde{x}) = \sum_{i=1}^{k} x_i \overline{H}_i(T, P, \tilde{x}) \tag{B-15}$$

where $\tilde{x}_i$ is the mole fraction of component i and $\overline{H}_i$ is the *partial molar enthalpy of component i*:

$$\overline{H}_i(T, P, \tilde{x}) \triangleq \left(\frac{\partial(n\tilde{H}(T, P, \tilde{x}))}{\partial n_i}\right)_{T, P, n_j \neq n_i} \tag{B-16}$$

In Eq. B-16, n_i is the number of moles of component i and n is the total number of moles, $n \triangleq \Sigma n_i$.

An important simplification occurs if the mixture can be considered to be an *ideal solution*. For an ideal solution, $\overline{H}_i(T, P, \tilde{x}) = \tilde{H}_i(T, P)$, where $\tilde{H}_i(T, P)$, is the molar enthalpy of pure component i. The mixture can be analyzed as a set of individual components, and (B-15) can be written as

$$\tilde{H}(T, P, x) = \sum_{i=1}^{k} \tilde{x}_i \tilde{H}_i(T, P) \quad \text{(for ideal solutions)} \tag{B-17}$$

Similarly, it can be shown that the enthalpy per unit mass of an ideal solution, $\hat{H}(T, P, x)$, can be expressed as

$$\hat{H}(T, P, x) = \sum_{i=1}^{k} x_i \hat{H}_i(T, P) \quad \text{(for ideal solutions)} \tag{B-18}$$

where x denotes the composition in mass units and x_i is the mass fraction of component i.

Equations B-17 and B-18 are very useful in developing dynamic models from the general energy balances in Eqs. 2-10 and 2-11. Similar expressions can be derived for $\hat{U}_{int}$ and $\tilde{U}_{int}$. Then the total internal energy of the system, U_{int}, can be expressed in terms of $\hat{U}_{int}$ and $\tilde{U}_{int}$ according to Eqs. B-7 and B-10 where ρ and V are now the density and volume of the mixture, respectively, and n is the total number of moles.

REFERENCE

Sandler, S. I., *Chemical, Biochemical, and Engineering Thermodynamics*, 4th ed., Wiley, New York, 2006.

Appendix C

Control Simulation Software

APPENDIX CONTENTS

MATLAB is a general-purpose software package for mathematical computations, analysis, and visualization available from The Mathworks (2010). This Appendix introduces the basic functionality of the MATLAB software and shows how to solve simple algebraic equations and ordinary differential equations (ODEs). Vector and matrix manipulations are considered first, and then solving simple linear algebraic equations is demonstrated using MATLAB. The basics of functions and scripts are presented next, and use of the ODE integration function, *ode45*, is described. Subsequent sections introduce the graphical modeling tools, Simulink and Lab VIEW, and their usage for computing responses for open-loop and closed-loop block diagrams. For more details on MATLAB usage, see Palma (2010), Bequette (1998), and Doyle III et al. (2000).

C.1 MATLAB OPERATIONS AND EQUATION SOLVING

In MATLAB statements, square brackets denote vectors and matrices. Elements in a row vector are separated by commas or spaces. For example, the row vector $\boldsymbol{v} = (1, 2, 3)$ can be represented by

$$v = [1\ 2\ 3] \quad \text{or} \quad v = [1, 2, 3]$$

The elements of a column vector are separated by semicolons. Thus, the column vector, $\boldsymbol{w} = col[4, 5, 6]$ is represented as $w = [4; 5; 6]$. A matrix $\boldsymbol{M}$,

$$\boldsymbol{M} = \begin{bmatrix} 3 & 7 & 9 \\ 2 & 6 & 8 \\ 1 & 0 & 4 \end{bmatrix}$$

has the MATLAB representation:

M = [3 7 9 ; 2 6 8 ; 1 0 4]

Note that MATLAB variables cannot be **boldface** or *italicized.* Similarly, subscripts, superscripts, and other accent marks are not allowed. Also, MATLAB is case-sensitive.

C.1.1 Matrix Operations

The transpose of a matrix M is calculated using the command M'. The inverse of a matrix M is calculated as inv(M). In MATLAB, the multiplication of matrices $\boldsymbol{A}$ and $\boldsymbol{B}$ is denoted by $A * B$, while their addition and subtraction are denoted by $A + B$ and $A - B$, respectively. Commands for element-by-element multiplication and division are also available. For more functions and help on any MATLAB operation, type *help*.

EXAMPLE C.1

Consider the following matrices:

$$A = \begin{bmatrix} 1 & -3 \\ 0 & 2 \end{bmatrix} \quad B = \begin{bmatrix} 5 & 2 \\ -1 & 4 \end{bmatrix}$$

$$C = \begin{bmatrix} 1 & 2 \\ -2 & -4 \end{bmatrix} \quad D = \begin{bmatrix} 2 & 3 \\ 1 & 4 \end{bmatrix}$$

Calculate the following:

(a) AB **(b)** AB^T **(c)** A^{-1} **(d)** DCD^T

(e) C^{-1} **(f)** $(ADA^T)^{-1}$ **(g)** $BC - D^{-1}$

SOLUTION

(a) $\begin{bmatrix} 8 & -10 \\ -2 & 8 \end{bmatrix}$ **(b)** $\begin{bmatrix} -1 & -13 \\ 4 & 8 \end{bmatrix}$ **(c)** $\begin{bmatrix} 1 & 1.5 \\ 0 & 0.5 \end{bmatrix}$

(d) $\begin{bmatrix} -32 & -36 \\ -56 & -63 \end{bmatrix}$ **(e)** $\begin{bmatrix} \text{Inf} & \text{Inf} \\ \text{Inf} & \text{Inf} \end{bmatrix}$

C is a singular matrix (not invertible).

(f) $\begin{bmatrix} 0.8 & 0.9 \\ 1.1 & 1.3 \end{bmatrix}$ **(g)** $\begin{bmatrix} -0.2 & 2.6 \\ -8.8 & -18.4 \end{bmatrix}$

Other matrix operations in MATLAB include:

- Eigenvalues and eigenvectors: *eig*
- Singular value decomposition: *svd*
- Pseudoinverse: *pinv*

C.1.2 Solution of Algebraic Linear or Nonlinear Equations

The solution to a set of linear algebraic equations, $Mx = b$, is given by $x = M^{-1}b$. The MATLAB solution can be written as either x = inv(M)*b or x = M\b, where the backslash operator (\) is used as a shortcut for the solution. The solution to a set of nonlinear algebraic equations can be obtained using the MATLAB routine *fsolve*.

EXAMPLE C.2

Solve the equation $Mx = b$ for x using the values of A, B, and D from Example C.1.

(a) $M = A$, $b = [1; 2]$ **(b)** $M = B$, $b = [1; 2]$

(c) $M = ADA^T$, $b = [5; 1]$.

SOLUTION

(a) $x = \begin{bmatrix} 4 \\ 1 \end{bmatrix}$ **(b)** $x = \begin{bmatrix} 0 \\ 0.5 \end{bmatrix}$ **(c)** $x = \begin{bmatrix} 4.9 \\ 5.8 \end{bmatrix}$

C.1.3 m-files

A MATLAB code, or *m-file*, is a collection of commands that are executed sequentially. Commands can be mathematical operations, function calls, flow control statements, and calls to the functions and scripts described in Section C.1.4. m-files are written using the MATLAB editor and have names such as *myfile.m*. They are executed from the MATLAB command window by typing the name of the m-file (without the .m). Saving an m-file will avoid many hours of retyping the same commands.

C.1.4 Functions and Scripts

There are two types of m-files, *functions* and *scripts*. A MATLAB function has variables that can be passed into and out of the function. Any other variables used inside the function are not saved in memory when the function is finished. Scripts, on the other hand, save all their variables in the MATLAB workspace. Functions and scripts have names like *myfunction.m*. The first line of a function must contain a function declaration, using the following format:

```
function [output1, output2, output3]
    = myfunction(input1, input2, input3)
```

Commented lines immediately following the function declaration comprise the help file for the function. To obtain information on any function, simply type *help function*. Some MATLAB functions that are useful for process control include:

- Unit step response of a transfer function: *step*
- Transfer function matrix derived from a state-space model: *ss2tf*
- State-space model derived from a transfer function matrix: *tf2ss*
- Transfer function multiplication: *series*
- Roots of the characteristic equation: *roots*
- Polynomial fitting of input-output data: *polyfit*
- Minimization of a multivariable function: *fminu*
- Frequency response of a linear, time-invariant system: *bode*

C.1.5 Solving a System of Differential Equations

MATLAB has several built-in functions for solving systems of differential equations. The basic use of the standard integration algorithm, *ode45*, is described in this section. First, a function containing the differential equations to be integrated must be created. This function must have at least two arguments, t and the state vector, y (see Section 6.5). The function returns a column vector containing the derivatives evaluated at the current time.

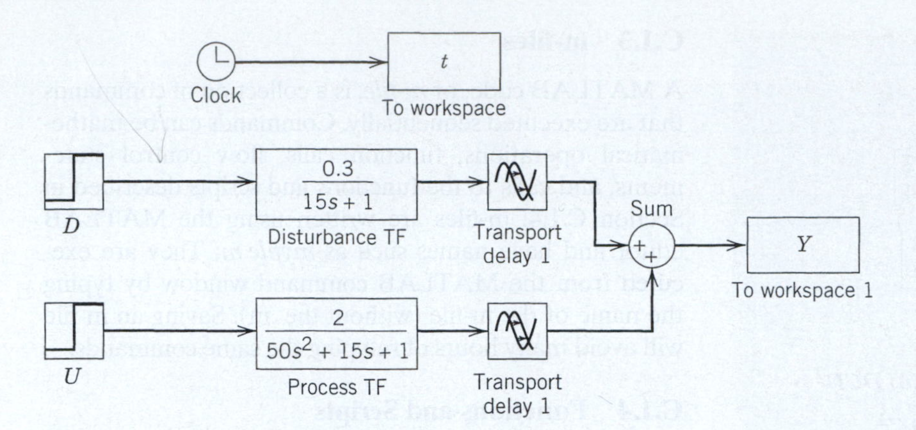

Figure C.1 Simulink block diagram for Equation C-1.

The commands inside the function calculate these derivatives. Additional arguments for the function are optional and can be used to pass parameter values from the script that calls the function.

Once the differential equation function is written and saved, a script (i.e., an m-file) containing the call to the integrator must be written. Here, parameter values, initial conditions, and options are specified, and the integration routine is called with the following command:

[t, y] = ode45(@myfunction, [*ti tf*], *y*0,
options, P1, P2, P3, . . .);

where *myfunction* is the function containing the differential equations as described above, *ti* and *tf* are the initial and final integration times, and *y*0 is the vector of initial conditions. *Options* is a parameter vector for *ode45*. More information is available in the help files. Empty brackets [ ] can be used in place of the options argument. P1, P2, . . . are additional parameter values that are passed to *myfunction*.

C.1.6 Plots

It is easy to display results in MATLAB graphically. The *plot* function is used to create simple plots. The command syntax is

plot(*x*1, *y*1, format1, *x*2, *y*2, format2, . . .)

*x*1 and *x*2 are independent variables (usually time), and *y*1 and *y*2 are dependent variables. The format1 and format2 arguments are short combinations of characters containing the plot-formatting commands. For example, a blue solid line is 'b-' (include the single quotes), a red dashed line is 'r--', and a green dotted line is 'g:'. More formats can be viewed by typing *help plot*. Axis labels, title, and legend can be created using *xlabel*, *ylabel*, *title*, and *legend* commands. These and other properties can also be edited directly on the figure by selecting the arrow icon and double-clicking on an object contained in the figure.

Additional plot commands in MATLAB are *loglog* for log-log plots, and *semilogx* and *semilogy* for semi-log plots, such as the Bode plots used in Chapter 14.

C.1.7 MATLAB Toolboxes

For advanced techniques in modeling, identification, and control, MATLAB has a variety of additional *toolboxes* that are licensed individually. Relevant toolboxes for process control include control system, fuzzy logic, system identification, model predictive control, neural networks, optimization, partial differential equations, robust control, and statistics.

C.2 COMPUTER SIMULATION WITH SIMULINK

Simulink, a companion software package to MATLAB, is an excellent interactive environment for simulation and analysis of control systems. Simulink enables the rapid creation of block diagrams based on transfer functions, followed by simulation for a given input signal. To facilitate model definitions, Simulink has a block diagram window in which blocks are created from the Simulink library browser and edited primarily by implementing drag-and-drop commands using a mouse. Blocks can be configured as additive transfer functions (see Fig. 4.1) or as multiplicative transfer functions (see Fig. 4.2), simply by connecting the output of one block to the input of another block. The coefficients of descending powers of *s* of the numerator and denominator polynomials in each block are entered as vectors. Time delays (called *transport delays* in Simulink) can be inserted in series with blocks for rational transfer functions. Input signals, called *sources*, include step, sinusoidal, and random inputs, but not the impulse function.[1] Clicking on the input block allows the user to specify the time when the input changes from an initial value of zero, and, for a step input, its initial and final values.

Consider a dynamic system consisting of a single output Y and two inputs U and D:

$$Y(s) = G_p(s)\, U(s) + G_d(s)\, D(s) \qquad \text{(C-1)}$$

[1]To obtain the unit impulse response of a single transfer function, use the function *impulse* from the MATLAB command window.

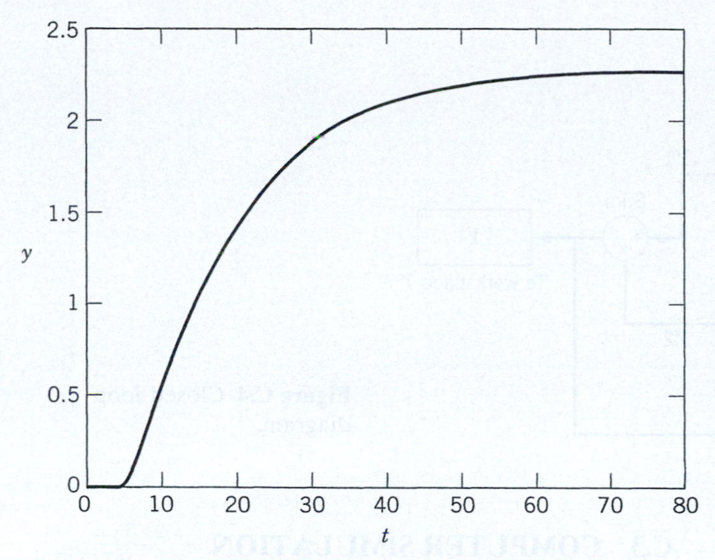

Figure C.2 Response for simultaneous unit step changes at $t = 0$ in U and D from the Simulink diagram in Fig. C.1.

where

$$G_p(s) = \frac{2e^{-5s}}{50s^2 + 15s + 1} \quad \text{(process transfer function)}$$

$$G_d(s) = \frac{0.3e^{-5s}}{15s + 1} \quad \text{(disturbance transfer function)}$$

Figure C.1 shows the Simulink diagram for Eq. C-1 (transport delay $1 = 5$ for both models). To generate a transient response, the simulation menu is selected to allow parameters for the simulation to be specified (start time, stop time, integration routine, maximum integration step size). Numerical values of time t are entered into the input-output data set via a *clock* block. After the simulation has been completed, the resulting data can be plotted (see Fig. C.2), manipulated, and analyzed from the MATLAB command window.

To simulate a closed-loop system, the procedure is somewhat more involved than for an open-loop system. Changing the previous example somewhat, start with Fig. C.1, but let $G_d = G_p$. Click on the connection between the U block and the *Process TF* block and delete it. Rename the U block, *Ysp*. This block will be used to produce a step change in the set point.

Place a copy of the *Sum* block to the right of *Ysp*. Double click above the *Sum* icon and label it *Sum1*. Open its dialog box and change the ++ sequence to +−. The *top* left input will have a + located to the right of it, while the bottom input will have a − located above it. Connect the output of *Ysp* to the left input of *Sum1*. Also, connect the output from *Sum* to the bottom input of *Sum1*. This can be done by clicking on the bottom input of *Sum1* and dragging the arrow to the line following the output of *Sum*. The output of *Sum1* is the error between the set point *Ysp* and the controlled variable *Y*. This leads to the block diagram in Fig. C.3.

To insert the controller, right-click the *Simulink Extras* block. Click on the *Additional Linear* block; then select the *PID Controller* and drag it to the right of the *Sum1* block. Connect the output of *Sum1* to the input of *PID controller* and the output of *PID controller* to the input of *Process TF*. Double-click on *PID controller* and use the following controller settings: $K_c = 1.65$, $\tau_I = 7.12$, $\tau_D = 1.80$. Note that Simulink PID controller settings are entered in the expanded form (see Eq. 8-16) as P, I, and D where $P = K_c$, $I = K_c/\tau_I$, and $D = K_c\tau_D$. Thus, the numerical values of P, I, and D should reflect these definitions. The model developed above represents the closed-loop system, as shown in Fig. C.4. Text can be added to the block diagram simply by double-clicking on a point in the diagram and typing the desired words (see Fig. C.4).

Now the closed-loop response of the system can be simulated. Starting with the set-point response, click on block D and set the *Final value* to 0 so that no step disturbance will occur. Create a step in the set point by

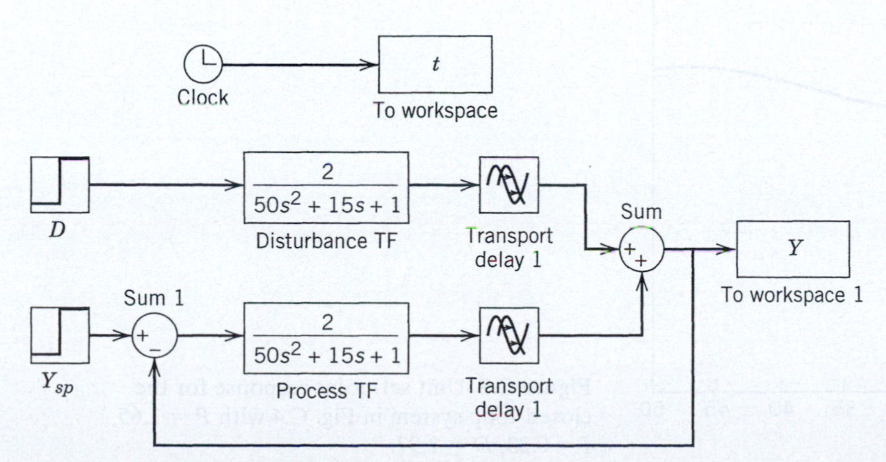

Figure C.3 Partially completed closed-loop diagram.

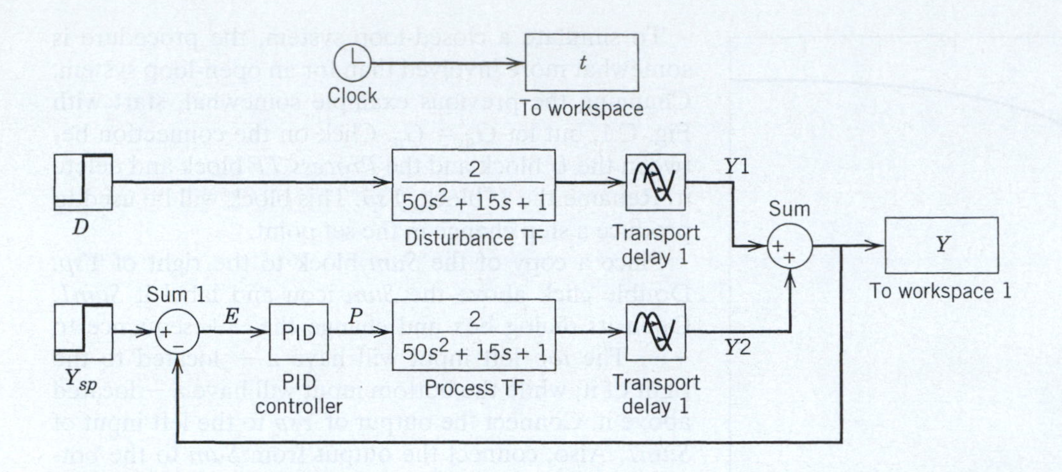

Figure C.4 Closed-loop diagram.

clicking on *Ysp* and setting the *Final value* to 1. In the *Simulation Parameters* menu, change the stop time to 50. Start the simulation by selecting *Start* from the *Simulation* menu. Because *D* (the disturbance) has been disabled, the resulting *Y* and *t* variables in the workspace will be for the unit set-point response, as shown in Fig. C.5.

Now simulate the unit response to a unit step disturbance. Double-click on *Ysp* and set *Final value* to 0. Double-click on *D* and set *Final value* to 1. Again, select *Start* from the *Simulation* menu to begin the simulation. Type *plot (t, Y)* to view the response. Figure C.6 shows the resulting disturbance response plot after modifying some of the labels. Simulink can be used to simulate the effects of different control strategies with realistic multivariable process models such as a distillation column or a furnace. See Appendix E and Doyle III et al. (2000) for a series of modules on such processes and various control strategies.

C.3 COMPUTER SIMULATION WITH LabVIEW

LabVIEW, which stands for *Laboratory Virtual Instrumentation Engineering Workbench*, is a graphical computing environment for instrumentation, system design, and signal processing. The *Control Design and Simulation (CDSim)* module for LabVIEW can be used to simulate dynamic systems. To facilitate model definition, CDSim has functions in the LabVIEW environment that resemble those found in Simulink. There is also the ability to use m-file syntax directly in LabVIEW through the new MathScript node.

A new program, called *VI* for *V*irtual *I*nstrument, can be created when LabVIEW is opened. Then one can right-click inside the block diagram to view the palette of functions used in creating programs. Select the Control Design & Simulation → Simulation palette to view the library of simulation functions. A dialog box opens, showing all the simulation parameters that

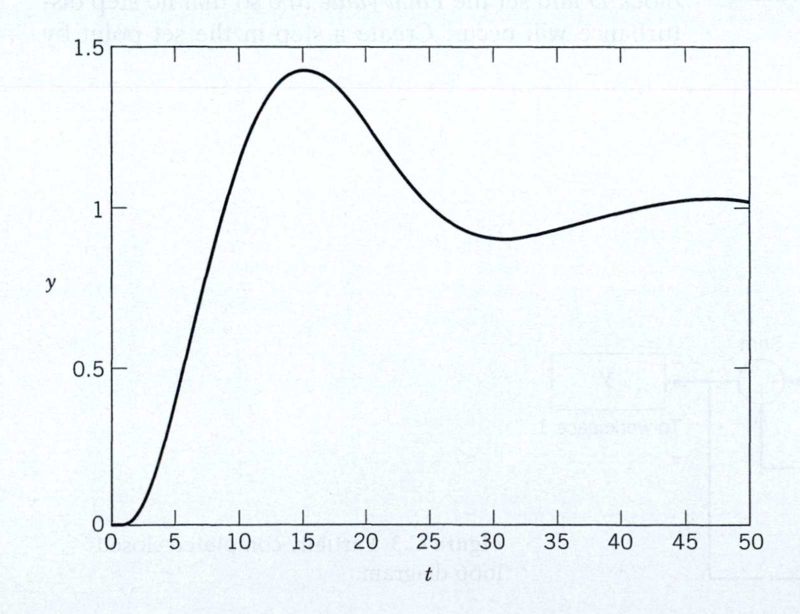

Figure C.5 Unit set-point response for the closed-loop system in Fig. C.4 with $P = 1.65$, $I = 0.23$, $D = 2.97$.

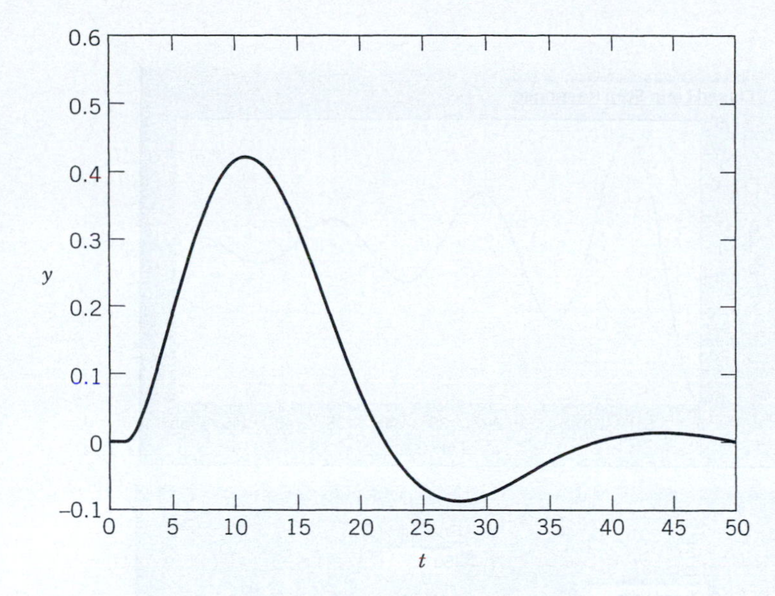

Figure C.6 Closed-loop response for a unit step disturbance.

can be modified, such as the final time and the maximum step size. Note that the LabVIEW Simulation loop includes an ODE solver. The maximum step size is used in LabVIEW for numerically integrating the ODE. A typical linear dynamic system is easy to integrate numerically, so a maximum step size of 1 usually result in a smooth curve. Larger step sizes produce more jagged curves. A typical block diagram of a closed-loop simulation is shown in Figure C.7.

An important feature of LabVIEW is interactivity. The PID controller parameters can be made interactive from the front panel, rather than editing them on the block diagram, as done in Simulink. By default, LabVIEW creates a standard numeric control faceplate, but this can easily be changed. Controller tuning parameters may be entered either from the slider or typed in to the numeric control box. A front panel for an interactive PID control tuner is shown in Figure C.8, which can depict open or closed-loop responses for any process transfer function.

Companion simulation courseware for selected examples in this book using National Instruments VIs is found on the book's student companion Web site at www.wiley.com/college/sebarg.

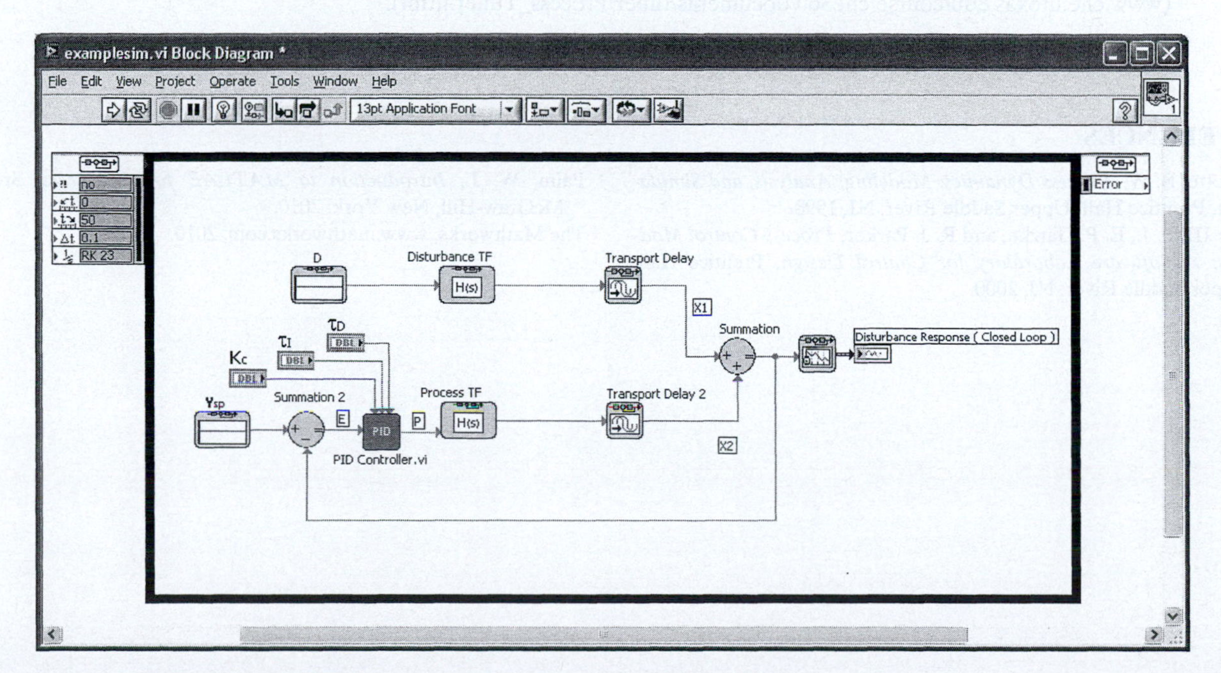

Figure C.7 Example LabVIEW block diagram of a closed-loop simulation.

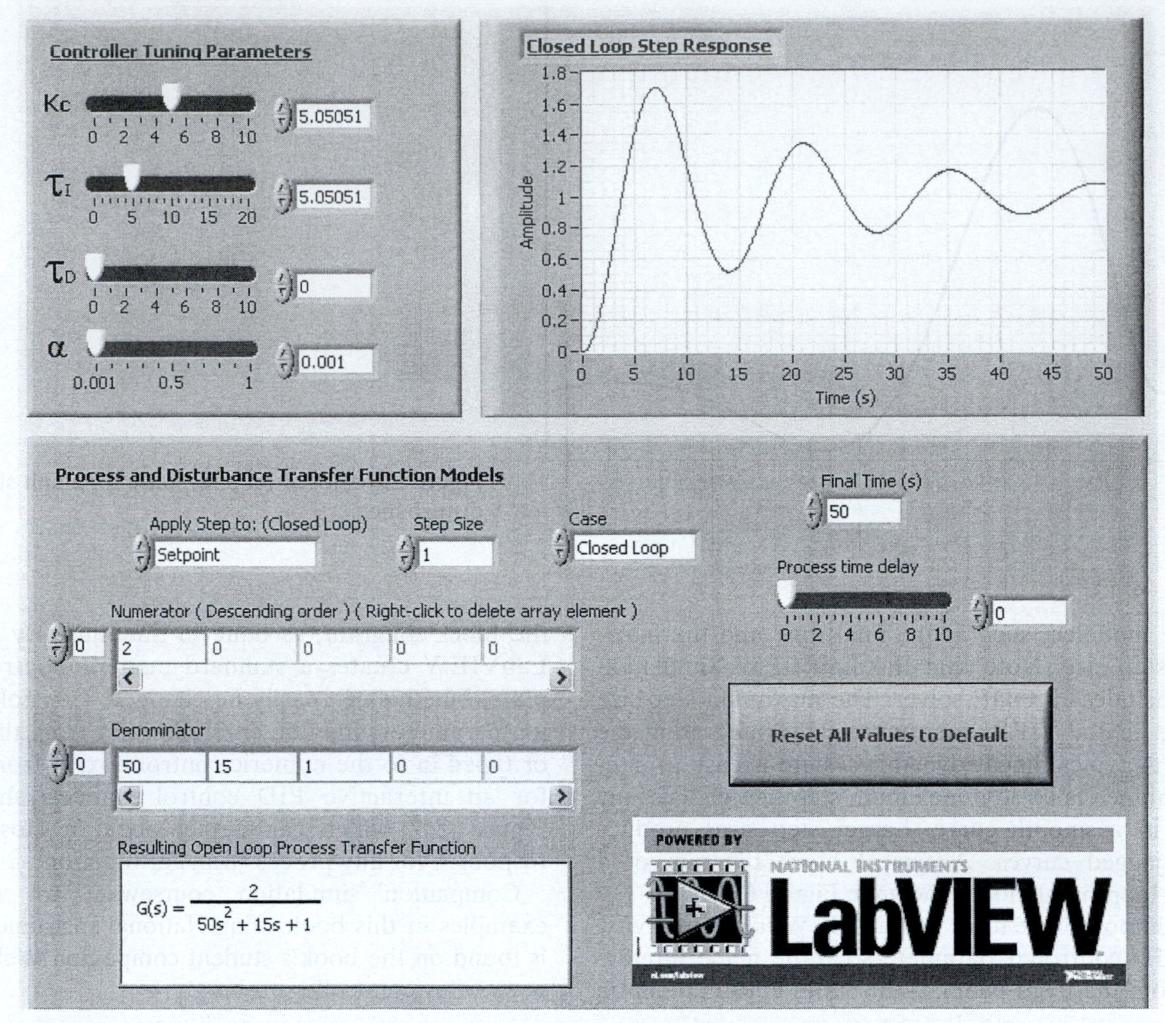

Figure C.8 The front panel of the Virtual Instruments (VI) for the interactive PID control tuner (www.che.utexas.edu/course/che360/documents/tuner/Process_Tuner.html).

REFERENCES

Bequette, B. W., *Process Dynamics: Modeling, Analysis, and Simulation*, Prentice Hall, Upper Saddle River, NJ, 1998.

Doyle III, F. J., E. P. Gatzke, and R. J. Parker, *Process Control Modules: A Software Laboratory for Control Design*, Prentice Hall, Upper Saddle River, NJ, 2000.

Palm, W. J., *Introduction to MATLAB for Engineers*, 3rd ed., McGraw-Hill, New York, 2010.

The Mathworks, www.mathworks.com, 2010.

Appendix D

Instrumentation Symbols

Process control systems and instrumentation can be described in several ways. Flowsheets show the process equipment, instruments, and control systems, as well as interconnections, such as piping and electrical and pneumatic transmission lines. More detailed flowsheets are referred to as *piping and instrumentation diagrams (P&IDs)*. They include additional information, such as valve characteristics, piping details (e.g., pipe sizes and fittings), and miscellaneous information, such as drains, vents, and sampling lines. Both types of diagrams are widely used in the process industries.

In order for flowsheets and P&IDs to be understood by people with different job responsibilities such as plant designers, process engineers, instrumentation specialists, and vendors, it is useful to use standardized symbols and conventions on the flowsheets. Standards concerning instrumentation symbols and flowsheet conventions have been developed by technical societies, such as the International Society of Automation (ISA). However, individual companies often use different or additional symbols for particular processes.

Figure D.1 lists some common instrument symbols and line designations. Instruments are usually shown as a circle with a letter designation and a number. The controller

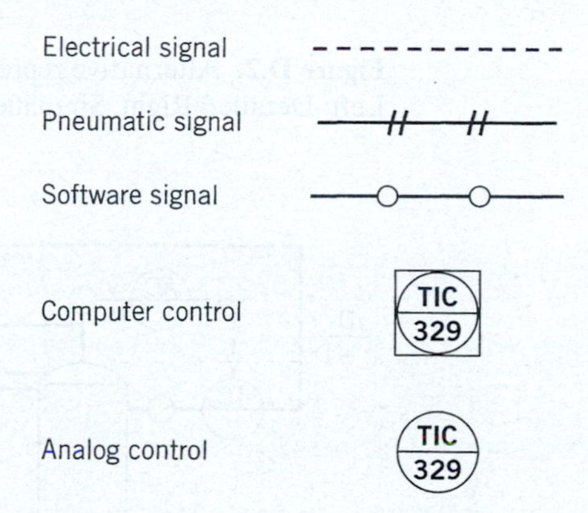

Figure D.1. Line and instrument symbols.

shown in Figure D.1 is a temperature-indicating controller (TIC). The square around the circle indicates that it is implemented via digital control. The *I* designation (for indicating) is an anachronism, because the vast majority of current analog and digital controllers display the measured value of the controlled variable. (Many decades ago, some controllers did not.) Each instrument in a control loop (e.g., sensor, control valve, controller) has the same identifying number, which is referred to as the tag number. Thus, in Figure D.1, the TIC is the temperature controller for control loop #329.

Figure D.2 shows alternative representations of a pressure control loop (Lipták, 2003). The simpler version would be used when the control strategy and its implementation are the main concerns. The more detailed version shows piping and instrumentation details. An example of a more complicated flowsheet is shown in Figure D.3 for a distillation column control strategy. In addition to the instrumentation and controllers, it includes special control calculations involving multiplication, addition, and subtraction.

Additional information concerning instrumentation symbols and flowsheets is available from ISA (1992) and the Instrument Engineer's Handbook (Lipták, 2003).

Table D.1 Some common letter symbols for instrumentation diagrams

Letter	Used as First Letter	Used as Succeeding Letters
A	Analysis	Alarm
C		Control
F	Flow rate	
G	User's choice	
H		High
I	Current	Indicate
J	Power	
L	Level	
P	Pressure	
R		Record
S	Speed	Switch
T	Temperature	

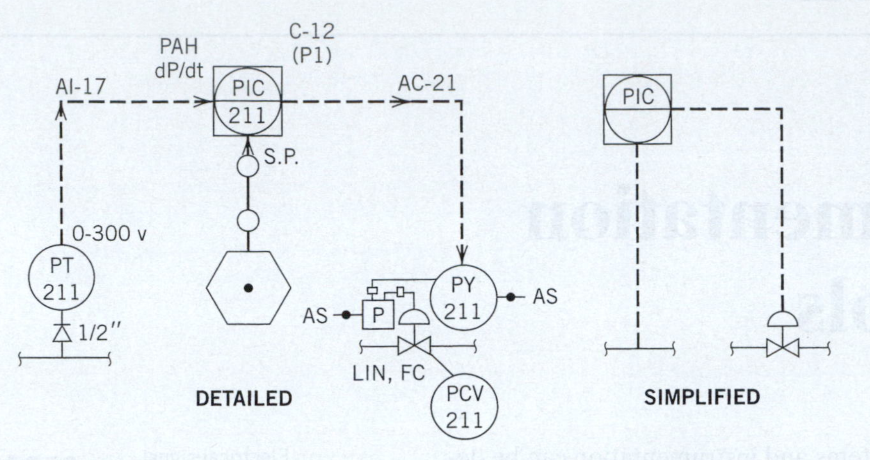

Figure D.2. Alternative representations of a pressure control loop:
Left: Detailed Right: Simplified for a process flow sheet (Liptak, 2003).

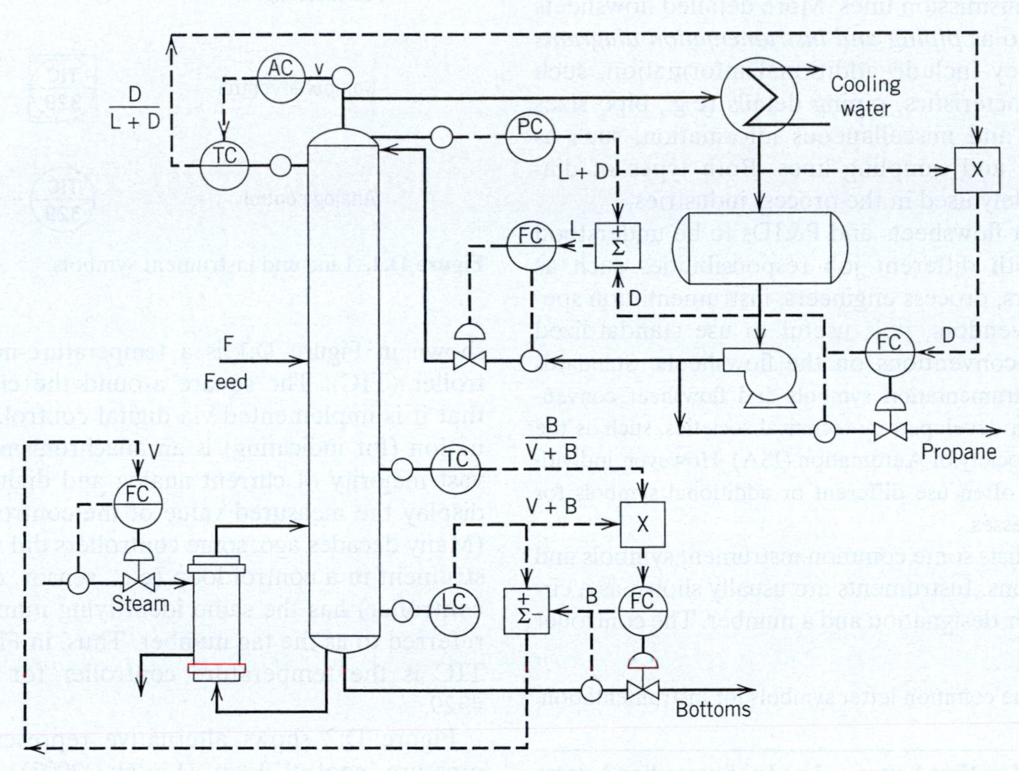

Figure D.3 A depropanizer control system (Perry et al., 2008).

REFERENCES

Instrumentation Symbols and Identification, Standard ISA-5.1-1984 (R1992), International Society of Automation (ISA), Research Triangle Park, NC (1992).

Lipták, B. (Ed.), *Instrument Engineers' Handbook*, 4th ed., *Vol. I, Process Measurement and Analysis*, Radnor, PA, 2003.

Perry, R. H. and D. W. Green (ed.), *Chemical Engineers' Handbook*, 8th ed., *Section 8, Process Control*, McGraw-Hill, NY, 2008.

Appendix E

Process Control Modules

APPENDIX CONTENTS

E.1. INTRODUCTION

The Process Control Modules (PCM), originally developed at the University of Delaware, have been designed to address the key engineering educational challenge of realistic problem solving within the constraints of a typical lecture course in process dynamics and control (Doyle III et al., 1998; Doyle III, 2001). These modules have been updated and adapted by Dr. Eyal Dassau at the University of California Santa Barbara, to be used in conjunction with the 3rd edition of *Process Dynamics and Control*. The primary objectives in creating these MATLAB® modules were to develop the following:

- Realistic computer simulation case studies, based on physical properties that exhibited nonlinear, high-order dynamic behavior in a rapid simulation environment
- A convenient graphical interface for students that allowed real-time interaction with the evolving virtual experiment
- A set of challenging exercises that reinforce the conventional lecture material through active learning and problem-based methods

E.2. MODULE ORGANIZATION

Eight distinct chemical and biological process applications, which range from simple single input-single output (SISO) processes to more complex 2×2 control loops, are formulated with a modular approach. The progression of the modules follows a typical undergraduate process dynamics and control course, starting with low-order dynamic system analysis and continuing through multivariable controller synthesis.

Table E.1 Organization of Process Control Modules (PCM)

Module	Modes						
Furnace	Operator Interface	PID	Feedforward	Multivariable	MPC		
Distillation Column	Operator Interface	PID	Feedforward	Multivariable	Decoupling	MPC	
Bioreactor	Operator Interface	PID	Feedforward	Multivariable	Decoupling		
Four Tanks	Operator Interface	PID	Feedforward	Multivariable	MPC		
Fermentor	Operator Interface	PID					
Diabetes	Operator Interface	PID	MPC				
First and Second Order Systems	First Order System	Second Order System	System Identification #1	System Identification #2			
Discrete	Aliasing	Model ID	PID-Furnace	PID-Column	PID-Four Tanks	IMC-Furnace	IMC-Column

E.3. HARDWARE AND SOFTWARE REQUIREMENTS

The Process Control Modules are a set of MATLAB/Simulink routines that require either a full license or the Student Version of MATLAB and Simulink. The current version of the modules has been tested with version 2007a of MATLAB and Simulink. The minimum recommended system configuration is a Windows (XP or Vista) PC with 1 GB RAM.

E.4. INSTALLATION

The Process Control Modules (PCM) software can be downloaded from www.wiley.com/college/seborg onto the user's computer. Then double-click on the PCM file, and follow the instructions on the installer to install the software. Note that MATLAB should be installed in order to use these modules. During the installation, users can create a shortcut icon to the software on their desktop (recommended).

E.5. RUNNING THE SOFTWARE

There are two ways to execute the software: the first is to double-click the PCM button on the desktop, which launches MATLAB and the PCM interface (Figure H.1), and the other way is to open MATLAB manually and to call the PCM software by pointing to the PCM installation folder and typing "PCM", followed by the Enter key.

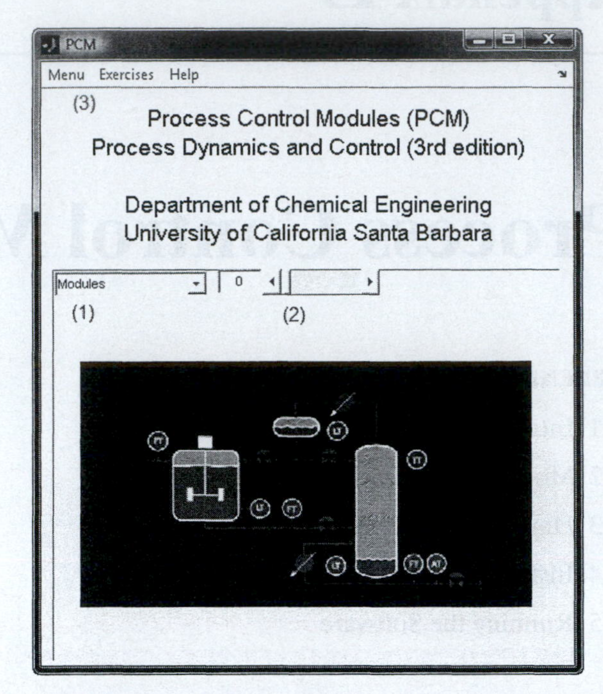

Figure E.1 PCM main interface.

The Web site for this textbook contains a more detailed tutorial on PCM, including case studies for the furnace and distillation column modules.

REFERENCES

Doyle III, F. J., E. P. Gatzke, and R.S. Parker, Practical Case Studies for Undergraduate Process Dynamics and Control Using Process Control Modules, *Comp. Appls. Eng. Educ.,* **6,** 181 (1998).

Doyle III, F. J. *Process Control Modules: A Software Laboratory for Control Design*, Prentice Hall PTR, Upper Saddle River, NJ, 2000.

Appendix F

Review of Basic Concepts From Probability and Statistics

APPENDIX CONTENTS

In this appendix, basic probability and statistics concepts are reviewed that are considered for the safety analysis of Chapter 10 and the quality control charts of Chapter 21.

F.1 PROBABILITY CONCEPTS

The term *probability* is used to quantify the likely outcome of a random event. For example, if a fair coin is flipped, the probability of a head is 0.5, and the probability of a tail is 0.5. Let $P(A)$ denote that probability that a random event A occurs. Then $P(A)$ is a number in the interval $0 \leq P(A) \leq 1$, such, that the larger $P(A)$ is, the more likely it is that A occurs. Let A' denote the *complement* of A, that is, the event that A does not occur. Then,

$$P(A') = 1 - P(A) \tag{F-1}$$

Now consider two events, A and B, with probabilities $P(A)$ and $P(B)$, respectively. The probability that one or both events occurs ($A \cup B$) can be expressed as

$$P(A \cup B) = P(A) + P(B) - P(A \cap B) \tag{F-2}$$

If A and B are *mutually exclusive*, this means that if one event occurs, the other cannot; consequently, their intersection is the null set $A \cap B = \varnothing$. Then $P(A \cap B) = 0$ and Eq. (F-2) becomes

$$P(A \cup B) = P(A) + P(B) \quad \text{(for mutually exclusive events)} \tag{F-3}$$

Analogous expressions are available for the union of more than two events (Montgomery and Runger, 2007).

If A and B are *independent*, then the probability that both occur is

$$P(A \cap B) = P(A)\,P(B) \quad \text{(for independent events)} \tag{F-4}$$

Similarly, the probability that n independent events, $E_1, E_2, \ldots, E_n$, occur is

$$P(E_1 \cap E_2 \cap \cdots \cap E_n) = P(E_1)\,P(E_2) \cdots P(E_n) \quad \text{(for independent events)} \tag{F-5}$$

These probability concepts are illustrated in two examples.

EXAMPLE F.1

A semiconductor processing operation consists of five independent batch steps where the probability of each step having its desired outcome is 0.95. What is the probability that the desired end product is actually produced?

SOLUTION

In order to make the product, each individual step must be successful. Because the steps are independent, the probability of a success, $P(S)$, can be calculated from Eq. (F-5):

$$P(S) = (0.95)^5 = 0.77$$

EXAMPLE F.2

In order to increase the reliability of a process, a critical process variable is measured on-line using two sensors. Sensor A is available 95% of the time while Sensor B is available 90% of the time. Suppose that the two sensors operate independently, and that their periods of unavailability occur randomly. What is the probability that neither sensor is available at any arbitrarily selected time?

SOLUTION

Let A denote the event that Sensor A is not available and B denote the event that Sensor B is not available. The event that neither Sensor is available can be expressed as $(A \cup B)'$. Then, from Eqs. (F-1) and (F-2),

$$P(A \cup B)' = 1 - P(A \cup B)$$
$$P(A \cup B)' = 1 - [P(A) + P(B) - P(A \cap B)]$$
$$P(A \cup B)' = 1 - [0.95 + 0.90 - (0.95)(0.90)] = 0.005$$

F.2 MEANS AND VARIANCES

Next, we consider two important statistical concepts, *means* and *variances*, and how they can be used to characterize both probability distributions and experimental data.

F.2.1 Means and Variances for Probability Distributions

In Section F.1, we considered the probability of one or more events occurring. The same probability concepts are also applicable for random variables such as temperatures or chemical compositions. For example, the product composition of a process could exhibit random fluctuations for several reasons, including feed disturbances and measurement errors. A temperature measurement could exhibit random variations due to turbulence near the sensor. Probability analysis can provide useful characterizations of such random phenomena.

Consider a continuous random variable, X, with an assumed probability distribution, $f(x)$, such as a Gaussian distribution. The probability that X has a numerical value in an interval [a, b] is given by (Montgomery and Runger, 2007),

$$P(a \leq X \leq b) \triangleq \int_a^b f(x)dx \tag{F-6}$$

where x denotes a numerical value of random variable, X. By definition, the *expected value* of X, μ_X, is defined as

$$\mu_X \triangleq E(X) \triangleq \int_{-\infty}^{\infty} x f(x)dx \tag{F-7}$$

The expected value is also called the *population mean* or *average*. It is an average over the expected range of values, weighted according to how likely each value is.

The *population variance* of X, σ_X^2, indicates the variability of X around its population mean. It is defined as:

$$\sigma_X^2 \triangleq E[(X - u_X)^2] \triangleq \int_{-\infty}^{\infty} (x - u_X)^2 f(x)dx \tag{F-8}$$

The positive square root of the variance is the *population standard deviation*, σ_X.

These calculations are illustrated in Example F.3.

EXAMPLE F.3

A mass fraction of an impurity X varies randomly between 0.3 and 0.5 with a uniform probability distribution:

$$f(x) = \frac{1}{0.2}$$

Determine its population mean and population standard deviation.

SOLUTION

Substituting $f(x)$ into Eq. F-7 gives:

$$\mu_X = \int_{-\infty}^{\infty} x f(x)\,dx = \int_{0.3}^{0.5} x\left(\frac{1}{0.2}\right)dx$$

$$\mu_X = \left(\frac{1}{0.2}\right)\left(\frac{1}{2}x^2\right)\Bigg|_{0.3}^{0.5} = 0.4$$

Thus μ_X is the midpoint of the [0.3, 0.5] interval for X. To determine σ_x, substitute $f(x)$ into Eq. F-8:

$$\sigma_X^2 = \int_{-\infty}^{\infty} (x - \mu_X)^2 f(x)\,dx = \int_{0.3}^{0.5} (x - 0.4)^2\left(\frac{1}{0.2}\right)dx$$

$$\sigma_X^2 = \left(\frac{1}{0.2}\right)\left(\frac{1}{3}(x - 0.4)^3\right)\Bigg|_{0.3}^{0.5} = 0.00333$$

$$\sigma_X = 0.0577$$

F.2.2 Means and Variances for Experimental Data

A set of experimental data can be characterized by its *sample mean* and *sample variance* (or simply, its *mean* and *variance*). Consider a set of N measurements, $\{x_1, x_2, \ldots, x_N\}$. Its mean, $\bar{x}$, and variance s^2 are defined as (Montgomery and Runger, 2007)

$$\bar{x} \triangleq \frac{1}{N}\sum_{i=1}^{N} x_i \tag{F-9}$$

$$s^2 \triangleq \frac{1}{N-1}\sum_{i=1}^{N}(x_i - x)^2 \tag{F-10}$$

The standard deviation s is the positive square root of the variance.

The mean is the average of the dataset while the variance and standard deviation characterize the variability in the data.

F.3 STANDARD NORMAL DISTRIBUTION

The normal (or Gaussian) probability distribution plays a central role in both the theory and application of statistics. It was introduced in Section 21.2.1. For probability calculations, it is convenient to use the *standard normal distribution*, $N(0, 1)$ which has a mean of zero and a variance of one. Suppose that a random variable X is normally distributed with a mean μ_X and variance $\sigma_X{}^2$. Then, the corresponding standard normal variable Z is

$$Z \triangleq \frac{X - \mu_X}{\sigma_X} \tag{F-11}$$

Statistics book contain tables of the *cumulative standard normal distribution*, $\Phi(z)$.

By definition, $\Phi(z)$ is the probability that Z is less than a specified numerical value, z (Montgomery and Runger, 2007; Ogunnaike, 2010):

$$\Phi(z) \triangleq P(Z \leq z) \tag{F-12}$$

Example 10.2 illustrates an application of $\Phi(z)$.

F.4 ERROR ANALYSIS

In engineering calculations, it can be important to determine how uncertainties in independent variables (or inputs) lead to even larger uncertainties in dependent variables (or outputs). This analysis is referred to as *error analysis*. Due to the uncertainties associated with input variables, they are considered to be random variables. The uncertainties can be attributed to imperfect measurements or uncertainties in unmeasured input variables. Error analysis is based on the statistical concepts of means and variances, considered in the previous section.

As an important example of error analysis, consider a linear combination of p variables,

$$Y = \sum_{i=1}^{p} c_i X_i \tag{F-13}$$

where X_i is an independent random variable with expected value μ_i and variance, σ_i^2. Then, Y has the following mean and variance (Montgomery and Runger, 2007):

$$\mu_Y = \sum_{i=1}^{p} c_i \mu_i \tag{F-14}$$

$$\sigma_Y^2 = \sum_{i=1}^{p} c_i^2 \sigma_i^2 \tag{F-15}$$

Equations F-14 and F-15 show how the variability of the individual X_i variables determines the variability of their linear combination, Y.

EXAMPLE F.3

Experimental tests are to be performed to determine whether a new catalyst A is superior to the current catalyst B, based on their yields for a chemical reaction. Denote the yields by X_A and X_B, and their standard deviations by 3% and 2%, respectively. What is the standard deviation for the difference in yields, $X_A - X_B$?

SOLUTION

Let $Y = X_A - X_B$, an expression in the form of (F-13) with $c_A = 1$ and $c_B = -1$. Thus Eq. (F-15) becomes

$$\sigma_Y^2 = \sigma_A^2 + \sigma_B^2$$

Thus,

$$\sigma_Y = \sqrt{\sigma_A^2 + \sigma_B^2} = \sqrt{(3\%)^2 + (2\%)^2} = 3.6\%$$

Thus, the standard deviation of the difference is larger than the individual standard deviations.

REFERENCES

Montgomery, D. C. and G. C. Runger, *Applied Statistics and Probability for Engineers, 4th ed.*, John Wiley & Sons, Hoboken, NJ, 2007.

Ogunnaike, B. A., *Random Phenomena: Fundamentals of Probability and Statistics for Engineers*, CRC Press, Boca Raton, FL, 2010.

Index

E

F